$X \times Y$ Cartesian product of $X$ and $Y$ [pairs $(x, y)$ with $x$ in $X$ and $y$ in $Y$]; page 9

$X_1 \times X_2 \times \cdots \times X_n$ Cartesian product of $X_1, X_2, \ldots, X_n$ ($n$-tuples with $x_i \in X_i$); page 10

$X \triangle Y$ symmetric difference of $X$ and $Y$; page 13

## RELATIONS

$xRy$ $(x, y)$ is in $R$ ($x$ is related to $y$ by the relation $R$); page 141

$[x]$ equivalence class containing $x$; page 153

$R^{-1}$ inverse relation [all $(y, x)$ with $(x, y)$ in $R$]; page 147

$R_2 \circ R_1$ composition of relations; page 147

$x \preceq y$ $xRy$; page 146

## FUNCTIONS

$f(x)$ value assigned to $x$; page 114

$f\colon X \to Y$ function from $X$ to $Y$; page 113

$f \circ g$ composition of $f$ and $g$; page 122

$f^{-1}$ inverse function [all $(y, x)$ with $(x, y)$ in $f$]; page 121

$f(n) = O(g(n))$ $|f(n)| \leq C|g(n)|$ for $n$ sufficiently large; page 186

$f(n) = \Omega(g(n))$ $c|g(n)| \leq |f(n)|$ for $n$ sufficiently large; page 186

$f(n) = \Theta(g(n))$ $c|g(n)| \leq |f(n)| \leq C|g(n)|$ for $n$ sufficiently large; page 186

## COUNTING

$C(n, r)$ number of $r$-combinations of an $n$-element set $(n!/[(n-r)!r!])$; page 273

$P(n, r)$ number of $r$-permutations of an $n$-element set $[n(n-1)\cdots(n-r+1)]$; page 271

## GRAPHS

$G = (V, E)$ graph $G$ with vertex set $V$ and edge set $E$; page 375

$(v, w)$ edge; page 375

$\delta(v)$ degree of vertex $v$; page 388

$(v_1, \ldots, v_n)$ path from $v_1$ to $v_n$; pages 384–385

$(v_1, \ldots, v_n), v_1 = v_n$ cycle; page 387

$K_n$ complete graph on $n$ vertices; page 380

$K_{m,n}$ complete bipartite graph on $m$ and $n$ vertices; page 381

$w(i, j)$ weight of edge $(i, j)$; page 405

$F_{ij}$ flow in edge $(i, j)$; page 507

$C_{ij}$ capacity of edge $(i, j)$; page 507

$(P, \overline{P})$ cut in a network; page 520

## PROBABILITY

$P(x)$ probability of outcome $x$; page 301

$P(E)$ probability of event $E$; page 302

$P(E \mid F)$ conditional probability of $E$ given $F$ $[P(E \cap F)/P(F)]$; page 306

# Discrete Mathematics

## Eighth Edition

**Richard Johnsonbaugh**

*DePaul University, Chicago*

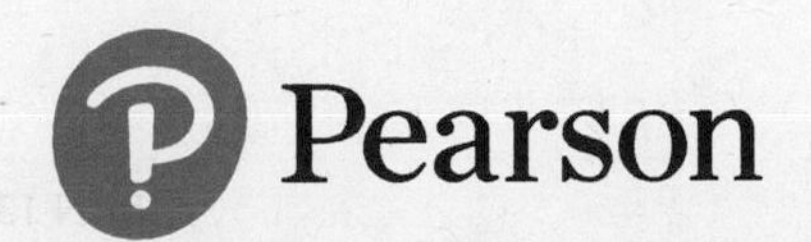


330 Hudson Street, NY, NY 10013

Director, Portfolio Management *Deirdre Lynch*
Executive Editor *Jeff Weidenaar*
Editorial Assistant *Jennifer Snyder*
Content Producer *Lauren Morse*
Managing Producer *Scott Disanno*
Media Producer *Nicholas Sweeney*
Product Marketing Manager *Yvonne Vannatta*
Field Marketing Manager *Evan St. Cyr*
Marketing Assistant *Jennifer Myers*
Senior Author Support/Technology Specialist *Joe Vetere*
Rights and Permissions Project Manager *Gina Cheselka*
Manufacturing Buyer *Carol Melville, LSC Communications*
Associate Director of Design *Blair Brown*
Text Design, Production Coordination, and Composition *SPi Global*
Cover Design *Laurie Entringer*
Cover and Chapter Opener *Helenecanada/iStock/Getty Images*



Johnsonbaugh, Richard, 1941-
Discrete mathematics / Richard Johnsonbaugh, DePaul University, Chicago. – Eighth edition.
pages cm
Includes bibliographical references and index.
ISBN 978-0-321-96468-7 – ISBN 0-321-96468-3
1. Mathematics. 2. Computer science–Mathematics. I. Title.
QA39.2.J65 2015
510–dc23

2014006017

13 2022

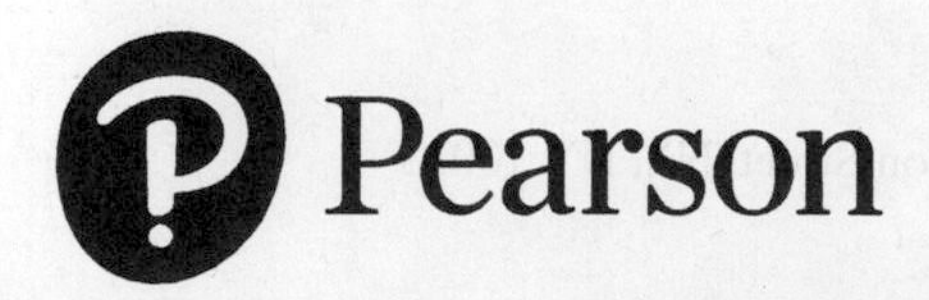

ISBN 13: 978-0-321-96468-7
ISBN 10: 0-321-96468-3

# Contents



†This section can be omitted without loss of continuity.

†This section can be omitted without loss of continuity.

## 6 Counting Methods and the Pigeonhole Principle 255



## 7 Recurrence Relations 327



## 8 Graph Theory 373



†This section can be omitted without loss of continuity.

†This section can be omitted without loss of continuity.

*Dedication*

To Pat, my wife, for her continuous support through my many book projects, for formally and informally copy-editing my books, for maintaining good cheer throughout, and for preventing all *egregious* mistakes that would have otherwise found their way into print. Her contributions are deeply appreciated.

# Preface

This updated edition is intended for a one- or two-term introductory course in discrete mathematics, based on my experience in teaching this course over many years and requests from users of previous editions. Formal mathematics prerequisites are minimal; calculus is not required. There are no computer science prerequisites. The book includes examples, exercises, figures, tables, sections on problem-solving, sections containing problem-solving tips, section reviews, notes, chapter reviews, self-tests, and computer exercises to help the reader master introductory discrete mathematics. In addition, an Instructor's Guide and website are available.

In the early 1980s there were few textbooks appropriate for an introductory course in discrete mathematics. However, there was a need for a course that extended students' mathematical maturity and ability to deal with abstraction, which also included useful topics such as combinatorics, algorithms, and graphs. The original edition of this book (1984) addressed this need and significantly influenced the development of discrete mathematics courses. Subsequently, discrete mathematics courses were endorsed by many groups for several different audiences, including mathematics and computer science majors. A panel of the Mathematical Association of America (MAA) endorsed a year-long course in discrete mathematics. The Educational Activities Board of the Institute of Electrical and Electronics Engineers (IEEE) recommended a freshman discrete mathematics course. The Association for Computing Machinery (ACM) and IEEE accreditation guidelines mandated a discrete mathematics course. This edition, like its predecessors, includes topics such as algorithms, combinatorics, sets, functions, and mathematical induction endorsed by these groups. It also addresses understanding and constructing proofs and, generally, expanding mathematical maturity.

## New to This Edition

The changes in this book, the eighth edition, result from comments and requests from numerous users and reviewers of previous editions of the book. This edition includes the following changes from the seventh edition:

- The web icons in the seventh edition have been replaced by short URLs, making it possible to quickly access the appropriate web page, for example, by using a hand-held device.
- The exercises in the chapter self-tests no longer identify the relevant sections making the self-test more like a real exam. (The hints to these exercises *do* identify the relevant sections.)

- Examples that are worked problems clearly identify where the solution begins and ends.
- The number of exercises in the first three chapters (Sets and Logic; Proofs; and Functions, Sequences, and Relations) has been increased from approximately 1640 worked examples and exercises in the seventh edition to over 1750 in the current edition.
- Many comments have been added to clarify potentially tricky concepts (e.g., "subset" and "element of," collection of sets, logical equivalence of a sequence of propositions, logarithmic scale on a graph).
- There are more examples illustrating diverse approaches to developing proofs and alternative ways to prove a particular result [see, e.g., Examples 2.2.4 and 2.2.8; Examples 6.1.3(c) and 6.1.12; Examples 6.7.7, 6.7.8, and 6.7.9; Examples 6.8.1 and 6.8.2].
- A number of definitions have been revised to allow them to be more directly applied in proofs [see, e.g., one-to-one function (Definition 3.1.22) and onto function (Definition 3.1.29)].
- Additional real-world examples (see descriptions in the following section) are included.
- The altered definition of sequence (Definition 3.2.1) provides more generality and makes subsequent discussion smoother (e.g., the discussion of subsequences).
- Exercises have been added (Exercises 40–49, Section 5.1) to give an example of an algebraic system in which prime factorization does not hold.
- An application of the binomial theorem is used to prove Fermat's little theorem (Exercises 40 and 41, Section 6.7).
- There is now a randomized algorithm to search for a Hamiltonian cycle in a graph (Algorithm 8.3.10).
- The Closest-Pair Problem (Section 13.1 in the seventh edition) has been integrated into Chapter 7 (Recurrence Relations) in the current edition. The algorithm to solve the closest-pair problem is based on merge sort, which is discussed and analyzed in Chapter 7. Chapter 13 in the seventh edition, which has now been removed, had only one additional section.
- A number of recent books and articles have been added to the list of references, and several book references have been updated to current editions.
- The number of exercises has been increased to nearly 4500. (There were approximately 4200 in the seventh edition.)

# Contents and Structure

## Content Overview

### Chapter 1 Sets and Logic

Coverage includes quantifiers and features practical examples such as using the Google search engine (Example 1.2.13). We cover translating between English and symbolic expressions as well as logic in programming languages. We also include a logic game (Example 1.6.15), which offers an alternative way to determine whether a quantified propositional function is true or false.

**Chapter 2 Proofs**

Proof techniques discussed include direct proofs, counterexamples, proof by contradiction, proof by contrapositive, proof by cases, proofs of equivalence, existence proofs (constructive and nonconstructive), and mathematical induction. We present loop invariants as a practical application of mathematical induction. We also include a brief, optional section on resolution proofs (a proof technique that can be automated).

**Chapter 3 Functions, Sequences, and Relations**

The chapter includes strings, sum and product notations, and motivating examples such as the Luhn algorithm for computing credit card check digits, which opens the chapter. Other examples include an introduction to hash functions (Example 3.1.15), pseudorandom number generators (Example 3.1.16). a real-world example of function composition showing its use in making a price comparison (Example 3.1.45), an application of partial orders to task scheduling (Section 3.3), and relational databases (Section 3.6).

**Chapter 4 Algorithms**

The chapter features a thorough discussion of algorithms, recursive algorithms, and the analysis of algorithms. We present a number of examples of algorithms before getting into big-oh and related notations (Sections 4.1 and 4.2), thus providing a gentle introduction and motivating the formalism that follows. We then continue with a full discussion of the "big oh," omega, and theta notations for the growth of functions (Section 4.3). Having all of these notations available makes it possible to make precise statements about the growth of functions and the time and space required by algorithms.

We use the algorithmic approach throughout the remainder of the book. We mention that many modern algorithms do not have all the properties of classical algorithms (e.g., many modern algorithms are not general, deterministic, or even finite). To illustrate the point, we give an example of a randomized algorithm (Example 4.2.4). Algorithms are written in a flexible form of pseudocode, which resembles currently popular languages such as C, C++, and Java. (The book does not assume any computer science prerequisites; the description of the pseudocode used is given in Appendix C.) Among the algorithms presented are:

- Tiling (Section 4.4)
- Euclidean algorithm for finding the greatest common divisor (Section 5.3)
- RSA public-key encryption algorithm (Section 5.4)
- Generating combinations and permutations (Section 6.4)
- Merge sort (Section 7.3)
- Finding a closest pair of points (Section 7.4)
- Dijkstra's shortest-path algorithm (Section 8.4)
- Backtracking algorithms (Section 9.3)
- Breadth-first and depth-first search (Section 9.3)
- Tree traversals (Section 9.6)
- Evaluating a game tree (Section 9.9)
- Finding a maximal flow in a network (Section 10.2)

**Chapter 5 Introduction to Number Theory**

The chapter includes classical results (e.g., divisibility, the infinitude of primes, fundamental theorem of arithmetic), as well as algorithmic number theory (e.g., the Euclidean algorithm to find the greatest common divisor, exponentiation using repeated squaring, computing $s$ and $t$ such that $\gcd(a, b) = sa + tb$, computing an inverse modulo an integer). The major application is the RSA public-key cryptosystem (Section 5.4). The calculations required by the RSA public-key cryptosystem are performed using the algorithms previously developed in the chapter.

**Chapter 6 Counting Methods and the Pigeonhole Principle**
Coverage includes combinations, permutations, discrete probability (optional Sections 6.5 and 6.6), and the Pigeonhole Principle. Applications include internet addressing (Section 6.1) and real-world pattern recognition problems in telemarketing (Example 6.6.21) and virus detection (Example 6.6.22) using Bayes' Theorem.

**Chapter 7 Recurrence Relations**
The chapter includes recurrence relations and their use in the analysis of algorithms.

**Chapter 8 Graph Theory**
Coverage includes graph models of parallel computers, the knight's tour, Hamiltonian cycles, graph isomorphisms, and planar graphs. Theorem 8.4.3 gives a simple, short, elegant proof of the correctness of Dijkstra's algorithm.

**Chapter 9 Trees**
Coverage includes binary trees, tree traversals, minimal spanning trees, decision trees, the minimum time for sorting, and tree isomorphisms.

**Chapter 10 Network Models**
Coverage includes the maximal flow algorithm and matching.

**Chapter 11 Boolean Algebras and Combinatorial Circuits**
Coverage emphasizes the relation of Boolean algebras to combinatorial circuits.

**Chapter 12 Automata, Grammars, and Languages**
Our approach emphasizes modeling and applications. We discuss the *SR* flip-flop circuit in Example 12.1.11, and we describe fractals, including the von Koch snowflake, which can be described by special kinds of grammars (Example 12.3.19).

**Book frontmatter and endmatter**
Appendixes include coverage of matrices, basic algebra, and pseudocode. A reference section provides more than 160 references to additional sources of information. Front and back endpapers summarize the mathematical and algorithm notation used in the book.

## Features of Content Coverage

- **A strong emphasis on the interplay among the various topics.** Examples of this include:
  - We closely tie mathematical induction to recursive algorithms (Section 4.4).
  - We use the Fibonacci sequence in the analysis of the Euclidean algorithm (Section 5.3).
  - Many exercises throughout the book require mathematical induction.
  - We show how to characterize the components of a graph by defining an equivalence relation on the set of vertices (see the discussion following Example 8.2.13).
  - We count the number of nonisomorphic $n$-vertex binary trees (Theorem 9.8.12).
- **A strong emphasis on reading and doing proofs.** We illustrate most proofs of theorems with annotated figures and/or motivate them by special Discussion sec-

tions. Separate sections (Problem-Solving Corners) show students how to attack and solve problems and how to do proofs. Special end-of-section Problem-Solving Tips highlight the main problem-solving techniques of the section.

- **A large number of applications, especially applications to computer science.**
- **Figures and tables** illustrate concepts, show how algorithms work, elucidate proofs, and motivate the material. Several figures illustrate proofs of theorems. The captions of these figures provide additional explanation and insight into the proofs.

## Textbook Structure

Each chapter is organized as follows:

Chapter X Overview
Section X.1
Section X.1 Review Exercises
Section X.1 Exercises
Section X.2
Section X.2 Review Exercises
Section X.2 Exercises
⋮
Chapter X Notes
Chapter X Review
Chapter X Self-Test
Chapter X Computer Exercises

In addition, most chapters have **Problem-Solving Corners** (see "Hallmark Features" for more information about this feature).

**Section review exercises** review the key concepts, definitions, theorems, techniques, and so on of the section. All section review exercises have answers in the back of the book. Although intended for reviews of the sections, section review exercises can also be used for placement and pretesting.

**Chapter notes** contain suggestions for further reading. **Chapter reviews** provide reference lists of the key concepts of the chapters. **Chapter self-tests** contain exercises based on material from throughout the chapter, with answers in the back of the book.

**Computer exercises** include projects, implementation of some of the algorithms, and other programming related activities. Although there is no programming prerequisite for this book and no programming is introduced in the book, these exercises are provided for those readers who want to explore discrete mathematics concepts with a computer.

# Hallmark Features

## Exercises

The book contains nearly 4500 exercises, approximately 150 of which are computer exercises. We use a star to label exercises felt to be more challenging than average. Exercise numbers in color (approximately one-third of the exercises) indicate that the exercise has a hint or solution in the back of the book. The solutions to most of the remaining exercises may be found in the Instructor's Guide. A handful of exercises are clearly identified as requiring calculus. No calculus concepts are used in the main body of the book and, except for these marked exercises, no calculus is needed to solve the exercises.

### Examples

The book contains almost 650 worked examples. These examples show students how to tackle problems in discrete mathematics, demonstrate applications of the theory, clarify proofs, and help motivate the material.

### Problem-Solving Corners

The Problem-Solving Corner sections help students attack and solve problems and show them how to do proofs. Written in an informal style, each is a self-contained section centered around a problem. The intent of these sections is to go beyond simply presenting a proof or a solution to the problem: we show alternative ways of attacking a problem, discuss what to look for in trying to obtain a solution to a problem, and present problem-solving and proof techniques.

Each Problem-Solving Corner begins with a statement of a problem. We then discuss ways to attack the problem, followed by techniques for finding a solution. After we present a solution, we show how to correctly write it up in a formal manner. Finally, we summarize the problem-solving techniques used in the section. Some sections include a Comments subsection, which discusses connections with other topics in mathematics and computer science, provides motivation for the problem, and lists references for further reading about the problem. Some Problem-Solving Corners conclude with a few exercises.

## Supplements and Technology

### Instructor's Solution Manual (downloadable)

ISBN-10: 0–321-98309-2 | ISBN-13: 978-0–321-98309-1

The Instructor's Guide, written by the author, provides worked-out solutions for most exercises in the text. It is available for download to qualified instructors from the Pearson Instructor Resource Center www.pearsonhighered.com/irc.

### Web Support

The short URLs in the margin of the text provide students with direct access to relevant content at point-of-use, including:

- Expanded explanations of difficult material and links to other sites for additional information about discrete mathematics topics.
- Computer programs (in C or C++).

The URL `bit.ly/2D1THYA` provides access to all of the above resources plus an errata list for the text.

**NOTE:**
When you enter URLs that appear in the text, take care to distinguish the following characters:
l = lowercase l
I = uppercase I
1 = one
O = uppercase O
0 = zero

## Acknowledgments

Special thanks go to reviewers of the text, who provided valuable input for this revision:

Venkata Dinavahi, *University of Findlay*
Matthew Elsey, *New York University*
Christophe Giraud-Carrier, *Brigham Young University*

Yevgeniy Kovchegov, *Oregon State University*
Filix Maisch, *Oregon State University*
Tyler McMillen, *California State University, Fullerton*
Christopher Storm, *Adelphi University*
Donald Vestal, *South Dakota State University*
Guanghua Zhao, *Fayetteville State University*

Thanks also to all of the users of the book for their helpful letters and e-mail.

I am grateful to my favorite consultant, Patricia Johnsonbaugh, for her careful reading of the manuscript, improving the exposition, catching miscues I wrote but should not have, and help with the index.

I have received consistent support from the staff at Pearson. Special thanks for their help go to Lauren Morse at Pearson, who managed production, Julie Kidd at SPi Global, who managed the design and typesetting, and Nick Fiala at St. Cloud State University, who accurately checked various stages of proof.

Finally, I thank editor Jeff Weidenaar who has been very helpful to me in preparing this edition. He paid close attention to details in the book, suggested several design enhancements, made many specific recommendations which improved the presentation and comprehension, and proposed changes which enhanced readability.

Richard Johnsonbaugh

# Chapter 1

# SETS AND LOGIC



Chapter 1 begins with sets. A **set** is a collection of objects; order is not taken into account. Discrete mathematics is concerned with objects such as graphs (sets of vertices and edges) and Boolean algebras (sets with certain operations defined on them). In this chapter, we introduce set terminology and notation. In Chapter 2, we treat sets more formally after discussing proof and proof techniques. However, in Section 1.1, we provide a taste of the logic and proofs to come in the remainder of Chapter 1 and in Chapter 2.

**Logic** is the study of reasoning; it is specifically concerned with whether reasoning is correct. Logic focuses on the relationship among statements as opposed to the content of any particular statement. Consider, for example, the following argument:

> All mathematicians wear sandals.
>
> Anyone who wears sandals is an algebraist.
>
> Therefore, all mathematicians are algebraists.

Technically, logic is of no help in determining whether any of these statements is true; however, if the first two statements are true, logic assures us that the statement,

> All mathematicians are algebraists,

is also true.

Logic is essential in reading and developing proofs, which we explore in detail in Chapter 2. An understanding of logic can also be useful in clarifying ordinary writing. For example, at one time, the following ordinance was in effect in Naperville, Illinois: "It shall be unlawful for any person to keep more than three dogs and three cats upon his property within the city." Was one of the citizens, who owned five dogs and no cats, in violation of the ordinance? Think about this question now; then analyze it (see Exercise 75, Section 1.2) after reading Section 1.2.

**Go Online**

For more on logic, see
`bit.ly/2JagVw5`

## 1.1 Sets

The concept of set is basic to all of mathematics and mathematical applications. A **set** is simply a collection of objects. The objects are sometimes referred to as elements or members. If a set is finite and not too large, we can describe it by listing the elements in it. For example, the equation

**Go Online**

For more on sets, see `bit.ly/2JagVw5`

$$A = \{1, 2, 3, 4\} \qquad \textbf{(1.1.1)}$$

describes a set $A$ made up of the four elements 1, 2, 3, and 4. A set is determined by its elements and not by any particular order in which the elements might be listed. Thus the set $A$ might just as well be specified as $A = \{1, 3, 4, 2\}$. The elements making up a set are assumed to be distinct, and although for some reason we may have duplicates in our list, only one occurrence of each element is in the set. For this reason we may also describe the set $A$ defined in (1.1.1) as $A = \{1, 2, 2, 3, 4\}$.

If a set is a large finite set or an infinite set, we can describe it by listing a property necessary for membership. For example, the equation

$$B = \{x \mid x \text{ is a positive, even integer}\} \qquad \textbf{(1.1.2)}$$

describes the set $B$ made up of all positive, even integers; that is, $B$ consists of the integers 2, 4, 6, and so on. The vertical bar "|" is read "such that." Equation (1.1.2) would be read "$B$ equals the set of all $x$ such that $x$ is a positive, even integer." Here the property necessary for membership is "is a positive, even integer." Note that the property appears after the vertical bar. The notation in (1.1.2) is called **set-builder notation**.

A set may contain *any* kind of elements whatsoever, and they need *not* be of the same "type." For example,

$$\{4.5, \text{Lady Gaga}, \pi, 14\}$$

is a perfectly fine set. It consists of four elements: the number 4.5, the person Lady Gaga, the number $\pi (= 3.1415\ldots)$, and the number 14.

A set may contain elements that are themselves sets. For example, the set

$$\{3, \{5, 1\}, 12, \{\pi, 4.5, 40, 16\}, \text{Henry Cavill}\}$$

consists of five elements: the number 3, the set $\{5, 1\}$, the number 12, the set $\{\pi, 4.5, 40, 16\}$, and the person Henry Cavill.

Some sets of numbers that occur frequently in mathematics generally, and in discrete mathematics in particular, are shown in Figure 1.1.1. The symbol **Z** comes from the German word, *Zahlen*, for *integer*. Rational numbers are quotients of integers, thus **Q** for *quotient*. The set of real numbers **R** can be depicted as consisting of all points on a straight line extending indefinitely in either direction (see Figure 1.1.2).[†]

| *Symbol* | *Set* | *Example of Members* |
|---|---|---|
| **Z** | Integers | $-3, 0, 2, 145$ |
| **Q** | Rational numbers | $-1/3, 0, 24/15$ |
| **R** | Real numbers | $-3, -1.766, 0, 4/15, \sqrt{2}, 2.666\ldots, \pi$ |

**Figure 1.1.1** Sets of numbers.

[†]The real numbers can be constructed by starting with a more primitive notion such as "set" or "integer," or they can be obtained by stating properties (axioms) they are assumed to obey. For our purposes, it suffices to think of the real numbers as points on a straight line. The construction of the real numbers and the axioms for the real numbers are beyond the scope of this book.

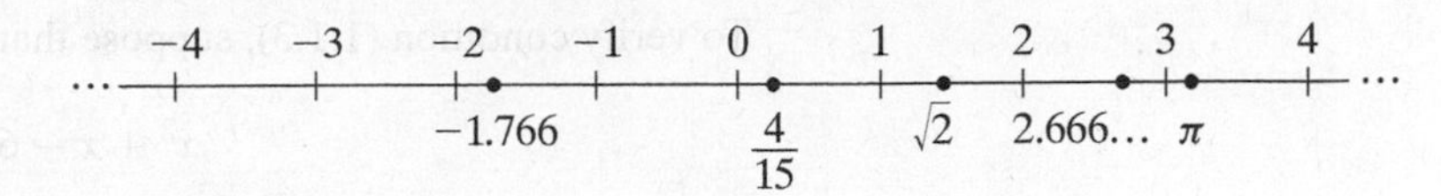


**Figure 1.1.2** The real number line.

To denote the negative numbers that belong to one of **Z**, **Q**, or **R**, we use the superscript minus. For example, $\mathbf{Z}^-$ denotes the set of negative integers, namely −1, −2, −3, .... Similarly, to denote the positive numbers that belong to one of the three sets, we use the superscript plus. For example, $\mathbf{Q}^+$ denotes the set of positive rational numbers. To denote the nonnegative numbers that belong to one of the three sets, we use the superscript *nonneg*. For example, $\mathbf{Z}^{nonneg}$ denotes the set of nonnegative integers, namely 0, 1, 2, 3, ....

If $X$ is a finite set, we let $|X| =$ number of elements in $X$. We call $|X|$ the **cardinality** of $X$. There is also a notion of cardinality of infinite sets, although we will not discuss it in this book. For example, the cardinality of the integers, Z, is denoted $\aleph_0$, read "aleph null." Aleph is the first letter of the Hebrew alphabet.

**Example 1.1.1** For the set $A$ in (1.1.1), we have $|A| = 4$, and the cardinality of $A$ is 4. The cardinality of the set $\{\mathbf{R}, \mathbf{Z}\}$ is 2 since it contains two elements, namely the two sets **R** and **Z**. ◀

Given a description of a set $X$ such as (1.1.1) or (1.1.2) and an element $x$, we can determine whether or not $x$ belongs to $X$. If the members of $X$ are listed as in (1.1.1), we simply look to see whether or not $x$ appears in the listing. In a description such as (1.1.2), we check to see whether the element $x$ has the property listed. If $x$ is in the set $X$, we write $x \in X$, and if $x$ is not in $X$, we write $x \notin X$. For example, $3 \in \{1, 2, 3, 4\}$, but $3 \notin \{x \mid x \text{ is a positive, even integer}\}$.

The set with no elements is called the **empty** (or **null** or **void**) **set** and is denoted $\varnothing$. Thus $\varnothing = \{\,\}$.

Two sets $X$ and $Y$ are **equal** and we write $X = Y$ if $X$ and $Y$ have the same elements. To put it another way, $X = Y$ if the following two conditions hold:

- For every $x$, if $x \in X$, then $x \in Y$,

and

- For every $x$, if $x \in Y$, then $x \in X$.

The first condition ensures that every element of $X$ is an element of $Y$, and the second condition ensures that every element of $Y$ is an element of $X$.

**Example 1.1.2** If $A = \{1, 3, 2\}$ and $B = \{2, 3, 2, 1\}$, by inspection, $A$ and $B$ have the same elements. Therefore $A = B$. ◀

**Example 1.1.3** Show that if $A = \{x \mid x^2 + x - 6 = 0\}$ and $B = \{2, -3\}$, then $A = B$.

**SOLUTION** According to the criteria in the paragraph immediately preceding Example 1.1.2, we must show that for every $x$,

$$\text{if } x \in A, \text{ then } x \in B, \tag{1.1.3}$$

and for every $x$,

$$\text{if } x \in B, \text{ then } x \in A. \tag{1.1.4}$$

To verify condition (1.1.3), suppose that $x \in A$. Then

$$x^2 + x - 6 = 0.$$

Solving for $x$, we find that $x = 2$ or $x = -3$. In either case, $x \in B$. Therefore, condition (1.1.3) holds.

To verify condition (1.1.4), suppose that $x \in B$. Then $x = 2$ or $x = -3$. If $x = 2$, then

$$x^2 + x - 6 = 2^2 + 2 - 6 = 0.$$

Therefore, $x \in A$. If $x = -3$, then

$$x^2 + x - 6 = (-3)^2 + (-3) - 6 = 0.$$

Again, $x \in A$. Therefore, condition (1.1.4) holds. We conclude that $A = B$. ◀

For a set $X$ to *not* be equal to a set $Y$ (written $X \neq Y$), $X$ and $Y$ must *not* have the same elements: There must be at least one element in $X$ that is not in $Y$ or at least one element in $Y$ that is not in $X$ (or both).

**Example 1.1.4** Let $A = \{1, 2, 3\}$ and $B = \{2, 4\}$. Then $A \neq B$ since there is at least one element in $A$ (1 for example) that is not in $B$. [Another way to see that $A \neq B$ is to note that there is at least one element in $B$ (namely 4) that is not in $A$.] ◀

Suppose that $X$ and $Y$ are sets. If every element of $X$ is an element of $Y$, we say that $X$ is a **subset** of $Y$ and write $X \subseteq Y$. In other words, $X$ is a subset of $Y$ if for every $x$, if $x \in X$, then $x \in Y$.

**Example 1.1.5** If $C = \{1, 3\}$ and $A = \{1, 2, 3, 4\}$, by inspection, every element of $C$ is an element of $A$. Therefore, $C$ is a subset of $A$ and we write $C \subseteq A$. ◀

**Example 1.1.6** Let $X = \{x \mid x^2 + x - 2 = 0\}$. Show that $X \subseteq \mathbf{Z}$.

**SOLUTION** We must show that for every $x$, if $x \in X$, then $x \in \mathbf{Z}$. If $x \in X$, then $x^2 + x - 2 = 0$. Solving for $x$, we obtain $x = 1$ or $x = -2$. In either case, $x \in \mathbf{Z}$. Therefore, for every $x$, if $x \in X$, then $x \in \mathbf{Z}$. We conclude that $X$ is a subset of $\mathbf{Z}$ and we write $X \subseteq \mathbf{Z}$. ◀

**Example 1.1.7** The set of integers $\mathbf{Z}$ is a subset of the set of rational numbers $\mathbf{Q}$. If $n \in \mathbf{Z}$, $n$ can be expressed as a quotient of integers, for example, $n = n/1$. Therefore $n \in \mathbf{Q}$ and $\mathbf{Z} \subseteq \mathbf{Q}$. ◀

**Example 1.1.8** The set of rational numbers $\mathbf{Q}$ is a subset of the set of real numbers $\mathbf{R}$. If $x \in \mathbf{Q}$, $x$ corresponds to a point on the number line (see Figure 1.1.2) so $x \in \mathbf{R}$. ◀

For $X$ to *not* be a subset of $Y$, there must be at least one member of $X$ that is not in $Y$.

**Example 1.1.9** Let $X = \{x \mid 3x^2 - x - 2 = 0\}$. Show that $X$ is not a subset of $\mathbf{Z}$.

**SOLUTION** If $x \in X$, then $3x^2 - x - 2 = 0$. Solving for $x$, we obtain $x = 1$ or $x = -2/3$. Taking $x = -2/3$, we have $x \in X$ but $x \notin \mathbf{Z}$. Therefore, $X$ is not a subset of $\mathbf{Z}$. ◀

Any set $X$ is a subset of itself, since any element in $X$ is in $X$. Also, the empty set is a subset of every set. If $\varnothing$ is *not* a subset of some set $Y$, according to the discussion preceding Example 1.1.9, there would have to be at least one member of $\varnothing$ that is not in $Y$. But this cannot happen because the empty set, by definition, has no members.

Notice the difference between the terms "subset" and "element of." The set $X$ is a *subset* of the set $Y (X \subseteq Y)$, if every element of $X$ is an element of $Y$; $x$ is an *element of* $X (x \in X)$, if $x$ is a member of the set $X$.

**Example 1.1.10** Let $X = \{1, 3, 5, 7\}$ and $Y = \{1, 2, 3, 4, 5, 6, 7\}$. Then $X \subseteq Y$ since every element of $X$ is an element of $Y$. But $X \notin Y$, since the *set* $X$ is not a member of $Y$. Also, $1 \in X$, but 1 is not a subset of $X$. Notice the difference between the number 1 and the *set* $\{1\}$. The set $\{1\}$ *is* a subset of $X$. ◀

If $X$ is a subset of $Y$ and $X$ does not equal $Y$, we say that $X$ is a **proper subset** of $Y$ and write $X \subset Y$.

**Example 1.1.11** Let $C = \{1, 3\}$ and $A = \{1, 2, 3, 4\}$. Then $C$ is a proper subset of $A$ since $C$ is a subset of $A$ but $C$ does not equal $A$. We write $C \subset A$. ◀

**Example 1.1.12** Example 1.1.7 showed that $\mathbf{Z}$ is a subset of $\mathbf{Q}$. In fact, $\mathbf{Z}$ is a proper subset of $\mathbf{Q}$ because, for example, $1/2 \in \mathbf{Q}$, but $1/2 \notin \mathbf{Z}$. ◀

**Example 1.1.13** Example 1.1.8 showed that $\mathbf{Q}$ is a subset of $\mathbf{R}$. In fact, $\mathbf{Q}$ is a proper subset of $\mathbf{R}$ because, for example, $\sqrt{2} \in \mathbf{R}$, but $\sqrt{2} \notin \mathbf{Q}$. (In Example 2.2.3, we will show that $\sqrt{2}$ is not the quotient of integers). ◀

The set of all subsets (proper or not) of a set $X$, denoted $\mathcal{P}(X)$, is called the **power set** of $X$.

**Example 1.1.14** If $A = \{a, b, c\}$, the members of $\mathcal{P}(A)$ are

$$\varnothing, \{a\}, \{b\}, \{c\}, \{a, b\}, \{a, c\}, \{b, c\}, \{a, b, c\}.$$

All but $\{a, b, c\}$ are proper subsets of $A$. ◀

In Example 1.1.14, $|A| = 3$ and $|\mathcal{P}(A)| = 2^3 = 8$. In Section 2.4 (Theorem 2.4.6), we will give a formal proof that this result holds in general; that is, the power set of a set with $n$ elements has $2^n$ elements.

Given two sets $X$ and $Y$, there are various set operations involving $X$ and $Y$ that can produce a new set. The set

$$X \cup Y = \{x \mid x \in X \text{ or } x \in Y\}$$

is called the **union** of $X$ and $Y$. The union consists of all elements belonging to either $X$ or $Y$ (or both).

The set

$$X \cap Y = \{x \mid x \in X \text{ and } x \in Y\}$$

is called the **intersection** of $X$ and $Y$. The intersection consists of all elements belonging to both $X$ and $Y$.

The set

$$X - Y = \{x \mid x \in X \text{ and } x \notin Y\}$$

is called the **difference** (or **relative complement**). The difference $X - Y$ consists of all elements in $X$ that are not in $Y$.

**Example 1.1.15** If $A = \{1, 3, 5\}$ and $B = \{4, 5, 6\}$, then

$$\begin{aligned} A \cup B &= \{1, 3, 4, 5, 6\} \\ A \cap B &= \{5\} \\ A - B &= \{1, 3\} \\ B - A &= \{4, 6\}. \end{aligned}$$

Notice that, in general, $A - B \neq B - A$. ◀

**Example 1.1.16** Since $\mathbf{Q} \subseteq \mathbf{R}$,

$$\begin{aligned} \mathbf{R} \cup \mathbf{Q} &= \mathbf{R} \\ \mathbf{R} \cap \mathbf{Q} &= \mathbf{Q} \\ \mathbf{Q} - \mathbf{R} &= \varnothing. \end{aligned}$$

The set $\mathbf{R} - \mathbf{Q}$, called the set of **irrational numbers,** consists of all real numbers that are not rational. ◀

We call a set $\mathcal{S}$, whose elements are sets, a **collection of sets** or a **family of sets.** For example, if

$$\mathcal{S} = \{\{1, 2\}, \{1, 3\}, \{1, 7, 10\}\},$$

then $\mathcal{S}$ is a collection or family of sets. The set $\mathcal{S}$ consists of the sets

$$\{1, 2\}, \{1, 3\}, \{1, 7, 10\}.$$

Sets $X$ and $Y$ are **disjoint** if $X \cap Y = \varnothing$. A collection of sets $\mathcal{S}$ is said to be **pairwise disjoint** if, whenever $X$ and $Y$ are distinct sets in $\mathcal{S}$, $X$ and $Y$ are disjoint.

**Example 1.1.17** The sets $\{1, 4, 5\}$ and $\{2, 6\}$ are disjoint. The collection of sets $\mathcal{S} = \{\{1, 4, 5\}, \{2, 6\}, \{3\}, \{7, 8\}\}$ is pairwise disjoint. ◀

Sometimes we are dealing with sets, all of which are subsets of a set $U$. This set $U$ is called a **universal set** or a **universe.** The set $U$ must be explicitly given or inferred from the context. Given a universal set $U$ and a subset $X$ of $U$, the set $U - X$ is called the **complement** of $X$ and is written $\overline{X}$.

**Example 1.1.18** Let $A = \{1, 3, 5\}$. If $U$, a universal set, is specified as $U = \{1, 2, 3, 4, 5\}$, then $\overline{A} = \{2, 4\}$. If, on the other hand, a universal set is specified as $U = \{1, 3, 5, 7, 9\}$, then $\overline{A} = \{7, 9\}$. The complement obviously depends on the universe in which we are working. ◀

**Example 1.1.19** Let the universal set be $\mathbf{Z}$. Then $\overline{\mathbf{Z}^-}$, the complement of the set of negative integers, is $\mathbf{Z}^{nonneg}$, the set of nonnegative integers. ◀

**Go Online**
For more on Venn diagrams, see `bit.ly/2JagVw5`

**Venn diagrams** provide pictorial views of sets. In a Venn diagram, a rectangle depicts a universal set (see Figure 1.1.3). Subsets of the universal set are drawn as circles. The inside of a circle represents the members of that set. In Figure 1.1.3 we see two sets $A$ and $B$ within the universal set $U$. Region 1 represents $\overline{(A \cup B)}$, the elements in neither $A$ nor $B$. Region 2 represents $A - B$, the elements in $A$ but not in $B$. Region 3 represents $A \cap B$, the elements in both $A$ and $B$. Region 4 represents $B - A$, the elements in $B$ but not in $A$.

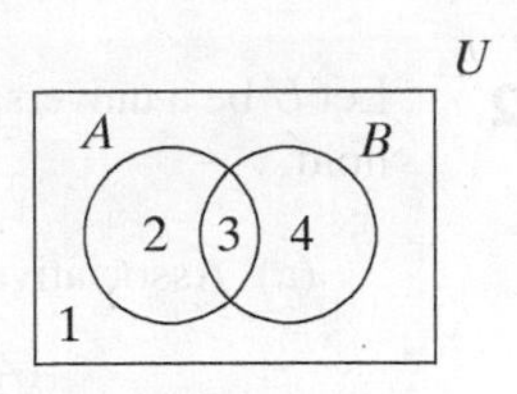


**Figure 1.1.3** A Venn diagram.

**Example 1.1.20** Particular regions in Venn diagrams are depicted by shading. The set $A \cup B$ is shown in Figure 1.1.4, and Figure 1.1.5 represents the set $A - B$. ◀

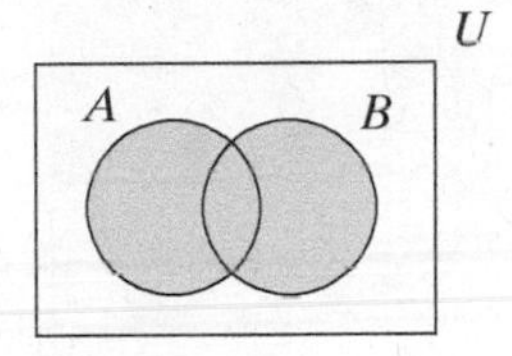


**Figure 1.1.4** A Venn diagram of $A \cup B$.

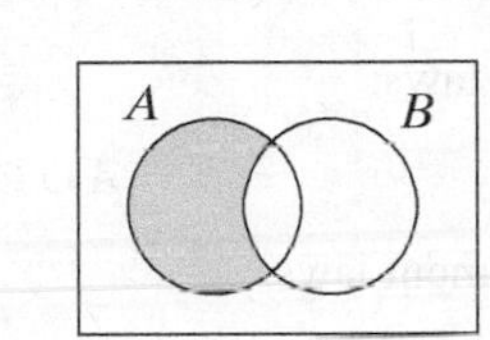


**Figure 1.1.5** A Venn diagram of $A - B$.

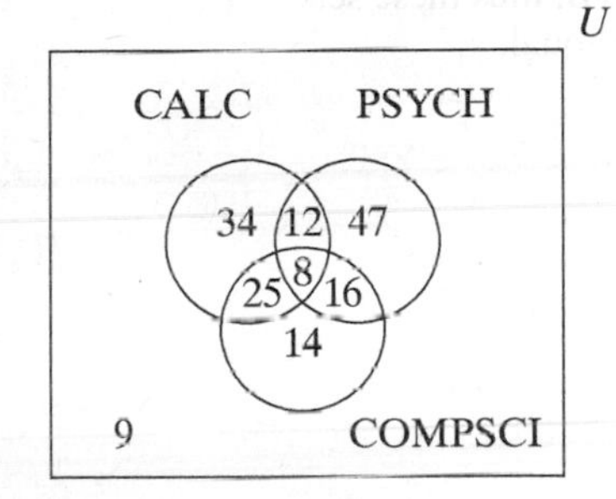


**Figure 1.1.6** A Venn diagram of three sets CALC, PSYCH, and COMPSCI. The numbers show how many students belong to the particular region depicted.

To represent three sets, we use three overlapping circles (see Figure 1.1.6).

**Example 1.1.21** Among a group of 165 students, 8 are taking calculus, psychology, and computer science; 33 are taking calculus and computer science; 20 are taking calculus and psychology; 24 are taking psychology and computer science; 79 are taking calculus; 83 are taking psychology; and 63 are taking computer science. How many are taking none of the three subjects?

**SOLUTION** Let CALC, PSYCH, and COMPSCI denote the sets of students taking calculus, psychology, and computer science, respectively. Let $U$ denote the set of all 165 students (see Figure 1.1.6). Since 8 students are taking calculus, psychology, and computer science, we write 8 in the region representing $\text{CALC} \cap \text{PSYCH} \cap \text{COMPSCI}$. Of the 33 students taking calculus and computer science, 8 are also taking psychology; thus 25 are taking calculus and computer science but not psychology. We write 25 in the region representing $\text{CALC} \cap \overline{\text{PSYCH}} \cap \text{COMPSCI}$. Similarly, we write 12 in the region representing $\text{CALC} \cap \text{PSYCH} \cap \overline{\text{COMPSCI}}$ and 16 in the region representing $\overline{\text{CALC}} \cap \text{PSYCH} \cap \text{COMPSCI}$. Of the 79 students taking calculus, 45 have now been accounted for. This leaves 34 students taking only calculus. We write 34 in the region representing $\text{CALC} \cap \overline{\text{PSYCH}} \cap \overline{\text{COMPSCI}}$. Similarly, we write 47 in the region representing $\overline{\text{CALC}} \cap \text{PSYCH} \cap \overline{\text{COMPSCI}}$ and 14 in the region representing

$\overline{\text{CALC}} \cap \overline{\text{PSYCH}} \cap \text{COMPSCI}$. At this point, 156 students have been accounted for. This leaves 9 students taking none of the three subjects. ◀

Venn diagrams can also be used to visualize certain properties of sets. For example, by sketching both $\overline{(A \cup B)}$ and $\overline{A} \cap \overline{B}$ (see Figure 1.1.7), we see that these sets are equal. A formal proof would show that for every $x$, if $x \in \overline{(A \cup B)}$, then $x \in \overline{A} \cap \overline{B}$, and if $x \in \overline{A} \cap \overline{B}$, then $x \in \overline{(A \cup B)}$. We state many useful properties of sets as Theorem 1.1.22.

**Theorem 1.1.22** Let $U$ be a universal set and let $A$, $B$, and $C$ be subsets of $U$. The following properties hold.

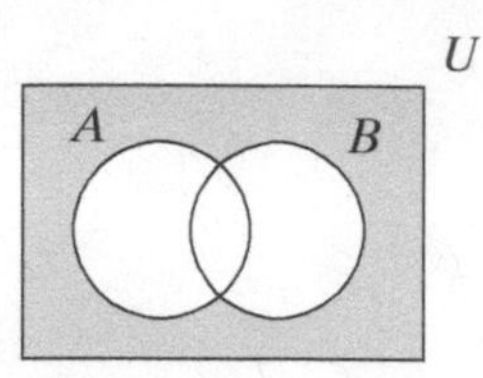


**Figure 1.1.7** The shaded region depicts both $\overline{(A \cup B)}$ and $\overline{A} \cap \overline{B}$; thus these sets are equal.

(a) Associative laws:

$$(A \cup B) \cup C = A \cup (B \cup C), \quad (A \cap B) \cap C = A \cap (B \cap C)$$

(b) Commutative laws:

$$A \cup B = B \cup A, \quad A \cap B = B \cap A$$

(c) Distributive laws:

$$A \cap (B \cup C) = (A \cap B) \cup (A \cap C), \quad A \cup (B \cap C) = (A \cup B) \cap (A \cup C)$$

(d) Identity laws:

$$A \cup \varnothing = A, \quad A \cap U = A$$

(e) Complement laws:

$$A \cup \overline{A} = U, \quad A \cap \overline{A} = \varnothing$$

(f) Idempotent laws:

$$A \cup A = A, \quad A \cap A = A$$

(g) Bound laws:

$$A \cup U = U, \quad A \cap \varnothing = \varnothing$$

(h) Absorption laws:

$$A \cup (A \cap B) = A, \quad A \cap (A \cup B) = A$$

(i) Involution law:

$$\overline{\overline{A}} = A^{\dagger}$$

(j) 0/1 laws:

$$\overline{\varnothing} = U, \quad \overline{U} = \varnothing$$

(k) De Morgan's laws for sets:

$$\overline{(A \cup B)} = \overline{A} \cap \overline{B}, \quad \overline{(A \cap B)} = \overline{A} \cup \overline{B}$$

**Go Online**
For a biography of De Morgan, see `bit.ly/2JagVw5`

**Proof** The proofs are left as exercises (Exercises 46–56, Section 2.1) to be done after more discussion of logic and proof techniques. ◀

We define the union of a collection of sets $\mathcal{S}$ to be those elements $x$ belonging to at least one set $X$ in $\mathcal{S}$. Formally,

$$\cup\mathcal{S} = \{x \mid x \in X \text{ for some } X \in \mathcal{S}\}.$$

† $\overline{\overline{A}}$ denotes the complement of the complement of $A$, that is, $\overline{\overline{A}} = \overline{(\overline{A})}$.

Similarly, we define the intersection of a collection of sets $\mathcal{S}$ to be those elements $x$ belonging to every set $X$ in $\mathcal{S}$. Formally,

$$\cap \mathcal{S} = \{x \mid x \in X \text{ for all } X \in \mathcal{S}\}.$$

**Example 1.1.23** Let $\mathcal{S} = \{\{1, 2\}, \{1, 3\}, \{1, 7, 10\}\}$. Then $\cup\mathcal{S} = \{1, 2, 3, 7, 10\}$ since each of the elements 1, 2, 3, 7, 10 belongs to at least one set in $\mathcal{S}$, and no other element belongs to any of the sets in $\mathcal{S}$. Also $\cap\mathcal{S} = \{1\}$ since only the element 1 belong to every set in $\mathcal{S}$. ◀

If

$$\mathcal{S} = \{A_1, A_2, \ldots, A_n\},$$

we write

$$\bigcup \mathcal{S} = \bigcup_{i=1}^{n} A_i, \qquad \bigcap \mathcal{S} = \bigcap_{i=1}^{n} A_i,$$

and if

$$\mathcal{S} = \{A_1, A_2, \ldots\},$$

we write

$$\bigcup \mathcal{S} = \bigcup_{i=1}^{\infty} A_i, \qquad \bigcap \mathcal{S} = \bigcap_{i=1}^{\infty} A_i.$$

**Example 1.1.24** For $i \geq 1$, define $A_i = \{i, i + 1, \ldots\}$ and $\mathcal{S} = \{A_1, A_2, \ldots\}$. As examples, $A_1 = \{1, 2, 3, \ldots\}$ and $A_2 = \{2, 3, 4, \ldots\}$. Then

$$\bigcup \mathcal{S} = \bigcup_{i=1}^{\infty} A_i = \{1, 2, \ldots\}, \qquad \bigcap \mathcal{S} = \bigcap_{i=1}^{\infty} A_i = \varnothing.$$ ◀

A partition of a set $X$ divides $X$ into nonoverlapping subsets. More formally, a collection $\mathcal{S}$ of nonempty subsets of $X$ is said to be a **partition** of the set $X$ if every element in $X$ belongs to exactly one member of $\mathcal{S}$. Notice that if $\mathcal{S}$ is a partition of $X$, $\mathcal{S}$ is pairwise disjoint and $\cup\mathcal{S} = X$.

**Example 1.1.25** Since each element of $X = \{1, 2, 3, 4, 5, 6, 7, 8\}$ is in exactly one member of $\mathcal{S} = \{\{1, 4, 5\}, \{2, 6\}, \{3\}, \{7, 8\}\}$, $\mathcal{S}$ is a partition of $X$. ◀

At the beginning of this section, we pointed out that a set is an unordered collection of elements; that is, a set is determined by its elements and not by any particular order in which the elements are listed. Sometimes, however, we do want to take order into account. An **ordered pair** of elements, written $(a, b)$, is considered distinct from the ordered pair $(b, a)$, unless, of course, $a = b$. To put it another way, $(a, b) = (c, d)$ precisely when $a = c$ and $b = d$. If $X$ and $Y$ are sets, we let $X \times Y$ denote the set of all ordered pairs $(x, y)$ where $x \in X$ and $y \in Y$. We call $X \times Y$ the **Cartesian product** of $X$ and $Y$.

**Example 1.1.26** If $X = \{1, 2, 3\}$ and $Y = \{a, b\}$, then

$$X \times Y = \{(1, a), (1, b), (2, a), (2, b), (3, a), (3, b)\}$$

$$Y \times X = \{(a, 1), (b, 1), (a, 2), (b, 2), (a, 3), (b, 3)\}$$

$$X \times X = \{(1, 1), (1, 2), (1, 3), (2, 1), (2, 2), (2, 3), (3, 1), (3, 2), (3, 3)\}$$

$$Y \times Y = \{(a, a), (a, b), (b, a), (b, b)\}.$$ ◀

Example 1.1.26 shows that, in general, $X \times Y \neq Y \times X$.

Notice that in Example 1.1.26, $|X \times Y| = |X| \cdot |Y|$ (both are equal to 6). The reason is that there are 3 ways to choose an element of $X$ for the first member of the ordered pair, there are 2 ways to choose an element of $Y$ for the second member of the ordered pair, and $3 \cdot 2 = 6$ (see Figure 1.1.8). The preceding argument holds for arbitrary finite sets $X$ and $Y$; it is always true that $|X \times Y| = |X| \cdot |Y|$.

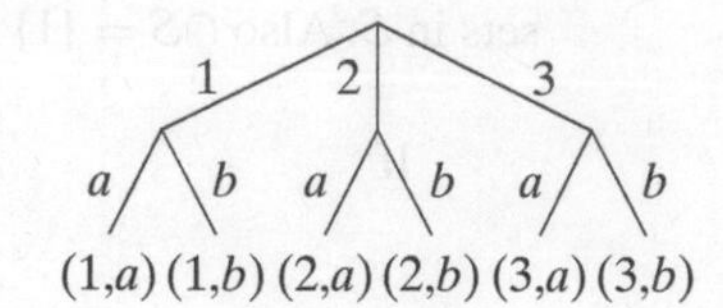


**Figure 1.1.8** $|X \times Y| = |X| \cdot |Y|$, where $X = \{1, 2, 3\}$ and $Y = \{a, b\}$. There are 3 ways to choose an element of $X$ for the first member of the ordered pair (shown at the top of the diagram) and, for each of these choices, there are 2 ways to choose an element of $Y$ for the second member of the ordered pair (shown at the bottom of the diagram). Since there are 3 groups of 2, there are $3 \cdot 2 = 6$ elements in $X \times Y$ (labeled at the bottom of the figure).

**Example 1.1.27** A restaurant serves four appetizers,

$$r = \text{ribs}, \quad n = \text{nachos}, \quad s = \text{shrimp}, \quad f = \text{fried cheese},$$

and three entrees,

$$c = \text{chicken}, \quad b = \text{beef}, \quad t = \text{trout}.$$

If we let $A = \{r, n, s, f\}$ and $E = \{c, b, t\}$, the Cartesian product $A \times E$ lists the 12 possible dinners consisting of one appetizer and one entree. ◀

Ordered lists need not be restricted to two elements. An ***n*-tuple,** written $(a_1, a_2, \ldots, a_n)$, takes order into account; that is,

$$(a_1, a_2, \ldots, a_n) = (b_1, b_2, \ldots, b_n)$$

precisely when

$$a_1 = b_1, a_2 = b_2, \ldots, a_n = b_n.$$

The Cartesian product of sets $X_1, X_2, \ldots, X_n$ is defined to be the set of all $n$-tuples $(x_1, x_2, \ldots, x_n)$ where $x_i \in X_i$ for $i = 1, \ldots, n$; it is denoted $X_1 \times X_2 \times \cdots \times X_n$.

**Example 1.1.28** If $X = \{1, 2\}$, $Y = \{a, b\}$, and $Z = \{\alpha, \beta\}$, then

$$X \times Y \times Z = \{(1, a, \alpha), (1, a, \beta), (1, b, \alpha), (1, b, \beta), (2, a, \alpha), (2, a, \beta), (2, b, \alpha), (2, b, \beta)\}.$$ ◀

Notice that in Example 1.1.28, $|X \times Y \times Z| = |X| \cdot |Y| \cdot |Z|$. In general,

$$|X_1 \times X_2 \times \cdots \times X_n| = |X_1| \cdot |X_2| \cdots |X_n|.$$

We leave the proof of this last statement as an exercise (see Exercise 27, Section 2.4).

**Example 1.1.29** If $A$ is a set of appetizers, $E$ is a set of entrees, and $D$ is a set of desserts, the Cartesian product $A \times E \times D$ lists all possible dinners consisting of one appetizer, one entree, and one dessert. ◀

## 1.1 Problem-Solving Tips

- To verify that two sets $A$ and $B$ are equal, written $A = B$, show that for every $x$, if $x \in A$, then $x \in B$, and if $x \in B$, then $x \in A$.
- To verify that two sets $A$ and $B$ are *not* equal, written $A \neq B$, find at least one element that is in $A$ but not in $B$, or find at least one element that is in $B$ but not in $A$. One or the other conditions suffices; you need not (and may not be able to) show both conditions.
- To verify that $A$ is a subset of $B$, written $A \subseteq B$, show that for every $x$, if $x \in A$, then $x \in B$. Notice that if $A$ is a subset of $B$, it is possible that $A = B$.
- To verify that $A$ is *not* a subset of $B$, find at least one element that is in $A$ but not in $B$.
- To verify that $A$ is a proper subset of $B$, written $A \subset B$, verify that $A$ is a subset of $B$ as described previously, and that $A \neq B$, that is, that there is at least one element that is in $B$ but not in $A$.
- To visualize relationships among sets, use a Venn diagram. A Venn diagram can suggest whether a statement about sets is true or false.
- A set of elements is determined by its members; order is irrelevant. On the other hand, ordered pairs and $n$-tuples take order into account.

## 1.1 Review Exercises

1.[†] What is a set?
2. What is set notation?
3. Describe the sets $\mathbf{Z}, \mathbf{Q}, \mathbf{R}, \mathbf{Z}^+, \mathbf{Q}^+, \mathbf{R}^+, \mathbf{Z}^-, \mathbf{Q}^-, \mathbf{R}^-, \mathbf{Z}^{nonneg}, \mathbf{Q}^{nonneg}$, and $\mathbf{R}^{nonneg}$, and give two examples of members of each set.
4. If $X$ is a finite set, what is $|X|$?
5. How do we denote *$x$ is an element of the set $X$*?
6. How do we denote *$x$ is not an element of the set $X$*?
7. How do we denote the empty set?
8. Define *set $X$ is equal to set $Y$*. How do we denote *$X$ is equal to $Y$*?
9. Explain a method of verifying that sets $X$ and $Y$ are equal.
10. Explain a method of verifying that sets $X$ and $Y$ are *not* equal.
11. Define *$X$ is a subset of $Y$*. How do we denote *$X$ is a subset of $Y$*?
12. Explain a method of verifying that $X$ is a subset of $Y$.
13. Explain a method of verifying that $X$ is *not* a subset of $Y$.
14. Define *$X$ is a proper subset of $Y$*. How do we denote *$X$ is a proper subset of $Y$*?
15. Explain a method of verifying that $X$ is a proper subset of $Y$.
16. What is the power set of $X$? How is it denoted?
17. Define *$X$ union $Y$*. How is the union of $X$ and $Y$ denoted?
18. If $\mathcal{S}$ is a family of sets, how do we define the union of $\mathcal{S}$? How is the union denoted?
19. Define *$X$ intersect $Y$*. How is the intersection of $X$ and $Y$ denoted?
20. If $\mathcal{S}$ is a family of sets, how do we define the intersection of $\mathcal{S}$? How is the intersection denoted?
21. Define *$X$ and $Y$ are disjoint sets*.
22. What is a pairwise disjoint family of sets?
23. Define the *difference* of sets $X$ and $Y$. How is the difference denoted?
24. What is a universal set?
25. What is the complement of the set $X$? How is it denoted?
26. What is a Venn diagram?
27. Draw a Venn diagram of three sets and identify the set represented by each region.

[†]Exercise numbers in color indicate that a hint or solution appears at the back of the book in the section following the References.

**28.** State the associative laws for sets.

**29.** State the commutative laws for sets.

**30.** State the distributive laws for sets.

**31.** State the identity laws for sets.

**32.** State the complement laws for sets.

**33.** State the idempotent laws for sets.

**34.** State the bound laws for sets.

**35.** State the absorption laws for sets.

**36.** State the involution law for sets.

**37.** State the 0/1 laws for sets.

**38.** State De Morgan's laws for sets.

**39.** What is a partition of a set $X$?

**40.** Define the *Cartesian product* of sets $X$ and $Y$. How is this Cartesian product denoted?

**41.** Define the *Cartesian product* of the sets $X_1, X_2, \ldots, X_n$. How is this Cartesian product denoted?

## 1.1 Exercises

*In Exercises 1–16, let the universe be the set $U = \{1, 2, 3, \ldots, 10\}$. Let $A = \{1, 4, 7, 10\}$, $B = \{1, 2, 3, 4, 5\}$, and $C = \{2, 4, 6, 8\}$. List the elements of each set.*

**1.** $A \cup B$ **2.** $B \cap C$

**3.** $A - B$ **4.** $B - A$

**5.** $\overline{A}$ **6.** $U - C$

**7.** $\overline{U}$ **8.** $A \cup \varnothing$

**9.** $B \cap \varnothing$ **10.** $A \cup U$

**11.** $B \cap U$ **12.** $A \cap (B \cup C)$

**13.** $\overline{B} \cap (C - A)$

**14.** $\overline{(A \cap B)} - C$

**15.** $\overline{A \cap B} \cup C$

**16.** $(A \cup B) - (C - B)$

*In Exercises 17–27, let the universe be the set $\mathbf{Z}^+$. Let $X = \{1, 2, 3, 4, 5\}$ and let $Y$ be the set of positive, even integers. In set-builder notation, $Y = \{2n \mid n \in \mathbf{Z}^+\}$. In Exercises 18–27, give a mathematical notation for the set by listing the elements if the set is finite, by using set-builder notation if the set is infinite, or by using a predefined set such as $\varnothing$.*

**17.** Describe $\overline{Y}$ in words.

**18.** $\overline{X}$ **19.** $\overline{Y}$

**20.** $X \cap Y$ **21.** $X \cup Y$

**22.** $\overline{X} \cap Y$ **23.** $\overline{X} \cup Y$

**24.** $X \cap \overline{Y}$ **25.** $X \cup \overline{Y}$

**26.** $\overline{X} \cap \overline{Y}$ **27.** $\overline{X} \cup \overline{Y}$

**28.** What is the cardinality of $\varnothing$?

**29.** What is the cardinality of $\{\varnothing\}$?

**30.** What is the cardinality of $\{a, b, a, c\}$?

**31.** What is the cardinality of $\{\{a\}, \{a, b\}, \{a, c\}, a, b\}$?

*In Exercises 32–35, show, as in Examples 1.1.2 and 1.1.3, that $A = B$.*

**32.** $A = \{3, 2, 1\}$, $B = \{1, 2, 3\}$

**33.** $C = \{1, 2, 3\}$, $D = \{2, 3, 4\}$, $A = \{2, 3\}$, $B = C \cap D$

**34.** $A = \{1, 2, 3\}$, $B = \{n \mid n \in \mathbf{Z}^+ \text{ and } n^2 < 10\}$

**35.** $A = \{x \mid x^2 - 4x + 4 = 1\}$, $B = \{1, 3\}$

*In Exercises 36–39, show, as in Example 1.1.4, that $A \neq B$.*

**36.** $A = \{1, 2, 3\}$, $B = \varnothing$

**37.** $A = \{1, 2\}$, $B = \{x \mid x^3 - 2x^2 - x + 2 = 0\}$

**38.** $A = \{1, 3, 5\}$, $B = \{n \mid n \in \mathbf{Z}^+ \text{ and } n^2 - 1 \leq n\}$

**39.** $B = \{1, 2, 3, 4\}$, $C = \{2, 4, 6, 8\}$, $A = B \cap C$

*In Exercises 40–43, determine whether each pair of sets is equal.*

**40.** $\{1, 2, 2, 3\}, \{1, 2, 3\}$

**41.** $\{1, 1, 3\}, \{3, 3, 1\}$ **42.** $\{x \mid x^2 + x = 2\}, \{1, -1\}$

**43.** $\{x \mid x \in \mathbf{R} \text{ and } 0 < x \leq 2\}, \{1, 2\}$

*In Exercises 44–47, show, as in Examples 1.1.5 and 1.1.6, that $A \subseteq B$.*

**44.** $A = \{1, 2\}$, $B = \{3, 2, 1\}$

**45.** $A = \{1, 2\}$, $B = \{x \mid x^3 - 6x^2 + 11x = 6\}$

**46.** $A = \{1\} \times \{1, 2\}$, $B = \{1\} \times \{1, 2, 3\}$

**47.** $A = \{2n \mid n \in \mathbf{Z}^+\}$, $B = \{n \mid n \in \mathbf{Z}^+\}$

*In Exercises 48–51, show, as in Example 1.1.9, that $A$ is not a subset of $B$.*

**48.** $A = \{1, 2, 3\}$, $B = \{1, 2\}$

**49.** $A = \{x \mid x^3 - 2x^2 - x + 2 = 0\}$, $B = \{1, 2\}$

**50.** $A = \{1, 2, 3, 4\}$, $C = \{5, 6, 7, 8\}$, $B = \{n \mid n \in A \text{ and } n + m = 8 \text{ for some } m \in C\}$

**51.** $A = \{1, 2, 3\}$, $B = \varnothing$

*In Exercises 52–59, draw a Venn diagram and shade the given set.*

**52.** $A \cap \overline{B}$ **53.** $\overline{A} - B$

**54.** $B \cup (B - A)$ **55.** $(A \cup B) - B$

**56.** $B \cap \overline{(C \cup A)}$ **57.** $(\overline{A} \cup B) \cap (\overline{C} - A)$

**58.** $((C \cap A) - \overline{(B - A)}) \cap C$

**59.** $(B - \overline{C}) \cup ((B - \overline{A}) \cap (C \cup B))$

**60.** A television commercial for a popular beverage showed the following Venn diagram

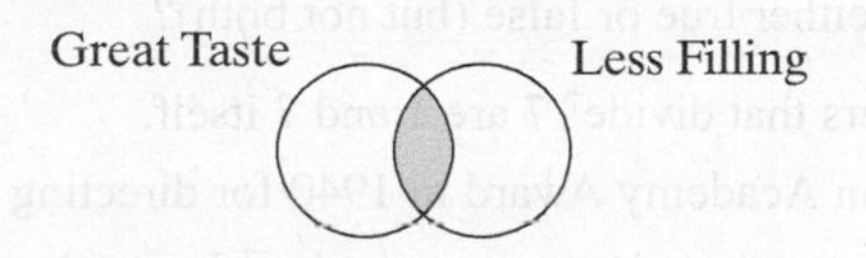


What does the shaded area represent?

*Exercises 61–65 refer to a group of 191 students, of which 10 are taking French, business, and music; 36 are taking French and business; 20 are taking French and music; 18 are taking business and music; 65 are taking French; 76 are taking business; and 63 are taking music.*

**61.** How many are taking French and music but not business?

**62.** How many are taking business and neither French nor music?

**63.** How many are taking French or business (or both)?

**64.** How many are taking music or French (or both) but not business?

**65.** How many are taking none of the three subjects?

**66.** A television poll of 151 persons found that 68 watched "Law and Disorder"; 61 watched "25"; 52 watched "The Tenors"; 16 watched both "Law and Disorder" and "25"; 25 watched both "Law and Disorder" and "The Tenors"; 19 watched both "25" and "The Tenors"; and 26 watched none of these shows. How many persons watched all three shows?

**67.** In a group of students, each student is taking a mathematics course or a computer science course or both. One-fifth of those taking a mathematics course are also taking a computer science course, and one-eighth of those taking a computer science course are also taking a mathematics course. Are more than one-third of the students taking a mathematics course?

*In Exercises 68–71, let $X = \{1, 2\}$ and $Y = \{a, b, c\}$. List the elements in each set.*

**68.** $X \times Y$

**69.** $Y \times X$

**70.** $X \times X$

**71.** $Y \times Y$

*In Exercises 72–75, let $X = \{1, 2\}$, $Y = \{a\}$, and $Z = \{\alpha, \beta\}$. List the elements of each set.*

**72.** $X \times Y \times Z$

**73.** $X \times Y \times Y$

**74.** $X \times X \times X$

**75.** $Y \times X \times Y \times Z$

*In Exercises 76–82, give a geometric description of each set in words. Consider the elements of the sets to be coordinates. For example, $\mathbf{R} \times \mathbf{Z}$ is the set $\{(x, n) \mid x \in \mathbf{R} \text{ and } n \in \mathbf{Z}\}$. Interpreting the ordered pairs $(x, n)$ as coordinates in the plane, the graph of all such ordered pairs is the set of all parallel horizontal lines spaced one unit apart, one of which passes through $(0,0)$.*

**76.** $\mathbf{R} \times \mathbf{R}$

**77.** $\mathbf{Z} \times \mathbf{R}$

**78.** $\mathbf{R} \times \mathbf{Z}^{nonneg}$

**79.** $\mathbf{Z} \times \mathbf{Z}$

**80.** $\mathbf{R} \times \mathbf{R} \times \mathbf{R}$

**81.** $\mathbf{R} \times \mathbf{R} \times \mathbf{Z}$

**82.** $\mathbf{R} \times \mathbf{Z} \times \mathbf{Z}$

*In Exercises 83–86, list all partitions of the set.*

**83.** $\{1\}$

**84.** $\{1, 2\}$

**85.** $\{a, b, c\}$

**86.** $\{a, b, c, d\}$

*In Exercises 87–92, answer true or false.*

**87.** $\{x\} \subseteq \{x\}$

**88.** $\{x\} \in \{x\}$

**89.** $\{x\} \in \{x, \{x\}\}$

**90.** $\{x\} \subseteq \{x, \{x\}\}$

**91.** $\{2\} \subseteq \mathcal{P}(\{1, 2\})$

**92.** $\{2\} \in \mathcal{P}(\{1, 2\})$

**93.** List the members of $\mathcal{P}(\{a, b\})$. Which are proper subsets of $\{a, b\}$?

**94.** List the members of $\mathcal{P}(\{a, b, c, d\})$. Which are proper subsets of $\{a, b, c, d\}$?

**95.** If $X$ has 10 members, how many members does $\mathcal{P}(X)$ have? How many proper subsets does $X$ have?

**96.** If $X$ has $n$ members, how many proper subsets does $X$ have?

*In Exercises 97–100, what relation must hold between sets A and B in order for the given condition to be true?*

**97.** $A \cap B = A$

**98.** $A \cup B = A$

**99.** $\overline{A} \cap U = \varnothing$

**100.** $\overline{A \cap B} = \overline{B}$

*The* symmetric difference *of two sets A and B is the set*

$$A \triangle B = (A \cup B) - (A \cap B).$$

**101.** If $A = \{1, 2, 3\}$ and $B = \{2, 3, 4, 5\}$, find $A \triangle B$.

**102.** Describe the symmetric difference of sets $A$ and $B$ in words.

**103.** Given a universe $U$, describe $A \triangle A$, $A \triangle \overline{A}$, $U \triangle A$, and $\varnothing \triangle A$.

**104.** Let $C$ be a circle and let $\mathcal{D}$ be the set of all diameters of $C$. What is $\cap \mathcal{D}$? (Here, by "diameter" we mean a line segment through the center of the circle with its endpoints on the circumference of the circle.)

[†]★**105.** Let $P$ denote the set of integers greater than 1. For $i \geq 2$, define

$$X_i = \{ik \mid k \in P\}.$$

Describe $P - \bigcup_{i=2}^{\infty} X_i$.

[†]A starred exercise indicates a problem of above-average difficulty.

## 1.2 Propositions

Which of sentences (a)–(f) are either true or false (but not both)?

(a) The only positive integers that divide† 7 are 1 and 7 itself.

(b) Alfred Hitchcock won an Academy Award in 1940 for directing *Rebecca*.

(c) For every positive integer $n$, there is a prime number‡ larger than $n$.

(d) Earth is the only planet in the universe that contains life.

(e) Buy two tickets to the "Unhinged Universe" rock concert for Friday.

(f) $x + 4 = 6$.

Sentence (a), which is another way to say that 7 is prime, is true.

Sentence (b) is false. Although *Rebecca* won the Academy Award for best picture in 1940, John Ford won the directing award for *The Grapes of Wrath*. It is a surprising fact that Alfred Hitchcock never won an Academy Award for directing.

Sentence (c), which is another way to say that the number of primes is infinite, is true.

Sentence (d) is either true or false (but not both), but no one knows which at this time.

Sentence (e) is neither true nor false [sentence (e) is a command].

The truth of equation (f) depends on the value of the variable $x$.

A sentence that is either true or false, but not both, is called a **proposition.** Sentences (a)–(d) are propositions, whereas sentences (e) and (f) are not propositions. A proposition is typically expressed as a declarative sentence (as opposed to a question, command, etc.). Propositions are the basic building blocks of any theory of logic.

We will use variables, such as $p$, $q$, and $r$, to represent propositions, much as we use letters in algebra to represent numbers. We will also use the notation

$$p\colon\ 1 + 1 = 3$$

to define $p$ to be the proposition $1 + 1 = 3$.

In ordinary speech and writing, we combine propositions using connectives such as *and* and *or*. For example, the propositions "It is raining" and "It is cold" can be combined to form the single proposition "It is raining and it is cold." The formal definitions of *and* and *or* follow.

**Definition 1.2.1 ▶** Let $p$ and $q$ be propositions.

The *conjunction* of $p$ and $q$, denoted $p \wedge q$, is the proposition

$$p \quad \text{and} \quad q.$$

The *disjunction* of $p$ and $q$, denoted $p \vee q$, is the proposition

$$p \quad \text{or} \quad q.$$

◀

**Example 1.2.2** If $p$: It is raining, and $q$: It is cold, then the conjunction of $p$ and $q$ is

$$p \wedge q\colon \text{ It is raining and it is cold.} \qquad (1.2.1)$$

The disjunction of $p$ and $q$ is

---

†"Divides" means "divides evenly." More formally, we say that a nonzero integer $d$ *divides* an integer $m$ if there is an integer $q$ such that $m = dq$. We call $q$ the *quotient*. We will explore the integers in detail in Chapter 5.

‡An integer $n > 1$ is *prime* if the only positive integers that divide $n$ are 1 and $n$ itself. For example, 2, 3, and 11 are prime numbers.

$$p \vee q\text{: It is raining or it is cold.}$$

◀

The truth value of the conjunction $p \wedge q$ is determined by the truth values of $p$ and $q$, and the definition is based upon the usual interpretation of "and." Consider the proposition (1.2.1) of Example 1.2.2. If it is raining (i.e., $p$ is true) and it is also cold (i.e., $q$ is also true), then we would consider the proposition (1.2.1) to be true. However, if it is not raining (i.e., $p$ is false) or it is not cold (i.e., $q$ is false) or both, then we would consider the proposition (1.2.1) to be false.

The truth values of propositions such as conjunctions and disjunctions can be described by **truth tables.** The truth table of a proposition $P$ made up of the individual propositions $p_1, \ldots, p_n$ lists all possible combinations of truth values for $p_1, \ldots, p_n$, T denoting true and F denoting false, and for each such combination lists the truth value of $P$. We use a truth table to formally define the truth value of $p \wedge q$.

A truth table for a proposition $P$ made up of $n$ propositions has $r = 2^n$ rows. Traditionally, for the first proposition, the first $r/2$ rows list T and the last $r/2$ rows list F. The next proposition has $r/4$ T's alternate with $r/4$ F's. The next proposition has $r/8$ T's alternate with $r/8$ F's, and so on. For example, a proposition $P$ made up of three propositions $p_1$, $p_2$, and $p_3$ has $8 = 2^3$ rows. Proposition $p_1$ will list $4 = 8/2$ T's followed by 4 F's. Proposition $p_2$ will list $2 = 8/4$ T's followed by 2 F's, followed by 2 T's, followed by 2 F's. Proposition $p_3$ will have one T, followed by one F, followed by one T, and so on. The truth table without the truth values of $P$ would be

| $p_1$ | $p_2$ | $p_3$ | $P$ |
|---|---|---|---|
| T | T | T | |
| T | T | F | |
| T | F | T | Here is where the truth values of $P$ go. |
| T | F | F | |
| F | T | T | |
| F | T | F | |
| F | F | T | |
| F | F | F | |

Notice that all possible combinations of truth values for $p_1, p_2, p_3$ are listed.

**Definition 1.2.3 ▶** The truth value of the proposition $p \wedge q$ is defined by the truth table

| $p$ | $q$ | $p \wedge q$ |
|---|---|---|
| T | T | T |
| T | F | F |
| F | T | F |
| F | F | F |

◀

Definition 1.2.3 states that the conjunction $p \wedge q$ is true provided that $p$ and $q$ are both true; $p \wedge q$ is false otherwise.

**Example 1.2.4** If $p$: A decade is 10 years, and $q$: A millennium is 100 years, then $p$ is true, $q$ is false (a millennium is 1000 years), and the conjunction,

$$p \wedge q\text{: A decade is 10 years and a millennium is 100 years,}$$

is false.

◀

**Example 1.2.5** Most programming languages define "and" exactly as in Definition 1.2.3. For example, in the Java programming language, (logical) "and" is denoted `&&`, and the expression

```
x < 10 && y > 4
```

is true precisely when the value of the variable `x` is less than 10 (i.e., `x < 10` is true) and the value of the variable `y` is greater than 4 (i.e., `y > 4` is also true). ◀

The truth value of the disjunction $p \vee q$ is also determined by the truth values of $p$ and $q$, and the definition is based upon the "inclusive" interpretation of "or." Consider the proposition,

$$p \vee q\text{: It is raining or it is cold,} \tag{1.2.2}$$

of Example 1.2.2. If it is raining (i.e., $p$ is true) or it is cold (i.e., $q$ is also true) *or both,* then we would consider the proposition (1.2.2) to be true (i.e., $p \vee q$ is true). If it is not raining (i.e., $p$ is false) and it is not cold (i.e., $q$ is also false), then we would consider the proposition (1.2.2) to be false (i.e., $p \vee q$ is false). The **inclusive-or** of propositions $p$ and $q$ is true if $p$ or $q$, *or both,* is true, and false otherwise. There is also an **exclusive-or** (see Exercise 67) that defines *p exor q* to be true if $p$ or $q$, *but not both,* is true, and false otherwise.

**Definition 1.2.6 ▶** The truth value of the proposition $p \vee q$, called the *inclusive-or* of $p$ and $q$, is defined by the truth table

| $p$ | $q$ | $p \vee q$ |
|---|---|---|
| T | T | T |
| T | F | T |
| F | T | T |
| F | F | F |

◀

**Example 1.2.7** If $p$: A millennium is 100 years, and $q$: A millennium is 1000 years, then $p$ is false, $q$ is true, and the disjunction,

$$p \vee q\text{: A millennium is 100 years or a millennium is 1000 years,}$$

is true. ◀

**Example 1.2.8** Most programming languages define (inclusive) "or" exactly as in Definition 1.2.6. For example, in the Java programming language, (logical) "or" is denoted `||`, and the expression

```
x < 10 || y > 4
```

is true precisely when the value of the variable `x` is less than 10 (i.e., `x < 10` is true) or the value of the variable `y` is greater than 4 (i.e., `y > 4` is true) or both. ◀

In ordinary language, propositions being combined (e.g., $p$ and $q$ combined to give the proposition $p \vee q$) are normally related; but in logic, these propositions are not required to refer to the same subject matter. For example, in logic, we permit propositions such as

$$3 < 5 \text{ or Paris is the capital of England.}$$

Logic is concerned with the form of propositions and the relation of propositions to each other and not with the subject matter itself. (The given proposition is true because $3 < 5$ is true.)

The final operator on a proposition $p$ that we discuss in this section is the **negation** of $p$.

**Definition 1.2.9 ▶** The *negation* of $p$, denoted $\neg p$, is the proposition

not $p$.

The truth value of the proposition $\neg p$ is defined by the truth table

| $p$ | $\neg p$ |
|---|---|
| T | F |
| F | T |

◀

In English, we sometimes write $\neg p$ as "It is not the case that $p$." For example, if

$p$: Paris is the capital of England,

the negation of $p$ could be written

$\neg p$: It is not the case that Paris is the capital of England,

or more simply as

$\neg p$: Paris is not the capital of England.

**Example 1.2.10** If

$p$: $\pi$ was calculated to 1,000,000 decimal digits in 1954,

the negation of $p$ is the proposition

$\neg p$: $\pi$ was not calculated to 1,000,000 decimal digits in 1954.

It was not until 1973 that 1,000,000 decimal digits of $\pi$ were computed; so, $p$ is false. (Since then over 12 trillion decimal digits of $\pi$ have been computed.) Since $p$ is false, $\neg p$ is true. ◀

**Example 1.2.11** Most programming languages define "not" exactly as in Definition 1.2.9. For example, in the Java programming language, "not" is denoted `!`, and the expression

```
!(x < 10)
```

is true precisely when the value of the variable x is not less than 10 (i.e., x is greater than or equal to 10). ◀

In expressions involving some or all of the operators $\neg$, $\wedge$, and $\vee$, in the absence of parentheses, we first evaluate $\neg$, then $\wedge$, and then $\vee$. We call such a convention **operator precedence.** In algebra, operator precedence tells us to evaluate $\cdot$ and $/$ before $+$ and $-$.

**Example 1.2.12** Given that proposition $p$ is false, proposition $q$ is true, and proposition $r$ is false, determine whether the proposition $\neg p \vee q \wedge r$ is true or false.

**SOLUTION** We first evaluate $\neg p$, which is true. We next evaluate $q \wedge r$, which is false. Finally, we evaluate $\neg p \vee q \wedge r$, which is true. ◀

**Example 1.2.13** **Searching the Web** A variety of Web search engines are available (e.g., Google, Yahoo!, Baidu) that allow the user to enter keywords that the search engine then tries to match with Web pages. For example, entering *mathematics* produces a (huge!) list of pages that contain the word "mathematics." Some search engines allow the user to use *and*, *or*, and *not* operators to combine keywords (see Figure 1.2.1), thus allowing more complex searches. In the Google search engine, *and* is the default operator so that, for example, entering *discrete mathematics* produces a list of pages containing both of the words "discrete" *and* "mathematics." The *or* operator is OR, and the *not* operator is the minus sign −. Furthermore, enclosing a phrase, typically with embedded spaces, in double quotation marks causes the phrase to be treated as a single word. For example, to search for pages containing the keywords

"*Shonda Rhimes*" and (*Grey's* or *Scandal*) and (not *Murder*),

the user could enter

"Shonda Rhimes" Grey's OR Scandal -Murder

◀

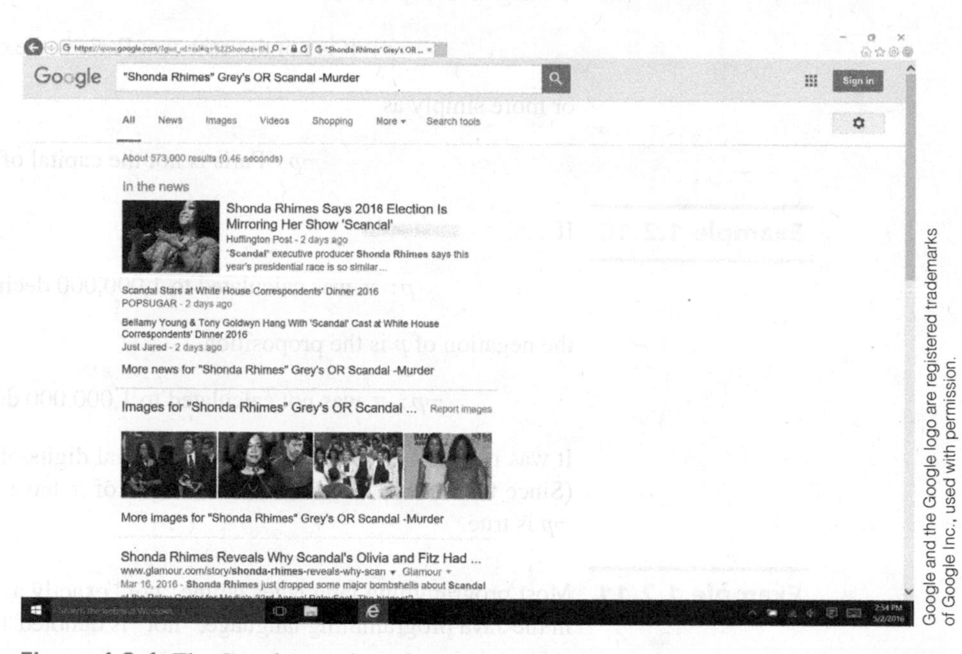




**Figure 1.2.1** The Google search engine, which allows the user to use *and* (space), *or* (OR), and *not* (−) operators to combine keywords. As shown, Google found about 573,000 Web pages containing "*Shonda Rhimes*" and (*Grey's* or *Scandal*) and (not *Murder*).

## 1.2 Problem-Solving Tips

Although there may be a shorter way to determine the truth values of a proposition $P$ formed by combining propositions $p_1, \ldots, p_n$ using operators such as $\neg$ and $\vee$, a truth table will always supply all possible truth values of $P$ for various truth values of the constituent propositions $p_1, \ldots, p_n$.

## 1.2 Review Exercises

1. What is a proposition?
2. What is a truth table?
3. What is the conjunction of $p$ and $q$? How is it denoted?
4. Give the truth table for the conjunction of $p$ and $q$.
5. What is the disjunction of $p$ and $q$? How is it denoted?
6. Give the truth table for the disjunction of $p$ and $q$.
7. What is the negation of $p$? How is it denoted?
8. Give the truth table for the negation of $p$.

## 1.2 Exercises

*Determine whether each sentence in Exercises 1–12 is a proposition. If the sentence is a proposition, write its negation. (You are* not *being asked for the truth values of the sentences that are propositions.)*

1. $2 + 5 = 19$.
2. $6 + 9 = 15$.
3. $x + 9 = 15$.
4. $\pi = 3.14$
5. Waiter, will you serve the nuts—I mean, would you serve the guests the nuts?
6. For some positive integer $n$, $19340 = n \cdot 17$.
7. Audrey Meadows was the original "Alice" in "The Honeymooners."
8. Peel me a grape.
9. The line "Play it again, Sam" occurs in the movie *Casablanca*.
10. Every even integer greater than 4 is the sum of two primes.
11. The difference of two primes.

★12. This statement is false.

*Exercises 13–16 refer to a coin that is flipped 10 times. Write the negation of the proposition.*

13. Ten heads were obtained.
14. Some heads were obtained.
15. Some heads and some tails were obtained.
16. At least one head was obtained.

*Given that proposition p is false, proposition q is true, and proposition r is false, determine whether each proposition in Exercises 17–22 is true or false.*

17. $p \vee q$
18. $\neg p \vee \neg q$
19. $\neg p \vee q$
20. $\neg p \vee \neg(q \wedge r)$
21. $\neg(p \vee q) \wedge (\neg p \vee r)$
22. $(p \vee \neg r) \wedge \neg((q \vee r) \vee \neg(r \vee p))$

*Write the truth table of each proposition in Exercises 23–30.*

23. $p \wedge \neg q$
24. $(\neg p \vee \neg q) \vee p$
25. $(p \vee q) \wedge \neg p$
26. $(p \wedge q) \wedge \neg p$
27. $(p \wedge q) \vee (\neg p \vee q)$
28. $\neg(p \wedge q) \vee (r \wedge \neg p)$
29. $(p \vee q) \wedge (\neg p \vee q) \wedge (p \vee \neg q) \wedge (\neg p \vee \neg q)$
30. $\neg(p \wedge q) \vee (\neg q \vee r)$

*In Exercises 31–33, represent the given proposition symbolically by letting*

$$p\colon\ 5 < 9, \quad q\colon\ 9 < 7, \quad r\colon\ 5 < 7.$$

*Determine whether each proposition is true or false.*

31. $5 < 9$ and $9 < 7$.
32. It is not the case that ($5 < 9$ and $9 < 7$).
33. $5 < 9$ or it is not the case that ($9 < 7$ and $5 < 7$).

*In Exercises 34–39, formulate the symbolic expression in words using*

*p : Lee takes computer science.*
*q : Lee takes mathematics.*

34. $\neg p$
35. $p \wedge q$
36. $p \vee q$
37. $p \vee \neg q$
38. $p \wedge \neg q$
39. $\neg p \wedge \neg q$

*In Exercises 40–44, formulate the symbolic expression in words using*

*p : You play football.*
*q : You miss the midterm exam.*
*r : You pass the course.*

40. $p \wedge q$
41. $\neg q \wedge r$
42. $p \vee q \vee r$
43. $\neg(p \vee q) \vee r$
44. $(p \wedge q) \vee (\neg q \wedge r)$

*In Exercises 45–49, formulate the symbolic expression in words using*

*p : Today is Monday.*
*q : It is raining.*
*r : It is hot.*

45. $p \vee q$
46. $\neg p \wedge (q \vee r)$
47. $\neg(p \vee q) \wedge r$
48. $(p \wedge q) \wedge \neg(r \vee p)$
49. $(p \wedge (q \vee r)) \wedge (r \vee (q \vee p))$

*In Exercises 50–55, represent the proposition symbolically by letting*

*p : There is a hurricane.*
*q : It is raining.*

50. There is no hurricane.

51. There is a hurricane and it is raining.

52. There is a hurricane, but it is not raining.

53. There is no hurricane and it is not raining.

54. Either there is a hurricane or it is raining (or both).

55. Either there is a hurricane or it is raining, but there is no hurricane.

*In Exercises 56–60, represent the proposition symbolically by letting*

$p$ : *You run 10 laps daily.*
$q$ : *You are healthy.*
$r$ : *You take multi-vitamins.*

56. You run 10 laps daily, but you are not healthy.

57. You run 10 laps daily, you take multi-vitamins, and you are healthy.

58. You run 10 laps daily or you take multi-vitamins, and you are healthy.

59. You do not run 10 laps daily, you do not take multi-vitamins, and you are not healthy.

60. Either you are healthy or you do not run 10 laps daily, and you do not take multi-vitamins.

*In Exercises 61–66, represent the proposition symbolically by letting*

$p$ : *You heard the* "Flying Pigs" *rock concert.*
$q$ : *You heard the* "Y2K" *rock concert.*
$r$ : *You have sore eardrums.*

61. You heard the "Flying Pigs" rock concert, and you have sore eardrums.

62. You heard the "Flying Pigs" rock concert, but you do not have sore eardrums.

63. You heard the "Flying Pigs" rock concert, you heard the "Y2K" rock concert, and you have sore eardrums.

64. You heard either the "Flying Pigs" rock concert or the "Y2K" rock concert, but you do not have sore eardrums.

65. You did not hear the "Flying Pigs" rock concert and you did not hear the "Y2K" rock concert, but you have sore eardrums.

66. It is not the case that: You heard the "Flying Pigs" rock concert or you heard the "Y2K" rock concert or you do not have sore eardrums.

67. Give the truth table for the exclusive-or of $p$ and $q$ in which $p$ *exor* $q$ is true if either $p$ or $q$, but not both, is true.

*In Exercises 68–74, state the meaning of each sentence if "or" is interpreted as the inclusive-or; then, state the meaning of each sentence if "or" is interpreted as the exclusive-or (see Exercise 67). In each case, which meaning do you think is intended?*

68. To enter Utopia, you must show a driver's license or a passport.

69. To enter Utopia, you must possess a driver's license or a passport.

70. The prerequisite to data structures is a course in Java or C++.

71. The car comes with a cupholder that heats or cools your drink.

72. We offer $1000 cash or 0 percent interest for two years.

73. Do you want fries or a salad with your burger?

74. The meeting will be canceled if fewer than 10 persons sign up or at least 3 inches of snow falls.

75. At one time, the following ordinance was in effect in Naperville, Illinois: "It shall be unlawful for any person to keep more than three [3] dogs and three [3] cats upon his property within the city." Was Charles Marko, who owned five dogs and no cats, in violation of the ordinance? Explain.

76. Write a command to search the Web for national parks in North or South Dakota.

77. Write a command to search the Web for information on lung disease other than cancer.

78. Write a command to search the Web for minor league baseball teams in Illinois that are not in the Midwest League.

## 1.3 Conditional Propositions and Logical Equivalence

The dean has announced that

If the Mathematics Department gets an additional $60,000,
then it will hire one new faculty member. **(1.3.1)**

Statement (1.3.1) states that on the condition that the Mathematics Department gets an additional $60,000, then the Mathematics Department will hire one new faculty member. A proposition such as (1.3.1) is called a **conditional proposition.**

**Definition 1.3.1 ▶** If $p$ and $q$ are propositions, the proposition

if $p$ then $q$ **(1.3.2)**

is called a *conditional proposition* and is denoted

$$p \to q.$$

The proposition $p$ is called the *hypothesis* (or *antecedent*), and the proposition $q$ is called the *conclusion* (or *consequent*). ◀

**Example 1.3.2** If we define

$p$: The Mathematics Department gets an additional \$60,000,
$q$: The Mathematics Department will hire one new faculty member,

then proposition (1.3.1) assumes the form (1.3.2). The hypothesis is the statement "The Mathematics Department gets an additional \$60,000," and the conclusion is the statement "The Mathematics Department will hire one new faculty member." ◀

What is the truth value of the dean's statement (1.3.1)? First, suppose that the Mathematics Department gets an additional \$60,000. If the Mathematics Department does hire an additional faculty member, surely the dean's statement is true. (Using the notation of Example 1.3.2, if $p$ and $q$ are both true, then $p \to q$ is true.) On the other hand, if the Mathematics Department gets an additional \$60,000 and does *not* hire an additional faculty member, the dean is wrong—statement (1.3.1) is false. (If $p$ is true and $q$ is false, then $p \to q$ is false.) Now, suppose that the Mathematics Department does *not* get an additional \$60,000. In this case, the Mathematics Department might or might not hire an additional faculty member. (Perhaps a member of the department retires and someone is hired to replace the retiree. On the other hand, the department might not hire anyone.) Surely we would not consider the dean's statement to be false. Thus, if the Mathematics Department does *not* get an additional \$60,000, the dean's statement must be true, regardless of whether the department hires an additional faculty member or not. (If $p$ is false, then $p \to q$ is true whether $q$ is true or false.) This discussion motivates the following definition.

**Definition 1.3.3 ▶** The truth value of the conditional proposition $p \to q$ is defined by the following truth table:

| $p$ | $q$ | $p \to q$ |
|---|---|---|
| T | T | T |
| T | F | F |
| F | T | T |
| F | F | T |

◀

For those who need additional evidence that we should define $p \to q$ to be true when $p$ is false, we offer further justification. Most people would agree that the proposition,

$$\text{For all real numbers } x, \text{ if } x > 0, \text{ then } x^2 > 0, \tag{1.3.3}$$

is true. (In Section 1.5, we will discuss such "for all" statements formally and in detail.) In the following discussion, we let $P(x)$ denote $x > 0$ and $Q(x)$ denote $x^2 > 0$. That proposition (1.3.3) is true means that no matter which real number we replace $x$ with, the proposition

$$\text{if } P(x) \text{ then } Q(x) \tag{1.3.4}$$

that results is true. For example, if $x = 3$, then $P(3)$ and $Q(3)$ are both true ($3 > 0$ and $3^2 > 0$ are both true), and, by Definition 1.3.3, (1.3.4) is true. Now let us consider the situation when $P(x)$ is false. If $x = -2$, then $P(-2)$ is false ($-2 > 0$ is false) and $Q(-2)$ is true [$(-2)^2 > 0$ is true]. In order for proposition (1.3.4) to be true in this case, we must define $p \to q$ to be true when $p$ is false and $q$ is true. This is exactly what occurs in the third line of the truth table of Definition 1.3.3. If $x = 0$, then $P(0)$ and $Q(0)$ are both false ($0 > 0$ and $0^2 > 0$ are both false). In order for proposition (1.3.4) to be true in this case, we must define $p \to q$ to be true when both $p$ and $q$ are false. This is exactly what occurs in the fourth line of the truth table of Definition 1.3.3. Even more motivation for defining $p \to q$ to be true when $p$ is false is given in Exercises 77 and 78.

**Example 1.3.4** Let $p : 1 > 2$ and $q : 4 < 8$. Then $p$ is false and $q$ is true. Therefore, $p \to q$ is true and $q \to p$ is false. ◀

In expressions that involve the logical operators $\wedge$, $\vee$, $\neg$, and $\to$, the conditional operator $\to$ is evaluated last. For example,

$$p \vee q \to \neg r$$

is interpreted as

$$(p \vee q) \to (\neg r).$$

**Example 1.3.5** Assuming that $p$ is true, $q$ is false, and $r$ is true, find the truth value of each proposition.

(a) $p \wedge q \to r$

(b) $p \vee q \to \neg r$

(c) $p \wedge (q \to r)$

(d) $p \to (q \to r)$

**SOLUTION**

(a) We first evaluate $p \wedge q$ because $\to$ is evaluated last. Since $p$ is true and $q$ is false, $p \wedge q$ is false. Therefore, $p \wedge q \to r$ is true (regardless of whether $r$ is true or false).

(b) We first evaluate $\neg r$. Since $r$ is true, $\neg r$ is false. We next evaluate $p \vee q$. Since $p$ is true and $q$ is false, $p \vee q$ is true. Therefore, $p \vee q \to \neg r$ is false.

(c) Since $q$ is false, $q \to r$ is true (regardless of whether $r$ is true or false). Since $p$ is true, $p \wedge (q \to r)$ is true.

(d) Since $q$ is false, $q \to r$ is true (regardless of whether $r$ is true or false). Thus, $p \to (q \to r)$ is true (regardless of whether $p$ is true or false). ◀

A conditional proposition that is true because the hypothesis is false is said to be **true by default** or **vacuously true.** For example, if the proposition,

If the Mathematics Department gets an additional \$60,000, then it will hire one new faculty member,

is true because the Mathematics Department did not get an additional \$60,000, we would say that the proposition is true by default or that it is vacuously true.

Some statements not of the form (1.3.2) may be rephrased as conditional propositions, as the next example illustrates.

**Example 1.3.6** Restate each proposition in the form (1.3.2) of a conditional proposition.

(a) Mary will be a good student if she studies hard.

(b) John takes calculus only if he has sophomore, junior, or senior standing.

(c) When you sing, my ears hurt.

(d) A necessary condition for the Cubs to win the World Series is that they sign a right-handed relief pitcher.

(e) A sufficient condition for Maria to visit France is that she goes to the Eiffel Tower.

**SOLUTION**

(a) The hypothesis is the clause following *if*; thus an equivalent formulation is

If Mary studies hard, then she will be a good student.

(b) The statement means that in order for John to take calculus, he must have sophomore, junior, or senior standing. In particular, if he is a freshman, he may *not* take calculus. Thus, we can conclude that if he takes calculus, then he has sophomore, junior, or senior standing. Therefore an equivalent formulation is

If John takes calculus, then he has sophomore, junior, or senior standing.

Notice that

If John has sophomore, junior, or senior standing, then he takes calculus,

is *not* an equivalent formulation. If John has sophomore, junior, or senior standing, he may or may *not* take calculus. (Although eligible to take calculus, he may have decided not to.)

The "if $p$ then $q$" formulation emphasizes the hypothesis, whereas the "$p$ only if $q$" formulation emphasizes the conclusion; the difference is only stylistic.

(c) *When* means the same as *if*; thus an equivalent formulation is

If you sing, then my ears hurt.

(d) A **necessary condition** is just that: a condition that is *necessary* for a particular outcome to be achieved. The condition does *not* guarantee the outcome; but, if the condition does not hold, the outcome will not be achieved. Here, the given statement means that if the Cubs win the World Series, we can be sure that they signed a right-handed relief pitcher since, without such a signing, they would not have won the World Series. Thus, an equivalent formulation of the given statement is

If the Cubs win the World Series, then they signed a right-handed relief pitcher.

The conclusion expresses a necessary condition.

Notice that

If the Cubs sign a right-handed relief pitcher, then they win the World Series,

is *not* an equivalent formulation. Signing a right-handed relief pitcher does not guarantee a World Series win. However, *not* signing a right-handed relief pitcher guarantees that they will not win the World Series.

(e) Similarly, a **sufficient condition** is a condition that *suffices* to guarantee a particular outcome. If the condition does not hold, the outcome might be achieved in other ways or it might not be achieved at all; but if the condition does hold, the outcome is guaranteed. Here, to be sure that Maria visits France, it suffices for her to go to the Eiffel Tower. (There are surely other ways to ensure that Maria visits France; for example, she could go to Lyon.) Thus, an equivalent formulation of the given statement is

If Maria goes to the Eiffel Tower, then she visits France.

The hypothesis expresses a sufficient condition.

Notice that

If Maria visits France, then she goes to the Eiffel Tower,

is *not* an equivalent formulation. As we have already noted, there are ways other than going to the Eiffel Tower to ensure that Maria visits France. ◀

Example 1.3.4 shows that the proposition $p \to q$ can be true while the proposition $q \to p$ is false. We call the proposition $q \to p$ the **converse** of the proposition $p \to q$. Thus a conditional proposition can be true while its converse is false.

**Example 1.3.7** Write the conditional proposition,

If Jerry receives a scholarship, then he will go to college,

and its converse symbolically and in words. Also, assuming that Jerry does not receive a scholarship, but wins the lottery and goes to college anyway, find the truth value of the original proposition and its converse.

**SOLUTION** Let $p$: Jerry receives a scholarship, and $q$: Jerry goes to college. The given proposition can be written symbolically as $p \to q$. Since the hypothesis $p$ is false, the conditional proposition is true.

The converse of the proposition is

If Jerry goes to college, then he receives a scholarship.

The converse can be written symbolically as $q \to p$. Since the hypothesis $q$ is true and the conclusion $p$ is false, the converse is false. ◀

Another useful proposition is

$p$ if and only if $q$,

which is considered to be true precisely when $p$ and $q$ have the same truth values (i.e., $p$ and $q$ are both true or $p$ and $q$ are both false).

**Definition 1.3.8** ▶ If $p$ and $q$ are propositions, the proposition

$p$ if and only if $q$

is called a *biconditional proposition* and is denoted

$$p \leftrightarrow q.$$

The truth value of the proposition $p \leftrightarrow q$ is defined by the following truth table:

| $p$ | $q$ | $p \leftrightarrow q$ |
|---|---|---|
| T | T | T |
| T | F | F |
| F | T | F |
| F | F | T |

◀

It is traditional in mathematical definitions to use "if" to mean "if and only if." Consider, for example, the definition of set equality: If sets $X$ and $Y$ have the same elements, then $X$ and $Y$ are equal. The meaning of this definition is that sets $X$ and $Y$ have the same elements *if and only if* $X$ and $Y$ are equal.

An alternative way to state "$p$ if and only if $q$" is "$p$ is a necessary and sufficient condition for $q$." The proposition "$p$ if and only if $q$" is sometimes written "$p$ iff $q$."

**Example 1.3.9** The proposition

$$1 < 5 \text{ if and only if } 2 < 8 \tag{1.3.5}$$

can be written symbolically as $p \leftrightarrow q$ if we define $p : 1 < 5$ and $q : 2 < 8$. Since both $p$ and $q$ are true, the proposition $p \leftrightarrow q$ is true. ◀

An alternative way to state (1.3.5) is: A necessary and sufficient condition for $1 < 5$ is that $2 < 8$.

In some cases, two different propositions have the same truth values no matter what truth values their constituent propositions have. Such propositions are said to be **logically equivalent.**

**Definition 1.3.10** ▶ Suppose that the propositions $P$ and $Q$ are made up of the propositions $p_1, \ldots, p_n$. We say that $P$ and $Q$ are *logically equivalent* and write

$$P \equiv Q,$$

provided that, given any truth values of $p_1, \ldots, p_n$, either $P$ and $Q$ are both true, or $P$ and $Q$ are both false. ◀

**Example 1.3.11** **De Morgan's Laws for Logic** We will verify the first of **De Morgan's laws**

$$\neg(p \vee q) \equiv \neg p \wedge \neg q, \quad \neg(p \wedge q) \equiv \neg p \vee \neg q,$$

and leave the second as an exercise (see Exercise 79).

By writing the truth tables for $P = \neg(p \vee q)$ and $Q = \neg p \wedge \neg q$, we can verify that, given any truth values of $p$ and $q$, either $P$ and $Q$ are both true or $P$ and $Q$ are both false:

| $p$ | $q$ | $\neg(p \vee q)$ | $\neg p \wedge \neg q$ |
|---|---|---|---|
| T | T | F | F |
| T | F | F | F |
| F | T | F | F |
| F | F | T | T |

Thus $P$ and $Q$ are logically equivalent. ◀

**Example 1.3.12** Show that, in Java, the expressions

```
x < 10 || x > 20
```

and

```
!(x >= 10 && x <= 20)
```

are equivalent. (In Java, >= means $\geq$, and <= means $\leq$.)

**SOLUTION** If we let $p$ denote the expression `x >= 10` and $q$ denote the expression `x <= 20`, the expression `(x >= 10 && x <= 20)` becomes $\neg(p \wedge q)$. By De Morgan's second law, $\neg(p \wedge q)$ is equivalent to $\neg p \vee \neg q$. Since $\neg p$ translates as `x < 10` and $\neg q$ translates as `x > 20`, $\neg p \vee \neg q$ translates as `x < 10 || x > 20`. Therefore, the expressions `x < 10 || x > 20` and `!(x >= 10 && x <= 20)` are equivalent. ◀

Our next example gives a logically equivalent form of the negation of $p \rightarrow q$.

**Example 1.3.13** Show that the negation of $p \rightarrow q$ is logically equivalent to $p \wedge \neg q$.

**SOLUTION** By writing the truth tables for $P = \neg(p \rightarrow q)$ and $Q = p \wedge \neg q$, we can verify that, given any truth values of $p$ and $q$, either $P$ and $Q$ are both true or $P$ and $Q$ are both false:

| $p$ | $q$ | $\neg(p \rightarrow q)$ | $p \wedge \neg q$ |
|---|---|---|---|
| T | T | F | F |
| T | F | T | T |
| F | T | F | F |
| F | F | F | F |

Thus $P$ and $Q$ are logically equivalent. ◀

**Example 1.3.14** Use the logical equivalence of $\neg(p \rightarrow q)$ and $p \wedge \neg q$ (see Example 1.3.13) to write the negation of

If Jerry receives a scholarship, then he goes to college,

symbolically and in words.

**SOLUTION** We let $p$: Jerry receives a scholarship, and $q$: Jerry goes to college. The given proposition can be written symbolically as $p \rightarrow q$. Its negation is logically equivalent to $p \wedge \neg q$. In words, this last expression is

Jerry receives a scholarship and he does not go to college. ◀

We now show that, according to our definitions, $p \leftrightarrow q$ is logically equivalent to $p \rightarrow q$ *and* $q \rightarrow p$. In words,

$p$ if and only if $q$

is logically equivalent to

if $p$ then $q$ and if $q$ then $p$.

**Example 1.3.15** The truth table shows that

$$p \leftrightarrow q \equiv (p \to q) \wedge (q \to p).$$

| $p$ | $q$ | $p \leftrightarrow q$ | $p \to q$ | $q \to p$ | $(p \to q) \wedge (q \to p)$ |
|---|---|---|---|---|---|
| T | T | T | T | T | T |
| T | F | F | F | T | F |
| F | T | F | T | F | F |
| F | F | T | T | T | T |

◀

Consider again the definition of set equality: If sets $X$ and $Y$ have the same elements, then $X$ and $Y$ are equal. We noted that the meaning of this definition is that sets $X$ and $Y$ have the same elements if and only if $X$ and $Y$ are equal. Example 1.3.15 shows that an equivalent formulation is: If sets $X$ and $Y$ have the same elements, then $X$ and $Y$ are equal, and if $X$ and $Y$ are equal, then $X$ and $Y$ have the same elements.

We conclude this section by defining the **contrapositive** of a conditional proposition. We will see (in Theorem 1.3.18) that the contrapositive is an alternative, logically equivalent form of the conditional proposition. Exercise 80 gives another logically equivalent form of the conditional proposition.

**Definition 1.3.16 ▶** The *contrapositive* (or *transposition*) of the conditional proposition $p \to q$ is the proposition $\neg q \to \neg p$. ◀

Notice the difference between the contrapositive and the converse. The converse of a conditional proposition merely reverses the roles of $p$ and $q$, whereas the contrapositive reverses the roles of $p$ and $q$ *and* negates each of them.

**Example 1.3.17** Write the conditional proposition,

If the network is down, then Dale cannot access the internet,

symbolically. Write the contrapositive and the converse symbolically and in words. Also, assuming that the network is not down and Dale can access the internet, find the truth value of the original proposition, its contrapositive, and its converse.

**SOLUTION** Let $p$: The network is down, and $q$: Dale cannot access the internet. The given proposition can be written symbolically as $p \to q$. Since the hypothesis $p$ is false, the conditional proposition is true.

The contrapositive can be written symbolically as $\neg q \to \neg p$ and, in words,

If Dale can access the internet, then the network is not down.

Since the hypothesis $\neg q$ and conclusion $\neg p$ are both true, the contrapositive is true. (Theorem 1.3.18 will show that the conditional proposition and its contrapositive are logically equivalent, that is, that they always have the same truth value.)

The converse of the given proposition can be written symbolically as $q \to p$ and, in words,

If Dale cannot access the internet, then the network is down.

Since the hypothesis $q$ is false, the converse is true. ◀

An important fact is that a conditional proposition and its contrapositive are logically equivalent.

**Theorem 1.3.18** The conditional proposition $p \to q$ and its contrapositive $\neg q \to \neg p$ are logically equivalent.

**Proof** The truth table

| $p$ | $q$ | $p \to q$ | $\neg q \to \neg p$ |
|---|---|---|---|
| T | T | T | T |
| T | F | F | F |
| F | T | T | T |
| F | F | T | T |

shows that $p \to q$ and $\neg q \to \neg p$ are logically equivalent. ◀

In ordinary language, "if" is often used to mean "if and only if." Consider the statement

If you fix my computer, then I'll pay you \$50.

The intended meaning is

If you fix my computer, then I'll pay you \$50, and
if you do not fix my computer, then I will not pay you \$50,

which is logically equivalent to (see Theorem 1.3.18)

If you fix my computer, then I'll pay you \$50, and
if I pay you \$50, then you fix my computer,

which, in turn, is logically equivalent to (see Example 1.3.15)

You fix my computer if and only if I pay you \$50.

In ordinary discourse, the intended meaning of statements involving logical operators can often (but, not always!) be inferred. However, in mathematics and science, precision is required. Only by carefully defining what we mean by terms such as "if" and "if and only if" can we obtain unambiguous and precise statements. In particular, logic carefully distinguishes among conditional, biconditional, converse, and contrapositive propositions.

## 1.3 Problem-Solving Tips

- In formal logic, "if" and "if and only if" are quite different. The conditional proposition $p \to q$ (if $p$ then $q$) is true except when $p$ is true and $q$ is false. On the other hand, the biconditional proposition $p \leftrightarrow q$ ($p$ if and only if $q$) is true precisely when $p$ and $q$ are both true or both false.
- To determine whether propositions $P$ and $Q$, made up of the propositions $p_1, \ldots,$ $p_n$, are logically equivalent, write the truth tables for $P$ and $Q$. If all of the entries for $P$ and $Q$ are always both true or both false, then $P$ and $Q$ *are* equivalent. If some entry is true for one of $P$ or $Q$ and false for the other, then $P$ and $Q$ are *not* equivalent.
- De Morgan's laws for logic

$$\neg(p \vee q) \equiv \neg p \wedge \neg q, \qquad \neg(p \wedge q) \equiv \neg p \vee \neg q$$

give formulas for negating "or" ($\vee$) and negating "and" ($\wedge$). Roughly speaking, negating "or" results in "and," and negating "and" results in "or."

■ Example 1.3.13 states a very important equivalence

$$\neg(p \to q) \equiv p \wedge \neg q,$$

which we will meet throughout this book. This equivalence shows that the negation of the conditional proposition can be written using the "and" ($\wedge$) operator. Notice that there is no conditional operator on the right-hand side of the equation.

## 1.3 Review Exercises

1. What is a conditional proposition? How is it denoted?
2. Give the truth table for the conditional proposition.
3. In a conditional proposition, what is the hypothesis?
4. In a conditional proposition, what is the conclusion?
5. What is a necessary condition?
6. What is a sufficient condition?
7. What is the converse of $p \to q$?
8. What is a biconditional proposition? How is it denoted?
9. Give the truth table for the biconditional proposition.
10. What does it mean for $P$ to be logically equivalent to $Q$?
11. State De Morgan's laws for logic.
12. What is the contrapositive of $p \to q$?

## 1.3 Exercises

*In Exercises 1–11, restate each proposition in the form (1.3.2) of a conditional proposition.*

1. Joey will pass the discrete mathematics exam if he studies hard.
2. Rosa may graduate if she has 160 quarter-hours of credits.
3. A necessary condition for Fernando to buy a computer is that he obtain $2000.
4. A sufficient condition for Katrina to take the algorithms course is that she pass discrete mathematics.
5. Getting that job requires knowing someone who knows the boss.
6. You can go to the Super Bowl unless you can't afford the ticket.
7. You may inspect the aircraft only if you have the proper security clearance.
8. When better cars are built, Buick will build them.
9. The audience will go to sleep if the chairperson gives the lecture.
10. The program is readable only if it is well structured.
11. A necessary condition for the switch to not be turned properly is that the light is not on.
12. Write the converse of each proposition in Exercises 1–11.
13. Write the contrapositive of each proposition in Exercises 1–11.

*Assuming that p and r are false and that q and s are true, find the truth value of each proposition in Exercises 14–22.*

14. $p \to q$
15. $\neg p \to \neg q$
16. $\neg(p \to q)$
17. $(p \to q) \wedge (q \to r)$
18. $(p \to q) \to r$
19. $p \to (q \to r)$
20. $(s \to (p \wedge \neg r)) \wedge ((p \to (r \vee q)) \wedge s)$
21. $((p \wedge \neg q) \to (q \wedge r)) \to (s \vee \neg q)$
22. $((p \vee q) \wedge (q \vee s)) \to ((\neg r \vee p) \wedge (q \vee s))$

*Exercises 23–32 refer to the propositions p, q, and r; p is true, q is false, and r's status is unknown at this time. Tell whether each proposition is true, is false, or has unknown status at this time.*

23. $p \vee r$
24. $p \wedge r$
25. $p \to r$
26. $q \to r$
27. $r \to p$
28. $r \to q$
29. $(p \wedge r) \leftrightarrow r$
30. $(p \vee r) \leftrightarrow r$
31. $(q \wedge r) \leftrightarrow r$
32. $(q \vee r) \leftrightarrow r$

*Determine the truth value of each proposition in Exercises 33–42.*

33. If $3 + 5 < 2$, then $1 + 3 = 4$.
34. If $3 + 5 < 2$, then $1 + 3 \neq 4$.
35. If $3 + 5 > 2$, then $1 + 3 = 4$.
36. If $3 + 5 > 2$, then $1 + 3 \neq 4$.
37. $3 + 5 > 2$ if and only if $1 + 3 = 4$.
38. $3 + 5 < 2$ if and only if $1 + 3 = 4$.
39. $3 + 5 < 2$ if and only if $1 + 3 \neq 4$.
40. If the earth has six moons, then $1 < 3$.
41. If $1 < 3$, then the earth has six moons.
42. If $\pi < 3.14$, then $\pi^2 < 9.85$.

*In Exercises 43–46, represent the given proposition symbolically by letting*

$$p:\ 4<2,\quad q:\ 7<10,\quad r:\ 6<6.$$

**43.** If $4 < 2$, then $7 < 10$.

**44.** If ($4 < 2$ and $6 < 6$), then $7 < 10$.

**45.** If it is not the case that ($6 < 6$ and 7 is not less than 10), then $6 < 6$.

**46.** $7 < 10$ if and only if ($4 < 2$ and 6 is not less than 6).

*In Exercises 47–52, represent the given proposition symbolically by letting*

$p$ : *You run 10 laps daily.*
$q$ : *You are healthy.*
$r$ : *You take multi-vitamins.*

**47.** If you run 10 laps daily, then you will be healthy.

**48.** If you do not run 10 laps daily or do not take multi-vitamins, then you will not be healthy.

**49.** Taking multi-vitamins is sufficient for being healthy.

**50.** You will be healthy if and only if you run 10 laps daily and take multi-vitamins.

**51.** If you are healthy, then you run 10 laps daily or you take multi-vitamins.

**52.** If you are healthy and run 10 laps daily, then you do not take multi-vitamins.

*In Exercises 53–58, formulate the symbolic expression in words using*

$p$: *Today is Monday,*
$q$: *It is raining,*
$r$: *It is hot.*

**53.** $p \to q$

**54.** $\neg q \to (r \wedge p)$

**55.** $\neg p \to (q \vee r)$

**56.** $\neg(p \vee q) \leftrightarrow r$

**57.** $(p \wedge (q \vee r)) \to (r \vee (q \vee p))$

**58.** $(p \vee (\neg p \wedge \neg(q \vee r))) \to (p \vee \neg(r \vee q))$

*In Exercises 59–62, write each conditional proposition symbolically. Write the converse and contrapositive of each proposition symbolically and in words. Also, find the truth value of each conditional proposition, its converse, and its contrapositive.*

**59.** If $4 < 6$, then $9 > 12$.

**60.** If $4 > 6$, then $9 > 12$.

**61.** $|1| < 3$ if $-3 < 1 < 3$.

**62.** $|4| < 3$ if $-3 < 4 < 3$.

*For each pair of propositions P and Q in Exercises 63–72, state whether or not $P \equiv Q$.*

**63.** $P = p,\ Q = p \vee q$

**64.** $P = p \wedge q,\ Q = \neg p \vee \neg q$

**65.** $P = p \to q,\ Q = \neg p \vee q$

**66.** $P = p \wedge (\neg q \vee r),\ Q = p \vee (q \wedge \neg r)$

**67.** $P = p \wedge (q \vee r),\ Q = (p \vee q) \wedge (p \vee r)$

**68.** $P = p \to q,\ Q = \neg q \to \neg p$

**69.** $P = p \to q,\ Q = p \leftrightarrow q$

**70.** $P = (p \to q) \wedge (q \to r),\ Q = p \to r$

**71.** $P = (p \to q) \to r,\ Q = p \to (q \to r)$

**72.** $P = (s \to (p \wedge \neg r)) \wedge ((p \to (r \vee q)) \wedge s),\ Q = p \vee t$

*Using De Morgan's laws for logic, write the negation of each proposition in Exercises 73–76.*

**73.** Pat will use the treadmill or lift weights.

**74.** Dale is smart and funny.

**75.** Shirley will either take the bus or catch a ride to school.

**76.** Red pepper and onions are required to make chili.

*Exercises 77 and 78 provide further motivation for defining $p \to q$ to be true when p is false. We consider changing the truth table for $p \to q$ when p is false. For the first change, we call the resulting operator impl (Exercise 77), and, for the second change, we call the resulting operator imp2 (Exercise 78). In both cases, we see that pathologies result.*

**77.** Define the truth table for *imp1* by

| $p$ | $q$ | $p$ *imp1* $q$ |
|---|---|---|
| T | T | T |
| T | F | F |
| F | T | F |
| F | F | T |

Show that $p$ *imp1* $q \equiv q$ *imp1* $p$.

**78.** Define the truth table for *imp2* by

| $p$ | $q$ | $p$ *imp2* $q$ |
|---|---|---|
| T | T | T |
| T | F | F |
| F | T | T |
| F | F | F |

(a) Show that

$$(p \text{ imp2 } q) \wedge (q \text{ imp2 } p) \not\equiv p \leftrightarrow q. \quad \textbf{(1.3.6)}$$

(b) Show that (1.3.6) remains true if we change the third row of *imp2*'s truth table to F T F.

**79.** Verify the second of De Morgan's laws, $\neg(p \wedge q) \equiv \neg p \vee \neg q$.

**80.** Show that $(p \to q) \equiv (\neg p \vee q)$.

## 1.4 Arguments and Rules of Inference

Consider the following sequence of propositions.

The bug is either in module 17 or in module 81.
The bug is a numerical error.
Module 81 has no numerical error. **(1.4.1)**

Assuming that these statements are true, it is reasonable to conclude

The bug is in module 17. **(1.4.2)**

This process of drawing a conclusion from a sequence of propositions is called **deductive reasoning.** The given propositions, such as (1.4.1), are called **hypotheses** or **premises,** and the proposition that follows from the hypotheses, such as (1.4.2), is called the **conclusion.** A **(deductive) argument** consists of hypotheses together with a conclusion. Many proofs in mathematics and computer science are deductive arguments.

Any argument has the form

$$\text{If } p_1 \text{ and } p_2 \text{ and } \cdots \text{ and } p_n, \text{ then } q. \qquad \textbf{(1.4.3)}$$

Argument (1.4.3) is said to be **valid** if the conclusion follows from the hypotheses; that is, if $p_1$ and $p_2$ and $\cdots$ and $p_n$ are true, then $q$ must also be true. This discussion motivates the following definition.

**Definition 1.4.1 ▶** An *argument* is a sequence of propositions written

$$\begin{array}{c} p_1 \\ p_2 \\ \vdots \\ p_n \\ \hline \therefore q \end{array}$$

or

$$p_1, p_2, \ldots, p_n / \therefore q.$$

The symbol $\therefore$ is read "therefore." The propositions $p_1, p_2, \ldots, p_n$ are called the *hypotheses* (or *premises*), and the proposition $q$ is called the *conclusion.* The argument is *valid* provided that if $p_1$ and $p_2$ and $\cdots$ and $p_n$ are all true, then $q$ must also be true; otherwise, the argument is *invalid* (or a *fallacy*). ◀

**Go Online**
For more on fallacies, see
bit.ly/2JagVw5

In a valid argument, we sometimes say that the conclusion follows from the hypotheses. Notice that we are not saying that the conclusion is true; we are only saying that if you grant the hypotheses, you must also grant the conclusion. An argument is valid because of its form, not because of its content.

Each step of an extended argument involves drawing intermediate conclusions. For the argument as a whole to be valid, each step of the argument must result in a valid, intermediate conclusion. **Rules of inference,** brief, valid arguments, are used within a larger argument.

**Example 1.4.2** Determine whether the argument

$$\begin{array}{l} p \rightarrow q \\ p \\ \hline \therefore q \end{array}$$

is valid.

**FIRST SOLUTION** We construct a truth table for all the propositions involved:

| $p$ | $q$ | $p \rightarrow q$ | $p$ | $q$ |
|---|---|---|---|---|
| T | T | T | T | T |
| T | F | F | T | F |
| F | T | T | F | T |
| F | F | T | F | F |

We observe that whenever the hypotheses $p \rightarrow q$ and $p$ are true, the conclusion $q$ is also true; therefore, the argument is valid.

**SECOND SOLUTION** We can avoid writing the truth table by directly verifying that whenever the hypotheses are true, the conclusion is also true.

Suppose that $p \rightarrow q$ and $p$ are true. Then $q$ must be true, for otherwise $p \rightarrow q$ would be false. Therefore, the argument is valid. ◀

The argument in Example 1.4.2 is used extensively and is known as the **modus ponens rule of inference** or **law of detachment.** Several useful rules of inference for propositions, which may be verified using truth tables (see Exercises 33–38), are listed in Table 1.4.1.

**TABLE 1.4.1** ■ Rules of Inference for Propositions

| *Rule of Inference* | *Name* | *Rule of Inference* | *Name* |
|---|---|---|---|
| $\begin{array}{l} p \rightarrow q \\ p \\ \hline \therefore q \end{array}$ | Modus ponens | $\begin{array}{l} p \\ q \\ \hline \therefore p \wedge q \end{array}$ | Conjunction |
| $\begin{array}{l} p \rightarrow q \\ \neg q \\ \hline \therefore \neg p \end{array}$ | Modus tollens | $\begin{array}{l} p \rightarrow q \\ q \rightarrow r \\ \hline \therefore p \rightarrow r \end{array}$ | Hypothetical syllogism |
| $\begin{array}{l} p \\ \hline \therefore p \vee q \end{array}$ | Addition | $\begin{array}{l} p \vee q \\ \neg p \\ \hline \therefore q \end{array}$ | Disjunctive syllogism |
| $\begin{array}{l} p \wedge q \\ \hline \therefore p \end{array}$ | Simplification | | |

**Example 1.4.3** Which rule of inference is used in the following argument?

If the computer has one gigabyte of memory, then it can run "Blast 'em." If the computer can run "Blast 'em," then the sonics will be impressive. Therefore, if the computer has one gigabyte of memory, then the sonics will be impressive.

**SOLUTION** Let $p$ denote the proposition "the computer has one gigabyte of memory," let $q$ denote the proposition "the computer can run 'Blast 'em,'" and let $r$ denote the proposition "the sonics will be impressive." The argument can be written symbolically as

$$\begin{array}{l} p \to q \\ q \to r \\ \hline \therefore p \to r \end{array}$$

Therefore, the argument uses the hypothetical syllogism rule of inference. ◀

**Example 1.4.4** Represent the argument

$$\begin{array}{l} \text{If } 2 = 3, \text{ then I ate my hat.} \\ \text{I ate my hat.} \\ \hline \therefore 2 = 3 \end{array}$$

symbolically and determine whether the argument is valid.

**SOLUTION** If we let $p$: $2 = 3$ and $q$: I ate my hat, the argument may be written

$$\begin{array}{l} p \to q \\ q \\ \hline \therefore p \end{array}$$

If the argument is valid, then whenever $p \to q$ and $q$ are both true, $p$ must also be true. Suppose that $p \to q$ and $q$ are true. This is possible if $p$ is false and $q$ is true. In this case, $p$ is not true; thus the argument is invalid. This fallacy is known as the **fallacy of affirming the conclusion.** ◀

We can also determine whether the argument in Example 1.4.4 is valid or not by examining the truth table of Example 1.4.2. In the third row of the table, the hypotheses are true and the conclusion is false; thus the argument is invalid.

**Example 1.4.5** Represent the argument

$$\begin{array}{l} \text{The bug is either in module 17 or in module 81.} \\ \text{The bug is a numerical error.} \\ \text{Module 81 has no numerical error.} \\ \hline \therefore \text{The bug is in module 17.} \end{array}$$

given at the beginning of this section symbolically and show that it is valid.

**SOLUTION** If we let

$p$ : The bug is in module 17.

$q$ : The bug is in module 81.

$r$ : The bug is a numerical error.

the argument may be written

$$\begin{array}{l} p \lor q \\ r \\ r \to \neg q \\ \hline \therefore p \end{array}$$

From $r \to \neg q$ and $r$, we may use modus ponens to conclude $\neg q$. From $p \lor q$ and $\neg q$, we may use the disjunctive syllogism to conclude $p$. Thus the conclusion $p$ follows from the hypotheses and the argument is valid. ◀

**Example 1.4.6** We are given the following hypotheses: If the Chargers get a good linebacker, then the Chargers can beat the Broncos. If the Chargers can beat the Broncos, then the Chargers can beat the Jets. If the Chargers can beat the Broncos, then the Chargers can beat the Dolphins. The Chargers get a good linebacker. Show by using the rules of inference (see Table 1.4.1) that the conclusion, the Chargers can beat the Jets and the Chargers can beat the Dolphins, follows from the hypotheses.

**SOLUTION** Let $p$ denote the proposition "the Chargers get a good linebacker," let $q$ denote the proposition "the Chargers can beat the Broncos," let $r$ denote the proposition "the Chargers can beat the Jets," and let $s$ denote the proposition "the Chargers can beat the Dolphins." Then the hypotheses are:

$$\begin{array}{l} p \to q \\ q \to r \\ q \to s \\ p. \end{array}$$

From $p \to q$ and $q \to r$, we may use the hypothetical syllogism to conclude $p \to r$. From $p \to r$ and $p$, we may use modus ponens to conclude $r$. From $p \to q$ and $q \to s$, we may use the hypothetical syllogism to conclude $p \to s$. From $p \to s$ and $p$, we may use modus ponens to conclude $s$. From $r$ and $s$, we may use conjunction to conclude $r \land s$. Since $r \land s$ represents the proposition "the Chargers can beat the Jets and the Chargers can beat the Dolphins," we conclude that the conclusion does follow from the hypotheses. ◀

### 1.4 Problem-Solving Tips

The validity of a very short argument or proof might be verified using a truth table. In practice, arguments and proofs use rules of inference.

## 1.4 Review Exercises

1. What is deductive reasoning?
2. What is a hypothesis in an argument?
3. What is a premise in an argument?
4. What is a conclusion in an argument?
5. What is a valid argument?
6. What is an invalid argument?
7. State the modus ponens rule of inference.
8. State the modus tollens rule of inference.
9. State the addition rule of inference.
10. State the simplification rule of inference.
11. State the conjunction rule of inference.
12. State the hypothetical syllogism rule of inference.
13. State the disjunctive syllogism rule of inference.

## 1.4 Exercises

*Formulate the arguments of Exercises 1–5 symbolically and determine whether each is valid. Let*

$p$: *I study hard.* $q$: *I get A's.* $r$: *I get rich.*

1. If I study hard, then I get A's.
   I study hard.
   ∴ I get A's.
2. If I study hard, then I get A's.
   If I don't get rich, then I don't get A's.
   ∴ I get rich.
3. I study hard if and only if I get rich.
   I get rich.
   ∴ I study hard.
4. If I study hard or I get rich, then I get A's.
   I get A's.
   ∴ If I don't study hard, then I get rich.
5. If I study hard, then I get A's or I get rich.
   I don't get A's and I don't get rich.
   ∴ I don't study hard.

*Formulate the arguments of Exercises 6–9 symbolically and determine whether each is valid.*

$p$: *The Democrats win.*
$q$: *The Republicans win.*
$r$: *Unemployment is up.*
$s$: *The economy is up.*

6. If the Democrats win, then the economy is up; and, if the Republicans win, then unemployment is up.
   The Democrats win or the Republicans win.
   ∴ Unemployment is up or the economy is up.
7. If the Democrats win or the Republicans win, then unemployment is up or the economy is up.
   The Democrats win and unemployment is not up.
   ∴ The economy is up.
8. If the Democrats win, then unemployment is up.
   If the Republicans win, then the economy is up
   The Republicans and Democrats do not both win.
   The Democrats do not win.
   ∴ The economy is up.
9. If the Democrats win, then unemployment is up or the economy is up.
   If the Republicans win, then unemployment is up.
   The economy is not up.
   The Democrats win
   ∴ Unemployment is up or the Republicans win.

*In Exercises 10–14, write the given argument in words and determine whether each argument is valid. Let*

$p$: *4 gigabytes is better than no memory at all.*
$q$: *We will buy more memory.*
$r$: *We will buy a new computer.*

10. $p \to r$
    $p \to q$
    $\therefore p \to (r \wedge q)$
11. $p \to (r \vee q)$
    $r \to \neg q$
    $\therefore p \to r$
12. $p \to r$
    $r \to q$
    $\therefore q$
13. $\neg r \to \neg p$
    $r$
    $\therefore p$
14. $p \to r$
    $r \to q$
    $p$
    $\therefore q$

*In Exercises 15–19, write the given argument in words and determine whether each argument is valid. Let*

$p$: *The for loop is faulty.*
$q$: *The while loop is faulty.*
$r$: *The hardware is unreliable.*
$s$: *The output is correct.*

15. $(p \vee q) \to (r \vee s)$
    $p$
    $\neg r$
    $\therefore s$
16. $(r \vee s) \to p$
    $s \to q$
    $p \vee s$
    $\therefore r$
17. $(p \to r) \to q$
    $(q \to s) \to p$
    $r \wedge s$
    $p \vee q$
    $\therefore p \wedge q$
18. $p \to q$
    $q \to r$
    $\neg r$
    $s \to r$
    $\therefore \neg s$
19. $p \to (q \vee r)$
    $q \to (p \vee s)$
    $p \vee \neg q$
    $\neg s$
    $\therefore p \vee r$

*Determine whether each argument in Exercises 20–24 is valid.*

20. $p \to q$
    $\neg p$
    $\therefore \neg q$
21. $p \to q$
    $\neg q$
    $\therefore \neg p$
22. $p \wedge \neg p$
    $\therefore q$

23. $p \to (q \to r)$
$q \to (p \to r)$
$\therefore (p \vee q) \to r$

24. $(p \to q) \wedge (r \to s)$
$p \vee r$
$\therefore q \vee s$

25. Show that if

$$p_1, p_2/\therefore p \quad \text{and} \quad p, p_3, \ldots, p_n/\therefore c$$

are valid arguments, the argument

$$p_1, p_2, \ldots, p_n/\therefore c$$

is also valid.

26. Comment on the following argument:

Hard disk drive storage is better than nothing.
Nothing is better than a solid state drive.
∴ Hard disk drive is better than a solid state drive.

*For each argument in Exercises 27–29, tell which rule of inference is used.*

27. Fishing is a popular sport. Therefore, fishing is a popular sport or lacrosse is wildly popular in California.

28. If fishing is a popular sport, then lacrosse is wildly popular in California. Fishing is a popular sport. Therefore, lacrosse is wildly popular in California.

29. Fishing is a popular sport or lacrosse is wildly popular in California. Lacrosse is not wildly popular in California. Therefore, fishing is a popular sport.

*In Exercises 30–32, give an argument using rules of inference to show that the conclusion follows from the hypotheses.*

30. Hypotheses: If there is gas in the car, then I will go to the store. If I go to the store, then I will get a soda. There is gas in the car. Conclusion: I will get a soda.

31. Hypotheses: If there is gas in the car, then I will go to the store. If I go to the store, then I will get a soda. I do not get a soda. Conclusion: There is not gas in the car, or the car transmission is defective.

32. Hypotheses: If Jill can sing or Dweezle can play, then I'll buy the compact disc. Jill can sing. I'll buy the compact disc player. Conclusion: I'll buy the compact disc and the compact disc player.

33. Show that modus tollens (see Table 1.4.1) is valid.

34. Show that addition (see Table 1.4.1) is valid.

35. Show that simplification (see Table 1.4.1) is valid.

36. Show that conjunction (see Table 1.4.1) is valid.

37. Show that hypothetical syllogism (see Table 1.4.1) is valid.

38. Show that disjunctive syllogism (see Table 1.4.1) is valid.

## 1.5 Quantifiers

**Go Online**
For more on quantifiers, see
`bit.ly/2JagVw5`

The logic in Sections 1.2 and 1.3 that deals with propositions is incapable of describing most of the statements in mathematics and computer science. Consider, for example, the statement

$$p\colon\ n \text{ is an odd integer.}$$

A proposition is a statement that is either true or false. The statement $p$ is not a proposition, because whether $p$ is true or false depends on the value of $n$. For example, $p$ is true if $n = 103$ and false if $n = 8$. Since most of the statements in mathematics and computer science use variables, we must extend the system of logic to include such statements.

**Definition 1.5.1 ▶** Let $P(x)$ be a statement involving the variable $x$ and let $D$ be a set. We call $P$ a *propositional function* or *predicate* (with respect to $D$) if for each $x \in D$, $P(x)$ is a proposition. We call $D$ the *domain of discourse* of $P$. ◀

In Definition 1.5.1, the domain of discourse specifies the allowable values for $x$.

**Example 1.5.2** Let $P(n)$ be the statement

$$n \text{ is an odd integer.}$$

Then $P$ is a propositional function with domain of discourse $\mathbf{Z}^+$ since for each $n \in \mathbf{Z}^+$, $P(n)$ is a proposition [i.e., for each $n \in \mathbf{Z}^+$, $P(n)$ is true or false but not both]. For example, if $n = 1$, we obtain the proposition

$$P(1): \text{ 1 is an odd integer}$$

(which is true). If $n = 2$, we obtain the proposition

$$P(2): \text{ 2 is an odd integer}$$

(which is false). ◀

A propositional function $P$, by itself, is neither true nor false. However, for each $x$ in the domain of discourse, $P(x)$ is a proposition and is, therefore, either true or false. We can think of a propositional function as defining a class of propositions, one for each element in the domain of discourse. For example, if $P$ is a propositional function with domain of discourse $\mathbf{Z}^+$, we obtain the class of propositions

$$P(1), P(2), \ldots .$$

Each of $P(1), P(2), \ldots$ is either true or false.

**Example 1.5.3** Explain why the following are propositional functions.

(a) $n^2 + 2n$ is an odd integer (domain of discourse $= \mathbf{Z}^+$).

(b) $x^2 - x - 6 = 0$ (domain of discourse $= \mathbf{R}$).

(c) The baseball player hit over .300 in 2015 (domain of discourse = set of baseball players).

(d) The film is rated over 20% by Rotten Tomatoes [the scale is 0% (awful) to 100% (terrific)]. The domain of discourse is the set of films rated by Rotten Tomatoes.

**SOLUTION** In statement (a), for each positive integer $n$, we obtain a proposition; therefore, statement (a) is a propositional function.

Similarly, in statement (b), for each real number $x$, we obtain a proposition; therefore, statement (b) is a propositional function.

We can regard the variable in statement (c) as "baseball player." Whenever we substitute a particular baseball player for the variable "baseball player," the statement is a proposition. For example, if we substitute "Joey Votto" for "baseball player," statement (c) is

Joey Votto hit over .300 in 2015,

which is true. If we substitute "Andrew McCutchen" for "baseball player," statement (c) is

Andrew McCutchen hit over .300 in 2015,

which is false. Thus statement (c) is a propositional function.

Statement (d) is similar in form to statement (c). Here the variable is "film." Whenever we substitute a film rated by Rotten Tomatoes for the variable "film," the statement is a proposition. For example if we substitute *Spectre* for "film," statement (d) is

*Spectre* is rated over 20% by Rotten Tomatoes,

which is true since *Spectre* is rated 64% by Rotten Tomatoes. If we substitute *Blended* for "film," statement (d) is

*Blended* is rated over 20% by Rotten Tomatoes,

which is false since *Blended* is rated 14% by Rotten Tomatoes. Thus statement (d) is a propositional function. ◀

Most of the statements in mathematics and computer science use terms such as "for every" and "for some." For example, in mathematics we have the following theorem:

For every triangle $T$, the sum of the angles of $T$ is equal to 180°.

In computer science, we have this theorem:

For some program $P$, the output of $P$ is $P$ itself.

We now extend the logical system of Sections 1.2 and 1.3 so that we can handle statements that include "for every" and "for some."

**Definition 1.5.4 ▶** Let $P$ be a propositional function with domain of discourse $D$. The statement

$$\text{for every } x, P(x)$$

is said to be a *universally quantified statement*. The symbol $\forall$ means "for every." Thus the statement

$$\text{for every } x, P(x)$$

may be written

$$\forall x\, P(x). \tag{1.5.1}$$

The symbol $\forall$ is called a *universal quantifier*.

The statement (1.5.1) is true if $P(x)$ is true for every $x$ in $D$. The statement (1.5.1) is false if $P(x)$ is false for at least one $x$ in $D$. ◀

**Example 1.5.5** Consider the universally quantified statement

$$\forall x(x^2 \geq 0).$$

The domain of discourse is **R**. The statement is true because, *for every* real number $x$, it is true that the square of $x$ is positive or zero. ◀

According to Definition 1.5.4, the universally quantified statement (1.5.1) is false if *for at least one* $x$ in the domain of discourse, the proposition $P(x)$ is false. A value $x$ in the domain of discourse that makes $P(x)$ false is called a **counterexample** to the statement (1.5.1).

**Example 1.5.6** Determine whether the universally quantified statement $\forall x(x^2 - 1 > 0)$ is true or false. The domain of discourse is **R**.

**SOLUTION** The statement is false since, if $x = 1$, the proposition $1^2 - 1 > 0$ is false. The value 1 is a counterexample to the statement $\forall x(x^2 - 1 > 0)$. Although there are values of $x$ that make the propositional function true, the counterexample provided shows that the universally quantified statement is false. ◀

**Example 1.5.7** Suppose that $P$ is a propositional function whose domain of discourse is the set $\{d_1, \ldots, d_n\}$. The following pseudocode[†] determines whether $\forall x\, P(x)$ is true or false:

```
for i = 1 to n
    if (¬P(d_i))
        return false
return true
```

The for loop examines the members $d_i$ of the domain of discourse one by one. If it finds a value $d_i$ for which $P(d_i)$ is false, the condition $\neg P(d_i)$ in the if statement is true; so the code returns false [to indicate that $\forall x\, P(x)$ is false] and terminates. In this case, $d_i$ is a counterexample. If $P(d_i)$ is true for every $d_i$, the condition $\neg P(d_i)$ in the if statement is always false. In this case, the for loop runs to completion, after which the code returns true [to indicate that $\forall x\, P(x)$ is true] and terminates.

Notice that if $\forall x\, P(x)$ is true, the for loop necessarily runs to completion so that *every* member of the domain of discourse is checked to ensure that $P(x)$ is true for every $x$. If $\forall x\, P(x)$ is false, the for loop terminates as soon as *one* element $x$ of the domain of discourse is found for which $P(x)$ is false. ◀

We call the variable $x$ in the propositional function $P(x)$ a *free variable*. (The idea is that $x$ is "free" to roam over the domain of discourse.) We call the variable $x$ in the universally quantified statement $\forall x\, P(x)$ a *bound variable*. (The idea is that $x$ is "bound" by the quantifier $\forall$.)

We previously pointed out that a propositional function does not have a truth value. On the other hand, Definition 1.5.4 assigns a truth value to the quantified statement $\forall x\, P(x)$. In sum, a statement with free (unquantified) variables is not a proposition, and a statement with no free variables (no unquantified variables) is a proposition.

Alternative ways to write $\forall x\, P(x)$ are

for all $x$, $P(x)$

and

for any $x$, $P(x)$.

The symbol $\forall$ may be read "for every," "for all," or "for any."

To prove that $\forall x\, P(x)$ is *true,* we must, in effect, examine *every* value of $x$ in the domain of discourse and show that for every $x$, $P(x)$ is true. One technique for proving that $\forall x\, P(x)$ is true is to let $x$ denote an *arbitrary* element of the domain of discourse $D$. The argument then proceeds using the symbol $x$. Whatever is claimed about $x$ must be true *no matter what value* $x$ might have in $D$. The argument must conclude by proving that $P(x)$ is true.

Sometimes to specify the domain of discourse $D$, we write a universally quantified statement as

for every $x$ in $D$, $P(x)$.

[†]The pseudocode used in this book is explained in Appendix C.

**Example 1.5.8** Verify that the universally quantified statement

$$\text{for every real number } x, \text{ if } x > 1, \text{ then } x + 1 > 1$$

is true.

**SOLUTION** This time we must verify that the statement

$$\text{if } x > 1, \text{ then } x + 1 > 1 \tag{1.5.2}$$

is true *for every* real number $x$.

Let $x$ be any real number whatsoever. It is true that for any real number $x$, either $x \leq 1$ or $x > 1$. If $x \leq 1$, the conditional proposition (1.5.2) is vacuously true. (The proposition is true because the hypothesis $x > 1$ is false. Recall that when the hypothesis is false, the conditional proposition is true regardless of whether the conclusion is true or false.) In most arguments, the vacuous case is omitted.

Now suppose that $x > 1$. Regardless of the specific value of $x$, $x + 1 > x$. Since $x + 1 > x$ and $x > 1$, we conclude that $x + 1 > 1$, so the conclusion is true. If $x > 1$, the hypothesis and conclusion are both true; hence the conditional proposition (1.5.2) is true.

We have shown that for every real number $x$, the proposition (1.5.2) is true. Therefore, the universally quantified statement

$$\text{for every real number } x, \text{ if } x > 1, \text{ then } x + 1 > 1$$

is true. ◀

The method of disproving the statement $\forall x\, P(x)$ is quite different from the method used to prove that the statement is true. To show that the universally quantified statement $\forall x\, P(x)$ is *false*, it is sufficient to find *one* value $x$ in the domain of discourse for which the proposition $P(x)$ is false. Such a value, we recall, is called a counterexample to the universally quantified statement.

We turn next to existentially quantified statements.

**Definition 1.5.9 ▶** Let $P$ be a propositional function with domain of discourse $D$. The statement

$$\text{there exists } x, P(x)$$

is said to be an *existentially quantified statement*. The symbol $\exists$ means "there exists." Thus the statement

$$\text{there exists } x, P(x)$$

may be written

$$\exists x\, P(x). \tag{1.5.3}$$

The symbol $\exists$ is called an *existential quantifier*.

The statement (1.5.3) is true if $P(x)$ is true for at least one $x$ in $D$. The statement (1.5.3) is false if $P(x)$ is false for every $x$ in $D$. ◀

**Example 1.5.10** Consider the existentially quantified statement

$$\exists x \left( \frac{x}{x^2 + 1} = \frac{2}{5} \right).$$

The domain of discourse is **R**. The statement is true because it is possible to find *at least one* real number $x$ for which the proposition

$$\frac{x}{x^2+1} = \frac{2}{5}$$

is true. For example, if $x = 2$, we obtain the true proposition

$$\frac{2}{2^2+1} = \frac{2}{5}.$$

It is not the case that *every* value of $x$ results in a true proposition. For example, if $x = 1$, the proposition

$$\frac{1}{1^2+1} = \frac{2}{5}$$

is false. ◀

According to Definition 1.5.9, the existentially quantified statement (1.5.3) is false if for every $x$ in the domain of discourse, the proposition $P(x)$ is false.

**Example 1.5.11** Verify that the existentially quantified statement

$$\exists x \in \mathbf{R} \left( \frac{1}{x^2+1} > 1 \right)$$

is false.

**SOLUTION** We must show that

$$\frac{1}{x^2+1} > 1$$

is false for every real number $x$. Now

$$\frac{1}{x^2+1} > 1$$

is false precisely when

$$\frac{1}{x^2+1} \le 1 \qquad \textbf{(1.5.4)}$$

is true. Thus, we must show that (1.5.4) is true for every real number $x$. To this end, let $x$ be any real number whatsoever. Since $0 \le x^2$, we may add 1 to both sides of this inequality to obtain $1 \le x^2 + 1$. If we divide both sides of this last inequality by $x^2 + 1$, we obtain (1.5.4) Therefore, the statement (1.5.4) is true for every real number $x$. Thus the statement

$$\frac{1}{x^2+1} > 1$$

is false for every real number $x$. We have shown that the existentially quantified statement

$$\exists x \left( \frac{1}{x^2+1} > 1 \right)$$

is false. ◀

**Example 1.5.12** Suppose that $P$ is a propositional function whose domain of discourse is the set $\{d_1, \ldots, d_n\}$. The following pseudocode determines whether $\exists x\, P(x)$ is true or false:

```
for i = 1 to n
  if (P(d_i))
    return true
return false
```

The for loop examines the members $d_i$ in the domain of discourse one by one. If it finds a value $d_i$ for which $P(d_i)$ is true, the condition $P(d_i)$ in the if statement is true; so the code returns true [to indicate that $\exists x\, P(x)$ is true] and terminates. In this case, the code found a value in the domain of discourse, namely $d_i$, for which $P(d_i)$ is true. If $P(d_i)$ is false for every $d_i$, the condition $P(d_i)$ in the if statement is always false. In this case, the for loop runs to completion, after which the code returns false [to indicate that $\exists x\, P(x)$ is true] and terminates.

Notice that if $\exists x\, P(x)$ is true, the for loop terminates as soon as *one* element $x$ in the domain of discourse is found for which $P(x)$ is true. If $\exists x\, P(x)$ is false, the for loop necessarily runs to completion so that *every* member in the domain of discourse is checked to ensure that $P(x)$ is false for every $x$. ◀

Alternative ways to write $\exists x\, P(x)$ are

there exists $x$ such that, $P(x)$

and

for some $x$, $P(x)$

and

for at least one $x$, $P(x)$.

The symbol $\exists$ may be read "there exists," "for some," or "for at least one."

**Example 1.5.13** Consider the existentially quantified statement

for some $n$, if $n$ is prime, then $n + 1$, $n + 2$, $n + 3$, and $n + 4$ are not prime.

The domain of discourse is $\mathbf{Z}^+$. This statement is true because we can find *at least one* positive integer $n$ that makes the conditional proposition

if $n$ is prime, then $n + 1$, $n + 2$, $n + 3$, and $n + 4$ are not prime

true. For example, if $n = 23$, we obtain the true proposition

if 23 is prime, then 24, 25, 26, and 27 are not prime.

(This conditional proposition is true because both the hypothesis "23 is prime" and the conclusion "24, 25, 26, and 27 are not prime" are true.) Some values of $n$ make the conditional proposition true (e.g., $n = 23$, $n = 4$, $n = 47$), while others make it false (e.g., $n = 2$, $n = 101$). The point is that we found *one* value that makes the conditional proposition

if $n$ is prime, then $n + 1$, $n + 2$, $n + 3$, and $n + 4$ are not prime true. For this reason, the existentially quantified statement

for some $n$, if $n$ is prime, then $n+1$, $n+2$, $n+3$, and $n+4$ are not prime

is true. ◀

In Example 1.5.11, we showed that an existentially quantified statement was false by proving that a related universally quantified statement was true. The following theorem makes this relationship precise. The theorem generalizes De Morgan's laws of logic (Example 1.3.11).

**Theorem 1.5.14** **Generalized De Morgan's Laws for Logic**

If $P$ is a propositional function, each pair of propositions in (a) and (b) has the same truth values (i.e., either both are true or both are false).

(a) $\neg(\forall x\, P(x));\ \exists x\, \neg P(x)$

(b) $\neg(\exists x\, P(x));\ \forall x\, \neg P(x)$

**Proof** We prove only part (a) and leave the proof of part (b) to the reader (Exercise 73).

Suppose that the proposition $\neg(\forall x\, P(x))$ is true. Then the proposition $\forall x\, P(x)$ is false. By Definition 1.5.4, the proposition $\forall x\, P(x)$ is false precisely when $P(x)$ is false for at least one $x$ in the domain of discourse. But if $P(x)$ is false for at least one $x$ in the domain of discourse, $\neg P(x)$ is true for at least one $x$ in the domain of discourse. By Definition 1.5.9, when $\neg P(x)$ is true for at least one $x$ in the domain of discourse, the proposition $\exists x\, \neg P(x)$ is true. Thus, if the proposition $\neg(\forall x\, P(x))$ is true, the proposition $\exists x\, \neg P(x)$ is true. Similarly, if the proposition $\neg(\forall x\, P(x))$ is false, the proposition $\exists x\, \neg P(x)$ is false.

Therefore, the pair of propositions in part (a) always has the same truth values. ◀

**Example 1.5.15** Let $P(x)$ be the statement

$$\frac{1}{x^2+1} > 1.$$

In Example 1.5.11 we showed that $\exists x\, P(x)$ is false by verifying that

$$\forall x\, \neg P(x) \tag{1.5.5}$$

is true.

The technique can be justified by appealing to Theorem 1.5.14. After we prove that proposition (1.5.5) is true, we may use Theorem 1.5.14, part (b), to conclude that $\neg(\exists x P(x))$ is also true. Thus $\neg\neg(\exists x P(x))$ or, equivalently, $\exists x\, P(x)$ is false. ◀

**Example 1.5.16** Write the statement

Every rock fan loves U2,

symbolically. Write its negation symbolically and in words.

**SOLUTION** Let $P(x)$ be the propositional function "$x$ loves U2." The given statement can be written symbolically as $\forall x\, P(x)$. The domain of discourse is the set of rock fans.

By Theorem 1.5.14, part (a), the negation of the preceding proposition $\neg(\forall x\, P(x))$ is equivalent to $\exists x\, \neg P(x)$. In words, this last proposition can be stated as: There exists a rock fan who does not love U2. ◀

**Example 1.5.17** Write the statement

Some birds cannot fly,

symbolically. Write its negation symbolically and in words.

**SOLUTION** Let $P(x)$ be the propositional function "$x$ flies." The given statement can be written symbolically as $\exists x\, \neg P(x)$. [The statement could also be written $\exists x\, Q(x)$, where $Q(x)$ is the propositional function "$x$ cannot fly." As in algebra, there are many ways to represent text symbolically.] The domain of discourse is the set of birds.

By Theorem 1.5.14, part (b), the negation $\neg(\exists x\, \neg P(x))$ of the preceding proposition is equivalent to $\forall x\, \neg\neg P(x)$ or, equivalently, $\forall x\, P(x)$. In words, this last proposition can be stated as: Every bird can fly. ◀

A universally quantified proposition generalizes the proposition

$$P_1 \wedge P_2 \wedge \cdots \wedge P_n \tag{1.5.6}$$

in the sense that the individual propositions $P_1, P_2, \ldots, P_n$ are replaced by an arbitrary family $P(x)$, where $x$ is in the domain of discourse, and (1.5.6) is replaced by

$$\forall x\, P(x). \tag{1.5.7}$$

The proposition (1.5.6) is true if and only if $P_i$ is true for every $i = 1, \ldots, n$. The truth value of proposition (1.5.7) is defined similarly: (1.5.7) is true if and only if $P(x)$ is true for every $x$ in the domain of discourse.

**Example 1.5.18** Suppose that the domain of discourse of the propositional function $P$ is $\{-1, 0, 1\}$. The propositional function $\forall x\, P(x)$ is equivalent to

$$P(-1) \wedge P(0) \wedge P(1).$$ ◀

Similarly, an existentially quantified proposition generalizes the proposition

$$P_1 \vee P_2 \vee \cdots \vee P_n \tag{1.5.8}$$

in the sense that the individual propositions $P_1, P_2, \ldots, P_n$ are replaced by an arbitrary family $P(x)$, where $x$ is in the domain of discourse, and (1.5.8) is replaced by $\exists x\, P(x)$.

**Example 1.5.19** Suppose that the domain of discourse of the propositional function $P$ is $\{1, 2, 3, 4\}$. The propositional function $\exists x\, P(x)$ is equivalent to

$$P(1) \vee P(2) \vee P(3) \vee P(4).$$ ◀

The preceding observations explain how Theorem 1.5.14 generalizes De Morgan's laws for logic (Example 1.3.11). Recall that the first of De Morgan's law for logic states that the propositions

$$\neg(P_1 \vee P_2 \vee \cdots \vee P_n) \quad \text{and} \quad \neg P_1 \wedge \neg P_2 \wedge \cdots \wedge \neg P_n$$

have the same truth values. In Theorem 1.5.14, part (b),

$$\neg P_1 \wedge \neg P_2 \wedge \cdots \wedge \neg P_n$$

is replaced by $\forall x \neg P(x)$ and

$$\neg(P_1 \vee P_2 \vee \cdots \vee P_n)$$

is replaced by $\neg(\exists x\, P(x))$.

**Example 1.5.20** Statements in words often have more than one possible interpretation. Consider the well-known quotation from Shakespeare's "The Merchant of Venice":

All that glitters is not gold.

One possible interpretation of this quotation is: Every object that glitters is not gold. However, this is surely not what Shakespeare intended. The correct interpretation is: Some object that glitters is not gold.

If we let $P(x)$ be the propositional function "$x$ glitters" and $Q(x)$ be the propositional function "$x$ is gold," the first interpretation becomes

$$\forall x(P(x) \rightarrow \neg Q(x)), \tag{1.5.9}$$

and the second interpretation becomes

$$\exists x(P(x) \wedge \neg Q(x)).$$

Using the result of Example 1.3.13, we see that the truth values of

$$\exists x(P(x) \wedge \neg Q(x))$$

and

$$\exists x \neg(P(x) \rightarrow Q(x))$$

are the same. By Theorem 1.5.14, the truth values of

$$\exists x \neg(P(x) \rightarrow Q(x))$$

and

$$\neg(\forall x\, P(x) \rightarrow Q(x))$$

are the same. Thus an equivalent way to represent the second interpretation is

$$\neg(\forall x\, P(x) \rightarrow Q(x)). \tag{1.5.10}$$

Comparing (1.5.9) and (1.5.10), we see that the ambiguity results from whether the negation applies to $Q(x)$ (the first interpretation) or to the entire statement

$$\forall x(P(x) \rightarrow Q(x))$$

(the second interpretation). The correct interpretation of the statement

All that glitters is not gold

results from negating the entire statement.

In positive statements, "any," "all," "each," and "every" have the same meaning. In negative statements, the situation changes:

Not all $x$ satisfy $P(x)$.

Not each $x$ satisfies $P(x)$.

Not every $x$ satisfies $P(x)$.

are considered to have the same meaning as

$$\text{For some } x, \neg P(x);$$

whereas

$$\text{Not any } x \text{ satisfies } P(x).$$
$$\text{No } x \text{ satisfies } P(x).$$

mean

$$\text{For all } x, \neg P(x).$$

See Exercises 61–71 for other examples. ◀

## Rules of Inference for Quantified Statements

We conclude this section by introducing some rules of inference for quantified statements and showing how they can be used with rules of inference for propositions (see Section 1.4).

Suppose that $\forall x P(x)$ is true. By Definition 1.5.4, $P(x)$ is true for every $x$ in $D$, the domain of discourse. In particular, if $d$ is in $D$, then $P(d)$ is true. We have shown that the argument

$$\frac{\forall x\, P(x)}{\therefore P(d) \text{ if } d \in D}$$

is valid. This rule of inference is called **universal instantiation.** Similar arguments (see Exercises 79–81) justify the other rules of inference listed in Table 1.5.1.

**TABLE 1.5.1** ■ Rules of Inference for Quantified Statements†

| *Rule of Inference* | *Name* |
|---|---|
| $\forall x\, P(x)$ / $\therefore P(d)$ if $d \in D$ | Universal instantiation |
| $P(d)$ for every $d \in D$ / $\therefore \forall x\, P(x)$ | Universal generalization |
| $\exists x\, P(x)$ / $\therefore P(d)$ for some $d \in D$ | Existential instantiation |
| $P(d)$ for some $d \in D$ / $\therefore \exists x\, P(x)$ | Existential generalization |

† The domain of discourse is $D$.

**Example 1.5.21** Given that

$$\text{for every positive integer } n,\ n^2 \geq n$$

is true, we may use universal instantiation to conclude that $54^2 \geq 54$ since 54 is a positive integer (i.e., a member of the domain of discourse). ◀

**Example 1.5.22** Let $P(x)$ denote the propositional function "$x$ owns a laptop computer," where the domain of discourse is the set of students taking MATH 201 (discrete mathematics). Suppose that Taylor, who is taking MATH 201, owns a laptop computer; in symbols, $P(\text{Taylor})$ is true. We may then use existential generalization to conclude that $\exists x\, P(x)$ is true. ◀

**Example 1.5.23** Write the following argument symbolically and then, using rules of inference, show that the argument is valid.

**SOLUTION** For every real number $x$, if $x$ is an integer, then $x$ is a rational number. The number $\sqrt{2}$ is not rational. Therefore, $\sqrt{2}$ is not an integer.

If we let $P(x)$ denote the propositional function "$x$ is an integer" and $Q(x)$ denote the propositional function "$x$ is rational," the argument becomes

$$\begin{array}{l} \forall x \in \mathbf{R}\, (P(x) \to Q(x)) \\ \neg Q(\sqrt{2}) \\ \hline \therefore \neg P(\sqrt{2}) \end{array}$$

Since $\sqrt{2} \in \mathbf{R}$, we may use universal instantiation to conclude $P(\sqrt{2}) \to Q(\sqrt{2})$. Combining $P(\sqrt{2}) \to Q(\sqrt{2})$ and $\neg Q(\sqrt{2})$, we may use modus tollens (see Table 1.4.1) to conclude $\neg P(\sqrt{2})$. Thus the argument is valid. ◀

The argument in Example 1.5.23 is called **universal modus tollens.**

**Example 1.5.24** We are given these hypotheses: Everyone loves either Microsoft or Apple. Lynn does not love Microsoft. Show that the conclusion, Lynn loves Apple, follows from the hypotheses.

**SOLUTION** Let $P(x)$ denote the propositional function "$x$ loves Microsoft," and let $Q(x)$ denote the propositional function "$x$ loves Apple." The first hypothesis is $\forall x(P(x) \vee Q(x))$. By universal instantiation, we have $P(\text{Lynn}) \vee Q(\text{Lynn})$. The second hypothesis is $\neg P(\text{Lynn})$. The disjunctive syllogism rule of inference (see Table 1.4.1) now gives $Q(\text{Lynn})$, which represents the proposition "Lynn loves Apple." We conclude that the conclusion does follow from the hypotheses. ◀

## 1.5 Problem-Solving Tips

- To prove that the universally quantified statement $\forall x\, P(x)$ is true, show that for *every* $x$ in the domain of discourse, the proposition $P(x)$ is true. Showing that $P(x)$ is true for a *particular* value $x$ does *not* prove that $\forall x\, P(x)$ is true.
- To prove that the existentially quantified statement $\exists x\, P(x)$ is true, find *one* value of $x$ in the domain of discourse for which the proposition $P(x)$ is true. *One* value suffices.
- To prove that the universally quantified statement $\forall x\, P(x)$ is false, find *one* value of $x$ (a counterexample) in the domain of discourse for which the proposition $P(x)$ is false.
- To prove that the existentially quantified statement $\exists x\, P(x)$ is false, show that for *every* $x$ in the domain of discourse, the proposition $P(x)$ is false. Showing that $P(x)$ is false for a *particular* value $x$ does *not* prove that $\exists x\, P(x)$ is false.

## 1.5 Review Exercises

1. What is a propositional function?
2. What is a domain of discourse?
3. What is a universally quantified statement?
4. What is a counterexample?
5. What is an existentially quantified statement?
6. State the generalized De Morgan's laws for logic.
7. Explain how to prove that a universally quantified statement is true.
8. Explain how to prove that an existentially quantified statement is true.
9. Explain how to prove that a universally quantified statement is false.
10. Explain how to prove that an existentially quantified statement is false.
11. State the universal instantiation rule of inference.
12. State the universal generalization rule of inference.
13. State the existential instantiation rule of inference.
14. State the existential generalization rule of inference.

## 1.5 Exercises

*In Exercises 1–6, tell whether the statement is a propositional function. For each statement that is a propositional function, give a domain of discourse.*

**1.** $(2n+1)^2$ is an odd integer.

**2.** Choose an integer between 1 and 10.

**3.** Let $x$ be a real number.

**4.** The movie won the Academy Award as the best picture of 1955.

**5.** $1+3=4$.

**6.** There exists $x$ such that $x < y$ ($x$, $y$ real numbers).

*Let $P(n)$ be the propositional function "$n$ divides 77." Write each proposition in Exercises 7–15 in words and tell whether it is true or false. The domain of discourse is $\mathbf{Z}^+$.*

**7.** $P(11)$ **8.** $P(1)$ **9.** $P(3)$

**10.** $\forall n\, P(n)$ **11.** $\exists n\, P(n)$ **12.** $\forall n \neg P(n)$

**13.** $\exists n \neg P(n)$ **14.** $\neg(\forall n\, P(n))$ **15.** $\neg(\exists n\, P(n))$

*Let $P(x)$ be the propositional function "$x \geq x^2$." Tell whether each proposition in Exercises 16–24 is true or false. The domain of discourse is $\mathbf{R}$.*

**16.** $P(1)$ **17.** $P(2)$ **18.** $P(1/2)$

**19.** $\forall x\, P(x)$ **20.** $\exists x\, P(x)$ **21.** $\neg(\forall x\, P(x))$

**22.** $\neg(\exists x\, P(x))$ **23.** $\forall x \neg P(x)$ **24.** $\exists x \neg P(x)$

*Suppose that the domain of discourse of the propositional function $P$ is $\{1, 2, 3, 4\}$. Rewrite each propositional function in Exercises 25–31 using only negation, disjunction, and conjunction.*

**25.** $\forall x\, P(x)$ **26.** $\forall x \neg P(x)$ **27.** $\neg(\forall x\, P(x))$

**28.** $\exists x\, P(x)$ **29.** $\exists x \neg P(x)$ **30.** $\neg(\exists x\, P(x))$

**31.** $\forall x((x \neq 1) \rightarrow P(x))$

*Let $P(x)$ denote the statement "$x$ is taking a math course." The domain of discourse is the set of all students. Write each proposition in Exercises 32–37 in words.*

**32.** $\forall x\, P(x)$ **33.** $\exists x\, P(x)$

**34.** $\forall x \neg P(x)$ **35.** $\exists x \neg P(x)$

**36.** $\neg(\forall x\, P(x))$ **37.** $\neg(\exists x\, P(x))$

**38.** Write the negation of each proposition in Exercises 32–37 symbolically and in words.

*Let $P(x)$ denote the statement "$x$ is a professional athlete," and let $Q(x)$ denote the statement "$x$ plays soccer." The domain of discourse is the set of all people. Write each proposition in Exercises 39–46 in words. Determine the truth value of each statement.*

**39.** $\forall x\, (P(x) \rightarrow Q(x))$ **40.** $\exists x\, (P(x) \rightarrow Q(x))$

**41.** $\forall x\, (Q(x) \rightarrow P(x))$ **42.** $\exists x\, (Q(x) \rightarrow P(x))$

**43.** $\forall x\, (P(x) \vee Q(x))$ **44.** $\exists x\, (P(x) \vee Q(x))$

**45.** $\forall x\, (P(x) \wedge Q(x))$ **46.** $\exists x\, (P(x) \wedge Q(x))$

**47.** Write the negation of each proposition in Exercises 39–46 symbolically and in words.

*Let $P(x)$ denote the statement "$x$ is an accountant," and let $Q(x)$ denote the statement "$x$ owns a Porsche." Write each statement in Exercises 48–51 symbolically.*

**48.** All accountants own Porsches.

**49.** Some accountant owns a Porsche.

**50.** All owners of Porsches are accountants.

**51.** Someone who owns a Porsche is an accountant.

**52.** Write the negation of each proposition in Exercises 48–51 symbolically and in words.

*Determine the truth value of each statement in Exercises 53–58. The domain of discourse is $\mathbf{R}$. Justify your answers.*

**53.** $\forall x(x^2 > x)$ **54.** $\exists x(x^2 > x)$

**55.** $\forall x(x > 1 \to x^2 > x)$

**56.** $\exists x(x > 1 \to x^2 > x)$

**57.** $\forall x(x > 1 \to x/(x^2 + 1) < 1/3)$

**58.** $\exists x(x > 1 \to x/(x^2 + 1) < 1/3)$

**59.** Write the negation of each proposition in Exercises 53–58 symbolically and in words.

**60.** Could the pseudocode of Example 1.5.7 be written as follows?

```
for i = 1 to n
    if (¬P(d_i))
        return false
    else
        return true
```

*What is the literal meaning of each statement in Exercises 61–71? What is the intended meaning? Clarify each statement by rephrasing it and writing it symbolically.*

**61.** From *Dear Abby:* All men do not cheat on their wives.

**62.** From the *San Antonio Express-News:* All old things don't covet twenty-somethings.

**63.** All 74 hospitals did not report every month.

**64.** Economist Robert J. Samuelson: Every environmental problem is not a tragedy.

**65.** Comment from a Door County alderman: This is still Door County and we all don't have a degree.

**66.** Headline over a Martha Stewart column: All lampshades can't be cleaned.

**67.** Headline in the *New York Times:* A World Where All Is Not Sweetness and Light.

**68.** Headline over a story about subsidized housing: Everyone can't afford home.

**69.** George W. Bush: I understand everybody in this country doesn't agree with the decisions I've made.

**70.** From *Newsweek:* Formal investigations are a sound practice in the right circumstances, but every circumstance is not right.

**71.** Joe Girardi (manager of the New York Yankees): Every move is not going to work out.

**72.** (a) Use a truth table to prove that if $p$ and $q$ are propositions, at least one of $p \to q$ or $q \to p$ is true.

(b) Let $I(x)$ be the propositional function "$x$ is an integer" and let $P(x)$ be the propositional function "$x$ is a positive number." The domain of discourse is $\mathbf{R}$. Determine whether or not the following proof that all integers are positive or all positive real numbers are integers is correct.

By part (a),

$$\forall x\,((I(x) \to P(x)) \vee (P(x) \to I(x)))$$

is true. In words: For all $x$, if $x$ is an integer, then $x$ is positive; or if $x$ is positive, then $x$ is an integer. Therefore, all integers are positive or all positive real numbers are integers.

**73.** Prove Theorem 1.5.14, part (b).

**74.** Analyze the following comments by film critic Roger Ebert: No good movie is too long. No bad movie is short enough. *Love Actually* is good, but it is too long.

**75.** Which rule of inference is used in the following argument? Every rational number is of the form $p/q$, where $p$ and $q$ are integers. Therefore, 9.345 is of the form $p/q$.

*In Exercises 76–78, give an argument using rules of inference to show that the conclusion follows from the hypotheses.*

**76.** Hypotheses: Everyone in the class has a graphing calculator. Everyone who has a graphing calculator understands the trigonometric functions. Conclusion: Ralphie, who is in the class, understands the trigonometric functions.

**77.** Hypotheses: Ken, a member of the Titans, can hit the ball a long way. Everyone who can hit the ball a long way can make a lot of money. Conclusion: Some member of the Titans can make a lot of money.

**78.** Hypotheses: Everyone in the discrete mathematics class loves proofs. Someone in the discrete mathematics class has never taken calculus. Conclusion: Someone who loves proofs has never taken calculus.

**79.** Show that universal generalization (see Table 1.5.1) is valid.

**80.** Show that existential instantiation (see Table 1.5.1) is valid.

**81.** Show that existential generalization (see Table 1.5.1) is valid.

## 1.6 Nested Quantifiers

Consider writing the statement

The sum of any two positive real numbers is positive,

symbolically. We first note that since two numbers are involved, we will need two variables, say $x$ and $y$. The assertion can be restated as: If $x > 0$ and $y > 0$, then $x + y > 0$. The given statement says that the sum of *any* two positive real numbers is positive, so we need two universal quantifiers. If we let $P(x, y)$ denote the expression $(x > 0) \wedge (y > 0) \to (x + y > 0)$, the given statement can be written symbolically as

$$\forall x \forall y\, P(x, y).$$

In words, for every $x$ and for every $y$, if $x > 0$ and $y > 0$, then $x + y > 0$. The domain of discourse of the two-variable propositional function $P$ is $\mathbf{R} \times \mathbf{R}$, which means that each variable $x$ and $y$ must belong to the set of real numbers. Multiple quantifiers such as $\forall x \forall y$ are said to be **nested quantifiers.** In this section we explore nested quantifiers in detail.

**Example 1.6.1** Restate $\forall m \exists n(m < n)$ in words. The domain of discourse is the set $\mathbf{Z} \times \mathbf{Z}$.

**SOLUTION** We may first rephrase this statement as: For every $m$, there exists $n$ such that $m < n$. Less formally, this means that if you take any integer $m$ whatsoever, there is an integer $n$ greater than $m$. Another restatement is then: There is no greatest integer. ◀

**Example 1.6.2** Write the assertion

Everybody loves somebody,

symbolically, letting $L(x, y)$ be the statement "$x$ loves $y$."

**SOLUTION** "Everybody" requires universal quantification and "somebody" requires existential quantification. Thus, the given statement may be written symbolically as

$$\forall x \exists y\, L(x, y).$$

In words, for every person $x$, there exists a person $y$ such that $x$ loves $y$.

Notice that

$$\exists x \forall y\, L(x, y)$$

is *not* a correct interpretation of the original statement. This latter statement is: There exists a person $x$ such that for all $y$, $x$ loves $y$. Less formally, someone loves everyone. The order of quantifiers is important; changing the order can change the meaning. ◀

By definition, the statement $\forall x \forall y\, P(x, y)$, with domain of discourse $X \times Y$, is true if, for *every* $x \in X$ and for *every* $y \in Y$, $P(x, y)$ is true. The statement $\forall x \forall y\, P(x, y)$ is false if there is *at least one* $x \in X$ and *at least one* $y \in Y$ such that $P(x, y)$ is false.

**Example 1.6.3** Consider the statement

$$\forall x \forall y((x > 0) \wedge (y > 0) \rightarrow (x + y > 0)).$$

The domain of discourse is $\mathbf{R} \times \mathbf{R}$. This statement is true because, for every real number $x$ and for every real number $y$, the conditional proposition

$$(x > 0) \wedge (y > 0) \rightarrow (x + y > 0)$$

is true. In words, for every real number $x$ and for every real number $y$, if $x$ and $y$ are positive, their sum is positive. ◀

**Example 1.6.4** Consider the statement

$$\forall x \forall y((x > 0) \wedge (y < 0) \rightarrow (x + y \neq 0)).$$

The domain of discourse is $\mathbf{R} \times \mathbf{R}$. This statement is false because if $x = 1$ and $y = -1$, the conditional proposition

$$(x > 0) \wedge (y < 0) \rightarrow (x + y \neq 0)$$

is false. We say that the pair $x = 1$ and $y = -1$ is a counterexample. ◀

**Example 1.6.5** Suppose that $P$ is a propositional function with domain of discourse $\{d_1, \ldots, d_n\} \times \{d_1, \ldots, d_n\}$. The following pseudocode determines whether $\forall x \forall y\, P(x, y)$ is true or false:

```
for i = 1 to n
    for j = 1 to n
        if (¬P(d_i, d_j))
            return false
return true
```

The for loops examine members of the domain of discourse. If they find a pair $d_i$, $d_j$ for which $P(d_i, d_j)$ is false, the condition $\neg P(d_i, d_j)$ in the if statement is true; so the code returns false [to indicate that $\forall x \forall y\, P(x, y)$ is false] and terminates. In this case, the pair $d_i$, $d_j$ is a counterexample. If $P(d_i, d_j)$ is true for every pair $d_i$, $d_j$, the condition $\neg P(d_i, d_j)$ in the if statement is always false. In this case, the for loops run to completion, after which the code returns true [to indicate that $\forall x \forall y\, P(x, y)$ is true] and terminates. ◀

By definition, the statement $\forall x \exists y\, P(x, y)$, with domain of discourse $X \times Y$, is true if, for *every* $x \in X$, there is *at least one* $y \in Y$ for which $P(x, y)$ is true. The statement $\forall x \exists y\, P(x, y)$ is false if there is *at least one* $x \in X$ such that $P(x, y)$ is false for *every* $y \in Y$.

**Example 1.6.6** Consider the statement

$$\forall x \exists y (x + y = 0).$$

The domain of discourse is $\mathbf{R} \times \mathbf{R}$. This statement is true because, for every real number $x$, there is at least one $y$ (namely $y = -x$) for which $x + y = 0$ is true. In words, for every real number $x$, there is a number that when added to $x$ makes the sum zero. ◀

**Example 1.6.7** Consider the statement

$$\forall x \exists y (x > y).$$

The domain of discourse is $\mathbf{Z}^+ \times \mathbf{Z}^+$. This statement is false because there is at least one $x$, namely $x = 1$, such that $x > y$ is false for every positive integer $y$. ◀

**Example 1.6.8** Suppose that $P$ is a propositional function with domain of discourse $\{d_1, \ldots, d_n\} \times \{d_1, \ldots, d_n\}$. The following pseudocode determines whether $\forall x \exists y\, P(x, y)$ is true or false:

```
for i = 1 to n
    if (¬ exists_dj(i))
        return false
return true
exists_dj(i) {
    for j = 1 to n
        if (P(d_i, d_j))
            return true
    return false
}
```

If for each $d_i$, there exists $d_j$ such that $P(d_i, d_j)$ is true, then for each $i$, $P(d_i, d_j)$ is true for some $j$. Thus, *exists_dj(i)* returns true for every $i$. Since $\neg$ *exists_dj(i)* is always false, the first for loop eventually terminates and true is returned to indicate that $\forall x \exists y\, P(x, y)$ is true.

If for some $d_i$, $P(d_i, d_j)$ is false for every $j$, then, for this $i$, $P(d_i, d_j)$ is false for every $j$. In this case, the for loop in *exists_dj(i)* runs to termination and false is returned. Since $\neg$ *exists_dj(i)* is true, false is returned to indicate that $\forall x \exists y\, P(x, y)$ is false. ◀

By definition, the statement $\exists x \forall y\, P(x, y)$, with domain of discourse $X \times Y$, is true if there is *at least one* $x \in X$ such that $P(x, y)$ is true for *every* $y \in Y$. The statement $\exists x \forall y\, P(x, y)$ is false if, for *every* $x \in X$, there is *at least one* $y \in Y$ such that $P(x, y)$ is false.

**Example 1.6.9** Consider the statement $\exists x \forall y (x \leq y)$. The domain of discourse is $\mathbf{Z}^+ \times \mathbf{Z}^+$. This statement is true because there is at least one positive integer $x$ (namely $x = 1$) for which $x \leq y$ is true for every positive integer $y$. In words, there is a smallest positive integer (namely 1). ◀

**Example 1.6.10** Consider the statement $\exists x \forall y (x \geq y)$. The domain of discourse is $\mathbf{Z}^+ \times \mathbf{Z}^+$. This statement is false because, for every positive integer $x$, there is at least one positive integer $y$, namely $y = x + 1$, such that $x \geq y$ is false. In words, there is no greatest positive integer. ◀

By definition, the statement $\exists x \exists y\, P(x, y)$, with domain of discourse $X \times Y$, is true if there is *at least one* $x \in X$ and *at least one* $y \in Y$ such that $P(x, y)$ is true. The statement $\exists x \exists y\, P(x, y)$ is false if, for *every* $x \in X$ and for *every* $y \in Y$, $P(x, y)$ is false.

**Example 1.6.11** Consider the statement

$$\exists x \exists y ((x > 1) \wedge (y > 1) \wedge (xy = 6)).$$

The domain of discourse is $\mathbf{Z}^+ \times \mathbf{Z}^+$. This statement is true because there is at least one integer $x > 1$ (namely $x = 2$) and at least one integer $y > 1$ (namely $y = 3$) such that $xy = 6$. In words, 6 is composite (i.e., not prime). ◀

**Example 1.6.12** Consider the statement

$$\exists x \exists y ((x > 1) \wedge (y > 1) \wedge (xy = 7)).$$

The domain of discourse is $\mathbf{Z}^+ \times \mathbf{Z}^+$. This statement is false because for every positive integer $x$ and for every positive integer $y$,

$$(x > 1) \wedge (y > 1) \wedge (xy = 7)$$

is false. In words, 7 is prime. ◀

The generalized De Morgan's laws for logic (Theorem 1.5.14) can be used to negate a proposition containing nested quantifiers.

**Example 1.6.13** Using the generalized De Morgan's laws for logic, we find that the negation of $\forall x \exists y\, P(x, y)$ is

$$\neg(\forall x \exists y\, P(x, y)) \equiv \exists x \neg(\exists y\, P(x, y)) \equiv \exists x \forall y \neg P(x, y).$$

Notice how in the negation, $\forall$ and $\exists$ are interchanged. ◀

**Example 1.6.14** Write the negation of $\exists x \forall y (xy < 1)$, where the domain of discourse is $\mathbf{R} \times \mathbf{R}$. Determine the truth value of the given statement and its negation.

**SOLUTION** Using the generalized De Morgan's laws for logic, we find that the negation is

$$\neg(\exists x \forall y(xy < 1)) \equiv \forall x \neg(\forall y(xy < 1)) \equiv \forall x \exists y \neg(xy < 1) \equiv \forall x \exists y(xy \geq 1).$$

The given statement $\exists x \forall y(xy < 1)$ is true because there is at least one $x$ (namely $x = 0$) such that $xy < 1$ for every $y$. Since the given statement is true, its negation is false. ◀

We conclude with a logic game, which presents an alternative way to determine whether a quantified propositional function is true or false. André Berthiaume contributed this example.

**Example 1.6.15** **The Logic Game** Given a quantified propositional function such as $\forall x \exists y P(x, y)$, you and your opponent, whom we call Farley, play a logic game. Your goal is to try to make $P(x, y)$ true, and Farley's goal is to try to make $P(x, y)$ false. The game begins with the first (left) quantifier. If the quantifier is $\forall$, Farley chooses a value for that variable; if the quantifier is $\exists$, you choose a value for that variable. The game continues with the second quantifier. After values are chosen for all the variables, if $P(x, y)$ is true, you win; if $P(x, y)$ is false, Farley wins. We will show that if you can always win regardless of how Farley chooses values for the variables, the quantified statement is true, but if Farley can choose values for the variables so that you cannot win, the quantified statement is false.

Consider the statement

$$\forall x \exists y(x + y = 0). \tag{1.6.1}$$

The domain of discourse is $\mathbf{R} \times \mathbf{R}$. Since the first quantifier is $\forall$, Farley goes first and chooses a value for $x$. Since the second quantifier is $\exists$, you go second. Regardless of what value Farley chose, you can choose $y = -x$, which makes the statement $x + y = 0$ true. You can always win the game, so the statement (1.6.1) is true.

Next, consider the statement

$$\exists x \forall y(x + y = 0). \tag{1.6.2}$$

Again, the domain of discourse is $\mathbf{R} \times \mathbf{R}$. Since the first quantifier is $\exists$, you go first and choose a value for $x$. Since the second quantifier is $\forall$, Farley goes second. Regardless of what value you chose, Farley can always choose a value for $y$, which makes the statement $x + y = 0$ false. (If you choose $x = 0$, Farley can choose $y = 1$. If you choose $x \neq 0$, Farley can choose $y = 0$.) Farley can always win the game, so the statement (1.6.2) is false.

We discuss why the game correctly determines the truth value of a quantified propositional function. Consider $\forall x \forall y P(x, y)$. If Farley can always win the game, this means that Farley can find values for $x$ and $y$ that make $P(x, y)$ false. In this case, the propositional function is false; the values Farley found provide a counterexample. If Farley cannot win the game, no counterexample exists; in this case, the propositional function is true.

Consider $\forall x \exists y P(x, y)$. Farley goes first and chooses a value for $x$. You choose second. If, no matter what value Farley chose, you can choose a value for $y$ that makes $P(x, y)$ true, you can always win the game and the propositional function is true. However, if Farley can choose a value for $x$ so that every value you choose for $y$ makes $P(x, y)$ false, then you will always lose the game and the propositional function is false.

An analysis of the other cases also shows that if you can always win the game, the propositional function is true; but if Farley can always win the game, the propositional function is false.

The logic game extends to propositional functions of more than two variables. The rules are the same and, again, if you can always win the game, the propositional function is true; but if Farley can always win the game, the propositional function is false. ◀

## 1.6 Problem-Solving Tips

- To prove that $\forall x \forall y\, P(x, y)$ is true, where the domain of discourse is $X \times Y$, you must show that $P(x, y)$ is true for all values of $x \in X$ and $y \in Y$. One technique is to argue that $P(x, y)$ is true using the symbols $x$ and $y$ to stand for *arbitrary* elements in $X$ and $Y$.
- To prove that $\forall x \forall y\, P(x, y)$ is false, where the domain of discourse is $X \times Y$, find one value of $x \in X$ and one value of $y \in Y$ (*two* values suffice—one for $x$ and one for $y$) that make $P(x, y)$ false.
- To prove that $\forall x \exists y\, P(x, y)$ is true, where the domain of discourse is $X \times Y$, you must show that for all $x \in X$, there is at least one $y \in Y$ such that $P(x, y)$ is true. One technique is to let $x$ stand for an arbitrary element in $X$ and then find a value for $y \in Y$ (*one* value suffices!) that makes $P(x, y)$ true.
- To prove that $\forall x \exists y\, P(x, y)$ is false, where the domain of discourse is $X \times Y$, you must show that for at least one $x \in X$, $P(x, y)$ is false for every $y \in Y$. One technique is to find a value of $x \in X$ (again *one* value suffices!) that has the property that $P(x, y)$ is false for every $y \in Y$. Having chosen a value for $x$, let $y$ stand for an arbitrary element of $Y$ and show that $P(x, y)$ is always false.
- To prove that $\exists x \forall y\, P(x, y)$ is true, where the domain of discourse is $X \times Y$, you must show that for at least one $x \in X$, $P(x, y)$ is true for every $y \in Y$. One technique is to find a value of $x \in X$ (again *one* value suffices!) that has the property that $P(x, y)$ is true for every $y \in Y$. Having chosen a value for $x$, let $y$ stand for an arbitrary element of $Y$ and show that $P(x, y)$ is always true.
- To prove that $\exists x \forall y\, P(x, y)$ is false, where the domain of discourse is $X \times Y$, you must show that for all $x \in X$, there is at least one $y \in Y$ such that $P(x, y)$ is false. One technique is to let $x$ stand for an arbitrary element in $X$ and then find a value for $y \in Y$ (*one* value suffices!) that makes $P(x, y)$ false.
- To prove that $\exists x \exists y\, P(x, y)$ is true, where the domain of discourse is $X \times Y$, find one value of $x \in X$ and one value of $y \in Y$ (*two* values suffice—one for $x$ and one for $y$) that make $P(x, y)$ true.
- To prove that $\exists x \exists y\, P(x, y)$ is false, where the domain of discourse is $X \times Y$, you must show that $P(x, y)$ is false for all values of $x \in X$ and $y \in Y$. One technique is to argue that $P(x, y)$ is false using the symbols $x$ and $y$ to stand for *arbitrary* elements in $X$ and $Y$.
- To negate an expression with nested quantifiers, use the generalized De Morgan's laws for logic. Loosely speaking, $\forall$ and $\exists$ are interchanged. Don't forget that the negation of $p \to q$ is equivalent to $p \wedge \neg q$.

## 1.6 Review Exercises

1. What is the interpretation of $\forall x \forall y P(x, y)$? When is this quantified expression true? When is it false?
2. What is the interpretation of $\forall x \exists y P(x, y)$? When is this quantified expression true? When is it false?
3. What is the interpretation of $\exists x \forall y P(x, y)$? When is this quantified expression true? When is it false?
4. What is the interpretation of $\exists x \exists y P(x, y)$? When is this quantified expression true? When is it false?
5. Give an example to show that, in general, $\forall x \exists y P(x, y)$ and $\exists x \forall y P(x, y)$ have different meanings.
6. Write the negation of $\forall x \forall y P(x, y)$ using the generalized De Morgan's laws for logic.

7. Write the negation of $\forall x \exists y P(x, y)$ using the generalized De Morgan's laws for logic.

8. Write the negation of $\exists x \forall y P(x, y)$ using the generalized De Morgan's laws for logic.

9. Write the negation of $\exists x \exists y P(x, y)$ using the generalized De Morgan's laws for logic.

10. Explain the rules for playing the logic game. How can the logic game be used to determine the truth value of a quantified expression?

## 1.6 Exercises

*In Exercises 1–33, the set $D_1$ consists of three students: Garth, who is 5 feet 11 inches tall; Erin, who is 5 feet 6 inches tall; and Marty, who is 6 feet tall. The set $D_2$ consists of four students: Dale, who is 6 feet tall; Garth, who is 5 feet 11 inches tall; Erin, who is 5 feet 6 inches tall; and Marty, who is 6 feet tall. The set $D_3$ consists of one student: Dale, who is 6 feet tall. The set $D_4$ consists of three students: Pat, Sandy, and Gale, each of whom is 5 feet 11 inches tall.*

*In Exercises 1–21, $T_1(x, y)$ is the propositional function "x is taller than y." Write each proposition in Exercises 1–4 in words.*

1. $\forall x \forall y\, T_1(x, y)$
2. $\forall x \exists y\, T_1(x, y)$
3. $\exists x \forall y\, T_1(x, y)$
4. $\exists x \exists y\, T_1(x, y)$
5. Write the negation of each proposition in Exercises 1–4 in words and symbolically.

*In Exercises 6–21, tell whether each proposition in Exercises 1–4 is true or false if the domain of discourse is $D_i \times D_j$ for the given values of i and j.*

6. $i = 1,\ j = 1$
7. $i = 1,\ j = 2$
8. $i = 1,\ j = 3$
9. $i = 1,\ j = 4$
10. $i = 2,\ j = 1$
11. $i = 2,\ j = 2$
12. $i = 2,\ j = 3$
13. $i = 2,\ j = 4$
14. $i = 3,\ j = 1$
15. $i = 3,\ j = 2$
16. $i = 3,\ j = 3$
17. $i = 3,\ j = 4$
18. $i = 4,\ j = 1$
19. $i = 4,\ j = 2$
20. $i = 4,\ j = 3$
21. $i = 4,\ j = 4$

*In Exercises 22–27, $T_2(x, y)$ is the propositional function "x is taller than or the same height as y." Write each proposition in Exercises 22–25 in words.*

22. $\forall x \forall y\, T_2(x, y)$
23. $\forall x \exists y\, T_2(x, y)$
24. $\exists x \forall y\, T_2(x, y)$
25. $\exists x \exists y\, T_2(x, y)$
26. Write the negation of each proposition in Exercises 22–25 in words and symbolically.
27. Tell whether each proposition in Exercises 22–25 is true or false if the domain of discourse is $D_i \times D_j$ for each pair of values $i, j$ given in Exercises 6–21. The sets $D_1, \ldots, D_4$ are defined before Exercise 1.

*In Exercises 28–33, $T_3(x, y)$ is the propositional function "if x and y are distinct persons, then x is taller than y." Write each proposition in Exercises 28–31 in words.*

28. $\forall x \forall y\, T_3(x, y)$
29. $\forall x \exists y\, T_3(x, y)$
30. $\exists x \forall y\, T_3(x, y)$
31. $\exists x \exists y\, T_3(x, y)$
32. Write the negation of each proposition in Exercises 28–31 in words and symbolically.
33. Tell whether each proposition in Exercises 28–31 is true or false if the domain of discourse is $D_i \times D_j$ for each pair of values $i, j$ given in Exercises 6–21. The sets $D_1, \ldots, D_4$ are defined before Exercise 1.

*Let $L(x, y)$ be the propositional function "x loves y." The domain of discourse is the Cartesian product of the set of all living people with itself (i.e., both x and y take on values in the set of all living people). Write each proposition in Exercises 34–37 symbolically. Which do you think are true?*

34. Someone loves everybody.
35. Everybody loves everybody.
36. Somebody loves somebody.
37. Everybody loves somebody.
38. Write the negation of each proposition in Exercises 34–37 in words and symbolically.

*Let $A(x, y)$ be the propositional function "x attended y's office hours" and let $E(x)$ be the propositional function "x is enrolled in a discrete math class." Let S be the set of students and let T denote the set of teachers—all at Hudson University. The domain of discourse of A is $S \times T$ and the domain of discourse of E is S. Write each proposition in Exercises 39–42 symbolically.*

39. Brit attended someone's office hours.
40. No one attended Professor Sandwich's office hours.
41. Every discrete math student attended someone's office hours.
42. All teachers had at least one student attend their office hours.

*Let $P(x, y)$ be the propositional function $x \geq y$. The domain of discourse is $\mathbf{Z}^+ \times \mathbf{Z}^+$. Tell whether each proposition in Exercises 43–46 is true or false.*

43. $\forall x \forall y\, P(x, y)$
44. $\forall x \exists y\, P(x, y)$
45. $\exists x \forall y\, P(x, y)$
46. $\exists x \exists y\, P(x, y)$
47. Write the negation of each proposition in Exercises 43–46.

*Determine the truth value of each statement in Exercises 48–65. The domain of discourse is $\mathbf{R} \times \mathbf{R}$. Justify your answers.*

48. $\forall x \forall y (x^2 < y + 1)$
49. $\forall x \exists y (x^2 < y + 1)$
50. $\exists x \forall y (x^2 < y + 1)$
51. $\exists x \exists y (x^2 < y + 1)$
52. $\exists y \forall x (x^2 < y + 1)$
53. $\forall y \exists x (x^2 < y + 1)$
54. $\forall x \forall y (x^2 + y^2 = 9)$
55. $\forall x \exists y (x^2 + y^2 = 9)$

**56.** $\exists x \forall y(x^2 + y^2 = 9)$ **57.** $\exists x \exists y(x^2 + y^2 = 9)$

**58.** $\forall x \forall y(x^2 + y^2 \geq 0)$ **59.** $\forall x \exists y(x^2 + y^2 \geq 0)$

**60.** $\exists x \forall y(x^2 + y^2 \geq 0)$ **61.** $\exists x \exists y(x^2 + y^2 \geq 0)$

**62.** $\forall x \forall y((x < y) \to (x^2 < y^2))$

**63.** $\forall x \exists y((x < y) \to (x^2 < y^2))$

**64.** $\exists x \forall y((x < y) \to (x^2 < y^2))$

**65.** $\exists x \exists y((x < y) \to (x^2 < y^2))$

**66.** Write the negation of each proposition in Exercises 48–65.

**67.** Suppose that $P$ is a propositional function with domain of discourse $\{d_1, \ldots, d_n\} \times \{d_1, \ldots, d_n\}$. Write pseudocode that determines whether

$$\exists x \forall y\, P(x, y)$$

is true or false.

**68.** Suppose that $P$ is a propositional function with domain of discourse $\{d_1, \ldots, d_n\} \times \{d_1, \ldots, d_n\}$. Write pseudocode that determines whether

$$\exists x \exists y\, P(x, y)$$

is true or false.

**69.** Explain how the logic game (Example 1.6.15) determines whether each proposition in Exercises 48–65 is true or false.

**70.** Use the logic game (Example 1.6.15) to determine whether the proposition

$$\forall x \forall y \exists z((z > x) \wedge (z < y))$$

is true or false. The domain of discourse is $\mathbf{Z} \times \mathbf{Z} \times \mathbf{Z}$.

**71.** Use the logic game (Example 1.6.15) to determine whether the proposition

$$\forall x \forall y \exists z((z < x) \wedge (z < y))$$

is true or false. The domain of discourse is $\mathbf{Z} \times \mathbf{Z} \times \mathbf{Z}$.

**72.** Use the logic game (Example 1.6.15) to determine whether the proposition

$$\forall x \forall y \exists z((x < y) \to ((z > x) \wedge (z < y)))$$

is true or false. The domain of discourse is $\mathbf{Z} \times \mathbf{Z} \times \mathbf{Z}$.

**73.** Use the logic game (Example 1.6.15) to determine whether the proposition

$$\forall x \forall y \exists z((x < y) \to ((z > x) \wedge (z < y)))$$

is true or false. The domain of discourse is $\mathbf{R} \times \mathbf{R} \times \mathbf{R}$.

*Assume that $\forall x \forall y\, P(x, y)$ is true and that the domain of discourse is nonempty. Which of Exercises 74–76 must also be true? Prove your answer.*

**74.** $\forall x \exists y\, P(x, y)$ **75.** $\exists x \forall y\, P(x, y)$ **76.** $\exists x \exists y\, P(x, y)$

*Assume that $\exists x \forall y\, P(x, y)$ is true and that the domain of discourse is nonempty. Which of Exercises 77–79 must also be true? Prove your answer.*

**77.** $\forall x \forall y\, P(x, y)$ **78.** $\forall x \exists y\, P(x, y)$ **79.** $\exists x \exists y\, P(x, y)$

*Assume that $\exists x \exists y\, P(x, y)$ is true and that the domain of discourse is nonempty. Which of Exercises 80–82 must also be true? Prove your answer.*

**80.** $\forall x \forall y\, P(x, y)$

**81.** $\forall x \exists y\, P(x, y)$

**82.** $\exists x \forall y\, P(x, y)$

*Assume that $\forall x \forall y\, P(x, y)$ is false and that the domain of discourse is nonempty. Which of Exercises 83–85 must also be false? Prove your answer.*

**83.** $\forall x \exists y\, P(x, y)$ **84.** $\exists x \forall y\, P(x, y)$ **85.** $\exists x \exists y\, P(x, y)$

*Assume that $\forall x \exists y\, P(x, y)$ is false and that the domain of discourse is nonempty. Which of Exercises 86–88 must also be false? Prove your answer.*

**86.** $\forall x \forall y\, P(x, y)$

**87.** $\exists x \forall y\, P(x, y)$

**88.** $\exists x \exists y\, P(x, y)$

*Assume that $\exists x \forall y\, P(x, y)$ is false and that the domain of discourse is nonempty. Which of Exercises 89–91 must also be false? Prove your answer.*

**89.** $\forall x \forall y\, P(x, y)$ **90.** $\forall x \exists y\, P(x, y)$ **91.** $\exists x \exists y\, P(x, y)$

*Assume that $\exists x \exists y\, P(x, y)$ is false and that the domain of discourse is nonempty. Which of Exercises 92–94 must also be false? Prove your answer.*

**92.** $\forall x \forall y\, P(x, y)$

**93.** $\forall x \exists y\, P(x, y)$

**94.** $\exists x \forall y\, P(x, y)$

*Which of Exercises 95–98 is logically equivalent to $\neg(\forall x \exists y\, P(x, y))$? Explain.*

**95.** $\exists x \neg(\forall y\, P(x, y))$ **96.** $\forall x \neg(\exists y\, P(x, y))$

**97.** $\exists x \forall y \neg P(x, y)$ **98.** $\exists x \exists y \neg P(x, y)$

**99.** [Requires calculus] The definition of

$$\lim_{x \to a} f(x) = L$$

is: For every $\varepsilon > 0$, there exists $\delta > 0$ such that for all $x$ if $0 < |x - a| < \delta$, then $|f(x) - L| < \varepsilon$. Write this definition symbolically using $\forall$ and $\exists$.

**100.** [Requires calculus] Write the negation of the definition of limit (see Exercise 99) in words and symbolically using $\forall$ and $\exists$ but not $\neg$.

★**101.** [Requires calculus] Write the definition of "$\lim_{x \to a} f(x)$ does not exist" (see Exercise 99) in words and symbolically using $\forall$ and $\exists$ but not $\neg$.

**102.** Consider the headline: Every school may not be right for every child. What is the literal meaning? What is the intended meaning? Clarify the headline by rephrasing it and writing it symbolically.

# Problem-Solving Corner — Quantifiers

## Problem

Assume that $\forall x \exists y\, P(x, y)$ is true and that the domain of discourse is nonempty. Which of the following must also be true? If the statement is true, explain; otherwise, give a counterexample.

(a) $\forall x \forall y\, P(x, y)$

(b) $\exists x \forall y\, P(x, y)$

(c) $\exists x \exists y\, P(x, y)$

## Attacking the Problem

Let's begin with part (a). We are given that $\forall x \exists y\, P(x, y)$ is true, which says, in words, for every $x$, there exists at least one $y$ for which $P(x, y)$ is true. If (a) is also true, then, in words, for every $x$, *for every* $y$, $P(x, y)$ is true. Let the words sink in. If for every $x$, $P(x, y)$ is true for *at least one* $y$, doesn't it seem unlikely that it would follow that $P(x, y)$ is true for *every* $y$? We suspect that (a) could be false. We'll need to come up with a counterexample.

Contrasting statement (b) with the given statement, we see that the quantifiers $\forall$ and $\exists$ have been swapped. There is a difference. In the given true statement $\forall x \exists y\, P(x, y)$, given *any* $x$, it's possible to find a $y$, which may depend on $x$, that makes $P(x, y)$ true. For statement (b), $\exists x \forall y\, P(x, y)$, to be true, for some $x$, $P(x, y)$ would need to be true for every $y$. Again, let the words sink in. These two statements seem quite different. We suspect that (b) also could be false. Again, we'll need to come up with a counterexample.

Now let's turn to part (c). We are given that $\forall x \exists y\, P(x, y)$ is true, which says, in words, for every $x$, there exists at least one $y$ for which $P(x, y)$ is true. For statement (c), $\exists x \exists y\, P(x, y)$, to be true, for some $x$ and for some $y$, $P(x, y)$ must be true. But the given statement says that for *every* $x$, there exists at least one $y$ for which $P(x, y)$ is true. So if we pick one $x$ (and we know we can since the domain of discourse is nonempty), the given statement assures us that there exists at least one $y$ for which $P(x, y)$ is true. Thus part (c) must be true. In fact, we have just given an explanation!

## Finding a Solution

As noted, we have already solved part (c). We need counterexamples for parts (a) and (b).

For part (a), we need the given statement, $\forall x \exists y\, P(x, y)$, to be true and $\forall x \forall y\, P(x, y)$ to be false. In order for the given statement to be true, we must find a propositional function $P(x, y)$ satisfying

for every $x$, there exists $y$ such that $P(x, y)$ is true. **(1)**

In order for (a) to be false, we must have

at least one value of $x$ and at least one value of $y$ such that $P(x, y)$ is false. **(2)**

We can arrange for (1) and (2) to hold simultaneously if we choose $P(x, y)$ so that for every $x$, $P(x, y)$ is true for some $y$, *but* for at least one $x$, $P(x, y)$ is also false for some other value of $y$. Upon reflection, many mathematical statements have this property. For example, $x > y$, $x, y \in \mathbf{R}$, suffices. For every $x$, there exists $y$ such that $x > y$ is true. Furthermore, for *every* $x$ (and, in particular, for at least one value of $x$), there exists $y$ such that $x > y$ is false.

For part (b), we again need the given statement, $\forall x \exists y\, P(x, y)$, to be true and $\exists x \forall y\, P(x, y)$ to be false. In order for the given statement to be true, we must find a propositional function $P(x, y)$ satisfying (1). In order for (b) to be false, we must have

for every $x$, there exists at least one value of $y$ such that $P(x, y)$ is false. **(3)**

We can arrange for (1) and (3) to hold simultaneously if we choose $P(x, y)$ so that for every $x$, $P(x, y)$ is true for some $y$ *and* false for some other value of $y$. We noted in the preceding paragraph that $x > y$, $x, y \in \mathbf{R}$, has this property.

## Formal Solution

(a) We give an example to show that statement (a) can be false while the given statement is true. Let $P(x, y)$ be the propositional function $x > y$ with domain of discourse $\mathbf{R} \times \mathbf{R}$. Then $\forall x \exists y\, P(x, y)$ is true since for any $x$, we may choose $y = x - 1$ to make $P(x, y)$ true. At the same time, $\forall x \forall y\, P(x, y)$ is false. A counterexample is $x = 0$, $y = 1$.

(b) We give an example to show that statement (b) can be false while the given statement is true. Let $P(x, y)$ be the propositional function $x > y$ with domain of discourse $\mathbf{R} \times \mathbf{R}$. As we showed in part (a), $\forall x \exists y\, P(x, y)$ is true. Now we show that $\exists x \forall y\, P(x, y)$ is false. Let $x$ be an arbitrary element in $\mathbf{R}$. We may choose $y = x + 1$ to make $x > y$ false. Thus for every $x$, there exists $y$ such that $P(x, y)$ is false. Therefore statement (b) is false.

(c) We show that if the given statement is true, statement (c) is necessarily true.

We are given that for every $x$, there exists $y$ such that $P(x, y)$ is true. We must show that there exist $x$ and $y$ such that $P(x, y)$ is true. Since the domain of discourse is nonempty, we may choose a value for $x$. For this chosen $x$, there exists $y$ such that $P(x, y)$ is true. We have found at least one value for $x$ and at least one value for $y$ that make $P(x, y)$ true. Therefore $\exists x \exists y\, P(x, y)$ is true.

## Summary of Problem-Solving Techniques

- When dealing with quantified statements, it is sometimes useful to write out the statements in words. For example, in this problem, it helped to write out exactly what $\forall x \exists y\, P(x, y)$ means. Take time to let the words sink in.
- If you have trouble finding examples, look at existing examples (e.g., examples in this book). To solve problems (a) and (b), we could have used the statement in Example 1.6.6. Sometimes, an existing example can be modified to solve a given problem.

## Exercises

1. Show that the statement in Example 1.6.6 solves problems (a) and (b) in this Problem-Solving Corner.
2. Could examples in Section 1.6 other than Example 1.6.6 have been used to solve problems (a) and (b) in this Problem-Solving Corner?

# Chapter 1 Notes

General references on discrete mathematics are [Graham, 1994; Liu, 1985; Tucker]. [Knuth, 1997, 1998a, 1998b] is the classic reference for much of this material.

[Halmos; Lipschutz; and Stoll] are recommended to the reader wanting to study set theory in more detail.

[Barker; Copi; Edgar] are introductory logic textbooks. A more advanced treatment is found in [Davis]. The first chapter of the geometry book by [Jacobs] is devoted to basic logic. For a history of logic, see [Kline]. The role of logic in reasoning about computer programs is discussed by [Gries].

# Chapter 1 Review

## Section 1.1

1. Set: any collection of objects
2. Notation for sets: $\{x \mid x \text{ has property } P\}$
3. $|X|$, the cardinality of $X$: the number of elements in the set $X$
4. $x \in X$: $x$ is an element of the set $X$
5. $x \notin X$: $x$ is not an element of the set $X$
6. Empty set: $\varnothing$ or $\{\,\}$
7. $X = Y$, where $X$ and $Y$ are sets: $X$ and $Y$ have the same elements
8. $X \subseteq Y$, $X$ is a subset of $Y$: every element in $X$ is also in $Y$
9. $X \subset Y$, $X$ is a proper subset of $Y$: $X \subseteq Y$ and $X \neq Y$
10. $\mathcal{P}(X)$, the power set of $X$: set of all subsets of $X$
11. $|\mathcal{P}(X)| = 2^{|X|}$
12. $X \cup Y$, $X$ union $Y$: set of elements in $X$ *or* $Y$ or both
13. Union of a family $\mathcal{S}$ of sets: $\cup\, \mathcal{S} = \{x \mid x \in X \text{ for some } X \in \mathcal{S}\}$
14. $X \cap Y$, $X$ intersect $Y$: set of elements in $X$ *and* $Y$
15. Intersection of a family $\mathcal{S}$ of sets: $\cap\, \mathcal{S} = \{x \mid x \in X \text{ for all } X \in \mathcal{S}\}$
16. Disjoint sets $X$ and $Y$: $X \cap Y = \varnothing$
17. Pairwise disjoint family of sets
18. $X - Y$, difference of $X$ and $Y$, relative complement: set of elements in $X$ but not in $Y$
19. Universal set, universe
20. $\overline{X}$, complement of $X$: $U - X$, where $U$ is a universal set
21. Venn diagram

22. Properties of sets (see Theorem 1.1.22)
23. De Morgan's laws for sets: $\overline{(A \cup B)} = \overline{A} \cap \overline{B}$, $\overline{(A \cap B)} = \overline{A} \cup \overline{B}$
24. Partition of $X$: a collection $\mathcal{S}$ of nonempty subsets of $X$ such that every element in $X$ belongs to exactly one member of $\mathcal{S}$
25. Ordered pair: $(x, y)$
26. Cartesian product of $X$ and $Y$: $X \times Y = \{(x, y) \mid x \in X, y \in Y\}$
27. Cartesian product of $X_1, X_2, \ldots, X_n$:

$$X_1 \times X_2 \times \cdots \times X_n = \{(a_1, a_2, \ldots, a_n) \mid a_i \in X_i\}$$

## Section 1.2

28. Logic
29. Proposition
30. Conjunction: $p$ and $q$, $p \wedge q$
31. Disjunction: $p$ or $q$, $p \vee q$
32. Negation: not $p$, $\neg p$
33. Truth table
34. Exclusive-or of propositions $p$, $q$: $p$ or $q$, but not both

## Section 1.3

35. Conditional proposition: if $p$, then $q$; $p \rightarrow q$
36. Hypothesis
37. Conclusion
38. Necessary condition
39. Sufficient condition
40. Converse of $p \rightarrow q$: $q \rightarrow p$
41. Biconditional proposition: $p$ if and only if $q$, $p \leftrightarrow q$
42. Logical equivalence: $P \equiv Q$
43. De Morgan's laws for logic: $\neg(p \vee q) \equiv \neg p \wedge \neg q$, $\neg(p \wedge q) \equiv \neg p \vee \neg q$
44. Contrapositive of $p \rightarrow q$: $\neg q \rightarrow \neg p$

## Section 1.4

45. Deductive reasoning
46. Hypothesis
47. Premises
48. Conclusion
49. Argument
50. Valid argument
51. Invalid argument
52. Rules of inference for propositions: modus ponens, modus tollens, addition, simplification, conjunction, hypothetical syllogism, disjunctive syllogism

## Section 1.5

53. Propositional function
54. Domain of discourse
55. Universal quantifier
56. Universally quantified statement
57. Counterexample
58. Existential quantifier
59. Existentially quantified statement
60. Generalized De Morgan's laws for logic:

    $\neg(\forall x P(x))$ and $\exists x \neg P(x)$ have the same truth values.
    $\neg(\exists x P(x))$ and $\forall x \neg P(x)$ have the same truth values.

61. To prove that the universally quantified statement $\forall x\, P(x)$ is true, show that for every $x$ in the domain of discourse, the proposition $P(x)$ is true.
62. To prove that the existentially quantified statement $\exists x\, P(x)$ is true, find one value of $x$ in the domain of discourse for which $P(x)$ is true.
63. To prove that the universally quantified statement $\forall x\, P(x)$ is false, find one value of $x$ (a counterexample) in the domain of discourse for which $P(x)$ is false.
64. To prove that the existentially quantified statement $\exists x\, P(x)$ is false, show that for every $x$ in the domain of discourse, the proposition $P(x)$ is false.
65. Rules of inference for quantified statements: universal instantiation, universal generalization, existential instantiation, existential generalization

## Section 1.6

66. To prove that $\forall x \forall y\, P(x, y)$ is true, show that $P(x, y)$ is true for all values of $x \in X$ and $y \in Y$, where the domain of discourse is $X \times Y$.
67. To prove that $\forall x \exists y\, P(x, y)$ is true, show that for all $x \in X$, there is at least one $y \in Y$ such that $P(x, y)$ is true, where the domain of discourse is $X \times Y$.
68. To prove that $\exists x \forall y\, P(x, y)$ is true, show that for at least one $x \in X$, $P(x, y)$ is true for every $y \in Y$, where the domain of discourse is $X \times Y$.
69. To prove that $\exists x \exists y\, P(x, y)$ is true, find one value of $x \in X$ and one value of $y \in Y$ that make $P(x, y)$ true, where the domain of discourse is $X \times Y$.
70. To prove that $\forall x \forall y\, P(x, y)$ is false, find one value of $x \in X$ and one value of $y \in Y$ that make $P(x, y)$ false, where the domain of discourse is $X \times Y$.
71. To prove that $\forall x \exists y\, P(x, y)$ is false, show that for at least one $x \in X$, $P(x, y)$ is false for every $y \in Y$, where the domain of discourse is $X \times Y$.
72. To prove that $\exists x \forall y\, P(x, y)$ is false, show that for all $x \in X$, there is at least one $y \in Y$ such that $P(x, y)$ is false, where the domain of discourse is $X \times Y$.
73. To prove that $\exists x \exists y\, P(x, y)$ is false, show that $P(x, y)$ is false for all values of $x \in X$ and $y \in Y$, where the domain of discourse is $X \times Y$.
74. To negate an expression with nested quantifiers, use the generalized De Morgan's laws for logic.
75. The logic game

## Chapter 1 Self-Test

1. If $A = \{1, 3, 4, 5, 6, 7\}$, $B = \{x \mid x \text{ is an even integer}\}$, $C = \{2, 3, 4, 5, 6\}$, find $(A \cap B) - C$.
2. If $p$, $q$, and $r$ are true, find the truth value of the proposition $(p \vee q) \wedge \neg((\neg p \wedge r) \vee q)$.
3. Restate the proposition "A necessary condition for Leah to get an A in discrete mathematics is to study hard" in the form of a conditional proposition.
4. Write the converse and contrapositive of the proposition of Exercise 3.
5. If $A \cup B = B$, what relation must hold between $A$ and $B$?
6. Are the sets

   $$\{3, 2, 2\}, \qquad \{x \mid x \text{ is an integer and } 1 < x \leq 3\}$$

   equal? Explain.
7. Determine whether the following argument is valid.

   $$\begin{array}{l} p \rightarrow q \vee r \\ p \vee \neg q \\ r \vee q \\ \hline \therefore q \end{array}$$

8. If $p$ is true and $q$ and $r$ are false, find the truth value of the proposition $(p \vee q) \rightarrow \neg r$.
9. Write the following argument symbolically and determine whether it is valid. If the Skyscrapers win, I'll eat my hat. If I eat my hat, I'll be quite full. Therefore, if I'm quite full, the Skyscrapers won.
10. Is the statement

    The team won the 2006 National Basketball Association championship

    a proposition? Explain.
11. Is the statement of Exercise 10 a propositional function? Explain.
12. Let $K(x, y)$ be the propositional function "$x$ knows $y$." The domain of discourse is the Cartesian product of the set of students taking discrete math with itself (i.e., both $x$ and $y$ take on values in the set of students taking discrete math). Represent the assertion "someone does not know anyone" symbolically.
13. Write the negation of the assertion of Exercise 12 symbolically and in words.
14. If $A = \{a, b, c\}$, how many elements are in $\mathcal{P}(A) \times A$?
15. Write the truth table of the proposition $\neg(p \wedge q) \vee (p \vee \neg r)$.
16. Formulate the proposition $p \wedge (\neg q \vee r)$ in words using

    $p$: I take hotel management.
    $q$: I take recreation supervision.
    $r$: I take popular culture.
17. Assume that $a$, $b$, and $c$ are real numbers. Represent the statement

    $$a < b \text{ or } (b < c \text{ and } a \geq c)$$

    symbolically, letting

    $$p\colon a < b, \qquad q\colon b < c, \qquad r\colon a < c.$$

*Let $P(n)$ be the statement $n$ and $n + 2$ are prime. In Exercises 18 and 19, write the statement in words and tell whether it is true or false.*

18. $\forall n\, P(n)$
19. $\exists n\, P(n)$
20. Which rule of inference is used in the following argument? If the Skyscrapers win, I'll eat my hat. If I eat my hat, I'll be quite full. Therefore, if the Skyscrapers win, I'll be quite full.
21. Give an argument using rules of inference to show that the conclusion follows from the hypotheses.

    Hypotheses: If the Council approves the funds, then New Atlantic will get the Olympic Games. If New Atlantic gets the Olympic Games, then New Atlantic will build a new stadium. New Atlantic does not build a new stadium. Conclusion: The Council does not approve the funds, or the Olympic Games are canceled.
22. Determine whether the statement $\forall x \exists y (x = y^3)$ is true or false. The domain of discourse is $\mathbf{R} \times \mathbf{R}$. Explain your answer. Explain, in words, the meaning of the statement.
23. Use the generalized De Morgan's laws for logic to write the negation of $\forall x \exists y \forall z\, P(x, y, z)$.
24. Represent the statement

    $$\text{If } (a \geq c \text{ or } b < c), \text{ then } b \geq c$$

    symbolically using the definitions of Exercise 17.

## Chapter 1 Computer Exercises

*In Exercises 1–6, assume that a set $X$ of $n$ elements is represented as an array $A$ of size at least $n + 1$. The elements of $X$ are listed consecutively in $A$ starting in the first position and terminating with 0. Assume further that no set contains 0.*

1. Write a program to represent the sets $X \cup Y$, $X \cap Y$, $X - Y$, and $X \,\Delta\, Y$, given the arrays representing $X$ and $Y$. (The symmetric difference is denoted $\Delta$.)

2. Write a program to determine whether $X \subseteq Y$, given arrays representing $X$ and $Y$.
3. Write a program to determine whether $X = Y$, given arrays representing $X$ and $Y$.
4. Assuming a universe represented as an array, write a program to represent the set $\overline{X}$, given the array representing $X$.
5. Given an element $E$ and the array $A$ that represents $X$, write a program that determines whether $E \in X$.
6. Given the array representing $X$, write a program that lists all subsets of $X$.
7. Write a program that reads a logical expression in $p$ and $q$ and prints the truth table of the expression.
8. Write a program that reads a logical expression in $p$, $q$, and $r$ and prints the truth table of the expression.
9. Write a program that tests whether two logical expressions in $p$ and $q$ are logically equivalent.
10. Write a program that tests whether two logical expressions in $p$, $q$, and $r$ are logically equivalent.

# Chapter 2

# PROOFS



**Go Online**
For more on proofs, see
`bit.ly/2ECHU4F`

This chapter uses the logic in Chapter 1 to discuss proofs. Logical methods are used in mathematics to prove theorems and in computer science to prove that programs do what they are alleged to do. Suppose, for example, that a student is assigned a program to compute shortest paths between cities. The program must be able to accept as input an arbitrary number of cities and the distances between cities directly connected by roads and produce as output the shortest paths (routes) between each distinct pair of cities. After the student writes the program, it is easy to test it for a small number of cities. Using paper and pencil, the student could simply list all possible paths between pairs of cities and find the shortest paths. This brute-force solution could then be compared with the output of the program. However, for a large number of cities, the brute-force technique would take too long. How can the student be sure that the program works properly for large input—almost surely the kind of input on which the instructor will test the program? The student will have to use logic to prove that the program is correct. The proof might be informal or formal using the techniques presented in this chapter, but a proof will be required.

After introducing some context and terminology in Section 2.1, we devote the remainder of this chapter to various proof techniques. Sections 2.1 and 2.2 introduce several proof techniques that build directly on the material in Chapter 1. Resolution, the topic of Section 2.3, is a special proof technique that can be automated. Sections 2.4 and 2.5 are concerned with mathematical induction, a proof technique especially useful in discrete mathematics and computer science.

†This section can be omitted without loss of continuity.

## 2.1 Mathematical Systems, Direct Proofs, and Counterexamples

A **mathematical system** consists of **axioms, definitions,** and **undefined terms.** Axioms are assumed to be true. Definitions are used to create new concepts in terms of existing ones. Some terms are not explicitly defined but rather are implicitly defined by the axioms. Within a mathematical system we can derive theorems. A **theorem** is a proposition that has been proved to be true. Special kinds of theorems are referred to as lemmas and corollaries. A **lemma** is a theorem that is usually not too interesting in its own right but is useful in proving another theorem. A **corollary** is a theorem that follows easily from another theorem.

An argument that establishes the truth of a theorem is called a **proof.** Logic is a tool for the analysis of proofs. In this section and the next, we introduce some general methods of proof. In Sections 2.3–2.5, we discuss resolution and mathematical induction, which are special proof techniques. We begin by giving some examples of mathematical systems.

**Example 2.1.1** Euclidean geometry furnishes an example of a mathematical system. Among the axioms are

- Given two distinct points, there is exactly one line that contains them.
- Given a line and a point not on the line, there is exactly one line parallel to the line through the point.

The terms *point* and *line* are undefined terms that are implicitly defined by the axioms that describe their properties.

Among the definitions are

- Two triangles are *congruent* if their vertices can be paired so that the corresponding sides and corresponding angles are equal.
- Two angles are *supplementary* if the sum of their measures is 180°. ◀

**Example 2.1.2** The real numbers furnish another example of a mathematical system. Among the axioms are

- For all real numbers $x$ and $y$, $xy = yx$.
- There is a subset **P** of real numbers satisfying

  (a) If $x$ and $y$ are in **P**, then $x + y$ and $xy$ are in **P**.

  (b) If $x$ is a real number, then exactly one of the following statements is true:

$$x \text{ is in } \mathbf{P}, \qquad x = 0, \qquad -x \text{ is in } \mathbf{P}.$$

Multiplication is implicitly defined by the first axiom and others that describe the properties multiplication is assumed to have.

Among the definitions are

- The elements in **P** (of the preceding axiom) are called *positive real numbers*.
- The *absolute value* $|x|$ of a real number $x$ is defined to be $x$ if $x$ is positive or 0 and $-x$ otherwise. ◀

We give several examples of theorems, corollaries, and lemmas in Euclidean geometry and in the system of real numbers.

**Example 2.1.3** Examples of theorems in Euclidean geometry are

- If two sides of a triangle are equal, then the angles opposite them are equal.
- If the diagonals of a quadrilateral bisect each other, then the quadrilateral is a parallelogram. ◀

**Example 2.1.4** An example of a corollary in Euclidean geometry is

- If a triangle is equilateral, then it is equiangular.

This corollary follows immediately from the first theorem of Example 2.1.3. ◀

**Example 2.1.5** Examples of theorems about real numbers are

- $x \cdot 0 = 0$ for every real number $x$.
- For all real numbers $x$, $y$, and $z$, if $x \le y$ and $y \le z$, then $x \le z$. ◀

**Example 2.1.6** An example of a lemma about real numbers is

- If $n$ is a positive integer, then either $n - 1$ is a positive integer or $n - 1 = 0$.

Surely this result is not that interesting in its own right, but it can be used to prove other results. ◀

## Direct Proofs

Theorems are often of the form

$$\text{For all } x_1, x_2, \ldots, x_n, \text{ if } p(x_1, x_2, \ldots, x_n), \text{ then } q(x_1, x_2, \ldots, x_n).$$

This universally quantified statement is true provided that the conditional proposition

$$\text{if } p(x_1, x_2, \ldots, x_n), \text{ then } q(x_1, x_2, \ldots, x_n) \tag{2.1.1}$$

is true for all $x_1, x_2, \ldots, x_n$ in the domain of discourse. To prove (2.1.1), we assume that $x_1, x_2, \ldots, x_n$ are arbitrary members of the domain of discourse. If $p(x_1, x_2, \ldots, x_n)$ is false, by Definition 1.3.3, (2.1.1) is vacuously true; thus, we need only consider the case that $p(x_1, x_2, \ldots, x_n)$ is true. A **direct proof** assumes that $p(x_1, x_2, \ldots, x_n)$ is true and then, using $p(x_1, x_2, \ldots, x_n)$ as well as other axioms, definitions, previously derived theorems, and rules of inference, shows directly that $q(x_1, x_2, \ldots, x_n)$ is true.

Everyone "knows" what an even or odd integer is, but the following definition makes these terms precise and provides a formal way to use the terms "even integer" and "odd integer" in proofs.

**Definition 2.1.7** ▶ An integer $n$ is *even* if there exists an integer $k$ such that $n = 2k$. An integer $n$ is *odd* if there exists an integer $k$ such that $n = 2k + 1$. ◀

**Example 2.1.8** The integer $n = 12$ is even because there exists an integer $k$ (namely $k = 6$) such that $n = 2k$; that is, $12 = 2 \cdot 6$. ◀

**Example 2.1.9** The integer $n = -21$ is odd because there exists an integer $k$ (namely $k = -11$) such that $n = 2k + 1$; that is, $-21 = 2(-11) + 1$. ◀

**Example 2.1.10** Give a direct proof of the following statement. For all integers $m$ and $n$, if $m$ is odd and $n$ is even, then $m + n$ is odd.

SOLUTION *Discussion* In a direct proof, we assume the hypotheses and derive the conclusion. A good start is achieved by writing out the hypotheses and conclusion so that we are clear where we start and where we are headed. In the case at hand, we have

> $m$ is odd and $n$ is even. (Hypotheses)
>
> ...
>
> $m + n$ is odd. (Conclusion)

The gap ($\cdots$) represents the part of the proof to be completed that leads from the hypotheses to the conclusion.

We can begin to fill in the gap by using the definitions of "odd" and "even" to obtain

> $m$ is odd and $n$ is even. (Hypotheses)
>
> There exists an integer, say $k_1$, such that $m = 2k_1 + 1$. (Because $m$ is odd)
>
> There exists an integer, say $k_2$, such that $n = 2k_2$. (Because $n$ is even)
>
> ...
>
> $m + n$ is odd. (Conclusion)

(Notice that we *cannot* assume that $k_1 = k_2$. For example if $m = 15$ and $n = 4$, then $k_1 = 7$ and $k_2 = 2$. That $k_1$ is not necessarily equal to $k_2$ is the reason that we *must* denote the two integers with different symbols.)

The missing part of our proof is the argument to show that $m + n$ is odd. How can we reach this conclusion? We can use the definition of "odd" again if we can show that $m + n$ is equal to

$$2 \times \text{some integer} + 1. \tag{2.1.2}$$

We already know that $m = 2k_1 + 1$ and $n = 2k_2$. How can we use these facts to reach our goal (2.1.2)? Since the goal involves $m + n$, we can add the equations $m = 2k_1 + 1$ and $n = 2k_2$ to obtain a fact about $m + n$, namely,

$$m + n = (2k_1 + 1) + 2k_2.$$

Now this expression is supposed to be of the form (2.1.2). We can use a little algebra to show that it is of the desired form:

$$m + n = (2k_1 + 1) + 2k_2 = 2(k_1 + k_2) + 1.$$

We have our proof.

**Proof** Let $m$ and $n$ be arbitrary integers, and suppose that $m$ is odd and $n$ is even. We prove that $m + n$ is odd. By definition, since $m$ is odd, there exists an integer $k_1$ such that $m = 2k_1 + 1$. Also, by definition, since $n$ is even, there exists an integer $k_2$ such that $n = 2k_2$. Now the sum is

$$m + n = (2k_1 + 1) + (2k_2) = 2(k_1 + k_2) + 1.$$

Thus, there exists an integer $k$ (namely $k = k_1 + k_2$) such that $m + n = 2k + 1$. Therefore, $m + n$ is odd. ◀ ◀

**Example 2.1.11** Give a direct proof of the following statement. For all sets $X$, $Y$, and $Z$, $X \cap (Y - Z) = (X \cap Y) - (X \cap Z)$.

**SOLUTION** ***Discussion*** The outline of the proof is

$X$, $Y$, and $Z$ are sets. (Hypothesis)

...

$X \cap (Y - Z) = (X \cap Y) - (X \cap Z)$ (Conclusion)

The conclusion asserts that the two sets $X \cap (Y - Z)$ and $(X \cap Y) - (X \cap Z)$ are equal. Recall (see Section 1.1) that to prove from the definition of set equality that these sets are equal, we must show that for all $x$,

$$\text{if } x \in X \cap (Y - Z), \text{ then } x \in (X \cap Y) - (X \cap Z) \tag{2.1.3}$$

and

$$\text{if } x \in (X \cap Y) - (X \cap Z), \text{ then } x \in X \cap (Y - Z). \tag{2.1.4}$$

Thus our proof outline becomes

$X$, $Y$, and $Z$ are sets. (Hypothesis)
If $x \in X \cap (Y - Z)$, then $x \in (X \cap Y) - (X \cap Z)$.
If $x \in (X \cap Y) - (X \cap Z)$, then $x \in X \cap (Y - Z)$.
$X \cap (Y - Z) = (X \cap Y) - (X \cap Z)$ (Conclusion)

We should be able to use the definitions of intersection ($\cap$) and set difference ($-$) to complete the proof.

To prove (2.1.3), we begin by assuming that (the arbitrary element) $x$ is in $X \cap (Y - Z)$. Because this latter set is an intersection, we immediately deduce that $x \in X$ and $x \in Y - Z$. The proof proceeds in this way. As one constructs the proof, it is essential to keep the goal in mind: $x \in (X \cap Y) - (X \cap Z)$. To help guide the construction of the proof, it may be helpful to translate the goal using the definition of set difference: $x \in (X \cap Y) - (X \cap Z)$ means $x \in X \cap Y$ and $x \notin X \cap Z$.

**Proof** Let $X$, $Y$, and $Z$ be arbitrary sets. We prove

$$X \cap (Y - Z) = (X \cap Y) - (X \cap Z)$$

by proving (2.1.3) and (2.1.4).

To prove equation (2.1.3), let $x \in X \cap (Y - Z)$. By the definition of intersection, $x \in X$ and $x \in Y - Z$. By the definition of set difference, since $x \in Y - Z$, $x \in Y$ and $x \notin Z$. By the definition of intersection, since $x \in X$ and $x \in Y$, $x \in X \cap Y$. Again by the definition of intersection, since $x \notin Z$, $x \notin X \cap Z$. By the definition of set difference, since $x \in X \cap Y$, but $x \notin X \cap Z$, $x \in (X \cap Y) - (X \cap Z)$. We have proved equation (2.1.3).

To prove equation (2.1.4), let $x \in (X \cap Y) - (X \cap Z)$. By the definition of set difference, $x \in X \cap Y$ and $x \notin X \cap Z$. By the definition of intersection, since $x \in X \cap Y$, $x \in X$ and $x \in Y$. Again, by the definition of intersection, since $x \notin X \cap Z$ and $x \in X$, $x \notin Z$. By the definition of set difference, since $x \in Y$ and $x \notin Z$, $x \in Y - Z$. Finally, by the definition of intersection, since $x \in X$ and $x \in Y - Z$, $x \in X \cap (Y - Z)$. We have proved equation (2.1.4).

Since we have proved both equations (2.1.3) and (2.1.4), it follows that

$$X \cap (Y - Z) = (X \cap Y) - (X \cap Z).$$ ◀ ◀

Our next example shows that in constructing a proof, we may find that we need some auxiliary results, at which point we pause, go off and prove these auxiliary results, and then return to the main proof. We call the proofs of auxiliary results **subproofs.** (For those familiar with programming, a subproof is similar to a subroutine.)

**Example 2.1.12** If $a$ and $b$ are real numbers, we define $\min\{a, b\}$ to be the *minimum* of $a$ and $b$ or the common value if they are equal. More precisely,

$$\min\{a, b\} = \begin{cases} a & \text{if } a < b \\ a & \text{if } a = b \\ b & \text{if } b < a. \end{cases}$$

Give a direct proof of the following statement. For all real numbers $d, d_1, d_2, x$,

$$\text{if } d = \min\{d_1, d_2\} \text{ and } x \le d, \text{ then } x \le d_1 \text{ and } x \le d_2.$$

**SOLUTION** ***Discussion*** The outline of the proof is

$d = \min\{d_1, d_2\}$ and $x \le d$ (Hypotheses)

...

$x \le d_1$ and $x \le d_2$ (Conclusion)

To help understand what is being asserted, let us look at a specific example. As we have remarked previously, when we are asked to prove a universally quantified statement, a specific example does not *prove* the statement. It may, however, help us to *understand* the statement.

Let us set $d_1 = 2$ and $d_2 = 4$. Then $d = \min\{d_1, d_2\} = 2$. The statement to be proved says that if $x \le d\ (= 2)$, then $x \le d_1\ (= 2)$ and $x \le d_2\ (= 4)$. Why is this true in general? The minimum $d$ of two numbers, $d_1$ and $d_2$, is equal to one of the two numbers (namely, the smallest) and less than or equal to the other one (namely, the largest)—in symbols, $d \le d_1$ and $d \le d_2$. If $x \le d$, then from $x \le d$ and $d \le d_1$, we may deduce $x \le d_1$. Similarly, from $x \le d$ and $d \le d_2$, we may deduce $x \le d_2$. Thus the outline of our proof becomes

$d = \min\{d_1, d_2\}$ and $x \le d$ (Hypotheses)

Subproof: Show that $d \le d_1$ and $d \le d_2$.

From $x \le d$ and $d \le d_1$, deduce $x \le d_1$.

From $x \le d$ and $d \le d_2$, deduce $x \le d_2$.

$x \le d_1$ and $x \le d_2$ (Conclusion—uses the conjunction inference rule)

At this point, the only part of the proof that is missing is the subproof to show that $d \le d_1$ and $d \le d_2$. Let us look at the definition of "minimum." If $d_1 \le d_2$, then $d = \min\{d_1, d_2\} = d_1$ and $d = d_1 \le d_2$. If $d_2 < d_1$, then $d = \min\{d_1, d_2\} = d_2$ and $d = d_2 < d_1$. In either case, $d \le d_1$ and $d \le d_2$.

**Proof** Let $d, d_1, d_2$, and $x$ be arbitrary real numbers, and suppose that

$$d = \min\{d_1, d_2\} \text{ and } x \le d.$$

We prove that $x \le d_1$ and $x \le d_2$.

We first show that $d \le d_1$ and $d \le d_2$. From the definition of "minimum," if $d_1 \le d_2$, then $d = \min\{d_1, d_2\} = d_1$ and $d = d_1 \le d_2$. If $d_2 < d_1$, then $d = \min\{d_1, d_2\} = d_2$ and $d = d_2 < d_1$. In either case, $d \le d_1$ and $d \le d_2$. From $x \le d$ and $d \le d_1$, it follows that $x \le d_1$ from a previous theorem (the second theorem of Example 2.1.5). From $x \le d$ and $d \le d_2$, we may derive $x \le d_2$ from the same previous theorem. Therefore, $x \le d_1$ and $x \le d_2$. ◀ ◀

**Example 2.1.13** There are frequently many different ways to prove a statement. We illustrate by giving two proofs of the statement

$$X \cup (Y - X) = X \cup Y \quad \text{for all sets } X \text{ and } Y.$$

**SOLUTION** ***Discussion*** We first give a direct proof like the proof in Example 2.1.11. We show that for all $x$, if $x \in X \cup (Y - X)$, then $x \in X \cup Y$, and if $x \in X \cup Y$, then $x \in X \cup (Y - X)$.

Our second proof uses Theorem 1.1.22, which gives laws of sets. The idea is to begin with $X \cup (Y - X)$ and use the laws of sets, which here we think of as rules to manipulate set equations, to obtain $X \cup Y$.

**Proof** [First proof] We show that for all $x$, if $x \in X \cup (Y - X)$, then $x \in X \cup Y$, and if $x \in X \cup Y$, then $x \in X \cup (Y - X)$.

Let $x \in X \cup (Y - X)$. Then $x \in X$ or $x \in Y - X$. If $x \in X$, then $x \in X \cup Y$. If $x \in Y - X$, then $x \in Y$, so again $x \in X \cup Y$. In either case, $x \in X \cup Y$.

Let $x \in X \cup Y$. Then $x \in X$ or $x \in Y$. If $x \in X$, then $x \in X \cup (Y - X)$. If $x \notin X$, then $x \in Y$. In this case, $x \in Y - X$. Therefore, $x \in X \cup (Y - X)$. In either case, $x \in X \cup (Y - X)$. The proof is complete. ◀

**Proof** [Second proof] We use Theorem 1.1.22, which gives laws of sets, and the fact that $Y - X = Y \cap \overline{X}$, which follows immediately from the definition of set difference. Letting $U$ denote the universal set, we obtain

$$\begin{aligned} X \cup (Y - X) &= X \cup (Y \cap \overline{X}) && [Y - X = Y \cap \overline{X}] \\ &= (X \cup Y) \cap (X \cup \overline{X}) && \text{[Distributive law; Theorem 1.1.22, part (c)]} \\ &= (X \cup Y) \cap U && \text{[Complement law; Theorem 1.1.22, part (e)]} \\ &= X \cup Y && \text{[Identity law; Theorem 1.1.22, part (d)].} \end{aligned}$$ ◀ ◀

## Disproving a Universally Quantified Statement

Recall (see Section 1.5) that to *disprove* $\forall x P(x)$ we simply need to find one member $x$ in the domain of discourse that makes $P(x)$ false. Such a value for $x$ is called a *counterexample*.

**Example 2.1.14** The statement $\forall n \in \mathbf{Z}^+ \ (2^n + 1 \text{ is prime})$ is false. A counterexample is $n = 3$ since $2^3 + 1 = 9$, which is not prime. ◀

**Example 2.1.15** If the statement

$$(A \cap B) \cup C = A \cap (B \cup C), \quad \text{for all sets } A, B, \text{ and } C$$

is true, prove it; otherwise, give a counterexample.

**SOLUTION** Let us begin by trying to prove the statement. We will first try to show that if $x \in (A \cap B) \cup C$, then $x \in A \cap (B \cup C)$. If $x \in (A \cap B) \cup C$, then

$$x \in A \cap B \quad \text{or} \quad x \in C. \tag{2.1.5}$$

We have to show that $x \in A \cap (B \cup C)$, that is,

$$x \in A \quad \text{and} \quad x \in B \cup C. \tag{2.1.6}$$

Statement (2.1.5) is true if $x$ is in $C$, and statement (2.1.6) is false if $x \notin A$. Thus the given statement is false; there is no direct proof (or any other proof!). If we choose sets $A$ and $C$ so that there is an element that is in $C$, but not in $A$, we will have a counterexample.

Let $A = \{1, 2, 3\}$, $B = \{2, 3, 4\}$, and $C = \{3, 4, 5\}$ so that there is an element that is in $C$, but not in $A$. Then $(A \cap B) \cup C = \{2, 3, 4, 5\}$, $A \cap (B \cup C) = \{2, 3\}$, and $(A \cap B) \cup C \neq A \cap (B \cup C)$. Thus $A$, $B$, and $C$ provide a counterexample that shows that the given statement is false. ◀

## 2.1 Problem-Solving Tips

- To construct a direct proof of a universally quantified statement, first write down the hypotheses (so you know what you are assuming), and then write down the conclusion (so you know what you must prove). The conclusion is what you will work toward—something like the answer in the back of the book to an exercise, except here it is essential to know the goal before proceeding. You must now give an argument that begins with the hypotheses and ends with the conclusion. To construct the argument, remind yourself what you know about the terms (e.g., "even," "odd"), symbols (e.g., $X \cap Y$, $\min\{d_1, d_2\}$), and so on. Look at relevant definitions and related results. For example, if a particular hypothesis refers to an even integer $n$, you know that $n$ is of the form $2k$ for some integer $k$. If you are to prove that two sets $X$ and $Y$ are equal from the definition of set equality, you know you must show that for every $x$, if $x \in X$ then $x \in Y$, and if $x \in Y$ then $x \in X$.
- To understand what is to be proved, look at some specific values in the domain of discourse. When we are asked to prove a universally quantified statement, showing that the statement is true for specific values does not *prove* the statement; it may, however, help to *understand* the statement.
- To *disprove* a universally quantified statement, find *one element* in the domain of discourse, called a *counterexample*, that makes the propositional function false. Here, your proof consists of presenting the counterexample together with justification that the propositional function is indeed false for your counterexample.
- When you write up your proof, begin by writing out the statement to be proved. Indicate clearly where your proof begins (e.g., by beginning a new paragraph or by writing "Proof."). Use complete sentences, which may include symbols. For example, it is perfectly acceptable to write: Thus $x \in X$. In words, this is the complete sentence: Thus $x$ is in $X$. End a direct proof by clearly stating the conclusion, and, perhaps, giving a reason to justify the conclusion. For example, Example 2.1.10 ends with:

  Thus, there exists an integer $k$ (namely $k = k_1 + k_2$) such that $m + n = 2k + 1$. Therefore, $m + n$ is odd.

  Here the conclusion ($m + n$ is odd) is clearly stated and justified by the statement $m + n = 2k + 1$.
- Alert the reader where you are headed. For example, if you are going to prove that $X = Y$, write "We will prove that $X = Y$" before launching into this part of the proof.
- Justify your steps. For example, if you conclude that $x \in X$ or $x \in Y$ because it is known that $x \in X \cup Y$, write "Since $x \in X \cup Y$, $x \in X$ or $x \in Y$," or perhaps even "Since $x \in X \cup Y$, by the definition of union $x \in X$ or $x \in Y$" if, like Richard Nixon, you want to be perfectly clear.
- If you are asked to prove or disprove a universally quantified statement, you can begin by trying to prove it. If you succeed, you are finished—the statement is true and you proved it! If your proof breaks down, look carefully at the point where it fails. The given statement may be false and your failed proof may give insight into how to construct a counterexample (see Example 2.1.15). On the other hand, if you have trouble constructing a counterexample, check where your proposed examples fail. This insight may show why the statement is true and guide construction of a proof.

### Some Common Errors

In Example 2.1.10, we pointed out that it is an error to use the same notation for two possibly distinct quantities. As an example, here is a faulty "proof" that for all $m$ and $n$, if $m$ and $n$ are even integers then $mn$ is a square (i.e., $mn = a^2$ for some integer $a$): Since $m$ and $n$ are even, $m = 2k$ and $n = 2k$. Now $mn = (2k)(2k) = (2k)^2$. If we let $a = 2k$, then $m = a^2$. The problem is that we *cannot* use $k$ for two potentially different quantities. If $m$ and $n$ are even, all we can conclude is that $m = 2k_1$ and $n = 2k_2$ for some integers $k_1$ and $k_2$. The integers $k_1$ and $k_2$ need *not* be equal. (In fact, it is false that for all $m$ and $n$, if $m$ and $n$ are even integers then $mn$ is a square. A counterexample is $m = 2$ and $n = 4$.)

Given a universally quantified propositional function, showing that the propositional function is true for specific values in the domain of discourse is *not* a proof that the propositional function is true for all values in the domain of discourse. (Such specific values may, however, *suggest* that the propositional function *might* be true for all values in the domain of discourse.) Example 2.1.10 is to prove that for *all* integers $m$ and $n$,

$$\text{if } m \text{ is odd and } n \text{ is even, then } m + n \text{ is odd.} \qquad \textbf{(2.1.7)}$$

Letting $m = 11$ and $n = 4$ and noting that $m + n = 15$ is odd does *not* constitute a proof that (2.1.7) is true for all integers $m$ and $n$, it merely proves that (2.1.7) is true for the specific values $m = 11$ and $n = 4$.

In constructing a proof, you cannot assume what you are supposed to prove. As an example, consider the *erroneous* "proof" that for all integers $m$ and $n$, if $m$ and $m + n$ are even, then $n$ is even: Let $m = 2k_1$ and $n = 2k_2$. Then $m + n = 2k_1 + 2k_2$. Therefore,

$$n = (m + n) - m = (2k_1 + 2k_2) - 2k_2 = 2(k_1 + k_2 - k_2).$$

Thus $n$ is even. The problem with the preceding "proof" is that we cannot write $n = 2k_2$ since this is true if and only if $n$ is even—which is what we are supposed to prove! This error is called **begging the question** or **circular reasoning.** [It is true that if $m$ and $m + n$ are even integers, then $n$ is even (see Exercise 12).]

## 2.1 Review Exercises

1. What is a mathematical system?
2. What is an axiom?
3. What is a definition?
4. What is an undefined term?
5. What is a theorem?
6. What is a proof?
7. What is a lemma?
8. What is a direct proof?
9. What is the formal definition of "even integer"?
10. What is the formal definition of "odd integer"?
11. What is a subproof?
12. How do you disprove a universally quantified statement?

## 2.1 Exercises

1. Give an example (different from those of Example 2.1.1) of an axiom in Euclidean geometry.
2. Give an example (different from those of Example 2.1.2) of an axiom in the system of real numbers.
3. Give an example (different from those of Example 2.1.1) of a definition in Euclidean geometry.
4. Give an example (different from those of Example 2.1.2) of a definition in the system of real numbers.
5. Give an example (different from those of Example 2.1.3) of a theorem in Euclidean geometry.
6. Give an example (different from those of Example 2.1.5) of a theorem in the system of real numbers.
7. Prove that for all integers $m$ and $n$, if $m$ and $n$ are even, then $m + n$ is even.
8. Prove that for all integers $m$ and $n$, if $m$ and $n$ are odd, then $m + n$ is even.

**9.** Prove that for all integers $m$ and $n$, if $m$ and $n$ are even, then $mn$ is even.

**10.** Prove that for all integers $m$ and $n$, if $m$ and $n$ are odd, then $mn$ is odd.

**11.** Prove that for all integers $m$ and $n$, if $m$ is odd and $n$ is even, then $mn$ is even.

**12.** Prove that for all integers $m$ and $n$, if $m$ and $m + n$ are even, then $n$ is even.

**13.** Prove that for all rational numbers $x$ and $y$, $x + y$ is rational.

**14.** Prove that for all rational numbers $x$ and $y$, $xy$ is rational.

**15.** Prove that for every rational number $x$, if $x \neq 0$, then $1/x$ is rational.

**16.** Prove that the product of two integers, one of the form $3k_1 + 1$ and the other of the form $3k_2 + 2$, where $k_1$ and $k_2$ are integers, is of the form $3k_3 + 2$ for some integer $k_3$.

**17.** Prove that the product of two integers, one of the form $3k_1 + 2$ and the other of the form $3k_2 + 2$, where $k_1$ and $k_2$ are integers, is of the form $3k_3 + 1$ for some integer $k_3$.

**18.** If $a$ and $b$ are real numbers, we define $\max\{a, b\}$ to be the *maximum* of $a$ and $b$ or the common value if they are equal. Prove that for all real numbers $d, d_1, d_2, x$,

$$\text{if } d = \max\{d_1, d_2\} \text{ and } x \geq d, \text{ then } x \geq d_1 \text{ and } x \geq d_2.$$

**19.** Justify each step of the following direct proof, which shows that if $x$ is a real number, then $x \cdot 0 = 0$. Assume that the following are previous theorems: If $a$, $b$, and $c$ are real numbers, then $b + 0 = b$ and $a(b + c) = ab + ac$. If $a + b = a + c$, then $b = c$.

**Proof** $x \cdot 0 + 0 = x \cdot 0 = x \cdot (0 + 0) = x \cdot 0 + x \cdot 0$; therefore, $x \cdot 0 = 0$. ◀

**20.** If $X$ and $Y$ are nonempty sets and $X \times Y = Y \times X$, what can we conclude about $X$ and $Y$? Prove your answer.

**21.** Prove that $X \cap Y \subseteq X$ for all sets $X$ and $Y$.

**22.** Prove that $X \subseteq X \cup Y$ for all sets $X$ and $Y$.

**23.** Prove that if $X \subseteq Y$, then $X \cup Z \subseteq Y \cup Z$ for all sets $X$, $Y$, and $Z$.

**24.** Prove that if $X \subseteq Y$, then $X \cap Z \subseteq Y \cap Z$ for all sets $X$, $Y$, and $Z$.

**25.** Prove that if $X \subseteq Y$, then $Z - Y \subseteq Z - X$ for all sets $X$, $Y$, and $Z$.

**26.** Prove that if $X \subseteq Y$, then $Y - (Y - X) = X$ for all sets $X$ and $Y$.

**27.** Prove that if $X \cap Y = X \cap Z$ and $X \cup Y = X \cup Z$, then $Y = Z$ for all sets $X$, $Y$, and $Z$.

**28.** Prove that $\mathcal{P}(X) \cup \mathcal{P}(Y) \subseteq \mathcal{P}(X \cup Y)$ for all sets $X$ and $Y$.

**29.** Prove that $\mathcal{P}(X \cap Y) = \mathcal{P}(X) \cap \mathcal{P}(Y)$ for all sets $X$ and $Y$.

**30.** Prove that if $\mathcal{P}(X) \subseteq \mathcal{P}(Y)$, then $X \subseteq Y$ for all sets $X$ and $Y$.

**31.** Disprove that $\mathcal{P}(X \cup Y) \subseteq \mathcal{P}(X) \cup \mathcal{P}(Y)$ for all sets $X$ and $Y$.

**32.** Give a direct proof along the lines of the second proof in Example 2.1.13 of the statement

$$X \cap (Y - Z) = (X \cap Y) - (X \cap Z) \quad \text{for all sets } X, Y, \text{ and } Z.$$

(In Example 2.1.11, we gave a direct proof of this statement using the definition of set equality.)

*In each of Exercises 33–45, if the statement is true, prove it; otherwise, give a counterexample. The sets X, Y, and Z are subsets of a universal set U. Assume that the universe for Cartesian products is U × U.*

**33.** For all sets $X$ and $Y$, either $X$ is a subset of $Y$ or $Y$ is a subset of $X$.

**34.** $X \cup (Y - Z) = (X \cup Y) - (X \cup Z)$ for all sets $X$, $Y$, and $Z$.

**35.** $\overline{Y - X} = X \cup \overline{Y}$ for all sets $X$ and $Y$.

**36.** $Y - Z = (X \cup Y) - (X \cup Z)$ for all sets $X$, $Y$, and $Z$.

**37.** $X - (Y \cup Z) = (X - Y) \cup Z$ for all sets $X$, $Y$, and $Z$.

**38.** $\overline{X - Y} = \overline{Y - X}$ for all sets $X$ and $Y$.

**39.** $\overline{X \cap Y} \subseteq X$ for all sets $X$ and $Y$.

**40.** $(X \cap Y) \cup (Y - X) = Y$ for all sets $X$ and $Y$.

**41.** $X \times (Y \cup Z) = (X \times Y) \cup (X \times Z)$ for all sets $X$, $Y$, and $Z$.

**42.** $\overline{X \times Y} = \overline{X} \times \overline{Y}$ for all sets $X$ and $Y$.

**43.** $X \times (Y - Z) = (X \times Y) - (X \times Z)$ for all sets $X$, $Y$, and $Z$.

**44.** $X - (Y \times Z) = (X - Y) \times (X - Z)$ for all sets $X$, $Y$, and $Z$.

**45.** $X \cap (Y \times Z) = (X \cap Y) \times (X \cap Z)$ for all sets $X$, $Y$, and $Z$.

**46.** Prove the associative laws for sets [Theorem 1.1.22, part (a)].

**47.** Prove the commutative laws for sets [Theorem 1.1.22, part (b)].

**48.** Prove the distributive laws for sets [Theorem 1.1.22, part (c)].

**49.** Prove the identity laws for sets [Theorem 1.1.22, part (d)].

**50.** Prove the complement laws for sets [Theorem 1.1.22, part (e)].

**51.** Prove the idempotent laws for sets [Theorem 1.1.22, part (f)].

**52.** Prove the bound laws for sets [Theorem 1.1.22, part (g)].

**53.** Prove the absorption laws for sets [Theorem 1.1.22, part (h)].

**54.** Prove the involution law for sets [Theorem 1.1.22, part (i)].

**55.** Prove the 0/1 laws for sets [Theorem 1.1.22, part (j)].

**56.** Prove De Morgan's laws for sets [Theorem 1.1.22, part (k)].

*In Exercises 57–65, △ denotes the symmetric difference operator defined as A △ B = (A ∪ B) − (A ∩ B), where A and B are sets.*

**57.** Prove that $A \triangle B = (A - B) \cup (B - A)$ for all sets $A$ and $B$.

**58.** Prove that $(A \triangle B) \triangle A = B$ for all sets $A$ and $B$.

**★59.** Prove or disprove: If $A$, $B$, and $C$ are sets satisfying $A \triangle C = B \triangle C$, then $A = B$.

**60.** Prove or disprove: $A \triangle (B \cup C) = (A \triangle B) \cup (A \triangle C)$ for all sets $A$, $B$, and $C$.

**61.** Prove or disprove: $A \triangle (B \cap C) = (A \triangle B) \cap (A \triangle C)$ for all sets $A$, $B$, and $C$.

**62.** Prove or disprove: $A \cup (B \triangle C) = (A \cup B) \triangle (A \cup C)$ for all sets $A$, $B$, and $C$.

**63.** Prove or disprove: $A \cap (B \triangle C) = (A \cap B) \triangle (A \cap C)$ for all sets $A$, $B$, and $C$.

**64.** Is △ commutative? If so, prove it; otherwise, give a counterexample.

**★65.** Is △ associative? If so, prove it; otherwise, give a counterexample.

## 2.2 More Methods of Proof

In this section, we discuss several more methods of proof: proof by contradiction, proof by contrapositive, proof by cases, proofs of equivalence, and existence proofs. We will find these proof techniques of use throughout this book.

### Proof by Contradiction

A **proof by contradiction** establishes $p \to q$ by assuming that the hypothesis $p$ is true and that the conclusion $q$ is false and then, using $p$ and $\neg q$ as well as other axioms, definitions, previously derived theorems, and rules of inference, derives a **contradiction.** A contradiction is a proposition of the form $r \wedge \neg r$ ($r$ may be any proposition whatever). A proof by contradiction is sometimes called an **indirect proof** since to establish $p \to q$ using proof by contradiction, we follow an indirect route: We derive $r \wedge \neg r$ and then conclude that $q$ is true.

The only difference between the assumptions in a direct proof and a proof by contradiction is the negated conclusion. In a direct proof the negated conclusion is *not* assumed, whereas in a proof by contradiction the negated conclusion *is* assumed.

Proof by contradiction may be justified by noting that the propositions $p \to q$ and $(p \wedge \neg q) \to (r \wedge \neg r)$ are equivalent. The equivalence is immediate from a truth table:

| $p$ | $q$ | $r$ | $p \to q$ | $p \wedge \neg q$ | $r \wedge \neg r$ | $(p \wedge \neg q) \to (r \wedge \neg r)$ |
|---|---|---|---|---|---|---|
| T | T | T | T | F | F | T |
| T | T | F | T | F | F | T |
| T | F | T | F | T | F | F |
| T | F | F | F | T | F | F |
| F | T | T | T | F | F | T |
| F | T | F | T | F | F | T |
| F | F | T | T | F | F | T |
| F | F | F | T | F | F | T |

**Example 2.2.1** Give a proof by contradiction of the following statement:

For every $n \in \mathbf{Z}$, if $n^2$ is even, then $n$ is even.

**SOLUTION** ***Discussion*** First, let us consider giving a *direct* proof of this statement. We would assume the hypothesis, that is, that $n^2$ is even. Then there exists an integer $k_1$ such that $n^2 = 2k_1$. To prove that $n$ is even, we must find an integer $k_2$ such that $n = 2k_2$. It is not clear how to get from $n^2 = 2k_1$ to $n = 2k_2$. (Taking the square root certainly does not work!) When a proof technique seems unpromising, try a different one.

In a proof by contradiction, we assume the hypothesis ($n^2$ is even) *and* the negation of the conclusion ($n$ is not even, i.e., $n$ is odd). Since $n$ is odd, there exists an integer $k$ such that $n = 2k + 1$. If we square both sides of this last equation, we obtain

$$n^2 = (2k+1)^2 = 4k^2 + 4k + 1 = 2(2k^2 + 2k) + 1.$$

But this last equation tells us that $n^2$ is odd. We have our contradiction: $n^2$ is even (hypothesis) and $n^2$ is odd. Formally, if $r$ is the statement "$n^2$ is even," we have deduced $r \wedge \neg r$.

**Proof** We give a proof by contradiction. Thus we assume the hypothesis $n^2$ is even and that the conclusion is false $n$ is odd. Since $n$ is odd, there exists an integer $k$ such that $n = 2k + 1$. Now

$$n^2 = (2k+1)^2 = 4k^2 + 4k + 1 = 2(2k^2 + 2k) + 1.$$

Thus $n^2$ is odd, which contradicts the hypothesis $n^2$ is even. The proof by contradiction is complete. We have proved that for every $n \in \mathbf{Z}$, if $n^2$ is even, then $n$ is even. ◀ ◀

**Example 2.2.2** Give a proof by contradiction of the following statement:

For all real numbers $x$ and $y$, if $x + y \geq 2$, then either $x \geq 1$ or $y \geq 1$.

SOLUTION ***Discussion*** As in the previous example, a direct proof seems unpromising—assuming only that $x + y \geq 2$ appears to be too little to get us started. We turn to a proof by contradiction.

**Proof** We begin by letting $x$ and $y$ be arbitrary real numbers. We then suppose that the conclusion is false, that is, that $\neg(x \geq 1 \vee y \geq 1)$ is true. By De Morgan's laws of logic (see Example 1.3.11),

$$\neg(x \geq 1 \vee y \geq 1) \equiv \neg(x \geq 1) \wedge \neg(y \geq 1) \equiv (x < 1) \wedge (y < 1).$$

In words, we are assuming that $x < 1$ *and* $y < 1$. Using a previous theorem, we may add these inequalities to obtain $x + y < 1 + 1 = 2$. At this point, we have derived a contradiction: $x + y \geq 2$ and $x + y < 2$. Thus we conclude that for all real numbers $x$ and $y$, if $x + y \geq 2$, then either $x \geq 1$ or $y \geq 1$. ◀ ◀

**Example 2.2.3** Prove that $\sqrt{2}$ is irrational using proof by contradiction.

SOLUTION ***Discussion*** Here a direct proof seems particularly bleak. It seems we have a blank slate with which to begin. However, if we use proof by contradiction, we may assume that $\sqrt{2}$ is rational. In this case, we know that there exist integers $p$ and $q$ such that $\sqrt{2} = p/q$. Now we have an entry on our slate. We can manipulate this equation and hope to obtain a contradiction.

**Proof** We use proof by contradiction and assume that $\sqrt{2}$ is rational. Then there exist integers $p$ and $q$ such that $\sqrt{2} = p/q$. We assume that the fraction $p/q$ is in lowest terms so that $p$ and $q$ are not both even. Squaring $\sqrt{2} = p/q$ gives $2 = p^2/q^2$, and multiplying by $q^2$ gives $2q^2 = p^2$. It follows that $p^2$ is even. Example 2.2.1 tells us that $p$ is even. Therefore, there exists an integer $k$ such that $p = 2k$. Substituting $p = 2k$ into $2q^2 = p^2$ gives $2q^2 = (2k)^2 = 4k^2$. Canceling 2 gives $q^2 = 2k^2$. Therefore $q^2$ is even, and Example 2.2.1 tells us the $q$ is even. Thus $p$ and $q$ are both even, which contradicts our assumption that $p$ and $q$ are not both even. Therefore, $\sqrt{2}$ is irrational. ◀ ◀

## Proof by Contrapositive

Suppose that we give a proof by contradiction of $p \to q$ in which, as in Examples 2.2.1 and 2.2.2, we deduce $\neg p$. In effect, we have proved $\neg q \to \neg p$. [Recall (see Theorem 1.3.18) that $p \to q$ and $\neg q \to \neg p$ are equivalent.] This special case of proof by contradiction is called **proof by contrapositive.**

**Example 2.2.4** Give a proof by contradiction that for any subset $S$ of 26 cards from an ordinary 52-card deck (composed of four suits of 13 cards each), there is a suit such that $S$ has at least 7 cards of that suit. (In the card game bridge, this result says that two partners, who each hold 13-card hands, will between them have a suit of at least 7 cards.)

**SOLUTION** ***Discussion*** The conclusion is: There is a suit such that $S$ has at least 7 cards of that suit. If the conclusion is false, it must be that $S$ has at most 6 cards of *every* suit. But, since there are four suits, we can account for at most $6 \cdot 4 = 24$ cards in $S$. But the hypothesis is that $S$ contains 26 cards. Contradiction! We have a proof.

We conclude the discussion by formally showing how De Morgan's laws of logic (see Example 1.3.1) are used to yield the negated conclusion. We take as our domain of discourse, the set of all subsets of cards from an ordinary 52-card deck. One piece of notation will be useful. If $Su$ is a suit, we let $S(Su)$ denote the number of cards of suit $Su$ in $S$. For example, the value of $S(Club)$ is the number of clubs in $S$.

We are to prove that

$$\forall S, \text{ if } |S| = 26, \text{ then } \exists \text{ a suit } Su \text{ such that } S(Su) \geq 7.$$

By De Morgan's laws of logic,

$$\neg(\exists \text{ a suit } Su \text{ such that } S(Su) \geq 7) \equiv \forall \text{ suits } Su,\ S(Su) < 7;$$

that is, the negation of the conclusion can be written: $S$ has less than 7 (i.e., at most 6) cards of every suit.

In Example 2.2.8, we will discuss an alternative approach to finding a proof of this result.

**Proof** We are given a subset $S$ of 26 cards from an ordinary 52-card deck. Suppose by way of contradiction that the conclusion is false; that is, suppose that $S$ has at most 6 cards of every suit. Since there are four suits, $S$ has at most $6 \cdot 4 = 24$ cards. This contradicts the fact that $S$ has 26 cards. Therefore, there is a suit such that $S$ has at least 7 cards of that suit. ◀ ◀

**Example 2.2.5** Give a proof by contrapositive to prove that

$$\text{for all } x \in \mathbf{R}, \text{ if } x^2 \text{ is irrational, then } x \text{ is irrational.}$$

**SOLUTION** ***Discussion*** For much the same reasons that a direct proof seemed unpromising in Examples 2.2.1 and 2.2.2, a direct proof in which we assume only that $x^2$ is irrational seems to be too little to get us started. A proof by contradiction can be devised (see Exercise 1), but here a proof by contrapositive is requested.

**Proof** We begin by letting $x$ be an arbitrary real number. We prove the contrapositive of the given statement, which is

$$\text{if } x \text{ is not irrational, then } x^2 \text{ is not irrational}$$

or, equivalently,

$$\text{if } x \text{ is rational, then } x^2 \text{ is rational.}$$

So suppose that $x$ is rational. Then $x = p/q$ for some integers $p$ and $q$. Now $x^2 = p^2/q^2$. Since $x^2$ is the quotient of integers, $x^2$ is rational. The proof is complete. ◀ ◀

## Proof by Cases

**Proof by cases** is used when the original hypothesis naturally divides itself into various cases. For example, the hypothesis "$x$ is a real number" can be divided into cases: (a) $x$

is a nonnegative real number and (b) $x$ is a negative real number. Suppose that the task is to prove $p \to q$ and that $p$ is equivalent to $p_1 \vee p_2 \vee \cdots \vee p_n$ ($p_1, \ldots, p_n$ are the cases). Instead of proving

$$(p_1 \vee p_2 \vee \cdots \vee p_n) \to q, \tag{2.2.1}$$

we prove

$$(p_1 \to q) \wedge (p_2 \to q) \wedge \cdots \wedge (p_n \to q). \tag{2.2.2}$$

As we will show, proof by cases is justified because the two statements are equivalent.

First suppose that $q$ is true. Then all the implications in (2.2.1) and (2.2.2) are true (regardless of the truth value of the hypotheses). Thus (2.2.1) and (2.2.2) are true.

Now suppose that $q$ is false. If all the $p_i$ are false, then all the implications in (2.2.1) and (2.2.2) are true, so (2.2.1) and (2.2.2) are true. If for some $j$, $p_j$ is true, then $p_1 \vee \cdots \vee p_n$ is true, so (2.2.1) is false. Since $p_j \to q$ is false, (2.2.2) is false. Thus (2.2.1) and (2.2.2) are false. Therefore, (2.2.1) and (2.2.2) are equivalent.

Sometimes the number of cases to prove is finite and not too large, so we can check them all one by one. We call this type of proof **exhaustive proof.**

**Example 2.2.6** Prove that $2m^2 + 3n^2 = 40$ has no solution in positive integers, that is, that $2m^2 + 3n^2 = 40$ is false for all positive integers $m$ and $n$.

**SOLUTION** ***Discussion*** We certainly *cannot* check $2m^2+3n^2$ for *all* positive integers $m$ and $n$, but we can rule out most positive integers because, if $2m^2+3n^2 = 40$, the sizes of $m$ and $n$ are restricted. In particular, we must have $2m^2 \le 40$ and $3n^2 \le 40$. (If, for example, $2m^2 > 40$, when we add $3n^2$ to $2m^2$, the sum $2m^2 + 3n^2$ will exceed 40.) If $2m^2 \le 40$, then $m^2 \le 20$ and $m$ can be at most 4. Similarly, if $3n^2 \le 40$, then $n^2 \le 40/3$ and $n$ can be at most 3. Thus it suffices to check the cases $m = 1, 2, 3, 4$ and $n = 1, 2, 3$.

**Proof** If $2m^2 + 3n^2 = 40$, we must have $2m^2 \le 40$. Thus $m^2 \le 20$ and $m \le 4$. Similarly, we must have $3n^2 \le 40$. Thus $n^2 \le 40/3$ and $n \le 3$. Therefore it suffices to check the cases $m = 1, 2, 3, 4$ and $n = 1, 2, 3$.

The entries in the table give the value of $2m^2 + 3n^2$ for the indicated values of $m$ and $n$.

| | | $m$ | | | |
|---|---|---|---|---|---|
| | | 1 | 2 | 3 | 4 |
| | 1 | 5 | 11 | 21 | 35 |
| $n$ | 2 | 14 | 20 | 30 | 44 |
| | 3 | 29 | 35 | 45 | 59 |

Since $2m^2+3n^2 \ne 40$ for $m = 1, 2, 3, 4$ and $n = 1, 2, 3$, and $2m^2+3n^2 > 40$ for $m > 4$ or $n > 3$, we conclude that $2m^2 + 3n^2 = 40$ has no solution in positive integers. ◀ ◀

Example 2.2.7 illustrates an important point: An *or* statement often leads itself to a proof by cases.

**Example 2.2.7** We prove that for every real number $x$, $x \le |x|$.

**SOLUTION** ***Discussion*** Since $x$ is a real number, either $x \ge 0$ or $x < 0$. We use this *or* statement to divide the proof into cases. We divide the proof into cases because the definition of absolute value is itself divided into cases $x \geq 0$ and $x < 0$ (see Example 2.1.2). Case 1 is $x \geq 0$ and case 2 is $x < 0$.

**Proof** If $x \geq 0$, by definition $|x| = x$. Thus $|x| \geq x$. If $x < 0$, by definition $|x| = -x$. Since $|x| = -x > 0$ and $0 > x$, $|x| \geq x$. In either case, $|x| \geq x$; so the proof is complete. ◀◀

Example 2.2.8 offers an alternative approach to developing a proof of the result in Example 2.2.4. There are often many approaches and proofs of a single result.

**Example 2.2.8** We revisit the problem (see Example 2.2.4) of proving that for any subset $S$ of 26 cards from an ordinary 52-card deck, there is a suit such that $S$ has at least 7 cards of that suit.

***Discussion*** Consider trying a direct proof. How about the club suit? If there are 7 or more clubs, we have the desired conclusion. What if there are 6 or fewer clubs? Try another suit! Let us try diamonds next. If there are 7 or more diamonds, we have the desired conclusion. What if there are 6 or fewer diamonds? Try another suit, and so on. If each of the four suits consists of 6 or fewer cards, we cannot deduce the conclusion; however, if each suit has 6 or fewer cards, we can account for only 24 cards. But there are 26 cards; this case cannot occur.

The preceding discussion shows that we can divide the proof into two cases. Case 1 is that some suit consists of 7 or more cards. (Proof complete!) Case 2 is that every suit consists of 6 or fewer cards. (We have shown that this case cannot occur.) Since only case 1 can occur, the proof is complete.

We started by thinking along the lines of a direct proof, which led us to a proof by cases. The proof by cases is essentially identical to the proof by contradiction in Example 2.2.4.

**Proof** Let $S$ be a subset of 26 cards from an ordinary 52-card deck. We consider two cases. Case 1 is that there is a suit such that $S$ has at least 7 or more cards of that suit. Case 2 is that $S$ has 6 or fewer cards of every suit.

If case 1 holds, the proof is complete. Suppose that case 2 holds. Since $S$ has 6 or fewer cards of each of the four suits, $S$ has at most 24 cards. But we are given that $S$ has 26 cards. Therefore, case 2 cannot hold. Since only case 1 holds, the proof is complete. ◀◀

## Proofs of Equivalence

Some theorems are of the form $p$ if and only if $q$. Such theorems are proved by using the equivalence (see Example 1.3.15)

$$p \leftrightarrow q \equiv (p \rightarrow q) \wedge (q \rightarrow p);$$

that is, to prove "$p$ if and only if $q$," prove "if $p$ then $q$" and "if $q$ then $p$."

**Example 2.2.9** Prove that for every integer $n$, $n$ is odd if and only if $n - 1$ is even.

**SOLUTION** ***Discussion*** We let $n$ be an arbitrary integer. We must prove that

$$\text{if } n \text{ is odd then } n - 1 \text{ is even} \tag{2.2.3}$$

and

$$\text{if } n - 1 \text{ is even then } n \text{ is odd.} \tag{2.2.4}$$

**Proof** We first prove (2.2.3). If $n$ is odd, then $n = 2k + 1$ for some integer $k$. Now $n - 1 = (2k + 1) - 1 = 2k$. Therefore, $n - 1$ is even.

Next we prove (2.2.4). If $n - 1$ is even, then $n - 1 = 2k$ for some integer $k$. Now $n = 2k + 1$. Therefore, $n$ is odd. The proof is complete. ◀ ◀

Some proofs of $p \leftrightarrow q$ combine the proofs of $p \rightarrow q$ and $q \rightarrow p$. For example, the proof in Example 2.2.9 could be written as follows:

> $n$ is odd if and only if $n = 2k + 1$ for some integer $k$ if and only if $n - 1 = 2k$ for some integer $k$ if and only if $n - 1$ is even.

For such a proof to be correct, it must be the case that each if-and-only-if statement is true. If such a proof of $p \leftrightarrow q$ is read in one direction, we obtain the proof of $p \rightarrow q$ and, if it is read in the other direction, we obtain a proof of $q \rightarrow p$. Reading the preceding proof in the if-then direction,

> if $n$ is odd, then $n = 2k + 1$ for some integer $k$; if $n = 2k + 1$ for some integer $k$, then $n - 1 = 2k$ for some integer $k$; if $n - 1 = 2k$ for some integer $k$, then $n - 1$ is even,

proves that if $n$ is odd, then $n - 1$ is even. Reversing the order,

> if $n - 1$ is even, then $n - 1 = 2k$ for some integer $k$; if $n - 1 = 2k$ for some integer $k$, then $n = 2k + 1$ for some integer $k$; if $n = 2k + 1$ for some integer $k$, then $n$ is odd,

proves that if $n - 1$ is even, then $n$ is odd.

**Example 2.2.10** Prove that for all real numbers $x$ and all positive real numbers $d$,

$$|x| < d \text{ if and only if } -d < x < d.$$

**SOLUTION** ***Discussion*** We let $x$ be an arbitrary real number and $d$ be an arbitrary positive real number. We must show

$$\text{if } |x| < d \text{ then } -d < x < d \qquad \textbf{(2.2.5)}$$

and

$$\text{if } -d < x < d \text{ then } |x| < d. \qquad \textbf{(2.2.6)}$$

Since $|x|$ is defined by cases, we expect to use proof by cases.

**Proof** To show (2.2.5), we use proof by cases. We assume that $|x| < d$. If $x \geq 0$, then $-d < 0 \leq x = |x| < d$. If $x < 0$, then $-d < 0 < -x = |x| < d$; that is, $-d < -x < d$. Multiplying by $-1$, we obtain $d > x > -d$. In either case, we have proved that $-d < x < d$.

To show (2.2.6), we also use proof by cases. We assume that $-d < x < d$. If $x \geq 0$, then $|x| = x < d$. If $x < 0$, then $|x| = -x$. Since $-d < x$, we may multiply by $-1$ to obtain $d > -x$. Combining $|x| = -x$ and $d > -x$ gives $|x| = -x < d$. In either case, we have proved that $|x| < d$. The proof is complete. ◀ ◀

In proving $p \leftrightarrow q$, we are proving that $p$ and $q$ are logically equivalent, that is, $p$ and $q$ are either both true or both false. Some theorems state that three or more statements are logically equivalent and, thus, have the form

The following are equivalent:

(a) —

(b) —

(c) —

⋮

Such a theorem asserts that (a), (b), (c), and so on are either all true or all false.

To prove that $p_1, p_2, \ldots, p_n$ are equivalent, the usual method is to prove

$$(p_1 \to p_2) \wedge (p_2 \to p_3) \wedge \cdots \wedge (p_{n-1} \to p_n) \wedge (p_n \to p_1). \quad \textbf{(2.2.7)}$$

We show that proving (2.2.7) shows that $p_1, p_2, \ldots, p_n$ are equivalent.

Suppose that we prove (2.2.7). We consider two cases: $p_1$ is true, $p_1$ is false. First, suppose that $p_1$ is true. Because $p_1$ and $p_1 \to p_2$ are true, $p_2$ is true; because $p_2$ and $p_2 \to p_3$ are true, $p_3$ is true; and so on. In this case, $p_1, p_2, \ldots, p_n$ have the same truth value: each is true.

Now suppose that $p_1$ is false. Because $p_1$ is false and $p_n \to p_1$ is true, $p_n$ is false; because $p_n$ is false and $p_{n-1} \to p_n$ is true, $p_{n-1}$ is false; and so on. In this case, $p_1, p_2, \ldots, p_n$ have the same truth value; each is false. Therefore, proving (2.2.7) shows that $p_1, p_2, \ldots, p_n$ are equivalent.

Not just any arrangement of implications along the lines of (2.3.7) will make $p_1, \ldots, p_n$ equivalent. For example, consider $p_1$: $2 = 3$, $p_2$: $4 = 6$, and $p_3$: $8 = 8$. For the arrangement

$$p_1 \to p_2, \quad p_2 \to p_3, \quad p_1 \to p_3,$$

$p_1 \to p_2$, $p_2 \to p_3$, and $p_1 \to p_3$ are all true, but $p_1, p_2, p_3$ are *not* equivalent.

**Example 2.2.11** Let $A$, $B$, and $C$ be sets. Prove that the following are equivalent:

$$\text{(a) } A \subseteq B \qquad \text{(b) } A \cap B = A \qquad \text{(c) } A \cup B = B.$$

**SOLUTION** ***Discussion*** According to the discussion preceding this example, we must prove

$$[\text{(a)} \to \text{(b)}] \wedge [\text{(b)} \to \text{(c)}] \wedge [\text{(c)} \to \text{(a)}].$$

**Proof** We prove (a) → (b), (b) → (c), and (c) → (a).

[(a) → (b).] We assume that $A \subseteq B$, and prove that $A \cap B = A$. Suppose that $x \in A \cap B$. We must show that $x \in A$. But if $x \in A \cap B$, $x \in A$ by the definition of intersection.

Now suppose that $x \in A$. We must show that $x \in A \cap B$. Since $A \subseteq B$, $x \in B$. Therefore $x \in A \cap B$. We have proved that $A \cap B = A$.

[(b) → (c).] We assume that $A \cap B = A$, and prove that $A \cup B = B$. Suppose that $x \in A \cup B$. We must show that $x \in B$. By assumption, either $x \in A$ or $x \in B$. If $x \in B$, we have the desired conclusion. If $x \in A$, since $A \cap B = A$, again $x \in B$.

Now suppose that $x \in B$. We must show that $x \in A \cup B$. But if $x \in B$, $x \in A \cup B$ by the definition of union. We have proved that $A \cup B = B$.

[(c) → (a).] We assume that $A \cup B = B$, and prove that $A \subseteq B$. Suppose that $x \in A$. We must show that $x \in B$. Since $x \in A$, $x \in A \cup B$ by the definition of union. Since $A \cup B = B$, $x \in B$. We have proved that $A \subseteq B$. The proof is complete. ◀ ◀

## Existence Proofs

A proof of

$$\exists x\, P(x) \tag{2.2.8}$$

is called an **existence proof.** In Section 1.5, we showed that one way to prove (2.2.8) is to exhibit one member $a$ in the domain of discourse that makes $P(a)$ true.

**Example 2.2.12** Let $a$ and $b$ be real numbers with $a < b$. Prove that there exists a real number $x$ satisfying $a < x < b$.

SOLUTION *Discussion* We will prove the statement by exhibiting a real number $x$ between $a$ and $b$. The point midway between $a$ and $b$ suffices.

**Proof** It suffices to find one real number $x$ satisfying $a < x < b$. The real number

$$x = \frac{a+b}{2},$$

halfway between $a$ and $b$, surely satisfies $a < x < b$. ◀ ◀

**Example 2.2.13** Prove that there exists a prime $p$ such that $2^p - 1$ is composite (i.e., *not* prime).

SOLUTION *Discussion* By trial and error, we find that $2^p - 1$ is prime for $p = 2, 3, 5, 7$ but not $p = 11$ since $2^{11} - 1 = 2048 - 1 = 2047 = 23 \cdot 89$. Thus $p = 11$ makes the given statement true.

**Proof** For the prime $p = 11$, $2^p - 1$ is composite:

$$2^{11} - 1 = 2048 - 1 = 2047 = 23 \cdot 89.$$

◀ ◀

**Go Online**
The Great Internet Mersenne Prime Search is at bit.ly/2ECHU4F

A prime number of the form $2^p - 1$, where $p$ is prime, is called a *Mersenne prime* [named for Marin Mersenne (1588–1648)]. It is not known whether the number of Mersenne primes is finite or infinite. The largest primes known are Mersenne primes. In January 2016, the 49th known Mersenne prime was found, $2^{74,207,281} - 1$, a number having 22,338,618 decimal digits. This number was found by the Great Internet Mersenne Prime Search (GIMPS). GIMPS is a computer program distributed over many personal computers maintained by volunteers. You can participate. Just check the web link. You may find the next Mersenne prime!

An existence proof of (2.2.8) that exhibits an element $a$ of the domain of discourse that makes $P(a)$ true is called a **constructive proof.** The proofs in Examples 2.2.12 and 2.2.13 are constructive proofs. A proof of (2.2.8) that does not exhibit an element $a$ of the domain of discourse that makes $P(a)$ true, but rather proves (2.2.8) some other way (e.g., using proof by contradiction), is called a **nonconstructive proof.**

**Example 2.2.14** Let

$$A = \frac{s_1 + s_2 + \cdots + s_n}{n}$$

be the average of the real numbers $s_1, \ldots, s_n$. Prove that there exists $i$ such that $s_i \geq A$.

SOLUTION *Discussion* It seems hopeless to choose an $i$ and prove that $s_i \geq A$; instead, we use proof by contradiction.

**Proof** We use proof by contradiction and assume the negation of the conclusion $\neg\exists i(s_i \geq A)$. By the generalized De Morgan's laws for logic (Theorem 1.5.14), the negation of the conclusion is equivalent to $\forall i \neg(s_i \geq A)$ or $\forall i(s_i < A)$. Thus we assume $s_1 < A, s_2 < A, \ldots, s_n < A$. Adding these inequalities yields

$$s_1 + s_2 + \cdots + s_n < nA.$$

Dividing by $n$ gives

$$\frac{s_1 + s_2 + \cdots + s_n}{n} < A,$$

which contradicts the hypothesis

$$A = \frac{s_1 + s_2 + \cdots + s_n}{n}.$$

Therefore, there exists $i$ such that $s_i \geq A$. ◀ ◀

The proof in Example 2.2.14 is nonconstructive; it does not exhibit an $i$ for which $s_i \geq A$. It does however *prove* indirectly using proof by contradiction that there is such an $i$. We could find such an $i$: We could check whether $s_1 \geq A$ and if true, stop. Otherwise, we could check whether $s_2 \geq A$ and if true, stop. We could continue in this manner until we find an $i$ for which $s_i \geq A$. Example 2.2.14 guarantees that there is such an $i$.

## 2.2 Problem-Solving Tips

It is worth reviewing the Problem-Solving Tips of Section 2.1. Tips specific to the present section follow.

- If you are trying to construct a direct proof of a statement of the form $p \rightarrow q$ and you seem to be getting stuck, try a proof by contradiction. You then have more to work with: Besides assuming $p$, you get to assume $\neg q$.
- When writing up a proof by contradiction, alert the reader by stating, "We give a proof by contradiction, thus we assume $\cdots$," where $\cdots$ is the negation of the conclusion. Another common introduction is: Assume by way of contradiction that $\cdots$.
- Proof by cases is useful if the hypotheses naturally break down into parts. For example, if the statement to prove involves the absolute value of $x$, you may want to consider the cases $x \geq 0$ and $x < 0$ because $|x|$ is itself defined by the cases $x \geq 0$ and $x < 0$. If the number of cases to prove is finite and not too large, the cases can be directly checked one by one.

  In writing up a proof by cases, it is sometimes helpful to the reader to indicate the cases, for example,

  [Case I: $x \geq 0$.] Proof of this case goes here.

  [Case II: $x < 0$.] Proof of this case goes here.
- To prove $p$ if and only if $q$, you must prove two statements: (1) if $p$ then $q$ and (2) if $q$ then $p$. It helps the reader if you state clearly what you are proving. You can write up the proof of (1) by beginning a new paragraph with a sentence that indicates that you are about to prove "if $p$ then $q$." You would then follow with a proof of (2) by beginning a new paragraph with a sentence that indicates that you are about to prove "if $q$ then $p$." Another common technique is to write

  [$p \rightarrow q$.] Proof of $p \rightarrow q$ goes here.

  [$q \rightarrow p$.] Proof of $q \rightarrow p$ goes here.

- To prove that several statements, say $p_1, \ldots, p_n$, are equivalent, prove $p_1 \to p_2$, $p_2 \to p_3, \ldots, p_{n-1} \to p_n, p_n \to p_1$. The statements can be ordered in any way and the proofs may be easier to construct for one ordering than another. For example, you could swap $p_2$ and $p_3$ and prove $p_1 \to p_3, p_3 \to p_2, p_2 \to p_4, p_4 \to p_5, \ldots, p_{n-1} \to p_n, p_n \to p_1$. You should indicate clearly what you are about to prove. One common form is

  [$p_1 \to p_2$.] Proof of $p_1 \to p_2$ goes here.
  [$p_2 \to p_3$.] Proof of $p_2 \to p_3$ goes here.
  And so forth.

- If the statement is existentially quantified (i.e., there exists $x \ldots$), the proof, called an existence proof, consists of showing that there exists at least one $x$ in the domain of discourse that makes the statement true. One type of existence proof exhibits a value of $x$ that makes the statement true (and proves that the statement is indeed true for the specific $x$). Another type of existence proof indirectly proves (e.g., using proof by contradiction) that a value of $x$ exists that makes the statement true without specifying any particular value of $x$ for which the statement is true.

## 2.2 Review Exercises

1. What is proof by contradiction?
2. Give an example of a proof by contradiction.
3. What is an indirect proof?
4. What is proof by contrapositive?
5. Give an example of a proof by contrapositive.
6. What is proof by cases?
7. Give an example of a proof by cases.
8. What is a proof of equivalence?
9. Give an example of a proof of equivalence.
10. How can we show that three statements, say (a), (b), and (c), are equivalent?
11. What is an existence proof?
12. What is a constructive existence proof?
13. Give an example of a constructive existence proof.
14. What is a nonconstructive existence proof?
15. Give an example of a nonconstructive existence proof.

## 2.2 Exercises

1. Use proof by contradiction to prove that for all $x \in \mathbf{R}$, if $x^2$ is irrational, then $x$ is irrational.
2. Is the converse of Exercise 1 true or false? Prove your answer.
3. Prove that for all $x \in \mathbf{R}$, if $x^3$ is irrational, then $x$ is irrational.
4. Prove that for every $n \in \mathbf{Z}$, if $n^2$ is odd, then $n$ is odd.
5. Prove that for all real numbers $x$, $y$, and $z$, if $x + y + z \geq 3$, then either $x \geq 1$ or $y \geq 1$ or $z \geq 1$.
6. Prove that for all real numbers $x$ and $y$, if $xy \leq 2$, then either $x \leq \sqrt{2}$ or $y \leq \sqrt{2}$.
7. Prove that $\sqrt[3]{2}$ is irrational.
8. Prove that for all $x, y \in \mathbf{R}$, if $x$ is rational and $y$ is irrational, then $x + y$ is irrational.
9. Prove or disprove: For all $x, y \in \mathbf{R}$, if $x$ is rational and $y$ is irrational, then $xy$ is irrational.
10. Prove that if $a$ and $b$ are real numbers with $a < b$, there exists a rational number $x$ satisfying $a < x < b$.
11. Prove that if $a$ and $b$ are real numbers with $a < b$, there exists an irrational number $x$ satisfying $a < x < b$.
12. Fill in the details of the following proof that there exist irrational numbers $a$ and $b$ such that $a^b$ is rational.

    **Proof** Let $x = y = \sqrt{2}$. If $x^y$ is rational, the proof is complete. (Explain.) Otherwise, suppose that $x^y$ is irrational. (Why?) Let $a = x^y$ and $b = \sqrt{2}$. Consider $a^b$. (How does this complete the proof?) ◀

    Is this proof constructive or nonconstructive?
13. Prove or disprove: There exist rational numbers $a$ and $b$ such that $a^b$ is rational. What kind of proof did you give?
14. Prove or disprove: There exist rational numbers $a$ and $b$ such that $a^b$ is irrational. What kind of proof did you give?
15. Let $x$ and $y$ be real numbers. Prove that if $x \leq y + \varepsilon$ for every positive real number $\varepsilon$, then $x \leq y$.
16. In American football, a safety counts two points, a field goal three points, a touchdown six points, a successful kick imme-

diately after a touchdown one point, and a successful run or pass immediately after a touchdown two points. What are the possible points a team can score? Prove your answer.

17. Suppose that a real number $a$ has the property that $a^n$ is irrational for every positive integer $n$. Prove that $a^r$ is irrational for every positive rational number $r$. (There are real numbers such as $a$. The real number $e = 2.718\ldots$, the base of the natural logarithm, has the property that $e^r$ is irrational for every rational number $r$. An elementary proof, using only calculus, may be found in [Aigner].)

18. Abby, Bosco, Cary, Dale, and Edie are considering going to hear the Flying Squirrels. If Abby goes, Cary will also go. If Bosco goes, Cary and Dale will also go. If Cary goes, Edie will also go. If Dale goes, Abby will also go. If Edie goes, Abby will also go. Exactly four of the five went to the concert. Who went? Prove your result.

19. Prove or disprove: $(X - Y) \cap (Y - X) = \varnothing$ for all sets $X$ and $Y$.

20. Prove or disprove: $X \times \varnothing = \varnothing$ for every set $X$.

21. Show, by giving a proof by contradiction, that if 100 balls are placed in nine boxes, some box contains 12 or more balls.

22. Show, by giving a proof by contradiction, that if 40 coins are distributed among nine bags so that each bag contains at least 1 coin, at least two bags contain the same number of coins.

23. Let $S$ be a subset of 26 cards from an ordinary 52-card deck. Suppose that there is a suit in which $S$ has exactly 7 cards. Prove that there is another suit in which $S$ has at least 7 cards.

24. Let $S_1$ be a subset of 26 cards from an ordinary 52-card deck, and let $S_2$ be the remaining 26 cards. Suppose that there is a suit in which $S_1$ has at least 9 cards. Prove that there is a suit in which $S_2$ has at least 8 cards. (In the card game bridge, two pairs compete against each other and initially hold 13 cards each. This result says a pair, who between them have a suit of at least 9 cards, will have opponents who between them have a suit of at least 8 cards.)

★25. Let $s_1, \ldots, s_n$ be a sequence[†] satisfying

(a) $s_1$ is a positive integer and $s_n$ is a negative integer,

(b) for all $i$, $1 \le i < n$, $s_{i+1} = s_i + 1$ or $s_{i+1} = s_i - 1$.

Prove that there exists $i$, $1 < i < n$, such that $s_i = 0$.

Calculus students will recognize this exercise as a discrete version of the calculus theorem: If $f$ is a continuous function on $[a, b]$ and $f(a) > 0$ and $f(b) < 0$, then $f(c) = 0$ for some $c$ in $(a, b)$. There are similar proofs of the two statements.

26. Disprove the statement: For every positive integer $n$, $n^2 \le 2^n$.

*In Exercises 27–31,*

$$A = \frac{s_1 + s_2 + \cdots + s_n}{n}$$

*is the average of the real numbers $s_1, \ldots, s_n$.*

27. Prove that there exists $i$ such that $s_i \le A$.

28. Prove or disprove: There exists $i$ such that $s_i > A$. What proof technique did you use?

29. Suppose that there exists $i$ such that $s_i < A$. Prove or disprove: There exists $j$ such that $s_j > A$. What proof technique did you use?

30. Suppose that there exist $i$ and $j$ such that $s_i \neq s_j$. Prove that there exists $k$ such that $s_k < A$.

31. Suppose that there exist $i$ and $j$ such that $s_i \neq s_j$. Prove that there exists $k$ such that $s_k > A$.

32. Prove that $2m + 5n^2 = 20$ has no solution in positive integers.

33. Prove that $m^3 + 2n^2 = 36$ has no solution in positive integers.

34. Prove that $2m^2 + 4n^2 - 1 = 2(m + n)$ has no solution in positive integers.

35. Prove that the product of two consecutive integers is even.

36. Prove that for every $n \in \mathbf{Z}$, $n^3 + n$ is even.

37. Use proof by cases to prove that $|xy| = |x||y|$ for all real numbers $x$ and $y$.

38. Use proof by cases to prove that $|x + y| \le |x| + |y|$ for all real numbers $x$ and $y$.

39. Define the *sign of the real number x*, sgn$(x)$, as

$$\operatorname{sgn}(x) = \begin{cases} 1 & \text{if } x > 0 \\ 0 & \text{if } x = 0 \\ -1 & \text{if } x < 0. \end{cases}$$

Use proof by cases to prove that $|x| = \operatorname{sgn}(x)x$ for every real number $x$.

40. Use proof by cases to prove that $\operatorname{sgn}(xy) = \operatorname{sgn}(x)\operatorname{sgn}(y)$ for all real numbers $x$ and $y$ (sgn is defined in Exercise 39).

41. Use Exercises 39 and 40 to give another proof that $|xy| = |x||y|$ for all real numbers $x$ and $y$.

42. Use proof by cases to prove that $\max\{x, y\} + \min\{x, y\} = x + y$ for all real numbers $x$ and $y$.

43. Use proof by cases to prove that

$$\max\{x, y\} = \frac{x + y + |x - y|}{2}$$

for all real numbers $x$ and $y$.

44. Use proof by cases to prove that

$$\min\{x, y\} = \frac{x + y - |x - y|}{2}$$

for all real numbers $x$ and $y$.

45. Use Exercises 43 and 44 to prove that $\max\{x, y\} + \min\{x, y\} = x + y$ for all real numbers $x$ and $y$.

46. Prove that for all $n \in \mathbf{Z}$, $n$ is even if and only if $n + 2$ is even.

47. Prove that for all $n \in \mathbf{Z}$, $n$ is odd if and only if $n + 2$ is odd.

48. Prove that for all sets $A$ and $B$, $A \subseteq B$ if and only if $\overline{B} \subseteq \overline{A}$.

[†]Informally, a *sequence* is a list of elements in which order is taken into account, so that $s_1$ is the first element, $s_2$ is the second element, and so on. We present the formal definition in Section 3.2.

**49.** Prove that for all sets $A$, $B$, and $C$, $A \subseteq C$ and $B \subseteq C$ if and only if $A \cup B \subseteq C$.

**50.** Prove that for all sets $A$, $B$, and $C$, $C \subseteq A$ and $C \subseteq B$ if and only if $C \subseteq A \cap B$.

★**51.** The ordered pair $(a, b)$ can be *defined* in terms of sets as

$$(a, b) = \{\{a\}, \{a, b\}\}.$$

Taking the preceding equation as the *definition* of ordered pair, prove that $(a, b) = (c, d)$ if and only if $a = c$ and $b = d$.

**52.** Prove that the following are equivalent for the integer $n$:
(a) $n$ is odd. (b) There exists $k \in \mathbf{Z}$ such that $n = 2k - 1$. (c) $n^2 + 1$ is even.

**53.** Prove that the following are equivalent for sets $A$, $B$, and $C$:

(a) $A \cap B = \varnothing$ (b) $B \subseteq \overline{A}$ (c) $A \triangle B = A \cup B$,

where $\triangle$ is the symmetric difference operator (see Exercise 101, Section 1.1).

**54.** Prove that the following are equivalent for sets $A$, $B$, and $C$:

(a) $A \cup B = U$ (b) $\overline{A} \cap \overline{B} = \varnothing$ (c) $\overline{A} \subseteq B$,

where $U$ is a universal set.

# Problem-Solving Corner Proving Some Properties of Real Numbers

## Problem

First some definitions:

(a) Let $X$ be a nonempty set of real numbers. An *upper bound for* $X$ is a real number $a$ having the property that $x \leq a$ for every $x \in X$.

(b) Let $a$ be an upper bound for a set $X$ of real numbers. If every upper bound $b$ for $X$ satisfies $b \geq a$, we call $a$ a *least upper bound for* $X$.

A fundamental property of the real numbers is that every nonempty subset of real numbers bounded above has a least upper bound.

Answer the following where $\mathbf{R}$ serves as a universal set:

1. Give an example of a set $X$ and three distinct upper bounds for $X$, one of which is a least upper bound for $X$.
2. Prove that if $a$ and $b$ are least upper bounds for a set $X$, then $a = b$. We say that the least upper bound for a set $X$ is *unique*. If $a$ is *the* least upper bound of a set $X$, we sometimes write $a = \operatorname{lub} X$.
3. Let $X$ be a set with least upper bound $a$. Prove that if $\varepsilon > 0$, then there exists $x \in X$ satisfying $a - \varepsilon < x \leq a$.
4. Let $X$ be a set with least upper bound $a$, and suppose that $t > 0$. Prove that $ta$ is the least upper bound of the set $\{tx \mid x \in X\}$.

## Attacking the Problem

To better understand the definitions, let's construct examples, write out the definitions in words, look at negations of the definitions, and draw pictures.

We'll start with definition (a) and construct a simple example—taking $X$ to be a small finite set, say $X = \{1, 2, 3, 4\}$. Now an upper bound $a$ for $X$ satisfies $x \leq a$ for every $x$ in $X$—here, we must have

$$1 \leq a, \quad 2 \leq a, \quad 3 \leq a, \quad 4 \leq a.$$

Examples of upper bounds for $X$ are 4, 6.9, $3\pi$, 9072.

In words, definition (a) says that $a$ is an upper bound for a set $X$ if every element in $X$ is less than or equal to $a$. We see that upper bounds $e$, $f$, and $g$ for a set $X$ (shown in color) look like



What would it mean that $a$ is *not* an upper bound for a set $X$? We would have to negate definition (a): $\neg\forall x(x \leq a)$ or, equivalently, $\exists x \neg(x \leq a)$ or $\exists x(x > a)$. In words, $a$ is not an upper bound for a set $X$ if there exists $x$ in $X$ such that $x > a$. Looking at the preceding picture, we see that any number less than $e$ is not an upper bound for $X$.

Let's turn to definition (b), which says, in words, that $a$ is a least upper bound for a set $X$ if, among all upper bounds for $X$, $a$ is smallest. Looking ahead, problem 2 is to show that there is only one (distinct) upper bound for a set $X$; thus, we usually say *the* least upper bound rather than *a* least upper bound. The least upper bound of our previous set $X = \{1, 2, 3, 4\}$ is 4. We have already noted that 4 is an upper bound for $X$. If $a$ is any upper bound for $X$, since $4 \in X$, $4 \leq a$. Therefore, 4 is the least upper bound for $X$. In the preceding figure, $e$ is a least upper bound for the set $X$.

## Finding a Solution

Now we consider the problems.

[Problem 1.] Our previous example, $X = \{1, 2, 3, 4\}$, will suffice. We have noted that 4 is the least upper bound for $X$. Any values greater than 4 serve as additional upper bounds.

[Problem 2.] One way to prove that two numbers $a$ and $b$ are equal is to show that $a \le b$ and $b \le a$. We'll try this first. Another possibility is proof by contradiction and assume that $a \ne b$.

[Problem 3.] Here we can use the fact that $a - \varepsilon$ is *not* an upper bound (since it's less than the least upper bound) and, as discussed previously, what it means for a value to *not* be an upper bound.

[Problem 4.] Here we are given a value, $ta$, and asked to prove that it is the least upper bound of the given set, which here we denote as $tX$. Going directly to the definitions, we must show that

(a) $z \le ta$ for every $z \in tX$ (i.e., $ta$ is an upper bound for $tX$),

(b) if $b$ is an upper bound for $tX$, then $b \ge ta$ (i.e., $ta$ is the least upper bound for $tX$).

For part (a), since $z = tx$ ($x \in X$), we must show that

$$tx \le ta \quad \text{for all } x \in X.$$

We are given that $a$ is a least upper bound for $X$. In particular, $a$ is an upper bound for $X$ so

$$x \le a \quad \text{for all } x \in X.$$

How do we deduce the first inequality from the second? Multiply by $t$! We hope that the proof of part (b) proceeds in a similar way.

## Formal Solution

[Problem 1.] Let $X = \{1, 2, 3, 4\}$. Upper bounds for $X$ are 4, 5, and 6 since $x \le 4$, $x \le 5$, and $x \le 6$ for every $x \in X$.

The least upper bound for $X$ is 4. We have already noted that 4 is an upper bound for $X$. If $a$ is any upper bound for $X$, since $4 \in X$, $4 \le a$. Therefore, 4 is the least upper bound for $X$.

[Problem 2.] Since $a$ is a least upper bound for $X$ and $b$ is an upper bound for $X$, $a \le b$. Since $b$ is a least upper bound for $X$ and $a$ is an upper bound for $X$, $b \le a$. Therefore, $a = b$.

[Problem 3.] Let $\varepsilon > 0$. Since $a$ is the least upper bound for $X$ and $a - \varepsilon < a$, $a - \varepsilon$ is *not* an upper bound for $X$. Therefore, by definition (a) there exists $x \in X$ such that $a - \varepsilon < x$. Since $a$ is an upper bound for $X$, $x \le a$. We have shown that there exists $x \in X$ such that $a - \varepsilon < x \le a$.

[Problem 4.] Let $tX$ denote the set $\{tx \mid x \in X\}$. We must prove that

(a) $z \le ta$ for every $z \in tX$ (i.e., $ta$ is an upper bound for $tX$),

(b) if $b$ is an upper bound for $tX$, then $b \ge ta$ (i.e., $ta$ is the least upper bound for $tX$).

We first prove part (a). Let $z \in tX$. Then $z = tx$ for some $x \in X$ (by the definition of the set $tX$). Since $a$ is an upper bound for $X$, $x \le a$. Multiplying by $t$ and noting that $t > 0$, we have $z = tx \le ta$. Therefore, $z \le ta$ for every $z \in tX$ and the proof of part (a) is complete.

Next we prove part (b). Let $b$ be an upper bound for $tX$. Then $tx \le b$ for every $x \in X$ (since an arbitrary element in $tX$ is of the form $tx$ for some $x \in X$). Dividing by $t$ and noting that $t > 0$, we have $x \le b/t$ for every $x \in X$. Therefore $b/t$ is an upper bound for $X$. Since $a$ is the least upper bound for $X$, $b/t \ge a$. Multiplying by $t$ and noting again that $t > 0$, we have $b \ge ta$. Therefore $ta$ is the least upper bound for $tX$. The proof is complete.

## Summary of Problem-Solving Techniques

- Before beginning a proof, familiarize yourself with relevant definitions, theorems, examples, and so on.
- Construct additional examples—especially small examples (e.g., for sets look at some small finite sets).
- Write out some of the technical statements in words.
- Look at negations of statements.
- Draw pictures.
- If one proof technique seems not to be working, try another. For example, if a direct proof seems unpromising, try a proof by contradiction.
- Review the Problem-Solving Tips sections in this chapter and the previous chapter.

## Comments

The fact that every nonempty set of real numbers that is bounded above has a least upper bound is called the **completeness property of the real numbers.** The real numbers are complete in the sense that there are no "holes" in the number line. Informally, if there was a hole in the line, the set of numbers to the left of the hole, although bounded above, would not have a least upper bound:

The set of rational numbers is *not* complete. The subset of rational numbers less than $\sqrt{2}$ is bounded above, but does not have a *rational* least upper bound. (The least upper bound of the subset of rational numbers less than $\sqrt{2}$ is the irrational number $\sqrt{2}$.)

### Exercises

1. What is the least upper bound of a nonempty finite set of real numbers?
2. What is the least upper bound of the set

$$\{1 - 1/n \mid n \text{ is a positive integer}\}?$$

Prove your answer.

3. Let $X$ and $Y$ be nonempty sets of real numbers such that $X \subseteq Y$ and $Y$ is bounded above. Prove that $X$ is bounded above and $\text{lub}\, X \leq \text{lub}\, Y$.
4. Let $X$ be a nonempty set. What is the least upper bound of the set $\{tx \mid x \in X\}$ if $t = 0$?
5. ★ Let $X$ be a set with least upper bound $a$, and let $Y$ be a set with least upper bound $b$. Prove that the set

$$\{x + y \mid x \in X \text{ and } y \in Y\}$$

is bounded above and its least upper bound is $a + b$.

*Let $X$ be a nonempty set of real numbers. A* lower bound *for $X$ is a real number $a$ having the property that $x \geq a$ for every $x \in X$. Let $a$ be a lower bound for a set $X$ of real numbers. If every lower bound $b$ for $X$ satisfies $b \leq a$, we call $a$ a* greatest lower bound *for $X$.*

6. Prove that if $a$ and $b$ are greatest lower bounds for a set $X$, then $a = b$.
7. Prove that every nonempty subset of real numbers bounded below has a greatest lower bound. *Hint*: If $X$ is a nonempty set of real numbers bounded below, let $Y$ denote the set of lower bounds. Prove that $Y$ has a least upper bound, say $a$. Prove that $a$ is the greatest lower bound for $X$.
8. Let $X$ be a set with greatest lower bound $a$. Prove that if $\varepsilon > 0$, then there exists $x \in X$ satisfying $a + \varepsilon > x \geq a$.
9. Let $X$ be a set with least upper bound $a$, and let $t < 0$. Prove that $ta$ is the greatest lower bound of the set $\{tx \mid x \in X\}$.

## 2.3 Resolution Proofs[†]

In this section, we will write $a \wedge b$ as $ab$.

**Resolution** is a proof technique proposed by J. A. Robinson in 1965 (see [Robinson]) that depends on a single rule:

$$\text{If } p \vee q \text{ and } \neg p \vee r \text{ are both true, then } q \vee r \text{ is true.} \qquad \textbf{(2.3.1)}$$

Statement (2.3.1) can be verified by writing the truth table (see Exercise 1). Because resolution depends on this single, simple rule, it is the basis of many computer programs that reason and prove theorems.

In a proof by resolution, the hypotheses and the conclusion are written as **clauses**. A clause consists of terms separated by *or*'s, where each term is a variable or the negation of a variable.

**Example 2.3.1** The expression $a \vee b \vee \neg c \vee d$ is a clause since the terms $a$, $b$, $\neg c$, and $d$ are separated by *or*'s and each term is a variable or the negation of a variable. ◀

**Example 2.3.2** The expression $xy \vee w \vee \neg z$ is *not* a clause even though the terms are separated by *or*'s, since the term $xy$ consists of two variables—not a single variable. ◀

**Example 2.3.3** The expression $p \rightarrow q$ is *not* a clause since the terms are separated by $\rightarrow$. Each term is, however, a variable. ◀

[†]This section can be omitted without loss of continuity.

A direct proof by resolution proceeds by repeatedly applying (2.3.1) to pairs of statements to derive new statements until the conclusion is derived. When we apply (2.3.1), $p$ must be a single variable, but $q$ and $r$ can be expressions. Notice that when (2.3.1) is applied to clauses, the result $q \vee r$ is a clause. (Since $q$ and $r$ each consist of terms separated by *or*'s, where each term is a variable or the negation of a variable, $q \vee r$ also consists of terms separated by *or*'s, where each term is a variable or the negation of a variable.)

**Example 2.3.4** Prove the following using resolution:

1. $a \vee b$
2. $\neg a \vee c$
3. $\neg c \vee d$

$\therefore b \vee d$

**SOLUTION** Applying (2.3.1) to expressions 1 and 2, we derive

4. $b \vee c$.

Applying (2.3.1) to expressions 3 and 4, we derive

5. $b \vee d$,

the desired conclusion. Given the hypotheses 1, 2, and 3, we have proved the conclusion $b \vee d$. ◀

Special cases of (2.3.1) are as follows:

$$\text{If } p \vee q \text{ and } \neg p \text{ are true, then } q \text{ is true.} \qquad \textbf{(2.3.2)}$$
$$\text{If } p \text{ and } \neg p \vee r \text{ are true, then } r \text{ is true.}$$

**Example 2.3.5** Prove the following using resolution:

1. $a$
2. $\neg a \vee c$
3. $\neg c \vee d$

$\therefore d$

**SOLUTION** Applying (2.3.2) to expressions 1 and 2, we derive

4. $c$.

Applying (2.3.2) to expressions 3 and 4, we derive

5. $d$,

the desired conclusion. Given the hypotheses 1, 2, and 3, we have proved the conclusion $d$. ◀

If a hypothesis is not a clause, it must be replaced by an equivalent expression that is either a clause or the *and* of clauses. For example, suppose that one of the hypotheses is $\neg(a \vee b)$. Since the negation applies to more than one term, we use the first of De Morgan's laws (see Example 1.3.11)

$$\neg(a \vee b) \equiv \neg a \neg b, \qquad \neg(ab) \equiv \neg a \vee \neg b \qquad \textbf{(2.3.3)}$$

to obtain an equivalent expression with the negation applying to single variables: $\neg(a \vee b) \equiv \neg a \neg b$. We then replace the original hypothesis $\neg(a \vee b)$ by the two hypotheses $\neg a$ and $\neg b$. This replacement is justified by recalling that individual hypotheses $h_1$ and $h_2$ are equivalent to $h_1 h_2$ (see Definition 1.4.1 and the discussion that precedes it). Repeated use of De Morgan's laws will result in each negation applying to only one variable.

An expression that consists of terms separated by *or*'s, where each term consists of the *and* of *several* variables, may be replaced by an equivalent expression that consists of the *and* of clauses by using the equivalence

$$a \vee bc \equiv (a \vee b)(a \vee c). \tag{2.3.4}$$

In this case, we may replace the single hypothesis $a \vee bc$ by the two hypotheses $a \vee b$ and $a \vee c$. By using first De Morgan's laws (2.3.3) and then (2.3.4), we can obtain equivalent hypotheses, each of which is a clause.

**Example 2.3.6** Prove the following using resolution:

1. $a \vee \neg bc$
2. $\neg(a \vee d)$

$\therefore \neg b$

**SOLUTION** We use (2.3.4) to replace hypothesis 1 with the two hypotheses $a \vee \neg b$ and $a \vee c$. We use the first of De Morgan's laws (2.3.3) to replace hypothesis 2 with the two hypotheses $\neg a$ and $\neg d$. The argument becomes

1. $a \vee \neg b$
2. $a \vee c$
3. $\neg a$
4. $\neg d$

$\therefore \neg b$

Applying (2.3.1) to expressions 1 and 3, we immediately derive the conclusion $\neg b$. ◀

In automated reasoning systems, proof by resolution is combined with proof by contradiction. We write the negated conclusion as clauses and add the clauses to the hypothesis. We then repeatedly apply (2.3.1) until we derive a contradiction.

**Example 2.3.7** Give another proof of Example 2.3.4 by combining resolution with proof by contradiction.

**SOLUTION** We first negate the conclusion and use the first of De Morgan's laws (2.3.3) to obtain $\neg(b \vee d) \equiv \neg b \neg d$. We then add the clauses $\neg b$ and $\neg d$ to the hypotheses to obtain

1. $a \vee b$
2. $\neg a \vee c$
3. $\neg c \vee d$
4. $\neg b$
5. $\neg d$

Applying (2.3.1) to expressions 1 and 2, we derive

6. $b \vee c$.

Applying (2.3.1) to expressions 3 and 6, we derive

7. $b \vee d$.

Applying (2.3.1) to expressions 4 and 7, we derive

8. $d$.

Now 5 and 8 combine to give a contradiction, and the proof is complete. ◀

It can be shown that resolution is *correct* and *refutation complete*. "Resolution is correct" means that if resolution derives a contradiction from a set of clauses, the clauses are inconsistent (i.e., the clauses are not all true). "Resolution is refutation complete" means that resolution will be able to derive a contradiction from a set of inconsistent clauses. Thus, if a conclusion follows from a set of hypotheses, resolution will be able to derive a contradiction from the hypotheses and the negation of the conclusion. Unfortunately, resolution does not tell us which clauses to combine in order to deduce the contradiction. A key challenge in automating a reasoning system is to help guide the search for clauses to combine. References on resolution and automated reasoning are [Gallier; Genesereth; and Wos].

### 2.3 Problem-Solving Tips

To construct a resolution proof, first replace any of the hypotheses or conclusion that is not a clause with one or more clauses. Then replace pairs of hypotheses of the form $p \vee q$ and $\neg p \vee r$ with $q \vee r$ until deriving the conclusion. Remember that resolution can be combined with proof by contradiction.

## 2.3 Review Exercises

**1.** What rule of logic does proof by resolution use?

**2.** What is a clause?

**3.** Explain how a proof by resolution proceeds.

## 2.3 Exercises

**1.** Write a truth table that proves (2.3.1).

*Use resolution to derive each conclusion in Exercises 2–6. Hint: In Exercises 5 and 6, replace → and ↔ with logically equivalent expressions that use or and and.*

**2.** $\neg p \vee q \vee r$
$\neg q$
$\neg r$
$\overline{\therefore \neg p}$

**3.** $\neg p \vee r$
$\neg r \vee q$
$p$
$\overline{\therefore q}$

**4.** $\neg p \vee t$
$\neg q \vee s$
$\neg r \vee st$
$p \vee q \vee r \vee u$
$\overline{\therefore s \vee t \vee u}$

**5.** $p \to q$
$p \vee q$
$\overline{\therefore q}$

**6.** $p \leftrightarrow r$
$r$
$\overline{\therefore p}$

**7.** Use resolution and proof by contradiction to re-prove Exercises 2–6.

**8.** Use resolution and proof by contradiction to re-prove Example 2.3.6.

# 2.4 Mathematical Induction

**Go Online**
For more on mathematical induction, see `bit.ly/2ECHU4F`

Suppose that a sequence of blocks numbered 1, 2, ... sits on an (infinitely) long table (see Figure 2.4.1) and that some blocks are marked with an "X." (All of the blocks visible in Figure 2.4.1 are marked.) Suppose that

The first block is marked. **(2.4.1)**

For all $n$, if block $n$ is marked, then block $n + 1$ is also marked. **(2.4.2)**

We claim that (2.4.1) and (2.4.2) imply that every block is marked.

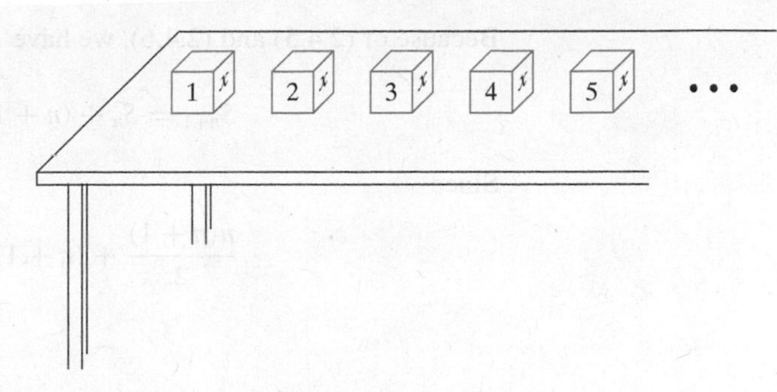


**Figure 2.4.1** Numbered blocks on a table.

We examine the blocks one by one. Statement (2.4.1) explicitly states that block 1 is marked. Consider block 2. Since block 1 is marked, by (2.4.2) (taking $n = 1$), block 2 is also marked. Consider block 3. Since block 2 is marked, by (2.4.2) (taking $n = 2$), block 3 is also marked. Continuing in this way, we can show that every block is marked. For example, suppose that we have verified that blocks 1–5 are marked, as shown in Figure 2.4.1. To show that block 6, which is not shown in Figure 2.4.1, is marked, we note that since block 5 is marked, by (2.4.2) (taking $n = 5$), block 6 is also marked.

The preceding example illustrates the **Principle of Mathematical Induction.** To show how mathematical induction can be used in a more profound way, let $S_n$ denote the sum of the first $n$ positive integers:

$$S_n = 1 + 2 + \cdots + n. \tag{2.4.3}$$

Suppose that someone claims that

$$S_n = \frac{n(n+1)}{2} \quad \text{for all } n \geq 1. \tag{2.4.4}$$

A sequence of statements is really being made, namely,

$$S_1 = \frac{1(2)}{2} = 1, \quad S_2 = \frac{2(3)}{2} = 3, \quad S_3 = \frac{3(4)}{2} = 6, \ldots.$$

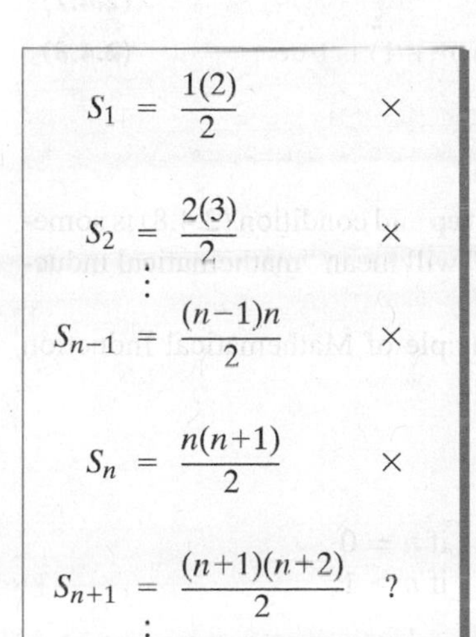


**Figure 2.4.2** A sequence of statements. True statements are marked with ×.

Suppose that each true equation has an "×" placed beside it (see Figure 2.4.2). Since the first equation is true, it is marked. Now suppose we can show that for all $n$, if equation $n$ is marked, then equation $n + 1$ is also marked. Then, as in the example involving the blocks, all of the equations are marked; that is, all the equations are true and the formula (2.4.4) is verified.

We must show that for all $n$, if equation $n$ is true, then equation $n + 1$ is also true. Equation $n$ is

$$S_n = \frac{n(n+1)}{2}. \tag{2.4.5}$$

Assuming that this equation is true, we must show that equation $n + 1$

$$S_{n+1} = \frac{(n+1)(n+2)}{2}$$

is true. According to definition (2.4.3),

$$S_{n+1} = 1 + 2 + \cdots + n + (n+1).$$

We note that $S_n$ is contained within $S_{n+1}$, in the sense that

$$S_{n+1} = 1 + 2 + \cdots + n + (n+1) = S_n + (n+1). \tag{2.4.6}$$

Because of (2.4.5) and (2.4.6), we have

$$S_{n+1} = S_n + (n+1) = \frac{n(n+1)}{2} + (n+1).$$

Since

$$\begin{aligned}\frac{n(n+1)}{2} + (n+1) &= \frac{n(n+1)}{2} + \frac{2(n+1)}{2}\\ &= \frac{n(n+1) + 2(n+1)}{2}\\ &= \frac{(n+1)(n+2)}{2},\end{aligned}$$

we have

$$S_{n+1} = \frac{(n+1)(n+2)}{2}.$$

Therefore, assuming that equation $n$ is true, we have proved that equation $n+1$ is true. We conclude that all of the equations are true.

Our proof using mathematical induction consisted of two steps. First, we verified that the statement corresponding to $n = 1$ was true. Second, we *assumed* that statement $n$ was true and then *proved* that statement $n+1$ was also true. In proving statement $n+1$, we were permitted to make use of statement $n$; indeed, the trick in constructing a proof using mathematical induction is to relate statement $n$ to statement $n+1$.

We next formally state the Principle of Mathematical Induction.

**Principle of Mathematical Induction**

Suppose that we have a propositional function $S(n)$ whose domain of discourse is the set of positive integers. Suppose that

$$S(1) \text{ is true;} \tag{2.4.7}$$

$$\text{for all } n \geq 1, \text{ if } S(n) \text{ is true, then } S(n+1) \text{ is true.} \tag{2.4.8}$$

Then $S(n)$ is true for every positive integer $n$.

Condition (2.4.7) is sometimes called the **Basis Step** and condition (2.4.8) is sometimes called the **Inductive Step.** Hereafter, "induction" will mean "mathematical induction."

After defining $n$ factorial, we illustrate the Principle of Mathematical Induction with another example.

**Definition 2.4.1** ▶ *n factorial* is defined as

$$n! = \begin{cases} 1 & \text{if } n = 0\\ n(n-1)(n-2)\cdots 2\cdot 1 & \text{if } n \geq 1.\end{cases}$$

That is, if $n \geq 1$, $n!$ is equal to the product of all the integers between 1 and $n$ inclusive. As a special case, $0!$ is defined to be 1. ◀

**Example 2.4.2**

$$0! = 1! = 1, \quad 3! = 3\cdot 2\cdot 1 = 6, \quad 6! = 6\cdot 5\cdot 4\cdot 3\cdot 2\cdot 1 = 720$$ ◀

**Example 2.4.3** Use induction to show that

$$n! \geq 2^{n-1} \quad \text{for all } n \geq 1. \tag{2.4.9}$$

**SOLUTION**

**Basis Step ($n = 1$)**

[Condition (2.4.7)] We must show that (2.4.9) is true if $n = 1$. This is easily accomplished, since $1! = 1 \geq 1 = 2^{1-1}$.

**Inductive Step**

[Condition (2.4.8)] We assume that the inequality is true for $n \geq 1$; that is, we assume that

$$n! \geq 2^{n-1} \tag{2.4.10}$$

is true. We must then prove that the inequality is true for $n + 1$; that is, we must prove that

$$(n+1)! \geq 2^n \tag{2.4.11}$$

is true. We can relate (2.4.10) and (2.4.11) by observing that $(n + 1)! = (n + 1)(n!)$. Now

$$\begin{aligned} (n+1)! &= (n+1)(n!) \\ &\geq (n+1)2^{n-1} && \text{by (2.4.10)} \\ &\geq 2 \cdot 2^{n-1} && \text{since } n+1 \geq 2 \\ &= 2^n. \end{aligned}$$

Therefore, (2.4.11) is true. We have completed the Inductive Step.

Since the Basis Step and the Inductive Step have been verified, the Principle of Mathematical Induction tells us that (2.4.9) is true for every positive integer $n$. ◀

If we want to verify that the statements $S(n_0), S(n_0 + 1), \ldots$, where $n_0 \neq 1$, are true, we must change the Basis Step to *$S(n_0)$ is true.* In words, the Basis Step is to prove that the propositional function $S(n)$ is true for the smallest value $n_0$ in the domain of discourse.

The Inductive Step then becomes

*for all $n \geq n_0$, if $S(n)$ is true, then $S(n + 1)$ is true.*

**Example 2.4.4** **Geometric Sum** Use induction to show that if $r \neq 1$,

$$a + ar^1 + ar^2 + \cdots + ar^n = \frac{a(r^{n+1} - 1)}{r - 1} \tag{2.4.12}$$

for all $n \geq 0$.

The sum on the left is called the **geometric sum.** In the geometric sum in which $a \neq 0$ and $r \neq 0$, the ratio of adjacent terms $[(ar^{i+1})/(ar^i) = r]$ is constant.

**SOLUTION**

**Basis Step ($n = 0$)**

Since the smallest value in the domain of discourse $\{n \mid n \geq 0\}$ is $n = 0$, the Basis Step is to prove that (2.4.12) is true for $n = 0$. For $n = 0$, (2.4.12) becomes

$$a = \frac{a(r^1 - 1)}{r - 1},$$

which is true.

### Inductive Step

Assume that statement (2.4.12) is true for $n$. Now

$$\begin{aligned} a + ar^1 + ar^2 + \cdots + ar^n + ar^{n+1} &= \frac{a(r^{n+1} - 1)}{r - 1} + ar^{n+1} \\ &= \frac{a(r^{n+1} - 1)}{r - 1} + \frac{ar^{n+1}(r - 1)}{r - 1} \\ &= \frac{a(r^{n+2} - 1)}{r - 1}. \end{aligned}$$

Since the Basis Step and the Inductive Step have been verified, the Principle of Mathematical Induction tells us that (2.4.12) is true for all $n \geq 0$. ◀

As an example of the use of the geometric sum, if we take $a = 1$ and $r = 2$ in (2.4.12), we obtain the formula

$$1 + 2 + 2^2 + 2^3 + \cdots + 2^n = \frac{2^{n+1} - 1}{2 - 1} = 2^{n+1} - 1.$$

The reader has surely noticed that in order to prove the previous formulas, one has to be given the correct formulas in advance. A reasonable question is: How does one come up with the formulas? There are many answers to this question. One technique to derive a formula is to experiment with small values and try to discover a pattern. (Another technique is discussed in Exercises 70–73). For example, consider the sum $1+3+\cdots+(2n-1)$. The following table gives the values of this sum for $n = 1, 2, 3, 4$.

| $n$ | $1 + 3 + \cdots + (2n - 1)$ |
|---|---|
| 1 | 1 |
| 2 | 4 |
| 3 | 9 |
| 4 | 16 |

Since the second column consists of squares, we conjecture that

$$1 + 3 + \cdots + (2n - 1) = n^2 \quad \text{for every positive integer } n.$$

The conjecture is correct and the formula can be proved by mathematical induction (see Exercise 1).

At this point, the reader may want to read the Problem-Solving Corner that follows this section. This Problem-Solving Corner gives an extended, detailed exposition of how to do proofs by mathematical induction.

Our final examples show that induction is not limited to proving formulas for sums and verifying inequalities.

**Example 2.4.5** Use induction to show that $5^n - 1$ is divisible by 4 for all $n \geq 1$.

### SOLUTION

### Basis Step ($n = 1$)

If $n = 1$, $5^n - 1 = 5^1 - 1 = 4$, which is divisible by 4.

### Inductive Step

We assume that $5^n - 1$ is divisible by 4. We must then show that $5^{n+1} - 1$ is divisible by 4. We use the fact that if $p$ and $q$ are each divisible by $k$, then $p + q$ is also divisible by $k$. In our case, $k = 4$. We leave the proof of this fact to the exercises (see Exercise 74).

We relate the $(n+1)$st case to the $n$th case by writing

$$5^{n+1} - 1 = 5^n - 1 + \text{to be determined.}$$

Now, by the inductive assumption, $5^n - 1$ is divisible by 4. If "to be determined" is also divisible by 4, then the preceding sum, which is equal to $5^{n+1} - 1$, will also be divisible by 4, and the Inductive Step will be complete. We must find the value of "to be determined."

Now

$$5^{n+1} - 1 = 5 \cdot 5^n - 1 = 4 \cdot 5^n + 1 \cdot 5^n - 1.$$

Thus, "to be determined" is $4 \cdot 5^n$, which is divisible by 4. Formally, we could write the Inductive Step as follows.

By the inductive assumption, $5^n - 1$ is divisible by 4 and, since $4 \cdot 5^n$ is divisible by 4, the sum

$$(5^n - 1) + 4 \cdot 5^n = 5^{n+1} - 1$$

is divisible by 4.

Since the Basis Step and the Inductive Step have been verified, the Principle of Mathematical Induction tells us that $5^n - 1$ is divisible by 4 for all $n \geq 1$. ◀

We next give the proof promised in Section 1.1 that if a set $X$ has $n$ elements, the power set of $X$, $\mathcal{P}(X)$, has $2^n$ elements.

**Theorem 2.4.6** If $|X| = n$, then

$$|\mathcal{P}(X)| = 2^n \tag{2.4.13}$$

for all $n \geq 0$.

**Proof** The proof is by induction on $n$.

**Basis Step ($n = 0$)**

If $n = 0$, $X$ is the empty set. The only subset of the empty set is the empty set itself; thus,

$$|\mathcal{P}(X)| = 1 = 2^0 = 2^n.$$

Thus, (2.4.13) is true for $n = 0$.

**Inductive Step**

Assume that (2.4.13) holds for $n$. Let $X$ be a set with $n+1$ elements. Choose $x \in X$. We claim that exactly half of the subsets of $X$ contain $x$, and exactly half of the subsets of $X$ do not contain $x$. To see this, notice that each subset $S$ of $X$ that contains $x$ can be paired uniquely with the subset obtained by removing $x$ from $S$ (see Figure 2.4.3). Thus exactly half of the subsets of $X$ contain $x$, and exactly half of the subsets of $X$ do not contain $x$.

| *Subsets of X that contain a* | *Subsets of X that do not contain a* |
|---|---|
| $\{a\}$ | $\emptyset$ |
| $\{a, b\}$ | $\{b\}$ |
| $\{a, c\}$ | $\{c\}$ |
| $\{a, b, c\}$ | $\{b, c\}$ |

**Figure 2.4.3** Subsets of $X = \{a, b, c\}$ divided into two classes: those that contain $a$ and those that do not contain $a$. Each subset in the right column is obtained from the corresponding subset in the left column by deleting the element $a$ from it.

If we let $Y$ be the set obtained from $X$ by removing $x$, $Y$ has $n$ elements. By the inductive assumption, $|\mathcal{P}(Y)| = 2^n$. But the subsets of $Y$ are precisely the subsets of $X$ that do not contain $x$. From the argument in the preceding paragraph, we conclude that

$$|\mathcal{P}(Y)| = \frac{|\mathcal{P}(X)|}{2}.$$

Therefore,

$$|\mathcal{P}(X)| = 2|\mathcal{P}(Y)| = 2 \cdot 2^n = 2^{n+1}.$$

Thus (2.4.13) holds for $n+1$ and the inductive step is complete. By the Principle of Mathematical Induction, (2.4.13) holds for all $n \geq 0$. ◀

**Example 2.4.7** **A Tiling Problem** A *right tromino,* hereafter called simply a *tromino,* is an object made up of three squares, as shown in Figure 2.4.4. A tromino is a type of polyomino. Since polyominoes were introduced by Solomon W. Golomb in 1954 (see [Golomb, 1954]), they have been a favorite topic in recreational mathematics. A *polyomino of order s* consists of $s$ squares joined at the edges. A tromino is a polyomino of order 3. Three squares in a row form the only other type of polyomino of order 3. (No one has yet found a simple formula for the number of polyominoes of order $s$.) Numerous problems using polyominoes have been devised (see [Martin]).

**Go Online**
For more on trominoes, see bit.ly/2ECHU4F

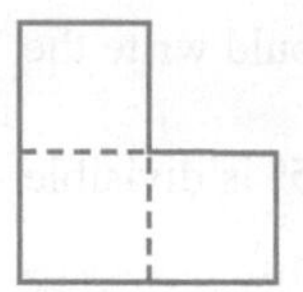

Figure 2.4.4 A tromino.

We give Golomb's inductive proof (see [Golomb, 1954]) that if we remove one square from an $n \times n$ board, where $n$ is a power of 2, we can tile the remaining squares with right trominoes (see Figure 2.4.5). By a *tiling* of a figure by trominoes, we mean an exact covering of the figure by trominoes without having any of the trominoes overlap each other or extend outside the figure. We call a board with one square missing a *deficient board.*

We now use induction on $k$ to prove that we can tile a $2^k \times 2^k$ deficient board with trominoes for all $k \geq 1$.

**Basis Step ($k = 1$)**

If $k = 1$, the $2 \times 2$ deficient board is itself a tromino and can therefore be tiled with one tromino.

**Inductive Step**

Assume that we can tile a $2^k \times 2^k$ deficient board. We show that we can tile a $2^{k+1} \times 2^{k+1}$ deficient board.

Consider a $2^{k+1} \times 2^{k+1}$ deficient board. Divide the board into four $2^k \times 2^k$ boards, as shown in Figure 2.4.6. Rotate the board so that the missing square is in the upper-left quadrant. By the inductive assumption, the upper-left $2^k \times 2^k$ board can be tiled. Place one tromino $T$ in the center, as shown in Figure 2.4.6, so that each square of $T$ is in each of the other quadrants. If we consider the squares covered by $T$ as missing, each of these quadrants is a $2^k \times 2^k$ deficient board. Again, by the inductive assumption, these boards can be tiled. We now have a tiling of the $2^{k+1} \times 2^{k+1}$ board. By the Principle of Mathematical Induction, it follows that any $2^k \times 2^k$ deficient board can be tiled with trominoes, $k = 1, 2, \ldots$.

If we can tile an $n \times n$ deficient board, where $n$ is not necessarily a power of 2, then the number of squares, $n^2 - 1$, must be divisible by 3. [Chu] showed that the converse is true, except when $n$ is 5. More precisely, if $n \neq 5$, any $n \times n$ deficient board can be tiled

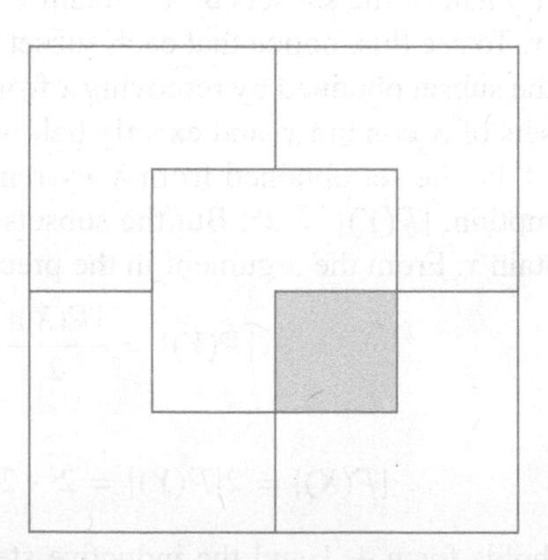

Figure 2.4.5 Tiling a $4 \times 4$ deficient board with trominoes.

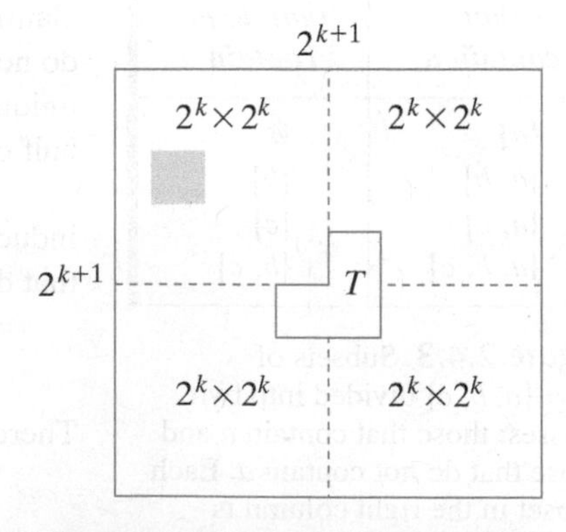


Figure 2.4.6 Using mathematical induction to tile a $2^{k+1} \times 2^{k+1}$ deficient board with trominoes.

with trominoes if and only if 3 divides $n^2 - 1$ (see Exercises 28 and 29, Section 2.5). [Some $5 \times 5$ deficient boards can be tiled and some cannot (see Exercises 33–35).]

Some real-world problems can be modeled as tiling problems. One example is the *VLSI layout problem*—the problem of packing many components on a computer chip (see [Wong]). (VLSI is short for very large scale integration.) The problem is to tile a rectangle of minimum area with the desired components. The components are sometimes modeled as rectangles and L-shaped figures similar to (right) trominoes. In practice, other constraints are imposed such as the proximity of various components that must be interconnected and restrictions on the ratios of width to height of the resulting rectangle. ◀

A **loop invariant** is a statement about program variables that is true just before a loop begins executing and is also true after each iteration of the loop. In particular, a loop invariant is true after the loop finishes, at which point the invariant tells us something about the state of the variables. Ideally, this statement tells us that the loop produces the expected result, that is, that the loop is correct. For example, a loop invariant for a while loop

```
while (condition)
    // loop body
```

is true just before *condition* is evaluated the first time, and it is also true each time the loop body is executed.

We can use mathematical induction to prove that an invariant has the desired behavior. The Basis Step proves that the invariant is true before the condition that controls looping is tested for the first time. The Inductive Step assumes that the invariant is true and then proves that if the condition that controls looping is true (so that the loop body is executed again), the invariant is true after the loop body executes. Since a loop iterates a finite number of times, the form of mathematical induction used here proves that a *finite* sequence of statements is true, rather than an infinite sequence of statements as in our previous examples. Whether the sequence of statements is finite or infinite, the steps needed for the proof by mathematical induction are the same. We illustrate a loop invariant with an example.

**Example 2.4.8** Use a loop invariant to prove that when the pseudocode

```
i − 1
fact = 1
while (i < n) {
    i = i + 1
    fact = fact * i
}
```

terminates, *fact* is equal to $n!$.

**SOLUTION** We prove that $fact = i!$ is an invariant for the while loop. Just before the while loop begins executing, $i = 1$ and $fact = 1$, so $fact = 1!$. We have proved the Basis Step.

Assume that $fact = i!$. If $i < n$ is true (so that the loop body executes again), $i$ becomes $i + 1$ and *fact* becomes

$$fact * (i + 1) = i! * (i + 1) = (i + 1)!.$$

We have proved the Inductive Step. Therefore, $fact = i!$ is an invariant for the while loop.

The while loop terminates when $i = n$. Because $fact = i!$ is an invariant, at this point, $fact = n!$. ◀

## 2.4 Problem-Solving Tips

To prove

$$a_1 + a_2 + \cdots + a_n = F(n) \qquad \text{for all } n \geq 1,$$

where $F(n)$ is the formula for the sum, first verify the equation for $n = 1$: $a_1 = F(1)$ (Basis Step). This is usually straightforward.

Now assume that the statement is true for $n$; that is, assume

$$a_1 + a_2 + \cdots + a_n = F(n).$$

Add $a_{n+1}$ to both sides to get

$$a_1 + a_2 + \cdots + a_n + a_{n+1} = F(n) + a_{n+1}.$$

Finally, show that

$$F(n) + a_{n+1} = F(n+1).$$

To verify the preceding equation, use algebra to manipulate the left-hand side of the equation $[F(n) + a_{n+1}]$ until you get $F(n+1)$. *Look at $F(n+1)$ so you know where you're headed.* (It's somewhat like looking up the answer in the back of the book!) You've shown that

$$a_1 + a_2 + \cdots + a_{n+1} = F(n+1),$$

which is the Inductive Step. Now the proof is complete.

Proving an inequality is handled in a similar fashion. The difference is that instead of obtaining equality $[F(n) + a_{n+1} = F(n+1)$ in the preceding discussion], you obtain an *inequality*.

In general, the key to devising a proof by induction is to find case $n$ "within" case $n+1$. Review the tiling problem (Example 2.4.7), which provides a striking example of case $n$ "within" case $n+1$.

## 2.4 Review Exercises

1. State the Principle of Mathematical Induction.
2. Explain how a proof by mathematical induction proceeds.
3. Give a formula for the sum $1 + 2 + \cdots + n$.
4. What is the geometric sum? Give a formula for it.

## 2.4 Exercises

*In Exercises 1–12, using induction, verify that each equation is true for every positive integer n.*

1. $1 + 3 + 5 + \cdots + (2n - 1) = n^2$
2. $1 \cdot 2 + 2 \cdot 3 + 3 \cdot 4 + \cdots + n(n+1) = \dfrac{n(n+1)(n+2)}{3}$
3. $1(1!) + 2(2!) + \cdots + n(n!) = (n+1)! - 1$
4. $1^2 + 2^2 + 3^2 + \cdots + n^2 = \dfrac{n(n+1)(2n+1)}{6}$
5. $1^2 - 2^2 + 3^2 - \cdots + (-1)^{n+1}n^2 = \dfrac{(-1)^{n+1}n(n+1)}{2}$
6. $1^3 + 2^3 + 3^3 + \cdots + n^3 = \left[\dfrac{n(n+1)}{2}\right]^2$
7. $\dfrac{1}{1 \cdot 3} + \dfrac{1}{3 \cdot 5} + \dfrac{1}{5 \cdot 7} + \cdots + \dfrac{1}{(2n-1)(2n+1)} = \dfrac{n}{2n+1}$

**8.** $$\frac{1}{2\cdot 4} + \frac{1\cdot 3}{2\cdot 4\cdot 6} + \frac{1\cdot 3\cdot 5}{2\cdot 4\cdot 6\cdot 8} + \cdots + \frac{1\cdot 3\cdot 5\cdots(2n-1)}{2\cdot 4\cdot 6\cdots(2n+2)} = \frac{1}{2} - \frac{1\cdot 3\cdot 5\cdots(2n+1)}{2\cdot 4\cdot 6\cdots(2n+2)}$$

**9.** $$\frac{1}{2^2-1} + \frac{1}{3^2-1} + \cdots + \frac{1}{(n+1)^2-1} = \frac{3}{4} - \frac{1}{2(n+1)} - \frac{1}{2(n+2)}$$

**10.** $$1\cdot 2^2 + 2\cdot 3^2 + \cdots + n(n+1)^2 = \frac{n(n+1)(n+2)(3n+5)}{12}$$

**★11.** $$\cos x + \cos 2x + \cdots + \cos nx = \frac{\cos[(x/2)(n+1)]\sin(nx/2)}{\sin(x/2)}$$
provided that $\sin(x/2) \neq 0$.

**★12.** $$1\sin x + 2\sin 2x + \cdots + n\sin nx = \frac{\sin[(n+1)x]}{4\sin^2(x/2)} - \frac{(n+1)\cos[(2n+1)x/2]}{2\sin(x/2)}$$
provided that $\sin(x/2) \neq 0$.

*In Exercises 13–18, using induction, verify the inequality.*

**13.** $\frac{1}{2n} \le \frac{1\cdot 3\cdot 5\cdots(2n-1)}{2\cdot 4\cdot 6\cdots(2n)}$, $n = 1, 2, \ldots$

**★14.** $\frac{1\cdot 3\cdot 5\cdots(2n-1)}{2\cdot 4\cdot 6\cdots(2n)} \le \frac{1}{\sqrt{n+1}}$, $n = 1, 2, \ldots$

**15.** $2n+1 \le 2^n$, $n = 3, 4, \ldots$

**★16.** $2^n \ge n^2$, $n = 4, 5, \ldots$

**★17.** $(a_1a_2\cdots a_{2^n})^{1/2^n} \le \frac{a_1+a_2+\cdots+a_{2^n}}{2^n}$, $n = 1, 2, \ldots,$ and the $a_i$ are positive numbers

**18.** $(1+x)^n \ge 1+nx$, for $x \ge -1$ and $n \ge 1$

**19.** Use the geometric sum to prove that
$$r^0 + r^1 + \cdots + r^n < \frac{1}{1-r}$$
for all $n \ge 0$ and $0 < r < 1$.

**★20.** Prove that
$$1\cdot r^1 + 2\cdot r^2 + \cdots + nr^n < \frac{r}{(1-r)^2}$$
for all $n \ge 1$ and $0 < r < 1$. *Hint:* Using the result of the previous exercise, compare the sum of the terms in

$$\begin{array}{llllll} r & r^2 & r^3 & r^4 & \cdots & r^n \\ r^2 & r^3 & r^4 & \cdots & r^n & \\ r^3 & r^4 & \cdots & r^n & & \\ r^4 & \cdots & & & & \\ \vdots & \vdots & & & & \\ r^{n-1} & r^n & & & & \\ r^n & & & & & \end{array}$$

in the diagonal direction (↙) with the sum of the terms by columns.

**21.** Prove that
$$\frac{1}{2^1} + \frac{2}{2^2} + \frac{3}{2^3} + \cdots + \frac{n}{2^n} < 2$$
for all $n \ge 1$.

*In Exercises 22–25, use induction to prove the statement.*

**22.** $7^n - 1$ is divisible by 6, for all $n \ge 1$.

**23.** $11^n - 6$ is divisible by 5, for all $n \ge 1$.

**24.** $6\cdot 7^n - 2\cdot 3^n$ is divisible by 4, for all $n \ge 1$.

**★25.** $3^n + 7^n - 2$ is divisible by 8, for all $n \ge 1$.

**26.** Use induction to prove that if $X_1, \ldots, X_n$ and $X$ are sets, then

(a) $X \cap (X_1 \cup X_2 \cup \cdots \cup X_n) = (X \cap X_1) \cup (X \cap X_2) \cup \cdots \cup (X \cap X_n)$.

(b) $\overline{X_1 \cap X_2 \cap \cdots \cap X_n} = \overline{X_1} \cup \overline{X_2} \cup \cdots \cup \overline{X_n}$.

**27.** Use induction to prove that if $X_1, \ldots, X_n$ are sets, then
$$|X_1 \times X_2 \times \cdots \times X_n| = |X_1|\cdot|X_2|\cdots|X_n|.$$

**28.** Prove that the number of subsets $S$ of $\{1, 2, \ldots, n\}$, with $|S|$ even, is $2^{n-1}$, $n \ge 1$.

**29.** By experimenting with small values of $n$, guess a formula for the given sum,
$$\frac{1}{1\cdot 2} + \frac{1}{2\cdot 3} + \cdots + \frac{1}{n(n+1)};$$
then use induction to verify your formula.

**30.** Use induction to show that $n$ straight lines in the plane divide the plane into $(n^2+n+2)/2$ regions. Assume that no two lines are parallel and that no three lines have a common point.

**31.** Show that the regions of the preceding exercise can be colored red and green so that no two regions that share an edge are the same color.

**32.** Given $n$ 0's and $n$ 1's distributed in any manner whatsoever around a circle (see the following figure), show, using induction on $n$, that it is possible to start at some number and proceed clockwise around the circle to the original starting position so that, at any point during the cycle, we have seen at least as many 0's as 1's. In the following figure, a possible starting point is marked with an arrow.

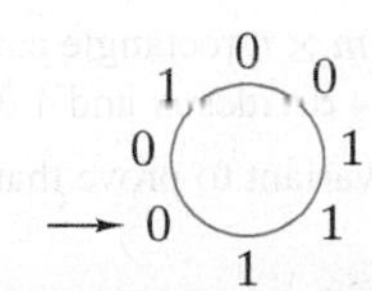

**33.** Give a tiling of a $5 \times 5$ board with trominoes in which the upper-left square is missing.

**34.** Show a $5 \times 5$ deficient board that is impossible to tile with trominoes. Explain why your board cannot be tiled with trominoes.

**35.** Which $5 \times 5$ deficient boards can be tiled?

**36.** Show that any $(2i) \times (3j)$ board, where $i$ and $j$ are positive integers, with no square missing, can be tiled with trominoes.

**★37.** Show that any $7 \times 7$ deficient board can be tiled with trominoes.

**38.** Show that any $11 \times 11$ deficient board can be tiled with trominoes. *Hint:* Subdivide the board into overlapping $7 \times 7$ and $5 \times 5$ boards and two $6 \times 4$ boards. Then, use Exercises 33, 36, and 37.

**39.** This exercise and the one that follows are due to Anthony Quas. A $2^n \times 2^n$ *L-shape,* $n \geq 0$, is a figure of the form

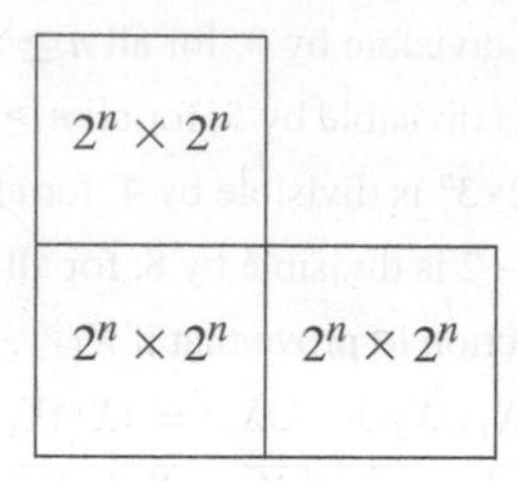


with no missing squares. Show that any $2^n \times 2^n$ L-shape can be tiled with trominoes.

**40.** Use the preceding exercise to give a different proof that any $2^n \times 2^n$ deficient board can be tiled with trominoes.

*A* straight tromino *is an object made up of three squares in a row:*

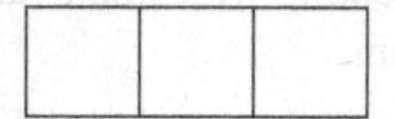

**41.** Which $4 \times 4$ deficient boards can be tiled with straight trominoes? *Hint:* Number the squares of the $4 \times 4$ board, left to right, top to bottom: 1, 2, 3, 1, 2, 3, and so on. Note that if there is a tiling, each straight tromino covers exactly one 2 and exactly one 3.

**42.** Which $5 \times 5$ deficient boards can be tiled with straight trominoes?

**43.** Which $8 \times 8$ deficient boards can be tiled with straight trominoes?

★**44.** A *T-tetromino* is an object made up of four squares

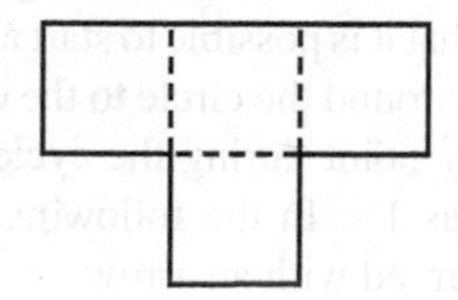

Prove that an $m \times n$ rectangle can be tiled with T-tetrominoes if and only if 4 divides $m$ and 4 divides $n$.

**45.** Use a loop invariant to prove that when the pseudocode

$i = 1$
$pow = 1$
while $(i \leq n)$ {
  $pow = pow * a$
  $i = i + 1$
}

terminates, $pow$ is equal to $a^n$.

**46.** Prove that, after the following pseudocode terminates, $a[h] = val$; for all $p$, $i \leq p < h$, $a[p] < val$; and for all $p$, $h < p \leq j$, $a[p] \geq val$. In particular, $val$ is in the position in the array $a[i], \ldots, a[j]$ where it would be if the array were sorted.

$val = a[i]$
$h = i$
for $k = i + 1$ to $j$
  if $(a[k] < val)$ {
    $h = h + 1$
    $swap(a[h], a[k])$
  }
$swap(a[i], a[h])$

*Hint:* Use the loop invariant: $h < k$; for all $p$, $i < p \leq h$, $a[p] < val$; and, for all $p$, $h < p < k$, $a[p] \geq val$. (A picture is helpful.)

This technique is called *partitioning*. This particular version is due to Nico Lomuto. Partitioning can be used to find the $k$th smallest element in an array and to construct a sorting algorithm called *quicksort*.

*A* 3D-septomino *is a three-dimensional* $2 \times 2 \times 2$ *cube with one* $1 \times 1 \times 1$ *corner cube removed. A* deficient cube *is a* $k \times k \times k$ *cube with one* $1 \times 1 \times 1$ *cube removed.*

**47.** Prove that a $2^n \times 2^n \times 2^n$ deficient cube can be tiled by 3D-septominoes.

**48.** Prove that if a $k \times k \times k$ deficient cube can be tiled by 3D-septominoes, then 7 divides one of $k - 1, k - 2, k - 4$.

**49.** Suppose that $S_n = (n + 2)(n - 1)$ is (incorrectly) proposed as a formula for

$$2 + 4 + \cdots + 2n.$$

(a) Show that the Inductive Step is satisfied but that the Basis Step fails.

★(b) If $S'_n$ is an arbitrary expression that satisfies the Inductive Step, what form must $S'_n$ assume?

★**50.** What is wrong with the following argument, which allegedly shows that any two positive integers are equal?

We use induction on $n$ to "prove" that if $a$ and $b$ are positive integers and $n = \max\{a, b\}$, then $a = b$.

### Basis Step ($n = 1$)

If $a$ and $b$ are positive integers and $1 = \max\{a, b\}$, we must have $a = b = 1$.

### Inductive Step

Assume that if $a'$ and $b'$ are positive integers and $n = \max\{a', b'\}$, then $a' = b'$. Suppose that $a$ and $b$ are positive integers and that $n + 1 = \max\{a, b\}$. Now $n = \max\{a - 1, b - 1\}$. By the inductive hypothesis, $a - 1 = b - 1$. Therefore, $a = b$.

Since we have verified the Basis Step and the Inductive Step, by the Principle of Mathematical Induction, any two positive integers are equal!

**51.** What is wrong with the following "proof" that

$$\frac{1}{2} + \frac{2}{3} + \cdots + \frac{n}{n+1} \neq \frac{n^2}{n+1}$$

for all $n \geq 2$?

Suppose by way of contradiction that

$$\frac{1}{2}+\frac{2}{3}+\cdots+\frac{n}{n+1}=\frac{n^2}{n+1}. \qquad \textbf{(2.4.14)}$$

Then also

$$\frac{1}{2}+\frac{2}{3}+\cdots+\frac{n}{n+1}+\frac{n+1}{n+2}=\frac{(n+1)^2}{n+2}.$$

We could prove statement (2.4.14) by induction. In particular, the Inductive Step would give

$$\left(\frac{1}{2}+\frac{2}{3}+\cdots+\frac{n}{n+1}\right)+\frac{n+1}{n+2}=\frac{n^2}{n+1}+\frac{n+1}{n+2}.$$

Therefore,

$$\frac{n^2}{n+1}+\frac{n+1}{n+2}=\frac{(n+1)^2}{n+2}.$$

Multiplying each side of this last equation by $(n+1)(n+2)$ gives

$$n^2(n+2)+(n+1)^2=(n+1)^3.$$

This last equation can be rewritten as

$$n^3+2n^2+n^2+2n+1=n^3+3n^2+3n+1$$

or

$$n^3+3n^2+2n+1=n^3+3n^2+3n+1,$$

which is a contradiction. Therefore,

$$\frac{1}{2}+\frac{2}{3}+\cdots+\frac{n}{n+1}\neq\frac{n^2}{n+1},$$

as claimed.

**52.** Use mathematical induction to prove that

$$\frac{1}{2}+\frac{2}{3}+\cdots+\frac{n}{n+1}<\frac{n^2}{n+1}$$

for all $n \geq 2$. This inequality gives a correct proof of the statement of the preceding exercise.

*In Exercises 53–57, suppose that $n > 1$ people are positioned in a field (Euclidean plane) so that each has a unique nearest neighbor. Suppose further that each person has a pie that is hurled at the nearest neighbor.* A survivor *is a person that is not hit by a pie.*

**53.** Give an example to show that if $n$ is even, there might be no survivor.

**54.** Give an example to show that there might be more than one survivor.

**★55.** [Carmony] Use induction on $n$ to show that if $n$ is odd, there is always at least one survivor.

**56.** Prove or disprove: If $n$ is odd, one of two persons farthest apart is a survivor.

**57.** Prove or disprove: If $n$ is odd, a person who throws a pie the greatest distance is a survivor.

*Exercises 58–61 deal with plane convex sets.* A plane convex set, *subsequently abbreviated to "convex set," is a nonempty set $X$ in the plane having the property that if $x$ and $y$ are any two points in $X$, the straight-line segment from $x$ to $y$ is also in $X$. The following figures illustrate.*

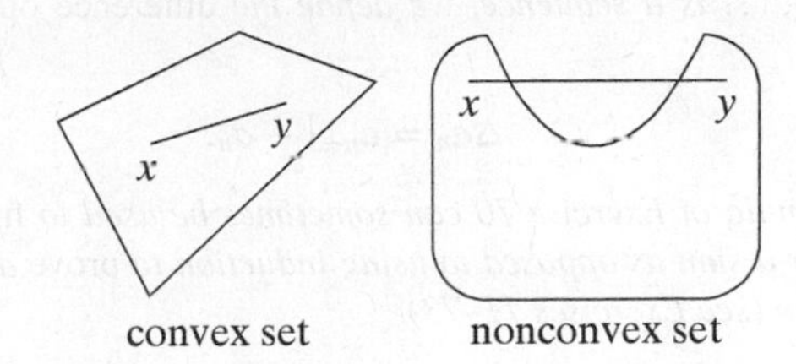


convex set    nonconvex set

**58.** Prove that if $X$ and $Y$ are convex sets and $X \cap Y$ is nonempty, $X \cap Y$ is a convex set.

**★59.** Suppose that $X_1, X_2, X_3, X_4$ are convex sets, each three of which have a common point. Prove that all four sets have a common point.

**★60.** Prove *Helly's Theorem:* Suppose that $X_1, X_2, \ldots, X_n$, $n \geq 4$, are convex sets, each three of which have a common point. Prove that all $n$ sets have a common point.

**61.** Suppose that $n \geq 3$ points in the plane have the property that each three of them are contained in a circle of radius 1. Prove that there is a circle of radius 1 that contains all of the points.

**62.** If $a$ and $b$ are real numbers with $a < b$, an *open interval* $(a, b)$ is the set of all real numbers $x$ such that $a < x < b$. Prove that if $I_1, \ldots, I_n$ is a set of $n \geq 2$ open intervals such that each pair has a nonempty intersection, then

$$I_1 \cap I_2 \cap \cdots \cap I_n$$

is nonempty.

*Flavius Josephus was a Jewish soldier and historian who lived in the first century (see [Graham, 1994; Schumer]). He was one of the leaders of a Jewish revolt against Rome in the year 66. The following year, he was among a group of trapped soldiers who decided to commit suicide rather than be captured. One version of the story is that, rather than being captured, they formed a circle and proceeded around the circle killing every third person. Josephus, being proficient in discrete math, figured out where he and a buddy should stand so they could avoid being killed.*

*Exercises 63–69 concern a variant of the* Josephus Problem *in which every second person is eliminated. We assume that $n$ people are arranged in a circle and numbered $1, 2, \ldots, n$ clockwise. Then, proceeding clockwise, 2 is eliminated, 4 is eliminated, and so on, until there is one survivor, denoted $J(n)$.*

**63.** Compute $J(4)$.

**64.** Compute $J(6)$.

**65.** Compute $J(10)$.

**66.** Use induction to show that $J(2^i) = 1$ for all $i \geq 1$.

**67.** Given a value of $n \geq 2$, let $2^i$ be the greatest power of 2 with $2^i \leq n$. (Examples: If $n = 10$, $i = 3$. If $n = 16$, $i = 4$.) Let $j = n - 2^i$. (After subtracting $2^i$, the greatest power of 2 less than or equal to $n$, from $n$, $j$ is what is left over.) By using the result of Exercise 66 or otherwise, prove that

$$J(n) = 2j + 1.$$

**68.** Use the result of Exercise 67 to compute $J(1000)$.

**69.** Use the result of Exercise 67 to compute $J(100{,}000)$.

*If $a_1, a_2, \ldots$ is a sequence, we define the* difference operator $\Delta$ *to be*

$$\Delta a_n = a_{n+1} - a_n.$$

*The formula of Exercise 70 can sometimes be used to* find *a formula for a sum as opposed to using induction to* prove *a formula for a sum (see Exercises 71–73).*

**70.** Suppose that $\Delta a_n = b_n$. Show that

$$b_1 + b_2 + \cdots + b_n = a_{n+1} - a_1.$$

This formula is analogous to the calculus formula $\int_c^d f(x)\,dx = g(d) - g(c)$, where $Dg = f$ ($D$ is the derivative operator). In the calculus formula, sum is replaced by integral, and $\Delta$ is replaced by derivative.

**71.** Let $a_n = n^2$, and compute $\Delta a_n$. Use Exercise 70 to find a formula for

$$1 + 2 + 3 + \cdots + n.$$

**72.** Use Exercise 70 to find a formula for

$$1(1!) + 2(2!) + \cdots + n(n!).$$

(Compare with Exercise 3.)

**73.** Use Exercise 70 to find a formula for

$$\frac{1}{1 \cdot 2} + \frac{1}{2 \cdot 3} + \cdots + \frac{1}{n(n+1)}.$$

(Compare with Exercise 29.)

**74.** Prove that if $p$ and $q$ are divisible by $k$, then $p + q$ is divisible by $k$.

# Problem-Solving Corner: Mathematical Induction

## Problem

Define

$$H_k = 1 + \frac{1}{2} + \frac{1}{3} + \cdots + \frac{1}{k} \tag{1}$$

for all $k \geq 1$. The numbers $H_1, H_2, \ldots$ are called the *harmonic numbers*. Prove that

$$H_{2^n} \geq 1 + \frac{n}{2} \tag{2}$$

for all $n \geq 0$.

## Attacking the Problem

It's often a good idea to begin attacking a problem by looking at some concrete examples of the expressions under consideration. Let's look at $H_k$ for some small values of $k$. The smallest value of $k$ for which $H_k$ is defined is $k = 1$. In this case, the last term $1/k$ in the definition of $H_k$ equals $1/1 = 1$. Since the first and last terms coincide, $H_1 = 1$. For $k = 2$, the last term $1/k$ in the definition of $H_k$ equals $1/2$, so

$$H_2 = 1 + \frac{1}{2}.$$

Similarly, we find that

$$H_3 = 1 + \frac{1}{2} + \frac{1}{3},$$

$$H_4 = 1 + \frac{1}{2} + \frac{1}{3} + \frac{1}{4}.$$

We observe that $H_1$ appears as the first term of $H_2$, $H_3$, and $H_4$, that $H_2$ appears as the first two terms of $H_3$ and $H_4$, and that $H_3$ appears as the first three terms of $H_4$. In general, $H_m$ appears as the first $m$ terms of $H_k$ if $m \leq k$. This observation will help us later because the Inductive Step in a proof by induction must relate smaller instances of a problem to larger instances of the problem.

In general, it's a good strategy to delay combining terms and simplifying until as late as possible, which is why, for example, we left $H_4$ as the sum of four terms rather than writing $H_4 = 25/12$. Since we left $H_4$ as the sum of four terms, we were able to see that each of $H_1$, $H_2$, and $H_3$ appears in the expression for $H_4$.

## Finding a Solution

The Basis Step is to prove the given statement for the smallest value of $n$, which here is $n = 0$. For $n = 0$, inequality (2) that we must prove becomes

$$H_{2^0} \geq 1 + \frac{0}{2} = 1.$$

We have already observed that $H_1 = 1$. Thus inequality (2) is true when $n = 0$; in fact, the inequality is an equality. (Recall that by definition, if $x = y$ is true, then $x \geq y$ is also true.)

Let's move to the Inductive Step. It's a good idea to write down what is assumed (here the case $n$),

$$H_{2^n} \geq 1 + \frac{n}{2}, \tag{3}$$

and what needs to be proved (here the case $n + 1$),

$$H_{2^{n+1}} \geq 1 + \frac{n+1}{2}. \tag{4}$$

It's also a good idea to write the formulas for any expressions that occur. Using equation (1), we may write

$$H_{2^n} = 1 + \frac{1}{2} + \cdots + \frac{1}{2^n} \tag{5}$$

and

$$H_{2^{n+1}} = 1 + \frac{1}{2} + \cdots + \frac{1}{2^{n+1}}.$$

It's not so evident from the last equation that $H_{2^n}$ appears as the first $2^n$ terms of $H_{2^{n+1}}$. Let's rewrite the last equation as

$$H_{2^{n+1}} = 1 + \frac{1}{2} + \cdots + \frac{1}{2^n} + \frac{1}{2^n+1} + \cdots + \frac{1}{2^{n+1}} \tag{6}$$

to make it clear that $H_{2^n}$ appears as the first $2^n$ terms of $H_{2^{n+1}}$.

For clarity, we have written the term that follows $1/2^n$. Notice that the denominators increase by one, so the term that follows $1/2^n$ is $1/(2^n+1)$. Also notice that there is a big difference between $1/(2^n+1)$, the term that follows $1/2^n$, and $1/2^{n+1}$, the last term in equation (6).

Using equations (5) and (6), we may relate $H_{2^n}$ to $H_{2^{n+1}}$ explicitly by writing

$$H_{2^{n+1}} = H_{2^n} + \frac{1}{2^n+1} + \cdots + \frac{1}{2^{n+1}}. \tag{7}$$

Combining (3) and (7), we obtain

$$H_{2^{n+1}} \geq 1 + \frac{n}{2} + \frac{1}{2^n+1} + \cdots + \frac{1}{2^{n+1}}. \tag{8}$$

This inequality shows that $H_{2^{n+1}}$ is greater than or equal to

$$1 + \frac{n}{2} + \frac{1}{2^n+1} + \cdots + \frac{1}{2^{n+1}},$$

but our goal (4) is to show that $H_{2^{n+1}}$ is greater than or equal to $1 + (n+1)/2$. We will achieve our goal if we show that

$$1 + \frac{n}{2} + \frac{1}{2^n+1} + \cdots + \frac{1}{2^{n+1}} \geq 1 + \frac{n+1}{2}.$$

In general, to prove an inequality, we replace terms in the larger expression with smaller terms so that the resulting expression equals the smaller expression; or we replace terms in the smaller expression with larger terms so that the resulting expression equals the larger expression. Here let's replace each of the terms in the sum

$$\frac{1}{2^n+1} + \cdots + \frac{1}{2^{n+1}}$$

by the smallest term $1/2^{n+1}$ in the sum. We obtain

$$\frac{1}{2^n+1} + \cdots + \frac{1}{2^{n+1}} \geq \frac{1}{2^{n+1}} + \cdots + \frac{1}{2^{n+1}}.$$

Since there are $2^n$ terms in the latter sum, each equal to $1/2^{n+1}$, we may rewrite the preceding inequality as

$$\frac{1}{2^n+1} + \cdots + \frac{1}{2^{n+1}} \geq \frac{1}{2^{n+1}} + \cdots + \frac{1}{2^{n+1}} = 2^n \frac{1}{2^{n+1}} = \frac{1}{2}. \tag{9}$$

Combining (8) and (9),

$$H_{2^{n+1}} \geq 1 + \frac{n}{2} + \frac{1}{2} = 1 + \frac{n+1}{2}.$$

We have the desired result, and the Inductive Step is complete.

## Formal Solution

The formal solution could be written as follows.

### Basis Step ($n = 0$)

$$H_{2^0} = 1 \geq 1 = 1 + \frac{0}{2}$$

### Inductive Step

We assume (2). Now

$$\begin{aligned} H_{2^{n+1}} &= 1 + \frac{1}{2} + \cdots + \frac{1}{2^n} + \frac{1}{2^n+1} + \cdots + \frac{1}{2^{n+1}} \\ &= H_{2^n} + \frac{1}{2^n+1} + \cdots + \frac{1}{2^{n+1}} \\ &\geq 1 + \frac{n}{2} + \frac{1}{2^{n+1}} + \cdots + \frac{1}{2^{n+1}} \\ &= 1 + \frac{n}{2} + 2^n \frac{1}{2^{n+1}} \\ &= 1 + \frac{n}{2} + \frac{1}{2} = 1 + \frac{n+1}{2}. \end{aligned}$$

## Summary of Problem-Solving Techniques

- Look at concrete examples of the expressions under consideration, typically for small values of the variables.
- Look for expressions for small values of $n$ to appear within expressions for larger values of $n$. In particular, the Inductive Step depends on relating case $n$ to case $n+1$.
- Delay combining terms and simplifying until as late as possible to help discover relationships among the expressions.
- Write out in full the specific cases to prove, specifically, the smallest value of $n$ for the Basis Step, the case $n$ that is assumed in the Inductive Step, and the case $n+1$ to prove in the Inductive Step. Write out the formulas for the various expressions that appear.

- To prove an inequality, replace terms in the larger expression with smaller terms so that the resulting expression equals the smaller expression, or replace terms in the smaller expression with larger terms so that the resulting expression equals the larger expression.

## Comments

The series

$$1 + \frac{1}{2} + \frac{1}{3} + \cdots,$$

which surfaces in calculus, is called the *harmonic series*. Inequality (2) shows that the harmonic numbers increase without bound. In calculus terminology, the harmonic series *diverges*.

## Exercises

1. Prove that $H_{2^n} \leq 1 + n$ for all $n \geq 0$.
2. Prove that
$$H_1 + H_2 + \cdots + H_n = (n+1)H_n - n$$
for all $n \geq 1$.
3. Prove that
$$H_n = H_{n+1} - \frac{1}{n+1}$$
for all $n \geq 1$.
4. Prove that
$$1 \cdot H_1 + 2 \cdot H_2 + \cdots + nH_n = \frac{n(n+1)}{2}H_{n+1} - \frac{n(n+1)}{4}$$
for all $n \geq 1$.
5. Prove that
$$\frac{H_1}{1} + \frac{H_2}{2} + \cdots + \frac{H_n}{n} = \frac{H_n^2}{2} + \frac{1}{2}\left[\frac{1}{1^2} + \frac{1}{2^2} + \cdots + \frac{1}{n^2}\right]$$
for all $n \geq 1$.

# 2.5 Strong Form of Induction and the Well-Ordering Property

In the Inductive Step of mathematical induction presented in Section 2.4, we assume that statement $n$ is true, and then prove that statement $n + 1$ is true. In other words, to prove that a statement is true (statement $n + 1$), we assume the truth of its immediate predecessor (statement $n$). In some cases in the Inductive Step, to prove a statement is true, it is helpful to assume the truth of *all* of the preceding statements (not just the immediate predecessor). The **Strong Form of Mathematical Induction** allows us to assume the truth of all of the preceding statements. Following the usual convention, the statement to prove is denoted $n$ rather than $n+1$. We next formally state the Strong Form of Mathematical Induction.

**Strong Form of Mathematical Induction.**

Suppose that we have a propositional function $S(n)$ whose domain of discourse is the set of integers greater than or equal to $n_0$. Suppose that

$S(n_0)$ is true;

for all $n > n_0$, if $S(k)$ is true for all $k$, $n_0 \leq k < n$, then $S(n)$ is true.

Then $S(n)$ is true for every integer $n \geq n_0$.

In the Inductive Step of the Strong Form of Mathematical Induction, we let $n$ denote an arbitrary integer, $n > n_0$. Then, assuming that $S(k)$ is true for all $k$ satisfying

$$n_0 \leq k < n, \qquad \textbf{(2.5.1)}$$

we prove that $S(n)$ is true. In inequality (2.5.1), $k$ indexes a statement $S(k)$ that is an *arbitrary* predecessor of the statement $S(n)$ (thus $k < n$), which we are to prove true. In inequality (2.5.1), $n_0 \leq k$ ensures that $k$ is in the domain of discourse

$$\{n_0, n_0 + 1, n_0 + 2, \ldots\}.$$

The two forms of mathematical induction are logically equivalent (see Exercise 38). We present several examples that illustrate the use of the Strong Form of Mathematical Induction.

**Example 2.5.1** Use mathematical induction to show that postage of 4 cents or more can be achieved by using only 2-cent and 5-cent stamps.

**SOLUTION** ***Discussion*** Consider the Inductive Step, where we want to prove that $n$-cents postage can be achieved using only 2-cent and 5-cent stamps. It would be particularly easy to prove this statement if we could assume that we can make postage of $n-2$ cents. We could then simply add a 2-cent stamp to make $n$-cents postage. How simple! If we use the Strong Form of Mathematical Induction, we *can* assume the truth of the statement for all $k < n$. In particular, we can assume the truth of the statement for $k = n-2$. Thus the Strong Form of Mathematical Induction allows us to give a correct proof based on our informal reasoning.

In this example, $n_0$ in inequality (2.5.1) is equal to 4. When we take $k = n-2$, to ensure that $n_0 \leq k$, that is $4 \leq n-2$, we must have $6 \leq n$. Now $n = 4$ is the Basis Step. What about $n = 5$? We explicitly prove this case. By convention, we add the case $n = 5$ to the Basis Step; thus the cases $n = 4$ and $n = 5$ become the Basis Steps. In general, if the Inductive Step assumes that the case $n - p$ is true (in this example, $p = 2$), there will be $p$ Basis Steps: $n = n_0, n = n_0 + 1, \ldots, n = n_0 + p - 1$.

**Basis Steps ($n = 4$, $n = 5$)**

We can make 4-cents postage by using two 2-cent stamps. We can make 5-cents postage by using one 5-cent stamp. The Basis Steps are verified.

**Inductive Step**

We assume that $n \geq 6$ and that postage of $k$ cents or more can be achieved by using only 2-cent and 5-cent stamps for $4 \leq k < n$.

By the inductive assumption, we can make postage of $n-2$ cents. We add a 2-cent stamp to make $n$-cents postage. The Inductive Step is complete. ◀

**Example 2.5.2** When an element of a sequence is defined in terms of some of its predecessors, the Strong Form of Mathematical Induction is sometimes useful to prove a property of the sequence. For example, suppose that the sequence $c_1, c_2, \ldots$ is defined by the equations[†]

$$c_1 = 0, \qquad c_n = c_{\lfloor n/2 \rfloor} + n \quad \text{for all } n > 1.$$

As examples,

$$\begin{aligned}
c_2 &= c_{\lfloor 2/2 \rfloor} + 2 = c_{\lfloor 1 \rfloor} + 2 = c_1 + 2 = 0 + 2 = 2,\\
c_3 &= c_{\lfloor 3/2 \rfloor} + 3 = c_{\lfloor 1.5 \rfloor} + 3 = c_1 + 3 = 0 + 3 = 3,\\
c_4 &= c_{\lfloor 4/2 \rfloor} + 4 = c_{\lfloor 2 \rfloor} + 4 = c_2 + 4 = 2 + 4 = 6,\\
c_5 &= c_{\lfloor 5/2 \rfloor} + 5 = c_{\lfloor 2.5 \rfloor} + 5 = c_2 + 5 = 2 + 5 = 7.
\end{aligned}$$

Use strong induction to prove that $c_n < 2n$, for all $n \geq 1$.

**SOLUTION** ***Discussion*** In this example, $n_0$ in inequality (2.5.1) is equal to 1 and (2.5.1) becomes $1 \leq k < n$. In particular, since $c_n$ is defined in terms of $c_{\lfloor n/2 \rfloor}$, in the Inductive Step we assume the truth of the statement for $k = \lfloor n/2 \rfloor$. Inequality (2.5.1)

[†]The *floor of* $x$, $\lfloor x \rfloor$, is the greatest integer less than or equal to $x$ (see Section 3.1). Informally, we are "rounding down." Examples: $\lfloor 2.3 \rfloor = 2$, $\lfloor 5 \rfloor = 5$, $\lfloor -2.7 \rfloor = -3$.

then becomes $1 \le \lfloor n/2 \rfloor < n$. Because $n/2 < n$, it follows that $\lfloor n/2 \rfloor < n$. If $n \ge 2$, then $1 \le n/2$ and so $1 \le \lfloor n/2 \rfloor$. Therefore if $n \ge 2$ and $k = \lfloor n/2 \rfloor$, inequality (2.5.1) is satisfied. Thus the Basis Step is $n = 1$.

**Basis Step ($n = 1$)**
Since $c_1 = 0 < 2 = 2 \cdot 1$, the Basis Step is verified.

**Inductive Step**
We assume that $c_k < 2k$, for all $k$, $1 \le k < n$, and prove that $c_n < 2n$, $n > 1$. Since $1 < n$, $2 \le n$. Thus $1 \le n/2 < n$. Therefore $1 \le \lfloor n/2 \rfloor < n$ and taking $k = \lfloor n/2 \rfloor$, we see that $k$ satisfies inequality (2.5.1). By the inductive assumption

$$c_{\lfloor n/2 \rfloor} = c_k < 2k = 2\lfloor n/2 \rfloor.$$

Now

$$c_n = c_{\lfloor n/2 \rfloor} + n < 2\lfloor n/2 \rfloor + n \le 2(n/2) + n = 2n.$$

The Inductive Step is complete. ◀

**Example 2.5.3** Define the sequence $c_1, c_2, \ldots$ by the equations

$$c_1 = 1, \qquad c_n = c_{\lfloor n/2 \rfloor} + n^2 \quad \text{for all } n > 1.$$

Suppose that we want to prove a statement for all $n \ge 2$ involving $c_n$. The Inductive Step will assume the truth of the statement involving $c_{\lfloor n/2 \rfloor}$. What are the Basis Steps?

**SOLUTION** In this example, $n_0$ in inequality (2.5.1) is equal to 2 and (2.5.1) becomes $2 \le k < n$. In the Inductive Step, we assume the truth of the statement for $k = \lfloor n/2 \rfloor$. Inequality (2.5.1) then becomes $2 \le \lfloor n/2 \rfloor < n$. Because $n/2 < n$, it follows that $\lfloor n/2 \rfloor < n$.

If $n = 3$, then $2 > \lfloor n/2 \rfloor$. Thus we must add $n = 3$ to the Basis Step ($n = 2$). If $n \ge 4$, then $2 \le n/2$ and so $2 \le \lfloor n/2 \rfloor$. Therefore if $n \ge 4$ and $k = \lfloor n/2 \rfloor$, inequality (2.5.1) is satisfied. Thus the Basis Steps are $n = 2$ and $n = 3$. ◀

**Example 2.5.4** Suppose that we insert parentheses and then multiply the $n$ numbers $a_1 a_2 \cdots a_n$. For example, if $n = 4$, we might insert the parentheses as shown:

$$(a_1 a_2)(a_3 a_4). \qquad \textbf{(2.5.2)}$$

Here we would first multiply $a_1$ by $a_2$ to obtain $a_1 a_2$ and $a_3$ by $a_4$ to obtain $a_3 a_4$. We would then multiply $a_1 a_2$ by $a_3 a_4$ to obtain $(a_1 a_2)(a_3 a_4)$. Notice that the number of multiplications is three. Use strong induction to prove that if we insert parentheses in any manner whatsoever and then multiply the $n$ numbers $a_1 a_2 \cdots a_n$, we perform $n - 1$ multiplications.

**SOLUTION**

**Basis Step ($n = 1$)**
We need 0 multiplications to compute $a_1$. The Basis Step is verified.

**Inductive Step**
We assume that for all $k$, $1 \le k < n$, it takes $k - 1$ multiplications to compute the product of $k$ numbers if parentheses are inserted in any manner whatsoever. We must prove that it takes $n$ multiplications to compute the product $a_1 a_2 \cdots a_n$ if parentheses are inserted in any manner whatsoever.

Suppose that parentheses are inserted in the product $a_1 a_2 \cdots a_n$. Consider the final multiplication, which looks like $(a_1 \cdots a_t)(a_{t+1} \cdots a_n)$, for some $t$, $1 \le t < n$. [For example, in equation (2.5.2), $t = 2$.] There are $t$ terms in the first set of parentheses, $1 \le t < n$, and $n - t$ terms in the second set of parentheses, $1 \le n - t < n$. By the inductive assumption, it takes $t - 1$ multiplications to compute $a_1 \cdots a_t$ and $n - t - 1$ multiplications to compute $a_{t+1} \cdots a_n$, regardless of how the parentheses are inserted. It takes one additional multiplication to multiply $a_1 \cdots a_t$ by $a_{t+1} \cdots a_n$. Thus the total number of multiplications is

$$(t - 1) + (n - t - 1) + 1 = n - 1.$$

The Inductive Step is complete. ◀

## Well-Ordering Property

The **Well-Ordering Property for nonnegative integers** states that every nonempty set of nonnegative integers has a least element. This property is equivalent to the two forms of induction (see Exercises 36–38). We use the Well-Ordering Property to prove something familiar from long division: When we divide an integer $n$ by a positive integer $d$, we obtain a quotient $q$ and a remainder $r$ satisfying $0 \le r < d$ so that $n = dq + r$.

**Example 2.5.5** When we divide $n = 74$ by $d = 13$

$$\begin{array}{r} 5 \\ 13\overline{)74} \\ \underline{65} \\ 9 \end{array}$$

we obtain the quotient $q = 5$ and the remainder $r = 9$. Notice that $r$ satisfies $0 \le r < d$; that is, $0 \le 9 < 13$. We have

$$n = 74 = 13 \cdot 5 + 9 = dq + r.$$ ◀

**Theorem 2.5.6** **Quotient-Remainder Theorem**

If $d$ and $n$ are integers, $d > 0$, there exist integers $q$ (quotient) and $r$ (remainder) satisfying

$$n = dq + r \qquad 0 \le r < d.$$

Furthermore, $q$ and $r$ are unique; that is, if

$$n = dq_1 + r_1 \qquad 0 \le r_1 < d$$

and

$$n = dq_2 + r_2 \qquad 0 \le r_2 < d,$$

then $q_1 = q_2$ and $r_1 = r_2$.

***Discussion*** We can devise a proof of Theorem 2.5.6 by looking carefully at the technique used in long division. Why is 5 the quotient in Example 2.5.5? Because $q = 5$ makes the remainder $n - dq$ nonnegative and as small as possible. If, for example, $q = 3$, the remainder would be $n - dq = 74 - 13 \cdot 3 = 35$, which is too large. As another example, if $q = 6$, the remainder would be $n - dq = 74 - 13 \cdot 6 = -4$, which is negative. The existence of a smallest, nonnegative remainder $n - dq$ is guaranteed by the Well-Ordering Property.

**Proof** Let

$$X = \{n - dk \mid n - dk \geq 0,\ k \in \mathbf{Z}\}.$$

We show that $X$ is nonempty using proof by cases. If $n \geq 0$, then $n - d \cdot 0 = n \geq 0$ so $n$ is in $X$. Suppose that $n < 0$. Since $d$ is a positive integer, $1 - d \leq 0$. Thus $n - dn = n(1 - d) \geq 0$. In this case, $n - dn$ is in $X$. Therefore $X$ is nonempty.

Since $X$ is a nonempty set of nonnegative integers, by the Well-Ordering Property, $X$ has a smallest element, which we denote $r$. We let $q$ denote the specific value of $k$ for which $r = n - dq$. Then $n = dq + r$.

Since $r$ is in $X$, $r \geq 0$. We use proof by contradiction to show that $r < d$. Suppose that $r \geq d$. Then

$$n - d(q + 1) = n - dq - d = r - d \geq 0.$$

Thus $n - d(q + 1)$ is in $X$. Also, $n - d(q + 1) = r - d < r$. But $r$ is the smallest integer in $X$. This contradiction shows that $r < d$.

We have shown that if $d$ and $n$ are integers, $d > 0$, there exist integers $q$ and $r$ satisfying

$$n = dq + r \qquad 0 \leq r < d.$$

We turn now to the uniqueness of $q$ and $r$. Suppose that

$$n = dq_1 + r_1 \qquad 0 \leq r_1 < d$$

and

$$n = dq_2 + r_2 \qquad 0 \leq r_2 < d.$$

We must show that $q_1 = q_2$ and $r_1 = r_2$. Subtracting the previous equations, we obtain

$$0 = n - n = (dq_1 + r_1) - (dq_2 + r_2) = d(q_1 - q_2) - (r_2 - r_1),$$

which can be rewritten

$$d(q_1 - q_2) = r_2 - r_1.$$

The preceding equation shows that $d$ divides $r_2 - r_1$. However, because $0 \leq r_1 < d$ and $0 \leq r_2 < d$,

$$-d < r_2 - r_1 < d.$$

But the only integer strictly between $-d$ and $d$ divisible by $d$ is 0. Therefore, $r_1 = r_2$. Thus, $d(q_1 - q_2) = 0$; hence, $q_1 = q_2$. The proof is complete. ◀

Notice that in Theorem 2.5.6 the remainder $r$ is zero if and only if $d$ divides $n$.

## 2.5 Problem-Solving Tips

In the Inductive Step of the Strong Form of Mathematical Induction, your goal is to prove case $n$. To do so, you can assume *all* preceding cases (not just the immediately preceding case as in Section 2.4). You could always use the Strong Form of Mathematical Induction. If it happens that you needed only the immediately preceding case in the Inductive Step, you merely used the form of mathematical induction of Section 2.4. However, assuming all previous cases potentially gives you more to work with in proving case $n$.

In the Inductive Step of the Strong Form of Mathematical Induction, when you assume that the statement $S(k)$ is true, you must be sure that $k$ is in the domain of discourse of the propositional function $S(n)$. In the terminology of this section, you must be sure that $n_0 \le k$ (see Examples 2.5.1 and 2.5.2).

In the Inductive Step of the Strong Form of Mathematical Induction, if you assume that case $n-p$ is true, there will be $p$ Basis Steps: $n = n_0, n = n_0+1, \ldots, n = n_0+p-1$.

In general, the key to devising a proof using the Strong Form of Mathematical Induction is to find smaller cases "within" case $n$. For example, the smaller cases in Example 2.5.4 are the parenthesized products $(a_1 \cdots a_t)$ and $(a_{t+1} \cdots a_n)$ for $1 \le t < n$.

## 2.5 Review Exercises

1. State the Strong Form of Mathematical Induction.
2. State the Well-Ordering Property.
3. State the Quotient-Remainder Theorem.

## 2.5 Exercises

1. Show that postage of 6 cents or more can be achieved by using only 2-cent and 7-cent stamps.
2. Show that postage of 24 cents or more can be achieved by using only 5-cent and 7-cent stamps.
3. Show that postage of 12 cents or more can be achieved by using only 3-cent and 7-cent stamps.

★4. Use the

If $S(n)$ is true, then $S(n+1)$ is true

form of the Inductive Step to prove the statement in Example 2.5.1.

★5. Use the

If $S(n)$ is true, then $S(n+1)$ is true

form of the Inductive Step to prove the statement in Exercise 1.

★6. Use the

If $S(n)$ is true, then $S(n+1)$ is true

form of the Inductive Step to prove the statement in Exercise 2.

*Exercises 7 and 8 refer to the sequence $c_1, c_2, \ldots$ defined by the equations*

$$c_1 = 0, \qquad c_n = c_{\lfloor n/2 \rfloor} + n^2 \text{ for all } n > 1.$$

7. Suppose that we want to prove a statement for all $n \ge 3$ involving $c_n$. The Inductive Step will assume the truth of the statement involving $c_{\lfloor n/2 \rfloor}$. What are the Basis Steps?
8. Suppose that we want to prove a statement for all $n \ge 4$ involving $c_n$. The Inductive Step will assume the truth of the statement involving $c_{\lfloor n/2 \rfloor}$. What are the Basis Steps?
9. Define the sequence $c_1, c_2, \ldots$ by the equations

$$c_1 = c_2 = 0, \qquad c_n = c_{\lfloor n/3 \rfloor} + n \text{ for all } n > 2.$$

Suppose that we want to prove a statement for all $n \ge 2$ involving $c_n$. The Inductive Step will assume the truth of the statement involving $c_{\lfloor n/3 \rfloor}$. What are the Basis Steps?

*Exercises 10 and 11 refer to the sequence $c_1, c_2, \ldots$ defined by the equations*

$$c_1 = 0, \qquad c_n = c_{\lfloor n/2 \rfloor} + n^2 \text{ for all } n > 1.$$

10. Compute $c_2$, $c_3$, $c_4$, and $c_5$.
11. Prove that $c_n < 4n^2$ for all $n \ge 1$.

*Exercises 12–14 refer to the sequence $c_1, c_2, \ldots$ defined by the equations*

$$c_1 = 0, \qquad c_n = 4c_{\lfloor n/2 \rfloor} + n \text{ for all } n > 1.$$

12. Compute $c_2$, $c_3$, $c_4$, and $c_5$.
13. Prove that $c_n \le 4(n-1)^2$ for all $n \ge 1$.
14. Prove that $(n+1)^2/8 < c_n$ for all $n \ge 2$. *Hint:* $\lfloor n/2 \rfloor \ge (n-1)/2$ for all $n$.
15. Define the sequence $c_0, c_1, \ldots$ by the equations

$$c_0 = 0, \qquad c_n = c_{\lfloor n/2 \rfloor} + 3 \text{ for all } n > 0.$$

What is wrong with the following "proof" that $c_n \le 2n$ for all $n \ge 3$? (You should verify that it is *false* that $c_n \le 2n$ for all $n \ge 3$.)

We use the Strong Form of Mathematical Induction.

### Basis Step ($n = 3$)

We have

$$c_3 = c_1 + 3 = (c_0 + 3) + 3 = 6 \le 2 \cdot 3.$$

The Basis Step is verified.

### Inductive Step

Assume that $c_k \le 2k$ for all $k < n$. Then

$$c_n = c_{\lfloor n/2 \rfloor} + 3 \le 2\lfloor n/2 \rfloor + 3 \le 2(n/2) + 3 = n + 3 < n + n = 2n.$$

(Since $3 < n$, $n + 3 < n + n$.) The Inductive Step is complete.

**16.** Suppose that we have two piles of cards each containing $n$ cards. Two players play a game as follows. Each player, in turn, chooses one pile and then removes any number of cards, but at least one, from the chosen pile. The player who removes the last card wins the game. Show that the second player can always win the game.

*In Exercises 17–22, find the quotient q and remainder r as in Theorem 2.5.6 when n is divided by d.*

**17.** $n = 47, d = 9$

**18.** $n = -47, d = 9$

**19.** $n = 7, d = 9$

**20.** $n = -7, d = 9$

**21.** $n = 0, d = 9$

**22.** $n = 47, d = 47$

*The Egyptians of antiquity expressed a fraction as a sum of fractions whose numerators were 1. For example, 5/6 might be expressed as*

$$\frac{5}{6} = \frac{1}{2} + \frac{1}{3}.$$

*We say that a fraction p/q, where p and q are positive integers, is in* Egyptian form *if*

$$\frac{p}{q} = \frac{1}{n_1} + \frac{1}{n_2} + \cdots + \frac{1}{n_k}, \tag{2.5.3}$$

*where $n_1, n_2, \ldots, n_k$ are positive integers satisfying $n_1 < n_2 < \cdots < n_k$.*

**23.** Show that the representation (2.5.3) need not be unique by representing 5/6 in two different ways.

**★24.** Show that the representation (2.5.3) is never unique.

**25.** By completing the following steps, give a proof by induction on $p$ to show that every fraction $p/q$ with $0 < p/q < 1$ may be expressed in Egyptian form.

(a) Verify the Basis Step ($p = 1$).

(b) Suppose that $0 < p/q < 1$ and that all fractions $i/q'$, with $1 \le i < p$ and $q'$ arbitrary, can be expressed in Egyptian form. Choose the smallest positive integer $n$ with $1/n \le p/q$. Show that

$$n > 1 \quad \text{and} \quad \frac{p}{q} < \frac{1}{n-1}.$$

(c) Show that if $p/q = 1/n$, the proof is complete.

(d) Assume that $1/n < p/q$. Let

$$p_1 = np - q \quad \text{and} \quad q_1 = nq.$$

Show that

$$\frac{p_1}{q_1} = \frac{p}{q} - \frac{1}{n}, \quad 0 < \frac{p_1}{q_1} < 1, \quad \text{and} \quad p_1 < p.$$

Conclude that

$$\frac{p_1}{q_1} = \frac{1}{n_1} + \frac{1}{n_2} + \cdots + \frac{1}{n_k}$$

with $n_1, n_2, \ldots, n_k$ distinct.

(e) Show that $p_1/q_1 < 1/n$.

(f) Show that

$$\frac{p}{q} = \frac{1}{n} + \frac{1}{n_1} + \cdots + \frac{1}{n_k}$$

and $n, n_1, \ldots, n_k$ are distinct.

**26.** Use the method of the preceding exercise to find Egyptian forms of 3/8, 5/7, and 13/19.

**★27.** Show that any fraction $p/q$, where $p$ and $q$ are positive integers, can be written in Egyptian form. (We are not assuming that $p/q < 1$.)

**★28.** Show that any $n \times n$ deficient board can be tiled with trominoes if $n$ is odd, $n > 5$, and 3 divides $n^2 - 1$. *Hint:* Use the ideas suggested in the hint for Exercise 38, Section 2.4.

**★29.** Show that any $n \times n$ deficient board can be tiled with trominoes if $n$ is even, $n > 8$, and 3 divides $n^2 - 1$. *Hint:* Use the fact that a $4 \times 4$ deficient board can be tiled with trominoes, Exercise 28, and Exercise 36, Section 2.4.

**★30.** Show that any $m \times n$ deficient rectangle, $2 \le m \le n$, can be tiled with trominoes if 3 divides $mn - 1$, neither side has length 2 unless both of them do, and $m \ne 5$.

**31.** Give an example of an $m \times n$ rectangle with two squares missing, where 3 divides $mn - 2$, that can be tiled with trominoes.

**32.** Give an example of an $m \times n$ rectangle with two squares missing, where 3 divides $mn - 2$, that cannot be tiled with trominoes.

**★33.** Which $m \times n$ rectangles with two squares missing, where 3 divides $mn - 2$, can be tiled with trominoes?

**34.** Give an alternative proof of the existence of $q$ and $r$ in Theorem 2.5.6 for the case $n \ge 0$ by first showing that the set $X$ consisting of all integers $k$ where $dk > n$ is a nonempty set of nonnegative integers, then showing that $X$ has a least element, and finally analyzing the least element of $X$.

**35.** Give an alternative proof of the existence of $q$ and $r$ in Theorem 2.5.6 using the form of mathematical induction where the Inductive Step is "if $S(n)$ is true, then $S(n + 1)$ is true." *Hint:* First assume that $n > 0$. Treat the case $n = 0$ separately. Reduce the case $n < 0$ to the case $n > 0$.

**★36.** Assume the form of mathematical induction where the Inductive Step is "if $S(n)$ is true, then $S(n + 1)$ is true." Prove the Well-Ordering Property.

**★37.** Assume the Well-Ordering Property. Prove the Strong Form of Mathematical Induction.

**★38.** Show that the Strong Form of Mathematical Induction and the form of mathematical induction where the Inductive Step is "if $S(n)$ is true, then $S(n + 1)$ is true" are equivalent. That is, assume the Strong Form of Mathematical Induction and prove the alternative form; then assume the alternative form and prove the Strong Form of Mathematical Induction.

## Chapter 2 Notes

[D'Angelo; Solow] address the problem of how to construct proofs. Tiling with polyominoes is the subject of the book by [Martin].

The "Fallacies, Flaws, and Flimflam" section of *The College Mathematics Journal*, published by the Mathematical Association of America, contains examples of mathematical mistakes, fallacious proofs, and faulty reasoning.

## Chapter 2 Review

### Section 2.1

1. Mathematical system
2. Axiom
3. Definition
4. Undefined term
5. Theorem
6. Proof
7. Lemma
8. Direct proof
9. Even integer
10. Odd integer
11. Subproof
12. Disproving a universally quantified statement
13. Begging the question
14. Circular reasoning

### Section 2.2

15. Proof by contradiction
16. Indirect proof
17. Proof by contrapositive
18. Proof by cases
19. Exhaustive proof
20. Proving an if-and-only-if statement
21. Proving several statements are equivalent
22. Existence proof
23. Constructive existence proof
24. Nonconstructive existence proof

### Section 2.3

25. Resolution proof; uses: if $p \vee q$ and $\neg p \vee r$ are both true, then $q \vee r$ is true.
26. Clause: consists of terms separated by *or*'s, where each term is a variable or a negation of a variable.

### Section 2.4

27. Principle of Mathematical Induction
28. Basis Step: prove true for the first instance.
29. Inductive Step: assume true for instance $n$; then prove true for instance $n+1$.
30. $n$ factorial: $n! = n(n-1)\cdots 1$, $0! = 1$
31. Formula for the sum of the first $n$ positive integers:
$$1 + 2 + \cdots + n = \frac{n(n+1)}{2}$$
32. Formula for the geometric sum:
$$ar^0 + ar^1 + \cdots + ar^n = \frac{a(r^{n+1}-1)}{r-1}, \qquad r \neq 1$$

### Section 2.5

33. Strong Form of Mathematical Induction
34. Basis Step for the Strong Form of Mathematical Induction: prove true for the first instance.
35. Inductive Step for the Strong Form of Mathematical Induction: assume true for all instances less than $n$; then prove true for instance $n$.
36. Well-Ordering Property: every nonempty set of nonnegative integers has a least element.
37. Quotient-Remainder Theorem: If $d$ and $n$ are integers, $d > 0$, there exist integers $q$ (quotient) and $r$ (remainder) satisfying $n = dq + r$, $0 \le r < d$. Furthermore, $q$ and $r$ are unique.

## Chapter 2 Self-Test

1. Distinguish between the terms *axiom* and *definition*.
2. What is the difference between a direct proof and a proof by contradiction?
3. Show, by giving a proof by contradiction, that if four teams play seven games, some pair of teams plays at least two times.
4. Prove that for all rational numbers $x$ and $y$, $y \neq 0$, $x/y$ is rational.
5. Use proof by cases to prove that
$$\min\{\min\{a, b\}, c\} = \min\{a, \min\{b, c\}\}$$
for all real numbers $a$, $b$, and $c$.

*Use mathematical induction to prove that the statements in Exercises 6–9 are true for every positive integer n.*

**6.** $2 + 4 + \cdots + 2n = n(n+1)$

**7.** $2^2 + 4^2 + \cdots + (2n)^2 = \frac{2n(n+1)(2n+1)}{3}$

**8.** $\frac{1}{2!} + \frac{2}{3!} + \cdots + \frac{n}{(n+1)!} = 1 - \frac{1}{(n+1)!}$

**9.** $2^{n+1} < 1 + (n+1)2^n$

**10.** Prove that for all integers $m$ and $n$, if $m$ and $m - n$ are odd, then $n$ is even.

**11.** Prove that the following are equivalent for sets $A$ and $B$:

(a) $A \subseteq B$ (b) $A \cap \overline{B} = \varnothing$ (c) $A \cup B = B$

**12.** Prove that for all sets $X$, $Y$, and $Z$, if $X \subseteq Y$ and $Y \subset Z$, then $X \subset Z$.

**13.** Find the quotient $q$ and remainder $r$ as in Theorem 2.5.6 when $n = 101$ is divided by $d = 11$.

*Exercises 14 and 15 refer to the sequence $c_1, c_2, \ldots$ defined by the equations*

$$c_1 = 0, \qquad c_n = 2c_{\lfloor n/2 \rfloor} + n \text{ for all } n > 1.$$

**14.** Compute $c_2$, $c_3$, $c_4$, and $c_5$.

**15.** Prove that $c_n \le n \lg n$ for all $n \ge 1$.

**16.** Use the Well-Ordering Property to show that any nonempty set $X$ of nonnegative integers that has an upper bound contains a largest element. *Hint:* Consider the set of integer upper bounds for $X$.

**17.** Find an expression, which is the *and* of clauses, equivalent to $(p \vee q) \to r$.

**18.** Find an expression, which is the *and* of clauses, equivalent to $(p \vee \neg q) \to \neg rs$.

**19.** Use resolution to prove

$$\begin{array}{l} \neg p \vee q \\ \neg q \vee \neg r \\ p \vee \neg r \\ \hline \therefore \neg r \end{array}$$

**20.** Reprove Exercise 19 using resolution and proof by contradiction.

## Chapter 2 Computer Exercises

**1.** Implement proof by resolution as a program.

**2.** Write a program that gives an Egyptian form of a fraction.

# Chapter 3

# FUNCTIONS, SEQUENCES, AND RELATIONS



All of mathematics, as well as subjects that rely on mathematics, such as computer science and engineering, make use of functions, sequences, and relations.

A function assigns to each member of a set $X$ exactly one member of a set $Y$. Functions are used extensively in discrete mathematics; for example, functions are used to analyze the time needed to execute algorithms.

A sequence is a special kind of function. A list of the letters as they appear in a word is an example of a sequence. Unlike a set, a sequence takes order into account. (Order is obviously important since, for example, *form* and *from* are different words.)

Relations generalize the notion of functions. A relation is a set of ordered pairs. The presence of the ordered pair $(a, b)$ in a relation is interpreted as indicating a relationship from $a$ to $b$. The relational database model that helps users access information in a database (a collection of records manipulated by a computer) is based on the concept of relation.

## 3.1 Functions

Credit card numbers typically consist of 13, 15, or 16 digits. For example,

$$4690\,3582\,1375\,4657 \tag{3.1.1}$$

is a hypothetical credit card number. The first digit designates the system. In (3.1.1), the first digit, 4, shows that the card would be a Visa card. The following digits specify other information such as the account number and the bank number. (The precise meaning depends on the type of card.) The last digit is special; it is computed from the preceding digits and is called a *check digit*. In (3.1.1), the check digit is 7 and is computed from the preceding digits 4690 3582 1375 465. Credit card check digits are used to identify certain erroneous card numbers. It is not a security measure, but rather it is used to help detect errors such as giving a credit card number over the phone and having it transcribed improperly or detecting an error in entering a credit card number while ordering a product online.

†This section can be omitted without loss of continuity.

The check digit is computed as follows. Starting from the right and skipping the check digit, double every other number. If the result of doubling is a two-digit number, add the digits; otherwise, use the original digit. The other digits are not modified.

| 4 | 6 | 9 | 0 | 3 | 5 | 8 | 2 | 1 | 3 | 7 | 5 | 4 | 6 | 5 | |
|---|---|---|---|---|---|---|---|---|---|---|---|---|---|---|---|
| ↓ | ↓ | ↓ | ↓ | ↓ | ↓ | ↓ | ↓ | ↓ | ↓ | ↓ | ↓ | ↓ | ↓ | ↓ | Double every other digit. |
| 8 | 6 | 18 | 0 | 6 | 5 | 16 | 2 | 2 | 3 | 14 | 5 | 8 | 6 | 10 | |
| ↓ | ↓ | ↓ | ↓ | ↓ | ↓ | ↓ | ↓ | ↓ | ↓ | ↓ | ↓ | ↓ | ↓ | ↓ | Add digits of two-digit numbers. |
| 8 | 6 | 9 | 0 | 6 | 5 | 7 | 2 | 2 | 3 | 5 | 5 | 8 | 6 | 1 | |

Sum the resulting digits

$$8+6+9+0+6+5+7+2+2+3+5+5+8+6+1=73.$$

If the last digit of the sum is 0, the check digit is 0. Otherwise, subtract the last digit of the sum from 10 to get the check digit, $10-3=7$. Verify the check digit on your favorite Visa, MasterCard, American Express, or Diners Club card. This method of calculating a check digit is called the *Luhn algorithm*. It is named after Hans Peter Luhn (1896–1964), who invented it while at IBM. Although originally patented, it is now in the public domain and is widely used.

One common error in copying a number is to change one digit. Each undoubled digit contributes a unique value to the sum ($0 \to 0$, $1 \to 1$, etc.). Each doubled digit also contributes a unique value to the sum ($0 \to 0$, $1 \to 2, \ldots$, $4 \to 8$, $5 \to 1$, $6 \to 3, \ldots$, $9 \to 9$). Thus if a single digit is changed in a credit card number, the sum used in the Luhn algorithm will change by an absolute amount less than 10, and the check digit will change. In the preceding example if 1 is changed to 7, the Luhn algorithm calculation becomes

| 4 | 6 | 9 | 0 | 3 | 5 | 8 | 2 | 7 | 3 | 7 | 5 | 4 | 6 | 5 | |
|---|---|---|---|---|---|---|---|---|---|---|---|---|---|---|---|
| ↓ | ↓ | ↓ | ↓ | ↓ | ↓ | ↓ | ↓ | ↓ | ↓ | ↓ | ↓ | ↓ | ↓ | ↓ | Double every other digit |
| 8 | 6 | 18 | 0 | 6 | 5 | 16 | 2 | 14 | 3 | 14 | 5 | 8 | 6 | 10 | |
| ↓ | ↓ | ↓ | ↓ | ↓ | ↓ | ↓ | ↓ | ↓ | ↓ | ↓ | ↓ | ↓ | ↓ | ↓ | Add digits of two-digit numbers. |
| 8 | 6 | 9 | 0 | 6 | 5 | 7 | 2 | 5 | 3 | 5 | 5 | 8 | 6 | 1 | |

and the sum becomes

$$8+6+9+0+6+5+7+2+5+3+5+5+8+6+1=76.$$

Therefore the check digit changes to 4. Thus, if 1 is inadvertently transcribed as 7, the error will be detected.

Another common error is transposition of adjacent digits. For example, if 82 is inadvertently written as 28, the error will be detected by the Luhn algorithm because the check digit will change (check this). In fact, the Luhn algorithm will detect every transposition of adjacent digits except for 90 and 09 (see Computer Exercise 4).

**Go Online**
For more on functions, see `bit.ly/2OFm8St`

The Luhn algorithm gives an example of a **function**. A function assigns to each member of a set $X$ exactly one member of a set $Y$. (The sets $X$ and $Y$ may or may not be the same.) The Luhn algorithm assigns to each integer 10 or greater (so there is a number available to compute a check digit) a single-digit integer, the check digit. In the preceding example, the integer 469035821375465 is assigned the value 7, and the integer 469035827375465 is assigned the value 4. We can represent these assignments as ordered pairs:

$$(469035821375465, 7) \quad \text{and} \quad (469035827375465, 4).$$

Formally, we *define* a function to be a particular kind of set of ordered pairs.

**Definition 3.1.1** ▶ Let $X$ and $Y$ be sets. A *function* $f$ from $X$ to $Y$ is a subset of the Cartesian product $X \times Y$ having the property that for each $x \in X$, there is exactly one $y \in Y$ with $(x, y) \in f$. We sometimes denote a function $f$ from $X$ to $Y$ as $f\colon X \to Y$.

The set $X$ is called the *domain of* $f$ and the set $Y$ is called the *codomain of* $f$. The set

$$\{y \mid (x, y) \in f\}$$

(which is a subset of the codomain $Y$) is called the *range of* $f$. ◀

**Example 3.1.2** For the check digit function, the domain is the set of positive integers 10 or greater and the range is the set of single-digit integers. We can take the codomain to be any set containing the set of single-digit integers, for example, the set of nonnegative integers. ◀

**Example 3.1.3** The set $f = \{(1, a), (2, b), (3, a)\}$ is a function from $X = \{1, 2, 3\}$ to $Y = \{a, b, c\}$. Each element of $X$ is assigned a unique value in $Y$: 1 is assigned the unique value $a$; 2 is assigned the unique value $b$; and 3 is assigned the unique value $a$. We can depict the situation as shown in Figure 3.1.1, where an arrow from $j$ to $x$ means that we assign the letter $x$ to the integer $j$. We call a picture such as Figure 3.1.1 an **arrow diagram.** For an arrow diagram to be a function, Definition 3.1.1 requires that there is exactly one arrow from each element in the domain. Notice that Figure 3.1.1 has this property.

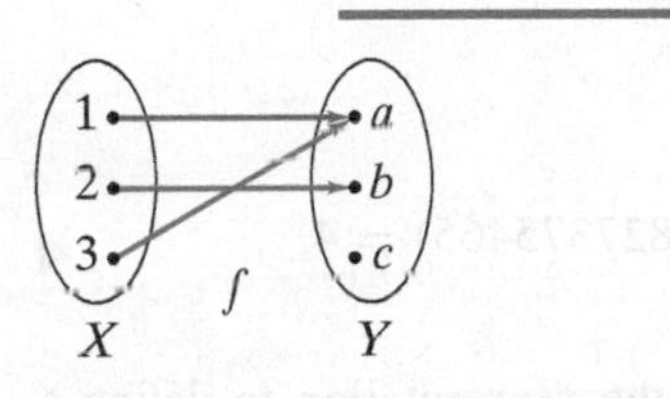

**Figure 3.1.1** The arrow diagram of the function of Example 3.1.3. There is exactly one arrow from each element in $X$.

Definition 3.1.1 allows us to reuse elements in $Y$. For the function $f$, the element $a$ in $Y$ is used twice. Further, Definition 3.1.1 does *not* require us to use all the elements in $Y$. No element in $X$ is assigned to the element $c$ in $Y$. The domain of $f$ is $X$, the codomain of $f$ is $Y$, and the range of $f$ is $\{a, b\}$. ◀

**Example 3.1.4** The set

$$\{(1, a), (2, a), (3, b)\} \tag{3.1.2}$$

is not a function from $X = \{1, 2, 3, 4\}$ to $Y = \{a, b, c\}$ because the element 4 in $X$ is not assigned to an element in $Y$. It is also apparent from the arrow diagram (see Figure 3.1.2) that this set is not a function because there is no arrow from 4. The set (3.1.2) *is* a function from $X' = \{1, 2, 3\}$ to $Y = \{a, b, c\}$.

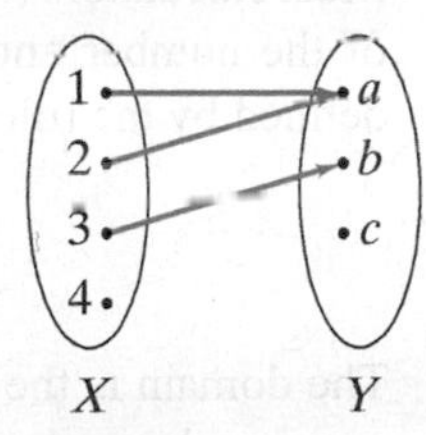

**Figure 3.1.2** The arrow diagram of the set in Example 3.1.4, which is not a function because there is no arrow from 4. ◀

**Example 3.1.5** The set $\{(1, a), (2, b), (3, c), (1, b)\}$ is not a function from $X = \{1, 2, 3\}$ to $Y = \{a, b, c\}$ because 1 is not assigned a *unique* element in $Y$ (1 is assigned the values $a$ *and* $b$). It is also apparent from the arrow diagram (see Figure 3.1.3) that this set is not a function because there are two arrows from 1.

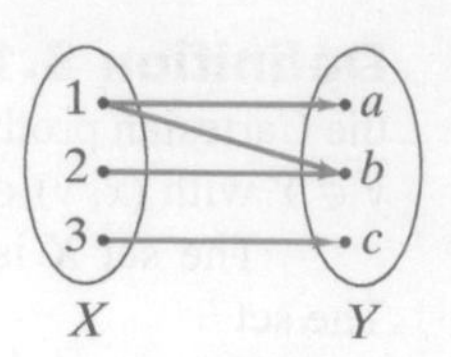


**Figure 3.1.3** The arrow diagram of the set in Example 3.1.5, which is not a function because there are two arrows from 1. ◀

Given a function $f$ from $X$ to $Y$, according to Definition 3.1.1, for each element $x$ in the domain $X$, there is exactly one $y$ in the codomain $Y$ with $(x, y) \in f$. This unique value $y$ is denoted $f(x)$. In other words, $f(x) = y$ is another way to write $(x, y) \in f$.

**Example 3.1.6** For the function $f$ of Example 3.1.3, we may write $f(1) = a$, $f(2) = b$, and $f(3) = a$. ◀

**Example 3.1.7** If we call the check digit function $L$, we may write

$$L(469035821375465) = 7 \quad \text{and} \quad L(469035827375465) = 4.$$

◀

The next example shows how we sometimes use the $f(x)$ notation to define a function.

**Example 3.1.8** Let $f$ be the function defined by the rule $f(x) = x^2$. For example, $f(2) = 4$, $f(-3.5) = 12.25$, and $f(0) = 0$. Although we frequently find functions defined in this way, the definition is incomplete since the domain and codomain are not specified. If we are told that the domain is the set of all real numbers and the codomain is the set of all nonnegative real numbers, in ordered-pair notation, we would have

$$f = \{(x, x^2) \mid x \text{ is a real number}\}.$$

The range of $f$ is the set of all nonnegative real numbers. ◀

**Example 3.1.9** Most calculators have a $1/x$ key. If you enter a number and hit the $1/x$ key, the reciprocal of the number entered (or an approximation to it) is displayed. This function can be defined by the rule

$$R(x) = \frac{1}{x}.$$

The domain is the set of all numbers that can be entered into the calculator and whose reciprocals can be computed and displayed by the calculator. The range is the set of all the reciprocals that can be computed and displayed. We could define the codomain also to be the set of all the reciprocals that can be computed and displayed. Notice that by the nature of the calculator, the domain and range are finite sets. ◀

Another way to visualize a function is to draw its graph. The **graph of a function** $f$ whose domain and codomain are subsets of the real numbers is obtained by plotting points in the plane that correspond to the elements in $f$. The domain is contained in the horizontal axis and the codomain is contained in the vertical axis.

**Example 3.1.10** The graph of the function $f(x) = x^2$ is shown in Figure 3.1.4. ◀

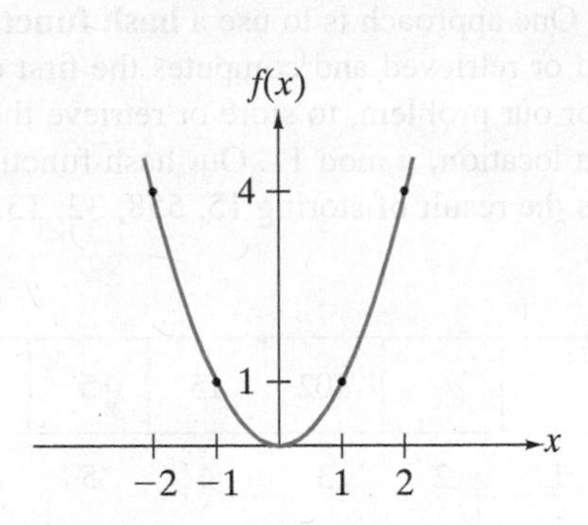


**Figure 3.1.4** The graph of $f(x) = x^2$.

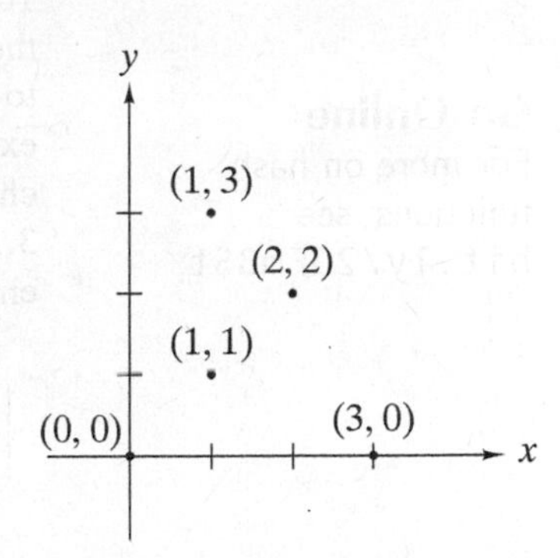


**Figure 3.1.5** A set that is not a function. The vertical line $x = 1$ intersects two points in the set.

We note that a set $S$ of points in the plane defines a function precisely when each vertical line intersects at most one point of $S$. If some vertical line contains two or more points of some set, the domain point does not assign a *unique* codomain point and the set does not define a function (see Figure 3.1.5).

Functions involving the **modulus operator** play an important role in mathematics and computer science.

**Definition 3.1.11 ▶** If $x$ is an integer and $y$ is a positive integer, we define $x$ mod $y$ to be the remainder when $x$ is divided by $y$. ◀

**Example 3.1.12** We have

$$6 \bmod 2 = 0, \quad 5 \bmod 1 = 0, \quad 8 \bmod 12 = 8, \quad 199673 \bmod 2 = 1. \quad ◀$$

**Example 3.1.13** The check digit calculated by the Luhn algorithm can be written

$$[10 - (S \bmod 10)] \bmod 10,$$

where $S$ is the sum used in the intermediate step of the calculation. The last digit in $S$ is given by $S$ mod 10. If this digit is 1 through 9, inclusive, $10 - (S \bmod 10)$ gives the check digit and the last "mod 10" is unnecessary, but harmless. However, if the last digit in $S$ is 0, $10 - (S \bmod 10) = 10$. In this case, adding the last "mod 10" gives the check digit as 0. ◀

**Example 3.1.14** What day of the week will it be 365 days from Wednesday?

**SOLUTION** Seven days after Wednesday, it is Wednesday again; 14 days after Wednesday, it is Wednesday again; and in general, if $n$ is a positive integer, $7n$ days after Wednesday, it is Wednesday again. Thus we need to subtract as many 7's as possible from 365 and see how many days are left, which is the same as computing 365 mod 7. Since 365 mod 7 = 1, 365 days from Wednesday, it will be one day later, namely Thursday. This explains why, except for leap year, when an extra day is added to February, the identical month and date in consecutive years move forward one day of the week. ◀

**Example 3.1.15** **Hash Functions** Suppose that we have cells in a computer memory indexed from 0 to 10 (see Figure 3.1.6). We wish to store and retrieve arbitrary nonnegative integers in these cells. One approach is to use a **hash function.** A hash function takes a data item to be stored or retrieved and computes the first choice for a location for the item. For example, for our problem, to store or retrieve the number $n$, we might take as the first choice for a location, $n \bmod 11$. Our hash function becomes $h(n) = n \bmod 11$. Figure 3.1.6 shows the result of storing 15, 558, 32, 132, 102, and 5, in this order, in initially empty cells.

**Go Online**
For more on hash functions, see
bit.ly/2OFm8St

| 132 | | | 102 | 15 | 5 | 257 | | 558 | | 32 |
|---|---|---|---|---|---|---|---|---|---|---|
| 0 | 1 | 2 | 3 | 4 | 5 | 6 | 7 | 8 | 9 | 10 |

**Figure 3.1.6** Cells in a computer memory.

Now suppose that we want to store 257. Since $h(257) = 4$, then 257 should be stored at location 4; however, this position is already occupied. In this case we say that a **collision** has occurred. More precisely, a collision occurs for a hash function $H$ if $H(x) = H(y)$, but $x \neq y$. To handle collisions, a **collision resolution policy** is required. One simple collision resolution policy is to find the next highest (with 0 assumed to follow 10) unoccupied cell. If we use this collision resolution policy, we would store 257 at location 6 (see Figure 3.1.6).

If we want to locate a stored value $n$, we compute $m = h(n)$ and begin looking at location $m$. If $n$ is not at this position, we look in the next-highest position (again, 0 is assumed to follow 10); if $n$ is not in this position, we proceed to the next-highest position, and so on. If we reach an empty cell or return to our original position, we conclude that $n$ is not present; otherwise, we obtain the position of $n$.

If collisions occur infrequently, and if when one does occur it is resolved quickly, then hashing provides a very fast method of storing and retrieving data. As an example, personnel data are frequently stored and retrieved by hashing on employee identification numbers. ◀

**Example 3.1.16** **Pseudorandom Numbers** Computers are often used to simulate random behavior. A game program might simulate rolling dice, and a client service program might simulate the arrival of customers at a bank. Such programs generate numbers that appear random and are called **pseudorandom numbers.** For example, the dice-rolling program would need pairs of pseudorandom numbers, each between 1 and 6, to simulate the outcome of rolling dice. Pseudorandom numbers are not truly random; if one knows the program that generates the numbers, one could predict what numbers would occur.

The method usually used to generate pseudorandom numbers is called the **linear congruential method.** This method requires four integers: the modulus $m$, the multiplier $a$, the increment $c$, and a seed $s$ satisfying $2 \leq a < m$, $0 \leq c < m$, and $0 \leq s < m$. We then set $x_0 = s$. The sequence of pseudorandom numbers generated, $x_1, x_2, \ldots$, is given by the formula

$$x_n = (ax_{n-1} + c) \bmod m.$$

The formula computes the next pseudorandom number using its immediate predecessor. For example, if $m = 11$, $a = 7$, $c = 5$, and $s = 3$, then

$$x_1 = (ax_0 + c) \bmod m = (7 \cdot 3 + 5) \bmod 11 = 4$$

and

$$x_2 = (ax_1 + c) \bmod m = (7 \cdot 4 + 5) \bmod 11 = 0.$$

Similar computations show that the sequence continues:

$$x_3 = 5,\ x_4 = 7,\ x_5 = 10,\ x_6 = 9,\ x_7 = 2,\ x_8 = 8,\ x_9 = 6,\ x_{10} = 3.$$

Since $x_{10} = 3$, which is the value of the seed, the sequence now repeats: 3, 4, 0, 5, 7, . . . .

Much effort has been invested in finding good values for a linear congruential method. Critical simulations such as those involving aircraft and nuclear research require "good" random numbers. In practice, large values are used for $m$ and $a$. Commonly used values are $m = 2^{31} - 1 = 2{,}147{,}483{,}647$; $a = 7^5 = 16{,}807$; and $c = 0$, which generate a sequence of $2^{31} - 1$ integers before repeating a value. ◀

In the 1990s, Daniel Corriveau of Quebec won three straight games of a computer keno game in Montreal, each time choosing 19 of 20 numbers correctly. The odds against this feat are 6 billion to 1. Suspicious officials at first refused to pay him. Although Corriveau attributed his success to chaos theory, what in fact happened was that whenever power was cut, the random number generator started with the same seed, thus generating the same sequence of numbers. The embarrassed casino finally paid Corriveau the $600,000 due him.

We next define the **floor** and **ceiling** of a real number.

**Definition 3.1.17** ▶ The *floor* of $x$, denoted $\lfloor x \rfloor$, is the greatest integer less than or equal to $x$. The *ceiling* of $x$, denoted $\lceil x \rceil$, is the least integer greater than or equal to $x$. ◀

**Example 3.1.18** $\lfloor 8.3 \rfloor = 8,\ \lceil 9.1 \rceil = 10,\ \lfloor -8.7 \rfloor = -9,\ \lceil -11.3 \rceil = -11,\ \lceil 6 \rceil = 6,\ \lceil -8 \rceil = -8$ ◀

The floor of $x$ "rounds $x$ down" while the ceiling of $x$ "rounds $x$ up." We will use the floor and ceiling functions throughout the book.

**Example 3.1.19** Figure 3.1.7 shows the graphs of the floor and ceiling functions. A bracket, [ or ], indicates that the point is to be included in the graph; a parenthesis, ( or ), indicates that the point is to be excluded from the graph.

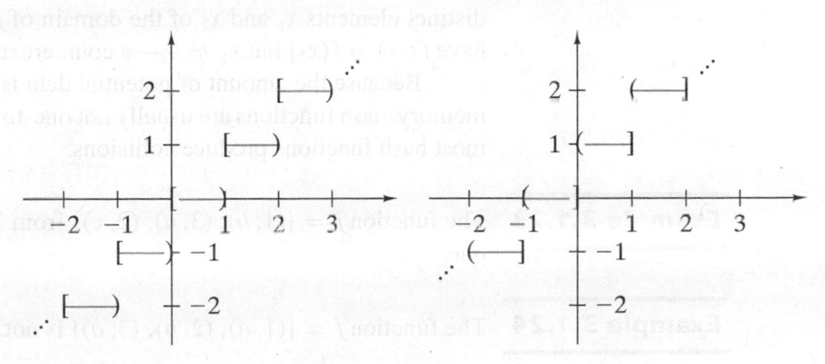


**Figure 3.1.7** The graphs of the floor (left graph) and ceiling (right graph) functions. ◀

**Example 3.1.20** The first-class postage rate for mail up to 13 ounces is 92 cents for the first ounce or fraction thereof and 20 cents for each additional ounce or fraction thereof. The postage $P(w)$ as a function of weight $w$ is given by the equation

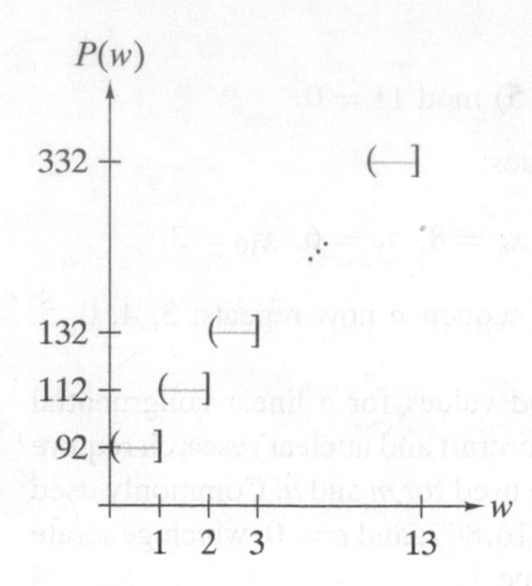


Figure 3.1.8 The graph of the postage function $P(w) = 92 + 20\lceil w - 1 \rceil$.

$$P(w) = 92 + 20\lceil w - 1 \rceil \quad 13 \geq w > 0.$$

The expression $\lceil w-1 \rceil$ counts the number of additional ounces beyond 1, with a fraction counting as one additional ounce. As examples,

$$P(3.7) = 92 + 20\lceil 3.7 - 1 \rceil = 92 + 20\lceil 2.7 \rceil = 92 + 20 \cdot 3 = 152,$$
$$P(2) = 92 + 20\lceil 2 - 1 \rceil = 92 + 20\lceil 1 \rceil = 92 + 20 \cdot 1 = 112.$$

The graph of the function $P$ is shown in Figure 3.1.8. ◀

The Quotient-Remainder Theorem (Theorem 2.5.6) states that if $d$ and $n$ are integers, $d > 0$, there exist integers $q$ (quotient) and $r$ (remainder) satisfying

$$n = dq + r \qquad 0 \leq r < d.$$

Dividing by $d$, we obtain

$$\frac{n}{d} = q + \frac{r}{d}.$$

Since $0 \leq r/d < 1$,

$$\left\lfloor \frac{n}{d} \right\rfloor = \left\lfloor q + \frac{r}{d} \right\rfloor = q.$$

Thus, we may compute the quotient $q$ as $\lfloor n/d \rfloor$. Having computed the quotient $q$, we may compute the remainder as $r = n - dq$. We previously introduced the notation $n$ mod $d$ for the remainder.

**Example 3.1.21** We have $36844/2427 = 15.18088\ldots$; thus the quotient is $q = \lfloor 36844/2427 \rfloor = 15$. Therefore, the remainder 36844 mod 2427 is $r = 36844 - 2427 \cdot 15 = 439$. We have $n = dq + r$ or $36844 = 2427 \cdot 15 + 439$. ◀

**Definition 3.1.22 ▶** A function $f$ from $X$ to $Y$ is said to be *one-to-one* (or *injective*) if for all $x_1, \ x_2 \in X$, if $f(x_1) = f(x_2)$ then $x_1 = x_2$. ◀

An equivalent way to state Definition 3.1.22 is: If $y$ is an element of the range of $f$, then there is *exactly one* $x$ in the domain of $f$ such that $f(x) = y$. If there were two distinct elements $x_1$ and $x_2$ of the domain of $f$ with $f(x_1) = y = f(x_2)$, then we would have $f(x_1) = f(x_2)$ but $x_1 \neq x_2$—a counterexample to the claim that $f$ is one-to-one.

Because the amount of potential data is usually so much larger than the available memory, hash functions are usually not one-to-one (see Example 3.1.15). In other words, most hash functions produce collisions.

**Example 3.1.23** The function $f = \{(1, b), (3, a), (2, c)\}$ from $X = \{1, 2, 3\}$ to $Y = \{a, b, c, d\}$ is one-to-one. ◀

**Example 3.1.24** The function $f = \{(1, a), (2, b), (3, a)\}$ is not one-to-one since $f(1) = a = f(3)$. ◀

**Example 3.1.25** If $X$ is the set of persons who have social security numbers and we assign each person $x \in X$ his or her social security number SS($x$), we obtain a one-to-one function since distinct persons are always assigned distinct social security numbers. It is because this correspondence is one-to-one that the government uses social security numbers as identifiers. ◀

**Example 3.1.26** If a function from $X$ to $Y$ is one-to-one, each element in $Y$ in its arrow diagram will have at most one arrow pointing to it (see Figure 3.1.9). If a function is not one-to-one, some element in $Y$ in its arrow diagram will have two or more arrows pointing to it (see Figure 3.1.10).

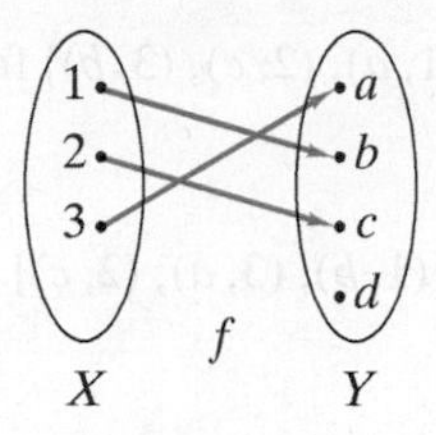


**Figure 3.1.9** The function of Example 3.1.23. This function is one-to-one because each element in $Y$ has at most one arrow pointing to it. This function is not onto $Y$ because there is no arrow pointing to $d$.

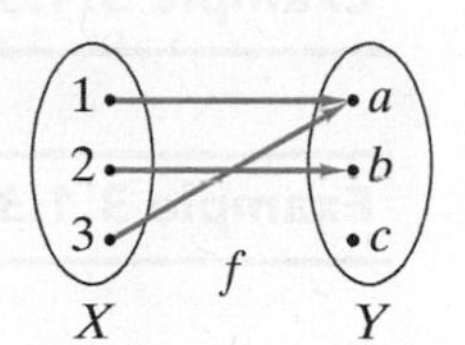


**Figure 3.1.10** A function that is not one-to-one. This function is not one-to-one because $a$ has two arrows pointing to it. This function is not onto $Y$ because there is no arrow pointing to $c$. ◀

**Example 3.1.27** Prove that the function $f(n) = 2n + 1$ from the set of positive integers to the set of positive integers is one-to-one.

**SOLUTION** We must show that for all positive integers $n_1$ and $n_2$, if $f(n_1) = f(n_2)$, then $n_1 = n_2$. So, suppose that $f(n_1) = f(n_2)$. Using the definition of $f$, this latter equation translates as $2n_1 + 1 = 2n_2 + 1$. Subtracting 1 from both sides of the equation and then dividing both sides of the equation by 2 yields $n_1 = n_2$. Therefore, $f$ is one-to-one. ◀

**Example 3.1.28** Prove that the function $f(n) = 2^n - n^2$ from the set of positive integers to the set of integers is *not* one-to-one.

**SOLUTION** We must find positive integers $n_1$ and $n_2$, $n_1 \neq n_2$, such that $f(n_1) = f(n_2)$. By checking the graph (see Figure 3.1.11) or otherwise, we find that $f(2) = f(4)$. Therefore, $f$ is not one-to-one.

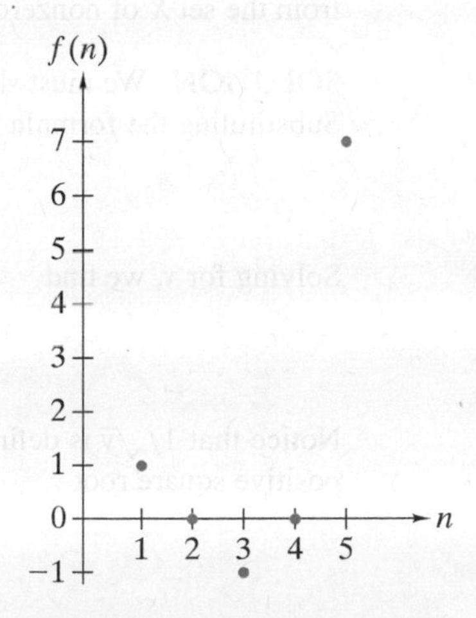


**Figure 3.1.11** The graph of $f(n) = 2^n - n^2$. ◀

If the range of a function $f$ is equal to its codomain $Y$, the function is said to be **onto** $Y$.

**Definition 3.1.29** ▶ A function $f$ from $X$ to $Y$ is said to be *onto* $Y$ (or *surjective*) if for every $y \in Y$, there exists $x \in X$ such that $f(x) = y$. ◀

**Example 3.1.30** The function $f = \{(1, a), (2, c), (3, b)\}$ from $X = \{1, 2, 3\}$ to $Y = \{a, b, c\}$ is one-to-one and onto $Y$. ◀

**Example 3.1.31** The function $f = \{(1, b), (3, a), (2, c)\}$ from $X = \{1, 2, 3\}$ to $Y = \{a, b, c, d\}$ is *not* onto $Y$. ◀

**Example 3.1.32** If a function from $X$ to $Y$ is onto, each element in $Y$ in its arrow diagram will have at least one arrow pointing to it (see Figure 3.1.12). If a function from $X$ to $Y$ is not onto, some element in $Y$ in its arrow diagram will fail to have an arrow pointing to it (see Figures 3.1.9 and 3.1.10).

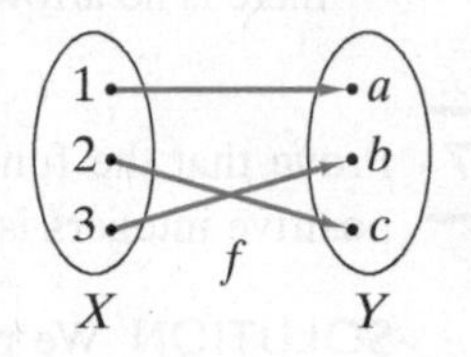


**Figure 3.1.12** The function of Example 3.1.30. This function is one-to-one because each element in $Y$ has at most one arrow. This function is onto because each element in $Y$ has at least one arrow pointing to it. ◀

**Example 3.1.33** Prove that the function

$$f(x) = \frac{1}{x^2}$$

from the set $X$ of nonzero real numbers to the set $Y$ of positive real numbers is onto $Y$.

SOLUTION We must show that for every $y \in Y$, there exists $x \in X$ such that $f(x) = y$. Substituting the formula for $f(x)$, this last equation becomes

$$\frac{1}{x^2} = y.$$

Solving for $x$, we find

$$x = \pm\frac{1}{\sqrt{y}}.$$

Notice that $1/\sqrt{y}$ is defined because $y$ is a positive real number. If we take $x$ to be the positive square root

$$x = \frac{1}{\sqrt{y}},$$

then $x \in X$. (We could just as well have taken $x = -1/\sqrt{y}$.) Thus, for every $y \in Y$, there exists $x$, namely, $x = 1/\sqrt{y}$ such that

$$f(x) = f(1/\sqrt{y}) = \frac{1}{(1/\sqrt{y})^2} = y.$$

Therefore, $f$ is onto $Y$. ◀

A function $f$ from $X$ to $Y$ is *not* onto $Y$ if for some $y \in Y$, for every $x \in X$, $f(x) \neq y$. In other words, $y$ is a counterexample to the claim that for every $y \in Y$, there exists $x \in X$ such that $f(x) = y$.

**Example 3.1.34** Prove that the function $f(n) = 2n - 1$ from the set $X$ of positive integers to the set $Y$ of positive integers is *not* onto $Y$.

**SOLUTION** We must find an element $m \in Y$ such that for all $n \in X$, $f(n) \neq m$. Since $f(n)$ is an odd integer for all $n$, we may choose for $y$ any positive, even integer, for example, $y = 2$. Then $y \in Y$ and $f(n) \neq y$ for all $n \in X$. Thus $f$ is not onto $Y$. ◀

**Definition 3.1.35 ▶** A function that is both one-to-one and onto is called a *bijection*. ◀

**Example 3.1.36** The function $f$ of Example 3.1.30 is a bijection. ◀

**Example 3.1.37** If $f$ is a bijection from a finite set $X$ to a finite set $Y$, then $|X| = |Y|$, that is, the sets have the same cardinality and are the same size. For example, $f = \{(1, a), (2, b), (3, c), (4, d)\}$ is a bijection from $X = \{1, 2, 3, 4\}$ to $Y = \{a, b, c, d\}$. Both sets have four elements. In effect, $f$ counts the elements in $Y$: $f(1) = a$ is the first element in $Y$; $f(2) = b$ is the second element in $Y$; and so on. ◀

Suppose that $f$ is a one-to-one, onto function from $X$ to $Y$. It can be shown (see Exercise 116) that $\{(y, x) \mid (x, y) \in f\}$ is a one-to-one, onto function from $Y$ to $X$. This new function, denoted $f^{-1}$, is called $f$ **inverse.**

**Example 3.1.38** For the function $f = \{(1, a), (2, c), (3, b)\}$, we have $f^{-1} = \{(a, 1), (c, 2), (b, 3)\}$. ◀

**Example 3.1.39** Given the arrow diagram for a one-to-one, onto function $f$ from $X$ to $Y$, we can obtain the arrow diagram for $f^{-1}$ simply by reversing the direction of each arrow (see Figure 3.1.13, which is the arrow diagram for $f^{-1}$, where $f$ is the function of Figure 3.1.12).

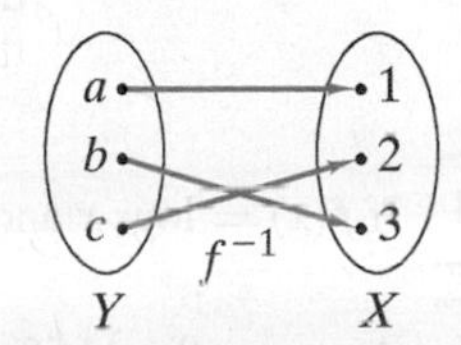


**Figure 3.1.13** The inverse of the function in Figure 3.1.12. The inverse is obtained by reversing all of the arrows in Figure 3.1.12. ◀

**Example 3.1.40** The function $f(x) = 2^x$ is a one-to-one function from the set $\mathbf{R}$ of all real numbers onto the set $\mathbf{R}^+$ of all positive real numbers. Derive a formula for $f^{-1}(y)$.

**SOLUTION** Suppose that $(y, x)$ is in $f^{-1}$; that is,

$$f^{-1}(y) = x. \tag{3.1.3}$$

Then $(x, y) \in f$. Thus, $y = 2^x$. By the definition of logarithm,

$$\log_2 y = x. \tag{3.1.4}$$

Combining (3.1.3) and (3.1.4), we have $f^{-1}(y) = x = \log_2 y$. That is, for each $y \in \mathbf{R}^+$, $f^{-1}(y)$ is the logarithm to the base 2 of $y$. We can summarize the situation by saying that the inverse of the exponential function is the logarithm function. ◀

Let $g$ be a function from $X$ to $Y$ and let $f$ be a function from $Y$ to $Z$. Given $x \in X$, we may apply $g$ to determine a unique element $y = g(x) \in Y$. We may then apply $f$ to determine a unique element $z = f(y) = f(g(x)) \in Z$. This compound action is called **composition.**

**Definition 3.1.41 ▶** Let $g$ be a function from $X$ to $Y$ and let $f$ be a function from $Y$ to $Z$. The *composition of $f$ with $g$*, denoted $f \circ g$, is the function

$$(f \circ g)(x) = f(g(x))$$

from $X$ to $Z$.

**Example 3.1.42** Given $g = \{(1, a), (2, a), (3, c)\}$, a function from $X = \{1, 2, 3\}$ to $Y = \{a, b, c\}$, and $f = \{(a, y), (b, x), (c, z)\}$, a function from $Y$ to $Z = \{x, y, z\}$, the composition function from $X$ to $Z$ is the function $f \circ g = \{(1, y), (2, y), (3, z)\}$. ◀

**Example 3.1.43** Given the arrow diagram for a function $g$ from $X$ to $Y$ and the arrow diagram for a function $f$ from $Y$ to $Z$, we can obtain the arrow diagram for the composition $f \circ g$ simply by "following the arrows" (see Figure 3.1.14).

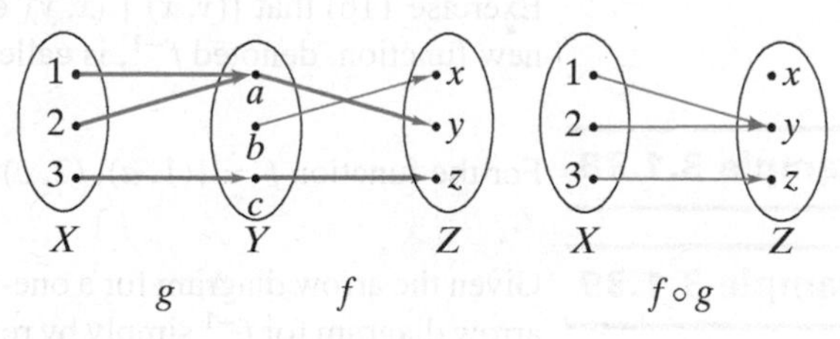


**Figure 3.1.14** The composition of the functions of Example 3.1.42. The composition is obtained by drawing an arrow from $x$ in $X$ to $z$ in $Z$ provided that there are arrows from $x$ to some $y$ in $Y$ and from $y$ to $z$. ◀

**Example 3.1.44** If $f(x) = \log_3 x$ and $g(x) = x^4$, then $f(g(x)) = \log_3(x^4)$, and $g(f(x)) = (\log_3 x)^4$. ◀

**Example 3.1.45** A store offers 15% off the price of certain items. A coupon is also available that offers \$20 off the price of the same items. The store will honor both discounts. The function $D(p) = 0.85p$ gives the cost with 15% off the price $p$. The function $C(p) = p - 20$ gives the cost using the \$20 coupon. The composition

$$(D \circ C)(p) = 0.85(p - 20) = 0.85p - 17$$

gives the cost using first the coupon and then the 15% discount. The composition $(C \circ D)(p) = 0.85p - 20$ gives the cost using first the 15% discount and then the coupon.

We see that regardless of the price of an item, it is always cheapest to use the discount first. ◀

**Example 3.1.46** Composition sometimes allows us to decompose complicated functions into simpler functions. For example, the function $f(x) = \sqrt{\sin 2x}$ can be decomposed into the functions

$$g(x) = \sqrt{x}, \qquad h(x) = \sin x, \qquad w(x) = 2x.$$

We can then write $f(x) = g(h(w(x)))$. This decomposition technique is important in differential calculus since there are rules for differentiating simple functions such as $g$, $h$, and $w$ and also rules about how to differentiate the composition of functions. Combining these rules, we can differentiate more complicated functions. ◀

A **binary operator** on a set $X$ associates with each ordered pair of elements in $X$ one element in $X$.

**Definition 3.1.47 ▶** A function from $X \times X$ to $X$ is called a *binary operator* on $X$. ◀

**Example 3.1.48** Let $X = \{1, 2, \ldots\}$. If we define $f(x, y) = x + y$, where $x, y \in X$, then $f$ is a binary operator on $X$. ◀

**Example 3.1.49** If $X$ is a set of propositions, $\wedge$, $\vee$, $\rightarrow$, and $\leftrightarrow$ are binary operators on $X$. ◀

A **unary operator** on a set $X$ associates with each single element of $X$ one element in $X$.

**Definition 3.1.50 ▶** A function from $X$ to $X$ is called a *unary operator* on $X$. ◀

**Example 3.1.51** Let $U$ be a universal set. If we define $f(X) = \overline{X}$, where $X \in \mathcal{P}(U)$, then $f$ is a unary operator on $\mathcal{P}(U)$. ◀

**Example 3.1.52** If $X$ is a set of propositions, $\neg$ is a unary operator on $X$. ◀

## 3.1 Problem-Solving Tips

The key to solving problems involving functions is clearly understanding the definition of function. A function $f$ from $X$ to $Y$ can be thought of in many ways. Formally, $f$ is a subset of $X \times Y$ having the property that for every $x \in X$, there is a unique $y \in Y$ such that $(x, y) \in X \times Y$. Informally, $f$ can be thought of as a *mapping* of elements from $X$ to $Y$. The arrow diagram emphasizes this view of a function. For an arrow diagram to be a function, there must be exactly one arrow from each element in $X$ to some element in $Y$.

A function is a very general concept. *Any* subset of $X \times Y$ having the property that for every $x \in X$, there is a unique $y \in Y$ such that $(x, y) \in X \times Y$ is a function. A function may be defined by listing its members; for example, $\{(a, 1), (b, 3), (c, 2), (d, 1)\}$ is a function from $\{a, b, c, d\}$ to $\{1, 2, 3\}$. Here, there is apparently no formula for membership; the definition just tells us which pairs make up the function.

On the other hand, a function may be defined by a formula. For example,

$$\{(n, n+2) \mid n \text{ is a positive integer}\}$$

defines a function from the set of positive integers to the set of positive integers. The "formula" for the mapping is "add 2."

The $f(x)$ notation may be used to indicate which element in the codomain is associated with an element $x$ in the domain or to define a function. For example, for the function $f = \{(a, 1), (b, 3), (c, 2), (d, 1)\}$, we could write $f(a) = 1, f(b) = 3$, and so on. Assuming that the domain of definition is the positive integers, the equation $g(n) = n+2$ defines the function $\{(n, n+2) \mid n \text{ is a positive integer}\}$ from the set of positive integers to the set of positive integers.

To prove that a function $f$ from $X$ to $Y$ is one-to-one, show that for all $x_1, x_2 \in X$, if $f(x_1) = f(x_2)$, then $x_1 = x_2$.

To prove that a function $f$ from $X$ to $Y$ is *not* one-to-one, find $x_1, x_2 \in X$, $x_1 \neq x_2$, such that $f(x_1) = f(x_2)$.

To prove that a function $f$ from $X$ to $Y$ is onto, show that for all $y \in Y$, there exists $x \in X$ such that $f(x) = y$.

To prove that a function $f$ from $X$ to $Y$ is *not* onto, find $y \in Y$ such that $f(x) \neq y$ for all $x \in X$.

## 3.1 Review Exercises

1. What is a function from $X$ to $Y$?
2. Explain how to use an arrow diagram to depict a function.
3. What is the graph of a function?
4. Given a set of points in the plane, how can we tell whether it is a function?
5. What is the value of $x$ mod $y$?
6. What is a hash function?
7. What is a collision for a hash function?
8. What is a collision resolution policy?
9. What are pseudorandom numbers?
10. Explain how a linear congruential random number generator works, and give an example of a linear congruential random number generator.
11. What is the floor of $x$? How is the floor denoted?
12. What is the ceiling of $x$? How is the ceiling denoted?
13. Define *one-to-one function*. Give an example of a one-to-one function. Explain how to use an arrow diagram to determine whether a function is one-to-one.
14. Define *onto function*. Give an example of an onto function. Explain how to use an arrow diagram to determine whether a function is onto.
15. What is a bijection? Give an example of a bijection. Explain how to use an arrow diagram to determine whether a function is a bijection.
16. Define *inverse function*. Give an example of a function and its inverse. Given the arrow diagram of a function, how can we find the arrow diagram of the inverse function?
17. Define *composition of functions*. How is the composition of $f$ and $g$ denoted? Give an example of functions $f$ and $g$ and their composition. Given the arrow diagrams of two functions, how can we find the arrow diagram of the composition of the functions?
18. What is a binary operator? Give an example of a binary operator.
19. What is a unary operator? Give an example of a unary operator.

## 3.1 Exercises

*In Exercises 1–6, determine which credit card numbers have correct check digits.*

1. 5366-2806-9965-4138
2. 5194-1132-8860-3905
3. 4004-6067-3429-0019
4. 3419-6888-7169-444
5. 3016-4773-7532-21
6. 4629-9521-3698-0203
7. Show that when 82 in the valid credit card number 4690-3582-1375-4657 is transposed to 28, the check digit changes.

*Determine whether each set in Exercises 8–12 is a function from* $X = \{1, 2, 3, 4\}$ *to* $Y = \{a, b, c, d\}$. *If it is a function, find its do-*

*main and range, draw its arrow diagram, and determine if it is one-to-one, onto, or both. If it is both one-to-one and onto, give the description of the inverse function as a set of ordered pairs, draw its arrow diagram, and give the domain and range of the inverse function.*

**8.** $\{(1, a), (2, a), (3, c), (4, b)\}$

**9.** $\{(1, c), (2, a), (3, b), (4, c), (2, d)\}$

**10.** $\{(1, c), (2, d), (3, a), (4, b)\}$

**11.** $\{(1, d), (2, d), (4, a)\}$

**12.** $\{(1, b), (2, b), (3, b), (4, b)\}$

*Draw the graphs of the functions in Exercises 13–16. The domain of each function is the set of real numbers. The codomain of each function is also the set of real numbers.*

**13.** $f(x) = \lceil x \rceil - \lfloor x \rfloor$ **14.** $f(x) = x - \lfloor x \rfloor$

**15.** $f(x) = \lceil x^2 \rceil$ **16.** $f(x) = \lfloor x^2 - x \rfloor$

*Determine whether each function in Exercises 17–22 is one-to-one, onto, or both. Prove your answers. The domain of each function is the set of all integers. The codomain of each function is also the set of all integers.*

**17.** $f(n) = n + 1$ **18.** $f(n) = n^2 - 1$

**19.** $f(n) = \lceil n/2 \rceil$ **20.** $f(n) = |n|$

**21.** $f(n) = 2n$ **22.** $f(n) = n^3$

*Determine whether each function in Exercises 23–28 is one-to-one, onto, or both. Prove your answers. The domain of each function is* $\mathbf{Z} \times \mathbf{Z}$. *The codomain of each function is* $\mathbf{Z}$.

**23.** $f(m, n) = m - n$ **24.** $f(m, n) = m$

**25.** $f(m, n) = mn$ **26.** $f(m, n) = m^2 + n^2$

**27.** $f(m, n) = n^2 + 1$ **28.** $f(m, n) = m + n + 2$

**29.** Prove that the function $f$ from $\mathbf{Z}^+ \times \mathbf{Z}^+$ to $\mathbf{Z}^+$ defined by $f(m, n) = 2^m 3^n$ is one-to-one but not onto.

*Determine whether each function in Exercises 30–35 is one-to-one, onto, or both. Prove your answers. The domain of each function is the set of all real numbers. The codomain of each function is also the set of all real numbers.*

**30.** $f(x) = 6x - 9$ **31.** $f(x) = 3x^2 - 3x + 1$

**32.** $f(x) = \sin x$ **33.** $f(x) = 2x^3 - 4$

**34.** $f(x) = 3^x - 2$ **35.** $f(x) = \dfrac{x}{1 + x^2}$

**36.** Give an example of a function different from those presented in the text that is one-to-one but not onto, and prove that your function has the required properties.

**37.** Give an example of a function different from those presented in the text that is onto but not one-to-one, and prove that your function has the required properties.

**38.** Give an example of a function different from those presented in the text that is neither one-to-one nor onto, and prove that your function has the required properties.

**39.** Write the definition of "one-to-one" using logical notation (i.e., use $\forall$, $\exists$, etc.).

**40.** Use De Morgan's laws of logic to negate the definition of "one-to-one."

**41.** Write the definition of "onto" using logical notation (i.e., use $\forall$, $\exists$, etc.).

**42.** Use De Morgan's laws of logic to negate the definition of "onto."

*Each function in Exercises 43–48 is one-to-one on the specified domain X. By letting Y = range of f, we obtain a bijection from X to Y. Find each inverse function.*

**43.** $f(x) = 4x + 2$, $X$ = set of real numbers

**44.** $f(x) = 3^x$, $X$ = set of real numbers

**45.** $f(x) = 3 \log_2 x$, $X$ = set of positive real numbers

**46.** $f(x) = 3 + \dfrac{1}{x}$, $X$ = set of nonzero real numbers

**47.** $f(x) = 4x^3 - 5$, $X$ = set of real numbers

**48.** $f(x) = 6 + 2^{7x-1}$, $X$ = set of real numbers

**49.** Given

$$g = \{(1, b), (2, c), (3, a)\},$$

a function from $X = \{1, 2, 3\}$ to $Y = \{a, b, c, d\}$, and

$$f = \{(a, x), (b, x), (c, z), (d, w)\},$$

a function from $Y$ to $Z = \{w, x, y, z\}$, write $f \circ g$ as a set of ordered pairs and draw the arrow diagram of $f \circ g$.

**50.** Let $f$ and $g$ be functions from the positive integers to the positive integers defined by the equations

$$f(n) = 2n + 1, \qquad g(n) = 3n - 1.$$

Find the compositions $f \circ f$, $g \circ g$, $f \circ g$, and $g \circ f$.

**51.** Let $f$ and $g$ be functions from the positive integers to the positive integers defined by the equations

$$f(n) = n^2, \qquad g(n) = 2^n.$$

Find the compositions $f \circ f$, $g \circ g$, $f \circ g$, and $g \circ f$.

**52.** Let $f$ and $g$ be functions from the nonnegative real numbers to the nonnegative real numbers defined by the equations

$$f(x) = \lfloor 2x \rfloor, \qquad g(x) = x^2.$$

Find the compositions $f \circ f$, $g \circ g$, $f \circ g$, and $g \circ f$.

**53.** A store offers a (fixed, nonzero) percentage off the price of certain items. A coupon is also available that offers a (fixed, nonzero) amount off the price of the same items. The store will honor both discounts. Show that regardless of the price of an item, the percentage off the price, and amount off the price, it is always cheapest to use the coupon first.

*In Exercises 54–59, decompose the function into simpler functions as in Example 3.1.46.*

**54.** $f(x) = \log_2(x^2 + 2)$

**55.** $f(x) = \dfrac{1}{2x^2}$

**56.** $f(x) = \sin 2x$

**57.** $f(x) = 2 \sin x$

**58.** $f(x) = (3 + \sin x)^4$

**59.** $f(x) = \dfrac{1}{(\cos 6x)^3}$

**60.** Given

$$f = \{(x, x^2) \mid x \in X\},$$

a function from $X = \{-5, -4, \ldots, 4, 5\}$ to the set of integers, write $f$ as a set of ordered pairs and draw the arrow diagram of $f$. Is $f$ one-to-one or onto?

**61.** How many functions are there from $\{1, 2\}$ to $\{a, b\}$? Which are one-to-one? Which are onto?

**62.** Given

$$f = \{(a, b), (b, a), (c, b)\},$$

a function from $X = \{a, b, c\}$ to $X$:

(a) Write $f \circ f$ and $f \circ f \circ f$ as sets of ordered pairs.

(b) Define

$$f^n = f \circ f \circ \cdots \circ f$$

to be the $n$-fold composition of $f$ with itself. Write $f^9$ and $f^{623}$ as sets of ordered pairs.

**63.** Let $f$ be the function from $X = \{0, 1, 2, 3, 4\}$ to $X$ defined by

$$f(x) = 4x \bmod 5.$$

Write $f$ as a set of ordered pairs and draw the arrow diagram of $f$. Is $f$ one-to-one? Is $f$ onto?

**64.** Let $f$ be the function from $X = \{0, 1, 2, 3, 4, 5\}$ to $X$ defined by

$$f(x) = 4x \bmod 6.$$

Write $f$ as a set of ordered pairs and draw the arrow diagram of $f$. Is $f$ one-to-one? Is $f$ onto?

**65.** An International Standard Book Number (ISBN) is a code of 13 characters separated by dashes, such as 978-1-59448-950-1. An ISBN consists of five parts: a product code, a group code, a publisher code, a code that uniquely identifies the book among those published by the particular publisher, and a check digit. For 978-1-59448-950-1, the product code 978 identifies the product as a book (the same scheme is used for other products). The group code is 1, which identifies the book as one from an English-speaking country. The publisher code 59448 identifies the book as one published by Riverhead Books, Penguin Group. The code 950 uniquely identifies the book among those published by Riverhead Books, Penguin Group (Hosseini: *A Thousand Splendid Suns*, in this case). The check digit is 1.

Let $S$ equal the sum of the first digit, plus three times the second digit, plus the third digit, plus three times the fourth digit, . . . , plus three times the twelfth digit. The check digit is equal to

$$[10 - (S \bmod 10)] \bmod 10.$$

Universal Product Codes (UPC), the bar codes that are scanned at the grocery store for example, use a similar method to compute the check digit.

Verify the check digit for this book.

*In Exercises 66–71, determine which ISBNs (see Exercise 65) have correct check digits.*

**66.** 978-1-61374-376-9

**67.** 978-0-8108-8139-2

**68.** 978-0-939460-91-5

**69.** 978-0-8174-3593-6

**70.** 978-1-4354-6028-7

**71.** 978-0-684-87018-0

**72.** Show that if a single digit of an ISBN is changed, the check digit will change. Thus, any single-digit error will be detected.

*For each hash function in Exercises 73–76, show how the data would be inserted in the order given in initially empty cells. Use the collision resolution policy of Example 3.1.15.*

**73.** $h(x) = x \bmod 11$; cells indexed 0 to 10; data: 53, 13, 281, 743, 377, 20, 10, 796

**74.** $h(x) = x \bmod 17$; cells indexed 0 to 16; data: 714, 631, 26, 373, 775, 906, 509, 2032, 42, 4, 136, 1028

**75.** $h(x) = x^2 \bmod 11$; cells and data as in Exercise 73

**76.** $h(x) = (x^2 + x) \bmod 17$; cells and data as in Exercise 74

**77.** Suppose that we store and retrieve data as described in Example 3.1.15. Will any problem arise if we delete data? Explain.

**78.** Suppose that we store data as described in Example 3.1.15 and that we never store more than 10 items. Will any problem arise when retrieving data if we stop searching when we encounter an empty cell? Explain.

**79.** Suppose that we store data as described in Example 3.1.15 and retrieve data as described in Exercise 78. Will any problem arise if we delete data? Explain.

*Let $g$ be a function from $X$ to $Y$ and let $f$ be a function from $Y$ to $Z$. For each statement in Exercises 80–87, if the statement is true, prove it; otherwise, give a counterexample.*

**80.** If $g$ is one-to-one, then $f \circ g$ is one-to-one.

**81.** If $f$ is onto, then $f \circ g$ is onto.

**82.** If $g$ is onto, then $f \circ g$ is onto.

**83.** If $f$ and $g$ are onto, then $f \circ g$ is onto.

**84.** If $f$ and $g$ are one-to-one and onto, then $f \circ g$ is one-to-one and onto.

**85.** If $f \circ g$ is one-to-one, then $f$ is one-to-one.

**86.** If $f \circ g$ is one-to-one, then $g$ is one-to-one.

**87.** If $f \circ g$ is onto, then $f$ is onto.

*If $f$ is a function from $X$ to $Y$ and $A \subseteq X$ and $B \subseteq Y$, we define*

$$f(A) = \{f(x) \mid x \in A\}, \qquad f^{-1}(B) = \{x \in X \mid f(x) \in B\}.$$

*We call $f^{-1}(B)$ the* inverse image of $B$ under $f$.

**88.** Let

$$g = \{(1, a), (2, c), (3, c)\}$$

be a function from $X = \{1, 2, 3\}$ to $Y = \{a, b, c, d\}$. Let $S = \{1\}$, $T = \{1, 3\}$, $U = \{a\}$, and $V = \{a, c\}$. Find $g(S)$, $g(T)$, $g^{-1}(U)$, and $g^{-1}(V)$.

★**89.** Let $f$ be a function from $X$ to $Y$. Prove that $f$ is one-to-one if and only if

$$f(A \cap B) = f(A) \cap f(B)$$

for all subsets $A$ and $B$ of $X$. [When $S$ is a set, we define $f(S) = \{f(x) \mid x \in S\}$.]

★**90.** Let $f$ be a function from $X$ to $Y$. Prove that $f$ is one-to-one if and only if whenever $g$ is a one-to-one function from any set $A$ to $X$, $f \circ g$ is one-to-one.

★**91.** Let $f$ be a function from $X$ to $Y$. Prove that $f$ is onto $Y$ if and only if whenever $g$ is a function from $Y$ onto any set $Z$, $g \circ f$ is onto $Z$.

**92.** Let $f$ be a function from $X$ onto $Y$. Let

$$\mathcal{S} = \{f^{-1}(\{y\}) \mid y \in Y\}.$$

Show that $\mathcal{S}$ is a partition of $X$.

*Let $\mathbf{R}^{\mathbf{R}}$ denote the set of functions from $\mathbf{R}$ to $\mathbf{R}$. We define the* evaluation function $E_a$, *where $a \in \mathbf{R}$, from $\mathbf{R}^{\mathbf{R}}$ to $\mathbf{R}$ as*

$$E_a(f) = f(a).$$

**93.** Is $E_1$ one-to-one? Prove your answer.

**94.** Is $E_1$ onto? Prove your answer.

**95.** Let $f$ be a function from $\mathbf{R}$ to $\mathbf{R}$ such that for some $r \in \mathbf{R}$, $f(rx) = rf(x)$ for all $x \in \mathbf{R}$. Prove that $f(r^n x) = r^n f(x)$ for all $x \in \mathbf{R}$, $n \in \mathbf{Z}^+$.

*Exercises 96–100 use the following definitions. Let $X = \{a, b, c\}$. Define a function $S$ from $\mathcal{P}(X)$ to the set of bit strings of length 3 as follows. Let $Y \subseteq X$. If $a \in Y$, set $s_1 = 1$; if $a \notin Y$, set $s_1 = 0$. If $b \in Y$, set $s_2 = 1$; if $b \notin Y$, set $s_2 = 0$. If $c \in Y$, set $s_3 = 1$; if $c \notin Y$, set $s_3 = 0$. Define $S(Y) = s_1 s_2 s_3$.*

**96.** What is the value of $S(\{a, c\})$?

**97.** What is the value of $S(\varnothing)$?

**98.** What is the value of $S(X)$?

**99.** Prove that $S$ is one-to-one.

**100.** Prove that $S$ is onto.

*Exercises 101–107 use the following definitions. Let $U$ be a universal set and let $X \subseteq U$. Define*

$$C_X(x) = \begin{cases} 1 & \text{if } x \in X \\ 0 & \text{if } x \notin X. \end{cases}$$

*We call $C_X$ the* characteristic function of $X$ (in $U$). *(A look ahead at the next Problem-Solving Corner may help in understanding the following exercises.)*

**101.** Prove that $C_{X \cap Y}(x) = C_X(x)C_Y(x)$ for all $x \in U$.

**102.** Prove that $C_{X \cup Y}(x) = C_X(x) + C_Y(x) - C_X(x)C_Y(x)$ for all $x \in U$.

**103.** Prove that $C_{\overline{X}}(x) = 1 - C_X(x)$ for all $x \in U$.

**104.** Prove that $C_{X-Y}(x) = C_X(x)[1 - C_Y(x)]$ for all $x \in U$.

**105.** Prove that if $X \subseteq Y$, then $C_X(x) \le C_Y(x)$ for all $x \in U$.

**106.** Find a formula for $C_{X \triangle Y}$. ($X \triangle Y$ is the symmetric difference of $X$ and $Y$. The definition is given before Exercise 101, Section 1.1.)

**107.** Prove that the function $f$ from $\mathcal{P}(U)$ to the set of characteristic functions in $U$ defined by

$$f(X) = C_X$$

is one-to-one and onto.

**108.** Let $X$ and $Y$ be sets. Prove that there is a one-to-one function from $X$ to $Y$ if and only if there is a function from $Y$ onto $X$.

*A binary operator $f$ on a set $X$ is* commutative *if $f(x, y) = f(y, x)$ for all $x, y \in X$. In Exercises 109–113, state whether the given function $f$ is a binary operator on the set $X$. If $f$ is not a binary operator, state why. State whether or not each binary operator is commutative.*

**109.** $f(x, y) = x + y, \quad X = \{1, 2, \ldots\}$

**110.** $f(x, y) = x - y, \quad X = \{1, 2, \ldots\}$

**111.** $f(x, y) = x \cup y, \quad X = \mathcal{P}(\{1, 2, 3, 4\})$

**112.** $f(x, y) = x/y, \quad X = \{0, 1, 2, \ldots\}$

**113.** $f(x, y) = x^2 + y^2 - xy, \quad X = \{1, 2, \ldots\}$

*In Exercises 114 and 115, give an example of a unary operator [different from $f(x) = x$, for all $x$] on the given set.*

**114.** $\{\ldots, -2, -1, 0, 1, 2, \ldots\}$

**115.** The set of all finite subsets of $\{1, 2, 3, \ldots\}$

**116.** Prove that if $f$ is a one-to-one, onto function from $X$ to $Y$, then

$$\{(y, x) \mid (x, y) \in f\}$$

is a one-to-one, onto function from $Y$ to $X$.

*In Exercises 117–119, if the statement is true for all real numbers, prove it; otherwise, give a counterexample.*

**117.** $\lceil x + 3 \rceil = \lceil x \rceil + 3$

**118.** $\lceil x + y \rceil = \lceil x \rceil + \lceil y \rceil$

**119.** $\lfloor x + y \rfloor = \lfloor x \rfloor + \lceil y \rceil$

**120.** Prove that if $n$ is an odd integer,

$$\left\lfloor \frac{n^2}{4} \right\rfloor = \left(\frac{n-1}{2}\right)\left(\frac{n+1}{2}\right).$$

**121.** Prove that if $n$ is an odd integer,

$$\left\lceil \frac{n^2}{4} \right\rceil = \frac{n^2+3}{4}.$$

**122.** Find a value for $x$ for which $\lceil 2x \rceil = 2\lceil x \rceil - 1$.

**123.** Prove that $2\lceil x \rceil - 1 \le \lceil 2x \rceil \le 2\lceil x \rceil$ for all real numbers $x$.

**124.** Prove that for all real numbers $x$ and integers $n$, $\lceil x \rceil = n$ if and only if there exists $\varepsilon$, $0 \le \varepsilon < 1$, such that $x + \varepsilon = n$.

**125.** State and prove a result analogous to Exercise 124 for $\lfloor x \rfloor$.

*The months with Friday the 13th in year x are found in row*

$$y = \left( x + \left\lfloor \frac{x-1}{4} \right\rfloor - \left\lfloor \frac{x-1}{100} \right\rfloor + \left\lfloor \frac{x-1}{400} \right\rfloor \right) \bmod 7$$

*in the appropriate column:*

| y | Non-Leap Year | Leap Year |
|---|---|---|
| 0 | January, October | January, April, July |
| 1 | April, July | September, December |
| 2 | September, December | June |
| 3 | June | March, November |
| 4 | February, March, November | February, August |
| 5 | August | May |
| 6 | May | October |

**126.** Find the months with Friday the 13th in 1945.

**127.** Find the months with Friday the 13th in the current year.

**128.** Find the months with Friday the 13th in 2040.

## Problem-Solving Corner: Functions

### Problem

Let $U$ be a universal set and let $X \subseteq U$. Define

$$C_X(x) = \begin{cases} 1 & \text{if } x \in X \\ 0 & \text{if } x \notin X. \end{cases}$$

[We call $C_X$ the *characteristic function* of $X$ (in $U$)]. Assume that $X$ and $Y$ are arbitrary subsets of the universal set $U$. Prove that $C_{X \cup Y}(x) = C_X(x) + C_Y(x)$ for all $x \in U$ if and only if $X \cap Y = \varnothing$.

### Attacking the Problem

First, let's be clear what we must do. Since the statement is of the form $p$ if and only if $q$, we have two tasks: (1) Prove if $p$ then $q$. (2) Prove if $q$ then $p$. It's a good idea to write out exactly what must be proved:

If $C_{X \cup Y}(x) = C_X(x) + C_Y(x)$ for all $x \in U$, then $X \cap Y = \varnothing$. **(1)**

If $X \cap Y = \varnothing$, then $C_{X \cup Y}(x) = C_X(x) + C_Y(x)$ for all $x \in U$. **(2)**

Consider the first statement in which we assume that $C_{X \cup Y}(x) = C_X(x) + C_Y(x)$ for all $x \in U$ and prove that $X \cap Y = \varnothing$. How do we prove that a set, $X \cap Y$ in this case, is the empty set? We have to show that $X \cap Y$ has no elements. How do we do that? There are several possibilities, but one thing that comes to mind is another question: What if $X \cap Y$ had an element? This suggests that we might prove the first statement by contradiction or by proving its contrapositive. If we let

$p$: $C_{X \cup Y}(x) = C_X(x) + C_Y(x)$ for all $x \in U$

$q$: $X \cap Y = \varnothing$,

the contrapositive is $\neg q \to \neg p$. Now the negation of $q$ is

$$\neg q : X \cap Y \neq \varnothing,$$

and, using De Morgan's law (roughly, negating $\forall$ results in $\exists$), the negation of $p$ is

$\neg p : C_{X \cup Y}(x) \neq C_X(x) + C_Y(x)$ for at least one $x \in U$.

Thus, the contrapositive is

If $X \cap Y \neq \varnothing$, then $C_{X \cup Y}(x) \neq C_X(x) + C_Y(x)$ for at least one $x \in U$. **(3)**

For the second statement, we assume that $X \cap Y = \varnothing$ and prove that $C_{X \cup Y}(x) = C_X(x) + C_Y(x)$ for all $x \in U$. Presumably, we can just use the definition of $C_X$ to compute both sides of the equation for all $x \in U$ and verify that the two sides are equal. The definition of $C_X$ suggests that we use proof by cases: $x \in X \cup Y$ (when $C_{X \cup Y}(x) = 1$) and $x \notin X \cup Y$ (when $C_{X \cup Y}(x) = 0$).

## Finding a Solution

We first consider proving the contrapositive (3) of statement (1). Since we assume that $X \cap Y \neq \varnothing$, there exists an element $x \in X \cap Y$. Now let's compare the values of the expressions $C_{X \cup Y}(x)$ and $C_X(x) + C_Y(x)$. Since $x \in X \cup Y$, $C_{X \cup Y}(x) = 1$. Since $x \in X \cap Y$, $x \in X$ and $x \in Y$. Therefore

$$C_X(x) + C_Y(x) = 1 + 1 = 2.$$

We have proved that

$$C_{X \cup Y}(x) \neq C_X(x) + C_Y(x) \text{ for at least one } x \in U.$$

Now consider proving the statement (2). This time we assume that $X \cap Y = \varnothing$. Let's compute each side of the equation

$$C_{X \cup Y}(x) = C_X(x) + C_Y(x) \qquad \textbf{(4)}$$

for each $x \in U$. As suggested earlier, we consider the cases: $x \in X \cup Y$ and $x \notin X \cup Y$. If $x \in X \cup Y$, then $C_{X \cup Y}(x) = 1$. Since $X \cap Y = \varnothing$, either $x \in X$ or $x \in Y$ *but not both*. Therefore,

$$C_X(x) + C_Y(x) = 1 + 0 = 1 = C_{X \cup Y}(x)$$

or

$$C_X(x) + C_Y(x) = 0 + 1 = 1 = C_{X \cup Y}(x).$$

Equation (4) is true if $x \in X \cup Y$.

If $x \notin X \cup Y$, then $C_{X \cup Y}(x) = 0$. But if $x \notin X \cup Y$, then $x \notin X$ *and* $x \notin Y$. Therefore,

$$C_X(x) + C_Y(x) = 0 + 0 = 0 = C_{X \cup Y}(x).$$

Equation (4) is true if $x \notin X \cup Y$. Thus it is true for all $x \in U$.

## Formal Solution

The formal proof could be written as follows.

CASE $\rightarrow$: If $C_{X \cup Y}(x) = C_X(x) + C_Y(x)$ for all $x \in U$, then $X \cap Y = \varnothing$.

We prove the equivalent contrapositive

If $X \cap Y \neq \varnothing$, then $C_{X \cup Y}(x) \neq C_X(x) + C_Y(x)$ for at least one $x \in U$.

Since $X \cap Y \neq \varnothing$, there exists $x \in X \cap Y$. Since $x \in X \cup Y$, $C_{X \cup Y}(x) = 1$. Since $x \in X \cap Y$, $x \in X$ and $x \in Y$. Therefore

$$C_X(x) + C_Y(x) = 1 + 1 = 2.$$

Thus,

$$C_{X \cup Y}(x) \neq C_X(x) + C_Y(x).$$

CASE $\leftarrow$: If $X \cap Y = \varnothing$, then $C_{X \cup Y}(x) = C_X(x) + C_Y(x)$ for all $x \in U$.

Suppose that $x \in X \cup Y$. Then $C_{X \cup Y}(x) = 1$. Since $X \cap Y = \varnothing$, either $x \in X$ or $x \in Y$ but not both. Therefore, $C_X(x) + C_Y(x) = 1$ and (4) holds.

If $x \notin X \cup Y$, then $C_{X \cup Y}(x) = 0$. If $x \notin X \cup Y$, then $x \notin X$ and $x \notin Y$. Therefore, $C_X(x) + C_Y(x) = 0$. Again, (4) holds. Therefore (4) holds for all $x \in U$.

## Summary of Problem-Solving Techniques

- Write out exactly what must be proved.
- Instead of proving $p \rightarrow q$ directly, consider proving its contrapositive $\neg q \rightarrow \neg p$ or a proof by contradiction.
- For statements involving negation, De Morgan's laws can be very helpful.
- Look for definitions and theorems relevant to the expressions mentioned in the statements to be proved.
- A definition that involves cases suggests a proof by cases.

# 3.2 Sequences and Strings

Blue Taxi Inc. charges \$1 for the first mile and 50 cents for each additional mile. The following table shows the cost of traveling from 1 to 10 miles. In general, the cost $C_n$ of traveling $n$ miles is 1.00 (the cost of traveling the first mile) plus 0.50 times the number $(n - 1)$ of additional miles. That is, $C_n = 1 + 0.5(n - 1)$. As examples,

$$C_1 = 1 + 0.5(1 - 1) = 1 + 0.5 \cdot 0 = 1,$$
$$C_5 = 1 + 0.5(5 - 1) = 1 + 0.5 \cdot 4 = 1 + 2 = 3.$$

| Mileage | Cost |
|---|---|
| 1 | \$1.00 |
| 2 | 1.50 |
| 3 | 2.00 |
| 4 | 2.50 |
| 5 | 3.00 |
| 6 | 3.50 |
| 7 | 4.00 |
| 8 | 4.50 |
| 9 | 5.00 |
| 10 | 5.50 |

The list of fares

$$C_1 = 1.00, \quad C_2 = 1.50, \quad C_3 = 2.00, \quad C_4 = 2.50, \quad C_5 = 3.00,$$
$$C_6 = 3.50, \quad C_7 = 4.00, \quad C_8 = 4.50, \quad C_9 = 5.00, \quad C_{10} = 5.50$$

furnishes an example of a **sequence,** which is a special type of function in which the domain consists of integers.

**Go Online**
For more on sequences, see
bit.ly/2OFm8St

**Definition 3.2.1 ▶** A *sequence s is a function whose domain D is a subset of integers. The notation $s_n$ is typically used instead of the more general function notation $s(n)$. The term n is called the index of the sequence. If D is a finite set, we call s a finite sequence; otherwise, s is an infinite sequence.* ◀

For the sequence of fares for Blue Taxi, Inc., the domain of the sequence $C$ is the subset of integers $\{1, 2, 3, 4, 5, 6, 7, 8, 9, 10\}$. The sequence $C$ is a finite sequence.

A sequence $s$ is denoted $s$ or $\{s_n\}$ if $n$ is the index of the sequence. If, for example, the domain is the set of positive integers $\mathbf{Z}^+$, $s$ or $\{s_n\}$ denotes the entire sequence $s_1, s_2, s_3, \ldots$. We use the notation $s_n$ to denote the single element of the sequence $s$ at index $n$. In this book, we will frequently use $\mathbf{Z}^+$ or $\mathbf{Z}^{nonneg}$ as the domain of a sequence.

**Example 3.2.2** Consider the sequence $s$: $2, 4, 6, \ldots, 2n, \ldots$. The first element of the sequence is 2, the second element of the sequence is 4, and so on. The $n$th element of the sequence is $2n$. If the domain of $s$ is $\mathbf{Z}^+$, we have $s_1 = 2,\ s_2 = 4,\ s_3 = 6, \ldots, s_n = 2n, \ldots$. The sequence $s$ is an infinite sequence. ◀

**Example 3.2.3** Consider the sequence $t$: $a, a, b, a, b$. The first element of the sequence is $a$, the second element of the sequence is $a$, and so on. If the domain of $t$ is $\{1, 2, 3, 4, 5\}$, we have $t_1 = a,\ t_2 = a,\ t_3 = b,\ t_4 = a,$ and $t_5 = b$. The sequence $t$ is a finite sequence. ◀

If the domain of a sequence $s$ is the (infinite) set of consecutive integers $\{k, k+1, k+2, \ldots\}$ and the index of $s$ is $n$, we can denote the sequence $s$ as $\{s_n\}_{n=k}^{\infty}$. For example, a sequence $s$ whose domain is $\mathbf{Z}^{nonneg}$ can be denoted $\{s_n\}_{n=0}^{\infty}$. If the domain of a sequence $s$ is the set of (finite) consecutive integers $\{i, i+1, \ldots, j\}$, we can denote the sequence $s$ as $\{s_n\}_{n=i}^{j}$. For example, a sequence $s$ whose domain is $\{-1, 0, 1, 2, 3\}$ can be denoted $\{s_n\}_{n=-1}^{3}$.

**Example 3.2.4** The sequence $\{u_n\}$ defined by the rule $u_n = n^2 - 1$, for all $n \geq 0$, can be denoted $\{u_n\}_{n=0}^{\infty}$. The name of the index can be chosen in any convenient way. For example, the sequence $u$ can also be denoted $\{u_m\}_{m=0}^{\infty}$. The formula for the term having index $m$ is $u_m = m^2 - 1$, for all $m \geq 0$. ◀

**Example 3.2.5** Define a sequence $b$ by the rule $b_n$ is the $n$th letter in the word *digital*. If the domain of $b$ is $\{1, 2, \ldots, 7\}$, then $b_1 = d$, $b_2 = b_4 = i$, and $b_7 = l$. ◀

**Example 3.2.6** If $x$ is the sequence defined by

$$x_n = \frac{1}{2^n} \qquad -1 \le n \le 4,$$

the elements of $x$ are 2, 1, 1/2, 1/4, 1/8, 1/16. ◀

**Example 3.2.7** Define a sequence $s$ as

$$s_n = 2^n + 4 \cdot 3^n \qquad n \ge 0. \tag{3.2.1}$$

(a) Find $s_0$.

(b) Find $s_1$.

(c) Find a formula for $s_i$.

(d) Find a formula for $s_{n-1}$.

(e) Find a formula for $s_{n-2}$.

(f) Prove that $\{s_n\}$ satisfies

$$s_n = 5s_{n-1} - 6s_{n-2} \qquad \text{for all } n \ge 2. \tag{3.2.2}$$

**SOLUTION**

(a) Replacing $n$ by 0 in Definition 3.2.1, we obtain

$$s_0 = 2^0 + 4 \cdot 3^0 = 5.$$

(b) Replacing $n$ by 1 in Definition 3.2.1, we obtain

$$s_1 = 2^1 + 4 \cdot 3^1 = 14.$$

(c) Replacing $n$ by $i$ in Definition 3.2.1, we obtain

$$s_i = 2^i + 4 \cdot 3^i.$$

(d) Replacing $n$ by $n - 1$ in Definition 3.2.1, we obtain

$$s_{n-1} = 2^{n-1} + 4 \cdot 3^{n-1}.$$

(e) Replacing $n$ by $n - 2$ in Definition 3.2.1, we obtain

$$s_{n-2} = 2^{n-2} + 4 \cdot 3^{n-2}.$$

(f) To prove equation (3.2.2), we will replace $s_{n-1}$ and $s_{n-2}$ in the right side of equation (3.2.2) by the formulas of parts (d) and (e). We will then use algebra to show that the result is equal to $s_n$. We obtain

$$\begin{aligned}
5s_{n-1} - 6s_{n-2} &= 5(2^{n-1} + 4 \cdot 3^{n-1}) - 6(2^{n-2} + 4 \cdot 3^{n-2}) \\
&= (5 \cdot 2 - 6)2^{n-2} + (5 \cdot 4 \cdot 3 - 6 \cdot 4)3^{n-2} \\
&= 4 \cdot 2^{n-2} + 36 \cdot 3^{n-2} \\
&= 2^2 2^{n-2} + (4 \cdot 3^2)3^{n-2} \\
&= 2^n + 4 \cdot 3^n = s_n.
\end{aligned}$$

The techniques shown in the example will be useful in checking solutions of recurrence relations in Chapter 7. ◀

Two important types of sequences are increasing sequences and decreasing sequences and their relatives, nonincreasing sequences and nondecreasing sequences. A sequence $s$ is **increasing** if for all $i$ and $j$ in the domain of $s$, if $i < j$, then $s_i < s_j$. A sequence $s$ is **decreasing** if for all $i$ and $j$ in the domain of $s$, if $i < j$, then $s_i > s_j$. A sequence $s$ is **nondecreasing** if for all $i$ and $j$ in the domain of $s$, if $i < j$, then $s_i \leq s_j$. (A nondecreasing sequence is like an increasing sequence except that, in the condition on $s$, "$<$" is replaced by "$\leq$.") A sequence $s$ is **nonincreasing** if for all $i$ and $j$ in the domain of $s$, if $i < j$, then $s_i \geq s_j$. (A nonincreasing sequence is like a decreasing sequence except that, in the condition on $s$, "$>$" is replaced by "$\geq$.") Notice that if the domain of a sequence $s$ is a set of consecutive integers, if $s_i < s_{i+1}$ for all $i$ for which $i$ and $i+1$ are in the domain of $s$, then $s$ is increasing. Similar remarks apply to decreasing, nondecreasing, and nonincreasing sequences. (We leave formal justification for this last comment to Exercise 4.)

**Example 3.2.8** The sequence 2, 5, 13, 104, 300 is increasing and nondecreasing. ◀

**Example 3.2.9** The sequence

$$a_i = \frac{1}{i} \qquad i \geq 1,$$

is decreasing and nonincreasing. ◀

**Example 3.2.10** The sequence 100, 90, 90, 74, 74, 74, 30 is nonincreasing, but it is *not* decreasing. ◀

**Example 3.2.11** The sequence 100 (consisting of a single element) is increasing, decreasing, nonincreasing, and nondecreasing since there are no distinct values of $i$ and $j$ for which both $i$ and $j$ are indexes. ◀

One way to form a new sequence from a given sequence is to retain only certain terms of the original sequence, maintaining the order of terms in the given sequence. The resulting sequence is called a **subsequence** of the original sequence.

**Definition 3.2.12 ▶** Let $s$ be a sequence. A *subsequence* of $s$ is a sequence obtained from $s$ by choosing certain terms of $s$ in the same order in which they appear in $s$. ◀

**Example 3.2.13** The sequence

$$b, c \qquad \textbf{(3.2.3)}$$

is a subsequence of the sequence

$$a, a, b, c, q. \qquad \textbf{(3.2.4)}$$

Subsequence (3.2.3) is obtained by choosing the third and fourth terms from sequence (3.2.4).

Notice that the sequence $c, b$ is *not* a subsequence of sequence (3.2.4) since the order of terms in the sequence (3.2.4) is not maintained. ◀

**Example 3.2.14** The sequence

$$2, 4, 8, 16, \ldots, 2^k, \ldots \tag{3.2.5}$$

is a subsequence of the sequence

$$2, 4, 6, 8, 10, 12, 14, 16, \ldots, 2n, \ldots. \tag{3.2.6}$$

Subsequence (3.2.5) is obtained by choosing the first, second, fourth, eighth, and so on, terms from sequence (3.2.6). ◀

We turn to the notation for a subsequence. A subsequence of a sequence $s$ is obtained by choosing certain terms from $s$. We let $n_1, n_2, n_3, \ldots$ denote the indexes (in order) of $s$ of the terms that are selected to obtain the subsequence. Thus the subsequence can be denoted $\{s_{n_k}\}$.

**Example 3.2.15** Let us return to the sequence (3.2.4) in Example 3.2.13, which we now call $t$:

$$t_1 = a, \quad t_2 = a, \quad t_3 = b, \quad t_4 = c, \quad t_5 = q.$$

The subsequence $b, c$ is obtained from $t$ by selecting the third and fourth terms of $t$. Thus $n_1 = 3$, $n_2 = 4$, and the subsequence $b, c$ is $t_3, t_4$, or $t_{n_1}, t_{n_2}$. ◀

**Example 3.2.16** Let us return to the sequence (3.2.6) in Example 3.2.14. We let $s_n = 2n$, so the sequence (3.2.6) can be denoted $\{s_n\}_{n=1}^{\infty}$. The subsequence (3.2.5) is obtained from $s$ by selecting the first, second, fourth, eighth, and so on, terms from $s$; thus

$$n_1 = 1, \quad n_2 = 2, \quad n_3 = 4, \quad n_4 = 8, \ldots.$$

We see that $n_k = 2^{k-1}$, $k = 1, 2, \ldots$. The subsequence (3.2.5) $\{s_{n_k}\}_{k=1}^{\infty}$ is defined by

$$s_{n_k} = s_{2^{k-1}} = 2 \cdot 2^{k-1} = 2^k.$$

In the notation for the original sequence $\{s_n\}_{n=1}^{\infty}$, $n$ is the index of the sequence $s$. In the notation for the subsequence $\{s_{n_k}\}_{k=1}^{\infty}$, $n$ is used in a totally different way; $n$ is itself a *sequence*, namely, the sequence of indexes selected to form the subsequence. Furthermore, $k$ is the index of the sequence $n$. It would have been nice to avoid having $n$ used in two different ways, but tradition demands that we continue this abuse of notation. ◀

Two important operations on numerical sequences are adding and multiplying terms.

**Definition 3.2.17 ▶** If $\{a_i\}_{i=m}^{n}$ is a sequence, we define

$$\sum_{i=m}^{n} a_i = a_m + a_{m+1} + \cdots + a_n, \qquad \prod_{i=m}^{n} a_i = a_m \cdot a_{m+1} \cdots a_n.$$

**Go Online**
For more on the sigma notation, see `bit.ly/2OFm8St`

The formalism

$$\sum_{i=m}^{n} a_i \tag{3.2.7}$$

is called the *sum* (or *sigma*) *notation* and

$$\prod_{i=m}^{n} a_i \tag{3.2.8}$$

is called the *product notation*.

In (3.2.7) or (3.2.8), $i$ is called the *index*, $m$ is called the *lower limit*, and $n$ is called the *upper limit*. ◀

**Example 3.2.18** Let $a$ be the sequence defined by $a_n = 2n$, $n \geq 1$. Then

$$\sum_{i=1}^{3} a_i = a_1 + a_2 + a_3 = 2 + 4 + 6 = 12,$$

$$\prod_{i=1}^{3} a_i = a_1 \cdot a_2 \cdot a_3 = 2 \cdot 4 \cdot 6 = 48.$$

◀

**Example 3.2.19** The geometric sum $a + ar + ar^2 + \cdots + ar^n$ can be rewritten compactly using the sum notation as $\sum_{i=0}^{n} ar^i$. ◀

It is sometimes useful to change not only the name of the index, but to change its limits as well. (The process is analogous to changing the variable in an integral in calculus.)

**Example 3.2.20** **Changing the Index and Limits in a Sum** Rewrite the sum $\sum_{i=0}^{n} ir^{n-i}$, replacing the index $i$ by $j$, where $i = j - 1$.

**SOLUTION** Since $i = j - 1$, the term $ir^{n-i}$ becomes

$$(j-1)r^{n-(j-1)} = (j-1)r^{n-j+1}.$$

Since $j = i + 1$, when $i = 0$, $j = 1$. Thus the lower limit for $j$ is 1. Similarly, when $i = n$, $j = n + 1$, and the upper limit for $j$ is $n + 1$. Therefore,

$$\sum_{i=0}^{n} ir^{n-i} = \sum_{j=1}^{n+1} (j-1)r^{n-j+1}.$$

◀

**Example 3.2.21** Let $a$ be the sequence defined by the rule $a_i = 2(-1)^i$, $i \geq 0$. Find a formula for the sequence $s$ defined by $s_n = \sum_{i=0}^{n} a_i$.

**SOLUTION** We find that

$$s_n = 2(-1)^0 + 2(-1)^1 + 2(-1)^2 + \cdots + 2(-1)^n$$

$$= 2 - 2 + 2 - \cdots \pm 2 = \begin{cases} 2 & \text{if } n \text{ is even} \\ 0 & \text{if } n \text{ is odd.} \end{cases}$$

◀

Sometimes the sum and product notations are modified to denote sums and products indexed over arbitrary sets of integers. Formally, if $S$ is a finite set of integers and $a$ is a sequence, $\sum_{i \in S} a_i$ denotes the sum of the elements $\{a_i \mid i \in S\}$. Similarly, $\prod_{i \in S} a_i$ denotes the product of the elements $\{a_i \mid i \in S\}$.

**Example 3.2.22** If $S$ denotes the set of prime numbers less than 20,

$$\sum_{i \in S} \frac{1}{i} = \frac{1}{2} + \frac{1}{3} + \frac{1}{5} + \frac{1}{7} + \frac{1}{11} + \frac{1}{13} + \frac{1}{17} + \frac{1}{19} = 1.455.$$ ◀

A *string* is a finite sequence of characters. In programming languages, strings can be used to denote text. For example, in Java

```
"Let's read Rolling Stone."
```

denotes the string consisting of the sequence of characters

Let's read Rolling Stone.

(The double quotes " mark the start and end of the string.)

Within a computer, *bit strings* (strings of 0's and 1's) represent data and instructions to execute. As we will see in Section 5.2, the bit string 101111 represents the number 47.

**Definition 3.2.23 ▶** A *string over* $X$, where $X$ is a finite set, is a finite sequence of elements from $X$. ◀

**Example 3.2.24** Let $X = \{a, b, c\}$. If we let

$$\beta_1 = b, \quad \beta_2 = a, \quad \beta_3 = a, \quad \beta_4 = c,$$

we obtain a string over $X$. This string is written $baac$. ◀

Since a string is a sequence, order is taken into account. For example, the string $baac$ is different from the string $acab$.

Repetitions in a string can be specified by superscripts. For example, the string $bbaaac$ may be written $b^2a^3c$.

The string with no elements is called the **null string** and is denoted $\lambda$. We let $X^*$ denote the set of all strings over $X$, including the null string, and we let $X^+$ denote the set of all nonnull strings over $X$.

**Example 3.2.25** Let $X = \{a, b\}$. Some elements in $X^*$ are $\lambda$, $a$, $b$, $abab$, and $b^{20}a^5ba$. ◀

The **length** of a string $\alpha$ is the number of elements in $\alpha$. The length of $\alpha$ is denoted $|\alpha|$.

**Example 3.2.26** If $\alpha = aabab$ and $\beta = a^3b^4a^{32}$, then $|\alpha| = 5$ and $|\beta| = 39$. ◀

If $\alpha$ and $\beta$ are two strings, the string consisting of $\alpha$ followed by $\beta$, written $\alpha\beta$, is called the **concatenation** of $\alpha$ and $\beta$.

**Example 3.2.27** If $\gamma = aab$ and $\theta = cabd$, then

$$\gamma\theta = aabcabd, \quad \theta\gamma = cabdaab, \quad \gamma\lambda = \gamma = aab, \quad \lambda\gamma = \gamma = aab.$$ ◀

**Example 3.2.28** Let $X = \{a, b, c\}$. If we define $f(\alpha, \beta) = \alpha\beta$, where $\alpha$ and $\beta$ are strings over $X$, then $f$ is a binary operator on $X^*$. ◀

A **substring** of a string $\alpha$ is obtained by selecting some or all consecutive elements of $\alpha$. The formal definition follows.

**Definition 3.2.29 ▶** A string $\beta$ is a *substring* of the string $\alpha$ if there are strings $\gamma$ and $\delta$ with $\alpha = \gamma\beta\delta$. ◀

**Example 3.2.30** The string $\beta = add$ is a substring of the string $\alpha = aaaddad$ since, if we take $\gamma = aa$ and $\delta = ad$, we have $\alpha = \gamma\beta\delta$. Note that if $\beta$ is a substring of $\alpha$, $\gamma$ is the part of $\alpha$ that precedes $\beta$ (in $\alpha$), and $\delta$ is the part of $\alpha$ that follows $\beta$ (in $\alpha$). ◀

**Example 3.2.31** Let $X = \{a, b\}$. If $\alpha \in X^*$, let $\alpha^R$ denote $\alpha$ written in reverse. For example, if $\alpha = abb$, $\alpha^R = bba$. Define a function from $X^*$ to $X^*$ as $f(\alpha) = \alpha^R$. Prove that $f$ is a bijection.

SOLUTION We must show that $f$ is one-to-one and onto $X^*$. We first show that $f$ is one-to-one. We must show that if $f(\alpha) = f(\beta)$, then $\alpha = \beta$. So suppose that $f(\alpha) = f(\beta)$. Using the definition of $f$, we have $\alpha^R = \beta^R$. Reversing each side, we find that $\alpha = \beta$. Therefore, $f$ is one-to-one.

Next we show that $f$ is onto $X^*$. We must show that if $\beta \in X^*$, there exists $\alpha \in X^*$ such that $f(\alpha) = \beta$. So suppose that $\beta \in X^*$. If we let $\alpha = \beta^R$, we have

$$f(\alpha) = \alpha^R = (\beta^R)^R = \beta$$

since if we twice reverse a string, we obtain the original string. Therefore, $f$ is onto $X^*$. We have proved that $f$ is a bijection. ◀

**Example 3.2.32** Let $X = \{a, b\}$. Define a function from $X^* \times X^*$ to $X^*$ as $f(\alpha, \beta) = \alpha\beta$. Is $f$ one-to-one? Is $f$ onto $X^*$?

SOLUTION We try to prove that $f$ is one-to-one. If we succeed, this part of the example is complete. If we fail, we may learn how to construct a counterexample. So suppose that $f(\alpha_1, \beta_1) = f(\alpha_2, \beta_2)$. We have to prove that $\alpha_1 = \beta_1$ and $\alpha_2 = \beta_2$. Using the definition of $f$, we have $\alpha_1\beta_1 = \alpha_2\beta_2$. Can we conclude that $\alpha_1 = \beta_1$ and $\alpha_2 = \beta_2$? No! It is possible to concatenate different strings and produce the same string. For example, $baa = \alpha_1\beta_1$ if we set $\alpha_1 = b$ and $\beta_1 = aa$. Also, $baa = \alpha_2\beta_2$ if we set $\alpha_2 = ba$ and $\beta_2 = a$. Therefore $f$ is *not* one-to-one. We could write up this part of the solution as follows.

If we set $\alpha_1 = b$, $\beta_1 = aa$, $\alpha_2 = ba$, and $\beta_2 = a$, then $f(\alpha_1, \beta_1) = baa = f(\alpha_2, \beta_2)$. Since $\alpha_1 \neq \alpha_2$, $f$ is not one-to-one.

The function $f$ is onto $X^*$ if given any string $\gamma \in X^*$, there exist $(\alpha, \beta) \in X^* \times X^*$ such that $f(\alpha, \beta) = \gamma$. In words, $f$ is onto $X^*$ if every string in $X^*$ is the concatenation of two strings, each in $X^*$. Since concatenating a string $\alpha$ with the null string $\lambda$ docs not change $\alpha$, every string in $X^*$ is the concatenation of two strings, each in $X^*$. This part of the solution could be written up as follows.

Let $\alpha \in X^*$. Then $f(\alpha, \lambda) = \alpha\lambda = \alpha$. Therefore $f$ is onto $X^*$. ◀

## 3.2 Problem-Solving Tips

A sequence is a special type of function; the domain is a set of integers. If $a_1, a_2, \ldots$ is a sequence, the numbers $1, 2, \ldots$ are called indexes. Index 1 identifies the first element of the sequence $a_1$; index 2 identifies the second element of the sequence $a_2$; and so on.

In this book, "increasing sequence" means *strictly* increasing; that is, the sequence $a$ is increasing if for all $i$ and $j$ in the domain of $a$, if $i < j$, then $a_i < a_j$. We require that $a_i$ is strictly less than (*not* less than or equal to) $a_j$. Allowing equality yields what we call in this book a "nondecreasing sequence." That is, the sequence $a$ is nondecreasing if for all $i$ and $j$ in the domain of $a$, if $i < j$, then $a_i \leq a_j$. Similar remarks apply to decreasing sequences and nonincreasing sequences.

## 3.2 Review Exercises

1. Define *sequence*.
2. What is an index in a sequence?
3. Define *increasing sequence*.
4. Define *decreasing sequence*.
5. Define *nonincreasing sequence*.
6. Define *nondecreasing sequence*.
7. Define *subsequence*.
8. What is $\sum_{i=m}^{n} a_i$?
9. What is $\prod_{i=m}^{n} a_i$?
10. Define *string*.
11. Define *null string*.
12. If $X$ is a finite set, what is $X^*$?
13. If $X$ is a finite set, what is $X^+$?
14. Define *length of a string*. How is the length of the string $\alpha$ denoted?
15. Define *concatenation of strings*. How is the concatenation of strings $\alpha$ and $\beta$ denoted?
16. Define *substring*.

## 3.2 Exercises

*Answer 1–3 for the sequence* $\{s_n\}_{n=1}^{6}$ *defined by*

$$c, d, d, c, d, c.$$

1. Find $s_1$.
2. Find $s_4$.
3. Write $s$ as a string.
4. Let $s$ be a sequence whose domain $D$ is a set of consecutive integers. Prove that if $s_i < s_{i+1}$ for all $i$ for which $i$ and $i+1$ are in $D$, then $s$ is increasing. *Hint:* Let $i \in D$. Use induction on $j$ to show that $s_i < s_j$ for all $j$, $i < j$ and $j \in D$.

*In Exercises 5–9, tell whether the sequence s defined by* $s_n = 2^n - n^2$ *is*

(a) *increasing*
(b) *decreasing*
(c) *nonincreasing*
(d) *nondecreasing*

*for the given domain D.*

5. $D = \{0, 1\}$
6. $D = \{0, 1, 2, 3\}$
7. $D = \{1, 2, 3\}$
8. $D = \{1, 2, 3, 4\}$
9. $D = \{n \mid n \in \mathbf{Z}, n \geq 3\}$

*Answer 10–22 for the sequence t defined by*

$$t_n = 2n - 1, \qquad n \geq 1.$$

10. Find $t_3$.
11. Find $t_7$.
12. Find $t_{100}$.
13. Find $t_{2077}$.
14. Find $\sum_{i=1}^{3} t_i$.
15. Find $\sum_{i=3}^{7} t_i$.
16. Find $\prod_{i=1}^{3} t_i$.
17. Find $\prod_{i=3}^{6} t_i$.
18. Find a formula that represents this sequence as a sequence whose lower index is 0.
19. Is $t$ increasing?
20. Is $t$ decreasing?
21. Is $t$ nonincreasing?
22. Is $t$ nondecreasing?

*Answer 23–30 for the sequence v defined by*

$$v_n = n! + 2, \qquad n \geq 1.$$

23. Find $v_3$.
24. Find $v_4$.
25. Find $\sum_{i=1}^{4} v_i$.
26. Find $\sum_{i=3}^{3} v_i$.
27. Is $v$ increasing?
28. Is $v$ decreasing?
29. Is $v$ nonincreasing?
30. Is $v$ nondecreasing?

*Answer 31–36 for the sequence*

$$q_1 = 8, \quad q_2 = 12, \quad q_3 = 12, \quad q_4 = 28, \quad q_5 = 33.$$

31. Find $\sum_{i=2}^{4} q_i$.
32. Find $\sum_{k=2}^{4} q_k$.
33. Is $q$ increasing?
34. Is $q$ decreasing?
35. Is $q$ nonincreasing?
36. Is $q$ nondecreasing?

*Answer 37–40 for the sequence*

$$\tau_0 = 5, \quad \tau_2 = 5.$$

37. Is $\tau$ increasing?
38. Is $\tau$ decreasing?
39. Is $\tau$ nonincreasing?
40. Is $\tau$ nondecreasing?

*Answer 41–44 for the sequence*

$$\Upsilon_2 = 5.$$

41. Is $\Upsilon$ increasing?
42. Is $\Upsilon$ decreasing?

**43.** Is $\Upsilon$ nonincreasing?

**44.** Is $\Upsilon$ nondecreasing?

*Answer 45–56 for the sequence a defined by*

$$a_n = n^2 - 3n + 3, \qquad n \geq 1.$$

**45.** Find $\sum_{i=1}^{4} a_i$.

**46.** Find $\sum_{j=3}^{5} a_j$.

**47.** Find $\sum_{i=4}^{4} a_i$.

**48.** Find $\sum_{k=1}^{6} a_k$.

**49.** Find $\prod_{i=1}^{2} a_i$.

**50.** Find $\prod_{i=1}^{3} a_i$.

**51.** Find $\prod_{n=2}^{3} a_n$.

**52.** Find $\prod_{x=3}^{4} a_x$.

**53.** Is $a$ increasing?

**54.** Is $a$ decreasing?

**55.** Is $a$ nonincreasing?

**56.** Is $a$ nondecreasing?

*Answer 57–64 for the sequence b defined by* $b_n = n(-1)^n$, $n \geq 1$.

**57.** Find $\sum_{i=1}^{4} b_i$.

**58.** Find $\sum_{i=1}^{10} b_i$.

**59.** Find a formula for the sequence $c$ defined by

$$c_n = \sum_{i=1}^{n} b_i.$$

**60.** Find a formula for the sequence $d$ defined by

$$d_n = \prod_{i=1}^{n} b_i.$$

**61.** Is $b$ increasing?

**62.** Is $b$ decreasing?

**63.** Is $b$ nonincreasing?

**64.** Is $b$ nondecreasing?

*Answer 65–72 for the sequence* $\Omega$ *defined by* $\Omega_n = 3$ *for all n.*

**65.** Find $\sum_{i=1}^{3} \Omega_i$.

**66.** Find $\sum_{i=1}^{10} \Omega_i$.

**67.** Find a formula for the sequence $c$ defined by

$$c_n = \sum_{i=1}^{n} \Omega_i.$$

**68.** Find a formula for the sequence $d$ defined by

$$d_n = \prod_{i=1}^{n} \Omega_i.$$

**69.** Is $\Omega$ increasing?

**70.** Is $\Omega$ decreasing?

**71.** Is $\Omega$ nonincreasing?

**72.** Is $\Omega$ nondecreasing?

*Answer 73–79 for the sequence x defined by*

$$x_1 = 2, \quad x_n = 3 + x_{n-1}, \qquad n \geq 2.$$

**73.** Find $\sum_{i=1}^{3} x_i$.

**74.** Find $\sum_{i=1}^{10} x_i$.

**75.** Find a formula for the sequence $c$ defined by

$$c_n = \sum_{i=1}^{n} x_i.$$

**76.** Is $x$ increasing?

**77.** Is $x$ decreasing?

**78.** Is $x$ nonincreasing?

**79.** Is $x$ nondecreasing?

*Answer 80–87 for the sequence w defined by*

$$w_n = \frac{1}{n} - \frac{1}{n+1}, \qquad n \geq 1.$$

**80.** Find $\sum_{i=1}^{3} w_i$.

**81.** Find $\sum_{i=1}^{10} w_i$.

**82.** Find a formula for the sequence $c$ defined by

$$c_n = \sum_{i=1}^{n} w_i.$$

**83.** Find a formula for the sequence $d$ defined by

$$d_n = \prod_{i=1}^{n} w_i.$$

**84.** Is $w$ increasing?

**85.** Is $w$ decreasing?

**86.** Is $w$ nonincreasing?

**87.** Is $w$ nondecreasing?

*Answer 88–100 for the sequence a defined by*

$$a_n = \frac{n-1}{n^2(n-2)^2}, \qquad n \geq 3$$

*and the sequence z defined by* $z_n = \sum_{i=3}^{n} a_i$.

**88.** Find $a_3$.

**89.** Find $a_4$.

**90.** Find $z_3$.

**91.** Find $z_4$.

★**92.** Find $z_{100}$. *Hint:* Show that

$$a_n = \frac{1}{4}\left[\frac{1}{(n-2)^2} - \frac{1}{n^2}\right]$$

and use this form in the sum. Write out $a_3 + a_4 + a_5 + a_6$ to see what is going on.

**93.** Is $a$ increasing?

★**94.** Is $a$ decreasing?

★**95.** Is $a$ nonincreasing?

**96.** Is $a$ nondecreasing?

**97.** Is $z$ increasing?

**98.** Is $z$ decreasing?

**99.** Is $z$ nonincreasing?

**100.** Is $z$ nondecreasing?

*Let X be the set of positive integers that are not perfect squares. (A perfect square m is an integer of the form* $m = i^2$ *where i is an integer.) Exercises 101–107 concern the sequence s from X to* **Z** *defined as follows. If* $n \in X$, *let* $s_n$ *be the least integer* $a_k$ *for* *which there exist integers $a_1, \ldots, a_k$ with $n < a_1 < a_2 < \cdots < a_k$ such that $n \cdot a_1 \cdots a_k$ is a perfect square. As an example, consider $s_2$. None of $2 \cdot 3, 2 \cdot 4, 2 \cdot 3 \cdot 4$ is a perfect square, so $s_2 \neq 3$ and $s_2 \neq 4$. If $s_2 = 5$, $2 \cdot a_1 \cdots a_k \cdot 5$ would be a perfect square for some $a_1, \ldots, a_k$, $2 < a_1 < a_2 < \cdots < a_k < 5$. But then one of the $a_i$ would be a multiple of 5, which is impossible. Therefore, $s_2 \neq 5$. However, $2 \cdot 3 \cdot 6$ is a perfect square, so $s_2 = 6$.*

**101.** Show that $s_n$ is defined for every $n \in X$, that is, that for any $n \in X$, there exist integers $a_1, \ldots, a_k$ with $n < a_1 < a_2 < \cdots < a_k$ such that $n \cdot a_1 \cdots a_k$ is a perfect square.

**102.** Show that $s_n \leq 4n$ for all $n \in X$.

**103.** Find $s_3$.

**104.** Find $s_5$.

**105.** Find $s_6$.

**★106.** Prove that if $p$ is a prime, $s_p = 2p$ for all $p \geq 5$.

**107.** Prove that $s$ is not increasing.

**108.** Let $u$ be the sequence defined by

$$u_1 = 3, \quad u_n = 3 + u_{n-1}, \qquad n \geq 2.$$

Find a formula for the sequence $d$ defined by

$$d_n = \prod_{i=1}^{n} u_i.$$

*Exercises 109–112 refer to the sequence $\{s_n\}$ defined by the rule*

$$s_n = 2n - 1, \qquad n \geq 1.$$

**109.** List the first seven terms of $s$.

*Answer 110–112 for the subsequence of s obtained by taking the first, third, fifth, ... terms.*

**110.** List the first seven terms of the subsequence.

**111.** Find a formula for the expression $n_k$ as described before Example 3.2.15.

**112.** Find a formula for the $k$th term of the subsequence.

*Exercises 113–116 refer to the sequence $\{t_n\}$ defined by the rule*

$$t_n = 2^n, \qquad n \geq 1.$$

**113.** List the first seven terms of $t$.

*Answer 114–116 for the subsequence of t obtained by taking the first, second, fourth, seventh, eleventh, ... terms.*

**114.** List the first seven terms of the subsequence.

**115.** Find a formula for the expression $n_k$ as described before Example 3.2.15.

**116.** Find a formula for the $k$th term of the subsequence.

*Answer 117–120 using the sequences y and z defined by*

$$y_n = 2^n - 1, \qquad z_n = n(n-1).$$

**117.** Find $\left(\sum_{i=1}^{3} y_i\right)\left(\sum_{i=1}^{3} z_i\right)$.

**118.** Find $\left(\sum_{i=1}^{5} y_i\right)\left(\sum_{i=1}^{4} z_i\right)$.

**119.** Find $\sum_{i=1}^{3} y_i z_i$.

**120.** Find $\left(\sum_{i=3}^{4} y_i\right)\left(\prod_{i=2}^{4} z_i\right)$.

*Answer 121–128 for the sequence r defined by*

$$r_n = 3 \cdot 2^n - 4 \cdot 5^n, \qquad n \geq 0.$$

**121.** Find $r_0$.

**122.** Find $r_1$.

**123.** Find $r_2$.

**124.** Find $r_3$.

**125.** Find a formula for $r_p$.

**126.** Find a formula for $r_{n-1}$.

**127.** Find a formula for $r_{n-2}$.

**128.** Prove that $\{r_n\}$ satisfies

$$r_n = 7r_{n-1} - 10r_{n-2}, \qquad n \geq 2.$$

*Answer 129–136 for the sequence z defined by*

$$z_n = (2+n)3^n, \qquad n \geq 0.$$

**129.** Find $z_0$.

**130.** Find $z_1$.

**131.** Find $z_2$.

**132.** Find $z_3$.

**133.** Find a formula for $z_i$.

**134.** Find a formula for $z_{n-1}$.

**135.** Find a formula for $z_{n-2}$.

**136.** Prove that $\{z_n\}$ satisfies

$$z_n = 6z_{n-1} - 9z_{n-2}, \qquad n \geq 2.$$

**137.** Find $b_n$, $n = 1, \ldots, 6$, where

$$b_n = n + (n-1)(n-2)(n-3)(n-4)(n-5).$$

**138.** Rewrite the sum

$$\sum_{i=1}^{n} i^2 r^{n-i},$$

replacing the index $i$ by $k$, where $i = k + 1$.

**139.** Rewrite the sum

$$\sum_{k=1}^{n} C_{k-1} C_{n-k},$$

replacing the index $k$ by $i$, where $k = i + 1$.

**140.** Let $a$ and $b$ be sequences, and let

$$s_k = \sum_{i=1}^{k} a_i.$$

Prove that

$$\sum_{k=1}^{n} a_k b_k = \sum_{k=1}^{n} s_k(b_k - b_{k+1}) + s_n b_{n+1}.$$

This equation, known as the *summation-by-parts formula*, is the discrete analog of the integration-by-parts formula in calculus.

**141.** Sometimes we generalize the notion of sequence as defined in this section by allowing more general indexing. Suppose that $\{a_{ij}\}$ is a sequence indexed over pairs of positive integers. Prove that

$$\sum_{i=1}^{n}\left(\sum_{j=i}^{n} a_{ij}\right) = \sum_{j=1}^{n}\left(\sum_{i=1}^{j} a_{ij}\right).$$

**142.** Compute the given quantity using the strings

$$\alpha = baab, \qquad \beta = caaba, \qquad \gamma = bbab.$$

(a) $\alpha\beta$ (b) $\beta\alpha$ (c) $\alpha\alpha$
(d) $\beta\beta$ (e) $|\alpha\beta|$ (f) $|\beta\alpha|$
(g) $|\alpha\alpha|$ (h) $|\beta\beta|$ (i) $\alpha\lambda$
(j) $\lambda\beta$ (k) $\alpha\beta\gamma$ (l) $\beta\beta\gamma\alpha$

**143.** List all strings over $X = \{0, 1\}$ of length 2.

**144.** List all strings over $X = \{0, 1\}$ of length 2 or less.

**145.** List all strings over $X = \{0, 1\}$ of length 3.

**146.** List all strings over $X = \{0, 1\}$ of length 3 or less.

**147.** Find all substrings of the string *babc*.

**148.** Find all substrings of the string *aabaabb*.

**149.** Use induction to prove that

$$\sum \frac{1}{n_1 \cdot n_2 \cdots n_k} = n,$$

for all $n \geq 1$, where the sum is taken over all nonempty subsets $\{n_1, n_2, \ldots, n_k\}$ of $\{1, 2, \ldots, n\}$.

**150.** Suppose that the sequence $\{a_n\}$ satisfies $a_1 = 0$, $a_2 = 1$, and

$$a_n = (n-1)(a_{n-1} + a_{n-2}) \quad \text{for all } n \geq 3.$$

Use induction to prove that

$$\frac{a_n}{n!} = \sum_{k=0}^{n} \frac{(-1)^k}{k!} \quad \text{for all } n \geq 1.$$

*In Exercises 151–153, $x_1, x_2, \ldots, x_n$, $n \geq 2$, are real numbers satisfying $x_1 < x_2 < \cdots < x_n$, and $x$ is an arbitrary real number.*

**151.** Prove that if $x_1 \leq x \leq x_n$, then

$$\sum_{i=1}^{n} |x - x_i| = \sum_{i=2}^{n-1} |x - x_i| + (x_n - x_1),$$

for all $n \geq 3$.

**152.** Prove that if $x < x_1$ or $x > x_n$, then

$$\sum_{i=1}^{n} |x - x_i| > \sum_{i=2}^{n-1} |x - x_i| + (x_n - x_1),$$

for all $n \geq 3$.

**153.** A *median* of $x_1, \ldots, x_n$ is the middle value of $x_1, \ldots, x_n$ when $n$ is odd, and any value between the two middle values of $x_1, \ldots, x_n$ when $n$ is even. For example, if $x_1 < x_2 < \cdots < x_5$, the median is $x_3$. If $x_1 < x_2 < x_3 < x_4$, a median is any value between $x_2$ and $x_3$, including $x_2$ and $x_3$.

Use Exercises 151 and 152 and mathematical induction to prove that the sum

$$\sum_{i=1}^{n} |x - x_i|, \tag{3.2.9}$$

$n \geq 1$, is minimized when $x$ is equal to a median of $x_1, \ldots, x_n$.

If we repeat an experiment $n$ times and observe the values $x_1, \ldots, x_n$, the sum (3.2.9) can be interpreted as a measure of the error in assuming that the correct value is $x$. This exercise shows that this error is minimized by choosing $x$ to be a median of the values $x_1, \ldots, x_n$. The requested inductive argument is attributed to J. Lancaster.

**154.** Prove that

$$\sum_{i=1}^{n}\sum_{j=1}^{n} (i-j)^2 = \frac{n^2(n^2-1)}{6}.$$

**155.** Let $X = \{a, b\}$. Define a function from $X^*$ to $X^*$ as $f(\alpha) = \alpha ab$. Is $f$ one-to-one? Is $f$ onto $X^*$? Prove your answers.

**156.** Let $X = \{a, b\}$. Define a function from $X^*$ to $X^*$ as $f(\alpha) = \alpha\alpha$. Is $f$ one-to-one? Is $f$ onto $X^*$? Prove your answers.

**157.** Let $X = \{a, b\}$. A *palindrome over X* is a string $\alpha$ for which $\alpha = \alpha^R$ (i.e., a string that reads the same forward and backward). An example of a palindrome over $X$ is *bbaabb*. Define a function from $X^*$ to the set of palindromes over $X$ as $f(\alpha) = \alpha\alpha^R$. Is $f$ one-to-one? Is $f$ onto? Prove your answers.

*Let L be the set of all strings, including the null string, that can be constructed by repeated application of the following rules:*

- *If $\alpha \in L$, then $a\alpha b \in L$ and $b\alpha a \in L$.*
- *If $\alpha \in L$ and $\beta \in L$, then $\alpha\beta \in L$.*

*For example, ab is in L, for if we take $\alpha = \lambda$, then $\alpha \in L$ and the first rule states that $ab = a\alpha b \in L$. Similarly, $ba \in L$. As another example, aabb is in L, for if we take $\alpha = ab$, then $\alpha \in L$; by the first rule, $aabb = a\alpha b \in L$. As a final example, aabbba is in L, for if we take $\alpha = aabb$ and $\beta = ba$, then $\alpha \in L$ and $\beta \in L$; by the second rule, $aabbba = \alpha\beta \in L$.*

**158.** Show that *aaabbb* is in $L$.

**159.** Show that *baabab* is in $L$.

**160.** Show that *aab* is *not* in $L$.

**161.** Prove that if $\alpha \in L$, $\alpha$ has equal numbers of $a$'s and $b$'s.

**★162.** Prove that if $\alpha$ has equal numbers of $a$'s and $b$'s, then $\alpha \in L$.

**163.** Let $\{a_n\}_{n=1}^{\infty}$ be a nondecreasing sequence, which is bounded above, and let $L$ be the least upper bound of the set $\{a_n \mid n = 1, 2, \ldots\}$. Prove that for every real number $\varepsilon > 0$, there exists a positive integer $N$ such that $L - \varepsilon < a_n \leq L$ for every $n \geq N$. In calculus terminology, a nondecreasing sequence, which is bounded above, converges to the limit $L$, where $L$ is the least upper bound of the set of elements of the sequence.

# 3.3 Relations

A **relation** from one set to another can be thought of as a table that lists which elements of the first set relate to which elements of the second set (see Table 3.3.1). Table 3.3.1 shows which students are taking which courses. For example, Bill is taking Computer Science and Art, and Mary is taking Mathematics. In the terminology of relations, we would say that Bill is related to Computer Science and Art, and that Mary is related to Mathematics.

**Go Online**

For more on relations, see bit.ly/2OFm8St

Of course, Table 3.3.1 is really just a set of ordered pairs. Abstractly, we *define* a relation to be a set of ordered pairs. In this setting, we consider the first element of the ordered pair to be related to the second element of the ordered pair.

**TABLE 3.3.1 ■ Relation of Students to Courses**

| *Student* | *Course* |
|---|---|
| Bill | CompSci |
| Mary | Math |
| Bill | Art |
| Beth | History |
| Beth | CompSci |
| Dave | Math |

**Definition 3.3.1 ▶** A *(binary) relation $R$ from a set $X$ to a set $Y$* is a subset of the Cartesian product $X \times Y$. If $(x, y) \in R$, we write $x R y$ and say that *$x$ is related to $y$*. If $X = Y$, we call $R$ a *(binary) relation on $X$*. ◀

A function (see Section 3.1) is a special type of relation. A function $f$ from $X$ to $Y$ is a relation from $X$ to $Y$ having the properties:

(a) The domain of $f$ is equal to $X$.

(b) For each $x \in X$, there is exactly one $y \in Y$ such that $(x, y) \in f$.

**Example 3.3.2** If we let $X = \{\text{Bill, Mary, Beth, Dave}\}$ and $Y = \{\text{CompSci, Math, Art, History}\}$, our relation $R$ of Table 3.3.1 can be written

$$R = \{(\text{Bill, CompSci}), (\text{Mary, Math}), (\text{Bill, Art}), (\text{Beth, History}), (\text{Beth, CompSci}), (\text{Dave, Math})\}.$$

Since $(\text{Beth, History}) \in R$, we may write Beth $R$ History. ◀

Example 3.3.2 shows that a relation can be given by simply specifying which ordered pairs belong to the relation. Our next example shows that sometimes it is possible to define a relation by giving a rule for membership in the relation.

**Example 3.3.3** Let $X = \{2, 3, 4\}$ and $Y = \{3, 4, 5, 6, 7\}$. If we define a relation $R$ from $X$ to $Y$ by

$$(x, y) \in R \qquad \text{if } x \text{ divides } y,$$

we obtain

$$R = \{(2, 4), (2, 6), (3, 3), (3, 6), (4, 4)\}.$$

If we rewrite $R$ as a table, we obtain

| $X$ | $Y$ |
|---|---|
| 2 | 4 |
| 2 | 6 |
| 3 | 3 |
| 3 | 6 |
| 4 | 4 |

◀

**Example 3.3.4** Let $R$ be the relation on $X = \{1, 2, 3, 4\}$ defined by $(x, y) \in R$ if $x \leq y$, $x, y \in X$. Then

$$R = \{(1, 1), (1, 2), (1, 3), (1, 4), (2, 2), (2, 3), (2, 4), (3, 3), (3, 4), (4, 4)\}.$$ ◀

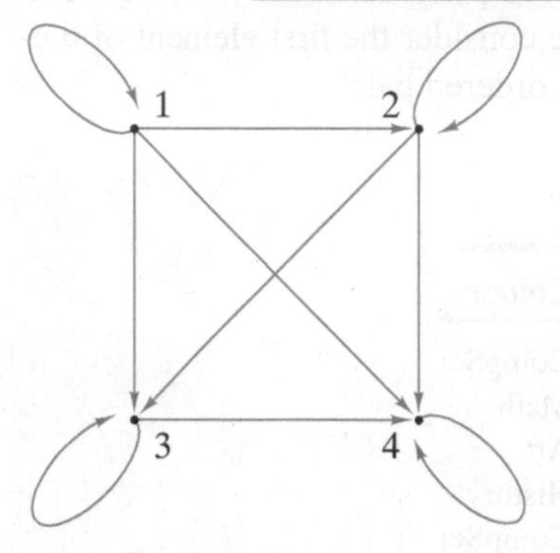


**Figure 3.3.1** The digraph of the relation of Example 3.3.4.

An informative way to picture a relation on a set is to draw its **digraph.** (Digraphs are discussed in more detail in Chapter 8. For now, we mention digraphs only in connection with relations.) To draw the digraph of a relation on a set $X$, we first draw dots or **vertices** to represent the elements of $X$. In Figure 3.3.1, we have drawn four vertices to represent the elements of the set $X$ of Example 3.3.4. Next, if the element $(x, y)$ is in the relation, we draw an arrow (called a **directed edge**) from $x$ to $y$. In Figure 3.3.1, we have drawn directed edges to represent the members of the relation $R$ of Example 3.3.4. Notice that an element of the form $(x, x)$ in a relation corresponds to a directed edge from $x$ to $x$. Such an edge is called a **loop.** There is a loop at every vertex in Figure 3.3.1.

**Example 3.3.5** The relation $R$ on $X = \{a, b, c, d\}$ given by the digraph of Figure 3.3.2 is $R = \{(a, a), (b, c), (c, b), (d, d)\}$.

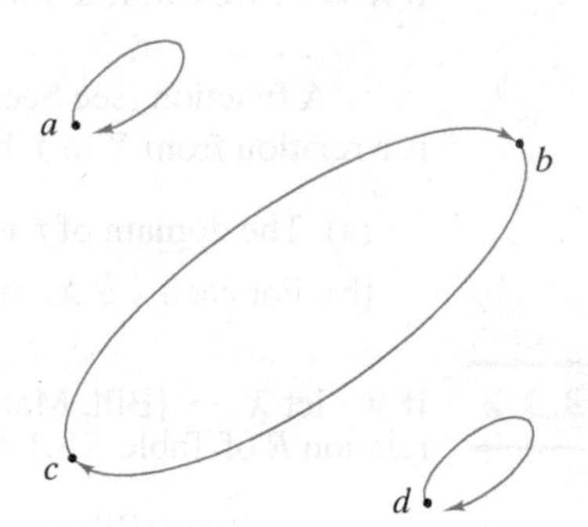


**Figure 3.3.2** The digraph of the relation of Example 3.3.5.

◀

We next define several properties that relations may have.

**Definition 3.3.6** ▶ A relation $R$ on a set $X$ is *reflexive* if $(x, x) \in R$ for every $x \in X$. ◀

**Example 3.3.7** The relation $R$ on $X = \{1, 2, 3, 4\}$ defined by $(x, y) \in R$ if $x \leq y$, $x, y \in X$, is reflexive because for each element $x \in X$, $(x, x) \in R$; specifically, $(1, 1)$, $(2, 2)$, $(3, 3)$, and $(4, 4)$ are each in $R$. The digraph of a reflexive relation has a loop at every vertex. Notice that the digraph of this relation (see Figure 3.3.1) has a loop at every vertex. ◀

By the generalized De Morgan's laws for logic (Theorem 1.5.14), a relation $R$ on $X$ is *not* reflexive if there exists $x \in X$ such that $(x, x) \notin R$.

**Example 3.3.8** The relation $R = \{(a, a), (b, c), (c, b), (d, d)\}$ on $X = \{a, b, c, d\}$ is not reflexive. For example, $b \in X$, but $(b, b) \notin R$. That this relation is not reflexive can also be seen by looking at its digraph (see Figure 3.3.2); vertex $b$ does not have a loop. ◀

**Definition 3.3.9 ▶** A relation $R$ on a set $X$ is *symmetric* if for all $x, y \in X$, if $(x, y) \in R$, then $(y, x) \in R$. ◀

**Example 3.3.10** The relation $R = \{(a, a), (b, c), (c, b), (d, d)\}$ on $X = \{a, b, c, d\}$ is symmetric because for all $x, y$, if $(x, y) \in R$, then $(y, x) \in R$. For example, $(b, c)$ is in $R$ and $(c, b)$ is also in $R$. The digraph of a symmetric relation has the property that whenever there is a directed edge from $v$ to $w$, there is also a directed edge from $w$ to $v$. Notice that the digraph of this relation (see Figure 3.3.2) has the property that for every directed edge from $v$ to $w$, there is also a directed edge from $w$ to $v$. ◀

In symbols, a relation $R$ is symmetric if

$$\forall x \forall y[(x, y) \in R] \rightarrow [(y, x) \in R].$$

Thus $R$ is *not* symmetric if

$$\neg[\forall x \forall y[(x, y) \in R] \rightarrow [(y, x) \in R]]. \tag{3.3.1}$$

Using the generalized De Morgan's laws for logic (Theorem 1.5.14) and the fact that $\neg(p \rightarrow q) \equiv p \wedge \neg q$ (see Example 1.3.13), we find that (3.3.1) is equivalent to

$$\exists x \exists y[[(x, y) \in R] \wedge \neg[(y, x) \in R]]$$

or, equivalently,

$$\exists x \exists y[[(x, y) \in R] \wedge [(y, x) \notin R]].$$

In words, a relation $R$ is not symmetric if there exist $x$ and $y$ such that $(x, y)$ is in $R$ and $(y, x)$ is not in $R$.

**Example 3.3.11** The relation $R$ on $X = \{1, 2, 3, 4\}$ defined by $(x, y) \in R$ if $x \leq y$, $x, y \in X$, is not symmetric. For example, $(2, 3) \in R$, but $(3, 2) \notin R$. The digraph of this relation (see Figure 3.3.1) has a directed edge from 2 to 3, but there is no directed edge from 3 to 2. ◀

**Definition 3.3.12 ▶** A relation $R$ on a set $X$ is *antisymmetric* if for all $x, y \in X$, if $(x, y) \in R$ and $(y, x) \in R$, then $x = y$. ◀

**Example 3.3.13** The relation $R$ on $X = \{1, 2, 3, 4\}$ defined by $(x, y) \in R$ if $x \leq y$, $x, y \in X$, is antisymmetric because for all $x, y$, if $(x, y) \in R$ (i.e., $x \leq y$) and $(y, x) \in R$ (i.e., $y \leq x$), then $x = y$. ◀

**Example 3.3.14** It is sometimes more convenient to replace

$$\text{if } (x, y) \in R \text{ and } (y, x) \in R, \text{ then } x = y$$

in the definition of "antisymmetric" (Definition 3.3.12) with its logically equivalent contrapositive (see Theorem 1.3.18)

$$\text{if } x \neq y, \text{ then } (x, y) \notin R \text{ or } (y, x) \notin R$$

to obtain a logically equivalent characterization of "antisymmetric": A relation $R$ on a set $X$ is antisymmetric if for all $x, y \in X$, if $x \neq y$, then $(x, y) \notin R$ or $(y, x) \notin R$.

Using this equivalent definition of "antisymmetric," we again see that the relation $R$ on $X = \{1, 2, 3, 4\}$ defined by $(x, y) \in R$ if $x \leq y$, $x, y \in X$, is antisymmetric because for all $x, y$, if $x \neq y$, $(x, y) \notin R$ (i.e., $x > y$) or $(y, x) \notin R$ (i.e., $y > x$).

The equivalent characterization of "antisymmetric" translates for digraphs as follows. The digraph of an antisymmetric relation has the property that between any two distinct vertices there is at most one directed edge. Notice that the digraph of the relation $R$ in the previous paragraph (see Figure 3.3.1) has at most one directed edge between each pair of vertices. ◀

**Example 3.3.15** If a relation has *no* members of the form $(x, y)$, $x \neq y$, we see that the equivalent characterization of "antisymmetric"

$$\text{for all } x, y \in X, \text{ if } x \neq y, \text{ then } (x, y) \notin R \text{ or } (y, x) \notin R$$

(see Example 3.3.14) is trivially true (since the hypothesis $x \neq y$ is always false). Thus if a relation $R$ has *no* members of the form $(x, y)$, $x \neq y$, $R$ is antisymmetric. For example, $R = \{(a, a), (b, b), (c, c)\}$ on $X = \{a, b, c\}$ is antisymmetric. The digraph of $R$ shown in Figure 3.3.3 has at most one directed edge between each pair of distinct vertices. Notice that $R$ is also reflexive and symmetric. This example shows that "antisymmetric" is not the same as "not symmetric" because this relation is in fact both symmetric and antisymmetric. ◀

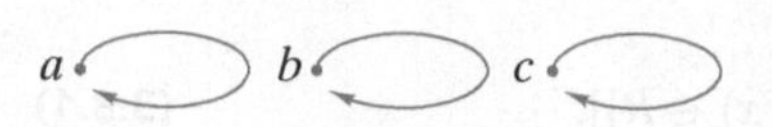


**Figure 3.3.3** The digraph of the relation of Example 3.3.15.

In symbols, a relation $R$ is antisymmetric if

$$\forall x \forall y [(x, y) \in R \wedge (y, x) \in R] \rightarrow [x = y].$$

Thus $R$ is *not* antisymmetric if

$$\neg[\forall x \forall y [(x, y) \in R \wedge (y, x) \in R] \rightarrow [x = y]]. \tag{3.3.2}$$

Using the generalized De Morgan's laws for logic (Theorem 1.5.14) and the fact that $\neg(p \rightarrow q) \equiv p \wedge \neg q$ (see Example 1.3.13), we find that (3.3.2) is equivalent to

$$\exists x \exists y [(x, y) \in R \wedge (y, x) \in R] \wedge \neg[x = y]]$$

which, in turn, is equivalent to

$$\exists x \exists y [(x, y) \in R \wedge (y, x) \in R \wedge (x \neq y)].$$

In words, a relation $R$ is not antisymmetric if there exist $x$ and $y$, $x \neq y$, such that $(x, y)$ and $(y, x)$ are both in $R$.

**Example 3.3.16** The relation $R = \{(a, a), (b, c), (c, b), (d, d)\}$ on $X = \{a, b, c, d\}$ is not antisymmetric because both $(b, c)$ and $(c, b)$ are in $R$. Notice that in the digraph of this relation (see Figure 3.3.2) there are two directed edges between $b$ and $c$. ◀

**Definition 3.3.17 ▶** A relation $R$ on a set $X$ is *transitive* if for all $x, y, z \in X$, if $(x, y)$ and $(y, z) \in R$, then $(x, z) \in R$. ◀

**Example 3.3.18** The relation $R$ on $X = \{1, 2, 3, 4\}$ defined by $(x, y) \in R$ if $x \leq y$, $x, y \in X$, is transitive because for all $x, y, z$, if $(x, y)$ and $(y, z) \in R$, then $(x, z) \in R$. To formally verify that this relation satisfies Definition 3.3.17, we can list all pairs of the form $(x, y)$ and $(y, z)$ in $R$ and then verify that in every case, $(x, z) \in R$:

| *Pairs of Form* | | | *Pairs of Form* | | |
|---|---|---|---|---|---|
| $(x, y)$ | $(y, z)$ | $(x, z)$ | $(x, y)$ | $(y, z)$ | $(x, z)$ |
| (1, 1) | (1, 1) | (1, 1) | (2, 2) | (2, 2) | (2, 2) |
| (1, 1) | (1, 2) | (1, 2) | (2, 2) | (2, 3) | (2, 3) |
| (1, 1) | (1, 3) | (1, 3) | (2, 2) | (2, 4) | (2, 4) |
| (1, 1) | (1, 4) | (1, 4) | (2, 3) | (3, 3) | (2, 3) |
| (1, 2) | (2, 2) | (1, 2) | (2, 3) | (3, 4) | (2, 4) |
| (1, 2) | (2, 3) | (1, 3) | (2, 4) | (4, 4) | (2, 4) |
| (1, 2) | (2, 4) | (1, 4) | (3, 3) | (3, 3) | (3, 3) |
| (1, 3) | (3, 3) | (1, 3) | (3, 3) | (3, 4) | (3, 4) |
| (1, 3) | (3, 4) | (1, 4) | (3, 4) | (4, 4) | (3, 4) |
| (1, 4) | (4, 4) | (1, 4) | (4, 4) | (4, 4) | (4, 4) |

Actually, some of the entries in the preceding table were unnecessary. If $x = y$ or $y = z$, we need not explicitly verify that the condition

$$\text{if } (x, y) \text{ and } (y, z) \in R, \text{ then } (x, z) \in R$$

is satisfied since it will automatically be true. Suppose, for example, that $x = y$ and $(x, y)$ and $(y, z)$ are in $R$. Since $x = y$, $(x, z) = (y, z)$ is in $R$ and the condition is satisfied. Eliminating the cases $x = y$ and $y = z$ leaves only the following to be explicitly checked to verify that the relation is transitive:

| *Pairs of Form* | | |
|---|---|---|
| $(x, y)$ | $(y, z)$ | $(x, z)$ |
| (1, 2) | (2, 3) | (1, 3) |
| (1, 2) | (2, 4) | (1, 4) |
| (1, 3) | (3, 4) | (1, 4) |
| (2, 3) | (3, 4) | (2, 4) |

The digraph of a transitive relation has the property that whenever there are directed edges from $x$ to $y$ and from $y$ to $z$, there is also a directed edge from $x$ to $z$. Notice that the digraph of this relation (see Figure 3.3.1) has this property. ◀

In symbols, a relation $R$ is transitive if

$$\forall x \forall y \forall z [(x, y) \in R \wedge (y, z) \in R] \rightarrow [(x, z) \in R].$$

Thus $R$ is *not* transitive if

$$\neg[\forall x \forall y \forall z [(x, y) \in R \wedge (y, z) \in R] \rightarrow [(x, z) \in R]]. \quad \textbf{(3.3.3)}$$

Using the generalized De Morgan's laws for logic (Theorem 1.5.14) and the fact that $\neg(p \rightarrow q) \equiv p \wedge \neg q$ (see Example 1.3.13), we find that (3.3.3) is equivalent to

$$\exists x \exists y \exists z [(x, y) \in R \wedge (y, z) \in R] \wedge \neg[(x, z) \in R]$$

or, equivalently,

$$\exists x \exists y \exists z [(x, y) \in R \wedge (y, z) \in R \wedge (x, z) \notin R].$$

In words, a relation $R$ is not transitive if there exist $x$, $y$, and $z$ such that $(x, y)$ and $(y, z)$ are in $R$, but $(x, z)$ is not in $R$.

**Example 3.3.19** The relation $R = \{(a, a), (b, c), (c, b), (d, d)\}$ on $X = \{a, b, c, d\}$ is not transitive. For example, $(b, c)$ and $(c, b)$ are in $R$, but $(b, b)$ is not in $R$. Notice that in the digraph of this relation (see Figure 3.3.2) there are directed edges from $b$ to $c$ and from $c$ to $b$, but there is no directed edge from $b$ to $b$. ◀

Relations can be used to order elements of a set. For example, the relation $R$ defined on the set of integers by

$$(x, y) \in R \qquad \text{if } x \leq y$$

orders the integers. Notice that the relation $R$ is reflexive, antisymmetric, and transitive. Such relations are called **partial orders.**

**Definition 3.3.20 ▶** A relation $R$ on a set $X$ is a *partial order* if $R$ is reflexive, antisymmetric, and transitive. ◀

**Example 3.3.21** Since the relation $R$ defined on the positive integers by

$$(x, y) \in R \qquad \text{if } x \text{ divides } y$$

is reflexive, antisymmetric, and transitive, $R$ is a partial order. ◀

If $R$ is a partial order on a set $X$, the notation $x \preceq y$ is sometimes used to indicate that $(x, y) \in R$. This notation suggests that we are interpreting the relation as an ordering of the elements in $X$.

Suppose that $R$ is a partial order on a set $X$. If $x, y \in X$ and either $x \preceq y$ or $y \preceq x$, we say that $x$ and $y$ are **comparable.** If $x, y \in X$ and $x \not\preceq y$ and $y \not\preceq x$, we say that $x$ and $y$ are **incomparable.** If every pair of elements in $X$ is comparable, we call $R$ a **total order.** The less than or equal to relation on the positive integers is a total order since, if $x$ and $y$ are integers, either $x \leq y$ or $y \leq x$. The reason for the term "partial order" is that in general some elements in $X$ may be incomparable. The "divides" relation on the positive integers (see Example 3.3.21) has both comparable and incomparable elements. For example, 2 and 3 are incomparable (since 2 does not divide 3 and 3 does not divide 2), but 3 and 6 are comparable (since 3 divides 6).

One application of partial orders is to task scheduling.

**Example 3.3.22** **Task Scheduling** Consider the set $T$ of tasks that must be completed in order to take an indoor flash picture with a particular camera.

1. Remove lens cap.
2. Focus camera.
3. Turn off safety lock.
4. Turn on flash unit.
5. Push photo button.

Some of these tasks must be done before others. For example, task 1 must be done before task 2. On the other hand, other tasks can be done in either order. For example, tasks 2 and 3 can be done in either order.

The relation $R$ defined on $T$ by

$$i\,R\,j \quad \text{if } i = j \text{ or task } i \text{ must be done before task } j$$

orders the tasks. We obtain

$$R = \{(1, 1), (2, 2), (3, 3), (4, 4), (5, 5), (1, 2), (1, 5), (2, 5), (3, 5), (4, 5)\}.$$

Since $R$ is reflexive, antisymmetric, and transitive, it is a partial order. A solution to the problem of scheduling the tasks so that we can take a picture is a total ordering of the tasks consistent with the partial order. More precisely, we require a total ordering of the tasks $t_1, t_2, t_3, t_4, t_5$ such that if $t_i \, R \, t_j$, then $i = j$ or $t_i$ precedes $t_j$ in the list. Among the solutions are 1, 2, 3, 4, 5 and 3, 4, 1, 2, 5. ◀

Given a relation $R$ from $X$ to $Y$, we may define a relation from $Y$ to $X$ by reversing the order of each ordered pair in $R$. The inverse relation generalizes the inverse function. The formal definition follows.

**Definition 3.3.23 ▶** Let $R$ be a relation from $X$ to $Y$. The *inverse* of $R$, denoted $R^{-1}$, is the relation from $Y$ to $X$ defined by

$$R^{-1} = \{(y, x) \mid (x, y) \in R\}.$$ ◀

**Example 3.3.24** If we define a relation $R$ from $X = \{2, 3, 4\}$ to $Y = \{3, 4, 5, 6, 7\}$ by

$$(x, y) \in R \quad \text{if } x \text{ divides } y,$$

we obtain

$$R = \{(2, 4), (2, 6), (3, 3), (3, 6), (4, 4)\}.$$

The inverse of this relation is

$$R^{-1} = \{(4, 2), (6, 2), (3, 3), (6, 3), (4, 4)\}.$$

In words, we might describe this relation as "is divisible by." ◀

If we have a relation $R_1$ from $X$ to $Y$ and a relation $R_2$ from $Y$ to $Z$, we can form the composition of the relations by applying first relation $R_1$ and then relation $R_2$. Composition of relations generalizes composition of functions. The formal definition follows.

**Definition 3.3.25 ▶** Let $R_1$ be a relation from $X$ to $Y$ and $R_2$ be a relation from $Y$ to $Z$. The *composition* of $R_1$ and $R_2$, denoted $R_2 \circ R_1$, is the relation from $X$ to $Z$ defined by

$$R_2 \circ R_1 = \{(x, z) \mid (x, y) \in R_1 \text{ and } (y, z) \in R_2 \text{ for some } y \in Y\}.$$ ◀

**Example 3.3.26** The composition of the relations

$$R_1 = \{(1, 2), (1, 6), (2, 4), (3, 4), (3, 6), (3, 8)\}$$

and

$$R_2 = \{(2, u), (4, s), (4, t), (6, t), (8, u)\}$$

is

$$R_2 \circ R_1 = \{(1, u), (1, t), (2, s), (2, t), (3, s), (3, t), (3, u)\}.$$

For example, $(1, u) \in R_2 \circ R_1$ because $(1, 2) \in R_1$ and $(2, u) \in R_2$. ◀

**Example 3.3.27** Suppose that $R$ and $S$ are transitive relations on a set $X$. Determine whether each of $R \cup S$, $R \cap S$, or $R \circ S$ must be transitive.

**SOLUTION** We try to prove each of the three statements. If we fail, we will try to determine where our proof fails and use this information to construct a counterexample.

To prove that $R \cup S$ is transitive, we must show that if $(x, y), (y, z) \in R \cup S$, then $(x, z) \in R \cup S$. Suppose that $(x, y), (y, z) \in R \cup S$. If $(x, y)$ and $(y, z)$ happen to both be in $R$, we could use the fact that $R$ is transitive to conclude that $(x, z) \in R$ and, therefore, $(x, z) \in R \cup S$. A similar argument shows that if $(x, y)$ and $(y, z)$ happen to both be in $S$, then $(x, z) \in R \cup S$. But what if $(x, y) \in R$ and $(y, z) \in S$? Now the fact that $R$ and $S$ are transitive seems to be of no help. We try to construct a counterexample in which $R$ and $S$ are transitive, but there exist $(x, y) \in R$ and $(y, z) \in S$ such that $(x, z) \notin R \cup S$.

We put $(1, 2)$ in $R$ and $(2, 3)$ in $S$ and ensure that $(1, 3)$ is not in $R \cup S$. In fact, if $R = \{(1, 2)\}$, $R$ is transitive. Similarly, if $S = \{(2, 3)\}$, $S$ is transitive. We have our counterexample. We could write up our solution as follows.

We show that $R \cup S$ need not be transitive. Let $R = \{(1, 2)\}$ and $S = \{(2, 3)\}$. Then $R$ and $S$ are transitive, but $R \cup S$ is not transitive; $(1, 2), (2, 3) \in R \cup S$, but $(1, 3) \notin R \cup S$.

Next we turn our attention to $R \cap S$. To prove that $R \cap S$ is transitive, we must show that if $(x, y), (y, z) \in R \cap S$, then $(x, z) \in R \cap S$. Suppose that $(x, y), (y, z) \in R \cap S$. Then $(x, y), (y, z) \in R$. Since $R$ is transitive, $(x, z) \in R$. Similarly, $(x, y), (y, z) \in S$, and since $S$ is transitive, $(x, z) \in S$. Therefore $(x, z) \in R \cap S$. We have proved that $R \cap S$ is transitive.

Finally, consider $R \circ S$. To prove that $R \circ S$ is transitive, we must show that if $(x, y), (y, z) \in R \circ S$, then $(x, z) \in R \circ S$. Suppose that $(x, y), (y, z) \in R \circ S$. Then there exists $a$ such that $(x, a) \in S$ and $(a, y) \in R$, and there exists $b$ such that $(y, b) \in S$ and $(b, z) \in R$. We now know that $(a, y), (b, z) \in R$, but the fact that $R$ is transitive does not allow us to infer anything from $(a, y), (b, z) \in R$. A similar statement applies to $S$. We try to construct a counterexample in which $R$ and $S$ are transitive but $R \circ S$ is not transitive.

We will arrange for $(1, 2), (2, 3) \in R \circ S$, but $(1, 3) \notin R \circ S$. In order for $(1, 2) \in R \circ S$, we must have $(1, a) \in S$ and $(a, 2) \in R$, for some $a$. We put $(1, 5)$ in $S$ and $(5, 2)$ in $R$. (We chose $a$ to be a number different from 1, 2, or 3 to avoid a clash with those numbers. Any number different from 1, 2, 3 would do.) So far, so good! In order for $(2, 3) \in R \circ S$, we must have $(2, b) \in S$ and $(b, 3) \in R$, for some $b$. We put $(2, 6)$ in $S$ and $(6, 3)$ in $R$. (Again, we chose $b = 6$ to avoid a clash with the other numbers already chosen.) Now $R = \{(5, 2), (6, 3)\}$ and $S = \{(1, 5), (2, 6)\}$. Notice that $R$ and $S$ are transitive. We have our counterexample. We could write up our solution as follows.

We show that $R \circ S$ need not be transitive. Let $R = \{(5, 2), (6, 3)\}$ and $S = \{(1, 5), (2, 6)\}$. Then $R$ and $S$ are transitive. Now $R \circ S = \{(1, 2), (2, 3)\}$ is not transitive; $(1, 2), (2, 3) \in R \circ S$, but $(1, 3) \notin R \circ S$. ◀

## 3.3 Problem-Solving Tips

- To prove that a relation is reflexive, show that $(x, x) \in R$ for every $x \in X$. In words, a relation is reflexive if every element in $X$ is related to itself. Given an arrow diagram, the relation is reflexive if there is a loop at every vertex.
- To prove that a relation $R$ on a set $X$ is *not* reflexive, find $x \in X$ such that $(x, x) \notin R$. Given an arrow diagram, the relation is not reflexive if some vertex has no loop.
- To prove that a relation $R$ on a set $X$ is symmetric, show that for all $x, y \in X$, if $(x, y) \in R$, then $(y, x) \in R$. In words, a relation is symmetric if whenever $x$ is related to $y$, then $y$ is related to $x$. Given an arrow diagram, the relation is symmetric

if whenever there is a directed edge from $x$ to $y$, there is also a directed edge from $y$ to $x$.

- To prove that a relation $R$ on a set $X$ is *not* symmetric, find $x, y \in X$ such that $(x, y) \in R$ and $(y, x) \notin R$. Given an arrow diagram, the relation is not symmetric if there are two distinct vertices $x$ and $y$ with a directed edge from $x$ to $y$ but no directed edge from $y$ to $x$.
- To prove that a relation $R$ on a set $X$ is antisymmetric, show that for all $x, y \in X$, if $(x, y) \in R$ and $(y, x) \in R$, then $x = y$. In words, a relation is antisymmetric if whenever $x$ is related to $y$ and $y$ is related to $x$, then $x = y$. An equivalent characterization of "antisymmetric" can also be used: Show that for all $x, y \in X$, if $x \neq y$, then $(x, y) \notin R$ or $(y, x) \notin R$. Given an arrow diagram, the relation is antisymmetric if between any two distinct vertices there is at most one directed edge. Note that "*not* symmetric" is not necessarily the same as "antisymmetric."
- To prove that a relation $R$ on a set $X$ is *not* antisymmetric, find $x, y \in X$, $x \neq y$, such that $(x, y) \in R$ and $(y, x) \in R$. Given an arrow diagram, the relation is not antisymmetric if there are two distinct vertices $x$ and $y$ and two directed edges, one from $x$ to $y$ and the other from $y$ to $x$.
- To prove that a relation $R$ on a set $X$ is transitive, show that for all $x, y, z \in X$, if $(x, y)$ and $(y, z)$ are in $R$, then $(x, z)$ is in $R$. [It suffices to check ordered pairs $(x, y)$ and $(y, z)$ with $x \neq y$ and $y \neq z$.] In words, a relation is transitive if whenever $x$ is related to $y$ and $y$ is related to $z$, then $x$ is related to $z$. Given an arrow diagram, the relation is transitive if whenever there are directed edges from $x$ to $y$ and from $y$ to $z$, there is also a directed edge from $x$ to $z$.
- To prove that a relation $R$ on a set $X$ is *not* transitive, find $x, y, z \in X$ such that $(x, y)$ and $(y, z)$ are in $R$, but $(x, z)$ is not in $R$. Given an arrow diagram, the relation is not transitive if there are three distinct vertices $x$, $y$, $z$ and directed edges from $x$ to $y$ and from $y$ to $z$, but no directed edge from $x$ to $z$.
- A *partial order* is a relation that is reflexive, antisymmetric, and transitive.
- The *inverse* $R^{-1}$ of the relation $R$ consists of the elements $(y, x)$, where $(x, y) \in R$. In words, $x$ is related to $y$ in $R$ if and only if $y$ is related to $x$ in $R^{-1}$.
- If $R_1$ is a relation from $X$ to $Y$ and $R_2$ is a relation from $Y$ to $Z$, the *composition* of $R_1$ and $R_2$, denoted $R_2 \circ R_1$, is the relation from $X$ to $Z$ defined by

$$R_2 \circ R_1 = \{(x, z) \mid (x, y) \in R_1 \text{ and } (y, z) \in R_2 \text{ for some } y \in Y\}.$$

To compute the composition, find all pairs of the form $(x, y) \in R_1$ and $(y, z) \in R_2$; then put $(x, z)$ in $R_2 \circ R_1$.

## 3.3 Review Exercises

1. What is a binary relation from $X$ to $Y$?
2. What is the digraph of a binary relation?
3. Define *reflexive relation*. Give an example of a reflexive relation. Give an example of a relation that is *not* reflexive.
4. Define *symmetric relation*. Give an example of a symmetric relation. Give an example of a relation that is *not* symmetric.
5. Define *antisymmetric relation*. Give an example of an antisymmetric relation. Give an example of a relation that is *not* antisymmetric.
6. Define *transitive relation*. Give an example of a transitive relation. Give an example of a relation that is *not* transitive.
7. Define *partial order* and give an example of a partial order.
8. Define *inverse relation* and give an example of an inverse relation.
9. Define *composition of relations* and give an example of the composition of relations.

## 3.3 Exercises

*In Exercises 1–4, write the relation as a set of ordered pairs.*

**1.**

| 8840 | Hammer |
|---|---|
| 9921 | Pliers |
| 452 | Paint |
| 2207 | Carpet |

**2.**

| Sally | Math |
|---|---|
| Ruth | Physics |
| Sam | Econ |

**3.**

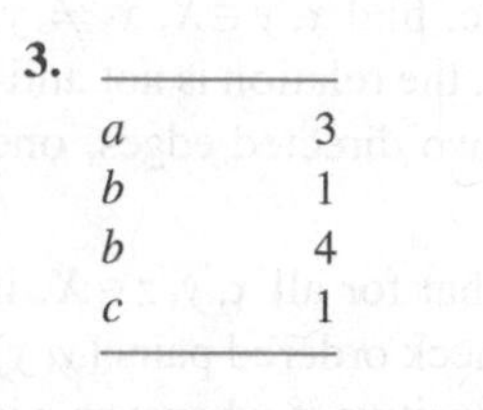

| $a$ | 3 |
|---|---|
| $b$ | 1 |
| $b$ | 4 |
| $c$ | 1 |

**4.**

| $a$ | $a$ |
|---|---|
| $b$ | $b$ |

*In Exercises 5–8, write the relation as a table.*

**5.** $R = \{(a, 6), (b, 2), (a, 1), (c, 1)\}$

**6.** The relation $R$ on $\{1, 2, 3, 4\}$ defined by $(x, y) \in R$ if $x^2 \geq y$

**7.** $R = \{$(Roger, Music), (Pat, History), (Ben, Math), (Pat, PolySci)$\}$

**8.** The relation $R$ from the set $X$ of planets to the set $Y$ of integers defined by $(x, y) \in R$ if $x$ is in position $y$ from the sun (nearest the sun being in position 1, second nearest the sun being in position 2, and so on)

*In Exercises 9–12, draw the digraph of the relation.*

**9.** The relation of Exercise 4 on $\{a, b, c\}$

**10.** The relation $R = \{(1, 2), (2, 3), (3, 4), (4, 1)\}$ on $\{1, 2, 3, 4\}$

**11.** The relation $R = \{(1, 2), (2, 1), (3, 3), (1, 1), (2, 2)\}$ on $X = \{1, 2, 3\}$

**12.** The relation of Exercise 6

*In Exercises 13–16, write the relation as a set of ordered pairs.*

**13.**

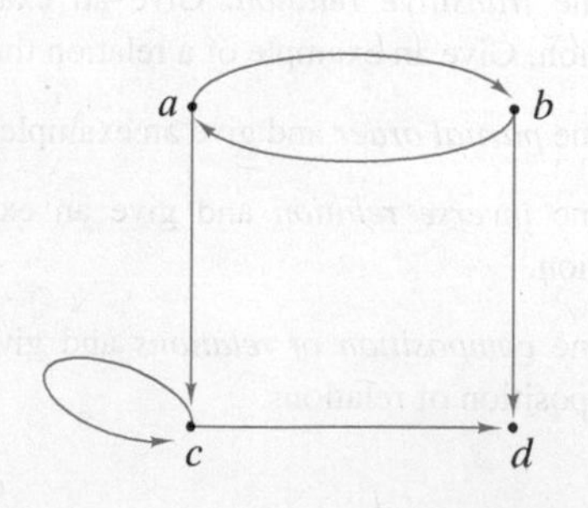


**14.**

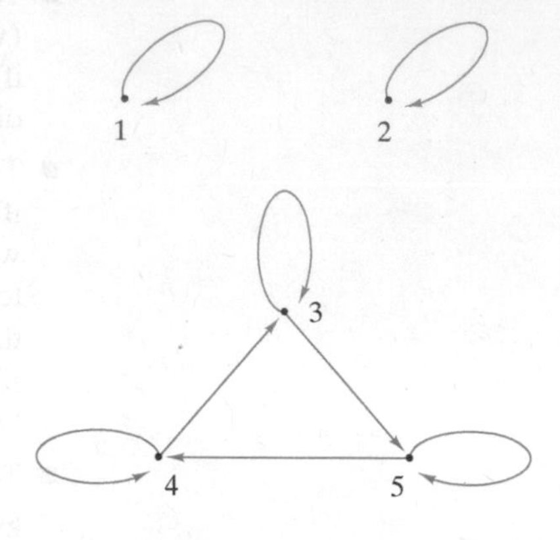


**15.**

**16.**

**17.** Find the inverse (as a set of ordered pairs) of each relation in Exercises 1–16.

*Exercises 18 and 19 refer to the relation $R$ on the set $\{1, 2, 3, 4, 5\}$ defined by the rule $(x, y) \in R$ if 3 divides $x - y$.*

**18.** List the elements of $R$.

**19.** List the elements of $R^{-1}$.

**20.** Repeat Exercises 18 and 19 for the relation $R$ on the set $\{1, 2, 3, 4, 5\}$ defined by the rule $(x, y) \in R$ if $x + y \leq 6$.

**21.** Repeat Exercises 18 and 19 for the relation $R$ on the set $\{1, 2, 3, 4, 5\}$ defined by the rule $(x, y) \in R$ if $x = y - 1$.

**22.** Is the relation of Exercise 20 reflexive, symmetric, antisymmetric, transitive, and/or a partial order?

**23.** Is the relation of Exercise 21 reflexive, symmetric, antisymmetric, transitive, and/or a partial order?

*In Exercises 24–34, determine whether each relation defined on the set of positive integers is reflexive, symmetric, antisymmetric, transitive, and/or a partial order.*

**24.** $(x, y) \in R$ if $xy = 1$.

**25.** $(x, y) \in R$ if $xy = 2$.

**26.** $(x, y) \in R$ if $xy \geq 1$.

**27.** $(x, y) \in R$ if $x = y^2$.

**28.** $(x, y) \in R$ if $x > y$.

**29.** $(x, y) \in R$ if $x \geq y$.

**30.** $(x, y) \in R$ if $x = y$.

**31.** $(x, y) \in R$ if 3 divides $x - y$.

**32.** $(x, y) \in R$ if 3 divides $x + 2y$.

**33.** $(x, y) \in R$ if $x - y = 2$.

**34.** $(x, y) \in R$ if $|x - y| = 2$.

**35.** Let $X$ be a nonempty set. Define a relation on $\mathcal{P}(X)$, the power set of $X$, as $(A, B) \in R$ if $A \subseteq B$. Is this relation reflexive, symmetric, antisymmetric, transitive, and/or a partial order?

**36.** Prove that a relation $R$ on a set $X$ is antisymmetric if and only if for all $x, y \in X$, if $(x, y) \in R$ and $x \neq y$, then $(y, x) \notin R$.

**37.** Let $X$ be the set of all four-bit strings (e.g., 0011, 0101, 1000). Define a relation $R$ on $X$ as $s_1 R s_2$ if some substring of $s_1$ of length 2 is equal to some substring of $s_2$ of length 2. Examples: 0111 $R$ 1010 (because both 0111 and 1010 contain 01). 1110 $\not R$ 0001 (because 1110 and 0001 do not share a common substring of length 2). Is this relation reflexive, symmetric, antisymmetric, transitive, and/or a partial order?

**38.** Suppose that $R_i$ is a partial order on $X_i$, $i = 1, 2$. Show that $R$ is a partial order on $X_1 \times X_2$ if we define

$$(x_1, x_2) R (x'_1, x'_2) \quad \text{if } x_1 R_1 x'_1 \text{ and } x_2 R_2 x'_2.$$

**39.** Let $R_1$ and $R_2$ be the relations on $\{1, 2, 3, 4\}$ given by

$$R_1 = \{(1, 1), (1, 2), (3, 4), (4, 2)\}$$
$$R_2 = \{(1, 1), (2, 1), (3, 1), (4, 4), (2, 2)\}.$$

List the elements of $R_1 \circ R_2$ and $R_2 \circ R_1$.

*Give examples of relations on {1, 2, 3, 4} having the properties specified in Exercises 40–44.*

**40.** Reflexive, symmetric, and not transitive

**41.** Reflexive, not symmetric, and not transitive

**42.** Reflexive, antisymmetric, and not transitive

**43.** Not reflexive, symmetric, not antisymmetric, and transitive

**44.** Not reflexive, not symmetric, and transitive

*Let R and S be relations on X. Determine whether each statement in Exercises 45–57 is true or false. If the statement is true, prove it; otherwise, give a counterexample.*

**45.** If $R$ is transitive, then $R^{-1}$ is transitive.

**46.** If $R$ and $S$ are reflexive, then $R \cup S$ is reflexive.

**47.** If $R$ and $S$ are reflexive, then $R \cap S$ is reflexive.

**48.** If $R$ and $S$ are reflexive, then $R \circ S$ is reflexive.

**49.** If $R$ is reflexive, then $R^{-1}$ is reflexive.

**50.** If $R$ and $S$ are symmetric, then $R \cup S$ is symmetric.

**51.** If $R$ and $S$ are symmetric, then $R \cap S$ is symmetric.

**52.** If $R$ and $S$ are symmetric, then $R \circ S$ is symmetric.

**53.** If $R$ is symmetric, then $R^{-1}$ is symmetric.

**54.** If $R$ and $S$ are antisymmetric, then $R \cup S$ is antisymmetric.

**55.** If $R$ and $S$ are antisymmetric, then $R \cap S$ is antisymmetric.

**56.** If $R$ and $S$ are antisymmetric, then $R \circ S$ is antisymmetric.

**57.** If $R$ is antisymmetric, then $R^{-1}$ is antisymmetric.

**58.** How many relations are there on an $n$-element set?

*In Exercises 59–61, determine whether each relation R defined on the collection of all nonempty subsets of real numbers is reflexive, symmetric, antisymmetric, transitive, and/or a partial order.*

**59.** $(A, B) \in R$ if for every $\varepsilon > 0$, there exists $a \in A$ and $b \in B$ with $|a - b| < \varepsilon$.

**60.** $(A, B) \in R$ if for every $a \in A$ and $\varepsilon > 0$, there exists $b \in B$ with $|a - b| < \varepsilon$.

**61.** $(A, B) \in R$ if for every $a \in A$, $b \in B$, and $\varepsilon > 0$, there exists $a' \in A$ and $b' \in B$ with $|a - b'| < \varepsilon$ and $|a' - b| < \varepsilon$.

**62.** What is wrong with the following argument, which supposedly shows that any relation $R$ on $X$ that is symmetric and transitive is reflexive?

Let $x \in X$. Using symmetry, we have $(x, y)$ and $(y, x)$ both in $R$. Since $(x, y), (y, x) \in R$, by transitivity we have $(x, x) \in R$. Therefore, $R$ is reflexive.

## 3.4 Equivalence Relations

**Go Online**
For more on equivalence relations, see
bit.ly/2OFm8St

Suppose that we have a set $X$ of 10 balls, each of which is either red, blue, or green (see Figure 3.4.1). If we divide the balls into sets $R$, $B$, and $G$ according to color, the family $\{R, B, G\}$ is a partition of $X$. (Recall that in Section 1.1, we defined a partition of a set $X$ to be a collection $\mathcal{S}$ of nonempty subsets of $X$ such that every element in $X$ belongs to exactly one member of $\mathcal{S}$.)

A partition can be used to define a relation. If $\mathcal{S}$ is a partition of $X$, we may define $x R y$ to mean that for some set $S \in \mathcal{S}$, both $x$ and $y$ belong to $S$. For the example of Figure 3.4.1, the relation obtained could be described as "is the same color as." The next theorem shows that such a relation is always reflexive, symmetric, and transitive.

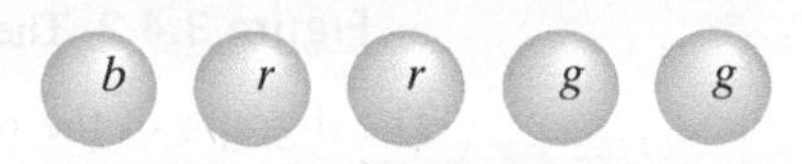


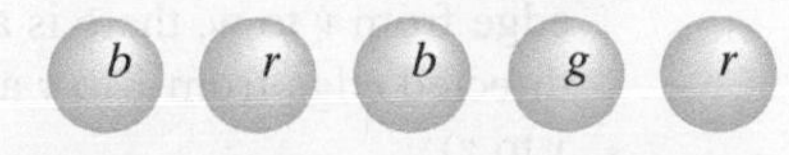


**Figure 3.4.1** A set of colored balls.

**Theorem 3.4.1** Let $\mathcal{S}$ be a partition of a set $X$. Define $x R y$ to mean that for some set $S$ in $\mathcal{S}$, both $x$ and $y$ belong to $S$. Then $R$ is reflexive, symmetric, and transitive.

**Proof** Let $x \in X$. By the definition of partition, $x$ belongs to some member $S$ of $\mathcal{S}$. Thus $x R x$ and $R$ is reflexive.

Suppose that $x R y$. Then both $x$ and $y$ belong to some set $S \in \mathcal{S}$. Since both $y$ and $x$ belong to $S$, $y R x$ and $R$ is symmetric.

Finally, suppose that $x R y$ and $y R z$. Then both $x$ and $y$ belong to some set $S \in \mathcal{S}$ and both $y$ and $z$ belong to some set $T \in \mathcal{S}$. Since $y$ belongs to exactly one member of $\mathcal{S}$, we must have $S = T$. Therefore, both $x$ and $z$ belong to $S$ and $x R z$. We have shown that $R$ is transitive. ◀

**Example 3.4.2** Consider the partition $\mathcal{S} = \{\{1, 3, 5\}, \{2, 6\}, \{4\}\}$ of $X = \{1, 2, 3, 4, 5, 6\}$. The relation $R$ on $X$ given by Theorem 3.4.1 contains the ordered pairs (1, 1), (1, 3), and (1, 5) because $\{1, 3, 5\}$ is in $\mathcal{S}$. The complete relation is

$$R = \{(1, 1), (1, 3), (1, 5), (3, 1), (3, 3), (3, 5), (5, 1), (5, 3), (5, 5), (2, 2), (2, 6), (6, 2), (6, 6), (4, 4)\}.$$
◀

Let $\mathcal{S}$ and $R$ be as in Theorem 3.4.1. If $S \in \mathcal{S}$, we can regard the members of $S$ as equivalent in the sense of the relation $R$, which motivates calling relations that are reflexive, symmetric, and transitive **equivalence relations.** In the example of Figure 3.4.1, the relation is "is the same color as"; hence *equivalent* means "is the same color as." Each set in the partition consists of all the balls of a particular color.

**Definition 3.4.3 ▶** A relation that is reflexive, symmetric, and transitive on a set $X$ is called an *equivalence relation on* $X$. ◀

**Example 3.4.4** The relation $R$ of Example 3.4.2 is an equivalence relation on $\{1, 2, 3, 4, 5, 6\}$ because of Theorem 3.4.1. We can also verify directly that $R$ is reflexive, symmetric, and transitive.

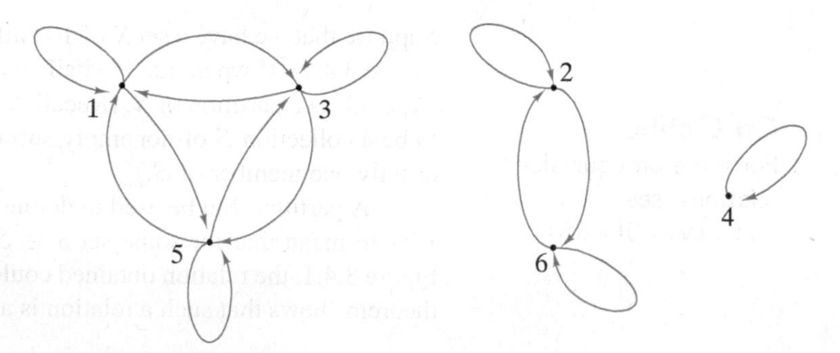


**Figure 3.4.2** The digraph of the relation of Example 3.4.2.

The digraph of the relation $R$ of Example 3.4.2 is shown in Figure 3.4.2. Again, we see that $R$ is reflexive (there is a loop at every vertex), symmetric (for every directed edge from $v$ to $w$, there is also a directed edge from $w$ to $v$), and transitive (if there is a directed edge from $x$ to $y$ and a directed edge from $y$ to $z$, there is a directed edge from $x$ to $z$). ◀

**Example 3.4.5** Consider the relation

$$R = \{(1, 1), (1, 3), (1, 5), (2, 2), (2, 4), (3, 1), (3, 3), (3, 5), (4, 2), (4, 4), (5, 1), (5, 3), (5, 5)\}$$

on $\{1, 2, 3, 4, 5\}$. The relation is reflexive because $(1, 1), (2, 2), (3, 3), (4, 4), (5, 5) \in R$. The relation is symmetric because whenever $(x, y)$ is in $R$, $(y, x)$ is also in $R$. Finally, the relation is transitive because whenever $(x, y)$ and $(y, z)$ are in $R$, $(x, z)$ is also in $R$. Since $R$ is reflexive, symmetric, and transitive, $R$ is an equivalence relation on $\{1, 2, 3, 4, 5\}$. ◀

**Example 3.4.6** The relation $R$ on $X = \{1, 2, 3, 4\}$ defined by $(x, y) \in R$ if $x \leq y$, $x, y \in X$, is not an equivalence relation because $R$ is not symmetric. [For example, $(2, 3) \in R$, but $(3, 2) \notin R$.] The relation $R$ is reflexive and transitive. ◀

**Example 3.4.7** The relation $R = \{(a, a), (b, c), (c, b), (d, d)\}$ on $X = \{a, b, c, d\}$ is not an equivalence relation because $R$ is neither reflexive nor transitive. [It is not reflexive because, for example, $(b, b) \notin R$. It is not transitive because, for example, $(b, c)$ and $(c, b)$ are in $R$, but $(b, b)$ is not in $R$.] ◀

Given an equivalence relation on a set $X$, we can partition $X$ by grouping related members of $X$. Elements related to one another may be thought of as equivalent. The next theorem gives the details.

**Theorem 3.4.8** Let $R$ be an equivalence relation on a set $X$. For each $a \in X$, let $[a] = \{x \in X \mid x R a\}$. (In words, $[a]$ is the set of all elements in $X$ that are related to $a$.) Then

$$\mathcal{S} = \{[a] \mid a \in X\}$$

is a partition of $X$.

**Proof** We must show that every element in $X$ belongs to exactly one member of $\mathcal{S}$.

Let $a \in X$. Since $a R a$, $a \in [a]$. Thus every element in $X$ belongs to *at least one* member of $\mathcal{S}$. It remains to show that every element in $X$ belongs to *exactly* one member of $\mathcal{S}$; that is,

$$\text{if } x \in X \text{ and } x \in [a] \cap [b], \text{ then } [a] = [b]. \qquad \textbf{(3.4.1)}$$

We first show that for all $c, d \in X$, if $c R d$, then $[c] = [d]$. Suppose that $c R d$. Let $x \in [c]$. Then $x R c$. Since $c R d$ and $R$ is transitive, $x R d$. Therefore, $x \in [d]$ and $[c] \subseteq [d]$. The argument that $[d] \subseteq [c]$ is the same as that just given, but with the roles of $c$ and $d$ interchanged. Thus $[c] = [d]$.

We now prove (3.4.1). Assume that $x \in X$ and $x \in [a] \cap [b]$. Then $x R a$ and $x R b$. Our preceding result shows that $[x] = [a]$ and $[x] = [b]$. Thus $[a] = [b]$. ◀

**Definition 3.4.9 ▶** Let $R$ be an equivalence relation on a set $X$. The sets $[a]$ defined in Theorem 3.4.8 are called the *equivalence classes of $X$ given by the relation $R$.* ◀

**Example 3.4.10** In Example 3.4.4, we showed that the relation

$$R = \{(1, 1), (1, 3), (1, 5), (3, 1), (3, 3), (3, 5), (5, 1), (5, 3), (5, 5), (2, 2), (2, 6), (6, 2), (6, 6), (4, 4)\}$$

on $X = \{1, 2, 3, 4, 5, 6\}$ is an equivalence relation. The equivalence class $[1]$ containing 1 consists of all $x$ such that $(x, 1) \in R$. Therefore, $[1] = \{1, 3, 5\}$. The remaining equivalence classes are found similarly:

$$[3] = [5] = \{1, 3, 5\}, \qquad [2] = [6] = \{2, 6\}, \qquad [4] = \{4\}.$$ ◀

**Example 3.4.11** The equivalence classes appear quite clearly in the digraph of an equivalence relation. The three equivalence classes of the relation $R$ of Example 3.4.10 appear in the digraph of $R$ (shown in Figure 3.4.2) as the three subgraphs whose vertices are $\{1, 3, 5\}$, $\{2, 6\}$, and $\{4\}$. A subgraph $G$ that represents an equivalence class is a largest subgraph of the original digraph having the property that for any vertices $v$ and $w$ in $G$, there is a directed edge from $v$ to $w$. For example, if $v, w \in \{1, 3, 5\}$, there is a directed edge from $v$ to $w$. Moreover, no additional vertices can be added to 1, 3, 5, and so the resulting vertex set has a directed edge between each pair of vertices. ◀

**Example 3.4.12** There are two equivalence classes for the equivalence relation

$$R = \{(1, 1), (1, 3), (1, 5), (2, 2), (2, 4), (3, 1), (3, 3), (3, 5), (4, 2), (4, 4), (5, 1), (5, 3), (5, 5)\}$$

on $\{1, 2, 3, 4, 5\}$ of Example 3.4.5, namely, $[1] = [3] = [5] = \{1, 3, 5\}$ and $[2] = [4] = \{2, 4\}$. ◀

**Example 3.4.13** We can readily verify that the relation $R = \{(a, a), (b, b), (c, c)\}$ on $X = \{a, b, c\}$ is reflexive, symmetric, and transitive. Thus $R$ is an equivalence relation. The equivalence classes are $[a] = \{a\}$, $[b] = \{b\}$, and $[c] = \{c\}$. ◀

**Example 3.4.14** Let $X = \{1, 2, \ldots, 10\}$. Define $x\,R\,y$ to mean that 3 divides $x - y$. We can readily verify that the relation $R$ is reflexive, symmetric, and transitive. Thus $R$ is an equivalence relation on $X$. Determine the members of the equivalence classes.

**SOLUTION** The equivalence class $[1]$ consists of all $x$ with $x\,R\,1$. Thus

$$[1] = \{x \in X \mid 3 \text{ divides } x - 1\} = \{1, 4, 7, 10\}.$$

Similarly, $[2] = \{2, 5, 8\}$ and $[3] = \{3, 6, 9\}$. These three sets partition $X$. Note that

$$[1] = [4] = [7] = [10], \qquad [2] = [5] = [8], \qquad [3] = [6] = [9].$$

For this relation, *equivalence* is "has the same remainder when divided by 3." ◀

**Example 3.4.15** Show that if a relation $R$ on a set $X$ is symmetric and transitive but not reflexive, the collection of sets $[a]$, $a \in X$, defined in Theorem 3.4.8 does *not* partition $X$ (see also Exercises 47–51).

**SOLUTION** Let $R$ be a relation on a set $X$ that is symmetric and transitive but not reflexive. We define "pseudo equivalence classes" as in Theorem 3.4.8:

$$[a] = \{x \in X \mid x\,R\,a\}.$$

Since $R$ is not reflexive, there exists $b \in X$ such that $(b, b) \notin R$. We show that $b$ is not in any pseudo equivalence class. Suppose, by way of contradiction, that $b \in [a]$ for some $a \in X$. Then $(b, a) \in R$. Since $R$ is symmetric, $(a, b) \in R$. Since $R$ is transitive, $(b, b) \in R$. But we assumed that $(b, b) \notin R$. This contradiction shows that $b$ is not in any pseudo equivalence class. Thus the collection of pseudo equivalence classes does not partition $X$. ◀

We close this section by proving a special result that we will need later (see Sections 6.2 and 6.3). The proof is illustrated in Figure 3.4.3.

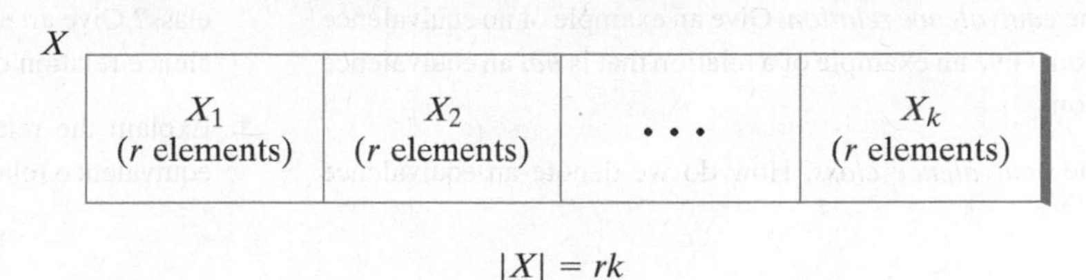


$$|X| = rk$$

**Figure 3.4.3** The proof of Theorem 3.4.16.

**Theorem 3.4.16** Let $R$ be an equivalence relation on a finite set $X$. If each equivalence class has $r$ elements, there are $|X|/r$ equivalence classes.

**Proof** Let $X_1, X_2, \ldots, X_k$ denote the distinct equivalence classes. Since these sets partition $X$,

$$|X| = |X_1| + |X_2| + \cdots + |X_k| = r + r + \cdots + r = kr$$

and the conclusion follows. ◀

## 3.4 Problem-Solving Tips

An *equivalence relation* is a relation that is reflexive, symmetric and transitive. To prove that a relation is an equivalence relation, you need to verify that these three properties hold (see *Problem-Solving Tips* for Section 3.3).

An equivalence relation on a set $X$ partitions $X$ into subsets. ("Partitions" means that every $x$ in $X$ belongs to exactly one of the subsets of the partition.) The subsets making up the partition can be determined in the following way. Choose $x_1 \in X$. Find the set, denoted $[x_1]$, of all elements related to $x_1$. Choose another element $x_2 \in X$ that is not related to $x_1$. Find the set $[x_2]$ of all elements related to $x_2$. Continue in this way until all the elements of $X$ have been assigned to a set. The sets $[x_i]$ are called the *equivalence classes*. The partition is $[x_1], [x_2], \ldots$. The elements of $[x_i]$ are *equivalent* in the sense that they are all related. For example, the relation $R$, defined by $x R y$ if $x$ and $y$ are the same color, partitions the set into subsets where each subset contains elements that are all the same color. Within a subset, the elements are equivalent in the sense that they are all the same color.

In the digraph of an equivalence relation, an equivalence class is a largest subgraph of the original digraph having the property that for any vertices $v$ and $w$ in $G$, there is a directed edge from $v$ to $w$.

A partition of a set gives rise to an equivalence relation. If $X_1, \ldots, X_n$ is a partition of a set $X$ and we define $x R y$ if for some $i$, $x$ and $y$ both belong to $X_i$, then $R$ is an equivalence relation on $X$. The equivalence classes turn out to be $X_1, \ldots, X_n$. Thus, "equivalence relation" and "partition of a set" are different views of the same situation. An equivalence relation on $X$ gives rise to a partition of $X$ (namely, the equivalence classes), and a partition of $X$ gives rise to an equivalence relation (namely, $x$ is related to $y$ if $x$ and $y$ are in the same set in the partition). This latter fact can be used to solve certain problems. If you are asked to find an equivalence relation, you can either find the equivalence relation directly or construct a partition and then use the associated equivalence relation. Similarly, if you are asked to find a partition, you can either find the partition directly or construct an equivalence relation and then take the equivalence classes as your partition.

## 3.4 Review Exercises

1. Define *equivalence relation*. Give an example of an equivalence relation. Give an example of a relation that is *not* an equivalence relation.
2. Define *equivalence class*. How do we denote an equivalence class? Give an example of an equivalence class for your equivalence relation of Exercise 1.
3. Explain the relationship between a partition of a set and an equivalence relation.

## 3.4 Exercises

*In Exercises 1–10, determine whether the given relation is an equivalence relation on* $\{1, 2, 3, 4, 5\}$. *If the relation is an equivalence relation, list the equivalence classes. (In Exercises 5–10,* $x, y \in \{1, 2, 3, 4, 5\}$.)

1. $\{(1, 1), (2, 2), (3, 3), (4, 4), (5, 5), (1, 3), (3, 1)\}$
2. $\{(1, 1), (2, 2), (3, 3), (4, 4), (5, 5), (1, 3), (3, 1), (3, 4), (4, 3)\}$
3. $\{(1, 1), (2, 2), (3, 3), (4, 4)\}$
4. $\{(1, 1), (2, 2), (3, 3), (4, 4), (5, 5), (1, 5), (5, 1), (3, 5), (5, 3), (1, 3), (3, 1)\}$
5. $\{(x, y) \mid 1 \leq x \leq 5 \text{ and } 1 \leq y \leq 5\}$
6. $\{(x, y) \mid 4 \text{ divides } x - y\}$
7. $\{(x, y) \mid 3 \text{ divides } x + y\}$
8. $\{(x, y) \mid x \text{ divides } 2 - y\}$
9. $\{(x, y) \mid x \text{ and } y \text{ are both even}\}$
10. $\{(x, y) \mid x \text{ and } y \text{ are both even or } x \text{ and } y \text{ are both odd}\}$

*In Exercises 11–16, determine whether the given relation is an equivalence relation on the set of all people.*

11. $\{(x, y) \mid x$ and $y$ are the same height$\}$
12. $\{(x, y) \mid x$ and $y$ have, at some time, lived in the same country$\}$
13. $\{(x, y) \mid x$ and $y$ have the same first name$\}$
14. $\{(x, y) \mid x$ is taller than $y\}$
15. $\{(x, y) \mid x$ and $y$ have the same parents$\}$
16. $\{(x, y) \mid x$ and $y$ have the same color hair$\}$

*In Exercises 17–22, list the members of the equivalence relation on* $\{1, 2, 3, 4\}$ *defined (as in Theorem 3.4.1) by the given partition. Also, find the equivalence classes* $[1]$, $[2]$, $[3]$, *and* $[4]$.

17. $\{\{1, 2\}, \{3, 4\}\}$
18. $\{\{1\}, \{2\}, \{3, 4\}\}$
19. $\{\{1\}, \{2\}, \{3\}, \{4\}\}$
20. $\{\{1, 2, 3\}, \{4\}\}$
21. $\{\{1, 2, 3, 4\}\}$
22. $\{\{1\}, \{2, 4\}, \{3\}\}$

*In Exercises 23–25, let* $X = \{1, 2, 3, 4, 5\}$, $Y = \{3, 4\}$, *and* $C = \{1, 3\}$. *Define the relation R on* $\mathcal{P}(X)$, *the set of all subsets of X, as*

$$A R B \quad \text{if } A \cup Y = B \cup Y.$$

23. Show that $R$ is an equivalence relation.
24. List the elements of $[C]$, the equivalence class containing $C$.
25. How many distinct equivalence classes are there?
26. Let

$$X = \{\text{San Francisco, Pittsburgh, Chicago, San Diego, Philadelphia, Los Angeles}\}.$$

Define a relation $R$ on $X$ as $x R y$ if $x$ and $y$ are in the same state.

(a) Show that $R$ is an equivalence relation.

(b) List the equivalence classes of $X$.

27. If an equivalence relation has only one equivalence class, what must the relation look like?
28. If $R$ is an equivalence relation on a finite set $X$ and $|X| = |R|$, what must the relation look like?
29. By listing ordered pairs, give an example of an equivalence relation on $\{1, 2, 3, 4, 5, 6\}$ having exactly four equivalence classes.
30. How many equivalence relations are there on the set $\{1, 2, 3\}$?
31. Let $R$ be a reflexive relation on $X$ satisfying: for all $x, y, z \in X$, if $x R y$ and $y R z$, then $z R x$. Prove that $R$ is an equivalence relation.
32. Define a relation $R$ on $\mathbf{R}^{\mathbf{R}}$, the set of functions from $\mathbf{R}$ to $\mathbf{R}$, by $f R g$ if $f(0) = g(0)$. Prove that $R$ is an equivalence relation on $\mathbf{R}^{\mathbf{R}}$. Let $f(x) = x$ for all $x \in \mathbf{R}$. Describe $[f]$.
33. Define a relation $R$ on $\mathbf{R}^{\mathbf{R}}$, the set of functions from $\mathbf{R}$ to $\mathbf{R}$ by $f R g$ if there exist $a, b \in \mathbf{R}$ such that $f(x) = g(x + a) + b$ for all $x \in \mathbf{R}$. Prove that $R$ is an equivalence relation on $\mathbf{R}^{\mathbf{R}}$. What property do all functions in an equivalence class share?
34. Let $X = \{1, 2, \ldots, 10\}$. Define a relation $R$ on $X \times X$ by $(a, b) R (c, d)$ if $a + d = b + c$.

(a) Show that $R$ is an equivalence relation on $X \times X$.

(b) List one member of each equivalence class of $X \times X$.

35. Let $X = \{1, 2, \ldots, 10\}$. Define a relation $R$ on $X \times X$ by $(a, b) R (c, d)$ if $ad = bc$.

(a) Show that $R$ is an equivalence relation on $X \times X$.

(b) List one member of each equivalence class of $X \times X$.

(c) Describe the relation $R$ in familiar terms.

36. Let $R$ be a reflexive and transitive relation on $X$. Show that $R \cap R^{-1}$ is an equivalence relation on $X$.

37. Let $R_1$ and $R_2$ be equivalence relations on $X$.

(a) Show that $R_1 \cap R_2$ is an equivalence relation on $X$.

(b) Describe the equivalence classes of $R_1 \cap R_2$ in terms of the equivalence classes of $R_1$ and the equivalence classes of $R_2$.

38. Suppose that $\mathcal{S}$ is a collection of subsets of a set $X$ and $X = \cup\mathcal{S}$. (It is not assumed that the family $\mathcal{S}$ is pairwise disjoint.) Define $x R y$ to mean that for some set $S \in \mathcal{S}$, both $x$ and $y$ are in $S$. Is $R$ necessarily reflexive, symmetric, or transitive?

39. Let $S$ be a unit square including the interior, as shown in the following figure.

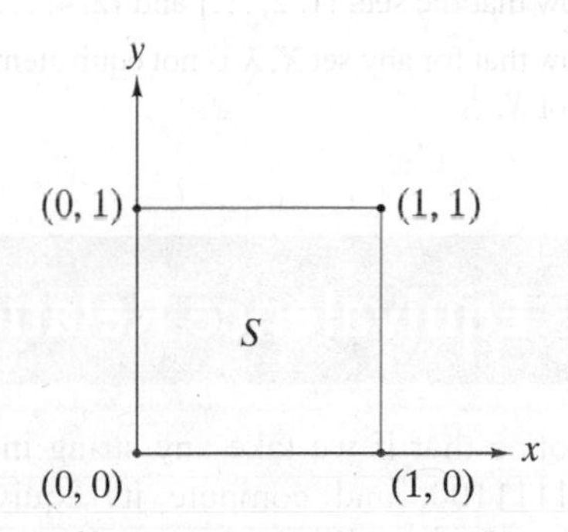


Define a relation $R$ on $S$ by $(x, y) R (x', y')$ if $(x = x'$ and $y = y')$, or $(y = y'$ and $x = 0$ and $x' = 1)$, or $(y = y'$ and $x = 1$ and $x' = 0)$.

(a) Show that $R$ is an equivalence relation on $S$.

(b) If points in the same equivalence class are glued together, how would you describe the figure formed?

40. Let $S$ be a unit square including the interior (as in Exercise 39). Define a relation $R'$ on $S$ by $(x, y) Rp(x', y')$ if $(x = x'$ and $y = y')$, or $(y = y'$ and $x = 0$ and $x' = 1)$, or $(y = y'$ and $x = 1$ and $x' = 0)$, or $(x = x'$ and $y = 0$ and $y' = 1)$, or $(x = x'$ and $y = 1$ and $y' = 0)$. Let

$$R = R' \cup \{((0, 0), (1, 1)), ((0, 1), (1, 0)), ((1, 0), (0, 1)), ((1, 1), (0, 0))\}.$$

(a) Show that $R$ is an equivalence relation on $S$.

(b) If points in the same equivalence class are glued together, how would you describe the figure formed?

41. Let $f$ be a function from $X$ to $Y$. Define a relation $R$ on $X$ by

$$x R y \quad \text{if } f(x) = f(y).$$

Show that $R$ is an equivalence relation on $X$.

42. Let $f$ be a characteristic function in $X$. ("Characteristic function" is defined before Exercise 101, Section 3.1.) Define a relation $R$ on $X$ by $x R y$ if $f(x) = f(y)$. According to the preceding exercise, $R$ is an equivalence relation. What are the equivalence classes?

43. Let $f$ be a function from $X$ onto $Y$. Let

$$\mathcal{S} = \{f^{-1}(\{y\}) \mid y \in Y\}.$$

[The definition of $f^{-1}(B)$, where $B$ is a set, precedes Exercise 88, Section 3.1.] Show that $\mathcal{S}$ is a partition of $X$. Describe an equivalence relation that gives rise to this partition.

44. Let $R$ be an equivalence relation on a set $A$. Define a function $f$ from $A$ to the set of equivalence classes of $A$ by the rule $f(x) = [x]$. When do we have $f(x) = f(y)$?

45. Let $R$ be an equivalence relation on a set $A$. Suppose that $g$ is a function from $A$ into a set $X$ having the property that if $x R y$, then $g(x) = g(y)$. Show that $h([x]) = g(x)$ defines a function from the set of equivalence classes of $A$ into $X$. [What needs to be shown is that $h$ *uniquely* assigns a value to $[x]$; that is, if $[x] = [y]$, then $g(x) = g(y)$.]

46. Suppose that a relation $R$ on a set $X$ is symmetric and transitive but not reflexive. Suppose, in particular, that $(b, b) \notin R$. Prove that the pseudo equivalence class $[b]$ (see Example 3.4.15) is empty.

47. Prove that if a relation $R$ on a set $X$ is not symmetric but transitive, the collection of pseudo equivalence classes (see Example 3.4.15) does not partition $X$.

48. Prove that if a relation $R$ on a set $X$ is reflexive but not symmetric, the collection of pseudo equivalence classes (see Example 3.4.15) does not partition $X$.

49. Prove that if a relation $R$ on a set $X$ is reflexive but not transitive, the collection of pseudo equivalence classes (see Example 3.4.15) does not partition $X$.

50. Give an example of a set $X$ and a relation $R$ on $X$ that is not reflexive, not symmetric, and not transitive, but for which the collection of pseudo equivalence classes (see Example 3.4.15) partitions $X$.

51. Give an example of a set $X$ and a relation $R$ on $X$ that is not reflexive, symmetric, and not transitive, but for which the collection of pseudo equivalence classes (see Example 3.4.15) partitions $X$.

52. Let $X$ denote the set of all sequences of real numbers. Define a relation $R$ on $X$ as $s R t$ if there exists an increasing function $f$ from the domain of $s$ onto the domain of $t$ such that $s_n = t_{f(n)}$ for all $n$ in the domain of $s$.

(a) Show that $R$ is an equivalence relation.

(b) Explain in words what it means for two sequences in $X$ to be equivalent under the relation $R$.

(c) Since a sequence is a function, a sequence is a set of ordered pairs. Two sequences are equal if the two sets of ordered pairs are equal. Contrast the difference between two equivalent sequences in $X$ and two equal sequences in $X$.

*Let $R$ be a relation on a set $X$. Define*

$$\rho(R) = R \cup \{(x, x) \mid x \in X\}$$

$$\sigma(R) = R \cup R^{-1}$$

$$R^n = R \circ R \circ R \circ \cdots \circ R \quad (nR\text{'s})$$

$$\tau(R) = \cup\{R^n \mid n = 1, 2, \ldots\}.$$

*The relation $\tau(R)$ is called the* transitive closure *of $R$.*

53. For the relations $R_1$ and $R_2$ of Exercise 39, Section 3.3, find $\rho(R_i)$, $\sigma(R_i)$, $\tau(R_i)$, and $\tau(\sigma(\rho(R_i)))$ for $i = 1, 2$.
54. Show that $\rho(R)$ is reflexive.
55. Show that $\sigma(R)$ is symmetric.
56. Show that $\tau(R)$ is transitive.
★57. Show that $\tau(\sigma(\rho(R)))$ is an equivalence relation containing $R$.
★58. Show that $\tau(\sigma(\rho(R)))$ is the smallest equivalence relation on $X$ containing $R$; that is, show that if $R'$ is an equivalence relation on $X$ and $R' \supseteq R$, then $R' \supseteq \tau(\sigma(\rho(R)))$.
★59. Show that $R$ is transitive if and only if $\tau(R) = R$.

*In Exercises 60–66, if the statement is true for all relations $R_1$ and $R_2$ on an arbitrary set $X$, prove it; otherwise, give a counterexample.*

60. $\rho(R_1 \cup R_2) = \rho(R_1) \cup \rho(R_2)$
61. $\sigma(R_1 \cap R_2) = \sigma(R_1) \cap \sigma(R_2)$
62. $\tau(R_1 \cup R_2) = \tau(R_1) \cup \tau(R_2)$
63. $\tau(R_1 \cap R_2) = \tau(R_1) \cap \tau(R_2)$
64. $\sigma(\tau(R_1)) = \tau(\sigma(R_1))$
65. $\sigma(\rho(R_1)) = \rho(\sigma(R_1))$
66. $\rho(\tau(R_1)) = \tau(\rho(R_1))$

*If $X$ and $Y$ are sets, we define $X$ to be* equivalent *to $Y$ if there is a one-to-one, onto function from $X$ to $Y$.*

67. Show that set equivalence is an equivalence relation.
68. Show that the sets $\{1, 2, \ldots\}$ and $\{2, 4, \ldots\}$ are equivalent.
★69. Show that for any set $X$, $X$ is not equivalent to $\mathcal{P}(X)$, the power set of $X$.

# Problem-Solving Corner: Equivalence Relations

## Problem

Answer the following questions for the relation $R$ defined on the set of eight-bit strings by $s_1 \, R \, s_2$, provided that the first four bits of $s_1$ and $s_2$ coincide.

(a) Show that $R$ is an equivalence relation.

(b) List one member of each equivalence class.

(c) How many equivalence classes are there?

## Attacking the Problem

Let's begin by looking at some specific eight-bit strings that are related according to the relation $R$. Let's take an arbitrary string 01111010 and find strings related to it. A string $s$ is related to 01111010 if the first four bits of 01111010 and $s$ coincide. This means that $s$ must begin 0111 and the last four bits can be anything. An example is $s = 01111000$.

Let's list all of the strings related to 01111010. In doing so, we must be careful to follow 0111 with every possible four-bit string:

01110000, 01110001, 01110010, 01110011,
01110100, 01110101, 01110110, 01110111,
01111000, 01111001, 01111010, 01111011,
01111100, 01111101, 01111110, 01111111.

Assuming for the moment that $R$ is an equivalence relation, the equivalence class containing 01111010, denoted [01111010], consists of all strings related to 01111010. Therefore, what we have just computed are the members of [01111010].

Notice that if we take any string in [01111010], say 01111100, and compute its equivalence class [01111100], we will obtain exactly the same set of strings—namely, the set of eight-bit strings that begin 0111.

To obtain a different example, we would have to start with a string whose first four bits are different from 0111, say 1011. As an example, the strings related to 10110100 are

10110000, 10110001, 10110010, 10110011,
10110100, 10110101, 10110110, 10110111,
10111000, 10111001, 10111010, 10111011,
10111100, 10111101, 10111110, 10111111.

What we have just computed are the members of [10110100]. We see that [01111010] and [10110100] have no members in common. It is always the case that two equivalence classes are identical or have no members in common (see Theorem 3.4.8).

Before reading on, compute the members of some other equivalence class.

## Finding a Solution

To show that $R$ is an equivalence relation, we must show that $R$ is reflexive, symmetric, and transitive (see Definition 3.4.3). For each property, we will go directly to the definition and check that the conditions specified in the definition hold.

For $R$ to be reflexive, we must have $s \, R \, s$ for every eight-bit string $s$. For $s \, R \, s$ to be true, the first four bits of $s$ and $s$ must coincide. This is certainly the case!

For $R$ to be symmetric, for all eight-bit strings $s_1$ and $s_2$, if $s_1 R s_2$, then $s_2 R s_1$. Using the definition of $R$, we may translate this condition to: If the first four bits of $s_1$ and $s_2$ coincide, then the first four bits of $s_2$ and $s_1$ coincide. This is also certainly the case!

For $R$ to be transitive, for all eight-bit strings $s_1$, $s_2$, and $s_3$, if $s_1 R s_2$ and $s_2 R s_3$, then $s_1 R s_3$. Again using the definition of $R$, we may translate this condition to: If the first four bits of $s_1$ and $s_2$ coincide and the first four bits of $s_2$ and $s_3$ coincide, then the first four bits of $s_1$ and $s_3$ coincide. This too is certainly the case! We have proved that $R$ is an equivalence relation.

In our earlier discussion, we found that each distinct four-bit string determines an equivalence class. For example, the string 0111 determines the equivalence class consisting of all eight-bit strings that begin 0111. Therefore, the number of equivalence classes is equal to the number of four-bit strings. We can simply list them all

0000, 0001, 0010, 0011,
0100, 0101, 0110, 0111,
1000, 1001, 1010, 1011,
1100, 1101, 1110, 1111

and then count them. There are 16 equivalence classes.

Consider the problem of listing one member of each equivalence class. The 16 four-bit strings listed previously determine the 16 equivalence classes. The first string 0000 determines the equivalence class consisting of all eight-bit strings that begin 0000; the second string 0001 determines the equivalence class consisting of all eight-bit strings that begin 0001; and so on. Thus to list one member of each equivalence class, we simply need to append some four-bit string to each of the strings in the previous list:

00000000, 00010000, 00100000, 00110000,
01000000, 01010000, 01100000, 01110000,
10000000, 10010000, 10100000, 10110000,
11000000, 11010000, 11100000, 11110000.

## Formal Solution

(a) We have already presented a formal proof that $R$ is an equivalence relation.

(b)

00000000, 00010000, 00100000, 00110000,
01000000, 01010000, 01100000, 01110000,
10000000, 10010000, 10100000, 10110000,
11000000, 11010000, 11100000, 11110000

lists one member of each equivalence class.

(c) There are 16 equivalence classes.

## Summary of Problem-Solving Techniques

- List elements that are related.
- Compute some equivalence classes; that is, list *all* elements related to a particular element.
- It may help to solve the parts of a problem in a different order than that given in the problem statement. In our example, it was helpful in looking at some concrete cases to *assume* that the relation was an equivalence relation before actually proving that it was an equivalence relation.
- To show that a particular relation $R$ is an equivalence relation, go directly to the definitions. Show that $R$ is reflexive, symmetric, and transitive by directly verifying that $R$ satisfies the definitions of reflexive, symmetric, and transitive.
- If the problem is to count the number of items satisfying some property (e.g., in our problem we were asked to count the number of equivalence classes) and the number is sufficiently small, just list all the items and count them directly.

## Comments

In programming languages, usually only some specified number of characters of the names of variables and special terms (technically these are called *identifiers*) are significant. For example, in the C programming language, only the first 31 characters of identifiers are significant. This means that if two identifiers begin with the same 31 characters, the system is allowed to consider them identical.

If we define a relation $R$ on the set of C identifiers by $s_1 R s_2$, provided that the first 31 characters of $s_1$ and $s_2$ coincide, then $R$ is an equivalence relation. An equivalence class consists of identifiers that the system is allowed to consider identical.

## 3.5 Matrices of Relations

**Go Online**
For more matrices of relations, see bit.ly/2OFm8St

A matrix is a convenient way to represent a relation $R$ from $X$ to $Y$. Such a representation can be used by a computer to analyze a relation. We label the rows with the elements of $X$ (in some arbitrary order), and we label the columns with the elements of $Y$ (again, in some arbitrary order). We then, set the entry in row $x$ and column $y$ to 1 if $x R y$ and to 0 otherwise. This matrix is called the **matrix of the relation** $R$ (relative to the orderings of $X$ and $Y$).

**Example 3.5.1** The matrix of the relation

$$R = \{(1, b), (1, d), (2, c), (3, c), (3, b), (4, a)\}$$

from $X = \{1, 2, 3, 4\}$ to $Y = \{a, b, c, d\}$ relative to the orderings 1, 2, 3, 4 and $a$, $b$, $c$, $d$ is

$$\begin{array}{c} \\ 1 \\ 2 \\ 3 \\ 4 \end{array} \begin{array}{c} \begin{array}{cccc} a & b & c & d \end{array} \\ \begin{pmatrix} 0 & 1 & 0 & 1 \\ 0 & 0 & 1 & 0 \\ 0 & 1 & 1 & 0 \\ 1 & 0 & 0 & 0 \end{pmatrix} \end{array}$$

◀

**Example 3.5.2** The matrix of the relation $R$ of Example 3.5.1 relative to the orderings 2, 3, 4, 1 and $d$, $b$, $a$, $c$ is

$$\begin{array}{c} \\ 2 \\ 3 \\ 4 \\ 1 \end{array} \begin{array}{c} \begin{array}{cccc} d & b & a & c \end{array} \\ \begin{pmatrix} 0 & 0 & 0 & 1 \\ 0 & 1 & 0 & 1 \\ 0 & 0 & 1 & 0 \\ 1 & 1 & 0 & 0 \end{pmatrix} \end{array}.$$

Obviously, the matrix of a relation from $X$ to $Y$ is dependent on the orderings of $X$ and $Y$. ◀

**Example 3.5.3** The matrix of the relation $R$ from $\{2, 3, 4\}$ to $\{5, 6, 7, 8\}$, relative to the orderings 2, 3, 4 and 5, 6, 7, 8, defined by

$$x R y \qquad \text{if } x \text{ divides } y$$

is

$$\begin{array}{c} \\ 2 \\ 3 \\ 4 \end{array} \begin{array}{c} \begin{array}{cccc} 5 & 6 & 7 & 8 \end{array} \\ \begin{pmatrix} 0 & 1 & 0 & 1 \\ 0 & 1 & 0 & 0 \\ 0 & 0 & 0 & 1 \end{pmatrix} \end{array}.$$

◀

When we write the matrix of a relation $R$ on a set $X$ (i.e., from $X$ to $X$), we use the same ordering for the rows as we do for the columns.

**Example 3.5.4** The matrix of the relation

$$R = \{(a, a), (b, b), (c, c), (d, d), (b, c), (c, b)\}$$

on $\{a, b, c, d\}$, relative to the ordering $a$, $b$, $c$, $d$, is

$$\begin{array}{c} \\ a \\ b \\ c \\ d \end{array} \begin{array}{c} \begin{array}{cccc} a & b & c & d \end{array} \\ \begin{pmatrix} 1 & 0 & 0 & 0 \\ 0 & 1 & 1 & 0 \\ 0 & 1 & 1 & 0 \\ 0 & 0 & 0 & 1 \end{pmatrix} \end{array}.$$

◀

Notice that the matrix of a relation on a set $X$ is always a square matrix.

We can quickly determine whether a relation $R$ on a set $X$ is reflexive by examining the matrix $A$ of $R$ (relative to some ordering). The relation $R$ is reflexive if and only if $A$ has 1's on the main diagonal. (The main diagonal of a square matrix consists of the entries on a line from the upper left to the lower right.) The relation $R$ is reflexive if and only if $(x, x) \in R$ for all $x \in X$. But this last condition holds precisely when the main diagonal consists of 1's. Notice that the relation $R$ of Example 3.5.4 is reflexive and that the main diagonal of the matrix of $R$ consists of 1's.

We can also quickly determine whether a relation $R$ on a set $X$ is symmetric by examining the matrix $A$ of $R$ (relative to some ordering). The relation $R$ is symmetric if and only if for all $i$ and $j$, the $ij$th entry of $A$ is equal to the $ji$th entry of $A$. (Less formally, $R$ is symmetric if and only if $A$ is symmetric about the main diagonal.) The reason is that $R$ is symmetric if and only if whenever $(x, y)$ is in $R$, $(y, x)$ is also in $R$. But this last condition holds precisely when $A$ is symmetric about the main diagonal. Notice that the relation $R$ of Example 3.5.4 is symmetric and that the matrix of $R$ is symmetric about the main diagonal.

We can also quickly determine whether a relation $R$ is antisymmetric by examining the matrix of $R$ (relative to some ordering) (see Exercise 11).

We conclude by showing how matrix multiplication relates to composition of relations and how we can use the matrix of a relation to test for transitivity.

**Example 3.5.5** Let $R_1$ be the relation from $X = \{1, 2, 3\}$ to $Y = \{a, b\}$ defined by

$$R_1 = \{(1, a), (2, b), (3, a), (3, b)\},$$

and let $R_2$ be the relation from $Y$ to $Z = \{x, y, z\}$ defined by

$$R_2 = \{(a, x), (a, y), (b, y), (b, z)\}.$$

The matrix of $R_1$ relative to the orderings 1, 2, 3 and $a, b$ is

$$A_1 = \begin{array}{c} \\ 1 \\ 2 \\ 3 \end{array} \begin{array}{c} \begin{array}{cc} a & b \end{array} \\ \begin{pmatrix} 1 & 0 \\ 0 & 1 \\ 1 & 1 \end{pmatrix} \end{array},$$

and the matrix of $R_2$ relative to the orderings $a, b$ and $x, y, z$ is

$$A_2 = \begin{array}{c} \\ a \\ b \end{array} \begin{array}{c} \begin{array}{ccc} x & y & z \end{array} \\ \begin{pmatrix} 1 & 1 & 0 \\ 0 & 1 & 1 \end{pmatrix} \end{array}.$$

The product of these matrices is

$$A_1 A_2 = \begin{pmatrix} 1 & 1 & 0 \\ 0 & 1 & 1 \\ 1 & 2 & 1 \end{pmatrix}.$$

Let us interpret this product.

The $ik$th entry in $A_1A_2$ is computed as

$$\begin{array}{c} \quad a \quad b \\ i\begin{pmatrix} s & t \end{pmatrix} \end{array} \begin{array}{c} k \\ \begin{pmatrix} u \\ v \end{pmatrix} \end{array} = su + tv.$$

If this value is nonzero, then either $su$ or $tv$ is nonzero. Suppose that $su \neq 0$. (The argument is similar if $tv \neq 0$.) Then $s \neq 0$ and $u \neq 0$. This means that $(i, a) \in R_1$ and $(a, k) \in R_2$. This implies that $(i, k) \in R_2 \circ R_1$. We have shown that if the $ik$th entry in $A_1A_2$ is nonzero, then $(i, k) \in R_2 \circ R_1$. The converse is also true, as we now show.

Assume that $(i, k) \in R_2 \circ R_1$. Then, either

1. $(i, a) \in R_1$ and $(a, k) \in R_2$

or

2. $(i, b) \in R_1$ and $(b, k) \in R_2$.

If 1 holds, then $s = 1$ and $u = 1$, so $su = 1$ and $su + tv$ is nonzero. Similarly, if 2 holds, $tv = 1$ and again we have $su + tv$ nonzero. We have shown that if $(i, k) \in R_2 \circ R_1$, then the $ik$th entry in $A_1A_2$ is nonzero.

We have shown that $(i, k) \in R_2 \circ R_1$ if and only if the $ik$th entry in $A_1A_2$ is nonzero; thus $A_1A_2$ is "almost" the matrix of the relation $R_2 \circ R_1$. To obtain the matrix of the relation $R_2 \circ R_1$, we need only change all nonzero entries in $A_1A_2$ to 1. Thus the matrix of the relation $R_2 \circ R_1$, relative to the previously chosen orderings 1, 2, 3 and $x, y, z$, is

$$\begin{array}{c} \\ 1 \\ 2 \\ 3 \end{array} \begin{array}{c} \;\; x \;\; y \;\; z \\ \begin{pmatrix} 1 & 1 & 0 \\ 0 & 1 & 1 \\ 1 & 1 & 1 \end{pmatrix} \end{array}.$$

◀

The argument given in Example 3.5.5 holds for any relations. We summarize this result as Theorem 3.5.6.

**Theorem 3.5.6** Let $R_1$ be a relation from $X$ to $Y$ and let $R_2$ be a relation from $Y$ to $Z$. Choose orderings of $X$, $Y$, and $Z$. Let $A_1$ be the matrix of $R_1$ and let $A_2$ be the matrix of $R_2$ with respect to the orderings selected. The matrix of the relation $R_2 \circ R_1$ with respect to the orderings selected is obtained by replacing each nonzero term in the matrix product $A_1A_2$ by 1.

**Proof** The proof is sketched before the statement of the theorem. ◀

Theorem 3.5.6 provides a quick test for determining whether a relation is transitive. If $A$ is the matrix of $R$ (relative to some ordering), we compute $A^2$. We then compare $A$ and $A^2$. The relation $R$ is transitive if and only if whenever entry $i, j$ in $A^2$ is nonzero, entry $i, j$ in $A$ is also nonzero. The reason is that entry $i, j$ in $A^2$ is nonzero if and only if there are elements $(i, k)$ and $(k, j)$ in $R$ (see the proof of Theorem 3.5.6). Now $R$ is transitive if and only if whenever $(i, k)$ and $(k, j)$ are in $R$, then $(i, j)$ is in $R$. But $(i, j)$ is in $R$ if and only if entry $i, j$ in $A$ is nonzero. Therefore, $R$ is transitive if and only if whenever entry $i, j$ in $A^2$ is nonzero, entry $i, j$ in $A$ is also nonzero.

**Example 3.5.7** The matrix of the relation

$$R = \{(a,a), (b,b), (c,c), (d,d), (b,c), (c,b)\}$$

on $\{a, b, c, d\}$, relative to the ordering $a, b, c, d$, is

$$A = \begin{pmatrix} 1 & 0 & 0 & 0 \\ 0 & 1 & 1 & 0 \\ 0 & 1 & 1 & 0 \\ 0 & 0 & 0 & 1 \end{pmatrix}.$$

Its square is

$$A^2 = \begin{pmatrix} 1 & 0 & 0 & 0 \\ 0 & 2 & 2 & 0 \\ 0 & 2 & 2 & 0 \\ 0 & 0 & 0 & 1 \end{pmatrix}.$$

We see that whenever entry $i, j$ in $A^2$ is nonzero, entry $i, j$ in $A$ is also nonzero. Therefore, $R$ is transitive. ◀

**Example 3.5.8** The matrix of the relation

$$R = \{(a,a), (b,b), (c,c), (d,d), (a,c), (c,b)\}$$

on $\{a, b, c, d\}$, relative to the ordering $a, b, c, d$, is

$$A = \begin{pmatrix} 1 & 0 & 1 & 0 \\ 0 & 1 & 0 & 0 \\ 0 & 1 & 1 & 0 \\ 0 & 0 & 0 & 1 \end{pmatrix}.$$

Its square is

$$A^2 = \begin{pmatrix} 1 & 1 & 2 & 0 \\ 0 & 1 & 0 & 0 \\ 0 & 2 & 1 & 0 \\ 0 & 0 & 0 & 1 \end{pmatrix}.$$

The entry in row 1, column 2 of $A^2$ is nonzero, but the corresponding entry in $A$ is zero. Therefore, $R$ is *not* transitive. ◀

## 3.5 Problem-Solving Tips

- The matrix of a relation $R$ is another way to represent or specify a relation from $X$ to $Y$. The entry in row $x$ and column $y$ is 1 if $x\,R\,y$, or 0 if $x \not R\, y$.
- A relation is reflexive if and only if the main diagonal of its matrix representation consists of all 1's.
- A relation is symmetric if and only if its matrix is symmetric (i.e., entry $i, j$ always equals entry $j, i$).
- Let $R_1$ be a relation from $X$ to $Y$ and let $R_2$ be a relation from $Y$ to $Z$. Let $A_1$ be the matrix of $R_1$ and let $A_2$ be the matrix of $R_2$. The matrix of the relation $R_2 \circ R_1$ is obtained by replacing each nonzero term in the matrix product $A_1A_2$ by 1.
- To test whether a relation is transitive, let $A$ be its matrix. Compute $A^2$. The relation is transitive if and only if whenever entry $i, j$ in $A^2$ is nonzero, entry $i, j$ in $A$ is also nonzero.

## 3.5 Review Exercises

1. What is the matrix of a relation?

2. Given the matrix of a relation, how can we determine whether the relation is reflexive?

3. Given the matrix of a relation, how can we determine whether the relation is symmetric?

4. Given the matrix of a relation, how can we determine whether the relation is transitive?

5. Given the matrix $A_1$ of the relation $R_1$ and the matrix $A_2$ of the relation $R_2$, explain how to obtain the matrix of the relation $R_2 \circ R_1$.

## 3.5 Exercises

*In Exercises 1–3, find the matrix of the relation R from X to Y relative to the orderings given.*

1. $R = \{(1, \delta), (2, \alpha), (2, \Sigma), (3, \beta), (3, \Sigma)\}$; ordering of $X$: 1, 2, 3; ordering of $Y$: $\alpha, \beta, \Sigma, \delta$

2. $R$ as in Exercise 1; ordering of $X$: 3, 2, 1; ordering of $Y$: $\Sigma, \beta, \alpha, \delta$

3. $R = \{(x, a), (x, c), (y, a), (y, b), (z, d)\}$; ordering of $X$: $x, y, z$; ordering of $Y$: $a, b, c, d$

*In Exercises 4–6, find the matrix of the relation R on X relative to the ordering given.*

4. $R = \{(1, 2), (2, 3), (3, 4), (4, 5)\}$; ordering of $X$: 1, 2, 3, 4, 5

5. $R$ as in Exercise 4; ordering of $X$: 5, 3, 1, 2, 4

6. $R = \{(x, y) \mid x < y\}$; ordering of $X$: 1, 2, 3, 4

7. Find matrices that represent the relations of Exercises 13–16, Section 3.3.

*In Exercises 8–10, write the relation R, given by the matrix, as a set of ordered pairs.*

8.
$$\begin{array}{c} \\ a \\ b \\ c \\ d \end{array} \begin{array}{c} \begin{matrix} w & x & y & z \end{matrix} \\ \begin{pmatrix} 1 & 0 & 1 & 0 \\ 0 & 0 & 0 & 0 \\ 0 & 0 & 1 & 0 \\ 1 & 1 & 1 & 1 \end{pmatrix} \end{array}$$

9.
$$\begin{array}{c} \\ 1 \\ 2 \end{array} \begin{array}{c} \begin{matrix} 1 & 2 & 3 & 4 \end{matrix} \\ \begin{pmatrix} 1 & 0 & 1 & 0 \\ 0 & 1 & 1 & 1 \end{pmatrix} \end{array}$$

10.
$$\begin{array}{c} \\ w \\ x \\ y \\ z \end{array} \begin{array}{c} \begin{matrix} w & x & y & z \end{matrix} \\ \begin{pmatrix} 1 & 0 & 1 & 0 \\ 0 & 0 & 0 & 0 \\ 1 & 0 & 1 & 0 \\ 0 & 0 & 0 & 1 \end{pmatrix} \end{array}$$

11. How can we quickly determine whether a relation $R$ is antisymmetric by examining the matrix of $R$ (relative to some ordering)?

12. Tell whether the relation of Exercise 10 is reflexive, symmetric, transitive, antisymmetric, a partial order, and/or an equivalence relation.

13. Given the matrix of a relation $R$ from $X$ to $Y$, how can we find the matrix of the inverse relation $R^{-1}$?

14. Find the matrix of the inverse of each of the relations of Exercises 8 and 9.

15. Use the matrix of the relation to test for transitivity (see Examples 3.5.7 and 3.5.8) for the relations of Exercises 4, 6, and 10.

*In Exercises 16–18, find*

*(a) The matrix $A_1$ of the relation $R_1$ (relative to the given orderings).*

*(b) The matrix $A_2$ of the relation $R_2$ (relative to the given orderings).*

*(c) The matrix product $A_1A_2$.*

*(d) Use the result of part (c) to find the matrix of the relation $R_2 \circ R_1$.*

*(e) Use the result of part (d) to find the relation $R_2 \circ R_1$ (as a set of ordered pairs).*

16. $R_1 = \{(1, x), (1, y), (2, x), (3, x)\}$; $R_2 = \{(x, b), (y, b), (y, a), (y, c)\}$; orderings: 1, 2, 3; $x, y$; $a, b, c$

17. $R_1 = \{(x, y) \mid x \text{ divides } y\}$; $R_1$ is from $X$ to $Y$; $R_2 = \{(y, z) \mid y > z\}$; $R_2$ is from $Y$ to $Z$; ordering of $X$ and $Y$: 2, 3, 4, 5; ordering of $Z$: 1, 2, 3, 4

18. $R_1 = \{(x, y) \mid x + y \le 6\}$; $R_1$ is from $X$ to $Y$; $R_2 = \{(y, z) \mid y = z + 1\}$; $R_2$ is from $Y$ to $Z$; ordering of $X$, $Y$, and $Z$: 1, 2, 3, 4, 5

19. Given the matrix of an equivalence relation $R$ on $X$, how can we easily find the equivalence class containing the element $x \in X$?

★20. Let $R_1$ be a relation from $X$ to $Y$ and let $R_2$ be a relation from $Y$ to $Z$. Choose orderings of $X$, $Y$, and $Z$. All matrices of relations are with respect to these orderings. Let $A_1$ be the matrix of $R_1$ and let $A_2$ be the matrix of $R_2$. Show that the $ik$th entry in the matrix product $A_1A_2$ is equal to the number of elements in the set

$$\{m \mid (i, m) \in R_1 \text{ and } (m, k) \in R_2\}.$$

21. Suppose that $R_1$ and $R_2$ are relations on a set $X$, $A_1$ is the matrix of $R_1$ relative to some ordering of $X$, and $A_2$ is the matrix of

$R_2$ relative to the same ordering of $X$. Let $A$ be a matrix whose $ij$th entry is 1 if the $ij$th entry of either $A_1$ or $A_2$ is 1. Prove that $A$ is the matrix of $R_1 \cup R_2$.

**22.** Suppose that $R_1$ and $R_2$ are relations on a set $X$, $A_1$ is the matrix of $R_1$ relative to some ordering of $X$, and $A_2$ is the matrix of $R_2$ relative to the same ordering of $X$. Let $A$ be a matrix whose $ij$th entry is 1 if the $ij$th entries of both $A_1$ and $A_2$ are 1. Prove that $A$ is the matrix of $R_1 \cap R_2$.

**23.** Suppose that the matrix of the relation $R_1$ on $\{1, 2, 3\}$ is

$$\begin{pmatrix} 1 & 0 & 0 \\ 0 & 1 & 1 \\ 1 & 0 & 1 \end{pmatrix}$$

relative to the ordering 1, 2, 3, and that the matrix of the relation $R_2$ on $\{1, 2, 3\}$ is

$$\begin{pmatrix} 0 & 1 & 0 \\ 0 & 1 & 0 \\ 1 & 0 & 1 \end{pmatrix}$$

relative to the ordering 1, 2, 3. Use Exercise 21 to find the matrix of the relation $R_1 \cup R_2$ relative to the ordering 1, 2, 3.

**24.** Use Exercise 22 to find the matrix of the relation $R_1 \cap R_2$ relative to the ordering 1, 2, 3 for the relations of Exercise 23.

**25.** How can we quickly determine whether a relation $R$ is a function by examining the matrix of $R$ (relative to some ordering)?

**26.** Let $A$ be the matrix of a function $f$ from $X$ to $Y$ (relative to some orderings of $X$ and $Y$). What conditions must $A$ satisfy for $f$ to be onto $Y$?

**27.** Let $A$ be the matrix of a function $f$ from $X$ to $Y$ (relative to some orderings of $X$ and $Y$). What conditions must $A$ satisfy for $f$ to be one-to-one?

## 3.6 Relational Databases†

The "bi" in a binary relation $R$ refers to the fact that $R$ has two columns when we write $R$ as a table. It is often useful to allow a table to have an arbitrary number of columns. If a table has $n$ columns, the corresponding relation is called an ***n*-ary relation.**

**Example 3.6.1** Table 3.6.1 represents a 4-ary relation. This table expresses the relationship among identification numbers, names, positions, and ages.

**TABLE 3.6.1 ■ PLAYER**

| *ID Number* | *Name* | *Position* | *Age* |
|---|---|---|---|
| 22012 | Johnsonbaugh | c | 22 |
| 93831 | Glover | of | 24 |
| 58199 | Battey | p | 18 |
| 84341 | Cage | c | 30 |
| 01180 | Homer | 1b | 37 |
| 26710 | Score | p | 22 |
| 61049 | Johnsonbaugh | of | 30 |
| 39826 | Singleton | 2b | 31 |

◀

We can also express an $n$-ary relation as a collection of $n$-tuples.

**Example 3.6.2** Table 3.6.1 can be expressed as the set

{(22012, Johnsonbaugh, c, 22), (93831, Glover, of, 24),
(58199, Battey, p, 18), (84341, Cage, c, 30),

†This section can be omitted without loss of continuity.

(01180, Homer, 1b, 37), (26710, Score, p, 22),
(61049, Johnsonbaugh, of, 30), (39826, Singleton, 2b, 31)}

of 4-tuples. ◀

A **database** is a collection of records that are manipulated by a computer. For example, an airline database might contain records of passengers' reservations, flight schedules, equipment, and so on. Computer systems are capable of storing large amounts of information in databases. The data are available to various applications. **Database management systems** are programs that help users access the information in databases. The **relational database model,** invented by E. F. Codd, is based on the concept of an $n$-ary relation. We will briefly introduce some of the fundamental ideas in the theory of relational databases. For more details on relational databases, the reader is referred to [Codd; Date; and Kroenke]. We begin with some of the terminology.

The columns of an $n$-ary relation are called **attributes.** The domain of an attribute is a set to which all the elements in that attribute belong. For example, in Table 3.6.1, the attribute Age might be taken to be the set of all positive integers less than 100. The attribute Name might be taken to be all strings over the alphabet having length 30 or less.

A single attribute or a combination of attributes for a relation is a **key** if the values of the attributes uniquely define an $n$-tuple. For example, in Table 3.6.1, we can take the attribute ID Number as a key. (It is assumed that each person has a unique identification number.) The attribute Name is not a key because different persons can have the same name. For the same reason, we cannot take the attribute Position or Age as a key. Name and Position, in combination, could be used as a key for Table 3.6.1, since in our example a player is uniquely defined by a name and a position.

A database management system responds to **queries.** A query is a request for information from the database. For example, "Find all persons who play outfield" is a meaningful query for the relation given by Table 3.6.1. We will discuss several operations on relations that are used to answer queries in the relational database model.

**Example 3.6.3** **Select** The **selection operator** chooses certain $n$-tuples from a relation. The choices are made by giving conditions on the attributes. For example, for the relation PLAYER given in Table 3.6.1, PLAYER [Position = c] will select the tuples

(22012, Johnsonbaugh, c, 22), (84341, Cage, c, 30). ◀

**Example 3.6.4** **Project** Whereas the selection operator chooses rows of a relation, the **projection operator** chooses columns. In addition, duplicates are eliminated. For example, for the relation PLAYER given by Table 3.6.1, PLAYER [Name, Position] will select the tuples

(Johnsonbaugh, c), (Glover, of), (Battey, p), (Cage, c),
(Homer, 1b), (Score, p), (Johnsonbaugh, of), (Singleton, 2b). ◀

**Example 3.6.5** **Join** The selection and projection operators manipulate a single relation; **join** manipulates two relations. The join operation on relations $R_1$ and $R_2$ begins by examining all pairs of tuples, one from $R_1$ and one from $R_2$. If the join condition is satisfied, the tuples are combined to form a new tuple. The join condition specifies a relationship between an attribute in $R_1$ and an attribute in $R_2$. For example, let us perform a join operation on Tables 3.6.1 and 3.6.2. As the condition we take ID Number = PID. We take a row from Table 3.6.1 and a row from Table 3.6.2 and if ID Number = PID, we combine the rows.

For example, the ID Number 01180 in the fifth row (01180, Homer, 1b, 37) of Table 3.6.1 matches the PID in the fourth row (01180, Mutts) of Table 3.6.2. These tuples are combined by first writing the tuple from Table 3.6.1, following it by the tuple from Table 3.6.2, and eliminating the equal entries in the specified attributes to give

(01180, Homer, 1b, 37, Mutts).

This operation is expressed as

PLAYER [ID Number = PID] ASSIGNMENT.

The relation obtained by executing this join is shown in Table 3.6.3.

**TABLE 3.6.2 ■ ASSIGNMENT**

| *PID* | *Team* |
|---|---|
| 39826 | Blue Sox |
| 26710 | Mutts |
| 58199 | Jackalopes |
| 01180 | Mutts |

**TABLE 3.6.3 ■ PLAYER [ID Number = PID] ASSIGNMENT**

| *ID Number* | *Name* | *Position* | *Age* | *Team* |
|---|---|---|---|---|
| 58199 | Battey | p | 18 | Jackalopes |
| 01180 | Homer | 1b | 37 | Mutts |
| 26710 | Score | p | 22 | Mutts |
| 39826 | Singleton | 2b | 31 | Blue Sox |

◀

Most queries to a relational database require several operations to provide the answer.

**Example 3.6.6** Describe operations that provide the answer to the query "Find the names of all persons who play for some team."

**SOLUTION** If we first join the relations given by Tables 3.6.1 and 3.6.2 subject to the condition ID Number = PID, we will obtain Table 3.6.3, which lists all persons who play for some team as well as other information. To obtain the names, we need only project on the attribute Name. We obtain the relation

| *Name* |
|---|
| Battey |
| Homer |
| Score |
| Singleton |

Formally, these operations would be specified as

TEMP := PLAYER [ID Number = PID] ASSIGNMENT
TEMP [Name]

◀

**Example 3.6.7** Describe operations that provide the answer to the query "Find the names of all persons who play for the Mutts."

**SOLUTION** If we first use the selection operator to pick the rows of Table 3.6.2 that reference Mutts' players, we obtain the relation

TEMP1

| *PID* | *Team* |
|---|---|
| 26710 | Mutts |
| 01180 | Mutts |

If we now join Table 3.6.1 and the relation TEMP1 subject to ID Number = PID, we obtain the relation

TEMP2

| *ID Number* | *Name* | *Position* | *Age* | *Team* |
|---|---|---|---|---|
| 01180 | Homer | 1b | 37 | Mutts |
| 26710 | Score | p | 22 | Mutts |

If we project the relation TEMP2 on the attribute Name, we obtain the relation

| *Name* |
|---|
| Homer |
| Score |

We would formally specify these operations as follows:

TEMP1 := ASSIGNMENT [Team = Mutts]
TEMP2 :=PLAYER [ID Number = PID] TEMP1
TEMP2 [Name]

Notice that the operations

TEMP1 := PLAYER [ID Number = PID] ASSIGNMENT
TEMP2 := TEMP1 [Team = Mutts]
TEMP2 [Name]

would also answer the query of Example 3.6.7.

## 3.6 Problem-Solving Tips

A relational database represents data as tables (*n*-ary relations). Information from the database is obtained by manipulating the tables. In this section, we discussed the operations *select* (choose rows specified by a given condition), *project* (choose columns specified by a given condition), and *join* (combine rows from two tables as specified by a given condition).

## 3.6 Review Exercises

1. What is an *n*-ary relation?
2. What is a database management system?
3. What is a relational database?
4. What is a key?
5. What is a query?
6. Explain how the selection operator works and give an example.
7. Explain how the project operator works and give an example.
8. Explain how the join operator works and give an example.

## 3.6 Exercises

**1.** Express the relation given by Table 3.6.4 as a set of $n$-tuples.

**TABLE 3.6.4 ■ EMPLOYEE**

| *ID* | *Name* | *Manager* |
|---|---|---|
| 1089 | Suzuki | Zamora |
| 5620 | Kaminski | Jones |
| 9354 | Jones | Yu |
| 9551 | Ryan | Washington |
| 3600 | Beaulieu | Yu |
| 0285 | Schmidt | Jones |
| 6684 | Manacotti | Jones |

**2.** Express the relation given by Table 3.6.5 as a set of $n$-tuples.

**TABLE 3.6.5 ■ DEPARTMENT**

| *Dept* | *Manager* |
|---|---|
| 23 | Jones |
| 04 | Yu |
| 96 | Zamora |
| 66 | Washington |

**3.** Express the relation given by Table 3.6.6 as a set of $n$-tuples.

**TABLE 3.6.6 ■ SUPPLIER**

| *Dept* | *Part No* | *Amount* |
|---|---|---|
| 04 | 335B2 | 220 |
| 23 | 2A | 14 |
| 04 | 8C200 | 302 |
| 66 | 42C | 3 |
| 04 | 900 | 7720 |
| 96 | 20A8 | 200 |
| 96 | 1199C | 296 |
| 23 | 772 | 39 |

**4.** Express the relation given by Table 3.6.7 as a set of $n$-tuples.

**TABLE 3.6.7 ■ BUYER**

| *Name* | *Part No* |
|---|---|
| United Supplies | 2A |
| ABC Unlimited | 8C200 |
| United Supplies | 1199C |
| JCN Electronics | 2A |
| United Supplies | 335B2 |
| ABC Unlimited | 772 |
| Danny's | 900 |
| United Supplies | 772 |
| Underhanded Sales | 20A8 |
| Danny's | 20A8 |
| DePaul University | 42C |
| ABC Unlimited | 20A8 |

*In Exercises 5–20, write a sequence of operations to answer the query. Also, provide an answer to the query. Use Tables 3.6.4–3.6.7.*

**5.** Find the names of all employees. (Do not include any managers.)

**6.** Find the names of all managers.

**7.** Find all part numbers.

**8.** Find the names of all buyers.

**9.** Find the names of all employees who are managed by Jones.

**10.** Find all part numbers supplied by department 96.

**11.** Find all buyers of part 20A8.

**12.** Find all employees in department 04.

**13.** Find the part numbers of parts of which there are at least 100 items on hand.

**14.** Find all department numbers of departments that supply parts to Danny's.

**15.** Find the part numbers and amounts of parts bought by United Supplies.

**16.** Find all managers of departments that produce parts for ABC Unlimited.

**17.** Find the names of all employees who work in departments that supply parts for JCN Electronics.

**18.** Find all buyers who buy parts in the department managed by Jones.

**19.** Find all buyers who buy parts that are produced by the department for which Suzuki works.

**20.** Find all part numbers and amounts for Zamora's department.

**21.** Make up at least three $n$-ary relations with artificial data that might be used in a medical database. Illustrate how your database would be used by posing and answering two queries. Also, write a sequence of operations that could be used to answer the queries.

**22.** Describe a *union* operation on a relational database. Illustrate how your operator works by answering the following query, using the relations of Tables 3.6.4–3.6.7: Find the names of all employees who work in either department 23 or department 96. Also, write a sequence of operations that could be used to answer the query.

**23.** Describe an *intersection* operation on a relational database. Illustrate how your operator works by answering the following query, using the relations of Tables 3.6.4–3.6.7: Find the names of all buyers who buy both parts 2A and 1199C. Also, write a sequence of operations that could be used to answer the query.

**24.** Describe a *difference* operation on a relational database. Illustrate how your operator works by answering the following query, using the relations of Tables 3.6.4–3.6.7: Find the names of all employees who do not work in department 04. Also, write a sequence of operations that could be used to answer the query.

## Chapter 3 Notes

Most general references on discrete mathematics address the topics of this chapter. [Halmos; Lipschutz; and Stoll] are recommended to the reader wanting to study functions in more detail. [Codd; Date; Kroenke; and Ullman] are recommended references on databases in general and the relational model in particular.

## Chapter 3 Review

### Section 3.1

1. Function from $X$ to $Y$, $f : X \to Y$: a subset $f$ of $X \times Y$ such that for each $x \in X$, there is exactly one $y \in Y$ with $(x, y) \in f$
2. $x$ mod $y$: remainder when $x$ is divided by $y$
3. Hash function
4. Collision for a hash function $H$: $H(x) = H(y)$
5. Collision resolution policy
6. Floor of $x$, $\lfloor x \rfloor$: greatest integer less than or equal to $x$
7. Ceiling of $x$, $\lceil x \rceil$: least integer greater than or equal to $x$
8. One-to-one function $f$: if $f(x) = f(x')$, then $x = x'$
9. Onto function $f$ from $X$ to $Y$: range of $f = Y$
10. Bijection: one-to-one and onto function
11. Inverse $f^{-1}$ of a one-to-one, onto function $f$: $\{(y, x) \mid (x, y) \in f\}$
12. Composition of functions: $f \circ g = \{(x, z) \mid (x, y) \in g$ and $(y, z) \in f\}$
13. Binary operator on $X$: function from $X \times X$ to $X$
14. Unary operator on $X$: function from $X$ to $X$

### Section 3.2

15. Sequence: function whose domain is a subset of integers
16. Index: in the sequence $\{s_n\}$, $n$ is the index
17. Increasing sequence: if $i < j$, then $s_i < s_j$
18. Decreasing sequence: if $i < j$, then $s_i > s_j$
19. Nonincreasing sequence: if $i < j$, then $s_i \geq s_j$
20. Nondecreasing sequence: if $i < j$, then $s_i \leq s_j$
21. Subsequence $s_{n_k}$ of the sequence $\{s_n\}$
22. Sum or sigma notation: $\sum_{i=m}^{n} a_i = a_m + a_{m+1} + \cdots + a_n$
23. Product notation: $\prod_{i=m}^{n} a_i = a_m \cdot a_{m+1} \cdots a_n$
24. Geometric sum: $\sum_{i=0}^{n} ar^i$
25. String: finite sequence
26. Null string, $\lambda$: string with no elements
27. $X^*$: set of all strings over $X$, including the null string
28. $X^+$: set of all nonnull strings over $X$
29. Length of string $\alpha$, $|\alpha|$: number of elements in $\alpha$
30. Concatenation of strings $\alpha$ and $\beta$, $\alpha\beta$: $\alpha$ followed by $\beta$
31. Substring of $\alpha$: a string $\beta$ for which there are strings $\gamma$ and $\delta$ with $\alpha = \gamma\beta\delta$

### Section 3.3

32. Binary relation from $X$ to $Y$: set of ordered pairs $(x, y)$, $x \in X, y \in Y$
33. Digraph of a binary relation
34. Reflexive relation $R$ on $X$: $(x, x) \in R$ for all $x \in X$
35. Symmetric relation $R$ on $X$: for all $x, y \in X$, if $(x, y) \in R$, then $(y, x) \in R$
36. Antisymmetric relation $R$ on $X$: for all $x, y \in X$, if $(x, y) \in R$ and $(y, x) \in R$, then $x = y$
37. Transitive relation $R$ on $X$: for all $x, y, z \in X$, if $(x, y)$ and $(y, z)$ are in $R$, then $(x, z) \in R$
38. Partial order: relation that is reflexive, antisymmetric, and transitive
39. Inverse relation $R^{-1}$: $\{(y, x) \mid (x, y) \in R\}$
40. Composition of relations $R_2 \circ R_1$: $\{(x, z) \mid (x, y) \in R_1$ and $(y, z) \in R_2\}$

### Section 3.4

41. Equivalence relation: relation that is reflexive, symmetric, and transitive
42. Equivalence class containing $a$ given by equivalence relation $R$: $[a] = \{x \mid xRa\}$
43. Equivalence classes partition the set (Theorem 3.4.8)

### Section 3.5

44. Matrix of a relation
45. $R$ is a reflexive relation if and only if the main diagonal of the matrix of $R$ consists of 1's.
46. $R$ is a symmetric relation if and only if the matrix of $R$ is symmetric about the main diagonal.
47. If $A_1$ is the matrix of the relation $R_1$ and $A_2$ is the matrix of the relation $R_2$, the matrix of the relation $R_2 \circ R_1$ is obtained by replacing each nonzero term in the matrix product $A_1A_2$ by 1.
48. If $A$ is the matrix of a relation $R$, $R$ is transitive if and only if whenever entry $i, j$ in $A^2$ is nonzero, entry $i, j$ in $A$ is also nonzero.

### Section 3.6

49. $n$-ary relation: Set of $n$-tuples
50. Database management system
51. Relational database
52. Key
53. Query
54. Select
55. Project
56. Join

## Chapter 3 Self-Test

1. Let $X$ be the set of strings over $\{a, b\}$ of length 4 and let $Y$ be the set of strings over $\{a, b\}$ of length 3. Define a function $f$ from $X$ to $Y$ by the rule

$$f(\alpha) = \text{string consisting of the first three characters of } \alpha.$$

Is $f$ one-to-one? Is $f$ onto?

2. Let $b_n = \sum_{i=1}^{n} (i+1)^2 - i^2$.
   (a) Find $b_5$ and $b_{10}$.
   (b) Find a formula for $b_n$.
   (c) Is $b$ increasing?
   (d) Is $b$ decreasing?

*In Exercises 3 and 4, determine whether the relation defined on the set of positive integers is reflexive, symmetric, antisymmetric, transitive, and/or a partial order.*

3. $(x, y) \in R$ if 2 divides $x + y$
4. $(x, y) \in R$ if 3 divides $x + y$
5. Is the relation

$$\{(1, 1), (1, 2), (2, 2), (4, 4), (2, 1), (3, 3)\}$$

an equivalence relation on $\{1, 2, 3, 4\}$? Explain.

6. Given that the relation

$$\{(1, 1), (2, 2), (3, 3), (4, 4), (1, 2), (2, 1), (3, 4), (4, 3)\}$$

is an equivalence relation on $\{1, 2, 3, 4\}$, find $[3]$, the equivalence class containing 3. How many (distinct) equivalence classes are there?

7. Find real numbers $x$ and $y$ satisfying $\lfloor x \rfloor \lfloor y \rfloor = \lfloor xy \rfloor - 1$.
8. For the sequence $a$ defined by $a_n = 2n + 2$, find
   (a) $a_6$
   (b) $\sum_{i=1}^{3} a_i$
   (c) $\prod_{i=1}^{3} a_i$
   (d) a formula for the subsequence of $a$ obtained by selecting every other term of $a$ starting with the first.
9. Give examples of functions $f$ and $g$ such that $f \circ g$ is onto, but $g$ is not onto.
10. Let $\alpha = ccddc$ and $\beta = c^3d^2$. Find
    (a) $\alpha\beta$ (b) $\beta\alpha$ (c) $|\alpha|$ (d) $|\alpha\alpha\beta\alpha|$
11. For the hash function $h(x) = x \bmod 13$, show how the data

$$784,\ 281,\ 1141,\ 18,\ 1,\ 329,\ 620,\ 43,\ 31,\ 684$$

would be inserted in the order given in initially empty cells indexed 0 to 12.

12. Find the equivalence relation (as a set of ordered pairs) on $\{a, b, c, d, e\}$ whose equivalence classes are

$$\{a\},\quad \{b, d, e\},\quad \{c\}.$$

13. Suppose that $R$ is a relation on $X$ that is symmetric and transitive but not reflexive. Suppose also that $|X| \geq 2$. Define the relation $\overline{R}$ on $X$ by $\overline{R} = X \times X - R$. Which of the following must be true? For each false statement, provide a counterexample.
    (a) $\overline{R}$ is reflexive.
    (b) $\overline{R}$ is symmetric.
    (c) $\overline{R}$ is not antisymmetric.
    (d) $\overline{R}$ is transitive.
14. Rewrite the sum $\sum_{i=1}^{n} (n-i)r^i$ replacing the index $i$ by $k$, where $i = k + 2$.
15. Give an example of a relation on $\{1, 2, 3, 4\}$ that is reflexive, not antisymmetric, and not transitive.
16. Let $R$ be the relation defined on the set of eight-bit strings by $s_1 R s_2$ provided that $s_1$ and $s_2$ have the same number of zeros.
    (a) Show that $R$ is an equivalence relation.
    (b) How many equivalence classes are there?
    (c) List one member of each equivalence class.

*Exercises 17–20 refer to the relations*

$$R_1 = \{(1, x), (2, x), (2, y), (3, y)\},$$
$$R_2 = \{(x, a), (x, b), (y, a), (y, c)\}.$$

17. Find the matrix $A_1$ of the relation $R_1$ relative to the orderings

$$1, 2, 3;\ x, y.$$

18. Find the matrix $A_2$ of the relation $R_2$ relative to the orderings

$$x, y;\ a, b, c.$$

19. Find the matrix product $A_1A_2$.
20. Use the result of Exercise 19 to find the matrix of the relation $R_2 \circ R_1$.

*In Exercises 21–24, write a sequence of operations to answer the query. Also, provide an answer to the query. Use Tables 3.6.1 and 3.6.2.*

21. Find all teams.
22. Find all players' names and ages.
23. Find the names of all teams that have a pitcher.
24. Find the names of all teams that have players aged 30 years or older.

## Chapter 3 Computer Exercises

1. Implement a hashing system for storing integers in an array.
2. Write a program that determines whether a credit card number has a valid check digit.
3. Write a program that determines whether an ISBN has a valid check digit.
4. Write a program that lists all transpositions of two distinct adjacent digits (e.g., 25 $\leftrightarrow$ 52) that are *not* detected by the Luhn algorithm (i.e., transpositions that produce equal check digits).
5. Write a program that lists all transpositions of two distinct adjacent digits (e.g., 25 $\leftrightarrow$ 52) that are *not* detected by the ISBN check digit (i.e., transpositions that produce equal check digits in their ISBNs).
6. Write a program that generates pseudorandom integers.

*In Exercises 7–12, assume that a sequence from* $\{1, \ldots, n\}$ *to the real numbers is represented as an array A, indexed from 1 to n.*

7. Write a program that tests whether $A$ is one-to-one.
8. Write a program that tests whether $A$ is onto a given set.
9. Write a program that tests whether $A$ is increasing.
10. Write a program that tests whether $A$ is decreasing.
11. Write a program that tests whether $A$ is nonincreasing.
12. Write a program that tests whether $A$ is nondecreasing.
13. Write a program to determine whether one sequence is a subsequence of another sequence.
14. Write a program to determine whether one string is a substring of another string.
15. Write a program to determine whether a relation is reflexive.
16. Write a program to determine whether a relation is antisymmetric.
17. Write a program to determine whether a relation is transitive.
18. Write a program that finds the inverse of a relation.
19. Write a program that finds the composition $R \circ S$ of relations $R$ and $S$.
20. Write a program that checks whether a relation $R$ is an equivalence relation. If $R$ is an equivalence relation, the program outputs the equivalence classes of $R$.
21. Write a program to determine whether a relation is a function from a set $X$ to a set $Y$.
22. [*Project*] Prepare a report on a commercial relational database such as Oracle or Access.

# Chapter 4

# ALGORITHMS



An **algorithm** is a step-by-step method of solving some problem. Such an approach to problem-solving is not new; indeed, the word "algorithm" derives from the name of the ninth-century Persian mathematician al-Khowārizmī. Today, "algorithm" typically refers to a solution that can be executed by a computer. In this book, we will be concerned primarily with algorithms that can be executed by a "traditional" computer, that is, a computer, such as a personal computer, with a single processor that executes instructions step-by-step.

After introducing algorithms and providing several examples, we turn to the analysis of algorithms, which refers to the time and space required to execute algorithms. We conclude by discussing recursive algorithms—algorithms that refer to themselves.

## 4.1 Introduction

Algorithms typically have the following characteristics:

- ***Input*** The algorithm receives *input*.
- ***Output*** The algorithm produces *output*.
- ***Precision*** The steps are precisely stated.
- ***Determinism*** The intermediate results of each step of execution are unique and are determined only by the inputs and the results of the preceding steps.
- ***Finiteness*** The algorithm *terminates;* that is, it stops after finitely many instructions have been executed.
- ***Correctness*** The output produced by the algorithm is correct; that is, the algorithm correctly solves the problem.
- ***Generality*** The algorithm applies to a set of inputs.

As an example, consider the following algorithm that finds the maximum of three numbers $a$, $b$, and $c$:

1. $large = a$.
2. If $b > large$, then $large = b$.
3. If $c > large$, then $large = c$.

(As explained in Appendix C, $=$ is the assignment operator.)

The idea of the algorithm is to inspect the numbers one by one and copy the largest value seen into a variable *large*. At the conclusion of the algorithm, *large* will then be equal to the largest of the three numbers.

We show how the preceding algorithm executes for some specific values of $a$, $b$, and $c$. Such a simulation is called a **trace.** First suppose that $a = 1$, $b = 5$, and $c = 3$. At line 1, we set *large* to $a$ (1). At line 2, $b > large$ $(5 > 1)$ is true, so we set *large* to $b$ (5). At line 3, $c > large$ $(3 > 5)$ is false, so we do nothing. At this point *large* is 5, the largest of $a$, $b$, and $c$.

Suppose that $a = 6$, $b = 1$, and $c = 9$. At line 1, we set *large* to $a$ (6). At line 2, $b > large$ $(1 > 6)$ is false, so we do nothing. At line 3, $c > large$ $(9 > 6)$ is true, so we set *large* to 9. At this point *large* is 9, the largest of $a$, $b$, and $c$.

We verify that our example algorithm has the properties set forth at the beginning of this section.

The algorithm receives the three values $a$, $b$, and $c$ as input and produces the value *large* as output.

The steps of the algorithm are stated sufficiently precisely so that the algorithm could be written in a programming language and executed by a computer.

Given values for the input, each intermediate step of an algorithm produces a unique result. For example, given the values $a = 1$, $b = 5$, and $c = 3$, at line 2, *large* will be set to 5 regardless of who executes the algorithm.

The algorithm terminates after finitely many steps (three steps) correctly answering the given question (find the largest of the three values input).

The algorithm is general; it can find the largest value of *any* three numbers.

Our description of what an algorithm is will suffice for our needs in this book. However, it should be noted that it is possible to give a precise, mathematical definition of "algorithm" (see the Notes for Chapter 12).

Although ordinary language is sometimes adequate to specify an algorithm, most mathematicians and computer scientists prefer **pseudocode** because of its precision, structure, and universality. Pseudocode is so named because it resembles the actual code of computer languages such as C++ and Java. There are many versions of pseudocode. Unlike actual computer languages, which must be concerned about semicolons, uppercase and lowercase letters, special words, and so on, any version of pseudocode is acceptable as long as its instructions are unambiguous. Our pseudocode is described in detail in Appendix C.

As our first example of an algorithm written in pseudocode, we rewrite the first algorithm in this section, which finds the maximum of three numbers.

**Algorithm 4.1.1** **Finding the Maximum of Three Numbers**

This algorithm finds the largest of the numbers $a$, $b$, and $c$.

Input: $a, b, c$

Output: *large* (the largest of $a$, $b$, and $c$)

```
1. max3(a, b, c) {
2.   large = a
3.   if (b > large) // if b is larger than large, update large
4.     large = b
5.   if (c > large) // if c is larger than large, update large
6.     large = c
7.   return large
8. }
```

Our algorithms consist of a title, a brief description of the algorithm, the input to and output from the algorithm, and the functions containing the instructions of the algorithm. Algorithm 4.1.1 consists of a single function. To make it convenient to refer to individual lines within a function, we sometimes number some of the lines. The function in Algorithm 4.1.1 has eight numbered lines.

When the function in Algorithm 4.1.1 executes, at line 2 we set *large* to $a$. At line 3, $b$ and *large* are compared. If $b$ is greater than *large,* we execute line 4

$$large = b$$

but if $b$ is not greater than *large,* we skip to line 5. At line 5, $c$ and *large* are compared. If $c$ is greater than *large,* we execute line 6

$$large = c$$

but if $c$ is not greater than *large,* we skip to line 7. Thus when we arrive at line 7, *large* will correctly hold the largest of $a$, $b$, and $c$.

At line 7 we return the value of *large,* which is equal to the largest of the numbers $a$, $b$, and $c$, to the invoker of the function and terminate the function. Algorithm 4.1.1 has correctly found the largest of three numbers.

The method of Algorithm 4.1.1 can be used to find the largest value in a sequence.

**Algorithm 4.1.2** **Finding the Maximum Value in a Sequence**

This algorithm finds the largest of the numbers $s_1, \ldots, s_n$.

Input: $s, n$

Output: *large* (the largest value in the sequence $s$)

```
max(s, n) {
    large = s_1
    for i = 2 to n
        if (s_i > large)
            large = s_i
    return large
}
```

We verify that Algorithm 4.1.2 is correct by proving that

$$large \text{ is the largest value in the subsequence } s_1, \ldots, s_i \tag{4.1.1}$$

is a loop invariant using induction on $i$.

For the Basis Step ($i = 1$), we note that just before the for loop begins executing, *large* is set to $s_1$; so *large* is surely the largest value in the subsequence $s_1$.

Assume that *large* is the largest value in the subsequence $s_1, \ldots, s_i$. If $i < n$ is true (so that the for loop body executes again), $i$ becomes $i + 1$. Suppose first that $s_{i+1} >$ *large*. It then follows that $s_{i+1}$ is the largest value in the subsequence $s_1, \ldots, s_i, s_{i+1}$. In this case, the algorithm assigns *large* the value $s_{i+1}$. Now *large* is equal to the largest value in the subsequence $s_1, \ldots, s_i, s_{i+1}$. Suppose next that $s_{i+1} \leq$ *large*. It then follows that *large* is the largest value in the subsequence $s_1, \ldots, s_i, s_{i+1}$. In this case, the algorithm does *not* change the value of *large;* thus, *large* is the largest value in the subsequence $s_1, \ldots, s_i, s_{i+1}$. We have proved the Inductive Step. Therefore, (4.1.1) is a loop invariant.

The for loop terminates when $i = n$. Because (4.1.1) is a loop invariant, at this point *large* is the largest value in the sequence $s_1, \ldots, s_n$. Therefore, Algorithm 4.1.2 is correct.

### 4.1 Problem-Solving Tips

To construct an algorithm, it is often helpful to assume that you are in the middle of the algorithm and part of the problem has been solved. For example, in finding the largest element in a sequence $s_1, \ldots, s_n$ (Algorithm 4.1.2), it was helpful to *assume* that we had already found the largest element *large* in the subsequence $s_1, \ldots, s_i$. Then, all we had to do was look at the next element $s_{i+1}$ and, if $s_{i+1}$ was larger than *large,* we simply updated *large.* If $s_{i+1}$ was not larger than *large,* we did not modify *large.* Iterating this procedure yields the algorithm. These observations also led to the loop invariant (4.1.1) which allowed us to *prove* that Algorithm 4.1.2 is correct.

## 4.1 Review Exercises

1. What is an algorithm?
2. Describe the following properties an algorithm typically has: input, output, precision, determinism, finiteness, correctness, and generality.
3. What is a trace of an algorithm?
4. What are the advantages of pseudocode over ordinary text in writing an algorithm?
5. How do algorithms relate to pseudocode functions?

## 4.1 Exercises

**1.** Consult the instructions for connecting a DVD or Blu-ray player to a TV. Which properties of an algorithm—input, output, precision, determinism, finiteness, correctness, generality—are present? Which properties are lacking?

**2.** Consult the instructions for adding a contact to a cell phone. Which properties of an algorithm—input, output, precision, determinism, finiteness, correctness, generality—are present? Which properties are lacking?

**3.** Goldbach's conjecture states that every even number greater than 2 is the sum of two prime numbers. Here is a proposed algorithm that checks whether Goldbach's conjecture is true:

1. Let $n = 4$.
2. If $n$ is not the sum of two primes, output "no" and stop.
3. Else increase $n$ by 2 and continue with step 2.
4. Output "yes" and stop.

Which properties of an algorithm—input, output, precision, determinism, finiteness, correctness, generality—does this proposed algorithm have? Do any of them depend on the truth of Goldbach's conjecture (which mathematicians have not yet settled)?

**4.** Write an algorithm that finds the smallest element among $a$, $b$, and $c$.

**5.** Write an algorithm that finds the second-smallest element among $a$, $b$, and $c$. Assume that the values of $a$, $b$, and $c$ are distinct.

**6.** Write an algorithm that returns the smallest value in the sequence $s_1, \ldots, s_n$.

**7.** Write an algorithm that returns the largest and second-largest values in the sequence $s_1, \ldots, s_n$. Assume that $n > 1$ and the values in the sequence are distinct.

**8.** Write an algorithm that returns the smallest and second-smallest values in the sequence $s_1, \ldots, s_n$. Assume that $n > 1$ and the values in the sequence are distinct.

**9.** Write an algorithm that outputs the smallest and largest values in the sequence $s_1, \ldots, s_n$.

**10.** Write an algorithm that returns the index of the first occurrence of the largest element in the sequence $s_1, \ldots, s_n$. *Example:* If the sequence is 6.2, 8.9, 4.2, 8.9, the algorithm returns the value 2.

**11.** Write an algorithm that returns the index of the last occurrence of the largest element in the sequence $s_1, \ldots, s_n$.

*Example:* If the sequence is 6.2, 8.9, 4.2, 8.9, the algorithm returns the value 4.

**12.** Write an algorithm that returns the sum of the sequence of numbers $s_1, \ldots, s_n$.

**13.** Write an algorithm that returns the index of the first item that is less than its predecessor in the sequence $s_1, \ldots, s_n$. If $s$ is in nondecreasing order, the algorithm returns the value 0. *Example:* If the sequence is

AMY BRUNO ELIE DAN ZEKE,

the algorithm returns the value 4.

**14.** Write an algorithm that returns the index of the first item that is greater than its predecessor in the sequence $s_1, \ldots, s_n$. If $s$ is in nonincreasing order, the algorithm returns the value 0. *Example:* If the sequence is

AMY BRUNO ELIE DAN ZEKE,

the algorithm returns the value 2.

**15.** Write an algorithm that reverses the sequence $s_1, \ldots, s_n$. *Example:* If the sequence is

AMY BRUNO ELIE,

the reversed sequence is

ELIE BRUNO AMY.

**16.** Write the standard method of adding two positive decimal integers, taught in elementary schools, as an algorithm.

**17.** Write an algorithm that receives as input the $n \times n$ matrix $A$ and outputs the transpose $A^T$.

**18.** Write an algorithm that receives as input the matrix of a relation $R$ and tests whether $R$ is reflexive.

**19.** Write an algorithm that receives as input the matrix of a relation $R$ and tests whether $R$ is symmetric.

**20.** Write an algorithm that receives as input the matrix of a relation $R$ and tests whether $R$ is transitive.

**21.** Write an algorithm that receives as input the matrix of a relation $R$ and tests whether $R$ is antisymmetric.

**22.** Write an algorithm that receives as input the matrix of a relation $R$ and tests whether $R$ is a function.

**23.** Write an algorithm that receives as input the matrix of a relation $R$ and produces as output the matrix of the inverse relation $R^{-1}$.

**24.** Write an algorithm that receives as input the matrices of relations $R_1$ and $R_2$ and produces as output the matrix of the composition $R_2 \circ R_1$.

**25.** Write an algorithm whose input is a sequence $s_1, \ldots, s_n$ and a value $x$. (Assume that all the values are real numbers.) The algorithm returns true if $s_i + s_j = x$, for some $i \neq j$, and false otherwise. *Example:* If the input sequence is 2, 12, 6, 14 and $x = 26$, the algorithm returns true because $12 + 14 = 26$. If the input sequence is 2, 12, 6, 14 and $x = 4$, the algorithm returns false because no *distinct* pair in the sequence sums to 4.

## 4.2 Examples of Algorithms

Algorithms have been devised to solve many problems. In this section, we give examples of several useful algorithms. Throughout the remainder of the book, we will investigate many additional algorithms.

### Searching

A large amount of computer time is devoted to searching. When a teller looks for a record in a bank, a computer program searches for the record. Looking for a solution to a puzzle or for an optimal move in a game can be stated as a searching problem. Using a search engine on the web is another example of a searching problem. Looking for specified text in a document when running a word processor is yet another example of a searching problem. We discuss an algorithm to solve the text-searching problem.

Suppose that we are given text $t$ (e.g., a word processor document) and we want to find the first occurrence of pattern $p$ in $t$ (e.g., we want to find the first occurrence of the string $p =$ "Nova Scotia" in $t$) or determine that $p$ does not occur in $t$. We index the characters in $t$ starting at 1. One approach to searching for $p$ is to check whether $p$ occurs at index 1 in $t$. If so, we stop, having found the first occurrence of $p$ in $t$. If not, we check whether $p$ occurs at index 2 in $t$. If so, we stop, having found the first occurrence of $p$ in $t$. If not, we next check whether $p$ occurs at index 3 in $t$, and so on.

We state the text-searching algorithm as Algorithm 4.2.1.

**Algorithm 4.2.1 Text Search**

This algorithm searches for an occurrence of the pattern $p$ in text $t$. It returns the smallest index $i$ such that $p$ occurs in $t$ starting at index $i$. If $p$ does not occur in $t$, it returns 0.

```
Input:  p (indexed from 1 to m), m, t (indexed from 1 to n), n
Output: i

text_search(p, m, t, n) {
  for i = 1 to n − m + 1 {
    j = 1

    // i is the index in t of the first character of the substring
    // to compare with p, and j is the index in p

    // the while loop compares t_i ··· t_{i+m−1} and p_1 ··· p_m
    while (t_{i+j−1} == p_j) {
      j = j + 1
      if (j > m)
        return i
    }
  }
  return 0
}
```

The variable $i$ marks the index in $t$ of the first character of the substring to compare with $p$. The algorithm first tries $i = 1$, then $i = 2$, and so on. Index $n - m + 1$ is the last possible value for $i$ since, at this point, the string $t_{n-m+1} t_{n-m+2} \cdots t_m$ has length exactly $m$.

After the value of $i$ is set, the while loop compares $t_i \cdots t_{i+m-1}$ and $p_1 \cdots p_m$. If the characters match,

$$t_{i+j-1} == p_j$$

$j$ is incremented

$$j = j + 1$$

and the next characters are compared. If $j$ is $m + 1$, all $m$ characters have matched and we have found $p$ at index $i$ in $t$. In this case, the algorithm returns $i$:

```
if (j > m)
  return i
```

If the for loop runs to completion, a match was never found; so the algorithm returns 0.

**Example 4.2.2** Figure 4.2.1 shows a trace of Algorithm 4.2.1 where we are searching for the pattern "001" in the text "010001".

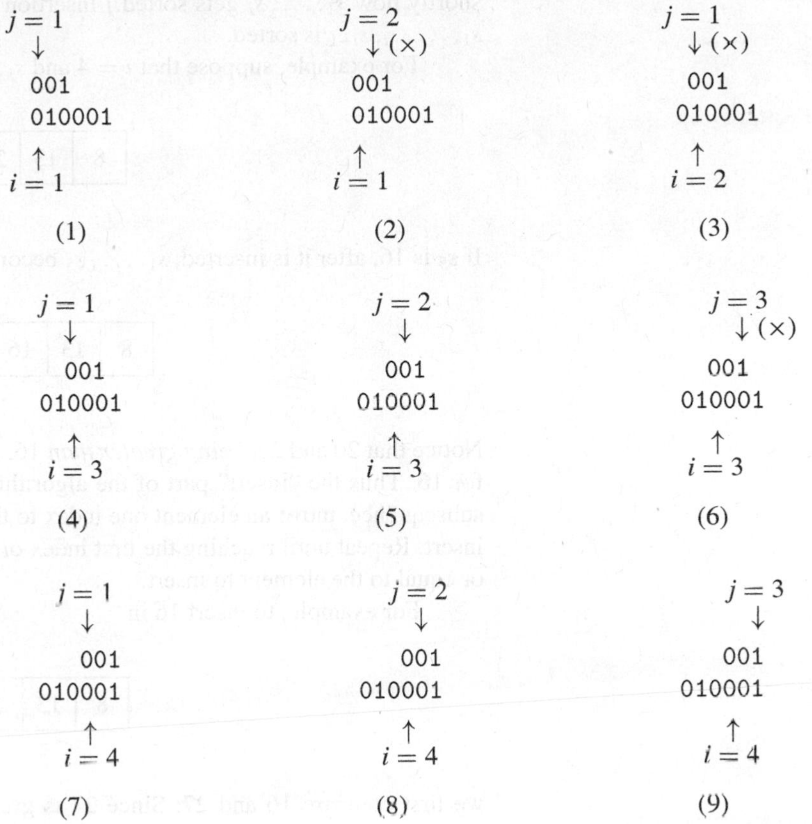

| | | |
|---|---|---|
| $j = 1$<br>↓<br>001<br>010001<br>↑<br>$i = 1$<br>(1) | $j = 2$<br>↓ (×)<br>001<br>010001<br>↑<br>$i = 1$<br>(2) | $j = 1$<br>↓ (×)<br>001<br>010001<br>↑<br>$i = 2$<br>(3) |
| $j = 1$<br>↓<br>001<br>010001<br>↑<br>$i = 3$<br>(4) | $j = 2$<br>↓<br>001<br>010001<br>↑<br>$i = 3$<br>(5) | $j = 3$<br>↓ (×)<br>001<br>010001<br>↑<br>$i = 3$<br>(6) |
| $j = 1$<br>↓<br>001<br>010001<br>↑<br>$i = 4$<br>(7) | $j = 2$<br>↓<br>001<br>010001<br>↑<br>$i = 4$<br>(8) | $j = 3$<br>↓<br>001<br>010001<br>↑<br>$i = 4$<br>(9) |

**Figure 4.2.1** Searching for "001" in "010001" using Algorithm 4.2.1. The cross (×) in steps (2), (3), and (6) marks a mismatch.

## Sorting

To **sort** a sequence is to put it in some specified order. If we have a sequence of names, we might want the sequence sorted in nondecreasing order according to dictionary order. For example, if the sequence is

Jones, Johnson, Appel, Zamora, Chu,

after sorting the sequence in nondecreasing order, we would obtain

Appel, Chu, Johnson, Jones, Zamora.

A major advantage of using a sorted sequence rather than an unsorted sequence is that it is much easier to find a particular item. Imagine trying to find the phone number of a particular individual in the New York City telephone book if the names were not sorted!

Many sorting algorithms have been devised (see, e.g., [Knuth, 1998b]). Which algorithm is preferred in a particular situation depends on factors such as the size of the data and how the data are represented. We discuss **insertion sort,** which is one of the fastest algorithms for sorting small sequences (less than 50 or so items).

We assume that the input to insertion sort is $s_1, \ldots, s_n$ and that the goal is to sort the data in nondecreasing order. At the $i$th iteration of insertion sort, the first part of the sequence $s_1, \ldots, s_i$ will have been rearranged so that it is sorted. (We will explain shortly how $s_1, \ldots, s_i$ gets sorted.) Insertion sort then *inserts* $s_{i+1}$ in $s_1, \ldots, s_i$ so that $s_1, \ldots, s_i, s_{i+1}$ is sorted.

For example, suppose that $i = 4$ and $s_1, \ldots, s_4$ is

| 8 | 13 | 20 | 27 |
|---|---|---|---|

If $s_5$ is 16, after it is inserted, $s_1, \ldots, s_5$ becomes

| 8 | 13 | 16 | 20 | 27 |
|---|---|---|---|---|

Notice that 20 and 27, being *greater than* 16, move one index to the right to make room for 16. Thus the "insert" part of the algorithm is: Beginning at the right of the sorted subsequence, move an element one index to the right if it is greater than the element to insert. Repeat until reaching the first index or encountering an element that is less than or equal to the element to insert.

For example, to insert 16 in

| 8 | 13 | 20 | 27 |
|---|---|---|---|

we first compare 16 and 27. Since 27 is greater than 16, 27 moves one index to the right:

| 8 | 13 | 20 |  | 27 |
|---|---|---|---|---|

We next compare 16 with 20. Since 20 is greater than 16, 20 moves one index to the right:

| 8 | 13 |  | 20 | 27 |
|---|---|---|---|---|

We next compare 16 with 13. Since 13 is less than or equal to 16, we insert (i.e., copy) 16 to the third index:

| 8 | 13 | 16 | 20 | 27 |
|---|---|---|---|---|

This subsequence is now sorted.

Having explained the key idea of insertion sort, we now complete the explanation of the algorithm. Insertion sort begins by inserting $s_2$ into the subsequence $s_1$. Note that $s_1$ by itself is sorted! Now $s_1, s_2$ is sorted. Next, insertion sort inserts $s_3$ into the now-sorted subsequence $s_1, s_2$. Now $s_1, s_2, s_3$ is sorted. This procedure continues until insertion sort inserts $s_n$ into the sorted subsequence $s_1, \ldots, s_{n-1}$. Now the entire sequence $s_1, \ldots, s_n$ is sorted. We obtain the following algorithm.

**Algorithm 4.2.3 Insertion Sort**

This algorithm sorts the sequence $s_1, \ldots, s_n$ in nondecreasing order.

Input: $s, n$

Output: $s$ (sorted)

```
insertion_sort(s, n) {
  for i = 2 to n {
    val = s_i // save s_i so it can be inserted into the correct place
    j = i − 1
    // if val < s_j, move s_j right to make room for s_i
    while (j ≥ 1 ∧ val < s_j) {
      s_{j+1} = s_j
      j = j − 1
    }
    s_{j+1} = val // insert val
  }
}
```

We leave proving that Algorithm 4.2.3 is correct as an exercise (see Exercise 14).

## Time and Space for Algorithms

It is important to know or be able to estimate the time (e.g., the number of steps) and space (e.g., the number of variables, length of the sequences involved) required by algorithms. Knowing the time and space required by algorithms allows us to compare algorithms that solve the same problem. For example, if one algorithm takes $n$ steps to solve a problem and another algorithm takes $n^2$ steps to solve the same problem, we would surely prefer the first algorithm, assuming that the space requirements are acceptable. In Section 4.3, we will give the technical definitions that allow us to make rigorous statements about the time and space required by algorithms.

The for loop in Algorithm 4.2.3 always executes $n - 1$ times, but the number of times that the while loop executes for a particular value of $i$ depends on the input. Thus, even for a fixed size $n$, the time required by Algorithm 4.2.3 depends on the input. For example, if the input sequence is already sorted in nondecreasing order,

$$val < s_j \tag{4.2.1}$$

will always be false and the body of the while loop will never be executed. We call this time the **best-case time.**

On the other hand, if the sequence is sorted in *decreasing* order, (4.2.1) will always be true and the while loop will execute the maximum number of times. (The while loop will execute $i - 1$ times during the $i$th iteration of the for loop.) We call this time the **worst-case time.**

## Randomized Algorithms

It is occasionally necessary to relax the requirements of an algorithm stated in Section 4.1. Many algorithms currently in use are not general, deterministic, or even finite. An operating system (e.g., Windows), for example, is better thought of as a program that never terminates rather than as a finite program with input and output. Algorithms written for more than one processor, whether for a multiprocessor machine or for a distributed environment (such as the internet), are rarely deterministic, for example, because of different execution speeds of the processors. Also, many practical problems are too difficult to be solved efficiently, and compromises either in generality or correctness are necessary. As an illustration, we present an example that shows the usefulness of allowing an algorithm to make random decisions, thereby violating the requirement of determinism.

A **randomized algorithm** does *not* require that the intermediate results of each step of execution be uniquely defined and depend only on the inputs and results of the preceding steps. By definition, when a randomized algorithm executes, at some points it makes *random* choices. In practice, a pseudorandom number generator is used (see Example 3.1.16).

We shall assume the existence of a function $rand(i, j)$, which returns a random integer between the integers $i$ and $j$, inclusive. As an example, we describe a randomized algorithm that shuffles a sequence of numbers. More precisely, it inputs a sequence $a_1, \ldots, a_n$ and moves the numbers to random positions. Major bridge tournaments use computer programs to shuffle the cards.

The algorithm first swaps (i.e., interchanges the values of) $a_1$ and $a_{rand(1,n)}$. At this point, the value of $a_1$ might be equal to *any* one of the original values in the sequence. Next, the algorithm swaps $a_2$ and $a_{rand(2,n)}$. Now the value of $a_2$ might be equal to *any* of the remaining values in the sequence. The algorithm continues in this manner until it swaps $a_{n-1}$ and $a_{rand(n-1,n)}$. Now the entire sequence is shuffled.

**Algorithm 4.2.4** **Shuffle**

This algorithm shuffles the values in the sequence

$$a_1, \ldots, a_n.$$

Input: $a, n$

Output: $a$ (shuffled)

*shuffle*$(a, n)$ {
  for $i = 1$ to $n - 1$
    $swap(a_i, a_{rand(i,n)})$
}

**Example 4.2.5** Suppose that the sequence $a$

| 17 | 9 | 5 | 23 | 21 |
|---|---|---|---|---|

is input to *shuffle*. We first swap $a_i$ and $a_j$, where $i = 1$ and $j = rand(1, 5)$. If $j = 3$, after the swap we have

| 5 | 9 | 17 | 23 | 21 |
|---|---|---|---|---|
| ↑ $i$ | | ↑ $j$ | | |

Next, $i = 2$. If $j = rand(2, 5) = 5$, after the swap we have

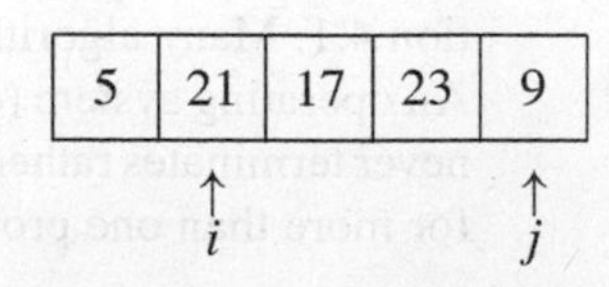

Next, $i = 3$. If $j = rand(3, 5) = 3$, the sequence does not change.
Finally, $i = 4$. If $j = rand(4, 5) = 5$, after the swap we have

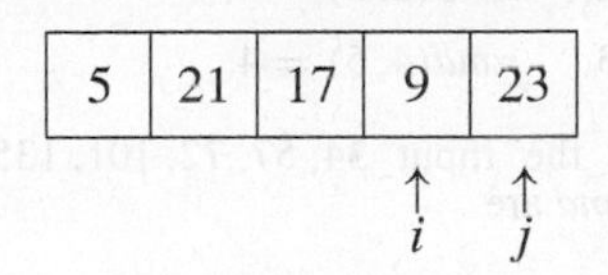

Notice that the output (i.e., the rearranged sequence) depends on the random choices made by the random number generator. ◀

Randomized algorithms can be used to search for nonrandom goals. For example, a person searching for the exit in a maze could randomly make a choice at each intersection. Of course, such an algorithm might not terminate (because of bad choices at the intersections). In Chapter 8, Graph Theory, we will present a randomized algorithm that searches for a particular structure in a graph (see Algorithm 8.3.10).

## 4.2 Problem-Solving Tips

Again, we emphasize that to construct an algorithm, it is often helpful to assume that you are in the middle of the algorithm and that part of the problem has been solved. In insertion sort (Algorithm 4.2.3), it was helpful to *assume* that the subsequence $s_1, \ldots, s_i$ was sorted. Then, all we had to do was insert the next element $s_{i+1}$ in the proper place. Iterating this procedure yields the algorithm. These observations lead to a loop invariant that can be used to prove that Algorithm 4.2.3 is correct (see Exercise 14).

## 4.2 Review Exercises

1. Give examples of searching problems.
2. What is text searching?
3. Describe, in words, an algorithm that solves the text-searching problem.
4. What does it mean to sort a sequence?
5. Give an example that illustrates why we might want to sort a sequence.
6. Describe insertion sort in words.
7. What do we mean by the time and space required by an algorithm?
8. Why is it useful to know or be able to estimate the time and space required by an algorithm?
9. Why is it sometimes necessary to relax the requirements of an algorithm as stated in Section 4.1?
10. What is a randomized algorithm?
11. Which requirements of an algorithm as stated in Section 4.1 does a randomized algorithm violate?
12. Describe the shuffle algorithm in words.
13. Give an application of the shuffle algorithm.

## 4.2 Exercises

1. Trace Algorithm 4.2.1 for the input $t$ = "balalaika" and $p$ = "bala".
2. Trace Algorithm 4.2.1 for the input $t$ = "balalaika" and $p$ = "lai".
3. Trace Algorithm 4.2.1 for the input $t$ = "000000000" and $p$ = "001".
4. Trace Algorithm 4.2.3 for the input 34, 20, 144, 55.
5. Trace Algorithm 4.2.3 for the input 34, 20, 19, 5.
6. Trace Algorithm 4.2.3 for the input 34, 55, 144, 259.
7. Trace Algorithm 4.2.3 for the input 34, 34, 34, 34.
8. Trace Algorithm 4.2.4 for the input 34, 57, 72, 101, 135. Assume that the values of *rand* are

$$rand(1, 5) = 5, \quad rand(2, 5) = 4,$$
$$rand(3, 5) = 3, \quad rand(4, 5) = 5.$$

9. Trace Algorithm 4.2.4 for the input 34, 57, 72, 101, 135.

Assume that the values of *rand* are

$$rand(1, 5) = 2, \quad rand(2, 5) = 5,$$
$$rand(3, 5) = 3, \quad rand(4, 5) = 4.$$

**10.** Trace Algorithm 4.2.4 for the input 34, 57, 72, 101, 135. Assume that the values of *rand* are

$$rand(1, 5) = 5, \quad rand(2, 5) = 5,$$
$$rand(3, 5) = 4, \quad rand(4, 5) = 4.$$

**11.** Is it possible that Algorithm 4.2.4 sorts the input 67, 32, 6, 89, 52 in increasing order? If so, show possible values for *rand* that perform the sort. It not, prove that Algorithm 4.2.4 cannot sort this input in increasing order.

**12.** Is it possible that Algorithm 4.2.4 sorts the input 67, 32, 6, 89, 52 in decreasing order? If so, show possible values for *rand* that perform the sort. It not, prove that Algorithm 4.2.4 cannot sort this input in decreasing order.

**13.** Prove that Algorithm 4.2.1 is correct.

**14.** Prove that Algorithm 4.2.3 is correct.

**15.** Write an algorithm that returns the index of the first occurrence of the value *key* in the sequence $s_1, \ldots, s_n$. If *key* is not in the sequence, the algorithm returns the value 0. *Example:* If the sequence is 12, 11, 12, 23 and *key* is 12, the algorithm returns the value 1.

**16.** Write an algorithm that returns the index of the last occurrence of the value *key* in the sequence $s_1, \ldots, s_n$. If *key* is not in the sequence, the algorithm returns the value 0. *Example:* If the sequence is 12, 11, 12, 23 and *key* is 12, the algorithm returns the value 3.

**17.** Write an algorithm whose input is a sequence $s_1, \ldots, s_n$ sorted in nondecreasing order and a value $x$. (Assume that all the values are real numbers.) The algorithm inserts $x$ into the sequence so that the resulting sequence is sorted in nondecreasing order. *Example:* If the input sequence is 2, 6, 12, 14 and $x = 5$, the resulting sequence is 2, 5, 6, 12, 14.

**18.** Modify Algorithm 4.2.1 so that it finds *all* occurrences of $p$ in $t$.

**19.** Describe best-case input for Algorithm 4.2.1.

**20.** Describe worst-case input for Algorithm 4.2.1.

**21.** Modify Algorithm 4.2.3 so that it sorts the sequence $s_1, \ldots, s_n$ in *nonincreasing* order.

**22.** The *selection sort* algorithm sorts the sequence $s_1, \ldots, s_n$ in nondecreasing order by first finding the smallest item, say $s_i$, and placing it first by swapping $s_1$ and $s_i$. It then finds the smallest item in $s_2, \ldots, s_n$, again say $s_i$, and places it second by swapping $s_2$ and $s_i$. It continues until the sequence is sorted. Write selection sort in pseudocode.

**23.** Trace selection sort (see Exercise 22) for the input of Exercises 4–7.

**24.** Show that the time for selection sort (see Exercise 22) is the same for all inputs of size $n$.

## 4.3 Analysis of Algorithms

A computer program, even though derived from a correct algorithm, might be useless for certain types of input because the time needed to run the program or the space needed to hold the data, program variables, and so on, is too great. **Analysis of an algorithm** refers to the process of deriving estimates for the time and space needed to execute the algorithm. In this section we deal with the problem of estimating the time required to execute algorithms.

Suppose that we are given a set $X$ of $n$ elements, some labeled "red" and some labeled "black," and we want to find the number of subsets of $X$ that contain at least one red item. Suppose we construct an algorithm that examines all subsets of $X$ and counts those that contain at least one red item and then implement this algorithm as a computer program. Since a set that has $n$ elements has $2^n$ subsets (see Theorem 2.4.6), the program would require at least $2^n$ units of time to execute. It does not matter what the units of time are—$2^n$ grows so fast as $n$ increases (see Table 4.3.1) that, except for small values of $n$, it would be impractical to run the program.

Determining the performance parameters of a computer program is a difficult task and depends on a number of factors such as the computer that is being used, the way the data are represented, and how the program is translated into machine instructions. Although precise estimates of the execution time of a program must take such factors into account, useful information can be obtained by analyzing the time of the underlying algorithm.

The time needed to execute an algorithm is a function of the input. Usually, it is difficult to obtain an explicit formula for this function, and we settle for less. Instead of dealing directly with the input, we use parameters that characterize the *size* of the input.

**TABLE 4.3.1** ■ Time to Execute an Algorithm if One Step Takes 1 Microsecond to Execute. $\lg n$ Denotes $\log_2 n$ (the logarithm of $n$ to base 2)

| *Number of Steps to Termination for Input of Size n* | *Time to Execute if n =* | | | | |
|---|---|---|---|---|---|
| | 3 | 6 | 9 | 12 | |
| 1 | $10^{-6}$ sec | $10^{-6}$ sec | $10^{-6}$ sec | $10^{-6}$ sec | |
| $\lg \lg n$ | $10^{-6}$ sec | $10^{-6}$ sec | $2 \times 10^{-6}$ sec | $2 \times 10^{-6}$ sec | |
| $\lg n$ | $2 \times 10^{-6}$ sec | $3 \times 10^{-6}$ sec | $3 \times 10^{-6}$ sec | $4 \times 10^{-6}$ sec | |
| $n$ | $3 \times 10^{-6}$ sec | $6 \times 10^{-6}$ sec | $9 \times 10^{-6}$ sec | $10^{-5}$ sec | |
| $n \lg n$ | $5 \times 10^{-6}$ sec | $2 \times 10^{-5}$ sec | $3 \times 10^{-5}$ sec | $4 \times 10^{-5}$ sec | |
| $n^2$ | $9 \times 10^{-6}$ sec | $4 \times 10^{-5}$ sec | $8 \times 10^{-5}$ sec | $10^{-4}$ sec | |
| $n^3$ | $3 \times 10^{-5}$ sec | $2 \times 10^{-4}$ sec | $7 \times 10^{-4}$ sec | $2 \times 10^{-3}$ sec | |
| $2^n$ | $8 \times 10^{-6}$ sec | $6 \times 10^{-5}$ sec | $5 \times 10^{-4}$ sec | $4 \times 10^{-3}$ sec | |
| | 50 | 100 | 1000 | $10^5$ | $10^6$ |
| 1 | $10^{-6}$ sec | $10^{-6}$ sec | $10^{-6}$ sec | $10^{-6}$ sec | $10^{-6}$ sec |
| $\lg \lg n$ | $2 \times 10^{-6}$ sec | $3 \times 10^{-6}$ sec | $3 \times 10^{-6}$ sec | $4 \times 10^{-6}$ sec | $4 \times 10^{-6}$ sec |
| $\lg n$ | $6 \times 10^{-6}$ sec | $7 \times 10^{-6}$ sec | $10^{-5}$ sec | $2 \times 10^{-5}$ sec | $2 \times 10^{-5}$ sec |
| $n$ | $5 \times 10^{-5}$ sec | $10^{-4}$ sec | $10^{-3}$ sec | 0.1 sec | 1 sec |
| $n \lg n$ | $3 \times 10^{-4}$ sec | $7 \times 10^{-4}$ sec | $10^{-2}$ sec | 2 sec | 20 sec |
| $n^2$ | $3 \times 10^{-3}$ sec | 0.01 sec | 1 sec | 3 hr | 12 days |
| $n^3$ | 0.13 sec | 1 sec | 16.7 min | 32 yr | 31,710 yr |
| $2^n$ | 36 yr | $4 \times 10^{16}$ yr | $3 \times 10^{287}$ yr | $3 \times 10^{30089}$ yr | $3 \times 10^{301016}$ yr |

For example, if the input is a set containing $n$ elements, we would say that the size of the input is $n$. We can ask for the minimum time needed to execute the algorithm among all inputs of size $n$. This time is called the **best-case time** for inputs of size $n$. We can also ask for the maximum time needed to execute the algorithm among all inputs of size $n$. This time is called the **worst-case time** for inputs of size $n$. Another important case is **average-case time**—the average time needed to execute the algorithm over some finite set of inputs all of size $n$.

Since we are primarily concerned with *estimating* the time of an algorithm rather than computing its exact time, as long as we count some fundamental, dominating steps of the algorithm, we will obtain a useful measure of the time. For example, if the principal activity of an algorithm is making comparisons, as might happen in a sorting routine, we might count the number of comparisons. As another example, if an algorithm consists of a single loop whose body executes in at most $C$ steps, for some constant $C$, we might count the number of iterations of the loop.

**Example 4.3.1** A reasonable definition of the size of input for Algorithm 4.1.2 that finds the largest value in a finite sequence is the number of elements in the input sequence. A reasonable definition of the execution time is the number of iterations of the while loop. With these definitions, the worst-case, best-case, and average-case times for Algorithm 4.1.2 for input of size $n$ are each $n - 1$ since the loop is always executed $n - 1$ times. ◀

Usually we are less interested in the exact best-case or worst-case time required for an algorithm to execute than we are in how the best-case or worst-case time grows as the size of the input increases. For example, suppose that the worst-case time of an algorithm is

$$t(n) = 60n^2 + 5n + 1$$

for input of size $n$. For large $n$, the term $60n^2$ is approximately equal to $t(n)$ (see Table 4.3.2). In this sense, $t(n)$ grows like $60n^2$.

**TABLE 4.3.2** ■ Comparing Growth of $t(n)$ with $60n^2$

| $n$ | $t(n) = 60n^2 + 5n + 1$ | $60n^2$ |
|---|---|---|
| 10 | 6,051 | 6,000 |
| 100 | 600,501 | 600,000 |
| 1,000 | 60,005,001 | 60,000,000 |
| 10,000 | 6,000,050,001 | 6,000,000,000 |

If $t(n)$ measures the worst-case time for input of size $n$ in seconds, then

$$T(n) = n^2 + \frac{5}{60}n + \frac{1}{60}$$

measures the worst-case time for input of size $n$ in minutes. Now this change of units does not affect how the worst-case time grows as the size of the input increases but only the units in which we measure the worst-case time for input of size $n$. Thus when we describe how the best-case or worst-case time grows as the size of the input increases, we not only seek the dominant term [e.g., $60n^2$ in the formula for $t(n)$], but we also may ignore constant coefficients. Under these assumptions, $t(n)$ grows like $n^2$ as $n$ increases. We say that $t(n)$ is of **order** $n^2$ and write $t(n) = \Theta(n^2)$, which is read "$t(n)$ is theta of $n^2$." The basic idea is to replace an expression, such as $t(n) = 60n^2 + 5n + 1$, with a simpler expression, such as $n^2$, that grows at the same rate as $t(n)$. The formal definitions follow.

**Definition 4.3.2** ▶ Let $f$ and $g$ be functions with domain $\{1, 2, 3, \ldots\}$.

We write

$$f(n) = O(g(n))$$

and say that $f(n)$ *is of order at most* $g(n)$ or $f(n)$ *is big oh of* $g(n)$ if there exists a positive constant $C_1$ such that

$$|f(n)| \leq C_1|g(n)|$$

for all but finitely many positive integers $n$.

**Go Online**
For more on these order notations, see
bit.ly/2OKWtYQ

We write

$$f(n) = \Omega(g(n))$$

and say that $f(n)$ *is of order at least* $g(n)$ or $f(n)$ *is omega of* $g(n)$ if there exists a positive constant $C_2$ such that

$$|f(n)| \geq C_2\,|g(n)|$$

for all but finitely many positive integers $n$.

We write

$$f(n) = \Theta(g(n))$$

and say that $f(n)$ *is of order* $g(n)$ or $f(n)$ *is theta of* $g(n)$ if $f(n) = O(g(n))$ and $f(n) = \Omega(g(n))$. ◀

Definition 4.3.2 can be loosely paraphrased as follows: $f(n) = O(g(n))$ if, except for a constant factor and a finite number of exceptions, $f$ is bounded above by $g$. We also say that $g$ is an **asymptotic upper bound** for $f$. Similarly, $f(n) = \Omega(g(n))$ if, except for a constant factor and a finite number of exceptions, $f$ is bounded below by $g$. We also say that $g$ is an **asymptotic lower bound** for $f$. Also, $f(n) = \Theta(g(n))$ if, except for constant factors and a finite number of exceptions, $f$ is bounded above and below by $g$. We also say that $g$ is an **asymptotic tight bound** for $f$.

According to Definition 4.3.2, if $f(n) = O(g(n))$, all that we can conclude is that, except for a constant factor and a finite number of exceptions, $f$ is bounded *above* by $g$, so $g$ grows at least as fast as $f$. For example, if $f(n) = n$ and $g(n) = 2^n$, then $f(n) = O(g(n))$, but $g$ grows considerably faster than $f$. The statement $f(n) = O(g(n))$ says nothing about a *lower* bound for $f$. On the other hand, if $f(n) = \Theta(g(n))$, we can draw the conclusion that, except for constant factors and a finite number of exceptions, $f$ is bounded *above* and *below* by $g$, so $f$ and $g$ grow at the same rate. Notice that $n = O(2^n)$, but $n \neq \Theta(2^n)$.

**Example 4.3.3** Since

$$60n^2 + 5n + 1 \leq 60n^2 + 5n^2 + n^2 = 66n^2 \qquad \text{for all } n \geq 1,$$

we may take $C_1 = 66$ in Definition 4.3.2 to obtain

$$60n^2 + 5n + 1 = O(n^2).$$

Since

$$60n^2 + 5n + 1 \geq 60n^2 \qquad \text{for all } n \geq 1,$$

we may take $C_2 = 60$ in Definition 4.3.2 to obtain

$$60n^2 + 5n + 1 = \Omega(n^2).$$

Since $60n^2 + 5n + 1 = O(n^2)$ and $60n^2 + 5n + 1 = \Omega(n^2)$,

$$60n^2 + 5n + 1 = \Theta(n^2).$$ ◀

The method of Example 4.3.3 can be used to show that a polynomial in $n$ of degree $k$ with nonnegative coefficients is $\Theta(n^k)$. [In fact, *any* polynomial in $n$ of degree $k$ is $\Theta(n^k)$, even if some of its coefficients are negative. To prove this more general result, the method of Example 4.3.3 has to be modified.]

**Theorem 4.3.4** Let

$$p(n) = a_k n^k + a_{k-1} n^{k-1} + \cdots + a_1 n + a_0$$

be a polynomial in $n$ of degree $k$, where each $a_i$ is nonnegative. Then

$$p(n) = \Theta(n^k).$$

**Proof** We first show that $p(n) = O(n^k)$. Let

$$C_1 = a_k + a_{k-1} + \cdots + a_1 + a_0.$$

Then, for all $n$,

$$\begin{aligned} p(n) &= a_k n^k + a_{k-1} n^{k-1} + \cdots + a_1 n + a_0 \\ &\le a_k n^k + a_{k-1} n^k + \cdots + a_1 n^k + a_0 n^k \\ &= (a_k + a_{k-1} + \cdots + a_1 + a_0) n^k = C_1 n^k. \end{aligned}$$

Therefore, $p(n) = O(n^k)$.

Next, we show that $p(n) = \Omega(n^k)$. For all $n$,

$$p(n) = a_k n^k + a_{k-1} n^{k-1} + \cdots + a_1 n + a_0 \ge a_k n^k = C_2 n^k,$$

where $C_2 = a_k$. Therefore, $p(n) = \Omega(n^k)$.

Since $p(n) = O(n^k)$ and $p(n) = \Omega(n^k)$, $p(n) = \Theta(n^k)$. ◀

**Example 4.3.5** In this book, we let $\lg n$ denote $\log_2 n$ (the logarithm of $n$ to the base 2). Since $\lg n < n$ for all $n \ge 1$ (see Figure 4.3.1[†]),

$$2n + 3\lg n < 2n + 3n = 5n \qquad \text{for all } n \ge 1.$$

Thus,

$$2n + 3\lg n = O(n).$$

Also, $2n + 3\lg n \ge 2n$, for all $n \ge 1$. Thus,

$$2n + 3\lg n = \Omega(n).$$

Therefore,

$$2n + 3\lg n = \Theta(n).$$ ◀

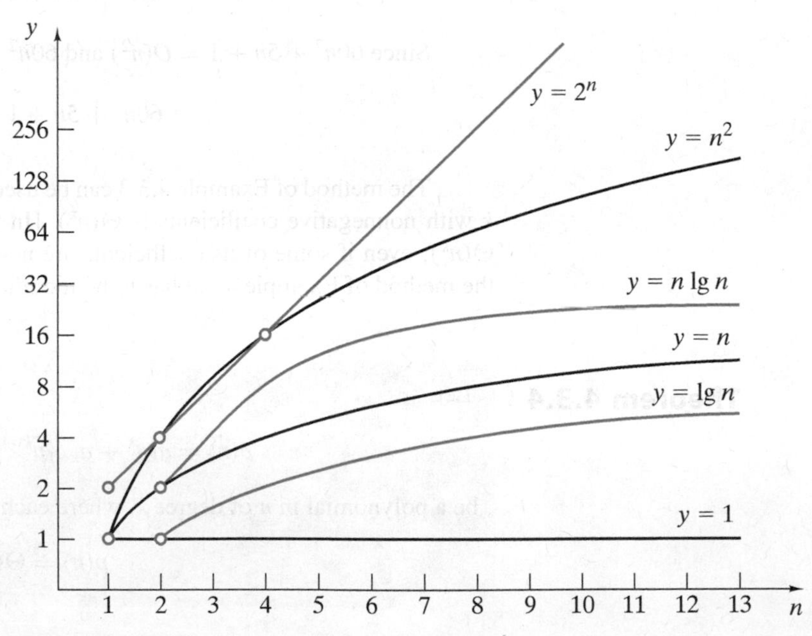


**Figure 4.3.1** Growth of some common functions.

[†]In Figure 4.3.1, the spacing on the $y$-axis is proportional to the logarithm of the number rather than to the number itself so that we can plot large $y$-values against smaller $x$-values. This $y$-axis scale is called a *logarithmic scale*. On the standard $xy$-graph, $y = n$ would be a straight line; here it is curved.

**Example 4.3.6** If $a > 1$ and $b > 1$ (to ensure that $\log_b a > 0$), by the change-of-base formula for logarithms [Theorem B.37(e)],

$$\log_b n = \log_b a \log_a n \qquad \text{for all } n \geq 1.$$

Therefore,

$$\log_b n \leq C \log_a n \qquad \text{for all } n \geq 1,$$

where $C = \log_b a$. Thus, $\log_b n = O(\log_a n)$.

Also,

$$\log_b n \geq C \log_a n \qquad \text{for all } n \geq 1;$$

so $\log_b n = \Omega(\log_a n)$. Since $\log_b n = O(\log_a n)$ and $\log_b n = \Omega(\log_a n)$, we conclude that $\log_b n = \Theta(\log_a n)$.

Because $\log_b n = \Theta(\log_a n)$, when using asymptotic notation we need not worry about which number is used as the base for the logarithm function (as long as the base is greater than 1). For this reason, we sometimes simply write log without specifying the base. ◀

**Example 4.3.7** If we replace each integer $1, 2, \ldots, n$ by $n$ in the sum $1 + 2 + \cdots + n$, the sum does not decrease and we have

$$1 + 2 + \cdots + n \leq n + n + \cdots + n = n \cdot n = n^2 \qquad \text{for all } n \geq 1. \tag{4.3.1}$$

It follows that

$$1 + 2 + \cdots + n = O(n^2).$$

To obtain a lower bound, we might imitate the preceding argument and replace each integer $1, 2, \ldots, n$ by 1 in the sum $1 + 2 + \cdots + n$ to obtain

$$1 + 2 + \cdots + n \geq 1 + 1 + \cdots + 1 = n \qquad \text{for all } n \geq 1.$$

In this case we conclude that

$$1 + 2 + \cdots + n = \Omega(n),$$

and while the preceding expression is true, we cannot deduce a $\Theta$-estimate for $1 + 2 + \cdots + n$, since the upper bound $n^2$ and lower bound $n$ are not equal. We must be craftier in deriving a lower bound.

One way to get a sharper lower bound is to argue as in the previous paragraph, but first throw away approximately the first half of the terms, to obtain

$$\begin{aligned} 1 + 2 + \cdots + n &\geq \lceil (n+1)/2 \rceil + \cdots + (n-1) + n \\ &\geq \lceil (n+1)/2 \rceil + \cdots + \lceil (n+1)/2 \rceil + \lceil (n+1)/2 \rceil \\ &= \lceil n/2 \rceil \lceil (n+1)/2 \rceil \geq (n/2)(n/2) = \frac{n^2}{4} \end{aligned} \tag{4.3.2}$$

for all $n \geq 1$. We can now conclude that

$$1 + 2 + \cdots + n = \Omega(n^2).$$

Therefore,

$$1 + 2 + \cdots + n = \Theta(n^2).$$

◀

**Example 4.3.8** If $k$ is a positive integer and, as in Example 4.3.7, we replace each integer $1, 2, \ldots, n$ by $n$, we have

$$1^k + 2^k + \cdots + n^k \le n^k + n^k + \cdots + n^k = n \cdot n^k = n^{k+1}$$

for all $n \ge 1$; hence

$$1^k + 2^k + \cdots + n^k = O(n^{k+1}).$$

We can also obtain a lower bound as in Example 4.3.7:

$$\begin{aligned} 1^k + 2^k + \cdots + n^k &\ge \lceil (n+1)/2 \rceil^k + \cdots + (n-1)^k + n^k \\ &\ge \lceil (n+1)/2 \rceil^k + \cdots + \lceil (n+1)/2 \rceil^k + \lceil (n+1)/2 \rceil^k \\ &= \lceil n/2 \rceil \lceil (n+1)/2 \rceil^k \ge (n/2)(n/2)^k = n^{k+1}/2^{k+1} \end{aligned}$$

for all $n \ge 1$. We conclude that

$$1^k + 2^k + \cdots + n^k = \Omega(n^{k+1}),$$

and hence

$$1^k + 2^k + \cdots + n^k = \Theta(n^{k+1}).$$

◀

Notice the difference between the polynomial

$$a_k n^k + a_{k-1} n^{k-1} + \cdots + a_1 n + a_0$$

in Theorem 4.3.4 and the expression

$$1^k + 2^k + \cdots + n^k$$

in Example 4.3.8. A polynomial has a fixed number of terms, whereas the number of terms in the expression in Example 4.3.8 is dependent on the value of $n$. Furthermore, the polynomial in Theorem 4.3.4 is $\Theta(n^k)$, but the expression in Example 4.3.8 is $\Theta(n^{k+1})$.

Our next example gives a theta notation for $\lg n!$.

**Example 4.3.9** Use an argument similar to that in Example 4.3.7, to show that $\lg n! = \Theta(n \lg n)$.

**SOLUTION** By properties of logarithms, we have

$$\lg n! = \lg n + \lg(n-1) + \cdots + \lg 2 + \lg 1$$

for all $n \ge 1$. Since lg is an increasing function,

$$\lg n + \lg(n-1) + \cdots + \lg 2 + \lg 1 \le \lg n + \lg n + \cdots + \lg n + \lg n = n \lg n$$

for all $n \ge 1$. We conclude that $\lg n! = O(n \lg n)$.

For all $n \ge 4$, we have

$$\begin{aligned} \lg n + \lg(n-1) + \cdots + \lg 2 + \lg 1 &\ge \lg n + \lg(n-1) + \cdots + \lg \lceil (n+1)/2 \rceil \\ &\ge \lg \lceil (n+1)/2 \rceil + \cdots + \lg \lceil (n+1)/2 \rceil \\ &= \lceil n/2 \rceil \lg \lceil (n+1)/2 \rceil \end{aligned}$$

$$
\begin{aligned}
&\geq (n/2)\lg(n/2) \\
&= (n/2)[\lg n - \lg 2] \\
&= (n/2)[(\lg n)/2 + ((\lg n)/2 - 1)] \\
&\geq (n/2)(\lg n)/2 \\
&= n \lg n/4
\end{aligned}
$$

[since $(\lg n)/2 \geq 1$ for all $n \geq 4$]. Therefore, $\lg n! = \Omega(n \lg n)$. It follows that $\lg n! = \Theta(n \lg n)$. ◀

**Example 4.3.10** Show that if $f(n) = \Theta(g(n))$ and $g(n) = \Theta(h(n))$, then $f(n) = \Theta(h(n))$.

**SOLUTION** Because $f(n) = \Theta(g(n))$, there are constants $C_1$ and $C_2$ such that

$$C_1|g(n)| \leq |f(n)| \leq C_2|g(n)|$$

for all but finitely many positive integers $n$. Because $g(n) = \Theta(h(n))$, there are constants $C_3$ and $C_4$ such that

$$C_3|h(n)| \leq |g(n)| \leq C_4|h(n)|$$

for all but finitely many positive integers $n$. Therefore,

$$C_1C_3|h(n)| \leq C_1|g(n)| \leq |f(n)| \leq C_2|g(n)| \leq C_2C_4|h(n)|$$

for all but finitely many positive integers $n$. It follows that $f(n) = \Theta(h(n))$. ◀

We next define what it means for the best-case, worst-case, or average-case time of an algorithm to be of order at most $g(n)$.

**Definition 4.3.11** ▶ If an algorithm requires $t(n)$ units of time to terminate in the best case for an input of size $n$ and

$$t(n) = O(g(n)),$$

we say that the *best-case time required by the algorithm is of order at most* $g(n)$ or that the *best-case time required by the algorithm* is $O(g(n))$.

If an algorithm requires $t(n)$ units of time to terminate in the worst case for an input of size $n$ and

$$t(n) = O(g(n)),$$

we say that the *worst-case time required by the algorithm is of order at most* $g(n)$ or that the *worst-case time required by the algorithm* is $O(g(n))$.

If an algorithm requires $t(n)$ units of time to terminate in the average case for an input of size $n$ and

$$t(n) = O(g(n)),$$

we say that the *average-case time required by the algorithm is of order at most* $g(n)$ or that the *average-case time required by the algorithm* is $O(g(n))$. ◀

By replacing $O$ by $\Omega$ and "at most" by "at least" in Definition 4.3.11, we obtain the definition of what it means for the best-case, worst-case, or average-case time of an algorithm to be of order at least $g(n)$. If the best-case time required by an algorithm is $O(g(n))$ and $\Omega(g(n))$, we say that the best-case time required by the algorithm is

$\Theta(g(n))$. An analogous definition applies to the worst-case and average-case times of an algorithm.

**Example 4.3.12** Suppose that an algorithm is known to take $60n^2+5n+1$ units of time to terminate in the worst case for inputs of size $n$. We showed in Example 4.3.3 that $60n^2+5n+1 = \Theta(n^2)$. Thus the worst-case time required by this algorithm is $\Theta(n^2)$. ◀

**Example 4.3.13** Find a theta notation in terms of $n$ for the number of times the statement $x = x + 1$ is executed.

```
1.  for i = 1 to n
2.      for j = 1 to i
3.          x = x + 1
```

**SOLUTION** First, $i$ is set to 1 and, as $j$ runs from 1 to 1, line 3 is executed one time. Next, $i$ is set to 2 and, as $j$ runs from 1 to 2, line 3 is executed two times, and so on. Thus the total number of times line 3 is executed is (see Example 4.3.7) $1+2+\cdots+n = \Theta(n^2)$. Thus a theta notation for the number of times the statement $x = x+1$ is executed is $\Theta(n^2)$. ◀

**Example 4.3.14** Find a theta notation in terms of $n$ for the number of times the statement $x = x + 1$ is executed:

```
1.  i = n
2.  while (i ≥ 1) {
3.      x = x + 1
4.      i = ⌊i/2⌋
5.  }
```

**SOLUTION** First, we examine some specific cases. Because of the floor function, the computations are simplified if $n$ is a power of 2. Consider, for example, the case $n = 8$. At line 1, $i$ is set to 8. At line 2, the condition $i \geq 1$ is true. At line 3, we execute the statement $x = x + 1$ the first time. At line 4, $i$ is reset to 4 and we return to line 2.

At line 2, the condition $i \geq 1$ is again true. At line 3, we execute the statement $x = x + 1$ the second time. At line 4, $i$ is reset to 2 and we return to line 2.

At line 2, the condition $i \geq 1$ is again true. At line 3, we execute the statement $x = x + 1$ the third time. At line 4, $i$ is reset to 1 and we return to line 2.

At line 2, the condition $i \geq 1$ is again true. At line 3, we execute the statement $x = x + 1$ the fourth time. At line 4, $i$ is reset to 0 and we return to line 2.

This time at line 2, the condition $i \geq 1$ is false. The statement $x = x + 1$ was executed four times.

Now suppose that $n$ is 16. At line 1, $i$ is set to 16. At line 2, the condition $i \geq 1$ is true. At line 3, we execute the statement $x = x + 1$ the first time. At line 4, $i$ is reset to 8 and we return to line 2. Now execution proceeds as before; the statement $x = x + 1$ is executed four more times, for a total of five times.

Similarly, if $n$ is 32, the statement $x = x + 1$ is executed a total of six times.

A pattern is emerging. Each time the initial value of $n$ is doubled, the statement $x = x + 1$ is executed one more time. More precisely, if $n = 2^k$, the statement $x = x + 1$ is executed $k + 1$ times. Since $k$ is the exponent for 2, $k = \lg n$. Thus if $n = 2^k$, the statement $x = x + 1$ is executed $1 + \lg n$ times.

If $n$ is an arbitrary positive integer (not necessarily a power of 2), it lies between two powers of 2; that is, for some $k \geq 1$,

$$2^{k-1} \leq n < 2^k.$$

We use induction on $k$ to show that in this case the statement $x = x + 1$ is executed $k$ times.

If $k = 1$, we have

$$1 = 2^{1-1} \leq n < 2^1 = 2.$$

Therefore, $n$ is 1. In this case, the statement $x = x + 1$ is executed once. Thus the Basis Step is proved.

Now suppose that if $n$ satisfies

$$2^{k-1} \leq n < 2^k, \tag{4.3.3}$$

the statement $x = x + 1$ is executed $k$ times. We must show that if $n$ satisfies

$$2^k \leq n < 2^{k+1}, \tag{4.3.4}$$

the statement $x = x + 1$ is executed $k + 1$ times.

Suppose that $n$ satisfies (4.3.4). At line 1, $i$ is set to $n$. At line 2, the condition $i \geq 1$ is true. At line 3, we execute the statement $x = x + 1$ the first time. At line 4, $i$ is reset to $\lfloor n/2 \rfloor$ and we return to line 2. Notice that

$$2^{k-1} \leq n/2 < 2^k.$$

Because $2^{k-1}$ is an integer, we must also have

$$2^{k-1} \leq \lfloor n/2 \rfloor < 2^k.$$

By the inductive assumption (4.3.3), the statement $x = x + 1$ is executed $k$ more times, for a total of $k+1$ times. The Inductive Step is complete. Therefore, if $n$ satisfies (4.3.3), the statement $x = x + 1$ is executed $k$ times.

Suppose that $n$ satisfies (4.3.3). Taking logarithms to the base 2, we have

$$k - 1 \leq \lg n < k.$$

Therefore, $k$, the number of times the statement $x = x + 1$ is executed, satisfies

$$\lg n < k \leq 1 + \lg n.$$

Because $k$ is an integer, we must have $k \leq 1 + \lfloor \lg n \rfloor$. Furthermore, $\lfloor \lg n \rfloor < k$. It follows from the last two inequalities that $k = 1 + \lfloor \lg n \rfloor$. Since $1 + \lfloor \lg n \rfloor = \Theta(\lg n)$, a theta notation for the number of times the statement $x = x + 1$ is executed is $\Theta(\lg n)$. ◀

Many algorithms are based on the idea of repeated halving. Example 4.3.14 shows that for size $n$, repeated halving takes time $\Theta(\lg n)$. Of course, the algorithm may do work in addition to the halving that will increase the overall time.

**Example 4.3.15** Find a theta notation in terms of $n$ for the number of times the statement $x = x + 1$ is executed.

```
1. j = n
2. while (j ≥ 1) {
3.     for i = 1 to j
4.         x = x + 1
5.     j = ⌊j/2⌋
6. }
```

**SOLUTION** Let $t(n)$ denote the number of times we execute the statement $x = x + 1$. The first time we arrive at the body of the while loop, the statement $x = x+1$ is executed $n$ times. Therefore $t(n) \geq n$ for all $n \geq 1$ and $t(n) = \Omega(n)$.

Next we derive a big oh notation for $t(n)$. After $j$ is set to $n$, we arrive at the while loop for the first time. The statement $x = x+1$ is executed $n$ times. At line 5, $j$ is replaced by $\lfloor n/2 \rfloor$; hence $j \leq n/2$. If $j \geq 1$, we will execute $x = x + 1$ at most $n/2$ additional times in the next iteration of the while loop, and so on. If we let $k$ denote the number of times we execute the body of the while loop, the number of times we execute $x = x + 1$ is at most

$$n + \frac{n}{2} + \frac{n}{4} + \cdots + \frac{n}{2^{k-1}}.$$

This geometric sum (see Example 2.4.4) is equal to

$$\frac{n\left(1 - \frac{1}{2^k}\right)}{1 - \frac{1}{2}}.$$

Now

$$t(n) \leq \frac{n\left(1 - \frac{1}{2^k}\right)}{1 - \frac{1}{2}} = 2n\left(1 - \frac{1}{2^k}\right) \leq 2n \qquad \text{for all } n \geq 1,$$

so $t(n) = O(n)$. Thus a theta notation for the number of times we execute $x = x + 1$ is $\Theta(n)$. ◀

**Example 4.3.16** Determine, in theta notation, the best-case, worst-case, and average-case times required to execute Algorithm 4.3.17, which follows. Assume that the input size is $n$ and that the run time of the algorithm is the number of comparisons made at line 3. Also, assume that the $n + 1$ possibilities of *key* being at any particular position in the sequence or not being in the sequence are equally likely.

**SOLUTION** The best-case time can be analyzed as follows. If $s_1 = key$, line 3 is executed once. Thus the best-case time of Algorithm 4.3.17 is $\Theta(1)$.

The worst-case time of Algorithm 4.3.17 is analyzed as follows. If *key* is not in the sequence, line 3 will be executed $n$ times, so the worst-case time of Algorithm 4.3.17 is $\Theta(n)$.

Finally, consider the average-case time of Algorithm 4.3.17. If *key* is found at the $i$th position, line 3 is executed $i$ times; if *key* is not in the sequence, line 3 is executed $n$ times. Thus the average number of times line 3 is executed is

$$\frac{(1 + 2 + \cdots + n) + n}{n + 1}.$$

Now

$$\begin{aligned} \frac{(1 + 2 + \cdots + n) + n}{n + 1} &\leq \frac{n^2 + n}{n + 1} && \text{by (4.3.1)} \\ &= \frac{n(n + 1)}{n + 1} = n. \end{aligned}$$

Therefore, the average-case time of Algorithm 4.3.17 is $O(n)$. Also,

$$\frac{(1+2+\cdots+n)+n}{n+1} \geq \frac{n^2/4+n}{n+1} \qquad \text{by (4.3.2)}$$

$$\geq \frac{n^2/4+n/4}{n+1} = \frac{n}{4}.$$

Therefore the average-case time of Algorithm 4.3.17 is $\Omega(n)$. Thus the average-case time of Algorithm 4.3.17 is $\Theta(n)$. For this algorithm, the worst-case and average-case times are both $\Theta(n)$. ◀

**Algorithm 4.3.17** **Searching an Unordered Sequence**

Given the sequence $s_1, \ldots, s_n$ and a value *key*, this algorithm returns the index of *key*. If *key* is not found, the algorithm returns 0.

Input: $s_1, s_2, \ldots, s_n$, $n$, and *key* (the value to search for)

Output: The index of *key*, or if *key* is not found, 0

```
1. linear_search(s, n, key) {
2.    for i = 1 to n
3.       if (key == s_i)
4.          return i // successful search
5.    return 0 // unsuccessful search
6. }
```

**Example 4.3.18** **Matrix Multiplication and Transitive Relations** If $A$ is a matrix, we let $A_{ij}$ denote the entry in row $i$, column $j$. The product of $n \times n$ matrices $A$ and $B$ (i.e., $A$ and $B$ have $n$ rows and $n$ columns) is defined as the $n \times n$ matrix $C$, where

$$C_{ij} = \sum_{k=1}^{n} A_{ik}B_{kj} \qquad 1 \leq i \leq n, \ 1 \leq j \leq n.$$

Algorithm 4.3.19, which computes the matrix product, is a direct translation of the preceding definition. Because of the nested loops, it runs in time $\Theta(n^3)$.

Recall (see the discussion following Theorem 3.5.6) that we can test whether a relation $R$ on an $n$-element set is transitive by squaring its adjacency matrix, say $A$, and then comparing $A^2$ with $A$. The relation $R$ is transitive if and only if, whenever the entry in row $i$, column $j$ in $A^2$ is nonzero, the corresponding entry in $A$ is also nonzero. Since there are $n^2$ entries in $A$ and $A^2$, the worst-case time to compare the entries is $\Theta(n^2)$. Using Algorithm 4.3.19 to compute $A^2$ requires time $\Theta(n^3)$. Therefore, the overall time to test whether a relation on an $n$-element set is transitive, using Algorithm 4.3.19 to compute $A^2$, is $\Theta(n^3)$.

For many years it was believed that the minimum time to multiply two $n \times n$ matrices was $\Theta(n^3)$; thus it was quite a surprise when a more efficient algorithm was discovered. Strassen's algorithm (see [Johnsonbaugh: Section 5.4]) to multiply two $n \times n$ matrices runs in time $\Theta(n^{\lg 7})$. Since lg 7 is approximately 2.807, Strassen's algorithm runs in time approximately $\Theta(n^{2.807})$, which is asymptotically faster than Algorithm 4.3.19. An algorithm by Coppersmith and Winograd (see [Coppersmith]) runs in time $\Theta(n^{2.376})$ and, so, is even asymptotically faster than Strassen's algorithm. Since the product of two $n \times n$ matrices contains $n^2$ terms, *any* algorithm that multiplies two $n \times n$ matrices requires time at least $\Omega(n^2)$. At the present time, no sharper lower bound is known. ◀

**Algorithm 4.3.19 Matrix Multiplication**

This algorithm computes the product $C$ of the $n \times n$ matrices $A$ and $B$ directly from the definition of matrix multiplication.

Input: $A, B, n$

Output: $C$, the product of $A$ and $B$

```
matrix_product(A, B, n) {
  for i = 1 to n
    for j = 1 to n {
      C_ij = 0
      for k = 1 to n
        C_ij = C_ij + A_ik * B_kj
    }
  return C
}
```

The constants that are suppressed in the theta notation may be important. Even if for any input of size $n$, algorithm $A$ requires exactly $C_1 n$ time units and algorithm $B$ requires exactly $C_2 n^2$ time units, for certain sizes of inputs algorithm $B$ may be superior. For example, suppose that for any input of size $n$, algorithm $A$ requires $300n$ units of time and algorithm $B$ requires $5n^2$ units of time. For an input size of $n = 5$, algorithm $A$ requires 1500 units of time and algorithm $B$ requires 125 units of time, and thus algorithm $B$ is faster. Of course, for sufficiently large inputs, algorithm $A$ is considerably faster than algorithm $B$.

A real-world example of the importance of constants in the theta notation is provided by matrix multiplication. Algorithm 4.3.19, which runs in time $\Theta(n^3)$, is typically used to multiply matrices even though the Strassen and Coppersmith-Winograd algorithms (see Example 4.3.18), which run in times $\Theta(n^{2.807})$ and $\Theta(n^{2.376})$, are asymptotically faster. The constants in the Strassen and Coppersmith-Winograd algorithms are so large that they are faster than Algorithm 4.3.19 only for very large matrices.

**TABLE 4.3.3 ■ Common Growth Functions**

| *Theta Form* | *Name* |
|---|---|
| $\Theta(1)$ | Constant |
| $\Theta(\lg \lg n)$ | Log log |
| $\Theta(\lg n)$ | Log |
| $\Theta(n)$ | Linear |
| $\Theta(n \lg n)$ | $n$ log $n$ |
| $\Theta(n^2)$ | Quadratic |
| $\Theta(n^3)$ | Cubic |
| $\Theta(n^k), k \geq 1$ | Polynomial |
| $\Theta(c^n), c > 1$ | Exponential |
| $\Theta(n!)$ | Factorial |

Certain growth functions occur so often that they are given special names, as shown in Table 4.3.3. The functions in Table 4.3.3, with the exception of $\Theta(n^k)$, are arranged so that if $\Theta(f(n))$ is above $\Theta(g(n))$, then $f(n) \leq g(n)$ for all but finitely many positive integers $n$. Thus, if algorithms $A$ and $B$ have run times that are $\Theta(f(n))$ and $\Theta(g(n))$, respectively, and $\Theta(f(n))$ is above $\Theta(g(n))$ in Table 4.3.3, then algorithm $A$ is more time-efficient than algorithm $B$ for sufficiently large inputs.

It is important to develop some feeling for the relative sizes of the functions in Table 4.3.3. In Figure 4.3.1 we have graphed some of these functions. Another way to develop some appreciation for the relative sizes of the functions $f(n)$ in Table 4.3.3 is to determine how long it would take an algorithm to terminate whose run time is exactly $f(n)$. For this purpose, let us assume that we have a computer that can execute one step in 1 microsecond ($10^{-6}$ sec). Table 4.3.1 shows the execution times, under this assumption, for various input sizes. Notice that it is practical to implement an algorithm that requires $2^n$ steps for an input of size $n$ only for very small input sizes. Algorithms requiring $n^2$ or $n^3$ steps also become impractical to implement, but for relatively larger input sizes. Also, notice the dramatic improvement that results when we move from $n^2$ steps to $n \lg n$ steps.

A problem that has a worst-case polynomial-time algorithm is considered to have a "good" algorithm; the interpretation is that such a problem has an efficient solution. Such problems are called **feasible** or **tractable.** Of course, if the worst-case time to solve the problem is proportional to a high-degree polynomial, the problem can still take a long time to solve. Fortunately, in many important cases, the polynomial bound has small degree.

A problem that does not have a worst-case polynomial-time algorithm is said to be **intractable.** Any algorithm, if there is one, that solves an intractable problem is guaranteed to take a long time to execute in the worst case, even for modest sizes of the input.

Certain problems are so hard that they have no algorithms at all. A problem for which there is no algorithm is said to be **unsolvable.** A large number of problems are known to be unsolvable, some of considerable practical importance. One of the earliest problems to be proved unsolvable is the **halting problem:** Given an arbitrary program and a set of inputs, will the program eventually halt?

A large number of solvable problems have an as yet undetermined status; they are thought to be intractable, but none of them has been proved to be intractable. (Most of these problems belong to the class of NP-complete problems; see [Johnsonbaugh] for details.) An example of an NP-complete problem is:

> Given a collection $\mathcal{C}$ of finite sets and a positive integer $k \le |\mathcal{C}|$, does $\mathcal{C}$ contain at least $k$ mutually disjoint sets?

Other NP-complete problems include the traveling-salesperson problem and the Hamiltonian-cycle problem (see Section 8.3).

NP-complete problems have efficient (i.e., polynomial-time) algorithms to check whether a proposed solution is, in fact, a solution. For example, given a collection $C$ of finite sets and $k$ sets in $C$, it is easy and fast to check whether the $k$ sets are mutually disjoint. (Just check each pair of sets!) On the other hand, no NP-complete problem is known to have an efficient algorithm. For example, given a collection $C$ of finite sets, *finding* $k$ mutually disjoint sets in $C$ is, in general, difficult and time consuming. NP-complete problems also have the property that if any one of them has a polynomial-time algorithm, then *all* NP-complete problems have polynomial-time algorithms.

## 4.3 Problem-Solving Tips

- To derive a big oh notation for an expression $f(n)$ directly, you must find a constant $C_1$ and a simple expression $g(n)$ (e.g., $n$, $n \lg n$, $n^2$) such that $|f(n)| \le C_1|g(n)|$ for all but finitely many $n$. Remember you're trying to derive an *inequality, not* an equality, so you can replace terms in $f(n)$ with other terms if the result is *larger* (see, e.g., Example 4.3.3).
- To derive an omega notation for an expression $f(n)$ directly, you must find a constant $C_2$ and a simple expression $g(n)$ such that $|f(n)| \ge C_2|g(n)|$ for all but finitely many $n$. Again, you're trying to derive an *inequality* so you can replace terms in $f(n)$ with other terms if the result is *smaller* (again, see Example 4.3.3).
- To derive a theta notation, you must derive both big oh and omega notations.
- Another way to derive big oh, omega, and theta estimates is to use known results.

| *Expression* | *Name* | *Estimate* | *Reference* |
|---|---|---|---|
| $a_k n^k + a_{k-1} n^{k-1} + \cdots + a_1 n + a_0$ | Polynomial | $\Theta(n^k)$ | Theorem 4.3.4 |
| $1 + 2 + \cdots + n$ | Arithmetic Sum (Case $k = 1$ for Next Entry) | $\Theta(n^2)$ | Example 4.3.7 |
| $1^k + 2^k + \cdots + n^k$ | Sum of Powers | $\Theta(n^{k+1})$ | Example 4.3.8 |
| $\lg n!$ | log $n$ Factorial | $\Theta(n \lg n)$ | Example 4.3.9 |

- To derive an asymptotic estimate for the time of an algorithm, count the number of steps $t(n)$ required by the algorithm, and then derive an estimate for $t(n)$ as described previously. Algorithms typically contain loops, in which case, deriving $t(n)$ requires counting the number of iterations of the loops.

## 4.3 Review Exercises

1. To what does "analysis of algorithms" refer?
2. What is the worst-case time of an algorithm?
3. What is the best-case time of an algorithm?
4. What is the average-case time of an algorithm?
5. Define $f(n) = O(g(n))$. What is this notation called?
6. Give an intuitive interpretation of how $f$ and $g$ are related if $f(n) = O(g(n))$.
7. Define $f(n) = \Omega(g(n))$. What is this notation called?
8. Give an intuitive interpretation of how $f$ and $g$ are related if $f(n) = \Omega(g(n))$.
9. Define $f(n) = \Theta(g(n))$. What is this notation called?
10. Give an intuitive interpretation of how $f$ and $g$ are related if $f(n) = \Theta(g(n))$.

## 4.3 Exercises

*Select a theta notation from Table 4.3.3 for each expression in Exercises 1–13.*

1. $6n + 1$
2. $2n^2 + 1$
3. $6n^3 + 12n^2 + 1$
4. $3n^2 + 2n \lg n$
5. $2 \lg n + 4n + 3n \lg n$
6. $6n^6 + n + 4$
7. $2 + 4 + 6 + \cdots + 2n$
8. $(6n + 1)^2$
9. $(6n + 4)(1 + \lg n)$
10. $\dfrac{(n+1)(n+3)}{n+2}$
11. $\dfrac{(n^2 + \lg n)(n+1)}{n + n^2}$
12. $2 + 4 + 8 + 16 + \cdots + 2^n$
13. $\lg[(2n)!]$

*In Exercises 14–16, select a theta notation for $f(n) + g(n)$.*

14. $f(n) = \Theta(1), \quad g(n) = \Theta(n^2)$
15. $f(n) = 6n^3 + 2n^2 + 4, \quad g(n) = \Theta(n \lg n)$
16. $f(n) = \Theta(n^{3/2}), \quad g(n) = \Theta(n^{5/2})$

*In Exercises 17–26, select a theta notation from among*

$$\Theta(1), \quad \Theta(\lg n), \quad \Theta(n), \quad \Theta(n \lg n), \quad \Theta(n^2), \quad \Theta(n^3), \quad \Theta(2^n), \quad or \quad \Theta(n!)$$

*for the number of times the statement $x = x + 1$ is executed.*

17.
```
for i = 1 to 2n
   x = x + 1
```

18.
```
i = 1
while (i ≤ 2n) {
   x = x + 1
   i = i + 2
}
```

19.
```
for i = 1 to n
   for j = 1 to n
      x = x + 1
```

20.
```
for i = 1 to 2n
   for j = 1 to n
      x = x + 1
```

21.
```
for i = 1 to n
   for j = 1 to ⌊i/2⌋
      x = x + 1
```

22.
```
for i = 1 to n
   for j = 1 to n
      for k = 1 to n
         x = x + 1
```

23.
```
for i = 1 to n
   for j = 1 to n
      for k = 1 to i
         x = x + 1
```

24.
```
for i = 1 to n
   for j = 1 to i
      for k = 1 to j
         x = x + 1
```

25.
```
j = n
while (j ≥ 1) {
   for i = 1 to j
      x = x + 1
   j = ⌊j/3⌋
}
```

26.
```
i = n
while (i ≥ 1) {
   for j = 1 to n
      x = x + 1
   i = ⌊i/2⌋
}
```

27. Find a theta notation for the number of times the statement $x = x + 1$ is executed.

```
i = 2
while (i < n) {
   i = i^2
   x = x + 1
}
```

28. Let $t(n)$ be the total number of times that $i$ is incremented and $j$ is decremented in the following pseudocode, where $a_1, a_2, \ldots$ is a sequence of real numbers.

```
    i = 1
    j = n
    while (i < j) {
      while (i < j ∧ a_i < 0)
        i = i + 1
      while (i < j ∧ a_j ≥ 0)
        j = j − 1
      if (i < j)
        swap(a_i, a_j)
    }
```

Find a theta notation for $t(n)$.

**29.** Find a theta notation for the worst-case time required by the following algorithm:

```
iskey(s, n, key) {
  for i = 1 to n − 1
    for j = i + 1 to n
      if (s_i + s_j == key)
        return 1
      else
        return 0
}
```

**30.** In addition to finding a theta notation in Exercises 1–29, prove that it is correct.

**31.** Find the exact number of comparisons (lines 10, 15, 17, 24, and 26) required by the following algorithm when $n$ is even and when $n$ is odd. Find a theta notation for this algorithm.

Input: $s_1, s_2, \ldots, s_n, n$

Output: *large* (the largest item in $s_1, s_2, \ldots, s_n$), *small* (the smallest item in $s_1, s_2, \ldots, s_n$)

```
 1. large_small(s, n, large, small) {
 2.   if (n == 1) {
 3.     large = s_1
 4.     small = s_1
 5.     return
 6.   }
 7.   m = 2⌊n/2⌋
 8.   i = 1
 9.   while (i ≤ m − 1) {
10.     if (s_i > s_{i+1})
11.       swap(s_i, s_{i+1})
12.     i = i + 2
13.   }
14.   if (n > m) {
15.     if (s_{m−1} > s_n)
16.       swap(s_{m−1}, s_n)
17.     if (s_n > s_m)
18.       swap(s_m, s_n)
19.   }
20.   small = s_1
21.   large = s_2
22.   i = 3
23.   while (i ≤ m − 1) {
24.     if (s_i < small)
25.       small = s_i
26.     if (s_{i+1} > large)
27.       large = s_{i+1}
28.     i = i + 2
29.   }
30. }
```

**32.** This exercise shows another way to guess a formula for $1 + 2 + \cdots + n$.

Example 4.3.7 suggests that

$$1 + 2 + \cdots + n = An^2 + Bn + C \qquad \text{for all } n,$$

for some constants $A$, $B$, and $C$. Assuming that this is true, plug in $n = 1, 2, 3$ to obtain three equations in the three unknowns $A$, $B$, and $C$. Now solve for $A$, $B$, and $C$. The resulting formula can now be proved using mathematical induction (see Section 2.4).

**33.** Suppose that $a > 1$ and that $f(n) = \Theta(\log_a n)$. Show that $f(n) = \Theta(\lg n)$.

**34.** Show that $n! = O(n^n)$.

**35.** Show that $2^n = O(n!)$.

**36.** By using an argument like the one shown in Examples 4.3.7–4.3.9 or otherwise, prove that $\sum_{i=1}^{n} i \lg i = \Theta(n^2 \lg n)$.

★**37.** Show that $n^{n+1} = O(2^{n^2})$.

**38.** Show that $\lg(n^k + c) = \Theta(\lg n)$ for every fixed $k > 0$ and $c > 0$.

**39.** Show that if $n$ is a power of 2, say $n = 2^k$, then

$$\sum_{i=0}^{k} \lg(n/2^i) = \Theta(\lg^2 n).$$

**40.** Suppose that $f(n) = O(g(n))$, and $f(n) \geq 0$ and $g(n) > 0$ for all $n \geq 1$. Show that for some constant $C$, $f(n) \leq Cg(n)$ for all $n \geq 1$.

**41.** State and prove a result for $\Omega$ similar to that for Exercise 40.

**42.** State and prove a result for $\Theta$ similar to that for Exercises 40 and 41.

*Determine whether each statement in Exercises 43–68 is true or false. If the statement is true, prove it. If the statement is false, give a counterexample. Assume that the functions $f$, $g$, and $h$ take on only positive values.*

**43.** $n^n = O(2^n)$

**44.** $2 + \sin n = O(2 + \cos n)$

**45.** $(2n)^2 = O(n^2)$

**46.** $(2n)^2 = \Omega(n^2)$

**47.** $(2n)^2 = \Theta(n^2)$

**48.** $2^{2n} = O(2^n)$

**49.** $2^{2n} = \Omega(2^n)$

**50.** $2^{2n} = \Theta(2^n)$

**51.** $n! = O((n+1)!)$

**52.** $n! = \Omega((n+1)!)$

**53.** $n! = \Theta((n+1)!)$

**54.** $\lg(2n)^2 = O(\lg n^2)$

**55.** $\lg(2n)^2 = \Omega(\lg n^2)$

**56.** $\lg(2n)^2 = \Theta(\lg n^2)$

**57.** $\lg 2^{2n} = O(\lg 2^n)$

**58.** $\lg 2^{2n} = \Omega(\lg 2^n)$

**59.** $\lg 2^{2n} = \Theta(\lg 2^n)$

**60.** If $f(n) = \Theta(h(n))$ and $g(n) = \Theta(h(n))$, then $f(n) + g(n) = \Theta(h(n))$.

**61.** If $f(n) = \Theta(g(n))$, then $cf(n) = \Theta(g(n))$ for any $c \neq 0$.

**62.** If $f(n) = \Theta(g(n))$, then $2^{f(n)} = \Theta(2^{g(n)})$.

**63.** If $f(n) = \Theta(g(n))$, then $\lg f(n) = \Theta(\lg g(n))$. Assume that $f(n) \geq 1$ and $g(n) \geq 1$ for all $n = 1, 2, \ldots$.

**64.** If $f(n) = O(g(n))$, then $g(n) = O(f(n))$.

**65.** If $f(n) = O(g(n))$, then $g(n) = \Omega(f(n))$.

**66.** If $f(n) = \Theta(g(n))$, then $g(n) = \Theta(f(n))$.

**67.** $f(n) + g(n) = \Theta(h(n))$, where $h(n) = \max\{f(n), g(n)\}$

**68.** $f(n) + g(n) = \Theta(h(n))$, where $h(n) = \min\{f(n), g(n)\}$

**69.** Write out exactly what $f(n) \neq O(g(n))$ means.

**70.** What is wrong with the following argument that purports to show that we cannot simultaneously have $f(n) \neq O(g(n))$ and $g(n) \neq O(f(n))$?

If $f(n) \neq O(g(n))$, then for every $C > 0$, $|f(n)| > C|g(n)|$. In particular, $|f(n)| > 2|g(n)|$. If $g(n) \neq O(f(n))$, then for every $C > 0$, $|g(n)| > C|f(n)|$. In particular, $|g(n)| > 2|f(n)|$. But now

$$|f(n)| > 2|g(n)| > 4|f(n)|.$$

Cancelling $|f(n)|$ gives $1 > 4$, which is a contradiction. Therefore, we cannot simultaneously have $f(n) \neq O(g(n))$ and $g(n) \neq O(f(n))$.

**★71.** Find functions $f$ and $g$ satisfying

$$f(n) \neq O(g(n)) \quad \text{and} \quad g(n) \neq O(f(n)).$$

**★72.** Give an example of increasing positive functions $f$ and $g$ defined on the positive integers for which

$$f(n) \neq O(g(n)) \quad \text{and} \quad g(n) \neq O(f(n)).$$

**★73.** Prove that $n^k = O(c^n)$ for all $k = 1, 2, \ldots$ and $c > 1$.

**74.** Find functions $f$, $g$, $h$, and $t$ satisfying

$$f(n) = \Theta(g(n)), \qquad h(n) = \Theta(t(n)),$$
$$f(n) - h(n) \neq \Theta(g(n) - t(n)).$$

**75.** Suppose that the worst-case time of an algorithm is $\Theta(n)$. What is the error in the following reasoning? Since $2n = \Theta(n)$, the worst-case time to run the algorithm with input of size $2n$ will be approximately the same as the worst-case time to run the algorithm with input of size $n$.

**76.** Does $f(n) = O(g(n))$ define an equivalence relation on the set of real-valued functions on $\{1, 2, \ldots\}$?

**77.** Does $f(n) = \Theta(g(n))$ define an equivalence relation on the set of real-valued functions on $\{1, 2, \ldots\}$?

**78.** [Requires the integral]

(a) Show, by consulting the figure, that

$$\frac{1}{2} + \frac{1}{3} + \cdots + \frac{1}{n} < \log_e n.$$

(b) Show, by consulting the figure, that

$$\log_e n < 1 + \frac{1}{2} + \cdots + \frac{1}{n-1}.$$

(c) Use parts (a) and (b) to show that

$$1 + \frac{1}{2} + \cdots + \frac{1}{n} = \Theta(\lg n).$$

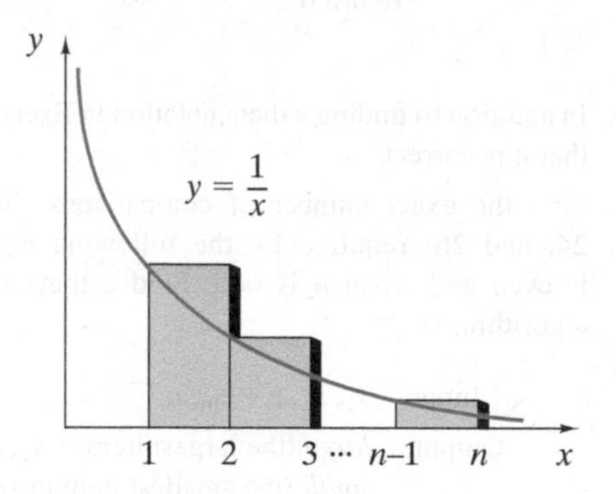


**79.** [Requires the integral] Use an argument like the one shown in Exercise 78 to show that

$$\frac{n^{m+1}}{m+1} < 1^m + 2^m + \cdots + n^m < \frac{(n+1)^{m+1}}{m+1},$$

where $m$ is a positive integer.

**80.** By using the formula

$$\frac{b^{n+1} - a^{n+1}}{b - a} = \sum_{i=0}^{n} a^i b^{n-i} \qquad 0 \leq a < b$$

or otherwise, prove that

$$\frac{b^{n+1} - a^{n+1}}{b - a} < (n+1)b^n \qquad 0 \leq a < b.$$

**81.** Take $a = 1 + 1/(n+1)$ and $b = 1 + 1/n$ in the inequality of Exercise 80 to prove that the sequence $\{(1 + 1/n)^n\}$ is increasing.

**82.** Take $a = 1$ and $b = 1 + 1/(2n)$ in the inequality of Exercise 80 to prove that

$$\left(1 + \frac{1}{2n}\right)^n < 2$$

for all $n \geq 1$. Use the preceding exercise to conclude that

$$\left(1 + \frac{1}{n}\right)^n < 4$$

for all $n \geq 1$.

The method used to prove the results of this exercise and its predecessor is apparently due to Fort in 1862 (see [Chrystal, vol. II, page 77]).

**83.** By using the preceding two exercises or otherwise, prove that

$$\frac{1}{n} \leq \lg(n+1) - \lg n < \frac{2}{n}$$

for all $n \geq 1$.

**84.** Use the preceding exercise to prove that

$$\sum_{i=1}^{n} \frac{1}{i} = \Theta(\lg n).$$

(Compare with Exercise 78.)

**85.** Prove that the sequence $\{n^{1/n}\}_{n=3}^{\infty}$ is decreasing.

**86.** Prove that if $0 \leq a < b$, then

$$\frac{b^{n+1} - a^{n+1}}{b - a} > (n+1)a^n.$$

**87.** Find appropriate values for $a$ and $b$ in the inequality in the preceding exercise to prove that the sequence $\{(1 - 1/n)^n\}_{n=1}^{\infty}$ is increasing and bounded above by 4/9.

**88.** By using the result of the preceding exercise, or otherwise, prove that the sequence $\{(1 + 1/n)^{n+1}\}_{n=1}^{\infty}$ is decreasing.

**89.** By using the result of the preceding exercise, or otherwise, prove that

$$\lg(n+1) - \lg n \leq \frac{2}{n+1}$$

for all $n \geq 1$.

**90.** What is wrong with the following "proof" that any algorithm has a run time that is $O(n)$?

We must show that the time required for an input of size $n$ is at most a constant times $n$.

**Basis Step**

Suppose that $n = 1$. If the algorithm takes $C$ units of time for an input of size 1, the algorithm takes at most $C \cdot 1$ units of time. Thus the assertion is true for $n = 1$.

**Inductive Step**

Assume that the time required for an input of size $n$ is at most $C'n$ and that the time for processing an additional item is $C''$. Let $C$ be the maximum of $C'$ and $C''$. Then the total time required for an input of size $n + 1$ is at most

$$C'n + C'' \leq Cn + C = C(n+1).$$

The Inductive Step has been verified.

By induction, for input of size $n$, the time required is at most a constant time $n$. Therefore, the run time is $O(n)$.

*In Exercises 91–96, determine whether the statement is true or false. If the statement is true, prove it. If the statement is false, give a counterexample. Assume that f and g are real-valued functions defined on the set of positive integers and that $g(n) \neq 0$ for $n \geq 1$. These exercises require calculus.*

**91.** If

$$\lim_{n\to\infty} \frac{f(n)}{g(n)} = 0,$$

then $f(n) = O(g(n))$.

**92.** If

$$\lim_{n\to\infty} \frac{f(n)}{g(n)} = 0,$$

then $f(n) = \Theta(g(n))$.

**93.** If

$$\lim_{n\to\infty} \frac{f(n)}{g(n)} = c \neq 0,$$

then $f(n) = O(g(n))$.

**94.** If

$$\lim_{n\to\infty} \frac{f(n)}{g(n)} = c \neq 0,$$

then $f(n) = \Theta(g(n))$.

**95.** If $f(n) = O(g(n))$, then

$$\lim_{n\to\infty} \frac{f(n)}{g(n)}$$

exists and is equal to some real number.

**96.** If $f(n) = \Theta(g(n))$, then

$$\lim_{n\to\infty} \frac{f(n)}{g(n)}$$

exists and is equal to some real number.

**★97.** Use induction to prove that

$$\lg n! \geq \frac{n}{2} \lg \frac{n}{2}.$$

**98.** [Requires calculus] Let $\ln x$ denote the natural logarithm ($\log_e x$) of $x$. Use the integral to obtain the estimate

$$n \ln n - n \leq \sum_{k=1}^{n} \ln k = \ln n!, \qquad n \geq 1.$$

**99.** Use the result of Exercise 98 and the change-of-base formula for logarithms to obtain the formula

$$n \lg n - n \lg e \leq \lg n!, \qquad n \geq 1.$$

**100.** Deduce

$$\lg n! \geq \frac{n}{2} \lg \frac{n}{2}$$

from the inequality of Exercise 99.

## Problem-Solving Corner Design and Analysis of an Algorithm

### Problem

Develop and analyze an algorithm that returns the maximum sum of consecutive values in the numerical sequence $s_1, \ldots, s_n$. In mathematical notation, the problem is to find the maximum sum of the form $s_j + s_{j+1} + \cdots + s_i$. *Example:* If the sequence is

$$27 \quad 6 \quad -50 \quad 21 \quad -3 \quad 14 \quad 16 \quad -8 \quad 42 \quad 33 \quad -21 \quad 9,$$

the algorithm returns 115—the sum of

$$21 \quad -3 \quad 14 \quad 16 \quad -8 \quad 42 \quad 33.$$

If all the numbers in a sequence are negative, the maximum sum of consecutive values is defined to be 0. (The idea is that the maximum of 0 is achieved by taking an "empty" sum.)

### Attacking the Problem

In developing an algorithm, a good way to start is to ask the question, "How would I solve this problem by hand?" At least initially, take a straightforward approach. Here we might just list the sums of *all* consecutive values and pick the largest. For the example sequence, the sums are as follows:

| | *j* | | | | | | | | | | | |
|---|---|---|---|---|---|---|---|---|---|---|---|---|
| *i* | 1 | 2 | 3 | 4 | 5 | 6 | 7 | 8 | 9 | 10 | 11 | 12 |
| 1 | 27 | | | | | | | | | | | |
| 2 | 33 | 6 | | | | | | | | | | |
| 3 | −17 | −44 | −50 | | | | | | | | | |
| 4 | 4 | −23 | −29 | 21 | | | | | | | | |
| 5 | 1 | −26 | −32 | 18 | −3 | | | | | | | |
| 6 | 15 | −12 | −18 | 32 | 11 | 14 | | | | | | |
| 7 | 31 | 4 | −2 | 48 | 27 | 30 | 16 | | | | | |
| 8 | 23 | −4 | −10 | 40 | 19 | 22 | 8 | −8 | | | | |
| 9 | 65 | 38 | 32 | 82 | 61 | 64 | 50 | 34 | 42 | | | |
| 10 | 98 | 71 | 65 | 115 | 94 | 97 | 83 | 67 | 75 | 33 | | |
| 11 | 77 | 50 | 44 | 94 | 73 | 76 | 62 | 46 | 54 | 12 | −21 | |
| 12 | 86 | 59 | 53 | 103 | 82 | 85 | 71 | 55 | 63 | 21 | −12 | 9 |

The entry in column $j$, row $i$, is the sum $s_j + \cdots + s_i$. For example, the entry in column 4, row 7, is 48—the sum

$$s_4 + s_5 + s_6 + s_7 = 21 + -3 + 14 + 16 = 48.$$

By inspection, we find that 115 is the largest sum.

### Finding a Solution

We begin by writing pseudocode for the straightforward algorithm that computes all consecutive sums and finds the largest:

Input: $s_1, \ldots, s_n$

Output: *max*

```
max_sum1(s, n) {
  // sum_ji is the sum s_j + ··· + s_i.
  for i = 1 to n {
    for j = 1 to i − 1
      sum_ji = sum_j,i−1 + s_i
    sum_ii = s_i
  }

  // step through sum_ji and find the maximum
  max = 0
  for i = 1 to n
    for j = 1 to i
      if (sum_ji > max)
        max = sum_ji
  return max
}
```

The first nested for loops compute the sums

$$sum_{ji} = s_j + \cdots + s_i.$$

The computation relies on the fact that

$$sum_{ji} = s_j + \cdots + s_i = s_j + \cdots + s_{i-1} + s_i$$
$$= sum_{j,i-1} + s_i.$$

The second nested for loops step through $sum_{ji}$ and find the largest value.

Since each of the nested for loops takes time $\Theta(n^2)$, *max_sum1*'s time is $\Theta(n^2)$.

We can improve the actual time, but not the asymptotic time, of the algorithm by computing the maximum within the same nested for loops in which we compute $sum_{ji}$:

```
Input:  s_1, ..., s_n
Output: max

max_sum2(s, n) {
  // sum_ji is the sum s_j + ··· + s_i.
  max = 0
  for i = 1 to n {
    for j = 1 to i − 1 {
      sum_ji = sum_j,i−1 + s_i
      if (sum_ji > max)
        max = sum_ji
    }
    sum_ii = s_i
    if (sum_ii > max)
      max = sum_ii
  }
  return max
}
```

Since the nested for loops take time $\Theta(n^2)$, *max_sum2*'s time is $\Theta(n^2)$. To reduce the asymptotic time, we need to take a hard look at the pseudocode to see where it can be improved.

Two key observations lead to improved time. First, since we are looking only for the *maximum* sum, there is no need to record all of the sums; we will store only the maximum sum that ends at index $i$. Second, the line

$$sum_{ji} = sum_{j,i-1} + s_i$$

shows how a consecutive sum that ends at index $i-1$ is related to a consecutive sum that ends at index $i$. The maximum can be computed by using a similar formula. If *sum* is the maximum consecutive sum that ends at index $i-1$, the maximum consecutive sum that ends at index $i$ is obtained by adding $s_i$ to *sum* provided that $sum + s_i$ is positive. (If some sum of consecutive terms that ends at index $i$ exceeds $sum + s_i$, we could remove $s_i$ and obtain a sum of consecutive terms ending at index $i-1$ that exceeds *sum*, which is impossible.) If $sum + s_i \leq 0$, the maximum consecutive sum that ends at index $i$ is obtained by taking no terms and has value 0. Thus we may compute the maximum consecutive sum that ends at index $i$ by executing

```
if (sum + s_i > 0)
  sum = sum + s_i
else
  sum = 0
```

## Formal Solution

```
Input:  s_1, ..., s_n
Output: max

max_sum3(s, n) {
  // max is the maximum sum seen so far.
  // After the ith iteration of the for
  // loop, sum is the largest consecutive
  // sum that ends at index i.
  max = 0
  sum = 0
  for i = 1 to n {
    if (sum + s_i > 0)
      sum = sum + s_i
    else
      sum = 0
    if (sum > max)
      max = sum
  }
  return max
}
```

Since this algorithm has a single for loop that runs from 1 to $n$, *max_sum3*'s time is $\Theta(n)$. The asymptotic time of this algorithm cannot be further improved. To find the maximum sum of consecutive values, we must at least look at each element in the sequence, which takes time $\Theta(n)$.

## Summary of Problem-Solving Techniques

- In developing an algorithm, a good way to start is to ask the question, "How would I solve this problem by hand?"
- In developing an algorithm, initially take a straightforward approach.
- After developing an algorithm, take a close look at the pseudocode to see where it can be improved. Look at the parts that perform key computations to gain insight into how to enhance the algorithm's efficiency.
- As in mathematical induction, extend a solution of a smaller problem to a larger problem. (In this problem, we extended a sum that ends at index $i-1$ to a sum that ends at index $i$.)

- Don't repeat computations. (In this problem, we extended a sum that ends at index $i-1$ to a sum that ends at index $i$ by adding an additional term rather than by computing the sum that ends at index $i$ from scratch. This latter method would have meant recomputing the sum that ends at index $i-1$.)

## Comments

According to [Bentley], the problem discussed in this section is the one-dimensional version of the original two-dimensional problem that dealt with pattern matching in digital images. The original problem was to find the maximum sum in a rectangular submatrix of an $n \times n$ matrix of real numbers.

## Exercises

1. Modify *max_sum3* so that it computes not only the maximum sum of consecutive values but also the indexes of the first and last terms of a maximum-sum subsequence. If there is no maximum-sum subsequence (which would happen, for example, if all of the values of the sequence were negative), the algorithm should set the first and last indexes to zero.

# 4.4 Recursive Algorithms

**Go Online**

For more on recursion, see bit.ly/2OKWtYQ

A **recursive function** (pseudocode) is a function that invokes itself. A **recursive algorithm** is an algorithm that contains a recursive function. Recursion is a powerful, elegant, and natural way to solve a large class of problems. A problem in this class can be solved using a *divide-and-conquer* technique in which the problem is decomposed into problems of the same type as the original problem. Each subproblem, in turn, is decomposed further until the process yields subproblems that can be solved in a straightforward way. Finally, solutions to the subproblems are combined to obtain a solution to the original problem.

**Example 4.4.1**

Recall that if $n \geq 1$, $n! = n(n-1)\cdots 2 \cdot 1$, and $0! = 1$. Notice that if $n \geq 2$, $n$ factorial can be written "in terms of itself" since, if we "peel off" $n$, the remaining product is simply $(n-1)!$; that is,

$$n! = n(n-1)(n-2)\cdots 2 \cdot 1 = n \cdot (n-1)!.$$

For example,

$$5! = 5 \cdot 4 \cdot 3 \cdot 2 \cdot 1 = 5 \cdot 4!.$$

The equation

$$n! = n \cdot (n-1)!, \qquad \textbf{(4.4.1)}$$

which happens to be true even when $n = 1$, shows how to decompose the original problem (compute $n!$) into increasingly simpler subproblems [compute $(n-1)!$, compute $(n-2)!, \ldots$] until the process reaches the straightforward problem of computing $0!$. The solutions to these subproblems can then be combined, by multiplying, to solve the original problem.

For example, the problem of computing 5! is reduced to computing 4!; the problem of computing 4! is reduced to computing 3!; and so on. Table 4.4.1 summarizes this process.

**TABLE 4.4.1 ■ Decomposing the Factorial Problem**

| *Problem* | *Simplified Problem* |
|---|---|
| 5! | $5 \cdot 4!$ |
| 4! | $4 \cdot 3!$ |
| 3! | $3 \cdot 2!$ |
| 2! | $2 \cdot 1!$ |
| 1! | $1 \cdot 0!$ |
| 0! | None |

Once the problem of computing 5! has been reduced to solving subproblems, the solution to the simplest subproblem can be used to solve the next simplest subproblem, and so on, until the original problem has been solved. Table 4.4.2 shows how the subproblems are combined to compute 5! ◀

**TABLE 4.4.2 ■ Combining Subproblems of the Factorial Problem**

| *Problem* | *Solution* |
|---|---|
| 0! | 1 |
| 1! | $1 \cdot 0! = 1$ |
| 2! | $2 \cdot 1! = 2$ |
| 3! | $3 \cdot 2! = 3 \cdot 2 = 6$ |
| 4! | $4 \cdot 3! = 4 \cdot 6 = 24$ |
| 5! | $5 \cdot 4! = 5 \cdot 24 = 120$ |

Next, we write a recursive algorithm that computes factorials. The algorithm is a direct translation of equation (4.4.1).

**Algorithm 4.4.2** **Computing $n$ Factorial**

This recursive algorithm computes $n!$.

Input: $n$, an integer greater than or equal to 0

Output: $n!$

```
1. factorial(n) {
2.    if (n == 0)
3.       return 1
4.    return n * factorial(n − 1)
5. }
```

We show how Algorithm 4.4.2 computes $n!$ for several values of $n$. If $n = 0$, at line 3 the function correctly returns the value 1.

If $n = 1$, we proceed to line 4 since $n \neq 0$. We use this function to compute $0!$. We have just observed that the function computes 1 as the value of $0!$. At line 4, the function correctly computes the value of $1!$:

$$n \cdot (n - 1)! = 1 \cdot 0! = 1 \cdot 1 = 1.$$

If $n = 2$, we proceed to line 4 since $n \neq 0$. We use this function to compute $1!$. We have just observed that the function computes 1 as the value of $1!$. At line 4, the function correctly computes the value of $2!$:

$$n \cdot (n - 1)! = 2 \cdot 1! = 2 \cdot 1 = 2.$$

If $n = 3$ we proceed to line 4 since $n \neq 0$. We use this function to compute $2!$. We have just observed that the function computes 2 as the value of $2!$. At line 4, the function correctly computes the value of $3!$:

$$n \cdot (n - 1)! = 3 \cdot 2! = 3 \cdot 2 = 6.$$

The preceding arguments may be generalized using mathematical induction to *prove* that Algorithm 4.4.2 correctly returns the value of $n!$ for any nonnegative integer $n$.

**Theorem 4.4.3** Algorithm 4.4.2 returns the value of $n!$, $n \geq 0$.

## Proof

### Basis Step ($n = 0$)

We have already observed that if $n = 0$, Algorithm 4.4.2 correctly returns the value of $0!$ (1).

### Inductive Step

Assume that Algorithm 4.4.2 correctly returns the value of $(n - 1)!$, $n > 0$. Now suppose that $n$ is input to Algorithm 4.4.2. Since $n \neq 0$, when we execute the function in Algorithm 4.4.2 we proceed to line 4. By the inductive assumption, the function correctly computes the value of $(n - 1)!$. At line 4, the function correctly computes the value $(n - 1)! \cdot n = n!$.

Therefore, Algorithm 4.4.2 correctly returns the value of $n!$ for every integer $n \geq 0$. ◀

If executed by a computer, Algorithm 4.4.2 would typically not be as efficient as a nonrecursive version because of the overhead of the recursive calls.

There must be some situations in which a recursive function does *not* invoke itself; otherwise, it would invoke itself forever. In Algorithm 4.4.2, if $n = 0$, the function does not invoke itself. We call the values for which a recursive function does not invoke itself the *base cases*. To summarize, every recursive function must have base cases.

We have shown how mathematical induction may be used to prove that a recursive algorithm computes the value it claims to compute. The link between mathematical induction and recursive algorithms runs deep. Often a proof by mathematical induction can be considered to be an algorithm to compute a value or to carry out a particular construction. The Basis Step of a proof by mathematical induction corresponds to the base cases of a recursive function, and the Inductive Step of a proof by mathematical induction corresponds to the part of a recursive function where the function calls itself.

In Example 2.4.7, we gave a proof using mathematical induction that, given an $n \times n$ deficient board (a board with one square removed), where $n$ is a power of 2, we can tile the board with right trominoes (three squares that form an "L"; see Figure 2.4.4). We now translate the inductive proof into a recursive algorithm to construct a tiling by right trominoes of an $n \times n$ deficient board where $n$ is a power of 2.

**Algorithm 4.4.4**

**Go Online**

For a C program implementing this algorithm, see bit.ly/2OKWtYQ

**Tiling a Deficient Board with Trominoes**

This algorithm constructs a tiling by right trominoes of an $n \times n$ deficient board where $n$ is a power of 2.

Input: $n$, a power of 2 (the board size); and the location $L$ of the missing square

Output: A tiling of an $n \times n$ deficient board

```
1.  tile(n, L) {
2.     if (n == 2) {
          // the board is a right tromino T
3.        tile with T
4.        return
5.     }
6.     divide the board into four (n/2) × (n/2) boards
7.     rotate the board so that the missing square is in the upper-left quadrant
8.     place one right tromino in the center // as in Figure 2.4.5
       // consider each of the squares covered by the center tromino as
       // missing, and denote the missing squares as m1, m2, m3, m4
9.     tile(n/2, m1)
10.    tile(n/2, m2)
11.    tile(n/2, m3)
12.    tile(n/2, m4)
13. }
```

Using the method of the proof of Theorem 4.4.3, we can prove that Algorithm 4.4.4 is correct (see Exercise 4).

We present one final example of a recursive algorithm.

**Example 4.4.5** A robot can take steps of 1 meter or 2 meters. We write an algorithm to calculate the number of ways the robot can walk $n$ meters. As examples:

| *Distance* | *Sequence of Steps* | *Number of Ways to Walk* |
|---|---|---|
| 1 | 1 | 1 |
| 2 | 1, 1 or 2 | 2 |
| 3 | 1, 1, 1 or 1, 2 or 2, 1 | 3 |
| 4 | 1, 1, 1, 1 or 1, 1, 2 or 1, 2, 1 or 2, 1, 1 or 2, 2 | 5 |

Let $walk(n)$ denote the number of ways the robot can walk $n$ meters. We have observed that $walk(1) = 1$ and $walk(2) = 2$. Now suppose that $n > 2$. The robot can begin by taking a step of 1 meter or a step of 2 meters. If the robot begins by taking a 1-meter step, a distance of $n - 1$ meters remains; but, by definition, the remainder of the walk can be completed in $walk(n - 1)$ ways. Similarly, if the robot begins by taking a 2-meter step, a distance of $n - 2$ meters remains and, in this case, the remainder of the walk can be completed in $walk(n - 2)$ ways. Since the walk must begin with either a 1-meter or a 2-meter step, all of the ways to walk $n$ meters are accounted for. We obtain the formula

$$walk(n) = walk(n-1) + walk(n-2). \tag{4.4.2}$$

For example,

$$walk(4) = walk(3) + walk(2) = 3 + 2 = 5.$$

We can write a recursive algorithm to compute $walk(n)$ by translating equation (4.4.2) directly into an algorithm. The base cases are $n = 1$ and $n = 2$. ◀

**Algorithm 4.4.6 Robot Walking**

This algorithm computes the function defined by

$$walk(n) = \begin{cases} 1, & n = 1 \\ 2, & n = 2 \\ walk(n-1) + walk(n-2) & n > 2. \end{cases}$$

Input: $n$
Output: $walk(n)$

```
walk(n) {
  if (n == 1 ∨ n == 2)
    return n
  return walk(n − 1) + walk(n − 2)
}
```

**Go Online**
For a C program implementing this algorithm, see bit.ly/2AmEpLO

Using the method of the proof of Theorem 4.4.3, we can prove that Algorithm 4.4.6 is correct (see Exercise 7).

The sequence $walk(1), walk(2), walk(3), \ldots,$ whose values begin 1, 2, 3, 5, 8, 13, ..., is related to the Fibonacci sequence. The **Fibonacci sequence** $\{f_n\}$ is defined by the equations

**Go Online**
For more on the Fibonacci sequence, see bit.ly/2OKWtYQ

$$f_1 = 1, \qquad f_2 = 1, \qquad f_n = f_{n-1} + f_{n-2} \qquad n \geq 3.$$

**Figure 4.4.1** A pine cone. There are 13 clockwise spirals (marked with white thread) and 8 counterclockwise spirals (marked with dark thread). [*Photo by the author; pine cone courtesy of André Berthiaume and Sigrid (Anne) Settle.*]

The Fibonacci sequence begins

$$1, 1, 2, 3, 5, 8, 13, \ldots.$$

Since $walk(1) = f_2$, $walk(2) = f_3$, and

$$walk(n) = walk(n-1) + walk(n-2), \qquad f_n = f_{n-1} + f_{n-2} \qquad \text{for all } n \geq 3,$$

it follows that

$$walk(n) = f_{n+1} \qquad \text{for all } n \geq 1.$$

(The argument can be formalized using mathematical induction; see Exercise 8.)

The Fibonacci sequence is named in honor of Leonardo Fibonacci (ca. 1170–1250), an Italian merchant and mathematician. The sequence originally arose in a puzzle about rabbits (see Exercises 18 and 19). After returning from the Orient in 1202, Fibonacci wrote his most famous work, *Liber Abaci* (available in an English translation by [Sigler]), which, in addition to containing what we now call the Fibonacci sequence, advocated the use of Hindu-Arabic numerals. This book was one of the main influences in bringing the decimal number system to Western Europe. Fibonacci signed much of his work "Leonardo Bigollo." *Bigollo* translates as "traveler" or "blockhead." There is some evidence that Fibonacci enjoyed having his contemporaries consider him a blockhead for advocating the new number system.

The Fibonacci sequence pops up in unexpected places. Figure 4.4.1 shows a pine cone with 13 clockwise spirals and 8 counterclockwise spirals. Many plants distribute their seeds as evenly as possible, thus maximizing the space available for each seed. The pattern in which the number of spirals is a Fibonacci number provides the most even distribution (see [Naylor, Mitchison]). In Section 5.3, the Fibonacci sequence appears in the analysis of the Euclidean algorithm.

**Example 4.4.7** Use mathematical induction to show that

$$\sum_{k=1}^{n} f_k = f_{n+2} - 1 \qquad \text{for all } n \geq 1.$$

**SOLUTION** For the basis step ($n = 1$), we must show that

$$\sum_{k=1}^{1} f_k = f_3 - 1.$$

Since $\sum_{k=1}^{1} f_k = f_1 = 1$ and $f_3 - 1 = 2 - 1 = 1$, the equation is verified.

For the inductive step, we assume case $n$

$$\sum_{k=1}^{n} f_k = f_{n+2} - 1$$

and prove case $n + 1$

$$\sum_{k=1}^{n+1} f_k = f_{n+3} - 1.$$

Now

$$\begin{aligned}\sum_{k=1}^{n+1} f_k &= \sum_{k=1}^{n} f_k + f_{n+1} \\ &= (f_{n+2} - 1) + f_{n+1} \qquad \text{by the inductive assumption} \\ &= f_{n+1} + f_{n+2} - 1 \\ &= f_{n+3} - 1.\end{aligned}$$

The last equality is true because of the definition of the Fibonacci numbers:

$$f_n = f_{n-1} + f_{n-2} \qquad \text{for all } n \geq 3.$$

Since the basis step and the inductive step have been verified, the given equation is true for all $n \geq 1$. ◀

## 4.4 Problem-Solving Tips

A recursive function is a function that invokes itself. The key to writing a recursive function is to find a smaller instance of the problem within the larger problem. For example, we can compute $n!$ recursively because $n! = n \cdot (n-1)!$ for all $n \geq 1$. The situation is analogous to the inductive step in mathematical induction when we must find a smaller case (e.g., case $n$) within the larger case (e.g., case $n + 1$).

As another example, tiling an $n \times n$ deficient board with trominoes when $n$ is a power of 2 can be done recursively because we can find four $(n/2) \times (n/2)$ subboards within the original $n \times n$ board. Note the similarity of the tiling algorithm to the inductive step of the proof that every $n \times n$ deficient board can be tiled with trominoes when $n$ is a power of 2.

To prove a statement about the Fibonacci numbers, use the equation

$$f_n = f_{n-1} + f_{n-2} \qquad \text{for all } n \geq 3.$$

The proof will often use mathematical induction *and* the previous equation (see Example 4.4.7).

## 4.4 Review Exercises

1. What is a recursive algorithm?
2. What is a recursive function?
3. Give an example of a recursive function.
4. Explain how the divide-and-conquer technique works.
5. What is a base case in a recursive function?
6. Why must every recursive function have a base case?
7. How is the Fibonacci sequence defined?
8. Give the first four values of the Fibonacci sequence.

## 4.4 Exercises

1. Trace Algorithm 4.4.2 for $n = 4$.
2. Trace Algorithm 4.4.4 when $n = 4$ and the missing square is the upper-left corner square.
3. Trace Algorithm 4.4.4 when $n = 8$ and the missing square is four from the left and six from the top.
4. Prove that Algorithm 4.4.4 is correct.
5. Trace Algorithm 4.4.6 for $n = 4$.
6. Trace Algorithm 4.4.6 for $n = 5$.
7. Prove that Algorithm 4.4.6 is correct.
8. Prove that

$$walk(n) = f_{n+1} \qquad \text{for all } n \geq 1.$$

9. (a) Use the formulas

$$s_1 = 1, \qquad s_n = s_{n-1} + n \qquad \text{for all } n \geq 2,$$

to write a recursive algorithm that computes

$$s_n = 1 + 2 + 3 + \cdots + n.$$

(b) Give a proof using mathematical induction that your algorithm for part (a) is correct.

10. (a) Use the formulas

$$s_1 = 2, \qquad s_n = s_{n-1} + 2n \qquad \text{for all } n \geq 2,$$

to write a recursive algorithm that computes

$$s_n = 2 + 4 + 6 + \cdots + 2n.$$

(b) Give a proof using mathematical induction that your algorithm for part (a) is correct.

11. (a) A robot can take steps of 1 meter, 2 meters, or 3 meters. Write a recursive algorithm to calculate the number of ways the robot can walk $n$ meters.

(b) Give a proof using mathematical induction that your algorithm for part (a) is correct.

12. Write a recursive algorithm to find the minimum of a finite sequence of numbers. Give a proof using mathematical induction that your algorithm is correct.
13. Write a recursive algorithm to find the maximum of a finite sequence of numbers. Give a proof using mathematical induction that your algorithm is correct.
14. Write a recursive algorithm that reverses a finite sequence. Give a proof using mathematical induction that your algorithm is correct.
15. Write a nonrecursive algorithm to compute $n!$.

★16. A robot can take steps of 1 meter or 2 meters. Write an algorithm to list all of the ways the robot can walk $n$ meters.

★17. A robot can take steps of 1 meter, 2 meters, or 3 meters. Write an algorithm to list all of the ways the robot can walk $n$ meters.

*Exercises 18–36 concern the Fibonacci sequence $\{f_n\}$.*

18. Suppose that at the beginning of the year, there is one pair of rabbits and that every month each pair produces a new pair that becomes productive after one month. Suppose further that no deaths occur. Let $a_n$ denote the number of pairs of rabbits at the end of the $n$th month. Show that $a_1 = 1, a_2 = 2$, and $a_n - a_{n-1} = a_{n-2}$. Prove that $a_n = f_{n+1}$ for all $n \geq 1$.
19. Fibonacci's original question was: Under the conditions of Exercise 18, how many pairs of rabbits are there after one year? Answer Fibonacci's question.
20. Show that the number of ways to tile a $2 \times n$ board with $1 \times 2$ rectangular pieces is $f_{n+1}$, the $(n + 1)$st Fibonacci number.
21. Use mathematical induction to show that

$$f_n^2 = f_{n-1}f_{n+1} + (-1)^{n+1} \qquad \text{for all } n \geq 2.$$

22. Show that

$$f_n^2 = f_{n-2}f_{n+2} + (-1)^n \qquad \text{for all } n \geq 3.$$

23. Show that

$$f_{n+2}^2 - f_{n+1}^2 = f_n f_{n+3} \qquad \text{for all } n \geq 1.$$

24. Use mathematical induction to show that

$$\sum_{k=1}^{n} f_k^2 = f_n f_{n+1} \qquad \text{for all } n \geq 1.$$

25. Use mathematical induction to show that for all $n \geq 1$, $f_n$ is even if and only if $n$ is divisible by 3.

★26. Use mathematical induction to show that

$$f_{2n} = f_{n+1}^2 - f_{n-1}^2 \quad \text{and} \quad f_{2n+1} = f_n^2 + f_{n+1}^2 \qquad \text{for all } n \geq 2.$$

27. Use mathematical induction to show that for all $n \geq 6$,

$$f_n > \left(\frac{3}{2}\right)^{n-1}.$$

28. Use mathematical induction to show that for all $n \geq 1$,

$$f_n \leq 2^{n-1}.$$

**29.** Use mathematical induction to show that for all $n \geq 1$,

$$\sum_{k=1}^{n} f_{2k-1} = f_{2n}, \qquad \sum_{k=1}^{n} f_{2k} = f_{2n+1} - 1.$$

**★30.** Use mathematical induction to show that every integer $n \geq 1$ can be expressed as the sum of distinct Fibonacci numbers, no two of which are consecutive.

**★31.** Show that the representation in Exercise 30 is unique if we do not allow $f_1$ as a summand.

**32.** Show that for all $n \geq 2$,

$$f_n = \frac{f_{n-1} + \sqrt{5f_{n-1}^2 + 4(-1)^{n+1}}}{2}.$$

Notice that this formula gives $f_n$ in terms of one predecessor rather than two predecessors as in the original definition.

**33.** Prove that

$$1 + \sum_{k=1}^{n} \frac{(-1)^{k+1}}{f_k f_{k+1}} = \frac{f_{n+2}}{f_{n+1}} \quad \text{for all } n \geq 1.$$

**34.** Define a sequence $\{g_n\}$ as $g_1 = c_1$ and $g_2 = c_2$ for constants $c_1$ and $c_2$, and

$$g_n = g_{n-1} + g_{n-2}$$

for $n \geq 3$. Prove that

$$g_n = g_1 f_{n-2} + g_2 f_{n-1}$$

for all $n \geq 3$.

**35.** Prove that

$$\sum_{k=1}^{n} (-1)^k f_k = (-1)^n f_{n-1} - 1 \quad \text{for all } n \geq 2.$$

**36.** Prove that

$$\sum_{k=1}^{n} (-1)^k k f_k = (-1)^n (n f_{n-1} + f_{n-3}) - 2 \quad \text{for all } n \geq 4.$$

**37.** [Requires calculus] Assume the formula for differentiating products:

$$\frac{d(fg)}{dx} = f\frac{dg}{dx} + g\frac{df}{dx}.$$

Use mathematical induction to prove that

$$\frac{dx^n}{dx} = nx^{n-1} \quad \text{for } n = 1, 2, \ldots.$$

**38.** [Requires calculus] Explain how the formula gives a recursive algorithm for integrating $\log^n |x|$:

$$\int \log^n |x|\, dx = x \log^n |x| - n \int \log^{n-1} |x|\, dx.$$

Give other examples of recursive integration formulas.

## Chapter 4 Notes

The first half of [Knuth, 1977] introduces the concept of an algorithm and various mathematical topics, including mathematical induction. The second half is devoted to data structures.

Most general references on computer science contain some discussion of algorithms. Books specifically on algorithms are [Aho; Baase; Brassard; Cormen; Johnsonbaugh; Knuth, 1997, 1998a, 1998b; Manber; Miller; Nievergelt; and Reingold]. [McNaughton] contains a very thorough discussion on an introductory level of what an algorithm is. Knuth's expository article about algorithms ([Knuth, 1977]) and his article about the role of algorithms in the mathematical sciences ([Knuth, 1985]) are also recommended. [Gardner, 1992] contains a chapter about the Fibonacci sequence.

## Chapter 4 Review

### Section 4.1

1. Algorithm
2. Properties of an algorithm: Input, output, precision, determinism, finiteness, correctness, generality
3. Trace
4. Pseudocode

### Section 4.2

5. Searching
6. Text search
7. Text-search algorithm
8. Sorting
9. Insertion sort
10. Time and space for algorithms
11. Best-case time
12. Worst-case time
13. Randomized algorithm
14. Shuffle algorithm

### Section 4.3

15. Analysis of algorithms
16. Worst-case time of an algorithm
17. Best-case time of an algorithm
18. Average-case time of an algorithm

19. Big oh notation: $f(n) = O(g(n))$
20. Omega notation: $f(n) = \Omega(g(n))$
21. Theta notation: $f(n) = \Theta(g(n))$

## Section 4.4

22. Recursive algorithm
23. Recursive function
24. Divide-and-conquer technique
25. Base cases: Situations where a recursive function does not invoke itself
26. Fibonacci sequence $\{f_n\}$ : $f_1 = 1, f_2 = 1, f_n = f_{n-1} + f_{n-2}, n \geq 3$

## Chapter 4 Self-Test

1. Trace Algorithm 4.1.1 for the values $a = 12$, $b = 3$, $c = 0$.
2. Which of the algorithm properties—input, output, precision, determinism, finiteness, correctness, generality—if any, are lacking in the following? Explain.

   Input: $S$ (a set of integers), $m$ (an integer)
   Output: All finite subsets of $S$ that sum to $m$

   1. List all finite subsets of $S$ and their sums.
   2. Step through the subsets listed in 1 and output each whose sum is $m$.

3. Trace Algorithm 4.2.1 for the input $t$ = "111011" and $p$ = "110".
4. Trace Algorithm 4.2.3 for the input 44, 64, 77, 15, 3.
5. Trace Algorithm 4.2.4 for the input 5, 51, 2, 44, 96. Assume that the values of *rand* are

   $$rand(1, 5) = 1, \quad rand(2, 5) = 3, \quad rand(3, 5) = 5, \quad rand(4, 5) = 5.$$

6. Write an algorithm that receives as input the distinct numbers $a$, $b$, and $c$ and assigns the values $a$, $b$, and $c$ to the variables $x$, $y$, and $z$ so that $x < y < z$.
7. Write an algorithm that receives as input the sequence $s_1, \ldots, s_n$ sorted in nondecreasing order and prints all values that appear more than once. *Example:* If the sequence is 1, 1, 1, 5, 8, 8, 9, 12, the output is 1 8.
8. Write an algorithm that returns true if the values of $a$, $b$, and $c$ are distinct, and false otherwise.
9. Write an algorithm that tests whether two $n \times n$ matrices are equal and find a theta notation for its worst-case time.

*Select a theta notation from among $\Theta(1)$, $\Theta(n)$, $\Theta(n^2)$, $\Theta(n^3)$, $\Theta(n^4)$, $\Theta(2^n)$, or $\Theta(n!)$ for each of the expressions in Exercises 10 and 11.*

10. $4n^3 + 2n - 5$
11. $1^3 + 2^3 + \cdots + n^3$
12. Select a theta notation from among $\Theta(1)$, $\Theta(n)$, $\Theta(n^2)$, $\Theta(n^3)$, $\Theta(2^n)$, or $\Theta(n!)$ for the number of times the line $x = x + 1$ is executed.

```
for i = 1 to n
  for j = 1 to n
    x = x + 1
```

13. Trace Algorithm 4.4.4 (the tromino tiling algorithm) when $n = 8$ and the missing square is four from the left and two from the top.

*Exercises 14–16 refer to the* tribonacci sequence $\{t_n\}$ *defined by the equations*

$$t_1 = t_2 = t_3 = 1, \quad t_n = t_{n-1} + t_{n-2} + t_{n-3} \quad \textit{for all } n \geq 4.$$

14. Find $t_4$ and $t_5$.
15. Write a recursive algorithm to compute $t_n$, $n \geq 1$.
16. Give a proof using mathematical induction that your algorithm for Exercise 15 is correct.

## Chapter 4 Computer Exercises

1. Implement Algorithm 4.1.2, finding the largest element in a sequence, as a program.
2. Implement Algorithm 4.2.1, text search, as a program.
3. Implement Algorithm 4.2.3, insertion sort, as a program.
4. Implement Algorithm 4.2.4, shuffle, as a program.
5. Run shuffle (Algorithm 4.2.4) many times for the same input sequence. How might the output be analyzed to determine if it is truly "random"?
6. Implement selection sort (see Exercise 22, Section 4.2) as a program.
7. Compare the running times of insertion sort (Algorithm 4.2.3) and selection sort (see Exercise 22, Section 4.2) for several inputs of different sizes. Include data sorted in nondecreasing order, data sorted in nonincreasing order, data containing many duplicates, and data in random order.
8. Write recursive and nonrecursive programs to compute $n!$. Compare the times required by the programs.

**9.** Write a program whose input is a $2^n \times 2^n$ board with one missing square and whose output is a tiling of the board by trominoes.

**10.** Write a program that uses a graphics display to show a tiling with trominoes of a $2^n \times 2^n$ board with one square missing.

**11.** Write a program that tiles with trominoes an $n \times n$ board with one square missing, provided that $n \neq 5$ and 3 does not divide $n$.

**12.** Write recursive and nonrecursive programs to compute the Fibonacci sequence. Compare the times required by the programs.

**13.** A robot can take steps of 1 meter or 2 meters. Write a program to list all of the ways the robot can walk $n$ meters.

**14.** A robot can take steps of 1, 2, or 3 meters. Write a program to list all of the ways the robot can walk $n$ meters.

# Chapter 5

# INTRODUCTION TO NUMBER THEORY



**Number theory** is the branch of mathematics concerned with the integers. Traditionally, number theory was a *pure* branch of mathematics—known for its abstract nature rather than its applications. The great English mathematician, G. H. Hardy (1877–1947), used number theory as an example of a beautiful, but impractical, branch of mathematics. However, in the late 1900s, number theory became extremely useful in cryptosystems—systems used for secure communications.

In the preceding chapters, we used some basic number theory definitions such as "divides" and "prime number." In Section 5.1, we review these basic definitions and extend the discussion to unique factorization, greatest common divisors, and least common multiples.

In Section 5.2, we discuss representations of integers and some algorithms for integer arithmetic.

The Euclidean algorithm for computing the greatest common divisor is the subject of Section 5.3. This is surely one of the oldest algorithms. Euclid lived about 295 B.C., and the algorithm probably predates him.

As an application of the number theory presented in Sections 5.1–5.3, we discuss the RSA system for secure communications in Section 5.4.

## 5.1 Divisors

In this section, we give the basic definitions and terminology. We begin by recalling the definition of "divides," and we introduce some related terminology.

**Definition 5.1.1 ▶** Let $n$ and $d$ be integers, $d \neq 0$. We say that $d$ *divides* $n$ if there exists an integer $q$ satisfying $n = dq$. We call $q$ the *quotient* and $d$ a *divisor* or *factor* of $n$. If $d$ divides $n$, we write $d \mid n$. If $d$ does not divide $n$, we write $d \nmid n$. ◀

**Example 5.1.2** Since $21 = 3 \cdot 7$, 3 divides 21 and we write $3 \mid 21$. The quotient is 7. We call 3 a divisor or factor of 21. ◀

We note that if $n$ and $d$ are positive integers and $d \mid n$, then $d \leq n$. (If $d \mid n$, there exists an integer $q$ such that $n = dq$. Since $n$ and $d$ are positive integers, $1 \leq q$. Therefore, $d \leq dq = n$.)

Whether an integer $d > 0$ divides an integer $n$ or not, we obtain a unique quotient $q$ and remainder $r$ as given by the Quotient-Remainder Theorem (Theorem 2.5.6): There exist unique integers $q$ (quotient) and $r$ (remainder) satisfying $n = dq + r, 0 \leq r < d$. The remainder $r$ equals zero if and only if $d$ divides $n$.

Some additional properties of divisors are given in the following theorem and will be useful in our subsequent work in this chapter.

**Theorem 5.1.3** Let $m$, $n$, and $d$ be integers.

(a) If $d \mid m$ and $d \mid n$, then $d \mid (m + n)$.

(b) If $d \mid m$ and $d \mid n$, then $d \mid (m - n)$.

(c) If $d \mid m$, then $d \mid mn$.

**Proof** (a) Suppose that $d \mid m$ and $d \mid n$. By Definition 5.1.1,

$$m = dq_1 \tag{5.1.1}$$

for some integer $q_1$ and

$$n = dq_2 \tag{5.1.2}$$

for some integer $q_2$. If we add equations (5.1.1) and (5.1.2), we obtain

$$m + n = dq_1 + dq_2 = d(q_1 + q_2).$$

Therefore, $d$ divides $m + n$ (with quotient $q_1 + q_2$). We have proved part (a).

The proofs of parts (b) and (c) are left as exercises (see Exercises 27 and 28). ◀

**Definition 5.1.4 ▶** An integer greater than 1 whose only positive divisors are itself and 1 is called *prime*. An integer greater than 1 that is not prime is called *composite*. ◀

**Example 5.1.5** The integer 23 is prime because its only divisors are itself and 1. The integer 34 is composite because it is divisible by 17, which is neither 1 nor 34. ◀

If an integer $n > 1$ is composite, then it has a positive divisor $d$ other than 1 and itself. Since $d$ is positive and $d \neq 1$, $d > 1$. Since $d$ is a divisor of $n$, $d \leq n$. Since $d \neq n$, $d < n$. Therefore, to determine if a positive integer $n$ is composite, it suffices to test whether any of the integers $2, 3, \ldots, n - 1$ divides $n$. If some integer in this list divides $n$, then $n$ is composite. If no integer in this list divides $n$, then $n$ is prime. (Actually, we can shorten this list considerably; see Theorem 5.1.7.)

**Example 5.1.6** By inspection, we find that none of $2, 3, 4, 5, \ldots, 41, 42$ divides 43; thus, 43 is prime.

Checking the list $2, 3, 4, 5, \ldots, 449, 450$ for potential divisors of 451, we find that 11 divides 451 ($451 = 11 \cdot 41$); thus, 451 is composite. ◀

In Example 5.1.6, to determine whether a positive integer $n > 1$ was prime, we checked the potential divisors $2, 3, \ldots, n - 1$. Actually, it suffices to check only

$$2, 3, \ldots, \lfloor\sqrt{n}\rfloor.$$

**Theorem 5.1.7** A positive integer $n$ greater than 1 is composite if and only if $n$ has a divisor $d$ satisfying $2 \leq d \leq \sqrt{n}$.

**Proof** We must prove

$$\text{If } n \text{ is composite, then } n \text{ has a divisor } d \text{ satisfying } 2 \leq d \leq \sqrt{n}, \tag{5.1.3}$$

and

$$\text{If } n \text{ has a divisor } d \text{ satisfying } 2 \leq d \leq \sqrt{n}, \text{ then } n \text{ is composite.} \tag{5.1.4}$$

We first prove (5.1.3). Suppose that $n$ is composite. The discussion following Example 5.1.5 shows that $n$ has a divisor $d'$ satisfying $2 \leq d' < n$. We now argue by cases. If $d' \leq \sqrt{n}$, then $n$ has a divisor $d$ (namely $d = d'$) satisfying $2 \leq d \leq \sqrt{n}$.

The other case is $d' > \sqrt{n}$. Since $d'$ divides $n$, by Definition 5.1.1 there exists an integer $q$ satisfying $n = d'q$. Thus $q$ is also a divisor of $n$. We claim that $q \leq \sqrt{n}$. To show that $q \leq \sqrt{n}$, we use proof by contradiction. Thus, suppose that $q > \sqrt{n}$. Multiplying $d' > \sqrt{n}$ and $q > \sqrt{n}$ gives

$$n = d'q > \sqrt{n}\sqrt{n} = n,$$

which is a contradiction. Thus, $q \leq \sqrt{n}$. Therefore, $n$ has a divisor $d$ (namely $d = q$) satisfying $2 \leq d \leq \sqrt{n}$.

It remains to prove (5.1.4). If $n$ has a divisor $d$ satisfying $2 \leq d \leq \sqrt{n}$, by Definition 5.1.4 $n$ is composite. The proof is complete. ◀

We may use Theorem 5.1.7 to construct the following algorithm that tests whether a positive integer $n > 1$ is prime.

**Algorithm 5.1.8** **Testing Whether an Integer Is Prime**

This algorithm determines whether the integer $n > 1$ is prime. If $n$ is prime, the algorithm returns 0. If $n$ is composite, the algorithm returns a divisor $d$ satisfying $2 \leq d \leq \sqrt{n}$. To test whether $d$ divides $n$, the algorithm checks whether the remainder when $n$ is divided by $d$, $n$ mod $d$, is zero.

Input: $n$
Output: $d$

```
is_prime(n) {
  for d = 2 to ⌊√n⌋
    if (n mod d == 0)
      return d
  return 0
}
```

**Go Online**
For a C++ program implementing this algorithm, see bit.ly/2q4cn1l

**Example 5.1.9** To determine whether 43 is prime, Algorithm 5.1.8 checks whether any of $2, 3, 4, 5, 6 = \lfloor\sqrt{43}\rfloor$ divides 43. Since none of these numbers divides 43, the condition

$$n \bmod d == 0 \tag{5.1.5}$$

is always false. Therefore, the algorithm returns 0 to indicate that 43 is prime.

To determine whether 451 is prime, Algorithm 5.1.8 checks whether any of $2, 3, \ldots, 21 = \lfloor\sqrt{451}\rfloor$ divides 451. For $d = 2, 3, \ldots, 10$, $d$ does not divide 451 and the condition (5.1.5) is false. However, when $d = 11$, $d$ does divide 451 and the condition

(5.1.5) is true. Therefore, the algorithm returns 11 to indicate that 451 is composite and 11 divides 451. ◀

In the worst case (when $n$ is prime and the for loop runs to completion), Algorithm 5.1.8 takes time $\Theta(\sqrt{n})$. Although Algorithm 5.1.8 runs in time polynomial in $n$ (since $\sqrt{n} \leq n$), it does *not* run in time polynomial in the *size* of the input (namely, $n$). [We can represent $n$ in considerably less space than $\Theta(n)$; see Example 5.2.1.] We say that Algorithm 5.1.8 is not a polynomial-time algorithm. It is not known whether there is a polynomial-time algorithm that can find a factor of a given integer; but most computer scientists think that there is no such algorithm. On the other hand, in 2002 Manindra Agarwal and two of his students, Nitin Saxena and Neeraj Kayal, discovered a polynomial-time algorithm that can determine whether or not a given integer is prime (see [Agarwal]). The question of whether there is a polynomial-time algorithm that can factor an integer is of more than academic interest since the security of certain encryption systems relies on the nonexistence of such an algorithm (see Section 5.4).

Notice that if a composite integer $n$ is input to Algorithm 5.1.8, the divisor returned is prime; that is, Algorithm 5.1.8 returns a prime factor of a composite integer. To prove this, we use proof by contradiction. If Algorithm 5.1.8 returns a composite divisor of $n$, say $a$, then $a$ has a divisor $a'$ less than $a$. Since $a'$ also divides $n$ and $a' < a$, when Algorithm 5.1.8 sets $d = a'$, it will return $a'$, not $a$. This contradiction shows that if a composite integer $n$ is input to Algorithm 5.1.8, the divisor returned is prime.

**Example 5.1.10** If the input to Algorithm 5.1.8 is $n = 1274$, the algorithm returns the prime 2 because 2 divides 1274, specifically $1274 = 2 \cdot 637$.

If we now input $n = 637$ to Algorithm 5.1.8, the algorithm returns the prime 7 because 7 divides 637, specifically $637 = 7 \cdot 91$.

If we now input $n = 91$ to Algorithm 5.1.8, the algorithm returns the prime 7 because 7 divides 91, specifically $91 = 7 \cdot 13$.

If we now input $n = 13$ to Algorithm 5.1.8, the algorithm returns 0 because 13 is prime.

Combining the previous equations gives 1274 as a product of primes

$$1274 = 2 \cdot 637 = 2 \cdot 7 \cdot 91 = 2 \cdot 7 \cdot 7 \cdot 13.$$

We have illustrated how to write any integer greater than 1 as a product of primes. It is also a fact (although we will not prove it in this book) that, except for the order of the prime factors, the prime factors are unique. This result is known as the **Fundamental Theorem of Arithmetic** or the **unique factorization theorem.** (Unique factorization does *not* hold in some systems; see Exercise 44.) ◀

**Theorem 5.1.11** **Fundamental Theorem of Arithmetic**

Any integer greater than 1 can be written as a product of primes. Moreover, if the primes are written in nondecreasing order, the factorization is unique. In symbols, if

$$n = p_1 p_2 \cdots p_i,$$

where the $p_k$ are primes and $p_1 \leq p_2 \leq \cdots \leq p_i$, and

$$n = p'_1 p'_2 \cdots p'_j,$$

where the $p'_k$ are primes and $p'_1 \leq p'_2 \leq \cdots \leq p'_j$, then $i = j$ and

$$p_k = p'_k \qquad \text{for all } k = 1, \ldots, i.$$

We next prove that the number of primes is infinite.

**Theorem 5.1.12** The number of primes is infinite.

**Proof** It suffices to show that if $p$ is a prime, there is a prime larger than $p$. To this end, we let $p_1, p_2, \ldots, p_n$ denote all of the distinct primes less than or equal to $p$. Consider the integer

$$m = p_1 p_2 \cdots p_n + 1.$$

Notice that when $m$ is divided by $p_i$, the remainder is 1:

$$m = p_i q + 1, \qquad q = p_1 p_2 \cdots p_{i-1} p_{i+1} \cdots p_n.$$

Therefore, for all $i = 1$ to $n$, $p_i$ does *not* divide $m$. Let $p'$ be a prime factor of $m$ ($m$ may or may not itself be prime; see Exercise 33). Then $p'$ is not equal to any of $p_i$, $i = 1$ to $n$. Since $p_1, p_2, \ldots, p_n$ is a list of all of the primes less than or equal to $p$, we must have $p' > p$. The proof is complete. ◀

**Example 5.1.13** Show how the proof of Theorem 5.1.12 produces a prime larger than 11.

**SOLUTION** We list the primes less than or equal to 11: 2, 3, 5, 7, 11. We let $m = 2 \cdot 3 \cdot 5 \cdot 7 \cdot 11 + 1 = 2311$. Using Algorithm 5.1.8, we find that 2311 is prime. We have found a prime, namely 2311, larger than each of 2, 3, 5, 7, 11. (If 2311 had turned out not to be prime, Algorithm 5.1.8 would have found a factor of 2311, which would necessarily be larger than each of 2, 3, 5, 7, 11.) ◀

The **greatest common divisor** of two integers $m$ and $n$ (not both zero) is the largest positive integer that divides both $m$ and $n$. For example, the greatest common divisor of 4 and 6 is 2, and the greatest common divisor of 3 and 8 is 1. We use the notion of greatest common divisor when we check to see if a fraction $m/n$, where $m$ and $n$ are integers, is in lowest terms. If the greatest common divisor of $m$ and $n$ is 1, $m/n$ is in lowest terms; otherwise, we can reduce $m/n$. For example, $4/6$ is not in lowest terms because the greatest common divisor of 4 and 6 is 2, not 1. (We can divide both 4 and 6 by 2.) The fraction $3/8$ is in lowest terms because the greatest common divisor of 3 and 8 is 1.

**Definition 5.1.14 ▶** Let $m$ and $n$ be integers with not both $m$ and $n$ zero. A *common divisor* of $m$ and $n$ is an integer that divides both $m$ and $n$. The *greatest common divisor,* written

$$\gcd(m, n),$$

is the largest common divisor of $m$ and $n$. ◀

**Example 5.1.15** The positive divisors of 30 are

$$1, 2, 3, 5, 6, 10, 15, 30,$$

and the positive divisors of 105 are

$$1, 3, 5, 7, 15, 21, 35, 105;$$

thus the positive common divisors of 30 and 105 are

$$1, 3, 5, 15.$$

It follows that the greatest common divisor of 30 and 105, $\gcd(30, 105)$, is 15. ◀

We can also find the greatest common divisor of two integers $m$ and $n$ by looking carefully at their prime factorizations. We illustrate with an example and then explain the technique in detail.

**Example 5.1.16** We find the greatest common divisor of 30 and 105 by looking at their prime factorizations

$$30 = 2 \cdot 3 \cdot 5 \qquad 105 = 3 \cdot 5 \cdot 7.$$

Notice that 3 is a common divisor of 30 and 105 since it occurs in the prime factorization of both numbers. For the same reason, 5 is also a common divisor of 30 and 105. Also, $3 \cdot 5 = 15$ is also a common divisor of 30 and 105. Since no larger product of primes is common to both 30 and 105, we conclude that 15 is the greatest common divisor of 30 and 105. ◀

We state the method of Example 5.1.16 as Theorem 5.1.17.

**Theorem 5.1.17** Let $m$ and $n$ be integers, $m > 1$, $n > 1$, with prime factorizations

$$m = p_1^{a_1} p_2^{a_2} \cdots p_k^{a_k}$$

and

$$n = p_1^{b_1} p_2^{b_2} \cdots p_k^{b_k}.$$

(If the prime $p_i$ is not a factor of $m$, we let $a_i = 0$. Similarly, if the prime $p_i$ is not a factor of $n$, we let $b_i = 0$.) Then

$$\gcd(m, n) = p_1^{\min(a_1,b_1)} p_2^{\min(a_2,b_2)} \cdots p_k^{\min(a_k,b_k)}.$$

**Proof** Let $g = \gcd(m, n)$. We note that if a prime $p$ appears in the prime factorization of $g$, $p$ must be equal to one of $p_1, \ldots, p_k$; otherwise, $g$ would not divide $m$ or $n$ (or both). Therefore $g = p_1^{c_1} \cdots p_k^{c_k}$ for some $c_1, \ldots, c_k$. Now

$$p_1^{\min(a_1,b_1)} p_2^{\min(a_2,b_2)} \cdots p_k^{\min(a_k,b_k)} \qquad \textbf{(5.1.6)}$$

divides both $m$ and $n$ and if any exponent, $\min(a_i, b_i)$, is increased, the resulting integer will fail to divide $m$ or $n$ (or both). Therefore (5.1.6) is the greatest common divisor of $m$ and $n$. ◀

**Example 5.1.18** Using the notation of Theorem 5.1.17, we have

$$82320 = 2^4 \cdot 3^1 \cdot 5^1 \cdot 7^3 \cdot 11^0$$

and

$$950796 = 2^2 \cdot 3^2 \cdot 5^0 \cdot 7^4 \cdot 11^1.$$

By Theorem 5.1.17,

$$\begin{aligned}\gcd(82320, 950796) &= 2^{\min(4,2)} \cdot 3^{\min(1,2)} \cdot 5^{\min(1,0)} \cdot 7^{\min(3,4)} \cdot 11^{\min(0,1)} \\ &= 2^2 \cdot 3^1 \cdot 5^0 \cdot 7^3 \cdot 11^0 \\ &= 4116.\end{aligned}$$

◀

Neither the "list all divisors" method of Example 5.1.15 nor use of prime factorization as in Example 5.1.18 is an efficient method of finding the greatest common divisor. The problem is that both methods require finding the prime factors of the numbers involved and no efficient algorithm is known to compute these prime factors. However, in Section 5.3, we will present the Euclidean algorithm, which does provide an efficient way to compute the greatest common divisor.

A companion to the greatest common divisor is the least common multiple.

**Definition 5.1.19 ▶** Let $m$ and $n$ be positive integers. A *common multiple* of $m$ and $n$ is an integer that is divisible by both $m$ and $n$. The *least common multiple,* written

$$\text{lcm}(m, n),$$

is the smallest positive common multiple of $m$ and $n$. ◀

**Example 5.1.20** The least common multiple of 30 and 105, lcm(30, 105), is 210 because 210 is divisible by both 30 and 105 and, by inspection, no positive integer smaller than 210 is divisible by both 30 and 105. ◀

**Example 5.1.21** We can find the least common multiple of 30 and 105 by looking at their prime factorizations

$$30 = 2 \cdot 3 \cdot 5 \qquad 105 = 3 \cdot 5 \cdot 7.$$

The prime factorization of lcm(30, 105) must contain 2, 3, and 5 as factors [so that 30 divides lcm(30, 105)]. It must also contain 3, 5, and 7 [so that 105 divides lcm(30, 105)]. The smallest number with this property is

$$2 \cdot 3 \cdot 5 \cdot 7 = 210.$$

Therefore, lcm(30, 105) = 210. ◀

We state the method of Example 5.1.21 as Theorem 5.1.22.

**Theorem 5.1.22** Let $m$ and $n$ be integers, $m > 1$, $n > 1$, with prime factorizations

$$m = p_1^{a_1} p_2^{a_2} \cdots p_k^{a_k}$$

and

$$n = p_1^{b_1} p_2^{b_2} \cdots p_k^{b_k}.$$

(If the prime $p_i$ is not a factor of $m$, we let $a_i = 0$. Similarly, if the prime $p_i$ is not a factor of $n$, we let $b_i = 0$.) Then

$$\text{lcm}(m, n) = p_1^{\max(a_1,b_1)} p_2^{\max(a_2,b_2)} \cdots p_k^{\max(a_k,b_k)}.$$

**Proof** Let $l = \text{lcm}(m, n)$. We note that if a prime $p$ appears in the prime factorization of $l$, $p$ must be equal to one of $p_1, \ldots, p_k$; otherwise, we could eliminate $p$ and obtain a smaller integer that is divisible by both $m$ and $n$. Therefore $l = p_1^{c_1} \cdots p_k^{c_k}$ for some $c_1, \ldots, c_k$. Now

$$p_1^{\max(a_1,b_1)} p_2^{\max(a_2,b_2)} \cdots p_k^{\max(a_k,b_k)} \tag{5.1.7}$$

is divisible by both $m$ and $n$ and if any exponent, $\max(a_i, b_i)$, is decreased, the resulting integer will fail to be divisible by $m$ or $n$ (or both). Therefore (5.1.7) is the least common multiple of $m$ and $n$. ◀

**Example 5.1.23** Using the notation of Theorem 5.1.22, we have

$$82320 = 2^4 \cdot 3^1 \cdot 5^1 \cdot 7^3 \cdot 11^0$$

and

$$950796 = 2^2 \cdot 3^2 \cdot 5^0 \cdot 7^4 \cdot 11^1.$$

By Theorem 5.1.22,

$$\begin{aligned}\text{lcm}(82320, 950796) &= 2^{\max(4,2)} \cdot 3^{\max(1,2)} \cdot 5^{\max(1,0)} \cdot 7^{\max(3,4)} \cdot 11^{\max(0,1)} \\ &= 2^4 \cdot 3^2 \cdot 5^1 \cdot 7^4 \cdot 11^1 \\ &= 19015920.\end{aligned}$$

◀

**Example 5.1.24** In Example 5.1.15, we found that $\gcd(30, 105) = 15$, and in Example 5.1.21, we found that $\text{lcm}(30, 105) = 210$. Notice that the product of the gcd and lcm is equal to the product of the pair of numbers; that is,

$$\gcd(30, 105) \cdot \text{lcm}(30, 105) = 15 \cdot 210 = 3150 = 30 \cdot 105.$$

This formula holds for any pair of numbers as we will show in Theorem 5.1.25. ◀

**Theorem 5.1.25** For any positive integers $m$ and $n$,

$$\gcd(m, n) \cdot \text{lcm}(m, n) = mn.$$

**Proof** If $m = 1$, then $\gcd(m, n) = 1$ and $\text{lcm}(m, n) = n$, so

$$\gcd(m, n) \cdot \text{lcm}(m, n) = 1 \cdot n = mn.$$

Similarly, if $n = 1$, then $\gcd(m, n) = 1$ and $\text{lcm}(m, n) = m$, so

$$\gcd(m, n) \cdot \text{lcm}(m, n) = 1 \cdot m = mn.$$

Thus, we may assume that $m > 1$ and $n > 1$.

The proof combines the formulas for the gcd (Theorem 5.1.17) and lcm (Theorem 5.1.22) (which require that $m > 1$ and $n > 1$) with the fact that

$$\min(x, y) + \max(x, y) = x + y \qquad \text{for all } x \text{ and } y.$$

This latter formula is true because one of $\{\min(x, y), \max(x, y)\}$ equals $x$ and the other equals $y$. We now put this all together to produce a proof.

Write the prime factorizations of $m$ and $n$ as

$$m = p_1^{a_1} p_2^{a_2} \cdots p_k^{a_k}$$

and

$$n = p_1^{b_1} p_2^{b_2} \cdots p_k^{b_k}.$$

(If the prime $p_i$ is not a factor of $m$, we let $a_i = 0$. Similarly, if the prime $p_i$ is not a factor of $n$, we let $b_i = 0$.) By Theorem 5.1.17,

$$\gcd(m, n) = p_1^{\min(a_1,b_1)} \cdots p_k^{\min(a_k,b_k)},$$

and by Theorem 5.1.22,

$$\text{lcm}(m, n) = p_1^{\max(a_1,b_1)} \cdots p_k^{\max(a_k,b_k)}.$$

Therefore,

$$\begin{aligned}\gcd(m, n) \cdot \text{lcm}(m, n) &= [p_1^{\min(a_1,b_1)} \cdots p_k^{\min(a_k,b_k)}] \cdot \\ &\quad [p_1^{\max(a_1,b_1)} \cdots p_k^{\max(a_k,b_k)}] \\ &= p_1^{\min(a_1,b_1)+\max(a_1,b_1)} \cdots p_k^{\min(a_k,b_k)+\max(a_k,b_k)} \\ &= p_1^{a_1+b_1} \cdots p_k^{a_k+b_k} \\ &= [p_1^{a_1} \cdots p_k^{a_k}][p_1^{b_1} \cdots p_k^{b_k}] = mn.\end{aligned}$$ ◀

If we have an algorithm to compute the greatest common divisor, we can compute the least common multiple by using Theorem 5.1.25:

$$\text{lcm}(m, n) = \frac{mn}{\gcd(m, n)}.$$

In particular, if we have an *efficient* algorithm to compute the greatest common divisor, we can efficiently compute the least common multiple as well.

## 5.1 Problem-Solving Tips

The straightforward way to determine whether an integer $n > 1$ is prime is to test whether any of $2, 3, \ldots, \lfloor\sqrt{n}\rfloor$ divides $n$. While this technique becomes too time-consuming as $n$ grows larger, it suffices for relatively small values of $n$. This technique can be iterated to find the prime factorization of $n$, again for relatively small values of $n$.

Two ways of finding the greatest common divisor of $a$ and $b$ were presented. The first way was to list all of the positive divisors of $a$ and all of the positive divisors of $b$ and then, among all of the common divisors, choose the largest. This technique is time consuming and was shown mainly to illustrate exactly what is meant by common divisors and the greatest common divisor.

The second technique was to compare the prime factorizations of $a$ and $b$. If $p^i$ appears in $a$ and $p^j$ appears in $b$, include $p^{\min(i,j)}$ in the prime factorization of the greatest common divisor. This technique works well if the numbers $a$ and $b$ are relatively small so that the prime factorizations of each can be found, or if the prime factorizations of each are given. In Section 5.3, we present the Euclidean algorithm that efficiently finds the greatest common divisor even for large values of $a$ and $b$.

If you compute the $\gcd(a, b)$, you can immediately compute the least common multiple using the formula

$$\text{lcm}(a, b) = \frac{ab}{\gcd(a, b)}.$$

The least common multiple can also be computed by comparing the prime factorizations of $a$ and $b$. If $p^i$ appears in $a$ and $p^j$ appears in $b$, include $p^{\max(i,j)}$ in the prime factorization of the least common multiple.

## 5.1 Review Exercises

1. Define *d divides n*.

2. Define *d is a divisor of n*.

3. Define *quotient*.

4. Define *n is prime*.

5. Define *n is composite*.

6. Explain why, when testing whether an integer $n > 1$ is prime by looking for divisors, we need only check whether any of 2 to $\lfloor\sqrt{n}\rfloor$ divides $n$.

7. Explain why Algorithm 5.1.8 is not considered to be a polynomial-time algorithm.

8. What is the Fundamental Theorem of Arithmetic?

9. Prove that the number of primes is infinite.

10. What is a common divisor?

11. What is the greatest common divisor?

12. Explain how to compute the greatest common divisor of $m$ and $n$, not both zero, given their prime factorizations.

13. What is a common multiple?

14. What is the least common multiple?

15. Explain how to compute the least common multiple of positive integers $m$ and $n$, given their prime factorizations.

16. How are the greatest common divisor and least common multiple related?

## 5.1 Exercises

*In Exercises 1–8, trace Algorithm 5.1.8 for the given input.*

1. $n = 9$
2. $n = 209$
3. $n = 47$
4. $n = 637$
5. $n = 4141$
6. $n = 1007$
7. $n = 3738$
8. $n = 1050703$

9. Which of the integers in Exercises 1–8 are prime?

10. Find the prime factorization of each integer in Exercises 1–8.

11. Find the prime factorization of 11!.

*Find the greatest common divisor of each pair of integers in Exercises 12–24.*

12. 0, 17
13. 60, 90
14. 5, 25
15. 110, 273
16. 315, 825
17. 220, 1400
18. 20, 40
19. 2091, 4807
20. 331, 993
21. 13, $13^2$
22. 15, $15^9$
23. $3^2 \cdot 7^3 \cdot 11, 2^3 \cdot 5 \cdot 7$
24. $3^2 \cdot 7^3 \cdot 11, 3^2 \cdot 7^3 \cdot 11$

25. Find the least common multiple of each pair of integers in Exercises 13–24.

26. For each pair of integers in Exercises 13–24, verify that $\gcd(m, n) \cdot \text{lcm}(m, n) = mn$.

27. Let $m$, $n$, and $d$ be integers. Show that if $d \mid m$ and $d \mid n$, then $d \mid (m - n)$.

28. Let $m$, $n$, and $d$ be integers. Show that if $d \mid m$, then $d \mid mn$.

29. Let $m$, $n$, $d_1$, and $d_2$ be integers. Show that if $d_1 \mid m$ and $d_2 \mid n$, then $d_1 d_2 \mid mn$.

30. Let $n$, $c$, and $d$ be integers. Show that if $dc \mid nc$, then $d \mid n$.

31. Let $a$, $b$, and $c$ be integers. Show that if $a \mid b$ and $b \mid c$, then $a \mid c$.

32. Suggest ways to make Algorithm 5.1.8 more efficient.

33. Give an example of consecutive primes $p_1 = 2, p_2, \ldots, p_n$ where

$$p_1 p_2 \cdots p_n + 1$$

is not prime.

*Exercises 34 and 35 use the following definition: A subset $\{a_1, \ldots, a_n\}$ of $\mathbf{Z}^+$ is a *-set of size $n$ if $(a_i - a_j) \mid a_i$ for all $i$ and $j$ where $i \neq j$, $1 \le i \le n$, and $1 \le j \le n$. These exercises are due to Martin Gilchrist.*

34. Prove that for all $n \ge 2$, there exists a *-set of size $n$. *Hint*. Use induction on $n$. For the Basis Step, consider the set $\{1, 2\}$. For the Inductive Step, let $b_0 = \prod_{k=1}^{n} a_k$ and $b_i = b_0 + a_i$ for $1 \le i \le n$.

35. Using the hint in Exercise 34, construct *-sets of sizes 3 and 4.

*The* Fermat numbers $F_0, F_1, \ldots$ *are defined as* $F_n = 2^{2^n} + 1$.

36. Prove that

$$\prod_{i=0}^{n-1} F_i = F_n - 2 \quad \text{for all } n, \; n \ge 1.$$

37. Using Exercise 36 or otherwise, prove that

$$\gcd(F_m, F_n) = 1 \quad \text{for all } m, n, \; 0 \le m < n.$$

38. Use Exercise 37 to prove that the number of primes is infinite.

39. Recall that a Mersenne prime (see the discussion before Example 2.2.14) is a prime of the form $2^p - 1$, where $p$ is prime. Prove that if $m$ is composite, $2^m - 1$ is also composite.

*Exercises 40–49 use the following notation and terminology. We let E denote the set of positive, even integers. If $n \in E$ can be written as a product of two or more elements in E, we say that n is* E-composite; *otherwise, we say that n is* E-prime. *As examples, 4 is E-composite and 6 is E-prime.*

**40.** Is 2 $E$-prime or $E$-composite?

**41.** Is 8 $E$-prime or $E$-composite?

**42.** Is 10 $E$-prime or $E$-composite?

**43.** Is 12 $E$-prime or $E$-composite?

**44.** Show that the number 36 can be written as a product of $E$-primes in two different ways, which shows that factoring into $E$-primes is *not* necessarily unique.

**45.** Find a necessary and sufficient condition for an integer to be an $E$-prime. Prove your statement.

**46.** Show that the set of $E$-primes is infinite.

**47.** Show that there are no twin $E$-primes, that is, two $E$-primes that differ by 2.

**48.** Show that there are infinitely many pairs of $E$-primes that differ by 4.

**49.** Give an example to show that the following is false: If an $E$-prime $p$ divides $mn \in E$, then $p$ divides $m$ or $p$ divides $n$. "Divides" means "divides in $E$." That is, if $p, q \in E$, we say that $p$ divides $q$ in $E$ if $q = pr$, where $r \in E$. (Compare this result with Exercise 27, Section 5.3.)

## 5.2 Representations of Integers and Integer Algorithms

**Go Online**
For more on representations of integers, see bit.ly/2NYUMkW

A **bit** is a *b*inary dig*it*, that is, a 0 or a 1. In a digital computer, data and instructions are encoded as bits. (The term *digital* refers to the use of the digits 0 and 1.) Technology determines how the bits are physically represented within a computer system. Today's hardware relies on the state of an electronic circuit to represent a bit. The circuit must be capable of being in two states—one representing 1, the other 0. In this section we discuss the **binary number system,** which represents integers using bits, and the **hexadecimal number system,** which represents integers using 16 symbols. The **octal number system,** which represents integers using eight symbols, is discussed before Exercise 42.

In the decimal number system, to represent integers we use the 10 symbols 0, 1, 2, 3, 4, 5, 6, 7, 8, and 9. In representing an integer, the symbol's position is significant; reading from the right, the first symbol represents the number of 1's, the next symbol the number of 10's, the next symbol the number of 100's, and so on. For example,

$$3854 = 3 \cdot 10^3 + 8 \cdot 10^2 + 5 \cdot 10^1 + 4 \cdot 10^0$$

(see Figure 5.2.1). In general, the symbol in position $n$ (with the rightmost symbol being in position 0) represents the number of $10^n$'s. Since $10^0 = 1$, the symbol in position 0 represents the number of $10^0$'s or 1's; since $10^1 = 10$, the symbol in position 1 represents the number of $10^1$'s or 10's; since $10^2 = 100$, the symbol in position 2 represents the number of $10^2$'s or 100's; and so on. We call the value on which the system is based (10 in the case of the decimal system) the **base** of the number system.

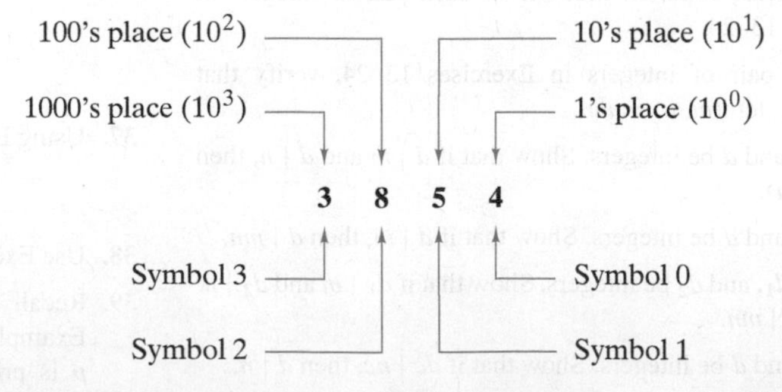


**Figure 5.2.1** The decimal number system.

In the binary (base 2) number system, to represent integers we need only two symbols, 0 and 1. In representing an integer, reading from the right, the first symbol represents the number of 1's, the next symbol the number of 2's, the next symbol the number of 4's, the next symbol the number of 8's, and so on. For example, in base 2,[†]

$$101101_2 = 1 \cdot 2^5 + 0 \cdot 2^4 + 1 \cdot 2^3 + 1 \cdot 2^2 + 0 \cdot 2^1 + 1 \cdot 2^0$$

(see Figure 5.2.2). In general, the symbol in position $n$ (with the rightmost symbol being in position 0) represents the number of $2^n$'s. Since $2^0 = 1$, the symbol in position 0 represents the number of $2^0$'s, or 1's; since $2^1 = 2$, the symbol in position 1 represents the number of $2^1$'s or 2's; since $2^2 = 4$, the symbol in position 2 represents the number of $2^2$'s or 4's; and so on.

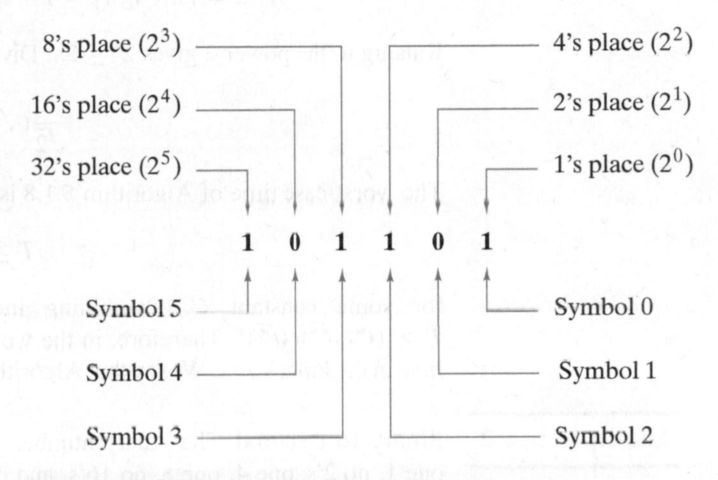


**Figure 5.2.2** The binary number system.

**Example 5.2.1** **Computer Representation of Integers** Computer systems represent integers in binary. Compute the number of bits necessary to represent a positive integer $n$. Deduce that Algorithm 5.1.8, which determines whether an integer is prime, is not a polynomial-time algorithm.

**SOLUTION** Suppose that the binary representation of the positive $k$-bit integer $n$ is

$$n = 1 \cdot 2^{k-1} + b_{k-2}2^{k-2} + \cdots + b_0 2^0.$$

Now

$$2^{k-1} \le n$$

and

$$\begin{aligned} n &= 1 \cdot 2^{k-1} + b_{k-2}2^{k-2} + \cdots + b_0 2^0 \\ &\le 1 \cdot 2^{k-1} + 1 \cdot 2^{k-2} + \cdots + 1 \cdot 2^0 = 2^k - 1 < 2^k. \end{aligned}$$

(The last equality follows from the formula for the geometric sum; see Example 2.4.4.) Therefore

$$2^{k-1} \le n < 2^k.$$

[†]Without knowing which number system is being used, a representation is ambiguous; for example, 101101 represents one number in decimal and quite a different number in binary. Often the context will make clear which number system is in effect; but when we want to be absolutely clear, we subscript the number to specify the base—the subscript 10 denotes the decimal system and the subscript 2 denotes the binary system.

Taking logs, we obtain

$$k - 1 \leq \lg n < k.$$

Adding one gives

$$k \leq 1 + \lg n < k + 1.$$

Therefore, $k = \lfloor 1 + \lg n \rfloor$. Since $k$ is the number of bits required to represent $n$, we have proved that the number of bits necessary to represent $n$ is $\lfloor 1 + \lg n \rfloor$.

The size $s$ of an integer $n$ input to Algorithm 5.1.8 is the number of bits necessary to represent $n$. Thus $s$ satisfies

$$s = \lfloor 1 + \lg n \rfloor \leq 1 + \lg n = \lg 2 + \lg n = \lg(2n).$$

Raising to the power 2 gives $2^s \leq 2n$. Dividing by 2 and taking square roots yields

$$\frac{1}{\sqrt{2}}(\sqrt{2})^s \leq \sqrt{n}. \tag{5.2.1}$$

The worst-case time of Algorithm 5.1.8 is $\Theta(\sqrt{n})$. Thus its worst-case time $T$ satisfies

$$T \geq C\sqrt{n} \tag{5.2.2}$$

for some constant $C$. Combining inequalities (5.2.1) and (5.2.2), we obtain $T \geq (C/\sqrt{2})(\sqrt{2})^s$. Therefore, in the worst case, Algorithm 5.1.8 runs in *exponential time* in the input size $s$. We say that Algorithm 5.1.8 is *not* a polynomial-time algorithm. ◀

**Example 5.2.2** **Binary to Decimal** The binary number $101101_2$ represents the number consisting of one 1, no 2's, one 4, one 8, no 16's, and one 32 (see Figure 5.2.2). This representation may be expressed

$$101101_2 = 1 \cdot 2^5 + 0 \cdot 2^4 + 1 \cdot 2^3 + 1 \cdot 2^2 + 0 \cdot 2^1 + 1 \cdot 2^0.$$

Computing the right-hand side in decimal, we find that

$$\begin{aligned} 101101_2 &= 1 \cdot 32 + 0 \cdot 16 + 1 \cdot 8 + 1 \cdot 4 + 0 \cdot 2 + 1 \cdot 1 \\ &= 32 + 8 + 4 + 1 = 45_{10}. \end{aligned}$$

◀

We turn the method of Example 5.2.2 into an algorithm. We generalize by allowing an arbitrary base $b$.

**Algorithm 5.2.3** **Converting an Integer from Base $b$ to Decimal**

This algorithm returns the decimal value of the base $b$ integer $c_n c_{n-1} \cdots c_1 c_0$.

Input: $c, n, b$

Output: *dec_val*

```
base_b_to_dec (c, n, b) {
  dec_val = 0
  b_to_the_i = 1
  for i = 0 to n {
    dec_val = dec_val + c_i * b_to_the_i
    b_to_the_i = b_to_the_i * b
  }
  return dec_val
}
```

Algorithm 5.2.3 runs in time $\Theta(n)$.

**Example 5.2.4** Show how Algorithm 5.2.3 converts the binary number 1101 to decimal.

SOLUTION Here $n = 3$, $b = 2$, and

$$c_3 = 1, \qquad c_2 = 1, \qquad c_1 = 0, \qquad c_0 = 1.$$

First, *dec_val* is set to 0, and *b_to_the_i* is set to 1. We then enter the for loop. Since $i = 0$ and $b\_to\_the\_i = 1$,

$$c_i * b\_to\_the\_i = 1 * 1 = 1.$$

Thus *dec_val* becomes 1. Executing

$$b\_to\_the\_i = b\_to\_the\_i * b$$

sets *b_to_the_i* to 2. We return to the top of the for loop. Since $i = 1$ and $b\_to\_the\_i = 2$,

$$c_i * b\_to\_the\_i = 0 * 2 = 0.$$

Thus *dec_val* remains 1. Executing

$$b\_to\_the\_i = b\_to\_the\_i * b$$

sets *b_to_the_i* to 4. We return to the top of the for loop. Since $i = 2$ and $b\_to\_the\_i = 4$,

$$c_i * b\_to\_the\_i = 1 * 4 = 4.$$

Thus *dec_val* becomes 5. Executing

$$b\_to\_the\_i = b\_to\_the\_i * b$$

sets *b_to_the_i* to 8. We return to the top of the for loop. Since $i = 3$ and $b\_to\_the\_i = 8$,

$$c_i * b\_to\_the\_i = 1 * 8 = 8.$$

Thus *dec_val* becomes 13. Executing

$$b\_to\_the\_i = b\_to\_the\_i * b$$

sets *b_to_the_i* to 16. The for loop terminates and the algorithm returns 13, the decimal value of the binary number 1101. ◀

Other important bases for number systems in computer science are base 8 or **octal** and base 16 or **hexadecimal** (sometimes shortened to **hex**). We will discuss the hexadecimal system and leave the octal system to the exercises (see Exercises 45–50).

In the hexadecimal number system, to represent integers we use the symbols 0, 1, 2, 3, 4, 5, 6, 7, 8, 9, A, B, C, D, E, and F. The symbols A–F are interpreted as decimal 10–15. (In general, in the base $N$ number system, $N$ distinct symbols, representing $0, 1, 2, \ldots, N - 1$ are required.) In representing an integer, reading from the right, the first symbol represents the number of 1's, the next symbol the number of 16's, the next symbol the number of $16^2$'s, and so on. For example, in base 16,

$$\text{B4F} = 11 \cdot 16^2 + 4 \cdot 16^1 + 15 \cdot 16^0$$

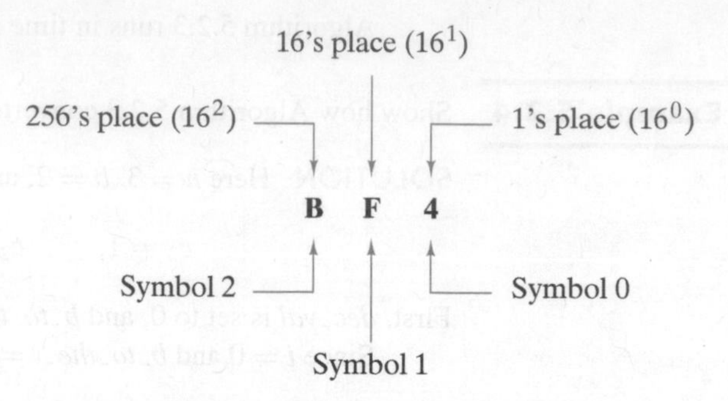


**Figure 5.2.3** The hexadecimal number system.

(see Figure 5.2.3). In general, the symbol in position $n$ (with the rightmost symbol being in position 0) represents the number of $16^n$'s.

**Example 5.2.5** **Hexadecimal to Decimal** Convert the hexadecimal number B4F to decimal.

**SOLUTION** We obtain

$$\begin{aligned} \text{B4F}_{16} &= 11 \cdot 16^2 + 4 \cdot 16^1 + 15 \cdot 16^0 \\ &= 11 \cdot 256 + 4 \cdot 16 + 15 = 2816 + 64 + 15 = 2895_{10}. \end{aligned}$$ ◀

Algorithm 5.2.3 shows how to convert an integer in base $b$ to decimal. Consider the reverse problem—converting a decimal number to base $b$. Suppose, for example, that we want to convert the decimal number 91 to binary. If we divide 91 by 2, we obtain

$$\begin{array}{r} 45 \\ 2\overline{)91} \\ \underline{8} \\ 11 \\ \underline{10} \\ 1 \end{array}$$

This computation shows that

$$91 = 2 \cdot 45 + 1. \tag{5.2.3}$$

We are beginning to express 91 in powers of 2. If we next divide 45 by 2, we find

$$45 = 2 \cdot 22 + 1. \tag{5.2.4}$$

Substituting this expression for 45 into (5.2.3), we obtain

$$\begin{aligned} 91 &= 2 \cdot 45 + 1 \\ &= 2 \cdot (2 \cdot 22 + 1) + 1 \\ &= 2^2 \cdot 22 + 2 + 1. \end{aligned} \tag{5.2.5}$$

If we next divide 22 by 2, we find

$$22 = 2 \cdot 11.$$

Substituting this expression for 22 into (5.2.5), we obtain

$$\begin{aligned} 91 &= 2^2 \cdot 22 + 2 + 1 \\ &= 2^2 \cdot (2 \cdot 11) + 2 + 1 \\ &= 2^3 \cdot 11 + 2 + 1. \end{aligned} \tag{5.2.6}$$

If we next divide 11 by 2, we find

$$11 = 2 \cdot 5 + 1.$$

Substituting this expression for 11 into (5.2.6), we obtain

$$91 = 2^4 \cdot 5 + 2^3 + 2 + 1. \tag{5.2.7}$$

If we next divide 5 by 2, we find

$$5 = 2 \cdot 2 + 1.$$

Substituting this expression for 5 into (5.2.7), we obtain

$$\begin{aligned} 91 &= 2^5 \cdot 2 + 2^4 + 2^3 + 2 + 1 \\ &= 2^6 + 2^4 + 2^3 + 2 + 1 \\ &= 1011011_2. \end{aligned}$$

The preceding computation shows that the *remainders,* as $N$ is successively divided by 2, give the bits in the binary representation of $N$. The first division by 2 in (5.2.3) gives the 1's bit; the second division by 2 in (5.2.4) gives the 2's bit; and so on. We illustrate with another example.

**Example 5.2.6** **Decimal to Binary** Write the decimal number 130 in binary.

**SOLUTION** The computation shows the successive divisions by 2 with the remainders recorded at the right.

| | | |
|---|---|---|
| 2)130 | remainder = 0 | 1's bit |
| 2)65 | remainder = 1 | 2's bit |
| 2)32 | remainder = 0 | 4's bit |
| 2)16 | remainder = 0 | 8's bit |
| 2)8 | remainder = 0 | 16's bit |
| 2)4 | remainder = 0 | 32's bit |
| 2)2 | remainder = 0 | 64's bit |
| 2)1 | remainder = 1 | 128's bit |
| 0 | | |

We may stop when the quotient is 0. Remembering that the first remainder gives the number of 1's, the second remainder gives the number of 2's, and so on, we obtain

$$130_{10} = 10000010_2.$$

We turn the method of Example 5.2.6 into an algorithm. We generalize by allowing an arbitrary base $b$.

**Algorithm 5.2.7** **Converting a Decimal Integer into Base $b$**

This algorithm converts the positive integer $m$ into the base $b$ integer $c_n c_{n-1} \cdots c_1 c_0$. The variable $n$ is used as an index in the sequence $c$. The value of $m \bmod b$ is the remainder when $m$ is divided by $b$. The value of $\lfloor m/b \rfloor$ is the quotient when $m$ is divided by $b$.

```
Input: m, b
Output: c, n

dec_to_base_b(m, b, c, n) {
  n = −1
  while (m > 0) {
    n = n + 1
    c_n = m mod b
    m = ⌊m/b⌋
  }
}
```

Just as a binary integer $m$ has $\lfloor 1 + \lg m \rfloor$ bits, a base $b$ integer $m$ has $\lfloor 1 + \log_b m \rfloor$ digits (see Exercise 55). Thus Algorithm 5.2.7 runs in time $\Theta(\log_b m)$.

**Example 5.2.8** Show how Algorithm 5.2.7 converts the decimal number $m = 11$ to binary.

**SOLUTION** The algorithm first sets $n$ to $-1$. The first time we arrive at the while loop, $m = 11$ and the condition $m > 0$ is true; so we execute the body of the while loop. The variable $n$ is incremented and becomes 0. Since $m \bmod b = 11 \bmod 2 = 1$, $c_0$ is set to 1. Since $\lfloor m/b \rfloor = \lfloor 11/2 \rfloor = 5$, $m$ is set to 5. We return to the top of the while loop.

Since $m = 5$, the condition $m > 0$ is true; so we execute the body of the while loop. The variable $n$ is incremented and becomes 1. Since $m \bmod b = 5 \bmod 2 = 1$, $c_1$ is set to 1. Since $\lfloor m/b \rfloor = \lfloor 5/2 \rfloor = 2$, $m$ is set to 2. We return to the top of the while loop.

Since $m = 2$, the condition $m > 0$ is true; so we execute the body of the while loop. The variable $n$ is incremented and becomes 2. Since $m \bmod b = 2 \bmod 2 = 0$, $c_2$ is set to 0. Since $\lfloor m/b \rfloor = \lfloor 2/2 \rfloor = 1$, $m$ is set to 1. We return to the top of the while loop.

Since $m = 1$, the condition $m > 0$ is true; so we execute the body of the while loop. The variable $n$ is incremented and becomes 3. Since $m \bmod b = 1 \bmod 2 = 1$, $c_3$ is set to 1. Since $\lfloor m/b \rfloor = \lfloor 1/2 \rfloor = 0$, $m$ is set to 0. We return to the top of the while loop.

Since $m = 0$, the algorithm terminates. The value 11 has been converted to the binary number $c_3c_2c_1c_0 = 1011$. ◀

**Example 5.2.9** **Decimal to Hexadecimal** Convert the decimal number 20385 to hexadecimal.

**SOLUTION** The computation shows the successive divisions by 16 with the remainders recorded at the right.

| | | |
|---|---|---|
| 16)20385 | remainder = 1 | 1's place |
| 16)1274 | remainder = 10 | 16's place |
| 16)79 | remainder = 15 | $16^2$'s place |
| 16)4 | remainder = 4 | $16^3$'s place |
| 0 | | |

We stop when the quotient is 0. The first remainder gives the number of 1's, the second remainder gives the number of 16's, and so on; thus we obtain $20385_{10} = 4FA1_{16}$. ◀

Next we turn our attention to addition of numbers in arbitrary bases. The same method that we use to add decimal numbers can be used to add binary numbers; however, we must replace the decimal addition table with the binary addition table

| + | 0 | 1 |
|---|---|---|
| 0 | 0 | 1 |
| 1 | 1 | 10 |

(In decimal, $1 + 1 = 2$, and $2_{10} = 10_2$; thus, in binary, $1 + 1 = 10$.)

**Example 5.2.10** **Binary Addition** Add the binary numbers 10011011 and 1011011.

**SOLUTION** We write the problem as

```
   10011011
 +  1011011
```

As in decimal addition, we begin from the right, adding 1 and 1. This sum is $10_2$; thus we write 0 and carry 1. At this point the computation is

```
          1
   10011011
 +  1011011
          0
```

Next, we add 1 and 1 and 1, which is $11_2$. We write 1 and carry 1. At this point, the computation is

```
         1
   10011011
 +  1011011
         10
```

Continuing in this way, we obtain

```
   10011011
 +  1011011
   11110110
```

**Example 5.2.11** The addition problem of Example 5.2.10, in decimal, is

```
   155
 +  91
   246
```

We turn the method of Example 5.2.10 into an algorithm. If the numbers to add are $b_n b_{n-1} \cdots b_1 b_0$ and $b'_n b'_{n-1} \cdots b'_1 b'_0$, at the iteration $i > 0$ the algorithm adds $b_i$, $b'_i$, and the carry bit from the previous iteration. When adding three bits, say $B_1$, $B_2$, and $B_3$, we obtain a two-bit binary number, say $cb$. For example, if we compute $1 + 0 + 1$, the result is $10_2$; in our notation, $c = 1$ and $b = 0$. By checking the various cases, we can verify that we can compute the binary sum $B_1 + B_2 + B_3$ by first computing the sum in decimal and then recovering $c$ and $b$ from the formulas

$$b = (B_1 + B_2 + B_3) \bmod 2, \qquad c = \lfloor (B_1 + B_2 + B_3)/2 \rfloor.$$

**Algorithm 5.2.12** **Adding Binary Numbers**

This algorithm adds the binary numbers $b_n b_{n-1} \cdots b_1 b_0$ and $b'_n b'_{n-1} \cdots b'_1 b'_0$ and stores the sum in $s_{n+1} s_n s_{n-1} \cdots s_1 s_0$. (Leading zeros can be appended to $b$ or $b'$ so that this algorithm can be used to add integers with different numbers of bits.)

Input: $b, b', n$

Output: $s$

```
binary_addition(b, b', n, s) {
  carry = 0
  for i = 0 to n {
    s_i = (b_i + b'_i + carry) mod 2
    carry = ⌊(b_i + b'_i + carry)/2⌋
  }
  s_{n+1} = carry
}
```

Algorithm 5.2.12 runs in time $\Theta(n)$.

Our next example shows that we can add hexadecimal numbers in the same way that we add decimal or binary numbers.

**Example 5.2.13** **Hexadecimal Addition** Add the hexadecimal numbers 84F and 42EA.

**SOLUTION** The problem may be written

```
   84F
+ 42EA
```

We begin in the rightmost column by adding F and A. Since F is $15_{10}$ and A is $10_{10}$, $\mathrm{F} + \mathrm{A} = 15_{10} + 10_{10} = 25_{10} = 19_{16}$. We write 9 and carry 1:

```
     1
   84F
+ 42EA
     9
```

Next, we add 1, 4, and E, obtaining $13_{16}$. We write 3 and carry 1:

```
    1
   84F
+ 42EA
    39
```

Continuing in this way, we obtain

```
   84F
+ 42EA
  4B39
```

◀

**Example 5.2.14** The addition problem of Example 5.2.13, in decimal, is

```
   2127
+ 17130
  19257
```

◀

We can multiply binary numbers by modifying the standard algorithm for multiplying decimal numbers (see Exercise 67).

We conclude by discussing a special algorithm, which we will need in Section 5.4, to compute powers mod $z$. We first discuss an algorithm to compute a power $a^n$ (without dealing with mod $z$). The straightforward way to compute this power is to repeatedly multiply by $a$

$$\underbrace{a \cdot a \cdots a}_{n\ a\text{'s}},$$

which uses $n - 1$ multiplications. We can do better using **repeated squaring.**

As a concrete example, consider computing $a^{29}$. We first compute $a^2 = a \cdot a$, which uses 1 multiplication. We next compute $a^4 = a^2 \cdot a^2$, which uses 1 additional multiplication. We next compute $a^8 = a^4 \cdot a^4$, which uses 1 additional multiplication. We next compute $a^{16} = a^8 \cdot a^8$, which uses 1 additional multiplication. So far, we have used only 4 multiplications. Noting that the expansion of 29 in powers of 2, that is the binary expansion, is

$$29 = 1 + 4 + 8 + 16,$$

we see that we can compute $a^{29}$ as

$$a^{29} = a^1 \cdot a^4 \cdot a^8 \cdot a^{16},$$

which uses 3 additional multiplications for a total of 7 multiplications. The straightforward technique uses 28 multiplications.

In Example 5.2.6, we saw that the remainders when $n$ is successively divided by 2 give the binary expansion of $n$. If the remainder is 1, the corresponding power of 2 is included; otherwise, it is not included. We can formalize the repeated squaring technique if, in addition to repeated squaring, we simultaneously determine the binary expansion of the exponent.

**Example 5.2.15** Figure 5.2.4 shows how $a^{29}$ is calculated using repeated squaring. Initially $x$ is set to $a$, and $n$ is set to the value of the exponent, 29 in this case. We then compute $n$ mod 2. Since this value is 1, we know that $1 = 2^0$ is included in the binary expansion of 29. Therefore $a^1$ is included in the product. We track the partial product in *Result;* so *Result* is set to $a$. We then compute the quotient when 29 is divided by 2. The quotient 14 becomes the new value of $n$. We then repeat this process.

| $x$ | *Current Value of n* | $n$ mod 2 | *Result* | *Quotient When n Divided by 2* |
|---|---|---|---|---|
| $a$ | 29 | 1 | $a$ | 14 |
| $a^2$ | 14 | 0 | Unchanged | 7 |
| $a^4$ | 7 | 1 | $a \cdot a^4 = a^5$ | 3 |
| $a^8$ | 3 | 1 | $a^5 \cdot a^8 = a^{13}$ | 1 |
| $a^{16}$ | 1 | 1 | $a^{13} \cdot a^{16} = a^{29}$ | 0 |

**Figure 5.2.4** Computing $a^{29}$ using repeated squaring.

We square $x$ to obtain $a^2$. We then compute $n$ mod 2. Since this value is 0, we know that $2 = 2^1$ is not included in the binary expansion of 29. Therefore $a^2$ is not included in the product, and *Result* is unchanged. We then compute the quotient when 14 is divided by 2. The quotient 7 becomes the new value of $n$. We then repeat this process.

We square $x$ to obtain $a^4$. We then compute $n \bmod 2$. Since this value is 1, we know that $4 = 2^2$ is included in the binary expansion of 29. Therefore $a^4$ is included in the product. *Result* becomes $a^5$. We then compute the quotient when 7 is divided by 2. The quotient 3 becomes the new value of $n$. The process continues until $n$ becomes 0. ◀

We state the method of repeated squaring as Algorithm 5.2.16.

**Algorithm 5.2.16** **Exponentiation by Repeated Squaring**

This algorithm computes $a^n$ using repeated squaring. The algorithm is explained in Example 5.2.15.

Input: $a, n$
Output: $a^n$

```
exp_via_repeated_squaring(a, n) {
  result = 1
  x = a
  while (n > 0) {
    if (n mod 2 == 1)
      result = result * x
    x = x * x
    n = ⌊n/2⌋
  }
  return result
}
```

The number of times that the while loop executes is determined by $n$. The variable $n$ is repeatedly halved

$$n = \lfloor n/2 \rfloor$$

and when $n$ becomes 0, the loop terminates. Example 4.3.14 shows that it takes time $\Theta(\lg n)$ to reduce $n$ to 0 by repeated halving. In the body of the while loop, at most two multiplications are performed. Thus, the number of multiplications is at most $\Theta(\lg n)$, an improvement over the straightforward algorithm that uses $\Theta(n)$ multiplications. The bottleneck in Algorithm 5.2.16 is the size of the numbers involved. The value returned $a^n$ requires $\lg a^n = n \lg a$ bits in its representation. Thus, simply to copy the final value into *Result* takes time, at least $\Omega(n \lg a)$, which is exponential in the size of $n$ (see Example 5.2.1).

In Section 5.4, we will need to compute $a^n \bmod z$ for large values of $a$ and $n$. In this case, $a^n$ will be huge; so it is impractical to compute $a^n$ and then compute the remainder when $a^n$ is divided by $z$. We can do much better. The key idea is to compute the remainder after each multiplication thereby keeping the numbers relatively small. The justification for this technique is given in our next theorem.

**Theorem 5.2.17** If $a$, $b$, and $z$ are positive integers,

$$ab \bmod z = [(a \bmod z)(b \bmod z)] \bmod z.$$

**Proof** Let $w = ab \bmod z$, $x = a \bmod z$, and $y = b \bmod z$. Since $w$ is the remainder when $ab$ is divided by $z$, by the quotient-remainder theorem, there exists $q_1$ such that

$$ab = q_1 z + w.$$

Thus

$$w = ab - q_1 z.$$

Similarly, there exists $q_2$ and $q_3$ such that

$$a = q_2 z + x, \qquad b = q_3 z + y.$$

Now

$$\begin{aligned} w &= ab - q_1 z \\ &= (q_2 z + x)(q_3 z + y) - q_1 z \\ &= (q_2 q_3 z + q_2 y + q_3 x - q_1) z + xy \\ &= qz + xy, \end{aligned}$$

where $q = q_2 q_3 z + q_2 y + q_3 x - q_1$. Therefore,

$$xy = -qz + w;$$

that is, $w$ is the remainder when $xy$ is divided by $z$. Thus, $w = xy \bmod z$, which translates to

$$ab \bmod z = [(a \bmod z)(b \bmod z)] \bmod z.$$ ◀

**Example 5.2.18** Show how to compute $572^{29} \bmod 713$ using Algorithm 5.2.16 and Theorem 5.2.17.

**SOLUTION** To compute $a^{29}$, we successively computed

$$a, \qquad a^5 = a \cdot a^4, \qquad a^{13} = a^5 \cdot a^8, \qquad a^{29} = a^{13} \cdot a^{16}$$

(see Example 5.2.15). To compute $a^{29} \bmod z$, we successively compute

$$a \bmod z, \qquad a^5 \bmod z, \qquad a^{13} \bmod z, \qquad a^{29} \bmod z.$$

Each multiplication is performed using Theorem 5.2.17. We compute $a^2$ using the formula

$$a^2 \bmod z = [(a \bmod z)(a \bmod z)] \bmod z.$$

We compute $a^4$ using the formula

$$a^4 \bmod z = a^2 a^2 \bmod z = [(a^2 \bmod z)(a^2 \bmod z)] \bmod z,$$

and so on.

We compute $a^5$ using the formula

$$a^5 \bmod z = aa^4 \bmod z = [(a \bmod z)(a^4 \bmod z)] \bmod z.$$

We compute $a^{13}$ using the formula

$$a^{13} \bmod z = a^5 a^8 \bmod z = [(a^5 \bmod z)(a^8 \bmod z)] \bmod z,$$

and so on.

The following shows the computation of $572^{29} \bmod 713$:

$$572^2 \bmod 713 = (572 \bmod 713)(572 \bmod 713) \bmod 713 = 572^2 \bmod 713 = 630$$

$$572^4 \bmod 713 = (572^2 \bmod 713)(572^2 \bmod 713) \bmod 713 = 630^2 \bmod 713 = 472$$

$$572^8 \bmod 713 = (572^4 \bmod 713)(572^4 \bmod 713) \bmod 713 = 472^2 \bmod 713 = 328$$

$$572^{16} \bmod 713 = (572^8 \bmod 713)(572^8 \bmod 713) \bmod 713 = 328^2 \bmod 713 = 634$$

$$572^5 \bmod 713 = (572 \bmod 713)(572^4 \bmod 713) \bmod 713 = 572 \cdot 472 \bmod 713 = 470$$

$$572^{13} \bmod 713 = (572^5 \bmod 713)(572^8 \bmod 713) \bmod 713 = 470 \cdot 328 \bmod 713 = 152$$

$$572^{29} \bmod 713 = (572^{13} \bmod 713)(572^{16} \bmod 713) \bmod 713 = 152 \cdot 634 \bmod 713 = 113.$$

The number $572^{29}$ has 80 digits, so Theorem 5.2.17 indeed simplifies the computation. ◀

The technique demonstrated in Example 5.2.18 is formalized as Algorithm 5.2.19.

**Algorithm 5.2.19** **Exponentiation Mod $z$ by Repeated Squaring**

This algorithm computes $a^n \bmod z$ using repeated squaring. The algorithm is explained in Example 5.2.18.

Input: $a, n, z$

Output: $a^n \bmod n$

```
exp_mod_z_via_repeated_squaring(a, n, z) {
  result = 1
  x = a mod z
  while (n > 0) {
    if (n mod 2 == 1)
      result = (result * x) mod z
    x = (x * x) mod z
    n = ⌊n/2⌋
  }
  return result
}
```

**Go Online**
For a C++ program implementing this algorithm, see bit.ly/2R4lVVJ

The key difference between Algorithms 5.2.16 and 5.2.19 is the size of the numbers that are multiplied. In Algorithm 5.2.19, the numbers multiplied are remainders after division by $z$ and so have magnitude less than $z$. If we modify the usual method of multiplying base 10 integers for base 2, it can be shown (see Exercise 68) that the time required to multiply $a$ and $b$ is $O(\lg a \lg b)$. Since the while loop in Algorithm 5.2.19 executes $\Theta(\lg n)$ times, the total time for Algorithm 5.2.19 is $O(\lg n \lg^2 z)$.

## 5.2 Problem-Solving Tips

- To convert the base $b$ number $c_n b^n + c_{n-1} b^{n-1} + \cdots + c_1 b^1 + c_0 b^0$ to decimal, carry out the indicated multiplications and additions in decimal.
- To convert the decimal number $n$ to base $b$, divide by $b$, divide the resulting quotient by $b$, divide the resulting quotient by $b$, and so on, until obtaining a zero quotient. The remainders give the base $b$ representation of $n$. The first remainder gives the 1's coefficient, the next remainder gives the $b$'s coefficient, and so on.
- When multiplying modulo $z$, compute the remainders as soon as possible to minimize the sizes of the numbers involved.

## 5.2 Review Exercises

1. What is the value of the decimal number $d_n d_{n-1} \ldots d_1 d_0$? (Each $d_i$ is one of 0–9.)
2. What is the value of the binary number $b_n b_{n-1} \ldots b_1 b_0$? (Each $b_i$ is 0 or 1.)
3. What is the value of the hexadecimal number $h_n h_{n-1} \ldots h_1 h_0$? (Each $h_i$ is one of 0–9 or A–F.)
4. How many bits are required to represent the positive integer $n$?
5. Explain how to convert from binary to decimal.
6. Explain how to convert from decimal to binary.
7. Explain how to convert from hexadecimal to decimal.
8. Explain how to convert from decimal to hexadecimal.
9. Explain how to add binary numbers.
10. Explain how to add hexadecimal numbers.
11. Explain how to compute $a^n$ using repeated squaring.
12. Explain how to compute $a^n \bmod z$ using repeated squaring.

## 5.2 Exercises

*How many bits are needed to represent each integer in Exercises 1–10?*

1. 60
2. 63
3. 64
4. 127
5. 128
6. $2^{1000}$
7. $3^{1000}$
8. $8^{1000}$
9. $3^{481}$
10. $6.87 \times 10^{48}$

*In Exercises 11–16, express each binary number in decimal.*

11. 1001
12. 11011
13. 11011011
14. 100000
15. 1111111
16. 110111011011

*In Exercises 17–22, express each decimal number in binary.*

17. 34
18. 61
19. 223
20. 400
21. 1024
22. 12,340

*In Exercises 23–28, add the binary numbers.*

23. 1001 + 1111
24. 11011 + 1101
25. 110110 + 101101
26. 101101 + 11011
27. 110110101 + 1101101
28. 1101 + 101100 + 11011011

*In Exercises 29–34, express each hexadecimal number in decimal.*

29. 3A
30. 1E9
31. 3E7C
32. A03
33. 209D
34. 4B07A
35. Express each binary number in Exercises 11–16 in hexadecimal.
36. Express each decimal number in Exercises 17–22 in hexadecimal.
37. Express each hexadecimal number in Exercises 29, 30, and 32 in binary.

*In Exercises 38–42, add the hexadecimal numbers.*

38. 4A + B4
39. 195 + 76E
40. 49F7 + C66
41. 349CC + 922D
42. 82054 + AEFA3
43. Does 2010 represent a number in binary? in decimal? in hexadecimal?
44. Does 1101010 represent a number in binary? in decimal? in hexadecimal?

*In the octal (base 8) number system, to represent integers we use the symbols 0, 1, 2, 3, 4, 5, 6, and 7. In representing an integer, reading from the right, the first symbol represents the number of 1's, the next symbol the number of 8's, the next symbol the number of $8^2$'s, and so on. In general, the symbol in position n (with the rightmost symbol being in position 0) represents the number of $8^n$'s. In Exercises 45–50, express each octal number in decimal.*

**45.** 63 **46.** 7643 **47.** 7711

**48.** 10732 **49.** 1007 **50.** 537261

**51.** Express each binary number in Exercises 11–16 in octal.

**52.** Express each decimal number in Exercises 17–22 in octal.

**53.** Express each hexadecimal number in Exercises 29–34 in octal.

**54.** Express each octal number in Exercises 45–50 in hexadecimal.

**55.** Does 1101010 represent a number in octal?

**56.** Does 30470 represent a number in binary? in octal? in decimal? in hexadecimal?

**57.** Does 9450 represent a number in binary? in octal? in decimal? in hexadecimal?

**58.** Prove that a base $b$ integer $m$ has $\lfloor 1 + \log_b m \rfloor$ digits.

*In Exercises 59–61, trace Algorithm 5.2.16 for the given value of n.*

**59.** $n = 16$ **60.** $n = 15$ **61.** $n = 80$

*In Exercises 62–64, trace Algorithm 5.2.19 for the given values of a, n, and z.*

**62.** $a = 5, n = 10, z = 21$

**63.** $a = 143, n = 10, z = 230$

**64.** $a = 143, n = 100, z = 230$

**65.** Let $T_n$ denote the highest power of 2 that divides $n$. Show that $T_{mn} = T_m + T_n$ for all $m, n \geq 1$.

**66.** Let $S_n$ denote the number of 1's in the binary representation of $n$. Use induction to prove that $T_{n!} = n - S_n$ for all $n \geq 1$. ($T_n$ is defined in the previous exercise.)

**67.** Modify the usual method of multiplying base 10 integers for base 2 to produce an algorithm to multiply binary numbers $b_m b_{m-1} \cdots b_1 b_0$ and $b'_n b'_{n-1} \cdots b'_1 b'_0$.

**68.** Show that the time required by the algorithm of Exercise 67 to multiply $a$ and $b$ is $O(\lg a \lg b)$.

## 5.3 The Euclidean Algorithm

**Go Online**

For more on the Euclidean algorithm, see bit.ly/2NYUMkW

In Section 5.1, we discussed some methods of computing the greatest common divisor of two integers that turned out to be inefficient. The **Euclidean algorithm** is an old, famous, and *efficient* algorithm for finding the greatest common divisor of two integers.

The Euclidean algorithm is based on the fact that if $r = a \bmod b$, then

$$\gcd(a, b) = \gcd(b, r). \tag{5.3.1}$$

Before proving (5.3.1), we illustrate how the Euclidean algorithm uses it to find the greatest common divisor.

**Example 5.3.1** Since 105 mod 30 = 15, by (5.3.1)

$$\gcd(105, 30) = \gcd(30, 15).$$

Since 30 mod 15 = 0, by (5.3.1)

$$\gcd(30, 15) = \gcd(15, 0).$$

By inspection, $\gcd(15, 0) = 15$. Therefore,

$$\gcd(105, 30) = \gcd(30, 15) = \gcd(15, 0) = 15.$$ ◀

We next prove equation (5.3.1).

**Theorem 5.3.2** If $a$ is a nonnegative integer, $b$ is a positive integer, and $r = a \bmod b$, then

$$\gcd(a, b) = \gcd(b, r).$$

**Proof** By the quotient-remainder theorem, there exist $q$ and $r$ satisfying

$$a = bq + r \qquad 0 \le r < b.$$

We show that the set of common divisors of $a$ and $b$ is equal to the set of common divisors of $b$ and $r$, thus proving the theorem.

Let $c$ be a common divisor of $a$ and $b$. By Theorem 5.1.3(c), $c \mid bq$. Since $c \mid a$ and $c \mid bq$, by Theorem 5.1.3(b), $c \mid a - bq\ (= r)$. Thus $c$ is a common divisor of $b$ and $r$. Conversely, if $c$ is a common divisor of $b$ and $r$, then $c \mid bq$ and $c \mid bq + r\ (= a)$ and $c$ is a common divisor of $a$ and $b$. Thus the set of common divisors of $a$ and $b$ is equal to the set of common divisors of $b$ and $r$. Therefore, $\gcd(a, b) = \gcd(b, r)$. ◀

We next formally state the Euclidean algorithm as Algorithm 5.3.3.

**Algorithm 5.3.3** **Euclidean Algorithm**

This algorithm finds the greatest common divisor of the nonnegative integers $a$ and $b$, where not both $a$ and $b$ are zero.

Input: $a$ and $b$ (nonnegative integers, not both zero)

Output: Greatest common divisor of $a$ and $b$

```
1.  gcd(a, b) {
2.     // make a largest
3.     if (a < b)
4.        swap(a, b)
5.     while (b ≠ 0) {
6.        r = a mod b
7.        a = b
8.        b = r
9.     }
10.    return a
11. }
```

We note that the while loop in the Euclidean algorithm (lines 5–9) always terminates since at the bottom of the loop (lines 7 and 8), the values of $a$ and $b$ are updated to *smaller* values. Since nonnegative integers cannot decrease indefinitely, eventually $b$ becomes zero and the loop terminates.

Let $G = \gcd(a, b)$, where $a$ and $b$ are the values input to Algorithm 5.3.3. We can prove that Algorithm 5.3.3 is correct by verifying that $G = \gcd(a, b)$ is a loop invariant, where now $a$ and $b$ denote the variables in the pseudocode.

By definition, the loop invariant is true the first time we arrive at line 5. Suppose that $G = \gcd(a, b)$ is true prior to another iteration of the loop and that $b \neq 0$. Theorem 5.3.2 then tells us that after line 6 executes, $\gcd(a, b) = \gcd(b, r)$. At lines 7 and 8, $a$ becomes $b$ and $b$ becomes $r$. Therefore, $G = \gcd(a, b)$ is true for the new values of $a$ and $b$. It follows that $G = \gcd(a, b)$ is a loop invariant. The while loop terminates when $b$ becomes 0. At this point, the loop invariant becomes $G = \gcd(a, 0)$. The algorithm then returns $a$ [$= \gcd(a, 0)$]. Thus the value that the algorithm returns is $G$, which by definition is the greatest common divisor of the input values. Therefore, Algorithm 5.3.3 is correct.

Algorithm 5.3.3 correctly finds the greatest common divisor if lines 3 and 4 are omitted (see Exercise 15). We include these lines because it simplifies the analysis of Algorithm 5.3.3 in the next subsection.

**Example 5.3.4** Show how Algorithm 5.3.3 finds gcd(504, 396).

**SOLUTION** Let $a = 504$ and $b = 396$. Since $a > b$, we move to line 5. Since $b \neq 0$, we proceed to line 6, where we set $r$ to

$$a \bmod b = 504 \bmod 396 = 108.$$

We then move to lines 7 and 8, where we set $a$ to 396 and $b$ to 108. We then return to line 5.

Since $b \neq 0$, we proceed to line 6, where we set $r$ to

$$a \bmod b = 396 \bmod 108 = 72.$$

We then move to lines 7 and 8, where we set $a$ to 108 and $b$ to 72. We then return to line 5.

Since $b \neq 0$, we proceed to line 6, where we set $r$ to

$$a \bmod b = 108 \bmod 72 = 36.$$

We then move to lines 7 and 8, where we set $a$ to 72 and $b$ to 36. We then return to line 5.

Since $b \neq 0$, we proceed to line 6, where we set $r$ to

$$a \bmod b = 72 \bmod 36 = 0.$$

We then move to lines 7 and 8, where we set $a$ to 36 and $b$ to 0. We then return to line 5.

This time $b = 0$, so we skip to line 10, where we return $a$ (36), the greatest common divisor of 396 and 504. ◀

## Analysis of the Euclidean Algorithm

We analyze the worst-case performance of Algorithm 5.3.3. We define the time required to be the number of modulus operations executed at line 6. Table 5.3.1 lists the number of modulus operations required for some small input values.

**TABLE 5.3.1** ■ Number of Modulus Operations Required by the Euclidean Algorithm for Various Values of the Input

| $a$ \ $b$ | 0 | 1 | 2 | 3 | 4 | 5 | 6 | 7 | 8 | 9 | 10 | 11 | 12 | 13 |
|---|---|---|---|---|---|---|---|---|---|---|---|---|---|---|
| 0 | — | 0 | 0 | 0 | 0 | 0 | 0 | 0 | 0 | 0 | 0 | 0 | 0 | 0 |
| 1 | 0 | 1 | 1 | 1 | 1 | 1 | 1 | 1 | 1 | 1 | 1 | 1 | 1 | 1 |
| 2 | 0 | 1 | 1 | 2 | 1 | 2 | 1 | 2 | 1 | 2 | 1 | 2 | 1 | 2 |
| 3 | 0 | 1 | 2 | 1 | 2 | 3 | 1 | 2 | 3 | 1 | 2 | 3 | 1 | 2 |
| 4 | 0 | 1 | 1 | 2 | 1 | 2 | 2 | 3 | 1 | 2 | 2 | 3 | 1 | 2 |
| 5 | 0 | 1 | 2 | 3 | 2 | 1 | 2 | 3 | 4 | 3 | 1 | 2 | 3 | 4 |
| 6 | 0 | 1 | 1 | 1 | 2 | 2 | 1 | 2 | 2 | 2 | 3 | 3 | 1 | 2 |
| 7 | 0 | 1 | 2 | 2 | 3 | 3 | 2 | 1 | 2 | 3 | 3 | 4 | 4 | 3 |
| 8 | 0 | 1 | 1 | 3 | 1 | 4 | 2 | 2 | 1 | 2 | 2 | 4 | 2 | 5 |
| 9 | 0 | 1 | 2 | 1 | 2 | 3 | 2 | 3 | 2 | 1 | 2 | 3 | 2 | 3 |
| 10 | 0 | 1 | 1 | 2 | 2 | 1 | 3 | 3 | 2 | 2 | 1 | 2 | 2 | 3 |
| 11 | 0 | 1 | 2 | 3 | 3 | 2 | 3 | 4 | 4 | 3 | 2 | 1 | 2 | 3 |
| 12 | 0 | 1 | 1 | 1 | 1 | 3 | 1 | 4 | 2 | 2 | 2 | 2 | 1 | 2 |
| 13 | 0 | 1 | 2 | 2 | 2 | 4 | 2 | 3 | 5 | 3 | 3 | 3 | 2 | 1 |

**TABLE 5.3.2** ■ Smallest Input Pair That Requires $n$ Modulus Operations in the Euclidean Algorithm

| $a$ | $b$ | $n$ (= *number of modulus operations*) |
|---|---|---|
| 1 | 0 | 0 |
| 2 | 1 | 1 |
| 3 | 2 | 2 |
| 5 | 3 | 3 |
| 8 | 5 | 4 |
| 13 | 8 | 5 |

The worst case for the Euclidean algorithm occurs when the number of modulus operations is as large as possible. By referring to Table 5.3.1, we can determine the input pair $a$, $b$, $a > b$, with $a$ as small as possible, that requires $n$ modulus operations for $n = 0, \ldots, 5$. The results are given in Table 5.3.2.

Recall that the Fibonacci sequence $\{f_n\}$ (see Section 4.4) is defined by the equations

$$f_1 = 1, \quad f_2 = 1, \quad f_n = f_{n-1} + f_{n-2} \quad n \geq 3.$$

The Fibonacci sequence begins $1, 1, 2, 3, 5, 8, \ldots$. A surprising pattern develops in Table 5.3.2: The $a$ column is the Fibonacci sequence starting with $f_2$ and, except for the first value, the $b$ column is also the Fibonacci sequence starting with $f_2$! We are led to conjecture that if the pair $a$, $b$, $a > b$, when input to the Euclidean algorithm requires $n \geq 1$ modulus operations, then $a \geq f_{n+2}$ and $b \geq f_{n+1}$. As further evidence of our conjecture, if we compute the smallest input pair that requires six modulus operations, we obtain $a = 21$ and $b = 13$. Our next theorem confirms that our conjecture is correct. The proof of this theorem is illustrated in Figure 5.3.1.

**Theorem 5.3.5** Suppose that the pair $a$, $b$, $a > b$, requires $n \geq 1$ modulus operations when input to the Euclidean algorithm. Then $a \geq f_{n+2}$ and $b \geq f_{n+1}$, where $\{f_n\}$ denotes the Fibonacci sequence.

**Proof** The proof is by induction on $n$.

**Basis Step (n = 1)**

We have already observed that the theorem is true if $n = 1$.

**Inductive Step**

Assume that the theorem is true for $n \geq 1$. We must show that the theorem is true for $n + 1$.

Suppose that the pair $a$, $b$, $a > b$, requires $n + 1$ modulus operations when input to the Euclidean algorithm. At line 6, we compute $r = a \bmod b$. Thus

$$a = bq + r \qquad 0 \leq r < b. \tag{5.3.2}$$

The algorithm then repeats using the values $b$ and $r$, $b > r$. These values require $n$ additional modulus operations. By the inductive assumption,

$$b \geq f_{n+2} \qquad \text{and} \qquad r \geq f_{n+1}. \tag{5.3.3}$$

Combining (5.3.2) and (5.3.3), we obtain

$$a = bq + r \geq b + r \geq f_{n+2} + f_{n+1} = f_{n+3}. \tag{5.3.4}$$

[The first inequality in (5.3.4) holds because $q > 0$; $q$ cannot equal 0, because $a > b$.] Inequalities (5.3.3) and (5.3.4) give

$$a \geq f_{n+3} \qquad \text{and} \qquad b \geq f_{n+2}.$$

The inductive step is finished and the proof is complete. ◀

We may use Theorem 5.3.5 to analyze the worst-case performance of the Euclidean algorithm.

| | |
|---|---|
| $34 = 91 \bmod 57$ | (1 modulus operation) |
| $57, 34$ requires 4 modulus operations | (to make a total of 5) |
| $57 \geq f_6$ and $34 \geq f_5$ | (by inductive assumption) |
| $\therefore\ 91 = 57 \cdot 1 + 34 \geq 57 + 34 \geq f_6 + f_5 = f_7$ | |

**Figure 5.3.1** The proof of Theorem 5.3.5. The pair 91, 57, which requires $n + 1 = 5$ modulus operations, is input to the Euclidean algorithm.

**Theorem 5.3.6** If integers in the range 0 to $m$, $m \geq 8$, not both zero, are input to the Euclidean algorithm, then at most

$$\log_{3/2} \frac{2m}{3}$$

modulus operations are required.

**Proof** Let $n$ be the maximum number of modulus operations required by the Euclidean algorithm for integers in the range 0 to $m$, $m \geq 8$. Let $a, b$ be an input pair in the range 0 to $m$ that requires $n$ modulus operations. Table 5.3.1 shows that $n \geq 4$ and that $a \neq b$. We may assume that $a > b$. (Interchanging the values of $a$ and $b$ does not alter the number of modulus operations required.) By Theorem 5.3.5, $a \geq f_{n+2}$. Thus

$$f_{n+2} \leq m.$$

By Exercise 27, Section 4.4, since $n + 2 \geq 6$,

$$\left(\frac{3}{2}\right)^{n+1} < f_{n+2}.$$

Combining these last inequalities, we obtain

$$\left(\frac{3}{2}\right)^{n+1} < m.$$

Taking the logarithm to the base 3/2, we obtain

$$n + 1 < \log_{3/2} m.$$

Therefore,

$$n < (\log_{3/2} m) - 1 = \log_{3/2} m - \log_{3/2} \frac{3}{2} = \log_{3/2} \frac{2m}{3}.$$

◀

Because the logarithm function grows so slowly, Theorem 5.3.6 tells us that the Euclidean algorithm is quite efficient, even for large values of the input. For example, since

$$\log_{3/2} \frac{2(1{,}000{,}000)}{3} = 33.07\ldots,$$

the Euclidean algorithm requires at most 33 modulus operations to compute the greatest common divisor of any pair of integers, not both zero, in the range 0 to 1,000,000.

## A Special Result

The following special result will be used to compute inverses modulo an integer (see the following subsection). Such inverses are used in the RSA cryptosystem (see Section 5.4). However, this special result is also useful in other ways (see Exercises 27 and 29 and the following Problem-Solving Corner).

**Theorem 5.3.7** If $a$ and $b$ are nonnegative integers, not both zero, there exist integers $s$ and $t$ such that

$$\gcd(a, b) = sa + tb.$$

The method of the Euclidean algorithm can be used to prove Theorem 5.3.7 and to compute $s$ and $t$. Before proving the theorem, we first illustrate the proof with a specific example.

**Example 5.3.8** Consider how the Euclidean algorithm computes gcd(273, 110). We begin with $a = 273$ and $b = 110$. The Euclidean algorithm first computes

$$r = 273 \bmod 110 = 53. \tag{5.3.5}$$

It then sets $a = 110$ and $b = 53$.

The Euclidean algorithm next computes

$$r = 110 \bmod 53 = 4. \tag{5.3.6}$$

It then sets $a = 53$ and $b = 4$.

The Euclidean algorithm next computes

$$r = 53 \bmod 4 = 1. \tag{5.3.7}$$

It then sets $a = 4$ and $b = 1$.

The Euclidean algorithm next computes

$$r = 4 \bmod 1 = 0.$$

Since $r = 0$, the algorithm terminates, having found the greatest common divisor of 273 and 110 to be 1.

To find $s$ and $t$, we work back, beginning with the last equation [equation (5.3.7)] in which $r \neq 0$. Equation (5.3.7) may be rewritten as

$$1 = 53 - 4 \cdot 13 \tag{5.3.8}$$

since the quotient when 53 is divided by 4 is 13.

Equation (5.3.6) may be rewritten as

$$4 = 110 - 53 \cdot 2.$$

We then substitute this formula for 4 into equation (5.3.8) to obtain

$$1 = 53 - 4 \cdot 13 = 53 - (110 - 53 \cdot 2)13 = 27 \cdot 53 - 13 \cdot 110. \tag{5.3.9}$$

Equation (5.3.5) may be rewritten as

$$53 = 273 - 110 \cdot 2.$$

We then substitute this formula for 53 into equation (5.3.9) to obtain

$$1 = 27 \cdot 53 - 13 \cdot 110 = 27(273 - 110 \cdot 2) - 13 \cdot 110 = 27 \cdot 273 - 67 \cdot 110.$$

Thus, if we take $s = 27$ and $t = -67$, we obtain

$$\gcd(273, 110) = 1 = s \cdot 273 + t \cdot 110.$$

◀

***Proof of Theorem 5.3.7*** Given $a > b \geq 0$, let $r_0 = a$, $r_1 = b$, and $r_i$ equal the value of $r$ after the $(i-1)$st time the while loop is executed in Algorithm 5.3.3 (e.g., $r_2 = a \bmod b$). Suppose that $r_n$ is the first $r$-value that is zero so that $\gcd(a, b) = r_{n-1}$.

In general,

$$r_i = r_{i+1}q_{i+2} + r_{i+2}. \tag{5.3.10}$$

Taking $i = n - 3$ in (5.3.10), we obtain

$$r_{n-3} = r_{n-2}q_{n-1} + r_{n-1},$$

which may be rewritten as

$$r_{n-1} = -q_{n-1}r_{n-2} + 1 \cdot r_{n-3}.$$

We may take $t_{n-3} = -q_{n-1}$ and $s_{n-3} = 1$ to obtain

$$r_{n-1} = t_{n-3}r_{n-2} + s_{n-3}r_{n-3}. \tag{5.3.11}$$

Taking $i = n - 4$ in (5.3.10), we obtain

$$r_{n-4} = r_{n-3}q_{n-2} + r_{n-2}$$

or

$$r_{n-2} = -q_{n-2}r_{n-3} + r_{n-4}. \tag{5.3.12}$$

Substituting (5.3.12) into (5.3.11), we obtain

$$\begin{aligned} r_{n-1} &= t_{n-3}[-q_{n-2}r_{n-3} + r_{n-4}] + s_{n-3}r_{n-3} \\ &= [-t_{n-3}q_{n-2} + s_{n-3}]r_{n-3} + t_{n-3}r_{n-4}. \end{aligned}$$

Setting $t_{n-4} = -t_{n-3}q_{n-2} + s_{n-3}$ and $s_{n-4} = t_{n-3}$, we obtain

$$r_{n-1} = t_{n-4}r_{n-3} + s_{n-4}r_{n-4}.$$

Continuing in this way, we ultimately obtain

$$\gcd(r_0, r_1) = r_{n-1} = t_0 r_1 + s_0 r_0 = t_0 b + s_0 a.$$

Taking $s = s_0$ and $t = t_0$, we have

$$\gcd(r_0, r_1) = sa + tb.$$

◀

We next give an algorithm to compute $s$ and $t$ satisfying $\gcd(a, b) = sa + tb$, where $a$ and $b$ are nonnegative integers not both zero. To compute $s$ and $t$, the proof of Theorem 5.3.7, which is illustrated in Example 5.3.8, first finds $\gcd(a, b)$ and then works backward from the last remainder obtained during the computation of $\gcd(a, b)$ to the first remainder obtained. Since recursion elegantly handles such a backward computation, we will write a recursive algorithm to compute $s$ and $t$. We begin by writing a recursive version of the Euclidean algorithm (Algorithm 5.3.9). We can then modify this recursive algorithm to obtain a recursive algorithm to compute $s$ and $t$.

**Algorithm 5.3.9** **Recursive Euclidean Algorithm**

This algorithm recursively finds the greatest common divisor of the nonnegative integers $a$ and $b$, where not both $a$ and $b$ are zero.

Input: $a$ and $b$ (nonnegative integers, not both zero)

Output: Greatest common divisor of $a$ and $b$

```
gcdr(a, b) {
   // make a largest
   if (a < b)
      swap(a, b)
   if (b == 0)
      return a
   r = a mod b
   return gcdr(b, r)
}
```

Algorithm 5.3.9 first makes $a$ largest. If $b$ is zero, it correctly returns $a$. Otherwise, Algorithm 5.3.9 computes $r = a \bmod b$ and returns the greatest common divisor of $b$ and $r$, which is correct since Theorem 5.3.2 tells us that $\gcd(b, r) = \gcd(a, b)$.

To compute $s$ and $t$, we modify Algorithm 5.3.9. We call the modification *STgcdr*. The idea is that whenever we compute the greatest common divisor, we also compute the values of $s$ and $t$. These values are stored in added parameters named $s$ and $t$.

Consider first the case when $b$ is zero. Then $\gcd(a, b) = a$. Here we must set $s = 1$. Since $b$ is zero, $t$ could be assigned any value; we choose $t = 0$. The first part of the modification of Algorithm 5.3.9 looks like:

```
STgcdr(a, b, s, t) {
   // make a largest
   if (a < b)
      swap(a, b)
   if (b == 0) {
      s = 1
      t = 0
      // now a = sa + tb
      return a
   }
```

Next, Algorithm 5.3.9 computes $r = a \bmod b$ and $gcdr(b, r)$. Our modified algorithm will compute $r = a \bmod b$ and $STgcdr(b, r, s', t')$. Thus $s'$ and $t'$ satisfy $g = s'b + t'r$, where $g = \gcd(b, r)$. We must compute $s$ and $t$ in terms of the available values. If we let $q$ be the quotient of $a$ divided by $b$, we have $a = bq + r$. Therefore, using the fact that $r = a - bq$, we have

$$\begin{aligned} g &= s'b + t'r \\ &= s'b + t'(a - bq) \\ &= t'a + (s' - t'q)b. \end{aligned}$$

Thus if we set $s = t'$ and $t = s' - t'q$, we have $g = sa + tb$. The formal algorithm follows.

**Algorithm 5.3.10 Computing $s$ and $t$ of Theorem 5.3.7**

This algorithm computes $s$ and $t$ satisfying $\gcd(a, b) = sa + tb$, where $a$ and $b$ are nonnegative integers not both zero, and returns $\gcd(a, b)$.

Input: $a$ and $b$ (nonnegative integers, not both zero)

Output: $s$ and $t$ of Theorem 5.3.7 (stored in parameters $s$ and $t$) and the greatest common divisor of $a$ and $b$ (which is returned)

```
STgcdr(a, b, s, t) {
  // make a largest
  if (a < b)
    swap(a, b)
  if (b == 0) {
    s = 1
    t = 0
    // now a = sa + tb
    return a
  }
  q = ⌊a/b⌋
  r = a mod b
  // a = bq + r
  g = STgcdr(b, r, s', t')
  // g = s'b + t'r
  // ∴ g = t'a + (s' − t'q)b
  s = t'
  t = s' − t' * q
  return g
}
```

## Computing an Inverse Modulo an Integer

Suppose that we have two integers $n > 0$ and $\phi > 1$ such that $\gcd(n, \phi) = 1$. We show how to efficiently compute an integer $s$, $0 < s < \phi$ such that $ns \bmod \phi = 1$. We call $s$ the **inverse of $n$ mod $\phi$**. Efficiently computing this inverse is required by the RSA cryptosystem in Section 5.4.

Since $\gcd(n, \phi) = 1$, we use the Euclidean algorithm, as explained previously, to find numbers $s'$ and $t'$ such that $s'n + t'\phi = 1$. Then $ns' = -t'\phi + 1$, and, since $\phi > 1$, 1 is the remainder. Thus

$$ns' \bmod \phi = 1. \tag{5.3.13}$$

Note that $s'$ is almost the desired value; the problem is that $s'$ may not satisfy $0 < s' < \phi$. However, we can convert $s'$ to the proper value by setting $s = s' \bmod \phi$. Now $0 \le s < \phi$. In fact $s \ne 0$ since, if $s = 0$, then $\phi \mid s'$, which contradicts (5.3.13). Since $s = s' \bmod \phi$, there exists $q$ such that $s' = q\phi + s$. Combining the previous equations, we have

$$ns = ns' - \phi nq = -t'\phi + 1 - \phi nq = \phi(-t' - nq) + 1.$$

Therefore

$$ns \bmod \phi = 1. \tag{5.3.14}$$

**Example 5.3.11** Let $n = 110$ and $\phi = 273$. In Example 5.3.8, we showed that $\gcd(n, \phi) = 1$ and that $s'n + t'\phi = 1$, where $s' = -67$ and $t' = 27$. Thus,

$$110(-67) \bmod 273 = ns' \bmod \phi = 1.$$

Here $s = s' \bmod \phi = -67 \bmod 273 = 206$. Therefore, the inverse of 110 modulo 273 is 206. ◀

We conclude by showing that the number $s$ in equation (5.3.14) is unique. Suppose that

$$ns \bmod \phi = 1 = ns' \bmod \phi, \qquad 0 < s < \phi, \qquad 0 < s' < \phi.$$

We must show that $s' = s$. Now

$$s' = (s' \bmod \phi)(ns \bmod \phi) = s'ns \bmod \phi = (s'n \bmod \phi)(s \bmod \phi) = s.$$

Therefore, the number $s$ in equation (5.3.14) is unique.

## 5.3 Problem-Solving Tips

The Euclidean algorithm for computing the greatest common divisor of nonnegative integers $a$ and $b$, not both zero, is based on the equation

$$\gcd(a, b) = \gcd(b, r),$$

where $r = a \bmod b$. We replace the original problem, compute $\gcd(a, b)$, with the problem, compute $\gcd(b, r)$. We then replace $a$ by $b$ and $b$ by $r$, and repeat. Eventually $r = 0$, so the solution is $\gcd(b, 0) = b$.

The Euclidean algorithm is quite efficient. If integers in the range 0 to $m$, $m \geq 8$, not both zero, are input to the Euclidean algorithm, then at most

$$\log_{3/2} \frac{2m}{3}$$

modulus operations are required.

If $a$ and $b$ are nonnegative integers, not both zero, there exist integers $s$ and $t$ such that

$$\gcd(a, b) = sa + tb.$$

To compute $s$ and $t$, use the Euclidean algorithm. In a problem that involves the greatest common divisor, the preceding equation may be helpful. (Try Exercises 27 and 29.)

Suppose that we have two integers $n > 0$ and $\phi > 1$ such that $\gcd(n, \phi) = 1$. To efficiently compute an integer $s$, $0 < s < \phi$ such that $ns \bmod \phi = 1$, first compute $s'$ and $t'$ satisfying

$$\gcd(n, \phi) = s'n + t'\phi$$

(see the subsection Computing an Inverse Modulo an Integer). Then set $s = s' \bmod \phi$.

## 5.3 Review Exercises

1. State the Euclidean algorithm.
2. What key theorem is the basis for the Euclidean algorithm?
3. If the pair $a, b, a > b$, requires $n \geq 1$ modulus operations when input to the Euclidean algorithm, how are $a$ and $b$ related to the Fibonacci sequence?

4. Integers in the range 0 to $m$, $m \geq 8$, not both zero, are input to the Euclidean algorithm. Give an upper bound for the number of modulus operations required.

5. Theorem 5.3.7 states that there exist integers $s$ and $t$ such that $\gcd(a, b) = sa + tb$. Explain how the Euclidean algorithm can be used to compute $s$ and $t$.

6. Explain what it means for $s$ to be the inverse of $n$ modulo $\phi$.

7. Suppose that $\gcd(n, \phi) = 1$. Explain how to compute the inverse of $n$ modulo $\phi$.

## 5.3 Exercises

*Use the Euclidean algorithm to find the greatest common divisor of each pair of integers in Exercises 1–12.*

1. 60, 90
2. 110, 273
3. 220, 1400
4. 315, 825
5. 20, 40
6. 331, 993
7. 2091, 4807
8. 2475, 32670
9. 67942, 4209
10. 490256, 337
11. 27, 27
12. 57853125, 555111200
13. For each number pair $a, b$ in Exercises 1–12, find integers $s$ and $t$ such that $sa + tb = \gcd(a, b)$.
14. Find two integers $a$ and $b$, each less than 100, that maximize the number of iterations of the while loop of Algorithm 5.3.3.
15. Show that Algorithm 5.3.3 correctly finds $\gcd(a, b)$ even if lines 3 and 4 are deleted.
16. Write a recursive version of the Euclidean algorithm that executes the check for $a < b$ and the call to the *swap* function one time. *Hint*: Use *two* functions.
17. If $a$ and $b$ are positive integers, show that $\gcd(a, b) = \gcd(a, a + b)$.
18. Show that if $a > b \geq 0$, then

$$\gcd(a, b) = \gcd(a - b, b).$$

19. Using Exercise 18, write an algorithm to compute the greatest common divisor of two nonnegative integers $a$ and $b$, not both zero, that uses subtraction but not the modulus operation.
20. How many subtractions are required by the algorithm of Exercise 19 in the worst case for numbers in the range 0 to $m$?
21. Extend Tables 5.3.1 and 5.3.2 to the range 0 to 21.
22. Exactly how many modulus operations are required by the Euclidean algorithm in the worst case for numbers in the range 0 to 1,000,000?
23. Prove that when the pair $f_{n+2}, f_{n+1}$ is input to the Euclidean algorithm, $n \geq 1$, exactly $n$ modulus operations are required.
24. Show that for any integer $k > 1$, the number of modulus operations required by the Euclidean algorithm to compute $\gcd(a, b)$ is the same as the number of modulus operations required to compute $\gcd(ka, kb)$.
25. Show that $\gcd(f_n, f_{n+1}) = 1$, $n \geq 1$.
26. Suppose that $d > 0$ is a common divisor of nonnegative integers $a$ and $b$, not both zero. Prove that $d \mid \gcd(a, b)$.
★27. Show that if $p$ is a prime number, $a$ and $b$ are positive integers, and $p \mid ab$, then $p \mid a$ or $p \mid b$.
28. Give an example of positive integers $p$, $a$, and $b$ where $p \mid ab$, $p \nmid a$, and $p \nmid b$.
29. Let $m$ and $n$ be positive integers. Let $f$ be the function from

$$X = \{0, 1, \ldots, m - 1\}$$

to $X$ defined by

$$f(x) = nx \bmod m.$$

Prove that $f$ is one-to-one and onto if and only if $\gcd(m, n) = 1$.

*Exercises 30–34 show another way to prove that if a and b are nonnegative integers, not both zero, there exist integers s and t such that*

$$\gcd(a, b) = sa + tb.$$

*However, unlike the Euclidean algorithm, this proof does not lead to a technique to compute s and t.*

30. Let

$$X = \{sa + tb \mid sa + tb > 0 \text{ and } s \text{ and } t \text{ are integers}\}.$$

Show that $X$ is nonempty.

31. Show that $X$ has a least element. Let $g$ denote the least element.
32. Show that if $c$ is a common divisor of $a$ and $b$, then $c$ divides $g$.
33. Show that $g$ is a common divisor of $a$ and $b$. *Hint:* Assume that $g$ does not divide $a$. Then $a = qg + r$, $0 < r < g$. Obtain a contradiction by showing that $r \in X$.
34. Show that $g$ is the greatest common divisor of $a$ and $b$.

*In Exercises 35–41, show that* $\gcd(n, \phi) = 1$, *and find the inverse s of n modulo* $\phi$ *satisfying* $0 < s < \phi$.

35. $n = 2, \phi = 3$
36. $n = 1, \phi = 47$
37. $n = 7, \phi = 20$
38. $n = 11, \phi = 47$
39. $n = 50, \phi = 231$
40. $n = 100, \phi = 231$
41. $n = 100, \phi = 243$
42. Show that 6 has no inverse modulo 15. Does this contradict the result preceding Example 5.3.11? Explain.
43. Show that $n > 0$ has an inverse modulo $\phi > 1$ if and only if $\gcd(n, \phi) = 1$.

# Problem-Solving Corner: Making Postage

## Problem

Let $p$ and $q$ be positive integers satisfying $\gcd(p, q) = 1$. Show that there exists $n$ such that for all $k \geq n$, postage of $k$ cents can be achieved by using only $p$-cent and $q$-cent stamps.

## Attacking the Problem

Does this type of problem sound familiar? Example 2.5.1 used mathematical induction to show that postage of four cents or more can be achieved by using only 2-cent and 5-cent stamps. This result illustrates the problem for $p = 2$ and $q = 5$. In this case, if we take $n = 4$, for all $k \geq 4$, postage of $k$ cents can be achieved by using only 2-cent and 5-cent stamps. At this point, you should review the induction proof of this result.

The induction proof of the $p = 2$, $q = 5$ problem can be summarized as follows. We first proved the base cases ($k = 4, 5$). In the inductive step, we assumed that we could make postage of $k - 2$ cents. We then added a 2-cent stamp to achieve $k$ cents postage. We will imitate this inductive proof to prove that, if $\gcd(p, q) = 1$ for arbitrary $p$ and $q$, there exists $n$ such that for all $k \geq n$, postage of $k$ cents can be achieved by using only $p$-cent and $q$-cent stamps.

## Finding a Solution

Let's first take care of a trivial case. If either $p$ or $q$ is 1, we can make $k$ cents postage for all $k \geq 1$ by using $k$ 1-cent stamps. Thus, we assume that $p > 1$ and $q > 1$.

Let's first develop some notation. For a particular amount of postage, we'll let $n_p$ denote the number of $p$-cent stamps used and $n_q$ denote the number of $q$-cent stamps used. Then the amount of postage is

$$n_p p + n_q q.$$

Does this expression remind you of anything? Theorem 5.3.7 states that there exist integers $s$ and $t$ such that

$$1 = \gcd(p, q) = sp + tq. \qquad \textbf{(1)}$$

This last equation suggests that we can make postage of 1 cent by using $s$ $p$-cent stamps and $t$ $q$-cent stamps. The problem is that one of $s$ or $t$ must be negative in order for the sum $sp + tq$ to be 1 (since $p$ and $q$ are greater than 1). In fact, since $p$ and $q$ are both greater than 1, one of $s$ or $t$ is positive and the other is negative. We assume that $s > 0$ and $t < 0$.

Let's see how the inductive step should work and then see what basis steps we need. Imitating the specific case discussed previously, we would like to assume that we can make $k - p$ cents postage and then add a $p$-cent stamp to make $k$ cents postage. Nothing to it! In order for this inductive step to work, our basis steps must be $n, n + 1, \ldots, n + p - 1$ for some $n$ that we get to choose.

Suppose that we can make $n$-cents postage:

$$n = n_p p + n_q q.$$

Because of equation (1), we can then make $(n + 1)$-cents postage

$$n + 1 = (n_p p + n_q q) + (sp + tq) = (n_p + s)p + (n_q + t)q$$

using $n_p + s$ $p$-cent stamps and $n_q + t$ $q$-cent stamps. Of course, this last statement is meaningful only if $n_p + s \geq 0$ and $n_q + t \geq 0$. However, $n_p + s \geq 0$ holds because $n_p \geq 0$ and $s > 0$. We can *arrange* for $n_q + t \geq 0$ to hold by choosing $n_q \geq -t$.

Similarly, we can make $(n + 2)$-cents postage

$$\begin{aligned} n + 2 &= (n_p p + n_q q) + 2(sp + tq) \\ &= (n_p + 2s)p + (n_q + 2t)q \end{aligned}$$

using $n_p + 2s$ $p$-cent stamps and $n_q + 2t$ $q$-cent stamps. As before, $n_p + 2s \geq 0$ holds because $n_p \geq 0$ and $s > 0$. We can *arrange* for $n_q + 2t \geq 0$ to hold by choosing $n_q \geq -2t$. Notice that for this choice of $n_q$, $n_q \geq -t$ also holds [so that we can *still* make $(n + 1)$ cents postage, too].

In general, we can make $(n + i)$-cents postage

$$\begin{aligned} n + i &= (n_p p + n_q q) + i(sp + tq) \\ &= (n_p + is)p + (n_q + it)q \end{aligned}$$

using $n_p + is$ $p$-cent stamps and $n_q + it$ $q$-cent stamps. As before, $n_p + is \geq 0$ holds because $n_p \geq 0$ and $s > 0$. We can *arrange* for $n_q + it \geq 0$ to hold by choosing $n_q \geq -it$. Notice that for this choice of $n_q$, $n_q \geq -jt$ also holds for $j = 0, \ldots, i - 1$ [so that we can *still* make $(n + j)$ cents postage, too].

It follows that we can make postage for $n, n + 1, \ldots, n + p - 1$ provided that we choose $n_q = -(p - 1)t$. Any value for $n_p \geq 0$ will do, so we take $n_p = 0$. This makes $n = n_q q = -(p - 1)tq$.

## Formal Solution

If either $p$ or $q$ equals 1, we may take $n = 1$; so assume that $p > 1$ and $q > 1$. By Theorem 5.3.7, there exist integers $s$ and $t$ such that $sp + tq = 1$. Since $p > 1$ and $q > 1$, $s \neq 0$ and $t \neq 0$. Furthermore, either $s$ or $t$ is negative. We may suppose that $t < 0$. Then $s > 0$.

Let $n = -t(p-1)q$. We next show that we can make postage for $n, n+1, \ldots, n+p-1$ using only $p$-cent and $q$-cent stamps.

Now

$$\begin{aligned} n+j &= -t(p-1)q + j(sp+tq) \\ &= (js)p + (-t(p-1)+jt)q. \end{aligned}$$

If $0 \le j \le p-1$, then

$$-t(p-1)+jt \ge -t(p-1)+t(p-1) = 0.$$

Therefore we can make postage for $n+j$, $0 \le j \le p-1$, by using $js$ $p$-cent stamps and $-t(p-1)+jt$ $q$-cent stamps.

Finally, we use induction to show that we can make postage of $n$ cents or more using only $p$-cent and $q$-cent stamps. The basis steps are $n, n+1, \ldots, n+p-1$. Suppose that $k \ge n+p$ and that we can make postage for $m$ satisfying $n \le m < k$. In particular, we can make postage for $k-p$. Add a $p$-cent stamp to make postage for $k$. The inductive step is complete.

## Summary of Problem-Solving Techniques

- Look for a similar problem. We already encountered a postage problem in Example 2.5.1.
- Try to use some of the ideas in a similar problem. To solve our problem, we were able to modify the induction proof in Example 2.5.1.
- Sometimes notation, setting, or context will suggest something useful. In our problem, the postage-amount equation $n_p p + n_q q$ was similar in form to the greatest common divisor formula $\gcd(p, q) = sp + tq$. Combining these equations was crucial to proving the base cases.
- Don't be afraid to make assumptions. If an assumption turns out to be unwarranted, it can sometimes be modified so that the modified assumption *is* correct. In our problem, we assumed that if we could make $n$-cents postage using $n_p$ $p$-cent stamps and $n_q$ $q$-cent stamps, we could then make $(n+1)$-cents postage by using $n_p + s$ $p$-cent stamps and $n_q + t$ $q$-cent stamps. We were able to force this latter statement to be true by choosing $n_q \ge -t$.

## Exercise

1. Show that if $\gcd(p, q) > 1$, it is false that there exists $n$ such that for all $k \ge n$, postage of $k$ cents can be achieved by using only $p$-cent and $q$-cent stamps.

# 5.4 The RSA Public-Key Cryptosystem

**Cryptology** is the study of systems, called **cryptosystems,** for secure communications. In a cryptosystem, the sender transforms the message before transmitting it, hoping that only authorized recipients can reconstruct the original message (i.e., the message before it was transformed). The sender is said to **encrypt** the message, and the recipient is said to **decrypt** the message. If the cryptosystem is secure, unauthorized persons will be unable to discover the decryption technique, so even if they read the encrypted message, they will be unable to decrypt it. Cryptosystems are important for large organizations (e.g., government and military), all internet-based businesses, and individuals. For example, if a credit card number is sent over the internet, it is important for the number to be read only by the intended recipient. In this section, we look at some algorithms that support secure communication.

In one of the oldest and simplest systems, the sender and receiver each have a key that defines a substitute character for each potential character to be sent. Moreover, the sender and receiver do not disclose the key. Such keys are said to be *private*.

**Example 5.4.1** If a key is defined as

character: `␣ABCDEFGHIJKLMNOPQRSTUVWXYZ`

replaced by: `EIJFUAXVHWP␣GSRKOBTQYDMLZNC`

the message `SEND␣MONEY` would be encrypted as `QARUESKRAN`. The encrypted message `SKRANEKRELIN` would be decrypted as `MONEY␣ON␣WAY`. ◀

**Go Online**
For more on the RSA public-key cryptosystem, see bit.ly/2NYUMkW

Simple systems such as that in Example 5.4.1 are easily broken since certain letters (e.g., E in English) and letter combinations (e.g., ER in English) appear more frequently than others. Also, a problem with private keys in general is that the keys have to be securely sent to the sender and recipient before messages can be sent. We devote the remainder of this section to the **RSA public-key cryptosystem,** named after its inventors, Ronald L. Rivest, Adi Shamir, and Leonard M. Adleman, that is believed to be secure. In the RSA system, each participant makes public an encryption key and hides a decryption key. To send a message, all one needs to do is look up the recipient's encryption key in a publicly distributed table. The recipient then decrypts the message using the hidden decryption key. The RSA system is used in all major secure internet transactions, for example, for secure financial transactions and secure e-mail exchanges.

In the RSA system, messages are represented as numbers. For example, each character might be represented as a number. If a blank space is represented as 1, A as 2, B as 3, and so on, the message SEND␣MONEY would be represented as 20, 6, 15, 5, 1, 14, 16, 15, 6, 26. If desired, the integers could be combined into the single integer

2006150501141615 0626

(note that leading zeros have been added to all single-digit numbers).

We next describe how the RSA system works, present a concrete example, and then discuss why it works. Each prospective recipient chooses two primes $p$ and $q$ and computes $z = pq$. Next, the prospective recipient computes $\phi = (p-1)(q-1)$ and chooses an integer $n$ such that $\gcd(n, \phi) = 1$. The pair $z, n$ is then made public. Finally, the prospective recipient computes the unique number $s$, $0 < s < \phi$, satisfying $ns \bmod \phi = 1$. (An efficient way to compute $s$ is given in Section 5.3.) The number $s$ is kept secret and used to decrypt messages.

To send the integer $a$, $0 \le a \le z-1$, to the holder of public key $z, n$, the sender computes $c = a^n \bmod z$ and sends $c$. (Algorithm 5.2.19 provides an efficient way to compute $a^n \bmod z$.) To decrypt the message, the recipient computes $c^s \bmod z$, which can be shown to be equal to $a$.

**Example 5.4.2** Suppose that we choose $p = 23$, $q = 31$, and $n = 29$. Then $z = pq = 713$ and $\phi = (p-1)(q-1) = 660$. Now $s = 569$ since $ns \bmod \phi = 29 \cdot 569 \bmod 660 = 16501 \bmod 660 = 1$. The pair $z, n = 713, 29$ is made publicly available.

To transmit $a = 572$ to the holder of public key 713, 29, the sender computes $c = a^n \bmod z = 572^{29} \bmod 713 = 113$ and sends 113. The receiver computes $c^s \bmod z = 113^{569} \bmod 713 = 572$ in order to decrypt the message. ◀

The main result that makes encryption and decryption work is that

$$a^u \bmod z = a \qquad \text{for all } 0 \le a < z \text{ and } u \bmod \phi = 1$$

(for a proof, see [Rivest]). Using this result and Theorem 5.2.17, we may show that decryption produces the correct result. Since $ns \bmod \phi = 1$,

$$c^s \bmod z = (a^n \bmod z)^s \bmod z = (a^n)^s \bmod z = a^{ns} \bmod z = a.$$

The security of the RSA encryption system relies mainly on the fact that currently there is no efficient algorithm known for factoring integers; that is, currently no algorithm is known for factoring $d$-bit integers in polynomial time, $O(d^k)$. Thus if the primes $p$ and $q$ are chosen large enough, it is impractical to compute the factorization $z = pq$. If the factorization could be found by a person who intercepts a message, the message could be decrypted just as the authorized recipient does. At this time, no efficient algorithm is known for factoring an arbitrary integer with 1024 or more bits. Thus, if $p$ and $p$ are chosen so that $z = pq$ is at least 1024 bits long and other technical requirements are implemented (e.g., $p$ and $q$ should not be chosen "too close" together), RSA would seem to be secure. (An integer having 1024 bits has over 300 decimal digits.)

The first description of the RSA encryption system was in Martin Gardner's February 1977 *Scientific American* column (see [Gardner, 1977]). Included in this column were an encoded message using the key $z, n$, where $z$ was the product of 64- and 65-digit primes, and $n = 9007$, and an offer of $100 to the first person to crack the code. At the time the article was written, it was estimated that it would take 40 quadrillion years to factor $z$. In fact, in April 1994, Arjen Lenstra, Paul Leyland, Michael Graff, and Derek Atkins, with the assistance of 600 volunteers from 25 countries using over 1600 computers, factored $z$ (see [Taubes]). The work was coordinated on the internet.

Another possible way a message could be intercepted and decrypted would be to compute the $n$th root of $c$ mod $z$, that is, compute an integer $a$ satisfying $c = a^n \bmod z$. This computation would give $a$, the decrypted value. Again, currently there is no efficient algorithm known for computing $n$th roots mod $z$. It is also conceivable that a message could be decrypted by some means other than factoring integers or taking $n$th roots mod $z$. For example, Paul Kocher proposed a way to break RSA based on the time it takes to decrypt messages (see [English]). The idea is that distinct secret keys require distinct amounts of time to decrypt messages and, by using this timing information, an unauthorized person might be able to unveil the secret key and thus decrypt the message. To thwart such attacks, implementors of RSA have taken steps to alter the observed time to decrypt messages.

## 5.4 Review Exercises

1. To what does "cryptology" refer?
2. What is a cryptosystem?
3. What does it mean to "encrypt a message"?
4. What does it mean to "decrypt a message"?
5. In the RSA public-key cryptosystem, how does one encrypt $a$ and send it to the holder of public key $z, n$?
6. In the RSA public-key cryptosystem, how does one decrypt $c$?
7. On what does the security of the RSA encryption system rest?

## 5.4 Exercises

1. Encrypt the message COOL_BEAVIS using the key of Example 5.4.1.
2. Decrypt the message UTWR_ENKDTEKMIGYWRA using the key of Example 5.4.1.
3. Encrypt the message I_AM_NOT_A_CROOK using the key of Example 5.4.1.
4. Decrypt the message JDQHLHIF_AU using the key of Example 5.4.1.
5. Encrypt 333 using the public key 713, 29 of Example 5.4.2.
6. Decrypt 411 using $s = 569$ as in Example 5.4.2.
7. Encrypt 241 using the public key 713, 29 of Example 5.4.2.
8. Decrypt 387 using $s = 569$ as in Example 5.4.2.

*In Exercises 9–13, assume that we choose primes* $p = 17$, $q = 23$, *and* $n = 31$.

9. Compute $z$.
10. Compute $\phi$.
11. Compute $s$.
12. Encrypt 101 using the public key $z, n$.
13. Decrypt 250.

*In Exercises 14–18, assume that we choose primes* $p = 59$, $q = 101$, *and* $n = 41$.

14. Compute $z$.
15. Compute $\phi$.
16. Compute $s$.
17. Encrypt 584 using the public key $z, n$.
18. Decrypt 250.

## Chapter 5 Notes

An accessible introduction to elementary number theory is [Niven, 1980]. An extended discussion of the greatest common divisor, including historical background, and other elementary number theory topics are in [Knuth, 1998a].

Full details of the RSA cryptosystem may be found in [Rivest]. [Pfleeger] is devoted to computer security.

## Chapter 5 Review

### Section 5.1

1. $d$ divides $n$: $d \mid n$
2. $d$ does not divide $n$: $d \nmid n$
3. $d$ is a divisor or factor of $n$
4. Prime
5. Composite
6. Fundamental theorem of arithmetic: any integer greater than 1 can be written as the product of primes
7. Common divisor
8. Greatest common divisor
9. Common multiple
10. Least common multiple

### Section 5.2

11. Bit
12. Decimal number system
13. Binary number system
14. Computer representation of integers: when represented in binary, the positive integer $n$ requires $\lfloor 1 + \lg n \rfloor$ bits
15. Hexadecimal number system
16. Base of a number system
17. Convert binary to decimal
18. Convert decimal to binary
19. Convert hexadecimal to decimal
20. Convert decimal to hexadecimal
21. Add binary numbers
22. Add hexadecimal numbers
23. Compute $a^n$ using repeated squaring
24. $ab \bmod z = [(a \bmod z)(b \bmod z)] \bmod z$
25. Compute $a^n \bmod z$ using repeated squaring

### Section 5.3

26. Euclidean algorithm
27. If the pair $a, b$, $a > b$, requires $n \geq 1$ modulus operations when input to the Euclidean algorithm, then $a \geq f_{n+2}$ and $b \geq f_{n+1}$, where $\{f_n\}$ denotes the Fibonacci sequence.
28. If integers in the range 0 to $m$, $m \geq 8$, not both zero, are input to the Euclidean algorithm, then at most

$$\log_{3/2} \frac{2m}{3}$$

modulus operations are required.
29. If $a$ and $b$ are nonnegative integers, not both zero, there exist integers $s$ and $t$ such that $\gcd(a, b) = sa + tb$.
30. Compute $s$ and $t$ such that $\gcd(a, b) = sa + tb$ using the Euclidean algorithm
31. Compute an inverse modulo an integer

### Section 5.4

32. Cryptology
33. Cryptosystem
34. Encrypt a message
35. Decrypt a message
36. RSA public key cryptosystem: To encrypt $a$ and send it to the holder of public key $z, n$, compute $c = a^n \bmod z$ and send $c$. To decrypt the message, compute $c^s \bmod z$, which can be shown to be equal to $a$.
37. The security of the RSA encryption system relies mainly on the fact that currently there is no efficient algorithm known for factoring integers.

## Chapter 5 Self-Test

1. Trace Algorithm 5.1.8 for the input $n = 539$.
2. Write the binary number 10010110 in decimal.
3. Find the prime factorization of 539.
4. Use the Euclidean algorithm to find the greatest common divisor of the integers 396 and 480.
5. Find $\gcd(2 \cdot 5^2 \cdot 7^2 \cdot 13^4, 7^4 \cdot 13^2 \cdot 17)$.
6. Find $\operatorname{lcm}(2 \cdot 5^2 \cdot 7^2 \cdot 13^4, 7^4 \cdot 13^2 \cdot 17)$.
7. Write the decimal number 430 in binary and hexadecimal.
8. Given that $\log_{3/2} 100 = 11.357747$, provide an upper bound for the number of modulus operations required by the Euclidean algorithm for integers in the range 0 to 100,000,000.
9. Trace Algorithm 5.2.16 for the value $n = 30$.
10. Trace Algorithm 5.2.19 for the values $a = 50$, $n = 30$, $z = 11$.
11. Use the Euclidean algorithm to find integers $s$ and $t$ satisfying $s \cdot 396 + t \cdot 480 = \gcd(396, 480)$.
12. Show that $\gcd(196, 425) = 1$ and find the inverse $s$ of 196 modulo 425 satisfying $0 < s < 425$.

*In Exercises 13–16, assume that we choose primes $p = 13$, $q = 17$, and $n = 19$.*

13. Compute $z$ and $\phi$.
14. Compute $s$.
15. Encrypt 144 using public key $z, n$.
16. Decrypt 28.

## Chapter 5 Computer Exercises

1. Implement Algorithm 5.1.8, testing whether a positive integer is prime, as a program.
2. Write a program that converts among decimal, hexadecimal, and octal.
3. Write a program that adds binary numbers.
4. Write a program that adds hexadecimal numbers.
5. Write a program that adds octal numbers.
6. Implement Algorithm 5.2.16, exponentiation by repeated squaring, as a program.
7. Implement Algorithm 5.2.19, exponentiation mod $z$, as a program.
8. Write recursive and nonrecursive programs to compute the greatest common divisor. Compare the times required by the programs.
9. Implement Algorithm 5.3.10, computing $s$ and $t$ satisfying $\gcd(a, b) = sa + tb$, as a program.
10. Write a program that, given integers $n > 0$ and $\phi > 1$, $\gcd(n, \phi) = 1$, computes the inverse of $n$ mod $\phi$.
11. Implement the RSA public-key cryptosystem.

# Chapter 6

# COUNTING METHODS AND THE PIGEONHOLE PRINCIPLE



In many discrete problems, we are confronted with the problem of counting. For example, in Section 4.3 we saw that in order to estimate the run time of an algorithm, we needed to count the number of times certain steps or loops were executed. Counting also plays a crucial role in probability theory. Because of the importance of counting, a variety of useful aids, some quite sophisticated, have been developed. In this chapter we develop several tools for counting. These techniques can be used to derive the binomial theorem. The chapter concludes with a discussion of the Pigeonhole Principle, which often allows us to prove the existence of an object with certain properties.

## 6.1 Basic Principles

**Go Online**
For more on basic principles, see
`bit.ly/2PNvtEt`

The menu for Kay's Quick Lunch is shown in Figure 6.1.1. As you can see, it features two appetizers, three main courses, and four beverages. How many different dinners consist of one main course and one beverage?

If we list all possible dinners consisting of one main course and one beverage,

$$\text{HT, HM, HC, HR, CT, CM, CC, CR, FT, FM, FC, FR,}$$

we see that there are 12 different dinners. (The dinner consisting of a main course whose first letter is $X$ and a beverage whose first letter is $Y$ is denoted $XY$. For example, CR refers to the dinner consisting of a cheeseburger and root beer.) Notice that there are three main courses and four beverages and $12 = 3 \cdot 4$.

There are 24 possible dinners consisting of one appetizer, one main course, and one beverage:

† These sections can be omitted without loss of continuity.

| APPETIZERS | |
|---|---|
| *Nachos* | 2.15 |
| *Salad* | 1.90 |
| **MAIN COURSES** | |
| *Hamburger* | 3.25 |
| *Cheeseburger* | 3.65 |
| *Fish Filet* | 3.15 |
| **BEVERAGES** | |
| *Tea* | .70 |
| *Milk* | .85 |
| *Cola* | .75 |
| *Root Beer* | .75 |

**Figure 6.1.1** Kay's Quick Lunch menu.

NHT, NHM, NHC, NHR, NCT, NCM, NCC, NCR,
NFT, NFM, NFC, NFR, SHT, SHM, SHC, SHR,
SCT, SCM, SCC, SCR, SFT, SFM, SFC, SFR.

(The dinner consisting of an appetizer whose first letter is $X$, a main course whose first letter is $Y$, and a beverage whose first letter is $Z$ is denoted $XYZ$.) Notice that there are two appetizers, three main courses, and four beverages and $24 = 2 \cdot 3 \cdot 4$.

In each of these examples, we found that the total number of dinners was equal to the product of numbers of each of the courses. These examples illustrate the **Multiplication Principle.**

**Multiplication Principle**

If an activity can be constructed in $t$ successive steps and step 1 can be done in $n_1$ ways, step 2 can then be done in $n_2$ ways, . . . , and step $t$ can then be done in $n_t$ ways, then the number of different possible activities is $n_1 \cdot n_2 \cdots n_t$.

In the problem of counting the number of dinners consisting of one main course and one beverage, the first step is "select the main course" and the second step is "select the beverage." Thus $n_1 = 3$ and $n_2 = 4$ and, by the Multiplication Principle, the total number of dinners is $3 \cdot 4 = 12$. Figure 6.1.2 shows why we multiply 3 times 4—we have three groups of four objects.

We may summarize the Multiplication Principle by saying that we multiply together the numbers of ways of doing each step when an activity is constructed in successive steps.

**Example 6.1.1**

How many dinners are available from Kay's Quick Lunch consisting of one main course and an *optional* beverage?

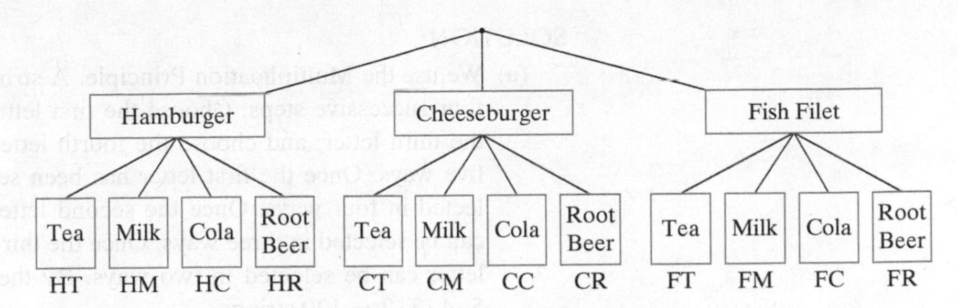


**Figure 6.1.2** An illustration of the Multiplication Principle.

**SOLUTION** We may construct a dinner consisting of one main course and an optional beverage by a two-step process. The first step is "select the main course" and the second step is "select an optional beverage." There are $n_1 = 3$ ways to select the main course (hamburger, cheeseburger, fish filet) and $n_2 = 5$ ways to select the optional beverage (tea, milk, cola, root beer, none). By the Multiplication Principle, there are $3 \cdot 5 = 15$ dinners. As confirmation, we list the 15 dinners (N = no beverage):

HT, HM, HC, HR, HN, CT, CM, CC, CR, CN, FT, FM, FC, FR, FN. ◀

**Example 6.1.2** **Melissa Virus** In the late 1990s, a computer virus named Melissa wreaked havoc by overwhelming system resources. The virus was spread by an e-mail message containing an attached word processor document with a malicious macro. When the word processor document was opened, the macro forwarded the e-mail message and the attached word processor document to the first 50 addresses obtained from the user's address book. When these forwarded copies were received and opened, the macro again forwarded the e-mail message and the attached word processor document, and so on. The virus caused problems by creating messages faster than they could be sent. The to-be-sent messages were temporarily stored on a disk. If the disk got full, the system could deadlock or even crash.

After the virus sent the e-mail to the first 50 addresses, each of those recipients then sent e-mail to 50 addresses. By the Multiplication Principle, there were then $50 \cdot 50 = 2500$ additional recipients. Each of these recipients, in turn, sent e-mail to 50 addresses. Again, by the Multiplication Principle, there were then $50 \cdot 50 \cdot 50 = 125{,}000$ additional recipients. After one more iteration, there were then $50 \cdot 50 \cdot 50 \cdot 50 = 6{,}250{,}000$ additional recipients. Thus after just four iterations,

$$6{,}250{,}000 + 125{,}000 + 2500 + 50 + 1 = 6{,}377{,}551$$

copies of the message had been sent.

Notice that we could have used the geometric sum (Example 2.4.4)

$$1 \cdot 50^4 + 1 \cdot 50^3 + 1 \cdot 50^2 + 1 \cdot 50 + 1 = \frac{1 \cdot (50^5 - 1)}{50 - 1} = 6{,}377{,}551$$

to calculate how many copies of the message were sent. ◀

**Example 6.1.3** (a) How many strings of length 4 can be formed using the letters *ABCDE* if repetitions are not allowed?

(b) How many strings of part (a) begin with the letter *B*?

(c) How many strings of part (a) do not begin with the letter *B*?

**SOLUTION**

(a) We use the Multiplication Principle. A string of length 4 can be constructed in four successive steps: Choose the first letter; choose the second letter; choose the third letter; and choose the fourth letter. The first letter can be selected in five ways. Once the first letter has been selected, the second letter can be selected in four ways. Once the second letter has been selected, the third letter can be selected in three ways. Once the third letter has been selected, the fourth letter can be selected in two ways. By the Multiplication Principle, there are $5 \cdot 4 \cdot 3 \cdot 2 = 120$ strings.

(b) The strings that begin with the letter $B$ can be constructed in four successive steps: Choose the first letter; choose the second letter; choose the third letter; and choose the fourth letter. The first letter ($B$) can be chosen in one way, the second letter in four ways, the third letter in three ways, and the fourth letter in two ways. Thus, by the Multiplication Principle, there are $1 \cdot 4 \cdot 3 \cdot 2 = 24$ strings that start with the letter $B$.

(c) Part (a) shows that there are 120 strings of length 4 that can be formed using the letters $ABCDE$, and part (b) shows that 24 of these start with the letter $B$. It follows that there are $120 - 24 = 96$ strings that do not begin with the letter $B$. ◀

**Example 6.1.4** In a digital picture, we wish to encode the amount of light at each point as an eight-bit string. How many values are possible at one point?

**SOLUTION** An eight-bit encoding can be constructed in eight successive steps: Select the first bit; select the second bit; . . . ; select the eighth bit. Since there are two ways to select each bit, by the Multiplication Principle the total number of eight-bit encodings is

$$2 \cdot 2 \cdot 2 \cdot 2 \cdot 2 \cdot 2 \cdot 2 \cdot 2 = 2^8 = 256.$$ ◀

We next give a proof using the Multiplication Principle that a set with $n$ elements has $2^n$ subsets. We previously gave a proof of this result using mathematical induction (Theorem 2.4.6).

**Example 6.1.5** Use the Multiplication Principle to prove that a set $\{x_1, \ldots, x_n\}$ containing $n$ elements has $2^n$ subsets; that is, the cardinality of the power set of an $n$-element set is $2^n$.

**SOLUTION** A subset can be constructed in $n$ successive steps: Pick or do not pick $x_1$; pick or do not pick $x_2$; . . . ; pick or do not pick $x_n$. Each step can be done in two ways. Thus the number of possible subsets is

$$\underbrace{2 \cdot 2 \cdots 2}_{n \text{ factors}} = 2^n.$$ ◀

**Example 6.1.6** Let $X$ be an $n$-element set. How many ordered pairs $(A, B)$ satisfy $A \subseteq B \subseteq X$?

**SOLUTION** Given an ordered pair $(A, B)$ satisfying $A \subseteq B \subseteq X$, we see that each element in $X$ is in exactly one of $A$, $B - A$, or $X - B$. Conversely, if we assign each element of $X$ to one of the three sets $A$ (and, by assumption, also to $B$ and $X$), $B - A$ (and, by assumption, also to $X$), or $X - B$, we obtain a unique ordered pair $(A, B)$ satisfying $A \subseteq B \subseteq X$. Thus the number of ordered pairs $(A, B)$ satisfying $A \subseteq B \subseteq X$ is equal to the number of ways to assign the elements of $X$ to the three sets $A$, $B-A$, and $X-B$. We can make such assignments by the following $n$-step process: Assign the first element of $X$ to one of $A, B-A, X-B$; assign the second element of $X$ to one of $A, B-A, X-B$; . . . ; assign the $n$th element of $X$ to one of $A$, $B-A$, $X-B$. Since each step can be done in three ways, the number of ordered pairs $(A, B)$ satisfying $A \subseteq B \subseteq X$ is

$$\underbrace{3 \cdot 3 \cdots 3}_{n \text{ factors}} = 3^n.$$

◀

**Example 6.1.7** How many reflexive relations are there on an $n$-element set?

**SOLUTION** We count the number of $n \times n$ matrices that represent reflexive relations on an $n$-element set $X$. Since $(x, x)$ is in the relation for all $x \in X$, the main diagonal of the matrix must consist of 1's. There is no restriction on the remaining entries; each can be 0 or 1. An $n \times n$ matrix has $n^2$ entries and the diagonal contains $n$ entries. Thus there are $n^2 - n$ off-diagonal entries. Since each can be assigned values in two ways, by the Multiplication Principle there are

$$\underbrace{2 \cdot 2 \cdots 2}_{n^2 - n \text{ factors}} = 2^{n^2-n}$$

matrices that represent reflexive relations on an $n$-element set. Therefore there are $2^{n^2-n}$ reflexive relations on an $n$-element set. ◀

**Example 6.1.8** **Internet Addresses** The internet is a network of interconnected computers. Each computer interface on the internet is identified by an internet address. In the IPv4 (Internet Protocol, Version 4) addressing scheme, the addresses are divided into five classes—Class A through Class E. Only Classes A, B, and C are used to identify computers on the internet. Class A addresses are used for large networks; Class B addresses are used for medium-size networks; and Class C addresses are used for small networks. In this example, we count the number of Class A addresses. Exercises 80–82 deal with Class B and C addresses.

A Class A address is a bit string of length 32. The first bit is 0 (to identify it as a Class A address). The next 7 bits, called the *netid*, identify the network. The remaining 24 bits, called the *hostid*, identify the computer interface. The netid must not consist of all 1's. The hostid must not consist of all 0's or all 1's. Arguing as in Example 6.1.4, we find that there are $2^7$ 7-bit strings. Since 1111111 is not allowed as a netid, there are $2^7 - 1$ netids. Again, arguing as in Example 6.1.4, we find that there are $2^{24}$ 24-bit strings. Since the two strings consisting of all 0's or all 1's are not allowed as a hostid, there are $2^{24} - 2$ hostids. By the Multiplication Principle, there are

$$(2^7 - 1)(2^{24} - 2) = 127 \cdot 16{,}777{,}214 = 2{,}130{,}706{,}178$$

Class A internet addresses. Because of the tremendous increase in the size of the internet, IPv6 (Internet Protocol, Version 6) uses 128-bit addresses rather than 32-bit addresses. ◀

Next, we illustrate the Addition Principle by an example and then present the principle.

**Example 6.1.9** How many eight-bit strings begin either 101 or 111?

**SOLUTION** An eight-bit string that begins 101 can be constructed in five successive steps: Select the fourth bit; select the fifth bit; . . . ; select the eighth bit. Since each of the five bits can be selected in two ways, by the Multiplication Principle, there are

$$2 \cdot 2 \cdot 2 \cdot 2 \cdot 2 = 2^5 = 32$$

eight-bit strings that begin 101. The same argument can be used to show that there are 32 eight-bit strings that begin 111. Since there are 32 eight-bit strings that begin 101 and 32 eight-bit strings that begin 111, there are $32 + 32 = 64$ eight-bit strings that begin either 101 or 111. ◀

In Example 6.1.9 we added the numbers of eight-bit strings (32 and 32) of each type to determine the final result. The **Addition Principle** tells us when to add to compute the total number of possibilities.

**Addition Principle**

Suppose that $X_1, \ldots, X_t$ are sets and that the ith set $X_i$ has $n_i$ elements. If $\{X_1, \ldots, X_t\}$ is a pairwise disjoint family (i.e., if $i \neq j$, $X_i \cap X_j = \varnothing$), the number of possible elements that can be selected from $X_1$ or $X_2$ or . . . or $X_t$ is

$$n_1 + n_2 + \cdots + n_t.$$

(Equivalently, the union $X_1 \cup X_2 \cup \cdots \cup X_t$ contains $n_1 + n_2 + \cdots + n_t$ elements.)

In Example 6.1.9 we could let $X_1$ denote the set of eight-bit strings that begin 101 and $X_2$ denote the set of eight-bit strings that begin 111. Since $X_1$ is disjoint from $X_2$, according to the Addition Principle, the number of eight-bit strings of either type, which is the number of elements in $X_1 \cup X_2$, is $32 + 32 = 64$.

We may summarize the Addition Principle by saying that we add the numbers of elements in each subset when the elements being counted can be decomposed into pairwise disjoint subsets.

If we are counting objects that are constructed in successive steps, we use the Multiplication Principle. If we have disjoint sets of objects and we want to know the total number of objects, we use the Addition Principle. It is important to recognize when to apply each principle. This skill comes from practice and careful thinking about each problem.

We close this section with examples that illustrate both counting principles.

**Example 6.1.10** In how many ways can we select two books from different subjects among five distinct computer science books, three distinct mathematics books, and two distinct art books?

**SOLUTION** Using the Multiplication Principle, we find that we can select two books, one from computer science and one from mathematics, in $5 \cdot 3 = 15$ ways. Similarly, we can select two books, one from computer science and one from art, in $5 \cdot 2 = 10$ ways, and we can select two books, one from mathematics and one from art, in $3 \cdot 2 = 6$ ways. Since these sets of selections are pairwise disjoint, we may use the Addition Principle to conclude that there are $15 + 10 + 6 = 31$ ways of selecting two books from different subjects among the computer science, mathematics, and art books. ◀

**Example 6.1.11** A six-person committee composed of Alice, Ben, Connie, Dolph, Egbert, and Francisco is to select a chairperson, secretary, and treasurer.

(a) In how many ways can this be done?

(b) In how many ways can this be done if either Alice or Ben must be chairperson?

(c) In how many ways can this be done if Egbert must hold one of the offices?

(d) In how many ways can this be done if both Dolph and Francisco must hold office?

**SOLUTION**

(a) We use the Multiplication Principle. The officers can be selected in three successive steps: Select the chairperson; select the secretary; select the treasurer. The chairperson can be selected in six ways. Once the chairperson has been selected, the secretary can be selected in five ways. After selection of the chairperson and secretary, the treasurer can be selected in four ways. Therefore, the total number of possibilities is $6 \cdot 5 \cdot 4 = 120$.

(b) Arguing as in part (a), if Alice is chairperson, we have $5 \cdot 4 = 20$ ways to select the remaining officers. Similarly, if Ben is chairperson, there are 20 ways to select the remaining officers. Since these cases are disjoint, by the Addition Principle, there are $20 + 20 = 40$ possibilities.

(c) [First solution] Arguing as in part (a), if Egbert is chairperson, we have 20 ways to select the remaining officers. Similarly, if Egbert is secretary, there are 20 possibilities, and if Egbert is treasurer, there are 20 possibilities. Since these three cases are pairwise disjoint, by the Addition Principle, there are $20+20+20 = 60$ possibilities.

[Second solution] Let us consider the activity of assigning Egbert and two others to offices to be made up of three successive steps: Assign Egbert an office; fill the highest remaining office; fill the last office. There are three ways to assign Egbert an office. Once Egbert has been assigned, there are five ways to fill the highest remaining office. Once Egbert has been assigned and the highest remaining office filled, there are four ways to fill the last office. By the Multiplication Principle, there are $3 \cdot 5 \cdot 4 = 60$ possibilities.

(d) Let us consider the activity of assigning Dolph, Francisco, and one other person to offices to be made up of three successive steps: Assign Dolph; assign Francisco; fill the remaining office. There are three ways to assign Dolph. Once Dolph has been assigned, there are two ways to assign Francisco. Once Dolph and Francisco have been assigned, there are four ways to fill the remaining office. By the Multiplication Principle, there are $3 \cdot 2 \cdot 4 = 24$ possibilities. ◀

**Example 6.1.12** We revisit Example 6.1.3(c), which asked how many strings of length 4 that do not begin with the letter $B$ can be formed using the letters $ABCDE$ (repetitions not allowed)?

**SOLUTION** We could have solved this problem using a direct approach by counting the number of strings that begin with $A$, with $C$, with $D$, and with $E$, and then, using the Addition Principle, summed these values. Since 24 strings begin with $A$ [see the solution to Example 6.1.3(b) with $B$ replaced by $A$], and 24 strings begin with $C$, and so on, we obtain

$$24 + 24 + 24 + 24 = 96.$$

However, the solution given in Example 6.1.3(c), which subtracted the number of strings that *do* start with $B$ from the total number of strings, was easier. It would have been *much* easier if more letters had been involved or if the numbers in the sum were not identical so that each involved a distinct calculation.

A variation of the direct approach might also have been used: Choose the first letter (4 ways—choose $A$, $C$, $D$, or $E$); choose the second letter (4 ways—any of the remaining letters); choose the third letter (3 ways—any of the remaining letters); and choose the fourth letter (2 ways—either of the remaining letters). By the Multiplication Principle, there are

$$4 \cdot 4 \cdot 3 \cdot 2 = 96$$

strings that do not begin with the letter $B$. ◀

## Inclusion-Exclusion Principle

Suppose that we want to count the number of eight-bit strings that start 10 or end 011 or both. Let $X$ denote the set of eight-bit strings that start 10 and $Y$ denote the set of eight-bit strings that end 011. The goal then is to compute $|X \cup Y|$. We *cannot* use the Addition Principle and add $|X|$ and $|Y|$ to compute $|X \cup Y|$ because the Addition Principle requires $X$ and $Y$ to be disjoint. Here $X$ and $Y$ are not disjoint; for example, $10111011 \in X \cap Y$. The **Inclusion-Exclusion Principle** generalizes the Addition Principle by giving a formula to compute the number of elements in a union without requiring the sets to be pairwise disjoint.

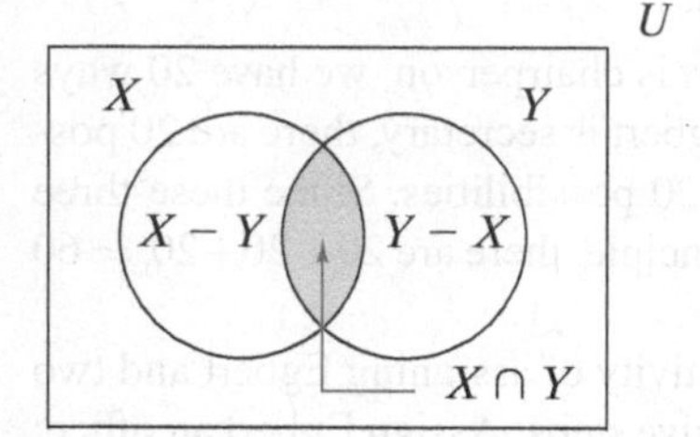


**Figure 6.1.3** $|X|$ counts the number of elements in $X - Y$ and $X \cap Y$, and $|Y|$ counts the number of elements in $Y - X$ and $X \cap Y$. Since $|X| + |Y|$ double-counts the elements in $X \cap Y$, $|X| + |Y| = |X \cup Y| + |X \cap Y|$.

Continuing the discussion in the previous paragraph, suppose that we compute $|X| + |Y|$. We have counted the elements in $X - Y$ (eight-bit strings that start 10 but do not end 011) once and the elements in $Y - X$ (eight-bit strings that end 011 but do not start 10) once, but we have counted the elements in $X \cap Y$ (eight-bit strings that start 10 and end 011) twice (see Figure 6.1.3). Thus if we subtract $|X \cap Y|$ from $|X| + |Y|$, to compensate for the double-counting, we will obtain the number of strings in $X \cup Y$; that is,

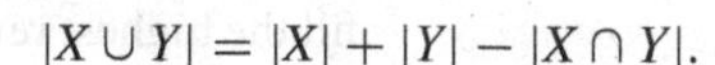

$$|X \cup Y| = |X| + |Y| - |X \cap Y|.$$

Arguing as in Example 6.1.7, we find that $|X| = 2^6$, $|Y| = 2^5$, and $|X \cap Y| = 2^3$. Therefore, the number of eight-bit strings that start 10 or end 011 or both is equal to

$$|X \cup Y| = |X| + |Y| - |X \cap Y| = 2^6 + 2^5 - 2^3.$$

We state the Inclusion-Exclusion Principle for two sets as Theorem 6.1.13.

**Theorem 6.1.13** **Inclusion-Exclusion Principle for Two Sets**

If $X$ and $Y$ are finite sets, then

$$|X \cup Y| = |X| + |Y| - |X \cap Y|.$$

**Proof** Since $X = (X - Y) \cup (X \cap Y)$ and $X - Y$ and $X \cap Y$ are disjoint, by the Addition Principle

$$|X| = |X - Y| + |X \cap Y|. \tag{6.1.1}$$

Similarly,

$$|Y| = |Y - X| + |X \cap Y|. \tag{6.1.2}$$

Since $X \cup Y = (X - Y) \cup (X \cap Y) \cup (Y - X)$ and $X - Y$, $X \cap Y$, and $Y - X$ are pairwise disjoint, by the Addition Principle

$$|X \cup Y| = |X - Y| + |X \cap Y| + |Y - X|. \tag{6.1.3}$$

Combining equations (6.1.1)–(6.1.3), we obtain

$$|X| + |Y| = |X - Y| + |X \cap Y| + |Y - X| + |X \cap Y| = |X \cup Y| + |X \cap Y|.$$

Subtracting $|X \cap Y|$ from both sides of the preceding equation gives the desired result. ◀

**Example 6.1.14** A committee composed of Alice, Ben, Connie, Dolph, Egbert, and Francisco is to select a chairperson, secretary, and treasurer. How many selections are there in which either Alice or Dolph or both are officers?

**SOLUTION** Let $X$ denote the set of selections in which Alice is an officer and let $Y$ denote the set of selections in which Dolph is an officer. We must compute $|X \cup Y|$. Since $X$ and $Y$ are *not* disjoint (both Alice and Dolph could be officers), we cannot use the Addition Principle. Instead we use the Inclusion-Exclusion Principle.

We first count the number of selections in which Alice is an officer. Alice can be assigned an office in three ways, the highest remaining office can be filled in five ways, and the last office can be filled in four ways. Thus the number of selections in which Alice is an officer is $3 \cdot 5 \cdot 4 = 60$, that is, $|X| = 60$. Similarly, the number of selections in which Dolph is an officer is 60, that is, $|Y| = 60$.

Now $X \cap Y$ is the set of selections in which both Alice and Dolph are officers. Alice can be assigned an office in three ways, Dolph can be assigned an office in two ways, and the last office can be filled in four ways. Thus the number of selections in which both Alice and Dolph are officers is $3 \cdot 2 \cdot 4 = 24$, that is, $|X \cap Y| = 24$.

The Inclusion-Exclusion Principle tells us that

$$|X \cup Y| = |X| + |Y| - |X \cap Y| = 60 + 60 - 24 = 96.$$

Thus there are 96 selections in which either Alice or Dolph or both are officers. ◀

The name "inclusion-exclusion" in Theorem 6.1.13 results from *including* $|X \cap Y|$ twice when computing $|X \cup Y|$ as $|X| + |Y|$ and then *excluding* it by subtracting $|X \cap Y|$ from $|X| + |Y|$.

We leave the Inclusion-Exclusion Principle for three or more sets to the exercises (see Exercises 98–105).

## 6.1 Problem-Solving Tips

The key to solving problems in this section is determining when to use the Multiplication Principle and when to use the Addition Principle. Use the Multiplication Principle when using a step-by-step process to *construct an activity*. For example, to construct a dinner from Kay's Quick Lunch menu (Figure 6.1.1) consisting of one appetizer, one main course, and one beverage, we use a three-step process:

1. Choose one appetizer.
2. Choose one main course.
3. Choose one beverage.

The number of different possible activities is the product of the number of ways each step can be done. Here we can select one appetizer in 2 ways, one main course in 3 ways, and one beverage in 4 ways. Thus, the number of dinners is $2 \cdot 3 \cdot 4 = 24$.

Use the Addition Principle when you want to count the number of elements in a set and you can divide the set into nonoverlapping subsets. Suppose, for example, that we want to count the total number of items available at Kay's Quick Lunch. Since there are 2 appetizers, 3 main course items, and 4 beverages, and no item belongs to two categories, the total number of items available is

$$2+3+4=9.$$

Notice the difference between the two examples. To construct a dinner consisting of one appetizer, one main course, and one beverage at Kay's Quick Lunch, we use a step-by-step process. The size of the set of dinners is *not* counted by dividing the set of dinners into nonoverlapping subsets. To count the number of dinners, we use the Multiplication Principle. To count the number of items available at Kay's Quick Lunch, we just sum the number of items in each category since dividing the items by category naturally splits them into nonoverlapping subsets. We are *not* counting the individual items available by constructing them using a step-by-step process. To count the total number of items available, we use the Addition Principle.

To count the number of items *not* having a given property, it is sometimes easier to count the number of items that *do* have the property and subtract that value from the total number of items [see Example 6.1.3(c)].

The Inclusion-Exclusion Principle (Theorem 6.1.13) is a variant of the Addition Principle that can be used when the sets involved are *not* pairwise disjoint.

## 6.1 Review Exercises

1. State the Multiplication Principle and give an example of its use.
2. State the Addition Principle and give an example of its use.
3. State the Inclusion-Exclusion Principle for two sets and give an example of its use.

## 6.1 Exercises

*Use the Multiplication Principle to solve Exercises 1–12.*

1. How many dinners at Kay's Quick Lunch (Figure 6.1.1) consist of one appetizer and one beverage?
2. How many dinners at Kay's Quick Lunch (Figure 6.1.1) consist of an optional appetizer, one main course, and an optional beverage?
3. How many dinners at Kay's Quick Lunch (Figure 6.1.1) consist of one appetizer, one main course, and an optional beverage?
4. A man has eight shirts, four pairs of pants, and five pairs of shoes. How many different outfits are possible?
5. The Braille system of representing characters was developed early in the nineteenth century by Louis Braille. The characters, used by the blind, consist of raised dots. The positions for the dots are selected from two vertical columns of three dots each. At least one raised dot must be present. How many distinct Braille characters are possible?
6. The options available on a particular model of a car are five interior colors, six exterior colors, two types of seats, three types of engines, and three types of radios. How many different possibilities are available to the consumer?
7. Two dice are rolled, one blue and one red. How many outcomes are possible?
8. A restaurant chain advertised a special in which a customer could choose one of five appetizers, one of 14 main dishes, and one of three desserts. The ad said that there were 210 possible dinners. Was the ad correct? Explain.
9. How many different car license plates can be constructed if the licenses contain three letters followed by two digits if repetitions are allowed? if repetitions are not allowed?

*Exercises 10–12 ask about strings of length 5 formed using the letters ABCDEFG without repetitions.*

10. How many strings begin with the letter $F$ and end with the letter $A$?
11. How many strings begin with the letter $F$ and do *not* end with $EB$ in that order?
12. How many strings contain $CEG$ together in that order?

*Use the Addition Principle to solve Exercises 13–24.*

13. Three departmental committees have 6, 12, and 9 members with no overlapping membership. In how many ways can these committees send one member to meet with the president?
14. In how many ways can a diner choose one item from among the appetizers and main courses at Kay's Quick Lunch (Figure 6.1.1)?

**15.** In how many ways can a diner choose one item from among the appetizers and beverages at Kay's Quick Lunch (Figure 6.1.1)?

**16.** How many times are the print statements executed?

for $i = 1$ to $m$
  $println(i)$
for $j = 1$ to $n$
  $println(j)$

**17.** How many times is the print statement executed?

for $i = 1$ to $m$
  for $j = 1$ to $n$
    $println(i, j)$

**18.** Given that there are 32 eight-bit strings that begin 101 and 16 eight-bit strings that begin 1101, how many eight-bit strings begin either 101 or 1101?

**19.** Two dice are rolled, one blue and one red. How many outcomes give the sum of 2 or the sum of 12?

**20.** A committee composed of Morgan, Tyler, Max, and Leslie is to select a president and secretary. How many selections are there in which Tyler is president or not an officer?

**21.** A committee composed of Morgan, Tyler, Max, and Leslie is to select a president and secretary. How many selections are there in which Max is president or secretary?

*Exercises 22–24 ask about strings of length 5 formed using the letters ABCDEFG without repetitions.*

**22.** How many strings begin with *AC* or *DB* in that order?

**23.** How many strings do *not* begin with *AB* (in that order) or *D*?

**24.** How many strings contain *B* and *D* together in either order (i.e., *BD* or *DB*)?

**25.** Comment on the following item from the *New York Times*:

> Big pickups also appeal because of the seemingly infinite ways they can be personalized; you need the math skills of Will Hunting to total the configurations. For starters, there are 32 combinations of cabs (standard, Club Cab, Quad Cab), cargo beds (6.5 or 8 feet), and engines (3.9-liter V6, 5.2-liter V8, 5.9-liter V8, 5.9-liter turbo-diesel inline 6, 8-liter V10).

*In Exercises 26–33, two dice are rolled, one blue and one red.*

**26.** How many outcomes give the sum of 4?

**27.** How many outcomes are doubles? (A double occurs when both dice show the same number.)

**28.** How many outcomes give the sum of 7 or the sum of 11?

**29.** How many outcomes have the blue die showing 2?

**30.** How many outcomes have exactly one die showing 2?

**31.** How many outcomes have at least one die showing 2?

**32.** How many outcomes have neither die showing 2?

**33.** How many outcomes give an even sum?

*In Exercises 34–36, suppose there are 10 roads from Oz to Mid Earth and five roads from Mid Earth to Fantasy Island.*

**34.** How many routes are there from Oz to Fantasy Island passing through Mid Earth?

**35.** How many round-trips are there of the form Oz–Mid Earth–Fantasy Island–Mid Earth–Oz?

**36.** How many round-trips are there of the form Oz–Mid Earth–Fantasy Island–Mid Earth–Oz in which on the return trip we do not reverse the original route from Oz to Fantasy Island?

**37.** How many eight-bit strings begin 1100?

**38.** How many eight-bit strings begin and end with 1?

**39.** How many eight-bit strings have either the second or the fourth bit 1 (or both)?

**40.** How many eight-bit strings have exactly one 1?

**41.** How many eight-bit strings have exactly two 1's?

**42.** How many eight-bit strings have at least one 1?

**43.** How many eight-bit strings read the same from either end? (An example of such an eight-bit string is 01111110. Such strings are called *palindromes*.)

*In Exercises 44–49, a six-person committee composed of Alice, Ben, Connie, Dolph, Egbert, and Francisco is to select a chairperson, secretary, and treasurer.*

**44.** How many selections exclude Connie?

**45.** How many selections are there in which neither Ben nor Francisco is an officer?

**46.** How many selections are there in which both Ben and Francisco are officers?

**47.** How many selections are there in which Dolph is an officer and Francisco is not an officer?

**48.** How many selections are there in which either Dolph is chairperson or he is not an officer?

**49.** How many selections are there in which Ben is either chairperson or treasurer?

*In Exercises 50–57, the letters ABCDE are to be used to form strings of length 3.*

**50.** How many strings can be formed if we allow repetitions?

**51.** How many strings can be formed if we do not allow repetitions?

**52.** How many strings begin with *A*, allowing repetitions?

**53.** How many strings begin with *A* if repetitions are not allowed?

**54.** How many strings do not contain the letter *A*, allowing repetitions?

**55.** How many strings do not contain the letter *A* if repetitions are not allowed?

**56.** How many strings contain the letter $A$, allowing repetitions?

**57.** How many strings contain the letter $A$ if repetitions are not allowed?

*Exercises 58–68 refer to the integers from 5 to 200, inclusive.*

**58.** How many numbers are there?

**59.** How many are even?

**60.** How many are odd?

**61.** How many are divisible by 5?

**62.** How many are greater than 72?

**63.** How many consist of distinct digits?

**64.** How many contain the digit 7?

**65.** How many do not contain the digit 0?

**66.** How many are greater than 101 and do not contain the digit 6?

**67.** How many have the digits in strictly increasing order? (Examples are 13, 147, 8.)

**68.** How many are of the form $xyz$, where $0 \neq x < y$ and $y > z$?

**69.** (a) In how many ways can the months of the birthdays of five people be distinct?

(b) How many possibilities are there for the months of the birthdays of five people?

(c) In how many ways can at least two people among five have their birthdays in the same month?

*Exercises 70–74 refer to a set of five distinct computer science books, three distinct mathematics books, and two distinct art books.*

**70.** In how many ways can these books be arranged on a shelf?

**71.** In how many ways can these books be arranged on a shelf if all five computer science books are on the left and both art books are on the right?

**72.** In how many ways can these books be arranged on a shelf if all five computer science books are on the left?

**73.** In how many ways can these books be arranged on a shelf if all books of the same discipline are grouped together?

**★74.** In how many ways can these books be arranged on a shelf if the two art books are not together?

**75.** In some versions of FORTRAN, an identifier consists of a string of one to six alphanumeric characters beginning with a letter. (An *alphanumeric character* is one of A to Z or 0 to 9.) How many valid FORTRAN identifiers are there?

**76.** If $X$ is an $n$-element set and $Y$ is an $m$-element set, how many functions are there from $X$ to $Y$?

**★77.** There are 10 copies of one book and one copy each of 10 other books. In how many ways can we select 10 books?

**78.** How many terms are there in the expansion of

$$(x + y)(a + b + c)(e + f + g)(h + i)?$$

**★79.** How many subsets of a $(2n+1)$-element set have $n$ elements or less?

**80.** A Class B internet address, used for medium-sized networks, is a bit string of length 32. The first bits are 10 (to identify it as a Class B address). The netid is given by the next 14 bits, which identifies the network. The hostid is given by the remaining 16 bits, which identifies the computer interface. The hostid must not consist of all 0's or all 1's. How many Class B addresses are available?

**81.** A Class C internet address, used for small networks, is a bit string of length 32. The first bits are 110 (to identify it as a Class C address). The netid is given by the next 21 bits, which identifies the network. The hostid is given by the remaining 8 bits, which identifies the computer interface. The hostid must not consist of all 0's or all 1's. How many Class C addresses are available?

**82.** Given that the IPv4 internet address of a computer interface is either Class A, Class B, or Class C, how many IPv4 Internet addresses are available?

**83.** How many symmetric relations are there on an $n$-element set?

**84.** How many antisymmetric relations are there on an $n$-element set?

**85.** How many reflexive and symmetric relations are there on an $n$-element set?

**86.** How many reflexive and antisymmetric relations are there on an $n$-element set?

**87.** How many symmetric and antisymmetric relations are there on an $n$-element set?

**88.** How many reflexive, symmetric, and antisymmetric relations are there on an $n$-element set?

**89.** How many truth tables are there for an $n$-variable function?

**90.** How many binary operators are there on $\{1, 2, \ldots, n\}$?

**91.** How many commutative binary operators are there on $\{1, 2, \ldots, n\}$?

*Use the Inclusion-Exclusion Principle (Theorem 6.1.13) to solve Exercises 92–97.*

**92.** How many eight-bit strings either begin with 100 or have the fourth bit 1 or both?

**93.** How many eight-bit strings either start with a 1 or end with a 1 or both?

*In Exercises 94 and 95, a six-person committee composed of Alice, Ben, Connie, Dolph, Egbert, and Francisco is to select a chairperson, secretary, and treasurer.*

**94.** How many selections are there in which either Ben is chairperson or Alice is secretary or both?

**95.** How many selections are there in which either Connie is chairperson or Alice is an officer or both?

**96.** Two dice are rolled, one blue and one red. How many outcomes have either the blue die 3 or an even sum or both?

**97.** How many integers from 1 to 10,000, inclusive, are multiples of 5 or 7 or both?

**98.** Prove the Inclusion-Exclusion Principle for three finite sets:

$$|X \cup Y \cup Z| = |X| + |Y| + |Z| - |X \cap Y| - |X \cap Z| - |Y \cap Z| + |X \cap Y \cap Z|.$$

*Hint*: Write the Inclusion-Exclusion Principle for two finite sets as

$$|A \cup B| = |A| + |B| - |A \cap B|$$

and let $A = X$ and $B = Y \cup Z$.

**99.** In a group of 191 students, 10 are taking French, business, and music; 36 are taking French and business; 20 are taking French and music; 18 are taking business and music; 65 are taking French; 76 are taking business; and 63 are taking music. Use the Inclusion-Exclusion Principle for three finite sets (see Exercise 98) to determine how many students are not taking any of the three courses.

**100.** Use the Inclusion-Exclusion Principle for three finite sets (see Exercise 98) to solve the problem in Example 1.1.21.

**101.** Use the Inclusion-Exclusion Principle for three finite sets (see Exercise 98) to solve Exercise 66, Section 1.1.

**102.** Use the Inclusion-Exclusion Principle for three finite sets (see Exercise 98) to compute the number of integers between 1 and 10,000, inclusive, that are multiples of 3 or 5 or 11 or any combination thereof.

**★103.** Use Mathematical Induction to prove the general Inclusion-Exclusion Principle for finite sets $X_1, X_2, \ldots, X_n$:

$$|X_1 \cup X_2 \cup \cdots \cup X_n| = \sum_{1 \le i \le n} |X_i| - \sum_{1 \le i < j \le n} |X_i \cap X_j| + \sum_{1 \le i < j < k \le n} |X_i \cap X_j \cap X_k| - \cdots + (-1)^{n+1} |X_1 \cap X_2 \cap \cdots \cap X_n|.$$

*Hint*: In the Inductive Step, refer to the hint in Exercise 98.

**104.** Using the previous exercise, write the Inclusion-Exclusion Principle for four finite sets.

**105.** How many integers between 1 and 10,000, inclusive, are multiples of 3 or 5 or 11 or 13 or any combination thereof?

# Problem-Solving Corner: Counting

## Problem

Find the number of ordered triples of sets $X_1, X_2, X_3$ satisfying

$$X_1 \cup X_2 \cup X_3 = \{1, 2, 3, 4, 5, 6, 7, 8\}$$
$$\text{and} \quad X_1 \cap X_2 \cap X_3 = \varnothing.$$

By *ordered triple,* we mean that the order of the sets $X_1$, $X_2, X_3$ is taken into account. For example, the triples

$$\{1, 2, 3\}, \quad \{1, 4, 8\}, \quad \{2, 5, 6, 7\}$$

and

$$\{1, 4, 8\}, \quad \{1, 2, 3\}, \quad \{2, 5, 6, 7\}$$

are considered distinct.

## Attacking the Problem

It would be nice to begin by enumerating triples, but there are so many it would be hard to gain much insight from staring at a few triples. Let's simplify the problem by making it smaller. Let's replace

$$\{1, 2, 3, 4, 5, 6, 7, 8\}$$

by $\{1\}$. What could be simpler than $\{1\}$? (Well, maybe $\varnothing$, but that's too simple!) We can now enumerate all ordered triples of sets $X_1, X_2, X_3$ satisfying $X_1 \cup X_2 \cup X_3 = \{1\}$ and $X_1 \cap X_2 \cap X_3 = \varnothing$. We must put 1 in at least one of the sets $X_1, X_2, X_3$ (so that the union will be $\{1\}$), but we must not put 1 in all three of the sets $X_1, X_2, X_3$ (otherwise, the intersection would not be empty). Thus 1 will be in exactly one or two of the sets $X_1, X_2, X_3$. The complete list of ordered triples is as follows:

$$\begin{aligned}
X_1 &= \{1\}, & X_2 &= \varnothing, & X_3 &= \varnothing;\\
X_1 &= \varnothing, & X_2 &= \{1\}, & X_3 &= \varnothing;\\
X_1 &= \varnothing, & X_2 &= \varnothing, & X_3 &= \{1\};\\
X_1 &= \{1\}, & X_2 &= \{1\}, & X_3 &= \varnothing;\\
X_1 &= \{1\}, & X_2 &= \varnothing, & X_3 &= \{1\};\\
X_1 &= \varnothing, & X_2 &= \{1\}, & X_3 &= \{1\}.
\end{aligned}$$

Thus there are six ordered triples of sets $X_1, X_2, X_3$ satisfying

$$X_1 \cup X_2 \cup X_3 = \{1\} \quad \text{and} \quad X_1 \cap X_2 \cap X_3 = \varnothing.$$

Let's step up one level and enumerate all ordered triples of sets $X_1, X_2, X_3$ satisfying $X_1 \cup X_2 \cup X_3 = \{1, 2\}$ and $X_1 \cap X_2 \cap X_3 = \varnothing$. As before, we must put 1 in at least one of the sets $X_1, X_2, X_3$ (so that 1 will be in the union), but we must not put 1 in all three of the sets $X_1, X_2, X_3$ (otherwise, the intersection would not be empty). This time we must also put 2 in at least one of the sets $X_1, X_2, X_3$ (so that 2 will also be in the union), but we must not put 2 in all three of the sets $X_1, X_2, X_3$ (otherwise, the intersection would not be empty). Thus each of 1 and 2 will be in exactly one or two of the sets

| *1 is in* | *2 is in* | *1 is in* | *2 is in* | *1 is in* | *2 is in* |
|---|---|---|---|---|---|
| $X_1$ | $X_1$ | $X_1$ | $X_2$ | $X_1$ | $X_3$ |
| $X_2$ | $X_1$ | $X_2$ | $X_2$ | $X_2$ | $X_3$ |
| $X_3$ | $X_1$ | $X_3$ | $X_2$ | $X_3$ | $X_3$ |
| $X_1, X_2$ | $X_1$ | $X_1, X_2$ | $X_2$ | $X_1, X_2$ | $X_3$ |
| $X_1, X_3$ | $X_1$ | $X_1, X_3$ | $X_2$ | $X_1, X_3$ | $X_3$ |
| $X_2, X_3$ | $X_1$ | $X_2, X_3$ | $X_2$ | $X_2, X_3$ | $X_3$ |
| $X_1$ | $X_1, X_2$ | $X_1$ | $X_1, X_3$ | $X_1$ | $X_2, X_3$ |
| $X_2$ | $X_1, X_2$ | $X_2$ | $X_1, X_3$ | $X_2$ | $X_2, X_3$ |
| $X_3$ | $X_1, X_2$ | $X_3$ | $X_1, X_3$ | $X_3$ | $X_2, X_3$ |
| $X_1, X_2$ | $X_1, X_2$ | $X_1, X_2$ | $X_1, X_3$ | $X_1, X_2$ | $X_2, X_3$ |
| $X_1, X_3$ | $X_1, X_2$ | $X_1, X_3$ | $X_1, X_3$ | $X_1, X_3$ | $X_2, X_3$ |
| $X_2, X_3$ | $X_1, X_2$ | $X_2, X_3$ | $X_1, X_3$ | $X_2, X_3$ | $X_2, X_3$ |

$X_1, X_2, X_3$. We enumerate the sets in a systematic way so that we can recognize any patterns that appear. The complete list of ordered triples is shown in the table at the top of this page. For example, the top left entry, $X_1\, X_1$, specifies that 1 is in $X_1$ and 2 is in $X_1$; therefore, this entry gives the ordered triple

$$X_1 = \{1, 2\}, \quad X_2 = \varnothing, \quad X_3 = \varnothing.$$

As shown, there are 36 ordered triples of sets $X_1, X_2, X_3$ satisfying

$$X_1 \cup X_2 \cup X_3 = \{1, 2\}$$
$$\text{and} \quad X_1 \cap X_2 \cap X_3 = \varnothing.$$

We see that there are six ways to assign 1 to the sets $X_1, X_2, X_3$, which accounts for six lines per block. Similarly, there are six ways to assign 2 to the sets $X_1, X_2, X_3$, which accounts for six blocks.

Before reading on, can you guess how many ordered triples of sets $X_1, X_2, X_3$ satisfy

$$X_1 \cup X_2 \cup X_3 = \{1, 2, 3\}$$
$$\text{and} \quad X_1 \cap X_2 \cap X_3 = \varnothing?$$

The pattern has emerged. If $X = \{1, 2, \ldots, n\}$, there are six ways to assign each of $1, 2, \ldots, n$ to the sets $X_1, X_2, X_3$. By the Multiplication Principle, the number of ordered triples is $6^n$.

## Finding Another Solution

We have just found a solution to the problem by starting with a simpler problem and then discovering and justifying the pattern that emerged.

Another approach is to look for a similar problem and imitate its solution. The problem of Example 6.1.6 is similar to the one at hand in that it also is a counting problem that deals with sets:

> Let $X$ be an $n$-element set. How many ordered pairs $(A, B)$ satisfy $A \subseteq B \subseteq X$?

(At this point, it would be a good idea to go back and reread Example 6.1.6.) The solution given in Example 6.1.6 counts the number of ways to assign elements of $X$ to exactly one of the sets $A$, $B - A$, or $X - B$.

We can solve our problem by taking a similar approach. Each element of $X$ is in exactly one of

$$\overline{X_1} \cap X_2 \cap X_3, \quad X_1 \cap \overline{X_2} \cap X_3, \quad X_1 \cap X_2 \cap \overline{X_3},$$
$$\overline{X_1} \cap \overline{X_2} \cap X_3, \quad \overline{X_1} \cap X_2 \cap \overline{X_3}, \quad X_1 \cap \overline{X_2} \cap \overline{X_3}.$$

Since each member of $X$ can be assigned to one of these sets in six ways, by the Multiplication Principle, the number of ordered triples is $6^8$.

Notice that while this approach to solving the problem is different than that of the preceding section, the final argument is essentially the same.

## Formal Solution

Each element in $X$ is in exactly one of

$$Y_1 = \overline{X_1} \cap X_2 \cap X_3, \quad Y_2 = X_1 \cap \overline{X_2} \cap X_3,$$
$$Y_3 = X_1 \cap X_2 \cap \overline{X_3}, \quad Y_4 = \overline{X_1} \cap \overline{X_2} \cap X_3,$$
$$Y_5 = \overline{X_1} \cap X_2 \cap \overline{X_3}, \quad Y_6 = X_1 \cap \overline{X_2} \cap \overline{X_3}.$$

We can construct an ordered triple by the following eight-step process: Choose $j$, $1 \le j \le 6$, and put 1 in $Y_j$; choose $j$, $1 \le j \le 6$, and put 2 in $Y_j$; ...; choose $j$, $1 \le j \le 6$, and put 8 in $Y_j$. For example, to construct the triple

$$\{1, 2, 3\}, \quad \{1, 4, 8\}, \quad \{2, 5, 6, 7\},$$

we first choose $j = 3$ and put 1 in $Y_3 = X_1 \cap X_2 \cap \overline{X_3}$. Next, we choose $j = 2$ and put 2 in $Y_2 = X_1 \cap \overline{X_2} \cap X_3$. The remaining choices for $j$ are $j = 6, 5, 4, 4, 4, 5$.

Each choice for $j$ can be made in six ways. By the Multiplication Principle, the number of ordered triples is

$$6 \cdot 6 \cdot 6 \cdot 6 \cdot 6 \cdot 6 \cdot 6 \cdot 6 = 6^8 = 1{,}679{,}616.$$

**Summary of Problem-Solving Techniques**

- Replace the original problem with a simpler problem. One way to do this is to reduce the size of the original problem.
- Directly enumerate the items to be counted.
- Enumerate items systematically so that patterns emerge.
- Look for patterns.
- Look for a similar problem and imitate its solution.

# 6.2 Permutations and Combinations

**Go Online**
For more on permutations and combinations, see bit.ly/2PNvtEt

Four candidates, Zeke, Yung, Xeno, and Wilma, are running for the same office. So that the positions of the names on the ballot will not influence the voters, it is necessary to print ballots with the names listed in every possible order. How many distinct ballots will there be?

We can use the Multiplication Principle. A ballot can be constructed in four successive steps: Select the first name to be listed; select the second name to be listed; select the third name to be listed; select the fourth name to be listed. The first name can be selected in four ways. Once the first name has been selected, the second name can be selected in three ways. Once the second name has been selected, the third name can be selected in two ways. Once the third name has been selected, the fourth name can be selected in one way. By the Multiplication Principle, the number of ballots is $4 \cdot 3 \cdot 2 \cdot 1 = 24$.

An ordering of objects, such as the names on the ballot, is called a **permutation.**

**Definition 6.2.1 ▶** A *permutation* of $n$ distinct elements $x_1, \ldots, x_n$ is an ordering of the $n$ elements $x_1, \ldots, x_n$. ◀

**Example 6.2.2** There are six permutations of three elements. If the elements are denoted $A$, $B$, $C$, the six permutations are

$$ABC, \quad ACB, \quad BAC, \quad BCA, \quad CAB, \quad CBA. \quad ◀$$

We found that there are 24 ways to order four candidates on a ballot; thus there are 24 permutations of four objects. The method that we used to count the number of distinct ballots containing four names may be used to derive a formula for the number of permutations of $n$ elements.

The proof of the following theorem for $n = 4$ is illustrated in Figure 6.2.1.

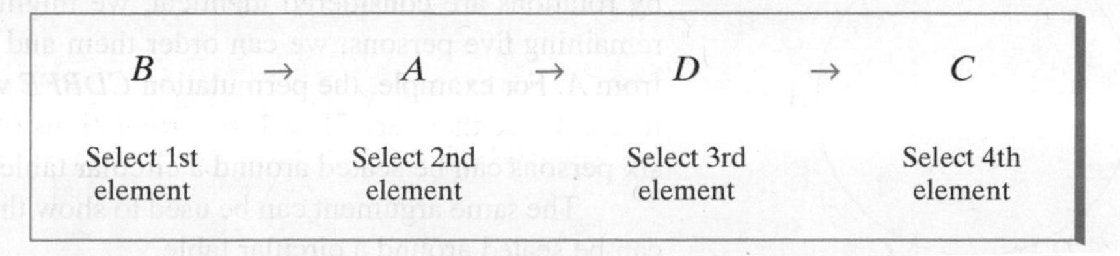


**Figure 6.2.1** The proof of Theorem 6.2.3 for $n = 4$. A permutation of $ABCD$ is constructed by successively selecting the first element, then the second element, then the third element, and, finally, the fourth element.

**Theorem 6.2.3** There are $n!$ permutations of $n$ elements.

**Proof** We use the Multiplication Principle. A permutation of $n$ elements can be constructed in $n$ successive steps: Select the first element; select the second element; . . . ; select the last element. The first element can be selected in $n$ ways. Once the first element has been selected, the second element can be selected in $n - 1$ ways. Once the second element has been selected, the third element can be selected in $n - 2$ ways, and so on. By the Multiplication Principle, there are

$$n(n-1)(n-2)\cdots 2\cdot 1 = n!$$

permutations of $n$ elements. ◀

**Example 6.2.4** There are

$$10! = 10\cdot 9\cdot 8\cdot 7\cdot 6\cdot 5\cdot 4\cdot 3\cdot 2\cdot 1 = 3{,}628{,}800$$

permutations of 10 elements. ◀

**Example 6.2.5** How many permutations of the letters *ABCDEF* contain the substring *DEF*?

*DEF* | *A* | *B* | *C*

**Figure 6.2.2** Four tokens to permute.

**SOLUTION** To guarantee the presence of the pattern *DEF* in the substring, these three letters must be kept together in this order. The remaining letters, *A*, *B*, and *C*, can be placed arbitrarily. We can think of constructing permutations of the letters *ABCDEF* that contain the pattern *DEF* by permuting four tokens—one labeled *DEF* and the others labeled *A*, *B*, and *C* (see Figure 6.2.2). By Theorem 6.2.3, there are 4! permutations of four objects. Thus the number of permutations of the letters *ABCDEF* that contain the substring *DEF* is $4! = 24$. ◀

**Example 6.2.6** How many permutations of the letters *ABCDEF* contain the letters *DEF* together in any order?

**SOLUTION** We can solve the problem by a two-step procedure: Select an ordering of the letters *DEF*; construct a permutation of *ABCDEF* containing the given ordering of the letters *DEF*. By Theorem 6.2.3, the first step can be done in $3! = 6$ ways and, according to Example 6.2.5, the second step can be done in 24 ways. By the Multiplication Principle, the number of permutations of the letters *ABCDEF* containing the letters *DEF* together in any order is $6\cdot 24 = 144$. ◀

**Example 6.2.7** In how many ways can six persons be seated around a circular table? If a seating is obtained from another seating by having everyone move $n$ seats clockwise, the seatings are considered identical.

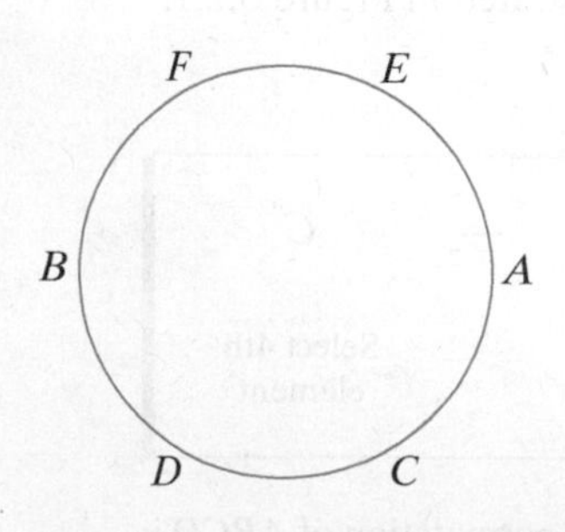


**SOLUTION** Let us denote the persons as *A*, *B*, *C*, *D*, *E*, and *F*. Since seatings obtained by rotations are considered identical, we might as well seat *A* arbitrarily. To seat the remaining five persons, we can order them and then seat them in this order clockwise from *A*. For example, the permutation *CDBFE* would define the seating in the adjacent figure. Since there are $5! = 120$ permutations of five elements, there are 120 ways that six persons can be seated around a circular table.

The same argument can be used to show that there are $(n-1)!$ ways that $n$ persons can be seated around a circular table. ◀

Sometimes we want to consider an ordering of $r$ elements selected from $n$ available elements. Such an ordering is called an ***r*-permutation.**

**Definition 6.2.8 ▶** An *r-permutation* of $n$ (distinct) elements $x_1, \ldots, x_n$ is an ordering of an $r$-element subset of $\{x_1, \ldots, x_n\}$. The number of $r$-permutations of a set of $n$ distinct elements is denoted $P(n, r)$. ◀

**Example 6.2.9** Examples of 2-permutations of $a$, $b$, $c$ are $ab$, $ba$, and $ca$. ◀

If $r = n$ in Definition 6.2.8, we obtain an ordering of all $n$ elements. Thus an $n$-permutation of $n$ elements is what we previously called simply a permutation. Theorem 6.2.3 tells us that $P(n, n) = n!$. The number $P(n, r)$ of $r$-permutations of a $n$-element set when $r < n$ may be derived as in the proof of Theorem 6.2.3. The proof of the theorem for $n = 6$ and $r = 3$ is illustrated in Figure 6.2.3.

| $C$ | $\to$ | $A$ | $\to$ | $E$ |
|---|---|---|---|---|
| Select 1st element | | Select 2nd element | | Select 3rd element |

**Figure 6.2.3** The proof of Theorem 6.2.10 for $n = 6$ and $r = 3$. An $r$-permutation of $ABCDEF$ is constructed by successively selecting the first element, then the second element, and, finally, the third element.

**Theorem 6.2.10** The number of $r$-permutations of a set of $n$ distinct objects is

$$P(n, r) = n(n-1)(n-2)\cdots(n-r+1) = \frac{n!}{(n-r)!} \qquad r \leq n.$$

**Proof** We are to count the number of ways to order $r$ elements selected from an $n$-element set. The first element can be selected in $n$ ways. Once the first element has been selected, the second element can be selected in $n - 1$ ways. We continue selecting elements until, having selected the $(r - 1)$st element, we select the $r$th element. This last element can be chosen in $n - r + 1$ ways. By the Multiplication Principle, the number of $r$-permutations of a set of $n$ distinct objects is

$$n(n-1)(n-2)\cdots(n-r+1) = \frac{[n(n-1)\cdots(n-r+1)][(n-r)(n-r-1)\cdots 2\cdot 1]}{(n-r)(n-r-1)\cdots 2\cdot 1} = \frac{n!}{(n-r)!}.$$ ◀

**Example 6.2.11** According to Theorem 6.2.10, the number of 2-permutations of $X = \{a, b, c\}$ is $P(3, 2) = 3 \cdot 2 = 6$. These six 2-permutations are $ab$, $ac$, $ba$, $bc$, $ca$, $cb$. ◀

**Example 6.2.12** In how many ways can we select a chairperson, vice-chairperson, secretary, and treasurer from a group of 10 persons?

**SOLUTION** We need to count the number of orderings of four persons selected from a group of 10, since an ordering picks (uniquely) a chairperson (first pick), a vice-chairperson (second pick), a secretary (third pick), and a treasurer (fourth pick). By Theorem 6.2.10, the solution is

$$P(10, 4) = 10 \cdot 9 \cdot 8 \cdot 7 = 5040$$

or

$$P(10, 4) = \frac{10!}{(10-4)!} = \frac{10!}{6!}.$$

◀

**Example 6.2.13** In how many ways can seven distinct Martians and five distinct Jovians wait in line if no two Jovians stand together?

**SOLUTION** We can line up the Martians and Jovians by a two-step process: Line up the Martians; line up the Jovians. The Martians can line up in $7! = 5040$ ways. Once we have lined up the Martians (e.g., in positions $M_1$–$M_7$), since no two Jovians can stand together, the Jovians have eight possible positions in which to stand (indicated by blanks):

$$\_\, M_1 \,\_\, M_2 \,\_\, M_3 \,\_\, M_4 \,\_\, M_5 \,\_\, M_6 \,\_\, M_7 \,\_\, .$$

Thus the Jovians can stand in $P(8, 5) = 8 \cdot 7 \cdot 6 \cdot 5 \cdot 4 = 6720$ ways. By the Multiplication Principle, the number of ways seven distinct Martians and five distinct Jovians can wait in line if no two Jovians stand together is $5040 \cdot 6720 = 33{,}868{,}800$. ◀

We turn next to combinations. A selection of objects without regard to order is called a **combination.**

**Definition 6.2.14 ▶** Given a set $X = \{x_1, \ldots, x_n\}$ containing $n$ (distinct) elements,

(a) An *r-combination* of $X$ is an unordered selection of $r$-elements of $X$ (i.e., an $r$-element subset of $X$).

(b) The number of $r$-combinations of a set of $n$ distinct elements is denoted $C(n, r)$ or $\binom{n}{r}$. ◀

**Example 6.2.15** A group of five students, Mary, Boris, Rosa, Ahmad, and Nguyen, has decided to talk with the Mathematics Department chairperson about having the Mathematics Department offer more courses in discrete mathematics. The chairperson has said that she will speak with three of the students. In how many ways can these five students choose three of their group to talk with the chairperson?

**SOLUTION** In solving this problem, we must *not* take order into account. (For example, it will make no difference whether the chairperson talks to Mary, Ahmad, and Nguyen or to Nguyen, Mary, and Ahmad.) By simply listing the possibilities, we see that there are 10 ways that the five students can choose three of their group to talk to the chairperson:

*MBR, MBA, MRA, BRA, MBN, MRN, BRN, MAN, BAN, RAN.*

In the terminology of Definition 6.2.14, the number of ways the five students can choose three of their group to talk with the chairperson is $C(5, 3)$, the number of 3-combinations of five elements. We have found that $C(5, 3) = 10$. ◀

We next derive a formula for $C(n, r)$ by counting the number of $r$-permutations of an $n$-element set in two ways. The first way simply uses the formula $P(n, r)$. The second way of counting the number of $r$-permutations of an $n$-element set involves $C(n, r)$. Equating the two values will enable us to derive a formula for $C(n, r)$.

We can construct $r$-permutations of an $n$-element set $X$ in two successive steps: First, select an $r$-combination of $X$ (an unordered subset of $r$ items); second, order it. For example, to construct a 2-permutation of $\{a, b, c, d\}$, we can first select a 2-combination and then order it. Figure 6.2.4 shows how all 2-permutations of $\{a, b, c, d\}$ are obtained

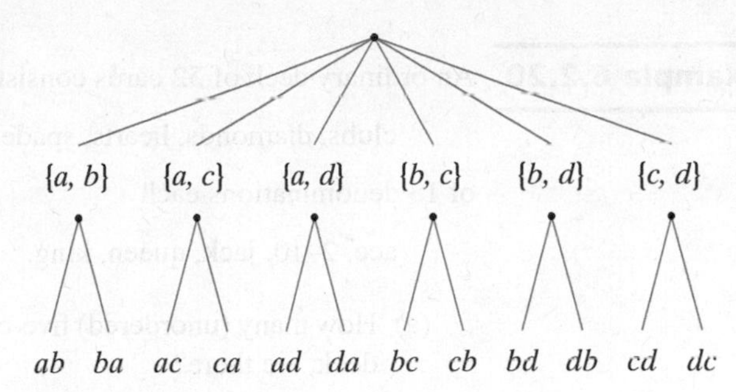


**Figure 6.2.4** 2-permutations of $\{a, b, c, d\}$.

in this way. The Multiplication Principle tells us that the number of $r$-permutations is the product of the number of $r$-combinations and the number of orderings of $r$ elements. That is,

$$P(n, r) = C(n, r)r!.$$

Therefore,

$$C(n, r) = \frac{P(n, r)}{r!}.$$

Our next theorem states this result and gives some alternative ways to write $C(n, r)$.

**Theorem 6.2.16** The number of $r$-combinations of a set of $n$ distinct objects is

$$C(n, r) = \frac{P(n, r)}{r!} = \frac{n(n-1)\cdots(n-r+1)}{r!} = \frac{n!}{(n-r)!\,r!} \qquad r \le n.$$

**Proof** The proof of the first equation is given before the statement of the theorem. The other forms of the equation follow from Theorem 6.2.10. ◀

**Example 6.2.17** In how many ways can we select a committee of three from a group of 10 distinct persons?

**SOLUTION** Since a committee is an unordered group of people, the answer is

$$C(10, 3) = \frac{10 \cdot 9 \cdot 8}{3!} = 120.$$

◀

**Example 6.2.18** In how many ways can we select a committee of two women and three men from a group of five distinct women and six distinct men?

**SOLUTION** As in Example 6.2.17, we find that the two women can be selected in $C(5, 2) = 10$ ways and that the three men can be selected in $C(6, 3) = 20$ ways. The committee can be constructed in two successive steps: Select the women; select the men. By the Multiplication Principle, the total number of committees is $10 \cdot 20 = 200$. ◀

**Example 6.2.19** How many eight-bit strings contain exactly four 1's?

**SOLUTION** An eight-bit string containing four 1's is uniquely determined once we tell which bits are 1. This can be done in $C(8, 4) = 70$ ways. ◀

**Example 6.2.20** An ordinary deck of 52 cards consists of four suits

clubs, diamonds, hearts, spades

of 13 denominations each

ace, 2–10, jack, queen, king.

(a) How many (unordered) five-card poker hands, selected from an ordinary 52-card deck, are there?

(b) How many poker hands contain cards all of the same suit?

(c) How many poker hands contain three cards of one denomination and two cards of a second denomination?

**SOLUTION**

(a) The answer is given by the combination formula $C(52, 5) = 2{,}598{,}960$.

(b) A hand containing cards all of the same suit can be constructed in two successive steps: Select a suit; select five cards from the chosen suit. The first step can be done in four ways, and the second step can be done in $C(13, 5)$ ways. By the Multiplication Principle, the answer is $4 \cdot C(13, 5) = 5148$.

(c) A hand containing three cards of one denomination and two cards of a second denomination can be constructed in four successive steps: Select the first denomination; select the second denomination; select three cards of the first denomination; select two cards of the second denomination. The first denomination can be chosen in 13 ways. Having selected the first denomination, we can choose the second denomination in 12 ways. We can select three cards of the first denomination in $C(4, 3)$ ways, and we can select two cards of the second denomination in $C(4, 2)$ ways. By the Multiplication Principle, the answer is

$$13 \cdot 12 \cdot C(4, 3) \cdot C(4, 2) = 3744.$$

◀

**Example 6.2.21** How many routes are there from the lower-left corner of an $n \times n$ square grid to the upper-right corner if we are restricted to traveling only to the right or upward? One such route is shown in a $4 \times 4$ grid in Figure 6.2.5(a).

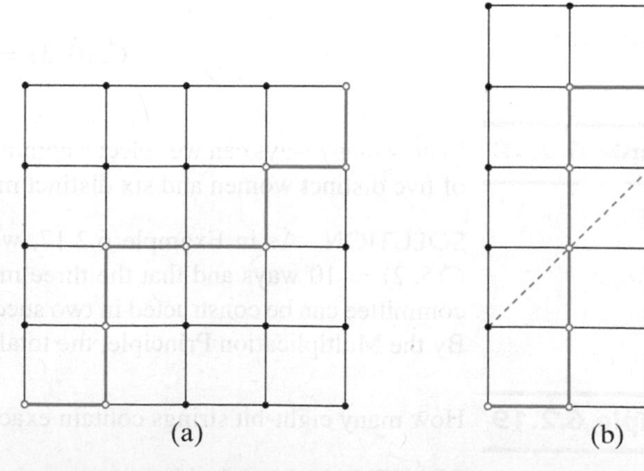


**Figure 6.2.5** (a) A $4 \times 4$ grid with a route from the lower-left corner to the upper-right corner. (b) The route in (a) transformed to a route in a $5 \times 3$ grid.

**SOLUTION** Each route can be described by a string of $n$ $R$'s (right) and $n$ $U$'s (up). For example, the route shown in Figure 6.2.5(a) can be described by the string *RUURRURU*. Any such string can be obtained by selecting $n$ positions for the $R$'s, without regard to the order of selection, among the $2n$ available positions in the string and then filling the remaining positions with $U$'s. Thus there are $C(2n, n)$ possible routes. ◀

**Example 6.2.22** How many routes are there from the lower-left corner of an $n \times n$ square grid to the upper-right corner if we are restricted to traveling only to the right or upward and if we are allowed to touch but not go above a diagonal line from the lower-left corner to the upper-right corner?

**SOLUTION** We call a route that touches but does not go above the diagonal a *good route*, and we call a route that goes above the diagonal a *bad route*. Our problem is to count the number of good routes. We let $G_n$ denote the number of good routes and $B_n$ denote the number of bad routes. In Example 6.2.21 we showed that $G_n+B_n = C(2n, n)$; thus it suffices to compute the number of bad routes.

We call a route from the lower-left corner of an $(n+1) \times (n-1)$ grid to the upper-right corner (with no restrictions) an $(n+1) \times (n-1)$ route. A $5 \times 3$ route is shown in Figure 6.2.5(b). We show that the number of bad routes is equal to the number of $(n+1) \times (n-1)$ routes by describing a one-to-one, onto function from the set of bad routes to the set of $(n+1) \times (n-1)$ routes.

Given a bad route, we find the first move (starting from the lower left) that takes it above the diagonal. Thereafter we replace each right move by an up move and each up move by a right move. For example, the route of Figure 6.2.5(a) is transformed to the route shown in Figure 6.2.5(b). This transformation can also be effected by rotating the portion of the route following the first move above the diagonal about the dashed line shown in Figure 6.2.5(b). We see that this transformation does indeed assign to each bad route an $(n+1) \times (n-1)$ route.

To show that our function is onto, consider any $(n+1) \times (n-1)$ route. Since this route ends above the diagonal, there is a first move where it goes above the diagonal. We may then rotate the remainder of the route about the dashed line shown in Figure 6.2.5(b) to obtain a bad route. The image of this bad route under our function is the $(n+1) \times (n-1)$ route with which we started. Therefore, our function is onto. Our function is also one-to-one, as we can readily verify that the function transforms distinct bad routes to distinct $(n+1) \times (n-1)$ routes. Therefore, the number of bad routes equals the number of $(n+1) \times (n-1)$ routes.

An argument like that in Example 6.2.21 shows that the number of $(n+1) \times (n-1)$ routes is equal to $C(2n, n-1)$. Thus the number of good routes is equal to

$$\begin{aligned}
C(2n, n) - B_n = C(2n, n) - C(2n, n-1) &= \frac{(2n)!}{n!\,n!} - \frac{(2n)!}{(n-1)!\,(n+1)!} \\
&= \frac{(2n)!}{n!\,(n-1)!}\left(\frac{1}{n} - \frac{1}{n+1}\right) = \frac{(2n)!}{n!\,(n-1)!} \cdot \frac{1}{n(n+1)} \\
&= \frac{(2n)!}{(n+1)n!\,n!} = \frac{C(2n, n)}{n+1}.
\end{aligned}$$

Exercises 75 and 76 outline an alternative proof using mathematical induction of the formula $C(2n, n)/(n+1)$ for the number of good routes. Exercises 27–33, Section 9.8, outline yet another way to derive the formula $C(2n, n)/(n+1)$. ◀

The numbers $C(2n, n)/(n + 1)$ are called **Catalan numbers** in honor of the Belgian mathematician Eugène-Charles Catalan (1814–1894), who discovered an elementary derivation of the formula $C(2n, n)/(n + 1)$. Catalan published numerous papers in analysis, combinatorics, algebra, geometry, probability, and number theory. In 1844, he conjectured that the only consecutive positive integers that are powers (i.e., $i^j$, where $j \geq 2$) are 8 and 9. Over 150 years later, Preda Mihailescu proved the result (in 2002).

**Go Online**
For more on Catalan numbers, see bit.ly/2PNvtEt

In this book, we denote the Catalan number $C(2n, n)/(n + 1)$ as $C_n$, $n \geq 1$, and we define $C_0$ to be 1. The first few Catalan numbers are

$$C_0 = 1, \quad C_1 = 1, \quad C_2 = 2, \quad C_3 = 5, \quad C_4 = 14, \quad C_5 = 42.$$

Like the Fibonacci numbers, the Catalan numbers have a way of appearing in unexpected places (e.g., Exercises 73 and 80–82, this section, and Exercises 30–32, Section 7.1).

Our next example illustrates a common error in counting—namely, counting some objects more than once.

**Example 6.2.23** What is wrong with the following argument, which purports to show that there are $C(8, 5)2^3$ bit strings of length 8 containing at least five 0's?

**SOLUTION** We can construct bit strings of length 8 by filling each of eight slots

$$\_\ \_\ \_\ \_\ \_\ \_\ \_\ \_$$

with either 0 or 1. To ensure that there are at least five 0's, we choose five slots and place a 0 in each of them. The five slots can be chosen in $C(8, 5)$ ways. We then fill the remaining three slots with either 0 or 1. Since each of the three remaining slots can be filled in two ways, the remaining slots can be filled in $2^3$ ways. Thus there are $C(8, 5)2^3$ bit strings of length 8 containing at least five 0's.

The problem is that some strings are counted more than one time. For example, suppose that we choose the first five slots and place 0's in them

$$\underline{0}\ \underline{0}\ \underline{0}\ \underline{0}\ \underline{0}\ \_\ \_\ \_$$

If we then place 0 1 0 in the last three slots, we obtain the string

$$\underline{0}\ \underline{0}\ \underline{0}\ \underline{0}\ \underline{0}\ \underline{0}\ \underline{1}\ \underline{0} \tag{6.2.1}$$

Now suppose that we choose the second through sixth slots and place 0's in them

$$\_\ \underline{0}\ \underline{0}\ \underline{0}\ \underline{0}\ \underline{0}\ \_\ \_$$

If we then place 0 in the first slot and 1 0 in the last two slots, we obtain the string

$$\underline{0}\ \underline{0}\ \underline{0}\ \underline{0}\ \underline{0}\ \underline{0}\ \underline{1}\ \underline{0} \tag{6.2.2}$$

In the argument given, strings (6.2.1) and (6.2.2) are counted as *distinct* strings.

A correct way to count the bit strings of length 8 containing at least five 0's is to count the number of strings containing *exactly* five 0's, the number of strings containing *exactly* six 0's, the number of strings containing *exactly* seven 0's, and the number of strings containing *exactly* eight 0's and sum these numbers. Notice that here each string is counted one time since no string can contain exactly $i$ 0's and exactly $j$ 0's when $i \neq j$.

To construct a bit string of length 8 containing exactly five 0's, we choose five slots for the 0's and put 1's in the other three slots. Since we can choose five slots in $C(8, 5)$ ways, there are $C(8, 5)$ bit strings of length 8 containing exactly five 0's. Similarly, there are $C(8, 6)$ bit strings of length 8 containing exactly six 0's, and so on. Therefore, there are

$$C(8, 5) + C(8, 6) + C(8, 7) + C(8, 8)$$

bit strings of length 8 containing at least five 0's. ◀

We close this section by providing another proof of Theorem 6.2.16 that gives a formula for the number of $r$-element subsets of an $n$-element set. The proof is illustrated in Figure 6.2.6. Let $X$ be an $n$-element set. We assume the formula $P(n, r) = n(n-1)\cdots(n-r+1)$ that counts the number of orderings of $r$-element subsets chosen from $X$. To count the number of $r$-element subsets of $X$, we do *not* want to take order into account—we want to consider permutations of the same subset *equivalent*. Formally, we define a relation $R$ on the set $S$ of $r$-permutations of $X$ by the following rule: $p_1 R p_2$ if $p_1$ and $p_2$ are permutations of the same $r$-element subset of $X$. It is straightforward to verify that $R$ is an equivalence relation on $S$.

If $p$ is an $r$-permutation of $X$, then $p$ is a permutation of some $r$-element subset $X_r$ of $X$; thus, the equivalence class containing $p$ consists of all permutations of $X_r$. We see that each equivalence class has $r!$ elements. An equivalence class is determined by the $r$-element subset of $X$ that is permuted to obtain its members. Therefore, there are $C(n, r)$ equivalence classes. Since the set $S$ has $P(n, r)$ elements, by Theorem 3.4.16, $C(n, r) = P(n, r)/r!$.

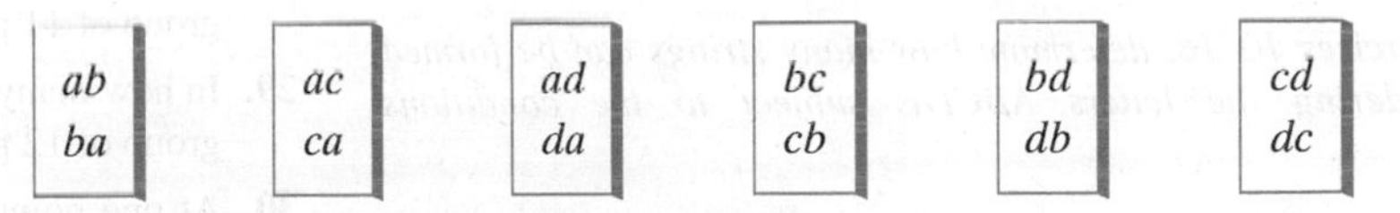


**Figure 6.2.6** The alternative proof of Theorem 6.2.16 for $n = 4$ and $r = 2$. Each box contains an equivalence class for the relation $R$ on the set of 2-permutations of $X = \{a, b, c, d\}$ defined by $p_1 R p_2$ if $p_1$ and $p_2$ are permutations of the same 2-element subset of $X$. There are $P(4, 2) = 12$ 2-permutations of $X$ and 2 ways to permute each 2-permutation. Since each equivalence class corresponds to a subset of $X$, $12/2 = C(4, 2)$.

In this section, we have presented techniques for counting objects where repetition is *not* allowed. In Section 6.3, we will count objects where repetition *is* allowed.

## 6.2 Problem-Solving Tips

The key points to remember in this section are that a permutation takes order into account and a combination does *not* take order into account. Thus, a key to solving counting problems is to determine whether we are counting ordered or unordered items. For example, a line of *distinct* persons is considered ordered. Thus six distinct persons can wait in line in 6! ways; the permutation formula is used. A committee is a typical example of an unordered group. For example, a committee of three can be selected from a set of six distinct persons in $C(6, 3)$ ways; the combination formula is used.

## 6.2 Review Exercises

1. What is a permutation of $x_1, \ldots, x_n$?
2. How many permutations are there of an $n$-element set? How is this formula derived?
3. What is an $r$-permutation of $x_1, \ldots, x_n$?
4. How many $r$-permutations are there of an $n$-element set? How is this formula derived?
5. How do we denote the number of $r$-permutations of an $n$-element set?
6. What is an $r$-combination of $\{x_1, \ldots, x_n\}$?
7. How many $r$-combinations are there of an $n$-element set? How is this formula derived?
8. How do we denote the number of $r$-combinations of an $n$-element set?

## 6.2 Exercises

1. How many permutations are there of $a, b, c, d$?
2. List the permutations of $a, b, c, d$.
3. How many 3-permutations are there of $a, b, c, d$?
4. List the 3-permutations of $a, b, c, d$.
5. How many permutations are there of 11 distinct objects?
6. How many 5-permutations are there of 11 distinct objects?
7. In how many ways can we select a chairperson, vice-chairperson, and recorder from a group of 11 persons?
8. In how many ways can we select a chairperson, vice-chairperson, secretary, and treasurer from a group of 12 persons?
9. In how many different ways can 12 horses finish in the order Win, Place, Show?

*In Exercises 10–18, determine how many strings can be formed by ordering the letters ABCDE subject to the conditions given.*

10. Contains the substring *ACE*
11. Contains the letters *ACE* together in any order
12. Contains the substrings *DB* and *AE*
13. Contains either the substring *AE* or the substring *EA* or both
14. *A* appears before *D*. *Examples*: *BCAED*, *BCADE*
15. Contains neither of the substrings *AB*, *CD*
16. Contains neither of the substrings *AB*, *BE*
17. *A* appears before *C* and *C* appears before *E*
18. Contains either the substring *DB* or the substring *BE* or both
19. In how many ways can five distinct Martians and eight distinct Jovians wait in line if no two Martians stand together?
20. In how many ways can five distinct Martians, ten distinct Vesuvians, and eight distinct Jovians wait in line if no two Martians stand together?
21. In how many ways can five distinct Martians and five distinct Jovians wait in line?
22. In how many ways can five distinct Martians and five distinct Jovians be seated at a circular table?
23. In how many ways can five distinct Martians and five distinct Jovians be seated at a circular table if no two Martians sit together?
24. In how many ways can five distinct Martians and eight distinct Jovians be seated at a circular table if no two Martians sit together?

*In Exercises 25–27, let* $X = \{a, b, c, d\}$.

25. Compute the number of 3-combinations of $X$.
26. List the 3-combinations of $X$.
27. Show the relationship between the 3-permutations and the 3-combinations of $X$ by drawing a picture like that in Figure 6.2.4.
28. In how many ways can we select a committee of three from a group of 11 persons?
29. In how many ways can we select a committee of four from a group of 12 persons?
30. At one point in the Illinois state lottery Lotto game, a person was required to choose six numbers (in any order) among 44 numbers. In how many ways can this be done? The state was considering changing the game so that a person would be required to choose six numbers among 48 numbers. In how many ways can this be done?
31. Suppose that a pizza parlor features four specialty pizzas and pizzas with three or fewer unique toppings (no choosing anchovies twice!) chosen from 17 available toppings. How many different pizzas are there?
32. Suppose that the pizza parlor of Exercise 31 has a special price for four pizzas. How many ways can four pizzas be selected?

*Exercises 33–38 refer to a club consisting of six distinct men and seven distinct women.*

33. In how many ways can we select a committee of five persons?
34. In how many ways can we select a committee of three men and four women?
35. In how many ways can we select a committee of four persons that has at least one woman?

**36.** In how many ways can we select a committee of four persons that has at most one man?

**37.** In how many ways can we select a committee of four persons that has persons of both sexes?

**38.** In how many ways can we select a committee of four persons so that Mabel and Ralph do not serve together?

**39.** In how many ways can we select a committee of four Republicans, three Democrats, and two Independents from a group of 10 distinct Republicans, 12 distinct Democrats, and four distinct Independents?

**40.** How many eight-bit strings contain exactly three 0's?

**41.** How many eight-bit strings contain three 0's in a row and five 1's?

★**42.** How many eight-bit strings contain at least two 0's in a row?

*In Exercises 43–51, find the number of (unordered) five-card poker hands, selected from an ordinary 52-card deck, having the properties indicated.*

**43.** Containing four aces

**44.** Containing four of a kind, that is, four cards of the same denomination

**45.** Containing all spades

**46.** Containing cards of exactly two suits

**47.** Containing cards of all suits

**48.** Of the form A2345 of the same suit

**49.** Consecutive and of the same suit (Assume that the ace is the lowest denomination.)

**50.** Consecutive (Assume that the ace is the lowest denomination.)

**51.** Containing two of one denomination, two of another denomination, and one of a third denomination

**52.** Find the number of (unordered) 13-card bridge hands selected from an ordinary 52-card deck.

**53.** How many bridge hands are all of the same suit?

**54.** How many bridge hands contain exactly two suits?

**55.** How many bridge hands contain all four aces?

**56.** How many bridge hands contain five spades, four hearts, three clubs, and one diamond?

**57.** How many bridge hands contain five of one suit, four of another suit, three of another suit, and one of another suit?

**58.** How many bridge hands contain four cards of three suits and one card of the fourth suit?

**59.** How many bridge hands contain no face cards? (A face card is one of 10, J, Q, K, A.)

*In Exercises 60–64, a coin is flipped 10 times.*

**60.** How many outcomes are possible? (An *outcome* is a list of 10 H's and T's that gives the result of each of 10 tosses. For example, the outcome

H H T H T H H H T H

represents 10 tosses, where a head was obtained on the first two tosses, a tail was obtained on the third toss, a head was obtained on the fourth toss, etc.)

**61.** How many outcomes have exactly three heads?

**62.** How many outcomes have at most three heads?

**63.** How many outcomes have a head on the fifth toss?

**64.** How many outcomes have as many heads as tails?

*Exercises 65–68 refer to a shipment of 50 microprocessors of which four are defective.*

**65.** In how many ways can we select a set of four microprocessors?

**66.** In how many ways can we select a set of four nondefective microprocessors?

**67.** In how many ways can we select a set of four microprocessors containing exactly two defective microprocessors?

**68.** In how many ways can we select a set of four microprocessors containing at least one defective microprocessor?

★**69.** Show that the number of bit strings of length $n \geq 4$ that contain exactly two occurrences of 10 is $C(n+1, 5)$.

★**70.** Show that the number of $n$-bit strings having exactly $k$ 0's, with no two 0's consecutive, is $C(n-k+1, k)$.

★**71.** Show that the product of any positive integer and its $k-1$ successors is divisible by $k!$.

**72.** Show that there are $(2n-1)(2n-3)\cdots 3\cdot 1$ ways to pick $n$ pairs from $2n$ distinct items.

*Exercises 73–77 refer to an election in which two candidates Wright and Upshaw ran for dogcatcher. After each vote was tabulated, Wright was never behind Upshaw. This problem is known as the* ballot problem.

**73.** Suppose that each candidate received exactly $r$ votes. Show that the number of ways the votes could be counted is $C_r$, the $r$th Catalan number.

**74.** Suppose that Wright received exactly $r$ votes and Upshaw received exactly $u$ votes, $r \geq u > 0$. Show that the number of ways the votes could be counted is $C(r+u, r) - C(r+u, r+1)$.

**75.** Suppose that Wright received exactly $r$ votes and Upshaw received exactly $u$ votes, $r \geq u \geq 0$. Use induction on $n = r+u$ to prove that the number of ways the votes could be counted is

$$\frac{r-u+1}{r+1}C(r+u, r).$$

**76.** Use Exercise 75 to derive the formula $C(2n, n)/(n+1)$ for the number of good routes as defined in Example 6.2.22.

**77.** Show that if exactly $n$ votes were cast, the number of ways the votes could be counted is $C(n, \lceil n/2 \rceil)$.

**78.** Suppose that we start at the origin in the $xy$-plane and take $n$ unit steps (i.e., each step is of length one), where each step is either vertical (up or down) or horizontal (left or right). How many such paths never go strictly below the $x$-axis?

**79.** Suppose that we start at the origin in the $xy$-plane and take $n$ unit steps (i.e., each step is of length one), where each step is either vertical (up or down) or horizontal (left or right). How many such paths stay in the first quadrant ($x \geq 0, y \geq 0$)?

**80.** Show that the number of ways that $2n$ persons, seated around a circular table, can shake hands in pairs without any arms crossing is $C_n$, the $n$th Catalan number.

**81.** A photo of $2n$ students, no two of which are the same height, is to be taken subject to the following rules:

(a) There will be two rows each containing $n$ students, all of whom are standing.

(b) Each student in the back row must be taller than the student standing directly in front of him.

(c) In each row, the students must be arranged in increasing order of height.

Show that the number of ways to arrange the students is $C_n$, the $n$th Catalan number.

**★82.** Show that the print statement in the pseudocode

```
for i_1 = 1 to n
  for i_2 = 1 to min(i_1, n − 1)
    for i_3 = 1 to min(i_2, n − 2)
      .
      .
      .
      for i_{n−1} = 1 to min(i_{n−2}, 2)
        for i_n = 1 to 1
          println(i_1, i_2, ..., i_n)
```

is executed $C_n$ times, where $C_n$ denotes the $n$th Catalan number.

**83.** Suppose that we have $n$ objects, $r$ distinct and $n - r$ identical. Give another derivation of the formula

$$P(n, r) = r!\,C(n, r)$$

by counting the number of orderings of the $n$ objects in two ways:

- Count the orderings by first choosing positions for the $r$ distinct objects.
- Count the orderings by first choosing positions for the $n - r$ identical objects.

**84.** What is wrong with the following argument, which purports to show that $4C(39, 13)$ bridge hands contain three or fewer suits?

There are $C(39, 13)$ hands that contain only clubs, diamonds, and spades. In fact, for any three suits, there are $C(39, 13)$ hands that contain only those three suits. Since there are four 3-combinations of the suits, the answer is $4C(39, 13)$.

**85.** What is wrong with the following argument, which purports to show that there are $13^4 \cdot 48$ (unordered) five-card poker hands containing cards of all suits?

Pick one card of each suit. This can be done in $13 \cdot 13 \cdot 13 \cdot 13 = 13^4$ ways. Since the fifth card can be chosen in 48 ways, the answer is $13^4 \cdot 48$.

**86.** What is wrong with the following argument, which purports to show that there are $P(n, m)m^{n-m}$ onto functions from the $n$-element set $X$ to the $m$-element set $Y$, $n > m$?

Let $Y = \{y_1, \ldots, y_m\}$. To ensure that a function from $X$ to $Y$ is onto $Y$, we select an $m$-permutation of $X$, say $x_1, \ldots, x_m$, and assign $x_1$ the value $y_1$, $x_2$ the value $y_2, \ldots$, and $x_m$ the value $y_m$. We can select the $m$-permutation in $P(n, m)$ ways. The remainder of the $n - m$ elements in $X$ may be assigned values in $Y$ arbitrarily. The first remaining element in $X$ can be assigned a value in $Y$ in $m$ ways. The next remaining element in $X$ can also be assigned a value in $Y$ in $m$ ways, and so on. Thus the remaining $n-m$ elements in $X$ can be assigned values in $Y$ in $m^{n-m}$ ways. Thus the number of functions from $X$ onto $Y$ is $P(n, m)m^{n-m}$.

**87.** How many times is the string 10100001 counted in the erroneous argument given in Example 6.2.23?

**88.** How many times is the string 10001000 counted in the erroneous argument given in Example 6.2.23?

**89.** How many times is the string 00000000 counted in the erroneous argument given in Example 6.2.23?

**90.** Let $s_{n,k}$ denote the number of ways to seat $n$ persons at $k$ round tables, with at least one person at each table. (The numbers $s_{n,k}$ are called *Stirling numbers of the first kind.*) The ordering of the tables is *not* taken into account. The seating arrangement at a table *is* taken into account except for rotations. *Examples:* The following pair is *not* distinct:

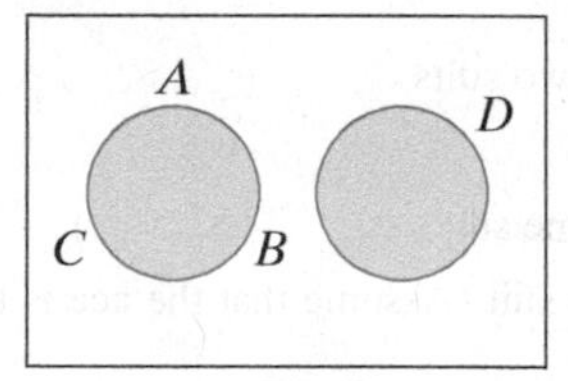

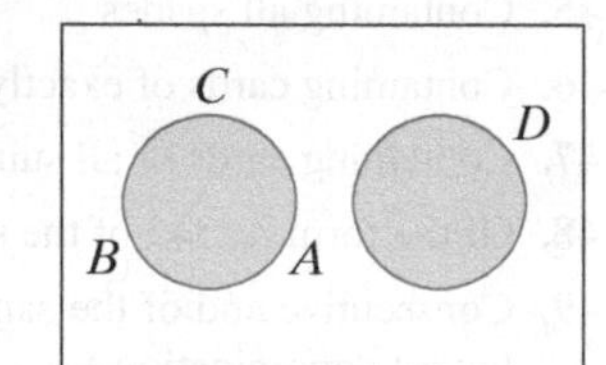


The following pair is *not* distinct:

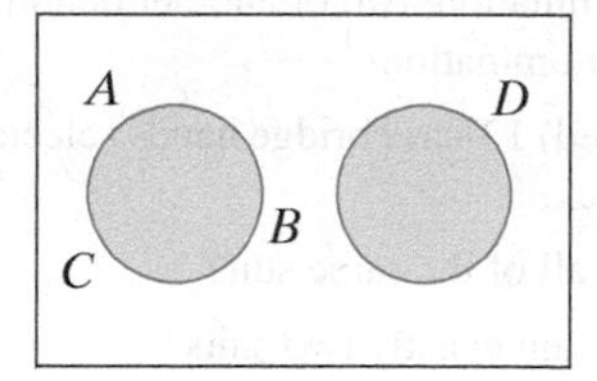

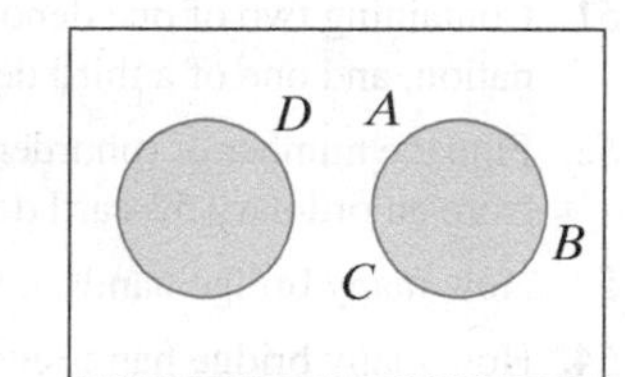


The following pair *is* distinct:

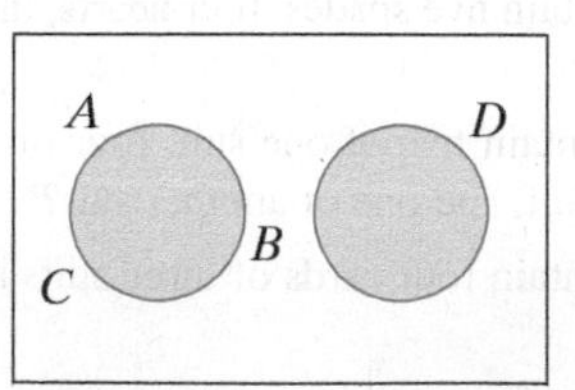

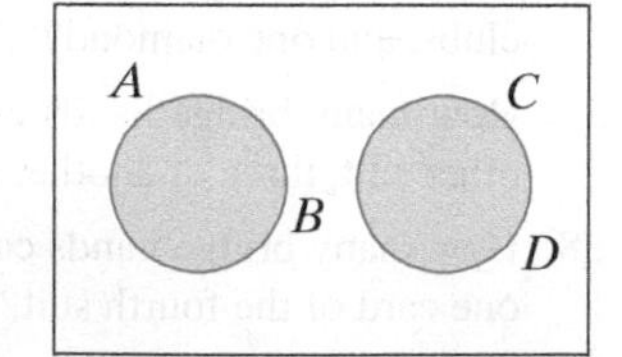


The following pair *is* distinct:

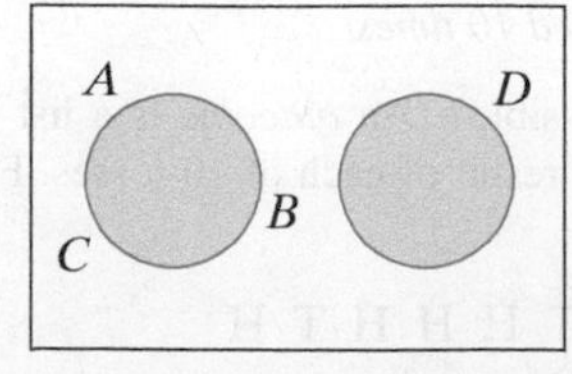

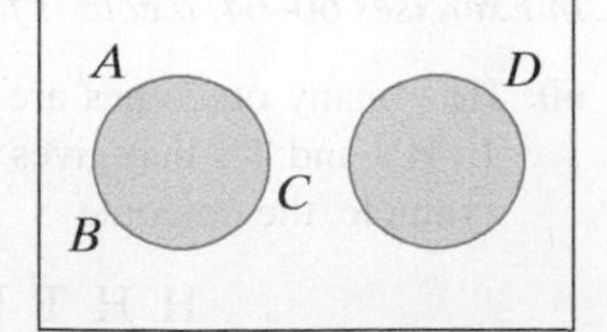

(a) Show that $s_{n,k} = 0$ if $k > n$.

(b) Show that $s_{n,n} = 1$ for all $n \geq 1$.

(c) Show that $s_{n,1} = (n-1)!$ for all $n \geq 1$.

(d) Show that $s_{n,n-1} = C(n, 2)$ for all $n \geq 2$.

(e) Show that

$$s_{n,2} = (n-1)!\left(1 + \frac{1}{2} + \frac{1}{3} + \cdots + \frac{1}{n-1}\right)$$

for all $n \geq 2$.

(f) Show that

$$\sum_{k=1}^{n} s_{n,k} = n! \qquad \text{for all } n \geq 1.$$

(g) Find a formula for $s_{n,n-2}$, $n \geq 3$, and prove it.

**91.** Let $S_{n,k}$ denote the number of ways to partition an $n$-element set into exactly $k$ nonempty subsets. The order of the subsets is not taken into account. (The numbers $S_{n,k}$ are called *Stirling numbers of the second kind.*)

(a) Show that $S_{n,k} = 0$ if $k > n$.

(b) Show that $S_{n,n} = 1$ for all $n \geq 1$.

(c) Show that $S_{n,1} = 1$ for all $n \geq 1$.

(d) Show that $S_{3,2} = 3$.

(e) Show that $S_{4,2} = 7$.

(f) Show that $S_{4,3} = 6$.

(g) Show that $S_{n,2} = 2^{n-1} - 1$ for all $n \geq 2$.

(h) Show that $S_{n,n-1} = C(n, 2)$ for all $n \geq 2$.

(i) Find a formula for $S_{n,n-2}$, $n \geq 3$, and prove it.

**92.** Show that there are

$$\sum_{k=1}^{n} S_{n,k}$$

equivalence relations on an $n$-element set. [The numbers $S_{n,k}$ are Stirling numbers of the second kind (see Exercise 91).]

**93.** If $X$ is an $n$-element set and $Y$ is an $m$-element set, $n \leq m$, how many one-to-one functions are there from $X$ to $Y$?

**94.** If $X$ and $Y$ are $n$-element sets, how many one-to-one, onto functions are there from $X$ to $Y$?

**95.** Show that $(n/k)^k \leq C(n, k) \leq n^k/k!$.

# Problem-Solving Corner Combinations

## Problem

(a) How many routes are there from the lower-left corner to the upper-right corner of an $m \times n$ grid in which we are restricted to traveling only to the right or upward? For example, the following figure is a $3 \times 5$ grid and one route is shown.

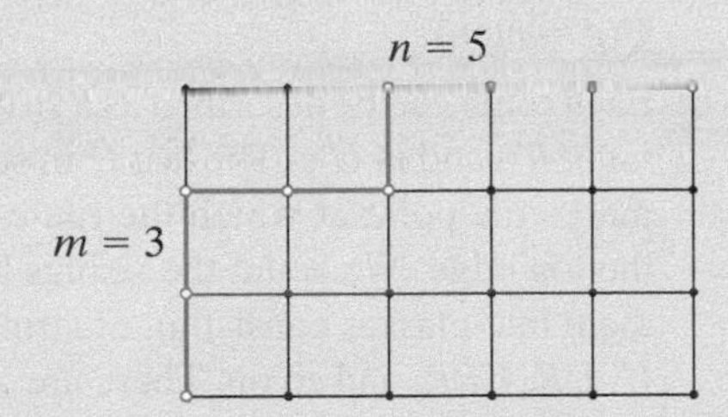


(b) Divide the routes into classes based on when the route first meets the top edge to derive the formula

$$\sum_{k=0}^{n} C(k+m-1, k) = C(m+n, m).$$

## Attacking the Problem

Example 6.2.21 asked how many paths there were from the lower-left corner to the upper-right corner of an $n \times n$ grid in which we are restricted to traveling only to the right or upward. The solution to that problem encoded each route as a string of $n$ $R$'s (right) and $n$ $U$'s (up). The problem then became one of counting the number of such strings. Any such string can be obtained by selecting $n$ positions for the $R$'s, without regard to the order of selection, among the $2n$ available positions in the string and then filling the remaining positions with $U$'s. Thus the number of strings and number of routes are equal to $C(2n, n)$.

In the present problem, we can encode each route as a string of $n$ $R$'s (right) and $m$ $U$'s (up). As in the previous problem, we must count the number of such strings. Any such string can be obtained by selecting $n$ positions for the $R$'s, without regard to the order of selection, among the $n + m$ available positions in the string and then filling the remaining positions with $U$'s. Thus the number of strings and number of routes are equal to $C(n + m, n)$. We have answered part (a).

In part (b) we are given a major hint: Divide the routes into classes based on when the route first meets the top edge. A route can first meet the top edge at any one of $n + 1$ positions. In the previous figure, the route shown first meets the top edge at the third position from the left. Before reading on, you might think about why we might divide the routes into classes.

Notice that when we divide the routes into classes based on when the route first meets the top edge:

- The classes are *disjoint.*

(A route cannot first meet the top edge in two or more distinct positions.) Notice also that every route meets the top edge somewhere, so

- Every route is in some class.

In the terminology of Section 1.1 (see Example 1.1.25 and the discussion that precedes it), the classes *partition* the set of routes. Because the classes partition the set of routes, the Addition Principle applies and the sum of the numbers of routes in each class is equal to the total number of routes. (No route is counted twice since the classes do not overlap, and every route is counted once since each route is in some class.) Evidently, the equation we're supposed to prove results from equating the sum of the number of routes in each class to the total number of routes.

## Finding a Solution

We have already solved part (a). For part (b), let's look at the $3 \times 5$ grid. There is exactly one route that first meets the top edge at the first position from the left. There are three routes that first meet the top edge at the second position from the left:

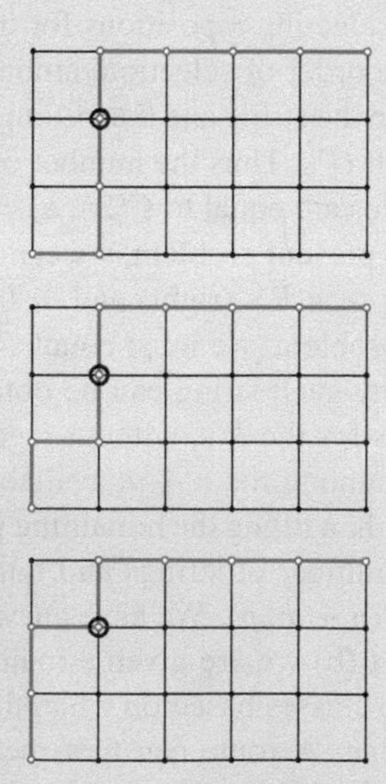

Notice that the only variation in the preceding figures occurs between the start and the circled dot. To put it another way, after a route meets the circled dot, there is only one way to finish the trip. Therefore, it suffices to count the number of routes from the lower-left corner to the upper-right corner of a $2 \times 1$ grid. But we already solved this problem in part (a)! The number of routes from the lower-left corner to the upper-right corner of a $2 \times 1$ grid is equal to $C(2+1, 1) = 3$. In a similar way, we find that the number of routes that first meet the top edge at the third position from the left is equal to the number of routes from the lower-left corner to the upper-right corner of a $2 \times 2$ grid—namely, $C(2+2, 2) = 6$. By summing we obtain all the routes:

$$\begin{aligned} C(5+3, 5) &= C(0+2, 0) + C(1+2, 1) \\ &\quad + C(2+2, 2) + C(3+2, 3) \\ &\quad + C(4+2, 4) + C(5+2, 5). \end{aligned}$$

If we replace each term $C(k+3-1, k)$ by its value, we obtain

$$56 = 1 + 3 + 6 + 10 + 15 + 21.$$

You should verify the preceding formula, find the six routes that first meet the top edge at the third position from the left, and see why the number of such routes is equal to the number of routes from the lower-left corner to the upper-right corner of a $2 \times 2$ grid.

## Formal Solution

(a) We can encode each route as a string of $n$ $R$'s (right) and $m$ $U$'s (up). Any such string can be obtained by selecting $n$ positions for the $R$'s, without regard to the order of selection, among the $n + m$ available positions in the string and then filling the remaining positions with $U$'s. Thus the number of routes is equal to $C(n+m, n)$.

(b) Each route can be described as a string containing $n$ $R$'s and $m$ $U$'s. The last $U$ in such a string marks the point at which the route first meets the top edge. We count the strings by dividing them into classes consisting of strings that end $U$, $UR$, $URR$, and so on. There are

$$C(n+m-1, n)$$

strings that end $U$, since we must choose $n$ slots among the first $n+m-1$ slots for the $n$ $R$'s. There are

$$C((n-1)+m-1, n-1)$$

strings that end $UR$, since we must choose $n-1$ slots from among the first $(n-1)+m-1$ slots for the $n-1$ $R$'s. In general, there are $C(k+m-1, k)$ strings that end $UR^{n-k}$. Since there are $C(m+n, m)$ strings altogether, the formula follows.

## Summary of Problem-Solving Techniques

- Look for a similar problem and imitate its solution.
- Counting the number of members of a set in two different ways leads to an equation. In particular, if $\{X_1, X_2, \ldots, X_n\}$ is a partition of $X$, the Addition Principle applies and

$$|X| = \sum_{i=1}^{n} |X_i|.$$

- Directly enumerate some of the items to be counted.
- Look for patterns.

## Comments

It's important to verify that an alleged partition is truly a partition before using the Addition Principle. If $X$ is the set of five-bit strings and $X_i$ is the set of five-bit strings that contain $i$ consecutive zeros, the Addition Principle does *not* apply; the sets $X_i$ are *not* pairwise disjoint. For example, $00001 \in X_2 \cap X_3$. As an example of a partition of $X$, we could let $X_i$ be the set of five-bit strings that contain exactly $i$ zeros.

## Exercises

**1.** Divide the routes into classes based on when the route first meets a vertical line, and use the Addition Principle to derive a formula like that proved in this section.

**2.** Divide the routes into classes based on when the route crosses the slanted line shown.

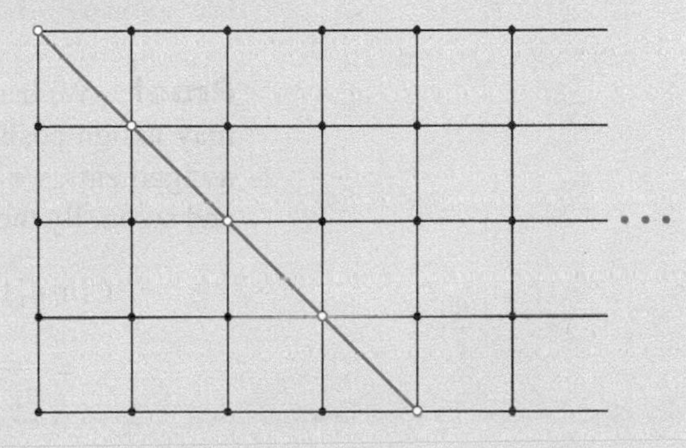

Use the Addition Principle to derive a formula like that proved in this section.

# 6.3 Generalized Permutations and Combinations

In Section 6.2, we dealt with orderings and selections without allowing repetitions. In this section we consider orderings of sequences containing repetitions and unordered selections in which repetitions are allowed.

**Example 6.3.1** How many strings can be formed using the following letters?

$$M\ I\ S\ S\ I\ S\ S\ I\ P\ P\ I$$

**Go Online**
For more generalized permutations and combinations, see bit.ly/2PNvtEt

**SOLUTION** Because of the duplication of letters, the answer is not 11!, but some number less than 11!.

Let us consider the problem of filling 11 blanks,

$$\_\ \_\ \_\ \_\ \_\ \_\ \_\ \_\ \_\ \_\ \_,$$

with the letters given. There are $C(11, 2)$ ways to choose positions for the two $P$'s. Once the positions for the $P$'s have been selected, there are $C(9, 4)$ ways to choose positions for the four $S$'s. Once the positions for the $S$'s have been selected, there are $C(5, 4)$ ways to choose positions for the four $I$'s. Once these selections have been made, there is one position left to be filled by the $M$. By the Multiplication Principle, the number of ways of ordering the letters is

$$C(11, 2)C(9, 4)C(5, 4) = \frac{11!}{2!\,9!}\frac{9!}{4!\,5!}\frac{5!}{4!\,1!} = \frac{11!}{2!\,4!\,4!\,1!} = 34{,}650. \qquad ◀$$

The solution to Example 6.3.1 assumes a nice form. The number 11 that appears in the numerator is the total number of letters. The values in the denominator give the numbers of duplicates of each letter. This method can be used to establish a general formula.

**Theorem 6.3.2** Suppose that a sequence $S$ of $n$ items has $n_1$ identical objects of type 1, $n_2$ identical objects of type 2, . . . , and $n_t$ identical objects of type $t$. Then the number of orderings of $S$ is

$$\frac{n!}{n_1!\, n_2! \cdots n_t!}.$$

**Proof** We assign positions to each of the $n$ items to create an ordering of $S$. We may assign positions to the $n_1$ items of type 1 in $C(n, n_1)$ ways. Having made these assignments, we may assign positions to the $n_2$ items of type 2 in $C(n - n_1, n_2)$ ways, and so on. By the Multiplication Principle, the number of orderings is

$$\begin{aligned}
&C(n, n_1)C(n - n_1, n_2)C(n - n_1 - n_2, n_3) \cdots C(n - n_1 - \cdots - n_{t-1}, n_t) \\
&= \frac{n!}{n_1!\,(n - n_1)!} \frac{(n - n_1)!}{n_2!\,(n - n_1 - n_2)!} \cdots \frac{(n - n_1 - \cdots - n_{t-1})!}{n_t!\,0!} \\
&= \frac{n!}{n_1!\, n_2! \cdots n_t!}.
\end{aligned}$$

◀

**Example 6.3.3** In how many ways can eight distinct books be divided among three students if Bill gets four books and Shizuo and Marian each get two books?

**SOLUTION** Put the books in some fixed order. Now consider orderings of four $B$'s, two $S$'s, and two $M$'s. An example is

$$B\ B\ B\ S\ M\ B\ M\ S.$$

Each such ordering determines a distribution of books. For the example ordering, Bill gets books 1, 2, 3, and 6, Shizuo gets books 4 and 8, and Marian gets books 5 and 7. Thus the number of ways of ordering *BBBBSSMM* is the number of ways to distribute the books. By Theorem 6.3.2, this number is

$$\frac{8!}{4!\,2!\,2!} = 420.$$

◀

We can give an alternate proof of Theorem 6.3.2 by using relations. Suppose that a sequence $S$ of $n$ items has $n_i$ identical objects of type $i$ for $i = 1, \ldots, t$. Let $X$ denote the set of $n$ elements obtained from $S$ by considering the $n_i$ objects of type $i$ *distinct* for $i = 1, \ldots, t$. For example, if $S$ is the sequence of letters

$$M\ I\ S\ S\ I\ S\ S\ I\ P\ P\ I,$$

$X$ would be the set

$$\{M, I_1, S_1, S_2, I_2, S_3, S_4, I_3, P_1, P_2, I_4\}.$$

We define a relation $R$ on the set of all permutations of $X$ by the rule: $p_1 R p_2$ if $p_2$ is obtained from $p_1$ by permuting the order of the objects of type 1 (but not changing their location) and/or permuting the order of the objects of type 2 (but not changing their location) ... and/or permuting the order of the objects of type $t$ (but not changing their location); for example,

$$(I_1S_1S_2I_2S_3S_4I_3P_1P_2I_4M)\, R\, (I_2S_3S_2I_1S_4S_1I_3P_1P_2I_4M).$$

It is straightforward to verify that $R$ is an equivalence relation on the set of all permutations of $X$.

The equivalence class containing the permutation $p$ consists of all permutations of $X$ that are identical if we consider the objects of type $i$ identical for $i = 1, \ldots, t$. Thus each equivalence class has $n_1!\, n_2! \cdots n_t!$ elements. Since an equivalence class is determined by an ordering of $S$, the number of orderings of $S$ is equal to the number of equivalence classes. There are $n!$ permutations of $X$, so by Theorem 3.4.16 the number of orderings of $S$ is

$$\frac{n!}{n_1!\, n_2! \cdots n_t!}.$$

Next, we turn to the problem of counting unordered selections where repetitions are allowed.

**Example 6.3.4** Consider three books: a computer science book, a physics book, and a history book. Suppose that the library has at least six copies of each of these books. In how many ways can we select six books?

**SOLUTION** The problem is to choose unordered, six-element selections from the set {computer science, physics, history}, repetitions allowed. A selection is uniquely determined by the number of each type of book selected. Let us denote a particular selection as

| CS | Physics | History |
|---|---|---|
| × × × | \| × × \| | × |

Here we have designated the selection consisting of three computer science books, two physics books, and one history book. Another example of a selection is

| CS | Physics | History |
|---|---|---|
|  | \| × × × × \| | × × |

which denotes the selection consisting of no computer science books, four physics books, and two history books. We see that each ordering of six ×'s and two |'s denotes a selection. Thus our problem is to count the number of such orderings. But this is just the number of ways $C(8, 2) = 28$ of selecting two positions for the |'s from eight possible positions. Thus there are 28 ways to select six books. ◀

The method used in Example 6.3.4 can be used to derive a general result.

**Theorem 6.3.5** If $X$ is a set containing $t$ elements, the number of unordered, $k$-element selections from $X$, repetitions allowed, is

$$C(k + t - 1, t - 1) = C(k + t - 1, k).$$

**Proof** Let $X = \{a_1, \ldots, a_t\}$. Consider the $k + t - 1$ slots

$$\_\_\ \_\_\ \_\_ \cdots \_\_\ \_\_$$

and $k+t-1$ symbols consisting of $k$ $\times$'s and $t-1$ |'s. Each placement of these symbols into the slots determines a selection. The number $n_1$ of $\times$'s up to the first | represents the selection of $n_1$ $a_1$'s; the number $n_2$ of $\times$'s between the first and second |'s represents the selection of $n_2$ $a_2$'s; and so on. Since there are $C(k+t-1, t-1)$ ways to select the positions for the |'s, there are also $C(k+t-1, t-1)$ selections. This is equal to $C(k+t-1, k)$, the number of ways to select the positions for the $\times$'s; hence there are

$$C(k+t-1, t-1) = C(k+t-1, k)$$

unordered, $k$-element selections from $X$, repetitions allowed. ◀

**Example 6.3.6** Suppose that there are piles of red, blue, and green balls and that each pile contains at least eight balls.

(a) In how many ways can we select eight balls?

(b) In how many ways can we select eight balls if we must have at least one ball of each color?

**SOLUTION**

(a) By Theorem 6.3.5, the number of ways of selecting eight balls is

$$C(8+3-1, 3-1) = C(10, 2) = 45.$$

(b) We can also use Theorem 6.3.5 to solve part (b) if we first select one ball of each color. To complete the selection, we must choose five additional balls. This can be done in

$$C(5+3-1, 3-1) = C(7, 2) = 21$$

ways. ◀

**Example 6.3.7** In how many ways can 12 identical mathematics books be distributed among the students Anna, Beth, Candy, and Dan?

**SOLUTION** We can use Theorem 6.3.5 to solve this problem if we consider the problem to be that of labeling each book with the name of the student who receives it. This is the same as selecting 12 items (the names of the students) from the set {Anna, Beth, Candy, Dan}, repetitions allowed. By Theorem 6.3.5, the number of ways to do this is

$$C(12+4-1, 4-1) = C(15, 3) = 455.$$ ◀

**Example 6.3.8**

(a) How many solutions in nonnegative integers are there to the equation

$$x_1 + x_2 + x_3 + x_4 = 29? \tag{6.3.1}$$

(b) How many solutions in integers are there to (6.3.1) satisfying $x_1 > 0$, $x_2 > 1$, $x_3 > 2$, $x_4 \geq 0$?

**SOLUTION**

(a) Each solution of (6.3.1) is equivalent to selecting 29 items, $x_i$ of type $i$, $i = 1, 2, 3, 4$. According to Theorem 6.3.5, the number of selections is

$$C(29+4-1, 4-1) = C(32, 3) = 4960.$$

(b) Each solution of (6.3.1) satisfying the given conditions is equivalent to selecting 29 items, $x_i$ of type $i$, $i = 1, 2, 3, 4$, where, in addition, we must have at least one item of type 1, at least two items of type 2, and at least three items of type 3. First, select one item of type 1, two items of type 2, and three items of type 3. Then, choose 23 additional items. By Theorem 6.3.5, this can be done in

$$C(23 + 4 - 1, 4 - 1) = C(26, 3) = 2600$$

ways. ◀

**Example 6.3.9** How many times is the print statement executed?

for $i_1 = 1$ to $n$
  for $i_2 = 1$ to $i_1$
    for $i_3 = 1$ to $i_2$
      ⋱
        for $i_k = 1$ to $i_{k-1}$
          $println(i_1, i_2, \ldots, i_k)$

**SOLUTION** Notice that each line of output consists of $k$ integers

$$i_1 i_2 \cdots i_k, \tag{6.3.2}$$

where

$$n \geq i_1 \geq i_2 \geq \cdots \geq i_k \geq 1, \tag{6.3.3}$$

and that every sequence (6.3.2) satisfying (6.3.3) occurs. Thus the problem is to count the number of ways of choosing $k$ integers, with repetitions allowed, from the set $\{1, 2, \ldots, n\}$. [Any such selection can be ordered to produce (6.3.3).] By Theorem 6.3.5, the total number of selections possible is $C(k + n - 1, k)$. ◀

## 6.3 Problem-Solving Tips

The formulas in Section 6.3 generalize the formulas of Section 6.2 by allowing repetitions. A *permutation* is an ordering of $s_1, \ldots, s_n$, where the $s_i$ are *distinct*. There are $n!$ permutations. Now suppose that we have $n$ items containing *duplicates*—specifically, $n_i$ identical objects of type $i$, for $i = 1, \ldots, t$. Then the number of orderings is

$$\frac{n!}{n_1! n_2! \cdots n_t!}. \tag{6.3.4}$$

To determine whether one of these formulas may be relevant to a particular problem, first be sure that the problem asks for *orderings*. If the items to be ordered are *distinct*, the permutation formula may be used. On the other hand, if there are *duplicates* among the items to be ordered, formula (6.3.4) may be used.

An *r-combination* is an unordered selection of $r$ elements, *repetitions not allowed*, from among $n$ items. There are $C(n, r)$ $r$-combinations. Now suppose that we want to count unordered selections of $k$ elements, *repetitions allowed*, from among $t$ items. The number of such selections is

$$C(k + t - 1, t - 1). \tag{6.3.5}$$

To determine whether one of these formulas may be relevant to a particular problem, first be sure that the problem asks for *unordered* selections. If the items are to be selected *without repetition,* the combination formula may be used. On the other hand, if the items are to be selected *with repetition,* formula (6.3.5) may be used.

The following table summarizes the various formulas:

| | *No Repetitions* | *Repetitions Allowed* |
|---|---|---|
| *Ordered Selections* | $n!$ | $n!/(n_1!\cdots n_t!)$ |
| *Unordered Selections* | $C(n, r)$ | $C(k+t-1, t-1)$ |

## 6.3 Review Exercises

1. How many orderings are there of $n$ items of $t$ types with $n_i$ identical objects of type $i$? How is this formula derived?
2. How many unordered, $k$-element selections are there from a $t$-element set, repetitions allowed? How is this formula derived?

## 6.3 Exercises

*In Exercises 1–6, determine the number of strings that can be formed by ordering the letters given.*

1. *GUIDE*
2. *SCHOOL*
3. *SALESPERSONS*
4. *GOOGOO*
5. *CLASSICS*
6. *SUGGESTS*
7. How many strings can be formed by ordering the letters *SALESPERSONS* if the four *S*'s must be consecutive?
8. How many strings can be formed by ordering the letters *SALESPERSONS* if no two *S*'s are consecutive?
9. How many strings can be formed by ordering the letters *SCHOOL* using some or all of the letters?

*Exercises 10–12 refer to selections among Action Comics, Superman, Captain Marvel, Archie, X-Man, and Nancy comics.*

10. How many ways are there to select six comics?
11. How many ways are there to select 10 comics?
12. How many ways are there to select 10 comics if we choose at least one of each book?
13. How many routes are there in the ordinary $xyz$-coordinate system from the origin to the point $(i, j, k)$, where $i$, $j$, and $k$ are positive integers, if we are limited to steps one unit in the positive $x$-direction, one unit in the positive $y$-direction, or one unit in the positive $z$-direction?
14. An exam has 12 problems. How many ways can (integer) points be assigned to the problems if the total of the points is 100 and each problem is worth at least five points?
15. A bicycle collector has 100 bikes. How many ways can the bikes be stored in four warehouses if the bikes and the warehouses are considered distinct?
16. A bicycle collector has 100 bikes. How many ways can the bikes be stored in four warehouses if the bikes are indistinguishable, but the warehouses are considered distinct?
17. In how many ways can 10 distinct books be divided among three students if the first student gets five books, the second three books, and the third two books?

*Exercises 18–24 refer to piles of identical red, blue, and green balls where each pile contains at least 10 balls.*

18. In how many ways can 10 balls be selected?
19. In how many ways can 10 balls be selected if at least one red ball must be selected?
20. In how many ways can 10 balls be selected if at least one red ball, at least two blue balls, and at least three green balls must be selected?
21. In how many ways can 10 balls be selected if exactly one red ball must be selected?
22. In how many ways can 10 balls be selected if exactly one red ball and at least one blue ball must be selected?
23. In how many ways can 10 balls be selected if at most one red ball is selected?
24. In how many ways can 10 balls be selected if twice as many red balls as green balls must be selected?

*In Exercises 25–30, find the number of integer solutions of*

$$x_1 + x_2 + x_3 = 15$$

*subject to the conditions given.*

25. $x_1 \geq 0, x_2 \geq 0, x_3 \geq 0$
26. $x_1 \geq 1, x_2 \geq 1, x_3 \geq 1$
27. $x_1 = 1, x_2 \geq 0, x_3 \geq 0$
28. $x_1 \geq 0, x_2 > 0, x_3 = 1$
29. $0 \leq x_1 \leq 6, x_2 \geq 0, x_3 \geq 0$

★30. $0 \leq x_1 < 6, 1 \leq x_2 < 9, x_3 \geq 0$

★31. Find the number of solutions in integers to

$$x_1 + x_2 + x_3 + x_4 = 12$$

satisfying $0 \le x_1 \le 4$, $0 \le x_2 \le 5$, $0 \le x_3 \le 8$, and $0 \le x_4 \le 9$.

32. Prove that the number of solutions to the equation

$$x_1 + x_2 + x_3 = n, \quad n \ge 3,$$

where $x_1, x_2$, and $x_3$ are positive integers, is $(n-1)(n-2)/2$.

33. Show that the number of solutions in nonnegative integers of the inequality

$$x_1 + x_2 + \cdots + x_n \le M,$$

where $M$ is a nonnegative integer, is $C(M+n, n)$.

34. How many integers between 1 and 1,000,000 have the sum of the digits equal to 15?

★35. How many integers between 1 and 1,000,000 have the sum of the digits equal to 20?

36. How many bridge deals are there? (A deal consists of partitioning a 52-card deck into four hands, each containing 13 cards.)

37. In how many ways can three teams containing four, two, and two persons be selected from a group of eight persons?

38. A *domino* is a rectangle divided into two squares with each square numbered one of 0, 1, ..., 6, repetitions allowed. How many distinct dominoes are there?

*Exercises 39–44 refer to a bag containing 20 balls—six red, six green, and eight purple.*

39. In how many ways can we select five balls if the balls are considered distinct?

40. In how many ways can we select five balls if balls of the same color are considered identical?

41. In how many ways can we draw two red, three green, and two purple balls if the balls are considered distinct?

42. We draw five balls, then replace the balls, and then draw five more balls. In how many ways can this be done if the balls are considered distinct?

43. We draw five balls without replacing them. We then draw five more balls. In how many ways can this be done if the balls are considered distinct?

44. We draw five balls where at least one is red, and then replace them. We then draw five balls and at most one is green. In how many ways can this be done if the balls are considered distinct?

45. In how many ways can 15 identical mathematics books be distributed among six students?

46. In how many ways can 15 identical computer science books and 10 identical psychology books be distributed among five students?

47. In how many ways can we place 10 identical balls in 12 boxes if each box can hold one ball?

48. In how many ways can we place 10 identical balls in 12 boxes if each box can hold 10 balls?

49. Show that $(kn)!$ is divisible by $(n!)^k$.

50. By considering

for $i_1 = 1$ to $n$
  for $i_2 = 1$ to $i_1$
    $println(i_1, i_2)$

and Example 6.3.9, deduce

$$1 + 2 + \cdots + n = \frac{n(n+1)}{2}.$$

★51. Use Example 6.3.9 to prove the formula

$$C(k-1, k-1) + C(k, k-1) + \cdots + C(n+k-2, k-1) = C(k+n-1, k).$$

52. Write an algorithm that lists all solutions in nonnegative integers to $x_1 + x_2 + x_3 = n$.

53. What is wrong with the following argument, which supposedly counts the number of partitions of a 10-element set into eight (nonempty) subsets?

List the elements of the set with blanks between them:

$$x_1 \_ x_2 \_ x_3 \_ x_4 \_ x_5 \_ x_6 \_ x_7 \_ x_8 \_ x_9 \_ x_{10}.$$

Every time we fill seven of the nine blanks with seven vertical bars, we obtain a partition of $\{x_1, \ldots, x_{10}\}$ into eight subsets. For example, the partition $\{x_1\}, \{x_2\}, \{x_3, x_4\}\ \{x_5\}, \{x_6\}, \{x_7, x_8\}\ \{x_9\}, \{x_{10}\}$ would be represented as

$$x_1 \mid x_2 \mid x_3\, x_4 \mid x_5 \mid x_6 \mid x_7\, x_8 \mid x_9 \mid x_{10}.$$

Thus the solution to the problem is $C(9, 7)$.

## 6.4 Algorithms for Generating Permutations and Combinations

**Go Online**
For more on generating permutations, see bit.ly/2PNvtEt

The rock group Unhinged Universe has recorded $n$ videos whose running times are $t_1, t_2, \ldots, t_n$ seconds. A DVD is to be released that can hold $C$ seconds. Since this is the first DVD by Unhinged Universe, the group wants to include as much material as possible. Thus the problem is to choose a subset $\{i_1, \ldots, i_k\}$ of $\{1, 2, \ldots, n\}$ such that the sum

$$\sum_{j=1}^{k} t_{i_j} \tag{6.4.1}$$

does not exceed $C$ and is as large as possible. A straightforward approach is to examine all subsets of $\{1, 2, \ldots, n\}$ and choose a subset so that the sum (6.4.1) does not exceed $C$ and is as large as possible. To implement this approach, we need an algorithm that generates all combinations of an $n$-element set. In this section we develop algorithms to generate combinations and permutations.

Since there are $2^n$ subsets of an $n$-element set, the running time of an algorithm that examines all subsets is $\Omega(2^n)$. As we saw in Section 4.3, such algorithms are impractical to run except for small values of $n$. Unfortunately, there are problems (an example of which is the DVD-filling problem described previously) for which no method much better than the "list all" approach is known.

Our algorithms list permutations and combinations in **lexicographic order.** Lexicographic order generalizes ordinary dictionary order.

Given two distinct words, to determine whether one precedes the other in the dictionary, we compare the letters in the words. There are two possibilities:

1. The words have different lengths, and each letter in the shorter word is identical to the corresponding letter in the longer word.
2. The words have the same or different lengths, and at some position, the letters in the words differ. **(6.4.2)**

If 1 holds, the shorter word precedes the longer. (For example, "dog" precedes "doghouse" in the dictionary.) If 2 holds, we locate the leftmost position $p$ at which the letters differ. The order of the words is determined by the order of the letters at position $p$. (For example, "gladiator" precedes "gladiolus" in the dictionary. At the leftmost position at which the letters differ, we find "a" in "gladiator" and "o" in "gladiolus"; "a" precedes "o" in the alphabet.)

Lexicographic order generalizes ordinary dictionary order by replacing the alphabet by any set of symbols on which an order has been defined. We will be concerned with strings of integers.

**Definition 6.4.1** ▶ Let $\alpha = s_1 s_2 \cdots s_p$ and $\beta = t_1 t_2 \cdots t_q$ be strings over $\{1, 2, \ldots, n\}$. We say that $\alpha$ is *lexicographically less than* $\beta$ and write $\alpha < \beta$ if either

(a) $p < q$ and $s_i = t_i$ for $i = 1, \ldots, p$,

or

(b) for some $i$, $s_i \neq t_i$, and for the smallest such $i$, we have $s_i < t_i$. ◀

In Definition 6.4.1, case (a) corresponds to possibility 1 of (6.4.2) and case (b) corresponds to possibility 2 of (6.4.2).

**Example 6.4.2** Let $\alpha = 132$ and $\beta = 1324$ be strings over $\{1, 2, 3, 4\}$. In the notation of Definition 6.4.1, $p = 3$, $q = 4$, $s_1 = 1$, $s_2 = 3$, $s_3 = 2$, $t_1 = 1$, $t_2 = 3$, $t_3 = 2$, and $t_4 = 4$. Since $p = 3 < 4 = q$ and $s_i = t_i$ for $i = 1, 2, 3$, condition (a) of Definition 6.4.1 is satisfied. Therefore, $\alpha < \beta$. ◀

**Example 6.4.3** Let $\alpha = 13246$ and $\beta = 1342$ be strings over $\{1, 2, 3, 4, 5, 6\}$. In the notation of Definition 6.4.1, $p = 5$, $q = 4$, $s_1 = 1$, $s_2 = 3$, $s_3 = 2$, $s_4 = 4$, $s_5 = 6$, $t_1 = 1$, $t_2 = 3$, $t_3 = 4$, and $t_4 = 2$. The smallest $i$ for which $s_i \neq t_i$ is $i = 3$. Since $s_3 < t_3$, by condition (b) of Definition 6.4.1, $\alpha < \beta$. ◀

**Example 6.4.4** Let $\alpha = 1324$ and $\beta = 1342$ be strings over $\{1, 2, 3, 4\}$. In the notation of Definition 6.4.1, $p = q = 4$, $s_1 = 1$, $s_2 = 3$, $s_3 = 2$, $s_4 = 4$, $t_1 = 1$, $t_2 = 3$, $t_3 = 4$, and $t_4 = 2$. The smallest $i$ for which $s_i \neq t_i$ is $i = 3$. Since $s_3 < t_3$, by condition (b) of Definition 6.4.1, $\alpha < \beta$. ◀

**Example 6.4.5** Let $\alpha = 13542$ and $\beta = 21354$ be strings over $\{1, 2, 3, 4, 5\}$. In the notation of Definition 6.4.1, $s_1 = 1$, $s_2 = 3$, $s_3 = 5$, $s_4 = 4$, $s_5 = 2$, $t_1 = 2$, $t_2 = 1$, $t_3 = 3$, $t_4 = 5$, and $t_5 = 4$. The smallest $i$ for which $s_i \neq t_i$ is $i = 1$. Since $s_1 < t_1$, by condition (b) of Definition 6.4.1, $\alpha < \beta$. ◀

For strings of the same length over $\{1, 2, \ldots, 9\}$, lexicographic order is the same as numerical order on the positive integers if we interpret the strings as decimal numbers (see Examples 6.4.4 and 6.4.5). For strings of unequal length, lexicographic order may be different than numerical order (see Example 6.4.3). Throughout the remainder of this section, *order* will refer to lexicographic order.

First we consider the problem of listing all $r$-combinations of $\{1, 2, \ldots, n\}$. In our algorithm, we will list the $r$-combination $\{x_1, \ldots, x_r\}$ as the string $s_1 \cdots s_r$, where $s_1 < s_2 < \cdots < s_r$ and $\{x_1, \ldots, x_r\} = \{s_1, \ldots, s_r\}$. For example, the 3-combination $\{6, 2, 4\}$ will be listed as 246.

We will list the $r$-combinations of $\{1, 2, \ldots, n\}$ in lexicographic order. Thus the first listed string will be $12 \cdots r$ and the last listed string will be $(n - r + 1) \cdots n$.

**Example 6.4.6** Consider the order in which the 5-combinations of $\{1, 2, 3, 4, 5, 6, 7\}$ will be listed. The first string is 12345, which is followed by 12346 and 12347. The next string is 12356, followed by 12357. The last string will be 34567. ◀

**Example 6.4.7** Find the string that follows 13467 when we list the 5-combinations of $X = \{1, 2, 3, 4, 5, 6, 7\}$.

**SOLUTION** No string that begins 134 and represents a 5-combination of $X$ exceeds 13467. Thus the string that follows 13467 must begin 135. Since 13567 is the smallest string that begins 135 and represents a 5-combination of $X$, the answer is 13567. ◀

**Example 6.4.8** Find the string that follows 2367 when we list the 4-combinations of $X = \{1, 2, 3, 4, 5, 6, 7\}$.

**SOLUTION** No string that begins 23 and represents a 4-combination of $X$ exceeds 2367. Thus the string that follows 2367 must begin 24. Since 2456 is the smallest string that begins 24 and represents a 4-combination of $X$, the answer is 2456. ◀

A pattern is developing. Given a string $\alpha = s_1 \cdots s_r$, which represents the $r$-combination $\{s_1, \ldots, s_r\}$, to find the next string $\beta = t_1 \cdots t_r$, we find the rightmost element $s_m$ that is not at its maximum value. ($s_r$ may have the maximum value $n$, $s_{r-1}$ may have the maximum value $n - 1$, etc.) Then

$$t_i = s_i \quad \text{for } i = 1, \ldots, m - 1.$$

The element $t_m$ is equal to $s_m + 1$. For the remainder of the string $\beta$ we have

$$t_{m+1} \cdots t_r = (s_m + 2)(s_m + 3) \cdots .$$

The algorithm follows.

**Algorithm 6.4.9 Generating Combinations**

This algorithm lists all $r$-combinations of $\{1, 2, \ldots, n\}$ in increasing lexicographic order.

Input: $r, n$

Output: All $r$-combinations of $\{1, 2, \ldots, n\}$ in increasing lexicographic order

```
 1. combination(r, n) {
 2.   for i = 1 to r
 3.     s_i = i
 4.   println(s_1, ..., s_r) // print the first r-combination
 5.   for i = 2 to C(n, r) {
 6.     m = r
 7.     max_val = n
 8.     while (s_m == max_val) {
 9.       // find the rightmost element not at its maximum value
10.       m = m − 1
11.       max_val = max_val − 1
12.     }
13.     // the rightmost element is incremented
14.     s_m = s_m + 1
15.     // the rest of the elements are the successors of s_m
16.     for j = m + 1 to r
17.       s_j = s_{j−1} + 1
18.     println(s_1, ..., s_r) // print the ith combination
19.   }
20. }
```

**Example 6.4.10** We will show how Algorithm 6.4.9 generates the 5-combination of $\{1, 2, 3, 4, 5, 6, 7\}$ that follows 23467. We are supposing that

$$s_1 = 2, \quad s_2 = 3, \quad s_3 = 4, \quad s_4 = 6, \quad s_5 = 7.$$

At line 13, we find that $s_3$ is the rightmost element not at its maximum value. At line 14, $s_3$ is set to 5. At lines 16 and 17, $s_4$ is set to 6 and $s_5$ is set to 7. At this point

$$s_1 = 2, \quad s_2 = 3, \quad s_3 = 5, \quad s_4 = 6, \quad s_5 = 7.$$

We have generated the 5-combination 23567, which follows 23467. ◀

We next prove that Algorithm 6.4.9 is correct, that is, that it generates all $r$-combinations of $\{1, 2, \ldots, n\}$ in increasing lexicographic order.

**Theorem 6.4.11** Algorithm 6.4.9 generates all r-combinations of $\{1, 2, \ldots, n\}$ in increasing lexicographic order. Furthermore, each $r$ combination output by Algorithm 6.4.9 lists the digits in increasing order.

**Proof** In this proof, "lexicographic order" means "increasing lexicographic order." We first show that Algorithm 6.4.9 generates $r$-combinations of $\{1, 2, \ldots, n\}$ in lexicographic order and that each $r$-combination lists the digits in increasing order. We use induction on $k$ to show that the $k$th $r$-combination output by Algorithm 6.4.9 lists the digits in increasing order, and that the $(k+1)$st $r$-combination output by Algorithm 6.4.9 is lexicographically the next $r$-combination.

For $k = 1$ (base case), the first $r$-combination output by Algorithm 6.4.9 is $12 \cdots r$, which has the digits in increasing order. If $r = n$, the proof is complete. Otherwise, in the for loop at lines 5–19, Algorithm 6.4.9 generates the next (i.e., second lexicographically) $r$-combination $12 \cdots (r-1)(r+1)$ and outputs it. This $r$-combination has the digits in increasing order.

If $k = C(n, r)$, the proof is complete. Otherwise, for the inductive step, assume that the $k$th $r$-combination output by Algorithm 6.4.9 has the digits in increasing order. In the for loop at lines 5–19, Algorithm 6.4.9 generates the $(k+1)$st (lexicographically, the next) $r$-combination, and, by construction, the digits are in increasing order. The inductive step is complete. Therefore, Algorithm 6.4.9 generates $r$-combinations of $\{1, 2, \ldots, n\}$ in lexicographic order, and each $r$-combination has the digits in increasing order.

Finally, we show that Algorithm 6.4.9 generates all $r$-combinations of $\{1, 2, \ldots, n\}$. Suppose, by way of contradiction, that some $r$-combination is not generated by Algorithm 6.4.9, and let $s$, with the digits listed in increasing order, be the least $r$-combination (lexicographically) that is not generated. Then $s$ is not $12 \cdots n$ because it is generated. Let $s'$ be the predecessor of $s$ lexicographically. Since $s$ is the least $r$-combination not generated, $s'$ is generated. But by construction Algorithm 6.4.9 generates the next $r$-combination, which is $s$. This contradiction shows that Algorithm 6.4.9 generates all $r$-combinations of $\{1, 2, \ldots, n\}$. ◀

**Example 6.4.12** The 4-combinations of $\{1, 2, 3, 4, 5, 6\}$ as listed by Algorithm 6.4.9 are

1234, 1235, 1236, 1245, 1246, 1256, 1345, 1346,
1356, 1456, 2345, 2346, 2356, 2456, 3456. ◀

**Example 6.4.13** Let $X = \{1, 2, 3, 4, 5, 6, 7, 8, 9\}$.

(a) What is the first 5-combination of $X$ listed by Algorithm 6.4.9 that begins 24?

(b) What is the last 5-combination of $X$ listed by Algorithm 6.4.9 that begins 13?

(c) How many 5-combinations of $X$ begin 13?

**SOLUTION**

(a) The digit following 4 must be larger than 4 since each combination listed by Algorithm 6.4.9 has the digits in increasing order. Since 246__, 247__, 248__, ... are lexicographically larger than 245__, and Algorithm 6.4.9 lists the 5-combinations in lexicographic order, the digit following 4 must be 5. Similarly the digit following 5 must be 6, and the final digit must be 7 giving us 24567. Therefore the first 5-combination of $X$ listed by Algorithm 6.4.9 that begins 24 is 24567.

(b) The digit following 3 must be as large as possible but smaller than the last two digits since Algorithm 6.4.9 lists 5-combinations with digits in increasing order. Thus the digit following 3 must be 7; hence, the desired 5-combination begins 137. Similarly the digit following 7 must be 8, and the digit following 8 must be 9. Therefore, the last 5-combination of $X$ listed by Algorithm 6.4.9 that begins 13 is 13789.

(c) Since Algorithm 6.4.9 generates all 5-combinations of $X$ with the digits in increasing order, the last three digits must be selected from

$$\{4, 5, 6, 7, 8, 9\},$$

and there are $C(6, 3) = 20$ of them. Therefore, there are twenty 5-combinations of $X$ that begin 13. ◀

Like the algorithm for generating $r$-combinations, the algorithm to generate permutations will list the permutations of $\{1, 2, \ldots, n\}$ in lexicographic order. (Exercise 26 asks for an algorithm that generates all $r$-permutations of an $n$-element set.)

**Example 6.4.14** To construct the permutation of $\{1, 2, 3, 4, 5, 6\}$ that follows 163542, we should keep as many digits as possible at the left the same.

Can the permutation following the given permutation have the form 1635__? Since the only permutation of the form 1635__ distinct from the given permutation is 163524, and 163524 is smaller than 163542, the permutation following the given permutation is not of the form 1635__.

Can the permutation following the given permutation have the form 163___? The last three digits must be a permutation of $\{2, 4, 5\}$. Since 542 is the largest permutation of $\{2, 4, 5\}$, any permutation that begins 163 is smaller than the given permutation. Thus the permutation following the given permutation is not of the form 163___.

The reason that the permutation following the given permutation cannot begin 1635 or 163 is that in either case the remaining digits in the given permutation (42 and 542, respectively) *decrease*. Therefore, working from the right, we must find the first digit $d$ whose right neighbor $r$ satisfies $d < r$. In our case, the third digit, 3, has this property. Thus the permutation following the given permutation will begin 16.

The digit following 16 must exceed 3. Since we want the next smallest permutation, the next digit is 4, the smallest available digit. Thus the desired permutation begins 164. The remaining digits 235 must be in increasing order to achieve the minimum value. Therefore, the permutation following the given permutation is 164235. ◀

We see that to generate all of the permutations of $\{1, 2, \ldots, n\}$, we can begin with the permutation $12 \cdots n$ and then repeatedly use the method of Example 6.4.12 to generate the next permutation. We will end when the permutation $n(n-1) \cdots 21$ is generated.

**Example 6.4.15** Using the method of Example 6.4.14, we can list the permutations of $\{1, 2, 3, 4\}$ in lexicographic order as

1234, 1243, 1324, 1342, 1423, 1432, 2134, 2143,
2314, 2341, 2413, 2431, 3124, 3142, 3214, 3241,
3412, 3421, 4123, 4132, 4213, 4231, 4312, 4321. ◀

The algorithm follows.

**Algorithm 6.4.16** **Generating Permutations**

This algorithm lists all permutations of $\{1, 2, \ldots, n\}$ in increasing lexicographic order.

**Go Online**
For a C program implementing this algorithm, see bit.ly/2yrcuZC

Input: $n$

Output: All permutations of $\{1, 2, \ldots, n\}$ in increasing lexicographic order

```
1.  permutation(n) {
2.    for i = 1 to n
3.      s_i = i
4.    println(s_1, ..., s_n) // print the first permutation
5.    for i = 2 to n! {
6.      m = n − 1
7.      while (s_m > s_{m+1})
8.        // find the first decrease working from the right
9.        m = m − 1
10.     k = n
11.     while (s_m > s_k)
12.       // find the rightmost element s_k with s_m < s_k
13.       k = k − 1
14.     swap(s_m, s_k)
15.     p = m + 1
16.     q = n
17.     while (p < q) {
18.       // swap s_{m+1} and s_n, swap s_{m+2} and s_{n−1}, and so on
19.       swap(s_p, s_q)
20.       p = p + 1
21.       q = q − 1
22.     }
23.     println(s_1, ..., s_n) // print the ith permutation
24.   }
25. }
```

We leave the proof that Algorithm 6.4.16 is correct to the exercises (see Exercise 33).

**Example 6.4.17** Show how Algorithm 6.4.16 generates the permutation that follows 163542.

**SOLUTION** Suppose that

$$s_1 = 1, \quad s_2 = 6, \quad s_3 = 3, \quad s_4 = 5, \quad s_5 = 4, \quad s_6 = 2$$

and that we are at line 6. The largest index $m$ satisfying $s_m < s_{m+1}$ is 3. At lines 10–13, we find that the largest index $k$ satisfying $s_k > s_m$ is 5. At line 14, we swap $s_m$ and $s_k$. At this point, we have $s = 164532$. At lines 15–22, we reverse the order of the elements $s_4s_5s_6 = 532$. We obtain the desired permutation, 164235. ◀

**Example 6.4.18** Let $X = \{1, 2, 3, 4, 5, 6, 7, 8, 9\}$.

(a) What is the first permutation of $X$ listed by Algorithm 6.4.16 that begins 48?

(b) What is the last permutation of $X$ listed by Algorithm 6.4.16 that begins 48?

(c) How many permutations of $X$ begin 48?

SOLUTION

(a) The digit following 8 must be 1 since $482\cdots, 483\cdots, \ldots$ are lexicographically larger than $481\cdots$, and Algorithm 6.4.16 lists the permutations in lexicographic order. Similarly the next digits must be 235679. Therefore the first permutation of $X$ listed by Algorithm 6.4.16 that begins 48 is 481235679.

(b) The digit following 8 must be as large as possible since Algorithm 6.4.16 lists the permutations in lexicographic order. Thus the digit following 8 is 9. Similarly, the next digits must be 765321. Therefore, the last permutation of $X$ listed by Algorithm 6.4.16 that begins 48 is 489765321.

(c) The permutations of $X$ that begin 48 are 48 followed by a permutation of

$$\{1, 2, 3, 5, 6, 7, 9\},$$

and there are $7! = 5040$ of them. Therefore, there are 5040 permutations of $X$ that begin 48. ◀

## 6.4 Review Exercises

1. Define *lexicographic order*.
2. Describe the algorithm for generating $r$-combinations.
3. Describe the algorithm for generating permutations.

## 6.4 Exercises

*In Exercises 1–3, find the r-combination that will be generated by Algorithm 6.4.9 with $n = 7$ after the r-combination given.*

**1.** 1356 **2.** 12367 **3.** 14567

*In Exercises 4–6, find the permutation that will be generated by Algorithm 6.4.16 after the permutation given.*

**4.** 12354 **5.** 625431 **6.** 12876543

**7.** For each string in Exercises 1–3, explain (as in Example 6.4.10) exactly how Algorithm 6.4.9 generates the next $r$-combination.

**8.** For each string in Exercises 4–6, explain (as in Example 6.4.17) exactly how Algorithm 6.4.16 generates the next permutation.

**9.** Show the output from Algorithm 6.4.9 when $n = 6$ and $r = 3$.

**10.** Show the output from Algorithm 6.4.9 when $n = 6$ and $r = 2$.

**11.** Show the output from Algorithm 6.4.9 when $n = 7$ and $r = 5$.

**12.** Show the output from Algorithm 6.4.16 when $n = 2$.

**13.** Show the output from Algorithm 6.4.16 when $n = 3$.

*In Exercises 14–23, let $X = \{1, 2, 3, 4, 5, 6, 7, 8, 9\}$.*

**14.** What is the first 5-combination of $X$ listed by Algorithm 6.4.9 that ends 79?

**15.** What is the last 5-combination of $X$ listed by Algorithm 6.4.9 that ends 68?

**16.** How many 5-combinations of $X$ listed by Algorithm 6.4.9 end 89?

**17.** How many 5-combinations of $X$ listed by Algorithm 6.4.9 end 53?

**18.** How many 6-combinations of $X$ listed by Algorithm 6.4.9 end 46?

**19.** How many 6-combinations of $X$ listed by Algorithm 6.4.9 start 2 and end 79?

**20.** What is the first permutation of $X$ listed by Algorithm 6.4.16 that ends 97?

**21.** What is the last permutation of $X$ listed by Algorithm 6.4.16 that ends 397?

**22.** How many permutations of $X$ listed by Algorithm 6.4.16 end 1938?

**23.** How many permutations of $X$ listed by Algorithm 6.4.16 end 1234?

**24.** Modify Algorithm 6.4.9 so that line 5

```
5.     for i = 2 to C(n, r) {
```

is eliminated. Base the terminating condition on the fact that the last $r$-combination has every element $s_i$ equal to its maximum value.

**25.** Modify Algorithm 6.4.16 so that line 5

```
5.     for i = 2 to n! {
```

is eliminated. Base the terminating condition on the fact that the last permutation has the elements $s_i$ in decreasing order.

**26.** Write an algorithm that generates all $r$-permutations of an $n$-element set.

**27.** Write an algorithm whose input is an $r$-combination of $\{1, 2, \ldots, n\}$. The output is the next (in lexicographic der) $r$-combination. The first $r$-combination follows the last $r$-combination.

**28.** Write an algorithm whose input is a permutation of $\{1, 2, \ldots, n\}$. The output is the next (in lexicographic order) permutation. The first permutation follows the last permutation.

**29.** Write an algorithm whose input is an $r$-combination of $\{1, 2, \ldots, n\}$. The output is the previous (in lexicographic order) $r$-combination. The last $r$-combination precedes the first $r$-combination.

**30.** Write an algorithm whose input is a permutation of $\{1, 2, \ldots, n\}$. The output is the previous (in lexicographic order) permutation. The last permutation precedes the first permutation.

★**31.** Write a recursive algorithm that generates all $r$-combinations of the set $\{s_1, s_2, \ldots, s_n\}$. Divide the problem into two subproblems:

- List the $r$-combinations containing $s_1$.
- List the $r$-combinations not containing $s_1$.

**32.** Write a recursive algorithm that generates all permutations of the set $\{s_1, s_2, \ldots, s_n\}$. Divide the problem into $n$ subproblems:

- List the permutations that begin with $s_1$.
- List the permutations that begin with $s_2$.

⋮

- List the permutations that begin with $s_n$.

**33.** Show that Algorithm 6.4.16 is correct.

## 6.5 Introduction to Discrete Probability†

**Go Online**
For more on probability, see `bit.ly/2PNvtEt`

**Probability** was developed in the seventeenth century to analyze games and, in this earliest form, directly involved counting. For example, suppose that a six-sided fair die whose sides are labeled 1, 2, 3, 4, 5, 6 is rolled (see Figure 6.5.1). "Fair" means that each number is equally likely to appear when the die is rolled. To compute the chance or *probability* that an even number appears, we first *count* how many ways an even number can appear (three: 2, 4, 6) and how may ways an arbitrary number can appear (six: 1, 2, 3, 4, 5, 6); then, the probability is the quotient: $3/6 = 1/2$. After introducing some terminology, we will give several examples of computing probabilities.

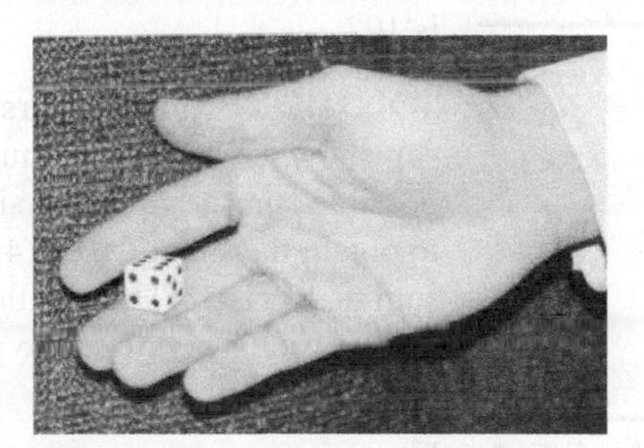

**Figure 6.5.1** Rolling a fair die. "Fair" means that each number is equally likely to appear when the die is rolled. [*Photo by the author. Hand courtesy of Ben Schneider.*]

An **experiment** is a process that yields an outcome. An **event** is an outcome or combination of outcomes from an experiment. The **sample space** is the event consisting of all possible outcomes.

**Example 6.5.1** Examples of experiments are

- Rolling a six-sided die.
- Randomly selecting 5 microprocessors from a lot of 1000 microprocessors.
- Selecting a newborn child at St. Rocco's Hospital.

†This section can be omitted without loss of continuity.

Examples of events that might occur when the previous experiments are performed are

- Obtaining a 4 when rolling a six-sided die.
- Finding no defective microprocessors out of 5 randomly chosen from a lot of 1000.
- Selecting a newborn female child at St. Rocco's Hospital.

The sample spaces for the previous experiments are

- The numbers 1, 2, 3, 4, 5, 6—all possible outcomes when a die is rolled.
- All possible combinations of 5 microprocessors selected from a lot of 1000 microprocessors.
- All newborn children at St. Rocco's Hospital. ◀

If all outcomes in a finite sample space are equally likely, the probability of an event is defined as the number of outcomes in the event divided by the number of outcomes in the sample space. In the following section, we will relax the assumption that all outcomes are equally likely.

**Definition 6.5.2** ▶ Let $S$ be a finite sample space in which all outcomes are equally likely. The *probability* $P(E)$ of an event $E$ from $S$ is

$$P(E) = \frac{|E|}{|S|}.^{\dagger}$$

**Example 6.5.3** Two fair dice are rolled. What is the probability that the sum of the numbers on the dice is 10?

**SOLUTION** Since the first die can show any one of six numbers and the second die can show any one of six numbers, by the Multiplication Principle there are $6 \cdot 6 = 36$ possible sums; that is, the size of the sample space is 36. There are three possible ways to obtain the sum of 10—(4, 6), (5, 5), (6, 4)—that is, the size of the event "obtaining a sum of 10" is 3. [The notation $(x, y)$ means that we obtain $x$ on the first die and $y$ on the second die.] Therefore, the probability is $3/36 = 1/12$. ◀

**Example 6.5.4** Five microprocessors are randomly selected from a lot of 1000 microprocessors among which 20 are defective. Find the probability of obtaining no defective microprocessors.

**SOLUTION** There are $C(1000, 5)$ ways to select 5 microprocessors among 1000. There are $C(980, 5)$ ways to select 5 good microprocessors since there are $1000 - 20 = 980$ good microprocessors. Therefore, the probability of obtaining no defective microprocessors is

$$\frac{C(980, 5)}{C(1000, 5)} = \frac{980 \cdot 979 \cdot 978 \cdot 977 \cdot 976}{1000 \cdot 999 \cdot 998 \cdot 997 \cdot 996} = 0.903735781.$$ ◀

**Example 6.5.5** In a state lottery game, to win the grand prize the contestant must match six distinct numbers, in any order, among the numbers 1 through 52 randomly drawn by a lottery representative. What is the probability of choosing the winning numbers?

---

†Recall that $|X|$ is the number of elements in a finite set $X$.

**SOLUTION** Six numbers among 52 can be selected in $C(52, 6)$ ways. Since there is one winning combination, the probability of choosing the winning numbers is

$$\frac{1}{C(52,6)} = \frac{6!}{52 \cdot 51 \cdot 50 \cdot 49 \cdot 48 \cdot 47} = 0.000000049.$$ ◀

**Example 6.5.6** A bridge hand consists of 13 cards from an ordinary 52-card deck. Find the probability of obtaining a 4–4–4–1 distribution, that is, four cards in each of three different suits and one card of a fourth suit.

**SOLUTION** There are $C(52, 13)$ bridge hands. The one-card suit can be chosen in 4 ways, and the card itself can be chosen in 13 ways. Having chosen this card, we must choose four cards from each of the three remaining suits, which can be done in $C(13, 4)^3$ ways. Thus there are $4 \cdot 13 \cdot C(13, 4)^3$ hands with a 4–4–4–1 distribution. Therefore, the probability of obtaining a 4–4–4–1 distribution is

$$\frac{4 \cdot 13 \cdot C(13,4)^3}{C(52,13)} = 0.03.$$ ◀

## 6.5 Review Exercises

1. What is an experiment?
2. What is an event?
3. What is a sample space?
4. If all outcomes in a finite sample space are equally likely, how is the probability of an event defined?

## 6.5 Exercises

*In Exercises 1–4, suppose that a coin is flipped and a die is rolled.*

1. List the members of the sample space.
2. List the members of the event "the coin shows a head and the die shows an even number."
3. List the members of the event "the die shows an odd number."
4. List the members of the event "the coin shows a head and the die shows a number less than 4."

*In Exercises 5–7, two dice are rolled.*

5. List the members of the event "the sum of the numbers on the dice is even."
6. List the members of the event "doubles occur" (i.e., the numbers are the same on both dice).
7. List the members of the event "4 appears on at least one die."
8. Give an example of an experiment different from those in this section.
9. Give an example of an event when the experiment of Exercise 8 is performed.
10. What is the sample space for the experiment of Exercise 8?
11. One fair die is rolled. What is the probability of getting a 5?
12. One fair die is rolled. What is the probability of getting an even number?
13. One fair die is rolled. What is the probability of not getting a 5?
14. A card is selected at random from an ordinary 52-card deck. What is the probability that it is the ace of spades?
15. A card is selected at random from an ordinary 52-card deck. What is the probability that it is a jack?
16. A card is selected at random from an ordinary 52-card deck. What is the probability that it is a heart?
17. Two fair dice are rolled. What is the probability that the sum of the numbers on the dice is 9?
18. Two fair dice are rolled. What is the probability that the sum of the numbers on the dice is odd?
19. Two fair dice are rolled. What is the probability of doubles?
20. Four microprocessors are randomly selected from a lot of 100 microprocessors among which 10 are defective. Find the probability of obtaining no defective microprocessors.
21. Four microprocessors are randomly selected from a lot of 100 microprocessors among which 10 are defective. Find the probability of obtaining exactly one defective microprocessor.
22. Four microprocessors are randomly selected from a lot of 100 microprocessors among which 10 are defective. Find the probability of obtaining at most one defective microprocessor.

**23.** In the California Daily 3 game, a contestant must select three numbers among 0 to 9, repetitions allowed. A "straight play" win requires that the numbers be matched in the exact order in which they are randomly drawn by a lottery representative. What is the probability of choosing the winning numbers?

**24.** In the California Daily 3 game, a contestant must select three numbers among 0 to 9. One type of "box play" win requires that three numbers match in any order those randomly drawn by a lottery representative, repetitions allowed. What is the probability of choosing the winning numbers, assuming that the contestant chooses three distinct numbers?

**25.** In the Maryland Lotto game, to win the grand prize the contestant must match six distinct numbers, in any order, among the numbers 1 through 49 randomly drawn by a lottery representative. What is the probability of choosing the winning numbers?

**26.** In the multi-state Big Game, to win the grand prize the contestant must match five distinct numbers, in any order, among the numbers 1 through 50, and one Big Money Ball number between 1 and 36, all randomly drawn by a lottery representative. What is the probability of choosing the winning numbers?

**27.** In the Maryland Cash In Hand game, to win the grand prize the contestant must match seven distinct numbers, in any order, among the numbers 1 through 31 randomly drawn by a lottery representative. What is the probability of choosing the winning numbers?

**28.** Find the probability of obtaining a bridge hand with 5–4–2–2 distribution, that is, five cards in one suit, four cards in another suit, and two cards in each of the other two suits.

**29.** Find the probability of obtaining a bridge hand consisting only of red cards, that is, no spades and no clubs.

*Exercises 30–33 concern an unprepared student who takes a 10-question true–false quiz and guesses at the answer to every question.*

**30.** What is the probability that the student answers every question correctly?

**31.** What is the probability that the student answers every question incorrectly?

**32.** What is the probability that the student answers exactly one question correctly?

**33.** What is the probability that the student answers exactly five questions correctly?

*Exercises 34–36 refer to a small consumer survey in which 10 people were asked to choose a cola among Coke, Pepsi, and RC.*

**34.** If each person chose a cola randomly, what is the probability that no one chose Coke?

**35.** If each person chose a cola randomly, what is the probability that at least one person did not choose Coke?

**36.** If each person chose a cola randomly, what is the probability that everyone chose Coke?

**37.** If five student records are chosen randomly, what is the probability that they are chosen so that the first record selected has the lowest grade point average (GPA), the second selected has the second-lowest GPA, and so on? Assume that the GPAs are distinct.

*Exercises 38–40 concern three persons who each randomly choose a locker among 12 consecutive lockers.*

**38.** What is the probability that the three lockers chosen are consecutive?

**39.** What is the probability that no two lockers are consecutive?

**40.** What is the probability that at least two of the lockers are consecutive?

*Exercises 41–44 deal with a roulette wheel that has 38 numbers: 18 red, 18 black, a 0, and a 00 (0 and 00 are neither red nor black). When the wheel is spun, all numbers are equally likely to be selected.*

**41.** What is the probability that the wheel lands on a black number?

**42.** What is the probability that the wheel lands on a black number twice in a row?

**43.** What is the probability that the wheel lands on 0?

**44.** What is the probability that the wheel lands on 0 or 00?

*Exercises 45–48 concern the Monty Hall problem, in which a contestant chooses one of three doors; behind one of the doors is a car, and behind the other two are goats. After the contestant chooses a door, the host opens one of the other two doors that hides a goat. (Because there are two goats, the host can open a door that hides a goat no matter which door the contestant first chooses.) The host then gives the contestant the option of abandoning the chosen door in favor of the still-closed, unchosen door. For each strategy, what is the probability of winning the car?*

**45.** Stay with the door initially chosen.

**46.** Make a random decision about whether to stay with the door initially chosen or switch to the unchosen, unopened door.

**47.** Switch to the unchosen, unopened door.

**48.** Suppose that the host forgets which door hides the car and, after the contestant chooses a door, picks a door at random. If the door hides a car, the game is over. Assuming that the host chooses a door that hides a goat, what is the probability of winning the car for each strategy in Exercises 45–47?

*Exercises 49–51 concern a variant of the Monty Hall problem, in which a contestant chooses one of four doors; behind one of the doors is a car, and behind the other three are goats. After the contestant chooses a door, the host opens one of the other three doors that hides a goat. The host then gives the contestant the option of abandoning the chosen door in favor of one of the two still-closed, unchosen doors. For each strategy, what is the probability of winning the car?*

**49.** Stay with the door initially chosen.

50. Make a random decision about whether to stay with the door initially chosen or switch to one of the unchosen, unopened doors.

51. Switch to one of the unchosen, unopened doors. The choice between the two unchosen, unopened doors is made randomly.

52. In a multiple-choice exam, one question has three choices: A, B, C. A student randomly chooses A. The teacher then states that choice C is incorrect. What is the probability of a correct answer if the student stays with choice A? What is the probability of a correct answer if the student switches to choice B?

53. Is the following reasoning correct? A county health inspector told a restaurant offering four-egg quiches that, because research by the FDA (Food and Drug Administration) shows that one in four eggs is contaminated with salmonella bacteria, the restaurant should use only three eggs in each quiche.

54. A two-person game is played in which a fair coin is tossed until either the sequence HT (heads, tails) or the sequence TT (tails, tails) appears. If HT appears, the first player wins; if TT appears, the second player wins. Would you rather be the first or second player? Explain.

*Exercises 55 and 56 refer to 10 identical compact discs that are randomly given to Mary, Ivan, and Juan.*

55. What is the probability that each person receives at least two compact discs?

56. What is the probability that Ivan receives exactly three compact discs?

## 6.6 Discrete Probability Theory†

In Section 6.5, we assume that all outcomes are equally likely; that is, if there are $n$ possible outcomes, the probability of each outcome is $1/n$. In general, the outcomes are not equally likely. For example, a "loaded" die is weighted so that certain numbers are more likely to appear than others. To handle the case of outcomes that are not equally likely, we assign a probability $P(x)$ to each outcome $x$. The values $P(x)$ need *not* all be the same. We call $P$ a *probability function*. Throughout this section, we assume that all sample spaces are finite.

**Definition 6.6.1** ▶ A *probability function* $P$ assigns to each outcome $x$ in a sample space $S$ a number $P(x)$ so that

$$0 \le P(x) \le 1 \qquad \text{for all } x \in S,$$

and

$$\sum_{x \in S} P(x) = 1.$$ ◀

The first condition guarantees that the probability of an outcome is nonnegative and at most 1, and the second condition guarantees that the sum of all the probabilities is 1—that is, that some outcome will occur when the experiment is performed.

**Example 6.6.2** Suppose that a die is loaded so that the numbers 2 through 6 are equally likely to appear, but that 1 is three times as likely as any other number to appear. To model this situation, we should have

$$P(2) = P(3) = P(4) = P(5) = P(6)$$

and

$$P(1) = 3P(2).$$

†This section can be omitted without loss of continuity.

Since

$$\begin{aligned} 1 &= P(1) + P(2) + P(3) + P(4) + P(5) + P(6) \\ &= 3P(2) + P(2) + P(2) + P(2) + P(2) + P(2) = 8P(2), \end{aligned}$$

we must have $P(2) = 1/8$. Therefore,

$$P(2) = P(3) = P(4) = P(5) = P(6) = \frac{1}{8}$$

and

$$P(1) = 3P(2) = \frac{3}{8}.$$ ◀

The **probability of an event** $E$ is defined as the sum of the probabilities of the outcomes in $E$.

**Definition 6.6.3 ▶** Let $E$ be an event. The *probability of* $E$, $P(E)$, is

$$P(E) = \sum_{x \in E} P(x).$$

◀

**Example 6.6.4** Given the assumptions of Example 6.6.2, the probability of an odd number is

$$P(1) + P(3) + P(5) = \frac{3}{8} + \frac{1}{8} + \frac{1}{8} = \frac{5}{8}.$$

Of course, for a fair die (with equally likely probabilities), the probability of an odd number is 1/2. ◀

## Formulas

We next develop some formulas that are useful in computing probabilities.

**Theorem 6.6.5** Let $E$ be an event. The probability of $\overline{E}$, the complement of $E$, satisfies

$$P(E) + P(\overline{E}) = 1.$$

**Proof** Suppose that $E = \{x_1, \ldots, x_k\}$ and $\overline{E} = \{x_{k+1}, \ldots, x_n\}$. Then

$$P(E) = \sum_{i=1}^{k} P(x_i) \quad \text{and} \quad P(\overline{E}) = \sum_{i=k+1}^{n} P(x_i).$$

Now

$$\begin{aligned} P(E) + P(\overline{E}) &= \sum_{i=1}^{k} P(x_i) + \sum_{i=k+1}^{n} P(x_i) \\ &= \sum_{i=1}^{n} P(x_i) = 1. \end{aligned}$$

The last equality follows from Definition 6.6.1, which states that the sum of the probabilities of all outcomes equals 1. ◀

Theorem 6.6.5 is often useful when it is easier to compute $P(\overline{E})$ than $P(E)$. After computing $P(\overline{E})$, we may obtain $P(E)$ by subtracting $P(\overline{E})$ from 1.

**Example 6.6.6** Five microprocessors are randomly selected from a lot of 1000 microprocessors among which 20 are defective. In Example 6.5.4, we found that the probability of obtaining no defective microprocessors is 0.903735781. By Theorem 6.6.5, the probability of obtaining at least one defective microprocessor is

$$1 - 0.903735781 = 0.096264219.$$

Notice how much more complex a direct approach would be to calculate the probability of obtaining at least one deficient microprocessor. We would have to calculate the probability of obtaining exactly one deficient microprocessor, then exactly two deficient microprocessors, then exactly three deficient microprocessors, then exactly four deficient microprocessors, and then exactly five deficient microprocessors; and then sum the resulting probabilities:

$$\frac{C(20,1)C(980,4) + C(20,2)C(980,3) + C(20,3)C(980,2) + C(20,4)C(980,1) + C(20,5)C(980,0)}{C(1000,5)}$$

◀

**Example 6.6.7** **Birthday Problem** Find the probability that among $n$ persons, at least two people have birthdays on the same month and date (but not necessarily in the same year). Assume that all months and dates are equally likely, and ignore February 29 birthdays.

**Go Online**
For more on the birthday problem, see `bit.ly/2PNvtEt`

SOLUTION We let $E$ denote the event "at least two persons have the same birthday." Then $\overline{E}$ is the event "no two persons have the same birthday." As we shall see, it is easier to compute $P(\overline{E})$ than $P(E)$. We can use Theorem 6.6.5 to obtain the desired probability.

Since all months and dates are equally likely and we are ignoring February 29 birthdays, the size of the sample space is $365^n$.

The first person's birthday can occur on any one of 365 days. If no two persons have the same birthday, the second person's birthday can occur on any day except the day of the first person's birthday. Therefore, the second person's birthday can occur on any one of 364 days. Similarly, the third person's birthday can occur on any one of 363 days. It follows that the size of the event "no two persons have the same birthday" is

$$365 \cdot 364 \cdots (365 - n + 1).$$

By Theorem 6.6.5, the probability that at least two persons have birthdays on the same month and date is

$$1 - \frac{365 \cdot 364 \cdots (365 - n + 1)}{365^n}.$$

For $n = 22$, the probability is 0.475695, and for $n = 23$, the probability is 0.507297. Thus if $n \geq 23$, the probability is greater than 1/2 that at least two persons have birthdays on the same month and date. Many persons would guess that $n$ would have to be *considerably* larger than 23 for the probability to be greater than 1/2. ◀

If $E_1$ and $E_2$ are events, the event $E_1 \cup E_2$ represents the event $E_1$ *or* $E_2$ (or both), and the event $E_1 \cap E_2$ represents the event $E_1$ *and* $E_2$.

**Example 6.6.8** Among a group of students, some take art and some take computer science. A student is selected at random. Let $A$ be the event "the student takes art," and let $C$ be the event "the student takes computer science." Then $A \cup C$ is the event "the student takes art or computer science or both," and $A \cap C$ is the event "the student takes art and computer science." ◀

The next theorem gives a formula for the probability of the union of two events. In the case of equally likely outcomes, it is an application of the Inclusion-Exclusion Principle for two sets (see Theorem 6.1.13).

**Theorem 6.6.9** Let $E_1$ and $E_2$ be events. Then

$$P(E_1 \cup E_2) = P(E_1) + P(E_2) - P(E_1 \cap E_2).$$

**Proof** Let

$$\begin{aligned} E_1 &= \{x_1, \ldots, x_i\} \\ E_2 &= \{y_1, \ldots, y_j\} \\ E_1 \cap E_2 &= \{z_1, \ldots, z_k\}, \end{aligned}$$

and assume that each set element is listed exactly one time per set (see Figure 6.6.1). Then in the list

$$x_1, \ldots, x_i, y_1, \ldots, y_j,$$

$z_1, \ldots, z_k$ occurs twice. It follows that

$$\begin{aligned} P(E_1 \cup E_2) &= \sum_{t=1}^{i} P(x_t) + \sum_{t=1}^{j} P(y_t) - \sum_{t=1}^{k} P(z_k) \\ &= P(E_1) + P(E_2) - P(E_1 \cap E_2). \end{aligned}$$

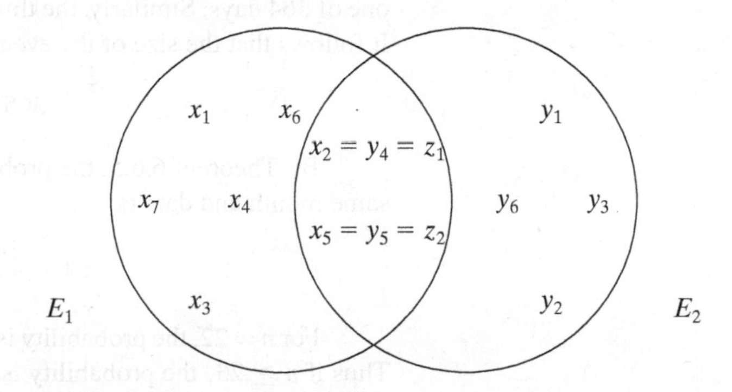


**Figure 6.6.1** Events $E_1$ and $E_2$. The $x$'s denote the elements in $E_1$, the $y$'s denote the elements in $E_2$, and the $z$'s denote the elements in $E_1 \cap E_2$. The $z$'s are thus found twice: once among the $x$'s and again among the $y$'s.

◀

**Example 6.6.10** Two fair dice are rolled. What is the probability of getting doubles (two dice showing the same number) or a sum of 6?

**SOLUTION** We let $E_1$ denote the event "get doubles" and $E_2$ denote the event "get a sum of 6." Since doubles can be obtained in six ways,

$$P(E_1) = \frac{6}{36} = \frac{1}{6}.$$

Since the sum of 6 can be obtained in five ways [(1, 5), (2, 4), (3, 3), (4, 2), (5, 1)],

$$P(E_2) = \frac{5}{36}.$$

The event $E_1 \cap E_2$ is "get doubles *and* get a sum of 6." Since this last event can occur only one way (by getting a pair of 3s),

$$P(E_1 \cap E_2) = \frac{1}{36}.$$

By Theorem 6.6.9, the probability of getting doubles or a sum of 6 is

$$\begin{aligned} P(E_1 \cup E_2) &= P(E_1) + P(E_2) - P(E_1 \cap E_2) \\ &= \frac{1}{6} + \frac{5}{36} - \frac{1}{36} = \frac{5}{18}. \end{aligned}$$

◀

Events $E_1$ and $E_2$ are **mutually exclusive** if $E_1 \cap E_2 = \varnothing$. It follows from Theorem 6.6.9 that if $E_1$ and $E_2$ are mutually exclusive,

$$P(E_1 \cup E_2) = P(E_1) + P(E_2).$$

**Corollary 6.6.11** If $E_1$ and $E_2$ are mutually exclusive events,

$$P(E_1 \cup E_2) = P(E_1) + P(E_2).$$

**Proof** Since $E_1$ and $E_2$ are mutually exclusive events, $E_1 \cap E_2 = \varnothing$. Therefore, $P(E_1 \cap E_2) = 0$. Theorem 6.6.9 now gives

$$P(E_1 \cup E_2) = P(E_1) + P(E_2) - P(E_1 \cap E_2) = P(E_1) + P(E_2).$$

◀

**Example 6.6.12** Two fair dice are rolled. Find the probability of getting doubles or the sum of 5.

**SOLUTION** We let $E_1$ denote the event "get doubles" and $E_2$ denote the event "get the sum of 5." Notice that $E_1$ and $E_2$ are mutually exclusive: You cannot get doubles and the sum of 5 simultaneously. Since doubles can be obtained in six ways,

$$P(E_1) = \frac{6}{36} = \frac{1}{6}.$$

Since the sum of 5 can be obtained in four ways [(1, 4), (2, 3), (3, 2), (4, 1)],

$$P(E_2) = \frac{4}{36} = \frac{1}{9}.$$

By Corollary 6.6.11,

$$P(E_1 \cup E_2) = P(E_1) + P(E_2) = \frac{1}{6} + \frac{1}{9} = \frac{5}{18}.$$

◀

| | | | | | |
|---|---|---|---|---|---|
| 1, 1 | 2, 1 | 3, 1 | 4, 1 | 5, 1 | 6, 1 |
| 1, 2 | 2, 2 | 3, 2 | 4, 2 | 5, 2 | 6, 2 |
| 1, 3 | 2, 3 | 3, 3 | 4, 3 | 5, 3 | 6, 3 |
| 1, 4 | 2, 4 | 3, 4 | 4, 4 | 5, 4 | 6, 4 |
| 1, 5 | 2, 5 | 3, 5 | 4, 5 | 5, 5 | 6, 5 |
| 1, 6 | 2, 6 | 3, 6 | 4, 6 | 5, 6 | 6, 6 |

**Figure 6.6.2** Rolling two fair dice. Since each outcome is assigned the value $1/36$, the probability of getting a sum of 10 is $1/12$. If at least one die shows a 5, one of the shaded outcomes occurs. The shaded outcomes become the new sample space, and each shaded outcome is reassigned the value $1/11$. The probability of getting a sum of 10 given that at least one 5 occurs is $1/11$.

## Conditional Probability

Suppose that we roll two fair dice. The sample space consists of all 36 possible outcomes with each outcome assigned the value $1/36$ (see Figure 6.6.2). The probability of getting a sum of 10 is $1/12$, the sum of the values of the outcomes that add up to 10.

Let us modify the example slightly. Suppose that we roll two dice and we are told that at least one die is 5. Now the probability of getting a sum of 10 is no longer $1/12$ because we know that one of the outcomes shown shaded in Figure 6.6.2 has occurred. Since the 11 outcomes shaded are equally likely, the probability of getting a sum of 10 given that at least one die is 5 is $1/11$. A probability given that some event occurred is called a **conditional probability.**

We now discuss conditional probabilities in general. We let $P(E \mid F)$ denote the probability of $E$ given $F$. In this situation, $F$ becomes the new sample space. Since the values of the outcomes in $F$ originally summed to $P(F)$, we change the value of each outcome in $F$ by dividing it by $P(F)$ so that the reassigned values sum to 1. The outcomes that satisfy $E$ given that $F$ occurred are precisely the outcomes in $E \cap F$. Summing the reassigned values of the outcomes in $E \cap F$, we obtain the value of $P(E \mid F)$:

$$\frac{P(E \cap F)}{P(F)}.$$

This discussion motivates the following definition.

**Definition 6.6.13 ▶** Let $E$ and $F$ be events, and assume that $P(F) > 0$. The *conditional probability of $E$ given $F$* is

$$P(E \mid F) = \frac{P(E \cap F)}{P(F)}.$$ ◀

**Example 6.6.14** Use Definition 6.6.13 to compute the probability of getting a sum of 10, given that at least one die shows 5, when two fair dice are rolled.

**SOLUTION** Let $E$ denote the event "getting a sum of 10," and let $F$ denote the event "at least one die shows 5." The event $E \cap F$ is "getting a sum of 10 *and* at least one die shows 5." Since only one outcome belongs to $E \cap F$,

$$P(E \cap F) = \frac{1}{36}.$$

Since 11 outcomes belong to $F$ (see Figure 6.6.2),

$$P(F) = \frac{11}{36}.$$

Therefore,

$$P(E \mid F) = \frac{P(E \cap F)}{P(F)} = \frac{\frac{1}{36}}{\frac{11}{36}} = \frac{1}{11}.$$

◀

**Example 6.6.15** Weather records show that the probability of high barometric pressure is 0.80, and the probability of rain and high barometric pressure is 0.10. Using Definition 6.6.13, the probability of rain given high barometric pressure is

$$P(R \mid H) = \frac{P(R \cap H)}{P(H)} = \frac{0.10}{0.80} = 0.125,$$

where $R$ denotes the event "rain," and $H$ denotes the event "high barometric pressure."

◀

## Independent Events

If the probability of event $E$ does not depend on event $F$ in the sense that $P(E \mid F) = P(E)$, we say that $E$ and $F$ are **independent events.** By Definition 6.6.13,

$$P(E \mid F) = \frac{P(E \cap F)}{P(F)}.$$

Thus if $E$ and $F$ are independent events,

$$P(E) = P(E \mid F) = \frac{P(E \cap F)}{P(F)}$$

or

$$P(E \cap F) = P(E)P(F).$$

We take this last equation as the formal definition of independent events.

**Definition 6.6.16 ▶** Events $E$ and $F$ are *independent* if

$$P(E \cap F) = P(E)P(F).$$

◀

**Example 6.6.17** Intuitively, if we flip a fair coin twice, the outcome of the second toss does not depend on the outcome of the first toss (after all, coins have no memory). For example, if $H$ is the event "head on first toss," and $T$ is the event "tail on second toss," we expect that events $H$ and $T$ are independent. Use Definition 6.6.16 to verify that $H$ and $T$ are indeed independent.

**SOLUTION** The event $H \cap T$ is the event "head on first toss and tail on second toss." Thus $P(H \cap T) = 1/4$. Since $P(H) = 1/2 = P(T)$, we have

$$P(H \cap T) = \frac{1}{4} = \left(\frac{1}{2}\right)\left(\frac{1}{2}\right) = P(H)P(T).$$

Therefore, events $H$ and $T$ are independent. ◀

**Example 6.6.18** Joe and Alicia take a final examination in discrete mathematics. The probability that Joe passes is 0.70, and the probability that Alicia passes is 0.95. Assuming that the events "Joe passes" and "Alicia passes" are independent, find the probability that Joe or Alicia, or both, passes the final exam.

**SOLUTION** We let $J$ denote the event "Joe passes the final exam" and $A$ denote the event "Alicia passes the final exam." We are asked to compute $P(J \cup A)$.

Theorem 6.6.9 says that

$$P(J \cup A) = P(J) + P(A) - P(J \cap A).$$

Since we are given $P(J)$ and $P(A)$, we need only compute $P(J \cap A)$. Because the events $J$ and $A$ are *independent,* Definition 6.6.16 says that

$$P(J \cap A) = P(J)P(A) = (0.70)(0.95) = 0.665.$$

Therefore,

$$P(J \cup A) = P(J) + P(A) - P(J \cap A) = 0.70 + 0.95 - 0.665 = 0.985.$$ ◀

## Pattern Recognition and Bayes' Theorem

**Pattern recognition** places items into various *classes* based on *features* of the items. For example, wine might be placed into the classes *premium, table wine,* and *swill* based on features such as acidity and bouquet. One way to perform such a classification uses probability theory. Given a set of features $F$, one computes the probability of a class given $F$ for each class and places the item into the most probable class; that is, the class $C$ chosen is the one for which $P(C \mid F)$ is greatest.

**Example 6.6.19** Let $R$ denote class *premium,* $T$ denote class *table wine,* and $S$ denote class *swill*. Suppose that a particular wine has feature set $F$ and

$$P(R \mid F) = 0.2, \qquad P(T \mid F) = 0.5, \qquad P(S \mid F) = 0.3.$$

Since class *table wine* has the greatest probability, this wine would be classified as *table wine.* ◀

**Bayes' Theorem** is useful in computing the probability of a class given a set of features.

**Theorem 6.6.20** **Bayes' Theorem**

Suppose that the possible classes are $C_1, \ldots, C_n$. Suppose further that each pair of classes is mutually exclusive and each item to be classified belongs to one of the classes. For a feature set $F$, we have

$$P(C_j \mid F) = \frac{P(F \mid C_j)P(C_j)}{\sum_{i=1}^{n} P(F \mid C_i)P(C_i)}.$$

**Proof** By Definition 6.6.13,

$$P(C_j \mid F) = \frac{P(C_j \cap F)}{P(F)},$$

and again by Definition 6.6.13,

$$P(F \mid C_j) = \frac{P(F \cap C_j)}{P(C_j)}.$$

Combining these equations, we obtain

$$P(C_j \mid F) = \frac{P(C_j \cap F)}{P(F)} = \frac{P(F \mid C_j)P(C_j)}{P(F)}.$$

To complete the proof of Bayes' Theorem, we need to show that

$$P(F) = \sum_{i=1}^{n} P(F \mid C_i)P(C_i).$$

Because each item to be classified belongs to one of the classes, we have

$$F = (F \cap C_1) \cup (F \cap C_2) \cup \cdots \cup (F \cap C_n).$$

Since the $C_i$ are pairwise mutually exclusive, the $F \cap C_i$ are also pairwise mutually exclusive. By Corollary 6.6.11,

$$P(F) = P(F \cap C_1) + P(F \cap C_2) + \cdots + P(F \cap C_n).$$

Again by Definition 6.6.13,

$$P(F \cap C_i) = P(F \mid C_i)P(C_i).$$

Therefore,

$$P(F) = \sum_{i=1}^{n} P(F \mid C_i)P(C_i),$$

and the proof is complete. ◀

**Example 6.6.21** **Telemarketing** At the telemarketing firm *SellPhone,* Dale, Rusty, and Lee make calls. The following table shows the percentage of calls each caller makes and the percentage of persons who are annoyed and hang up on each caller:

| | *Caller* | | |
|---|---|---|---|
| | *Dale* | *Rusty* | *Lee* |
| *Percent of calls* | 40 | 25 | 35 |
| *Percent of hang-ups* | 20 | 55 | 30 |

Let $D$ denote the event "Dale made the call," let $R$ denote the event "Rusty made the call," let $L$ denote the event "Lee made the call," and let $H$ denote the event "the caller hung up." Find $P(D)$, $P(R)$, $P(L)$, $P(H \mid D)$, $P(H \mid R)$, $P(H \mid L)$, $P(D \mid H)$, $P(R \mid H)$, $P(L \mid H)$, and $P(H)$.

**SOLUTION** Since Dale made 40 percent of the calls, $P(D) = 0.4$. Similarly, from the table we obtain $P(R) = 0.25$ and $P(L) = 0.35$.

Given that Dale made the call, the table shows that 20 percent of the persons hung up; therefore, $P(H \mid D) = 0.2$. Similarly,

$$P(H \mid R) = 0.55 \text{ and } P(H \mid L) = 0.3.$$

To compute $P(D \mid H)$, we use Bayes' Theorem:

$$\begin{aligned} P(D \mid H) &= \frac{P(H \mid D)P(D)}{P(H \mid D)P(D) + P(H \mid R)P(R) + P(H \mid L)P(L)} \\ &= \frac{(0.2)(0.4)}{(0.2)(0.4) + (0.55)(0.25) + (0.3)(0.35)} = 0.248. \end{aligned}$$

A similar computation using Bayes' Theorem gives $P(R \mid H) = 0.426$. Again using Bayes' Theorem or noting that

$$P(D \mid H) + P(R \mid H) + P(L \mid H) = 1,$$

we obtain $P(L \mid H) = 0.326$.

Finally, the proof of Bayes' Theorem shows that

$$\begin{aligned} P(H) &= P(H \mid D)P(D) + P(H \mid R)P(R) + P(H \mid L)P(L) \\ &= (0.2)(0.4) + (0.55)(0.25) + (0.3)(0.35) = 0.3225. \end{aligned}$$

◀

**Example 6.6.22** **Detecting the HIV Virus** The enzyme-linked immunosorbent assay (ELISA) test is used to detect antibodies in blood and can indicate the presence of the HIV virus. Approximately 15 percent of the patients at one clinic have the HIV virus. Furthermore, among those that have the HIV virus, approximately 95 percent test positive on the ELISA test. Among those that do not have the HIV virus, approximately 2 percent test positive on the ELISA test. Find the probability that a patient has the HIV virus if the ELISA test is positive.

**SOLUTION** The classes are "has the HIV virus," which we denote $H$, and "does not have the HIV virus" ($\overline{H}$). The feature is "tests positive," which we denote $Pos$. Using this notation, the given information may be written

$$P(H) = 0.15, \quad P(\overline{H}) = 0.85, \quad P(Pos \mid H) = 0.95, \quad P(Pos \mid \overline{H}) = 0.02.$$

Bayes' Theorem gives the desired probability:

$$\begin{aligned} P(H \mid Pos) &= \frac{P(Pos \mid H)P(H)}{P(Pos \mid H)P(H) + P(Pos \mid \overline{H})P(\overline{H})} \\ &= \frac{(0.95)(0.15)}{(0.95)(0.15) + (0.02)(0.85)} = 0.893. \end{aligned}$$

◀

## 6.6 Review Exercises

1. What is a probability function?
2. If $P$ is a probability function and all outcomes are equally likely, what is the value of $P(x)$, where $x$ is an outcome?
3. How is the probability of an event defined?
4. If $E$ is an event, how are $P(E)$ and $P(\overline{E})$ related? Explain how the formula is derived.
5. If $E_1$ and $E_2$ are events, what does the event $E_1 \cup E_2$ represent?
6. If $E_1$ and $E_2$ are events, what does the event $E_1 \cap E_2$ represent?
7. If $E_1$ and $E_2$ are events, how are $P(E_1 \cup E_2)$, $P(E_1 \cap E_2)$, $P(E_1)$, and $P(E_2)$ related? Explain how the formula is derived.
8. Explain what it means for two events to be "mutually exclusive."
9. Give an example of mutually exclusive events.
10. If $E_1$ and $E_2$ are mutually exclusive events, how are $P(E_1 \cup E_2)$, $P(E_1)$, and $P(E_2)$ related? Explain how the formula is derived.
11. Give an informal, intuitive description of the meaning of the event *E given F*.
12. How is the event *E given F* denoted?
13. Give a formula for the probability of *E given F*.
14. Explain what it means for two events to be "independent."
15. Give an example of independent events.
16. What is pattern recognition?
17. State Bayes' Theorem. Explain how the formula is derived.

## 6.6 Exercises

*Exercises 1–3 refer to Example 6.6.2 in which a die is loaded so that the numbers 2 through 6 are equally likely to appear, but 1 is three times as likely as any other number to appear.*

1. One die is rolled. What is the probability of getting a 5?
2. One die is rolled. What is the probability of getting an even number?
3. One die is rolled. What is the probability of not getting a 5?

*Exercises 4–9 refer to a die that is loaded so that the numbers 1 and 3 are equally likely to appear and 2, 4, 5, and 6 are equally likely to appear. However, 1 is three times as likely to appear as 2.*

4. One die is rolled. Assign probabilities to the outcomes that accurately model the likelihood of the various numbers to appear.
5. One die is rolled. What is the probability of getting a 3?
6. One die is rolled. What is the probability of getting a 4?
7. One die is rolled. What is the probability of not getting a 3?
8. One die is rolled. What is the probability of getting a 1 or a 4?
9. One die is rolled. What is the probability of getting a 1 or a 4 or a 6?

*Exercises 10–19 refer to dice that are loaded so that the numbers 2, 4, and 6 are equally likely to appear. 1, 3, and 5 are also equally likely to appear, but 1 is three times as likely as 2 is to appear.*

10. One die is rolled. Assign probabilities to the outcomes that accurately model the likelihood of the various numbers to appear.
11. One die is rolled. What is the probability of getting a 5?
12. One die is rolled. What is the probability of getting an even number?
13. One die is rolled. What is the probability of not getting a 5?
14. Two dice are rolled. What is the probability of getting doubles?
15. Two dice are rolled. What is the probability of getting a sum of 7?
16. Two dice are rolled. What is the probability of getting doubles or a sum of 6?
17. Two dice are rolled. What is the probability of getting a sum of 6 given that at least one die shows 2?
18. Two dice are rolled. What is the probability of getting a sum of 6 or doubles given that at least one die shows 2?
19. Two dice are rolled. What is the probability of getting a sum of 6 or a sum of 8 given that at least one die shows 2?

*In Exercises 20–24, suppose that a coin is flipped and a die is rolled. Let $E_1$ denote the event "the coin shows a tail," let $E_2$ denote the event "the die shows a 3," and let $E_3$ denote the event "the coin shows heads and the die shows an odd number."*

20. List the elements of the event $E_1$ *or* $E_2$.
21. List the elements of the event $E_2$ *and* $E_3$.
22. Are $E_1$ and $E_2$ mutually exclusive?
23. Are $E_1$ and $E_3$ mutually exclusive?
24. Are $E_2$ and $E_3$ mutually exclusive?
25. Six microprocessors are randomly selected from a lot of 100 microprocessors among which 10 are defective. Find the probability of obtaining no defective microprocessors.
26. Six microprocessors are randomly selected from a lot of 100 microprocessors among which 10 are defective. Find the probability of obtaining at least one defective microprocessor.
27. Six microprocessors are randomly selected from a lot of 100 microprocessors among which 10 are defective. Find the probability of obtaining at least three defective microprocessors.

*Exercises 28–35 refer to a family with four children. Assume that it is equally probable for a boy or a girl to be born.*

**28.** What is the probability of all girls?

**29.** What is the probability of exactly two girls?

**30.** What is the probability of at least one boy and at least one girl?

**31.** What is the probability of all girls given that there is at least one girl?

**32.** What is the probability of exactly two girls given that there is at least one girl?

**33.** What is the probability of at least one boy and at least one girl given that there is at least one girl?

**34.** Are the events "there are children of both sexes" and "there is at most one boy" independent?

**35.** Are the events "there is at most one boy" and "there is at most one girl" independent?

**36.** A family has $n$ children. Assume that it is equally probable for a boy or a girl to be born. For which values of $n$ are the events "there are children of both sexes" and "there is at most one girl" independent?

*Exercises 37–45 refer to a fair coin that is repeatedly flipped.*

**37.** If the coin is flipped 10 times, what is the probability of no heads?

**38.** If the coin is flipped 10 times, what is the probability of exactly five heads?

**39.** If the coin is flipped 10 times, what is the probability of exactly four or five or six heads?

**40.** If the coin is flipped 10 times, what is the probability of at least one head?

**41.** If the coin is flipped 10 times, what is the probability of at most five heads?

**42.** If the coin is flipped 10 times, what is the probability of exactly five heads given at least one head?

**43.** If the coin is flipped 10 times, what is the probability of exactly four or five or six heads given at least one head?

**44.** If the coin is flipped 10 times, what is the probability of at least one head given at least one tail?

**45.** If the coin is flipped 10 times, what is the probability of at most five heads given at least one head?

**46.** Suppose that you buy a lottery ticket containing $k$ distinct numbers from among $\{1, 2, \ldots, n\}$, $1 \leq k \leq n$. To determine the winning ticket, $k$ balls are randomly drawn without replacement from a bin containing $n$ balls numbered $1, 2, \ldots, n$. What is the probability that at least one of the numbers on your lottery ticket is among those drawn from the bin?

*Exercises 47–50 ask about the following situation. In a small charity fundraiser, 70 tickets are sold numbered 1 through 70. Each person buys one ticket. Later in the evening, 20 numbers are randomly drawn from among 1 through 70, and those holding these numbers win modest prizes. Among those buying the tickets are Maya and Chloe.*

**47.** What is the probability that either Maya or Chloe (or both) wins a prize?

**48.** What is the probability that both Maya and Chloe win prizes?

**49.** What is the probability that either Maya or Chloe, but not both, wins a prize?

**50.** What is the probability that Maya, but not Chloe, wins a prize?

**51.** Prove that Algorithm 4.2.4, which generates random permutations, can potentially output any permutation of the input and that all outcomes are equally likely.

**52.** Mr. Wizard suggests changing the third line of Algorithm 4.2.4, which generates random permutations, to

$$swap(a_i, a_{rand(1,n)})$$

Show that in Mr. Wizard's version, the algorithm can potentially output any permutation of the input, but that the outcomes are *not* equally likely.

**53.** Find the probability that among $n$ persons, at least two people have birthdays on April 1 (but not necessarily in the same year). Assume that all months and dates are equally likely, and ignore February 29 birthdays.

**54.** Find the least $n$ such that among $n$ persons, the probability that at least two persons have birthdays on April 1 (but not necessarily in the same year) is greater than 1/2. Assume that all months and dates are equally likely, and ignore February 29 birthdays.

**55.** Find the probability that among $n \geq 3$ persons, at least three people have birthdays on the same month and date (but not necessarily in the same year). Assume that all months and dates are equally likely, and ignore February 29 birthdays.

**56.** Under the conditions of Exercise 55, find the minimum value of $n$ for which the probability of at least three people having birthdays on the same month and date is greater than or equal to 1/2.

**57.** Suppose that a professional wrestler is selected at random among 90 wrestlers, where 35 are over 350 pounds, 20 are bad guys, and 15 are over 350 pounds and bad guys. What is the probability that the wrestler selected is over 350 pounds or a bad guy?

**58.** Suppose that the probability of a person having a headache is 0.01, that the probability of a person having a fever given that the person has a headache is 0.4, and that the probability of a person having a fever is 0.02. Find the probability that a person has a headache given that the person has a fever.

*Exercises 59–62 refer to a company that buys computers from three vendors and tracks the number of defective machines. The following table shows the results.*

| | *Vendor* | | |
|---|---|---|---|
| | *Acme* | *DotCom* | *Nuclear* |
| *Percent purchased* | 55 | 10 | 35 |
| *Percent defective* | 1 | 3 | 3 |

*Let A denote the event "the computer was purchased from Acme," let D denote the event "the computer was purchased from Dot-Com," let N denote the event "the computer was purchased from Nuclear," and let B denote the event "the computer was defective."*

**59.** Find $P(A)$, $P(D)$, and $P(N)$.

**60.** Find $P(B \mid A)$, $P(B \mid D)$, and $P(B \mid N)$.

**61.** Find $P(A \mid B)$, $P(D \mid B)$, and $P(N \mid B)$.

**62.** Find $P(B)$.

**63.** In Example 6.6.22, how small would $P(H)$ have to be so that the conclusion would be "no HIV" even if the result of the test is positive?

**64.** Show that for any events $E_1$ and $E_2$,

$$P(E_1 \cap E_2) \ge P(E_1) + P(E_2) - 1.$$

**65.** Use mathematical induction to show that if $E_1, E_2, \ldots, E_n$ are events, then

$$P(E_1 \cup E_2 \cup \cdots \cup E_n) \le \sum_{i=1}^{n} P(E_i).$$

**66.** If $E$ and $F$ are independent events, are $\overline{E}$ and $\overline{F}$ independent?

**67.** If $E$ and $F$ are independent events, are $E$ and $\overline{F}$ independent?

**68.** Is the following reasoning correct? Explain.

A person, concerned about the possibility of a bomb on a plane, estimates the probability of a bomb on a plane to be 0.000001. Not satisfied with the chances, the person computes the probability of *two* bombs on a plane to be

$$0.000001^2 = 0.000000000001.$$

Satisfied with the chances now, the person always carries a bomb on a plane so that the probability of someone else carrying a bomb, and thus there being two bombs on the plane, is 0.000000000001—small enough to be safe.

**69.** A track enthusiast decides to try to complete the East-South-East Marathon. The runner will stop if the marathon is completed or after three attempts. The probability of completing the marathon in one attempt is $1/3$. Analyze the following argument that, assuming independence, purportedly shows that the runner is almost, but not quite, certain to complete the marathon.

Since the probability of each attempt is $1/3 = 0.3333$, after three attempts the probability of completing the marathon is 0.9999, which means that the runner is almost, but not quite, certain to complete the marathon.

# 6.7 Binomial Coefficients and Combinatorial Identities

At first glance the expression $(a+b)^n$ does not have much to do with combinations; but as we will see in this section, we can obtain the formula for the expansion of $(a+b)^n$ by using the formula for the number of $r$-combinations of $n$ objects. Frequently, we can relate an algebraic expression to some counting process. Several advanced counting techniques use such methods (see [Riordan; and Tucker]).

**Go Online**
For more on the binomial theorem, see bit.ly/2PNvtEt

The **Binomial Theorem** gives a formula for the coefficients in the expansion of $(a+b)^n$. Since

$$(a+b)^n = \underbrace{(a+b)(a+b)\cdots(a+b)}_{n \text{ factors}}, \tag{6.7.1}$$

the expansion results from selecting either $a$ or $b$ from each of the $n$ factors, multiplying the selections together, and then summing all such products obtained. For example, in the expansion of $(a+b)^3$, we select either $a$ or $b$ from the first factor $(a+b)$; either $a$ or $b$ from the second factor $(a+b)$; and either $a$ or $b$ from the third factor $(a+b)$; multiply the selections together; and then sum the products obtained. If we select $a$ from all factors and multiply, we obtain the term $aaa$. If we select $a$ from the first factor, $b$ from the second factor, and $a$ from the third factor and multiply, we obtain the term $aba$. Table 6.7.1 shows all the possibilities. If we sum the products of all the selections, we obtain

$$\begin{aligned}(a+b)^3 &= (a+b)(a+b)(a+b)\\ &= aaa + aab + aba + abb + baa + bab + bba + bbb\\ &= a^3 + a^2b + a^2b + ab^2 + a^2b + ab^2 + ab^2 + b^3\\ &= a^3 + 3a^2b + 3ab^2 + b^3.\end{aligned}$$

**TABLE 6.7.1** ■ Computing $(a+b)^3$

| *Selection from First Factor* $(a+b)$ | *Selection from Second Factor* $(a+b)$ | *Selection from Third Factor* $(a+b)$ | *Product of Selections* |
|---|---|---|---|
| $a$ | $a$ | $a$ | $aaa = a^3$ |
| $a$ | $a$ | $b$ | $aab = a^2b$ |
| $a$ | $b$ | $a$ | $aba = a^2b$ |
| $a$ | $b$ | $b$ | $abb = ab^2$ |
| $b$ | $a$ | $a$ | $baa = a^2b$ |
| $b$ | $a$ | $b$ | $bab = ab^2$ |
| $b$ | $b$ | $a$ | $bba = ab^2$ |
| $b$ | $b$ | $b$ | $bbb = b^3$ |

In (6.7.1), a term of the form $a^{n-k}b^k$ arises from choosing $b$ from $k$ factors and $a$ from the other $n-k$ factors. But this can be done in $C(n, k)$ ways, since $C(n, k)$ counts the number of ways of selecting $k$ things from $n$ items. Thus $a^{n-k}b^k$ appears $C(n, k)$ times. It follows that

$$(a+b)^n = C(n,0)a^nb^0 + C(n,1)a^{n-1}b^1 + C(n,2)a^{n-2}b^2 + \cdots + C(n,n-1)a^1b^{n-1} + C(n,n)a^0b^n. \tag{6.7.2}$$

This result is known as the *Binomial Theorem*.

**Theorem 6.7.1** **Binomial Theorem**

If $a$ and $b$ are real numbers and $n$ is a positive integer, then

$$(a+b)^n = \sum_{k=0}^{n} C(n,k)a^{n-k}b^k.$$

**Proof** The proof precedes the statement of the theorem. ◀

The Binomial Theorem can also be proved using induction on $n$ (see Exercise 17).

The numbers $C(n, r)$ are known as **binomial coefficients** because they appear in the expansion (6.7.2) of the binomial $a+b$ raised to a power.

**Example 6.7.2** Taking $n = 3$ in Theorem 6.7.1, we obtain

$$\begin{aligned}(a+b)^3 &= C(3,0)a^3b^0 + C(3,1)a^2b^1 + C(3,2)a^1b^2 + C(3,3)a^0b^3\\ &= a^3 + 3a^2b + 3ab^2 + b^3.\end{aligned}$$ ◀

**Example 6.7.3** Expand $(3x-2y)^4$ using the Binomial Theorem.

**SOLUTION** If we take $a = 3x$, $b = -2y$, and $n = 4$ in Theorem 6.7.1, we obtain

$$\begin{aligned}(3x-2y)^4 &= (a+b)^4\\ &= C(4,0)a^4b^0 + C(4,1)a^3b^1 + C(4,2)a^2b^2\\ &\quad + C(4,3)a^1b^3 + C(4,4)a^0b^4\\ &= C(4,0)(3x)^4(-2y)^0 + C(4,1)(3x)^3(-2y)^1\\ &\quad + C(4,2)(3x)^2(-2y)^2 + C(4,3)(3x)^1(-2y)^3\end{aligned}$$

$$+ C(4, 4)(3x)^0(-2y)^4$$
$$= 3^4x^4 + 4 \cdot 3^3x^3(-2y) + 6 \cdot 3^2x^2(-2)^2y^2$$
$$+ 4(3x)(-2)^3y^3 + (-2)^4y^4$$
$$= 81x^4 - 216x^3y + 216x^2y^2 - 96xy^3 + 16y^4.$$

◀

**Example 6.7.4** Find the coefficient of $a^5b^4$ in the expansion of $(a + b)^9$.

**SOLUTION** The term involving $a^5b^4$ arises in the Binomial Theorem by taking $n = 9$ and $k = 4$:

$$C(n, k)a^{n-k}b^k = C(9, 4)a^5b^4 = 126a^5b^4.$$

Thus the coefficient of $a^5b^4$ is 126. ◀

**Example 6.7.5** Find the coefficient of $x^2y^3z^4$ in the expansion of $(x + y + z)^9$.

**SOLUTION** Since

$$(x + y + z)^9 = (x + y + z)(x + y + z) \cdots (x + y + z) \quad \text{(nine terms)},$$

we obtain $x^2y^3z^4$ each time we multiply together $x$ chosen from two of the nine terms, $y$ chosen from three of the nine terms, and $z$ chosen from four of the nine terms. We can choose two terms for the $x$'s in $C(9, 2)$ ways. Having made this selection, we can choose three terms for the $y$'s in $C(7, 3)$ ways. This leaves the remaining four terms for the $z$'s. Thus the coefficient of $x^2y^3z^4$ in the expansion of $(x + y + z)^9$ is

$$C(9, 2)C(7, 3) = \frac{9!}{2!\,7!}\frac{7!}{3!\,4!} = \frac{9!}{2!\,3!\,4!} = 1260.$$

◀

```
          1
        1   1
      1   2   1
    1   3   3   1
  1   4   6   4   1
1   5  10  10   5   1
.         .         .
```

Figure 6.7.1 Pascal's triangle.

**Go Online**
A biography of Pascal is at bit.ly/2PNvtEt

We can write the binomial coefficients in a triangular form known as **Pascal's triangle** (see Figure 6.7.1). The border consists of 1's, and any interior value is the sum of the two numbers above it. This relationship is stated formally in the next theorem. The proof is a combinatorial argument. An identity that results from some counting process is called a **combinatorial identity** and the argument that leads to its formulation is called a **combinatorial argument.**

**Theorem 6.7.6**

$$C(n + 1, k) = C(n, k - 1) + C(n, k)$$

for $1 \le k \le n$.

**Proof** Let $X$ be a set with $n$ elements. Choose $a \notin X$. Then $C(n + 1, k)$ is the number of $k$-element subsets of $Y = X \cup \{a\}$. Now the $k$-element subsets of $Y$ can be divided into two disjoint classes:

1. Subsets of $Y$ not containing $a$.
2. Subsets of $Y$ containing $a$.

The subsets of class 1 are just $k$-element subsets of $X$ and there are $C(n, k)$ of these. Each subset of class 2 consists of a $(k - 1)$-element subset of $X$ together with $a$ and there are $C(n, k - 1)$ of these. Therefore,

$$C(n + 1, k) = C(n, k - 1) + C(n, k).$$

◀

Theorem 6.7.6 can also be proved using Theorem 6.2.17 (Exercise 18 of this section).

We conclude by showing how the Binomial Theorem (Theorem 6.7.1) and Theorem 6.7.6 can be used to derive other combinatorial identities.

**Example 6.7.7** Use the Binomial Theorem to derive the equation

$$\sum_{k=0}^{n} C(n,k) = 2^n. \tag{6.7.3}$$

**SOLUTION** The sum is the same as the sum in the Binomial Theorem,

$$\sum_{k=0}^{n} C(n,k)a^{n-k}b^k,$$

except that the expression $a^{n-k}b^k$ is missing. One way to "eliminate" this expression is to take $a = b = 1$, in which case the Binomial Theorem becomes

$$2^n = (1+1)^n = \sum_{k=0}^{n} C(n,k)1^{n-k}1^k = \sum_{k=0}^{n} C(n,k).$$

There are often many different ways to prove a result. In Example 6.7.7, we used an algebraic technique (the Binomial Theorem) to prove equation (6.7.3). In the next two examples, we first give a combinatorial proof and then a proof using mathematical induction of equation (6.7.3).

**Example 6.7.8** Prove equation (6.7.3) using a combinatorial argument.

**SOLUTION** Given an $n$-element set $X$, $C(n,k)$ counts the number of $k$-element subsets of $X$. Thus the left side of equation (6.7.3) counts the total number of subsets of $X$. But the number of subsets of $X$ is $2^n$ (see Theorem 2.4.6), and equation (6.7.3) follows immediately.

**Example 6.7.9** Prove equation (6.7.3) using mathematical induction.

**SOLUTION** The Basis Step ($n = 0$) is readily verified.

For the Inductive Step, we first assume that equation (6.7.3) is true for $n$. Using Theorem 6.7.6, we have

$$\begin{aligned}
\sum_{k=0}^{n+1} C(n+1,k) &= C(n+1,0) + \sum_{k=1}^{n} C(n+1,k) + C(n+1,n+1) \\
&= 1 + \sum_{k=1}^{n} [C(n,k-1) + C(n,k)] + 1 \\
&= 2 + \sum_{k=1}^{n} C(n,k-1) + \sum_{k=1}^{n} C(n,k) \\
&= 2 + \left[\sum_{k=0}^{n} C(n,k) - C(n,n)\right] + \left[\sum_{k=0}^{n} C(n,k) - C(n,0)\right] \\
&= 2 + [2^n - 1] + [2^n - 1] = 2 \cdot 2^n = 2^{n+1}.
\end{aligned}$$

**Example 6.7.10** Use Theorem 6.7.6 to show that

$$\sum_{i=k}^{n} C(i, k) = C(n+1, k+1). \tag{6.7.4}$$

**SOLUTION** We use Theorem 6.7.6 in the form

$$C(i, k) = C(i+1, k+1) - C(i, k+1)$$

to obtain

$$\begin{aligned} C(k, k) &+ C(k+1, k) + C(k+2, k) + \cdots + C(n, k) \\ &= 1 + C(k+2, k+1) - C(k+1, k+1) + C(k+3, k+1) \\ &\quad - C(k+2, k+1) + \cdots + C(n+1, k+1) - C(n, k+1) \\ &= C(n+1, k+1). \end{aligned}$$

◀

Exercise 51, Section 6.3, shows another way to prove equation (6.7.4).

**Example 6.7.11** Use equation (6.7.4) to find the sum $1 + 2 + \cdots + n$.

**SOLUTION** We may write

$$\begin{aligned} 1 + 2 + \cdots + n &= C(1, 1) + C(2, 1) + \cdots + C(n, 1) \\ &= C(n+1, 2) \qquad \text{by equation (6.7.4)} \\ &= \frac{(n+1)n}{2}. \end{aligned}$$

◀

## 6.7 Review Exercises

1. State the Binomial Theorem.
2. Explain how the Binomial Theorem is derived.
3. What is Pascal's triangle?
4. State the formulas that can be used to generate Pascal's triangle.

## 6.7 Exercises

1. Expand $(x + y)^4$ using the Binomial Theorem.
2. Expand $(2c - 3d)^5$ using the Binomial Theorem.

*In Exercises 3–9, find the coefficient of the term when the expression is expanded.*

3. $x^4y^7$; $(x + y)^{11}$
4. $s^6t^6$; $(2s - t)^{12}$
5. $x^2y^3z^5$; $(x + y + z)^{10}$
6. $w^2x^3y^2z^5$; $(2w + x + 3y + z)^{12}$
7. $a^2x^3$; $(a + x + c)^2(a + x + d)^3$
8. $a^2x^3$; $(a + ax + x)(a + x)^4$
9. $a^3x^4$; $(a + \sqrt{ax} + x)^2(a + x)^5$

*In Exercises 10–12, find the number of terms in the expansion of each expression.*

10. $(x + y + z)^{10}$
11. $(w + x + y + z)^{12}$
★12. $(x + y + z)^{10}(w + x + y + z)^2$
13. Find the next row of Pascal's triangle given the row

$$1 \quad 7 \quad 21 \quad 35 \quad 35 \quad 21 \quad 7 \quad 1.$$

14. (a) Show that $C(n, k) < C(n, k+1)$ if and only if $k < (n-1)/2$.
    (b) Use part (a) to deduce that the maximum of $C(n, k)$ for $k = 0, 1, \ldots, n$ is $C(n, \lfloor n/2 \rfloor)$.
15. Prove that

$$\sum_{k=0}^{m} (-1)^k C(n, k) = (-1)^m C(n-1, m)$$

for all $m$, $0 \le m \le n - 1$.

**16.** Use the Binomial Theorem to show that

$$0 = \sum_{k=0}^{n} (-1)^k C(n, k).$$

**17.** Use induction on $n$ to prove the Binomial Theorem.

**18.** Prove Theorem 6.7.6 by using Theorem 6.2.16.

**19.** Give a combinatorial argument to show that

$$C(n, k) = C(n, n - k).$$

**★20.** Prove equation (6.7.4) by giving a combinatorial argument.

**21.** Find the sum

$$1 \cdot 2 + 2 \cdot 3 + \cdots + (n - 1)n.$$

**★22.** Use equation (6.7.4) to derive a formula for

$$1^2 + 2^2 + \cdots + n^2.$$

**23.** Use the Binomial Theorem to show that

$$\sum_{k=0}^{n} 2^k C(n, k) = 3^n.$$

**24.** Suppose that $n$ is even. Prove that

$$\sum_{k=0}^{n/2} C(n, 2k) = 2^{n-1} = \sum_{k=1}^{n/2} C(n, 2k - 1).$$

**25.** Prove

$$(a + b + c)^n = \sum_{0 \le i+j \le n} \frac{n!}{i!\, j!\, (n - i - j)!} a^i b^j c^{n-i-j}.$$

**26.** Use Exercise 25 to write the expansion of $(x + y + z)^3$.

**27.** Prove

$$3^n = \sum_{0 \le i+j \le n} \frac{n!}{i!\, j!\, (n - i - j)!}.$$

**28.** Prove that $2C(2n - 1, n) = C(2n, n)$ for all $n \ge 1$.

**★29.** Prove that

$$\sum_{i=0}^{n-1} C(n - 1, i)C(n, n - i) = C(2n - 1, n)$$

for all $n \ge 1$.

**★30.** Give a combinatorial argument to prove that

$$\sum_{k=0}^{n} C(n, k)^2 = C(2n, n).$$

**★31.** Prove that for all $n$, $q$, and $k$, $0 < q < n$ and $0 < k < n$,

$$C(n, q) = \sum_{i=0}^{q} C(k, i)C(n - k, q - i).$$

If $j \ge i$, define $C(i, j)$ to be 0. Notice that Exercise 30 is a special case of this result with $n$ replaced by $2n$, and $k$ and $q$ replaced by $n$. *Hint*: Use a combinatorial argument. Consider $k$ red balls and $n-k$ blue balls to be distributed into containers of size $q$ and $n - q$.

**32.** Prove

$$n(1 + x)^{n-1} = \sum_{k=1}^{n} C(n, k)kx^{k-1}.$$

**33.** Use the result of Exercise 32 to show that

$$n2^{n-1} = \sum_{k=1}^{n} kC(n, k). \qquad \textbf{(6.7.5)}$$

**★34.** Prove equation (6.7.5) by induction.

**35.** A *smoothing sequence* $b_0, \ldots, b_{k-1}$ is a (finite) sequence satisfying $b_i \ge 0$ for $i = 0, \ldots, k - 1$, and $\sum_{i=0}^{k-1} b_i = 1$. A *smoothing of the (infinite) sequence* $a_1, a_2, \ldots$ *by the smoothing sequence* $b_0, \ldots, b_{k-1}$ is the sequence $\{a'_j\}$ defined by

$$a'_j = \sum_{i=0}^{k-1} a_{i+j} b_i.$$

The idea is that averaging smooths noisy data.

The *binomial smoother of size* $k$ is the sequence

$$\frac{B_0}{2^n}, \ldots, \frac{B_{k-1}}{2^n},$$

where $B_0, \ldots, B_{k-1}$ is row $n$ of Pascal's triangle (row 0 being the top row).

Let $c_0, c_1$ be the smoothing sequence defined by $c_0 = c_1 = 1/2$. Show that if a sequence $a$ is smoothed by $c$, the resulting sequence is smoothed by $c$, and so on $k$ times; then, the sequence that results can be obtained by one smoothing of $a$ by the binomial smoother of size $k + 1$.

**36.** In Example 6.1.6 we showed that there are $3^n$ ordered pairs $(A, B)$ satisfying $A \subseteq B \subseteq X$, where $X$ is an $n$-element set. Derive this result by considering the cases $|A| = 0$, $|A| = 1, \ldots, |A| = n$, and then using the Binomial Theorem.

**37.** Show that

$$\sum_{k=m}^{n} C(k, m)H_k = C(n + 1, m + 1)\left(H_{n+1} - \frac{1}{m + 1}\right)$$

for all $n \ge m$, where $H_k$, the $k$th harmonic number, is defined as

$$H_k = \sum_{i=1}^{k} \frac{1}{i}.$$

**38.** Prove that

$$\sum_{i=1}^{n} \frac{1}{C(n, i)} = \frac{n + 1}{2^n} \sum_{i=0}^{n-1} \frac{2^i}{i + 1},$$

for all $n \in \mathbf{Z}^+$.

**★39.** Prove that

$$\sum_{k=0}^{\lfloor n/2 \rfloor} C(n - k, k) = f_n$$

for all $n \ge 1$, where $f_n$ is the $n$th Fibonacci number.

**40.** Prove that if $p$ is prime, $C(p, i)$ is divisible by $p$ for all $i$, $1 \le i \le p - 1$.

**41.** Prove Fermat's little theorem, which states that if $p$ is a prime and $a$ is a positive integer, $p$ divides $a^p - a$. *Hint*: $(j+1)^p - j^p - 1 = \sum_{i=1}^{p-1} C(p,i)j^i$ for all $j$, $1 \le j \le a-1$.

**42.** Prove that

$$\left(\frac{m}{m+n}\right)^m \left(\frac{n}{m+n}\right)^n C(m+n, m) < 1$$

for all $m, n \in \mathbf{Z}^+$. *Hint*: Consider the term for $k = m$ in the binomial theorem expansion of $(x+y)^{m+n}$ for appropriate $x$ and $y$.

**★43.** Prove that for each $k \in \mathbf{Z}^{nonneg}$, there exist constants $C_0, C_1, \ldots, C_k$, depending on $k$, such that

$$\sum_{i=1}^{n} i^k = \frac{n^{k+1}}{k+1} + C_k n^k + C_{k-1} n^{k-1} + \cdots + C_1 n + C_0,$$

for all $n \in \mathbf{Z}^+$. *Hint*: Use Strong Induction on $k$; Exercise 70, Section 2.4, with $a_k = i^{k+1}$; and the Binomial Theorem.

## 6.8 The Pigeonhole Principle

**Go Online**
For more on the Pigeonhole Principle, see
bit.ly/2PNvtEt

**Go Online**
A biography of Dirichlet is at
bit.ly/2PNvtEt

The **Pigeonhole Principle** (also known as the *Dirichlet Drawer Principle* or the *Shoe Box Principle*) is sometimes useful in answering the question: Is there an item having a given property? When the Pigeonhole Principle is successfully applied, the principle tells us only that the object exists; the principle will not tell us how to find the object or how many there are.

The first version of the Pigeonhole Principle that we will discuss asserts that if $n$ pigeons fly into $k$ pigeonholes and $k < n$, some pigeonhole contains at least two pigeons (see Figure 6.8.1). The reason this statement is true can be seen by arguing by contradiction. If the conclusion is false, each pigeonhole contains at most one pigeon and, in this case, we can account for at most $k$ pigeons. Since there are $n$ pigeons and $n > k$, we have a contradiction.

**Figure 6.8.1** $n = 6$ pigeons in $k = 4$ pigeonholes. Some pigeonhole contains at least two pigeons.

**Pigeonhole Principle (First Form)**

If $n$ pigeons fly into $k$ pigeonholes and $k < n$, some pigeonhole contains at least two pigeons.

We note that the Pigeonhole Principle tells us nothing about how to locate the pigeonhole that contains two or more pigeons. It only asserts the *existence* of a pigeonhole containing two or more pigeons.

To apply the Pigeonhole Principle, we must decide which objects will play the roles of the pigeons and which objects will play the roles of the pigeonholes. Our first example illustrates one possibility.

**Example 6.8.1** Ten persons have first names Alice, Bernard, and Charles and last names Lee, McDuff, and Ng. Show that at least two persons have the same first and last names.

**SOLUTION** There are nine possible names for the 10 persons. If we think of the persons as pigeons and the names as pigeonholes, we can consider the assignment of names to people to be that of assigning pigeonholes to the pigeons. By the Pigeonhole Principle, some name (pigeonhole) is assigned to at least two persons (pigeons). ◀

A proof using the Pigeonhole Principle can often be recast as a proof by contradiction. (After all, the validity of the Pigeonhole Principle was verified using proof by contradiction!) Although the proofs are essentially the same, one may be easier than the other for a particular individual to construct.

**Example 6.8.2** Give another proof of the result in Example 6.8.1 using proof by contradiction.

**SOLUTION** Suppose, by way of contradiction, that no two of the 10 people in Example 6.8.1 have the same first and last names. Since there are three first names and three last names, there are at most nine people. This contradiction shows that there are at least two people having the same first and last names. ◀

We next restate the Pigeonhole Principle in an alternative form.

**Pigeonhole Principle (Second Form)** If $f$ is a function from a finite set $X$ to a finite set $Y$ and $|X| > |Y|$, then $f(x_1) = f(x_2)$ for some $x_1, x_2 \in X$, $x_1 \neq x_2$.

The second form of the Pigeonhole Principle can be reduced to the first form by letting $X$ be the set of pigeons and $Y$ be the set of pigeonholes. We assign pigeon $x$ to pigeonhole $f(x)$. By the first form of the Pigeonhole Principle, at least two pigeons, $x_1, x_2 \in X$, are assigned to the same pigeonhole; that is, $f(x_1) = f(x_2)$ for some $x_1, x_2 \in X$, $x_1 \neq x_2$.

Our next examples illustrate the use of the second form of the Pigeonhole Principle.

**Example 6.8.3** If 20 processors are interconnected, show that at least 2 processors are directly connected to the same number of processors.

**SOLUTION** Designate the processors 1, 2, . . . , 20. Let $a_i$ be the number of processors to which processor $i$ is directly connected. We are to show that $a_i = a_j$, for some $i \neq j$. The domain of the function $a$ is $X = \{1, 2, \ldots, 20\}$ and the range $Y$ is some subset of $\{0, 1, \ldots, 19\}$. Unfortunately, $|X| = |\{0, 1, \ldots, 19\}|$ and we cannot immediately use the second form of the Pigeonhole Principle.

Let us examine the situation more closely. Notice that we cannot have $a_i = 0$, for some $i$, *and* $a_j = 19$, for some $j$, for then we would have one processor (the $i$th processor) not connected to any other processor while, at the same time, some other processor (the $j$th processor) is connected to all the other processors (including the $i$th processor). Thus the range $Y$ is a subset of either $\{0, 1, \ldots, 18\}$ or $\{1, 2, \ldots, 19\}$. In either case, $|Y| < 20 = |X|$. By the second form of the Pigeonhole Principle, $a_i = a_j$, for some $i \neq j$, as desired. ◀

**Example 6.8.4** Show that if we select 151 distinct computer science courses numbered between 1 and 300 inclusive, at least two are consecutively numbered.

**SOLUTION** Let the selected course numbers be

$$c_1, \quad c_2, \quad \ldots, \quad c_{151}. \tag{6.8.1}$$

The 302 numbers consisting of (6.8.1) together with

$$c_1 + 1, \quad c_2 + 1, \quad \ldots, \quad c_{151} + 1 \tag{6.8.2}$$

range in value between 1 and 301. By the second form of the Pigeonhole Principle, at least two of these values coincide. The numbers (6.8.1) are all distinct and hence the numbers (6.8.2) are also distinct. It must then be that one of (6.8.1) and one of (6.8.2) are equal. Thus we have $c_i = c_j + 1$ and course $c_i$ follows course $c_j$. ◀

**Example 6.8.5** An inventory consists of a list of 89 items, each marked "available" or "unavailable." There are 50 available items. Show that there are at least two available items in the list exactly nine items apart. (For example, available items at positions 13 and 22 or positions 69 and 78 satisfy the condition.)

**SOLUTION** Let $a_i$ denote the position of the $i$th available item. We must show that $a_i - a_j = 9$ for some $i$ and $j$. Consider the numbers

$$a_1, \quad a_2, \quad \ldots, \quad a_{50} \tag{6.8.3}$$

and

$$a_1 + 9, \quad a_2 + 9, \quad \ldots, \quad a_{50} + 9. \tag{6.8.4}$$

The 100 numbers in (6.8.3) and (6.8.4) have possible values from 1 to 98. By the second form of the Pigeonhole Principle, two of the numbers must coincide. We cannot have two of (6.8.3) or two of (6.8.4) identical; thus some number in (6.8.3) is equal to some number in (6.8.4). Therefore, $a_i - a_j = 9$ for some $i$ and $j$, as desired. ◀

We next state yet another form of the Pigeonhole Principle.

**Pigeonhole Principle (Third Form)** Let $f$ be a function from a finite set $X$ into a finite set $Y$. Suppose that $|X| = n$ and $|Y| = m$. Let $k = \lceil n/m \rceil$. Then there are at least $k$ distinct values $a_1, \ldots, a_k \in X$ such that

$$f(a_1) = f(a_2) = \cdots = f(a_k).$$

To prove the third form of the Pigeonhole Principle, we argue by contradiction. Let $Y = \{y_1, \ldots, y_m\}$. Suppose that the conclusion is false. Then there are at most $k - 1$ distinct values $x \in X$ with $f(x) = y_1$; there are at most $k - 1$ distinct values $x \in X$ with $f(x) = y_2$; ...; there are at most $k - 1$ distinct values $x \in X$ with $f(x) = y_m$. Thus there are at most $m(k - 1)$ members in the domain of $f$. But

$$m(k - 1) < m\frac{n}{m} = n,$$

which is a contradiction. Therefore, there are at least $k$ distinct values, $a_1, \ldots, a_k \in X$, such that

$$f(a_1) = f(a_2) = \cdots = f(a_k).$$

Our last example illustrates the use of the third form of the Pigeonhole Principle.

**Example 6.8.6** A useful feature of black-and-white pictures is the average brightness of the picture. Let us say that two pictures are similar if their average brightness differs by no more than some fixed value. Show that among six pictures, there are either three that are mutually similar or three that are mutually dissimilar.

**SOLUTION** Denote the pictures $P_1, P_2, \ldots, P_6$. Each of the five pairs

$$(P_1, P_2), \quad (P_1, P_3), \quad (P_1, P_4), \quad (P_1, P_5), \quad (P_1, P_6),$$

has the value "similar" or "dissimilar." By the third form of the Pigeonhole Principle, there are at least $\lceil 5/2 \rceil = 3$ pairs with the same value; that is, there are three pairs

$$(P_1, P_i), \quad (P_1, P_j), \quad (P_1, P_k)$$

all similar or all dissimilar. Suppose that each pair is similar. (The case that each pair is dissimilar is Exercise 14.) If any pair

$$(P_i, P_j), \quad (P_i, P_k), \quad (P_j, P_k) \tag{6.8.5}$$

is similar, then these two pictures together with $P_1$ are mutually similar and we have found three mutually similar pictures. Otherwise, each of the pairs (6.8.5) is dissimilar and we have found three mutually dissimilar pictures. ◀

## 6.8 Review Exercises

1. State three forms of the Pigeonhole Principle.

2. Give an example of the use of each form of the Pigeonhole Principle.

## 6.8 Exercises

1. Prove that if five cards are chosen from an ordinary 52-card deck, at least two cards are of the same suit.

2. Prove that among a group of six students, at least two received the same grade on the final exam. (The grades assigned were chosen from A, B, C, D, F.)

3. Suppose that each person in a group of 32 people receives a check in January. Prove that at least two people receive checks on the same day.

4. Prove that among 35 students in a class, at least two have first names that start with the same letter.

5. Prove that if $f$ is a function from the finite set $X$ to the finite set $Y$ and $|X| > |Y|$, then $f$ is not one-to-one.

6. Suppose that six distinct integers are selected from the set $\{1, 2, 3, 4, 5, 6, 7, 8, 9, 10\}$. Prove that at least two of the six have a sum equal to 11. *Hint*: Consider the partition $\{1, 10\}$, $\{2, 9\}$, $\{3, 8\}$, $\{4, 7\}$, $\{5, 6\}$.

7. Thirteen persons have first names Dennis, Evita, and Ferdinand and last names Oh, Pietro, Quine, and Rostenkowski. Show that at least two persons have the same first and last names.

8. Eighteen persons have first names Alfie, Ben, and Cissi and last names Dumont and Elm. Show that at least three persons have the same first and last names.

9. Professor Euclid is paid every other week on Friday. Show that in some month she is paid three times.

10. Is it possible to interconnect five processors so that exactly two processors are directly connected to an identical number of processors? Explain.

11. An inventory consists of a list of 115 items, each marked "available" or "unavailable." There are 60 available items. Show that there are at least two available items in the list exactly four items apart.

12. An inventory consists of a list of 100 items, each marked "available" or "unavailable." There are 55 available items. Show that there are at least two available items in the list exactly nine items apart.

★13. An inventory consists of a list of 80 items, each marked "available" or "unavailable." There are 50 available items. Show that there are at least two unavailable items in the list either three or six items apart.

14. Complete Example 6.8.6 by showing that if the pairs $(P_1, P_i)$, $(P_1, P_j)$, $(P_1, P_k)$ are dissimilar, there are three pictures that are mutually similar or mutually dissimilar.

15. Does the conclusion to Example 6.8.6 necessarily follow if there are fewer than six pictures? Explain.

16. Does the conclusion to Example 6.8.6 necessarily follow if there are more than six pictures? Explain.

17. Prove that for any positive integer $n$, there exists a positive integer which, when expressed in decimal, consists of at most $n$ 0s and 1s and is a multiple of $n$. *Hint*: Consider the set of the

$n$ integers, $\{1, 11, 111, \ldots\}$, using only 1s, and the remainders of these numbers when divided by $n$.

*Answer Exercises 18–21 to give an argument that shows that if $X$ is any $(n+2)$-element subset of $\{1, 2, \ldots, 2n+1\}$ and $m$ is the greatest element in $X$, there exist distinct $i$ and $j$ in $X$ with $m = i+j$.*

*For each element $k \in X - \{m\}$, let*

$$a_k = \begin{cases} k & \text{if } k \leq \frac{m}{2} \\ m - k & \text{if } k > \frac{m}{2}. \end{cases}$$

**18.** How many elements are in the domain of $a$?

**19.** Show that the range of $a$ is contained in $\{1, 2, \ldots, n\}$.

**20.** Explain why Exercises 18 and 19 imply that $a_i = a_j$ for some $i \neq j$.

**21.** Explain why Exercise 20 implies that there exist distinct $i$ and $j$ in $X$ with $m = i+j$.

**22.** Give an example of an $(n+1)$-element subset $X$ of $\{1, 2, \ldots, 2n+1\}$ having the property: For no distinct $i, j \in X$ do we have $i+j \in X$.

*Answer Exercises 23–26 to give an argument that proves the following result.*

> *A sequence $a_1, a_2, \ldots, a_{n^2+1}$ of $n^2+1$ distinct numbers contains either an increasing subsequence of length $n+1$ or a decreasing subsequence of length $n+1$.*

*Suppose by way of contradiction that every increasing or decreasing subsequence has length $n$ or less. Let $b_i$ be the length of a longest increasing subsequence starting at $a_i$, and let $c_i$ be the length of a longest decreasing subsequence starting at $a_i$.*

**23.** Show that the ordered pairs $(b_i, c_i)$, $i = 1, \ldots, n^2+1$, are distinct.

**24.** How many ordered pairs $(b_i, c_i)$ are there?

**25.** Explain why $1 \leq b_i \leq n$ and $1 \leq c_i \leq n$.

**26.** What is the contradiction?

*Answer Exercises 27–30 to give an argument that shows that in a group of 10 persons there are at least two such that either the difference or sum of their ages is divisible by 16. Assume that the ages are given as whole numbers.*

*Let $a_1, \ldots, a_{10}$ denote the ages. Let $r_i = a_i \bmod 16$ and let*

$$s_i = \begin{cases} r_i & \text{if } r_i \leq 8 \\ 16 - r_i & \text{if } r_i > 8. \end{cases}$$

**27.** Show that $s_1, \ldots, s_{10}$ range in value from 0 to 8.

**28.** Explain why $s_j = s_k$ for some $j \neq k$.

**29.** Suppose that $s_j = s_k$ for some $j \neq k$. Explain why if $s_j = r_j$ and $s_k = r_k$ or $s_j = 16 - r_j$ and $s_k = 16 - r_k$, then 16 divides $a_j - a_k$.

**30.** Show that if the conditions in Exercise 29 fail, then 16 divides $a_j + a_k$.

**31.** Show that in the decimal expansion of the quotient of two integers, eventually some block of digits repeats. *Examples:*

$$\frac{1}{6} = 0.1\underline{66}6\ldots, \qquad \frac{217}{660} = 0.32\underline{87}8787\ldots.$$

**★32.** Twelve basketball players, whose uniforms are numbered 1 through 12, stand around the center ring on the court in an arbitrary arrangement. Show that some three consecutive players have the sum of their numbers at least 20.

**★33.** For the situation of Exercise 32, find and prove an estimate for how large the sum of some four consecutive players' numbers must be.

**★34.** Let $f$ be a one-to-one function from $X = \{1, 2, \ldots, n\}$ onto $X$. Let $f^k = f \circ f \circ \cdots \circ f$ denote the $k$-fold composition of $f$ with itself. Show that there are distinct positive integers $i$ and $j$ such that $f^i(x) = f^j(x)$ for all $x \in X$. Show that for some positive integer $k$, $f^k(x) = x$ for all $x \in X$.

**★35.** A $3 \times 7$ rectangle is divided into 21 squares each of which is colored red or black. Prove that the board contains a nontrivial rectangle (not $1 \times k$ or $k \times 1$) whose four corner squares are all black or all red.

**★36.** Prove that if $p$ ones and $q$ zeros are placed around a circle in an arbitrary manner, where $p$, $q$, and $k$ are positive integers satisfying $p \geq kq$, the arrangement must contain at least $k$ consecutive ones.

**★37.** Write an algorithm that, given a sequence $a$, finds the length of a longest increasing subsequence of $a$.

**38.** A $2k \times 2k$ grid is divided into $4k^2$ squares and four $k \times k$ subgrids. The following figure shows the grid for $k = 4$:

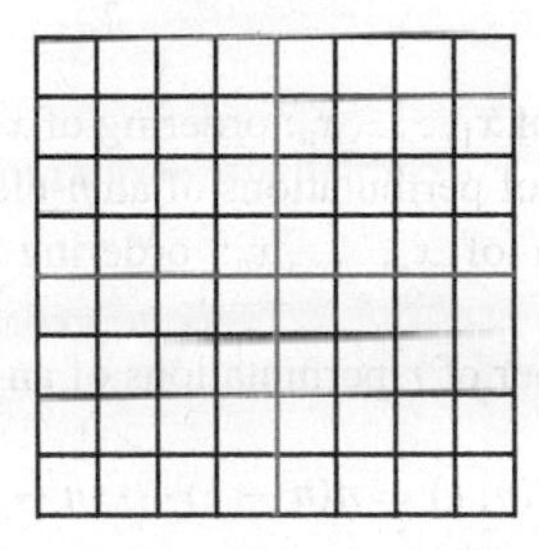

Show that it is impossible to mark $k$ squares in the upper-left, $k \times k$ subgrid and $k$ squares in the lower-right, $k \times k$ subgrid so that no two marked squares are in the same row, column, or diagonal of the $2k \times 2k$ grid.

This is a variant of the *n-queens problem,* which we discuss in detail in Section 9.3.

**39.** Prove that every $(n+1)$-element subset of $\{1, 2, \ldots, 2n\}$ contains two distinct integers $p$ and $q$ such that $\gcd(p, q) = 1$. *Hint:* Let $\{a_1, \ldots, a_{n+1}\}$ be an $(n+1)$-element subset of $\{1, 2, \ldots, 2n\}$. Consider the list $a_1, \ldots, a_{n+1}, a_1 + 1, \ldots, a_{n+1} + 1$.

**40.** Show that if "$(n+1)$-element subset" is changed to "$n$-element subset" in Exercise 39, the statement is false.

**41.** Prove that every $(n+1)$-element subset of $\{1, 2, \ldots, 2n\}$ contains two distinct integers $p$ and $q$ such that $p$ divides $q$. *Hint:*

If $X$ is an $(n+1)$-element subset of $\{1, 2, \ldots, 2n\}$, write the numbers in $X$ in the form $2^m k$, where $k$ is odd.

★**42.** Prove Exercise 41 using mathematical induction.

**43.** Show that if "$(n+1)$-element subset" is changed to "$n$-element subset" in Exercise 41, the statement is false.

**44.** Prove that a planar polygon with $n$ sides, $n \geq 3$, has at least three interior angles each less than 180 degrees. Assume no 0-degree interior angles. As an example, in the following figure angles $A$, $C$, and $E$ are each less than 180 degrees.

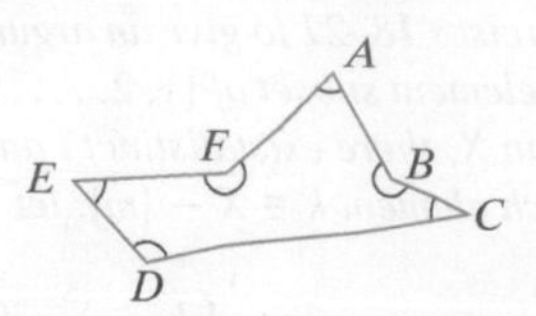


## Chapter 6 Notes

An elementary book concerning counting methods is [Niven, 1965]. References on combinatorics are [Brualdi; Even, 1973; Liu, 1968; Riordan; and Roberts]. [Vilenkin] contains many worked-out combinatorial examples. [Benjamin] contains an outstanding collection of combinatorial proofs. The general discrete mathematical references [Liu, 1985; and Tucker] devote several sections to the topics of Chapter 6. [Even, 1973; Hu; and Reingold] treat combinatorial algorithms. References on probability are [Billingsley; Ghahramani; Kelly; Ross; and Rozanov]. [Fukunaga; Gose; and Nadler] are texts on pattern recognition.

Several proofs of Fermat's little theorem can be found at http://en.wikipedia.org/wiki/Proofs_of_Fermat's_little_theorem.

## Chapter 6 Review

### Section 6.1

1. Multiplication Principle
2. Addition Principle
3. Inclusion-Exclusion Principle

### Section 6.2

4. Permutation of $x_1, \ldots, x_n$: ordering of $x_1, \ldots, x_n$
5. $n!$ = number of permutations of an $n$-element set
6. $r$-permutation of $x_1, \ldots, x_n$: ordering of $r$ elements of $x_1, \ldots, x_n$
7. $P(n, r)$: number of $r$-permutations of an $n$-element set;

$$P(n, r) = n(n-1)\cdots(n-r+1)$$

8. $r$-combination of $\{x_1, \ldots, x_n\}$: (unordered) subset of $\{x_1, \ldots, x_n\}$ containing $r$ elements
9. $C(n, r)$: number of $r$-combinations of an $n$-element set; $C(n, r) = P(n, r)/r! = n!/[(n-r)!\,r!]$

### Section 6.3

10. Number of orderings of $n$ items of $t$ types with $n_i$ identical objects of type $i = n!/[n_1! \cdots n_t!]$
11. Number of unordered, $k$-element selections from a $t$-element set, repetitions allowed $= C(k+t-1, k)$

### Section 6.4

12. Lexicographic order
13. Algorithm for generating $r$-combinations: Algorithm 6.4.9
14. Algorithm for generating permutations: Algorithm 6.4.16

### Section 6.5

15. Experiment
16. Event
17. Sample space
18. Probability of an event when all outcomes are equally likely

### Section 6.6

19. Probability function
20. Probability of an event
21. If $E$ is an event, $P(E) + P(\overline{E}) = 1$.
22. If $E_1$ and $E_2$ are events, $P(E_1 \cup E_2) = P(E_1) + P(E_2) - P(E_1 \cap E_2)$.
23. Events $E_1$ and $E_2$ are mutually exclusive if $E_1 \cap E_2 = \varnothing$.
24. If events $E_1$ and $E_2$ are mutually exclusive, $P(E_1 \cup E_2) = P(E_1) + P(E_2)$.
25. If $E$ and $F$ are events and $P(F) > 0$, the conditional probability of $E$ given $F$ is $P(E \mid F) = P(E \cap F)/P(F)$.
26. Events $E$ and $F$ are independent if $P(E \cap F) = P(E)P(F)$.
27. Bayes' Theorem: If the possible classes are $C_1, \ldots, C_n$, each pair of these classes is mutually exclusive, and each item to be classified belongs to one of these classes, for a feature set $F$ we have

$$P(C_j \mid F) = \frac{P(F \mid C_j)}{\sum_{i=1}^{n} P(F \mid C_i)P(C_i)}.$$

### Section 6.7

28. Binomial Theorem: $(a+b)^n = \sum_{k=0}^{n} C(n,k)a^{n-k}b^k$
29. Pascal's triangle: $C(n+1,k) = C(n,k-1) + C(n,k)$

### Section 6.8

30. Pigeonhole Principle (three forms)

## Chapter 6 Self-Test

1. How many eight-bit strings begin with 0 and end with 101?
2. How many 3-combinations are there of six objects?
3. How many strings can be formed by ordering the letters *ABCDEF* if *A* appears before *C* and *E* appears before *C*?
4. How many strings can be formed by ordering the letters *ILLINOIS*?
5. How many strings can be formed by ordering the letters *ILLINOIS* if some *I* appears before some *L*?
6. How many ways can we select three books each from a different subject from a set of six distinct history books, nine distinct classics books, seven distinct law books, and four distinct education books?
7. How many functions are there from an $n$-element set onto $\{0, 1\}$?
8. How many six-card hands chosen from an ordinary 52-card deck contain three cards of one suit and three cards of another suit?
9. Expand the expression $(s-r)^4$ using the Binomial Theorem.
10. Show that every set of 15 socks chosen among 14 pairs of socks contains at least one matched pair.
11. A seven-person committee composed of Greg, Hwang, Isaac, Jasmine, Kirk, Lynn, and Manuel is to select a chairperson, vice-chairperson, social events chairperson, secretary, and treasurer. How many ways can the officers be chosen if either Greg is secretary or he is not an officer?
12. A shipment of 100 compact discs contains five defective discs. In how many ways can we select a set of four compact discs that contains more defective than nondefective discs?
13. In how many ways can 12 distinct books be divided among four students if each student gets three books?
14. Find the coefficient of $x^3yz^4$ in the expansion of $(2x + y + z)^8$.
15. Use the Binomial Theorem to prove that
$$\sum_{k=0}^{n} 2^{n-k}(-1)^k C(n,k) = 1.$$
16. How many integer solutions of
$$x_1 + x_2 + x_3 + x_4 = 17$$
satisfy $x_1 \geq 0$, $x_2 \geq 1$, $x_3 \geq 2$, $x_4 \geq 3$?
17. Nineteen persons have first names Zeke, Wally, and Linda; middle names Lee and David; and last names Yu, Zamora, and Smith. Show that at least two persons have the same first, middle, and last names.
18. Rotate Pascal's triangle counterclockwise so that the top row consists of 1's. Explain why the second row lists the positive integers in order 1, 2, ....
19. An inventory consists of a list of 200 items, each marked "available" or "unavailable." There are 110 available items. Show that there are at least two available items in the list exactly 19 items apart.
20. Let $P = \{p_1, p_2, p_3, p_4, p_5\}$ be a set of five (distinct) points in the ordinary Euclidean plane each of which has integer coordinates. Show that some pair has a midpoint that has integer coordinates.
21. Find the 5-combination that will be generated by Algorithm 6.4.9 after 12467 if $n = 7$.
22. Find the 6-combination that will be generated by Algorithm 6.4.9 after 145678 if $n = 8$.
23. Find the permutation that will be generated by Algorithm 6.4.16 after 6427135.
24. Find the permutation that will be generated by Algorithm 6.4.16 after 625431.
25. A card is selected at random from an ordinary 52-card deck. What is the probability that it is a heart?
26. A coin is loaded so that a head is five times as likely to occur as a tail. Assign probabilities to the outcomes that accurately model the likelihood of the outcomes to occur.
27. A family has three children. Assume that it is equally probable for a boy or a girl to be born. Are the events "there are children of both sexes" and "there is at most one girl" independent? Explain.
28. Two fair dice are rolled. What is the probability that the sum of the numbers on the dice is 8?
29. In the Maryland Cash In Hand game, the contestant chooses seven distinct numbers among the numbers 1 through 31. The contestant wins a modest amount ($40) if exactly five numbers, in any order, match those among the seven distinct numbers randomly drawn by a lottery representative. What is the probability of winning $40?
30. Joe and Alicia take a final examination in C++. The probability that Joe passes is 0.75, and the probability that Alicia passes is 0.80. Assume that the events "Joe passes the final examination" and "Alicia passes the final examination" are independent. Find the probability that Joe does not pass. Find the probability that both pass. Find the probability that both fail. Find the probability that at least one passes.

**31.** Trisha, Roosevelt, and José write programs that schedule tasks for manufacturing dog toys. The following table shows the percentage of code written by each person and the percentage of buggy code for each person.

| | *Coder* | | |
|---|---|---|---|
| | *Trisha* | *Roosevelt* | *José* |
| *Percent of code* | 30 | 45 | 25 |
| *Percent of bugs* | 3 | 2 | 5 |

Given that a bug was found, find the probability that it was in the code written by José.

**32.** Find the probability of obtaining a bridge hand with 6–5–2–0 distribution, that is, six cards in one suit, five cards in another suit, two cards in another suit, and no cards in the fourth suit.

## Chapter 6 Computer Exercises

**1.** Write a program that generates all $r$-combinations of the elements $\{1, \ldots, n\}$.

**2.** Write a program that generates all permutations of the elements $\{1, \ldots, n\}$.

**3.** Write a program that generates all $r$-permutations of the elements $\{1, \ldots, n\}$.

**4.** [*Project*] Report on algorithms different from those presented in this chapter for generating combinations and permutations. Implement some of these algorithms as programs.

**5.** Write a program that lists all permutations of *ABCDEF* in which *A* appears before *D*.

**6.** Write a program that lists all permutations of *ABCDEF* in which *C* and *E* are side by side in either order.

**7.** Write a program that lists all the ways that $m$ distinct Martians and $n$ distinct Jovians can wait in line if no two Jovians stand together.

**8.** Write a program to compute the Catalan numbers.

**9.** Write a program that generates Pascal's triangle to level $n$, for arbitrary $n$.

**10.** Write a program that finds an increasing or decreasing subsequence of length $n + 1$ of a sequence of $n^2 + 1$ distinct numbers.

# Chapter 7

# RECURRENCE RELATIONS



This chapter offers an introduction to recurrence relations. Recurrence relations are useful in certain counting problems. A recurrence relation relates the $n$th element of a sequence to its predecessors. Because recurrence relations are closely related to recursive algorithms, recurrence relations arise naturally in the analysis of recursive algorithms.

## 7.1 Introduction

**Go Online**

For more on recurrence relations, see `bit.ly/2yYdlQL`

Consider the following instructions for generating a sequence:

1. Start with 5.
2. Given any term, add 3 to get the next term.

If we list the terms of the sequence, we obtain

$$5, 8, 11, 14, 17, \ldots. \tag{7.1.1}$$

The first term is 5 because of instruction 1. The second term is 8 because instruction 2 says to add 3 to 5 to get the next term, 8. The third term is 11 because instruction 2 says to add 3 to 8 to get the next term, 11. By following instructions 1 and 2, we can compute any term in the sequence. Instructions 1 and 2 do not give an explicit formula for the $n$th term of the sequence in the sense of providing a formula that we can "plug $n$ into" to obtain the value of the $n$th term, but by computing term by term we can eventually compute any term of the sequence.

If we denote the sequence (7.1.1) as $a_1, a_2, \ldots$, we may rephrase instruction 1 as

$$a_1 = 5 \tag{7.1.2}$$

and we may rephrase instruction 2 as

$$a_n = a_{n-1} + 3, \qquad n \geq 2. \tag{7.1.3}$$

[†]This section can be omitted without loss of continuity.

Taking $n = 2$ in (7.1.3), we obtain $a_2 = a_1 + 3$. By (7.1.2), $a_1 = 5$; thus

$$a_2 = a_1 + 3 = 5 + 3 = 8.$$

Taking $n = 3$ in (7.1.3), we obtain $a_3 = a_2 + 3$. Since $a_2 = 8$,

$$a_3 = a_2 + 3 = 8 + 3 = 11.$$

By using (7.1.2) and (7.1.3), we can compute any term in the sequence just as we did using instructions 1 and 2. We see that (7.1.2) and (7.1.3) are equivalent to instructions 1 and 2.

Equation (7.1.3) furnishes an example of a **recurrence relation.** A recurrence relation defines a sequence by giving the $n$th value in terms of certain of its predecessors. In (7.1.3) the $n$th value is given in terms of the immediately preceding value. In order for a recurrence relation such as (7.1.3) to define a sequence, a "start-up" value or values, such as (7.1.2), must be given. These start-up values are called **initial conditions.** The formal definitions follow.

**Definition 7.1.1 ▶** A *recurrence relation* for the sequence $a_0, a_1, \ldots$ is an equation that relates $a_n$ to certain of its predecessors $a_0, a_1, \ldots, a_{n-1}$.

*Initial conditions* for the sequence $a_0, a_1, \ldots$ are explicitly given values for a finite number of the terms of the sequence. ◀

We have seen that it is possible to define a sequence by a recurrence relation together with certain initial conditions. Next we give several examples of recurrence relations.

**Example 7.1.2** The Fibonacci sequence (see the discussion following Algorithm 4.4.6) is defined by the recurrence relation $f_n = f_{n-1} + f_{n-2}$, $n \geq 3$, and initial conditions $f_1 = 1$, $f_2 = 1$. ◀

**Example 7.1.3** A person invests \$1000 at 12 percent interest compounded annually. If $A_n$ represents the amount at the end of $n$ years, find a recurrence relation and initial conditions that define the sequence $\{A_n\}$.

**Go Online**
For more on compound interest, see `bit.ly/2yYdlQL`

SOLUTION At the end of $n - 1$ years, the amount is $A_{n-1}$. After one more year, we will have the amount $A_{n-1}$ plus the interest. Thus

$$A_n = A_{n-1} + (0.12)A_{n-1} = (1.12)A_{n-1}, \qquad n \geq 1. \tag{7.1.4}$$

To apply this recurrence relation for $n = 1$, we need to know the value of $A_0$. Since $A_0$ is the beginning amount, we have the initial condition

$$A_0 = 1000. \tag{7.1.5}$$

◀

The initial condition (7.1.5) and the recurrence relation (7.1.4) allow us to compute the value of $A_n$ for any $n$. For example,

$$\begin{aligned} A_3 &= (1.12)A_2 = (1.12)(1.12)A_1 \\ &= (1.12)(1.12)(1.12)A_0 = (1.12)^3(1000) = 1404.93. \end{aligned} \tag{7.1.6}$$

Thus, at the end of the third year, the amount is \$1404.93.

The computation (7.1.6) can be carried out for an arbitrary value of $n$ to obtain

$$A_n = (1.12)A_{n-1}$$
$$\vdots$$
$$= (1.12)^n(1000).$$

We see that sometimes an explicit formula can be derived from a recurrence relation and initial conditions. Finding explicit formulas from recurrence relations is the topic of Section 7.2.

Although it is easy to obtain an explicit formula from the recurrence relation and initial condition for the sequence of Example 7.1.3, it is not immediately apparent how to obtain an explicit formula for the Fibonacci sequence. In Section 7.2 we give a method that yields an explicit formula for the Fibonacci sequence.

Recurrence relations, recursive algorithms, and mathematical induction are closely related. In all three, prior instances of the current case are assumed known. A recurrence relation uses prior values in a sequence to compute the current value. A recursive algorithm uses smaller instances of the current input to process the current input. The inductive step in a proof by mathematical induction assumes the truth of prior instances of the statement to prove the truth of the current statement.

A recurrence relation that defines a sequence can be directly converted to an algorithm to compute the sequence. For example, Algorithm 7.1.4, derived from recurrence relation (7.1.4) and initial condition (7.1.5), computes the sequence of Example 7.1.3.

**Algorithm 7.1.4** **Computing Compound Interest**

This recursive algorithm computes the amount of money at the end of $n$ years assuming an initial amount of \$1000 and an interest rate of 12 percent compounded annually.

Input: $n$, the number of years

Output: The amount of money at the end of $n$ years

```
1. compound_interest(n) {
2.     if (n == 0)
3.         return 1000
4.     return 1.12 * compound_interest(n − 1)
5. }
```

Algorithm 7.1.4 is a direct translation of equations (7.1.4) and (7.1.5), which define the sequence $A_0, A_1, \ldots$. Lines 2 and 3 correspond to initial condition (7.1.5), and line 4 corresponds to recurrence relation (7.1.4).

**Example 7.1.5** Let $S_n$ denote the number of subsets of an $n$-element set. Since going from an $(n-1)$-element set to an $n$-element set doubles the number of subsets (see Theorem 2.4.6), we obtain the recurrence relation $S_n = 2S_{n-1}$. The initial condition is $S_0 = 1$. ◀

One of the main reasons for using recurrence relations is that sometimes it is easier to determine the $n$th term of a sequence in terms of its predecessors than it is to find an explicit formula for the $n$th term in terms of $n$. The next examples are intended to illustrate this thesis.

**Example 7.1.6** Let $S_n$ denote the number of $n$-bit strings that do not contain the pattern 111. Develop a recurrence relation for $S_1, S_2, \ldots$ and initial conditions that define the sequence $S$.

**SOLUTION** We will count the number of $n$-bit strings that do not contain the pattern 111

(a) that begin with 0;
(b) that begin with 10;
(c) that begin with 11.

Since the sets of strings of types (a), (b), and (c) are disjoint, by the Addition Principle $S_n$ will equal the sum of the numbers of strings of types (a), (b), and (c). Suppose that an $n$-bit string begins with 0 and does not contain the pattern 111. Then the $(n-1)$-bit string following the initial 0 does not contain the pattern 111. Since any $(n-1)$-bit string not containing 111 can follow the initial 0, there are $S_{n-1}$ strings of type (a). If an $n$-bit string begins with 10 and does not contain the pattern 111, then the $(n-2)$-bit string following the initial 10 cannot contain the pattern 111; therefore, there are $S_{n-2}$ strings of type (b). If an $n$-bit string begins with 11 and does not contain the pattern 111, then the third bit must be 0. The $(n-3)$-bit string following the initial 110 cannot contain the pattern 111; therefore, there are $S_{n-3}$ strings of type (c). Thus

$$S_n = S_{n-1} + S_{n-2} + S_{n-3} \qquad n \geq 4.$$

By inspection, we find the initial conditions $S_1 = 2$, $S_2 = 4$, $S_3 = 7$. ◀

**Example 7.1.7** Recall (see Example 6.2.22) that the Catalan number $C_n$ is equal to the number of routes from the lower-left corner of an $n \times n$ square grid to the upper-right corner if we are restricted to traveling only to the right or upward and if we are allowed to touch but not go above a diagonal line from the lower-left corner to the upper-right corner. We call such a route a *good route*. We give a recurrence relation for the Catalan numbers.

We divide the good routes into classes based on when they first meet the diagonal after leaving the lower-left corner. For example, the route in Figure 7.1.1 first meets the diagonal at point (3, 3). We regard the routes that first meet the diagonal at $(k, k)$ as

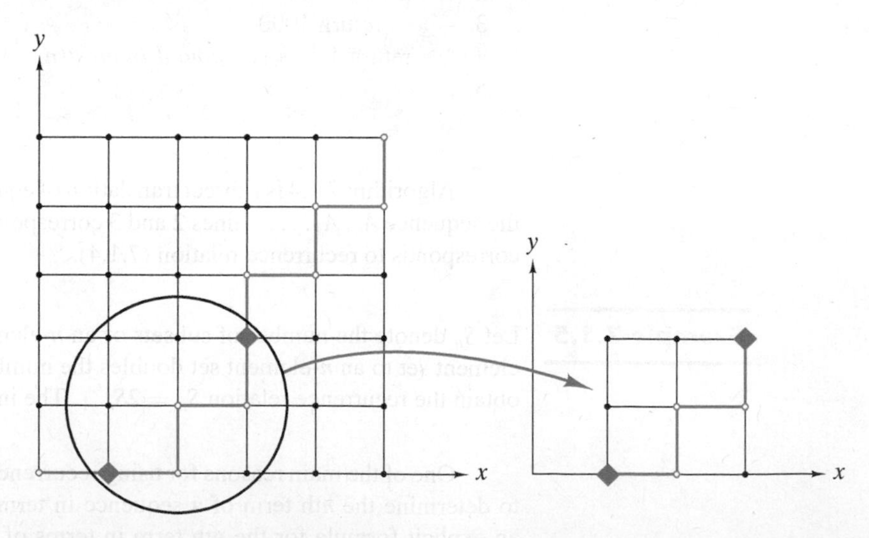


**Figure 7.1.1** Decomposition of a good route.

constructed by a two-step process: First, construct the part from $(0, 0)$ to $(k, k)$. Second, construct the part from $(k, k)$ to $(n, n)$. A good route always leaves $(0, 0)$ by moving right to $(1, 0)$ and it always arrives at $(k, k)$ by moving up from $(k, k-1)$. The moves from $(1, 0)$ to $(k, k-1)$ give a good route in the $(k-1) \times (k-1)$ grid with corners at $(1, 0)$, $(1, k-1)$, $(k, k-1)$, and $(k, 0)$. [In Figure 7.1.1, we marked the points $(1, 0)$ and $(k, k-1)$, $k = 3$, with diamonds, and we isolated the $(k-1) \times (k-1)$ subgrid.] Thus there are $C_{k-1}$ routes from $(0, 0)$ to $(k, k)$ that first meet the diagonal at $(k, k)$. The part from $(k, k)$ to $(n, n)$ is a good route in the $(n-k) \times (n-k)$ grid with corners at $(k, k)$, $(k, n)$, $(n, n)$, and $(n, k)$ (see Figure 7.1.1). There are $C_{n-k}$ such routes. By the Multiplication Principle, there are $C_{k-1}C_{n-k}$ good routes in an $n \times n$ grid that first meet the diagonal at $(k, k)$. The good routes that first meet the diagonal at $(k, k)$ are distinct from those that first meet the diagonal at $(k', k')$, $k \neq k'$. Thus we may use the Addition Principle to obtain a recurrence relation for the total number of good routes in an $n \times n$ grid:

$$C_n = \sum_{k=1}^{n} C_{k-1}C_{n-k}.$$

◀

**Example 7.1.8** **Tower of Hanoi** The Tower of Hanoi is a puzzle consisting of three pegs mounted on a board and $n$ disks of various sizes with holes in their centers (see Figure 7.1.2). It is assumed that if a disk is on a peg, only a disk of smaller diameter can be placed on top of the first disk. Given all the disks stacked on one peg as in Figure 7.1.2, the problem is to transfer the disks to another peg by moving one disk at a time.

**Go Online**
For more on the Tower of Hanoi, see `bit.ly/2yYd1QL`

We will provide a solution and then find a recurrence relation and an initial condition for the sequence $c_1, c_2, \ldots,$ where $c_n$ denotes the number of moves our solution takes to solve the $n$-disk puzzle. We will then show that our solution is optimal; that is, we will show that no other solution uses fewer moves.

We give a recursive algorithm. If there is only one disk, we simply move it to the desired peg. If we have $n > 1$ disks on peg 1 as in Figure 7.1.2, we begin by recursively invoking our algorithm to move the top $n - 1$ disks to peg 2 (see Figure 7.1.3). During these moves, the bottom disk on peg 1 stays fixed. Next, we move the remaining disk on peg 1 to peg 3. Finally, we again recursively invoke our algorithm to move the $n - 1$ disks on peg 2 to peg 3. We have succeeded in moving $n$ disks from peg 1 to peg 3.

If $n > 1$, we solve the $(n-1)$-disk problem twice and we explicitly move one disk. Therefore, $c_n = 2c_{n-1} + 1$, $n > 1$. The initial condition is $c_1 = 1$. In Section 7.2, we will show that $c_n = 2^n - 1$.

We next show that our solution is optimal. We let $d_n$ be the number of moves an optimal solution requires. We use mathematical induction to show that

$$c_n = d_n \qquad n \geq 1. \tag{7.1.7}$$

**Go Online**
For a C program to solve the Tower of Hanoi puzzle, see `bit.ly/2R5vgfW`

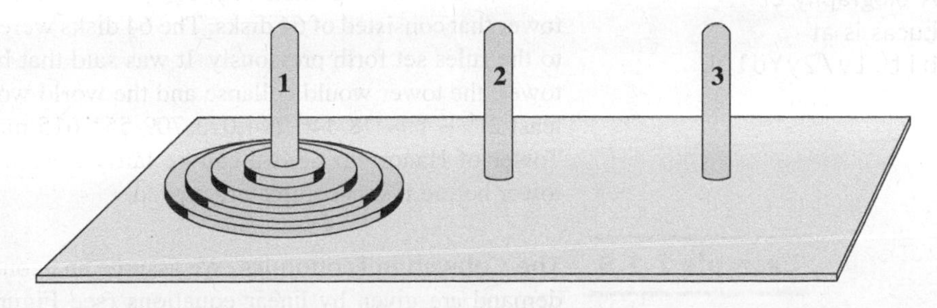


**Figure 7.1.2** Tower of Hanoi.

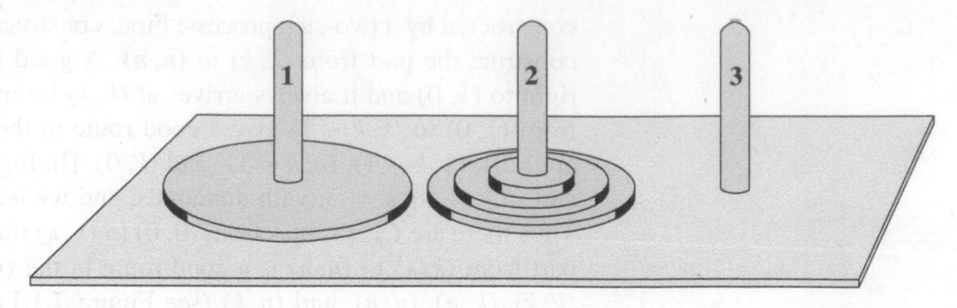


Figure 7.1.3 After recursively moving the top $n-1$ disks from peg 1 to peg 2 in the Tower of Hanoi.

**Basis Step ($n=1$)**

By inspection, $c_1 = 1 = d_1$; thus (7.1.7) is true when $n = 1$.

**Inductive Step**

Assume that (7.1.7) is true for $n-1$. Consider the point in an optimal solution to the $n$-disk problem when the largest disk is moved for the first time. The largest disk must be on a peg by itself (so that it can be removed from the peg) and another peg must be empty (so that this peg can receive the largest disk). Thus the $n-1$ smaller disks must be stacked on a third peg (see Figure 7.1.3). In other words, the $(n-1)$ disk problem must have been solved, which required at least $d_{n-1}$ moves. The largest disk was then moved, which required one additional move. Finally, at some point the $n-1$ disks were moved on top of the largest disk, which required at least $d_{n-1}$ additional moves. It follows that

$$d_n \geq 2d_{n-1} + 1.$$

By the inductive assumption, $c_{n-1} = d_{n-1}$. Thus

$$d_n \geq 2d_{n-1} + 1 = 2c_{n-1} + 1 = c_n. \tag{7.1.8}$$

The last equality follows from the recurrence relation for the sequence $c_1, c_2, \ldots$. By definition, no solution can take fewer moves than an optimal solution, so

$$c_n \geq d_n. \tag{7.1.9}$$

Inequalities (7.1.8) and (7.1.9) combine to give $c_n = d_n$. The inductive step is complete. Therefore, our solution is optimal. ◀

The Tower of Hanoi puzzle was invented by the French mathematician Édouard Lucas in the late nineteenth century. [Lucas was the first person to call the sequence 1, 1, 2, 3, 5, ... the Fibonacci sequence. The Lucas sequence (see Exercise 68) is also named after him.] The following myth was also created to accompany the puzzle (and, one assumes, to help market it). The puzzle was said to be derived from a mythical gold tower that consisted of 64 disks. The 64 disks were to be transferred by monks according to the rules set forth previously. It was said that before the monks finished moving the tower, the tower would collapse and the world would end in a clap of thunder. Since at least $2^{64} - 1 = 18{,}446{,}744{,}073{,}709{,}551{,}615$ moves are required to solve the 64-disk Tower of Hanoi puzzle, we can be fairly certain that something would happen to the tower before it was completely moved.

**Go Online**
A biography of Lucas is at bit.ly/2yYdlQL

**Example 7.1.9** **The Cobweb in Economics** We assume an economics model in which the supply and demand are given by linear equations (see Figure 7.1.4). Specifically, the demand is given by the equation $p = a - bq$, where $p$ is the price, $q$ is the quantity, and $a$ and $b$

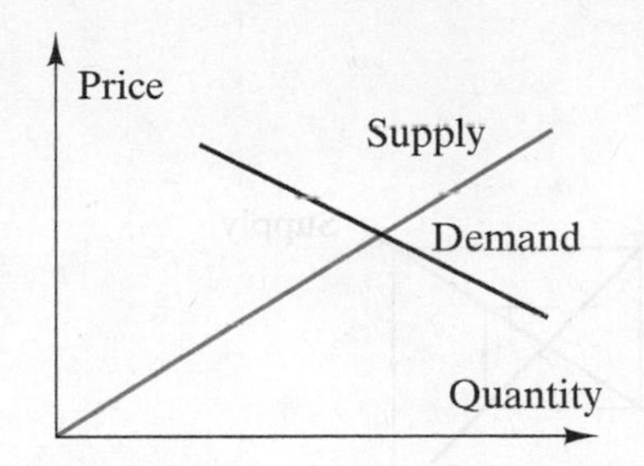


**Figure 7.1.4** An economics model.

are positive parameters. The idea is that as the price increases, the consumers demand less of the product. The supply is given by the equation $p = kq$, where $p$ is the price, $q$ is the quantity, and $k$ is a positive parameter. The idea is that as the price increases, the manufacturer is willing to supply greater quantities.

We assume further that there is a time lag as the supply reacts to changes. (For example, it takes time to manufacture goods and it takes time to grow crops.) We denote the discrete time intervals as $n = 0, 1, \ldots$. We assume that the demand is given by the equation

$$p_n = a - bq_n; \tag{7.1.10}$$

that is, at time $n$, the quantity $q_n$ of the product will be sold at price $p_n$. We assume that the supply is given by the equation

$$p_n = kq_{n+1}; \tag{7.1.11}$$

that is, one unit of time is required for the manufacturer to adjust the quantity $q_{n+1}$, at time $n + 1$, to the price $p_n$, at the prior time $n$.

If we solve equation (7.1.11) for $q_{n+1}$ and substitute into the demand equation (7.1.10) for time $n + 1$, $p_{n+1} = a - bq_{n+1}$, we obtain the recurrence relation

$$p_{n+1} = a - \frac{b}{k}p_n$$

for the price. We will solve this recurrence relation in Section 7.2.

The price changes through time may be viewed graphically. If the initial price is $p_0$, the manufacturer will be willing to supply the quantity $q_1$, at time $n = 1$. We locate this quantity by moving horizontally to the supply curve (see Figure 7.1.5). However, the market forces drive the price down to $p_1$, as we can see by moving vertically to the demand curve. At price $p_1$, the manufacturer will be willing to supply the quantity $q_2$ at time $n = 2$, as we can see by moving horizontally to the supply curve. Now the market forces drive the price up to $p_2$, as we can see by moving vertically to the demand curve. By continuing this process, we obtain the "cobweb" shown in Figure 7.1.5.

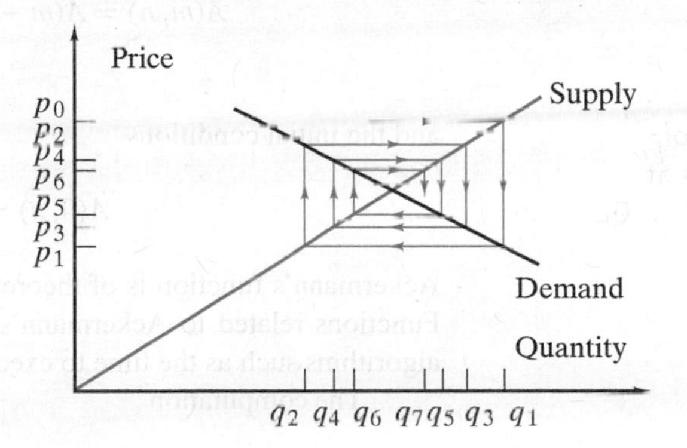


**Figure 7.1.5** A cobweb with a stabilizing price.

For the supply and demand functions of Figure 7.1.5, the price approaches that given by the intersection of the supply and demand curves. This is not always the case, however. For example, in Figure 7.1.6, the price fluctuates between $p_0$ and $p_1$, whereas in Figure 7.1.7, the price swings become more and more pronounced. The behavior is determined by the slopes of the supply and demand lines. To produce the fluctuating behavior of Figure 7.1.6, the angles $\alpha$ and $\beta$ must add to 180°. The slopes of the supply and demand curves are $\tan\alpha$ and $\tan\beta$, respectively; thus in Figure 7.1.6, we have

$$k = \tan\alpha = -\tan\beta = b.$$

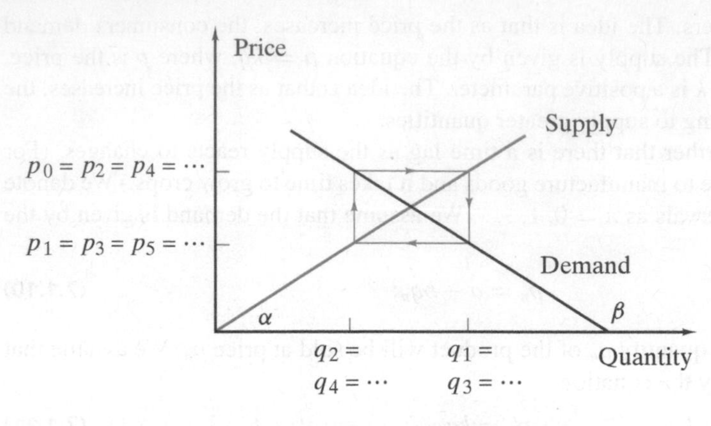


**Figure 7.1.6** A cobweb with a fluctuating price.

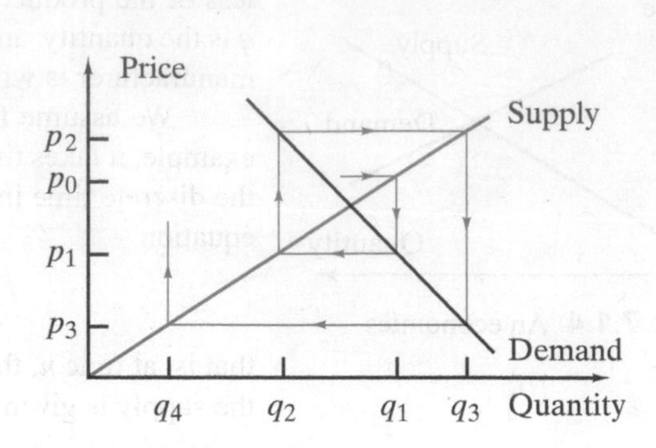


**Figure 7.1.7** A cobweb with increasing price swings.

We have shown that the price fluctuates between two values when $k = b$. A similar analysis shows that the price tends to that given by the intersection of the supply and demand curves (Figure 7.1.5) when $b < k$; the increasing price-swings case (Figure 7.1.7) occurs when $b > k$ (see Exercises 38 and 39). In Section 7.2 we discuss the behavior of the price through time by analyzing an explicit formula for the price $p_n$. ◀

It is possible to extend the definition of recurrence relation to include functions indexed over $n$-tuples of positive integers. Our last example is of this form.

**Example 7.1.10** **Ackermann's Function** Ackermann's function can be defined by the recurrence relations

$$A(m, 0) = A(m-1, 1) \qquad m = 1, 2, \ldots, \tag{7.1.12}$$

$$A(m, n) = A(m-1, A(m, n-1)) \qquad m = 1, 2, \ldots, \quad n = 1, 2, \ldots, \tag{7.1.13}$$

**Go Online**
A biography of Ackermann is at `bit.ly/2yYdlQL`

and the initial conditions

$$A(0, n) = n + 1 \qquad n = 0, 1, \ldots. \tag{7.1.14}$$

Ackermann's function is of theoretical importance because of its rapid rate of growth. Functions related to Ackermann's function appear in the time complexity of certain algorithms such as the time to execute union/find algorithms (see [Tarjan, pp. 22–29]).

The computation

$$\begin{aligned} A(1, 1) &= A(0, A(1, 0)) && \text{by (7.1.13)} \\ &= A(0, A(0, 1)) && \text{by (7.1.12)} \\ &= A(0, 2) && \text{by (7.1.14)} \\ &= 3 && \text{by (7.1.14)} \end{aligned}$$

illustrates the use of equations (7.1.12)–(7.1.14).

What we and others have called Ackermann's function is actually derived from Ackermann's original function. Ackermann's original function defines increasing orders of arithmetic: Addition is defined in terms of the "add-one" function, multiplication is defined in terms of addition, exponentiation is defined in terms of multiplication, and so on. (See Exercises 47–52.) ◀

### 7.1 Problem-Solving Tips

To set up a recurrence relation, first *define*, say $A_n$, to be the desired quantity. For example, let $A_n$ denote the amount of money at the end of $n$ years if \$1000 is invested at 12 percent interest compounded annually. Next, just as in proof by induction, *look for smaller instances within instance n.* Continuing the compound interest example, instance $n-1$ is "within" instance $n$ in the sense that the amount of money at the end of $n$ years is the amount at the end of $n-1$ years plus interest. We obtain the recurrence relation

$$A_n = A_{n-1} + (0.12)A_{n-1}.$$

## 7.1 Review Exercises

1. What is a recurrence relation?
2. What is an initial condition?
3. What is compound interest and how can it be described by a recurrence relation?
4. What is the Tower of Hanoi puzzle?
5. Give a solution of the Tower of Hanoi puzzle.
6. Describe the cobweb in economics.
7. Define *Ackermann's function.*

## 7.1 Exercises

*In Exercises 1–3, find a recurrence relation and initial conditions that generate a sequence that begins with the given terms.*

1. 3, 7, 11, 15, ...
2. 3, 6, 9, 15, 24, 39, ...
3. 1, 1, 2, 4, 16, 128, 4096, ...

*In Exercises 4–8, assume that a person invests \$2000 at 14 percent interest compounded annually. Let $A_n$ represent the amount at the end of n years.*

4. Find a recurrence relation for the sequence $A_0, A_1, \ldots$.
5. Find an initial condition for the sequence $A_0, A_1, \ldots$.
6. Find $A_1$, $A_2$, and $A_3$.
7. Find an explicit formula for $A_n$.
8. How long will it take for a person to double the initial investment?

*If a person invests in a tax-sheltered annuity, the money invested, as well as the interest earned, is not subject to taxation until withdrawn from the account. In Exercises 9–12, assume that a person invests \$2000 each year in a tax-sheltered annuity at 10 percent interest compounded annually. Let $A_n$ represent the amount at the end of n years.*

9. Find a recurrence relation for the sequence $A_0, A_1, \ldots$.
10. Find an initial condition for the sequence $A_0, A_1, \ldots$.
11. Find $A_1$, $A_2$, and $A_3$.
12. Find an explicit formula for $A_n$.

*In Exercises 13–17, assume that a person invests \$3000 at 12 percent annual interest compounded quarterly. Let $A_n$ represent the amount at the end of n years.*

13. Find a recurrence relation for the sequence $A_0, A_1, \ldots$.
14. Find an initial condition for the sequence $A_0, A_1, \ldots$.
15. Find $A_1$, $A_2$, and $A_3$.
16. Find an explicit formula for $A_n$.
17. How long will it take for a person to double the initial investment?
18. Let $S_n$ denote the number of $n$ bit strings that do not contain the pattern 000. Find a recurrence relation and initial conditions for the sequence $\{S_n\}$.

*Exercises 19–21 refer to the sequence S where $S_n$ denotes the number of n-bit strings that do not contain the pattern 00.*

19. Find a recurrence relation and initial conditions for the sequence $\{S_n\}$.
20. Show that $S_n = f_{n+2}$, $n = 1, 2, \ldots$, where $f$ denotes the Fibonacci sequence.
21. By considering the number of $n$-bit strings with exactly $i$ 0's and Exercise 20, show that

$$f_{n+2} = \sum_{i=0}^{\lfloor (n+1)/2 \rfloor} C(n+1-i, i) \qquad n = 1, 2, \ldots,$$

where $f$ denotes the Fibonacci sequence.

*Exercises 22–24 refer to the sequence $S_1, S_2, \ldots$, where $S_n$ denotes the number of n-bit strings that do not contain the pattern 010.*

**22.** Compute $S_1$, $S_2$, $S_3$, and $S_4$.

**23.** By considering the number of $n$-bit strings that do not contain the pattern 010 that have no leading 0's (i.e., that begin with 1); that have one leading 0 (i.e., that begin 01); that have two leading 0's; and so on, derive the recurrence relation

$$S_n = S_{n-1} + S_{n-3} + S_{n-4} + S_{n-5} + \cdots + S_1 + 3. \quad (7.1.15)$$

**24.** By replacing $n$ by $n - 1$ in (7.1.15), write a formula for $S_{n-1}$. Subtract the formula for $S_{n-1}$ from the formula for $S_n$ and use the result to derive the recurrence relation

$$S_n = 2S_{n-1} - S_{n-2} + S_{n-3}.$$

*In Exercises 25–33, $C_0, C_1, C_2, \ldots$ denotes the sequence of Catalan numbers.*

**25.** Given that $C_0 = C_1 = 1$ and $C_2 = 2$, compute $C_3$, $C_4$, and $C_5$ by using the recurrence relation of Example 7.1.7.

**26.** Show that the Catalan numbers are given by the recurrence relation

$$(n + 2)C_{n+1} = (4n + 2)C_n \qquad n \geq 0,$$

and initial condition $C_0 = 1$.

**27.** Prove that $n + 2 < C_n$ for all $n \geq 4$.

**★28.** Prove that $C_n$ is prime if and only if $n = 2$ or $n = 3$. *Hint:* First, use proof by contradiction to prove that if $n \geq 5$, $C_n$ is not prime. Use Exercises 26 and 27, this section, and Exercise 27, Section 5.3, to derive a contradiction. Now examine the cases $n = 0, 1, 2, 3, 4$.

**29.** Prove that

$$C_n \geq \frac{4^{n-1}}{n^2} \qquad \text{for all } n \geq 1.$$

**30.** Derive a recurrence relation and an initial condition for the number of ways to parenthesize the product

$$a_1 * a_2 * \cdots * a_n \qquad n \geq 2.$$

*Examples:* There is one way to parenthesize $a_1 * a_2$, namely, $(a_1 * a_2)$. There are two ways to parenthesize $a_1 * a_2 * a_3$, namely, $((a_1 * a_2) * a_3)$ and $(a_1 * (a_2 * a_3))$. Deduce that the number of ways to parenthesize the product of $n$ elements is $C_{n-1}$, $n \geq 2$.

**★31.** Derive a recurrence relation and an initial condition for the number of ways to divide a convex $(n + 2)$-sided polygon, $n \geq 1$, into triangles by drawing $n - 1$ lines through the corners that do not intersect in the interior of the polygon. (A polygon is *convex* if any line joining two points in the polygon lies wholly in the polygon.) For example, there are five ways to divide a convex pentagon into triangles by drawing two nonintersecting lines through the corners:

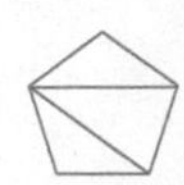 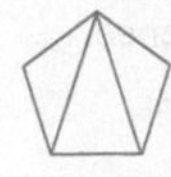 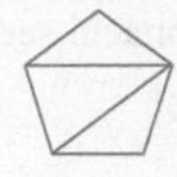  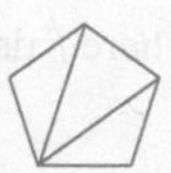

Deduce that the number of ways to divide a convex $(n + 2)$ sided polygon into triangles by drawing $n - 1$ nonintersecting lines through the corners is $C_n$, $n \geq 1$.

**32.** How many parenthesized expressions are there containing $n$ distinct binary operators, $n + 1$ distinct variables, and $n - 1$ pairs of parentheses? For example, if $n = 2$, and we choose $*$ and $+$ as the operators and $x$, $y$, and $z$ as the variables, some of the expressions are

$$(x * y) + z, \quad x * (y + z), \quad x * (z + y), \quad x + (z * y), \quad z * (y + x).$$

**33.** Consider routes from the lower-left corner to the upper-right corner in an $(n + 1) \times (n + 1)$ grid in which we are restricted to traveling only to the right or upward. By dividing the routes into classes based on when, after leaving the lower-left corner, the route first meets the diagonal line from the lower-left corner to the upper-right corner, derive the recurrence relation

$$C_n = \frac{1}{2}C(2(n + 1), n + 1) - \sum_{k=0}^{n-1} C_k C(2(n - k), n - k).$$

*In Exercises 34 and 35, let $S_n$ denote the number of routes from the lower-left corner of an $n \times n$ grid to the upper-right corner in which we are restricted to traveling to the right, upward, or diagonally northeast [i.e., from $(i, j)$ to $(i+1, j+1)$] and in which we are allowed to touch but not go above a diagonal line from the lower-left corner to the upper-right corner. The numbers $S_0, S_1, \ldots$ are called the* Schröder numbers.

**34.** Show that $S_0 = 1$, $S_1 = 2$, $S_2 = 6$, and $S_3 = 22$.

**35.** Derive a recurrence relation for the sequence of Schröder numbers.

**36.** Write explicit solutions for the Tower of Hanoi puzzle for $n = 3, 4$.

**37.** To what values do the price and quantity tend in Example 7.1.9 when $b < k$?

**38.** Show that when $b < k$ in Example 7.1.9, the price tends to that given by the intersection of the supply and demand curves.

**39.** Show that when $b > k$ in Example 7.1.9, the differences between successive prices increase.

*Exercises 40–46 refer to Ackermann's function $A(m, n)$.*

**40.** Compute $A(2, 2)$ and $A(2, 3)$.

**41.** Use induction to show that

$$A(1, n) = n + 2 \qquad n = 0, 1, \ldots.$$

**42.** Use induction to show that

$$A(2, n) = 3 + 2n \qquad n = 0, 1 \ldots.$$

**43.** Guess a formula for $A(3, n)$ and prove it by using induction.

**★44.** Prove that $A(m, n) > n$ for all $m \geq 0, n \geq 0$ by induction on $m$. The inductive step will use induction on $n$.

**45.** By using Exercise 44 or otherwise, prove that $A(m, n) > 1$ for all $m \geq 1, n \geq 0$.

**46.** By using Exercise 44 or otherwise, prove that $A(m, n) < A(m, n + 1)$ for all $m \geq 0, n \geq 0$.

*In Exercises 47–52, Ackermann's original function AO is defined by*

$$
\begin{aligned}
AO(0, y, z) &= z + 1 && y, z \geq 0,\\
AO(1, y, z) &= y + z && y, z \geq 0,\\
AO(2, y, z) &= yz && y, z \geq 0,\\
AO(x + 3, y, 0) &= 1 && x \geq 0, y \geq 1,\\
AO(x + 3, y, z + 1) &= \\
AO(x + 2, y, &AO(x + 3, y, z)) && x \geq 0, y \geq 1, z \geq 0.
\end{aligned}
$$

*Exercises 47–52 refer to the function AO and to Ackermann's function A.*

**47.** Prove that $AO(3, y, z) = y^z$ for $y \geq 1, z \geq 0$.

**48.** Find a formula for $AO(4, y, z)$ for $y \geq 1, z \geq 0$ and prove it.

**49.** Show that $A(x, y) = AO(x, 2, y + 3) - 3$ for $y \geq 0$ and $x = 0, 1, 2$.

**50.** Show that $AO(x, 2, 1) = 2$ for $x \geq 2$.

**51.** Show that $AO(x, 2, 2) = 4$ for $x \geq 2$.

**★52.** Show that $A(x, y) = AO(x, 2, y + 3) - 3$ for $x, y \geq 0$.

**53.** A network consists of $n$ nodes. Each node has communications facilities and local storage. Periodically, all files must be shared. A *link* consists of two nodes sharing files. Specifically, when nodes $A$ and $B$ are linked, $A$ transmits all its files to $B$ and $B$ transmits all its files to $A$. Only one link exists at a time, and after a link is established and the files are shared, the link is deleted. Let $a_n$ be the minimum number of links required by $n$ nodes so that all files are known to all nodes.

(a) Show that $a_2 = 1, a_3 \leq 3, a_4 \leq 4$.

(b) Show that $a_n \leq a_{n-1} + 2, n \geq 3$.

**54.** If $P_n$ denotes the number of permutations of $n$ distinct objects, find a recurrence relation and an initial condition for the sequence $P_1, P_2, \ldots$.

**55.** Suppose that we have $n$ dollars and that each day we buy either orange juice ($1), milk ($2), or beer ($2). If $R_n$ is the number of ways of spending all the money, show that

$$R_n = R_{n-1} + 2R_{n-2}.$$

Order is taken into account. For example, there are 11 ways to spend $4: *MB, BM, OOM, OOB, OMO, OBO, MOO, BOO, OOOO, MM, BB*.

**56.** Suppose that we have $n$ dollars and that each day we buy either tape ($1), paper ($1), pens ($2), pencils ($2), or binders ($3). If $R_n$ is the number of ways of spending all the money, derive a recurrence relation for the sequence $R_1, R_2, \ldots$.

**57.** Let $R_n$ denote the number of regions into which the plane is divided by $n$ lines. Assume that each pair of lines meets in a point, but that no three lines meet in a point. Derive a recurrence relation for the sequence $R_1, R_2, \ldots$.

*Exercises 58 and 59 refer to the sequence $S_n$ defined by*

$$S_1 = 0, \quad S_2 = 1, \quad S_n = \frac{S_{n-1} + S_{n-2}}{2} \qquad n = 3, 4, \ldots.$$

**58.** Compute $S_3$ and $S_4$.

**★59.** Guess a formula for $S_n$ and use induction to show that it is correct.

**★60.** Let $F_n$ denote the number of functions $f$ from $X = \{1, \ldots, n\}$ into $X$ having the property that if $i$ is in the range of $f$, then $1, 2, \ldots, i - 1$ are also in the range of $f$. (Set $F_0 = 1$.) Show that the sequence $F_0, F_1, \ldots$ satisfies the recurrence relation

$$F_n = \sum_{j=0}^{n-1} C(n, j)F_j.$$

**61.** If $\alpha$ is a bit string, let $C(\alpha)$ be the maximum number of consecutive 0's in $\alpha$. [*Examples:* $C(10010) = 2$, $C(00110001) = 3$.] Let $S_n$ be the number of $n$-bit strings $\alpha$ with $C(\alpha) \leq 2$. Develop a recurrence relation for $S_1, S_2, \ldots$.

**62.** The sequence $g_1, g_2, \ldots$ is defined by the recurrence relation

$$g_n = g_{n-1} + g_{n-2} + 1 \qquad n \geq 3,$$

and initial conditions $g_1 = 1$, $g_2 = 3$. By using mathematical induction or otherwise, show that

$$g_n = 2f_{n+1} - 1 \qquad n \geq 1,$$

where $f_1, f_2, \ldots$ is the Fibonacci sequence.

**63.** Consider the formula

$$u_n = \begin{cases} u_{3n+1} & \text{if } n \text{ is odd and greater than 1} \\ u_{n/2} & \text{if } n \text{ is even and greater than 1} \end{cases}$$

and initial condition $u_1 = 1$. Explain why the formula is *not* a recurrence relation. A longstanding, open conjecture is that for every positive integer $n$, $u_n$ is defined and equal to 1. Compute $u_n$ for $n = 2, \ldots, 7$.

**64.** Define the sequence $t_1, t_2, \ldots$ by the recurrence relation

$$t_n = t_{n-1}t_{n-2} \qquad n \geq 3,$$

and initial conditions $t_1 = 1$, $t_2 = 2$. What is wrong with the following "proof" that $t_n = 1$ for all $n \geq 1$?

**Basis Step** For $n = 1$, we have $t_1 = 1$; thus, the Basis Step is verified.

**Inductive Step** Assume that $t_k = 1$ for $k < n$. We must prove that $t_n = 1$. Now

$$
\begin{aligned}
t_n &= t_{n-1}t_{n-2}\\
&= 1 \cdot 1 && \text{by the inductive assumption}\\
&= 1.
\end{aligned}
$$

The Inductive Step is complete.

**65.** Derive a recurrence relation for $C(n, k)$, the number of $k$-element subsets of an $n$-element set. Specifically, write $C(n + 1, k)$ in terms of $C(n, i)$ for appropriate $i$.

**66.** Derive a recurrence relation for $S(k, n)$, the number of ways of choosing $k$ items, allowing repetitions, from $n$ available types. Specifically, write $S(k, n)$ in terms of $S(k - 1, i)$ for appropriate $i$.

**67.** Let $S(n, k)$ denote the number of functions from $\{1, \ldots, n\}$ onto $\{1, \ldots, k\}$. Show that $S(n, k)$ satisfies the recurrence relation

$$S(n, k) = k^n - \sum_{i=1}^{k-1} C(k, i)S(n, i).$$

**68.** The *Lucas sequence* $L_1, L_2, \ldots$ (named after Édouard Lucas, the inventor of the Tower of Hanoi puzzle) is defined by the recurrence relation

$$L_n = L_{n-1} + L_{n-2} \qquad n \geq 3,$$

and the initial conditions $L_1 = 1$, $L_2 = 3$.

(a) Find the values of $L_3$, $L_4$, and $L_5$.

(b) Show that

$$L_{n+2} = f_{n+1} + f_{n+3} \qquad n \geq 1,$$

where $f_1, f_2, \ldots$ denotes the Fibonacci sequence.

**69.** Establish the recurrence relation

$$s_{n+1,k} = s_{n,k-1} + ns_{n,k}$$

for Stirling numbers of the first kind (see Exercise 90, Section 6.2).

**70.** Establish the recurrence relation

$$S_{n+1,k} = S_{n,k-1} + kS_{n,k}$$

for Stirling numbers of the second kind (see Exercise 91, Section 6.2).

★**71.** Show that

$$S_{n,k} = \frac{1}{k!}\sum_{i=0}^{k}(-1)^i(k-i)^n C(k,i),$$

where $S_{n,k}$ denotes a Stirling number of the second kind (see Exercise 91, Section 6.2).

**72.** Assume that a person invests a sum of money at $r$ percent compounded annually. Explain the rule of thumb: To estimate the time to double the investment, divide 70 by $r$.

**73.** Derive a recurrence relation for the number of multiplications needed to evaluate an $n \times n$ determinant by the cofactor method.

*A rise/fall permutation is a permutation p of* $1, 2, \ldots, n$ *satisfying*

$$p(i) < p(i+1) \qquad \text{for } i = 1, 3, 5, \ldots$$

*and*

$$p(i) > p(i+1) \qquad \text{for } i = 2, 4, 6, \ldots.$$

*For example, there are five rise/fall permutations of 1, 2, 3, 4:*

$$1, 3, 2, 4; \quad 1, 4, 2, 3; \quad 2, 3, 1, 4;$$
$$2, 4, 1, 3; \quad 3, 4, 1, 2.$$

*Let* $E_n$ *denote the number of rise/fall permutations of* $1, 2, \ldots, n$. *(Define* $E_0 = 1$.*) The numbers* $E_0, E_1, E_2, \ldots$ *are called the* Euler numbers.

**74.** List all rise/fall permutations of 1, 2, 3. What is the value of $E_3$?

**75.** List all rise/fall permutations of 1, 2, 3, 4, 5. What is the value of $E_5$?

**76.** Show that in a rise/fall permutation of $1, 2, \ldots, n$, $n$ must occur in position $2i$, for some $i$.

★**77.** Use Exercise 76 to derive the recurrence relation

$$E_n = \sum_{j=1}^{\lfloor n/2 \rfloor} C(n-1, 2j-1)E_{2j-1}E_{n-2j}.$$

★**78.** By considering where 1 must occur in a rise/fall permutation, derive the recurrence relation

$$E_n = \sum_{j=0}^{\lfloor (n-1)/2 \rfloor} C(n-1, 2j)E_{2j}E_{n-2j-1}.$$

★**79.** Prove that

$$E_n = \frac{1}{2}\sum_{j=1}^{n-1} C(n-1, j)E_jE_{n-j-1}.$$

## 7.2 Solving Recurrence Relations

**Go Online**
For more on solving recurrence relations, see bit.ly/2yYdlQL

To solve a recurrence relation involving the sequence $a_0, a_1, \ldots$ is to find an explicit formula for the general term $a_n$. In this section we discuss two methods of solving recurrence relations: **iteration** and a special method that applies to **linear homogeneous recurrence relations with constant coefficients.** For more powerful methods, such as methods that make use of generating functions, consult [Brualdi].

To solve a recurrence relation involving the sequence $a_0, a_1, \ldots$ by iteration, we use the recurrence relation to write the $n$th term $a_n$ in terms of certain of its predecessors $a_{n-1}, \ldots, a_0$. We then successively use the recurrence relation to replace each of $a_{n-1}, \ldots$ by certain of their predecessors. We continue until an explicit formula is obtained. The iterative method was used to solve the recurrence relation of Example 7.1.3.

**Example 7.2.1** Solve the recurrence relation

$$a_n = a_{n-1} + 3, \qquad (7.2.1)$$

subject to the initial condition, $a_1 = 2$, by iteration.

**SOLUTION** Replacing $n$ by $n-1$ in (7.2.1), we obtain

$$a_{n-1} = a_{n-2} + 3.$$

If we substitute this expression for $a_{n-1}$ into (7.2.1), we obtain

$$\begin{aligned} a_n &= \boxed{a_{n-1}} + 3 \\ &\quad\downarrow \\ &= \boxed{a_{n-2} + 3} + 3 \\ &= a_{n-2} + 2 \cdot 3. \end{aligned} \tag{7.2.2}$$

Replacing $n$ by $n-2$ in (7.2.1), we obtain

$$a_{n-2} = a_{n-3} + 3.$$

If we substitute this expression for $a_{n-2}$ into (7.2.2), we obtain

$$\begin{aligned} a_n &= \boxed{a_{n-2}} + 2 \cdot 3 \\ &\quad\downarrow \\ &= \boxed{a_{n-3} + 3} + 2 \cdot 3 \\ &= a_{n-3} + 3 \cdot 3. \end{aligned}$$

In general, we have

$$a_n = a_{n-k} + k \cdot 3.$$

If we set $k = n - 1$ in this last expression, we have

$$a_n = a_1 + (n-1) \cdot 3.$$

Since $a_1 = 2$, we obtain the explicit formula

$$a_n = 2 + 3(n-1)$$

for the sequence $a$. ◀

**Example 7.2.2** Solve the recurrence relation $S_n = 2S_{n-1}$ of Example 7.1.5, subject to the initial condition, $S_0 = 1$, by iteration.

**SOLUTION** $S_n = 2S_{n-1} = 2(2S_{n-2}) = \cdots = 2^n S_0 = 2^n$. ◀

**Example 7.2.3** **Population Growth** Assume that the deer population of Rustic County is 1000 at time $n = 0$ and that the increase from time $n-1$ to time $n$ is 10 percent of the size at time $n-1$. Write a recurrence relation and an initial condition that define the deer population at time $n$ and then solve the recurrence relation.

**SOLUTION** Let $d_n$ denote the deer population at time $n$. We have the initial condition $d_0 = 1000$. The increase from time $n-1$ to time $n$ is $d_n - d_{n-1}$. Since this increase is 10 percent of the size at time $n-1$, we obtain the recurrence relation $d_n - d_{n-1} = 0.1d_{n-1}$, which may be rewritten

$$d_n = 1.1d_{n-1}.$$

The recurrence relation may be solved by iteration:

$$d_n = 1.1d_{n-1} = 1.1(1.1d_{n-2}) = (1.1)^2(d_{n-2})$$
$$= \cdots = (1.1)^n d_0 = (1.1)^n 1000.$$

The assumptions imply exponential population growth. ◀

**Example 7.2.4** Find an explicit formula for $c_n$, the minimum number of moves in which the $n$-disk Tower of Hanoi puzzle can be solved (see Example 7.1.8).

**SOLUTION** In Example 7.1.8 we obtained the recurrence relation

$$c_n = 2c_{n-1} + 1 \tag{7.2.3}$$

and initial condition $c_1 = 1$. Applying the iterative method to (7.2.3), we obtain

$$\begin{aligned} c_n &= 2c_{n-1} + 1 \\ &= 2(2c_{n-2} + 1) + 1 \\ &= 2^2 c_{n-2} + 2 + 1 \\ &= 2^2(2c_{n-3} + 1) + 2 + 1 \\ &= 2^3 c_{n-3} + 2^2 + 2 + 1 \\ &\ \vdots \\ &= 2^{n-1}c_1 + 2^{n-2} + 2^{n-3} + \cdots + 2 + 1 \\ &= 2^{n-1} + 2^{n-2} + 2^{n-3} + \cdots + 2 + 1 \\ &= 2^n - 1. \end{aligned}$$

The last step results from the formula for the geometric sum (see Example 2.4.4). ◀

**Example 7.2.5** Solve the recurrence relation

$$p_n = a - \frac{b}{k}p_{n-1}$$

for the price $p_n$ in the economics model of Example 7.1.9 by iteration.

**SOLUTION** To simplify the notation, we set $s = -b/k$.

$$\begin{aligned} p_n &= a + sp_{n-1} \\ &= a + s(a + sp_{n-2}) \\ &= a + as + s^2 p_{n-2} \\ &= a + as + s^2(a + sp_{n-3}) \\ &= a + as + as^2 + s^3 p_{n-3} \\ &\ \vdots \\ &= a + as + as^2 + \cdots + as^{n-1} + s^n p_0 \\ &= \frac{a - as^n}{1 - s} + s^n p_0 \\ &= s^n\left(\frac{-a}{1-s} + p_0\right) + \frac{a}{1-s} \\ &= \left(-\frac{b}{k}\right)^n \left(\frac{-ak}{k+b} + p_0\right) + \frac{ak}{k+b}. \end{aligned} \tag{7.2.4}$$

We see that if $b/k < 1$, the term

$$\left(-\frac{b}{k}\right)^n \left(\frac{-ak}{k+b} + p_0\right)$$

becomes small as $n$ gets large so that the price tends to stabilize at approximately $ak/(k+b)$. If $b/k = 1$, (7.2.4) shows that $p_n$ oscillates between $p_0$ and $p_1$. If $b/k > 1$, (7.2.4) shows that the differences between successive prices increase. Previously, we observed these properties graphically (see Example 7.1.9). ◀

We turn next to a special class of recurrence relations.

**Definition 7.2.6** ▶ A *linear homogeneous recurrence relation of order k with constant coefficients* is a recurrence relation of the form

$$a_n = c_1 a_{n-1} + c_2 a_{n-2} + \cdots + c_k a_{n-k}, \quad c_k \neq 0. \tag{7.2.5}$$

◀

Notice that a linear homogeneous recurrence relation of order $k$ with constant coefficients (7.2.5), together with the $k$ initial conditions

$$a_0 = C_0, \quad a_1 = C_1, \ldots, \quad a_{k-1} = C_{k-1},$$

uniquely defines a sequence $a_0, a_1, \ldots$.

**Example 7.2.7** The recurrence relations

$$S_n = 2S_{n-1} \tag{7.2.6}$$

of Example 7.2.2 and

$$f_n = f_{n-1} + f_{n-2}, \tag{7.2.7}$$

which defines the Fibonacci sequence, are both linear homogeneous recurrence relations with constant coefficients. The recurrence relation (7.2.6) is of order 1 and (7.2.7) is of order 2. ◀

**Example 7.2.8** The recurrence relation

$$a_n = 3a_{n-1}a_{n-2} \tag{7.2.8}$$

is not a linear homogeneous recurrence relation with constant coefficients. In a linear homogeneous recurrence relation with constant coefficients, each term is of the form $ca_k$. Terms such as $a_{n-1}a_{n-2}$ are not permitted. Recurrence relations such as (7.2.8) are said to be *nonlinear*. ◀

**Example 7.2.9** The recurrence relation $a_n - a_{n-1} = 2n$ is not a linear homogeneous recurrence relation with constant coefficients because the expression on the right side of the equation is not zero. (Such an equation is said to be *inhomogeneous*. Linear inhomogeneous recurrence relations with constant coefficients are discussed in Exercises 42–48.) ◀

**Example 7.2.10** The recurrence relation $a_n = 3na_{n-1}$ is not a linear homogeneous recurrence relation with constant coefficients because the coefficient $3n$ is not constant. It is a linear homogeneous recurrence relation with nonconstant coefficients. ◀

We will illustrate the general method of solving linear homogeneous recurrence relations with constant coefficients by finding an explicit formula for the sequence defined by the recurrence relation

$$a_n = 5a_{n-1} - 6a_{n-2} \tag{7.2.9}$$

and initial conditions

$$a_0 = 7, \qquad a_1 = 16. \tag{7.2.10}$$

Often in mathematics, when trying to solve a more difficult instance of some problem, we begin with an expression that solved a simpler version. For the first-order recurrence relation (7.2.6), we found in Example 7.2.2 that the solution was of the form $S_n = t^n$; thus for our first attempt at finding a solution of the second-order recurrence relation (7.2.9), we will search for a solution of the form $V_n = t^n$.

If $V_n = t^n$ is to solve (7.2.9), we must have

$$V_n = 5V_{n-1} - 6V_{n-2}$$

or

$$t^n = 5t^{n-1} - 6t^{n-2}$$

or

$$t^n - 5t^{n-1} + 6t^{n-2} = 0.$$

Dividing by $t^{n-2}$, we obtain the equivalent equation

$$t^2 - 5t + 6 = 0. \tag{7.2.11}$$

Solving (7.2.11), we find the solutions

$$t = 2, \qquad t = 3.$$

At this point, we have two solutions $S$ and $T$ of (7.2.9), given by

$$S_n = 2^n, \qquad T_n = 3^n. \tag{7.2.12}$$

We can verify (see Theorem 7.2.11) that if $S$ and $T$ are solutions of (7.2.9), then $bS + dT$, where $b$ and $d$ are any numbers whatever, is also a solution of (7.2.9). In our case, if we define the sequence $U$ by the equation

$$U_n = bS_n + dT_n = b2^n + d3^n,$$

$U$ is a solution of (7.2.9).

To satisfy the initial conditions (7.2.10), we must have

$$7 = U_0 = b2^0 + d3^0 = b + d, \qquad 16 = U_1 = b2^1 + d3^1 = 2b + 3d.$$

Solving these equations for $b$ and $d$, we obtain $b = 5$ and $d = 2$. Therefore, the sequence $U$ defined by

$$U_n = 5 \cdot 2^n + 2 \cdot 3^n$$

satisfies the recurrence relation (7.2.9) and the initial conditions (7.2.10). We conclude that

$$a_n = U_n = 5 \cdot 2^n + 2 \cdot 3^n \qquad \text{for } n = 0, 1, \ldots.$$

At this point we will summarize and justify the techniques used to solve the preceding recurrence relation.

**Theorem 7.2.11** Let

$$a_n = c_1 a_{n-1} + c_2 a_{n-2} \tag{7.2.13}$$

be a second-order, linear homogeneous recurrence relation with constant coefficients.

If $S$ and $T$ are solutions of (7.2.13), then $U = bS + dT$ is also a solution of (7.2.13).

If $r$ is a root of

$$t^2 - c_1 t - c_2 = 0, \tag{7.2.14}$$

then the sequence $r^n$, $n = 0, 1, \ldots$, is a solution of (7.2.13).

If $a$ is the sequence defined by (7.2.13),

$$a_0 = C_0, \qquad a_1 = C_1, \tag{7.2.15}$$

and $r_1$ and $r_2$ are roots of (7.2.14) with $r_1 \neq r_2$, then there exist constants $b$ and $d$ such that

$$a_n = br_1^n + dr_2^n \qquad n = 0, 1, \ldots.$$

**Proof** Since $S$ and $T$ are solutions of (7.2.13),

$$S_n = c_1 S_{n-1} + c_2 S_{n-2}, \qquad T_n = c_1 T_{n-1} + c_2 T_{n-2}.$$

If we multiply the first equation by $b$ and the second by $d$ and add, we obtain

$$\begin{aligned} U_n = bS_n + dT_n &= c_1(bS_{n-1} + dT_{n-1}) + c_2(bS_{n-2} + dT_{n-2}) \\ &= c_1 U_{n-1} + c_2 U_{n-2}. \end{aligned}$$

Therefore, $U$ is a solution of (7.2.13).

Since $r$ is a root of (7.2.14),

$$r^2 = c_1 r + c_2.$$

Now

$$c_1 r^{n-1} + c_2 r^{n-2} = r^{n-2}(c_1 r + c_2) = r^{n-2} r^2 = r^n;$$

thus the sequence $r^n$, $n = 0, 1, \ldots$, is a solution of (7.2.13).

If we set $U_n = br_1^n + dr_2^n$, then $U$ is a solution of (7.2.13). To meet the initial conditions (7.2.15), we must have

$$U_0 = b + d = C_0, \qquad U_1 = br_1 + dr_2 = C_1.$$

If we multiply the first equation by $r_1$ and subtract, we obtain

$$d(r_1 - r_2) = r_1 C_0 - C_1.$$

Since $r_1 - r_2 \neq 0$, we can solve for $d$. Similarly, we can solve for $b$. With these choices for $b$ and $d$, we have

$$U_0 = C_0, \qquad U_1 = C_1.$$

Let $a$ be the sequence defined by (7.2.13) and (7.2.15). Since $U$ also satisfies (7.2.13) and (7.2.15), it follows that $U_n = a_n$, $n = 0, 1, \ldots$. ◀

**Example 7.2.12** **More Population Growth** Assume that the deer population of Rustic County is 200 at time $n = 0$ and 220 at time $n = 1$ and that the increase from time $n - 1$ to time $n$ is twice the increase from time $n - 2$ to time $n - 1$. Write a recurrence relation and an initial condition that define the deer population at time $n$ and then solve the recurrence relation.

SOLUTION Let $d_n$ denote the deer population at time $n$. We have the initial conditions

$$d_0 = 200, \qquad d_1 = 220.$$

The increase from time $n - 1$ to time $n$ is $d_n - d_{n-1}$, and the increase from time $n - 2$ to time $n - 1$ is $d_{n-1} - d_{n-2}$. Thus we obtain the recurrence relation

$$d_n - d_{n-1} = 2\left(d_{n-1} - d_{n-2}\right),$$

which may be rewritten

$$d_n = 3d_{n-1} - 2d_{n-2}.$$

To solve this recurrence relation, we first solve the quadratic equation $t^2 - 3t + 2 = 0$ to obtain roots 1 and 2. The sequence $d$ is of the form

$$d_n = b \cdot 1^n + c \cdot 2^n = b + c2^n.$$

To meet the initial conditions, we must have

$$200 = d_0 = b + c, \qquad 220 = d_1 = b + 2c.$$

Solving for $b$ and $c$, we find that $b = 180$ and $c = 20$. Thus $d_n$ is given by $d_n = 180 + 20 \cdot 2^n$. As in Example 7.2.3, the growth is exponential. ◀

**Example 7.2.13** Find an explicit formula for the Fibonacci sequence.

SOLUTION The Fibonacci sequence is defined by the linear homogeneous, second-order recurrence relation

$$f_n - f_{n-1} - f_{n-2} = 0 \qquad n \geq 3,$$

and initial conditions $f_1 = 1$ and $f_2 = 1$. We begin by using the quadratic formula to solve $t^2 - t - 1 = 0$. The solutions are

$$t = \frac{1 \pm \sqrt{5}}{2}.$$

Thus the solution is of the form

$$f_n = b\left(\frac{1 + \sqrt{5}}{2}\right)^n + d\left(\frac{1 - \sqrt{5}}{2}\right)^n.$$

To satisfy the initial conditions, we must have

$$b\left(\frac{1+\sqrt{5}}{2}\right)+d\left(\frac{1-\sqrt{5}}{2}\right)=1$$

$$b\left(\frac{1+\sqrt{5}}{2}\right)^2+d\left(\frac{1-\sqrt{5}}{2}\right)^2=1.$$

Solving these equations for $b$ and $d$, we obtain

$$b=\frac{1}{\sqrt{5}}, \qquad d=-\frac{1}{\sqrt{5}}.$$

Therefore, an explicit formula for the Fibonacci sequence is

$$f_n=\frac{1}{\sqrt{5}}\left(\frac{1+\sqrt{5}}{2}\right)^n-\frac{1}{\sqrt{5}}\left(\frac{1-\sqrt{5}}{2}\right)^n.$$

Surprisingly, even though $f_n$ is an integer, the preceding formula involves the irrational number $\sqrt{5}$. ◀

In Example 7.2.13, the number $(1+\sqrt{5})/2$ that appears in the formula for the $n$th Fibonacci number is known as the *golden ratio*. Calculus fans may enjoy showing that $\lim_{n\to\infty} f_{n+1}/f_n = (1+\sqrt{5})/2$. Historically, a ratio $a$:$b$, where $a/b=(1+\sqrt{5})/2$, was considered to be esthetically pleasing. For example, pictures and paintings whose dimensions were in the golden ratio were admired. The resolution $1920 \times 1200$ of some computer monitors is close to the golden ratio. The ratio 5:3, which is also close to the golden ratio, is a film standard in several European countries.

Theorem 7.2.11 states that any solution of (7.2.13) may be given in terms of two basic solutions $r_1^n$ and $r_2^n$. However, in case (7.2.14) has two equal roots $r$, we obtain only one basic solution $r^n$. The next theorem shows that in this case, $nr^n$ furnishes the other basic solution.

**Theorem 7.2.14** Let

$$a_n = c_1 a_{n-1} + c_2 a_{n-2} \tag{7.2.16}$$

be a second-order linear homogeneous recurrence relation with constant coefficients. Let $a$ be the sequence satisfying (7.2.16) and

$$a_0 = C_0, \qquad a_1 = C_1.$$

If both roots of

$$t^2 - c_1 t - c_2 = 0 \tag{7.2.17}$$

are equal to $r$, then there exist constants $b$ and $d$ such that

$$a_n = br^n + dnr^n \qquad n = 0, 1, \ldots.$$

**Proof** The proof of Theorem 7.2.11 shows that the sequence $r^n$, $n = 0, 1, \ldots$, is a solution of (7.2.16). We show that the sequence $nr^n$, $n = 0, 1, \ldots$, is also a solution of (7.2.16).

Since $r$ is the only solution of (7.2.17), we must have

$$t^2 - c_1 t - c_2 = (t - r)^2.$$

It follows that $c_1 = 2r$ and $c_2 = -r^2$. Now

$$\begin{aligned} c_1\left[(n-1)r^{n-1}\right] + c_2\left[(n-2)r^{n-2}\right] &= 2r(n-1)r^{n-1} - r^2(n-2)r^{n-2} \\ &= r^n\left[2(n-1) - (n-2)\right] = nr^n. \end{aligned}$$

Therefore, the sequence $nr^n$, $n = 0, 1, \ldots$, is a solution of (7.2.16).

By Theorem 7.2.11, the sequence $U$ defined by $U_n = br^n + dnr^n$ is a solution of (7.2.16).

The proof that there are constants $b$ and $d$ such that $U_0 = C_0$ and $U_1 = C_1$ is similar to the argument given in Theorem 7.2.11 and is left as an exercise (Exercise 50). It follows that $U_n = a_n$, $n = 0, 1, \ldots$. ◀

**Example 7.2.15** Solve the recurrence relation

$$d_n = 4(d_{n-1} - d_{n-2}) \tag{7.2.18}$$

subject to the initial conditions $d_0 = 1 = d_1$.

**SOLUTION** According to Theorem 7.2.11, $S_n = r^n$ is a solution of (7.2.18), where $r$ is a solution of

$$t^2 - 4t + 4 = 0. \tag{7.2.19}$$

Thus we obtain the solution $S_n = 2^n$ of (7.2.18). Since 2 is the only solution of (7.2.19), by Theorem 7.2.14, $T_n = n2^n$ is also a solution of (7.2.18). Thus the general solution of (7.2.18) is of the form $U = aS + bT$. We must have $U_0 = 1 = U_1$. These last equations become

$$aS_0 + bT_0 = a + 0b = 1, \qquad aS_1 + bT_1 = 2a + 2b = 1.$$

Solving for $a$ and $b$, we obtain

$$a = 1, \qquad b = -\frac{1}{2}.$$

Therefore, the solution of (7.2.18) is

$$d_n = 2^n - n2^{n-1}.$$ ◀

For the general linear homogeneous recurrence relation of order $k$ with constant coefficients (7.2.5), if $r$ is a root of

$$t^k - c_1 t^{k-1} - c_2 t^{k-2} - \cdots - c_k = 0$$

of multiplicity $m$, it can be shown that $r^n, nr^n, \ldots, n^{m-1}r^n$ are solutions of (7.2.5). This fact can be used, just as in the previous examples for recurrence relations of order 2, to solve a linear homogeneous recurrence relation of order $k$ with constant coefficients. For a precise statement and a proof of the general result, see [Brualdi].

## 7.2 Problem-Solving Tips

To solve a recurrence relation in which $a_n$ is defined in terms of its immediate predecessor $a_{n-1}$

$$a_n = \ldots a_{n-1} \ldots, \tag{7.2.20}$$

use *iteration*. Start with the equation (7.2.20). Substitute for $a_{n-1}$, which will yield an expression involving $a_{n-2}$, to obtain $a_n = \ldots a_{n-2} \ldots$. Continue until you obtain $a_k$ (e.g., $a_0$), whose value is given explicitly as the initial condition $a_n = \ldots a_k \ldots$. Plug in the value for $a_k$ and you have solved the recurrence relation.

To solve the recurrence relation

$$a_n = c_1 a_{n-1} + c_2 a_{n-2},$$

first solve the equation

$$t^2 - c_1 t - c_2 = 0. \tag{7.2.21}$$

If the roots are $r_1$ and $r_2$, $r_1 \neq r_2$, the solution is

$$a_n = b r_1^n + d r_2^n \tag{7.2.22}$$

for some constants $b$ and $d$. To determine these constants, use the initial conditions, say $a_0 = C_0,\ a_1 = C_1$. Set $n = 0$ and then $n = 1$ in (7.2.22) to obtain

$$C_0 = a_0 = b + d,\ C_1 = a_1 = br_1 + dr_2.$$

Now solve

$$C_0 = b + d,\ C_1 = br_1 + dr_2$$

for $b$ and $d$ to obtain the solution of the recurrence relation.

If the roots of (7.2.21) are $r_1$ and $r_2$, $r_1 = r_2$, the solution is

$$a_n = br^n + dnr^n \tag{7.2.23}$$

for some constants $b$ and $d$, where $r = r_1 = r_2$. To determine these constants, use the initial conditions, say $a_0 = C_0,\ a_1 = C_1$. Set $n = 0$ and then $n = 1$ in (7.2.23) to obtain

$$C_0 = a_0 = b,\ C_1 = a_1 = br + dr.$$

Now solve

$$C_0 = b,\ C_1 = br + dr$$

for $b$ and $d$ to obtain the solution of the recurrence relation.

## 7.2 Review Exercises

1. Explain how to solve a recurrence relation by iteration.
2. What is an $n$th-order, linear homogeneous recurrence relation with constant coefficients?
3. Give an example of a second-order, linear homogeneous recurrence relation with constant coefficients.
4. Explain how to solve a second-order, linear homogeneous recurrence relation with constant coefficients.

## 7.2 Exercises

*Tell whether or not each recurrence relation in Exercises 1–10 is a linear homogeneous recurrence relation with constant coefficients. Give the order of each linear homogeneous recurrence relation with constant coefficients.*

1. $a_n = -3a_{n-1}$
2. $a_n = 2na_{n-2} - a_{n-1}$
3. $a_n = 2na_{n-1}$

4. $a_n = a_{n-1} + n$
5. $a_n = a_{n-1} + 1 + 2^{n-1}$
6. $a_n = 7a_{n-2} - 6a_{n-3}$
7. $a_n = (\lg 2n)a_{n-1} - [\lg(n-1)]a_{n-2}$
8. $a_n = -a_{n-1} - a_{n-2}$
9. $a_n = 6a_{n-1} - 9a_{n-2}$
10. $a_n = -a_{n-1} + 5a_{n-2} - 3a_{n-3}$

*In Exercises 11–26, solve the given recurrence relation for the initial conditions given.*

11. Exercise 1; $a_0 = 2$
12. Exercise 3; $a_0 = 1$
13. Exercise 4; $a_0 = 0$
14. $a_n = 2^n a_{n-1}$; $a_0 = 1$
15. $a_n = 6a_{n-1} - 8a_{n-2}$; $a_0 = 1$, $a_1 = 0$
16. $a_n = 2a_{n-1} + 8a_{n-2}$; $a_0 = 4$, $a_1 = 10$
17. $a_n = 7a_{n-1} - 10a_{n-2}$; $a_0 = 5$, $a_1 = 16$
18. $2a_n = 7a_{n-1} - 3a_{n-2}$; $a_0 = a_1 = 1$
19. Exercise 5; $a_0 = 0$
20. Exercise 9; $a_0 = a_1 = 1$
21. $a_n = -8a_{n-1} - 16a_{n-2}$; $a_0 = 2$, $a_1 = -20$
22. The Lucas sequence

$$L_n = L_{n-1} + L_{n-2}, \quad n \geq 3; \qquad L_1 = 1, \quad L_2 = 3$$

23. $9a_n = 6a_{n-1} - a_{n-2}$; $a_0 = 6$, $a_1 = 5$
24. Exercise 55, Section 7.1
25. Exercise 57, Section 7.1
26. The recurrence relation preceding Exercise 58, Section 7.1
27. The population of Utopia increases 5 percent per year. In 2000 the population was 10,000. What was the population in 1970?
28. Assume that the deer population of Rustic County is 0 at time $n = 0$. Suppose that at time $n$, $100n$ deer are introduced into Rustic County and that the population increases 20 percent each year. Write a recurrence relation and an initial condition that define the deer population at time $n$ and then solve the recurrence relation. The following formula may be of use:

$$\sum_{i=1}^{n-1} ix^{i-1} = \frac{(n-1)x^n - nx^{n-1} + 1}{(x-1)^2}.$$

*Exercises 29–35 concern Toots and Sly, who flip fair pennies. If the pennies are both heads or both tails, Toots wins one of Sly's pennies. If one penny is a head and the other is a tail, Sly wins one of Toots' pennies. The game ends when one person has all of the pennies. There are C total pennies.*

*Let $E_n$ denote the event "Toots starts with n coins and gets all of Sly's coins"; let W denote the event "Toots wins the first toss"; and let $p_n = P(E_n)$.*

29. Find $P(W)$.
30. Find $P(\overline{W})$.
31. Show that $P(E_n) = P(E_n \mid W) + P(E_n \mid \overline{W})$.
32. Find $p_0$.
33. Find $p_C$.
34. Find a recurrence relation for $p_n$.
35. If Toots starts with $n$ pennies, what is the probability that Toots wins all of Sly's pennies?

*Sometimes a recurrence relation that is not a linear homogeneous equation with constant coefficients can be transformed into a linear homogeneous equation with constant coefficients. In Exercises 36 and 37, make the given substitution and solve the resulting recurrence relation; then find the solution to the original recurrence relation.*

36. Solve the recurrence relation

$$\sqrt{a_n} = \sqrt{a_{n-1}} + 2\sqrt{a_{n-2}}$$

with initial conditions $a_0 = a_1 = 1$ by making the substitution $b_n = \sqrt{a_n}$.

37. Solve the recurrence relation

$$a_n = \sqrt{\frac{a_{n-2}}{a_{n-1}}}$$

with initial conditions $a_0 = 8, a_1 = 1/(2\sqrt{2})$ by taking the logarithm of both sides and making the substitution $b_n = \lg a_n$.

*In Exercises 38–40, solve the recurrence relation for the initial conditions given.*

38. $a_n = -2na_{n-1} + 3n(n-1)a_{n-2}$; $a_0 = 1$, $a_1 = 2$

★39. $c_n = 2 + \sum_{i=1}^{n-1} c_i$, $n \geq 2$; $c_1 = 1$

★40. $A(n, m) = 1 + A(n-1, m-1) + A(n-1, m)$, $n-1 \geq m \geq 1$, $n \geq 2$; $A(n, 0) = A(n, n) = 1$, $n \geq 0$

41. Show that

$$f_{n+1} \geq \left(\frac{1+\sqrt{5}}{2}\right)^{n-1} \qquad n \geq 1,$$

where $f$ denotes the Fibonacci sequence.

42. The equation

$$a_n = c_1 a_{n-1} + c_2 a_{n-2} + f(n) \qquad (7.2.24)$$

is called a **second-order, linear inhomogeneous recurrence relation with constant coefficients.**

Let $g(n)$ be a solution of (7.2.24). Show that any solution $U$ of (7.2.24) is of the form

$$U_n = V_n + g(n), \tag{7.2.25}$$

where $V$ is a solution of the homogeneous equation (7.2.13).

*If $f(n) = C$ in (7.2.24), it can be shown that $g(n) = C'$ in (7.2.25) if 1 is not a root of*

$$t^2 - c_1 t - c_2 = 0, \tag{7.2.26}$$

*$g(n) = C'n$ if 1 is a root of (7.2.26) of multiplicity one, and $g(n) = C'n^2$ if 1 is a root of (7.2.26) of multiplicity two. Similarly, if $f(n) = Cn$, it can be shown that $g(n) = C_1'n + C_0'$ if 1 is not a root of (7.2.26), $g(n) = C_1'n^2 + C_0'n$ if 1 is a root of multiplicity one, and $g(n) = C_1'n^3 + C_0'n^2$ if 1 is a root of multiplicity two. If $f(n) = Cn^2$, it can be shown that $g(n) = C_2'n^2 + C_1'n + C_0'$ if 1 is not a root of (7.2.26), $g(n) = C_2'n^3 + C_1'n^2 + C_0'$ if 1 is a root of multiplicity one, and $g(n) = C_2'n^4 + C_1'n^3 + C_0'n^2$ if 1 is a root of multiplicity two. If $f(n) = C^n$, it can be shown that $g(n) = C'C^n$ if $C$ is not a root of (7.2.26), $g(n) = C'nC^n$ if $C$ is a root of multiplicity one, and $g(n) = C'n^2C^n$ if $C$ is a root of multiplicity two. The constants can be determined by substituting $g(n)$ into the recurrence relation and equating coefficients on the two sides of the resulting equation. As examples, the constant terms on the two sides of the equation must be equal, and the coefficient of $n$ on the left side of the equation must equal the coefficient of $n$ on the right side of the equation. Use these facts together with Exercise 42 to find the general solutions of the recurrence relations of Exercises 43–48.*

**43.** $a_n = 6a_{n-1} - 8a_{n-2} + 3$

**44.** $a_n = 7a_{n-1} - 10a_{n-2} + 16n$

**45.** $a_n = 2a_{n-1} + 8a_{n-2} + 81n^2$

**46.** $2a_n = 7a_{n-1} - 3a_{n-2} + 2^n$

**47.** $a_n = -8a_{n-1} - 16a_{n-2} + 3n$

**48.** $9a_n = 6a_{n-1} - a_{n-2} + 5n^2$

**49.** The equation

$$a_n = f(n)a_{n-1} + g(n)a_{n-2} \tag{7.2.27}$$

is called a **second-order, linear homogeneous recurrence relation.** The coefficients $f(n)$ and $g(n)$ are not necessarily constant. Show that if $S$ and $T$ are solutions of (7.2.27), then $bS + dT$ is also a solution of (7.2.27).

**50.** Suppose that both roots of

$$t^2 - c_1 t - c^2 = 0$$

are equal to $r$, and suppose that $a_n$ satisfies

$$a_n = c_1 a_{n-1} + c_2 a_{n-2}, \qquad a_0 = C_0, \qquad a_1 = C_1.$$

Show that there exist constants $b$ and $d$ such that

$$a_n = br^n + dnr^n \qquad n = 0, 1, \ldots,$$

thus completing the proof of Theorem 7.2.14.

**51.** Let $a_n$ be the minimum number of links required to solve the $n$-node communication problem (see Exercise 53, Section 7.1). Use iteration to show that $a_n \le 2n - 4$, $n \ge 4$.

*The n-disk, four-peg Tower of Hanoi puzzle has the same rules as the three-peg puzzle; the only difference is that there is an extra peg. Exercises 52–55 refer to the following algorithm to solve the n-disk, four-peg Tower of Hanoi puzzle.*

*Assume that the pegs are numbered 1, 2, 3, 4 and that the problem is to move the disks, which are initially stacked on peg 1, to peg 4. If $n = 1$, move the disk to peg 4 and stop. If $n > 1$, let $k_n$ be the largest integer satisfying*

$$\sum_{i=1}^{k_n} i \le n.$$

*Fix $k_n$ disks at the bottom of peg 1. Recursively invoke this algorithm to move the $n - k_n$ disks at the top of peg 1 to peg 2. During this part of the algorithm, the $k_n$ bottom disks on peg 1 remain fixed. Next, move the $k_n$ disks on peg 1 to peg 4 by invoking the optimal three-peg algorithm (see Example 7.1.8) and using only pegs 1, 3, and 4. Finally, again recursively invoke this algorithm to move the $n - k_n$ disks on peg 2 to peg 4. During this part of the algorithm, the $k_n$ disks on peg 4 remain fixed. Let $T(n)$ denote the number of moves required by this algorithm.*

*This algorithm, although not known to be optimal, uses as few moves as any other algorithm that has been proposed for the four-peg problem.*

**52.** Derive the recurrence relation

$$T(n) = 2T(n - k_n) + 2^{k_n} - 1.$$

**53.** Compute $T(n)$ for $n = 1, \ldots, 10$. Compare these values with the optimal number of moves to solve the three-peg problem.

**★54.** Let

$$r_n = n - \frac{k_n(k_n + 1)}{2}.$$

Using induction or otherwise, prove that

$$T(n) = (k_n + r_n - 1)2^{k_n} + 1.$$

**★55.** Show that $T(n) = O(4^{\sqrt{n}})$.

**★56.** Give an optimal algorithm to solve a variant of the Tower of Hanoi puzzle in which a stack of disks, except for the beginning and ending stacks, is allowed as long as the largest disk in the stack is on the bottom (the disks above the largest disk in the stack can be ordered in any way whatsoever). The problem is to transfer the disks to another peg by moving one disk at a time beginning with all the disks stacked on one peg in order from largest (bottom) to smallest as in the original puzzle. The end position will be the same as in the original puzzle—the disks will be stacked in order from largest (bottom) to smallest. Prove that your algorithm is optimal. This problem is due to John McCarthy.

## Problem-Solving Corner — Recurrence Relations

### Problem

(a) Several persons order compact discs. Their names, together with their order codes, are entered into a spreadsheet:

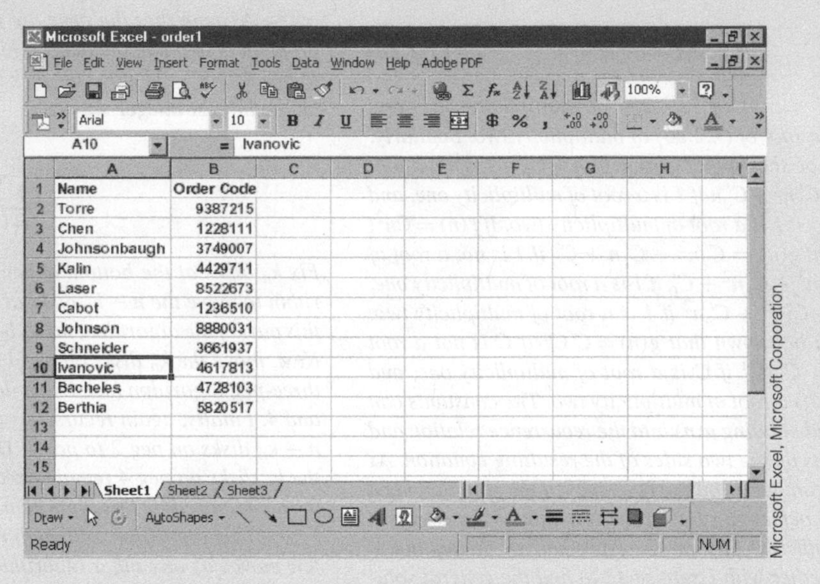

| | Name | Order Code |
|---|---|---|
| 2 | Torre | 9387215 |
| 3 | Chen | 1228111 |
| 4 | Johnsonbaugh | 3749007 |
| 5 | Kalin | 4429711 |
| 6 | Laser | 8522673 |
| 7 | Cabot | 1236510 |
| 8 | Johnson | 8880031 |
| 9 | Schneider | 3661937 |
| 10 | Ivanovic | 4617813 |
| 11 | Bacheles | 4728103 |
| 12 | Berthia | 5820517 |

Microsoft Excel, Microsoft Corporation.

The records are to be sorted in alphabetical order, but an error occurs when the *names* are sorted but the order codes are unchanged:

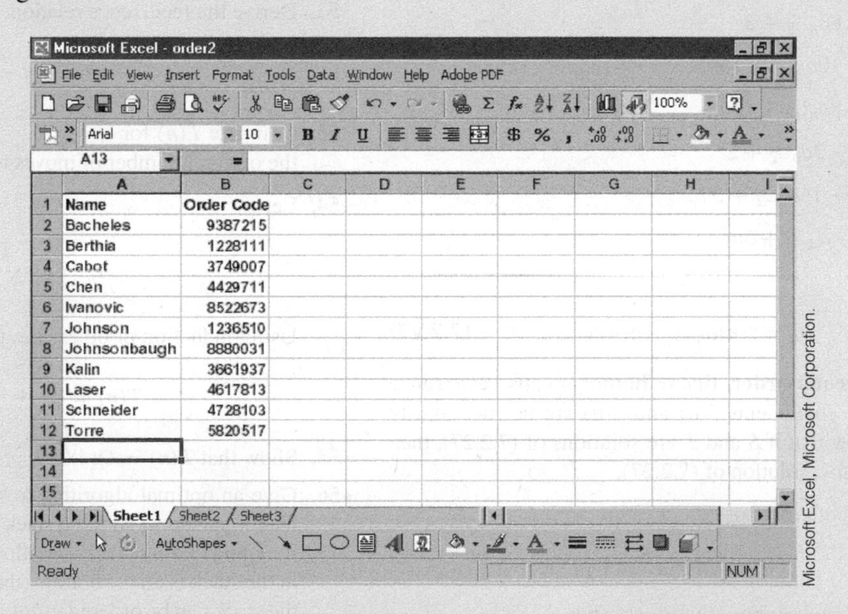

| | Name | Order Code |
|---|---|---|
| 2 | Bacheles | 9387215 |
| 3 | Berthia | 1228111 |
| 4 | Cabot | 3749007 |
| 5 | Chen | 4429711 |
| 6 | Ivanovic | 8522673 |
| 7 | Johnson | 1236510 |
| 8 | Johnsonbaugh | 8880031 |
| 9 | Kalin | 3661937 |
| 10 | Laser | 4617813 |
| 11 | Schneider | 4728103 |
| 12 | Torre | 5820517 |

Microsoft Excel, Microsoft Corporation.

As a result, no one receives the correct compact discs. Let $D_n$ denote the number of ways that $n$ persons can all receive the wrong orders. Show that the sequence $D_1, D_2, \ldots$ satisfies the recurrence relation

$$D_n = (n-1)(D_{n-1} + D_{n-2}).$$

(b) Solve the recurrence relation of part (a) by making the substitution $C_n = D_n - nD_{n-1}$.

## Attacking the Problem

Before attacking the problem, consider what is required for part (a). To prove the recurrence relation, we must reduce the $n$-person problem to the $(n-1)$- and $(n-2)$-person problems (since in the formula $D_n$ is given in terms of $D_{n-1}$ and $D_{n-2}$). Thus as we look systematically at some examples, we should try to see how the case for $n$ persons relates to the cases for $n-1$ and $n-2$ persons. The situation is similar to that of mathematical induction and recursive algorithms in which a given instance of a problem is related to smaller instances of the same problem.

Now for some examples. The smallest case is $n=1$. A single person must get the correct order, so $D_1 = 0$. For $n = 2$, there is one way for everyone to get the wrong orders: Person 1 gets order 2, and person 2 gets order 1. Thus $D_2 = 1$. Before continuing, let's develop some notation for the distribution of the orders. Carefully chosen notation can help in solving a problem.

We'll write

$$c_1, c_2, \ldots, c_n$$

to mean that person 1 got order $c_1$, person 2 got order $c_2$, and so on. The one way for two persons to get the wrong orders is denoted 2,1.

If $n = 3$, person 1 gets either order 2 or 3, so the possibilities are 2,?,? and 3,?,?. Let's fill in the missing numbers. Suppose that person 1 gets order 2. Person 2 can't get order 1 (for then person 3 would get the correct order); thus, person 2 gets order 3. This leaves order 1 for person 3. Thus if person 1 gets order 2, the only possibility is

$$2, 3, 1.$$

Suppose that person 1 gets order 3. Person 3 can't get order 1 (for then person 2 would get the correct order); thus, person 3 gets order 2. This leaves order 1 for person 2. Thus if person 1 gets order 3, the only possibility is

$$3, 1, 2.$$

Thus $D_3 = 2$.

Let's check that the recurrence relation holds for $n = 3$:

$$D_3 = 2 = 2(1 + 0) = (3 - 1)(D_2 + D_1).$$

If $n = 4$, person 1 gets either order 2, 3, or 4, so the possibilities are 2,?,?,?; 3,?,?,?; and 4,?,?,?. (If there are $n$ persons, person 1 gets either order 2 or 3 or ... or $n$. These $n-1$ possibilities account for the leading $n-1$ factor in the recurrence relation.) Let's fill in the missing numbers. Suppose that person 1 gets order 2. If person 2 gets order 1, then person 3 gets order 4 and person 4 gets order 3, which gives 2,1,4,3. If person 2 does not get order 1, the possibilities are 2,3,4,1 and 2,4,1,3. Thus if person 1 gets order 2, there are three possibilities. Similarly, if person 1 gets order 3, there are three possibilities, and if person 1 gets order 4, there are three possibilities. You should list the possibilities to confirm this last statement. Thus $D_4 = 9$.

Let's check that the recurrence relation also holds for $n = 4$:

$$D_4 = 9 = 3(2 + 1) = (4 - 1)(D_3 + D_2).$$

Before reading on, work through the case $n = 5$. List only the possibilities when person 1 gets order 2. (There are too many possibilities to list them all.) Also verify the recurrence relation for $n = 5$.

Notice that if person 1 gets order 2 and person 2 gets order 1, the number of ways for the other persons to get the wrong orders is $D_{n-2}$ (the remaining $n-2$ persons must all get the wrong orders). This accounts for the presence of $D_{n-2}$ in the recurrence relation. We'll have a solution, provided that the $D_{n-1}$ term appears in the remaining case: Person 1 gets order 2, but person 2 does not get order 1.

## Finding a Solution

Suppose there are $n$ persons. Let's summarize the argument we've developed through our examples. Person 1 gets either order 2, or 3, ..., or $n$; so there are $n-1$ possible ways for person 1 to get the wrong order. Suppose that person 1 gets order 2. There are two possibilities: Person 2 gets order 1, person 2 does not get order 1. If person 2 gets order 1, the number of ways for the other persons to get the wrong orders is $D_{n-2}$. The remaining case is that person 2 does not get order 1.

Let's write out carefully what it is we have to count. Persons 2, 3, ..., $n$ have among themselves orders 1, 3, 4, ..., $n$ (order 2 is missing because person 1 has it). We want to find the number of ways for persons 2, 3, ..., $n$ each to get the wrong order *and* for person 2 not to get order 1. This is almost the problem of $n-1$ persons all getting the wrong orders. We can turn it into this problem if we tell persons 2, 3, ..., $n$ that order 1 is order 2. Now person 2 will not get order 1 because person 2 thinks it's order 2. Since there are $n-1$ persons, there are $D_{n-1}$ ways for persons 2, 3, ..., $n$ each to get the wrong order *and* for person 2 not to get order 1. It follows that there are $D_{n-1} + D_{n-2}$ ways for person 1 to get order 2 and for all the others to get the wrong orders. Since there are $n-1$ possible ways for person 1 to get the wrong order, the recurrence relation now follows.

The recurrence relation defines $D_n$ in terms of $D_{n-1}$ and $D_{n-2}$, so it can't be solved using iteration. Also, the recurrence relation does not have constant coefficients (although it is linear), so it can't be solved using Theorem 7.2.11 or 7.2.14. This explains the need to make the substitution in part (b). Evidently, after making the substitution, the recurrence relation for $C_n$ can be solved by the methods of Section 7.2.

If we expand

$$D_n = (n-1)(D_{n-1} + D_{n-2}),$$

we obtain

$$D_n = nD_{n-1} - D_{n-1} + (n-1)D_{n-2}.$$

If we then move $nD_{n-1}$ to the left side of the equation (to obtain an expression equal to $C_n$), we obtain

$$D_n - nD_{n-1} = -D_{n-1} + (n-1)D_{n-2}.$$

Now the left side of the equation is equal to $C_n$ and the right side is equal to $-C_{n-1}$. Thus we obtain the recurrence relation

$$C_n = -C_{n-1}.$$

This equation may be solved by iteration.

## Formal Solution

Part (a): Suppose that $n$ persons all have the wrong orders. We consider the order that one person $p$ has. Suppose that $p$ has $q$'s order. We consider two cases: $q$ has $p$'s order, and $q$ does not have $p$'s order.

There are $D_{n-2}$ distributions in which $q$ has $p$'s order since the remaining $n-2$ orders are in possession of the remaining $n-2$ persons, but each has the wrong order.

We show that there are $D_{n-1}$ distributions in which $q$ does not have $p$'s order. Note that the set of orders $S$ that the $n-1$ persons (excluding $p$) possess includes all except $q$'s (which $p$ has). Temporarily assign ownership of $p$'s order to $q$. Then any distribution of $S$ among the $n-1$ persons in which no one has his or her own order yields a distribution in which $q$ does not have the order that is really $p$'s. Since there are $D_{n-1}$ such distributions, there are $D_{n-1}$ distributions in which $q$ does not have $p$'s order.

It follows that there are $D_{n-1} + D_{n-2}$ distributions in which $p$ has $q$'s order. Since $p$ can have any one of $n-1$ orders, the recurrence relation follows.

Part (b): Making the given substitution, we obtain

$$C_n = -C_{n-1}.$$

Using iteration, we obtain

$$\begin{aligned} C_n &= (-1)^1 C_{n-1} \\ &= (-1)^2 C_{n-2} = \cdots \\ &= (-1)^{n-2} C_2 \\ &= (-1)^n C_2 \\ &= (-1)^n (D_2 - 2D_1) = (-1)^n. \end{aligned}$$

Therefore,

$$D_n - nD_{n-1} = (-1)^n.$$

Solving this last recurrence relation by iteration, we obtain

$$\begin{aligned} D_n &= (-1)^n + nD_{n-1} \\ &= (-1)^n + n[(-1)^{n-1} + (n-1)D_{n-2}] \\ &= (-1)^n + n(-1)^{n-1} \\ &\quad + n(n-1)[(-1)^{n-2} + (n-2)D_{n-3}] \\ &\ \ \vdots \\ &= (-1)^n + n(-1)^{n-1} + n(n-1)(-1)^{n-2} + \cdots \\ &\quad - [n(n-1)\cdots 4] + [n(n-1)\cdots 3]. \end{aligned}$$

## Summary of Problem-Solving Techniques

- When examples get too wordy, develop a notation for concisely describing the examples. Carefully chosen notation can help enormously in solving a problem.
- When looking at examples, try to see how the current problem relates to smaller instances of the same problem.
- It often helps to write out carefully what is to be counted.
- It is sometimes possible to convert a recurrence relation that is not a linear homogeneous equation with constant coefficients into a linear homogeneous equation with constant coefficients. Such a recurrence relation can then be solved by the methods of Section 7.2.

## Comment

The technical name of a permutation in which no element is in its original position is a **derangement.**

## Probability

We can compute the probability that no one receives the correct order if we assume that all permutations are equally likely. In this case, since there are $n!$ permutations, the probability that no one receives the correct order is

$$\frac{D_n}{n!} = \frac{1}{n!}\{(-1)^n + n(-1)^{n-1}$$
$$+ n(n-1)(-1)^{n-2} + \cdots$$
$$- [n(n-1)\cdots 4] + [n(n-1)\cdots 3]\}$$
$$= \frac{(-1)^n}{n!} + \frac{(-1)^{n-1}}{(n-1)!} + \frac{(-1)^{n-2}}{(n-2)!} + \cdots$$
$$- \frac{1}{3!} + \frac{1}{2!}.$$

As $n$ increases, the additional terms are so small that the probability changes very little. In other words, for sufficiently large $n$, the probability that no one receives the correct order is essentially insensitive to the number of orders!

Calculus tells us that

$$e^x = \sum_{i=0}^{\infty} \frac{x^i}{i!}.$$

In particular,

$$e^{-1} = \sum_{i=0}^{\infty} \frac{(-1)^i}{i!} = \frac{1}{2!} - \frac{1}{3!} + \frac{1}{4!} \cdots.$$

Thus for large $n$, the probability that no one receives the correct order is approximately $e^{-1} = 0.368$.

## 7.3 Applications to the Analysis of Algorithms

In this section we use recurrence relations to analyze the time algorithms require. The technique is to develop a recurrence relation and initial conditions that define a sequence $a_1, a_2, \ldots,$ where $a_n$ is the time (best-case, average-case, or worst-case time) required for an algorithm to execute an input of size $n$. By solving the recurrence relation, we can determine the time needed by the algorithm.

Our first algorithm is a version of the selection sorting algorithm. This algorithm selects the largest item and places it last, then recursively repeats this process.

**Algorithm 7.3.1 Selection Sort**

**Go Online**
For more on selection sort, see bit.ly/2yYdlQL

This algorithm sorts the sequence

$$s_1, s_2, \ldots, s_n$$

in nondecreasing order by first selecting the largest item and placing it last and then recursively sorting the remaining elements.

Input: $s_1, s_2, \ldots, s_n$ and the length $n$ of the sequence
Output: $s_1, s_2, \ldots, s_n$, arranged in nondecreasing order

```
1.  selection_sort(s, n) {
2.     // base case
3.     if (n == 1)
4.        return
5.     // find largest
6.     max_index = 1 // assume initially that s_1 is largest
7.     for i = 2 to n
8.        if (s_i > s_max_index) // found larger, so update
9.           max_index = i
10.    // move largest to end
11.    swap(s_n, s_max_index)
12.    selection_sort(s, n − 1)
13. }
```

As a measure of the time required by this algorithm, we count the number of comparisons $b_n$ at line 8 required to sort $n$ items. (Notice that the best-case, average-case, and worst-case times are all the same for this algorithm.) We immediately obtain the initial condition $b_1 = 0$. To obtain a recurrence relation for the sequence $b_1, b_2, \ldots,$ we simulate the execution of the algorithm for arbitrary input of size $n > 1$. We count the number of comparisons at each line and then sum these numbers to obtain the total number of comparisons $b_n$. At lines 1–7, there are zero comparisons (of the type we are counting). At line 8, there are $n - 1$ comparisons (since line 7 causes line 8 to be executed $n - 1$ times). There are zero comparisons at lines 9–11. The recursive call occurs at line 12, where we invoke this algorithm with input of size $n-1$. But, by definition, this algorithm requires $b_{n-1}$ comparisons for input of size $n - 1$. Thus there are $b_{n-1}$ comparisons at line 12. Therefore, the total number of comparisons is $b_n = n - 1 + b_{n-1}$, which yields the desired recurrence relation.

Our recurrence relation can be solved by iteration:

$$
\begin{aligned}
b_n &= b_{n-1} + n - 1 \\
&= (b_{n-2} + n - 2) + (n - 1) \\
&= (b_{n-3} + n - 3) + (n - 2) + (n - 1) \\
&\vdots \\
&= b_1 + 1 + 2 + \cdots + (n - 2) + (n - 1) \\
&= 0 + 1 + 2 + \cdots + (n - 1) = \frac{(n - 1)n}{2} = \Theta(n^2).
\end{aligned}
$$

Thus the time required by Algorithm 7.3.1 is $\Theta(n^2)$.

Our next algorithm (Algorithm 7.3.2) is **binary search.** Binary search looks for a value in a *sorted* sequence and returns the index of the value if it is found or 0 if it is not found. The algorithm uses the divide-and-conquer approach. The sequence is divided into two nearly equal parts (line 4). If the item is found at the dividing point (line 5), the algorithm terminates. If the item is not found, because the sequence is sorted, an additional comparison (line 7) will locate the half of the sequence in which the item appears if it is present. We then recursively invoke binary search (line 11) to continue the search.

**Algorithm 7.3.2** Binary Search

This algorithm looks for a value in a nondecreasing sequence and returns the index of the value if it is found or 0 if it is not found.

Input: A sequence $s_i, s_{i+1}, \ldots, s_j, i \geq 1$, sorted in nondecreasing order, a value *key*, $i$, and $j$

Output: The output is an index $k$ for which $s_k = key$, or if *key* is not in the sequence, the output is the value 0.

```
1. binary_search(s, i, j, key) {
2.    if (i > j) // not found
3.       return 0
4.    k = ⌊(i + j)/2⌋
5.    if (key == s_k) // found
6.       return k
7.    if (key < s_k) // search left half
8.       j = k − 1
```

```
9.     else // search right half
10.        i = k + 1
11.    return binary_search(s, i, j, key)
12. }
```

**Example 7.3.3** We illustrate Algorithm 7.3.2 for the input

$$s_1 = \text{'}B\text{'},\quad s_2 = \text{'}D\text{'},\quad s_3 = \text{'}F\text{'},\quad s_4 = \text{'}S\text{'},$$

and *key* = '*S*'. At line 2, since $i > j$ $(1 > 4)$ is false, we proceed to line 4, where we set $k$ to 2. At line 5, since *key* ('*S*') is not equal to $s_2$ ('*D*'), we proceed to line 7. At line 7, $key < s_k$ ('*S*' < '*D*') is false, so at line 10, we set $i$ to 3. We then invoke this algorithm with $i = 3, j = 4$ to search for *key* in

$$s_3 = \text{'}F\text{'},\quad s_4 = \text{'}S\text{'}.$$

At line 2, since $i > j$ $(3 > 4)$ is false, we proceed to line 4, where we set $k$ to 3. At line 5, since *key* ('*S*') is not equal to $s_3$ ('*F*'), we proceed to line 7. At line 7, $key < s_k$ ('*S*' < '*F*') is false, so at line 10, we set $i$ to 4. We then invoke this algorithm with $i = j = 4$ to search for *key* in

$$s_4 = \text{'}S\text{'}.$$

At line 2, since $i > j$ $(4 > 4)$ is false, we proceed to line 4, where we set $k$ to 4. At line 5, since *key* ('*S*') is equal to $s_4$ ('*S*'), we return 4, the index of *key* in the sequence $s$. ◀

Next we turn to the worst-case analysis of binary search. We define the worst-case time required by binary search to be the number of times the algorithm is invoked in the worst case for a sequence containing $n$ items. We let $a_n$ denote the worst-case time.

Suppose that $n$ is 1; that is, suppose that the sequence consists of one element $s_i$ and $i = j$. In the worst case, the item will not be found at line 5, so the algorithm will be invoked a second time at line 11. However, at the second invocation we will have $i > j$ and the algorithm will terminate unsuccessfully at line 3. We have shown that if $n$ is 1, the algorithm is invoked twice. We obtain the initial condition

$$a_1 = 2. \tag{7.3.1}$$

Now suppose that $n > 1$. In this case, $i < j$, so the condition in line 2 is false. In the worst case, the item will not be found at line 5, so the algorithm will be invoked at line 11. By definition, the invocation at line 11 will require a total of $a_m$ invocations, where $m$ is the size of the sequence that is input at line 11. Since the sizes of the left and right sides of the original sequence are $\lfloor (n-1)/2 \rfloor$ and $\lfloor n/2 \rfloor$ and the worst case occurs with the larger sequence, the total number of invocations at line 11 will be $a_{\lfloor n/2 \rfloor}$. The original invocation together with the invocations at line 11 gives all the invocations; thus we obtain the recurrence relation

$$a_n = 1 + a_{\lfloor n/2 \rfloor}. \tag{7.3.2}$$

The recurrence relation (7.3.2) is typical of those that result from divide-and-conquer algorithms. Such recurrence relations are usually not easily solved explicitly (see, however, Exercise 6). Rather, one estimates the growth of the sequence involved using theta notation. Our method of deriving a theta notation for the sequence defined by (7.3.1) and (7.3.2) illustrates a general method of handling such recurrence relations. First we *explicitly solve* (7.3.2) in case $n$ is a power of 2. When $n$ is not a power of 2, $n$ lies between two powers of 2, say $2^{k-1}$ and $2^k$, and $a_n$ lies between $a_{2^{k-1}}$ and $a_{2^k}$. Since explicit formulas are known for $a_{2^{k-1}}$ and $a_{2^k}$, we can *estimate* $a_n$ and thereby derive a theta notation for $a_n$.

First we solve the recurrence relation (7.3.2) in case $n$ is a power of 2. If $n = 2^k$, (7.3.2) becomes

$$a_{2^k} = 1 + a_{2^{k-1}} \qquad k = 1, 2, \ldots.$$

If we let $b_k = a_{2^k}$, we obtain the recurrence relation

$$b_k = 1 + b_{k-1} \qquad k = 1, 2, \ldots, \tag{7.3.3}$$

and the initial condition $b_0 = 2$. The recurrence relation (7.3.3) can be solved by the iterative method:

$$b_k = 1 + b_{k-1} = 2 + b_{k-2} = \cdots = k + b_0 = k + 2.$$

Thus, if $n = 2^k$,

$$a_n = 2 + \lg n. \tag{7.3.4}$$

An arbitrary value of $n$ falls between two powers of 2, say

$$2^{k-1} < n \leq 2^k. \tag{7.3.5}$$

Since the sequence $a$ is nondecreasing (a fact that can be *proved* using induction—see Exercise 5),

$$a_{2^{k-1}} \leq a_n \leq a_{2^k}. \tag{7.3.6}$$

Notice that (7.3.5) gives

$$k - 1 < \lg n \leq k. \tag{7.3.7}$$

From (7.3.4), (7.3.6), and (7.3.7), we deduce that

$$\lg n < 1 + k = a_{2^{k-1}} \leq a_n \leq a_{2^k} = 2 + k < 3 + \lg n = O(\lg n).$$

Therefore, $a_n = \Theta(\lg n)$, so binary search is $\Theta(\lg n)$ in the worst case. This result is important enough to highlight as a theorem.

**Theorem 7.3.4** The worst-case time for binary search for input of size $n$ is $\Theta(\lg n)$.

**Proof** The proof precedes the statement of the theorem. ◀

For our last example, we present and analyze another sorting algorithm known as **merge sort** (Algorithm 7.3.8). We will show that merge sort has worst-case run time $\Theta(n \lg n)$, so for large input, merge sort is much faster than selection sort (Algorithm 7.3.1), which has worst-case run time $\Theta(n^2)$. In Section 9.7 we will show that *any* sorting algorithm that compares elements and, based on the result of a comparison, moves items around in an array is $\Omega(n \lg n)$ in the worst case; thus merge sort is optimal within this class of sorting algorithms.

In merge sort the sequence to be sorted,

$$s_i, \ldots, s_j,$$

is divided into two nearly equal sequences,

$$s_i, \ldots, s_m, \qquad s_{m+1}, \ldots, s_j,$$

where $m = \lfloor (i + j)/2 \rfloor$. Each of these sequences is recursively sorted, after which they are combined to produce a sorted arrangement of the original sequence. The process of combining two sorted sequences is called **merging.**

**Algorithm 7.3.5** **Merging Two Sequences**

This algorithm combines two nondecreasing sequences into a single nondecreasing sequence.

Input: Two nondecreasing sequences: $s_i, \ldots, s_m$ and $s_{m+1}, \ldots, s_j$, and indexes $i$, $m$, and $j$

Output: The sequence $c_i, \ldots, c_j$ consisting of the elements $s_i, \ldots, s_m$ and $s_{m+1}, \ldots, s_j$ combined into one nondecreasing sequence

```
1.  merge(s, i, m, j, c) {
2.     // p is the position in the sequence s_i, ..., s_m
3.     // q is the position in the sequence s_{m+1}, ..., s_j
4.     // r is the position in the sequence c_i, ..., c_j
5.     p = i
6.     q = m + 1
7.     r = i
8.     // copy smaller of s_p and s_q
9.     while (p ≤ m ∧ q ≤ j) {
10.       if (s_p < s_q) {
11.          c_r = s_p
12.          p = p + 1
13.       }
14.       else {
15.          c_r = s_q
16.          q = q + 1
17.       }
18.       r = r + 1
19.    }
20.    // copy remainder of first sequence
21.    while (p ≤ m) {
22.       c_r = s_p
23.       p = p + 1
24.       r = r + 1
25.    }
26.    // copy remainder of second sequence
27.    while (q ≤ j) {
28.       c_r = s_q
29.       q = q + 1
30.       r = r + 1
31.    }
32. }
```

**Example 7.3.6** Figure 7.3.1 shows how Algorithm 7.3.5 merges the sequences

$$1, 3, 4 \quad \text{and} \quad 2, 4, 5, 6.$$

| | $p$ | $p$ | $p$ | $p$ | $p$ | $p$ | $p$ |
|---|---|---|---|---|---|---|---|
| $s_i, \ldots, s_m$: | 134 | 134 | 134 | 134 | 134 | 134 | 134 |
| | $q$ | $q$ | $q$ | $q$ | $q$ | $q$ | $q$ |
| $s_{m+1}, \ldots, s_j$: | 2456 | 2456 | 2456 | 2456 | 2456 | 2456 | 2456 |
| | $r$ | $r$ | $r$ | $r$ | $r$ | $r$ | $r$ |
| $c_i, \ldots, c_j$: | 1 | 12 | 123 | 1234 | 12344 | 123445 | 1234456 |

**Figure 7.3.1** Merging $s_i, \ldots, s_m$ and $s_{m+1}, \ldots, s_j$. The result is $c_i, \ldots, c_j$. ◀

Theorem 7.3.7 shows that in the worst case, $n-1$ comparisons are needed to merge two sequences the sum of whose lengths is $n$.

**Theorem 7.3.7** In the worst case, Algorithm 7.3.5 requires $j - i$ comparisons. In particular, in the worst case, $n - 1$ comparisons are needed to merge two sequences the sum of whose lengths is $n$.

**Proof** In Algorithm 7.3.5, the comparison of elements in the sequences occurs in the while loop at line 10. The while loop will execute as long as $p \leq m$ and $q \leq j$. Thus, in the worst case, Algorithm 7.3.5 requires $j - i$ comparisons. ◀

We next use Algorithm 7.3.5 (merging) to construct merge sort.

**Algorithm 7.3.8** **Merge Sort**

This recursive algorithm sorts a sequence into nondecreasing order by using Algorithm 7.3.5, which merges two nondecreasing sequences.

**Go Online**
For more on merge sort, see bit.ly/2yYdlQL

Input: $s_i, \ldots, s_j$, $i$, and $j$

Output: $s_i, \ldots, s_j$, arranged in nondecreasing order

```
1.  merge_sort(s, i, j) {
2.     // base case: i == j
3.     if (i == j)
4.        return
5.     // divide sequence and sort
6.     m = ⌊(i + j)/2⌋
7.     merge_sort(s, i, m)
8.     merge_sort(s, m + 1, j)
9.     // merge
10.    merge(s, i, m, j, c)
11.    // copy c, the output of merge, into s
12.    for k = i to j
13.       s_k = c_k
14. }
```

**Example 7.3.9** Figure 7.3.2 shows how Algorithm 7.3.8 sorts the sequence

$$12,\quad 30,\quad 21,\quad 8,\quad 6,\quad 9,\quad 1,\quad 7.$$

| | Merge one-element arrays | Merge two-element arrays | Merge four-element arrays |
|---|---|---|---|
| 12 | 12 | 8 | 1 |
| 30 | 30 | 12 | 6 |
| 21 | 8 | 21 | 7 |
| 8 | 21 | 30 | 8 |
| 6 | 6 | 1 | 9 |
| 9 | 9 | 6 | 12 |
| 1 | 1 | 7 | 21 |
| 7 | 7 | 9 | 30 |

**Figure 7.3.2** Sorting by merge sort.

We conclude by showing that merge sort (Algorithm 7.3.8) is $\Theta(n \lg n)$ in the worst case. The method of proof is the same as we used to show that binary search is $\Theta(\lg n)$ in the worst case.

**Theorem 7.3.10** Merge sort (Algorithm 7.3.8) is $\Theta(n \lg n)$ in the worst case.

**Proof** Let $a_n$ be the number of comparisons required by Algorithm 7.3.8 to sort $n$ items in the worst case. Then $a_1 = 0$. If $n > 1$, $a_n$ is at most the sum of the numbers of comparisons in the worst case resulting from the recursive calls at lines 7 and 8, and the number of comparisons in the worst case required by merge at line 10. That is,

$$a_n \le a_{\lfloor n/2 \rfloor} + a_{\lfloor (n+1)/2 \rfloor} + n - 1.$$

In fact, this upper bound is achievable (see Exercise 22), so that

$$a_n = a_{\lfloor n/2 \rfloor} + a_{\lfloor (n+1)/2 \rfloor} + n - 1.$$

First we solve the preceding recurrence relation in case $n$ is a power of 2, say $n = 2^k$. The equation becomes

$$a_{2^k} = 2a_{2^{k-1}} + 2^k - 1.$$

We may solve this last equation by using iteration (see Section 7.2):

$$\begin{aligned}
a_{2^k} &= 2a_{2^{k-1}} + 2^k - 1 \\
&= 2[2a_{2^{k-2}} + 2^{k-1} - 1] + 2^k - 1 \\
&= 2^2 a_{2^{k-2}} + 2 \cdot 2^k - 1 - 2 \\
&= 2^2[2a_{2^{k-3}} + 2^{k-2} - 1] + 2 \cdot 2^k - 1 - 2
\end{aligned}$$

$$
\begin{aligned}
&= 2^3 a_{2^{k-3}} + 3 \cdot 2^k - 1 - 2 - 2^2 \\
&\vdots \\
&= 2^k a_{2^0} + k \cdot 2^k - 1 - 2 - 2^2 - \cdots - 2^{k-1} \\
&= k \cdot 2^k - (2^k - 1) \\
&= (k-1)2^k + 1. \qquad (7.3.8)
\end{aligned}
$$

An arbitrary value of $n$ falls between two powers of 2, say,

$$2^{k-1} < n \le 2^k. \qquad (7.3.9)$$

Since the sequence $a$ is nondecreasing (see Exercise 25),

$$a_{2^{k-1}} \le a_n \le a_{2^k}. \qquad (7.3.10)$$

Notice that (7.3.9) gives

$$k - 1 < \lg n \le k. \qquad (7.3.11)$$

From (7.3.8), (7.3.10), and (7.3.11), we deduce that

$$
\begin{aligned}
\Omega(n \lg n) = (-2 + \lg n)\frac{n}{2} &< (k-2)2^{k-1} + 1 = a_{2^{k-1}} \\
&\le a_n \le a_{2^k} \le k2^k + 1 \le (1 + \lg n)2n + 1 = O(n \lg n).
\end{aligned}
$$

Therefore, $a_n = \Theta(n \lg n)$, so merge sort is $\Theta(n \lg n)$ in the worst case. ◀

As remarked previously, in Section 9.7 we will show that any comparison-based sorting algorithm is $\Omega(n \lg n)$ in the worst case. This result implies, in particular, that merge sort is $\Omega(n \lg n)$ in the worst case. If we had already proved this result, to prove that merge sort is $\Theta(n \lg n)$ in the worst case, it would have been sufficient to prove that merge sort is $O(n \lg n)$ in the worst case.

Even though merge sort, Algorithm 7.3.8, is optimal, it may not be the algorithm of choice for a particular sorting problem. Factors such as the average-case time of the algorithm, the number of items to be sorted, available memory, the data structures to be used, whether the items to be sorted are in memory or reside on peripheral storage devices such as disks or tapes, whether the items to be sorted are already "nearly" sorted, and the hardware to be used must be taken into account.

## 7.3 Review Exercises

1. Explain how to find a recurrence relation that describes the time a recursive algorithm requires.
2. How does selection sort work?
3. What is the time required by selection sort?
4. How does binary search work? What properties must the input have?
5. Write a recurrence relation that describes the worst-case time required by binary search.
6. What is the worst-case time of binary search?
7. How does the merge algorithm work? What properties must the input have?
8. What is the worst-case time of the merge algorithm?
9. Explain how merge sort works.
10. Write a recurrence relation that describes the worst-case time required by merge sort.
11. Why is it easy to solve for the worst-case time of merge sort if the input size is a power of two?
12. Explain, in words, how we can obtain bounds on the worst-case time of merge sort for input of arbitrary size if we know the worst-case time when the input size is a power of two.
13. What is the worst-case time of merge sort?

## 7.3 Exercises

*Exercises 1–4 refer to the sequence*

$$s_1 = \text{'C'}, \quad s_2 = \text{'G'}, \quad s_3 = \text{'J'}, \quad s_4 = \text{'M'}, \quad s_5 = \text{'X'}.$$

**1.** Show how Algorithm 7.3.2 executes in case $key$ = 'G'.

**2.** Show how Algorithm 7.3.2 executes in case $key$ = 'P'.

**3.** Show how Algorithm 7.3.2 executes in case $key$ = 'C'.

**4.** Show how Algorithm 7.3.2 executes in case $key$ = 'Z'.

**5.** Let $a_n$ denote the worst-case time of binary search (Algorithm 7.3.2). Prove that $a_n \le a_{n+1}$ for $n \ge 1$.

**★6.** Prove that if $a_n$ is the number of times the binary search algorithm (Algorithm 7.3.2) is invoked in the worst case for a sequence containing $n$ items, then

$$a_n = 2 + \lfloor \lg n \rfloor$$

for every positive integer $n$.

**7.** Give an example to show that if the input to Algorithm 7.3.2 is *not* in nondecreasing order, Algorithm 7.3.2 may not find $key$ even if it is present.

**8.** Suppose that the input to Algorithm 7.3.2 is *not* in nondecreasing order. Could Algorithm 7.3.2 erroneously find $key$ even though it is not present?

**9.** Write a nonrecursive version of the binary search algorithm.

**★10.** Write a nonrecursive version of the binary search algorithm that uses at most $1 + \lfloor \lg n \rfloor$ array comparisons. The input is a sequence $s_1, \ldots, s_n$, sorted in nondecreasing order, a value $key$ to search for, and $n$. Assume that the only array comparisons allowed are $s_i < key$ and $s_i == key$.

**11.** Assume the conditions of Exercise 10 and assume further that one incorrect response to an array comparison is allowed. [An "incorrect response to an array comparison" occurs in one of four ways: (1) $s_i < key$ is true, but in the algorithm, the relation $s_i < key$ evaluates to false. (2) $s_i < key$ is false, but in the algorithm, the relation $s_i < key$ evaluates to true. (3) $s_i == key$ is true, but in the algorithm, the relation $s_i == key$ evaluates to false. (4) $s_i == key$ is false, but in the algorithm, the relation $s_i == key$ evaluates to true.] Write an algorithm that uses at most $3 + 2\lceil \lg n \rceil$ array comparisons to determine whether $key$ is in the sequence. (The problem can be solved in fewer than $3 + 2\lceil \lg n \rceil$ array comparisons in the worst case.)

**12.** Professor T. R. S. Eighty proposes the following version of binary search:

```
binary_search2(s, i, j, key) {
    if (i > j)
        return 0
    k = ⌊(i + j)/2⌋
    if (key == s_k)
        return k
    k1 = binary_search2(s, i, k − 1, key)
    k2 = binary_search2(s, k + 1, j, key)
    return k1 + k2
}
```

(a) Show that *binary_search2* is correct; that is, if $key$ is present, the algorithm returns its index, but if $key$ is not present, it returns 0.

(b) Find the worst-case running time of *binary_search2*.

**13.** Professor Larry proposes the following version of binary search:

```
binary_search3(s, i, j, key) {
    while (i ≤ j) {
        k = ⌊(i + j)/2⌋
        if (key == s_k)
            return k
        if (key < s_k)
            j = k
        else
            i = k
    }
    return 0
}
```

Is the Professor's version correct (i.e., does it find $key$ if it is present and return 0 if it is not present)? If the Professor's version is correct, what is the worst-case time?

**14.** Professor Curly proposes the following version of binary search:

```
binary_search4(s, i, j, key) {
    if (i > j)
        return 0
    k = ⌊(i + j)/2⌋
    if (key == s_k)
        return k
    flag = binary_search4(s, i, k − 1, key)
    if (flag == 0)
        return binary_search4(s, k + 1, j, key)
    else
        return flag
}
```

Is the Professor's version correct (i.e., does it find $key$ if it is present and return 0 if it is not present)? If the Professor's version is correct, what is the worst-case time?

**15.** Professor Moe proposes the following version of binary search:

```
binary_search5(s, i, j, key) {
    if (i > j)
        return 0
    k = ⌊(i + j)/2⌋
    if (key == s_k)
        return k
    if (key < s_k)
        return binary_search5(s, i, k, key)
    else
        return binary_search5(s, k + 1, j, key)
}
```

Is the Professor's version correct (i.e., does it find *key* if it is present and return 0 if it is not present)? If the Professor's version is correct, what is the worst-case time?

16. Suppose that we replace the line

$$k = \lfloor (i + j)/2 \rfloor$$

with

$$k = \lfloor i + (j - i)/3 \rfloor$$

in Algorithm 7.3.2. Is the resulting algorithm still correct (i.e., does it find *key* if it is present and return 0 if it is not present)? If it is correct, what is the worst-case time?

17. Suppose that we replace the line

$$k = \lfloor (i + j)/2 \rfloor$$

with

$$k = \lfloor j - 2 \rfloor$$

in Algorithm 7.3.2. Is the resulting algorithm still correct (i.e., does it find *key* if it is present and return 0 if it is not present)? If it is correct, what is the worst-case time?

18. Suppose that algorithm *A* requires $\lceil n \lg n \rceil$ comparisons to sort $n$ items and algorithm *B* requires $\lceil n^2/4 \rceil$ comparisons to sort $n$ items. For which $n$ is algorithm *B* superior to algorithm *A*?

19. Show how merge sort (Algorithm 7.3.8) sorts the sequence 1, 9, 7, 3.

20. Show how merge sort (Algorithm 7.3.8) sorts the sequence 2, 3, 7, 2, 8, 9, 7, 5, 4.

21. Suppose that we have two sequences each of size $n$ sorted in nondecreasing order.
    (a) Under what conditions does the maximum number of comparisons occur in Algorithm 7.3.5?
    (b) Under what conditions does the minimum number of comparisons occur in Algorithm 7.3.5?

22. Let $a_n$ be as in the proof of Theorem 7.3.10. Describe input for which

$$a_n = a_{\lfloor n/2 \rfloor} + a_{\lfloor (n+1)/2 \rfloor} + n - 1.$$

23. What is the minimum number of comparisons required by Algorithm 7.3.8 to sort an array of size 6?

24. What is the maximum number of comparisons required by Algorithm 7.3.8 to sort an array of size 6?

25. Let $a_n$ be as in the proof of Theorem 7.3.10. Show that $a_n \le a_{n+1}$ for all $n \ge 1$.

26. Let $a_n$ denote the number of comparisons required by merge sort in the worst case. Show that $a_n \le 3n \lg n$ for $n = 1, 2, 3, \ldots$.

27. Show that in the best case, merge sort requires $\Theta(n \lg n)$ comparisons.

*Let $p_1, \ldots, p_n$ be a permutation of $\{1, \ldots, n\}$. An inversion in $p$ is a pair $p_i, p_j$ where $i < j$ and $p_i > p_j$. Informally, the pair $p_i, p_j$ is an inversion if $p_i$ and $p_j$ are out of order, where order means increasing order. Exercises 28–38 are concerned with inversions.*

28. List all inversions in the permutation 4, 1, 3, 2.

29. List all inversions in the permutation 3, 2, 5, 4, 1.

30. List all inversions in the permutation 1, 2, 3, 4.

31. Show a permutation of $\{1, 2, 3, 4\}$ having the maximum number of inversions.

32. What is the maximum number of inversions in a permutation of $\{1, \ldots, n\}$?

33. Show that any sorting algorithm that moves data by swapping only adjacent elements has worst-case time $\Omega(n^2)$.

34. Show that the total number of inversions in all of the permutations of $\{1, \ldots, n\}$ is $n!n(n-1)/4$. *Hint*: If $p_1, \ldots, p_n$ is a permutation of $\{1, \ldots, n\}$, let $p^R = p_n, \ldots, p_1$ denote the reverse of $p$. Consider the total number of permutations in $p$ and $p^R$.

35. Show that the average number of inversions in permutations of $\{1, \ldots, n\}$ is $n(n-1)/4$. Assume that all permutations are equally likely.

36. Write an algorithm whose input is a permutation $p$ of $\{1, \ldots, n\}$ and $n$. The algorithm returns the number of inversions in $p$. What is the worst-case time of your algorithm?

★37. Write an algorithm whose worst-case time is $O(n \lg n)$ that returns the number of inversions in a permutation of $\{1, \ldots, n\}$. The algorithm's input is a permutation of $\{1, \ldots, n\}$ and $n$.

38. What is wrong with the following argument that purports to show that any algorithm that computes the number of inversions in a permutation of $n$ elements has worst-case time $\Omega(n^2)$? Since any algorithm that computes the number of inversions in a permutation of $n$ elements must check each pair of elements and there are $\Omega(n^2)$ pairs, any such algorithm has worst-case time $\Omega(n^2)$. (Compare this exercise with Exercise 37.)

*Exercises 39–43 refer to Algorithm 7.3.11.*

### Algorithm 7.3.11
### *Computing an Exponential*

This algorithm computes $a^n$ recursively, where $a$ is a real number and $n$ is a positive integer.

Input: $a$ (a real number), $n$ (a positive integer)
Output: $a^n$

```
1.  expl(a, n) {
2.     if (n == 1)
3.        return a
4.     m = ⌊n/2⌋
5.     return expl(a, m) * expl(a, n − m)
6.  }
```

*Let $b_n$ be the number of multiplications (line 5) required to compute $a^n$.*

39. Explain how Algorithm 7.3.11 computes $a^n$.

40. Find a recurrence relation and initial conditions for the sequence $\{b_n\}$.

41. Compute $b_2$, $b_3$, and $b_4$.

42. Solve the recurrence relation of Exercise 40 in case $n$ is a power of 2.

43. Prove that $b_n = n - 1$ for every positive integer $n$.

*Exercises 44–49, refer to Algorithm 7.3.12.*

## Algorithm 7.3.12
### Computing an Exponential

This algorithm computes $a^n$ recursively, where $a$ is a real number and $n$ is a positive integer.

Input: $a$ (a real number), $n$ (a positive integer)
Output: $a^n$

```
1.  exp2(a, n) {
2.     if (n == 1)
3.        return a
4.     m = ⌊n/2⌋
5.     power = exp2(a, m)
6.     power = power * power
7.     if (n is even)
8.        return power
9.     else
10.       return power * a
11. }
```

*Let $b_n$ be the number of multiplications (lines 6 and 10) required to compute $a^n$.*

44. Explain how Algorithm 7.3.12 computes $a^n$.

45. Show that

$$b_n = \begin{cases} b_{(n-1)/2} + 2 & \text{if } n \text{ is odd} \\ b_{n/2} + 1 & \text{if } n \text{ is even.} \end{cases}$$

46. Find $b_1$, $b_2$, $b_3$, and $b_4$.

47. Solve the recurrence relation of Exercise 45 in case $n$ is a power of 2.

48. Show, by an example, that $b$ is not nondecreasing.

★49. Prove that $b_n = \Theta(\lg n)$.

*Exercises 50–55 refer to Algorithm 7.3.13.*

## Algorithm 7.3.13
### Finding the Largest and Smallest Elements in a Sequence

This recursive algorithm finds the largest and smallest elements in a sequence.

Input: $s_i, \ldots, s_j$, $i$, and $j$
Output: *large* (the largest element in the sequence), *small* (the smallest element in the sequence)

```
1.  large_small(s, i, j, large, small) {
2.     if (i == j) {
3.        large = s_i
4.        small = s_i
5.        return
6.     }
7.     m = ⌊(i + j)/2⌋
8.     large_small(s, i, m, large_left, small_left)
9.     large_small(s, m + 1, j, large_right, small_right)
10.    if (large_left > large_right)
11.       large = large_left
12.    else
13.       large = large_right
14.    if (small_left > small_right)
15.       small = small_right
16.    else
17.       small = small_left
18. }
```

*Let $b_n$ be the number of comparisons (lines 10 and 14) required for an input of size $n$.*

50. Explain how Algorithm 7.3.13 finds the largest and smallest elements.

51. Show that $b_1 = 0$ and $b_2 = 2$.

52. Find $b_3$.

53. Establish the recurrence relation

$$b_n = b_{\lfloor n/2 \rfloor} + b_{\lfloor (n+1)/2 \rfloor} + 2 \qquad (7.3.12)$$

for $n > 1$.

54. Solve the recurrence relation (7.3.12) in case $n$ is a power of 2 to obtain

$$b_n = 2n - 2 \qquad n = 1, 2, 4, \ldots.$$

55. Use induction to show that

$$b_n = 2n - 2$$

for every positive integer $n$.

*Exercises 56–59 refer to Algorithm 7.3.13, with the following inserted after line 6.*

```
6a.  if (j == i + 1) {
6b.     if (s_i > s_j) {
6c.        large = s_i
6d.        small = s_j
6e.     }
6f.     else {
6g.        small = s_i
6h.        large = s_j
6i.     }
6j.     return
6k.  }
```

*Let $b_n$ be the number of comparisons (lines 6b, 10, and 14) for an input of size n.*

**56.** Show that $b_1 = 0$ and $b_2 = 1$.

**57.** Compute $b_3$ and $b_4$.

**58.** Show that the recurrence relation (7.3.12) holds for $n > 2$.

**59.** Solve the recurrence relation (7.3.12) in case $n$ is a power of 2 to obtain

$$b_n = \frac{3n}{2} - 2 \qquad n = 2, 4, 8, \ldots.$$

**★60.** Modify Algorithm 7.3.13 by inserting the lines preceding Exercise 56 after line 6 and replacing line 7 with the following.

```
7a.  if (j − i is odd ∧ (1 + j − i)/2 is odd)
7b.      m = ⌊(i + j)/2⌋ − 1
7c.  else
7d.      m = ⌊(i + j)/2⌋
```

Show that in the worse case, this modified algorithm requires at most $\lceil (3n/2) - 2 \rceil$ comparisons to find the largest and smallest elements in an array of size $n$.

*Exercises 61–65 refer to Algorithm 7.3.14.*

### Algorithm 7.3.14

*Insertion Sort (Recursive Version)*

This algorithm sorts the sequence

$$s_1, s_2, \ldots, s_n$$

in nondecreasing order by recursively sorting the first $n - 1$ elements and then inserting $s_n$ in the correct position.

Input: $s_1, s_2, \ldots, s_n$ and the length $n$ of the sequence

Output: $s_1, s_2, \ldots, s_n$ arranged in nondecreasing order

```
1.  insertion_sort(s, n) {
2.      if (n == 1)
3.          return
4.      insertion_sort(s, n − 1)
5.      i = n − 1
6.      temp = s_n
7.      while (i ≥ 1 ∧ s_i > temp) {
8.          s_{i+1} = s_i
9.          i = i − 1
10.     }
11.     s_{i+1} = temp
12. }
```

*Let $b_n$ be the number of times the comparison $s_i > temp$ in line 7 is made in the worst case. Assume that if $i < 1$, the comparison $s_i > temp$ is not made.*

**61.** Explain how Algorithm 7.3.14 sorts the sequence.

**62.** Which input produces the worst-case behavior for Algorithm 7.3.14?

**63.** Find $b_1$, $b_2$, and $b_3$.

**64.** Find a recurrence relation for the sequence $\{b_n\}$.

**65.** Solve the recurrence relation of Exercise 64.

*Exercises 66–68 refer to Algorithm 7.3.15.*

### Algorithm 7.3.15

Input: $s_1, \ldots, s_n, n$

Output: $s_1, \ldots, s_n$

```
algor1(s, n) {
    i = n
    while (i ≥ 1) {
        s_i = s_i + 1
        i = ⌊i/2⌋
    }
    n = ⌊n/2⌋
    if (n ≥ 1)
        algor1(s, n)
}
```

*Let $b_n$ be the number of times the statement $s_i = s_i + 1$ is executed.*

**66.** Find a recurrence relation for the sequence $\{b_n\}$ and compute $b_1$, $b_2$, and $b_3$.

**67.** Solve the recurrence relation of Exercise 66 in case $n$ is a power of 2.

**68.** Prove that $b_n = \Theta((\lg n)^2)$.

**69.** Find a theta notation in terms of $n$ for the number of times *algor2* is called when invoked as *algor2*$(1, n)$.

```
algor2(i, j) {
    if (i == j)
        return
    k = ⌊(i + j)/2⌋
    algor2(i, k)
    algor2(k + 1, j)
}
```

*Exercises 70–74 refer to Algorithm 7.3.16.*

### Algorithm 7.3.16

Input: A sequence $s_i, \ldots, s_j$ of zeros and ones

Output: $s_i, \ldots, s_j$ where all the zeros precede all the ones

```
sort(s, i, j) {
    if (i == j)
        return
    if (s_i == 1) {
        swap(s_i, s_j)
        sort(s, i, j − 1)
    }
    else
        sort(s, i + 1, j)
}
```

**70.** Use mathematical induction on $n$, the number of items in the input sequence, to prove that *sort* does produce as output a rearranged version of the input sequence in which all of the zeros precede all of the ones. (The Basis Step is $n = 1$.)

*Let $b_n$ denote the number of times sort is called when the input sequence contains n items.*

**71.** Find $b_1$, $b_2$, and $b_3$.

**72.** Write a recurrence relation for $b_n$.

**73.** Solve your recurrence relation of Exercise 72 for $b_n$.

**74.** In theta notation, what is the running time of *sort* as a function of $n$, the number of items in the input sequence?

**75.** Solve the recurrence relation

$$a_n = 3a_{\lfloor n/2 \rfloor} + n \qquad n > 1,$$

in case $n$ is a power of 2. Assume that $a_1 = 1$.

**76.** Show that $a_n = \Theta(n^{\lg 3})$, where $a_n$ is as in Exercise 75.

*Exercises 77–84 refer to an algorithm that accepts as input the sequence*

$$s_i, \ldots, s_j.$$

*If $j > i$, the subproblems*

$$s_i, \ldots, s_{\lfloor (i+j)/2 \rfloor} \quad \textit{and} \quad s_{\lfloor (i+j)/2+1 \rfloor}, \ldots, s_j$$

*are solved recursively. Solutions to subproblems of sizes m and k can be combined in time $c_{m,k}$ to solve the original problem. Let $b_n$ be the time required by the algorithm for an input of size n.*

**77.** Write a recurrence relation for $b_n$ assuming that $c_{m,k} = 3$.

**78.** Write a recurrence relation for $b_n$ assuming that $c_{m,k} = m+k$.

**79.** Solve the recurrence relation of Exercise 77 in case $n$ is a power of 2, assuming that $b_1 = 0$.

**80.** Solve the recurrence relation of Exercise 77 in case $n$ is a power of 2, assuming that $b_1 = 1$.

**81.** Solve the recurrence relation of Exercise 78 in case $n$ is a power of 2, assuming that $b_1 = 0$.

**82.** Solve the recurrence relation of Exercise 78 in case $n$ is a power of 2, assuming that $b_1 = 1$.

★**83.** Assume that if $m_1 \geq m_2$ and $k_1 \geq k_2$, then $c_{m_1,k_1} \geq c_{m_2,k_2}$. Show that the sequence $b_1, b_2, \ldots$ is nondecreasing.

★**84.** Assuming that $c_{m,k} = m+k$ and $b_1 = 0$, show that $b_n \leq 4n \lg n$.

*Exercises 85–90 refer to the following situation. We let $P_n$ denote a particular problem of size n. If $P_n$ is divided into subproblems of sizes i and j, there is an algorithm that combines the solutions of these two subproblems into a solution to $P_n$ in time at most $2 + \lg(ij)$. Assume that a problem of size 1 is already solved.*

**85.** Write a recursive algorithm to solve $P_n$ similar to Algorithm 7.3.8.

**86.** Let $a_n$ be the worst-case time to solve $P_n$ by the algorithm of Exercise 85. Show that

$$a_n \leq a_{\lfloor n/2 \rfloor} + a_{\lfloor (n+1)/2 \rfloor} + 2 \lg n.$$

**87.** Let $b_n$ be the recurrence relation obtained from Exercise 86 by replacing "$\leq$" by "$=$." Assume that $b_1 = a_1 = 0$. Show that if $n$ is a power of 2,

$$b_n = 4n - 2 \lg n - 4.$$

**88.** Show that $a_n \leq b_n$ for $n = 1, 2, 3, \ldots$.

**89.** Show that $b_n \leq b_{n+1}$ for $n = 1, 2, 3, \ldots$.

**90.** Show that $a_n \leq 8n$ for $n = 1, 2, 3, \ldots$.

**91.** Suppose that $\{a_n\}$ is a nondecreasing sequence and that whenever $m$ divides $n$,

$$a_n = a_{n/m} + d,$$

where $d$ is a positive real number and $m$ is an integer satisfying $m > 1$. Show that $a_n = \Theta(\lg n)$.

★**92.** Suppose that $\{a_n\}$ is a nondecreasing sequence and that whenever $m$ divides $n$,

$$a_n = c a_{n/m} + d,$$

where $c$ and $d$ are real numbers satisfying $c > 1$ and $d > 0$, and $m$ is an integer satisfying $m > 1$. Show that $a_n = \Theta(n^{\log_m c})$.

**93.** [Project] Investigate other sorting algorithms. Consider specifically complexity, empirical studies, and special features of the algorithms (see [Knuth, 1998b]).

## 7.4 The Closest-Pair Problem[†]

**Go Online**
For more on computational geometry, see bit.ly/2yYdlQL

**Computational geometry** is concerned with the design and analysis of algorithms to solve geometry problems. Efficient geometric algorithms are useful in fields such as computer graphics, statistics, image processing, and very-large-scale-integration (VLSI) design.

The **closest-pair problem** furnishes an example of a problem from computational geometry: Given $n$ points in the plane, find a closest pair. (We say *a* closest pair since it is possible that several pairs achieve the same minimum distance.) Our distance measure is ordinary Euclidean distance.

---

[†]This section can be omitted without loss of continuity.

One way to solve this problem is to list the distance between each pair and choose the minimum in this list of distances. Since there are $C(n, 2) = n(n-1)/2 = \Theta(n^2)$ pairs, this "list-all" algorithm's time is $\Theta(n^2)$. We can do better; we will give a divide-and-conquer closest-pair algorithm whose worst-case time is $\Theta(n \lg n)$. We first discuss the algorithm and then give a more precise description using pseudocode.

Our algorithm begins by finding a vertical line $l$ that divides the points into two nearly equal parts (see Figure 7.4.1). [If $n$ is even, we divide the points into parts each having $n/2$ points. If $n$ is odd, we divide the points into parts, one having $(n+1)/2$ points and the other having $(n-1)/2$ points.]

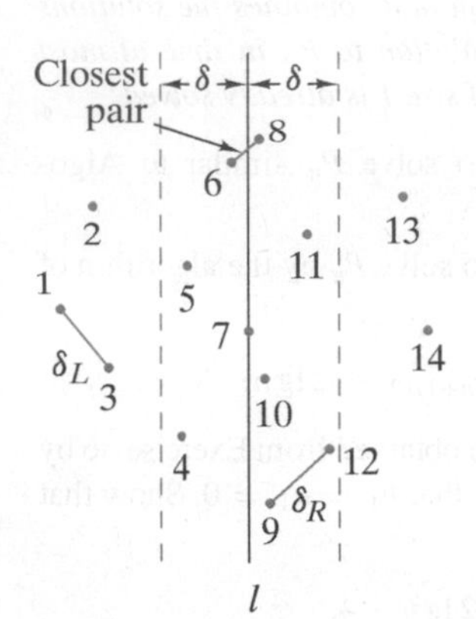


**Figure 7.4.1** $n$ points in the plane. The problem is to find a closest pair. For this set, the closest pair is 6 and 8. Line $l$ divides the points into two approximately equal parts. The closest pair in the left half is 1 and 3, which is $\delta_L$ apart. The closest pair in the right half is 9 and 12, which is $\delta_R$ apart. Any pair (e.g., 6 and 8) closer together than $\delta = \min\{\delta_L, \delta_R\}$ must lie in the vertical strip of width $2\delta$ centered at $l$.

We then recursively solve the problem for each of the parts. We let $\delta_L$ be the distance between a closest pair in the left part; we let $\delta_R$ be the distance between a closest pair in the right part; and we let $\delta = \min\{\delta_L, \delta_R\}$. Unfortunately, $\delta$ may not be the distance between a closest pair from the original set of points because a pair of points, one from the left part and the other from the right part, might be closer than $\delta$ (see Figure 7.4.1). Thus we must consider distances between points on opposite sides of the line $l$.

We first note that if the distance between a pair of points is less than $\delta$, the points must lie in the vertical strip of width $2\delta$ centered at $l$ (see Figure 7.4.1). (Any point not in this strip is at least $\delta$ away from every point on the other side of $l$.) Thus we can restrict our search for a pair closer than $\delta$ to points in this strip.

If there are $n$ points in the strip and we check *all* pairs in the strip, the worst-case time to process the points in the strip is $\Theta(n^2)$. In this case the worst-case time of our algorithm will be $\Omega(n^2)$, which is at least as bad as the exhaustive search; thus we must avoid checking all pairs in the strip.

We order the points in the strip in nondecreasing order of their $y$-coordinates. We then examine the points in this order. When we examine a point $p$ in the strip, any point $q$ following $p$ whose distance to $p$ is less than $\delta$ must lie strictly within or on the base of the rectangle of height $\delta$ whose base contains $p$ and whose vertical sides are at a distance $\delta$ from $l$ (see Figure 7.4.2). (We need not compute the distance between $p$ and points below $p$. These distances would already have been considered since we are examining the points in nondecreasing order of their $y$-coordinates.) We will show that this rectangle contains at most eight points, including $p$ itself, so if we compute the distances between $p$ and the next seven points in the strip, we can be sure that we will compute the distance between $p$ and all the points in the rectangle. Of course, if fewer than seven points follow $p$ in the list, we compute the distances between $p$ and all of the remaining points. By restricting the search in the strip in this way, the time spent processing the points in the strip is $O(n)$. (Since there are at most $n$ points in the strip, the time spent processing the points in the strip is at most $7n$.)

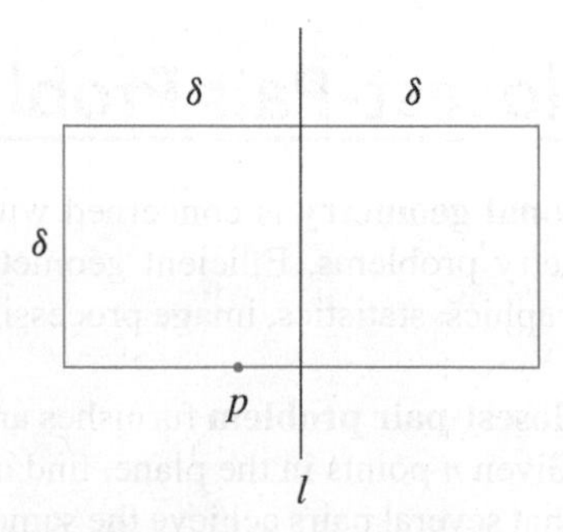


**Figure 7.4.2** Any point $q$ following $p$ whose distance to $p$ is less than $\delta$ must lie within the rectangle.

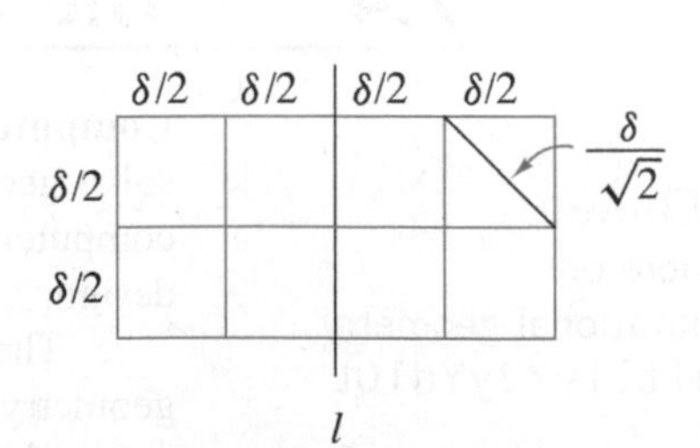


**Figure 7.4.3** The large rectangle contains at most eight points because each square contains at most one point.

We show that the rectangle of Figure 7.4.2 contains at most eight points. Figure 7.4.3 shows the rectangle of Figure 7.4.2 divided into eight equal squares. Notice that the length of a diagonal of a square is

$$\left(\left(\frac{\delta}{2}\right)^2 + \left(\frac{\delta}{2}\right)^2\right)^{1/2} = \frac{\delta}{\sqrt{2}} < \delta;$$

thus each square contains at most one point. Therefore, the $2\delta \times \delta$ rectangle contains at most eight points.

**Example 7.4.1** Show how the closest-pair algorithm finds a closest pair for the input of Figure 7.4.1.

**SOLUTION** The algorithm begins by finding a vertical line $l$ that divides the points into two equal parts,

$$S_1 = \{1, 2, 3, 4, 5, 6, 7\}, \qquad S_2 = \{8, 9, 10, 11, 12, 13, 14\}.$$

For this input there are many possible choices for the dividing line. The particular line chosen here happens to go through point 7.

Next we recursively solve the problem for $S_1$ and $S_2$. The closest pair of points in $S_1$ is 1 and 3. We let $\delta_L$ denote the distance between points 1 and 3. The closest pair of points in $S_2$ is 9 and 12. We let $\delta_R$ denote the distance between points 9 and 12. We let

$$\delta = \min\{\delta_L, \delta_R\} = \delta_L.$$

We next order the points in the vertical strip of width $2\delta$ centered at $l$ in nondecreasing order of their $y$-coordinates:

$$9, 12, 4, 10, 7, 5, 11, 6, 8.$$

We then examine the points in this order. We compute the distances between each point and the following seven points, or between each point and the remaining points if fewer than seven points follow it.

We first compute the distances from 9 to each of 12, 4, 10, 7, 5, 11, and 6. Since each of these distances exceeds $\delta$, at this point we have not found a closer pair.

We next compute the distances from 12 to each of 4, 10, 7, 5, 11, 6, and 8. Since each of these distances exceeds $\delta$, at this point we still have not found a closer pair.

We next compute the distances from 4 to each of 10, 7, 5, 11, 6, and 8. Since each of these distances exceeds $\delta$, at this point we still have not found a closer pair.

We next compute the distances from 10 to each of 7, 5, 11, 6, and 8. Since the distance between points 10 and 7 is less than $\delta$, we have discovered a closer pair. We update $\delta$ to the distance between points 10 and 7.

We next compute the distances from 7 to each of 5, 11, 6, and 8. Since each of these distances exceeds $\delta$, we have not found a closer pair.

We next compute the distances from 5 to each of 11, 6, and 8. Since each of these distances exceeds $\delta$, we have not found a closer pair.

We next compute the distances from 11 to each of 6 and 8. Since each of these distances exceeds $\delta$, we have not found a closer pair.

We next compute the distance from 6 to 8. Since the distance between points 6 and 8 is less than $\delta$, we have discovered a closer pair. We update $\delta$ to the distance between points 6 and 8. Since there are no more points in the strip to consider, the algorithm terminates. The closest pair is 6 and 8 and the distance between them is $\delta$. ◀

Before we give a formal statement of the closest-pair algorithm, there are several technical points to resolve.

In order to terminate the recursion, we check the number of points in the input and if there are three or fewer points, we find a closest pair directly. Dividing the input and using recursion only if there are four or more points ensures that each of the two parts contains at least one pair of points and, therefore, that there is a closest pair in each part.

Before invoking the recursive procedure, we sort the entire set of points by $x$-coordinate. This makes it easy to divide the points into two nearly equal parts.

We use merge sort (see Section 7.3) to sort by $y$-coordinate. However, instead of sorting each time we examine points in the vertical strip, we assume as in merge sort that each half is sorted by $y$-coordinate. Then we simply merge the two halves to sort all of the points by $y$-coordinate.

We can now formally state the closest-pair algorithm. To simplify the description, our version outputs only the distance between a closest pair, not a closest pair. We leave this enhancement as an exercise (Exercise 5).

**Go Online**
For more on the closest-pair algorithm, see bit.ly/2yYdlQL

**Go Online**
For a C++ program implementation of this algorithm, see bit.ly/2ytRD86

### Algorithm 7.4.2 Finding the Distance Between a Closest Pair of Points

Input: $p_1, \ldots, p_n$ ($n \geq 2$ points in the plane)

Output: $\delta$, the distance between a closest pair of points

```
closest_pair(p, n) {
  sort p_1, ..., p_n by x-coordinate
  return rec_cl_pair(p, 1, n)
}
rec_cl_pair(p, i, j) {
  // the input is the sequence p_i, ..., p_j of points in the plane
  // sorted by x-coordinate

  // at termination of rec_cl_pair, the sequence is sorted by
  // y-coordinate

  // rec_cl_pair returns the distance between a closest pair
  // in the input

  // denote the x-coordinate of point p by p.x

  // trivial case (3 or fewer points)
  if (j − i < 3) {
     sort p_i, ..., p_j by y-coordinate
     directly find the distance δ between a closest pair
     return δ
  }
  // divide
  k = ⌊(i + j)/2⌋
  l = p_k.x
  δ_L = rec_cl_pair(p, i, k)
  δ_R = rec_cl_pair(p, k + 1, j)
  δ = min{δ_L, δ_R}

  // p_i, ..., p_k is now sorted by y-coordinate
  // p_{k+1}, ..., p_j is now sorted by y-coordinate
  merge p_i, ..., p_k and p_{k+1}, ..., p_j by y-coordinate
  // assume that the result of the merge is stored back in p_i, ..., p_j
```

```
      // p_i, ..., p_j is now sorted by y-coordinate

      // store points in the vertical strip in v
      t = 0
      for k = i to j
        if (p_k.x > l - δ ∧ p_k.x < l + δ) {
          t = t + 1
          v_t = p_k
        }
      // points in strip are v_1, ..., v_t
      // look for closer pairs in strip
      // compare each to next seven points
      for k = 1 to t - 1
        for s = k + 1 to min{t, k + 7}
          δ = min{δ, dist(v_k, v_s)}
      return δ
   }
```

We now consider the worst-case time required by the closest-pair algorithm. The procedure *closest_pair* begins by sorting the points by $x$-coordinate. If we use an optimal sort (e.g., merge sort), the worst-case sorting time will be $\Theta(n \lg n)$. Next *closest_pair* invokes *rec_cl_pair*. We let $a_n$ denote the worst-case time of *rec_cl_pair* for input of size $n$. If $n > 3$, *rec_cl_pair* first invokes itself with input size $\lfloor n/2 \rfloor$ and $\lfloor (n+1)/2 \rfloor$. Merging the points, extracting the points in the strip, and checking the distances in the strip each takes time $O(n)$. Thus we obtain the recurrence

$$a_n \le a_{\lfloor n/2 \rfloor} + a_{\lfloor (n+1)/2 \rfloor} + cn \qquad n > 3.$$

This is essentially the same recurrence that merge sort satisfies, so we expect that *rec_cl_pair* has the same worst-case time $O(n \lg n)$ as merge sort. Since the worst-case time of sorting the points by $x$-coordinate is $\Theta(n \lg n)$, we expect that the worst-case time of *closest_pair* is $\Theta(n \lg n)$. While this is true, the gap in the argument is that in the proof that merge sort has worst-case time $\Theta(n \lg n)$, we used the fact that, in the worst case, the recurrence for merge sort is actually $a_n = a_{\lfloor n/2 \rfloor} + a_{\lfloor (n+1)/2 \rfloor} + n - 1$. Notice that in the recurrence, "$=$" has replaced "$\le$." We leave the technical details of filling the gap to the exercises (see Exercises 18–21).

Making reasonable assumptions about what kinds of computations are allowed and using a decision tree model together with advanced methods, Preparata (see [Preparata, 1985: Theorem 5.2, page 188]) shows that any algorithm that solves the closest-pair problem is $\Omega(n \lg n)$; thus Algorithm 7.4.2 is asymptotically optimal.

It can be shown (Exercise 10) that there are at most six points in the rectangle of Figure 7.4.2 when the base is included and the other sides are excluded. This result is the best possible since it is possible to place six points in the rectangle (Exercise 8). By considering the possible locations of the points in the rectangle, D. Lerner and R. Johnsonbaugh have shown that it suffices to compare each point in the strip with the next three points (rather than the next seven). This result is the best possible since checking the next two points does not lead to a correct algorithm (Exercise 7).

## 7.4 Review Exercises

1. What is computational geometry?
2. State the closest-pair problem.
3. Describe a brute-force closest-pair algorithm.
4. Describe the divide-and-conquer closest-pair algorithm.
5. Compare the worst-case times of the brute-force and divide-and-conquer closest-pair algorithms.

## 7.4 Exercises

**1.** Describe how the closest-pair algorithm finds the closest pair of points if the input is (8, 4), (3, 11), (12, 10), (5, 4), (1, 2), (17, 10), (8, 7), (8, 9), (11, 3), (1, 5), (11, 7), (5, 9), (1, 9), (7, 6), (3, 7), (14, 7).

**2.** What can you conclude about input to the closest-pair algorithm when the output is zero for the distance between a closest pair?

**3.** Give an example of input for which the closest-pair algorithm puts some points on the dividing line $l$ into the left half and other points on $l$ into the right half.

**4.** Explain why in some cases, when dividing a set of points by a vertical line into two nearly equal parts, it is necessary for the line to contain some of the points.

**5.** Write a closest-pair algorithm that finds one closest pair as well as the distance between the pair of points.

**6.** Write an algorithm that finds the distance between a closest pair of points on a (straight) line.

**7.** Give an example of input for which comparing each point in the strip with the next two points gives incorrect output.

**8.** Give an example to show that it is possible to place six points in the rectangle of Figure 7.4.2 when the base is included and the other sides are excluded.

**9.** When we compute the distances between a point $p$ in the strip and points following it, can we stop computing distances from $p$ if we find a point $q$ such that the distance between $p$ and $q$ exceeds $\delta$? Explain.

**★10.** Show that there are at most six points in the rectangle of Figure 7.4.2 when the base is included and the other sides are excluded.

**11.** Write a $\Theta(n \lg n)$ algorithm that finds the distance $\delta$ between a closest pair and, if $\delta > 0$, also finds *all* pairs $\delta$ apart.

**12.** Write a $\Theta(n \lg n)$ algorithm that finds the distance $\delta$ between a closest pair and *all* pairs less than $2\delta$ apart.

**13.** Write a $\Theta(n \lg n)$ algorithm that finds the distance $\delta$ between a closest pair and all pairs less than $2\delta$ apart, where each pair less than $2\delta$ apart is listed exactly one time.

**14.** Explain why the following algorithm does not find the distance $\delta$ between a closest pair and all pairs less than $2\delta$ apart.

```
exercise14(p, n) {
  δ = closest_pair(p, n)
  for i = 1 to n − 1
    for j = i + 1 to min{n, i + 31}
      if (dist(p_i, p_j) < 2 * δ)
        println(i + " " + j)
}
```

**15.** Write an algorithm that combines finding points in the strip and searching in the strip. In this way, the strip size changes dynamically so that, in general, this version checks fewer points in the strip than does Algorithm 7.4.2.

**16.** Suppose that the distance between a closest pair in a set of $n$ points in the plane is $\delta > 0$. Show that the number of pairs exactly $\delta$ apart is $O(n)$.

**17.** Suppose that the distance between a closest pair in a set of $n$ points in the plane is $\delta > 0$. Show that the number of pairs less than $2\delta$ apart is $O(n)$.

*Let $\{a_n\}_{n=2}^{\infty}$ be a sequence satisfying the recurrence*

$$a_n \le a_{\lfloor n/2 \rfloor} + a_{\lfloor (n+1)/2 \rfloor} + cn,$$

*for some positive constant c and for all $n > 3$.*

*Define the sequence $\{t_n\}_{n=2}^{\infty}$ by the rules $t_2 = a_2$, $t_3 = \max\{a_2, a_3\}$, and*

$$t_n = t_{\lfloor n/2 \rfloor} + t_{\lfloor (n+1)/2 \rfloor} + cn \qquad n > 3.$$

*Exercises 18–21 deal with these recurrences.*

**18.** Prove that $a_n \le t_n$ for all $n \ge 2$.

**19.** Prove that the sequence $\{t_n\}_{n=2}^{\infty}$ is nondecreasing.

**20.** Prove that $t_n = \Theta(n \lg n)$.

**21.** Deduce that *rec_cl_pair* has worst-case time $O(n \lg n)$.

## Chapter 7 Notes

Recurrence relations are treated more fully in [Liu, 1985; Roberts; and Tucker]. Several applications to the analysis of algorithms are presented in [Johnsonbaugh].

[Cull] gives algorithms for solving certain Tower of Hanoi problems with minimum space and time complexity. [Hinz] is a comprehensive discussion of the Tower of Hanoi with 50 references.

The cobweb in economics first appeared in [Ezekiel].

All data structures and algorithms books have extended discussions of searching and sorting (see, e.g., [Brassard; Cormen; Johnsonbaugh; Knuth, 1998b; Kruse; and Nyhoff]).

[de Berg; Preparata, 1985; and Edelsbrunner] are books on computational geometry. The closest-pair algorithm in Section 7.4 is due to M. I. Shamos and appears in [Preparata, 1985]. [Preparata, 1985] also gives a $\Theta(n \lg n)$ algorithm to find the closest pair in an arbitrary number of dimensions.

Recurrence relations are also called *difference equations*. [Goldberg] contains a discussion of difference equations and applications.

## Chapter 7 Review

### Section 7.1

1. Recurrence relation
2. Initial condition
3. Compound interest
4. Tower of Hanoi
5. Cobweb in economics
6. Ackermann's function

### Section 7.2

7. Solving a recurrence relation by iteration
8. $n$th-order, linear homogeneous recurrence relation with constant coefficients and how to solve a second-order recurrence relation
9. Population growth

### Section 7.3

10. How to find a recurrence relation that describes the time required by a recursive algorithm
11. Selection sort
12. Binary search
13. Merging sequences
14. Merge sort

### Section 7.4

15. Computational geometry
16. Closest-pair problem
17. Closest-pair algorithm

## Chapter 7 Self-Test

1. Answer parts (a)–(c) for the sequence defined by the rules:

   1. The first term is 3.
   2. The $n$th term is $n$ plus the previous term.

   (a) Write the first four terms of the sequence.

   (b) Find an initial condition for the sequence.

   (c) Find a recurrence relation for the sequence.

2. Is the recurrence relation

$$a_n = a_{n-1} + a_{n-3}$$

   a linear homogeneous recurrence relation with constant coefficients?

3. Assume that a person invests \$4000 at 17 percent interest compounded annually. Let $A_n$ represent the amount at the end of $n$ years. Find a recurrence relation and an initial condition for the sequence $A_0, A_1, \ldots$.

4. Let $P_n$ be the number of partitions of an $n$-element set. Show that the sequence $P_0, P_1, \ldots$ satisfies the recurrence relation

$$P_n = \sum_{k=0}^{n-1} C(n-1, k)P_k.$$

*In Exercises 5 and 6, solve the recurrence relation subject to the initial conditions.*

5. $a_n = -4a_{n-1} - 4a_{n-2}$;  $a_0 = 2$,  $a_1 = 4$
6. $a_n = 3a_{n-1} + 10a_{n-2}$;  $a_0 = 4$,  $a_1 = 13$
7. Suppose that we have a $2 \times n$ rectangular board divided into $2n$ squares. Let $a_n$ denote the number of ways to exactly cover this board by $1 \times 2$ dominoes. Show that the sequence $\{a_n\}$ satisfies the recurrence relation

$$a_n = a_{n-1} + a_{n-2}.$$

   Show that $a_n = f_{n+1}$, where $\{f_n\}$ is the Fibonacci sequence.

8. Let $c_n$ denote the number of strings over $\{0, 1, 2\}$ of length $n$ that contain an even number of 1's. Write a recurrence relation and initial condition that define the sequence $c_1, c_2, \ldots$. Solve the recurrence relation to obtain an explicit formula for $c_n$.

*Exercises 9–12 refer to the following algorithm.*

**Algorithm**
*Polynomial Evaluation*
This algorithm evaluates the polynomial

$$p(x) = \sum_{k=0}^{n} c_k x^{n-k}$$

at the point $t$.

Input: The sequence of coefficients $c_0, c_1, \ldots, c_n$, the value $t$, and $n$

Output: $p(t)$

```
poly(c, n, t) {
  if (n == 0)
    return c_0
  return t * poly(c, n − 1, t) + c_n
}
```

*Let $b_n$ be the number of multiplications required to compute $p(t)$.*

**9.** Find a recurrence relation and an initial condition for the sequence $\{b_n\}$.

**10.** Compute $b_1$, $b_2$, and $b_3$.

**11.** Solve the recurrence relation of Exercise 9.

**12.** Suppose that we compute $p(t)$ by a straightforward technique that requires $n - k$ multiplications to compute $c_k t^{n-k}$. How many multiplications would be required to compute $p(t)$? Would you prefer this method or the preceding algorithm? Explain.

**13.** Describe how the closest-pair algorithm finds the closest pair of points for the input (10, 1), (7, 7), (3, 13), (6, 10), (16, 4), (10, 5), (7, 13), (13, 8), (4, 4), (2, 2), (1, 8), (10, 13), (7, 1), (4, 8), (12, 3), (16, 10), (14, 5), (10, 9).

**14.** In order to terminate the recursion in the closest-pair algorithm, if there are three or fewer points in the input we find the closest pair directly. Why can't we replace "three" by "two"?

**15.** Show that there are at most four points in the lower half of the rectangle of Figure 7.4.3.

**16.** What would the worst-case time of the closest-pair algorithm be if instead of merging $p_i, \ldots, p_k$ and $p_{k+1}, \ldots, p_j$, we used merge sort to sort $p_i, \ldots, p_j$?

## Chapter 7 Computer Exercises

**1.** Write a program that prints the amount accumulated yearly if a person invests $n$ dollars at $p$ percent compounded annually.

**2.** Write a program that prints the amount accumulated yearly if a person invests $n$ dollars at $p$ percent annual interest compounded $m$ times yearly.

**3.** Write a program that solves the three-peg Tower of Hanoi puzzle.

**4.** Write a program that solves the four-peg Tower of Hanoi puzzle in fewer moves than does the solution to the three-peg puzzle.

**5.** Write a program to display the cobweb of economics.

**6.** Write a program to compute Ackermann's function.

**7.** Write a program that prints a solution to the $n$-node communication program (see Exercise 53, Section 7.1).

**8.** Write a program that prints all the ways to spend $n$ dollars under the conditions of Exercise 55, Section 7.1.

**9.** Write a program that prints all $n$-bit strings that do not contain the pattern 010.

**10.** Write a program to compute the Lucas sequence (see Exercise 68, Section 7.1).

**11.** Write a program to list all rise/fall permutations of length $n$ (see the definition before Exercise 74, Section 7.1).

**12.** Implement Algorithms 7.3.1 (selection sort) and 7.3.8 (merge sort) and other sorting algorithms as programs and compare the times needed to sort $n$ items.

**13.** Implement binary search (Algorithm 7.3.2) as a program. Measure the time used by the program for various keys and for various values of $n$. Compare these results with the theoretical estimate for the worst-case time $\Theta(\lg n)$.

**14.** Write a program to merge two sequences.

**15.** Implement a computation of $a^n$ nonrecursively that uses repeated multiplication and Algorithms 7.3.11 and 7.3.12 as computer programs, and compare the times needed to execute each.

**16.** Implement the methods of computing the largest and smallest elements in an array (see Exercises 50–60, Section 7.3) and compare the times needed to execute each.

**17.** Implement the closest-pair algorithm (Algorithm 7.4.2) as a program. The program should find not only the closest distance but also a closest pair.

**18.** Write a program to find all closest pairs in the plane.

**19.** Write a program to find the closest-pair in three-dimensional space. An algorithm is given in [Preparata].

# Chapter 8

# GRAPH THEORY



Although the first paper in graph theory goes back to 1736 (see Example 8.2.16) and several important results in graph theory were obtained in the nineteenth century, it is only since the 1920s that there has been a sustained, widespread, intense interest in graph theory. Indeed, the first text on graph theory ([König]) appeared in 1936. Undoubtedly, one of the reasons for the recent interest in graph theory is its applicability in many diverse fields, including computer science, chemistry, operations research, electrical engineering, linguistics, and economics.

We begin with some basic graph terminology and examples. We then discuss some important concepts in graph theory, including paths and cycles. Two classical graph problems, the existence of Hamiltonian cycles and the traveling salesperson problem, are then considered. A shortest-path algorithm is presented that efficiently finds a shortest path between two given points. After presenting ways of representing graphs, we study the question of when two graphs are essentially the same (i.e., when two graphs are isomorphic) and when a graph can be drawn in the plane without having any of its edges cross. We conclude by presenting a solution based on a graph model to the Instant Insanity puzzle.

## 8.1 Introduction

**Go Online**
For more on graph theory, see
bit.ly/2yUuHxX

Figure 8.1.1 shows the highway system in Wyoming that a particular person is responsible for inspecting. Specifically, this road inspector must travel all of these roads and file reports on road conditions, visibility of lines on the roads, status of traffic signs, and so on. Since the road inspector lives in Greybull, the most economical way to inspect all of the roads would be to start in Greybull, travel each of the roads exactly once, and return to Greybull. Is this possible? See if you can decide before reading on.

The problem can be modeled as a **graph.** In fact, since graphs are drawn with dots and lines, they look like road maps. In Figure 8.1.2, we have drawn a graph $G$ that models the map of Figure 8.1.1. The dots in Figure 8.1.2 are called **vertices,** and the lines that connect the vertices are called **edges.** (Later in this section we will define all of these terms carefully.) We have labeled each vertex with the first three letters of the city to which it corresponds. We have labeled the edges $e_1, \ldots, e_{13}$. When we draw a graph,

†This section can be omitted without loss of continuity.

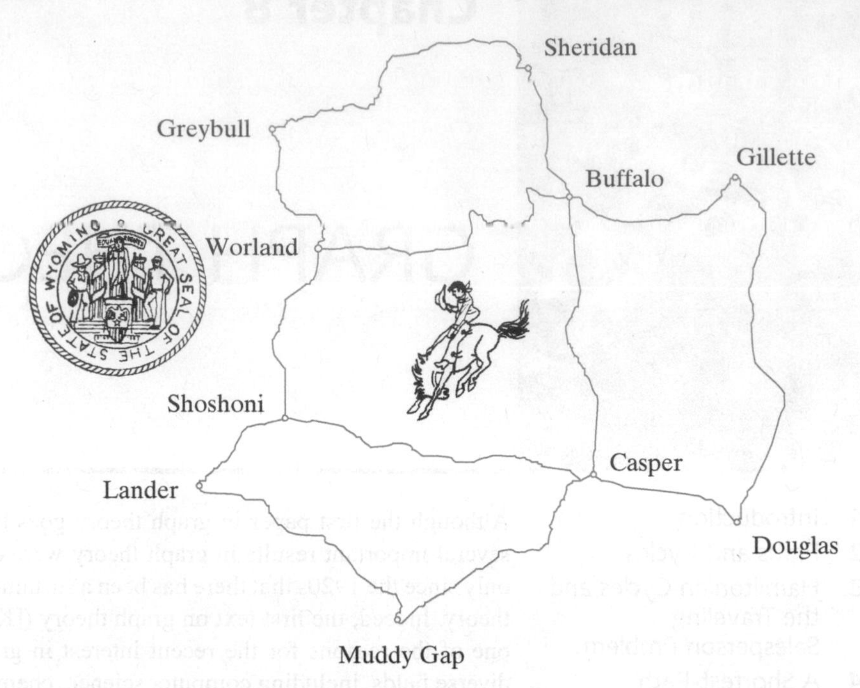


**Figure 8.1.1** Part of the Wyoming highway system.

the only information of importance is which vertices are connected by which edges. For this reason, the graph of Figure 8.1.2 could just as well be drawn as in Figure 8.1.3.

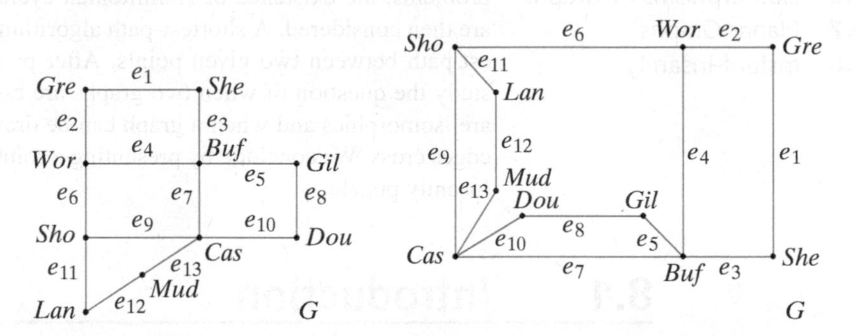


**Figure 8.1.2** A graph model of the highway system shown in Figure 8.1.1.

**Figure 8.1.3** An alternative, but equivalent, graph model of the highway system shown in Figure 8.1.1.

If we start at a vertex $v_0$, travel along an edge to vertex $v_1$, travel along another edge to vertex $v_2$, and so on, and eventually arrive at vertex $v_n$, we call the complete tour a **path** from $v_0$ to $v_n$. The path that starts at *She*, then goes to *Buf*, and ends at *Gil* corresponds to a trip on the map of Figure 8.1.1 that begins in Sheridan, goes to Buffalo, and ends at Gillette. The road inspector's problem can be rephrased for the graph model $G$ in the following way: Is there a path from vertex *Gre* to vertex *Gre* that traverses every edge exactly once?

We can show that the road inspector cannot start in Greybull, travel each of the roads exactly once, and return to Greybull. To put the answer in graph terms, there is no path from vertex *Gre* to vertex *Gre* in Figure 8.1.2 that traverses every edge exactly once. To see this, suppose that there is such a path and consider vertex *Wor*. Each time we arrive at *Wor* on some edge, we must leave *Wor* on a different edge. Furthermore, every edge that touches *Wor* must be used. Thus the edges at *Wor* occur in pairs. It follows that an even number of edges must touch *Wor*. Since three edges touch *Wor*, we have a contradiction. Therefore, there is no path from vertex *Gre* to vertex *Gre* in Figure 8.1.2 that traverses every edge exactly once. The argument applies to an arbitrary graph $G$. If $G$ has a path from vertex $v$ to $v$ that traverses every edge exactly once, an even number of edges must touch each vertex. We discuss this problem in greater detail in Section 8.2.

At this point we give some formal definitions.

**Definition 8.1.1 ▶** A *graph* (or *undirected graph*) $G$ consists of a set $V$ of *vertices* (or *nodes*) and a set $E$ of *edges* (or *arcs*) such that each edge $e \in E$ is associated with an unordered pair of vertices. If there is a unique edge $e$ associated with the vertices $v$ and $w$, we write $e = (v, w)$ or $e = (w, v)$. In this context, $(v, w)$ denotes an edge between $v$ and $w$ in an undirected graph and *not* an ordered pair.

A *directed graph* (or *digraph*) $G$ consists of a set $V$ of *vertices* (or *nodes*) and a set $E$ of *edges* (or *arcs*) such that each edge $e \in E$ is associated with an ordered pair of vertices. If there is a unique edge $e$ associated with the ordered pair $(v, w)$ of vertices, we write $e = (v, w)$, which denotes an edge from $v$ to $w$.

An edge $e$ in a graph (undirected or directed) that is associated with the pair of vertices $v$ and $w$ is said to be *incident on* $v$ *and* $w$, and $v$ and $w$ are said to be *incident on* $e$ and to be *adjacent vertices*.

If $G$ is a graph (undirected or directed) with vertices $V$ and edges $E$, we write $G = (V, E)$.

Unless specified otherwise, the sets $E$ and $V$ are assumed to be finite and $V$ is assumed to be nonempty. ◀

**Go Online**

For a C program to generate random graphs with specified properties, see bit.ly/2PenwuY

**Example 8.1.2** In Figure 8.1.2 the (undirected) graph $G$ consists of the set

$$V = \{Gre, She, Wor, Buf, Gil, Sho, Cas, Dou, Lan, Mud\}$$

of vertices and the set

$$E = \{e_1, e_2, \ldots, e_{13}\}$$

of edges. Edge $e_1$ is associated with the unordered pair {*Gre, She*} of vertices, and edge $e_{10}$ is associated with the unordered pair {*Cas, Dou*} of vertices. Edge $e_1$ is denoted (*Gre, She*) or (*She, Gre*), and edge $e_{10}$ is denoted (*Cas, Dou*) or (*Dou, Cas*). Edge $e_4$ is incident on *Wor* and *Buf*, and the vertices *Wor* and *Buf* are adjacent. ◀

**Example 8.1.3** A directed graph is shown in Figure 8.1.4. The directed edges are indicated by arrows. Edge $e_1$ is associated with the ordered pair $(v_2, v_1)$ of vertices, and edge $e_7$ is associated with the ordered pair $(v_6, v_6)$ of vertices. Edge $e_1$ is denoted $(v_2, v_1)$, and edge $e_7$ is denoted $(v_6, v_6)$. ◀

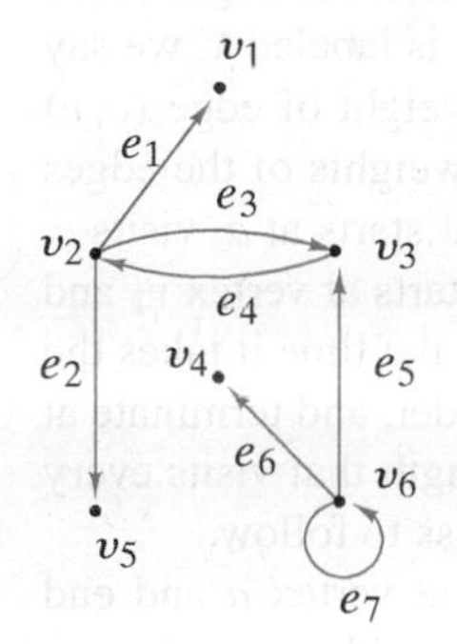

**Figure 8.1.4** A directed graph.

Definition 8.1.1 allows distinct edges to be associated with the same pair of vertices. For example, in Figure 8.1.5, edges $e_1$ and $e_2$ are both associated with the vertex pair $\{v_1, v_2\}$. Such edges are called **parallel edges.** An edge incident on a single vertex

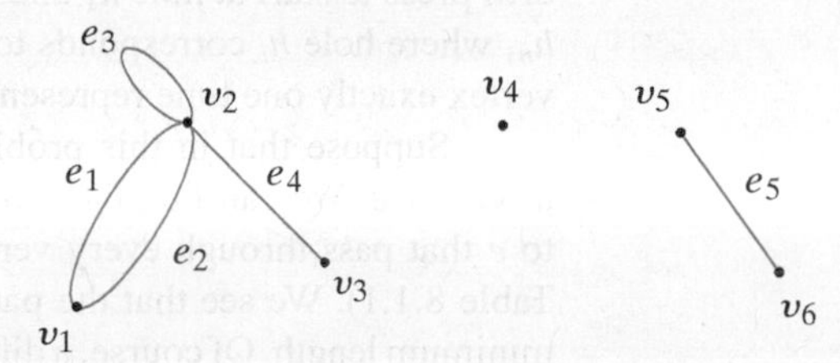

**Figure 8.1.5** A graph with parallel edges and loops.

is called a **loop.** For example, in Figure 8.1.5, edge $e_3 = (v_2, v_2)$ is a loop. A vertex, such as vertex $v_4$ in Figure 8.1.5, that is not incident on any edge is called an **isolated vertex.** A graph with neither loops nor parallel edges is called a **simple graph.**

**Example 8.1.4** Since the graph of Figure 8.1.2 has neither parallel edges nor loops, it is a simple graph. ◀

Some authors do not permit loops and parallel edges when they define graphs. One would expect that if agreement has not been reached on the definition of "graph," most other terms in graph theory would also not have standard definitions. This is indeed the case. In reading articles and books about graphs, it is necessary to check on the definitions being used.

We turn next to an example that shows how a graph model can be used to analyze a manufacturing problem.

**Example 8.1.5** Frequently in manufacturing, it is necessary to bore many holes in sheets of metal (see Figure 8.1.6). Components can then be bolted to these sheets of metal. The holes can be drilled using a drill press under the control of a computer. To save time and money, the drill press should be moved as quickly as possible. Model the situation as a graph.

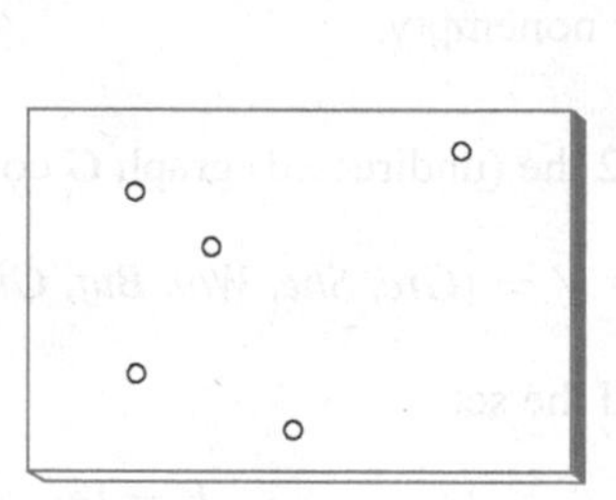

**Figure 8.1.6** A sheet of metal with holes for bolts.

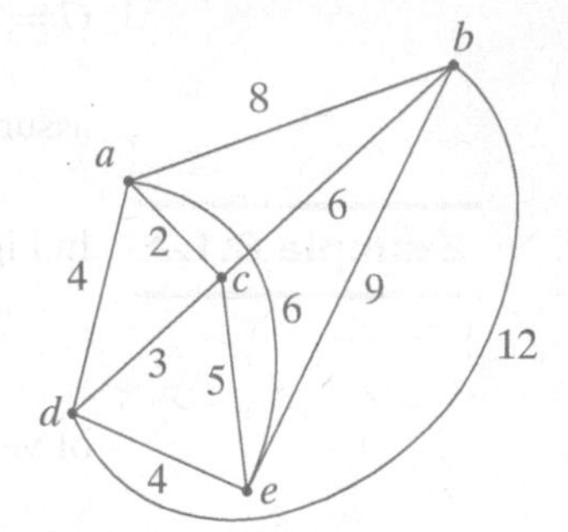


**Figure 8.1.7** A graph model of sheet metal in Figure 8.1.6. The edge weight is the time to move the drill press.

**SOLUTION** The vertices of the graph correspond to the holes (see Figure 8.1.7). Every pair of vertices is connected by an edge. We write on each edge the time to move the drill press between the corresponding holes. A graph with numbers on the edges (such as the graph of Figure 8.1.7) is called a **weighted graph.** If edge $e$ is labeled $k$, we say that the **weight of edge** $e$ is $k$. For example, in Figure 8.1.7 the weight of edge $(c, e)$ is 5. In a weighted graph, the **length of a path** is the sum of the weights of the edges in the path. For example, in Figure 8.1.7 the length of the path that starts at $a$, visits $c$, and terminates at $b$ is 8. In this problem, the length of a path that starts at vertex $v_1$ and then visits $v_2, v_3, \ldots,$ in this order, and terminates at $v_n$ represents the time it takes the drill press to start at hole $h_1$ and then move to $h_2, h_3, \ldots,$ in this order, and terminate at $h_n$, where hole $h_i$ corresponds to vertex $v_i$. A path of minimum length that visits every vertex exactly one time represents the optimal path for the drill press to follow.

Suppose that in this problem the path is required to begin at vertex $a$ and end at vertex $e$. We can find the minimum-length path by listing all possible paths from $a$ to $e$ that pass through every vertex exactly one time and choose the shortest one (see Table 8.1.1). We see that the path that visits the vertices $a, b, c, d, e$, in this order, has minimum length. Of course, a different pair of starting and ending vertices might produce an even shorter path.

**TABLE 8.1.1** ■ Paths in the Graph of Figure 8.1.7 from $a$ to $e$ That Pass Through Every Vertex Exactly One Time, and Their Lengths

| *Path* | *Length* |
|---|---|
| $a, b, c, d, e$ | 21 |
| $a, b, d, c, e$ | 28 |
| $a, c, b, d, e$ | 24 |
| $a, c, d, b, e$ | 26 |
| $a, d, b, c, e$ | 27 |
| $a, d, c, b, e$ | 22 |

Listing all paths from vertex $v$ to vertex $w$, as we did in Example 8.1.5, is a rather time-consuming way to find a minimum-length path from $v$ to $w$ that visits every vertex exactly one time. Unfortunately, no one knows a method that is much more practical for arbitrary graphs. This problem is a version of the **traveling salesperson problem.** We discuss that problem in Section 8.3.

**Example 8.1.6** **Bacon Numbers** Actor Kevin Bacon has appeared in numerous films including *Diner* and *Apollo 13.* Actors who have appeared in a film with Bacon are said to have *Bacon number one*. For example, Ellen Barkin has Bacon number one because she appeared with Bacon in *Diner*. Actors who did not appear in a film with Bacon but who appeared in a film with an actor whose Bacon number is one are said to have *Bacon number two*. Higher Bacon numbers are defined similarly. For example, Bela Lugosi has Bacon number three. Lugosi was in *Black Friday* with Emmett Vogan, Vogan was in *With a Song in My Heart* with Robert Wagner, and Wagner was in *Wild Things* with Bacon. Develop a graph model for Bacon numbers.

SOLUTION We let vertices denote actors, and we place one edge between two distinct actors if they appeared in at least one film together (see Figure 8.1.8). In an unweighted

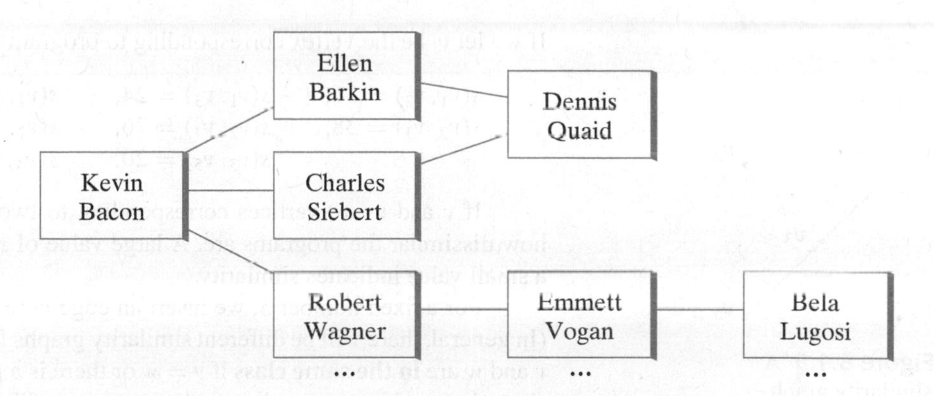


**Figure 8.1.8** Part of a graph that models Bacon numbers. The vertices denote actors. There is an edge between two distinct actors if they appeared in at least one film together. For example, there is an edge between Ellen Barkin and Dennis Quaid because they both appeared in *The Big Easy*. An actor's Bacon number is the length of a shortest path from that actor to Bacon. For example, Bela Lugosi's Bacon number is three because the length of a shortest path from Lugosi to Bacon is three.

graph, the length of a path is the number of edges in the path. Thus an actor's Bacon number is the length of a shortest path from the vertex corresponding to that actor to the vertex corresponding to Bacon. In Section 8.4, we discuss the general problem of finding shortest paths in graphs. Unlike the situation in Example 8.1.5, there are efficient algorithms for finding shortest paths.

**Go Online**
For more on Bacon numbers, see
`bit.ly/2yUuHxX`

It is interesting that *most* actors, even actors who died many years ago, have Bacon numbers of three or less. See Exercise 30 for a similar graph model. ◀

**Example 8.1.7** **Similarity Graphs** This example deals with the problem of grouping "like" objects into classes based on properties of the objects. For example, suppose that a particular algorithm is implemented in C++ by a number of persons and we want to group "like" programs into classes based on certain properties of the programs (see Table 8.1.2). Suppose that we select as properties

1. The number of lines in the program
2. The number of `return` statements in the program
3. The number of function calls in the program

**TABLE 8.1.2 ■** C++ Programs That Implement the Same Algorithm

| *Program* | *Number of Program Lines* | *Number of* `return` *Statements* | *Number of Function Calls* |
|---|---|---|---|
| 1 | 66 | 20 | 1 |
| 2 | 41 | 10 | 2 |
| 3 | 68 | 5 | 8 |
| 4 | 90 | 34 | 5 |
| 5 | 75 | 12 | 14 |

A **similarity graph** $G$ is constructed as follows. The vertices correspond to programs. A vertex is denoted $(p_1, p_2, p_3)$, where $p_i$ is the value of property $i$. We define a **dissimilarity function** $s$ as follows. For each pair of vertices $v = (p_1, p_2, p_3)$ and $w = (q_1, q_2, q_3)$, we set

$$s(v, w) = |p_1 - q_1| + |p_2 - q_2| + |p_3 - q_3|.$$

If we let $v_i$ be the vertex corresponding to program $i$, we obtain

$$\begin{array}{llll} s(v_1, v_2) = 36, & s(v_1, v_3) = 24, & s(v_1, v_4) = 42, & s(v_1, v_5) = 30, \\ s(v_2, v_3) = 38, & s(v_2, v_4) = 76, & s(v_2, v_5) = 48, & s(v_3, v_4) = 54, \\ & s(v_3, v_5) = 20, & s(v_4, v_5) = 46. & \end{array}$$

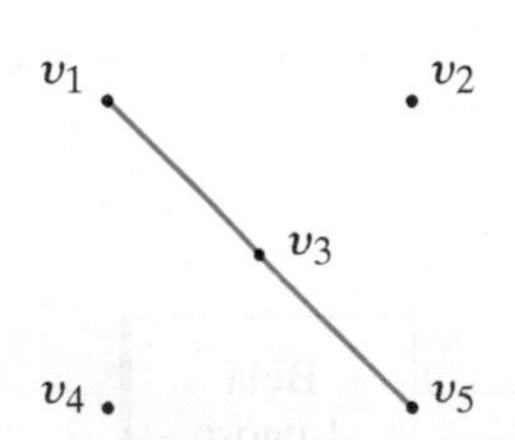

**Figure 8.1.9** A similarity graph corresponding to the programs of Table 8.1.2 with $S = 25$.

If $v$ and $w$ are vertices corresponding to two programs, $s(v, w)$ is a measure of how dissimilar the programs are. A large value of $s(v, w)$ indicates dissimilarity, while a small value indicates similarity.

For a fixed number $S$, we insert an edge between vertices $v$ and $w$ if $s(v, w) < S$. (In general, there will be different similarity graphs for different values of $S$.) We say that $v$ and $w$ are **in the same class** if $v = w$ or there is a path from $v$ to $w$. In Figure 8.1.9 we show the graph corresponding to the programs of Table 8.1.2 with $S = 25$. In this graph, the programs are grouped into three classes: {1, 3, 5}, {2}, and {4}. In a real problem, an appropriate value for $S$ might be selected by trial and error or the value of $S$ might be selected automatically according to some predetermined criteria. ◀

**Example 8.1.8** **The $n$-Cube (Hypercube)** The traditional computer, often called a **serial computer,** executes one instruction at a time. Our definition of "algorithm" also assumes that one instruction is executed at a time. Such algorithms are called **serial algorithms.** As hardware costs have declined, it has become feasible to build **parallel computers** with many processors that are capable of executing several instructions at a time. Graphs are often convenient models to describe these machines. The associated algorithms are known as **parallel algorithms.** Many problems can be solved much faster using parallel computers rather than serial computers. We discuss one model for parallel computation known as the ***n*-cube** or **hypercube.**

**Go Online**
For more on the hypercube, see
bit.ly/2yUuHxX

The $n$-cube has $2^n$ processors, $n \geq 1$, which are represented by vertices (see Figure 8.1.10) labeled $0, 1, \ldots, 2^n - 1$. Each processor has its own local memory. An edge connects two vertices if the binary representation of their labels differs in exactly one bit. During one time unit, all processors in the $n$-cube may execute an instruction simultaneously and then communicate with an adjacent processor. If a processor needs to communicate with a nonadjacent processor, the first processor sends a message that includes the route to, and ultimate destination of, the recipient. It may take several time units for a processor to communicate with a nonadjacent processor.

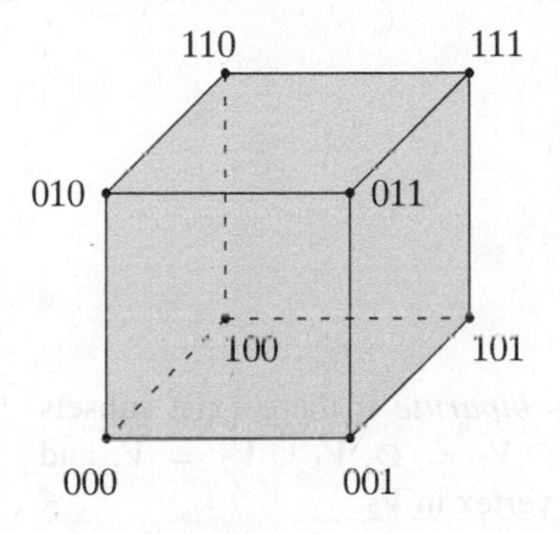

**Figure 8.1.10** The 3-cube.

The $n$-cube may also be described recursively. The 1-cube has two processors, labeled 0 and 1, and one edge. Let $H_1$ and $H_2$ be two $(n-1)$-cubes whose vertices are labeled in binary $0, \ldots, 2^{n-1} - 1$ (see Figure 8.1.11). We place an edge between each pair of vertices, one from $H_1$ and one from $H_2$, provided that the vertices have identical labels. We then change the label $L$ on each vertex in $H_1$ to $0L$ and we change the label $L$ on each vertex in $H_2$ to $1L$. We obtain an $n$-cube (Exercise 39). See Exercises 43–45 for an alternative way to construct the $n$-cube.

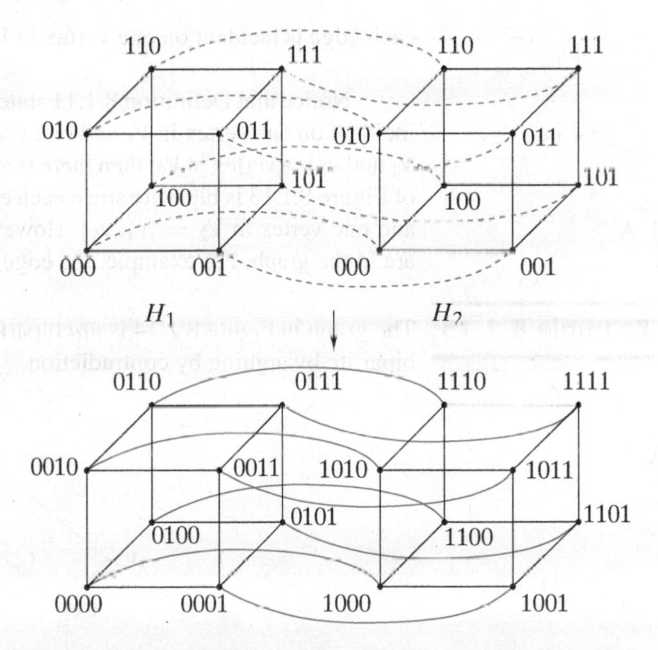

**Figure 8.1.11** Combining two 3-cubes to obtain a 4-cube.

The $n$-cube is an important model of computation because several such machines have been built and are running. Furthermore, several other parallel computation models can be simulated by the hypercube. The latter point is considered in more detail in Examples 8.3.5 and 8.6.3. ◀

We conclude this introductory section by defining some special graphs that appear frequently in graph theory.

**Definition 8.1.9 ▶** The *complete graph on n vertices,* denoted $K_n$, is the simple graph with $n$ vertices in which there is an edge between every pair of distinct vertices. ◀

**Example 8.1.10** The complete graph on four vertices, $K_4$, is shown in Figure 8.1.12.

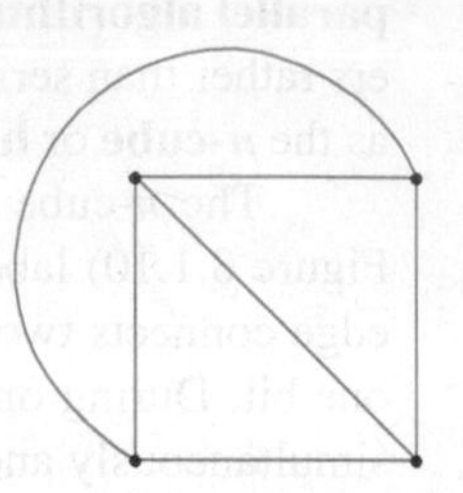

**Figure 8.1.12** The complete graph $K_4$. ◀

**Definition 8.1.11 ▶** A graph $G = (V, E)$ is *bipartite* if there exist subsets $V_1$ and $V_2$ (either possibly empty) of $V$ such that $V_1 \cap V_2 = \varnothing$, $V_1 \cup V_2 = V$, and each edge in $E$ is incident on one vertex in $V_1$ and one vertex in $V_2$. ◀

**Example 8.1.12** The graph in Figure 8.1.13 is bipartite since if we let

$$V_1 = \{v_1, v_2, v_3\} \quad \text{and} \quad V_2 = \{v_4, v_5\},$$

each edge is incident on one vertex in $V_1$ and one vertex in $V_2$. ◀

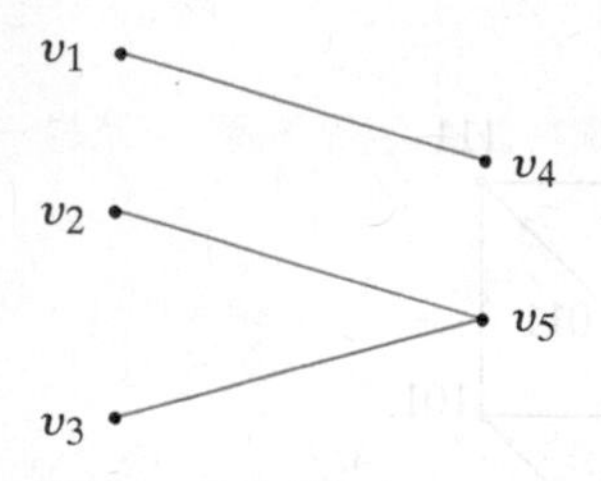


**Figure 8.1.13** A bipartite graph.

Notice that Definition 8.1.11 states that if $e$ is an edge in a bipartite graph, then $e$ is incident on one vertex in $V_1$ and one vertex in $V_2$. It does *not* state that if $v_1$ is a vertex in $V_1$ and $v_2$ is a vertex in $V_2$, then there is an edge between $v_1$ and $v_2$. For example, the graph of Figure 8.1.13 is bipartite since each edge is incident on one vertex in $V_1 = \{v_1, v_2, v_3\}$ and one vertex in $V_2 = \{v_4, v_5\}$. However, not all edges between vertices in $V_1$ and $V_2$ are in the graph. For example, the edge $(v_1, v_5)$ is absent.

**Example 8.1.13** The graph in Figure 8.1.14 is *not* bipartite. It is often easiest to prove that a graph is not bipartite by arguing by contradiction.

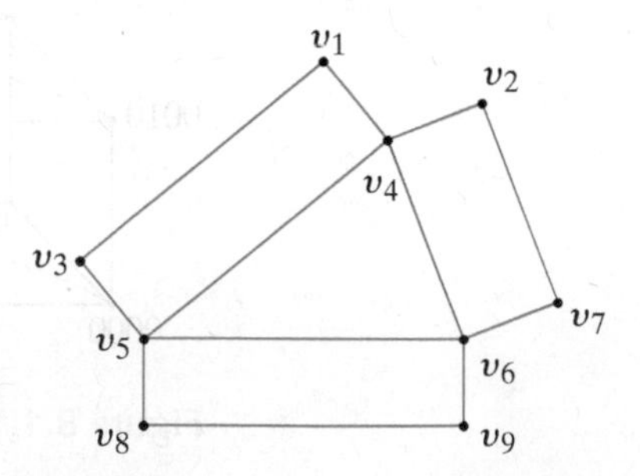


**Figure 8.1.14** A graph that is not bipartite.

Suppose that the graph in Figure 8.1.14 is bipartite. Then the vertex set can be partitioned into two subsets $V_1$ and $V_2$ such that each edge is incident on one vertex in $V_1$ and one vertex in $V_2$. Now consider the vertices $v_4$, $v_5$, and $v_6$. Since $v_4$ and $v_5$ are adjacent, one is in $V_1$ and the other in $V_2$. We may assume that $v_4$ is in $V_1$ and that $v_5$ is in $V_2$. Since $v_5$ and $v_6$ are adjacent and $v_5$ is in $V_2$, $v_6$ is in $V_1$. Since $v_4$ and $v_6$ are adjacent and $v_4$ is in $V_1$, $v_6$ is in $V_2$. But now $v_6$ is in both $V_1$ and $V_2$, which is a contradiction since $V_1$ and $V_2$ are disjoint. Therefore, the graph in Figure 8.1.14 is not bipartite. ◀

**Example 8.1.14** The complete graph $K_1$ on one vertex is bipartite. We may let $V_1$ be the set containing the one vertex and $V_2$ be the empty set. Then each edge (namely none!) is incident on one vertex in $V_1$ and one vertex in $V_2$. ◀

**Definition 8.1.15 ▶** The *complete bipartite graph on m and n vertices,* denoted $K_{m,n}$, is the simple graph whose vertex set is partitioned into sets $V_1$ with $m$ vertices and $V_2$ with $n$ vertices in which the edge set consists of all edges of the form $(v_1, v_2)$ with $v_1 \in V_1$ and $v_2 \in V_2$. ◀

**Example 8.1.16** The complete bipartite graph on two and four vertices, $K_{2,4}$, is shown in Figure 8.1.15. ◀

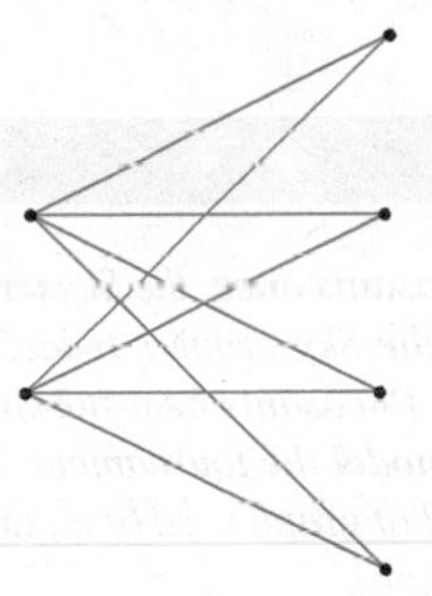

**Figure 8.1.15** The complete bipartite graph $K_{2,4}$.

## 8.1 Problem-Solving Tips

To model a given situation as a graph, you must first decide what the vertices represent. Then an edge between two vertices represents some kind of relation. For example, if several teams play soccer games, we could let the vertices represent the teams. We could then put an edge between two vertices (teams) if the two teams represented by the two vertices played at least one game. The graph would then show which teams have played each other.

To determine whether a graph is bipartite, try to separate the vertices into two disjoint sets $V_1$ and $V_2$ so that each edge is incident on one vertex in one set and one vertex in the other set. If you succeed, the graph is bipartite and you have discovered the sets $V_1$ and $V_2$. If you fail, the graph is not bipartite. To try to separate the vertices into two disjoint sets, pick a start vertex $v$. Put $v \in V_1$. Put all vertices adjacent to $v$ in $V_2$. Pick a vertex $w \in V_2$. Put all vertices adjacent to $w$ in $V_1$. Pick a vertex $v' \in V_2$, $v' \neq v$. Put all vertices adjacent to $v'$ in $V_2$. Continue in this way. If you can put each vertex into either $V_1$ or $V_2$, but not both, the graph is bipartite. If at some point, you are forced to put a vertex into both $V_1$ *and* $V_2$, the graph is not bipartite.

## 8.1 Review Exercises

1. Define *undirected graph.*
2. Give an example of something in the real world that can be modeled by an undirected graph.
3. Define *directed graph.*
4. Give an example of something in the real world that can be modeled by a directed graph.

5. What does it mean for an edge to be *incident on a vertex*?
6. What does it mean for a vertex to be *incident on an edge*?
7. What does it mean for $v$ and $w$ to be *adjacent vertices*?
8. What are parallel edges?
9. What is a loop?
10. What is an isolated vertex?
11. What is a simple graph?
12. What is a weighted graph?
13. Give an example of something in the real world that can be modeled by a weighted graph.
14. Define *length of path* in a weighted graph.
15. What is a similarity graph?
16. Define *n-cube*.
17. What is a serial computer?
18. What is a serial algorithm?
19. What is a parallel computer?
20. What is a parallel algorithm?
21. What is the complete graph on $n$ vertices? How is it denoted?
22. Define *bipartite graph*.
23. What is the complete bipartite graph on $m$ and $n$ vertices? How is it denoted?

## 8.1 Exercises

*In a tournament, the Snow beat the Pheasants once, the Skyscrapers beat the Tuna once, the Snow beat the Skyscrapers twice, the Pheasants beat the Tuna once, and the Pheasants beat the Snow once. In Exercises 1–4, use a graph to model the tournament. The teams are the vertices. Describe the kind of graph used (e.g., undirected graph, directed graph, simple graph).*

1. There is an edge between teams if the teams played.
2. There is an edge from team $t_i$ to team $t_j$ if $t_i$ beat $t_j$ at least one time.
3. There is an edge between teams for each game played.
4. There is an edge from team $t_i$ to team $t_j$ for each victory of $t_i$ over $t_j$.

*Explain why none of the graphs in Exercises 5–7 has a path from a to a that passes through each edge exactly one time.*

5.

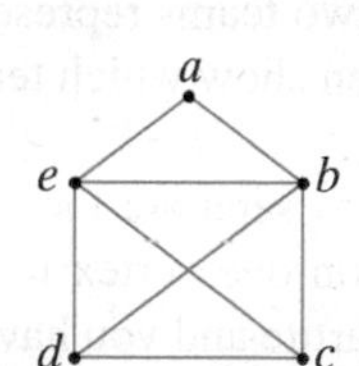


6.

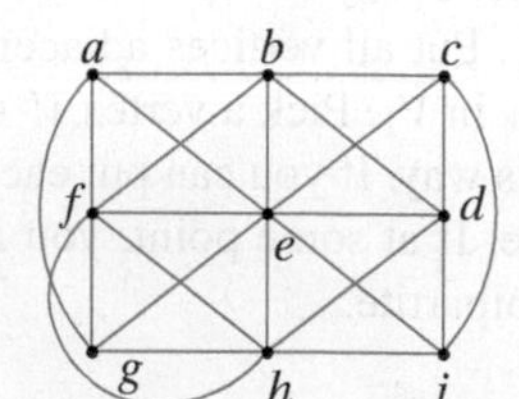


7.

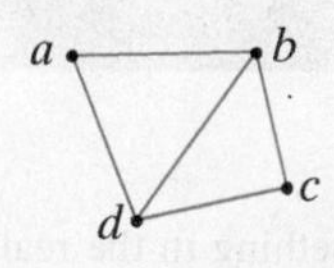


*Show that each graph in Exercises 8–10 has a path from a to a that passes through each edge exactly one time by finding such a path by inspection.*

8.

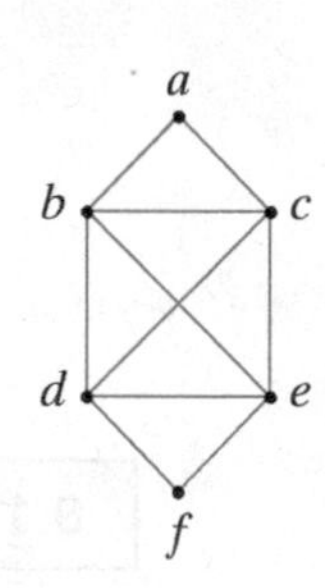


9.

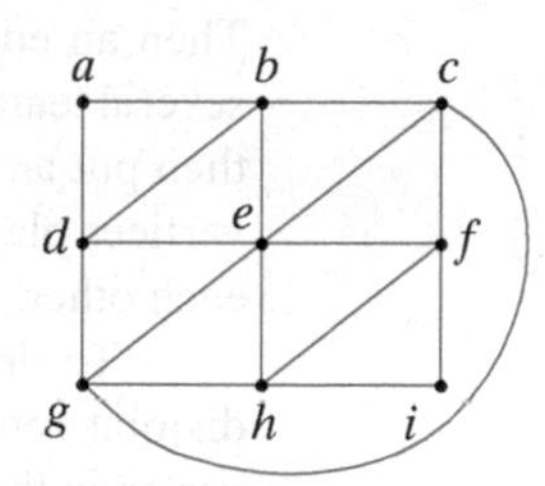


10.

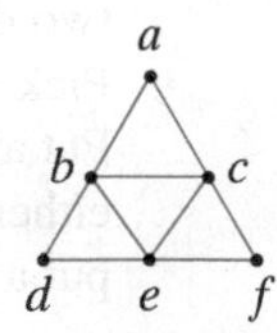


*For each graph $G = (V, E)$ in Exercises 11–13, find V, E, all parallel edges, all loops, all isolated vertices, and tell whether G is a simple graph. Also, tell on which vertices edge $e_1$ is incident.*

**11.**

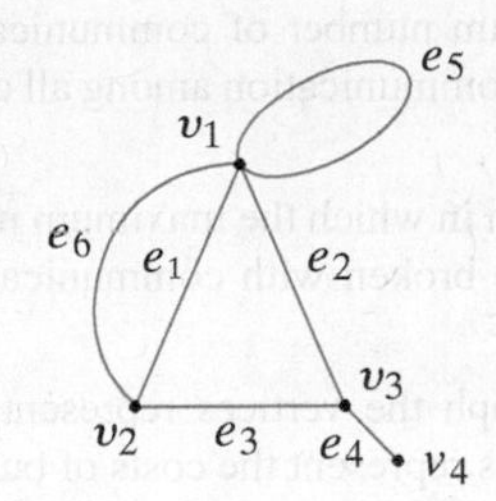


**12.**

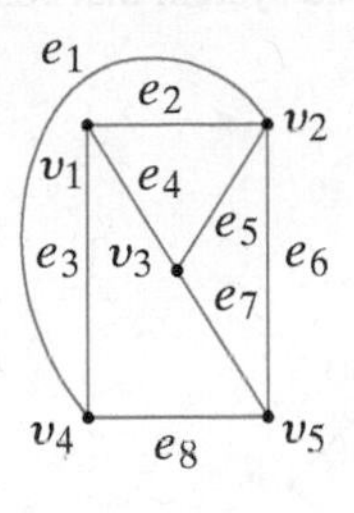


**Go Online**
For more on Paul Erdős and Erdős numbers, see `bit.ly/2yUuHxX`

**13.**

**14.** Draw $K_3$ and $K_5$.

**15.** Give an example of a bipartite graph different from those in the examples of this section. Specify the disjoint vertex sets.

**16.** Find a formula for the number of edges in $K_n$.

*State which graphs in Exercises 17–23 are bipartite graphs. If the graph is bipartite, specify the disjoint vertex sets.*

**17.**

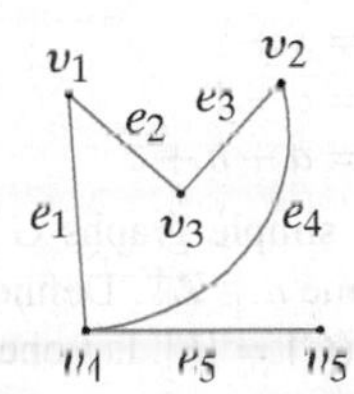


**18.**

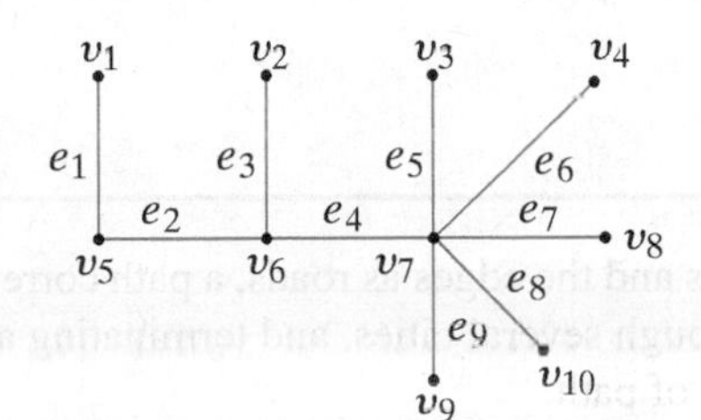


**19.** Figure 8.1.2

**20.** Figure 8.1.5

**21.** Exercise 11

**22.** Exercise 12

**23.** Exercise 13

**24.** Draw $K_{2,3}$ and $K_{3,3}$.

**25.** Find a formula for the number of edges in $K_{m,n}$.

**26.** Most authors require that $V_1$ and $V_2$ be nonempty in Definition 8.1.11. According to these authors, which of the graphs in Examples 8.1.12–8.1.14 are bipartite?

*In Exercises 27–29, find a path of minimum length from $v$ to $w$ in the graph of Figure 8.1.7 that passes through each vertex exactly one time.*

**27.** $v = b, w = e$

**28.** $v = a, w = b$

**29.** $v = c, w = d$

**30.** Paul Erdős (1913–1996) was one of the most prolific mathematicians of all time. He was the author or co-author of nearly 1500 papers. Mathematicians who co-authored a paper with Erdős are said to have *Erdős number one*. Mathematicians who did not co-author a paper with Erdős but who co-authored a paper with a mathematician whose Erdős number is one are said to have *Erdős number two*. Higher Erdős numbers are defined similarly. For example, the author of this book has Erdős number five. Johnsonbaugh co-authored a paper with Tadao Murata, Murata co-authored a paper with A. T. Amin, Amin co-authored a paper with Peter J. Slater, Slater co-authored a paper with Frank Harary, and Harary co-authored a paper with Erdős. Develop a graph model for Erdős numbers. In your model, what is an Erdős number?

**31.** Is the graph model for Bacon numbers (see Example 8.1.6) a simple graph?

**32.** Draw the similarity graph that results from setting $S = 40$ in Example 8.1.7. How many classes are there?

**33.** Draw the similarity graph that results from setting $S = 50$ in Example 8.1.7. How many classes are there?

**34.** In general, is "is similar to" an equivalence relation?

**35.** Suggest additional properties for Example 8.1.7 that might be useful in comparing programs.

**36.** How might one automate the selection of $S$ to group data into classes using a similarity graph?

**37.** Draw a 2-cube.

**38.** Draw a picture like that in Figure 8.1.11 to show how a 3-cube may be constructed from two 2-cubes.

**39.** Prove that the recursive construction in Example 8.1.8 actually yields an $n$-cube.

**40.** How many edges are incident on a vertex in an $n$-cube?

**41.** How many edges are in an $n$-cube?

**★42.** In how many ways can the vertices of an $n$-cube be labeled $0, \ldots, 2^n - 1$ so that there is an edge between two vertices if and only if the binary representation of their labels differs in exactly one bit?

*[Bain] invented an algorithm to draw the n-cube in the plane. In the algorithm, all vertices are on the unit circle in the xy-plane. The angle of a point is the angle from the positive x-axis counterclockwise to the ray from the origin to the point. The input is n.*

1. *If $n = 0$, put one unlabeled vertex at $(-1, 0)$ and stop.*
2. *Recursively invoke this algorithm with input $n - 1$.*
3. *Move each vertex so that its new angle is half the current angle, maintaining edge connections.*
4. *Reflect each vertex and edge in the x-axis.*

5. *Connect each vertex above the x-axis to its mirror image below the x-axis.*
6. *Prefix 0 to the label of each vertex above the x-axis, and prefix 1 to the label of each vertex below the x-axis.*

*The following figures show how the algorithm draws the 2-cube and 3-cube.*

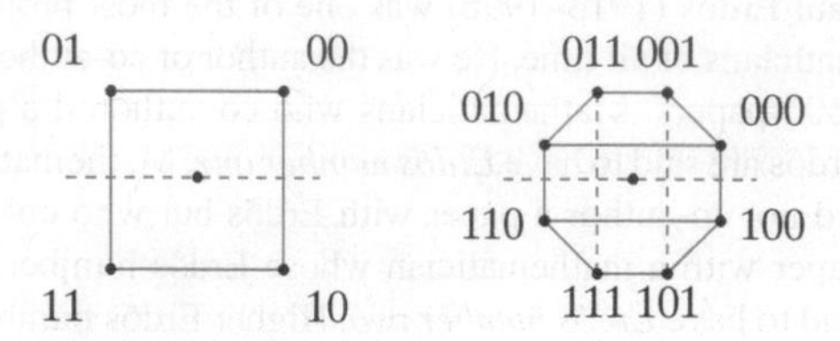


**43.** Show how the algorithm constructs the 2-cube from the 1-cube.

**44.** Show how the algorithm constructs the 3-cube from the 2-cube.

**45.** Show how the algorithm constructs the 4-cube from the 3-cube.

*Exercises 46–48 refer to the following graph. The vertices represent offices. An edge connects two offices if there is a communication link between the two. Notice that any office can communicate with any other either directly through a communication link or by having others relay the message.*

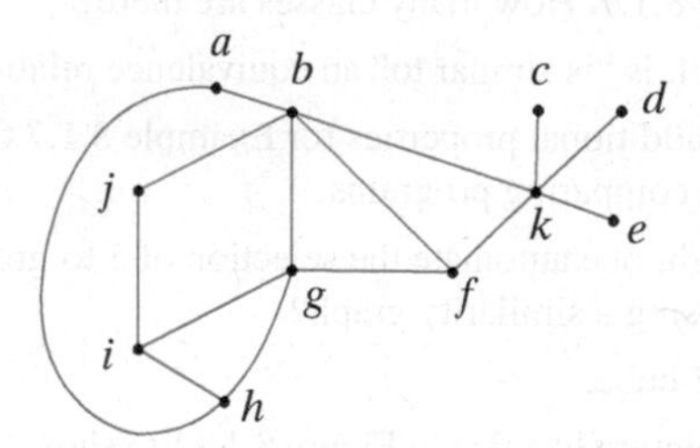


**46.** Show, by giving an example, that communication among all offices is still possible even if some communication links are broken.

**47.** What is the maximum number of communication links that can be broken with communication among all offices still possible?

**48.** Show a configuration in which the maximum number of communication links are broken with communication among all offices still possible.

**49.** In the following graph the vertices represent cities and the numbers on the edges represent the costs of building the indicated roads. Find a least-expensive road system that connects all the cities.

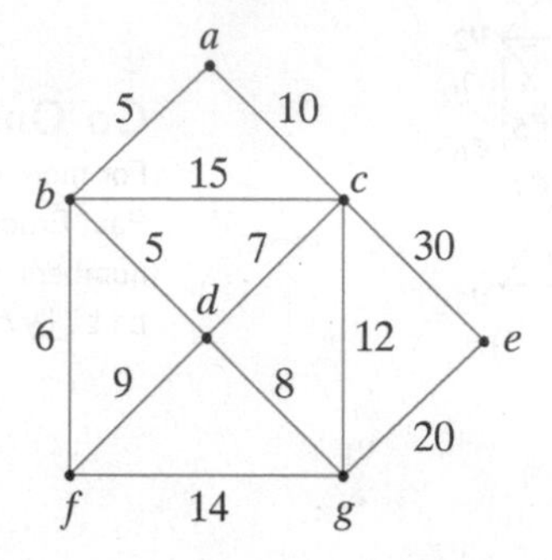


*In a precedence graph, the vertices model certain actions. For example, a vertex might model a statement in a computer program. There is an edge from vertex v to vertex w if the action modeled by v must occur before the action modeled by w. Draw a precedence graph for each computer program in Exercises 50–52.*

**50.**
$x = 1$
$y = 2$
$z = x + y$
$z = z + 1$

**51.**
$x = 1$
$y = 2$
$z = 3$
$a = x + y$
$b = y + z$
$c = x + z$
$c = c + 1$
$x = a + b + c$

**52.**
$x = 1$
$y = 2$
$z = y + 2$
$w = x + 5$
$x = z + w$

**53.** Let $\mathcal{G}$ denote the set of simple graphs $G = (V, E)$, where $V = \{1, 2, \ldots, n\}$ for some $n \in \mathbf{Z}^+$. Define a function $f$ from $\mathcal{G}$ to $\mathbf{Z}^{nonneg}$ by the rule $f(G) = |E|$. Is $f$ one-to-one? Is $f$ onto? Explain.

## 8.2 Paths and Cycles

If we think of the vertices in a graph as cities and the edges as roads, a path corresponds to a trip beginning at some city, passing through several cities, and terminating at some city. We begin by giving a formal definition of path.

**Definition 8.2.1 ▶** Let $v_0$ and $v_n$ be vertices in a graph. A *path* from $v_0$ to $v_n$ of length $n$ is an alternating sequence of $n + 1$ vertices and $n$ edges beginning with vertex $v_0$ and ending with vertex $v_n$,

$$(v_0, e_1, v_1, e_2, v_2, \ldots, v_{n-1}, e_n, v_n),$$

in which edge $e_i$ is incident on vertices $v_{i-1}$ and $v_i$ for $i = 1, \ldots, n$. ◀

The formalism in Definition 8.2.1 means: Start at vertex $v_0$; go along edge $e_1$ to $v_1$; go along edge $e_2$ to $v_2$; and so on.

**Example 8.2.2** In the graph of Figure 8.2.1,

$$(1, e_1, 2, e_2, 3, e_3, 4, e_4, 2) \tag{8.2.1}$$

is a path of length 4 from vertex 1 to vertex 2. ◀

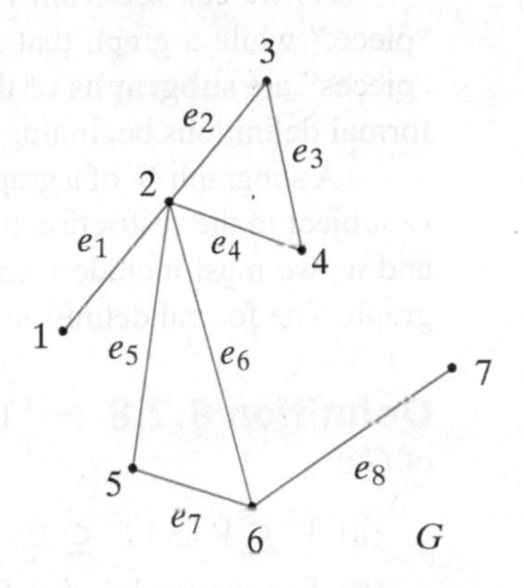


**Figure 8.2.1** A connected graph with paths $(1, e_1, 2, e_2, 3, e_3, 4, e_4, 2)$ of length 4 and (6) of length 0.

**Example 8.2.3** In the graph of Figure 8.2.1, the path (6) consisting solely of vertex 6 is a path of length 0 from vertex 6 to vertex 6. ◀

In the absence of parallel edges, in denoting a path we may suppress the edges. For example, the path (8.2.1) may also be written (1, 2, 3, 4, 2).

A **connected graph** is a graph in which we can get from any vertex to any other vertex on a path. The formal definition follows.

**Definition 8.2.4 ▶** A graph $G$ is *connected* if given any vertices $v$ and $w$ in $G$, there is a path from $v$ to $w$. ◀

**Example 8.2.5** The graph $G$ of Figure 8.2.1 is connected since, given any vertices $v$ and $w$ in $G$, there is a path from $v$ to $w$. ◀

**Example 8.2.6** The graph $G$ of Figure 8.2.2 is not connected since, for example, there is no path from vertex $v_2$ to vertex $v_5$.

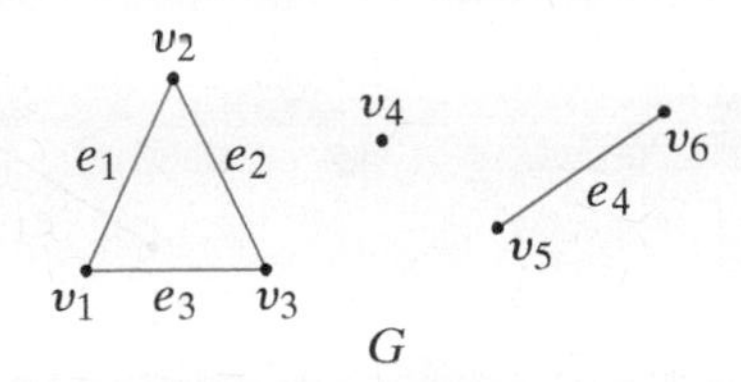


**Figure 8.2.2** A graph that is not connected. ◀

**Example 8.2.7** Let $G$ be the graph whose vertex set consists of the 50 states of the United States. Put an edge between states $v$ and $w$ if $v$ and $w$ share a border. For example, there is an edge between California and Oregon and between Illinois and Missouri. There is no edge between Georgia and New York, nor is there an edge between Utah and New Mexico. (Touching does not count; the states must share a border.) The graph $G$ is not connected because there is no path from Hawaii to California (or from Hawaii to any other state). ◀

As we can see from Figures 8.2.1 and 8.2.2, a connected graph consists of one "piece," while a graph that is not connected consists of two or more "pieces." These "pieces" are **subgraphs** of the original graph and are called **components.** We give the formal definitions beginning with subgraph.

A subgraph $G'$ of a graph $G$ is obtained by selecting certain edges and vertices from $G$ subject to the restriction that if we select an edge $e$ in $G$ that is incident on vertices $v$ and $w$, we must include $v$ and $w$ in $G'$. The restriction is to ensure that $G'$ is actually a graph. The formal definition follows.

**Definition 8.2.8 ▶** Let $G = (V, E)$ be a graph. We call $(V', E')$ a *subgraph* of $G$ if

(a) $V' \subseteq V$ and $E' \subseteq E$.

(b) For every edge $e' \in E'$, if $e'$ is incident on $v'$ and $w'$, then $v', w' \in V'$. ◀

**Example 8.2.9** The graph $G' = (V', E')$ of Figure 8.2.3 is a subgraph of the graph $G = (V, E)$ of Figure 8.2.4 since $V' \subseteq V$ and $E' \subseteq E$.

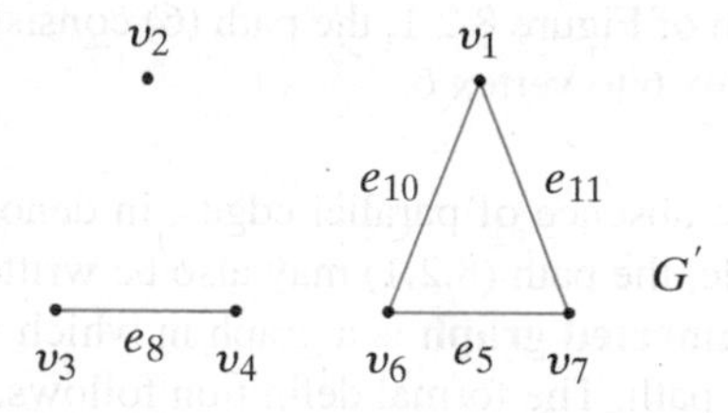


**Figure 8.2.3** A subgraph of the graph of Figure 8.2.4.

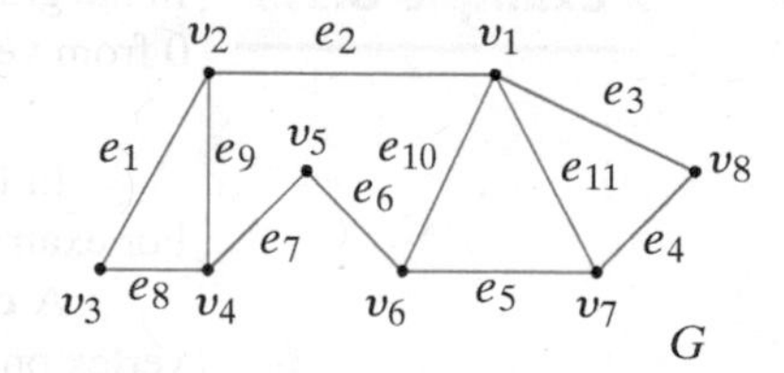


**Figure 8.2.4** A graph, one of whose subgraphs is shown in Figure 8.2.3. ◀

**Example 8.2.10** Find all subgraphs of the graph $G$ of Figure 8.2.5 having at least one vertex.

**SOLUTION** If we select no edges, we may select one or both vertices yielding the subgraphs $G_1$, $G_2$, and $G_3$ shown in Figure 8.2.6. If we select the one available edge $e_1$, we must select the two vertices on which $e_1$ is incident. In this case, we obtain the subgraph $G_4$ shown in Figure 8.2.6. Thus $G$ has the four subgraphs shown in Figure 8.2.6.

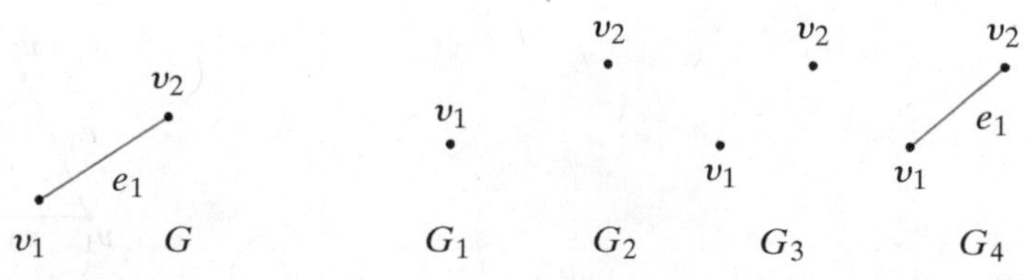


**Figure 8.2.5** The graph for Example 8.2.10.

**Figure 8.2.6** The four subgraphs of the graph of Figure 8.2.5. ◀

We can now define "component."

**Definition 8.2.11 ▶** Let $G$ be a graph and let $v$ be a vertex in $G$. The subgraph $G'$ of $G$ consisting of all edges and vertices in $G$ that are contained in some path beginning at $v$ is called the *component* of $G$ containing $v$. ◀

**Example 8.2.12** The graph $G$ of Figure 8.2.1 has one component, namely itself. Indeed, a graph is connected if and only if it has exactly one component. ◀

**Example 8.2.13** Let $G$ be the graph of Figure 8.2.2. The component of $G$ containing $v_3$ is the subgraph

$$G_1 = (V_1, E_1), \qquad V_1 = \{v_1, v_2, v_3\}, \qquad E_1 = \{e_1, e_2, e_3\}.$$

The component of $G$ containing $v_4$ is the subgraph

$$G_2 = (V_2, E_2), \qquad V_2 = \{v_4\}, \qquad E_2 = \varnothing.$$

The component of $G$ containing $v_5$ is the subgraph

$$G_3 = (V_3, E_3), \qquad V_3 = \{v_5, v_6\}, \qquad E_3 = \{e_4\}.$$ ◀

Another characterization of the components of a graph $G = (V, E)$ is obtained by defining a relation $R$ on the set of vertices $V$ by the rule

$$v_1 R v_2 \quad \text{if there is a path from } v_1 \text{ to } v_2.$$

It can be shown (Exercise 69) that $R$ is an equivalence relation on $V$ and that if $v \in V$, the set of vertices in the component containing $v$ is the equivalence class

$$[v] = \{w \in V \mid wRv\}.$$

Notice that the definition of "path" allows repetitions of vertices or edges or both. In the path (8.2.1), vertex 2 appears twice.

Subclasses of paths are obtained by prohibiting duplicate vertices or edges or by making the vertices $v_0$ and $v_n$ of Definition 8.2.1 identical.

**Definition 8.2.14 ▶** Let $v$ and $w$ be vertices in a graph $G$.

A *simple path* from $v$ to $w$ is a path from $v$ to $w$ with no repeated vertices.

A *cycle* (or *circuit*) is a path of nonzero length from $v$ to $v$ with no repeated edges.

A *simple cycle* is a cycle from $v$ to $v$ in which, except for the beginning and ending vertices that are both equal to $v$, there are no repeated vertices. ◀

**Example 8.2.15** For the graph of Figure 8.2.1, we have the following information.

| *Path* | *Simple Path?* | *Cycle?* | *Simple Cycle?* |
|---|---|---|---|
| (6, 5, 2, 4, 3, 2, 1) | No | No | No |
| (6, 5, 2, 4) | Yes | No | No |
| (2, 6, 5, 2, 4, 3, 2) | No | Yes | No |
| (5, 6, 2, 5) | No | Yes | Yes |
| (7) | Yes | No | No |

◀

We next reexamine the problem introduced in Section 8.1 of finding a cycle in a graph that traverses each edge exactly one time.

**Example 8.2.16** **Königsberg Bridge Problem** The first paper in graph theory was Leonhard Euler's in 1736. The paper presented a general theory that included a solution to what is now called the Königsberg bridge problem.

**Go Online**
For more on the Königsberg Bridge Problem, see bit.ly/2yUuHxX

Two islands lying in the Pregel River in Königsberg (now Kaliningrad in Russia) were connected to each other and the river banks by bridges, as shown in Figure 8.2.7. The problem is to start at any location—$A$, $B$, $C$, or $D$; walk over each bridge exactly once; then return to the starting location.

The bridge configuration can be modeled as a graph, as shown in Figure 8.2.8. The vertices represent the locations and the edges represent the bridges. The Königsberg bridge problem is now reduced to finding a cycle in the graph of Figure 8.2.8 that includes all of the edges and all of the vertices. In honor of Euler, a cycle in a graph $G$ that includes all of the edges and all of the vertices of $G$ is called an **Euler cycle.**† From the discussion of Section 8.1, we see that there is no Euler cycle in the graph of Figure 8.2.8 because the number of edges incident on vertex $A$ is odd. (In fact, in the graph of Figure 8.2.8, every vertex is incident on an odd number of edges.)

**Go Online**
A biography of Euler is at bit.ly/2yUuHxX

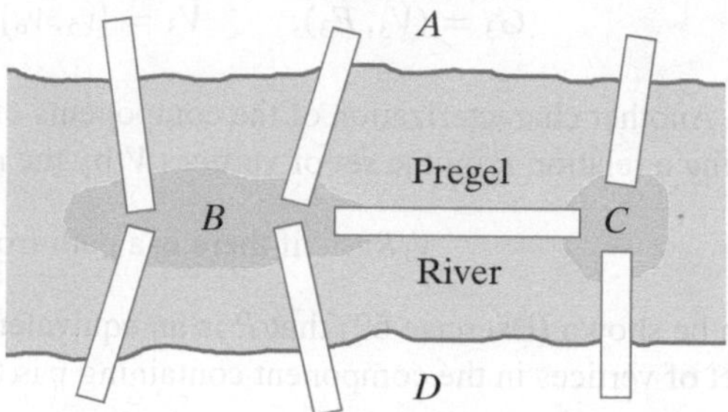


Figure 8.2.7 The bridges of Königsberg.

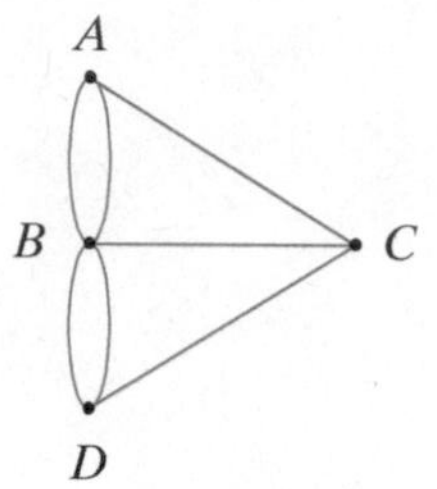

Figure 8.2.8 A graph model of the bridges of Königsberg. ◀

The solution to the existence of Euler cycles is nicely stated by introducing the degree of a vertex. The **degree of a vertex** $v$, $\delta(v)$, is the number of edges incident on $v$. (By definition, each loop on $v$ contributes 2 to the degree of $v$.) In Section 8.1 we found that if a graph $G$ has an Euler cycle, then every vertex in $G$ has even degree. We can also prove that $G$ is connected.

**Theorem 8.2.17** If a graph $G$ has an Euler cycle, then $G$ is connected and every vertex has even degree.

**Proof** Suppose that $G$ has an Euler cycle. We argued in Section 8.1 that every vertex in $G$ has even degree. If $v$ and $w$ are vertices in $G$, the portion of the Euler cycle that takes us from $v$ to $w$ serves as a path from $v$ to $w$. Therefore, $G$ is connected. ◀

The converse of Theorem 8.2.17 is also true. We give a proof by mathematical induction due to [Fowler].

**Theorem 8.2.18** If $G$ is a connected graph and every vertex has even degree, then $G$ has an Euler cycle.

**Proof** The proof is by induction on the number $n$ of edges in $G$.

†For technical reasons, if $G$ consists of one vertex $v$ and no edges, we call the path $(v)$ an Euler cycle for $G$.

**Basis Step ($n=0$)**

Since $G$ is connected, if $G$ has no edges, $G$ consists of a single vertex. An Euler cycle consists of the single vertex and no edges.

**Inductive Step**

Suppose that $G$ has $n$ edges, $n > 0$, and that any connected graph with $k$ edges, $k < n$, in which every vertex has even degree, has an Euler cycle.

It is straightforward to verify that a connected graph with one or two vertices, each of which has even degree, has an Euler cycle (see Exercise 70); thus we assume that the graph has at least three vertices.

Since $G$ is connected, there are vertices $v_1$, $v_2$, and $v_3$ in $G$ with edge $e_1$ incident on $v_1$ and $v_2$ and edge $e_2$ incident on $v_2$ and $v_3$. We delete edges $e_1$ and $e_2$, but no vertices, and add an edge $e$ incident on $v_1$ and $v_3$ to obtain the graph $G'$ [see Figure 8.2.9(a)]. Notice that each component of the graph $G'$ has less than $n$ edges and that in each component of the graph $G'$, every vertex has even degree. We show that $G'$ has either one or two components.

Let $v$ be a vertex. Since $G$ is connected, there is a path $P$ in $G$ from $v$ to $v_1$. Let $P'$ be the portion of the path $P$ starting at $v$ whose edges are also in $G'$. Now $P'$ ends at either $v_1$, $v_2$, or $v_3$ because the only way that $P$ could fail to be a path in $G'$ is that $P$ contains one of the deleted edges $e_1$ or $e_2$. If $P'$ ends at $v_1$, then $v$ is in the same component as $v_1$ in $G'$. If $P'$ ends at $v_3$ [see Figure 8.2.9(b)], then $v$ is in the same component as $v_3$ in $G'$, which is in the same component as $v_1$ in $G'$ (since edge $e$ in $G'$ is incident on $v_1$ and $v_3$). If $P'$ ends at $v_2$, then $v_2$ is in the same component as $v$. Therefore, any vertex in $G'$ is in the same component as either $v_1$ or $v_2$. Thus $G'$ has one or two components.

If $G'$ has one component, that is, if $G'$ is connected, we may apply the inductive hypothesis to conclude that $G'$ has an Euler cycle $C'$. This Euler cycle may be modified to produce an Euler cycle in $G$: We simply replace the occurrence of edge $e$ in $C'$ by edges $e_1$ and $e_2$.

Suppose that $G'$ has two components [see Figure 8.2.9(c)]. By the inductive hypothesis, the component containing $v_1$ has an Euler cycle $C'$ and the component containing $v_2$ has an Euler cycle $C''$ beginning and ending at $v_2$. An Euler cycle in $G$ is obtained by modifying $C'$ by replacing $(v_1, v_3)$ in $C'$ by $(v_1, v_2)$ followed by $C''$ followed

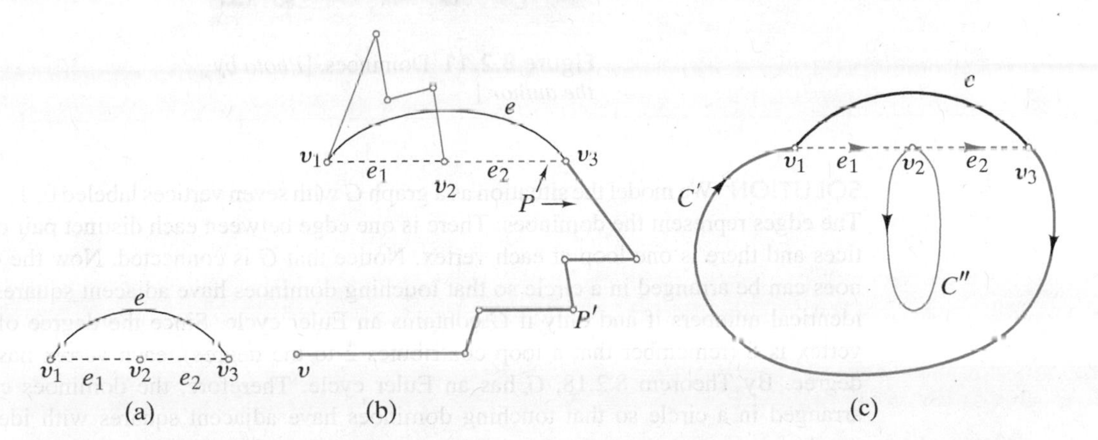


**Figure 8.2.9** The proof of Theorem 8.2.18. In (a), edges $e_1$ and $e_2$ are deleted and edge $e$ is added. In (b), $P$ (shown in color) is a path in $G$ from $v$ to $v_1$, and $P'$ (shown in heavy color) is the portion of $P$ starting at $v$ whose edges are also in $G'$. As shown, $P'$ ends at $v_3$. Since edge $e$ is in $G'$, there is a path in $G'$ from $v$ to $v_1$. Thus $v$ and $v_1$ are in the same component. In (c), $C'$ (shown with a heavy line) is an Euler cycle for one component, and $C''$ (shown with a light, solid line) is an Euler cycle for the other component. If we replace $e$ in $C'$ by $e_1$, $C''$, $e_2$, we obtain an Euler cycle (shown in color) for $G$.

by $(v_2, v_3)$ or by replacing $(v_3, v_1)$ in $C'$ by $(v_3, v_2)$ followed by $C''$ followed by $(v_2, v_1)$. The Inductive Step is complete; $G$ has an Euler cycle. ◀

If $G$ is a connected graph and every vertex has even degree and $G$ has only a few edges, we can usually find an Euler cycle by inspection.

**Example 8.2.19** Let $G$ be the graph of Figure 8.2.10. Use Theorem 8.2.18 to verify that $G$ has an Euler cycle. Find an Euler cycle for $G$.

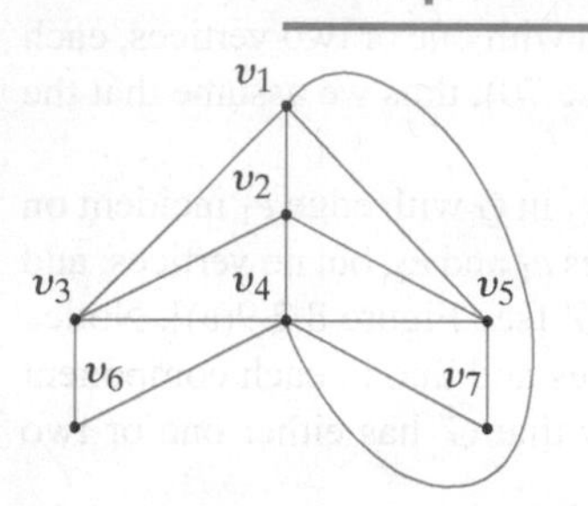


**Figure 8.2.10** The graph for Example 8.2.19.

**SOLUTION** We observe that $G$ is connected and that

$$\delta(v_1) = \delta(v_2) = \delta(v_3) = \delta(v_5) = 4, \qquad \delta(v_4) = 6, \qquad \delta(v_6) = \delta(v_7) = 2.$$

Since the degree of every vertex is even, by Theorem 8.2.18, $G$ has an Euler cycle. By inspection, we find the Euler cycle

$$(v_6, v_4, v_7, v_5, v_1, v_3, v_4, v_1, v_2, v_5, v_4, v_2, v_3, v_6).$$ ◀

**Example 8.2.20** A domino is a rectangle divided into two squares with each square numbered one of 0, 1, . . . , 6 (see Figure 8.2.11). Two squares on a single domino can have the same number. Show that the distinct dominoes can be arranged in a circle so that touching dominoes have adjacent squares with identical numbers.

**Figure 8.2.11** Dominoes. [*Photo by the author.*]

**SOLUTION** We model the situation as a graph $G$ with seven vertices labeled 0, 1, . . . , 6. The edges represent the dominoes: There is one edge between each distinct pair of vertices and there is one loop at each vertex. Notice that $G$ is connected. Now the dominoes can be arranged in a circle so that touching dominoes have adjacent squares with identical numbers if and only if $G$ contains an Euler cycle. Since the degree of each vertex is 8 (remember that a loop contributes 2 to the degree), each vertex has even degree. By Theorem 8.2.18, $G$ has an Euler cycle. Therefore, the dominoes can be arranged in a circle so that touching dominoes have adjacent squares with identical numbers. ◀

What can be said about a connected graph in which not all the vertices have even degree? The first observation (Corollary 8.2.22) is that the number of vertices of odd degree is even. This follows from the fact (Theorem 8.2.21) that the sum of all of the degrees in a graph is an even number.

**Theorem 8.2.21** If $G$ is a graph with $m$ edges and vertices $\{v_1, v_2, \ldots, v_n\}$, then

$$\sum_{i=1}^{n} \delta(v_i) = 2m.$$

In particular, the sum of the degrees of all the vertices in a graph is even.

**Proof** When we sum over the degrees of all the vertices, we count each edge $(v_i, v_j)$ twice—once when we count it as $(v_i, v_j)$ in the degree of $v_i$ and again when we count it as $(v_j, v_i)$ in the degree of $v_j$. The conclusion follows. ◀

**Corollary 8.2.22** In any graph, the number of vertices of odd degree is even.

**Proof** Let us divide the vertices into two groups: those with even degree $x_1, \ldots, x_m$ and those with odd degree $y_1, \ldots, y_n$. Let

$$S = \delta(x_1) + \delta(x_2) + \cdots + \delta(x_m), \qquad T = \delta(y_1) + \delta(y_2) + \cdots + \delta(y_n).$$

By Theorem 8.2.21, $S + T$ is even. Since $S$ is the sum of even numbers, $S$ is even. Thus $T$ is even. But $T$ is the sum of $n$ odd numbers, and therefore $n$ is even. ◀

Suppose that a connected graph $G$ has exactly two vertices $v$ and $w$ of odd degree. Let us temporarily insert an edge $e$ from $v$ to $w$. The resulting graph $G'$ is connected and every vertex has even degree. By Theorem 8.2.18, $G'$ has an Euler cycle. If we delete $e$ from this Euler cycle, we obtain a path with no repeated edges from $v$ to $w$ containing all the edges and vertices of $G$. We have shown that if a graph has exactly two vertices $v$ and $w$ of odd degree, there is a path with no repeated edges containing all the edges and vertices from $v$ to $w$. The converse can be proved similarly.

**Theorem 8.2.23** A graph has a path with no repeated edges from $v$ to $w$ ($v \neq w$) containing all the edges and vertices if and only if it is connected and $v$ and $w$ are the only vertices having odd degree.

**Proof** Suppose that a graph has a path $P$ with no repeated edges from $v$ to $w$ containing all the edges and vertices. The graph is surely connected. If we add an edge from $v$ to $w$, the resulting graph has an Euler cycle, namely, the path $P$ together with the added edge. By Theorem 8.2.17, every vertex has even degree. Removing the added edge affects only the degrees of $v$ and $w$, which are each reduced by 1. Thus in the original graph, $v$ and $w$ have odd degree and all other vertices have even degree.

The converse was discussed just before the statement of the theorem. ◀

Generalizations of Theorem 8.2.23 are given as Exercises 43 and 45.

We conclude by proving a rather special result that we will use in Section 9.2.

**Theorem 8.2.24** If a graph $G$ contains a cycle from $v$ to $v$, $G$ contains a simple cycle from $v$ to $v$.

**Proof** Let

$$C = (v_0, e_1, v_1, \ldots, e_i, v_i, e_{i+1}, \ldots, e_j, v_j, e_{j+1}, v_{j+1}, \ldots, e_n, v_n)$$

be a cycle from $v$ to $v$ where $v = v_0 = v_n$ (see Figure 8.2.12). If $C$ is not a simple cycle, then $v_i = v_j$, for some $i < j < n$. We can replace $C$ by the cycle

$$C' = (v_0, e_1, v_1, \ldots, e_i, v_i, e_{j+1}, v_{j+1}, \ldots, e_n, v_n).$$

If $C'$ is not a simple cycle from $v$ to $v$, we repeat the previous procedure. Eventually we obtain a simple cycle from $v$ to $v$.

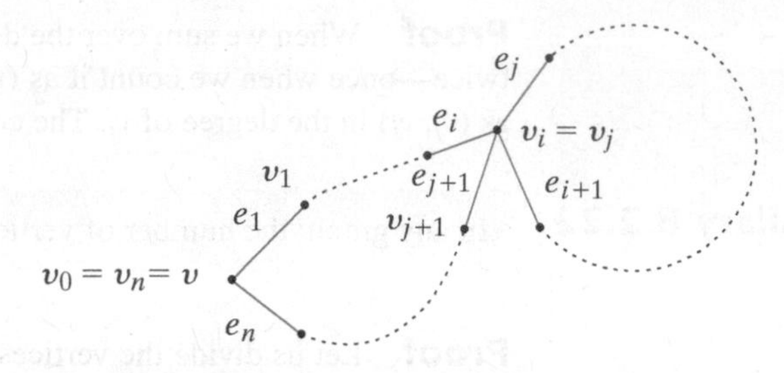


**Figure 8.2.12** A cycle that either is a simple cycle or can be reduced to a simple cycle. ◀

## 8.2 Review Exercises

1. What is a path?
2. What is a simple path?
3. Give an example of a path that is not a simple path.
4. What is a cycle?
5. What is a simple cycle?
6. Give an example of a cycle that is not a simple cycle.
7. Define *connected graph*.
8. Give an example of a connected graph.
9. Give an example of a graph that is not connected.
10. What is a subgraph?
11. Give an example of a subgraph.
12. What is a component of a graph?
13. Give an example of a component of a graph.
14. If a graph is connected, how many components does it have?
15. Define *degree of vertex v*.
16. What is an Euler cycle?
17. State a necessary and sufficient condition that a graph have an Euler cycle.
18. Give an example of a graph that has an Euler cycle. Specify the Euler cycle.
19. Give an example of a graph that does *not* have an Euler cycle. Prove that it does not have an Euler cycle.
20. What is the relationship between the sum of the degrees of the vertices in a graph and the number of edges in the graph?
21. In any graph, must the number of vertices of odd degree be even?
22. State a necessary and sufficient condition that a graph have a path with no repeated edges from $v$ to $w$ ($v \neq w$) containing all the edges and vertices.
23. If a graph $G$ contains a cycle from $v$ to $v$, must $G$ contain a simple cycle from $v$ to $v$?

## 8.2 Exercises

*In Exercises 1–9, tell whether the given path in the graph is*

*(a) A simple path*

*(b) A cycle*

*(c) A simple cycle*

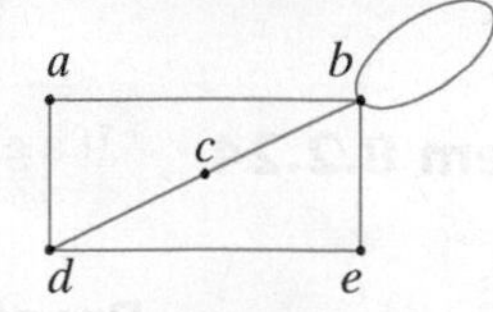

**1.** $(b, b)$ **2.** $(a, d, c, d, e)$

**3.** $(e, d, c, b)$ **4.** $(d, c, b, e, d)$

**5.** $(b, c, d, e, b, b)$ **6.** $(b, c, d, a, b, e, d, c, b)$

**7.** $(a, d, c, b, e)$ **8.** $(d, c, b)$

**9.** $(d)$

*In Exercises 10–18, draw a graph having the given properties or explain why no such graph exists.*

**10.** Six vertices each of degree 3

**11.** Four vertices each of degree 1

**12.** Five vertices each of degree 3

**13.** Six vertices; four edges

**14.** Four vertices having degrees 1, 2, 3, 4

**15.** Four edges; four vertices having degrees 1, 2, 3, 4

**16.** Simple graph; six vertices having degrees 1, 2, 3, 4, 5, 5

**17.** Simple graph; five vertices having degrees 2, 2, 4, 4, 4

**18.** Simple graph; five vertices having degrees 2, 3, 3, 4, 4

**19.** Find all the simple cycles in the following graph.

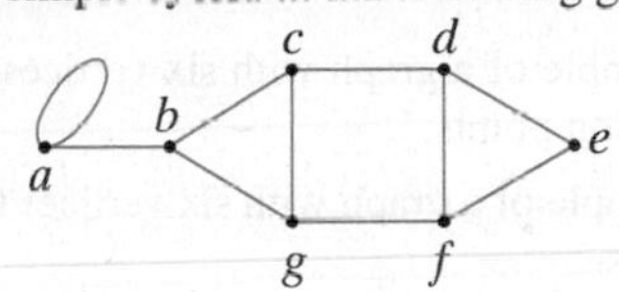


**20.** Find all simple paths from $a$ to $e$ in the graph of Exercise 19.

**21.** Find all connected subgraphs of the following graph containing all of the vertices of the original graph and having as few edges as possible. Which are simple paths? Which are cycles? Which are simple cycles?

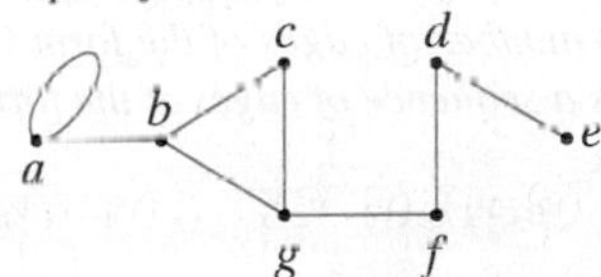


*Find the degree of each vertex for the following graphs.*

**22.**

**23.**

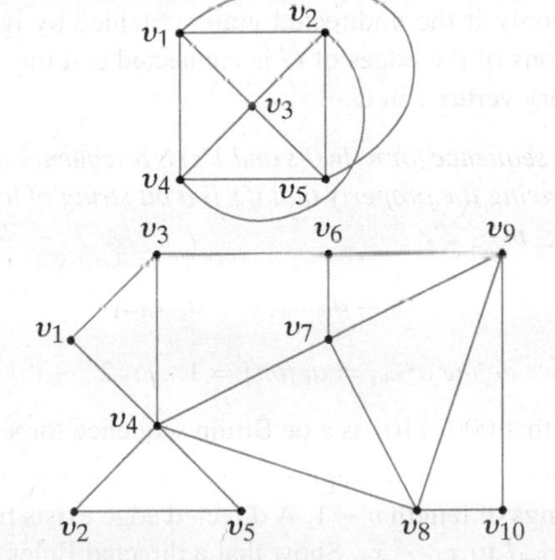


*In Exercises 24–27, find all subgraphs having at least one vertex of the graph given.*

**24.**

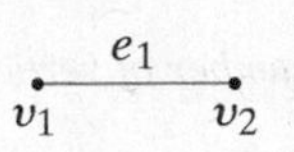


**25.**

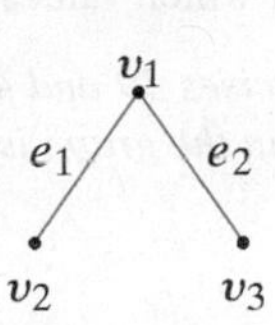


**26.**

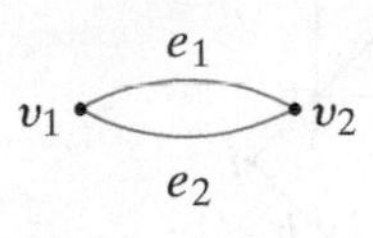


**★27.**

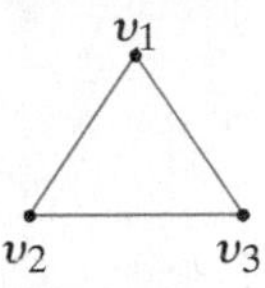


*In Exercises 28–33, decide whether the graph has an Euler cycle. If the graph has an Euler cycle, exhibit one.*

**28.** Exercise 21 **29.** Exercise 22

**30.** Exercise 23 **31.** Figure 8.2.4

**32.**

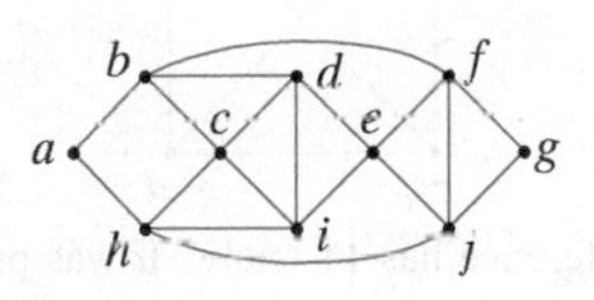


**33.**

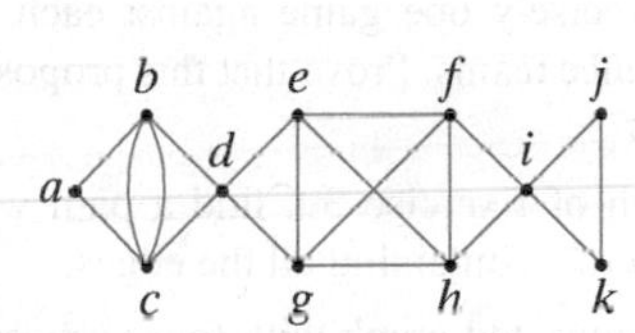


**34.** The following graph is continued to an arbitrary, finite depth. Does the graph contain an Euler cycle? If the answer is yes, describe one.

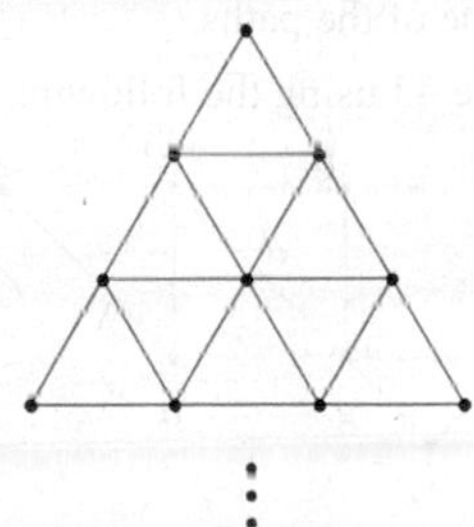

**35.** When does the complete graph $K_n$ contain an Euler cycle?

**36.** When does the complete bipartite graph $K_{m,n}$ contain an Euler cycle?

**37.** For which values of $m$ and $n$ does the graph contain an Euler cycle?

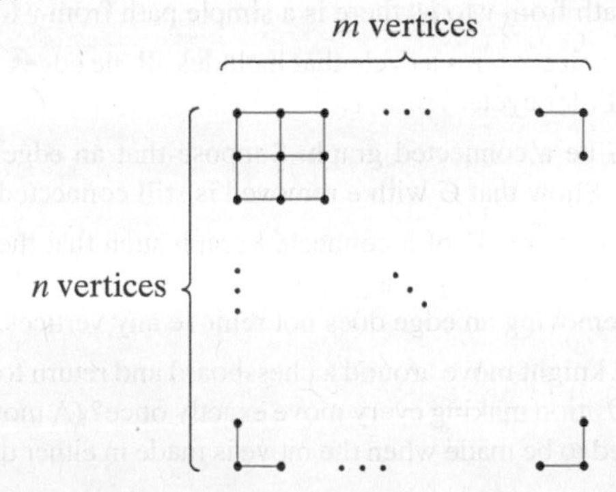

**38.** For which values of $n$ does the $n$-cube contain an Euler cycle?

*In Exercises 39 and 40, verify that the number of vertices of odd degree in the graph is even.*

**39.**

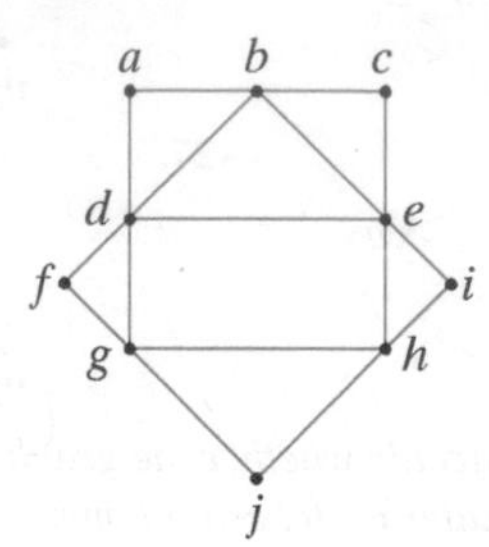


**40.**

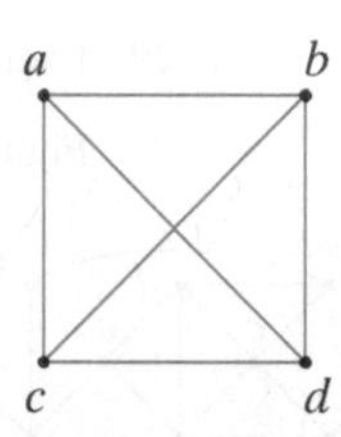


**41.** A sports conference has 11 teams. It was proposed that each team play precisely one game against each of exactly nine other conference teams. Prove that this proposal is impossible to implement.

**42.** For the graph of Exercise 39, find a path with no repeated edges from $d$ to $e$ containing all the edges.

**43.** Let $G$ be a connected graph with four vertices $v_1$, $v_2$, $v_3$, and $v_4$ of odd degree. Show that there are paths with no repeated edges from $v_1$ to $v_2$ and from $v_3$ to $v_4$ such that every edge in $G$ is in exactly one of the paths.

**44.** Illustrate Exercise 43 using the following graph.

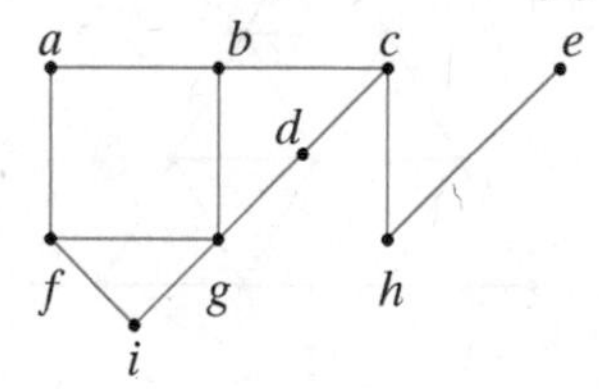


**45.** State and prove a generalization of Exercise 43 where the number of vertices of odd degree is arbitrary.

*In Exercises 46 and 47, tell whether each assertion is true or false. If false, give a counterexample and if true, prove it.*

**46.** Let $G$ be a graph and let $v$ and $w$ be distinct vertices. If there is a path from $v$ to $w$, there is a simple path from $v$ to $w$.

**47.** If a graph contains a cycle that includes all the edges, the cycle is an Euler cycle.

**48.** Let $G$ be a connected graph. Suppose that an edge $e$ is in a cycle. Show that $G$ with $e$ removed is still connected.

**49.** Give an example of a connected graph such that the removal of any edge results in a graph that is not connected. (Assume that removing an edge does not remove any vertices.)

★**50.** Can a knight move around a chessboard and return to its original position making every move exactly once? (A move is considered to be made when the move is made in either direction.)

**51.** Show that if $G'$ is a connected subgraph of a graph $G$, then $G'$ is contained in a component.

**52.** Show that if a graph $G$ is partitioned into connected subgraphs so that each edge and each vertex in $G$ belong to one of the subgraphs, the subgraphs are components.

**53.** Let $G$ be a directed graph and let $G'$ be the undirected graph obtained from $G$ by ignoring the direction of edges in $G$. Assume that $G$ is connected. If $v$ is a vertex in $G$, we say the *parity of* $v$ is *even* if the number of edges of the form $(v, w)$ is even; *odd parity* is defined similarly. Prove that if $v$ and $w$ are vertices in $G$ having odd parity, it is possible to change the orientation of certain edges in $G$ so that $v$ and $w$ have even parity and the parity of all other vertices in $G$ is unchanged.

★**54.** Show that the maximum number of edges in a simple, disconnected graph with $n$ vertices is $(n - 1)(n - 2)/2$.

★**55.** Show that the maximum number of edges in a simple, bipartite graph with $n$ vertices is $\lfloor n^2/4 \rfloor$.

*A vertex v in a connected graph G is an articulation point if the removal of v and all edges incident on v disconnects G.*

**56.** Give an example of a graph with six vertices that has exactly two articulation points.

**57.** Give an example of a graph with six vertices that has no articulation points.

**58.** Show that a vertex $v$ in a connected graph $G$ is an articulation point if and only if there are vertices $w$ and $x$ in $G$ having the property that every path from $w$ to $x$ passes through $v$.

*Let G be a directed graph and let v be a vertex in G. The indegree of v, in(v), is the number of edges of the form (w, v). The outdegree of v, out(v), is the number of edges of the form (v, w). A directed Euler cycle in G is a sequence of edges of the form*

$$(v_0, v_1), (v_1, v_2), \ldots, (v_{n-1}, v_n),$$

*where* $v_0 = v_n$, *every edge in G occurs exactly one time, and all vertices appear.*

**59.** Show that a directed graph $G$ contains a directed Euler cycle if and only if the undirected graph obtained by ignoring the directions of the edges of $G$ is connected and $\text{in}(v) = \text{out}(v)$ for every vertex $v$ in $G$.

*A de Bruijn sequence for n (in 0's and 1's) is a sequence* $a_1, \ldots, a_{2^n}$ *of* $2^n$ *bits having the property that if s is a bit string of length n, for some m,* $1 \le m \le 2^n$,

$$s = a_m a_{m+1} \cdots a_{m+n-1}. \qquad (8.2.2)$$

*In (8.2.2), we define* $a_{2^n+i} = a_i$ *for* $i = 1, \ldots, 2^n - 1$.

**60.** Verify that 00011101 is a de Bruijn sequence for $n = 3$.

**61.** Let $G$ be a directed graph with vertices corresponding to all bit strings of length $n - 1$. A directed edge exists from vertex $x_1 \cdots x_{n-1}$ to $x_2 \cdots x_n$. Show that a directed Euler cycle in $G$ corresponds to a de Bruijn sequence.

★**62.** Show that there is a de Bruijn sequence for every $n = 1, 2, \ldots$.

★**63.** A *closed path* is a path from $v$ to $v$. Show that a connected graph $G$ is bipartite if and only if every closed path in $G$ has even length.

**64.** How many paths of length $k \geq 1$ are there in $K_n$?

**65.** Show that there are

$$\frac{n(n-1)[(n-1)^k - 1]}{n-2}$$

paths whose lengths are between 1 and $k$, inclusive, in $K_n$, $n > 2$.

**66.** Let $v$ and $w$ be distinct vertices in $K_n$. Let $p_m$ denote the number of paths of length $m$ from $v$ to $w$ in $K_n$, $1 \leq m \leq n$.

(a) Derive a recurrence relation for $p_m$.

(b) Find an explicit formula for $p_m$.

**67.** Let $v$ and $w$ be distinct vertices in $K_n$, $n \geq 2$. Show that the number of simple paths from $v$ to $w$ is

$$(n-2)! \sum_{k=0}^{n-2} \frac{1}{k!}.$$

★**68.** [Requires calculus] Show that there are $\lfloor n!e - 1 \rfloor$ simple paths in $K_n$. ($e = 2.71828\ldots$ is the base of the natural logarithm.)

**69.** Let $G$ be a graph. Define a relation $R$ on the set $V$ of vertices of $G$ as $vRw$ if there is a path from $v$ to $w$. Prove that $R$ is an equivalence relation on $V$.

**70.** Prove that a connected graph with one or two vertices, each of which has even degree, has an Euler cycle.

*Let $G$ be a connected graph. The distance between vertices $v$ and $w$ in $G$, $dist(v, w)$, is the length of a shortest path from $v$ to $w$. The diameter of $G$ is*

$$d(G) = \max\{dist(v, w) \mid v \text{ and } w \text{ are vertices in } G\}.$$

**71.** Find the diameter of the graph of Figure 8.2.10.

**72.** Find the diameter of the $n$-cube. In the context of parallel computation, what is the meaning of this value?

**73.** Find the diameter of $K_n$, the complete graph on $n$ vertices.

**74.** Show that the number of paths in the following graph from $v_1$ to $v_1$ of length $n$ is equal to the $(n+1)$st Fibonacci number $f_{n+1}$.

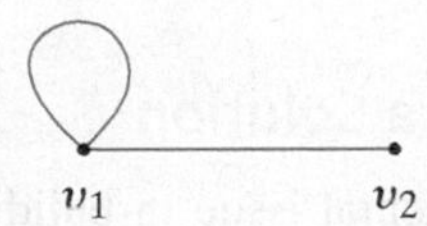


**75.** Let $G$ be a simple graph with $n$ vertices in which every vertex has degree $k$ and

$$k \geq \frac{n-3}{2} \quad \text{if } n \bmod 4 = 1$$
$$k \geq \frac{n-1}{2} \quad \text{if } n \bmod 4 \neq 1.$$

Show that $G$ is connected.

*A cycle in a simple directed graph [i.e., a directed graph in which there is at most one edge of the form $(v, w)$ and no edges of the form $(v, v)$] is a sequence of three or more vertices*

$$(v_0, v_1, \ldots, v_n)$$

*in which $(v_{i-1}, v_i)$ is an edge for $i = 1, \ldots, n$ and $v_0 = v_n$. A directed acyclic graph (dag) is a simple directed graph with no cycles.*

**76.** Show that a dag has at least one vertex with no out edges [i.e., there is at least one vertex $v$ such that there are no edges of the form $(v, w)$].

**77.** Show that the maximum number of edges in an $n$-vertex dag is $n(n-1)/2$.

**78.** An *independent set* in a graph $G$ is a subset $S$ of the vertices of $G$ having the property that no two vertices in $S$ are adjacent. (Note that $\varnothing$ is an independent set for any graph.) Prove the following result due to [Prodinger].

Let $P_n$ be the graph that is a simple path with $n$ vertices. Prove that the number of independent sets in $P_n$ is equal to $f_{n+2}$, $n = 1, 2, \ldots$, where $\{f_n\}$ is the Fibonacci sequence.

**79.** Let $G$ be a graph. Suppose that for every pair of distinct vertices $v_1$ and $v_2$ in $G$, there is a unique vertex $w$ in $G$ such that $v_1$ and $w$ are adjacent and $v_2$ and $w$ are adjacent.

(a) Prove that if $v$ and $w$ are nonadjacent vertices in $G$, then $\delta(v) = \delta(w)$.

(b) Prove that if there is a vertex of degree $k > 1$ and no vertex is adjacent to all other vertices, then the degree of every vertex is $k$.

# Problem-Solving Corner Graphs

## Problem

Is it possible in a department of 25 persons, racked by dissension, for each person to get along with exactly five others?

## Attacking the Problem

Where do we start? Since this problem is in Chapter 8, which deals with graphs, it would probably be a good idea to try to model the problem as a graph. If this problem were not associated with a particular section or chapter in the book, we might try several approaches—one of which might be to model the problem as a graph. Many discrete problems can be solved by modeling them using graphs. This is not to say that this is the only approach possible. Most of the time by taking different approaches, we can solve a single problem in many ways. (A nice example is [Wagon].)

### Finding a Solution

A fundamental issue in building a graph model is to figure out what the graph is—what are the vertices, and what are the edges? In this problem, there's not much choice; we have persons and dissension. Let's try letting the vertices be the people. It's very common in a graph model for the edges to indicate a *relationship* between the vertices. Here the relationship is "gets along with," so we'll put an edge between two vertices (people) if they get along.

Now suppose that each person gets along with exactly five others. For example, in the figure that follows, which shows part of our graph, Jeremy gets along with Samantha, Alexandra, Lance, Bret, and Tiffany, and no others.

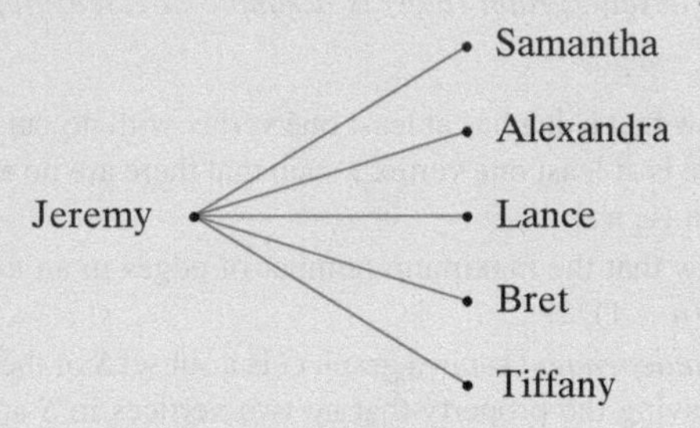


It follows that the degree of every vertex is 5. Now let's take stock of the situation: We have 25 vertices and each vertex has degree 5. Before reading on, try to determine whether this is possible.

Corollary 8.2.22 says that the number of vertices of odd degree is even. We have a contradiction because the number of vertices of odd degree is *odd*. Therefore, it is not possible in a department of 25 persons racked by dissension for each person to get along with exactly five others.

### Formal Solution

No. It is not possible in a department of 25 persons racked by dissension for each person to get along with exactly five others. Suppose by way of contradiction that it is possible. Consider a graph where the vertices are the persons and an edge connects two vertices (people) if the people get along. Since every vertex has odd degree, the number of vertices of odd degree is odd, which is a contradiction.

### Summary of Problem-Solving Techniques

- Many discrete problems can be solved by modeling them using graphs.
- To build a graph model, determine what the vertices represent and what the edges represent.
- It's very common in a graph model for the edges to indicate a relationship between the vertices.

## 8.3 Hamiltonian Cycles and the Traveling Salesperson Problem

**Go Online**
For more on Hamilton, Hamiltonian cycles, and the traveling salesperson problem, see bit.ly/2yUuHxX

Sir William Rowan Hamilton marketed a puzzle in the mid-1800s in the form of a dodecahedron (see Figure 8.3.1). Each corner bore the name of a city and the problem was to start at any city, travel along the edges, visit each city exactly one time, and return to the initial city. The graph of the edges of the dodecahedron is given in Figure 8.3.2. We can solve Hamilton's puzzle if we can find a cycle in the graph of Figure 8.3.2 that contains each vertex exactly once (except for the starting and ending vertex that appears twice). See if you can find a solution before looking at a solution given in Figure 8.3.3.

In honor of Hamilton, we call a cycle in a graph $G$ that contains each vertex in $G$ exactly once, except for the starting and ending vertex that appears twice, a **Hamiltonian cycle.**

**Go Online**
For a C program that determines whether a graph has a Hamiltonian cycle, see bit.ly/2EDLJ9Z

Hamilton (1805–1865) was one of Ireland's greatest scholars. He was professor of astronomy at the University of Dublin, where he published articles in physics and mathematics. In mathematics, Hamilton is most famous for inventing the quaternions, a generalization of the complex number system. The quaternions provided inspiration for the development of modern abstract algebra. In this connection, Hamilton introduced the term *vector*.

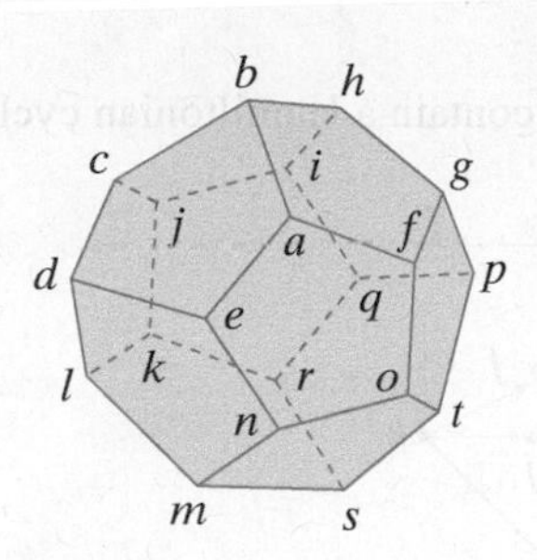


Figure 8.3.1 Hamilton's puzzle.

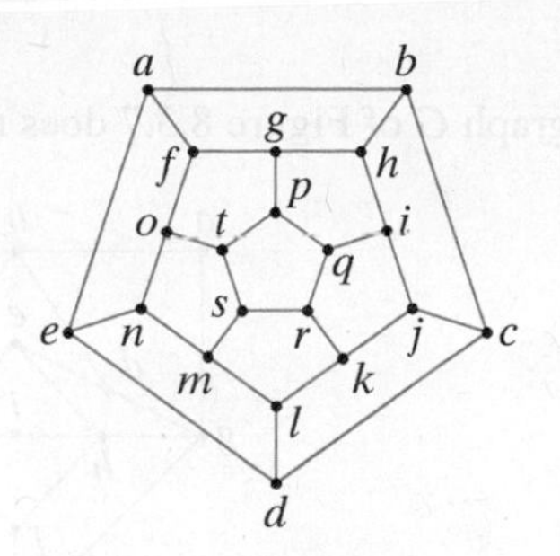


Figure 8.3.2 The graph of Hamilton's puzzle.

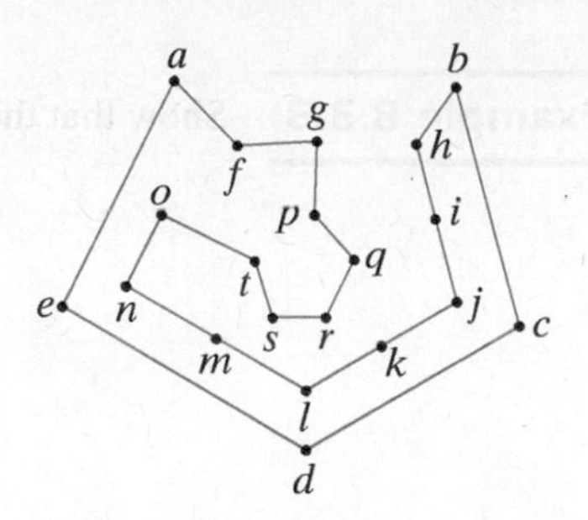


Figure 8.3.3 Visiting each vertex once in the graph of Figure 8.3.2.

**Example 8.3.1** The cycle $(a, b, c, d, e, f, g, a)$ is a Hamiltonian cycle for the graph of Figure 8.3.4. ◀

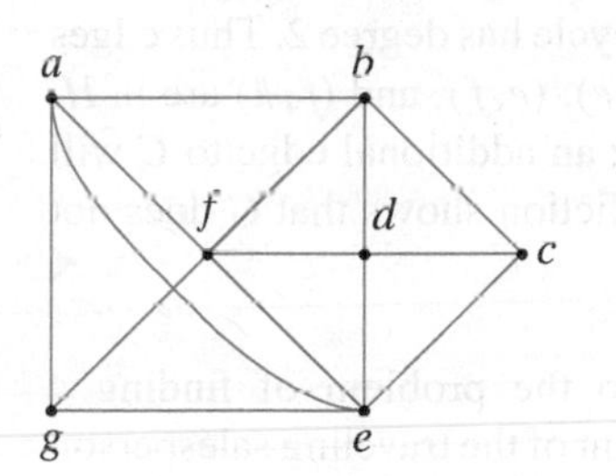


Figure 8.3.4 A graph with a Hamiltonian cycle.

The problem of finding a Hamiltonian cycle in a graph sounds similar to the problem of finding an Euler cycle in a graph. An Euler cycle visits each edge once, whereas a Hamiltonian cycle visits each vertex once; however, the problems are actually quite distinct. For example, the graph $G$ of Figure 8.3.4 does not have an Euler cycle since there are vertices of odd degree, yet Example 8.3.1 showed that $G$ has a Hamiltonian cycle. Furthermore, unlike the situation for Euler cycles (see Theorems 8.2.17 and 8.2.18), no easily verified necessary and sufficient conditions are known for the existence of a Hamiltonian cycle in a graph.

The following examples show that sometimes we can argue that a graph does not contain a Hamiltonian cycle.

**Example 8.3.2** Show that the graph of Figure 8.3.5 does not contain a Hamiltonian cycle.

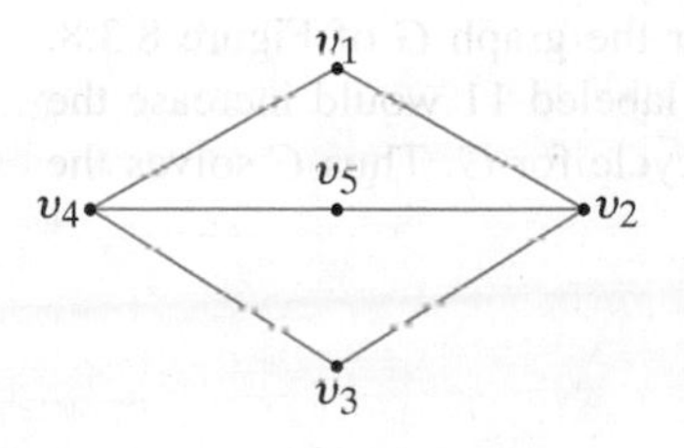


Figure 8.3.5 A graph with no Hamiltonian cycle.

**SOLUTION** Since there are five vertices, a Hamiltonian cycle must have five edges. Suppose that we could eliminate edges from the graph, leaving just a Hamiltonian cycle. We would have to eliminate one edge incident at $v_2$ and one edge incident at $v_4$, since each vertex in a Hamiltonian cycle has degree 2. But this leaves only four edges—not enough for a Hamiltonian cycle of length 5. Therefore, the graph of Figure 8.3.5 does not contain a Hamiltonian cycle. ◀

We must be careful not to count an eliminated edge more than once when using an argument like that in Example 8.3.2 to show that a graph does not have a Hamiltonian cycle. Notice in Example 8.3.2 (which refers to Figure 8.3.5) that if we eliminate one edge incident at $v_2$ and one edge incident at $v_4$, these edges are distinct. Therefore, we are correct in reasoning that we must eliminate two edges from the graph of Figure 8.3.5 to produce a Hamiltonian cycle.

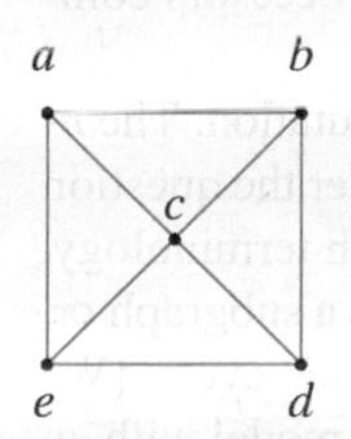


Figure 8.3.6 A graph with a Hamiltonian cycle.

As an example of double counting, consider the following *faulty* argument that purports to show that the graph of Figure 8.3.6 has no Hamiltonian cycle. Since there are five vertices, a Hamiltonian cycle must have five edges. Suppose that we could eliminate edges from the graph to produce a Hamiltonian cycle. We would have to eliminate two edges incident at $c$ and one edge incident at each of $a$, $b$, $d$, and $e$. This leaves two edges—not enough for a Hamiltonian cycle. Therefore, the graph of Figure 8.3.6 does not contain a Hamiltonian cycle. The error in this argument is that if we eliminate two edges incident at $c$ (as we must do), we also eliminate edges incident at two of $a$, $b$, $d$, or $e$. We must not count the two eliminated edges incident at the two vertices again. Notice that the graph of Figure 8.3.6 does have a Hamiltonian cycle.

**Example 8.3.3** Show that the graph $G$ of Figure 8.3.7 does not contain a Hamiltonian cycle.

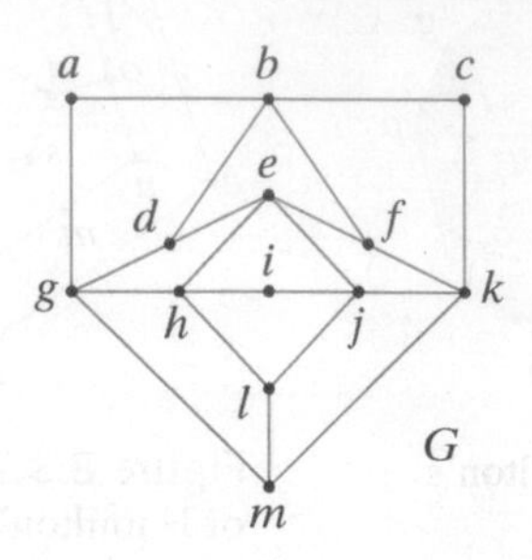


Figure 8.3.7 A graph with no Hamiltonian cycle.

**SOLUTION** Suppose that $G$ has a Hamiltonian cycle $H$. The edges $(a, b)$, $(a, g)$, $(b, c)$, and $(c, k)$ must be in $H$ since each vertex in a Hamiltonian cycle has degree 2. Thus edges $(b, d)$ and $(b, f)$ are not in $H$. Therefore, edges $(g, d)$, $(d, e)$, $(e, f)$, and $(f, k)$ are in $H$. The edges now known to be in $H$ form a cycle $C$. Adding an additional edge to $C$ will give some vertex in $H$ degree greater than 2. This contradiction shows that $G$ does not have a Hamiltonian cycle. ◀

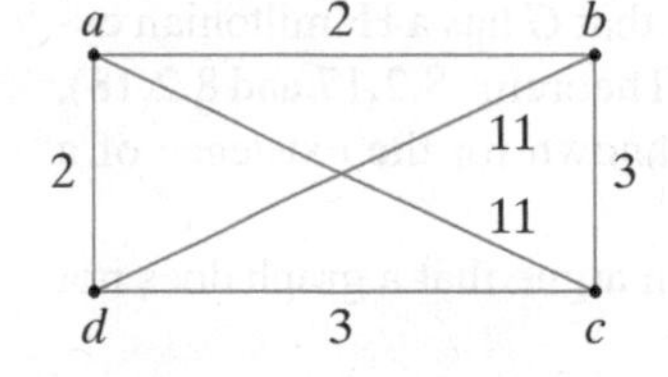


Figure 8.3.8 A graph for the traveling salesperson problem.

The **traveling salesperson problem** is related to the problem of finding a Hamiltonian cycle in a graph. (We referred briefly to a variant of the traveling salesperson problem in Section 8.1.) The problem is: Given a weighted graph $G$, find a minimum-length Hamiltonian cycle in $G$. If we think of the vertices in a weighted graph as cities and the edge weights as distances, the traveling salesperson problem is to find a shortest route in which the salesperson can visit each city one time, starting and ending at the same city.

**Example 8.3.4** The cycle $C = (a, b, c, d, a)$ is a Hamiltonian cycle for the graph $G$ of Figure 8.3.8. Replacing any of the edges in $C$ by either of the edges labeled 11 would increase the length of $C$; thus $C$ is a minimum-length Hamiltonian cycle for $G$. Thus $C$ solves the traveling salesperson problem for $G$. ◀

We next look at Hamiltonian cycles in the $n$-cube.

**Example 8.3.5** **Gray Codes and Hamiltonian Cycles in the $n$-Cube** Consider a **ring model** for parallel computation that, when represented as a graph, is a simple cycle (see Figure 8.3.9). The vertices represent processors. An edge between processors $p$ and $q$ indicates that $p$ and $q$ can communicate directly with one another. We see that each processor can communicate directly with exactly two other processors. Nonadjacent processors communicate by sending messages.

**Go Online**
For more on Gray codes, see
`bit.ly/2yUuHxX`

The $n$-cube (see Example 8.1.7) is another model for parallel computation. The $n$-cube has a greater degree of connectivity among its processors. We consider the question of when an $n$-cube can simulate a ring model with $2^n$ processors. In graph terminology, we are asking when the $n$-cube contains a simple cycle with $2^n$ vertices as a subgraph or, since the $n$-cube has $2^n$ processors, when the $n$-cube contains a Hamiltonian cycle. [We leave to the exercises the question of when an $n$-cube can simulate a ring model with an arbitrary number of processors (see Exercise 18).]

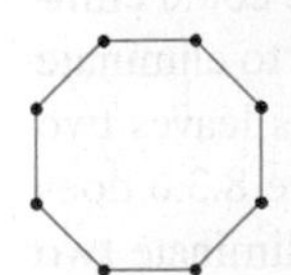

Figure 8.3.9 The ring model for parallel computation.

We first observe that if the $n$-cube contains a Hamiltonian cycle, we must have $n \geq 2$ since the 1-cube has no cycles at all.

Recall (see Example 8.1.7) that we may label the vertices of the $n$-cube $0, 1, \ldots, 2^n - 1$ in such a way that an edge connects two vertices if and only if the binary representation of their labels differs in exactly one bit. Thus the $n$-cube has a Hamiltonian cycle if and only if $n \geq 2$ and there is a sequence,

$$s_1, s_2, \ldots, s_{2^n} \tag{8.3.1}$$

where each $s_i$ is a string of $n$ bits, satisfying:

- Every $n$-bit string appears somewhere in the sequence.
- $s_i$ and $s_{i+1}$ differ in exactly one bit, $i = 1, \ldots, 2^n - 1$.
- $s_{2^n}$ and $s_1$ differ in exactly one bit.

A sequence (8.3.1) is called a **Gray code.** When $n \geq 2$, a Gray code (8.3.1) corresponds to the Hamiltonian cycle $s_1, s_2, \ldots, s_{2^n}, s_1$ since every vertex appears and the edges $(s_i, s_{i+1})$, $i = 1, \ldots, 2^n - 1$, and $(s_{2^n}, s_1)$ are distinct. When $n = 1$, the Gray code 0, 1 corresponds to the path (0, 1, 0), which is not a cycle because the edge (0, 1) is repeated.

Gray codes have been extensively studied in other contexts. For example, Gray codes have been used in converting analog information to digital form (see [Deo]). We show how to construct a Gray code for each positive integer $n$, thus proving that the $n$-cube has a Hamiltonian cycle for every positive integer $n \geq 2$. ◀

**Theorem 8.3.6** Let $G_1$ denote the sequence 0, 1. We define $G_n$ in terms of $G_{n-1}$ by the following rules:

(a) Let $G_{n-1}^R$ denote the sequence $G_{n-1}$ written in reverse.

(b) Let $G'_{n-1}$ denote the sequence obtained by prefixing each member of $G_{n-1}$ with 0.

(c) Let $G''_{n-1}$ denote the sequence obtained by prefixing each member of $G_{n-1}^R$ with 1.

(d) Let $G_n$ be the sequence consisting of $G'_{n-1}$ followed by $G''_{n-1}$.

Then $G_n$ is a Gray code for every positive integer $n$.

**Proof** We prove the theorem by induction on $n$.

**Basis Step ($n = 1$)**

Since the sequence 0, 1 is a Gray code, the theorem is true when $n$ is 1.

**Inductive Step**

Assume that $G_{n-1}$ is a Gray code. Each string in $G'_{n-1}$ begins with 0, so any difference between consecutive strings must result from differing bits in the corresponding strings in $G_{n-1}$. But since $G_{n-1}$ is a Gray code, each consecutive pair of strings in $G_{n-1}$ differs in exactly one bit. Therefore, each consecutive pair of strings in $G'_{n-1}$ differs in exactly one bit. Similarly, each consecutive pair of strings in $G''_{n-1}$ differs in exactly one bit.

Let $\alpha$ denote the last string in $G'_{n-1}$, and let $\beta$ denote the first string in $G''_{n-1}$. If we delete the first bit from $\alpha$ and the first bit from $\beta$, the resulting strings are identical. Since the first bit in $\alpha$ is 0 and the first bit in $\beta$ is 1, the last string in $G'_{n-1}$ and the first string in $G''_{n-1}$ differ in exactly one bit. Similarly, the first string in $G'_{n-1}$ and the last string in $G''_{n-1}$ differ in exactly one bit. Therefore, $G_n$ is a Gray code. ◀

**Corollary 8.3.7** The $n$-cube has a Hamiltonian cycle for every positive integer $n \geq 2$.

**Example 8.3.8** We use Theorem 8.3.6 to construct the Gray code $G_3$ beginning with $G_1$.

| | | | | | | | | |
|---|---|---|---|---|---|---|---|---|
| $G_1$: | 0 | 1 | | | | | | |
| $G_1^R$: | 1 | 0 | | | | | | |
| $G_1'$: | 00 | 01 | | | | | | |
| $G_1''$: | 11 | 10 | | | | | | |
| $G_2$: | 00 | 01 | 11 | 10 | | | | |
| $G_2^R$: | 10 | 11 | 01 | 00 | | | | |
| $G_2'$: | 000 | 001 | 011 | 010 | | | | |
| $G_2''$: | 110 | 111 | 101 | 100 | | | | |
| $G_3$: | 000 | 001 | 011 | 010 | 110 | 111 | 101 | 100 |

◀

We next examine a problem that goes back some 200 years.

**Example 8.3.9** **The Knight's Tour** In chess, the knight's move consists of moving two squares horizontally or vertically and then moving one square in the perpendicular direction. For example, in Figure 8.3.10 a knight on the square marked $K$ can move to any of the squares marked $X$. A **knight's tour of an $n \times n$ board** begins at some square, visits each square exactly once making legal moves, and returns to the initial square. The problem is to determine for which $n$ a knight's tour exists.

| | | | | |
|---|---|---|---|---|
| | $X$ | | $X$ | |
| $X$ | | | | $X$ |
| | | $K$ | | |
| $X$ | | | | $X$ |
| | $X$ | | $X$ | |

**Figure 8.3.10** The knight's legal moves in chess.

**Go Online**

For more on the knight's tour, see bit.ly/2yUuHxX

We can use a graph to model this problem. We let the squares of the board, alternately colored black and white in the usual way, be the vertices of the graph, and we place an edge between two vertices if the corresponding squares on the board represent a legal move for the knight (see Figure 8.3.11). We denote the graph as $GK_n$. Then there is a knight's tour on the $n \times n$ board if and only if $GK_n$ has a Hamiltonian cycle.

We show that if $GK_n$ has a Hamiltonian cycle, $n$ is even. To see this, note that $GK_n$ is bipartite. We can partition the vertices into sets $V_1$, those corresponding to the white squares, and $V_2$, those corresponding to the black squares; each edge is incident on a vertex in $V_1$ and $V_2$. Since any cycle must alternate between a vertex in $V_1$ and one in $V_2$, any cycle in $GK_n$ must have even length. But since a Hamiltonian cycle must visit each vertex exactly once, a Hamiltonian cycle in $GK_n$ must have length $n^2$. Thus $n$ must be even.

In view of the preceding result, the smallest possible board that might have a knight's tour is the $2 \times 2$ board, but it does not have a knight's tour because the board is so small the knight has no legal moves. The next smallest board that might have a knight's tour is the $4 \times 4$ board, although, as we shall show, it too does not have a knight's tour.

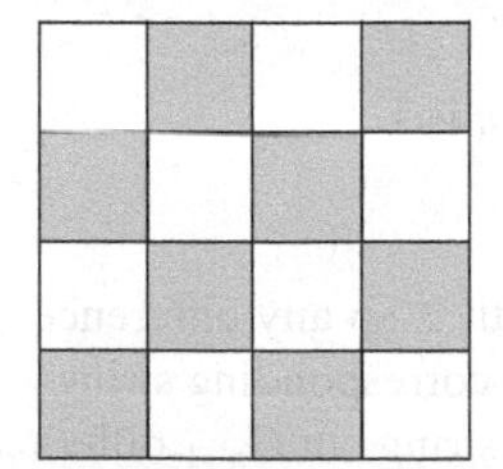

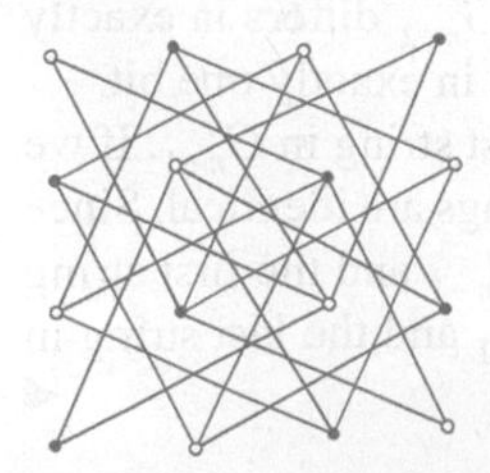

**Figure 8.3.11** A $4 \times 4$ chessboard and the graph $GK_4$.

We argue by contradiction to show that $GK_4$ does not have a Hamiltonian cycle. Suppose that $GK_4$ has a Hamiltonian cycle $C = (v_1, v_2, \ldots, v_{17})$. We assume that $v_1$ corresponds to the upper-left square. We call the eight squares across the top and bottom *outside squares,* and we call the other eight squares *inside squares*. Notice that the knight must arrive at an outside square from an inside square and that the knight must move from an outside square to an inside square. Thus in the cycle $C$, each vertex corresponding to an outside square must be preceded and followed by a vertex corresponding to an inside square. Since there are equal numbers of outside and inside squares, vertices $v_i$ where $i$ is odd correspond to outside squares, and vertices $v_i$ where $i$ is even correspond to inside squares. But looking at the moves the knight makes, we see that vertices $v_i$ where $i$ is odd correspond to white squares, and vertices $v_i$ where $i$ is even correspond to black squares.

Therefore, the only outside squares visited are white and the only inside squares visited are black. Thus $C$ is not a Hamiltonian cycle. This contradiction completes the proof that $GK_4$ has no Hamiltonian cycle. This argument was given by Louis Pósa when he was a teenager.

The graph $GK_6$ has a Hamiltonian cycle. This fact can be proved by simply exhibiting one (see Exercise 21). It can be shown using elementary methods that $GK_n$ has a Hamiltonian cycle for all even $n \geq 6$ (see [Schwenk]). The proof explicitly constructs Hamiltonian cycles for certain smaller boards and then pastes smaller boards together to obtain Hamiltonian cycles for the larger boards. ◀

Although there are algorithms (see e.g., [Even, 1979]) for finding an Euler cycle in an $n$-edge graph if there is one in time $\Theta(n)$, there are no known polynomial-time algorithms for finding a Hamiltonian cycle in a graph. Indeed, it is known that the Hamiltonian cycle problem is NP-complete (see Section 4.3), which means that if someone discovers a polynomial-time algorithm to solve the Hamiltonian-cycle problem, *all* NP-complete problems will then have polynomial-time algorithms. The traveling salesperson problem is also NP-complete.

Because of the lack of a polynomial-time algorithm, other approaches to the traveling salesperson problem have been proposed, for example, finding an *approximate* solution (i.e., a near-minimum-length Hamiltonian cycle). Other approaches to finding a Hamiltonian cycle have also been suggested. We discuss a randomized algorithm (see Section 4.2) that searches for a Hamiltonian cycle.

Our randomized algorithm is a version of an algorithm due to Pósa (see [Pósa]). The idea is to build a Hamiltonian cycle vertex by vertex. We assume that our graph $G$ is a simple graph with $n$ vertices. We begin with an arbitrary vertex, which we call $v_1$. Our initial path consists of the vertex $v_1$ and no edges. At each successive step, we try to extend the current path by adding a new edge and vertex to the end of the current path. Thus, at the first step, we select a random, adjacent vertex of $v_1$, which we call $v_2$. Our path is now $(v_1, v_2)$. At the $i$th step, we have the path $(v_1, v_2, \ldots, v_i)$. If the path contains all of the vertices of $G$, that is, if $i = n$, we check whether $v_n$ is adjacent to $v_1$. If so, we have found a Hamiltonian cycle, namely, $(v_1, v_2, \ldots, v_n, v_1)$. In this case, we return true. If $i \neq n$, we select a random, adjacent vertex of $v_i$, which is not on the path, and continue with the new path $(v_1, v_2, \ldots, v_i, v_{i+1})$. If all of the vertices adjacent to $v_i$ are already on the path, we randomly select one of them different from $v_{i-1}$, which we call $v_j$, and change the path by reversing the order of the vertices following $v_j$, that is, the path $(v_1, v_2, \ldots, v_j, v_{j+1}, \ldots, v_i)$ becomes $(v_1, v_2, \ldots, v_j, v_i, v_{i-1}, \ldots, v_{j+1})$ (see Figure 8.3.12). If the only vertex on the path adjacent to $v_i$ is $v_{i-1}$, the preceding adjustment of path would not change the path. In this case, we have found a vertex of degree 1. Since there is no Hamiltonian cycle, we return false.

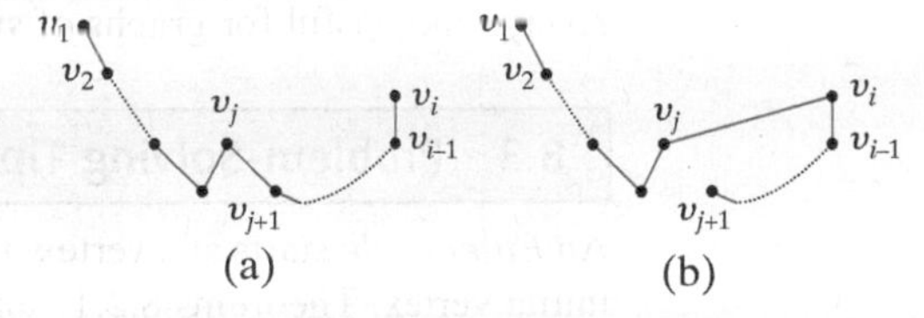


**Figure 8.3.12** Given a path $(v_1, \ldots, v_j, v_{j+1}, \ldots, v_i)$, as shown in (a), where $v_i$ is adjacent to $v_j$, we change the path to $(v_1, \ldots, v_j, v_i, \ldots, v_{j+1})$, as shown in (b).

**Algorithm 8.3.10 Randomized Hamiltonian Cycle**

This algorithm takes as an input a simple graph $G = (V, E)$ and searches for a Hamiltonian cycle. If the algorithm returns true, it has found a Hamiltonian cycle. If the algorithm returns false, there is no Hamiltonian cycle (in fact, there is a vertex of degree 1). If the algorithm does not terminate, there may or may not be a Hamiltonian cycle. If $v$ is a vertex, $N(v)$ is the set of vertices adjacent to $v$.

Input: A simple graph $G = (V, E)$ with $n$ vertices

Output: If the algorithm terminates, it returns true if it finds a Hamiltonian cycle and false otherwise.

```
randomized_hamiltonian_cycle (G, n) {
  if (n == 1 ∨ n == 2) // trivial cases
     return false
  v_1 = random vertex in G
  i = 1
  while (i ¬= n ∨ v_1 ∉ N(v_i)) {
     N = N(v_i) − {v_1, ..., v_{i−1}} // if i is 1, {v_1, ..., v_{i−1}} is ∅
     // N contains the vertices adjacent to v_i (the current last vertex
     // of the path) that are not already on the path
     if (N ≠ ∅) {
        i = i + 1
        v_i = random vertex in N
     }
     else if (v_j ∈ N(v_i) for some j, 1 ≤ j < i − 1)
        (v_1, ..., v_i) = (v_1, ..., v_j, v_i, ..., v_{j+1})
     else
        return false
  }
  return true
}
```

If *randomized_hamiltonian_cycle* returns true, it has indeed found a Hamiltonian cycle, which is stored in variables $v_1, \ldots, v_n, v_1$. If the graph input to *randomized_hamiltonian_cycle* has Hamiltonian cycles, it may fail to find one, and, in this case, it will not terminate (see Exercise 25). If there are vertices of degree 1, the algorithm may not find one and, in this case, it will not terminate (see Exercise 28). If there is no Hamiltonian cycle and all vertices have degree at least 2, the algorithm will not terminate (see Exercise 29). Failure to terminate is common for randomized algorithms. One way to guarantee termination is to simply stop the algorithm after some specified number of iterations. Experiments with the algorithm show that it is very good at finding Hamiltonian cycles in large graphs, and there are theoretical results proving that it is almost always successful for graphs of sufficiently large minimum degree.

## 8.3 Problem-Solving Tips

An *Euler cycle* starts at a vertex, traverses each *edge* exactly one time, and returns to the initial vertex. Theorems 8.2.17 and 8.2.18 allow us to easily determine whether a graph has an Euler cycle: A graph $G$ has an Euler cycle if and only if $G$ is connected and every vertex has even degree.

A *Hamiltonian cycle* starts at a vertex, visits each *vertex* exactly one time (except for the initial vertex, which is visited twice: at the beginning and end of the cycle), and returns to the initial vertex. Unlike Theorems 8.2.17 and 8.2.18, no easily verified necessary and sufficient condition is known for a graph to have a Hamiltonian cycle. If a relatively small graph has a Hamiltonian cycle, trial and error will discover one. If a graph does not have a Hamiltonian cycle, you can sometimes use the fact that a Hamiltonian cycle in a graph containing $n$ vertices has length $n$ together with proof by contradiction to prove that it does not have a Hamiltonian cycle. Two proof-by-contradiction techniques were shown in Section 8.3. In the first, we assume that the graph has a Hamiltonian cycle. Certain edges cannot appear in the Hamiltonian cycle: If a graph has a vertex $v$ of degree greater than 2, only two edges incident on $v$ can appear in the Hamiltonian cycle. We can sometimes obtain a contradiction by showing that so many edges must be eliminated that the graph cannot have a Hamiltonian cycle (see Example 8.3.2).

In the second proof-by-contradiction technique shown in Section 8.3, we again assume that the graph with $n$ vertices has a Hamiltonian cycle. We then argue that certain edges *must* be in the Hamiltonian cycle. For example, if a vertex $v$ has degree 2, both edges incident on $v$ must be in the cycle. We can sometimes obtain a contradiction by showing that edges that must be in the Hamiltonian cycle form a cycle of length less than $n$ (see Example 8.3.3).

## 8.3 Review Exercises

1. What is a Hamiltonian cycle?
2. Give an example of a graph that has a Hamiltonian cycle and an Euler cycle. Prove that the graph has the specified properties.
3. Give an example of a graph that has a Hamiltonian cycle but not an Euler cycle. Prove that the graph has the specified properties.
4. Give an example of a graph that does not have a Hamiltonian cycle but does have an Euler cycle. Prove that the graph has the specified properties.
5. Give an example of a graph that has neither a Hamiltonian cycle nor an Euler cycle. Prove that the graph has the specified properties.
6. What is the traveling salesperson problem? How is it related to the Hamiltonian cycle problem?
7. What is the ring model for parallel computation?
8. What is a Gray code?
9. Explain how to construct a Gray code.
10. Explain how the randomized Hamiltonian algorithm (Algorithm 8.3.10) works.

## 8.3 Exercises

*Find a Hamiltonian cycle in each graph.*

1.

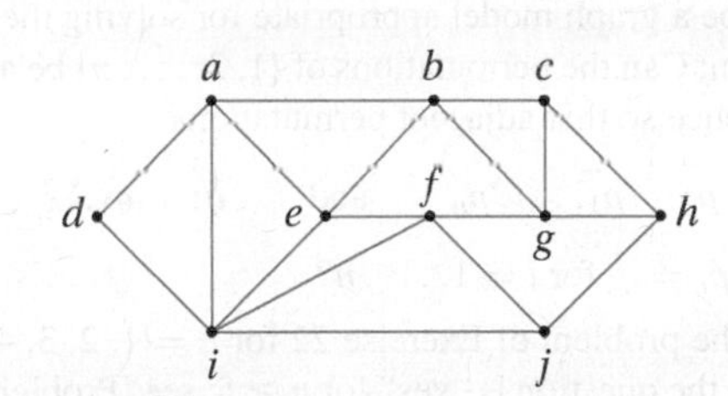


2.

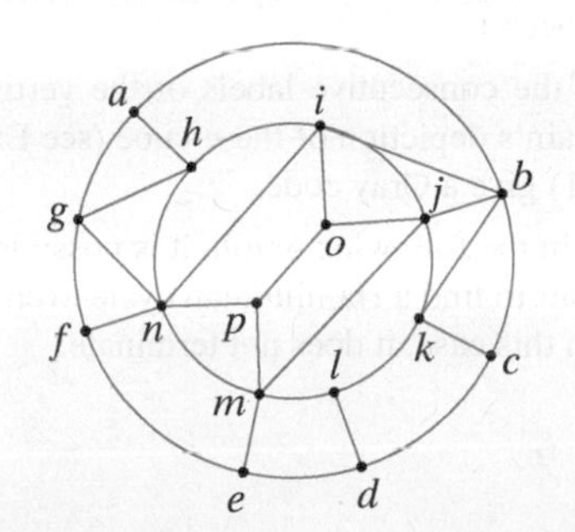


*Show that none of the graphs contains a Hamiltonian cycle.*

3.

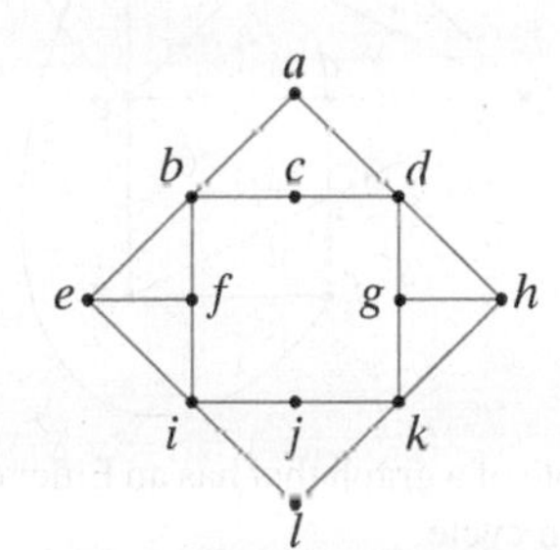


4.

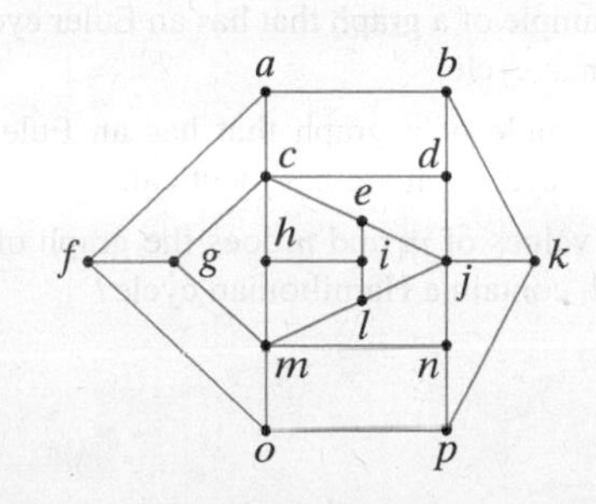

**5.**

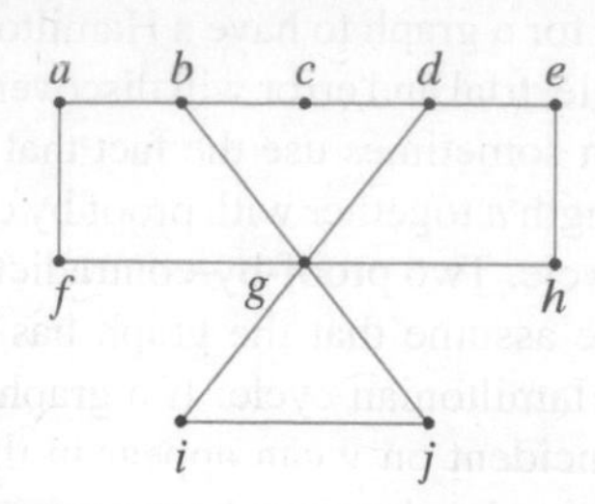


*Determine whether or not each graph contains a Hamiltonian cycle. If there is a Hamiltonian cycle, exhibit it; otherwise, give an argument that shows there is no Hamiltonian cycle.*

**6.**

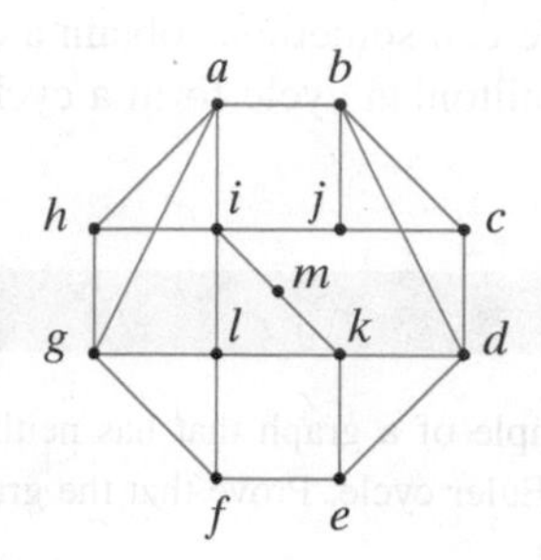


**7.**

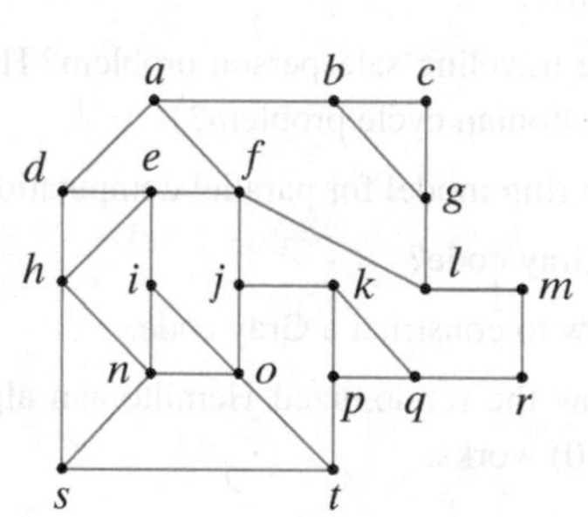


**8.**

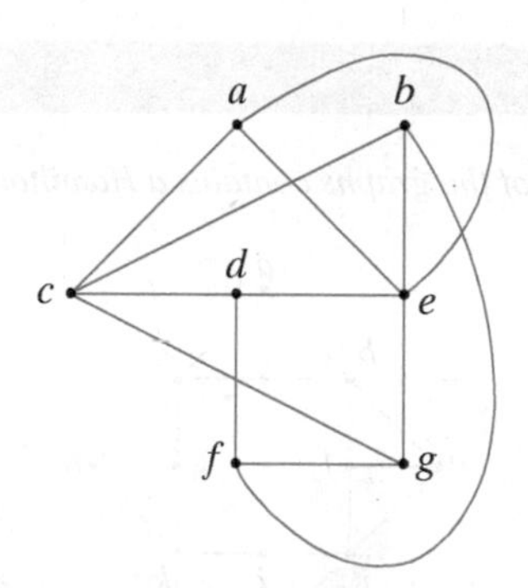


**9.** Give an example of a graph that has an Euler cycle but contains no Hamiltonian cycle.

**10.** Give an example of a graph that has an Euler cycle that is also a Hamiltonian cycle.

**11.** Give an example of a graph that has an Euler cycle and a Hamiltonian cycle that are not identical.

★**12.** For which values of $m$ and $n$ does the graph of Exercise 37, Section 8.2, contain a Hamiltonian cycle?

**13.** Modify the graph of Exercise 37, Section 8.2, by inserting an edge between the vertex in row $i$, column 1, and the vertex in row $i$, column $m$, for $i = 1, \ldots, n$. Show that the resulting graph always has a Hamiltonian cycle.

**14.** Show that if $n \geq 3$, the complete graph on $n$ vertices $K_n$ contains a Hamiltonian cycle.

**15.** When does the complete bipartite graph $K_{m,n}$ contain a Hamiltonian cycle?

**16.** Show that the cycle $(e, b, a, c, d, e)$ provides a solution to the traveling salesperson problem for the graph shown.

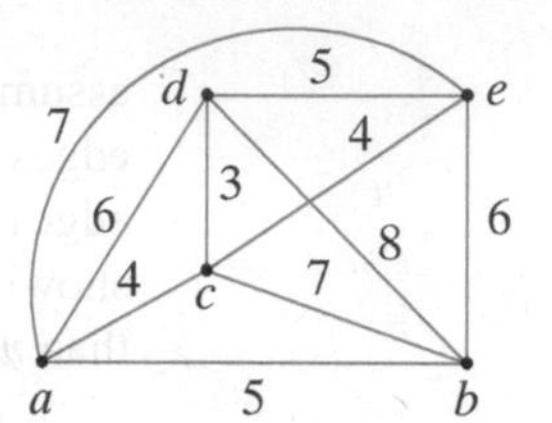


**17.** Solve the traveling salesperson problem for the graph shown.

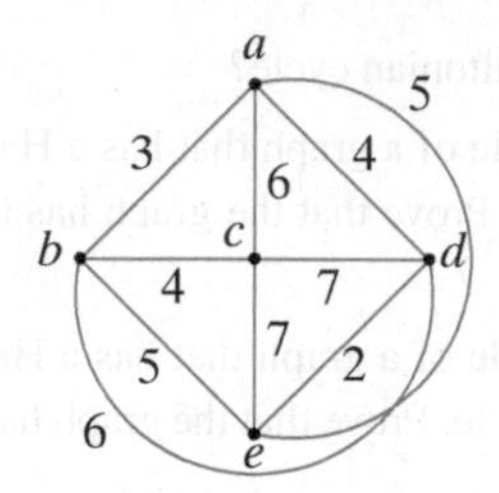


★**18.** Let $m$ and $n$ be integers satisfying $1 \leq m \leq 2^n$. Prove that the $n$-cube has a simple cycle of length $m$ if and only if $m \geq 4$ and $m$ is even.

**19.** Use Theorem 8.3.6 to compute the Gray code $G_4$.

**20.** Let $G$ be a bipartite graph with disjoint vertex sets $V_1$ and $V_2$, as in Definition 8.1.11. Show that if $G$ has a Hamiltonian cycle, $V_1$ and $V_2$ have the same number of elements.

**21.** Find a Hamiltonian cycle in $GK_6$ (see Example 8.3.9).

**22.** Describe a graph model appropriate for solving the following problem: Can the permutations of $\{1, 2, \ldots, n\}$ be arranged in a sequence so that adjacent permutations

$$p: \quad p_1, \ldots, p_n \qquad \text{and} \qquad q: \quad q_1, \ldots, q_n$$

satisfy $p_i \neq q_i$ for $i = 1, \ldots, n$?

**23.** Solve the problem of Exercise 22 for $n = 1, 2, 3, 4$. (The answer to the question is "yes" for $n \geq 5$; see [Problem 1186] in the References.)

**24.** Show that the consecutive labels of the vertices on the unit circle in Bain's depiction of the $n$-cube (see Exercises 43–45, Section 8.1) give a Gray code.

**25.** Show that in the following graph, it is possible for Algorithm 8.3.10 to fail to find a Hamiltonian cycle even though there is one and, in this case, it does not terminate.

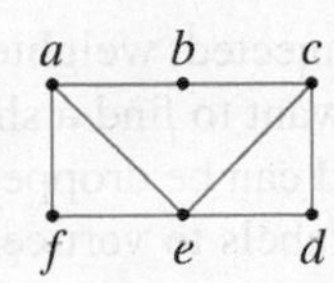


26. Show that in the graph of Exercise 25, if the first two choices of vertices for Algorithm 8.3.10 are $a$ and then $f$, Algorithm 8.3.10 will always find a Hamiltonian cycle.
27. Show that in the following graph, Algorithm 8.3.10 will always find a Hamiltonian cycle.

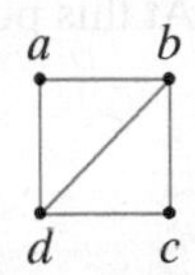


28. Show that in the following graph, Algorithm 8.3.10 may not find the vertex of degree 1, and, in this case, it will not terminate.

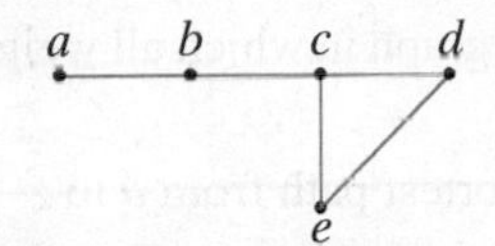


29. Prove that if a graph with no Hamiltonian cycle and all vertices of degree at least 2 is input to Algorithm 8.3.10, the algorithm will not terminate.
30. Answer true or false, and prove your response. In Algorithm 8.3.10, when a vertex is added to the path, it is never removed from the path.
31. Does Algorithm 8.3.10 always terminate when the input is $K_n$, the complete graph on $n$ vertices? Prove your answer.
32. For which $n$ does Algorithm 8.3.10 always find a Hamiltonian cycle when the input is $K_n$, the complete graph on $n$ vertices? Prove your answer.
33. Does Algorithm 8.3.10 always terminate when the input is $K_{m,n}$, the complete bipartite graph on $m$ and $n$ vertices? Prove your answer.
34. Does Algorithm 8.3.10 always find a Hamiltonian cycle if there is one when the input is $K_{m,n}$, the complete bipartite graph on $m$ and $n$ vertices? Prove your answer.
35. Is there a graph containing a Hamiltonian cycle for which Algorithm 8.3.10 always fails; that is, for which whatever the sequence of random guesses made by the algorithm, the Hamiltonian cycle is not found? Prove your answer.
36. Suggest ways to improve Algorithm 8.3.10.

*A Hamiltonian path in a graph G is a simple path that contains every vertex in G exactly once. (A Hamiltonian path begins and ends at different vertices.)*

37. If a graph has a Hamiltonian cycle, must it have a Hamiltonian path? Explain.
38. If a graph has a Hamiltonian path, must it have a Hamiltonian cycle? Explain.
39. Does the graph of Figure 8.3.5 have a Hamiltonian path?
40. Does the graph of Figure 8.3.7 have a Hamiltonian path?
41. Does the graph of Exercise 3 have a Hamiltonian path?
42. Does the graph of Exercise 4 have a Hamiltonian path?
43. Does the graph of Exercise 5 have a Hamiltonian path?
44. Does the graph of Exercise 6 have a Hamiltonian path?
45. Does the graph of Exercise 7 have a Hamiltonian path?
46. Does the graph of Exercise 8 have a Hamiltonian path?
47. For which values of $m$ and $n$ does the graph of Exercise 37, Section 8.2, have a Hamiltonian path?
48. For which $n$ does the complete graph on $n$ vertices have a Hamiltonian path?
49. Modify Algorithm 8.3.10 for finding a Hamiltonian cycle to search for a Hamiltonian path.

## 8.4 A Shortest-Path Algorithm

Recall (see Section 8.1) that a weighted graph is a graph in which values are assigned to the edges and that the length of a path in a weighted graph is the sum of the weights of the edges in the path. We let $w(i, j)$ denote the weight of edge $(i, j)$. In weighted graphs, we often want to find a **shortest path** (i.e., a path having minimum length) between two given vertices. Algorithm 8.4.1, due to E. W. Dijkstra, which efficiently solves this problem, is the topic of this section.

Edsger W. Dijkstra (1930–2002) was born in The Netherlands. He was an early proponent of programming as a science. So dedicated to programming was he that when he was married in 1957, he listed his profession as a programmer. However, the Dutch authorities said that there was no such profession, and he had to change the entry to "theoretical physicist." He won the prestigious Turing Award from the Association for Computing Machinery in 1972. He was appointed to the Schlumberger Centennial Chair in Computer Science at the University of Texas at Austin in 1984 and retired as Professor Emeritus in 1999.

Throughout this section, $G$ denotes a connected, weighted graph. We assume that the weights are positive numbers and that we want to find a shortest path from vertex $a$ to vertex $z$. The assumption that $G$ is connected can be dropped (see Exercise 9).

**Go Online**
For more on the shortest-path algorithm, see
`bit.ly/2yUuHxX`

Dijkstra's algorithm involves assigning labels to vertices. We let $L(v)$ denote the label of vertex $v$. At any point, some vertices have temporary labels and the rest have permanent labels. We let $T$ denote the set of vertices having temporary labels. In illustrating the algorithm, we will circle vertices having permanent labels. We will show later that if $L(v)$ is the permanent label of vertex $v$, then $L(v)$ is the length of a shortest path from $a$ to $v$. Initially, all vertices have temporary labels. Each iteration of the algorithm changes the status of one label from temporary to permanent; thus we may terminate the algorithm when $z$ receives a permanent label. At this point $L(z)$ gives the length of a shortest path from $a$ to $z$.

**Algorithm 8.4.1**

### Dijkstra's Shortest-Path Algorithm

This algorithm finds the length of a shortest path from vertex $a$ to vertex $z$ in a connected, weighted graph. The weight of edge $(i, j)$ is $w(i, j) > 0$ and the label of vertex $x$ is $L(x)$. At termination, $L(z)$ is the length of a shortest path from $a$ to $z$.

Input: A connected, weighted graph in which all weights are positive; vertices $a$ and $z$

Output: $L(z)$, the length of a shortest path from $a$ to $z$

```
 1. dijkstra(w, a, z, L) {
 2.    L(a) = 0
 3.    for all vertices x ≠ a
 4.       L(x) = ∞
 5.    T = set of all vertices
 6.    // T is the set of vertices whose shortest distance from a has
 7.    // not been found
 8.    while (z ∈ T) {
 9.       choose v ∈ T with minimum L(v)
10.       T = T − {v}
11.       for each x ∈ T adjacent to v
12.          L(x) = min{L(x), L(v) + w(v, x)}
13.    }
14. }
```

**Example 8.4.2** We show how Algorithm 8.4.1 finds a shortest path from $a$ to $z$ in the graph of Figure 8.4.1. (The vertices in $T$ are uncircled and have temporary labels. The circled vertices have permanent labels.) Figure 8.4.2 shows the result of executing lines 2–5. At line 8, $z$ is not circled. We proceed to line 9, where we select vertex $a$, the uncircled vertex with the smallest label, and circle it (see Figure 8.4.3). At lines 11 and 12 we update each of the uncircled vertices, $b$ and $f$, adjacent to $a$. We obtain the new labels

$$L(b) = \min\{\infty, 0 + 2\} = 2, \qquad L(f) = \min\{\infty, 0 + 1\} = 1$$

(see Figure 8.4.3). At this point, we return to line 8.

Since $z$ is not circled, we proceed to line 9, where we select vertex $f$, the uncircled vertex with the smallest label, and circle it (see Figure 8.4.4). At lines 11 and 12 we update each label of the uncircled vertices, $d$ and $g$, adjacent to $f$. We obtain the labels shown in Figure 8.4.4.

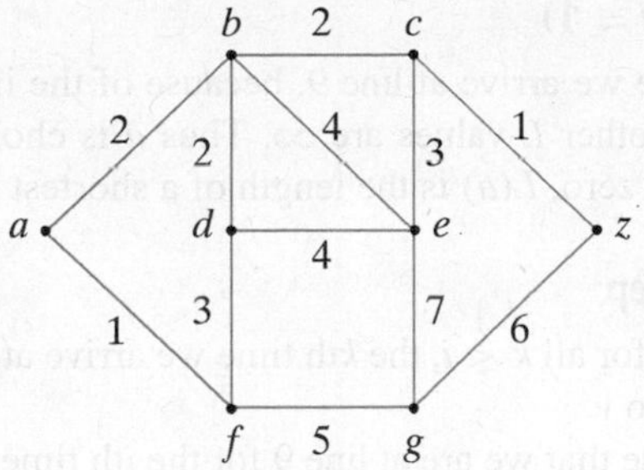

Figure 8.4.1 The graph for Example 8.4.2.

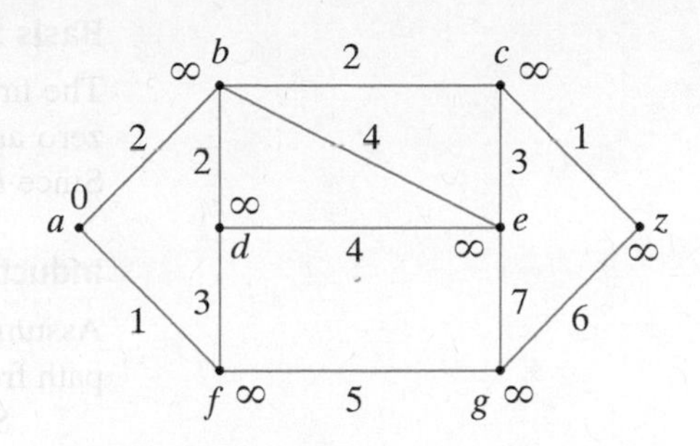

Figure 8.4.2 Initialization in Dijkstra's shortest-path algorithm.

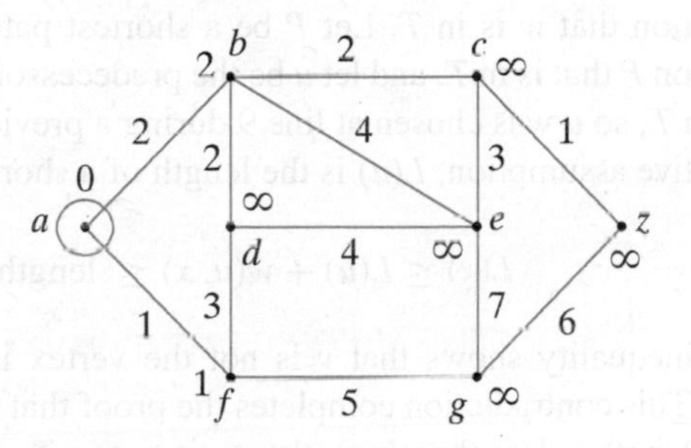

Figure 8.4.3 The first iteration of Dijkstra's shortest-path algorithm.

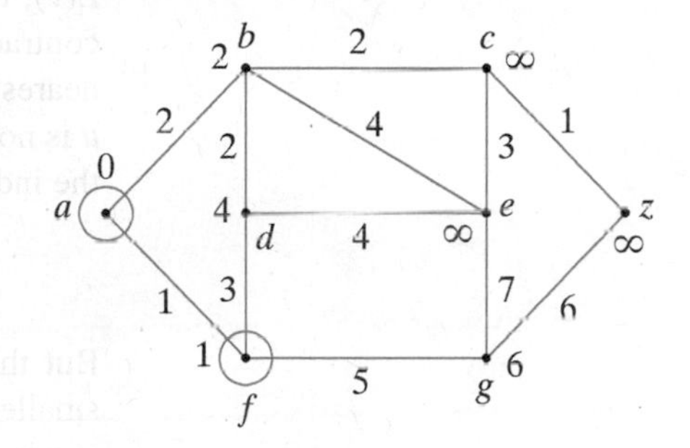

Figure 8.4.4 The second iteration of Dijkstra's shortest-path algorithm.

You should verify that the next iteration of the algorithm produces the labeling shown in Figure 8.4.5 and that at the termination of the algorithm, $z$ is labeled 5, indicating that the length of a shortest path from $a$ to $z$ is 5. A shortest path is given by $(a, b, c, z)$.

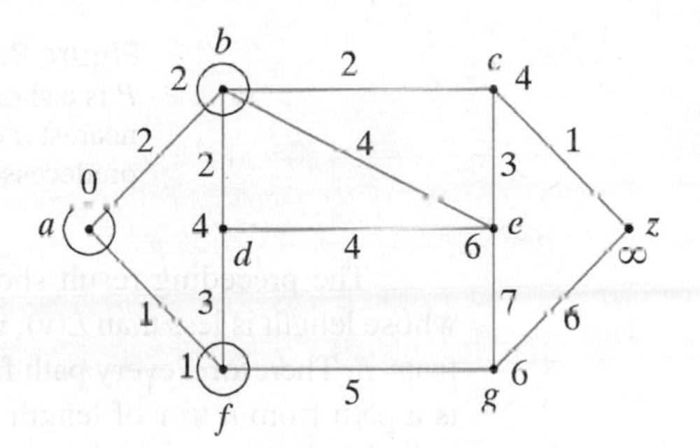

Figure 8.4.5 The third iteration of Dijkstra's shortest-path algorithm.

◀

We next show that Algorithm 8.4.1 is correct. The proof hinges on the fact that Dijkstra's algorithm finds the lengths of shortest paths from $a$ in nondecreasing order.

**Theorem 8.4.3** Dijkstra's shortest-path algorithm (Algorithm 8.4.1) correctly finds the length of a shortest path from $a$ to $z$.

**Proof** We use mathematical induction on $i$ to prove that the $i$th time we arrive at line 9, $L(v)$ is the length of a shortest path from $a$ to $v$. When this is proved, correctness of the algorithm follows since when $z$ is chosen at line 9, $L(z)$ will give the length of a shortest path from $a$ to $z$.

**Basis Step ($i = 1$)**

The first time we arrive at line 9, because of the initialization steps (lines 2–4), $L(a)$ is zero and all other $L$-values are $\infty$. Thus $a$ is chosen the first time we arrive at line 9. Since $L(a)$ is zero, $L(a)$ is the length of a shortest path from $a$ to $a$.

**Inductive Step**

Assume that for all $k < i$, the $k$th time we arrive at line 9, $L(v)$ is the length of a shortest path from $a$ to $v$.

Suppose that we are at line 9 for the $i$th time and we choose $v$ in $T$ with minimum value $L(v)$.

First we show that if there is a path from $a$ to a vertex $w$ whose length is less than $L(v)$, then $w$ is not in $T$ (i.e., $w$ was previously selected at line 9). Suppose by way of contradiction that $w$ is in $T$. Let $P$ be a shortest path from $a$ to $w$, let $x$ be the vertex nearest $a$ on $P$ that is in $T$, and let $u$ be the predecessor of $x$ on $P$ (see Figure 8.4.6). Then $u$ is not in $T$, so $u$ was chosen at line 9 during a previous iteration of the while loop. By the inductive assumption, $L(u)$ is the length of a shortest path from $a$ to $u$. Now

$$L(x) \leq L(u) + w(u, x) \leq \text{ length of } P < L(v).$$

But this inequality shows that $v$ is not the vertex in $T$ with minimum $L(v)$ [$L(x)$ is smaller]. This contradiction completes the proof that if there is a path from $a$ to a vertex $w$ whose length is less than $L(v)$, then $w$ is not in $T$.

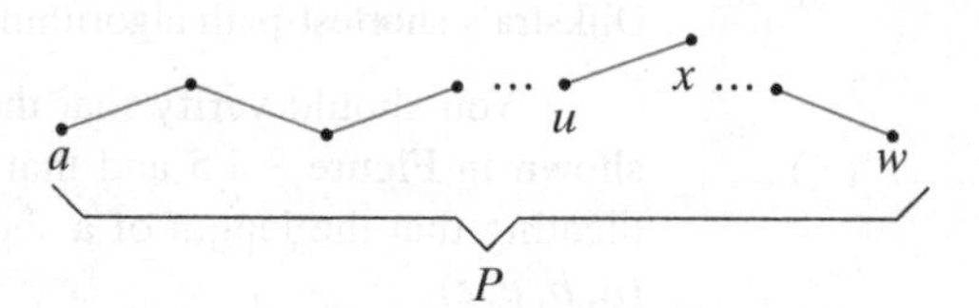


**Figure 8.4.6** The proof of Theorem 8.4.3. $P$ is a shortest path from $a$ to $w$, $x$ is the vertex nearest $a$ on $P$ that is in $T$, and $u$ is the predecessor of $x$ on $P$.

The preceding result shows, in particular, that if there were a path from $a$ to $v$ whose length is less than $L(v)$, $v$ would already have been selected at line 9 and removed from $T$. Therefore, every path from $a$ to $v$ has length at least $L(v)$. By construction, there is a path from $a$ to $v$ of length $L(v)$, so this is a shortest path from $a$ to $v$. The proof is complete. ◀

Algorithm 8.4.1 finds the length of a shortest path from $a$ to $z$. In most applications, we would also want to identify a shortest path. A slight modification of Algorithm 8.4.1 allows us to find a shortest path.

**Example 8.4.4** Find a shortest path from $a$ to $z$ and its length for the graph of Figure 8.4.7.

SOLUTION We will apply Algorithm 8.4.1 with a slight modification. In addition to circling a vertex, we will also label it with the name of the vertex from which it was labeled.

Figure 8.4.7 shows the result of executing lines 2–4 of Algorithm 8.4.1. First, we circle $a$ (see Figure 8.4.8). Next, we label the vertices $b$ and $d$ adjacent to $a$. Vertex $b$ is labeled "$a$, 2" to indicate its value and the fact that it was labeled from $a$. Similarly, vertex $d$ is labeled "$a$, 1."

Next, we circle vertex $d$ and update the label of the vertex $e$ adjacent to $d$ (see Figure 8.4.9). Then we circle vertex $b$ and update the labels of vertices $c$ and $e$ (see Figure 8.4.10). Next, we circle vertex $e$ and update the label of vertex $z$ (see Figure 8.4.11).

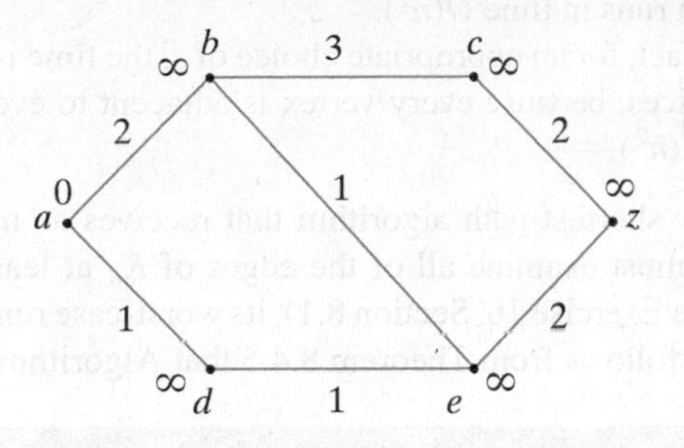


**Figure 8.4.7** Initialization in Dijkstra's shortest-path algorithm.

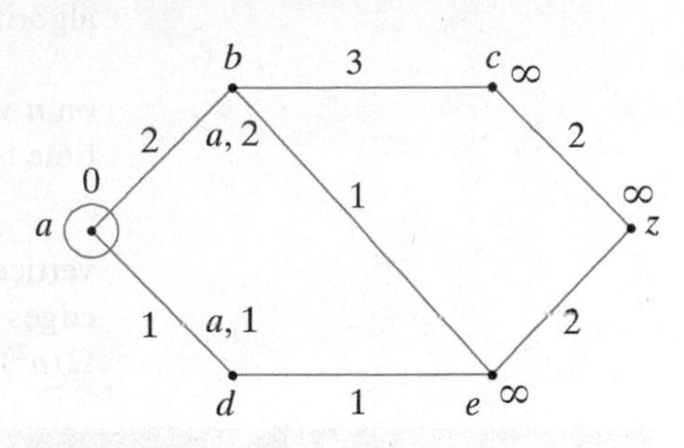


**Figure 8.4.8** The first iteration of Dijkstra's shortest-path algorithm.

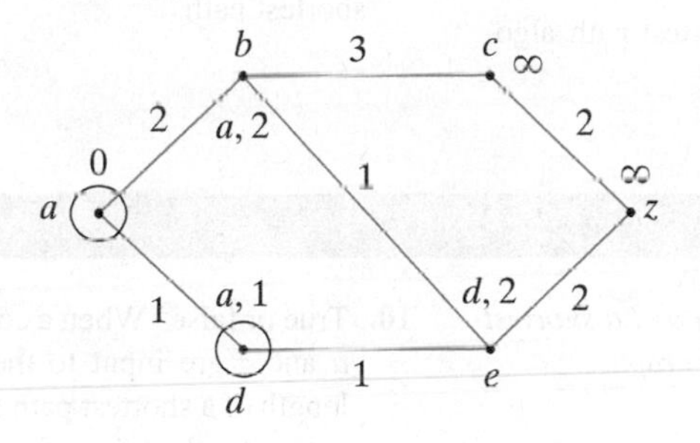


**Figure 8.4.9** The second iteration of Dijkstra's shortest-path algorithm.

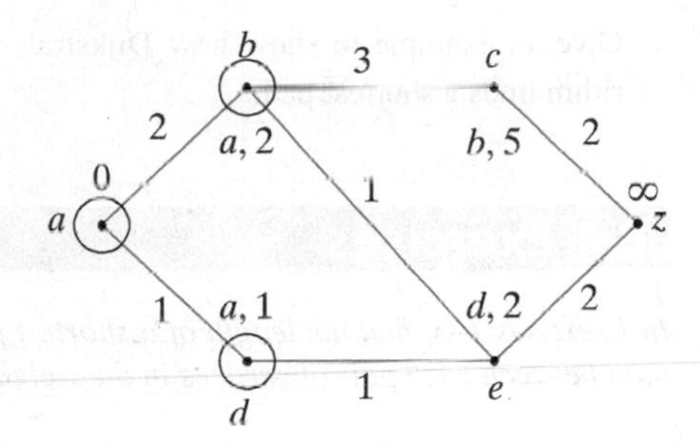


**Figure 8.4.10** The third iteration of Dijkstra's shortest-path algorithm.

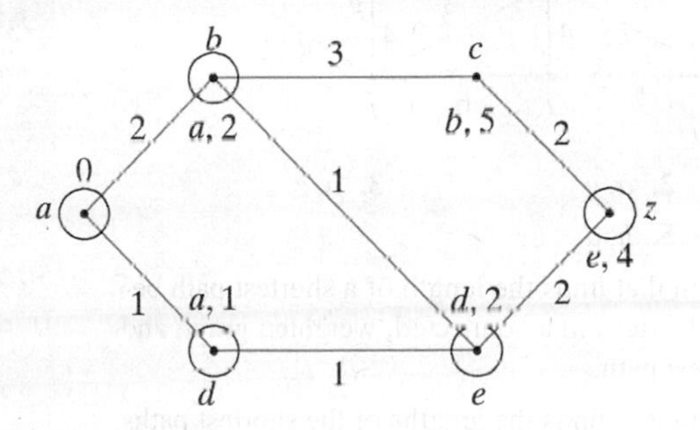


**Figure 8.4.11** The conclusion of Dijkstra's shortest-path algorithm.

At this point, we may circle $z$, so the algorithm terminates. The length of a shortest path from $a$ to $z$ is 4. Starting at $z$, we can retrace the labels to find the shortest path

$$(a, d, e, z).$$

Our next theorem shows that Dijkstra's algorithm is $\Theta(n^2)$ in the worst case.

**Theorem 8.4.5** For input consisting of an $n$-vertex, simple, connected, weighted graph, Dijkstra's algorithm (Algorithm 8.4.1) has worst-case run time $\Theta(n^2)$.

**Proof** We consider the time spent in the loops, which provides an upper bound on the total time. Line 4 is executed $O(n)$ times. Within the while loop, line 9 takes time

$O(n)$ [we could find the minimum $L(v)$ by examining all the vertices in $T$]. The body of the for loop (line 12) takes time $O(n)$. Since lines 9 and 12 are nested within a while loop, which takes time $O(n)$, the total time for lines 9 and 12 is $O(n^2)$. Thus Dijkstra's algorithm runs in time $O(n^2)$.

In fact, for an appropriate choice of $z$, the time is $\Omega(n^2)$ for $K_n$, the complete graph on $n$ vertices, because every vertex is adjacent to every other. Thus the worst-case run time is $\Theta(n^2)$. ◀

*Any* shortest-path algorithm that receives as input $K_n$, the complete graph on $n$ vertices, must examine all of the edges of $K_n$ at least once. Since $K_n$ has $n(n-1)/2$ edges (see Exercise 16, Section 8.1), its worst-case run time must be at least $n(n-1)/2 = \Omega(n^2)$. It follows from Theorem 8.4.5 that Algorithm 8.4.1 is optimal.

## 8.4 Review Exercises

1. Describe Dijkstra's shortest-path algorithm.
2. Give an example to show how Dijkstra's shortest-path algorithm finds a shortest path.
3. Prove that Dijkstra's shortest-path algorithm correctly finds a shortest path.

## 8.4 Exercises

*In Exercises 1–5, find the length of a shortest path and a shortest path between each pair of vertices in the weighted graph.*

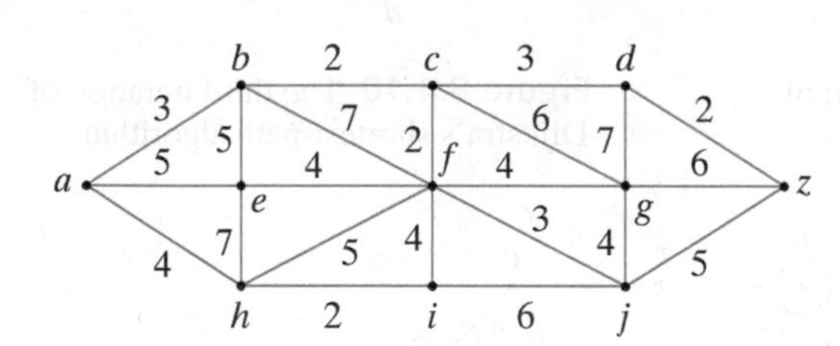


1. $a, f$
2. $a, g$
3. $a, z$
4. $b, j$
5. $h, d$
6. Write an algorithm that finds the length of a shortest path between two given vertices in a connected, weighted graph and also finds a shortest path.
7. Write an algorithm that finds the lengths of the shortest paths from a given vertex to every other vertex in a connected, weighted graph $G$.

★8. Write an algorithm that finds the lengths of the shortest paths between all vertex pairs in a simple, connected, weighted graph having $n$ vertices in time $O(n^3)$.

9. Modify Algorithm 8.4.1 so that it accepts a weighted graph that is not necessarily connected. At termination, what is $L(z)$ if there is no path from $a$ to $z$?
10. True or false? When a connected, weighted graph and vertices $a$ and $z$ are input to the following algorithm, it returns the length of a shortest path from $a$ to $z$. If the algorithm is correct, prove it; otherwise, give an example of a connected, weighted graph and vertices $a$ and $z$ for which it fails.

**Algorithm 8.4.6**

```
algor(w, a, z) {
   length = 0
   v = a
   T = set of all vertices
   while (v ¬= z) {
      T = T − {v}
      choose x ∈ T with minimum w(v, x)
      length = length + w(v, x)
      v = x
   }
   return length
}
```

11. True or false? Algorithm 8.4.1 finds the length of a shortest path in a connected, weighted graph even if some weights are negative. If true, prove it; otherwise, provide a counterexample.

# 8.5 Representations of Graphs

In the preceding sections we represented a graph by drawing it. Sometimes, as for example in using a computer to analyze a graph, we need a more formal representation. Our first method of representing a graph uses the **adjacency matrix.**

**Example 8.5.1** **Adjacency Matrix** Consider the graph of Figure 8.5.1. To obtain the adjacency matrix of this graph, we first select an ordering of the vertices, say $a, b, c, d, e$. Next, we label the rows and columns of a matrix with the ordered vertices. The entry in this matrix in row $i$, column $j$, $i \neq j$, is the number of edges incident on $i$ and $j$. If $i = j$, the entry is twice the number of loops incident on $i$. The adjacency matrix for this graph is

**Go Online**
For more on the adjacency matrix, see bit.ly/2yUuHxX

$$\begin{array}{c} \\ a \\ b \\ c \\ d \\ e \end{array} \begin{array}{c} \begin{array}{ccccc} a & b & c & d & e \end{array} \\ \begin{pmatrix} 0 & 1 & 0 & 0 & 1 \\ 1 & 0 & 1 & 0 & 1 \\ 0 & 1 & 2 & 0 & 1 \\ 0 & 0 & 0 & 0 & 2 \\ 1 & 1 & 1 & 2 & 0 \end{pmatrix} \end{array}.$$

◀

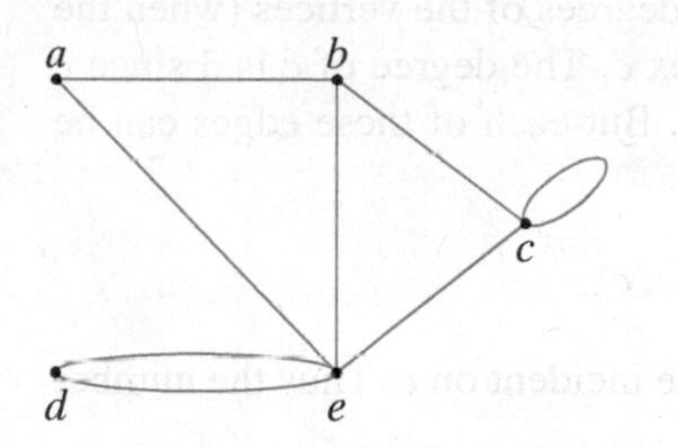


Figure 8.5.1 The graph for Example 8.5.1.

Notice that we can obtain the degree of a vertex $v$ in a graph $G$ by summing row $v$ or column $v$ in $G$'s adjacency matrix.

The adjacency matrix is not a very efficient way to represent a graph. Since the matrix is symmetric about the main diagonal (the elements on a line from the upper-left corner to the lower-right corner), the information, except that on the main diagonal, appears twice.

**Example 8.5.2** The adjacency matrix of the simple graph of Figure 8.5.2 is

$$A = \begin{array}{c} \\ a \\ b \\ c \\ d \\ e \end{array} \begin{array}{c} \begin{array}{ccccc} a & b & c & d & e \end{array} \\ \begin{pmatrix} 0 & 1 & 0 & 1 & 0 \\ 1 & 0 & 1 & 0 & 1 \\ 0 & 1 & 0 & 1 & 1 \\ 1 & 0 & 1 & 0 & 0 \\ 0 & 1 & 1 & 0 & 0 \end{pmatrix} \end{array}.$$

◀

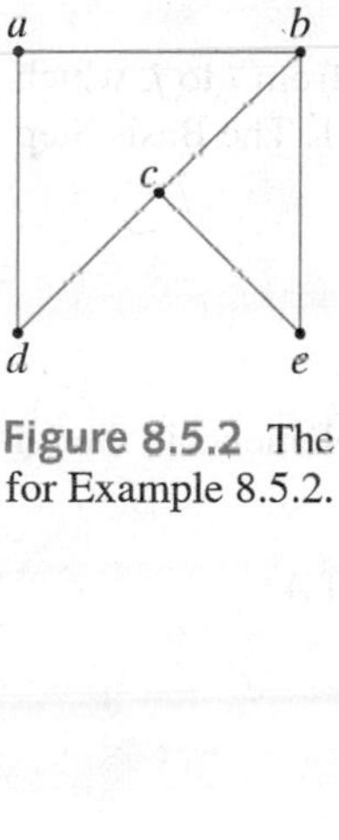

Figure 8.5.2 The graph for Example 8.5.2.

We will show that if $A$ is the adjacency matrix of a simple graph $G$, the powers of $A$,

$$A, A^2, A^3, \ldots,$$

count the number of paths of various lengths. More precisely, if the vertices of $G$ are labeled $1, 2, \ldots,$ the $ij$ th entry in the matrix $A^n$ is equal to the number of paths from $i$ to $j$ of length $n$. For example, suppose that we square the matrix $A$ of Example 8.5.2 to obtain

$$A^2 = \begin{pmatrix} 0 & 1 & 0 & 1 & 0 \\ 1 & 0 & 1 & 0 & 1 \\ 0 & 1 & 0 & 1 & 1 \\ 1 & 0 & 1 & 0 & 0 \\ 0 & 1 & 1 & 0 & 0 \end{pmatrix} \begin{pmatrix} 0 & 1 & 0 & 1 & 0 \\ 1 & 0 & 1 & 0 & 1 \\ 0 & 1 & 0 & 1 & 1 \\ 1 & 0 & 1 & 0 & 0 \\ 0 & 1 & 1 & 0 & 0 \end{pmatrix} = \begin{array}{c} \\ a \\ b \\ c \\ d \\ e \end{array} \begin{array}{c} \begin{array}{ccccc} a & b & c & d & e \end{array} \\ \begin{pmatrix} 2 & 0 & 2 & 0 & 1 \\ 0 & 3 & 1 & 2 & 1 \\ 2 & 1 & 3 & 0 & 1 \\ 0 & 2 & 0 & 2 & 1 \\ 1 & 1 & 1 & 1 & 2 \end{pmatrix} \end{array}.$$

Consider the entry for row $a$, column $c$ in $A^2$, obtained by multiplying pairwise the entries in row $a$ by the entries in column $c$ of the matrix $A$ and summing:

$$a \overset{\phantom{b}}{(}0 \quad \overset{b}{1} \quad 0 \quad \overset{d}{1} \quad 0) \begin{pmatrix} 0 \\ 1 \\ 0 \\ 1 \\ 1 \end{pmatrix}^{c} \begin{array}{c} \\ b \\ \\ d \\ \\ \end{array} = 0 \cdot 0 + 1 \cdot 1 + 0 \cdot 0 + 1 \cdot 1 + 0 \cdot 1 = 2.$$

The only way a nonzero product appears in this sum is if both entries to be multiplied are 1. This happens if there is a vertex $v$ whose entry in row $a$ is 1 and whose entry in column $c$ is 1. In other words, there must be edges of the form $(a, v)$ and $(v, c)$. Such edges form a path $(a, v, c)$ of length 2 from $a$ to $c$, and each path increases the sum by 1. In this example, the sum is 2 because there are two paths

$$(a, b, c), \quad (a, d, c)$$

of length 2 from $a$ to $c$. In general, the entry in row $x$ and column $y$ of the matrix $A^2$ is the number of paths of length 2 from vertex $x$ to vertex $y$.

The entries on the main diagonal of $A^2$ give the degrees of the vertices (when the graph is a simple graph). Consider, for example, vertex $c$. The degree of $c$ is 3 since $c$ is incident on the three edges $(c, b)$, $(c, d)$, and $(c, e)$. But each of these edges can be converted to a path of length 2 from $c$ to $c$:

$$(c, b, c), \quad (c, d, c), \quad (c, e, c).$$

Similarly, a path of length 2 from $c$ to $c$ defines an edge incident on $c$. Thus the number of paths of length 2 from $c$ to $c$ is 3, the degree of $c$.

We now use induction to show that the entries in the $n$th power of an adjacency matrix give the number of paths of length $n$.

**Theorem 8.5.3** If $A$ is the adjacency matrix of a simple graph, the $ij$th entry of $A^n$ is equal to the number of paths of length $n$ from vertex $i$ to vertex $j$, $n = 1, 2, \ldots$ .

**Proof** We will use induction on $n$.

If $n = 1$, $A^1$ is simply $A$. The $ij$th entry is 1 if there is an edge from $i$ to $j$, which is a path of length 1, and 0 otherwise. Thus the theorem is true if $n = 1$. The Basis Step has been verified.

Assume that the theorem is true for $n$. Now

$$A^{n+1} = A^n A$$

so that the $ik$th entry in $A^{n+1}$ is obtained by multiplying pairwise the elements in the $i$th row of $A^n$ by the elements in the $k$th column of $A$ and summing:

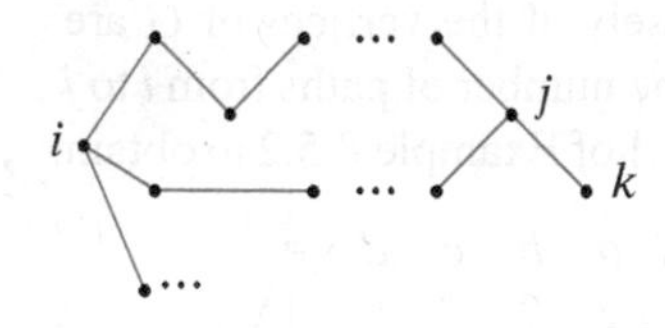


Figure 8.5.3 The proof of Theorem 8.5.3. A path from $i$ to $k$ of length $n+1$ whose next-to-last vertex is $j$ consists of a path of length $n$ from $i$ to $j$ followed by edge $(j, k)$. If there are $s_j$ paths of length $n$ from $i$ to $j$ and $t_j$ is 1 if edge $(j, k)$ exists and 0 otherwise, the sum of $s_j t_j$ over all $j$ gives the number of paths of length $n+1$ from $i$ to $k$.

$$\begin{array}{c} \\ i\text{th row of } A^n \end{array} \quad (s_1, \; s_2, \; \ldots, \; s_j, \; \ldots, \; s_m) \begin{array}{c} k\text{th column of } A \\ \begin{pmatrix} t_1 \\ t_2 \\ \vdots \\ t_j \\ \vdots \\ t_m \end{pmatrix} \end{array}$$

$$= s_1 t_1 + s_2 t_2 + \cdots + s_j t_j + \cdots + s_m t_m$$
$$= ik\text{th entry in } A^{n+1}.$$

By induction, $s_j$ gives the number of paths of length $n$ from $i$ to $j$ in the graph $G$. Now $t_j$ is either 0 or 1. If $t_j$ is 0, there is no edge from $j$ to $k$, so there are $s_j t_j = 0$ paths of length $n+1$ from $i$ to $k$, where the last edge is $(j, k)$. If $t_j$ is 1, there is an edge from vertex $j$ to vertex $k$ (see Figure 8.5.3). Since there are $s_j$ paths of length $n$ from vertex $i$ to vertex $j$, there are $s_j t_j = s_j$ paths of length $n+1$ from $i$ to $k$, where the last edge is $(j, k)$ (see Figure 8.5.3). Summing over all $j$, we will count all paths of length $n+1$ from $i$ to $k$. Thus the $ik$th entry in $A^{n+1}$ gives the number of paths of length $n+1$ from $i$ to $k$, and the Inductive Step is verified.

By the Principle of Mathematical Induction, the theorem is established. ◀

**Example 8.5.4** After Example 8.5.2, we showed that if $A$ is the matrix of the graph of Figure 8.5.2, then

$$A^2 = \begin{array}{c} \\ a \\ b \\ c \\ d \\ e \end{array} \begin{array}{c} \begin{array}{ccccc} a & b & c & d & e \end{array} \\ \begin{pmatrix} 2 & 0 & 2 & 0 & 1 \\ 0 & 3 & 1 & 2 & 1 \\ 2 & 1 & 3 & 0 & 1 \\ 0 & 2 & 0 & 2 & 1 \\ 1 & 1 & 1 & 1 & 2 \end{pmatrix} \end{array}.$$

By multiplying,

$$A^4 = A^2A^2 = \begin{pmatrix} 2 & 0 & 2 & 0 & 1 \\ 0 & 3 & 1 & 2 & 1 \\ 2 & 1 & 3 & 0 & 1 \\ 0 & 2 & 0 & 2 & 1 \\ 1 & 1 & 1 & 1 & 2 \end{pmatrix} \begin{pmatrix} 2 & 0 & 2 & 0 & 1 \\ 0 & 3 & 1 & 2 & 1 \\ 2 & 1 & 3 & 0 & 1 \\ 0 & 2 & 0 & 2 & 1 \\ 1 & 1 & 1 & 1 & 2 \end{pmatrix},$$

we find that

$$A^4 = \begin{array}{c} \\ a \\ b \\ c \\ d \\ e \end{array} \begin{array}{c} \begin{array}{ccccc} a & b & c & d & e \end{array} \\ \begin{pmatrix} 9 & 3 & 11 & 1 & 6 \\ 3 & 15 & 7 & 11 & 8 \\ 11 & 7 & 15 & 3 & 8 \\ 1 & 11 & 3 & 9 & 6 \\ 6 & 8 & 8 & 6 & 8 \end{pmatrix} \end{array}.$$

The entry from row $d$, column $e$ is 6, which means that there are six paths of length 4 from $d$ to $e$. By inspection, we find them to be

$$(d, a, d, c, e), \quad (d, c, d, c, e), \quad (d, a, b, c, e),$$
$$(d, c, e, c, e), \quad (d, c, e, b, e), \quad (d, c, b, c, e).$$

◀

Another useful matrix representation of a graph is known as the **incidence matrix.**

**Example 8.5.5** **Incidence Matrix** To obtain the incidence matrix of the graph in Figure 8.5.4, we label the rows with the vertices and the columns with the edges (in some arbitrary order). The entry for row $v$ and column $e$ is 1 if $e$ is incident on $v$ and 0 otherwise. Thus the incidence matrix for the graph of Figure 8.5.4 is

$$\begin{array}{c} \\ v_1 \\ v_2 \\ v_3 \\ v_4 \\ v_5 \end{array} \begin{array}{c} \begin{array}{ccccccc} e_1 & e_2 & e_3 & e_4 & e_5 & e_6 & e_7 \end{array} \\ \begin{pmatrix} 1 & 1 & 1 & 0 & 0 & 0 & 0 \\ 0 & 0 & 1 & 1 & 0 & 1 & 1 \\ 0 & 0 & 0 & 0 & 0 & 1 & 0 \\ 1 & 1 & 0 & 1 & 0 & 0 & 0 \\ 0 & 0 & 0 & 0 & 1 & 1 & 0 \end{pmatrix} \end{array}.$$

A column such as $e_7$ is understood to represent a loop. ◀

**Go Online**
For more on the incidence matrix, see
bit.ly/2yUuHxX

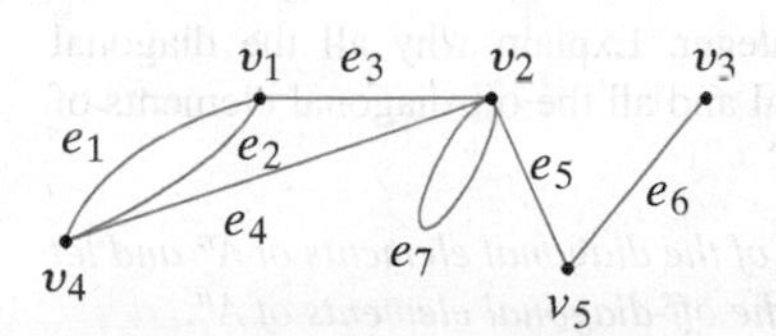


**Figure 8.5.4** The graph for Example 8.5.5.

Notice that in a graph without loops each column has two 1's and that the sum of a row gives the degree of the vertex identified with that row.

## 8.5 Review Exercises

1. What is an adjacency matrix?
2. If $A$ is the adjacency matrix of a simple graph, what are the values of the entries in $A^n$?
3. What is an incidence matrix?

## 8.5 Exercises

*In Exercises 1–6, write the adjacency matrix of each graph.*

1.

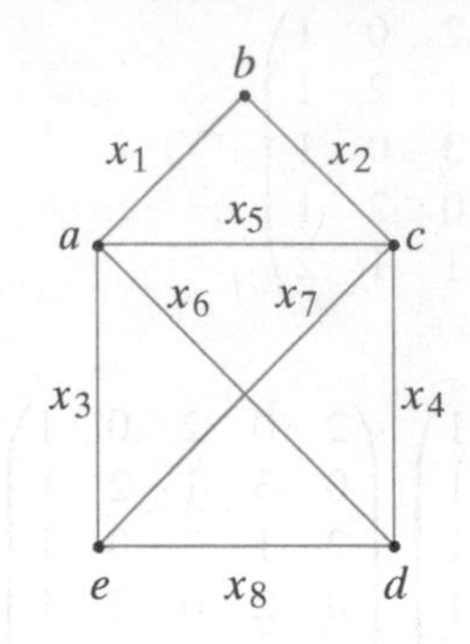


2.

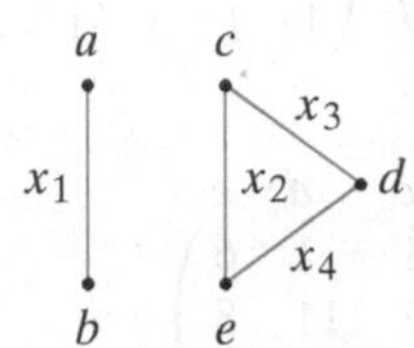


3.

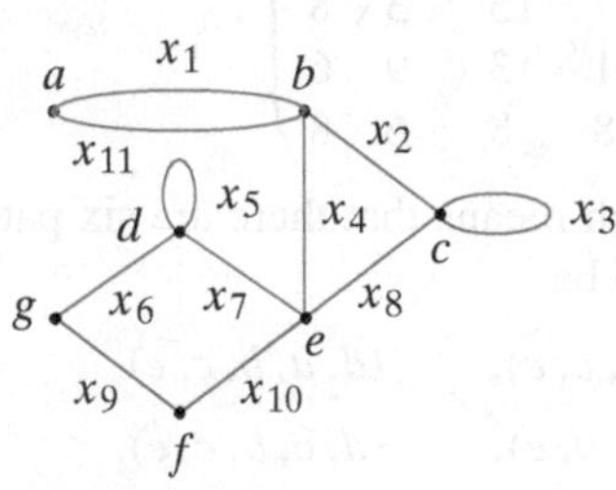


4. The graph of Figure 8.2.2
5. The complete bipartite graph $K_{2,3}$
6. The complete graph on five vertices $K_5$

*In Exercises 7–12, write the incidence matrix of each graph.*

7. The graph of Exercise 1
8. The graph of Exercise 2
9. The graph of Exercise 3
10. The graph of Figure 8.2.1
11. The complete bipartite graph $K_{2,3}$
12. The complete graph on five vertices $K_5$

*In Exercises 13–17, draw the graph represented by each adjacency matrix.*

13.
$$\begin{array}{c} \\ a \\ b \\ c \\ d \\ e \end{array} \begin{array}{c} \begin{array}{ccccc} a & b & c & d & e \end{array} \\ \begin{pmatrix} 2 & 0 & 0 & 1 & 0 \\ 0 & 0 & 1 & 0 & 1 \\ 0 & 1 & 2 & 1 & 1 \\ 1 & 0 & 1 & 0 & 0 \\ 0 & 1 & 1 & 0 & 0 \end{pmatrix} \end{array}$$

14.
$$\begin{array}{c} \\ a \\ b \\ c \\ d \\ e \\ f \end{array} \begin{array}{c} \begin{array}{cccccc} a & b & c & d & e & f \end{array} \\ \begin{pmatrix} 0 & 0 & 1 & 0 & 0 & 1 \\ 0 & 2 & 0 & 1 & 2 & 0 \\ 1 & 0 & 0 & 0 & 0 & 1 \\ 0 & 1 & 0 & 0 & 1 & 0 \\ 0 & 2 & 0 & 1 & 0 & 0 \\ 1 & 0 & 1 & 0 & 0 & 0 \end{pmatrix} \end{array}$$

15.
$$\begin{array}{c} \\ a \\ b \\ c \\ d \\ e \end{array} \begin{array}{c} \begin{array}{ccccc} a & b & c & d & e \end{array} \\ \begin{pmatrix} 0 & 1 & 0 & 0 & 0 \\ 1 & 0 & 0 & 0 & 0 \\ 0 & 0 & 0 & 1 & 1 \\ 0 & 0 & 1 & 0 & 1 \\ 0 & 0 & 1 & 1 & 2 \end{pmatrix} \end{array}$$

16.
$$\begin{array}{c} \\ a \\ b \\ c \\ d \\ e \\ f \end{array} \begin{array}{c} \begin{array}{cccccc} a & b & c & d & e & f \end{array} \\ \begin{pmatrix} 4 & 1 & 1 & 1 & 0 & 2 \\ 1 & 0 & 1 & 1 & 1 & 0 \\ 1 & 1 & 0 & 1 & 1 & 3 \\ 1 & 1 & 1 & 0 & 1 & 1 \\ 0 & 1 & 1 & 1 & 0 & 1 \\ 2 & 0 & 3 & 1 & 1 & 0 \end{pmatrix} \end{array}$$

17. The $7 \times 7$ matrix whose $ij$ th entry is 1 if $i + 1$ divides $j + 1$ or $j + 1$ divides $i + 1$, $i \neq j$; whose $ij$ th entry is 2 if $i = j$; and whose $ij$ th entry is 0 otherwise

18. Write the adjacency matrices of the components of the graphs given by the adjacency matrices of Exercises 13–17.

19. Compute the squares of the adjacency matrices of $K_5$ and the graphs of Exercises 1 and 2.

20. Let $A$ be the adjacency matrix for the graph of Exercise 1. What is the entry in row $a$, column $d$ of $A^5$?

21. Suppose that a graph has an adjacency matrix of the form

$$A = \left( \begin{array}{c|c} & A' \\ \hline A'' & \end{array} \right),$$

where all entries of the submatrices $A'$ and $A''$ are 0. What must the graph look like?

22. Repeat Exercise 21 with "adjacency" replaced by "incidence."

23. Let $A$ be an adjacency matrix of a graph. Why is $A^n$ symmetric about the main diagonal for every positive integer $n$?

*In Exercises 24 and 25, draw the graphs represented by the incidence matrices.*

24.
$$\begin{array}{c} a \\ b \\ c \\ d \\ e \end{array} \begin{pmatrix} 1 & 0 & 0 & 0 & 0 & 1 \\ 0 & 1 & 1 & 0 & 1 & 0 \\ 1 & 0 & 0 & 1 & 0 & 0 \\ 0 & 1 & 0 & 1 & 0 & 0 \\ 0 & 0 & 1 & 0 & 1 & 1 \end{pmatrix}$$

25.
$$\begin{array}{c} a \\ b \\ c \\ d \\ e \end{array} \begin{pmatrix} 0 & 1 & 0 & 0 & 1 & 1 \\ 0 & 1 & 1 & 0 & 1 & 0 \\ 0 & 0 & 0 & 0 & 0 & 1 \\ 1 & 0 & 0 & 1 & 0 & 0 \\ 1 & 0 & 0 & 1 & 0 & 0 \end{pmatrix}$$

26. What must a graph look like if some row of its incidence matrix consists only of 0's?

27. Let $A$ be the adjacency matrix of a graph $G$ with $n$ vertices. Let

$$Y = A + A^2 + \cdots + A^{n-1}.$$

If some off-diagonal entry in the matrix $Y$ is zero, what can you say about the graph $G$?

*Exercises 28–31 refer to the adjacency matrix $A$ of $K_5$.*

28. Let $n$ be a positive integer. Explain why all the diagonal elements of $A^n$ are equal and all the off-diagonal elements of $A^n$ are equal.

*Let $d_n$ be the common value of the diagonal elements of $A^n$ and let $a_n$ be the common value of the off-diagonal elements of $A^n$.*

★29. Show that

$$d_{n+1} = 4a_n; \quad a_{n+1} = d_n + 3a_n; \quad a_{n+1} = 3a_n + 4a_{n-1}.$$

★30. Show that

$$a_n = \frac{1}{5}[4^n + (-1)^{n+1}].$$

31. Show that

$$d_n = \frac{4}{5}[4^{n-1} + (-1)^n].$$

⋆**32.** Derive results similar to those of Exercises 29–31 for the adjacency matrix $A$ of the graph $K_m$.

⋆**33.** Let $A$ be the adjacency matrix of the graph $K_{m,n}$. Find a formula for the entries in $A^j$.

# 8.6 Isomorphisms of Graphs

The following instructions are given to two persons who cannot see each other's paper: "Draw and label five vertices $a$, $b$, $c$, $d$, and $e$. Connect $a$ and $b$, $b$ and $c$, $c$ and $d$, $d$ and $e$, and $a$ and $e$." The graphs produced are shown in Figure 8.6.1. Surely these figures define the same graph even though they appear dissimilar. Such graphs are said to be **isomorphic.**

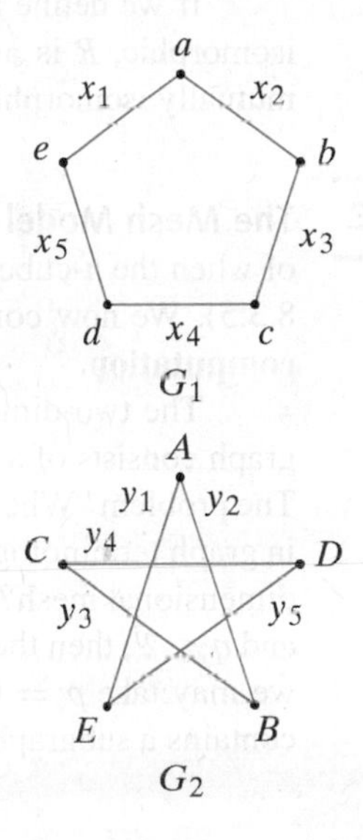


**Figure 8.6.1**
Isomorphic graphs.

Graphs are isomorphic if they are the same as *graphs*, even though the names of the vertices and edges may be different. As drawn, the graphs $G_1$ and $G_2$ in Figure 8.6.1 do not look the same; but, except for the names of the vertices and edges, they are the same graph. To put it another way, if we change the name of vertex $a$ in $G_1$ to $A$, and, similarly, $b$ to $B$, $c$ to $C$, $d$ to $D$, and $e$ to $E$; and change the name of edge $x_i$ in $G_1$ to $y_i$ for $i = 1$ to 5, we obtain graph $G_2$. [Edge $x_1$, which was incident on $(e, a)$, becomes edge $y_1$, which is now incident on $(E, A)$; edge $x_2$, which was incident on $(a, b)$, becomes edge $y_2$, which is now incident on $(A, B)$; and so on.] In general, in mathematics, computer science, biology, chemistry, and so on, structures are "isomorphic" if they are the same *in form,* but are not necessarily identical.

**Go Online**
For a C program that determines whether two graphs are isomorphic, see bit.ly/2AmJIuB

There are many practical applications of graph isomorphism. Fingerprints can be modeled using graphs in which vertices represent local geometric patterns and edges represent relations between the patterns (see [Nandi]). In this way, a fingerprint graph can be potentially identified by checking it against a database of fingerprint graphs. A similar application of graph isomorphism represents chemical and biological structures as graphs and compares them to databases of such representations (see [Willett]). This latter work is particularly useful in pharmaceutical research.

**Definition 8.6.1 ▶** Graphs $G_1$ and $G_2$ are *isomorphic* if there is a one-to-one, onto function $f$ from the vertices of $G_1$ to the vertices of $G_2$ and a one-to-one, onto function $g$ from the edges of $G_1$ to the edges of $G_2$, so that an edge $e$ is incident on

$v$ and $w$ in $G_1$ if and only if the edge $g(e)$ is incident on $f(v)$ and $f(w)$ in $G_2$. The pair of functions $f$ and $g$ is called an *isomorphism* of $G_1$ onto $G_2$. ◀

**Example 8.6.2** An isomorphism for the graphs $G_1$ and $G_2$ of Figure 8.6.1 is defined by

$$f(a) = A, \quad f(b) = B, \quad f(c) = C, \quad f(d) = D, \quad f(e) = E,$$
$$g(x_i) = y_i, \quad i = 1, \ldots, 5.$$

We can think of functions $f$ and $g$ as "renaming functions." ◀

If we define a relation $R$ on a set of graphs by the rule $G_1RG_2$ if $G_1$ and $G_2$ are isomorphic, $R$ is an equivalence relation. Each equivalence class consists of a set of mutually isomorphic graphs.

**Example 8.6.3** **The Mesh Model for Parallel Computation** Previously, we considered the problem of when the $n$-cube could simulate a ring model for parallel computation (see Example 8.3.5). We now consider when the $n$-cube can simulate the **mesh model for parallel computation.**

The two-dimensional mesh model for parallel computation when described as a graph consists of a rectangular array of vertices connected as shown (see Figure 8.6.2). The problem "When can an $n$-cube simulate a two-dimensional mesh?" can be rephrased in graph terminology as "When does an $n$-cube contain a subgraph isomorphic to a two-dimensional mesh?" We show that if $M$ is a mesh $p$ vertices by $q$ vertices, where $p \leq 2^i$ and $q \leq 2^j$, then the $(i+j)$-cube contains a subgraph isomorphic to $M$. (In Figure 8.6.2, we may take $p = 6$, $q = 4$, $i = 3$, and $j = 2$. Thus our result shows that the 5-cube contains a subgraph isomorphic to the graph in Figure 8.6.2.)

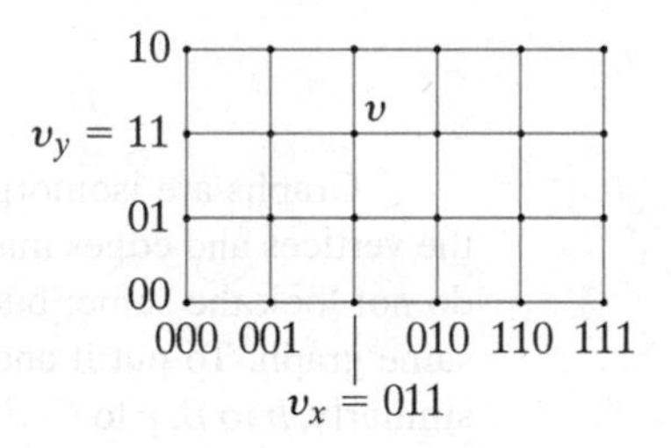


**Figure 8.6.2** Mesh model for parallel computation.

Let $M$ be a mesh $p$ vertices by $q$ vertices, where $p \leq 2^i$ and $q \leq 2^j$. We consider $M$ to be a rectangular array in ordinary 2-space with $p$ vertices in the horizontal direction and $q$ vertices in the vertical direction (see Figure 8.6.2). As coordinates for the vertices, we use elements of Gray codes. (Gray codes are discussed in Example 8.3.5.) The coordinates in the horizontal direction are the first $p$ members of an $i$-bit Gray code, and the coordinates in the vertical direction are the first $q$ members of a $j$-bit Gray code (see Figure 8.6.2). If a vertex $v$ is in the mesh, we let $v_x$ denote the horizontal coordinate of $v$ and $v_y$ denote the vertical coordinate of $v$. We then define a function $f$ on the vertices of $M$ by $f(v) = v_xv_y$. (The string $v_xv_y$ is the string $v_x$ followed by the string $v_y$.) Notice that $f$ is one-to-one.

If $(v, w)$ is an edge in $M$, the bit strings $v_xv_y$ and $w_xw_y$ differ in exactly one bit. Thus $(v_xv_y, w_xw_y)$ is an edge in the $(i+j)$-cube. We define a function $g$ on the edges of

$M$ by $g((v, w)) = (v_x v_y, w_x w_y)$. Notice that $g$ is one-to-one. The pair $f, g$ of functions is an isomorphism of $M$ onto the subgraph $(V, E)$ of the $(i + j)$-cube where

$$V = \{f(v) \mid v \text{ is a vertex in } M\}, \qquad E = \{g(e) \mid e \text{ is an edge in } M\}.$$

Therefore, if $M$ is a mesh $p$ vertices by $q$ vertices, where $p \le 2^i$ and $q \le 2^j$, the $(i + j)$-cube contains a subgraph isomorphic to $M$.

The argument given extends to an arbitrary number of dimensions (see Exercise 12); that is, if $M$ is a $p_1 \times p_2 \times \cdots \times p_k$ mesh, where $p_i \le 2^{t_i}$ for $i = 1, \ldots, k$, then the $(t_1 + t_2 + \cdots + t_k)$-cube contains a subgraph isomorphic to $M$. ◀

In general, the adjacency matrix of a graph changes when the ordering of its vertices is changed. We can show that graphs $G_1$ and $G_2$ are isomorphic if and only if *for some* ordering of their vertices, their adjacency matrices are equal.

**Theorem 8.6.4** Graphs $G_1$ and $G_2$ are isomorphic if and only if for some ordering of their vertices, their adjacency matrices are equal.

**Proof** Suppose that $G_1$ and $G_2$ are isomorphic. Then there is a one-to-one, onto function $f$ from the vertices of $G_1$ to the vertices of $G_2$ and a one-to-one, onto function $g$ from the edges of $G_1$ to the edges of $G_2$, so that an edge $e$ is incident on $v$ and $w$ if and only if the edge $g(e)$ is incident on $f(v)$ and $f(w)$ in $G_2$.

Let $v_1, \ldots, v_n$ be an ordering of the vertices of $G_1$. Let $A_1$ be the adjacency matrix of $G_1$ relative to the ordering $v_1, \ldots, v_n$, and let $A_2$ be the adjacency matrix of $G_2$ relative to the ordering $f(v_1), \ldots, f(v_n)$. Suppose that the entry in row $i$, column $j$, $i \ne j$, of $A_1$ is equal to $k$. Then there are $k$ edges, say $e_1, \ldots, e_k$, incident on $v_i$ and $v_j$. Therefore, there are exactly $k$ edges $g(e_1), \ldots, g(e_k)$ incident on $f(v_i)$ and $f(v_j)$ in $G_2$. Thus the entry in row $i$, column $j$ in $A_2$, which counts the number of edges incident on $f(v_i)$ and $f(v_j)$, is also equal to $k$. A similar argument shows that the entries on the diagonals in $A_1$ and $A_2$ are also equal. Therefore, $A_1 = A_2$.

The converse is similar and is left as an exercise (see Exercise 35). ◀

**Corollary 8.6.5** Let $G_1$ and $G_2$ be simple graphs. The following are equivalent:

(a) $G_1$ and $G_2$ are isomorphic.

(b) There is a one-to-one, onto function $f$ from the vertex set of $G_1$ to the vertex set of $G_2$ satisfying the following: Vertices $v$ and $w$ are adjacent in $G_1$ if and only if the vertices $f(v)$ and $f(w)$ are adjacent in $G_2$.

**Proof** It follows immediately from Definition 8.6.1 that (a) implies (b).

We prove that (b) implies (a). Suppose there is a one-to-one, onto function $f$ from the vertex set of $G_1$ to the vertex set of $G_2$ satisfying the following: Vertices $v$ and $w$ are adjacent in $G_1$ if and only if the vertices $f(v)$ and $f(w)$ are adjacent in $G_2$.

Let $v_1, \ldots, v_n$ be an ordering of the vertices of $G_1$. Let $A_1$ be the adjacency matrix of $G_1$ relative to the ordering $v_1, \ldots, v_n$, and let $A_2$ be the adjacency matrix of $G_2$ relative to the ordering $f(v_1), \ldots, f(v_n)$. Since $G_1$ and $G_2$ are simple graphs, the entries in the adjacency matrices are either 1 (to indicate that vertices are adjacent) or 0 (to indicate that vertices are not adjacent). Because vertices $v$ and $w$ are adjacent in $G_1$ if and only if the vertices $f(v)$ and $f(w)$ are adjacent in $G_2$, it follows that $A_1 = A_2$. By Theorem 8.6.4, $G_1$ and $G_2$ are isomorphic. ◀

**Example 8.6.6** The adjacency matrix of graph $G_1$ in Figure 8.6.1 relative to the vertex ordering $a, b, c, d, e$,

$$\begin{array}{c} \\ a \\ b \\ c \\ d \\ e \end{array} \begin{array}{c} \begin{array}{ccccc} a & b & c & d & e \end{array} \\ \begin{pmatrix} 0 & 1 & 0 & 0 & 1 \\ 1 & 0 & 1 & 0 & 0 \\ 0 & 1 & 0 & 1 & 0 \\ 0 & 0 & 1 & 0 & 1 \\ 1 & 0 & 0 & 1 & 0 \end{pmatrix} \end{array},$$

is equal to the adjacency matrix of graph $G_2$ in Figure 8.6.1 relative to the vertex ordering $A, B, C, D, E$,

$$\begin{array}{c} \\ A \\ B \\ C \\ D \\ E \end{array} \begin{array}{c} \begin{array}{ccccc} A & B & C & D & E \end{array} \\ \begin{pmatrix} 0 & 1 & 0 & 0 & 1 \\ 1 & 0 & 1 & 0 & 0 \\ 0 & 1 & 0 & 1 & 0 \\ 0 & 0 & 1 & 0 & 1 \\ 1 & 0 & 0 & 1 & 0 \end{pmatrix} \end{array}.$$

We see again that $G_1$ and $G_2$ are isomorphic. ◀

Although the graph isomorphism problem is not known to be NP-complete, there are no known worst-case, polynomial-time algorithms to test whether two graphs are isomorphic. There are, however, average-case, linear-time algorithms for the graph isomorphism problem (see [Read] and [Babai]).

The following is one way to show that two simple graphs $G_1$ and $G_2$ are *not* isomorphic. Find a property of $G_1$ that $G_2$ does *not* have but that $G_2$ *would* have if $G_1$ and $G_2$ were isomorphic. Such a property is called an **invariant.** More precisely, a property $P$ is an invariant if whenever $G_1$ and $G_2$ are isomorphic graphs:

If $G_1$ has property $P$, $G_2$ also has property $P$.

By Definition 8.6.1, if graphs $G_1$ and $G_2$ are isomorphic, there are one-to-one, onto functions from the edges (respectively, vertices) of $G_1$ to the edges (respectively, vertices) of $G_2$. Thus, if $G_1$ and $G_2$ are isomorphic, then $G_1$ and $G_2$ have the same number of edges and the same number of vertices. Therefore, if $e$ and $n$ are nonnegative integers, the properties "has $e$ edges" and "has $n$ vertices" are invariants.

**Example 8.6.7** The graphs $G_1$ and $G_2$ in Figure 8.6.3 are not isomorphic, since $G_1$ has seven edges and $G_2$ has six edges and "has seven edges" is an invariant.

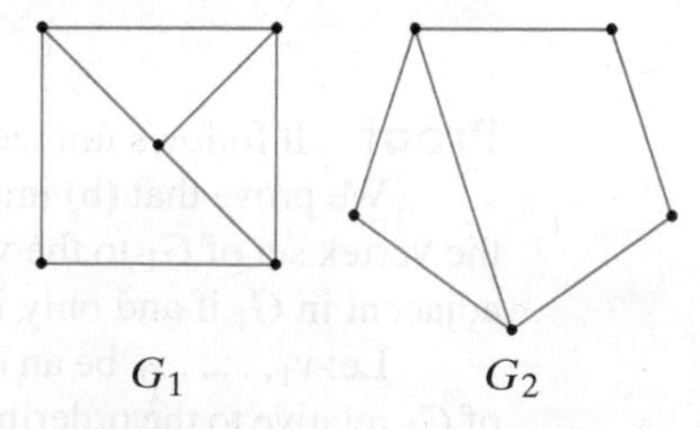


**Figure 8.6.3** Nonisomorphic graphs. $G_1$ has seven edges and $G_2$ has six edges. ◀

**Example 8.6.8** Show that if $k$ is a positive integer, "has a vertex of degree $k$" is an invariant.

**SOLUTION** Suppose $G_1$ and $G_2$ are isomorphic graphs and $f$ (respectively, $g$) is a one-to-one, onto function from the vertices (respectively, edges) of $G_1$ onto the vertices (respectively, edges) of $G_2$. Suppose that $G_1$ has a vertex $v$ of degree $k$. Then there are $k$ edges $e_1, \ldots, e_k$ incident on $v$. By Definition 8.6.1, $g(e_1), \ldots, g(e_k)$ are incident on $f(v)$. Because $g$ is one-to-one, $\delta(f(v)) \geq k$.

Let $E$ be an edge that is incident on $f(v)$ in $G_2$. Since $g$ is onto, there is an edge $e$ in $G_1$ with $g(e) = E$. Since $g(e)$ is incident on $f(v)$ in $G_2$, by Definition 8.6.1, $e$ is incident on $v$ in $G_1$. Since $e_1, \ldots, e_k$ are the only edges in $G_1$ incident on $v$, $e = e_i$ for some $i \in \{1, \ldots, k\}$. Now $g(e_i) = g(e) = E$. Thus $\delta(f(v)) = k$, so $G_2$ has a vertex, namely $f(v)$, of degree $k$. ◀

**Example 8.6.9** Since "has a vertex of degree 3" is an invariant, the graphs $G_1$ and $G_2$ of Figure 8.6.4 are not isomorphic; $G_1$ has vertices ($a$ and $f$) of degree 3, but $G_2$ does not have a vertex of degree 3. Notice that $G_1$ and $G_2$ have the same numbers of edges and vertices.

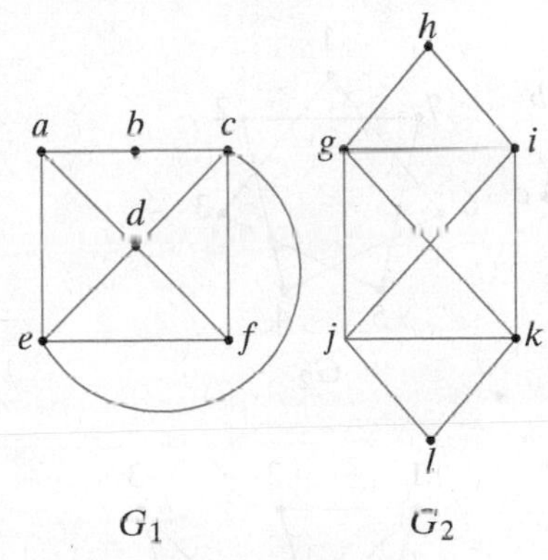


**Figure 8.6.4** Nonisomorphic graphs. $G_1$ has vertices of degree 3, but $G_2$ has no vertices of degree 3. ◀

Another invariant that is sometimes useful is "has a simple cycle of length $k$." We leave the proof that this property is an invariant to the exercises (Exercise 22).

**Example 8.6.10** Since "has a simple cycle of length 3" is an invariant, the graphs $G_1$ and $G_2$ of Figure 8.6.5 are not isomorphic; the graph $G_2$ has a simple cycle of length 3, but all simple cycles in $G_1$ have length at least 4. Notice that $G_1$ and $G_2$ have the same numbers of edges and vertices and that every vertex in $G_1$ or $G_2$ has degree 4.

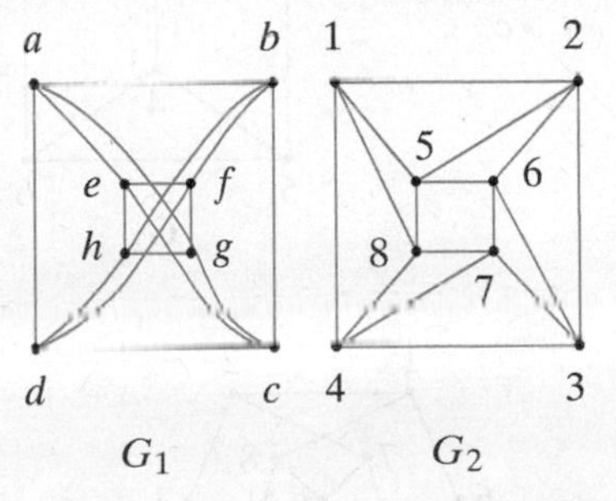


**Figure 8.6.5** Nonisomorphic graphs. $G_2$ has a simple cycle of length 3, but $G_1$ has no simple cycles of length 3. ◀

It would be easy to test whether a pair of graphs is isomorphic if we could find a small number of easily checked invariants that isomorphic graphs and *only* isomorphic graphs share. Unfortunately, no one has succeeded in finding such a set of invariants.

## 8.6 Review Exercises

1. Define what it means for two graphs to be isomorphic.
2. Give an example of isomorphic, nonidentical graphs. Explain why they are isomorphic.
3. Give an example of two graphs that are *not* isomorphic. Explain why they are not isomorphic.
4. What is an invariant in a graph?
5. How is "invariant" related to isomorphism?
6. How can one determine whether graphs are isomorphic from their adjacency matrices?
7. What is the mesh model for parallel computation?

## 8.6 Exercises

*In Exercises 1–4, prove that the graphs $G_1$ and $G_2$ are isomorphic.*

**1.**

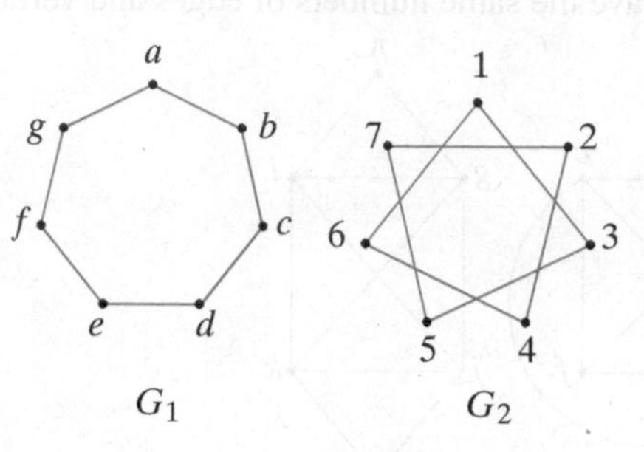


**2.**

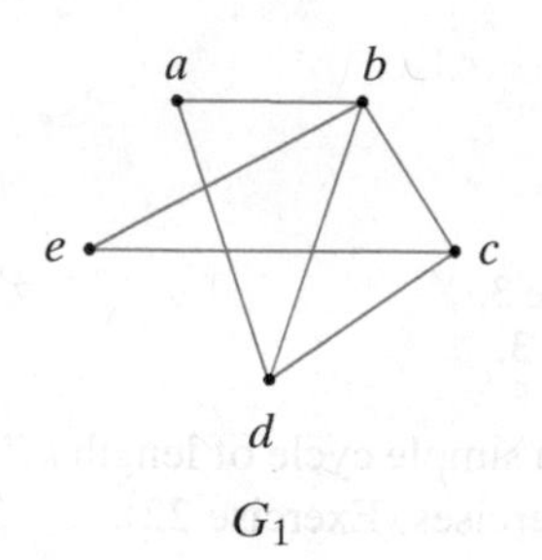


1 2 3 4 5 $G_2$

**3.**

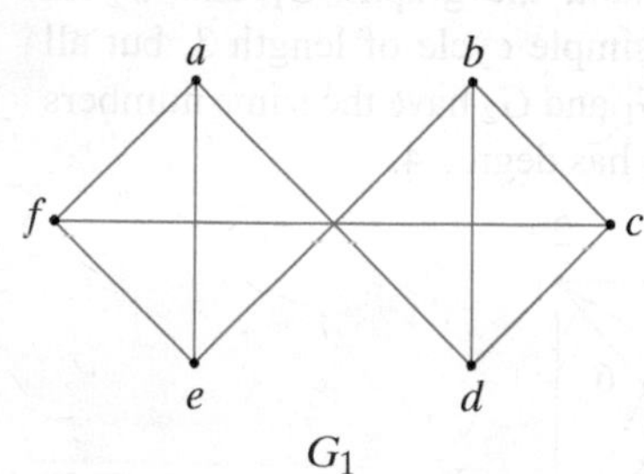


1 2 3 4 5 6 $G_2$

**4.**

a b c d e f g h i j $G_1$

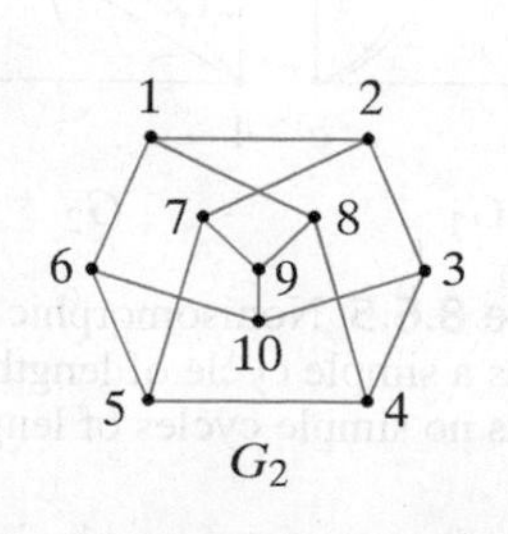


*Any graph isomorphic to $G_1$ and $G_2$ is called the* Petersen graph*. The Petersen graph is much used as an example; in fact, D. A. Holton and J. Sheehan wrote an entire book about it (see [Holton]).*

**5.** Prove that the following graph is the Petersen graph; that is, prove that it is isomorphic to the graphs in Exercise 4.

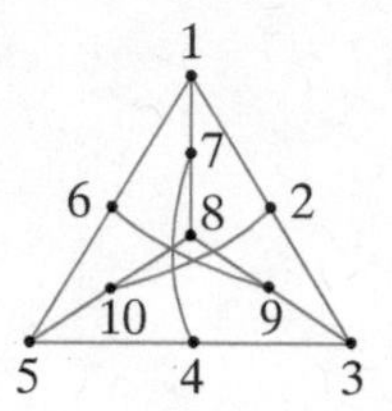


**6.** Draw a graph with 10 vertices. Label each vertex with one of the 10 distinct two-element subsets of $\{1, 2, 3, 4, 5\}$. Put an edge between two vertices if their labels (i.e., subsets) have no elements in common. Prove that your graph is the Petersen graph; that is, prove that it is isomorphic to the graphs in Exercise 4.

*In Exercises 7–9, prove that the graphs $G_1$ and $G_2$ are not isomorphic.*

**7.**

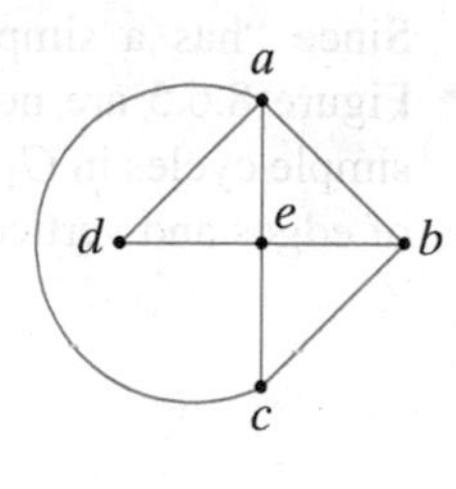


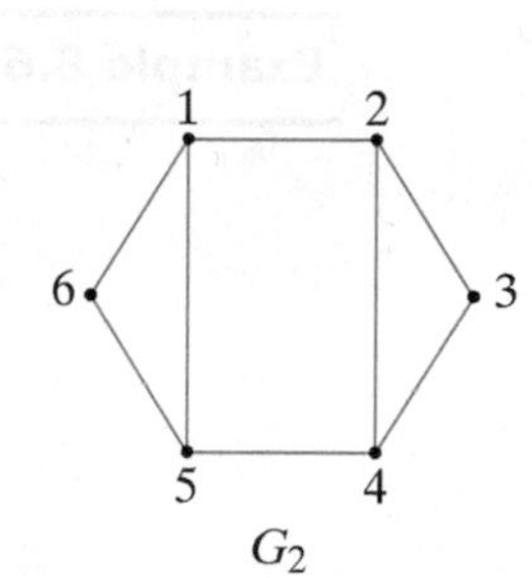


**8.**

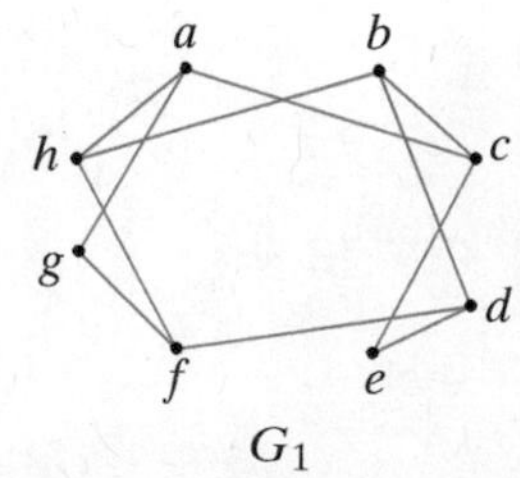


1 2 3 4 5 6 7 8 $G_2$

**9.**

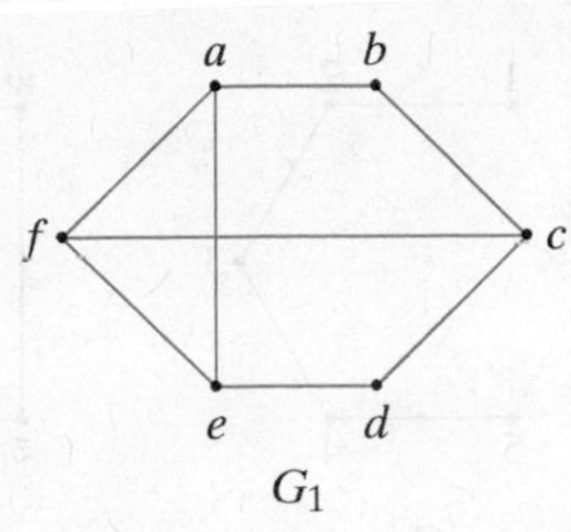


$G_1$

The second graph is $G_2$ of Exercise 3.

*In Exercises 10–15, determine whether the graphs $G_1$ and $G_2$ are isomorphic. Prove your answer.*

**10.**

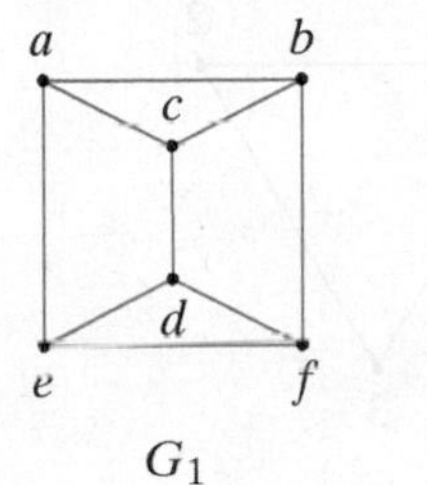


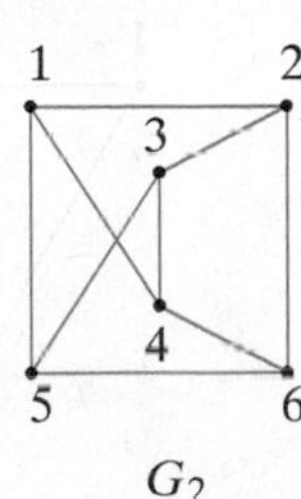


$G_1$ $G_2$

**11.**

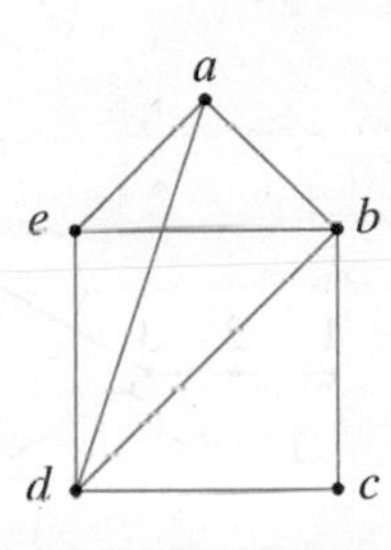


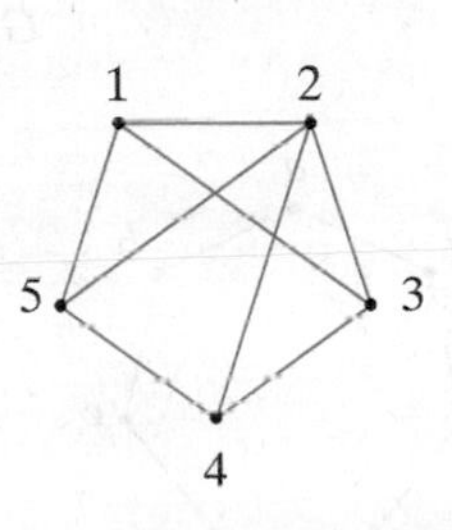


$G_1$ $G_2$

**12.**

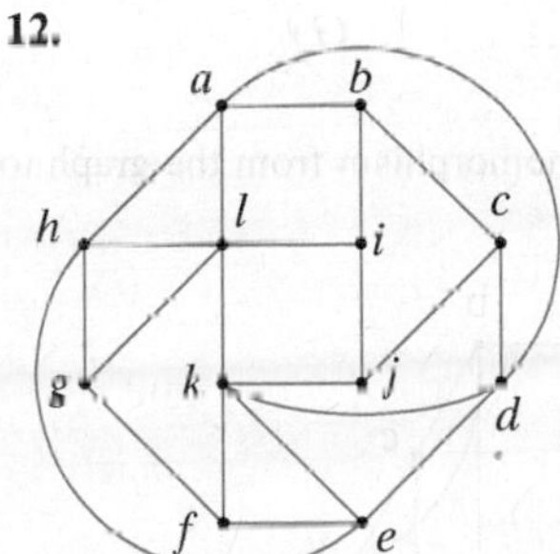


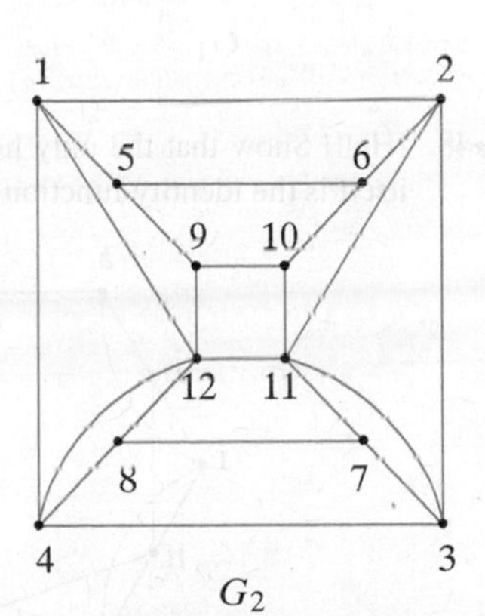


$G_1$ $G_2$

**★13.**

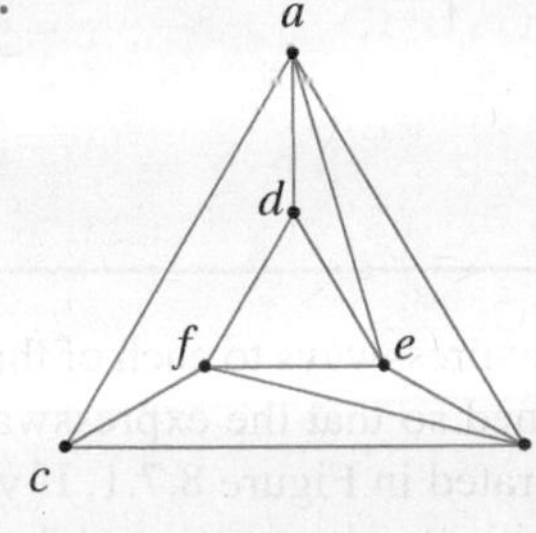


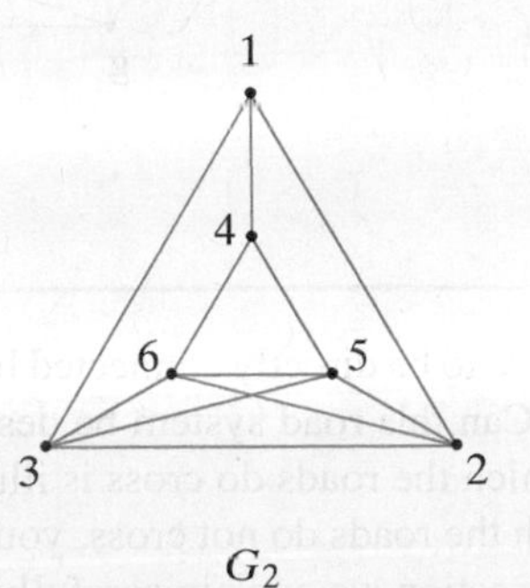


$G_1$ $G_2$

**★14.**

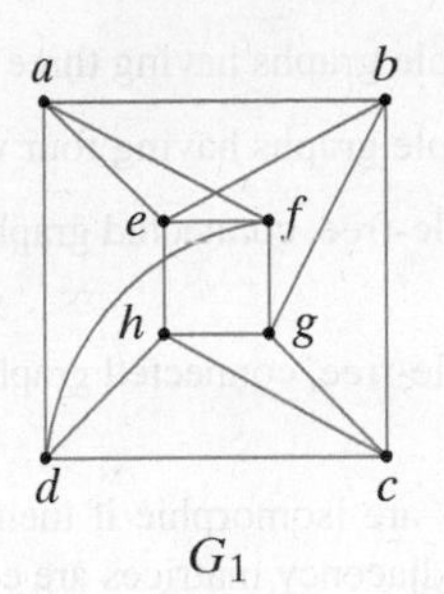


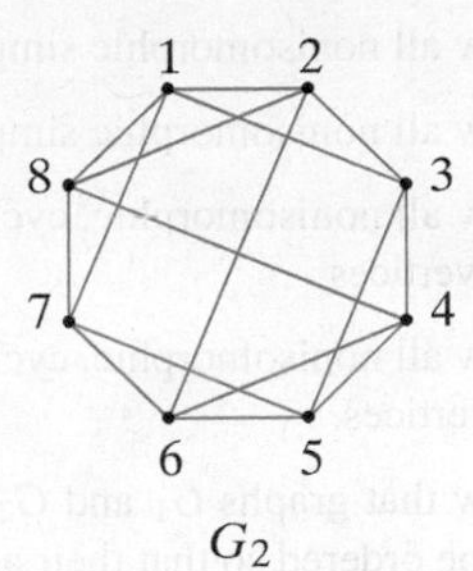


$G_1$ $G_2$

**★15.**

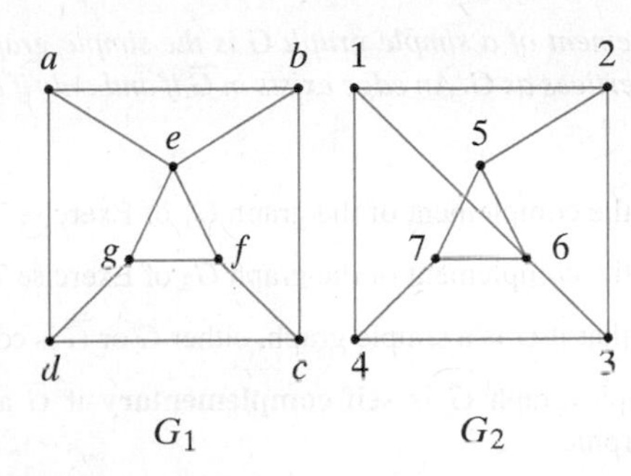


$G_1$ $G_2$

**16.** Show that if $M$ is a $p_1 \times p_2 \times \cdots \times p_k$ mesh, where $p_i \leq 2^{t_i}$ for $i = 1, \ldots, k$, then the $(t_1 + t_2 + \cdots + t_k)$-cube contains a subgraph isomorphic to $M$.

*An r-*regular graph *is a graph in which all vertices have degree r. A* regular graph *is a graph which is regular for some r.*

**17.** Show that for each $r$, all connected, simple, 2-vertex, $r$-regular graphs are isomorphic.

**18.** Show that for each $r$, all connected, simple, 3-vertex, $r$-regular graphs are isomorphic.

**19.** Show that for each $r$, all connected, simple, 4-vertex, $r$-regular graphs are isomorphic.

**20.** Show that for each $r$, all connected, simple, 5-vertex, $r$-regular graphs are isomorphic.

**21.** Find a value $r$ and two nonisomorphic, connected, simple, 6-vertex, $r$-regular graphs.

*In Exercises 22–26, show that the property given is an invariant.*

**22.** Has a simple cycle of length $k$

**23.** Has $n$ vertices of degree $k$

**24.** Is connected

**25.** Has $n$ simple cycles of length $k$

**26.** Has an edge $(v, w)$, where $\delta(v) = i$ and $\delta(w) = j$

**27.** Find an invariant not given in this section or in Exercises 22–26. Prove that your property is an invariant.

*In Exercises 28–30, tell whether or not each property is an invariant. If the property is an invariant, prove that it is; otherwise, give a counterexample.*

**28.** Has an Euler cycle

**29.** Has a vertex inside some simple cycle

**30.** Is bipartite

31. Draw all nonisomorphic simple graphs having three vertices.

32. Draw all nonisomorphic simple graphs having four vertices.

33. Draw all nonisomorphic, cycle-free, connected graphs having five vertices.

34. Draw all nonisomorphic, cycle-free, connected graphs having six vertices.

35. Show that graphs $G_1$ and $G_2$ are isomorphic if their vertices can be ordered so that their adjacency matrices are equal.

*The complement of a simple graph $G$ is the simple graph $\overline{G}$ with the same vertices as $G$. An edge exists in $\overline{G}$ if and only if it does not exist in $G$.*

36. Draw the complement of the graph $G_1$ of Exercise 7.

37. Draw the complement of the graph $G_2$ of Exercise 7.

★38. Show that if $G$ is a simple graph, either $G$ or $\overline{G}$ is connected.

39. A simple graph $G$ is **self-complementary** if $G$ and $\overline{G}$ are isomorphic.

(a) Find a self-complementary graph having five vertices.

(b) Find another self-complementary graph.

40. Let $G_1$ and $G_2$ be simple graphs. Show that $G_1$ and $G_2$ are isomorphic if and only if $\overline{G_1}$ and $\overline{G_2}$ are isomorphic.

41. Given two graphs $G_1$ and $G_2$, suppose that there is a one-to-one, onto function $f$ from the vertices of $G_1$ to the vertices of $G_2$ and a one-to-one, onto function $g$ from the edges of $G_1$ to the edges of $G_2$, so that if an edge $e$ is incident on $v$ and $w$ in $G_1$, the edge $g(e)$ is incident on $f(v)$ and $f(w)$ in $G_2$. Are $G_1$ and $G_2$ isomorphic?

*A homomorphism from a graph $G_1$ to a graph $G_2$ is a function $f$ from the vertex set of $G_1$ to the vertex set of $G_2$ with the property that if $v$ and $w$ are adjacent in $G_1$, then $f(v)$ and $f(w)$ are adjacent in $G_2$.*

42. Suppose that $G_1$ and $G_2$ are simple graphs. Show that if $f$ is a homomorphism of $G_1$ to $G_2$ and $f$ is one-to-one and onto, $G_1$ and $G_2$ are isomorphic.

*In Exercises 43–47, for each pair of graphs, give an example of a homomorphism from $G_1$ to $G_2$.*

43.

a b c d      1 2 3

$G_1$      $G_2$

44.

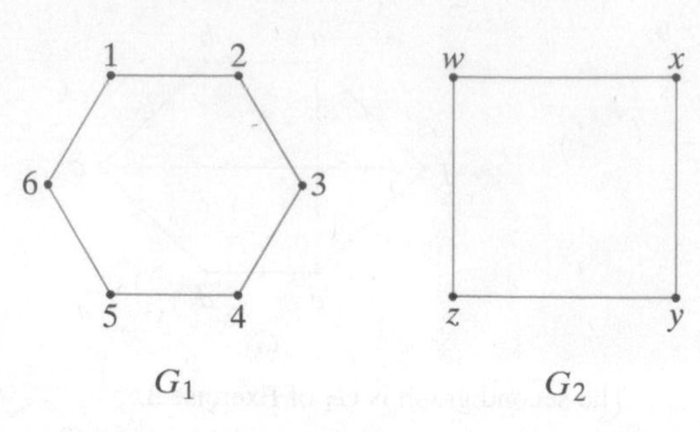


45. $G_1 = G_1$ of Exercise 44; $G_2 = G_1$ of Exercise 43

46. $G_1 = G_1$ of Exercise 43

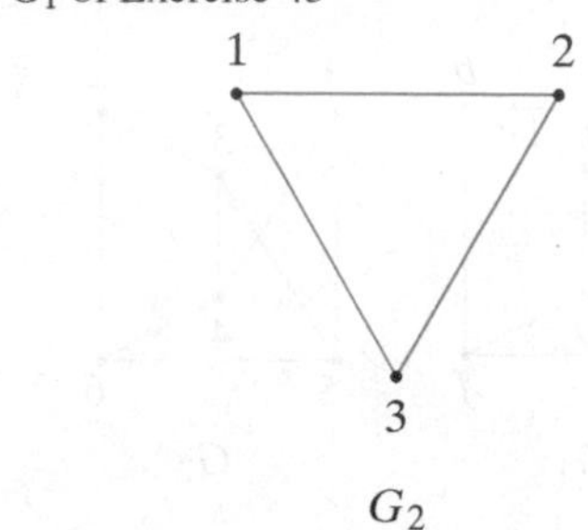


47.

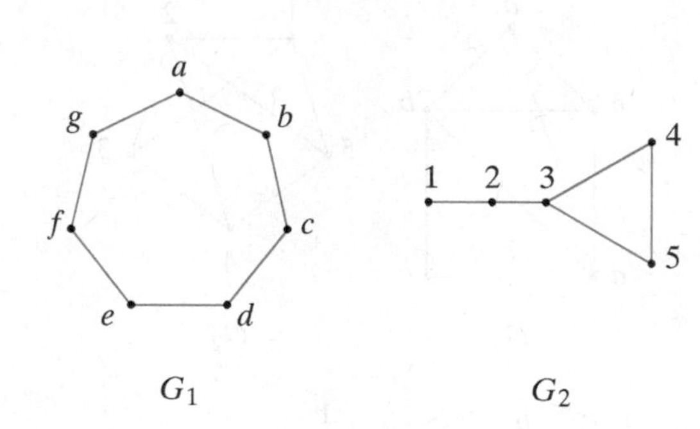


★48. [Hell] Show that the only homomorphism from the graph to itself is the identity function.

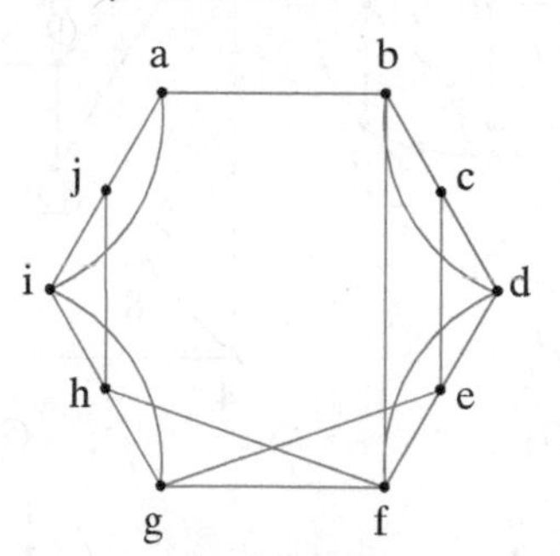


## 8.7 Planar Graphs

**Go Online**

For a planar graph algorithm, see `bit.ly/2yUuHxX`

Three cities, $C_1$, $C_2$, and $C_3$, are to be directly connected by expressways to each of three other cities, $C_4$, $C_5$, and $C_6$. Can this road system be designed so that the expressways do not cross? A system in which the roads do cross is illustrated in Figure 8.7.1. If you try drawing a system in which the roads do not cross, you will soon be convinced that it cannot be done. Later in this section we explain carefully why it cannot be done.

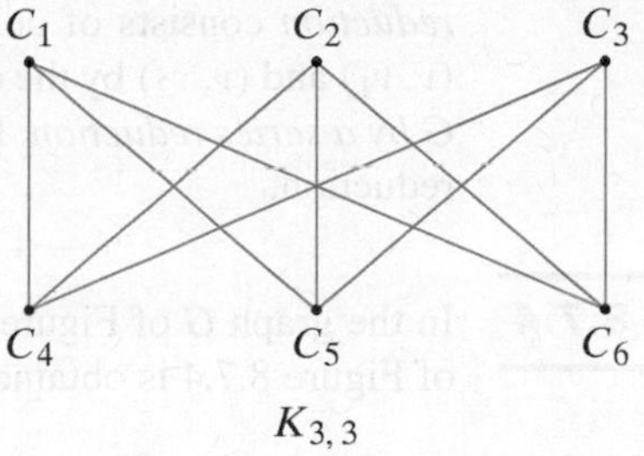

$C_1$ $C_2$ $C_3$

$C_4$ $C_5$ $C_6$

$K_{3,3}$

**Figure 8.7.1** Cities connected by expressways.

**Definition 8.7.1 ▶** A graph is *planar* if it can be drawn in the plane without its edges crossing. ◀

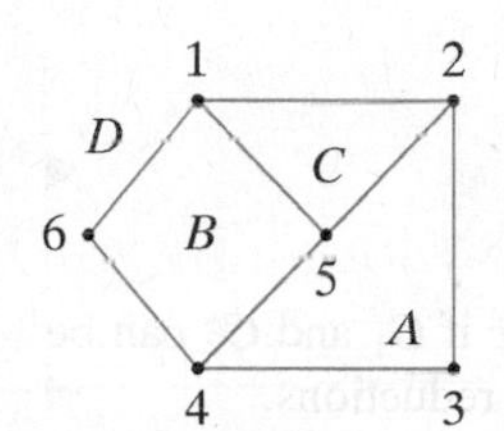


**Figure 8.7.2** A connected, planar graph with $f = 4$ faces ($A, B, C, D$), $e = 8$ edges, and $v = 6$ vertices; $f = e - v + 2$.

In designing printed circuits it is desirable to have as few lines cross as possible; thus the designer of printed circuits faces the problem of planarity.

If a connected, planar graph is drawn in the plane, the plane is divided into contiguous regions called **faces.** A face is characterized by the cycle that forms its boundary. For example, in the graph of Figure 8.7.2, face $A$ is bounded by the cycle (5, 2, 3, 4, 5) and face $C$ is bounded by the cycle (1, 2, 5, 1). The outer face $D$ is considered to be bounded by the cycle (1, 2, 3, 4, 6, 1). The graph of Figure 8.7.2 has $f = 4$ faces, $e = 8$ edges, and $v = 6$ vertices. Notice that $f$, $e$, and $v$ satisfy the equation

$$f = e - v + 2. \tag{8.7.1}$$

In 1752, Euler proved that equation (8.7.1) holds for any connected, planar graph. At the end of this section we will show how to prove (8.7.1), but for now let us show how (8.7.1) can be used to show that certain graphs are not planar.

**Example 8.7.2** Show that the graph $K_{3,3}$ of Figure 8.7.1 is not planar.

**SOLUTION** Suppose that $K_{3,3}$ is planar. Since every cycle has at least four edges, each face is bounded by at least four edges. Thus the number of edges that bound faces is at least $4f$. In a planar graph, each edge belongs to at most two bounding cycles. Therefore, $2e \geq 4f$. Using (8.7.1), we find that

$$2e \geq 4(e - v + 2). \tag{8.7.2}$$

For the graph of Figure 8.7.1, $e = 9$ and $v = 6$, so (8.7.2) becomes

$$18 = 2 \cdot 9 \geq 4(9 - 6 + 2) = 20,$$

which is a contradiction. Therefore, $K_{3,3}$ is not planar. ◀

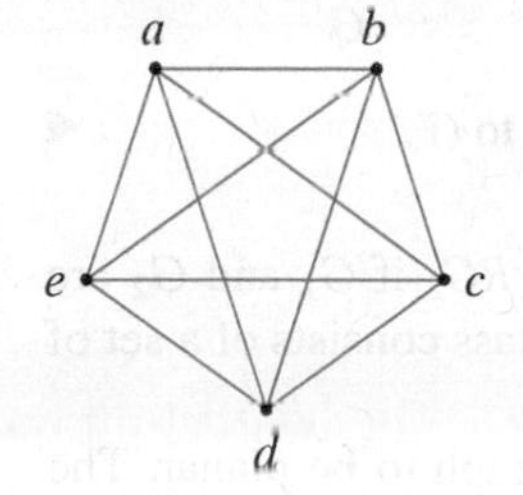


**Figure 8.7.3** The nonplanar graph $K_5$.

By a similar kind of argument (see Exercise 15), we can show that the graph $K_5$ of Figure 8.7.3 is not planar.

Obviously, if a graph contains $K_{3,3}$ or $K_5$ as a subgraph, it cannot be planar. The converse is not true; however, if we introduce the concept of "homeomorphic graphs," we can obtain a true statement similar to the converse (see Theorem 8.7.7).

**Definition 8.7.3 ▶** If a graph $G$ has a vertex $v$ of degree 2 and edges $(v, v_1)$ and $(v, v_2)$ with $v_1 \neq v_2$, we say that the edges $(v, v_1)$ and $(v, v_2)$ are in *series*. A *series* *reduction* consists of deleting the vertex $v$ from the graph $G$ and replacing the edges $(v, v_1)$ and $(v, v_2)$ by the edge $(v_1, v_2)$. The resulting graph $G'$ is said to be *obtained from G by a series reduction*. By convention, $G$ is said to be obtainable from itself by a series reduction. ◀

**Example 8.7.4** In the graph $G$ of Figure 8.7.4, the edges $(v, v_1)$ and $(v, v_2)$ are in series. The graph $G'$ of Figure 8.7.4 is obtained from $G$ by a series reduction.

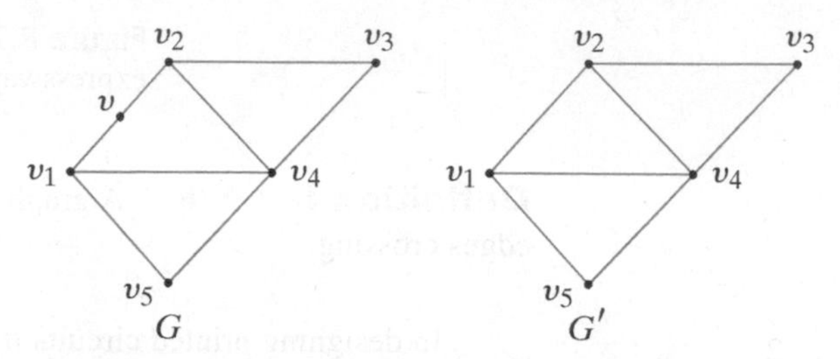


**Figure 8.7.4** $G'$ is obtained from $G$ by a series reduction. ◀

**Definition 8.7.5 ▶** Graphs $G_1$ and $G_2$ are *homeomorphic* if $G_1$ and $G_2$ can be reduced to isomorphic graphs by performing a sequence of series reductions. ◀

According to Definitions 8.7.3 and 8.7.5, any graph is homeomorphic to itself. Also, graphs $G_1$ and $G_2$ are homeomorphic if $G_1$ can be reduced to a graph isomorphic to $G_2$ or if $G_2$ can be reduced to a graph isomorphic to $G_1$.

**Example 8.7.6** The graphs $G_1$ and $G_2$ of Figure 8.7.5 are homeomorphic since they can both be reduced to the graph $G'$ of Figure 8.7.5 by a sequence of series reductions.

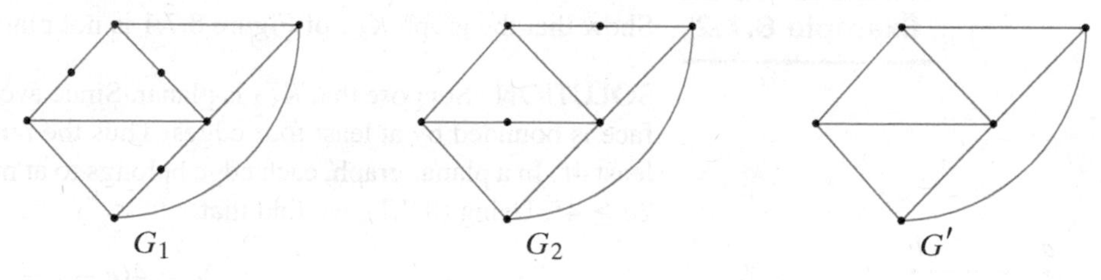


**Figure 8.7.5** $G_1$ and $G_2$ are homeomorphic; each can be reduced to $G'$. ◀

If we define a relation $R$ on a set of graphs by the rule $G_1 R G_2$ if $G_1$ and $G_2$ are homeomorphic, $R$ is an equivalence relation. Each equivalence class consists of a set of mutually homeomorphic graphs.

**Go Online**
A biography of Kuratowski is at `bit.ly/2yUuHxX`

We now state a necessary and sufficient condition for a graph to be planar. The theorem was first stated and proved by Kuratowski in 1930. The proof appears in [Even, 1979].

**Theorem 8.7.7** **Kuratowski's Theorem**

A graph $G$ is planar if and only if $G$ does not contain a subgraph homeomorphic to $K_5$ or $K_{3,3}$.

**Example 8.7.8** Show that the graph $G$ of Figure 8.7.6 is not planar by using Kuratowski's Theorem.

**SOLUTION** Let us try to find $K_{3,3}$ in the graph $G$ of Figure 8.7.6. We first note that the vertices $a, b, f$, and $e$ each have degree 4. In $K_{3,3}$ each vertex has degree 3, so let us eliminate the edges $(a, b)$ and $(f, e)$ so that all vertices have degree 3 (see Figure 8.7.6). We note that if we eliminate one more edge, we will obtain two vertices of degree 2 and we can then carry out two series reductions. The resulting graph will have nine edges; since $K_{3,3}$ has nine edges, this approach looks promising. Using trial and error, we finally see that if we eliminate edge $(g, h)$ and carry out the series reductions, we obtain an isomorphic copy of $K_{3,3}$ (see Figure 8.7.7). Therefore, the graph $G$ of Figure 8.7.6 is not planar, since it contains a subgraph homeomorphic to $K_{3,3}$.

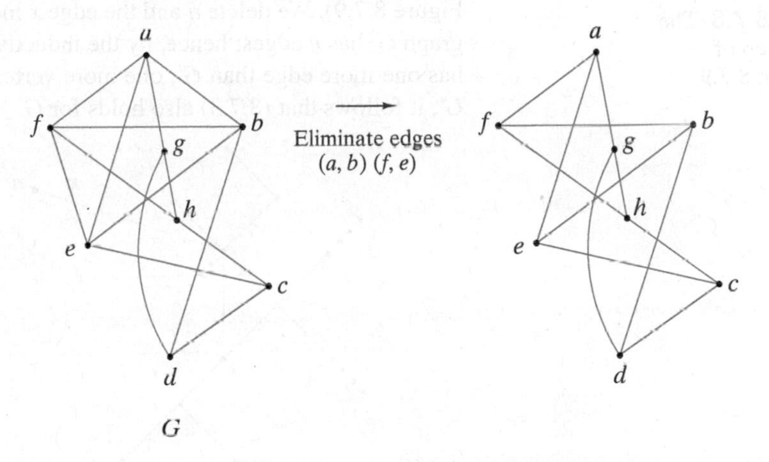


**Figure 8.7.6** Eliminating edges to obtain a subgraph.

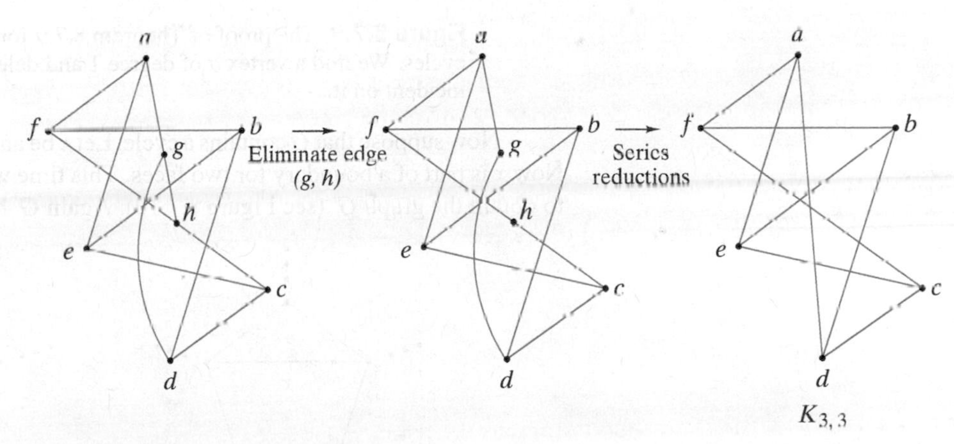


**Figure 8.7.7** Elimination of an edge to obtain a subgraph, followed by series reductions. ◀

Although Theorem 8.7.7 does give an elegant characterization of planar graphs, it does not lead to an efficient algorithm for recognizing planar graphs. However, algorithms are known that can determine whether a graph having $n$ vertices is planar in time $O(n)$ (see [Even, 1979]).

We will conclude this section by proving Euler's formula.

**Theorem 8.7.9** **Euler's Formula for Graphs**

If $G$ is a connected, planar graph with $e$ edges, $v$ vertices, and $f$ faces, then

$$f = e - v + 2. \tag{8.7.3}$$

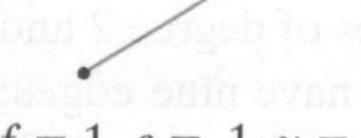

$f = 1, e = 1, v = 2$

$f = 2, e = 1, v = 1$

**Figure 8.7.8** The Basis Step of Theorem 8.7.9.

**Proof** We will use induction on the number of edges.

Suppose that $e = 1$. Then $G$ is one of the two graphs shown in Figure 8.7.8. In either case, the formula holds. We have verified the Basis Step.

Suppose that the formula holds for connected, planar graphs with $n$ edges. Let $G$ be a graph with $n+1$ edges. First, suppose that $G$ contains no cycles. Pick a vertex $v$ and trace a path starting at $v$. Since $G$ is cycle-free, every time we trace an edge, we arrive at a new vertex. Eventually, we will reach a vertex $a$, with degree 1, that we cannot leave (see Figure 8.7.9). We delete $a$ and the edge $x$ incident on $a$ from the graph $G$. The resulting graph $G'$ has $n$ edges; hence, by the inductive assumption, (8.7.3) holds for $G'$. Since $G$ has one more edge than $G'$, one more vertex than $G'$, and the same number of faces as $G'$, it follows that (8.7.3) also holds for $G$.

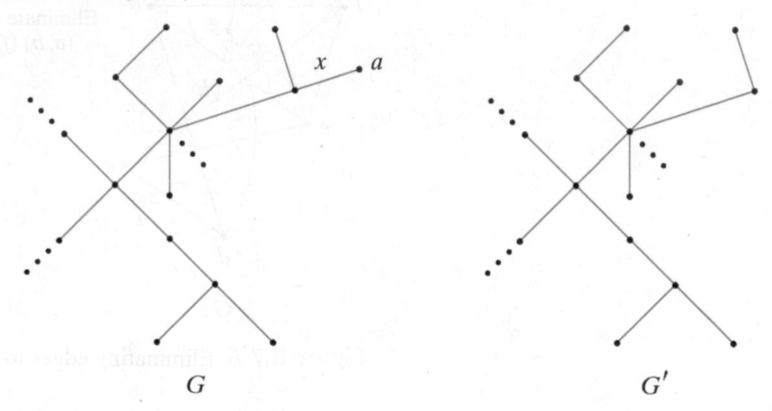


**Figure 8.7.9** The proof of Theorem 8.7.9 for the case that $G$ has no cycles. We find a vertex $a$ of degree 1 and delete $a$ and the edge $x$ incident on it.

Now suppose that $G$ contains a cycle. Let $x$ be an edge in a cycle (see Figure 8.7.10). Now $x$ is part of a boundary for two faces. This time we delete the edge $x$ but no vertices to obtain the graph $G'$ (see Figure 8.7.10). Again $G'$ has $n$ edges; hence,

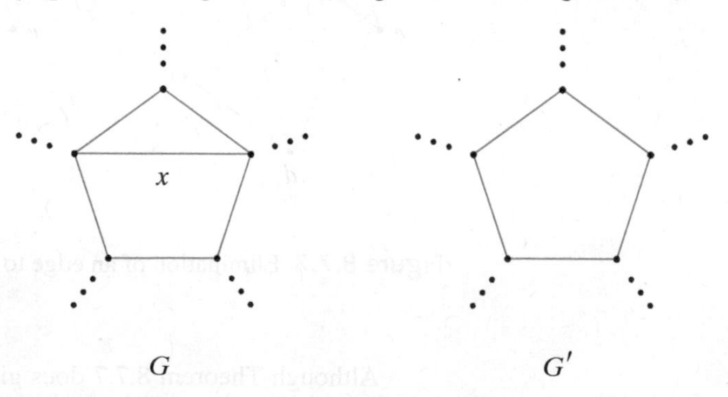


**Figure 8.7.10** The proof of Theorem 8.7.9 for the case that $G$ has a cycle. We delete edge $x$ in a cycle.

by the inductive assumption, (8.7.3) holds for $G'$. Since $G$ has one more face than $G'$, one more edge than $G'$, and the same number of vertices as $G'$, it follows that (8.7.3) also holds for $G$.

Since we have verified the Inductive Step, by the Principle of Mathematical Induction, the theorem is proved. ◀

## 8.7 Review Exercises

1. What is a planar graph?
2. What is a face?
3. State Euler's equation for a connected, planar graph.
4. What are series edges?
5. What is a series reduction?
6. Define *homeomorphic graphs.*
7. State Kuratowski's theorem.

## 8.7 Exercises

*In Exercises 1–3, show that each graph is planar by redrawing it so that no edges cross.*

1.

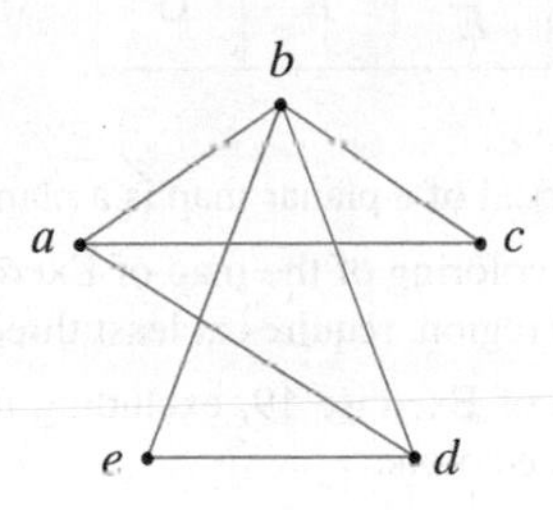


2.

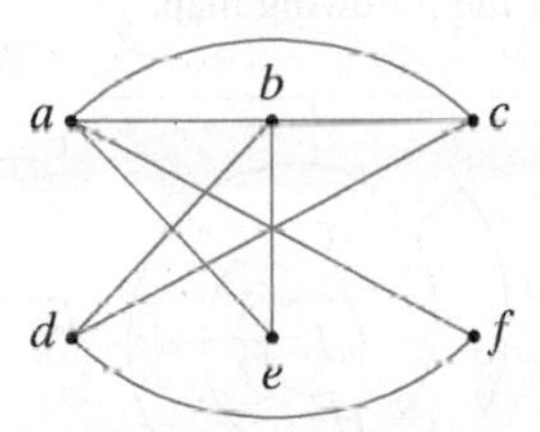


3.

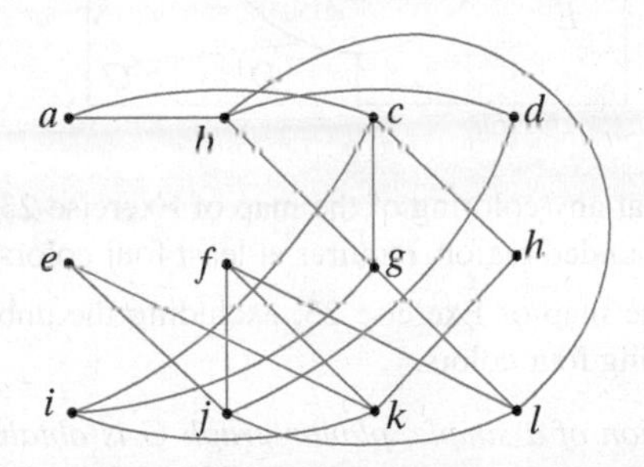


*In Exercises 4 and 5, show that each graph is not planar by finding a subgraph homeomorphic to either $K_5$ or $K_{3,3}$.*

4.

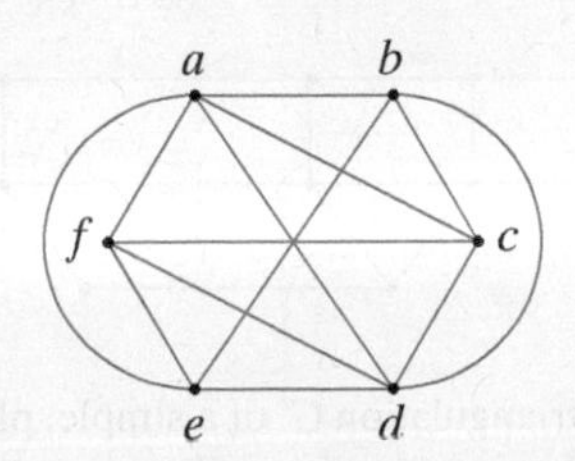


5.

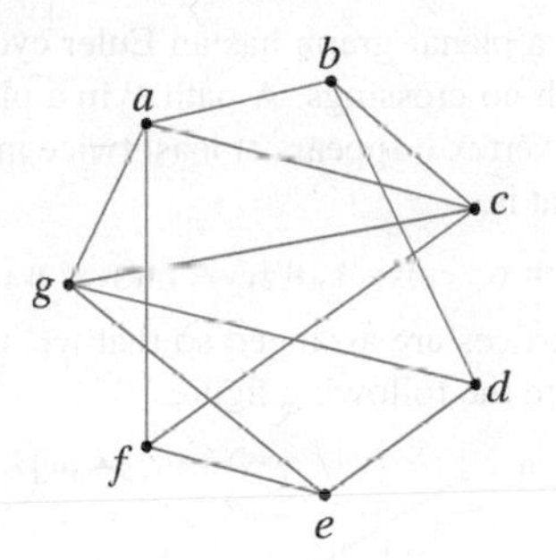


*In Exercises 6–8, determine whether each graph is planar. If the graph is planar, redraw it so that no edges cross; otherwise, find a subgraph homeomorphic to either $K_5$ or $K_{3,3}$.*

6.

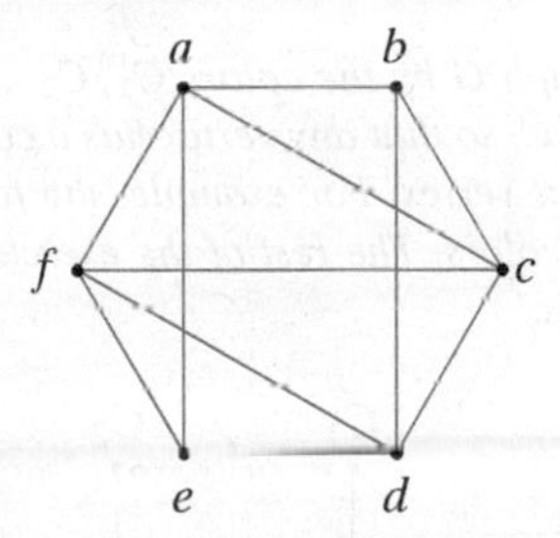


7.

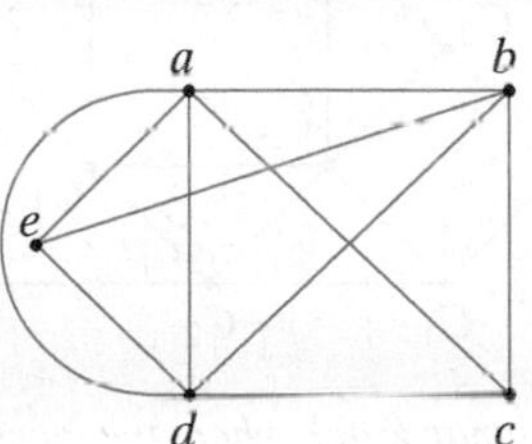


8.

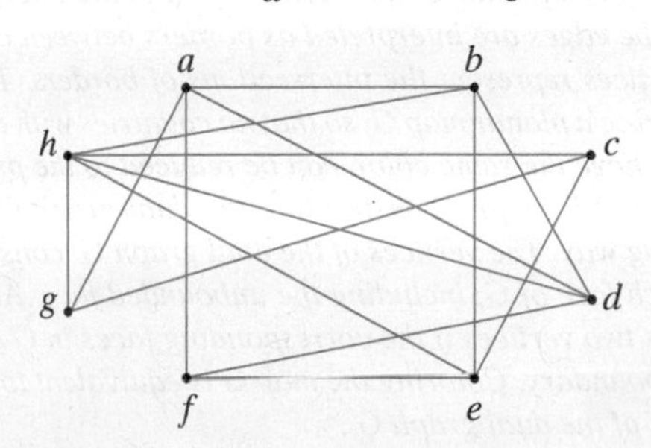

9. A connected, planar graph has nine vertices having degrees 2, 2, 2, 3, 3, 3, 4, 4, and 5. How many edges are there? How many faces are there?
10. Show that adding or deleting loops, parallel edges, or edges in series does not affect the planarity of a graph.
11. Show that any graph having four or fewer vertices is planar.
12. Show that any graph having five or fewer vertices and a vertex of degree 2 is planar.
13. Show that in any simple, connected, planar graph, $e \le 3v - 6$.
14. Give an example of a simple, connected, nonplanar graph for which $e \le 3v - 6$.
15. Use Exercise 13 to show that $K_5$ is not planar.

★16. Show that if a simple graph $G$ has 11 or more vertices, then either $G$ or its complement $\overline{G}$ is not planar.

★17. Prove that if a planar graph has an Euler cycle, it has an Euler cycle with no crossings. A path $P$ in a planar graph has a *crossing* if a vertex $v$ appears at least twice in $P$ and $P$ crosses itself at $v$; that is,

$$P = (\ldots, w_1, v, w_2, \ldots, w_3, v, w_4, \ldots),$$

where the vertices are arranged so that $w_1$, $v$, $w_2$ crosses $w_3$, $v$, $w_4$ at $v$ as in the following figure.

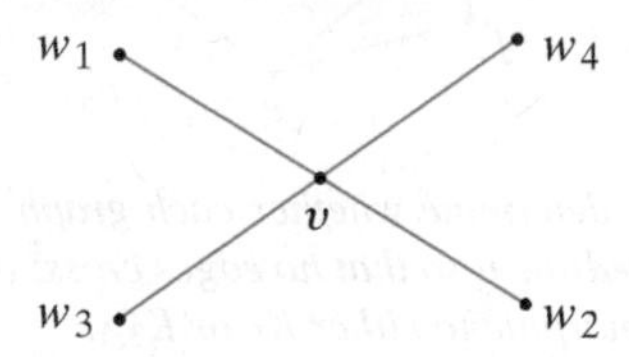


*A coloring of a graph G by the colors $C_1, C_2, \ldots, C_n$ assigns to each vertex a color $C_i$ so that any vertex has a color different from that of any adjacent vertex. For example, the following graph is colored with three colors. The rest of the exercises deal with coloring planar graphs.*

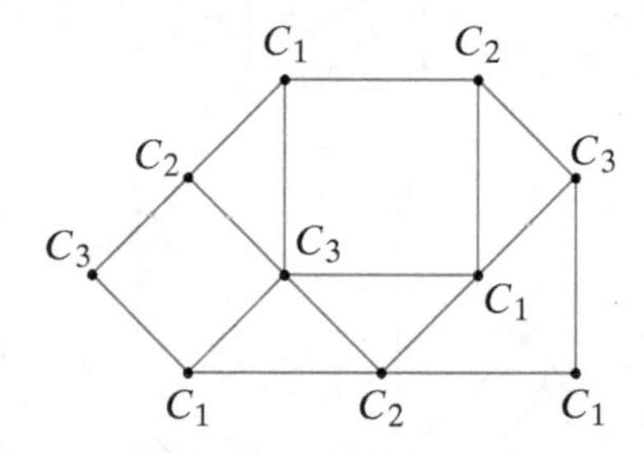


*A planar map is a planar graph where the faces are interpreted as countries, the edges are interpreted as borders between countries, and the vertices represent the intersections of borders. The problem of coloring a planar map G, so that no countries with adjoining boundaries have the same color, can be reduced to the problem of coloring a graph by first constructing the dual graph G′ of G in the following way. The vertices of the dual graph G′ consist of one point in each face of G, including the unbounded face. An edge in G′ connects two vertices if the corresponding faces in G are separated by a boundary. Coloring the map G is equivalent to coloring the vertices of the dual graph G′.*

18. Prove that any planar graph whose boundary is a polygon, all of whose vertices lie on the boundary, and whose interior consists of triangles (bounded by straight-line edges) can be colored using three colors. The following figure shows such a graph.

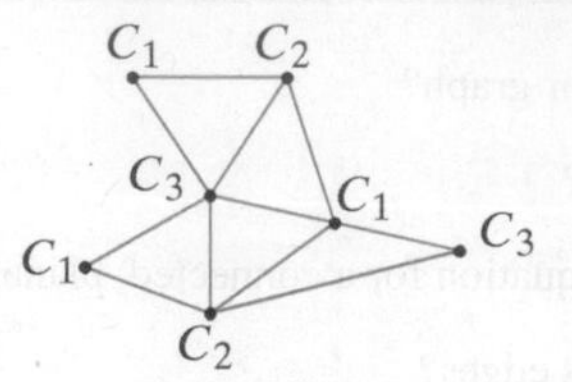


19. Find the dual of the following map.

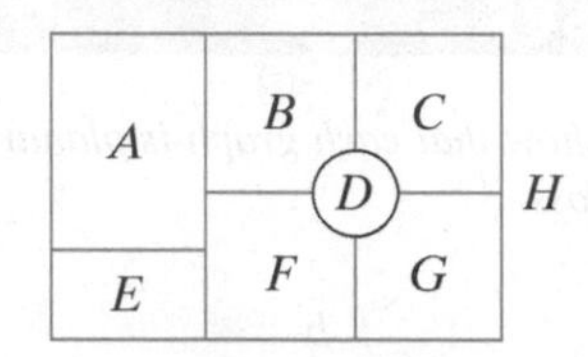


20. Show that the dual of a planar map is a planar graph.
21. Show that any coloring of the map of Exercise 19, excluding the unbounded region, requires at least three colors.
22. Color the map of Exercise 19, excluding the unbounded region, using three colors.
23. Find the dual of the following map.

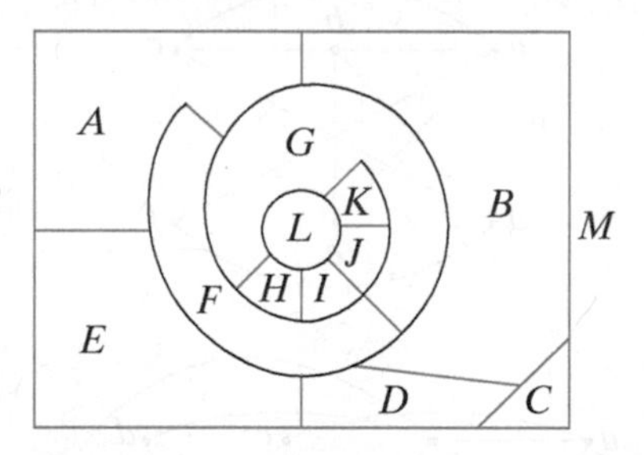


24. Show that any coloring of the map of Exercise 23, excluding the unbounded region, requires at least four colors.
25. Color the map of Exercise 23, excluding the unbounded region, using four colors.

*A triangulation of a simple, planar graph G is obtained from G by connecting as many vertices as possible while maintaining planarity and not introducing loops or parallel edges.*

26. Find a triangulation of the following graph.

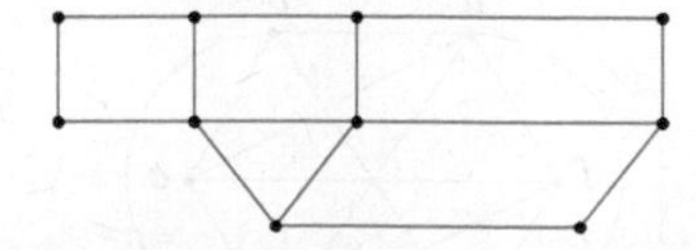

27. Show that if a triangulation $G'$ of a simple, planar graph $G$ can be colored with $n$ colors, so can $G$.

**28.** Show that in a triangulation of a simple, planar graph, $3f = 2e$.

*Appel and Haken proved (see [Appel]) that every simple, planar graph can be colored with four colors. The problem had been posed in the mid-1800s and for years no one had succeeded in giving a proof. Those working on the four-color problem in recent years had one advantage their predecessors did not—the use of fast electronic computers. The following exercises show how the proof begins.*

*Suppose there is a simple, planar graph that requires more than four colors to color. Among all such graphs, there is one with the fewest number of vertices. Let G be a triangulation of this graph. Then G also has a minimal number of vertices and by Exercise 27, G requires more than four colors to color.*

**29.** If the dual of a map has a vertex of degree 3, what must the original map look like?

**30.** Show that $G$ cannot have a vertex of degree 3.

★**31.** Show that $G$ cannot have a vertex of degree 4.

★**32.** Show that $G$ has a vertex of degree 5.

*The contribution of Appel and Haken was to show that only a finite number of cases involving the vertex of degree 5 needed to be considered and to analyze all of these cases and show that all could be colored using four colors. The reduction to a finite number of cases was facilitated by using the computer to help find the cases to be analyzed. The computer was then used again to analyze the resulting cases.*

★**33.** Show that any simple, planar graph can be colored using five colors.

## 8.8 Instant Insanity[†]

**Instant Insanity** is a puzzle consisting of four cubes each of whose faces is painted one of four colors: red, white, blue, or green (see Figure 8.8.1). (There are different versions of the puzzle, depending on which faces are painted which colors.) The problem is to stack the cubes, one on top of the other, so that whether the cubes are viewed from front, back, left, or right, one sees all four colors (see Figure 8.8.2). Since 331,776 different stacks are possible (see Exercise 12), a solution by hand by trial and error is impractical. We present a solution, using a graph model, that makes it possible to discover a solution, if there is one, in a few minutes!

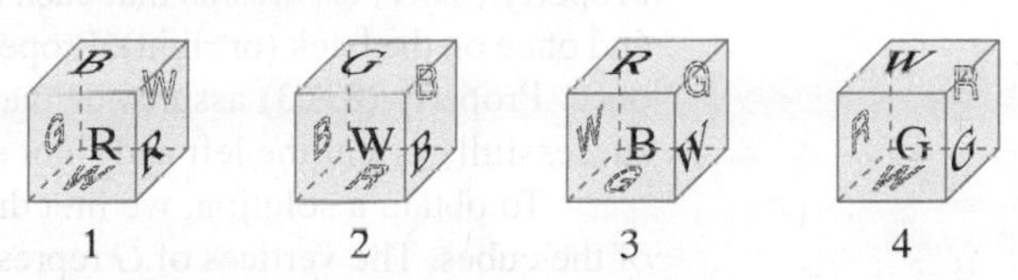

**Figure 8.8.1** An Instant Insanity puzzle.

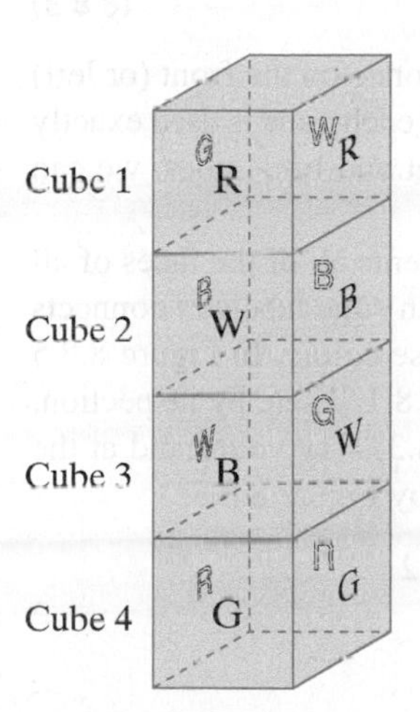


**Figure 8.8.2** A solution to the Instant Insanity puzzle of Figure 8.8.1.

First, notice that any particular stacking can be represented by two graphs, one representing the front/back colors and the other representing the left/right colors. For example, in Figure 8.8.3 we represent the stacking of Figure 8.8.2. The vertices represent the colors, and an edge connects two vertices if the opposite faces have those colors. For example, in the front/back graph, the edge labeled 1 connects $R$ and $W$, since the front and back faces of cube 1 are red and white. As another example, in the left/right graph, $W$ has a loop, since both the left and right faces of cube 3 are white.

We can also construct a stacking from a pair of graphs such as those in Figure 8.8.3, which represent a solution of the Instant Insanity puzzle. Begin with the front/back graph. Cube 1 is to have red and white opposing faces. Arbitrarily assign one of these colors, say red, to the front. Then cube 1 has a white back face. The other edge incident on $W$ is 2, so make cube 2's front face white. This gives cube 2 a blue back face. The other edge incident on $B$ is 3, so make cube 3's front face blue. This gives cube 3 a green back face. The other edge incident on $G$ is 4. Cube 4 then gets a green front face and a red back face. The front and back faces are now properly aligned. At this point, the left and right faces are randomly arranged; however, we will show how to correctly orient the left and right faces without altering the colors of the front and back faces.

[†]This section can be omitted without loss of continuity.

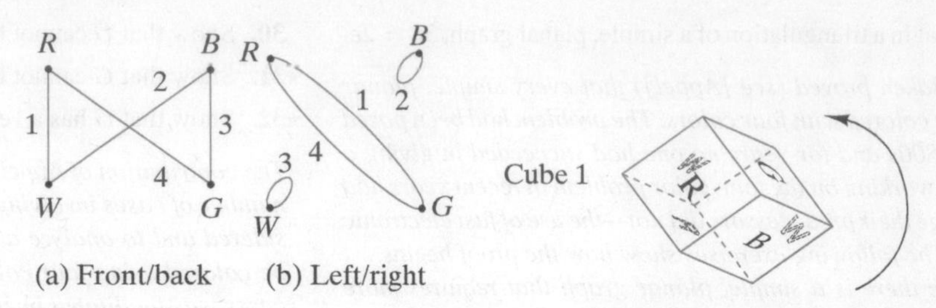


**Figure 8.8.3** Graphs that represent the stacking of Figure 8.8.2.

**Figure 8.8.4** Rotating a cube to obtain a left/right orientation, without changing the front/back colors.

Cube 1 is to have red and green opposing left and right faces. Arbitrarily assign one of these colors, say green, to the left. Then cube 1 has a red right face. Notice that by rotating the cube, we can obtain this left/right orientation without changing the colors of the front and back (see Figure 8.8.4). We can similarly orient cubes 2, 3, and 4. Notice that cubes 2 and 3 have the same colors on opposing sides. The stacking of Figure 8.8.2 has been reconstructed.

It is apparent from the preceding discussion that a solution to the Instant Insanity puzzle can be obtained if we can find two graphs like those of Figure 8.8.3. The properties needed are

- Each vertex should have degree 2. **(8.8.1)**
- Each cube should be represented by an edge exactly once in each graph. **(8.8.2)**
- The two graphs should not have any edges in common. **(8.8.3)**

Property (8.8.1) assures us that each color can be used twice, once on the front (or left) and once on the back (or right). Property (8.8.2) assures us that each cube is used exactly once. Property (8.8.3) assures us that, after orienting the front and back sides, we can successfully orient the left and right sides.

To obtain a solution, we first draw a graph $G$ that represents all of the faces of all of the cubes. The vertices of $G$ represent the four colors, and an edge labeled $i$ connects two vertices (colors) if the opposing faces of cube $i$ have those colors. In Figure 8.8.5 we have drawn the graph that represents the cubes of Figure 8.8.1. Then, by inspection, we find two subgraphs of $G$ satisfying properties (8.8.1)–(8.8.3). Try your hand at the method by finding another solution to the puzzle represented by Figure 8.8.5.

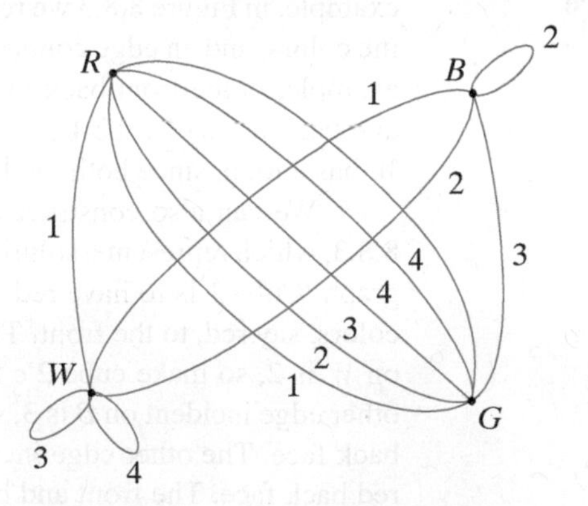


**Figure 8.8.5** A graph representation of the Instant Insanity puzzle of Figure 8.8.1.

**Example 8.8.1** Find a solution to the Instant Insanity puzzle of Figure 8.8.6.

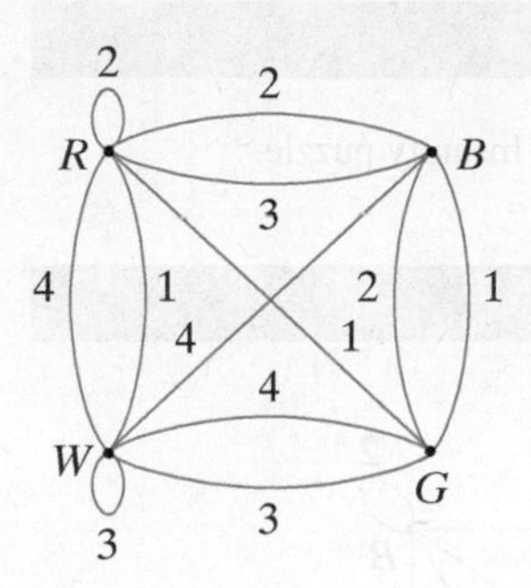


Figure 8.8.6 The Instant Insanity puzzle for Example 8.8.1.

SOLUTION We begin by trying to construct one subgraph having properties (8.8.1) and (8.8.2). We arbitrarily choose a vertex, say $B$, and choose two edges incident on vertex $B$. Suppose that we select the two edges shown as solid lines in Figure 8.8.7. Now consider the problem of picking two edges incident on vertex $R$. We cannot select any edges incident on $B$ or $G$ since $B$ and $G$ must each have degree 2. Since each cube must appear in each subgraph exactly once, we cannot select any of the edges labeled 1 or 2 since we already have selected edges with these labels. Edges incident on $R$ that cannot be selected are shown dashed in Figure 8.8.7. This leaves only the edge labeled 4. Since we need two edges incident on $R$, our initial selection of edges incident on $B$ must be revised.

For our next attempt at choosing two edges incident on vertex $B$, let us select the edges labeled 2 and 3, as shown in Figure 8.8.8. Since this choice includes one edge incident on $R$, we must choose one additional edge incident on $R$. We have three possibilities for selecting the additional edge (shown in color in Figure 8.8.8). (The loop incident at $R$ counts as two edges and so cannot be chosen.) If we select the edge labeled 1 incident on $R$ and $G$, we would need a loop at $W$ labeled 4. Since there is no such loop, we do not select this edge. If we select the edge labeled 1 incident on $R$ and $W$, we can then select the edge labeled 4 incident on $W$ and $G$ (see Figure 8.8.9). We have obtained one of the graphs.

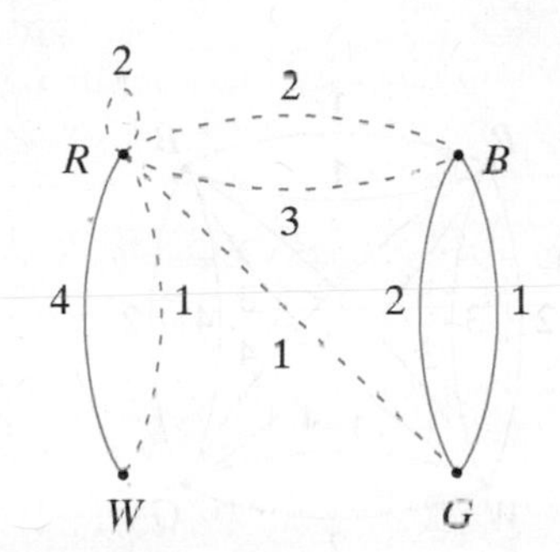


Figure 8.8.7 Trying to find a subgraph of Figure 8.8.6 satisfying (8.8.1) and (8.8.2).

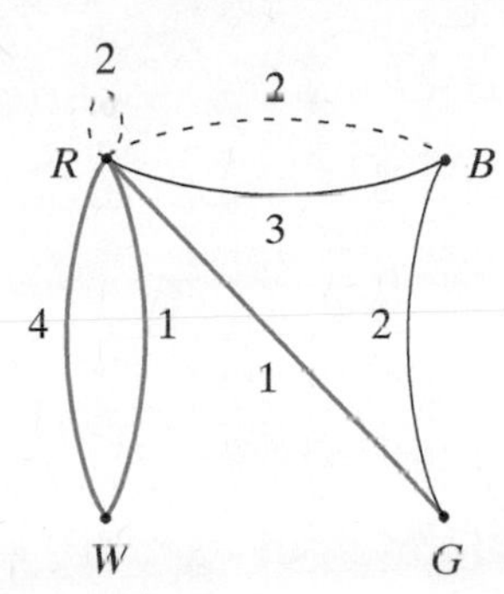


Figure 8.8.8 Another attempt to find a subgraph of Figure 8.8.6 satisfying (8.8.1) and (8.8.2).

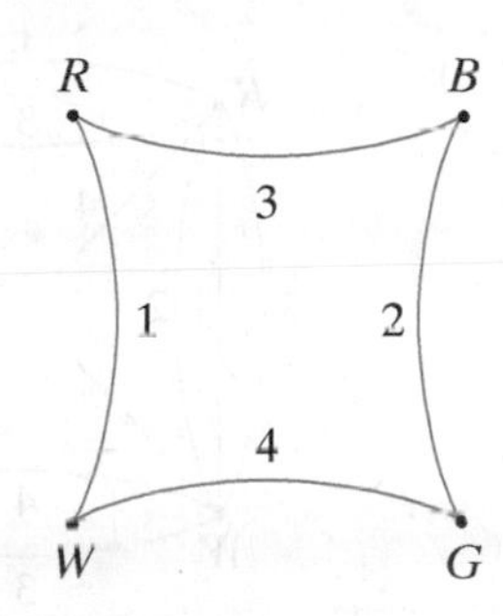


Figure 8.8.9 A subgraph of Figure 8.8.6 satisfying (8.8.1) and (8.8.2).

We now look for a second graph having no edges in common with the graph just chosen. Let us again begin by picking two edges incident on $B$. Because we cannot reuse edges, our choices are limited to three edges (see Figure 8.8.10). Choosing the edges labeled 1 and 4 leads to the graph of Figure 8.8.11. The graphs of Figures 8.8.9 and 8.8.11 solve the Instant Insanity puzzle of Figure 8.8.6.

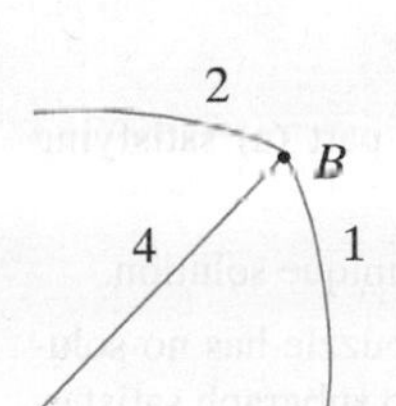


Figure 8.8.10 Edges incident on $B$ not used in Figure 8.8.9.

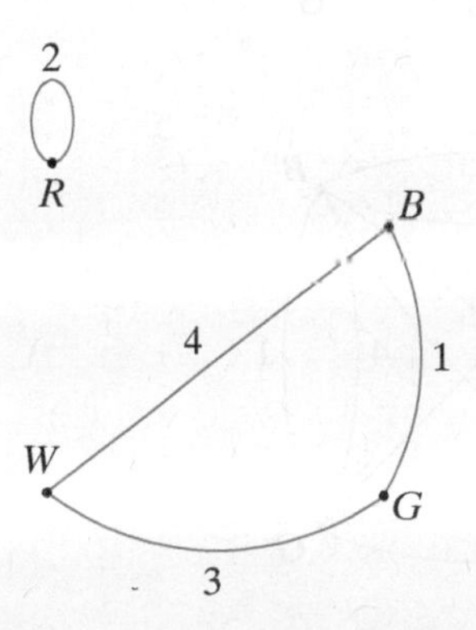


Figure 8.8.11 Another subgraph of Figure 8.8.6, with no edges in common with Figure 8.8.9, satisfying (8.8.1) and (8.8.2). This graph and that of Figure 8.8.9 solve the Instant Insanity puzzle of Figure 8.8.6. ◀

## 8.8 Review Exercises

1. Describe the Instant Insanity puzzle.

2. Describe how to solve the Instant Insanity puzzle.

## 8.8 Exercises

*Find solutions to the following Instant Insanity puzzles.*

**1.**

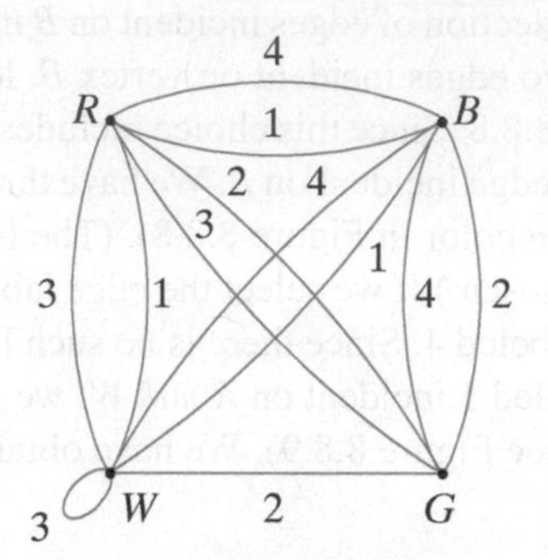

**2.**

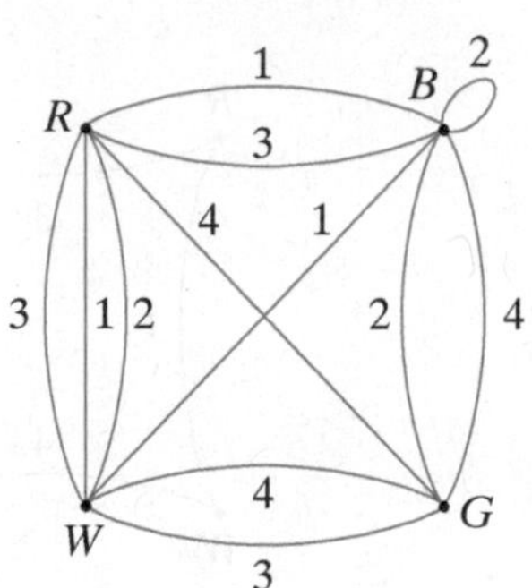

**3.**

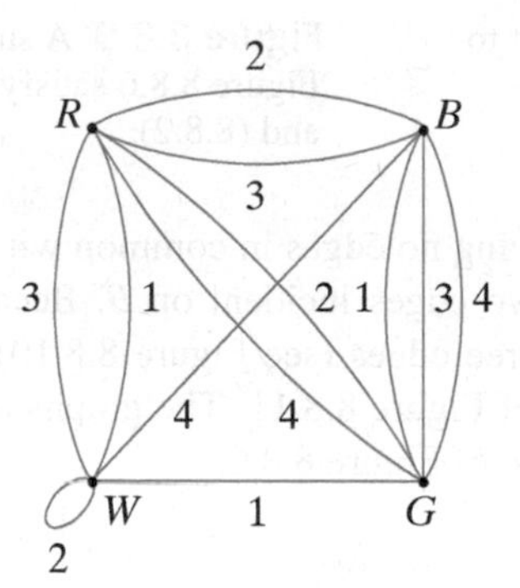

**4.**

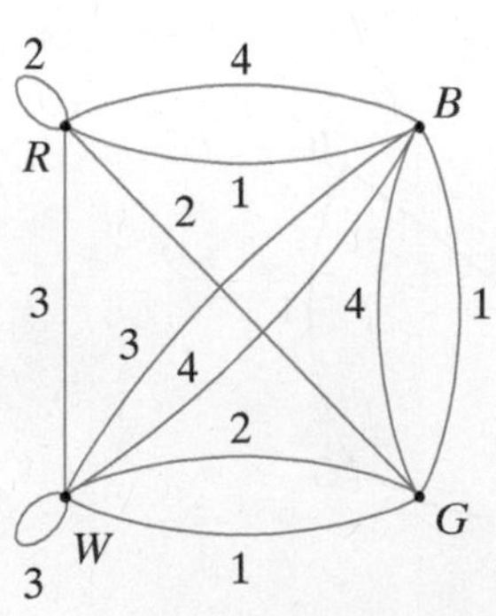

**5.**

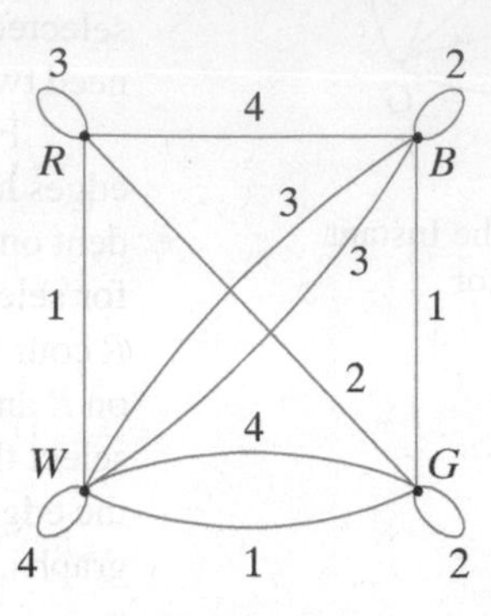

**6.**

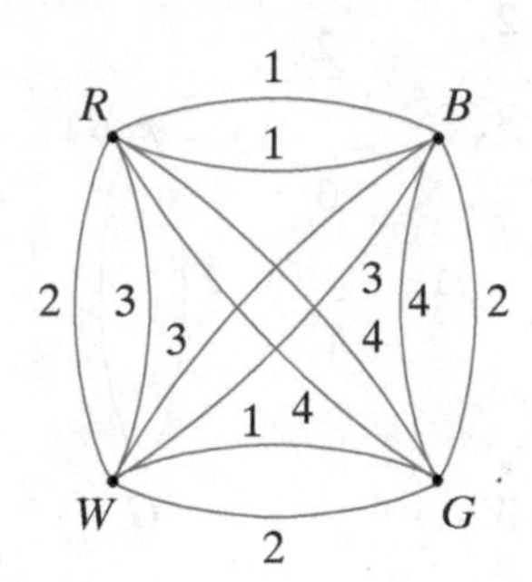

**7.** (a) Find all subgraphs of Figure 8.8.5 satisfying properties (8.8.1) and (8.8.2).

(b) Find all the solutions to the Instant Insanity puzzle of Figure 8.8.5.

**8.** (a) Represent the Instant Insanity puzzle by a graph.

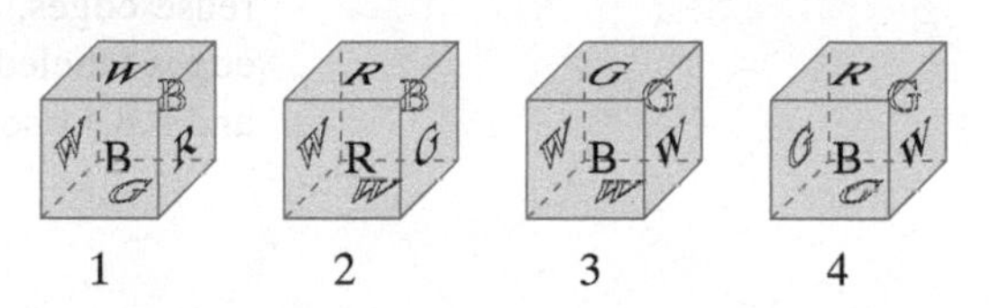

(b) Find a solution to the puzzle.

(c) Find all subgraphs of your graph of part (a) satisfying properties (8.8.1) and (8.8.2).

(d) Use (c) to show that the puzzle has a unique solution.

**9.** Show that the following Instant Insanity puzzle has no solution by giving an argument to show that no subgraph satisfies properties (8.8.1) and (8.8.2). Notice that there is no solution even though each cube contains all four colors.

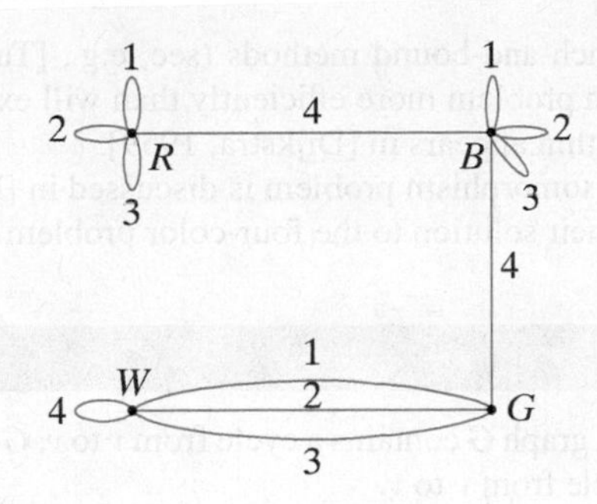


**★10.** Give an example of an Instant Insanity puzzle satisfying:

(a) There is no solution.

(b) Each cube contains all four colors.

(c) There is a subgraph satisfying properties (8.8.1) and (8.8.2).

**11.** Show that there are 24 orientations of a cube.

**12.** Number the cubes of an Instant Insanity puzzle 1, 2, 3, and 4. Show that the number of stackings in which the cubes are stacked 1, 2, 3, and 4, reading from bottom to top, is 331,776.

**★13.** How many Instant Insanity graphs are there; that is, how many graphs are there with four vertices and 12 edges—three of each of four types?

*Exercises 14–21 refer to a modified version of Instant Insanity where a solution is defined to be a stacking that, when viewed from the front, back, left, or right, shows one color. (The front, back, left, and right are of different colors.)*

**14.** Give an argument that shows that if we graph the puzzle as in regular Instant Insanity, a solution to modified Instant Insanity consists of two subgraphs of the form shown in the figure, with no edges or vertices in common.

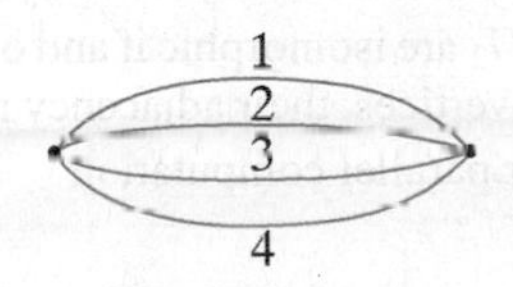


*Find solutions to modified Instant Insanity for the following puzzles.*

**15.**

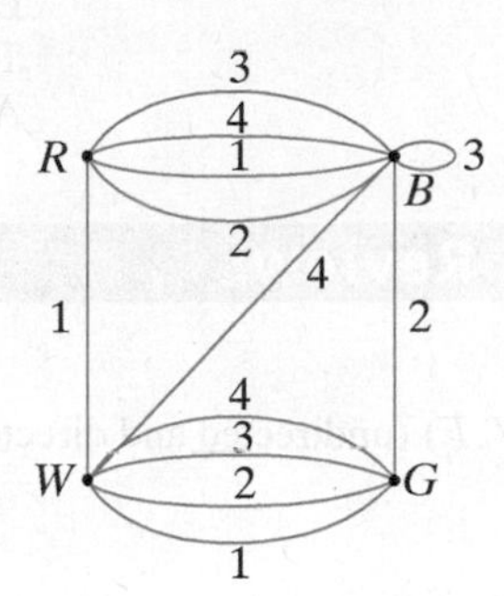


**16.**

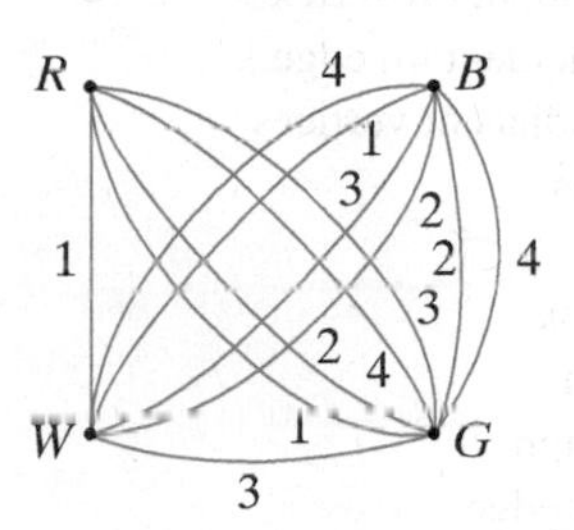


**17.**

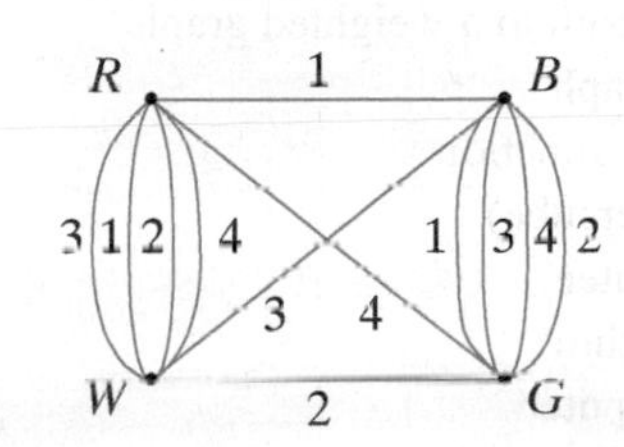


**18.** Graph of Exercise 6

**19.** Show that for Figure 8.8.5, Instant Insanity, as given in the text, has a solution, but the modified version does not have a solution.

**20.** Show that if modified Instant Insanity has a solution, the version given in the text must also have a solution.

**21.** Is it possible for neither version of Instant Insanity to have a solution even if each cube contains all four colors? If the answer is yes, prove it; otherwise, give a counterexample.

## Chapter 8 Notes

Virtually any reference on discrete mathematics contains one or more chapters on graph theory. Books specifically on graph theory are [Berge; Bondy; Chartrand; Deo; Even, 1979; Gibbons; Harary; König; Ore; West; and Wilson]. [Deo; Even, 1979; and Gibbons] emphasize graph algorithms. [Brassard, Cormen, and Johnsonbaugh] also treat graphs and graph algorithms.

[Akl; Leighton; Lester; Lewis; Miller; and Quinn] discuss parallel computers and algorithms for parallel computers. Our results on subgraphs of the hypercube are from [Saad].

Euler's original paper on the Königsberg bridges, edited by J. R. Newman, was reprinted as [Newman].

In [Gardner, 1959], Hamiltonian cycles are related to the Tower of Hanoi puzzle.

Concerning the traveling salesperson problem, [Applegate] is accurately self-described as "the definitive book on the subject."

In many cases, so called branch-and-bound methods (see, e.g., [Tucker]) often give solutions to the traveling salesperson problem more efficiently than will exhaustive search.

Dijkstra's shortest-path algorithm appears in [Dijkstra, 1959].

The complexity of the graph isomorphism problem is discussed in [Köbler].

Appel and Haken published their solution to the four-color problem in [Appel].

## Chapter 8 Review

### Section 8.1

1. Graph $G = (V, E)$ (undirected and directed)
2. Vertex
3. Edge
4. Edge $e$ is incident on vertex $v$
5. Vertex $v$ is incident on edge $e$
6. $v$ and $w$ are adjacent vertices
7. Parallel edges
8. Loop
9. Isolated vertex
10. Simple graph
11. Weighted graph
12. Weight of an edge
13. Length of a path in a weighted graph
14. Similarity graph
15. Dissimilarity function
16. $n$-cube (hypercube)
17. Serial computer
18. Serial algorithm
19. Parallel computer
20. Parallel algorithm
21. Complete graph on $n$ vertices, $K_n$
22. Bipartite graph
23. Complete bipartite graph on $m$ and $n$ vertices, $K_{m,n}$

### Section 8.2

24. Path
25. Simple path
26. Cycle
27. Simple cycle
28. Connected graph
29. Subgraph
30. Component of a graph
31. Degree of a vertex $\delta(v)$
32. Königsberg bridge problem
33. Euler cycle
34. A graph $G$ has an Euler cycle if and only if $G$ is connected and every vertex has even degree.
35. The sum of the degrees of all vertices in a graph is equal to twice the number of edges.
36. In any graph, the number of vertices of odd degree is even.
37. A graph has a path with no repeated edges from $v$ to $w$ ($v \neq w$) containing all the edges and vertices if and only if it is connected and $v$ and $w$ are the only vertices having odd degree.
38. If a graph $G$ contains a cycle from $v$ to $v$, $G$ contains a simple cycle from $v$ to $v$.

### Section 8.3

39. Hamiltonian cycle
40. Traveling salesperson problem
41. Ring model for parallel computation
42. Gray code

### Section 8.4

43. Dijkstra's shortest-path algorithm

### Section 8.5

44. Adjacency matrix
45. If $A$ is the adjacency matrix of a simple graph, the $ij$ th entry of $A^n$ is equal to the number of paths of length $n$ from vertex $i$ to vertex $j$.
46. Incidence matrix

### Section 8.6

47. Graphs $G_1 = (V_1, E_1)$ and $G_2 = (V_2, E_2)$ are isomorphic if there are one-to-one, onto functions $f\colon V_1 \to V_2$ and $g\colon E_1 \to E_2$ such that $e \in E_1$ is incident on $v, w \in V_1$ if and only if $g(e)$ is incident on $f(v)$ and $f(w)$.
48. Graphs $G_1$ and $G_2$ are isomorphic if and only if, for some orderings of their vertices, their adjacency matrices are equal.
49. Mesh model for parallel computation
50. Invariant

### Section 8.7

51. Planar graph
52. Face
53. Euler's equation for connected, planar graphs: $f = e - v + 2$
54. Edges in series
55. Series reduction
56. Homeomorphic graphs
57. Kuratowski's theorem: A graph is planar if and only if it does not contain a subgraph homeomorphic to $K_5$ or $K_{3,3}$.

### Section 8.8

58. Instant Insanity
59. How to solve an Instant Insanity puzzle

## Chapter 8 Self-Test

1. For the graph $G = (V, E)$, find $V$, $E$, all parallel edges, all loops, and all isolated vertices, and state whether $G$ is a simple graph. Also state on which vertices edge $e_3$ is incident and on which edges vertex $v_2$ is incident.

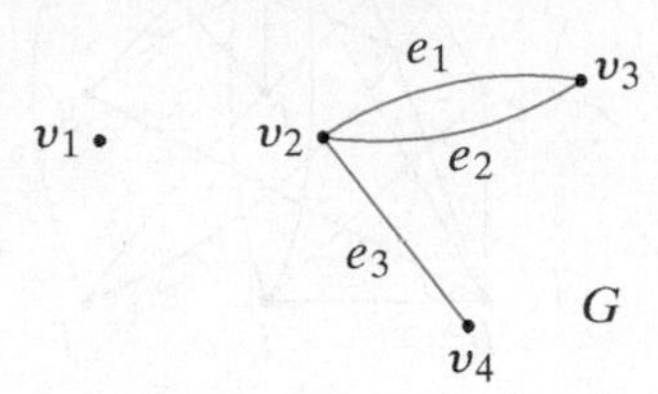


2. Tell whether the path $(v_2, v_3, v_4, v_2, v_6, v_1, v_2)$ in the graph is a simple path, a cycle, a simple cycle, or none of these.

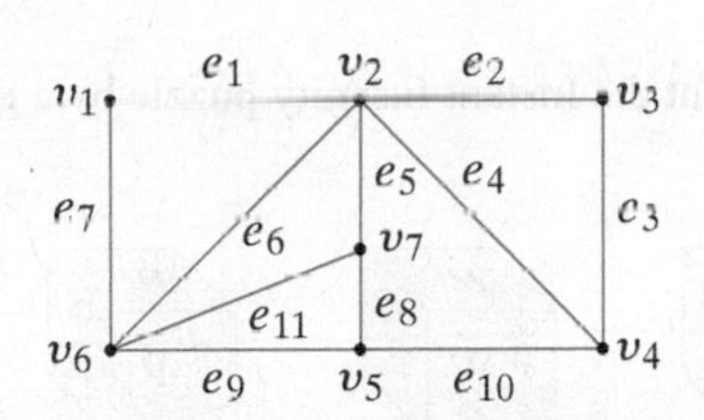


3. Find a Hamiltonian cycle in the graph of Exercise 2.
4. Write the adjacency matrix of the graph of Exercise 2.
5. Write the incidence matrix of the graph of Exercise 2.
6. If $A$ is the adjacency matrix of the graph of Exercise 2, what does the entry in row $v_2$ and column $v_3$ of $A^3$ represent?

*In Exercises 7 and 8, determine whether the graphs $G_1$ and $G_2$ are isomorphic. Prove your answer.*

7.

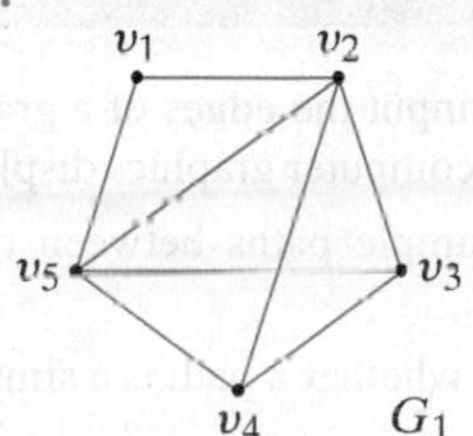


w1 w2 w3 w4 w5 $G_2$

8.

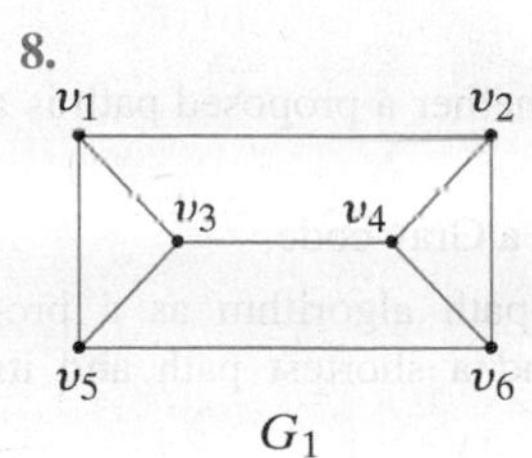


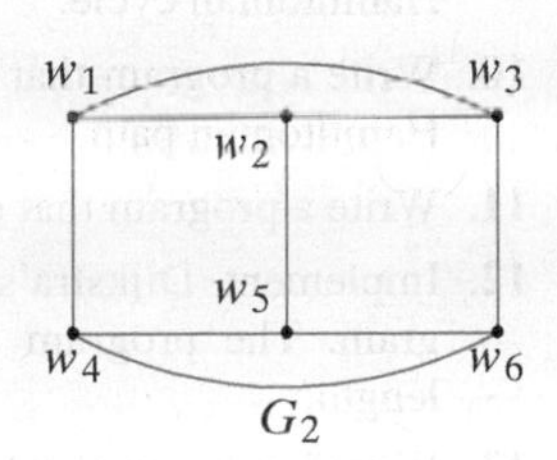


9. Explain why the following graph does not have a path from $a$ to $a$ that passes through each edge exactly one time.

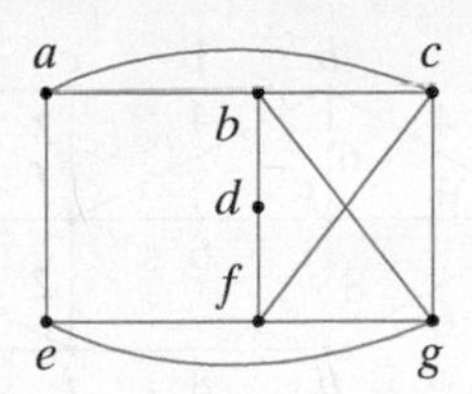


10. Draw $K_{2,5}$, the complete bipartite graph on 2 and 5 vertices.
11. Draw all subgraphs containing exactly two edges of the following graph.

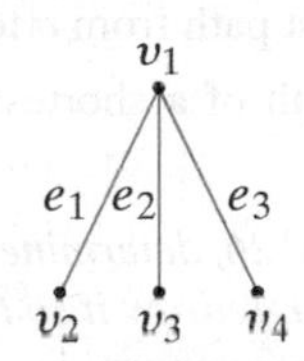


12. Find a connected subgraph of the graph of Exercise 2 containing all of the vertices of the original graph and having as few edges as possible.
13. Find a Hamiltonian cycle in the 3-cube.
14. Show that the graph has no Hamiltonian cycle.

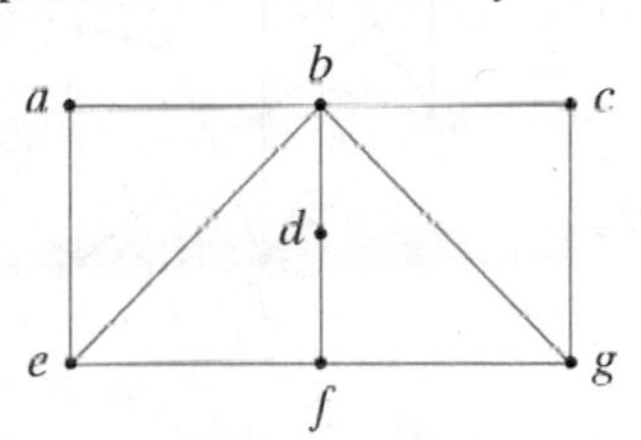


15. Can a column of an incidence matrix consist only of zeros? Explain.
16. Prove that the $n$-cube is bipartite for all $n \geq 1$.
17. Does the graph of Exercise 2 contain an Euler cycle? Explain.
18. Show that the cycle $(a, b, c, d, e, f, g, h, i, j, a)$ provides a solution to the traveling salesperson problem for the graph shown.

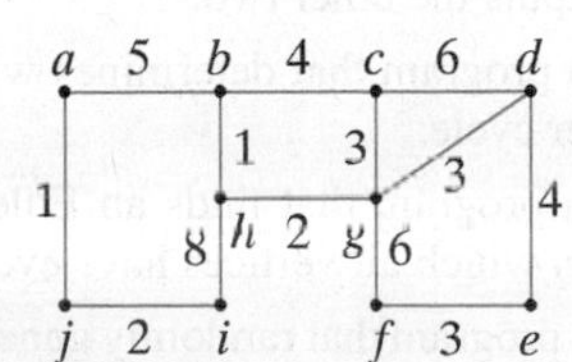


19. Draw all nonisomorphic, simple graphs having exactly five vertices and two edges.
20. Draw all nonisomorphic, simple graphs having exactly five vertices, two components, and no cycles.

*Exercises 21–24 refer to the following graph.*

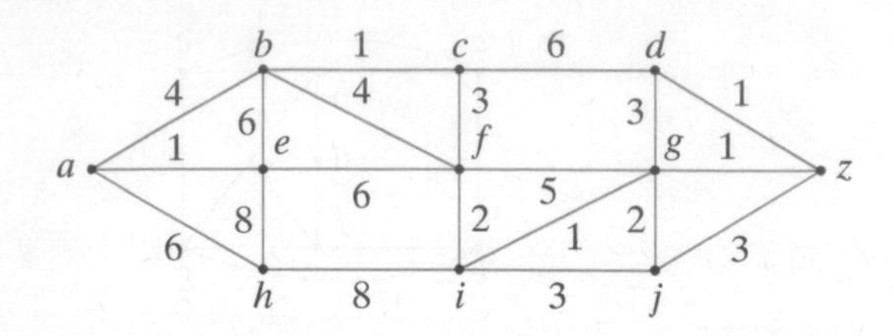


**21.** Find the length of a shortest path from $a$ to $i$.

**22.** Find the length of a shortest path from $a$ to $z$.

**23.** Find a shortest path from $a$ to $z$.

**24.** Find the length of a shortest path from $a$ to $z$ that passes through $c$.

*In Exercises 25 and 26, determine whether the graph is planar. If the graph is planar, redraw it so that no edges cross; otherwise, find a subgraph homeomorphic to either $K_5$ or $K_{3,3}$.*

**25.**

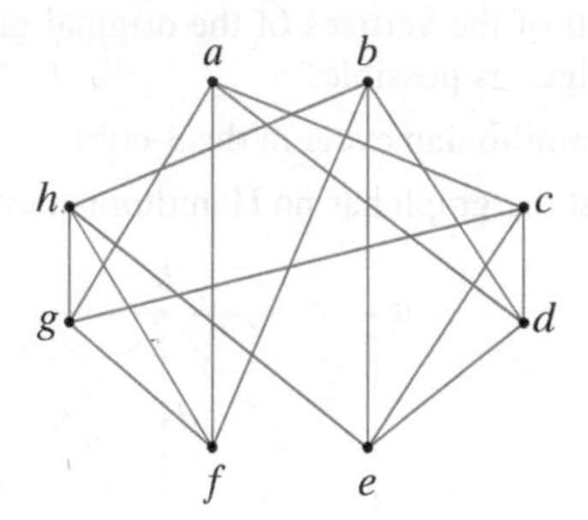


**26.**

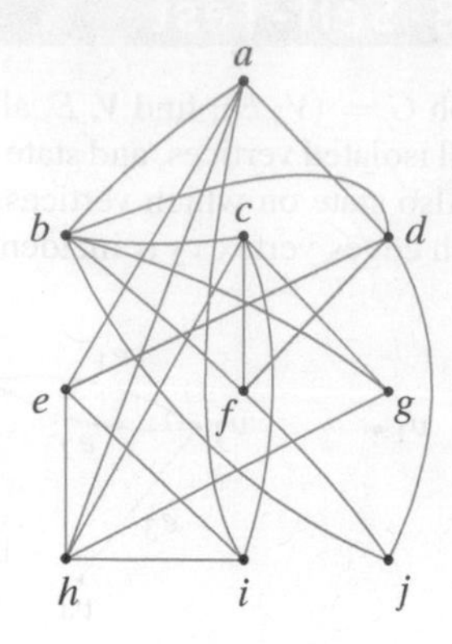


**27.** Show that any simple, connected graph with 31 edges and 12 vertices is not planar.

**28.** Show that the $n$-cube is planar if $n \leq 3$ and not planar if $n > 3$.

**29.** Represent the Instant Insanity puzzle by a graph.

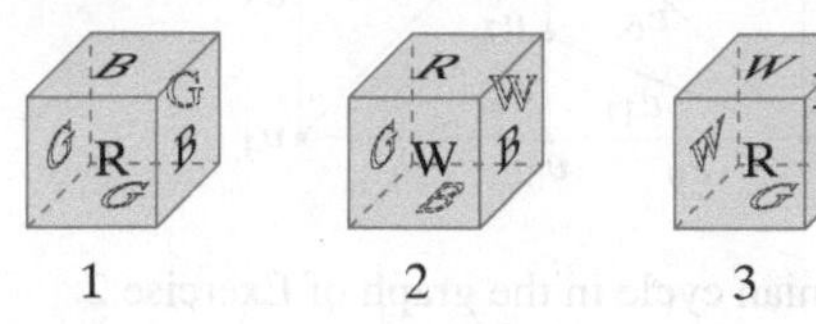


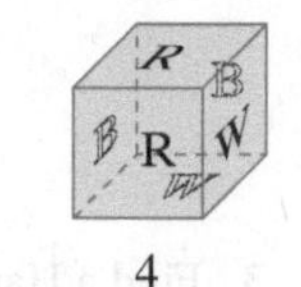


**30.** Find a solution to the puzzle of Exercise 29.

**31.** Find all subgraphs of the graph of Exercise 29 satisfying properties (8.8.1) and (8.8.2).

**32.** Use Exercise 31 to determine how many solutions the puzzle of Exercise 29 has.

## Chapter 8 Computer Exercises

**1.** Write a program that accepts as input any of the following:
- a listing of edges of a graph given as pairs of positive integers
- the adjacency matrix
- the incidence matrix

and outputs the other two.

**2.** Write a program that determines whether a graph contains an Euler cycle.

**3.** Write a program that finds an Euler cycle in a connected graph in which all vertices have even degree.

**4.** Write a program that randomly generates an $n \times n$ adjacency matrix. Have your program print the adjacency matrix, the number of edges, the number of loops, and the degree of each vertex.

**5.** Write a program that determines whether a graph is a bipartite graph. If it is a bipartite graph, the program should list the disjoint sets of vertices.

**6.** Write a program that accepts as input the edges of a graph and then draws the graph using a computer graphics display.

**7.** Write a program that lists all simple paths between two given vertices.

**8.** Write a program that determines whether a path is a simple path, a cycle, or a simple cycle.

**9.** Write a program that checks whether a proposed cycle is a Hamiltonian cycle.

**10.** Write a program that checks whether a proposed path is a Hamiltonian path.

**11.** Write a program that constructs a Gray code.

**12.** Implement Dijkstra's shortest-path algorithm as a program. The program should find a shortest path and its length.

**13.** Write a program that tests whether a proposed isomorphism is an isomorphism.

**14.** Write a program to determine whether a graph is planar.

**15.** Write a program that solves an arbitrary Instant Insanity puzzle.

**16.** Implement *randomized_hamiltonian_cycle* (Algorithm 8.3.10) as a computer program. Add a condition to stop the algorithm after some number of iterations to guarantee that it will always terminate. Try your program out on several sample graphs. (There is a random graph generator at the website that accompanies this book.)

# Chapter 9

# TREES



Trees form one of the most widely used subclasses of graphs. Computer science, in particular, makes extensive use of trees. In computer science, trees are useful in organizing and relating data in a database (see Example 9.1.7). Trees also arise in theoretical problems such as the optimal time for sorting (see Section 9.7).

In this chapter we begin by giving the requisite terminology. We look at subclasses of trees (e.g., rooted trees and binary trees) and many applications of trees (e.g., spanning trees, decision trees, and game trees). Our discussion of tree isomorphisms extends the discussion of Section 8.6 on graph isomorphisms.

## 9.1 Introduction

Figure 9.1.1 shows the results of the semifinals and finals of a classic tennis competition at Wimbledon, which featured four of the greatest players in the history of tennis. At Wimbledon, when a player loses, she is out of the tournament. Winners continue to play until only one person, the champion, remains. (Such a competition is called a *single-elimination tournament.*) Figure 9.1.1 shows that in the semifinals Monica Seles defeated Martina Navratilova and Steffi Graf defeated Gabriela Sabatini. The winners, Seles and Graf, then played, and Graf defeated Seles. Steffi Graf, being the sole undefeated player, became Wimbledon champion.

**Go Online**
For more on trees, see `bit.ly/2q58dGw`

If we regard the single-elimination tournament of Figure 9.1.1 as a graph (see Figure 9.1.2), we obtain a **tree.** If we rotate Figure 9.1.2, it looks like a natural tree (see Figure 9.1.3). The formal definition follows.

**Definition 9.1.1 ▶** A (*free*) *tree* $T$ is a simple graph satisfying the following: If $v$ and $w$ are vertices in $T$, there is a unique simple path from $v$ to $w$.

A *rooted tree* is a tree in which a particular vertex is designated the root. ◀

†This section can be omitted without loss of continuity.

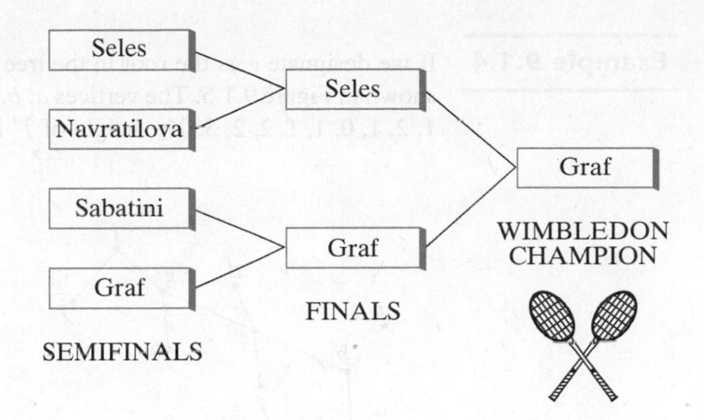


**Figure 9.1.1** Semifinals and finals at Wimbledon.

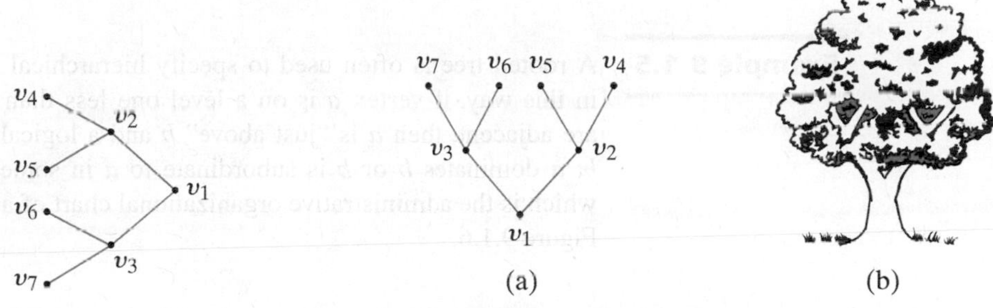


**Figure 9.1.2** The tournament of Figure 9.1.1 as a tree.

**Figure 9.1.3** The tree of Figure 9.1.2 rotated (a) compared with a natural tree (b).

**Example 9.1.2** If we designate the winner as the root, the single-elimination tournament of Figure 9.1.1 (or Figure 9.1.2) is a rooted tree. Notice that if $v$ and $w$ are vertices in this graph, there is a unique simple path from $v$ to $w$. For example, the unique simple path from $v_2$ to $v_7$ is $(v_2, v_1, v_3, v_7)$. ◀

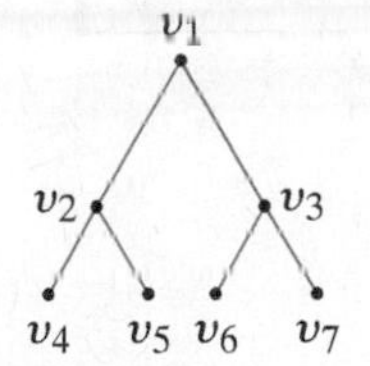


**Figure 9.1.4** The tree Figure 9.1.3(a) with the root at the top.

In contrast to natural trees, which have their roots at the bottom, in graph theory rooted trees are typically drawn with their roots at the top. Figure 9.1.4 shows the way the tree of Figure 9.1.2 would be drawn (with $v_1$ as root). First, we place the root $v_1$ at the top. Under the root and on the same level, we place the vertices, $v_2$ and $v_3$, that can be reached from the root on a simple path of length 1. Under each of these vertices and on the same level, we place the vertices, $v_4$, $v_5$, $v_6$, and $v_7$, that can be reached from the root on a simple path of length 2. We continue in this way until the entire tree is drawn. Since the simple path from the root to any given vertex is unique, each vertex is on a uniquely determined level. We call the level of the root level 0. The vertices under the root are said to be on level 1, and so on. Thus the **level of a vertex** $v$ is the length of the simple path from the root to $v$. The **height** of a rooted tree is the maximum level number that occurs.

**Example 9.1.3** The vertices $v_1$, $v_2$, $v_3$, $v_4$, $v_5$, $v_6$, $v_7$ in the rooted tree of Figure 9.1.4 are on (respectively) levels 0, 1, 1, 2, 2, 2, 2. The height of the tree is 2. ◀

**Example 9.1.4** If we designate $e$ as the root in the tree $T$ of Figure 9.1.5, we obtain the rooted tree $T'$ shown in Figure 9.1.5. The vertices $a, b, c, d, e, f, g, h, i, j$ are on (respectively) levels 2, 1, 2, 1, 0, 1, 1, 2, 2, 3. The height of $T'$ is 3.

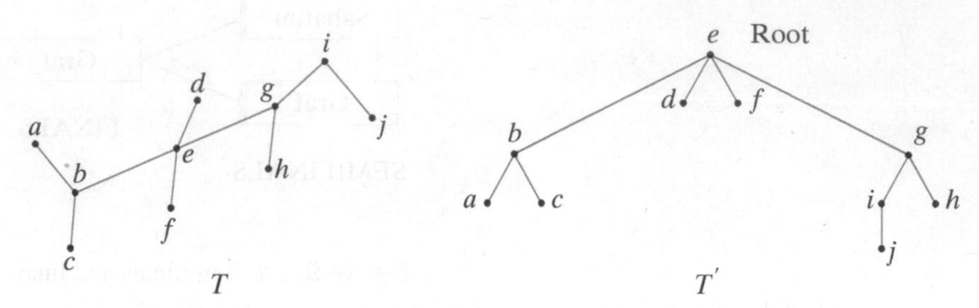


**Figure 9.1.5** A tree $T$ and a rooted tree $T'$. $T'$ is obtained from $T$ by designating $e$ as the root. ◀

**Example 9.1.5** A rooted tree is often used to specify hierarchical relationships. When a tree is used in this way, if vertex $a$ is on a level one less than the level of vertex $b$ and $a$ and $b$ are adjacent, then $a$ is "just above" $b$ and a logical relationship exists between $a$ and $b$: $a$ dominates $b$ or $b$ is subordinate to $a$ in some way. An example of such a tree, which is the administrative organizational chart of a hypothetical university, is given in Figure 9.1.6.

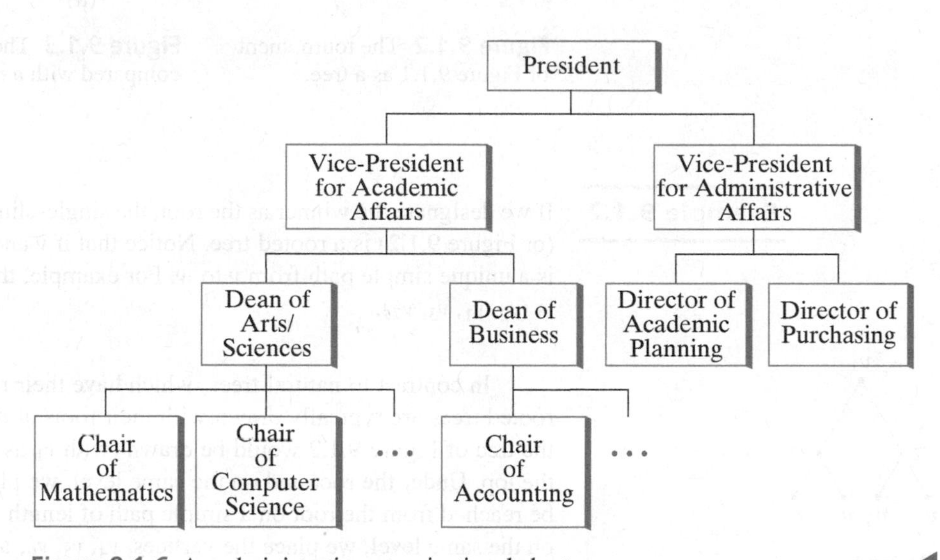


**Figure 9.1.6** An administrative organizational chart. ◀

**Example 9.1.6** **Computer File Systems** Modern computer operating systems organize folders and files using a tree structure. A *folder* contains other folders and files. Figure 9.1.7 shows the Windows Explorer display of folders on the left and files on the right on a particular computer. Figure 9.1.8 shows the same structure as a rooted tree. The root is called *Desktop*. Under *Desktop* are *My Documents*, *My Computer*, and others. Under *My Documents* are *Fax*, *My Data Sources*, *My Pictures*, and others. Under *My Pictures* are *archived*, *basement water*, and *My eBooks*. Under *basement water*, which is highlighted, are the files *11-18-03*, *DSC01007 11-18-03*, and others, which appear on the right of Figure 9.1.7.

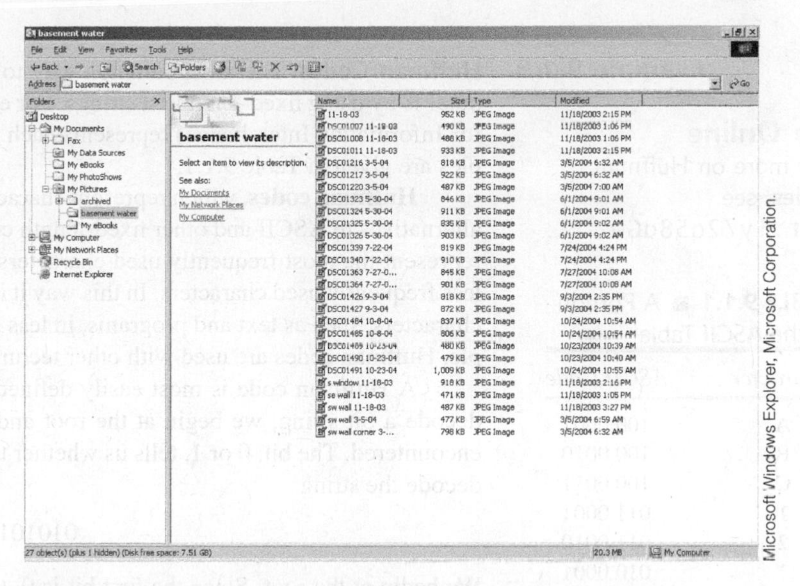




**Figure 9.1.7** Windows Explorer display of folders and files on a particular computer. The folders appear on the left and files appear on the right. Under *basement water*, which is highlighted, are the files *11-18-03*, *DSC01007 11-18-03*, and others, which appear on the right.

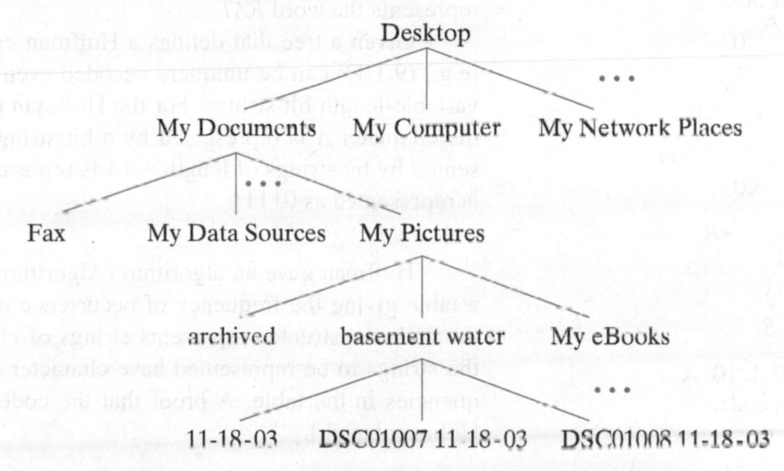


**Figure 9.1.8** The structure of Figure 9.1.7 shown as a rooted tree.

**Example 9.1.7** **Hierarchical Definition Trees** Figure 9.1.9 is an example of a **hierarchical definition tree.** Such trees are used to show logical relationships among records in a database. [Recall (see Section 3.6) that a database is a collection of records that are manipulated by a computer.] The tree of Figure 9.1.9 might be used as a model for setting up a database to maintain records about books housed in several libraries.

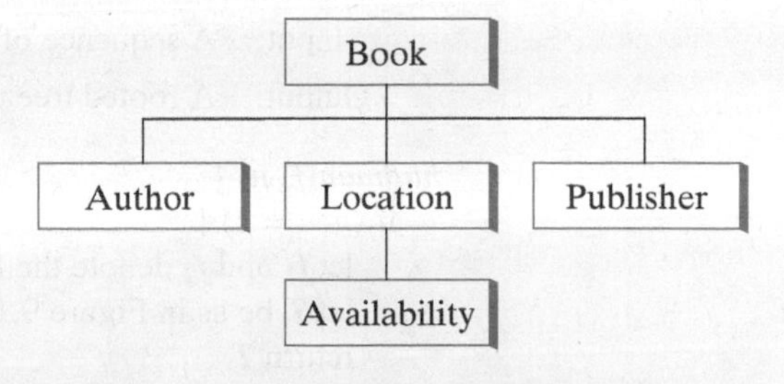


**Figure 9.1.9** A hierarchical definition tree.

**Example 9.1.8** **Huffman Codes** The most common way to represent characters internally in a computer is by using fixed-length bit strings. For example, ASCII (American Standard Code for Information Interchange) represents each character by a string of seven bits. Examples are given in Table 9.1.1.

**Go Online**
For more on Huffman codes, see bit.ly/2q58dGw

TABLE 9.1.1 ■ A Portion of the ASCII Table

| *Character* | *ASCII Code* |
|---|---|
| A | 100 0001 |
| B | 100 0010 |
| C | 100 0011 |
| 1 | 011 0001 |
| 2 | 011 0010 |
| ! | 010 0001 |
| * | 010 1010 |

**Huffman codes,** which represent characters by variable-length bit strings, provide alternatives to ASCII and other fixed-length codes. The idea is to use short bit strings to represent the most frequently used characters and to use longer bit strings to represent less frequently used characters. In this way it is generally possible to represent strings of characters, such as text and programs, in less space than if ASCII were used. For example, Huffman codes are used with other techniques to compress data for fax machines.

A Huffman code is most easily defined by a rooted tree (see Figure 9.1.10). To decode a bit string, we begin at the root and move down the tree until a character is encountered. The bit, 0 or 1, tells us whether to move right or left. As an example, let us decode the string

$$01010111. \quad \textbf{(9.1.1)}$$

We begin at the root. Since the first bit is 0, the first move is right. Next, we move left and then right. At this point, we encounter the first character $R$. To decode the next character, we begin again at the root. The next bit is 1, so we move left and encounter the next character $A$. The last bits 0111 decode as $T$. Therefore, the bit string (9.1.1) represents the word *RAT*.

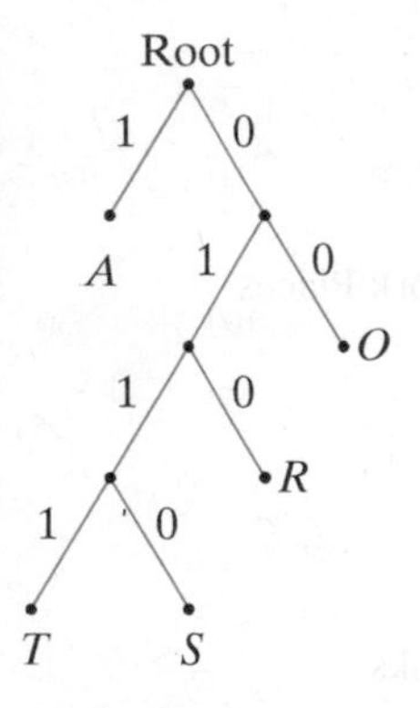

**Figure 9.1.10** A Huffman code.

Given a tree that defines a Huffman code, such as Figure 9.1.10, any bit string [e.g., (9.1.1)] can be uniquely decoded even though the characters are represented by variable-length bit strings. For the Huffman code defined by the tree of Figure 9.1.10, the character $A$ is represented by a bit string of length 1, whereas $S$ and $T$ are represented by bit strings of length 4. ($A$ is represented as 1, $S$ is represented as 0110, and $T$ is represented as 0111.) ◀

Huffman gave an algorithm (Algorithm 9.1.9) to construct a Huffman code from a table giving the frequency of occurrence of the characters to be represented so that the code constructed represents strings of characters in minimal space, provided that the strings to be represented have character frequencies identical to the character frequencies in the table. A proof that the code constructed is optimal may be found in [Johnsonbaugh].

**Algorithm 9.1.9** **Constructing an Optimal Huffman Code**

This algorithm constructs an optimal Huffman code from a table giving the frequency of occurrence of the characters to be represented. The output is a rooted tree with the vertices at the lowest levels labeled with the frequencies and with the edges labeled with bits as in Figure 9.1.10. The coding tree is obtained by replacing each frequency with a character having that frequency.

Input: A sequence of $n$ frequencies, $n \geq 2$

Output: A rooted tree that defines an optimal Huffman code

```
huffman(f, n) {
   if (n == 2) {
      let f1 and f2 denote the frequencies
      let T be as in Figure 9.1.11
      return T
   }
```

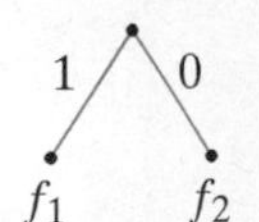

**Figure 9.1.11** The case $n = 2$ for Algorithm 9.1.9.

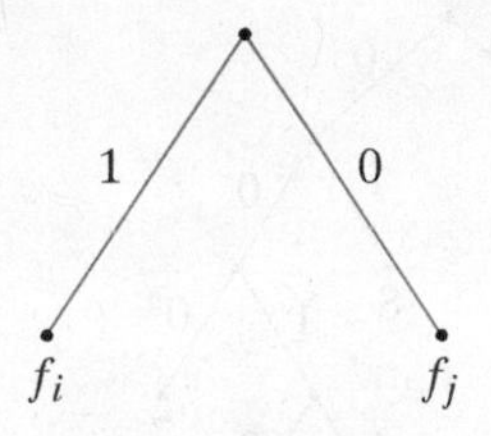

**Figure 9.1.12** The case $n > 2$ for Algorithm 9.1.9.

```
        let f_i and f_j denote the smallest frequencies
        replace f_i and f_j in the list f by f_i + f_j
        T' = huffman(f, n - 1)
        replace a vertex in T' labeled f_i + f_j by the tree shown in Figure 9.1.12
            to obtain the tree T
        return T
}
```

**Example 9.1.10** Show how Algorithm 9.1.9 constructs an optimal Huffman code using Table 9.1.2.

**TABLE 9.1.2** ■ Input for Example 9.1.10

| *Character* | *Frequency* |
|---|---|
| ! | 2 |
| @ | 3 |
| # | 7 |
| $ | 8 |
| % | 12 |

**SOLUTION** The algorithm begins by repeatedly replacing the smallest two frequencies with the sum until a two-element sequence is obtained:

$$2, 3, 7, 8, 12 \rightarrow 2 + 3, 7, 8, 12$$
$$5, 7, 8, 12 \rightarrow 5 + 7, 8, 12$$
$$8, 12, 12 \rightarrow 8 + 12, 12$$
$$12, 20$$

The algorithm then constructs trees working backward beginning with the two-element sequence 12, 20 as shown in Figure 9.1.13. For example, the second tree is obtained from the first by replacing the vertex labeled 20 by the tree of Figure 9.1.14 since 20 arose as the sum of 8 and 12. Finally, to obtain the optimal Huffman coding tree, we replace each frequency by a character having that frequency (see Figure 9.1.15).

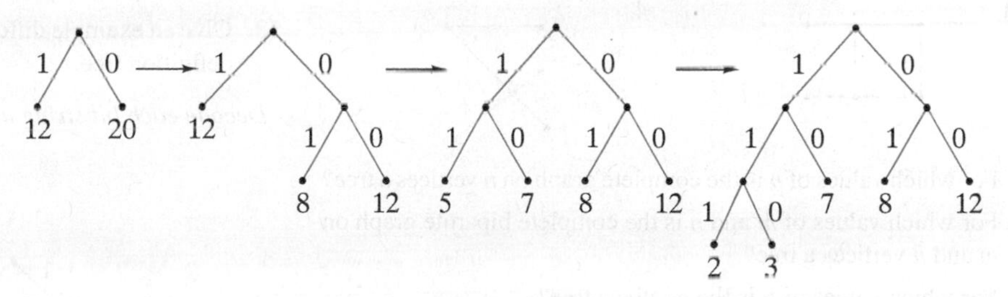

**Figure 9.1.13** Constructing an optimal Huffman code.

Notice that the Huffman tree for Table 9.1.2 is not unique. When 12 is replaced by 5, 7, because there are two vertices labeled 12, there is a choice. In Figure 9.1.13, we arbitrarily chose one of the vertices labeled 12. If we choose the other vertex labeled 12, we will obtain the tree of Figure 9.1.16. Either of the Huffman trees gives an optimal code; that is, either will encode text having the frequencies of Table 9.1.2 in exactly the same (optimal) space.

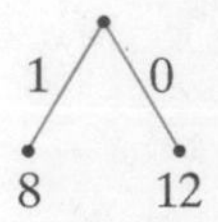


**Figure 9.1.14** The tree that replaces the vertex labeled 20 in Figure 9.1.13.

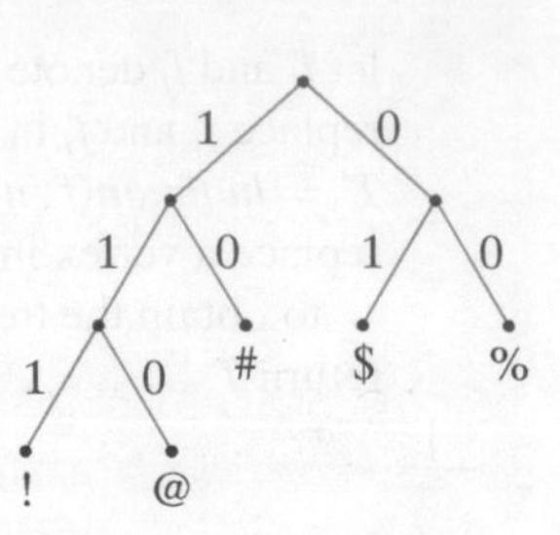


**Figure 9.1.15** The final tree of Figure 9.1.13 with each frequency replaced by a character having that frequency.

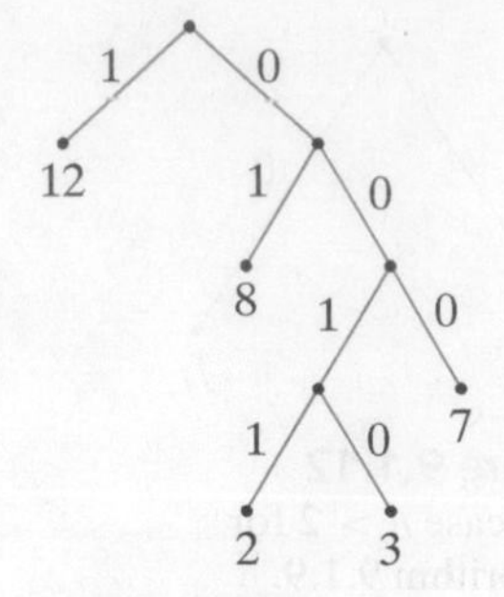


**Figure 9.1.16** Another optimal Huffman tree for Example 9.1.10.

## 9.1 Review Exercises

1. Define *free tree*.
2. Define *rooted tree*.
3. What is the level of a vertex in a rooted tree?
4. What is the height of a rooted tree?
5. Give an example of a hierarchical definition tree.
6. Explain how files and folders in a computer system are organized into a rooted tree structure.
7. What is a Huffman code?
8. Explain how to construct an optimal Huffman code.

## 9.1 Exercises

*Which of the graphs in Exercises 1–4 are trees? Explain.*

**1.**

**2.**

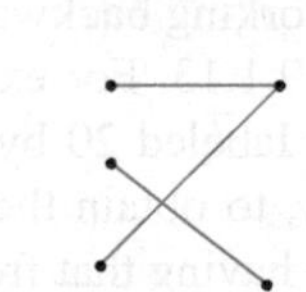

**3.**

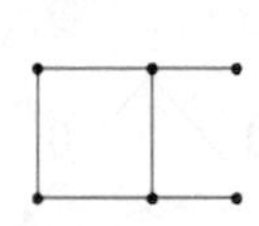

**4.**

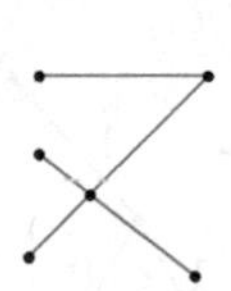

5. For which values of $n$ is the complete graph on $n$ vertices a tree?
6. For which values of $m$ and $n$ is the complete bipartite graph on $m$ and $n$ vertices a tree?
7. For which values of $n$ is the $n$-cube a tree?
8. Find the level of each vertex in the tree shown.

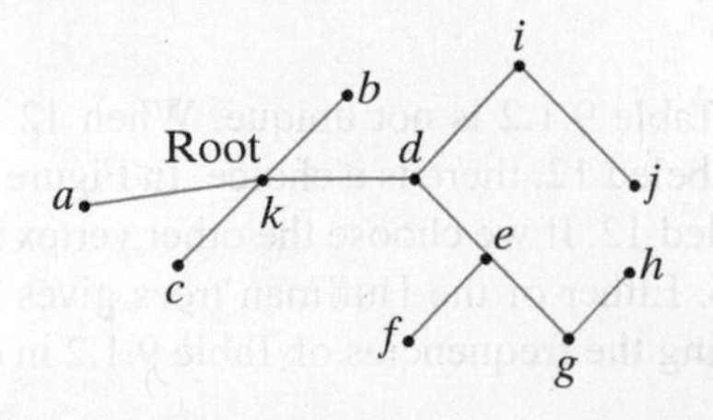


9. Find the height of the tree of Exercise 8.
10. Draw the tree $T$ of Figure 9.1.5 as a rooted tree with $a$ as root. What is the height of the resulting tree?
11. Draw the tree $T$ of Figure 9.1.5 as a rooted tree with $b$ as root. What is the height of the resulting tree?
12. Give an example similar to Example 9.1.5 of a tree that is used to specify hierarchical relationships.
13. Give an example different from Example 9.1.7 of a hierarchical definition tree.

*Decode each bit string using the Huffman code given.*

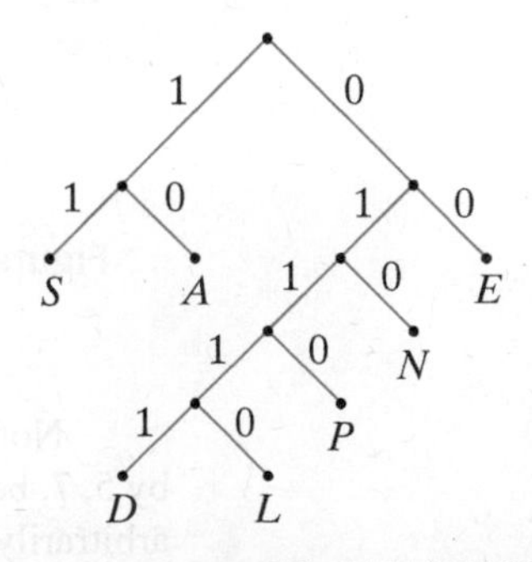


14. 011000010
15. 01111001001110
16. 01110100110
17. 111001110100111

*Encode each word using the preceding Huffman code.*

**18.** DEN **19.** LEADEN

**20.** NEED **21.** PENNED

**22.** What factors in addition to the amount of memory used should be considered when choosing a code, such as ASCII or a Huffman code, to represent characters in a computer?

**23.** What techniques in addition to the use of Huffman codes might be used to save memory when storing text?

**24.** Construct an optimal Huffman code for the set of letters in the table.

| *Letter* | *Frequency* | *Letter* | *Frequency* |
|---|---|---|---|
| $\alpha$ | 5 | $\delta$ | 11 |
| $\beta$ | 6 | $\varepsilon$ | 20 |
| $\gamma$ | 6 | | |

**25.** Construct an optimal Huffman code for the set of letters in the table.

| *Letter* | *Frequency* | *Letter* | *Frequency* |
|---|---|---|---|
| I | 7.5 | C | 5.0 |
| U | 20.0 | H | 10.0 |
| B | 2.5 | M | 2.5 |
| S | 27.5 | P | 25.0 |

**26.** Use the code developed in Exercise 25 to encode the following words (which have frequencies consistent with the table of Exercise 25):

BUS, CUPS, MUSH, PUSS, SIP, PUSH,

CUSS, HIP, PUP, PUPS, HIPS.

**27.** Construct two optimal Huffman coding trees for the table of Exercise 24 of different heights.

**28.** Construct an optimal Huffman code for the set of letters in the table.

| *Letter* | *Frequency* | *Letter* | *Frequency* |
|---|---|---|---|
| *a* | 2 | *d* | 8 |
| *b* | 3 | *e* | 13 |
| *c* | 5 | *f* | 21 |

**29.** Professor Ter A. Byte needs to store text made up of the characters A, B, C, D, E, which occur with the following frequencies:

| *Character* | *Frequency* | *Character* | *Frequency* |
|---|---|---|---|
| A | 6 | D | 2 |
| B | 2 | E | 8 |
| C | 3 | | |

Professor Byte suggests using the variable-length codes

| *Character* | *Code* |
|---|---|
| A | 1 |
| B | 00 |
| C | 01 |
| D | 10 |
| E | 0 |

which, he argues, store the text in less space than that used by an optimal Huffman code. Is the professor correct? Explain.

**30.** Show that a tree is a planar graph.

**31.** Show that any tree with two or more vertices has a vertex of degree 1.

**32.** Show that a tree is a bipartite graph.

**33.** Show that the vertices of a tree can be colored with two colors so that each edge is incident on vertices of different colors.

*The* eccentricity *of a vertex $v$ in a tree $T$ is the maximum length of a simple path that begins at $v$.*

**34.** Find the eccentricity of each vertex in the tree of Figure 9.1.5.

*A vertex $v$ in a tree $T$ is a* center *for $T$ if the eccentricity of $v$ is minimal.*

**35.** Find the center(s) of the tree of Figure 9.1.5.

**★36.** Show that a tree has either one or two centers.

**★37.** Show that if a tree has two centers they are adjacent.

**38.** Define the radius $r$ of a tree using the concepts of eccentricity and center. The diameter $d$ of any graph was defined before Exercise 71, Section 8.2. Is it always true, according to your definition of radius, that $2r = d$? Explain.

**39.** Give an example of a tree $T$ that does not satisfy the following property: If $v$ and $w$ are vertices in $T$, there is a unique path from $v$ to $w$.

## 9.2 Terminology and Characterizations of Trees

A portion of the family tree of the ancient Greek gods is shown in Figure 9.2.1. (Not all children are listed.) As shown, we can regard a family tree as a rooted tree. The vertices adjacent to a vertex $v$ and on the next-lower level are the children of $v$. For example,

Kronos's children are Zeus, Poseidon, Hades, and Ares. The terminology adapted from a family tree is used routinely for any rooted tree. The formal definitions follow.

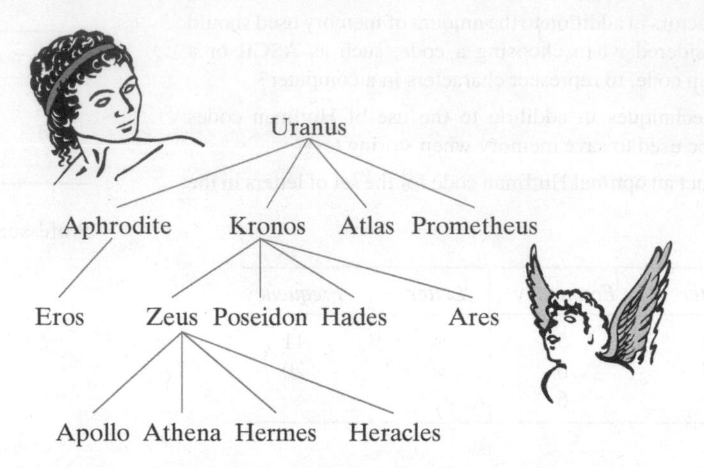


**Figure 9.2.1** A portion of the family tree of ancient Greek gods.

**Definition 9.2.1** ▶ Let $T$ be a tree with root $v_0$. Suppose that $x$, $y$, and $z$ are vertices in $T$ and that $(v_0, v_1, \ldots, v_n)$ is a simple path in $T$. Then

(a) $v_{n-1}$ is the *parent* of $v_n$.

(b) $v_0, \ldots, v_{n-1}$ are *ancestors* of $v_n$.

(c) $v_n$ is a *child* of $v_{n-1}$.

(d) If $x$ is an ancestor of $y$, $y$ is a *descendant* of $x$.

(e) If $x$ and $y$ are children of $z$, $x$ and $y$ are *siblings*.

(f) If $x$ has no children, $x$ is a *terminal vertex* (or a *leaf*).

(g) If $x$ is not a terminal vertex, $x$ is an *internal* (or *branch*) *vertex*.

(h) The *subtree of T rooted at x* is the graph with vertex set $V$ and edge set $E$, where $V$ is $x$ together with the descendants of $x$ and

$$E = \{e \mid e \text{ is an edge on a simple path from } x \text{ to some vertex in } V\}.$$ ◀

**Example 9.2.2** In the rooted tree of Figure 9.2.1,

(a) The parent of Eros is Aphrodite.

(b) The ancestors of Hermes are Zeus, Kronos, and Uranus.

(c) The children of Zeus are Apollo, Athena, Hermes, and Heracles.

(d) The descendants of Kronos are Zeus, Poseidon, Hades, Ares, Apollo, Athena, Hermes, and Heracles.

(e) Aphrodite and Prometheus are siblings.

(f) The terminal vertices are Eros, Apollo, Athena, Hermes, Heracles, Poseidon, Hades, Ares, Atlas, and Prometheus.

(g) The internal vertices are Uranus, Aphrodite, Kronos, and Zeus.

(h) The subtree rooted at Kronos is shown in Figure 9.2.2.

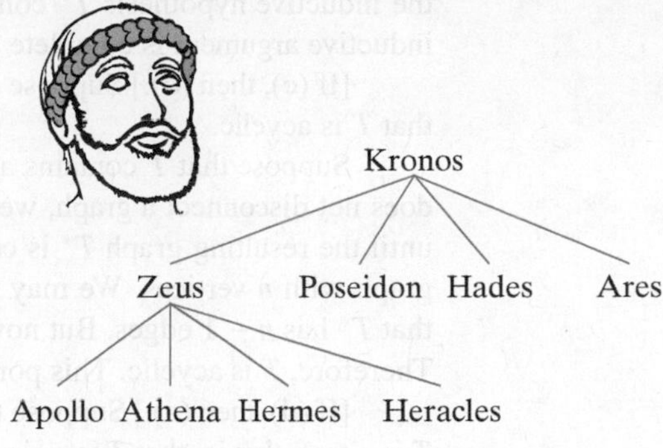


**Figure 9.2.2** The subtree rooted at Kronos of the tree of Figure 9.2.1.

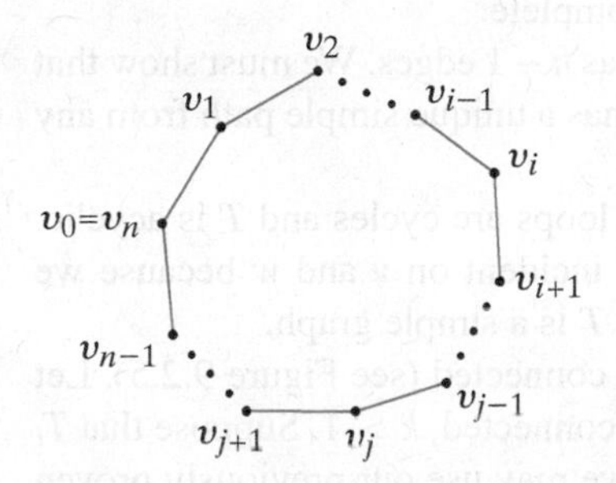


**Figure 9.2.3** A simple cycle.

The remainder of this section is devoted to providing alternative characterizations of trees. Let $T$ be a tree. We note that $T$ is connected since there is a simple path from any vertex to any other vertex. Further, we can show that $T$ does not contain a cycle. To see this, suppose that $T$ contains a cycle $C'$. By Theorem 8.2.24, $T$ contains a simple cycle (see Figure 9.2.3)

$$C = (v_0, \ldots, v_n),$$

$v_0 = v_n$. Since $T$ is a simple graph, $C$ cannot be a loop; so $C$ contains at least two distinct vertices $v_i$ and $v_j$, $i < j$. Now

$$(v_i, v_{i+1}, \ldots, v_j), \qquad (v_i, v_{i-1}, \ldots, v_0, v_{n-1}, \ldots, v_j)$$

are distinct simple paths from $v_i$ to $v_j$, which contradicts the definition of a tree. Therefore, a tree cannot contain a cycle.

A graph with no cycles is called an **acyclic graph.** We just showed that a tree is a connected, acyclic graph. The converse is also true; every connected, acyclic graph is a tree. The next theorem gives this characterization of trees as well as others.

**Theorem 9.2.3** Let $T$ be a graph with $n$ vertices. The following are equivalent.

(*a*) $T$ is a tree.

(*b*) $T$ is connected and acyclic.

(*c*) $T$ is connected and has $n-1$ edges.

(*d*) $T$ is acyclic and has $n-1$ edges.

**Go Online**

For more on equivalent definitions of trees, see `bit.ly/2q58dGw`

**Proof** To show that (a)–(d) are equivalent, we will prove four results: If (a), then (b); if (b), then (c); if (c), then (d); and if (d), then (a).

[If (a), then (b).] The proof of this result was given before the statement of the theorem.

[If (b), then (c).] Suppose that $T$ is connected and acyclic. We will prove that $T$ has $n-1$ edges by induction on $n$.

If $n = 1$, $T$ consists of one vertex and zero edges, so the result is true if $n = 1$.

Now suppose that the result holds for a connected, acyclic graph with $n$ vertices. Let $T$ be a connected, acyclic graph with $n+1$ vertices. Choose a path $P$ with no repeated edges of maximum length. Since $T$ is acyclic, $P$ contains no cycles. Therefore, $P$ contains a vertex $v$ of degree 1 (see Figure 9.2.4). Let $T^*$ be $T$ with $v$ and the edge incident on $v$ removed. Then $T^*$ is connected and acyclic, and because $T^*$ contains $n$ vertices, by

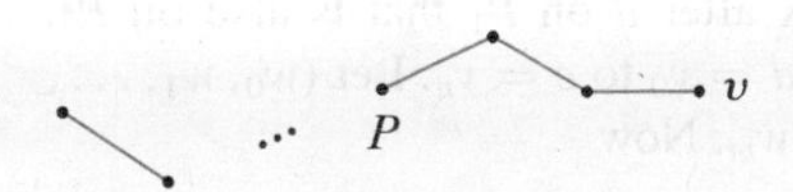


**Figure 9.2.4** The proof of Theorem 9.2.3 [if (b), then (c)]. $P$ is a simple path. $v$ and the edge incident on $v$ are removed so that the inductive hypothesis can be invoked.

the inductive hypothesis $T^*$ contains $n - 1$ edges. Therefore, $T$ contains $n$ edges. The inductive argument is complete and this portion of the proof is complete.

[If (c), then (d).] Suppose that $T$ is connected and has $n - 1$ edges. We must show that $T$ is acyclic.

Suppose that $T$ contains at least one cycle. Since removing an edge from a cycle does not disconnect a graph, we may remove edges, but no vertices, from cycle(s) in $T$ until the resulting graph $T^*$ is connected and acyclic. Now $T^*$ is an acyclic, connected graph with $n$ vertices. We may use our just proven result, (b) implies (c), to conclude that $T^*$ has $n - 1$ edges. But now $T$ has more than $n - 1$ edges. This is a contradiction. Therefore, $T$ is acyclic. This portion of the proof is complete.

[If (d), then (a).] Suppose that $T$ is acyclic and has $n - 1$ edges. We must show that $T$ is a tree, that is, that $T$ is a simple graph and that $T$ has a unique simple path from any vertex to any other vertex.

The graph $T$ cannot contain any loops because loops are cycles and $T$ is acyclic. Similarly, $T$ cannot contain distinct edges $e_1$ and $e_2$ incident on $v$ and $w$ because we would then have the cycle $(v, e_1, w, e_2, v)$. Therefore, $T$ is a simple graph.

Suppose, by way of contradiction, that $T$ is not connected (see Figure 9.2.5). Let $T_1, T_2, \ldots, T_k$ be the components of $T$. Since $T$ is not connected, $k > 1$. Suppose that $T_i$ has $n_i$ vertices. Each $T_i$ is connected and acyclic, so we may use our previously proven result, (b) implies (c), to conclude that $T_i$ has $n_i - 1$ edges. Now

$$\begin{aligned} n - 1 &= (n_1 - 1) + (n_2 - 1) + \cdots + (n_k - 1) && \text{(counting edges)} \\ &< (n_1 + n_2 + \cdots + n_k) - 1 && \text{(since } k > 1\text{)} \\ &= n - 1, && \text{(counting vertices)} \end{aligned}$$

which is impossible. Therefore, $T$ is connected.

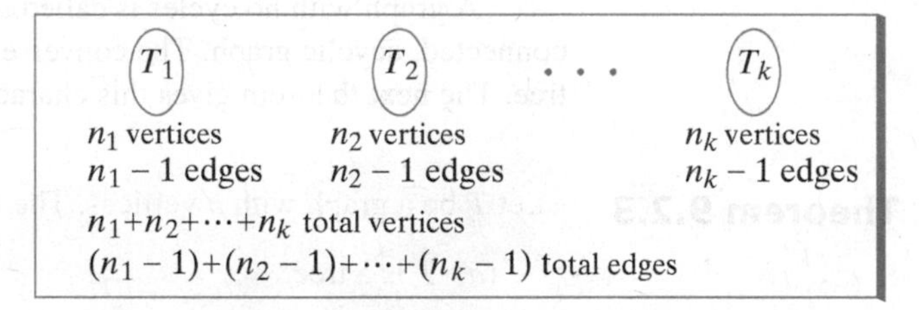


**Figure 9.2.5** The proof of Theorem 9.2.3 [if (d), then (a)]. The $T_i$ are components of $T$. $T_i$ has $n_i$ vertices and $n_i - 1$ edges. A contradiction results from the fact that the total number of edges must equal $n - 1$.

Suppose that there are distinct simple paths $P_1$ and $P_2$ from $a$ to $b$ in $T$ (see Figure 9.2.6). Let $c$ be the first vertex after $a$ on $P_1$ that is not in $P_2$; let $d$ be the vertex preceding $c$ on $P_1$; and let $e$ be the first vertex after $d$ on $P_1$ that is also on $P_2$. Let $(v_0, v_1, \ldots, v_{n-1}, v_n)$ be the portion of $P_1$ from $d = v_0$ to $e = v_n$. Let $(w_0, w_1, \ldots, w_{m-1}, w_m)$ be the portion of $P_2$ from $d = w_0$ to $e = w_m$. Now

$$(v_0, \ldots, v_n = w_m, w_{m-1}, \ldots, w_1, w_0) \qquad \textbf{(9.2.1)}$$

is a cycle in $T$, which is a contradiction. [In fact, (9.2.1) is a simple cycle since no vertices are repeated except for $v_0$ and $w_0$.] Thus there is a unique simple path from any vertex to any other vertex in $T$. Therefore, $T$ is a tree. This completes the proof.

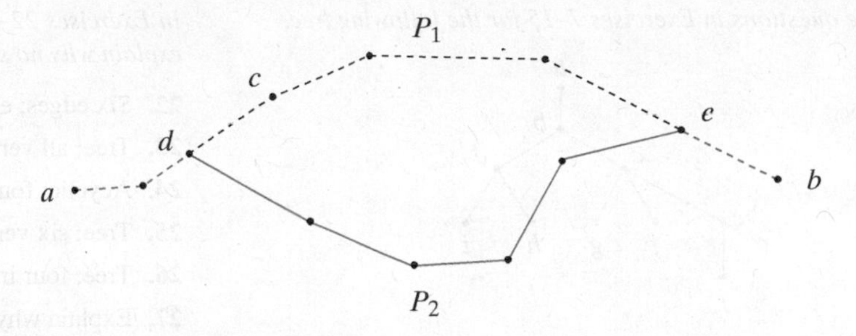


**Figure 9.2.6** The proof of Theorem 9.2.3 [if (d), then (a)]. $P_1$ (shown dashed) and $P_2$ (shown in color) are distinct simple paths from $a$ to $b$. $c$ is the first vertex after $a$ on $P_1$ not in $P_2$. $d$ is the vertex preceding $c$ on $P_1$. $e$ is the first vertex after $d$ on $P_1$ that is also on $P_2$. As shown, a cycle results, which gives a contradiction. ◀

## 9.2 Problem-Solving Tips

This section introduces some useful terminology and provides several different characterizations of trees. If $T$ is a graph with $n$ vertices, the following are equivalent (Theorem 9.2.3):

(a) $T$ is a tree.

(b) If $v$ and $w$ are vertices in $T$, there is a unique simple path from $v$ to $w$ (definition of tree).

(c) $T$ is connected and acyclic.

(d) $T$ is connected and has $n - 1$ edges.

(e) $T$ is acyclic and has $n - 1$ edges.

You can use the preceding characterizations in various ways. For example, a graph with four edges and six vertices cannot be a tree because it violates parts (d) and (e). Furthermore, the graph is either not connected or contains a cycle. (If it is both connected and acyclic, it would be a tree and thus have five edges.) A connected graph with $n$ vertices and more than $n-1$ edges must contain a cycle. (If it were acyclic, it would be a tree and, therefore, have $n - 1$ edges.)

## 9.2 Review Exercises

1. Define *parent* in a rooted tree.
2. Define *descendant* in a rooted tree.
3. Define *sibling* in a rooted tree.
4. Define *terminal vertex* in a rooted tree.
5. Define *internal vertex* in a rooted tree.
6. Define *acyclic graph*.
7. Give alternative characterizations of trees.

## 9.2 Exercises

*Answer the questions in Exercises 1–6 for the tree in Figure 9.2.1.*

1. Find the parent of Poseidon.
2. Find the children of Uranus.
3. Find the ancestors of Eros.
4. Find the descendants of Zeus.
5. Find the siblings of Ares.
6. Draw the subtree rooted at Aphrodite.

*Answer the questions in Exercises 7–15 for the following tree.*

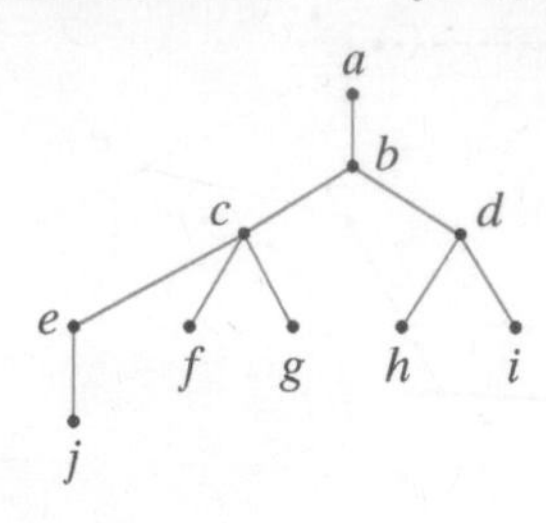


**7.** Find the parents of $c$ and of $h$.

**8.** Find the children of $d$ and of $e$.

**9.** Find the ancestors of $c$ and of $j$.

**10.** Find the descendants of $c$ and of $e$.

**11.** Find the terminal vertices.

**12.** Find the siblings of $f$ and of $h$.

**13.** Find the internal vertices.

**14.** Draw the subtree rooted at $e$.

**15.** Draw the subtree rooted at $j$.

**16.** Answer the questions in Exercises 7–15 for the following tree.

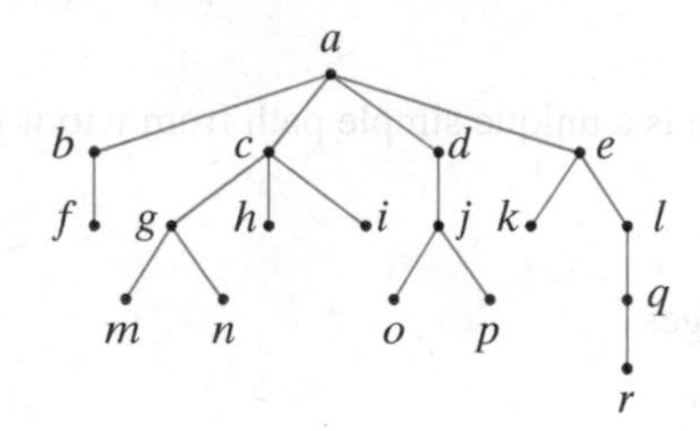


**17.** What can you say about two vertices in a rooted tree that have the same parent?

**18.** What can you say about a vertex in a rooted tree that has no ancestors?

**19.** What can you say about two vertices in a rooted tree that have the same ancestors?

**20.** What can you say about a vertex in a rooted tree that has no descendants?

**21.** What can you say about two vertices in a rooted tree that have a descendant in common?

*In Exercises 22–26, draw a graph having the given properties or explain why no such graph exists.*

**22.** Six edges; eight vertices

**23.** Tree; all vertices of degree 2

**24.** Acyclic; four edges, six vertices

**25.** Tree; six vertices having degrees 1, 1, 1, 1, 3, 3

**26.** Tree; four internal vertices; six terminal vertices

**27.** Explain why if we allow cycles of length 0, a graph consisting of a single vertex and no edges is not acyclic.

**28.** The connected graph shown has a unique simple path from any vertex to any other vertex, but it is not a tree. Explain.

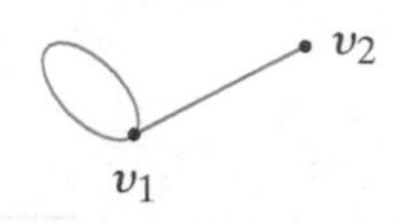


**29.** Explain why if we allow cycles to repeat edges, a graph consisting of a single edge and two vertices is not acyclic.

*A forest is a simple graph with no cycles.*

**30.** Explain why a forest is a union of trees.

**31.** If a forest $F$ consists of $m$ trees and has $n$ vertices, how many edges does $F$ have?

**32.** If $P_1 = (v_0, \ldots, v_n)$ and $P_2 = (w_0, \ldots, w_m)$ are distinct simple paths from $a$ to $b$ in a simple graph $G$, is

$$(v_0, \ldots, v_n = w_m, w_{m-1}, \ldots, w_1, w_0)$$

necessarily a cycle? Explain. (This exercise is relevant to the last paragraph of the proof of Theorem 9.2.3.)

**33.** Show that a graph $G$ with $n$ vertices and fewer than $n-1$ edges is not connected.

★**34.** Prove that $T$ is a tree if and only if $T$ is connected and when an edge is added between any two vertices, exactly one cycle is created.

**35.** Show that if $G$ is a tree, every vertex of degree 2 or more is an articulation point. ("Articulation point" is defined before Exercise 56, Section 8.2.)

**36.** Give an example to show that the converse of Exercise 35 is false, even if $G$ is assumed to be connected.

## Problem-Solving Corner — Trees

### Problem

Let $T$ be a simple graph. Prove that $T$ is a tree if and only if $T$ is connected but the removal of any edge (but no vertices) from $T$ disconnects $T$.

### Attacking the Problem

Let's be clear about what we have to prove. Since the statement is "if and only if," we must prove two statements:

If $T$ is a tree, then $T$ is connected but the removal of any edge (but no vertices) from $T$ disconnects $T$. **(1)**

If $T$ is connected but the removal of any edge (but no vertices) from $T$ disconnects $T$, then $T$ is a tree. **(2)**

In (1), from the assumption that $T$ is a tree, we must deduce that $T$ is connected but the removal of any edge (but no vertices) from $T$ disconnects $T$. In (2), from the assumption that $T$ is connected but the removal of any edge (but no vertices) from $T$ disconnects $T$, we must deduce that $T$ is a tree.

In developing a proof, it is usually helpful to review definitions and other results related to the statement to be proved. Here of direct relevance are the definition of a tree and Theorem 9.2.3, which gives equivalent conditions for a graph to be a tree.

Definition 9.1.1 states:

A tree $T$ is a simple graph satisfying the following: If $v$ and $w$ are vertices in $T$, there is a unique simple path from $v$ to $w$. **(3)**

Theorem 9.2.3 states that the following are equivalent for an $n$-vertex graph $T$:

$T$ is a tree. **(4)**

$T$ is connected and acyclic. **(5)**

$T$ is connected and has $n - 1$ edges. **(6)**

$T$ is acyclic and has $n - 1$ edges. **(7)**

## Finding a Solution

Let's first try to prove (1). We assume that $T$ is a tree. We must prove two things: $T$ is connected, and the removal of any edge (but no vertices) from $T$ disconnects $T$.

Statements (5) and (6) immediately tell us that $T$ is connected. None of the statements (3) through (7) directly tells us anything about removal of edges or about a graph that's not connected. However, if we argue by contradiction and assume that the removal of some edge (but no vertices) from $T$ does not disconnect $T$, then when we remove that edge from $T$, the graph $T'$ that results is connected. In this case, for the graph $T'$, (5) is true, but (6) and (7) are false, which is a contradiction since either (5), (6), and (7) are all true (and the graph is a tree), or (5) , (6) , and (7) are all false (and the graph is a not a tree).

Now consider proving (2). We assume that $T$ is connected and that the removal of any edge (but no vertices) from $T$ disconnects $T$. We must show that $T$ is a tree. Let's try to show that $T$ is connected and acyclic. We can then appeal to (5) to conclude that $T$ is a tree.

Since $T$ is connected, all we have to do is show that $T$ is acyclic. Again, we'll approach this by contradiction. Suppose that $T$ has a cycle. Keeping in mind what we're assuming (the removal of any edge from $T$ disconnects $T$), try to figure out how to deduce a contradiction from the assumption that $T$ contains a cycle before reading on.

If we delete an edge from $T$'s cycle, $T$ will remain connected. This contradiction shows that $T$ is acyclic. By (5), $T$ is a tree.

## Formal Solution

Suppose that $T$ has $n$ vertices.

Suppose that $T$ is a tree. Then by Theorem 9.2.3, $T$ is connected and has $n - 1$ edges. Suppose that we can remove an edge from $T$ to obtain $T'$ so that $T'$ is connected. Since $T$ contains no cycles, $T'$ also contains no cycles. By Theorem 9.2.3, $T'$ is a tree. Again by Theorem 9.2.3, $T'$ has $n - 1$ edges. This is a contradiction. Therefore, $T$ is connected, but the removal of any edge (but no vertices) from $T$ disconnects $T$.

If $T$ is connected and the removal of any edge (but no vertices) from $T$ disconnects $T$, then $T$ contains no cycles. By Theorem 9.2.3, $T$ is a tree.

## Summary of Problem-Solving Techniques

- When trying to construct a proof, write out carefully what is assumed and what is to be proved.
- When trying to construct a proof, consider using closely related definitions and theorems.
- When trying to construct a proof, review the proofs of similar and related theorems.
- If none of the conditions of potentially useful definitions and theorems applies, try proof by contradiction. When you assume the negation of the hypotheses, additional statements become available that might make some of the conditions of the definitions and theorems apply.

## 9.3 Spanning Trees

In this section we consider the problem of finding a subgraph $T$ of a graph $G$ such that $T$ is a tree containing all of the vertices of $G$. Such a tree is called a **spanning tree.** We will see that the methods of finding spanning trees may be applied to other problems as well.

**Go Online**
For more on spanning trees, see
bit.ly/2q58dGw

**Definition 9.3.1** ▶ A tree $T$ is a *spanning tree* of a graph $G$ if $T$ is a subgraph of $G$ that contains all of the vertices of $G$. ◀

**Example 9.3.2** A spanning tree of the graph $G$ of Figure 9.3.1 is shown in black. ◀

**Example 9.3.3** In general, a graph will have several spanning trees. Another spanning tree of the graph $G$ of Figure 9.3.1 is shown in Figure 9.3.2.

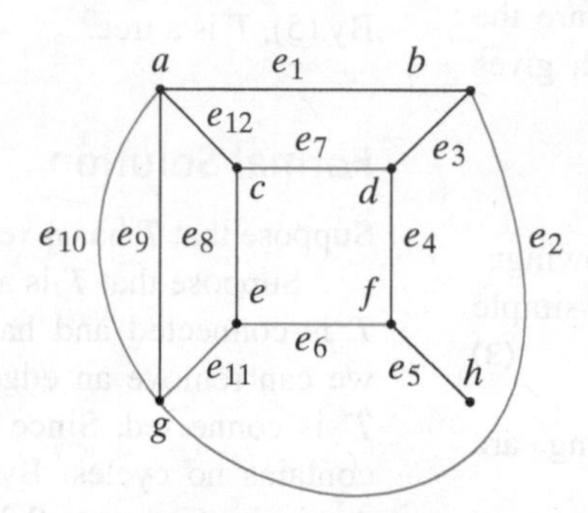


**Figure 9.3.1** A graph and a spanning tree shown in black.

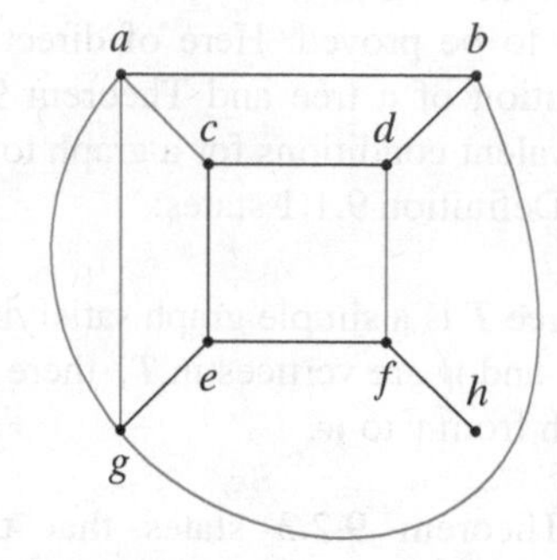


**Figure 9.3.2** Another spanning tree (in black) of the graph of Figure 9.3.1. ◀

Suppose that a graph $G$ has a spanning tree $T$. Let $a$ and $b$ be vertices of $G$. Since $a$ and $b$ are also vertices in $T$ and $T$ is a tree, there is a path $P$ from $a$ to $b$. However, $P$ also serves as a path from $a$ to $b$ in $G$; thus $G$ is connected. The converse is also true.

**Theorem 9.3.4** A graph $G$ has a spanning tree if and only if $G$ is connected.

**Proof** We have already shown that if $G$ has a spanning tree, then $G$ is connected. Suppose that $G$ is connected. If $G$ is acyclic, by Theorem 9.2.3, $G$ is a tree.

Suppose that $G$ contains a cycle. We remove an edge (but no vertices) from this cycle. The graph produced is still connected. If it is acyclic, we stop. If it contains a cycle, we remove an edge from this cycle. Continuing in this way, we eventually produce an acyclic, connected subgraph $T$. By Theorem 9.2.3, $T$ is a tree. Since $T$ contains all the vertices of $G$, $T$ is a spanning tree of $G$. ◀

An algorithm for finding a spanning tree based on the proof of Theorem 9.3.4 would not be very efficient; it would involve the time-consuming process of finding cycles. We can do much better. We shall illustrate the first algorithm for finding a spanning tree by an example and then we will state the algorithm.

**Example 9.3.5** Find a spanning tree for the graph $G$ of Figure 9.3.1.

**SOLUTION** We will use a method called **breadth-first search** (Algorithm 9.3.6). The idea of breadth-first search is to process all the vertices on a given level before moving to the next-higher level.

**Go Online**
For more on breadth-first search, see
bit.ly/2q58dGw

First, select an ordering, say *abcdefgh*, of the vertices of $G$. Select the first vertex $a$ and label it the root. Let $T$ consist of the single vertex $a$ and no edges. Add to $T$ all edges $(a, x)$ and vertices on which they are incident, for $x = b$ to $h$, that do not produce a cycle when added to $T$. We would add to $T$ edges $(a, b)$, $(a, c)$, and $(a, g)$. (We could use either of the parallel edges incident on $a$ and $g$.) Repeat this procedure with the vertices on level 1 by examining each in order:

$b$: Include $(b, d)$.
$c$: Include $(c, e)$.
$g$: None

Repeat this procedure with the vertices on level 2:

$d$: Include $(d, f)$.
$e$: None

Repeat this procedure with the vertices on level 3:

$f$: Include $(f, h)$.

Since no edges can be added to the single vertex $h$ on level 4, the procedure ends. We have found the spanning tree shown in Figure 9.3.1. ◀

We formalize the method of Example 9.3.5 as Algorithm 9.3.6.

**Algorithm 9.3.6** **Breadth-First Search for a Spanning Tree**

This algorithm finds a spanning tree using the breadth-first search method.

Input: A connected graph $G$ with vertices ordered

$$v_1, v_2, \ldots, v_n$$

Output: A spanning tree $T$

```
bfs(V, E) {
  // V = vertices ordered v1, ..., vn; E = edges
  // V′ = vertices of spanning tree T; E′ = edges of spanning tree T
  // v1 is the root of the spanning tree
  // S is an ordered list
  S = (v1)
  V′ = {v1}
  E′ = ∅
  while (true) {
    for each x ∈ S, in order,
      for each y ∈ V − V′, in order,
        if ((x, y) is an edge)
          add edge (x, y) to E′ and y to V′
    if (no edges were added)
      return T
    S = children of S ordered consistently with the original vertex ordering
  }
}
```

Exercise 20 is to prove that Algorithm 9.3.6 finds a spanning tree.

Breadth-first search can be used to test whether an arbitrary graph $G$ with $n$ vertices is connected (see Exercise 30). We use the method of Algorithm 9.3.6 to produce a tree $T$. Then $G$ is connected if and only if $T$ has $n$ vertices.

Breadth-first search can also be used to find minimum-length paths in an unweighted graph from a fixed vertex $v$ to all other vertices (see Exercise 24). We use the method of Algorithm 9.3.6 to generate a spanning tree rooted at $v$. We note that the length of a shortest path from $v$ to a vertex on level $i$ of the spanning tree is $i$. Dijkstra's shortest-path algorithm for weighted graphs (Algorithm 8.4.1) can be considered as a generalization of breadth-first search (see Exercise 25).

**Go Online**

For more on depth-first search, see bit.ly/2q58dGw

An alternative to breadth-first search is **depth-first search,** which proceeds to successive levels in a tree at the earliest possible opportunity.

**Algorithm 9.3.7** **Depth-First Search for a Spanning Tree**

This algorithm finds a spanning tree using the depth-first search method.

Input: A connected graph $G$ with vertices ordered

$$v_1, v_2, \ldots, v_n$$

Output: A spanning tree $T$

```
dfs(V, E) {
  // V' = vertices of spanning tree T; E' = edges of spanning tree T
  // v1 is the root of the spanning tree
  V' = {v1}
  E' = ∅
  w = v1
  while (true) {
    while (there is an edge (w, v) that when added to T does not create a cycle
      in T) {
      choose the edge (w, vk) with minimum k that when added to T
        does not create a cycle in T
      add (w, vk) to E'
      add vk to V'
      w = vk
    }
    if (w == v1)
      return T
    w = parent of w in T // backtrack
  }
}
```

Exercise 21 is to prove that Algorithm 9.3.7 finds a spanning tree.

**Example 9.3.8** Use depth-first search (Algorithm 9.3.7) to find a spanning tree for the graph of Figure 9.3.2 with the vertex ordering *abcdefgh*.

SOLUTION We select the first vertex $a$ and call it the root (see Figure 9.3.2). Next, we add the edge $(a, x)$, with minimal $x$, to our tree. In our case we add the edge $(a, b)$.

We repeat this process. We add the edges $(b, d)$, $(d, c)$, $(c, e)$, $(e, f)$, and $(f, h)$. At this point, we cannot add an edge of the form $(h, x)$, so we backtrack to the parent $f$ of $h$ and try to add an edge of the form $(f, x)$. Again, we cannot add an edge of the form $(f, x)$, so we backtrack to the parent $e$ of $f$. This time we succeed in adding the edge

$(e, g)$. At this point, no more edges can be added, so we finally backtrack to the root and the procedure ends. ◀

Because of the line in Algorithm 9.3.7 where we retreat along an edge toward the initially chosen root, depth-first search is also called **backtracking.** In the following example, we use backtracking to solve a puzzle.

**Example 9.3.9** **Four-Queens Problem** The four-queens problem is to place four tokens on a $4 \times 4$ grid so that no two tokens are on the same row, column, or diagonal. Construct a backtracking algorithm to solve the four-queens problem. (To use chess terminology, this is the problem of placing four queens on a $4 \times 4$ board so that no queen attacks another queen.)

**Go Online**
For more on the $n$-queens problem, see bit.ly/2q58dGw

**SOLUTION** The idea of the algorithm is to place tokens successively in the columns. When it is impossible to place a token in a column, we backtrack and adjust the token in the preceding column. ◀

**Algorithm 9.3.10** **Solving the Four-Queens Problem Using Backtracking**

**Go Online**
For a C program implementing this algorithm, see bit.ly/2D1THry

This algorithm uses backtracking to search for an arrangement of four tokens on a $4 \times 4$ grid so that no two tokens are on the same row, column, or diagonal.

Input: An array *row* of size 4

Output: true, if there is a solution
false, if there is no solution
[If there is a solution, the $k$th queen is in column $k$, row $row(k)$.]

```
four_queens(row) {
   k = 1 // start in column 1

   // start in row 1
   // since row(k) is incremented prior to use, set row(1) to 0
   row(1) = 0
   while (k > 0) {
      row(k) = row(k) + 1
      // look for a legal move in column k
      while (row(k) ≤ 4 ∧ column k, row(k) conflicts)
         // try next row
         row(k) = row(k) + 1
      if (row(k) ≤ 4)
         if (k == 4)
            return true
         else { // next column
            k = k + 1
            row(k) = 0
         }
      else // backtrack to previous column
         k = k − 1
   }
   return false // no solution
}
```

The tree that Algorithm 9.3.10 generates is shown in Figure 9.3.3. The numbering indicates the order in which the vertices are generated. The solution is found at vertex 8.

The $n$-queens problem is to place $n$ tokens on an $n \times n$ grid so that no two tokens are on the same row, column, or diagonal. Checking that there is no solution to the

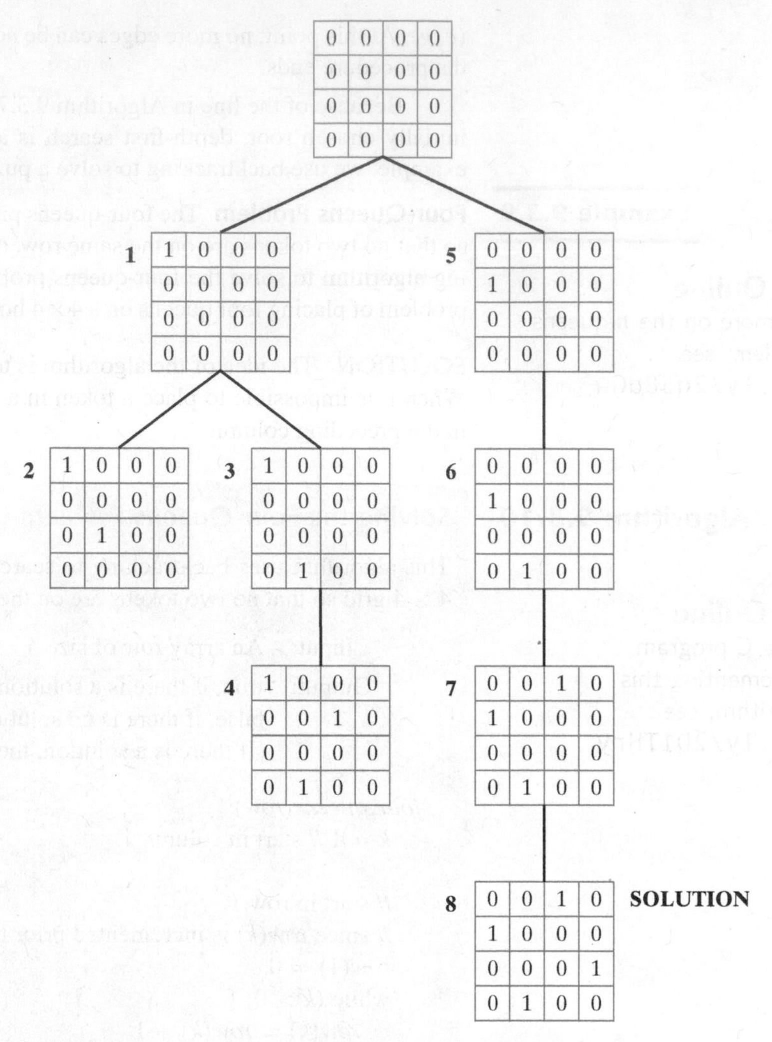


**Figure 9.3.3** The tree generated by the backtracking algorithm (Algorithm 9.3.10) in the search for a solution to the four-queens problem.

two- or three-queens problem (see Exercise 10) is straightforward. We have just seen that Algorithm 9.3.10 generates a solution to the four-queens problem. Many constructions have been given to generate solutions to the $n$-queens problem for all $n \geq 4$ (see, e.g., [Johnsonbaugh]).

Backtracking or depth-first search is especially attractive in a problem such as that in Example 9.3.9, where all that is desired is one solution. Since a solution, if one exists, is found at a terminal vertex, by moving to the terminal vertices as rapidly as possible, in general we can avoid generating some unnecessary vertices.

## 9.3 Problem-Solving Tips

Depth-first search and breadth-first search are the basis of many graph algorithms. For example, either can be used to determine whether a graph is connected: If we can visit *every* vertex in a graph from an initial vertex, the graph is connected; otherwise, it is not connected (see Exercises 30 and 31). Depth-first search can be used as a searching algorithm, in which case it is called backtracking. In Algorithm 9.3.10, backtracking is used to search for solutions to the four-queens problem. Backtracking can also be used to search for Hamiltonian cycles in a graph, to generate permutations, to solve Sudoku puzzles, and to determine whether two graphs are isomorphic.

## 9.3 Review Exercises

1. What is a spanning tree?
2. State a necessary and sufficient condition for a graph to have a spanning tree.
3. Explain how breadth-first search works.
4. Explain how depth-first search works.
5. What is backtracking?

## 9.3 Exercises

1. Use breadth-first search (Algorithm 9.3.6) with the vertex ordering *hgfedcba* to find a spanning tree for graph $G$ of Figure 9.3.1.
2. Use breadth-first search (Algorithm 9.3.6) with the vertex ordering *chbgadfe* to find a spanning tree for graph $G$ of Figure 9.3.1.
3. Use breadth-first search (Algorithm 9.3.6) with the vertex ordering *hfdbgeca* to find a spanning tree for graph $G$ of Figure 9.3.1.
4. Use depth-first search (Algorithm 9.3.7) with the vertex ordering *hgfedcba* to find a spanning tree for graph $G$ of Figure 9.3.1.
5. Use depth-first search (Algorithm 9.3.7) with the vertex ordering *dhcbefag* to find a spanning tree for graph $G$ of Figure 9.3.1.
6. Use depth-first search (Algorithm 9.3.7) with the vertex ordering *hfdbgeca* to find a spanning tree for graph $G$ of Figure 9.3.1.

*In Exercises 7–9, find a spanning tree for each graph.*

7.

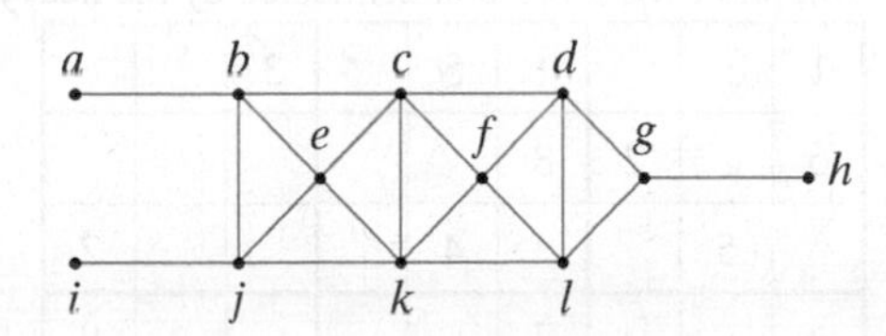


8.

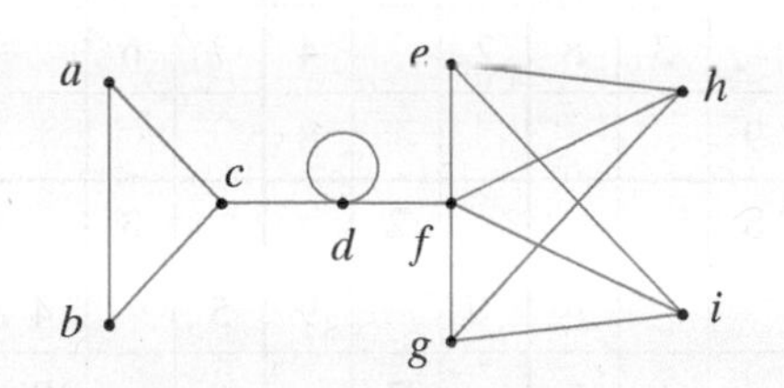


9.

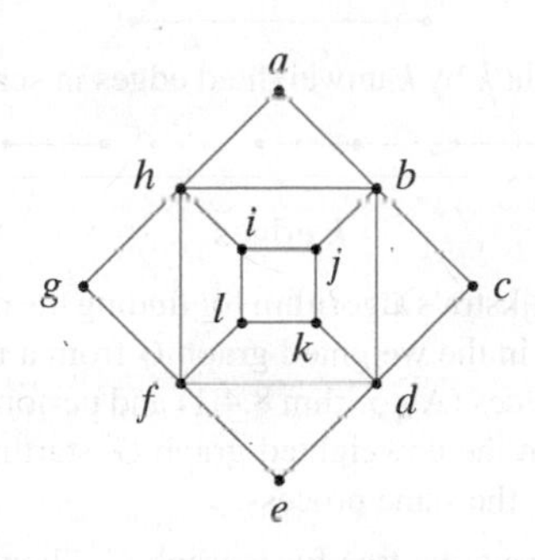


10. Show that there is no solution to the two-queens or the three-queens problem.
11. Show all solutions to the four-queens problem.
12. Find a solution to the five-queens and six-queens problems.
13. Show all solutions to the five-queens problem in which one queen is in the first column, second row.
14. How many solutions are there to the five-queens problem?
15. Show all solutions to the six-queens problem in which one queen is in row 2, column 1, and a second queen is in row 4, column 2.
16. True or false? If $G$ is a connected graph and $T$ is a spanning tree for $G$, there is an ordering of the vertices of $G$ such that Algorithm 9.3.6 produces $T$ as a spanning tree. If true, prove it; otherwise, give a counterexample.
17. True or false? If $G$ is a connected graph and $T$ is a spanning tree for $G$, there is an ordering of the vertices of $G$ such that Algorithm 9.3.7 produces $T$ as a spanning tree. If true, prove it; otherwise, give a counterexample.
18. Show, by an example, that Algorithm 9.3.6 can produce identical spanning trees for a connected graph $G$ from two distinct vertex orderings of $G$.
19. Show, by an example, that Algorithm 9.3.7 can produce identical spanning trees for a connected graph $G$ from two distinct vertex orderings of $G$.

**20.** Prove that Algorithm 9.3.6 is correct.

**21.** Prove that Algorithm 9.3.7 is correct.

**22.** Under what conditions is an edge in a connected graph $G$ contained in every spanning tree of $G$?

**23.** Let $T$ and $T'$ be two spanning trees of a connected graph $G$. Suppose that an edge $x$ is in $T$ but not in $T'$. Show that there is an edge $y$ in $T'$ but not in $T$ such that $(T - \{x\}) \cup \{y\}$ and $(T' - \{y\}) \cup \{x\}$ are spanning trees of $G$.

**24.** Write an algorithm based on breadth-first search that finds the minimum length of each path in an unweighted graph from a fixed vertex $v$ to all other vertices.

**25.** Let $G$ be a weighted graph in which the weight of each edge is a positive integer. Let $G'$ be the graph obtained from $G$ by replacing each edge

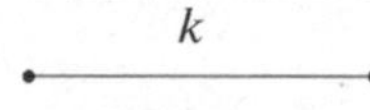


in $G$ of weight $k$ by $k$ unweighted edges in series:

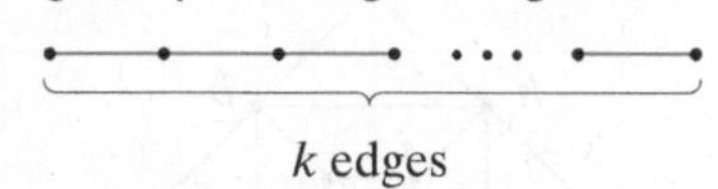


Show that Dijkstra's algorithm for finding the minimum length of each path in the weighted graph $G$ from a fixed vertex $v$ to all other vertices (Algorithm 8.4.1) and performing a breadth-first search in the unweighted graph $G'$ starting with vertex $v$ are, in effect, the same process.

**26.** Let $T$ be a spanning tree for a graph $G$. Show that if an edge in $G$, but not in $T$, is added to $T$, a unique cycle is produced.

*A cycle as described in Exercise 26 is called a* fundamental cycle. *The* fundamental cycle matrix *of a graph G has its rows indexed by the fundamental cycles of G relative to a spanning tree T for G and its columns indexed by the edges of G. The ijth entry is 1 if edge j is in the ith fundamental cycle and 0 otherwise. For example, the fundamental cycle matrix of the graph G of Figure 9.3.1 relative to the spanning tree shown in Figure 9.3.1 is*

$$
\begin{array}{c} \\ (abdca) \\ (efdbace) \\ (ageca) \\ (aga) \\ (abga) \end{array}
\begin{array}{c} \begin{array}{cccccccccccc} e_7 & e_6 & e_{11} & e_{10} & e_2 & e_1 & e_3 & e_4 & e_5 & e_8 & e_9 & e_{12} \end{array} \\
\begin{pmatrix}
1 & 0 & 0 & 0 & 0 & 1 & 1 & 0 & 0 & 0 & 0 & 1 \\
0 & 1 & 0 & 0 & 0 & 1 & 1 & 1 & 0 & 1 & 0 & 1 \\
0 & 0 & 1 & 0 & 0 & 0 & 0 & 0 & 0 & 1 & 1 & 1 \\
0 & 0 & 0 & 1 & 0 & 0 & 0 & 0 & 0 & 0 & 1 & 0 \\
0 & 0 & 0 & 0 & 1 & 1 & 0 & 0 & 0 & 0 & 1 & 0
\end{pmatrix} \end{array}.
$$

*Find the fundamental cycle matrix of each graph. The spanning tree to be used is drawn in black.*

**27.**

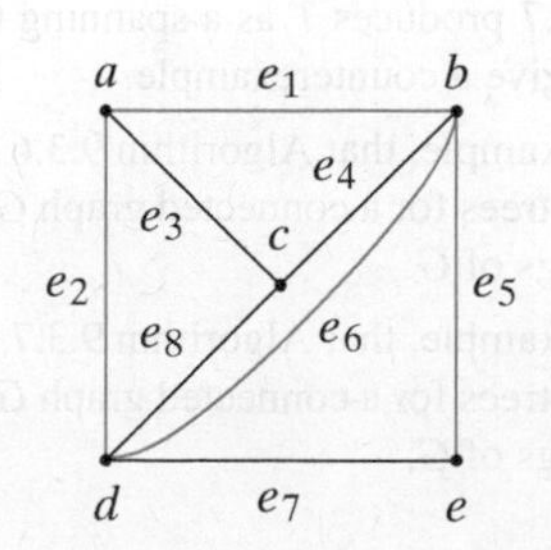


**28.**

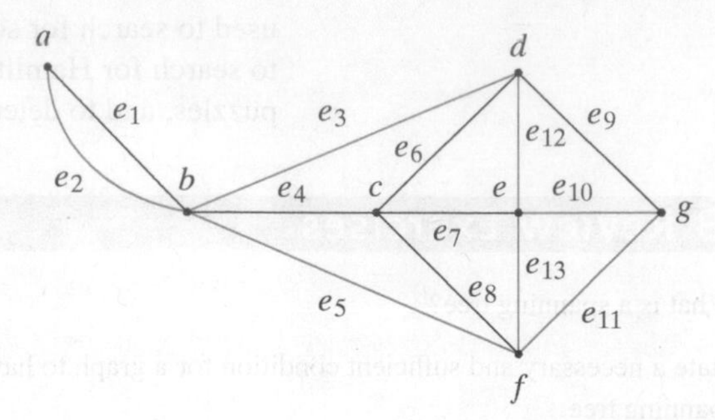


**29.**

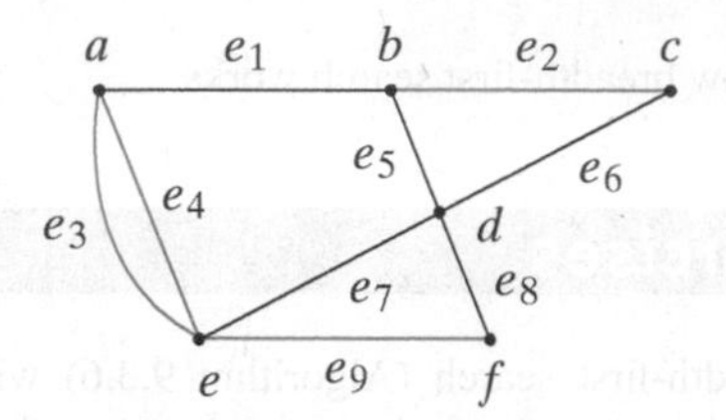


**30.** Write a breadth-first search algorithm to test whether a graph is connected.

**31.** Write a depth-first search algorithm to test whether a graph is connected.

**32.** Write a depth-first search algorithm that finds all solutions to the four-queens problem.

**33.** Modify Algorithm 9.3.6 so that it tracks the parent $p$ of a vertex $c$ ($p$ is the *parent* of $c$ if $c$ was visited from $p$).

**34.** Write an algorithm that uses the output of your algorithm in Exercise 33 to print each vertex and its parent.

**35.** Modify Algorithm 9.3.7 so that it tracks the parent $p$ of a vertex $c$ ($p$ is the *parent* of $c$ if $c$ was visited from $p$).

**36.** Write an algorithm that uses the output of your algorithm in Exercise 35 to print each vertex and its parent.

**37.** Write a backtracking algorithm that outputs all permutations of $1, 2, \ldots, n$.

**38.** Write a backtracking algorithm that outputs all subsets of $\{1, 2, \ldots, n\}$.

**39.** Sudoku is a puzzle in which the goal is to fill in a $9 \times 9$ grid so that each of the numbers 1 through 9 appears in each column, each row, and each $3 \times 3$ box delineated by the heavy lines:

| 1 |  |  |  | 5 |  | 3 |  |  |
|---|---|---|---|---|---|---|---|---|
| 6 |  | 2 | 8 |  |  |  |  |  |
|  | 5 |  |  | 4 |  |  |  | 2 |
|  |  |  | 7 |  |  |  |  | 8 |
|  | 3 | 8 | 2 |  | 4 | 7 | 6 |  |
| 9 |  |  |  |  | 8 |  |  |  |
| 8 |  |  |  | 2 |  |  | 3 |  |
|  |  |  |  |  | 9 | 5 |  | 4 |
|  |  | 6 |  | 7 |  |  |  | 9 |

As shown, in each puzzle some numbers are given. Solve the preceding Sudoku puzzle.

**40.** Write a backtracking algorithm that solves an arbitrary Sudoku puzzle.

**41.** The *minimum-queens problem* asks for the minimum number of queens that can attack all of the squares of an $n \times n$ board (i.e., the minimum number of queens such that each row, column, and diagonal contains at least one queen). Write a backtracking algorithm that determines whether $k$ queens can attack all squares of an $n \times n$ board.

**42.** The *subset-sum problem* is: Given a set $\{c_1, \ldots, c_n\}$ of positive integers and a positive integer $M$, find all subsets $\{c_{k_1}, \ldots, c_{k_j}\}$ of $\{c_1, \ldots, c_n\}$ satisfying

$$\sum_{i=1}^{j} c_{k_i} = M.$$

Write a backtracking algorithm to solve the subset-sum problem.

# 9.4 Minimal Spanning Trees

**Go Online**
For more on minimal spanning trees, see `bit.ly/2q58dGw`

The weighted graph $G$ of Figure 9.4.1 shows six cities and the costs of building roads between certain pairs of cities. We want to build the lowest-cost road system that will connect the six cities. The solution can be represented by a subgraph. This subgraph must be a spanning tree since it must contain all the vertices (so that each city is in the road system), it must be connected (so that any city can be reached from any other), and it must have a unique simple path between each pair of vertices (since a graph containing multiple simple paths between a vertex pair could not represent a minimum-cost system). Thus what is needed is a spanning tree the sum of whose weights is a minimum. Such a tree is called a **minimal spanning tree.**

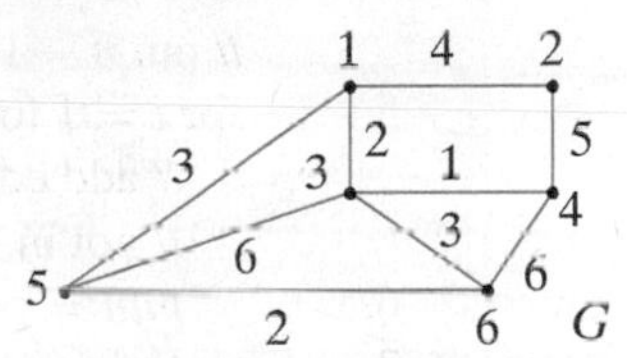


**Figure 9.4.1** Six cities 1–6 and the costs of building roads between certain pairs of them.

**Definition 9.4.1 ▶** Let $G$ be a weighted graph. A *minimal spanning tree* of $G$ is a spanning tree of $G$ with minimum weight. ◀

**Example 9.4.2** The tree $T'$ shown in Figure 9.4.2 is a spanning tree for graph $G$ of Figure 9.4.1. The weight of $T'$ is 20. This tree is not a minimal spanning tree since spanning tree $T$ shown in Figure 9.4.3 has weight 12. We will see later that $T$ is a minimal spanning tree for $G$.

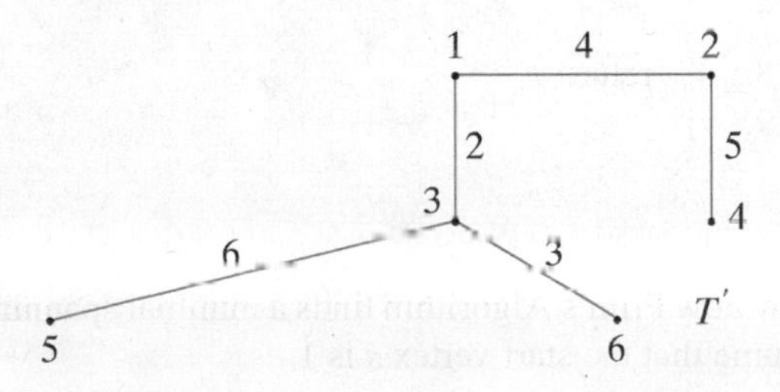


**Figure 9.4.2** A spanning tree of weight 20 of the graph of Figure 9.4.1.

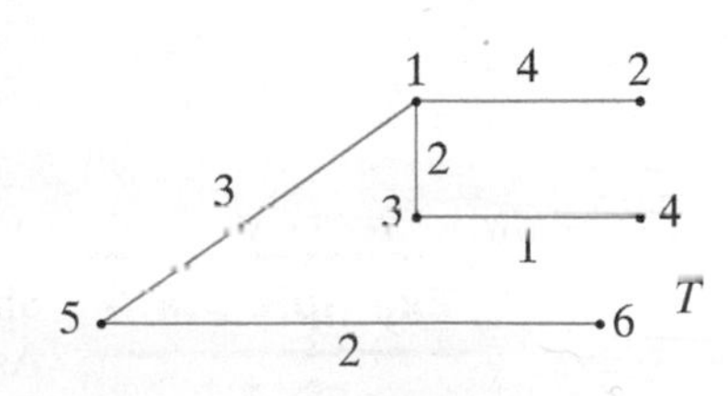


**Figure 9.4.3** A spanning tree of weight 12 of the graph of Figure 9.4.1. ◀

**Go Online**
For more on Prim's Algorithm, see `bit.ly/2q58dGw`

The algorithm to find a minimal spanning tree that we will discuss is known as **Prim's Algorithm** (Algorithm 9.4.3). This algorithm builds a tree by iteratively adding edges until a minimal spanning tree is obtained. The algorithm begins with a single vertex. Then at each iteration, it adds to the current tree a minimum-weight edge that does not complete a cycle. Another algorithm to find a minimal spanning tree, known as **Kruskal's Algorithm,** is presented in the exercises (see Exercises 20–22).

**Algorithm 9.4.3** **Prim's Algorithm**

**Go Online**
For a C program implementing this algorithm, see `bit.ly/2PinbXS`

This algorithm finds a minimal spanning tree in a connected, weighted graph.

Input: A connected, weighted graph with vertices $1, \ldots, n$ and start vertex $s$. If $(i, j)$ is an edge, $w(i, j)$ is equal to the weight of $(i, j)$; if $(i, j)$ is not an edge, $w(i, j)$ is equal to $\infty$ (a value greater than any actual weight).

Output: The set of edges $E$ in a minimal spanning tree (mst)

```
     prim(w, n, s) {
        // v(i) = 1 if vertex i has been added to mst
        // v(i) = 0 if vertex i has not been added to mst
 1.     for i = 1 to n
 2.        v(i) = 0
        // add start vertex to mst
 3.     v(s) = 1
        // begin with an empty edge set
 4.     E = ∅
        // put n − 1 edges in the minimal spanning tree
 5.     for i = 1 to n − 1 {
           // add edge of minimum weight with one vertex in mst and one vertex
           // not in mst
 6.        min = ∞
 7.        for j = 1 to n
 8.           if (v(j) == 1) // if j is a vertex in mst
 9.              for k = 1 to n
10.                 if (v(k) == 0 ∧ w(j, k) < min) {
11.                    add_vertex = k
12.                    e = (j, k)
13.                    min = w(j, k)
14.                 }
           // put vertex and edge in mst
15.        v(add_vertex) = 1
16.        E = E ∪ {e}
17.     }
18.     return E
19.  }
```

**Example 9.4.4** Show how Prim's Algorithm finds a minimal spanning tree for the graph of Figure 9.4.1. Assume that the start vertex $s$ is 1.

**SOLUTION** At line 3 we add vertex 1 to the minimal spanning tree. The first time we execute the for loop in lines 7–14, the edges with one vertex in the tree and one vertex not in the tree are

| *Edge* | *Weight* |
|---|---|
| (1, 2) | 4 |
| (1, 3) | 2 |
| (1, 5) | 3 |

The edge (1, 3) with minimum weight is selected. At lines 15 and 16, vertex 3 is added to the minimal spanning tree and edge (1, 3) is added to $E$.

The next time we execute the for loop in lines 7–14, the edges with one vertex in the tree and one vertex not in the tree are

| *Edge* | *Weight* |
|---|---|
| (1, 2) | 4 |
| (1, 5) | 3 |
| (3, 4) | 1 |
| (3, 5) | 6 |
| (3, 6) | 3 |

The edge (3, 4) with minimum weight is selected. At lines 15 and 16, vertex 4 is added to the minimal spanning tree and edge (3, 4) is added to $E$.

The next time we execute the for loop in lines 7–14, the edges with one vertex in the tree and one vertex not in the tree are

| *Edge* | *Weight* |
|---|---|
| (1, 2) | 4 |
| (1, 5) | 3 |
| (2, 4) | 5 |
| (3, 5) | 6 |
| (3, 6) | 3 |
| (4, 6) | 6 |

This time two edges have minimum weight 3. A minimal spanning tree will be constructed when either edge is selected. In this version, edge (1, 5) is selected. At lines 15 and 16, vertex 5 is added to the minimal spanning tree and edge (1, 5) is added to $E$.

The next time we execute the for loop in lines 7–14, the edges with one vertex in the tree and one vertex not in the tree are

| *Edge* | *Weight* |
|---|---|
| (1, 2) | 4 |
| (2, 4) | 5 |
| (3, 6) | 3 |
| (4, 6) | 6 |
| (5, 6) | 2 |

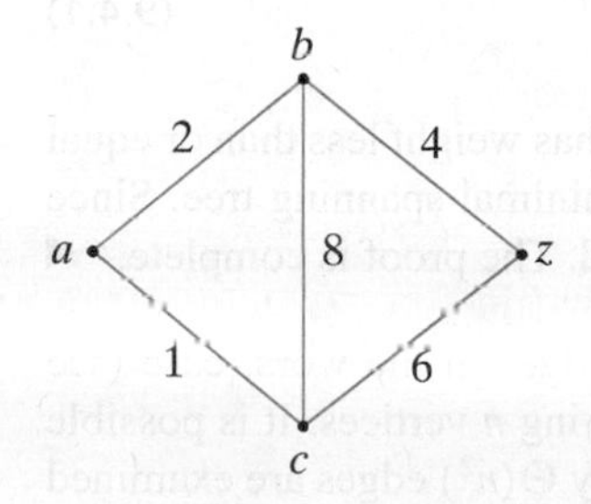


**Figure 9.4.4** A graph that shows that selecting an edge having minimum weight incident on the most recently added vertex does *not* necessarily yield a shortest path. Starting at $a$, we obtain $(a, c, z)$, but the shortest path from $a$ to $z$ is $(a, b, z)$.

The edge (5, 6) with minimum weight is selected. At lines 15 and 16, vertex 6 is added to the minimal spanning tree and edge (5, 6) is added to $E$.

The last time we execute the for loop in lines 7–14, the edges with one vertex in the tree and one vertex not in the tree are

| *Edge* | *Weight* |
|---|---|
| (1, 2) | 4 |
| (2, 4) | 5 |

The edge (1, 2) with minimum weight is selected. At lines 15 and 16, vertex 2 is added to the minimal spanning tree and edge (1, 2) is added to $E$. The minimal spanning tree constructed is shown in Figure 9.4.3. ◀

Prim's Algorithm furnishes an example of a **greedy algorithm.** A greedy algorithm is an algorithm that optimizes the choice at each iteration. The principle can be summarized as "doing the best locally." In Prim's Algorithm, since we want a minimal spanning tree, at each iteration we simply add an available edge with minimum weight.

Optimizing at each iteration does not necessarily give an optimal solution to the original problem. We will show shortly (Theorem 9.4.5) that Prim's Algorithm is correct; that is, we do obtain a minimal spanning tree. As an example of a greedy algorithm that does not lead to an optimal solution, consider a "shortest-path algorithm" in which at each step we select an available edge having minimum weight incident on the most recently added vertex. If we apply this algorithm to the weighted graph of Figure 9.4.4 to find a shortest path from $a$ to $z$, we would select the edge $(a, c)$ and then the edge $(c, z)$. Unfortunately, this is not the shortest path from $a$ to $z$.

We next show that Prim's Algorithm is correct.

**Theorem 9.4.5** Prim's Algorithm (Algorithm 9.4.3) is correct; that is, at the termination of Algorithm 9.4.3, $T$ is a minimal spanning tree.

**Proof** We let $T_i$ denote the graph constructed by Algorithm 9.4.3 after the $i$th iteration of the for loop, lines 5–17. More precisely, the edge set of $T_i$ is the set $E$ constructed after the $i$th iteration of the for loop, lines 5–17, and the vertex set of $T_i$ is the set of vertices on which the edges in $E$ are incident. We let $T_0$ be the graph constructed by Algorithm 9.4.3 just before the for loop at line 5 is entered for the first time; $T_0$ consists of the single vertex $s$ and no edges. Subsequently in this proof, we suppress the vertex set and refer to a graph by specifying its edge set.

By construction, at the termination of Algorithm 9.4.3, the resulting graph, $T_{n-1}$, is a connected, acyclic subgraph of the given graph $G$ containing all the vertices of $G$; hence $T_{n-1}$ is a spanning tree of $G$.

We use induction to show that for all $i = 0, \ldots, n-1$, $T_i$ is contained in a minimal spanning tree. It then follows that at termination, $T_{n-1}$ is a minimal spanning tree.

If $i = 0$, $T_0$ consists of a single vertex. In this case $T_0$ is contained in every minimal spanning tree. We have verified the Basis Step.

Next, assume that $T_i$ is contained in a minimal spanning tree $T'$. Let $V$ be the set of vertices in $T_i$. Algorithm 9.4.3 selects an edge $(j, k)$ of minimum weight, where $j \in V$ and $k \notin V$, and adds it to $T_i$ to produce $T_{i+1}$. If $(j, k)$ is in $T'$, then $T_{i+1}$ is contained in the minimal spanning tree $T'$. If $(j, k)$ is not in $T'$, $T' \cup \{(j, k)\}$ contains a cycle $C$. Choose an edge $(x, y)$ in $C$, different from $(j, k)$, with $x \in V$ and $y \notin V$. Then

$$w(x, y) \geq w(j, k). \tag{9.4.1}$$

Because of (9.4.1), the graph $T'' = [T' \cup \{(j, k)\}] - \{(x, y)\}$ has weight less than or equal to the weight of $T'$. Since $T''$ is a spanning tree, $T''$ is a minimal spanning tree. Since $T_{i+1}$ is contained in $T''$, the Inductive Step has been verified. The proof is complete. ◀

Our version of Prim's Algorithm examines $\Theta(n^3)$ edges in the worst case (see Exercise 6) to find a minimal spanning tree for a graph having $n$ vertices. It is possible (see Exercise 8) to implement Prim's Algorithm so that only $\Theta(n^2)$ edges are examined in the worst case. Since $K_n$ has $\Theta(n^2)$ edges, the latter version is optimal.

## 9.4 Review Exercises

1. What is a minimal spanning tree?
2. Explain how Prim's Algorithm finds a minimal spanning tree.
3. What is a greedy algorithm?

## 9.4 Exercises

*In Exercises 1–5, find the minimal spanning tree given by Algorithm 9.4.3 for each graph.*

**1.**

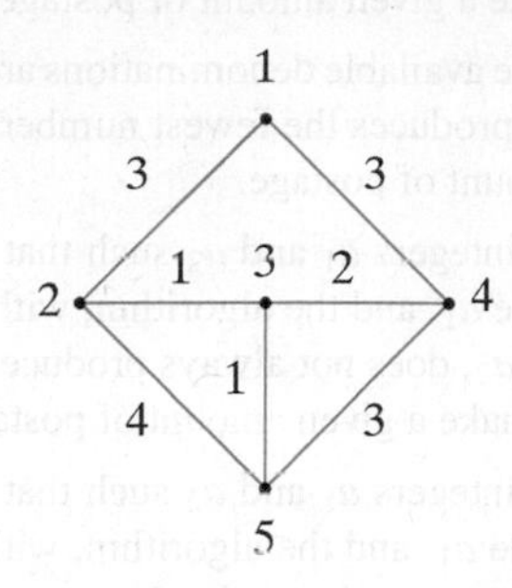


**2.**

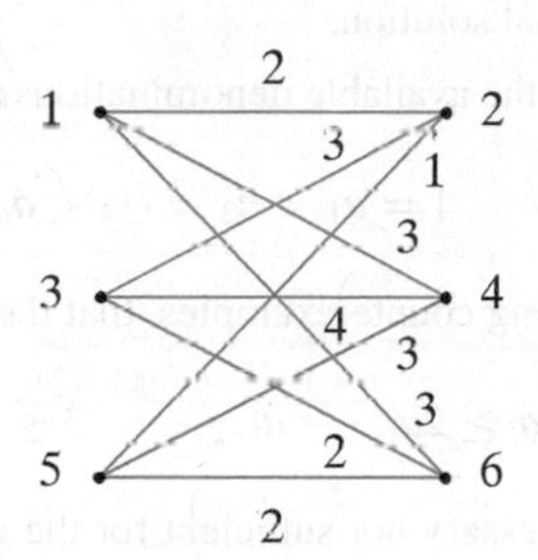


**3.**

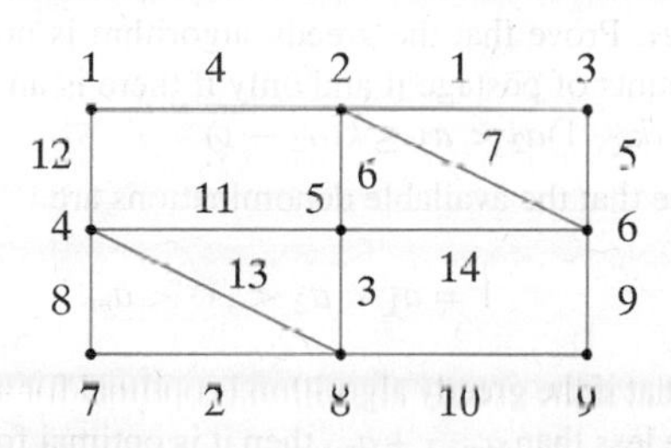


**4.**

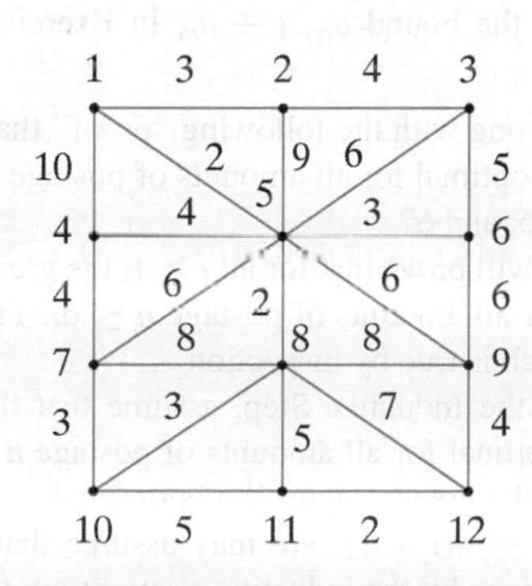


**5.**

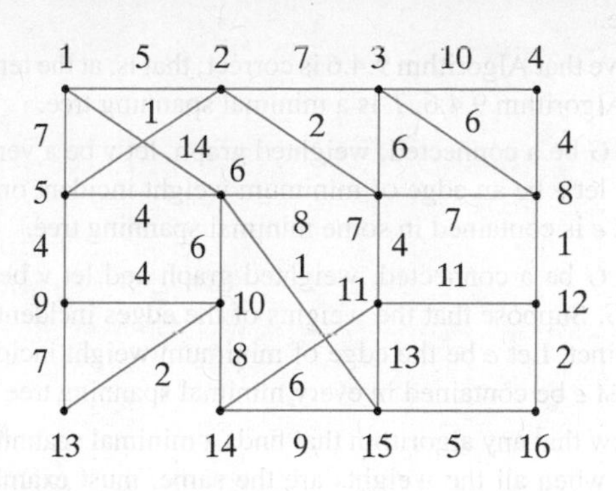


**6.** Show that Algorithm 9.4.3 examines $\Theta(n^3)$ edges in the worst case.

*Exercises 7–9 refer to an alternate version of Prim's Algorithm (Algorithm 9.4.6).*

### Algorithm 9.4.6

#### *Alternate Version of Prim's Algorithm*

This algorithm finds a minimal spanning tree in a connected, weighted graph $G$. At each step, some vertices have temporary labels and some have permanent labels. The label of vertex $i$ is denoted $L_i$.

Input: A connected, weighted graph with vertices $1, \ldots, n$ and start vertex $s$. If $(i, j)$ is an edge, $w(i, j)$ is equal to the weight of $(i, j)$; if $(i, j)$ is not an edge, $w(i, j)$ is equal to $\infty$ (a value greater than any actual weight).

Output: A minimal spanning tree $T$

```
prim_alternate(w, n, s) {
   let T be the graph with vertex s and no edges
   for j = 1 to n {
      L_j = w(s, j) // these labels are temporary
      back(j) = s
   }
   L_s = 0
   make L_s permanent
   while (temporary labels remain) {
      choose the smallest temporary label L_i
      make L_i permanent
      add edge (i, back(i)) to T
      add vertex i to T
      for each temporary label L_k
         if (w(i, k) < L_k) {
            L_k = w(i, k)
            back(k) = i
         }
   }
   return T
}
```

7. Show how Algorithm 9.4.6 finds a minimal spanning tree for the graphs of Exercises 1–5.

8. Show that Algorithm 9.4.6 examines $\Theta(n^2)$ edges in the worst case.

9. Prove that Algorithm 9.4.6 is correct; that is, at the termination of Algorithm 9.4.6, $T$ is a minimal spanning tree.

10. Let $G$ be a connected, weighted graph, let $v$ be a vertex in $G$, and let $e$ be an edge of minimum weight incident on $v$. Show that $e$ is contained in some minimal spanning tree.

11. Let $G$ be a connected, weighted graph and let $v$ be a vertex in $G$. Suppose that the weights of the edges incident on $v$ are distinct. Let $e$ be the edge of minimum weight incident on $v$. Must $e$ be contained in every minimal spanning tree?

12. Show that any algorithm that finds a minimal spanning tree in $K_n$, when all the weights are the same, must examine every edge in $K_n$.

13. Show that if all weights in a connected graph $G$ are distinct, $G$ has a unique minimal spanning tree.

*In Exercises 14–16, decide if the statement is true or false. If the statement is true, prove it; otherwise, give a counterexample. In each exercise, G is a connected, weighted graph.*

14. If all the weights in $G$ are distinct, then distinct spanning trees of $G$ have distinct weights.

15. If $e$ is an edge in $G$ whose weight is less than the weight of every other edge, $e$ is in every minimal spanning tree of $G$.

16. If $T$ is a minimal spanning tree of $G$, there is a labeling of the vertices of $G$ so that Algorithm 9.4.3 produces $T$.

17. Let $G$ be a connected, weighted graph. Show that if, as long as possible, we remove an edge from $G$ having maximum weight whose removal does not disconnect $G$, the result is a minimal spanning tree for $G$.

★18. Write an algorithm that finds a maximal spanning tree in a connected, weighted graph.

19. Prove that your algorithm in Exercise 18 is correct.

**Go Online**
For more on Kruskal's Algorithm, see
bit.ly/2q58dGw

*Kruskal's Algorithm finds a minimal spanning tree in a connected, weighted graph G having n vertices as follows. The graph T initially consists of the vertices of G and no edges. At each iteration, we add an edge e to T having minimum weight that does not complete a cycle in T. When T has n − 1 edges, we stop.*

20. Formally state Kruskal's Algorithm.

21. Show how Kruskal's Algorithm finds minimal spanning trees for the graphs of Exercises 1–5.

22. Show that Kruskal's Algorithm is correct; that is, at the termination of Kruskal's Algorithm, $T$ is a minimal spanning tree.

23. Let $V$ be a set of $n$ vertices and let $s$ be a "dissimilarity function" on $V \times V$ (see Example 8.1.7). Let $G$ be the complete, weighted graph having vertices $V$ and weights $w(v_i, v_j) = s(v_i, v_j)$. Modify Kruskal's Algorithm so that it groups data into classes. This modification is known as the **method of nearest neighbors** (see [Gose]).

*Exercises 24–31 refer to the following situation. Suppose that we have stamps of various denominations and that we want to choose the minimum number of stamps to make a given amount of postage. Consider a greedy algorithm that selects stamps by choosing as many of the largest denomination as possible, then as many of the second-largest denomination as possible, and so on.*

24. Show that if the available denominations are 1, 8, and 10 cents, the algorithm does not always produce the fewest number of stamps to make a given amount of postage.

★25. Show that if the available denominations are 1, 5, and 25 cents, the algorithm produces the fewest number of stamps to make any given amount of postage.

26. Find positive integers $a_1$ and $a_2$ such that $a_1 > 2a_2 > 1$, $a_2$ does not divide $a_1$, and the algorithm, with available denominations 1, $a_1$, $a_2$, does not always produce the fewest number of stamps to make a given amount of postage.

★27. Find positive integers $a_1$ and $a_2$ such that $a_1 > 2a_2 > 1$, $a_2$ does not divide $a_1$, and the algorithm, with available denominations 1, $a_1$, $a_2$, produces the fewest number of stamps to make any given amount of postage. Prove that your values do give an optimal solution.

★28. Suppose that the available denominations are

$$1 = a_1 < a_2 < \cdots < a_n.$$

Show, by giving counterexamples, that the condition

$$a_i \geq 2a_{i-1} - a_{i-2}, \qquad 3 \leq i \leq n,$$

is neither necessary nor sufficient for the greedy algorithm to be optimal for all amounts of postage.

★29. Suppose that the available denominations are $1 = a_1 < a_2 < a_3$. Prove that the greedy algorithm is not optimal for all amounts of postage if and only if there is an integer $k$ satisfying $(k-1)a_2 < a_3 \leq k(a_2 - 1)$.

★30. Suppose that the available denominations are

$$1 = a_1 < a_2 < \cdots < a_m.$$

Prove that if the greedy algorithm is optimal for all amounts of postage less than $a_{m-1} + a_m$, then it is optimal for all amounts of postage.

31. Show that the bound $a_{m-1} + a_m$ in Exercise 30 cannot be lowered.

32. What is wrong with the following "proof" that the greedy algorithm is optimal for all amounts of postage for the denominations 1, 5, and 6?

We will prove that for all $i \geq 1$, the greedy algorithm is optimal for all amounts of postage $n \leq 6i$. The Basis Step is $i = 1$, which is true by inspection.

For the Inductive Step, assume that the greedy algorithm is optimal for all amounts of postage $n \leq 6i$. We must show that the greedy algorithm is optimal for all amounts of postage $n \leq 6(i+1)$. We may assume that $n > 6i$. Now $n - 6 \leq 6i$, so by the inductive assumption, the greedy algorithm is optimal for $n - 6$. Suppose that the greedy algorithm chooses $k$ stamps for $n - 6$. In determining the postage for the amount $n$, the greedy algorithm will first choose a 6-cent stamp and then stamps for $n - 6$ for a total of $k + 1$ stamps. These $k + 1$ stamps must be optimal or otherwise $n - 6$ would use less than $k$ stamps. The Inductive Step is complete.

## 9.5 Binary Trees

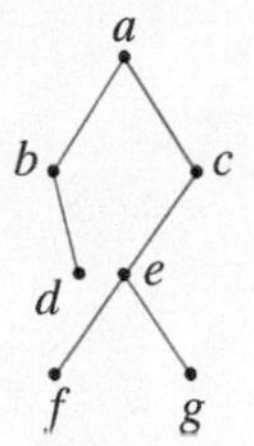

Figure 9.5.1 A binary tree.

**Binary trees** are among the most important special types of rooted trees. Every vertex in a binary tree has at most two children (see Figure 9.5.1). Moreover, each child is designated as either a **left child** or a **right child.** When a binary tree is drawn, a left child is drawn to the left and a right child is drawn to the right. The formal definition follows.

**Definition 9.5.1 ▶** A *binary tree* is a rooted tree in which each vertex has either no children, one child, or two children. If a vertex has one child, that child is designated as either a left child or a right child (but not both). If a vertex has two children, one child is designated a left child and the other child is designated a right child. ◀

**Example 9.5.2** In the binary tree of Figure 9.5.1, vertex $b$ is the left child of vertex $a$ and vertex $c$ is the right child of vertex $a$. Vertex $d$ is the right child of vertex $b$; vertex $b$ has no left child. Vertex $e$ is the left child of vertex $c$; vertex $c$ has no right child. ◀

**Example 9.5.3** A tree that defines a Huffman code is a binary tree. For example, in the Huffman coding tree of Figure 9.1.10, moving from a vertex to a left child corresponds to using the bit 1, and moving from a vertex to a right child corresponds to using the bit 0. ◀

A **full binary tree** is a binary tree in which each vertex has either two children or zero children. A fundamental result about full binary trees is our next theorem.

**Theorem 9.5.4** If $T$ is a full binary tree with $i$ internal vertices, then $T$ has $i + 1$ terminal vertices and $2i + 1$ total vertices.

**Proof** The vertices of $T$ consist of the vertices that are children (of some parent) and the vertices that are not children (of any parent). There is one nonchild—the root. Since there are $i$ internal vertices, each having two children, there are $2i$ children. Thus the total number of vertices of $T$ is $2i + 1$, and the number of terminal vertices is

$$(2i + 1) - i = i + 1.$$ ◀

**Example 9.5.5** A single-elimination tournament is a tournament in which a contestant is eliminated after one loss. The graph of a single-elimination tournament is a full binary tree (see Figure 9.5.2). The contestants' names are listed on the left. Winners progress to the right. Eventually, there is a single winner at the root. If the number of contestants is not a power of 2, some contestants receive byes. In Figure 9.5.2, contestant 7 has a first-round bye.

Show that if there are $n$ contestants in a single-elimination tournament, a total of $n - 1$ matches are played.

**SOLUTION** The number of contestants is the same as the number of terminal vertices, and the number of matches $i$ is the same as the number of internal vertices. Thus, by Theorem 9.5.4,

$$n + i = 2i + 1,$$

so that $i = n - 1$.

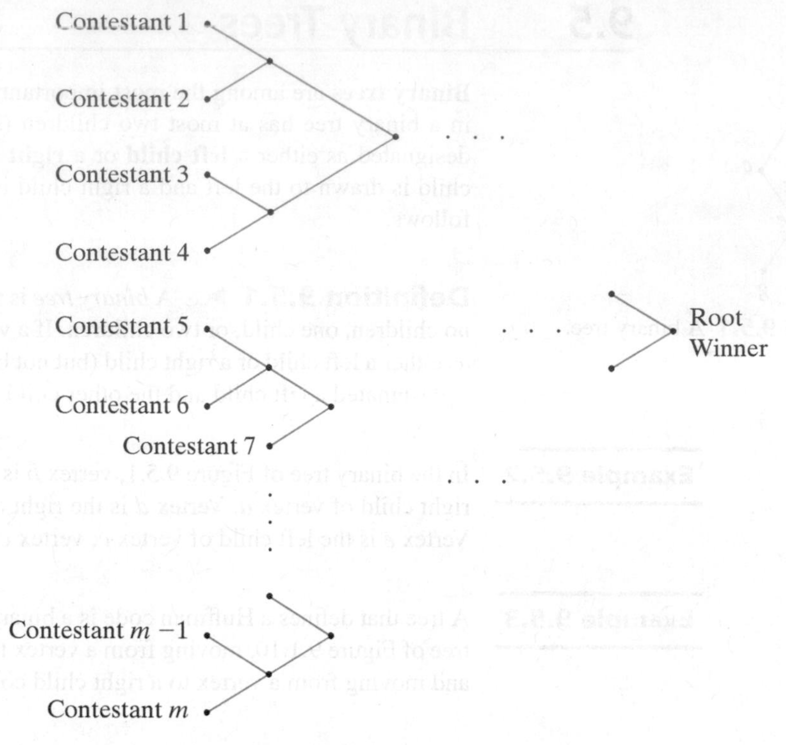


**Figure 9.5.2** The graph (full binary tree) of a single-elimination tournament.

Our next result about binary trees relates the number of terminal vertices to the height.

**Theorem 9.5.6** If a binary tree of height $h$ has $t$ terminal vertices, then

$$\lg t \leq h. \tag{9.5.1}$$

**Proof** We will prove the equivalent inequality

$$t \leq 2^h \tag{9.5.2}$$

by induction on $h$. Inequality (9.5.1) is obtained from (9.5.2) by taking the logarithm to the base 2 of both sides of (9.5.2).

If $h = 0$, the binary tree consists of a single vertex. In this case, $t = 1$ and thus (9.5.2) is true.

Assume that the result holds for a binary tree whose height is less than $h$. Let $T$ be a binary tree of height $h > 0$ with $t$ terminal vertices. Suppose first that the root of $T$ has only one child. If we eliminate the root and the edge incident on the root, the resulting tree has height $h-1$ and the same number of terminals as $T$. By induction, $t \leq 2^{h-1}$. Since $2^{h-1} < 2^h$, (9.5.2) is established for this case.

Now suppose that the root of $T$ has children $v_1$ and $v_2$. Let $T_i$ be the subtree rooted at $v_i$ and suppose that $T_i$ has height $h_i$ and $t_i$ terminal vertices, $i = 1, 2$. By induction,

$$t_i \leq 2^{h_i}, \qquad i = 1, 2. \tag{9.5.3}$$

The terminal vertices of $T$ consist of the terminal vertices of $T_1$ and $T_2$. Hence

$$t = t_1 + t_2. \tag{9.5.4}$$

Combining (9.5.3) and (9.5.4), we obtain

$$t = t_1 + t_2 \leq 2^{h_1} + 2^{h_2} \leq 2^{h-1} + 2^{h-1} = 2^h.$$

The inductive step has been verified and the proof is complete. ◀

**Example 9.5.7** The binary tree in Figure 9.5.3 has height $h = 3$ and the number of terminals $t = 8$. For this tree, the inequality (9.5.1) becomes an equality.

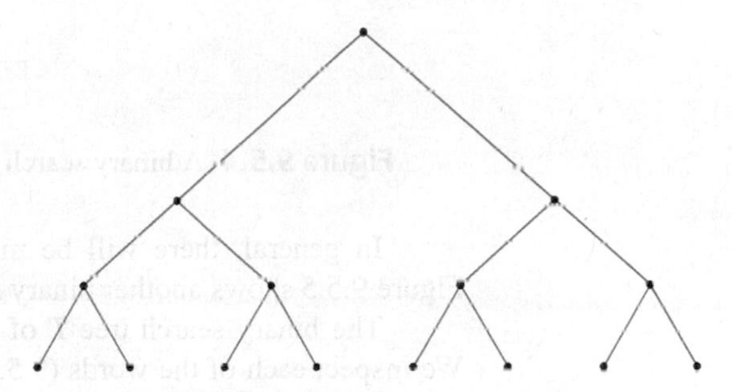

**Figure 9.5.3** A binary tree of height $h = 3$ with $t = 8$ terminals. For this binary tree, $\lg t = h$.

◀

Suppose that we have a set $S$ whose elements can be ordered. For example, if $S$ consists of numbers, we can use ordinary ordering defined on numbers, and if $S$ consists of strings of alphabetic characters, we can use lexicographic order. Binary trees are used extensively in computer science to store elements from an ordered set such as a set of numbers or a set of strings. If data item $d(v)$ is stored in vertex $v$ and data item $d(w)$ is stored in vertex $w$, then if $v$ is a left child (or right child) of $w$, some ordering relationship will be guaranteed to exist between $d(v)$ and $d(w)$. One example is a **binary search tree.**

**Definition 9.5.8 ▶** A *binary search tree* is a binary tree $T$ in which data are associated with the vertices. The data are arranged so that, for *each* vertex $v$ in $T$, each data item in the left subtree of $v$ is less than the data item in $v$, and each data item in the right subtree of $v$ is greater than the data item in $v$. ◀

**Example 9.5.9** The words

OLD PROGRAMMERS NEVER DIE
THEY JUST LOSE THEIR MEMORIES (9.5.5)

may be placed in a binary search tree as shown in Figure 9.5.4. Notice that for any vertex $v$, each data item in the left subtree of $v$ is less than (i.e., precedes alphabetically) the data item in $v$ and each data item in the right subtree of $v$ is greater than the data item in $v$.

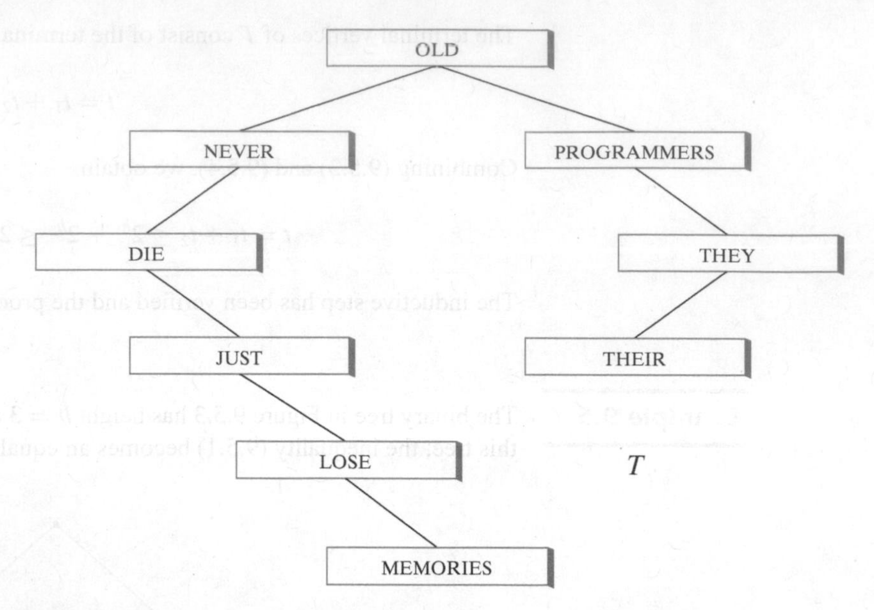


**Figure 9.5.4** A binary search tree. ◀

In general, there will be many ways to place data into a binary search tree. Figure 9.5.5 shows another binary search tree that stores the words (9.5.5).

The binary search tree $T$ of Figure 9.5.4 was constructed in the following way. We inspect each of the words (9.5.5) *in the order in which they appear,* OLD first, then PROGRAMMERS, then NEVER, and so on. To start, we create a vertex and place the first word OLD in this vertex. We designate this vertex the root. Thereafter, given a word in the list (9.5.5), we add a vertex $v$ and an edge to the tree and place the word in the vertex $v$. To decide where to add the vertex and edge, we begin at the root. If the word to be added is less than (using lexicographic order) the word at the root, we move to the left child; if the word to be added is greater than the word at the root, we move to the right child. If there is no child, we create one, put in an edge incident on the root and new vertex, and place the word in the new vertex. If there is a child $v$, we repeat this process. That is, we compare the word to be added with the word at $v$ and move to the left child of $v$ if the word to be added is less than the word at $v$; otherwise, we move to the right child of $v$. If there is no child to move to, we create one, put in an edge incident on $v$ and the new vertex, and place the word in the new vertex. If there is a child to move to, we repeat this process. Eventually, we place the word in the tree. We then get the

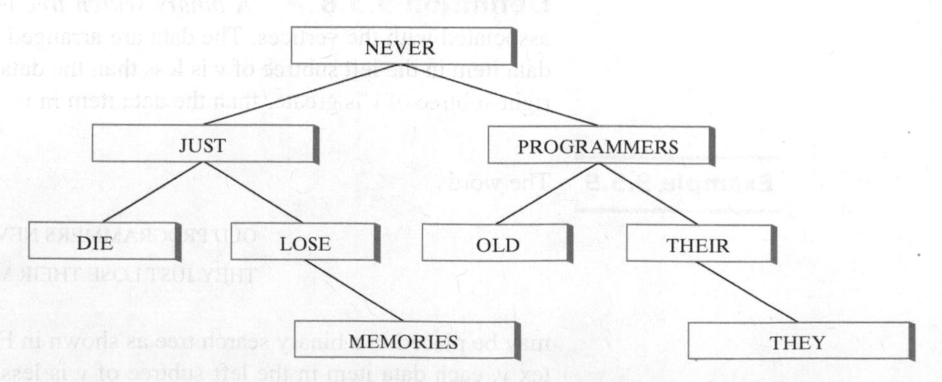


**Figure 9.5.5** Another binary search tree that stores the same words as the tree in Figure 9.5.4.

next word in the list, compare it with the root, move left or right, compare it with the new vertex, move left or right, and so on, and eventually store it in the tree. In this way, we store all of the words in the tree and thus create a binary search tree. We formally state this method of constructing a binary search tree as Algorithm 9.5.10.

**Algorithm 9.5.10** **Constructing a Binary Search Tree**

This algorithm constructs a binary search tree. The input is read in the order submitted. After each word is read, it is inserted into the tree.

Input: A sequence $w_1, \ldots, w_n$ of distinct words and the length $n$ of the sequence

Output: A binary search tree $T$

```
make_bin_search_tree(w, n) {
  let T be the tree with one vertex, root
  store w_1 in root
  for i = 2 to n {
    v = root
    search = true // find spot for w_i
    while (search) {
      s = word in v
      if (w_i < s)
        if (v has no left child) {
          add a left child l to v
          store w_i in l
          search = false // end search
        }
        else
          v = left child of v
      else // w_i > s
        if (v has no right child) {
          add a right child r to v
          store w_i in r
          search = false // end search
        }
        else
          v = right child of v
    } // end while
  } // end for
  return T
}
```

Binary search trees are useful for locating data. That is, given a data item $D$, we can easily determine if $D$ is in a binary search tree and, if it is present, where it is located. To determine if a data item $D$ is in a binary search tree, we would begin at the root. We would then repeatedly compare $D$ with the data item at the current vertex. If $D$ is equal to the data item at the current vertex, we have found $D$, so we stop. If $D$ is less than the data item at the current vertex $v$, we move to $v$'s left child and repeat this process. If $D$ is greater than the data item at the current vertex $v$, we move to $v$'s right child and repeat this process. If at any point the child to move to is missing, we conclude that $D$ is not in the tree. (Exercise 6 asks for a formal statement of this process.)

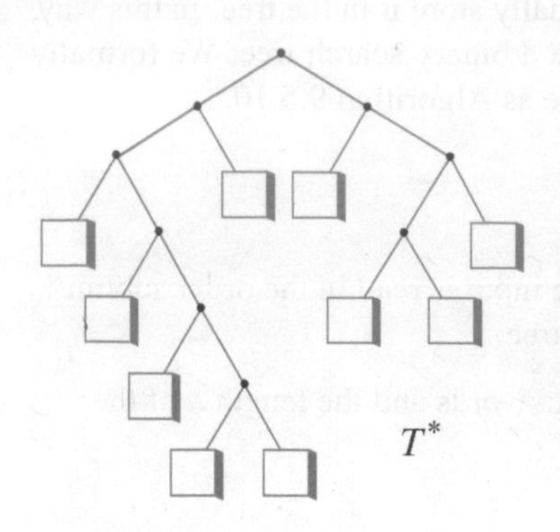


**Figure 9.5.6** Expanding a binary search tree to a full binary tree.

The time spent searching for an item in a binary search tree is longest when the item is not present and we follow a longest path from the root. Thus the maximum time to search for an item in a binary search tree is approximately proportional to the height of the tree. Therefore, if the height of a binary search tree is small, searching the tree will always be very fast (see Exercise 25). Many ways are known to minimize the height of a binary search tree (see, e.g., [Cormen]).

We make more precise statements about worst-case searching in a binary search tree. Let $T$ be a binary search tree with $n$ vertices and let $T^*$ be the full binary tree obtained from $T$ by adding left and right children to existing vertices in $T$ wherever possible. In Figure 9.5.6, we show the full binary tree that results from modifying the binary search tree of Figure 9.5.4. The added vertices are drawn as boxes. An unsuccessful search in $T$ corresponds to arriving at an added (box) vertex in $T^*$. Let us define the worst-case time needed to execute the search procedure as the height $h$ of the tree $T^*$. By Theorem 9.5.6, $\lg t \leq h$, where $t$ is the number of terminal vertices in $T^*$. The full binary tree $T^*$ has $n$ internal vertices, so by Theorem 9.5.4, $t = n + 1$. Thus in the worst case, the time will be equal to at least $\lg t = \lg(n+1)$. Exercise 7 shows that if the height of $T$ is minimized, the worst case requires time equal to $\lceil \lg(n+1) \rceil$. For example, since

$$\lceil \lg(2{,}000{,}000 + 1) \rceil = 21,$$

it is possible to store 2 million items in a binary search tree and find an item, or determine that it is not present, in at most 21 steps.

## 9.5 Review Exercises

1. Define *binary tree.*
2. What is a left child in a binary tree?
3. What is a right child in a binary tree?
4. What is a full binary tree?
5. If $T$ is a full binary tree with $i$ internal vertices, how many terminal vertices does $T$ have?
6. If $T$ is a full binary tree with $i$ internal vertices, how many total vertices does $T$ have?
7. How is the height of a binary tree related to the number of its terminal vertices?
8. What is a binary search tree?
9. Give an example of a binary search tree.
10. Give an algorithm to construct a binary search tree.

## 9.5 Exercises

*Exercises 1–4 concern n teams that play a single-elimination tournament.*

1. After the teams are assigned, in how many ways can the tournament unfold? For example, if there are three teams, Scientists, Whales, Pilots, assigned as

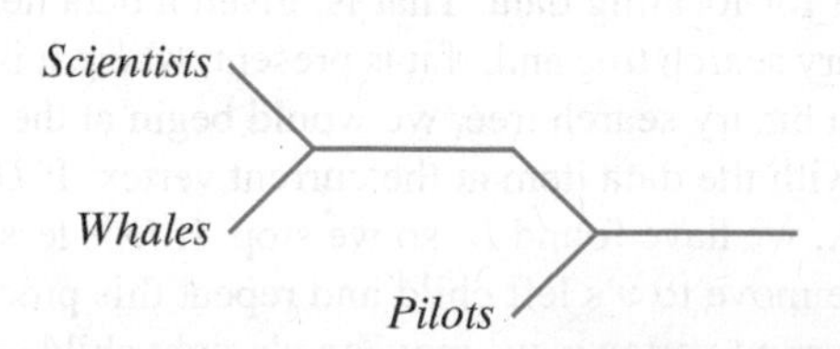


one way the tournament can unfold is

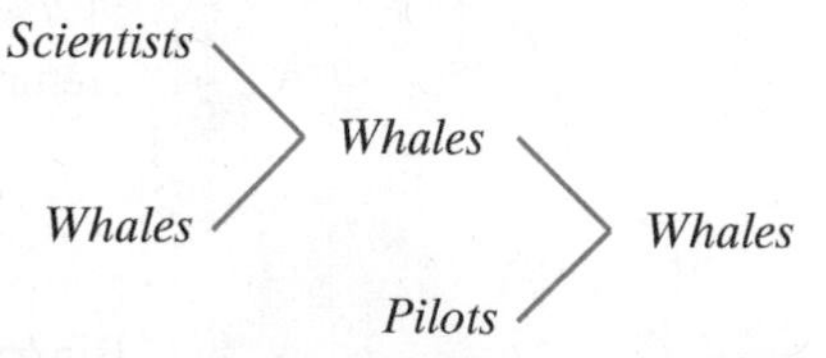


There are three other ways that the tournament can unfold:

(a) Whales defeat Scientists; Pilots defeat Whales.

(b) Scientist defeat Whales; Scientists defeat Pilots.

(c) Scientist defeat Whales; Pilots defeat Scientists.

Thus, if three teams play a single-elimination tournament, after the teams are assigned, the tournament can unfold in four ways.

2. After so-called play-in games, the NCAA men's basketball tournament is a 64-team single-elimination tournament. After the teams are assigned, in how many ways can the tournament unfold? How many (base-10) digits does this number have?

3. Suppose that after the teams are assigned in the NCAA men's basketball tournament, someone randomly guesses how the tournament will unfold. What is the probability that the guess is correct?

4. Is the value in Exercise 3 a good estimate of the chance that someone knowledgeable about basketball will successfully predict how the tournament will unfold?

5. Place the words FOUR SCORE AND SEVEN YEARS AGO OUR FOREFATHERS BROUGHT FORTH, in the order in which they appear, in a binary search tree.

6. Write a formal algorithm for searching in a binary search tree.

7. Write an algorithm that stores $n$ distinct words in a binary search tree $T$ of minimal height. Show that the derived tree $T^*$, as described in the text, has height $\lceil \lg(n+1) \rceil$.

8. True or false? Let $T$ be a binary tree. If for every vertex $v$ in $T$ the data item in $v$ is greater than the data item in the left child of $v$ and the data item in $v$ is less than the data item in the right child of $v$, then $T$ is a binary search tree. Explain.

*In Exercises 9–11, draw a graph having the given properties or explain why no such graph exists.*

9. Full binary tree; four internal vertices; five terminal vertices

10. Full binary tree; height = 4; nine terminal vertices

11. Full binary tree; height = 3; nine terminal vertices

12. A **full $m$-ary tree** is a rooted tree such that every parent has $m$ ordered children. If $T$ is a full $m$-ary tree with $i$ internal vertices, how many vertices does $T$ have? How many terminal vertices does $T$ have? Prove your results.

13. Give an algorithm for constructing a full binary tree with $n > 1$ terminal vertices.

14. Give a recursive algorithm to insert a word in a binary search tree.

15. Find the maximum height of a full binary tree having $t$ terminal vertices.

16. Write an algorithm that tests whether a binary tree in which data are stored in the vertices is a binary search tree.

17. Let $T$ be a full binary tree. Let $I$ be the sum of the lengths of the simple paths from the root to the internal vertices. We call $I$ the *internal path length*. Let $E$ be the sum of the lengths of the simple paths from the root to the terminal vertices. We call $E$ the *external path length*. Prove that if $T$ has $n$ internal vertices, then $E = I + 2n$.

*A binary tree $T$ is balanced if for every vertex $v$ in $T$, the heights of the left and right subtrees of $v$ differ by at most 1. (Here the height of a "missing subtree" is defined to be $-1$.)*

*State whether each tree in Exercises 18–21 is balanced or not.*

18.

a
b c
d

19.

a
b c
d e f
g h

20. 21.

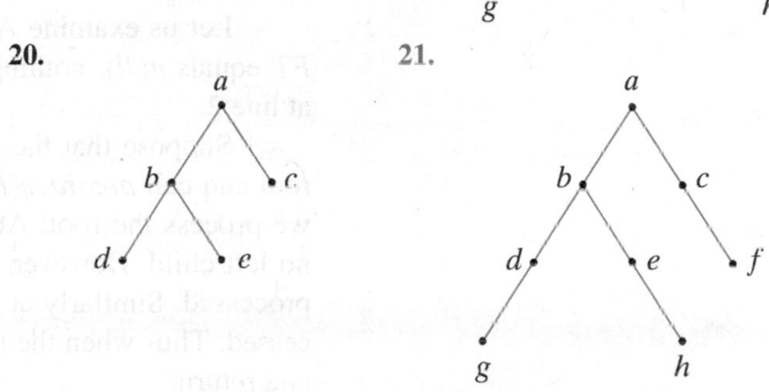


*In Exercises 22–24, $N_h$ is defined as the minimum number of vertices in a balanced binary tree of height $h$ and $f_1, f_2, \ldots$ denotes the Fibonacci sequence.*

22. Show that $N_0 = 1$, $N_1 = 2$, and $N_2 = 4$.

23. Show that $N_h = 1 + N_{h-1} + N_{h-2}$, for $h \geq 0$.

24. Show that $N_h = f_{h+3} - 1$, for $h \geq 0$.

★25. Show that the height $h$ of an $n$-vertex balanced binary tree satisfies $h = O(\lg n)$. This result shows that the worst-case time to search in an $n$-vertex balanced binary search tree is $O(\lg n)$.

★26. Prove that if a binary tree of height $h$ has $n \geq 1$ vertices, then $\lg n < h+1$. This result, together with Exercise 25, shows that the worst-case time to search in an $n$-vertex balanced binary search tree is $\Theta(\lg n)$.

## 9.6 Tree Traversals

Breadth-first search and depth-first search provide ways to "walk" a tree, that is, to traverse a tree in a systematic way so that each vertex is visited exactly once. In this section we consider three additional tree-traversal methods. We define these traversals recursively.

**Algorithm 9.6.1 Preorder Traversal**

This recursive algorithm processes the vertices of a binary tree using preorder traversal.

Input: *PT*, the root of a binary tree, or the special value *null* to indicate that no tree is input

Output: Dependent on how "process" is interpreted in line 3

```
   preorder(PT) {
1.   if (PT == null)
2.     return
3.   process PT
4.   l = left child of PT
5.   preorder(l)
6.   r = right child of PT
7.   preorder(r)
   }
```

Let us examine Algorithm 9.6.1 for some simple cases. If no tree is input (i.e., *PT* equals *null*), nothing is processed since, in this case, the algorithm simply returns at line 2.

Suppose that the input consists of a tree with a single vertex. We set *PT* to the root and call *preorder*(*PT*). Since *PT* is not equal to *null*, we proceed to line 3, where we process the root. At line 5, we call *preorder* with *PT* equal to *null* since there is no left child. However, we just saw that when no tree is input to *preorder*, nothing is processed. Similarly at line 7, when no tree is input to *preorder*, again nothing is processed. Thus when the input consists of a tree with a single vertex, we process the root and return.

Now suppose that the input is the tree of Figure 9.6.1. We set *PT* to the root and call *preorder*(*PT*). Since *PT* is not equal to *null*, we proceed to line 3, where we process the root. At line 5 we call *preorder* with *PT* equal to the left child of the root (see Figure 9.6.2). We just saw that if the tree input to *preorder* consists of a single vertex, *preorder* processes that vertex. Thus we next process vertex *B*. Similarly, at line 7, we process vertex *C*. Thus the vertices are processed in the order *ABC*.

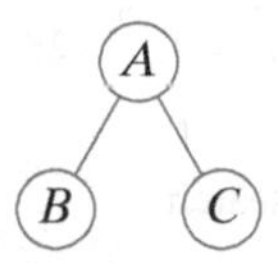


**Figure 9.6.1** Input for Algorithm 9.6.1.

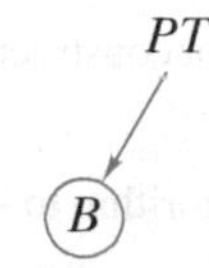


**Figure 9.6.2** At line 5 of Algorithm 9.6.1, where the input is the tree of Figure 9.6.1.

**Example 9.6.2** In what order are the vertices of the tree of Figure 9.6.3 processed if preorder traversal is used?

**SOLUTION** Following lines 3–7 (root/left/right) of Algorithm 9.6.1, the traversal proceeds as shown in Figure 9.6.4. Thus the order of processing is *ABCDEFGHIJ*.

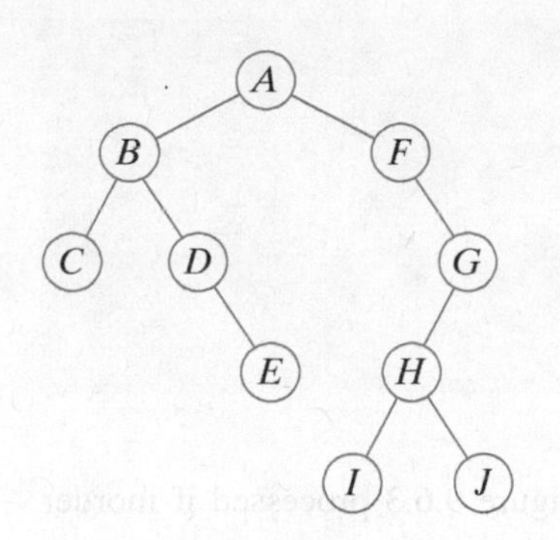


Figure 9.6.3 A binary tree.
Preorder is *ABCDEFGHIJ*.
Inorder is *CBDEAFIHJG*.
Postorder is *CEDBIJHGFA*.

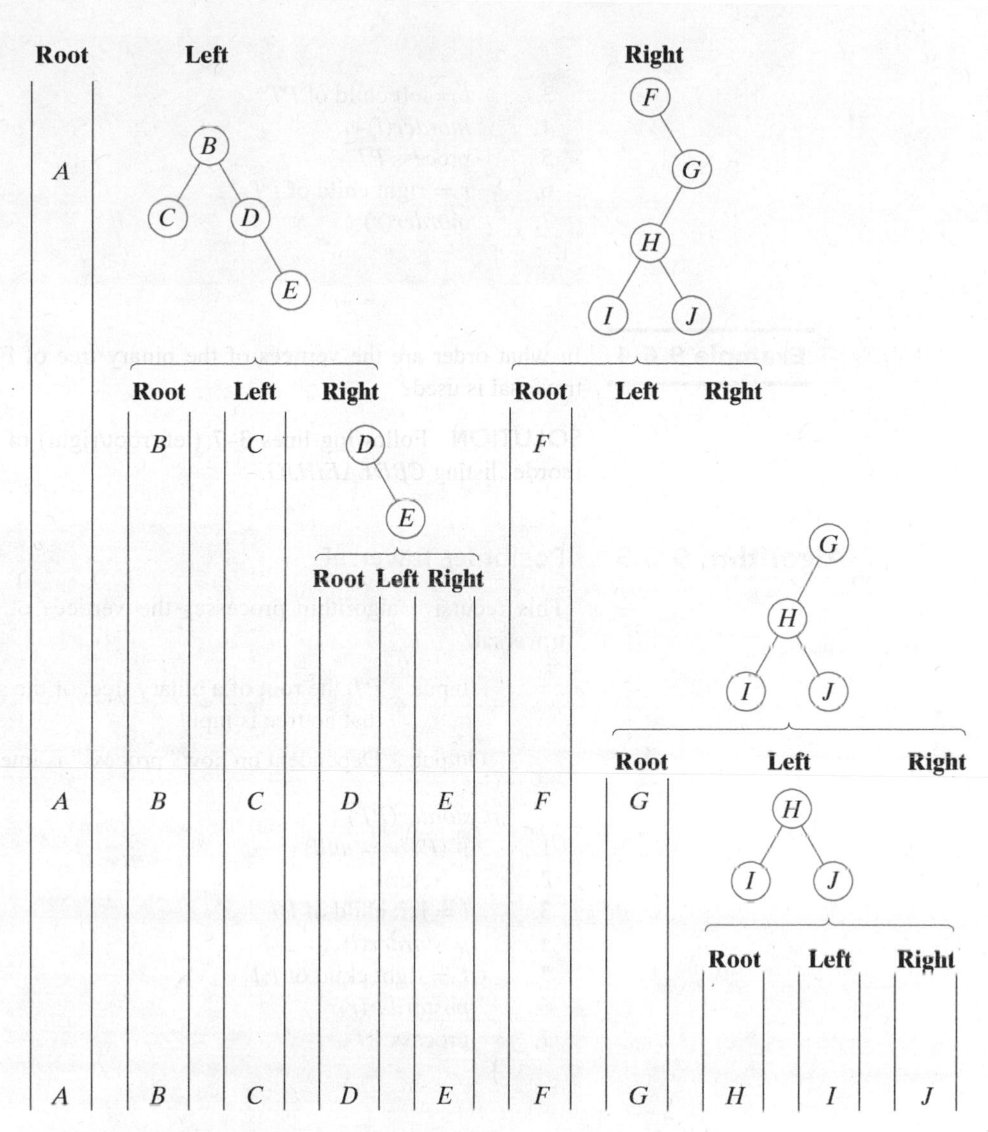


Figure 9.6.4 Preorder traversal of the tree in Figure 9.6.3.

Inorder traversal and postorder traversal are obtained by changing the position of line 3 (root) in Algorithm 9.6.1. "Pre," "in," and "post" refer to the position of the root in the traversal; that is, "preorder" means root first, "inorder" means root second, and "postorder" means root last.

**Algorithm 9.6.3 Inorder Traversal**

This recursive algorithm processes the vertices of a binary tree using inorder traversal.

Input: *PT*, the root of a binary tree, or the special value *null* to indicate that no tree is input

Output: Dependent on how "process" is interpreted in line 5

```
   inorder(PT) {
1.    if (PT == null)
2.       return
```

```
3.       l = left child of PT
4.       inorder(l)
5.       process PT
6.       r = right child of PT
7.       inorder(r)
     }
```

**Example 9.6.4** In what order are the vertices of the binary tree of Figure 9.6.3 processed if inorder traversal is used?

**SOLUTION** Following lines 3–7 (left/root/right) of Algorithm 9.6.3, we obtain the inorder listing *CBDEAFIHJG*. ◀

**Algorithm 9.6.5**

**Postorder Traversal**

This recursive algorithm processes the vertices of a binary tree using postorder traversal.

Input: *PT*, the root of a binary tree, or the special value *null* to indicate that no tree is input

Output: Dependent on how "process" is interpreted in line 7

```
     postorder(PT) {
1.       if (PT == null)
2.           return
3.       l = left child of PT
4.       postorder(l)
5.       r = right child of PT
6.       postorder(r)
7.       process PT
     }
```

**Example 9.6.6** In what order are the vertices of the binary tree of Figure 9.6.3 processed if postorder traversal is used?

**SOLUTION** Following lines 3–7 (left/right/root) of Algorithm 9.6.5, we obtain the postorder listing *CEDBIJHGFA*. ◀

Notice that preorder traversal may be obtained by following the route shown in Figure 9.6.5, and that reverse postorder traversal may be obtained by following the route shown in Figure 9.6.6.

If data are stored in a binary search tree, as described in Section 9.5, inorder traversal will process the data in order, since the sequence left/root/right agrees with the ordering of the data in the tree.

In the remainder of this section we consider binary tree representations of arithmetic expressions. Such representations facilitate the computer evaluation of expressions.

We will restrict our operators to $+$, $-$, $*$, and $/$. An example of an expression involving these operators is

$$(A + B) * C - D/E. \tag{9.6.1}$$

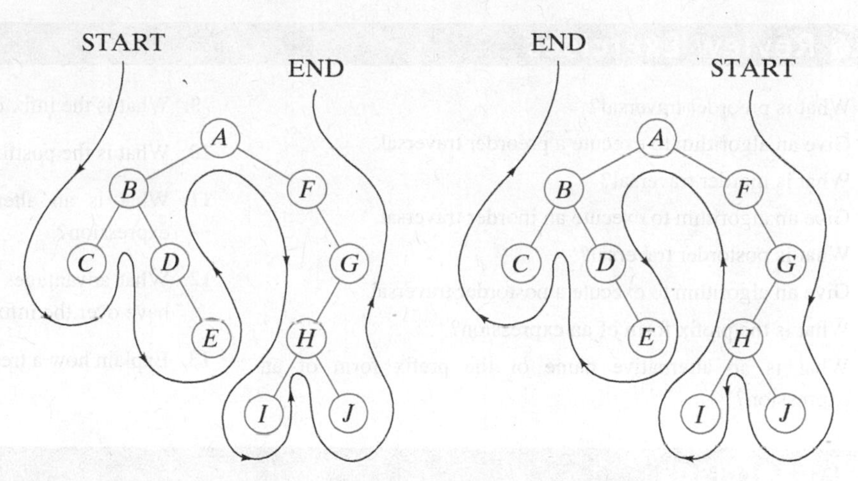


**Figure 9.6.5** Preorder traversal.

**Figure 9.6.6** Reverse postorder traversal.

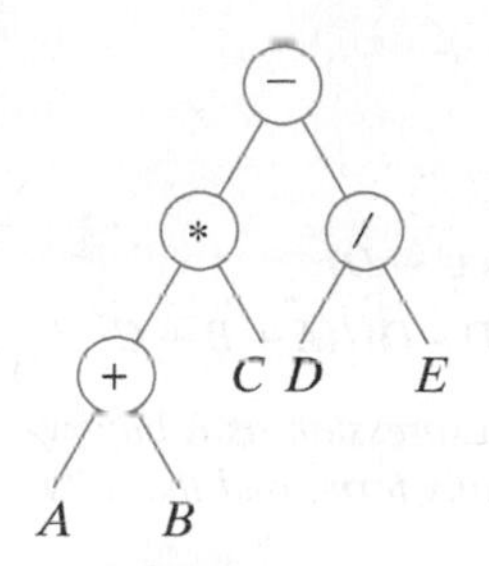


**Figure 9.6.7** The binary tree representation of the expression $(A+B)*C-D/E$.

**Go Online**
For more on reverse Polish notation, see bit.ly/2q58dGw

This standard way of representing expressions is called the **infix form of an expression.** The variables $A$, $B$, $C$, $D$, and $E$ are referred to as **operands.** The **operators** $+$, $-$, $*$, and $/$ operate on pairs of operands or expressions. In the infix form of an expression, an operator appears between its operands.

An expression such as (9.6.1) can be represented as a binary tree. The terminal vertices correspond to the operands, and the internal vertices correspond to the operators. The expression (9.6.1) would be represented as shown in Figure 9.6.7. In the binary tree representation of an expression, an operator operates on its left and right subtrees. For example, in the subtree whose root is / in Figure 9.6.7, the divide operator operates on the operands $D$ and $E$; that is, $D$ is to be divided by $E$. In the subtree whose root is $*$ in Figure 9.6.7, the multiplication operator operates on the subtree headed by $+$, which itself represents an expression, and $C$.

In a binary tree we distinguish the left and right subtrees of a vertex. The left and right subtrees of a vertex correspond to the left and right operands or expressions. This left/right distinction is important in expressions. For example, 4–6 and 6–4 are different.

If we traverse the binary tree of Figure 9.6.7 using inorder, and insert a pair of parentheses for each operation, we obtain $(((A+B)*C)-(D/E))$. This form of an expression is called the **fully parenthesized form of the expression.** In this form we do not need to specify which operations (such as multiplication) are to be performed before others (such as addition), since the parentheses unambiguously dictate the order of operations.

If we traverse the tree of Figure 9.6.7 using postorder, we obtain $AB+C*DE/-$. This form of the expression is called the **postfix form of the expression** (or **reverse Polish notation**). In postfix, the operator follows its operands. For example, the first three symbols $AB+$ indicate that $A$ and $B$ are to be added. Advantages of the postfix form over the infix form are that in postfix no parentheses are needed and no conventions are necessary regarding the order of operations. The expression will be unambiguously evaluated. For these reasons and others, many compilers translate infix expressions to postfix form. Also, some calculators require expressions to be entered in postfix form.

A third form of an expression can be obtained by applying preorder traversal to a binary tree representation of an expression. In this case, the result is called the **prefix form of the expression** (or **Polish notation**). As in postfix, no parentheses are needed and no conventions are necessary regarding the order of operations. The prefix form of (9.6.1), obtained by applying preorder traversal to the tree of Figure 9.6.7, is $-*+ABC/DE$.

## 9.6 Review Exercises

1. What is preorder traversal?
2. Give an algorithm to execute a preorder traversal.
3. What is inorder traversal?
4. Give an algorithm to execute an inorder traversal.
5. What is postorder traversal?
6. Give an algorithm to execute a postorder traversal.
7. What is the prefix form of an expression?
8. What is an alternative name of the prefix form of an expression?
9. What is the infix form of an expression?
10. What is the postfix form of an expression?
11. What is an alternative name of the postfix form of an expression?
12. What advantages do prefix and postfix forms of expressions have over the infix form?
13. Explain how a tree can be used to represent an expression.

## 9.6 Exercises

*In Exercises 1–5, list the order in which the vertices are processed using preorder, inorder, and postorder traversal.*

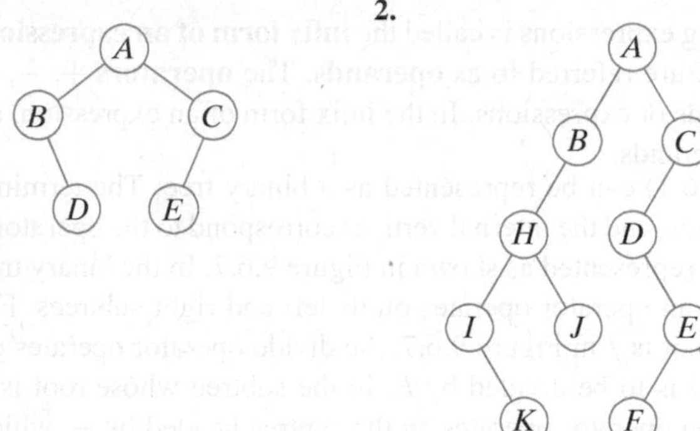


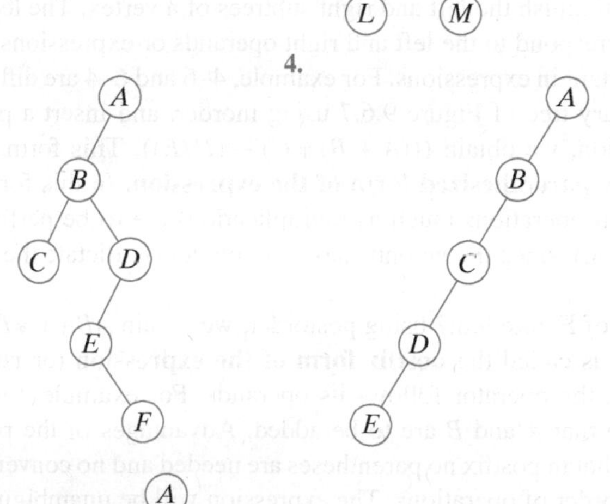


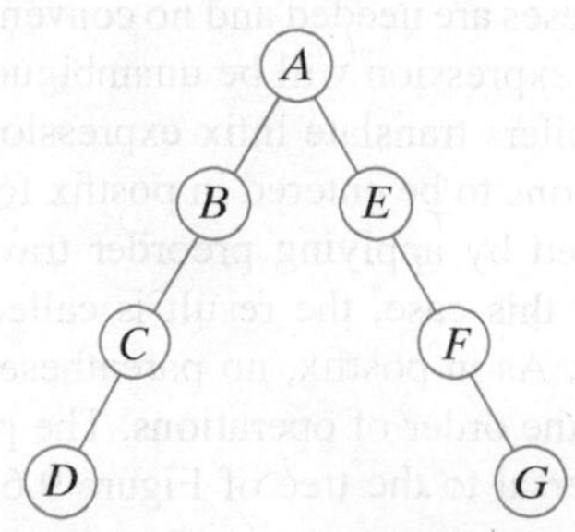


*In Exercises 6–10, represent the expression as a binary tree and write the prefix and postfix forms of the expression.*

6. $(A + B) * (C - D)$
7. $(A * B + C * D) - (A/B - (D + E))$
8. $((A - C) * D) / (A + (B + D))$
9. $(((A + B) * C + D) * E) - ((A + B) * C - D)$
10. $(A * B - C/D + E) + (A - B - C - D * D)/(A + B + C)$

*In Exercises 11–15, represent the postfix expression as a binary tree and write the prefix form, the usual infix form, and the fully parenthesized infix form of the expression.*

11. $AB{+}C{-}$
12. $ABCD{+}{*}/E{-}$
13. $ABC{+}{-}$
14. $ABC{*}{*}CDE{+}/{-}$
15. $AB{+}CD{*}EF/{-}{-}A{*}$

*In Exercises 16–21, find the value of the postfix expression if $A = 1$, $B = 2$, $C = 3$, and $D = 4$.*

16. $ABC{+}{-}$
17. $AB{+}CD{*}AA/{-}{-}B{*}$
18. $AB{+}C{-}$
19. $ABC{*}{*}ABC{+}{+}{-}$
20. $ADBCD{*}{-}{+}{*}$
21. $ABAB{*}{+}{*}D{*}$
22. Show, by example, that distinct binary trees with vertices $A$, $B$, and $C$ can have the same preorder listing $ABC$.
23. Show that there is a unique binary tree with six vertices whose preorder vertex listing is $ABCEFD$ and whose inorder vertex listing is $ACFEBD$.

★24. Write an algorithm that reconstructs the binary tree given its preorder and inorder vertex orderings.

25. Give examples of distinct binary trees, $B_1$ and $B_2$, each with two vertices, with the preorder vertex listing of $B_1$ equal to the preorder listing of $B_2$ and the postorder vertex listing of $B_1$ equal to the postorder listing of $B_2$.
26. Let $P_1$ and $P_2$ be permutations of $ABCDEF$. Is there a binary tree with vertices $A$, $B$, $C$, $D$, $E$, and $F$ whose preorder listing is $P_1$ and whose inorder listing is $P_2$? Explain.
27. Write a recursive algorithm that prints the contents of the terminal vertices of a binary tree from left to right.

28. Write a recursive algorithm that interchanges all left and right children of a binary tree.
29. Write an algorithm that returns the number of terminal nodes in a binary tree.
30. Write a recursive algorithm that initializes each vertex of a binary tree to the number of its descendants.
31. Prove that the algorithm

```
funnyorder(PT) {
  if (PT == null)
    return
  process PT
  r = right child of PT
  funnyorder(r)
  l = left child of PT
  funnyorder(l)
}
```

visits the nodes in the reverse order of postorder.

*In Exercises 32 and 33, every expression involves only the operands $A, B, \ldots, Z$ and the operators $+, -, *, /$.*

★32. Give a necessary and sufficient condition for a string of symbols to be a valid postfix expression.
33. Write an algorithm that, given the binary tree representation of an expression, outputs the fully parenthesized infix form of the expression.
34. Write an algorithm that prints the characters and their codes given a Huffman coding tree (see Example 9.1.8). Assume that each terminal vertex stores a character and its frequency.

*Use the following definitions in Exercises 35–40.*

*Let $G = (V, E)$ be a simple undirected graph. A* vertex cover *of $G$ is a subset $V'$ of $V$ such that for each edge $(v, w) \in E$, either $v \in V'$ or $w \in V'$. The size of a vertex cover $V'$ is the number of vertices in $V'$. An* optimal vertex cover *is a vertex cover of minimum size.*

*An* edge disjoint set *for $G$ is a subset $E'$ of $E$ such that for every pair of distinct edges $e_1 = (v_1, w_1)$ and $e_2 = (v_2, w_2)$ in $E'$, we have $\{v_1, w_1\} \cap \{v_2, w_2\} = \varnothing$.*

35. Prove that for every $n$, there is a connected graph with $n$ vertices that has a vertex cover of size 1.
36. Could the size of an optimal vertex cover of a graph with $n$ vertices equal $n$? Explain.
37. Show that the size of an optimal vertex cover of the complete graph on $n$ vertices is $n - 1$.
★38. Write an algorithm that finds an optimal vertex cover of a tree $T = (V, E)$ whose worst-case time is $\Theta(|E|)$.
39. Show that if $V'$ is any vertex cover of a graph $G$ and $E'$ is any edge disjoint set for $G$, then $|E'| \leq |V'|$.
40. Give an example of a connected graph in which, for every vertex cover $V'$ and every edge disjoint set $E'$, we have $|E'| < |V'|$. Prove that your example has the required property.
41. Show how a binary tree with $n$ edges can be encoded as a string of $n + 1$ ones and $n + 1$ zeros where, reading from left to right, the number of zeros never exceeds the number of ones. Show that each such string represents a binary tree. *Hint:* Consider a preorder traversal of the binary tree in which a one means that an edge is present, and a zero means that an edge is absent. Add an extra one to the beginning of the string, and delete the last zero.

## 9.7 Decision Trees and the Minimum Time for Sorting

The binary tree of Figure 9.7.1 gives an algorithm for choosing a restaurant. Each internal vertex asks a question. If we begin at the root, answer each question, and follow the appropriate edge, we will eventually arrive at a terminal vertex that chooses a restaurant. Such a tree is called a **decision tree.** In this section we use decision trees to specify algorithms and to obtain lower bounds on the worst-case time for sorting as well as solving certain coin puzzles. We begin with coin puzzles.

**Example 9.7.1** **Five-Coins Puzzle** Five coins are identical in appearance, but one coin is either heavier or lighter than the others, which all weigh the same. The problem is to identify the bad coin and determine whether it is heavier or lighter than the others using only a pan balance (see Figure 9.7.2), which compares the weights of two sets of coins.

An algorithm to solve the puzzle is given in Figure 9.7.3 as a decision tree. The coins are labeled $C_1, C_2, C_3, C_4, C_5$. As shown, we begin at the root and place coin $C_1$ in the left pan and coin $C_2$ in the right pan. An edge labeled ⟋ means that the left side of the pan balance is heavier than the right side. Similarly, an edge labeled ⟍ means that the right side of the pan balance is heavier than the left side, and an edge labeled –•– means that the two sides balance. For example, at the root when we compare $C_1$ with

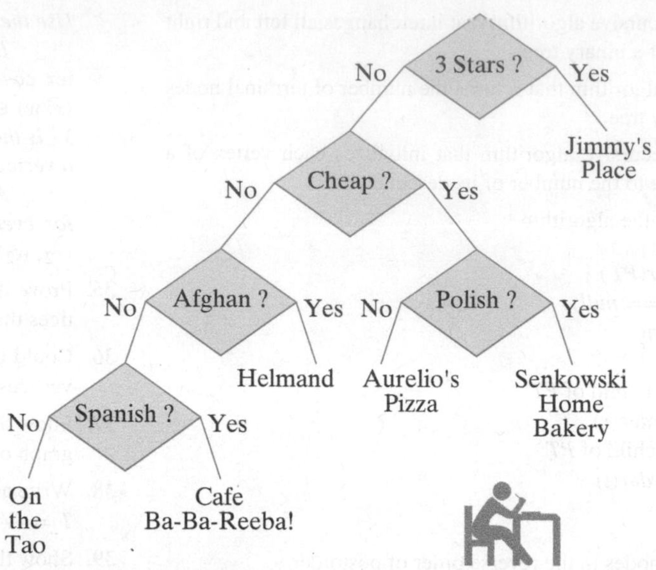


**Figure 9.7.1** A decision tree.

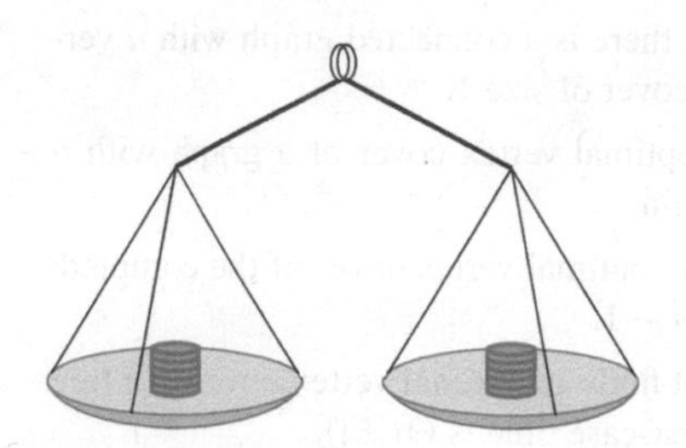

**Figure 9.7.2** A pan balance for comparing weights of coins.

$C_2$, if the left side is heavier than the right side, we know that either $C_1$ is the heavy coin or $C_2$ is the light coin. In this case, as shown in the decision tree, we next compare $C_1$ with $C_5$ (which is known to be a good coin) and immediately determine whether the bad coin is $C_1$ or $C_2$ and whether it is heavy or light. The terminal vertices give the solution. For example, when we compare $C_1$ with $C_5$ and the pans balance, we follow the edge to the terminal vertex labeled $C_2, L$, which tells us that the bad coin is $C_2$ and that it is lighter than the others.

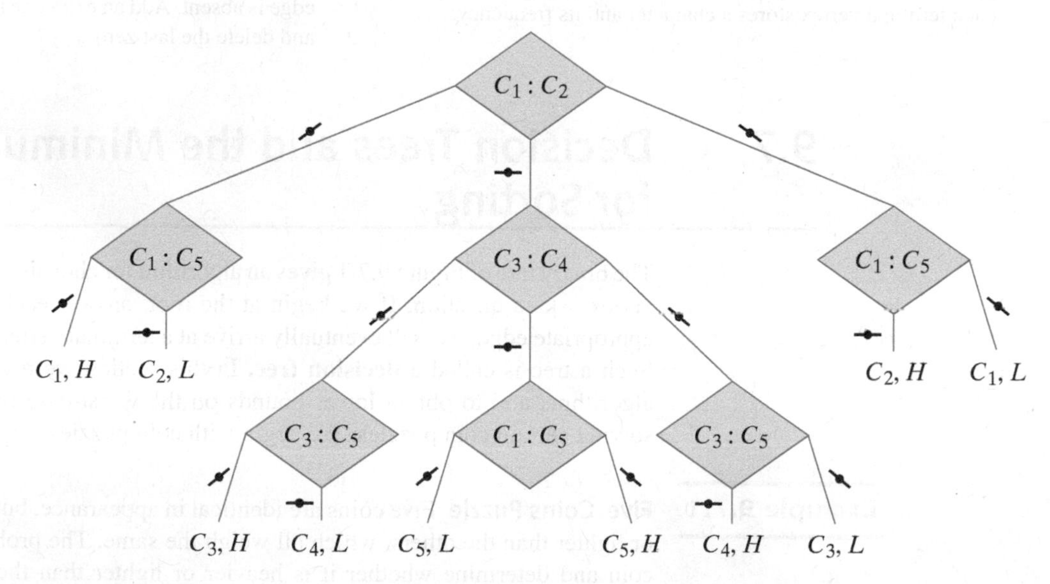


**Figure 9.7.3** An algorithm to solve the five-coins puzzle. ◀

If we define the worst-case time to solve a coin-weighing problem to be the number of weighings required in the worst case, it is easy to determine the worst-case time from the decision tree; the worst-case time is equal to the height of the tree. For example, the height of the decision tree of Figure 9.7.3 is 3, so the worst-case time for this algorithm is equal to 3.

We can use decision trees to show that the algorithm given in Figure 9.7.3 to solve the five-coins puzzle is optimal, that is, that *no* algorithm that solves the five-coins puzzle has a worst-case time less than 3.

We argue by contradiction to show that no algorithm that solves the five-coins puzzle has a worst-case time less than 3. Suppose that there is an algorithm that solves the five-coins puzzle in the worst case in two or fewer weighings. The algorithm can be described by a decision tree, and since the worst-case time is 2 or less, the height of the decision tree is 2 or less. Since each internal vertex has at most three children, such a tree can have at most nine terminal vertices (see Figure 9.7.4). Now the terminal vertices correspond to possible outcomes. Thus a decision tree of height 2 or less can account for at most nine outcomes. But the five-coins puzzle has 10 outcomes:

$$C_1, L, \quad C_1, H, \quad C_2, L, \quad C_2, H, \quad C_3, L,$$
$$C_3, H, \quad C_4, L, \quad C_4, H, \quad C_5, L, \quad C_5, H.$$

This is a contradiction. Therefore, no algorithm that solves the five-coins puzzle has a worst-case time less than 3, and the algorithm of Figure 9.7.3 is optimal.

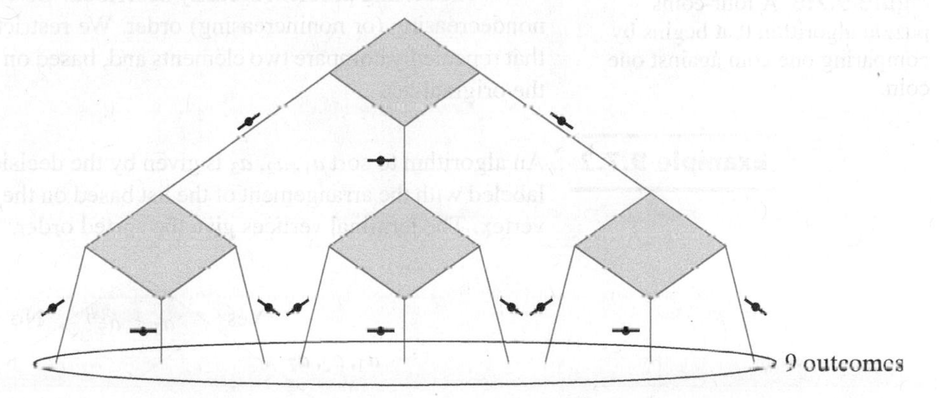


**Figure 9.7.4** A five-coins puzzle algorithm that uses at most two weighings.

We have seen how a decision tree can be used to give a lower bound for the worst-case time to solve a problem. Sometimes, the lower bound is unattainable.

Consider the four-coins puzzle (all the rules are the same as for the five-coins puzzle except that the number of coins is reduced by one). Since there are now eight outcomes rather than 10, we can conclude that any algorithm to solve the four-coins puzzle requires at least two weighings in the worst case. (This time we *cannot* conclude that at least three weighings are required in the worst case.) However, closer inspection shows that, in fact, three weighings are required.

The first weighing either compares two coins against two coins or one coin against one coin. Figure 9.7.5 shows that if we begin by comparing two coins against two coins, the decision tree can account for at most six outcomes. Since there are eight outcomes, no algorithm that begins by comparing two coins against two coins can solve the problem in two weighings or less in the worst case. Similarly, Figure 9.7.6 shows that if we begin by comparing one coin against one coin and the coins balance, the decision tree can account for only three outcomes. Since four outcomes are possible after identifying two good coins, no algorithm that begins by comparing one coin against one coin can solve the problem in two weighings or less in the worst case. Therefore, any algorithm that solves the four-coins puzzle requires at least three weighings in the worst case.

If we modify the four-coins puzzle by requiring only that we identify the bad coin (without determining whether it is heavy or light), we can solve the puzzle in two weighings in the worst case (see Exercise 1).

We turn now to sorting. We can use decision trees to estimate the worst-case time to sort.

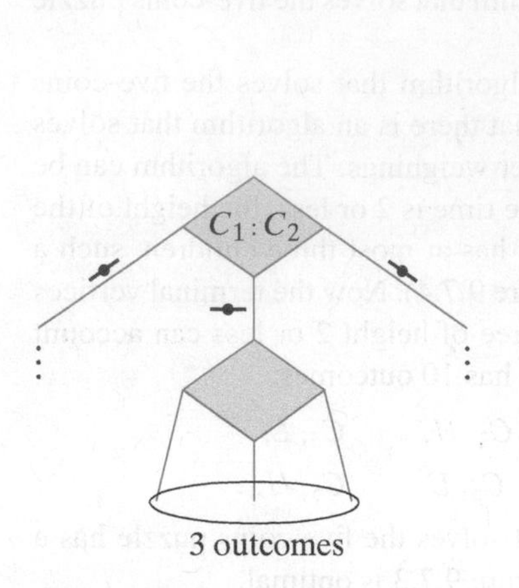


Figure 9.7.6 A four-coins puzzle algorithm that begins by comparing one coin against one coin.

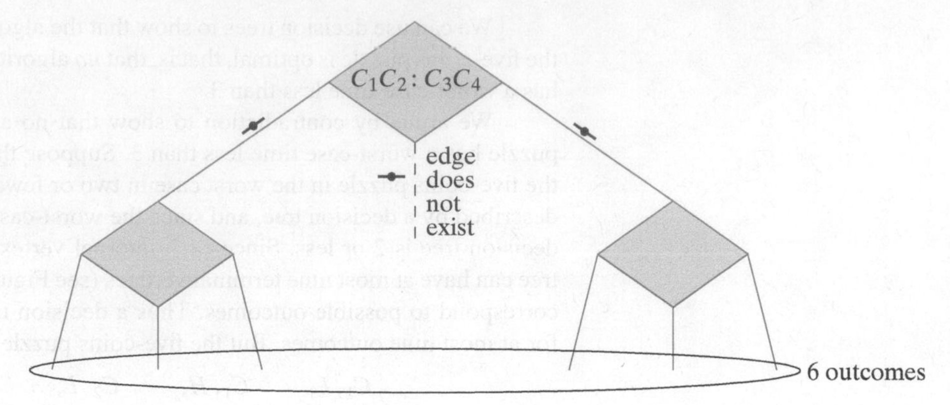


Figure 9.7.5 A four-coins puzzle algorithm that begins by comparing two coins against two coins.

The sorting problem is easily described: Given $n$ items $x_1, \ldots, x_n$, arrange them in nondecreasing (or nonincreasing) order. We restrict our attention to sorting algorithms that repeatedly compare two elements and, based on the result of the comparison, modify the original list.

**Example 9.7.2** An algorithm to sort $a_1, a_2, a_3$ is given by the decision tree of Figure 9.7.7. Each edge is labeled with the arrangement of the list based on the answer to the question at an internal vertex. The terminal vertices give the sorted order.

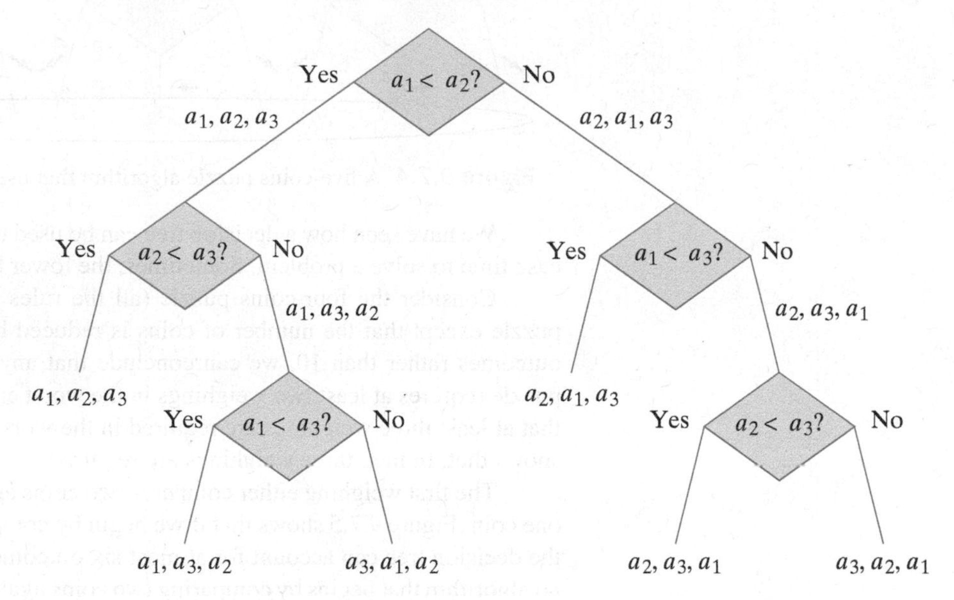


Figure 9.7.7 An algorithm to sort $a_1, a_2, a_3$.

Let us define the worst-case time to sort to be the number of comparisons in the worst case. Just as in the case of the decision trees that solve coin puzzle problems, the height of a decision tree that solves a sorting problem is equal to the worst-case time. For example, the worst-case time for the algorithm given by the decision tree of Figure 9.7.7 is equal to 3. We show that this algorithm is optimal, that is, that *no* algorithm that sorts three items has a worst-case time less than 3.

We argue by contradiction to show that no algorithm that sorts three items has a worst-case time less than 3. Suppose that there is an algorithm that sorts three items in the worst case in two or fewer comparisons. The algorithm can be described by a

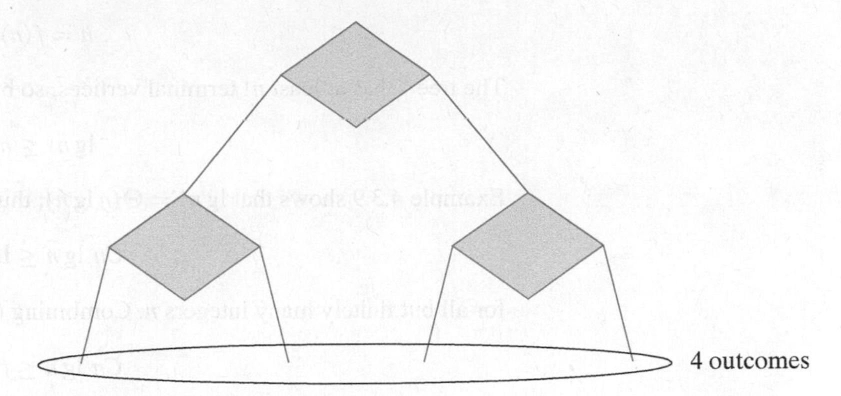


**Figure 9.7.8** A sorting algorithm that makes at most two comparisons.

decision tree, and since the worst-case time is 2 or less, the height of the decision tree is 2 or less. Since each internal vertex has at most two children, such a tree can have at most four terminal vertices (see Figure 9.7.8). Now the terminal vertices correspond to possible outcomes. Thus a decision tree of height 2 or less can account for at most four outcomes. But the problem of sorting three items has six possible outcomes (when the items are distinct), corresponding to the $3! = 6$ ways that three items can be arranged:

$$s_1, s_2, s_3, \quad s_1, s_3, s_2, \quad s_2, s_1, s_3, \quad s_2, s_3, s_1, \quad s_3, s_1, s_2, \quad s_3, s_2, s_1.$$

This is a contradiction. Therefore, no algorithm that sorts three items has a worst-case time less than 3, and the algorithm of Figure 9.7.7 is optimal. ◀

Since $4! = 24$, there are 24 possible outcomes to the problem of sorting four items (when the items are distinct). To accommodate 24 terminal vertices, we must have a tree of height at least 5 (see Figure 9.7.9). Therefore, any algorithm that sorts four items requires at least five comparisons in the worst case. Exercise 9 is to give an algorithm that sorts four items using five comparisons in the worst case.

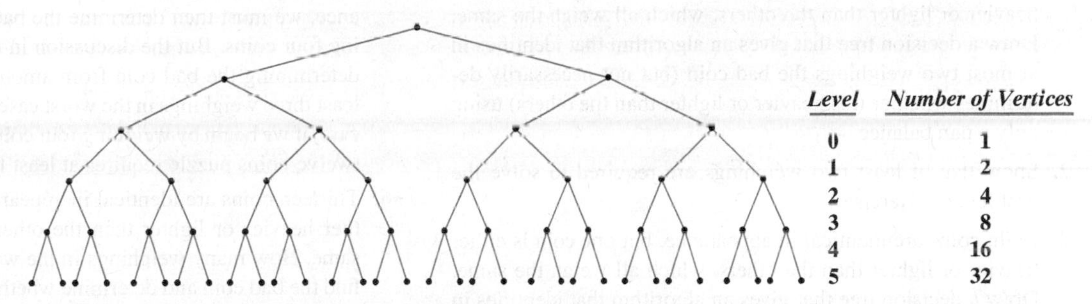

| *Level* | *Number of Vertices* |
|---|---|
| 0 | 1 |
| 1 | 2 |
| 2 | 4 |
| 3 | 8 |
| 4 | 16 |
| 5 | 32 |

**Figure 9.7.9** Level compared with the maximum number of vertices in that level in a binary tree.

The method of Example 9.7.2 can be used to give a lower bound on the number of comparisons required in the worst case to sort an arbitrary number of items.

**Theorem 9.7.3** If $f(n)$ is the number of comparisons needed to sort $n$ items in the worst case by a sorting algorithm, then $f(n) = \Omega(n \lg n)$.

**Proof** Let $T$ be the decision tree that represents the algorithm for input of size $n$ and let $h$ denote the height of $T$. Then the algorithm requires $h$ comparisons in the worst case, so

$$h = f(n). \tag{9.7.1}$$

The tree $T$ has at least $n!$ terminal vertices, so by Theorem 9.5.6,

$$\lg n! \leq h. \tag{9.7.2}$$

Example 4.3.9 shows that $\lg n! = \Theta(n \lg n)$; thus, for some positive constant $C$,

$$Cn \lg n \leq \lg n! \tag{9.7.3}$$

for all but finitely many integers $n$. Combining (9.7.1) through (9.7.3), we obtain

$$Cn \lg n \leq f(n)$$

for all but finitely many integers $n$. Therefore,

$$f(n) = \Omega(n \lg n).$$ ◀

Theorem 7.3.10 states that merge sort (Algorithm 7.3.8) uses $\Theta(n \lg n)$ comparisons in the worst case and is, by Theorem 9.7.3, optimal. Many other sorting algorithms are known that also attain the optimal number $\Theta(n \lg n)$ of comparisons; one, tournament sort, is described before Exercise 14.

## 9.7 Review Exercises

1. What is a decision tree?
2. How is the height of a decision tree that represents an algorithm related to the worst-case time of the algorithm?
3. Use decision trees to explain why worst-case sorting requires at least $\Omega(n \lg n)$ comparisons.

## 9.7 Exercises

1. Four coins are identical in appearance, but one coin is either heavier or lighter than the others, which all weigh the same. Draw a decision tree that gives an algorithm that identifies in at most two weighings the bad coin (but not necessarily determines whether it is heavier or lighter than the others) using only a pan balance.
2. Show that at least two weighings are required to solve the problem of Exercise 1.
3. Eight coins are identical in appearance, but one coin is either heavier or lighter than the others, which all weigh the same. Draw a decision tree that gives an algorithm that identifies in at most three weighings the bad coin and determines whether it is heavier or lighter than the others using only a pan balance.
4. Twelve coins are identical in appearance, but one coin is either heavier or lighter than the others, which all weigh the same. Draw a decision tree that gives an algorithm that identifies in at most three weighings the bad coin and determines whether it is heavier or lighter than the others using only a pan balance.
5. What is wrong with the following argument, which supposedly shows that the twelve-coins puzzle requires at least four weighings in the worst case if we begin by weighing four coins against four coins?

   If we weigh four coins against four coins and they balance, we must then determine the bad coin from the remaining four coins. But the discussion in this section showed that determining the bad coin from among four coins requires at least three weighings in the worst case. Therefore, in the worst case, if we begin by weighing four coins against four coins, the twelve-coins puzzle requires at least four weighings.

★6. Thirteen coins are identical in appearance, but one coin is either heavier or lighter than the others, which all weigh the same. How many weighings in the worst case are required to find the bad coin and determine whether it is heavier or lighter than the others using only a pan balance? Prove your answer.

7. Solve Exercise 6 for the fourteen-coins puzzle.
8. $(3^n - 3)/2$, $n \geq 2$, coins are identical in appearance, but one coin is either heavier or lighter than the others, which all weigh the same. [Kurosaka] gave an algorithm to find the bad coin and determine whether it is heavier or lighter than the others using only a pan balance in $n$ weighings in the worst case. Prove that the coin cannot be found and identified as heavy or light in fewer than $n$ weighings.

*Exercises 9 and 10 concern the following variant of the coin-weighing problem. We are given n coins, some of which are bad, but are otherwise identical in appearance. All of the good coins have the same weight. All of the bad coins also have the same weight, but they are lighter than the good coins. We assume that there is at least one bad coin and at least one good coin among the n coins. The task is to determine the number of bad coins.*

**9.** Show that at least $\log_3(n-1)$ weighings are necessary to determine the number of bad coins.

**10.** Show how to determine the number of bad coins in at most $n-1$ weighings.

**11.** Give an algorithm that sorts four items using five comparisons in the worst case.

**12.** Use decision trees to find a lower bound on the number of comparisons required to sort five items in the worst case. Give an algorithm that uses this number of comparisons to sort five items in the worst case.

**13.** Use decision trees to find a lower bound on the number of comparisons required to sort six items in the worst case. Give an algorithm that uses this number of comparisons to sort six items in the worst case.

*Exercises 14–20 refer to tournament sort.*

*Tournament Sort. We are given a sequence $s_1, \ldots, s_{2^k}$ to sort in nondecreasing order.*

*We will build a binary tree with terminal vertices labeled $s_1, \ldots, s_{2^k}$. An example is shown.*

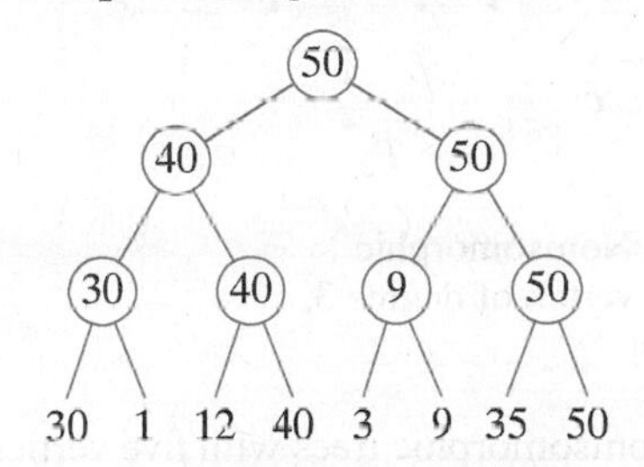


*Working left to right, create a parent for each pair and label it with the maximum of the children. Continue in this way until you reach the root. At this point, the largest value, m, has been found.*

*To find the second-largest value, first pick a value v less than all the items in the sequence. Replace the terminal vertex w containing m with v. Relabel the vertices by following the path from w to the root, as shown. At this point, the second-largest value is found. Continue until the sequence is ordered.*

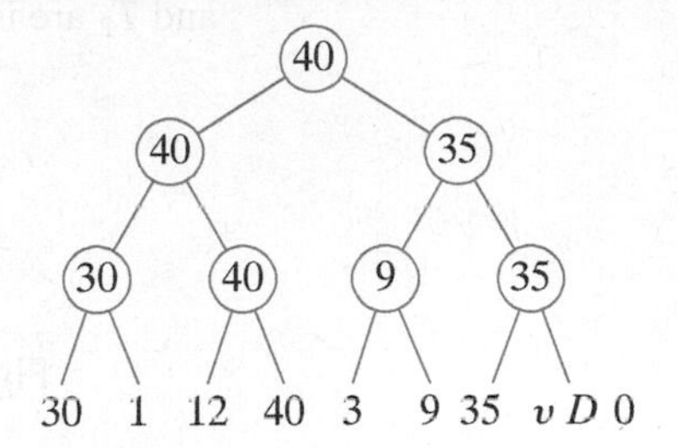


**14.** Why is the name "tournament" appropriate?

**15.** Draw the two trees that would be created after the preceding tree when tournament sort is applied.

**16.** How many comparisons does tournament sort require to find the largest element?

**17.** Show that any algorithm that finds the largest value among $n$ items requires at least $n-1$ comparisons.

**18.** How many comparisons does tournament sort require to find the second-largest element?

**19.** Write tournament sort as a formal algorithm.

**20.** Show that if $n$ is a power of 2, tournament sort requires $\Theta(n \lg n)$ comparisons.

**21.** Give an example of a real situation (like that of Figure 9.7.1) that can be modeled as a decision tree. Draw the decision tree.

**22.** Draw a decision tree that can be used to determine who must file a federal tax return.

**23.** Draw a decision tree that gives a reasonable strategy for playing blackjack (see, e.g., [Ainslie]).

## 9.8 Isomorphisms of Trees

In Section 8.6 we defined what it means for two graphs to be isomorphic. (You might want to review Section 8.6 before continuing.) In this section we discuss isomorphic trees, isomorphic rooted trees, and isomorphic binary trees.

Corollary 8.6.5 states that simple graphs $G_1$ and $G_2$ are isomorphic if and only if there is a one-to-one, onto function $f$ from the vertex set of $G_1$ to the vertex set of $G_2$ that preserves the adjacency relation in the sense that vertices $v_i$ and $v_j$ are adjacent in $G_1$ if and only if the vertices $f(v_i)$ and $f(v_j)$ are adjacent in $G_2$. Since a (free) tree is a simple graph, trees $T_1$ and $T_2$ are isomorphic if and only if there is a one-to-one, onto function $f$ from the vertex set of $T_1$ to the vertex set of $T_2$ that preserves the adjacency relation; that is, vertices $v_i$ and $v_j$ are adjacent in $T_1$ if and only if the vertices $f(v_i)$ and $f(v_j)$ are adjacent in $T_2$.

**Example 9.8.1** The function $f$ from the vertex set of the tree $T_1$ shown in Figure 9.8.1 to the vertex set of the tree $T_2$ shown in Figure 9.8.2 defined by

$$f(a) = 1, \quad f(b) = 3, \quad f(c) = 2, \quad f(d) = 4, \quad f(e) = 5$$

is a one-to-one, onto function that preserves the adjacency relation. Thus the trees $T_1$ and $T_2$ are isomorphic.

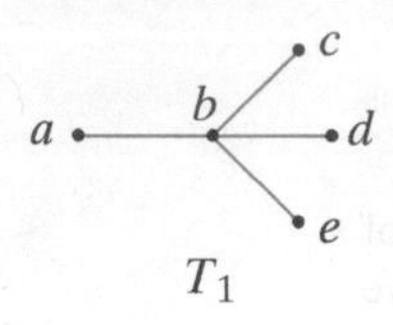


Figure 9.8.1 A tree.

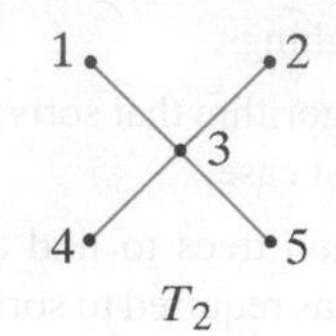


Figure 9.8.2 A tree isomorphic to the tree in Figure 9.8.1.

As in the case of graphs, we can show that two trees are not isomorphic if we can exhibit an invariant that the trees do not share.

**Example 9.8.2** The trees $T_1$ and $T_2$ of Figure 9.8.3 are not isomorphic because $T_2$ has a vertex ($x$) of degree 3, but $T_1$ does not have a vertex of degree 3.

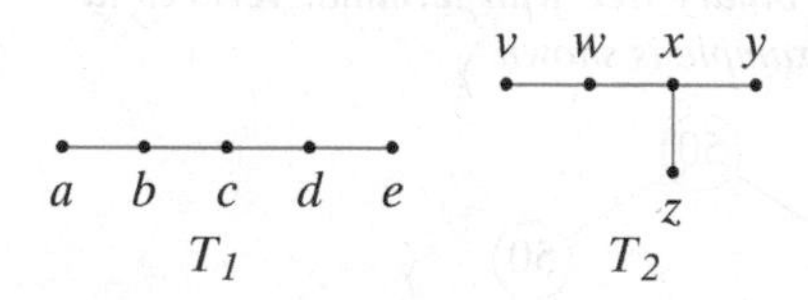


Figure 9.8.3 Nonisomorphic trees. $T_2$ has a vertex of degree 3, but $T_2$ does not.

We can show that there are three nonisomorphic trees with five vertices. The three nonisomorphic trees are shown in Figures 9.8.1 and 9.8.3.

**Theorem 9.8.3** There are three nonisomorphic trees with five vertices.

**Proof** We will give an argument to show that any tree with five vertices is isomorphic to one of the trees in Figure 9.8.1 or 9.8.3.

If $T$ is a tree with five vertices, by Theorem 9.2.3 $T$ has four edges. If $T$ had a vertex $v$ of degree greater than 4, $v$ would be incident on more than four edges. It follows that each vertex in $T$ has degree at most 4.

We will first find all nonisomorphic trees with five vertices in which the maximum vertex degree that occurs is 4. We will next find all nonisomorphic trees with five vertices in which the maximum vertex degree that occurs is 3, and so on.

Let $T$ be a tree with five vertices and suppose that $T$ has a vertex $v$ of degree 4. Then there are four edges incident on $v$ and, because of Theorem 9.2.3, these are all the edges. It follows that in this case $T$ is isomorphic to the tree in Figure 9.8.1.

Suppose that $T$ is a tree with five vertices and the maximum vertex degree that occurs is 3. Let $v$ be a vertex of degree 3. Then $v$ is incident on three edges, as shown in Figure 9.8.4. The fourth edge cannot be incident on $v$ since then $v$ would have degree 4. Thus the fourth edge is incident on one of $v_1$, $v_2$, or $v_3$. Adding an edge incident on any of $v_1$, $v_2$, or $v_3$ gives a tree isomorphic to the tree $T_2$ of Figure 9.8.3.

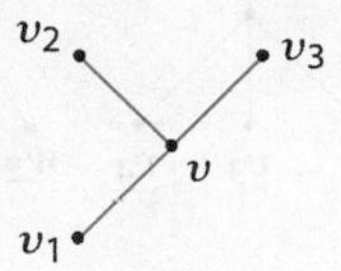

Figure 9.8.4 Vertex $v$ has degree 3.

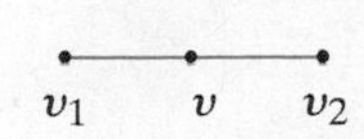


Figure 9.8.5 Vertex $v$ has degree 2.

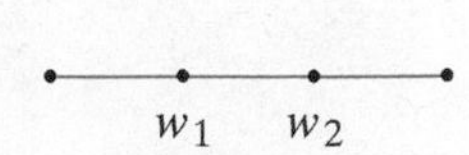


Figure 9.8.6 Adding a third edge to the graph of Figure 9.8.5.

Now suppose that $T$ is a tree with five vertices and the maximum vertex degree that occurs is 2. Let $v$ be a vertex of degree 2. Then $v$ is incident on two edges, as shown in Figure 9.8.5. A third edge cannot be incident on $v$; thus it must be incident on either $v_1$ or $v_2$. Adding the third edge gives the graph of Figure 9.8.6. For the same reason, the fourth edge cannot be incident on either of the vertices $w_1$ or $w_2$ of Figure 9.8.6. Adding the last edge gives a tree isomorphic to the tree $T_1$ of Figure 9.8.3.

Since a tree with five vertices must have a vertex of degree 2, we have found all nonisomorphic trees with five vertices. ◀

For two *rooted* trees $T_1$ and $T_2$ to be isomorphic, there must be a one-to-one, onto function $f$ from $T_1$ to $T_2$ that preserves the adjacency relation and that preserves the root. The latter condition means that $f(\text{root of } T_1) = \text{root of } T_2$. The formal definition follows.

**Definition 9.8.4 ▶** Let $T_1$ be a rooted tree with root $r_1$ and let $T_2$ be a rooted tree with root $r_2$. The rooted trees $T_1$ and $T_2$ are *isomorphic* if there is a one-to-one, onto function $f$ from the vertex set of $T_1$ to the vertex set of $T_2$ satisfying the following:

(a) Vertices $v_i$ and $v_j$ are adjacent in $T_1$ if and only if the vertices $f(v_i)$ and $f(v_j)$ are adjacent in $T_2$.

(b) $f(r_1) = r_2$.

We call the function $f$ an *isomorphism*. ◀

**Example 9.8.5** The rooted trees $T_1$ and $T_2$ in Figure 9.8.7 are isomorphic. An isomorphism is

$$f(v_1) = w_1, \quad f(v_2) = w_3, \quad f(v_3) = w_4, \quad f(v_4) = w_2,$$
$$f(v_5) = w_7, \quad f(v_6) = w_6, \quad f(v_7) = w_5.$$

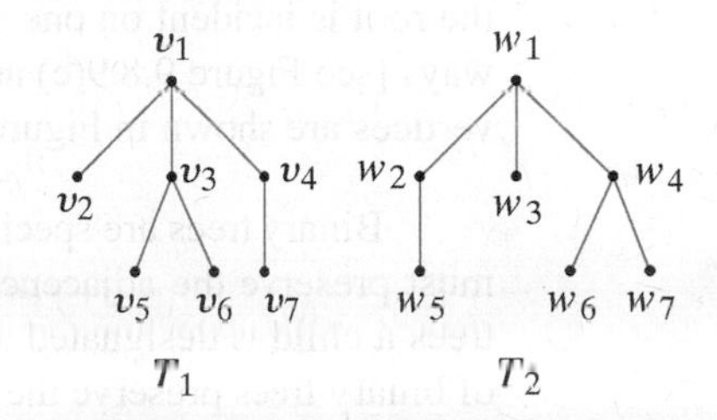


Figure 9.8.7 Isomorphic rooted trees. ◀

The isomorphism of Example 9.8.5 is not unique. Can you find another isomorphism of the rooted trees of Figure 9.8.7?

**Example 9.8.6** The rooted trees $T_1$ and $T_2$ of Figure 9.8.8 are not isomorphic since the root of $T_1$ has degree 3 but the root of $T_2$ has degree 2. These trees are isomorphic as *free* trees. Each is isomorphic to the tree $T_2$ of Figure 9.8.3.

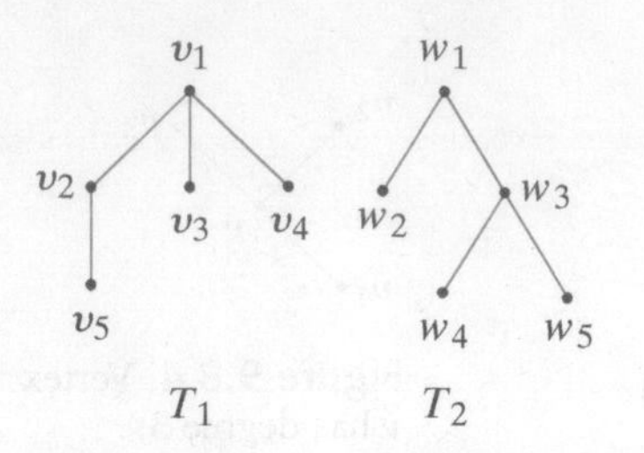


**Figure 9.8.8** Nonisomorphic rooted trees. (The trees are isomorphic as *free* trees.) ◀

Arguing as in the proof of Theorem 9.8.3, we can show that there are four nonisomorphic rooted trees with four vertices.

**Theorem 9.8.7** There are four nonisomorphic rooted trees with four vertices. These four rooted trees are shown in Figure 9.8.9.

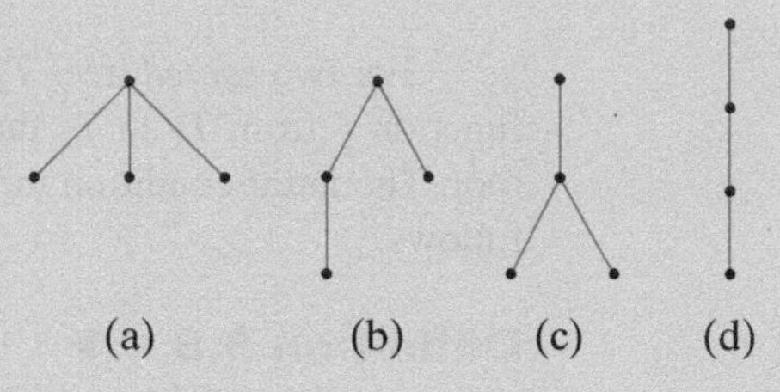


**Figure 9.8.9** The four nonisomorphic rooted trees with four vertices.

**Proof** We first find all nonisomorphic rooted trees with four vertices in which the root has degree 3; we then find all nonisomorphic rooted trees with four vertices in which the root has degree 2; and so on. We note that the root of a rooted tree with four vertices cannot have degree greater than 3.

A rooted tree with four vertices in which the root has degree 3 must be isomorphic to the tree in Figure 9.8.9(a).

A rooted tree with four vertices in which the root has degree 2 must be isomorphic to the tree in Figure 9.8.9(b).

Let $T$ be a rooted tree with four vertices in which the root has degree 1. Then the root is incident on one edge. The two remaining edges may be added in one of two ways [see Figure 9.8.9(c) and (d)]. Therefore, all nonisomorphic rooted trees with four vertices are shown in Figure 9.8.9. ◀

Binary trees are special kinds of rooted trees; thus an isomorphism of binary trees must preserve the adjacency relation and must preserve the roots. However, in binary trees a child is designated a left child or a right child. We require that an isomorphism of binary trees preserve the left and right children. The formal definition follows.

**Definition 9.8.8 ▶** Let $T_1$ be a binary tree with root $r_1$ and let $T_2$ be a binary tree with root $r_2$. The binary trees $T_1$ and $T_2$ are *isomorphic* if there is a one-to-one, onto function $f$ from the vertex set of $T_1$ to the vertex set of $T_2$ satisfying the following:

(a) Vertices $v_i$ and $v_j$ are adjacent in $T_1$ if and only if the vertices $f(v_i)$ and $f(v_j)$ are adjacent in $T_2$.

(b) $f(r_1) = r_2$.

(c) $v$ is a left child of $w$ in $T_1$ if and only if $f(v)$ is a left child of $f(w)$ in $T_2$.

(d) $v$ is a right child of $w$ in $T_1$ if and only if $f(v)$ is a right child of $f(w)$ in $T_2$.

We call the function $f$ an *isomorphism*.

**Example 9.8.9** The binary trees $T_1$ and $T_2$ in Figure 9.8.10 are isomorphic. The isomorphism is $f(v_i) = w_i$ for $i = 1, \ldots, 4$.

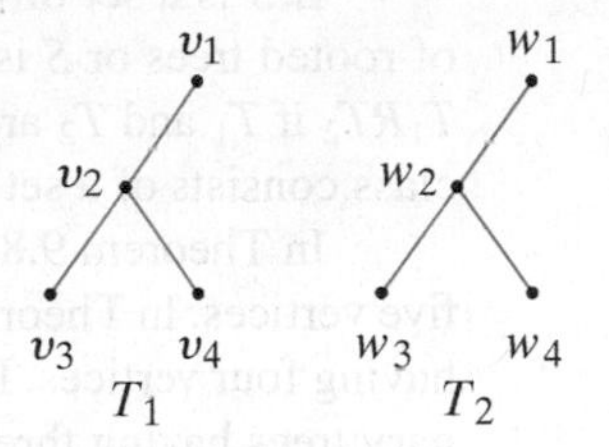


**Figure 9.8.10** Isomorphic binary trees.

**Example 9.8.10** The binary trees $T_1$ and $T_2$ in Figure 9.8.11 are not isomorphic. The root $v_1$ in $T_1$ has a right child, but the root $w_1$ in $T_2$ has no right child.

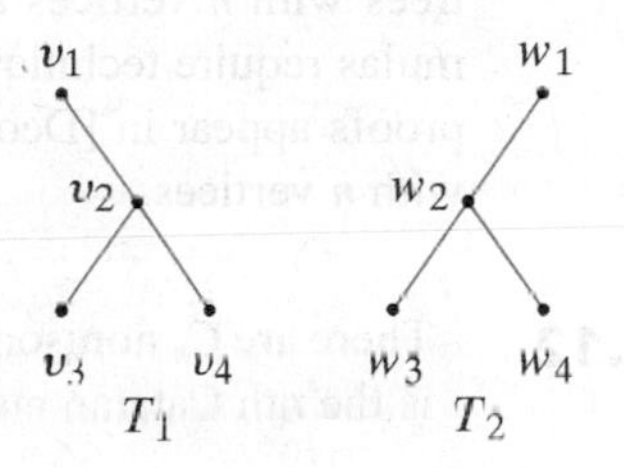


**Figure 9.8.11** Nonisomorphic binary trees. (The trees are isomorphic as *rooted* trees and as *free* trees.)

The trees $T_1$ and $T_2$ in Figure 9.8.11 *are* isomorphic as rooted trees and as free trees. As *rooted* trees, either of the trees of Figure 9.8.11 is isomorphic to the rooted tree $T$ of Figure 9.8.9(c).

Arguing as in the proofs of Theorems 9.8.3 and 9.8.7, we can show that there are five nonisomorphic binary trees with three vertices.

**Theorem 9.8.11** There are five nonisomorphic binary trees with three vertices. These five binary trees are shown in Figure 9.8.12.

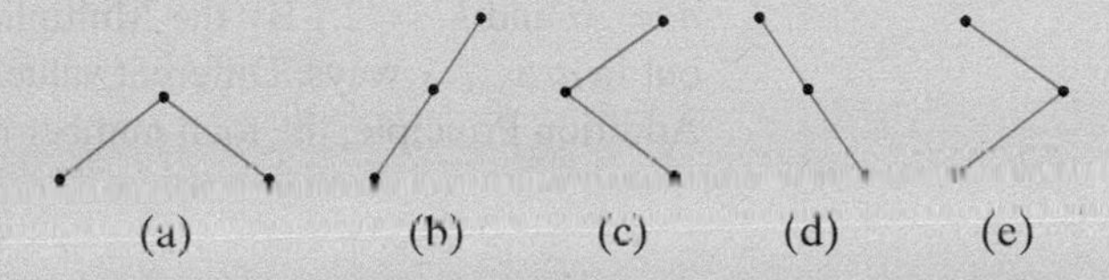


**Figure 9.8.12** The five nonisomorphic binary trees with three vertices.

**Proof** We first find all nonisomorphic binary trees with three vertices in which the root has degree 2. We then find all nonisomorphic binary trees with three vertices in which the root has degree 1. We note that the root of any binary tree cannot have degree greater than 2.

A binary tree with three vertices in which the root has degree 2 must be isomorphic to the tree in Figure 9.8.12(a). In a binary tree with three vertices in which the root has degree 1, the root either has a left child and no right child or has a right child and no left child. If the root has a left child, the child itself has either a left or a right child. We obtain the two binary trees in Figure 9.8.12(b) and (c). Similarly, if the root has a right child, the child itself has either a left or a right child. We obtain the two binary trees in Figure 9.8.12(d) and (e). Therefore, all nonisomorphic binary trees with three vertices are shown in Figure 9.8.12. ◀

If $S$ is a set of trees of a particular type (e.g., $S$ is a set of free trees or $S$ is a set of rooted trees or $S$ is a set of binary trees) and we define a relation $R$ on $S$ by the rule $T_1 R T_2$ if $T_1$ and $T_2$ are isomorphic, then $R$ is an equivalence relation. Each equivalence class consists of a set of mutually isomorphic trees.

In Theorem 9.8.3 we showed that there are three nonisomorphic free trees having five vertices. In Theorem 9.8.7 we showed that there are four nonisomorphic rooted trees having four vertices. In Theorem 9.8.11 we showed that there are five nonisomorphic binary trees having three vertices. You might have wondered if there are formulas for the number of nonisomorphic $n$-vertex trees of a particular type. There are formulas for the number of nonisomorphic $n$-vertex free trees, for the number of nonisomorphic $n$-vertex rooted trees, and for the number of nonisomorphic $n$-vertex binary trees. The formulas for the number of nonisomorphic free trees and for the number of nonisomorphic rooted trees with $n$ vertices are quite complicated. Furthermore, the derivations of these formulas require techniques beyond those that we develop in this book. The formulas and proofs appear in [Deo, Sec. 10-3]. We derive a formula for the number of binary trees with $n$ vertices.

**Theorem 9.8.12** There are $C_n$ nonisomorphic binary trees with $n$ vertices where $C_n = C(2n, n)/(n+1)$ is the $n$th Catalan number.

Figure 9.8.13 The two nonisomorphic binary trees with two vertices.

**Proof** Let $a_n$ denote the number of binary trees with $n$ vertices. For example, $a_0 = 1$ since there is one binary tree having no vertices; $a_1 = 1$ since there is one binary tree having one vertex; $a_2 = 2$ since there are two binary trees having two vertices (see Figure 9.8.13); and $a_3 = 5$ since there are five binary trees having three vertices (see Figure 9.8.12).

We derive a recurrence relation for the sequence $a_0, a_1, \ldots$. Consider the construction of a binary tree with $n$ vertices, $n > 0$. One vertex must be the root. Since there are $n - 1$ vertices remaining, if the left subtree has $k$ vertices, the right subtree must have $n - k - 1$ vertices. We construct an $n$-vertex binary tree whose left subtree has $k$ vertices and whose right subtree has $n - k - 1$ vertices by a two-step process: Construct the left subtree, construct the right subtree. (Figure 9.8.14 shows this construction for $n = 6$ and $k = 2$.) By the Multiplication Principle, this construction can be carried out in $a_k a_{n-k-1}$ ways. Different values of $k$ give distinct $n$-vertex binary trees, so by the Addition Principle, the total number of $n$-vertex binary trees is

$$\sum_{k=0}^{n-1} a_k a_{n-k-1}.$$

We obtain the recurrence relation

$$a_n = \sum_{k=0}^{n-1} a_k a_{n-k-1}, \qquad n \geq 1.$$

But this recurrence relation and initial condition $a_0 = 1$ define the sequence of Catalan numbers (see Examples 6.2.22 and 7.1.7). Thus $a_n$ is equal to the Catalan number $C(2n, n)/(n + 1)$. ◀

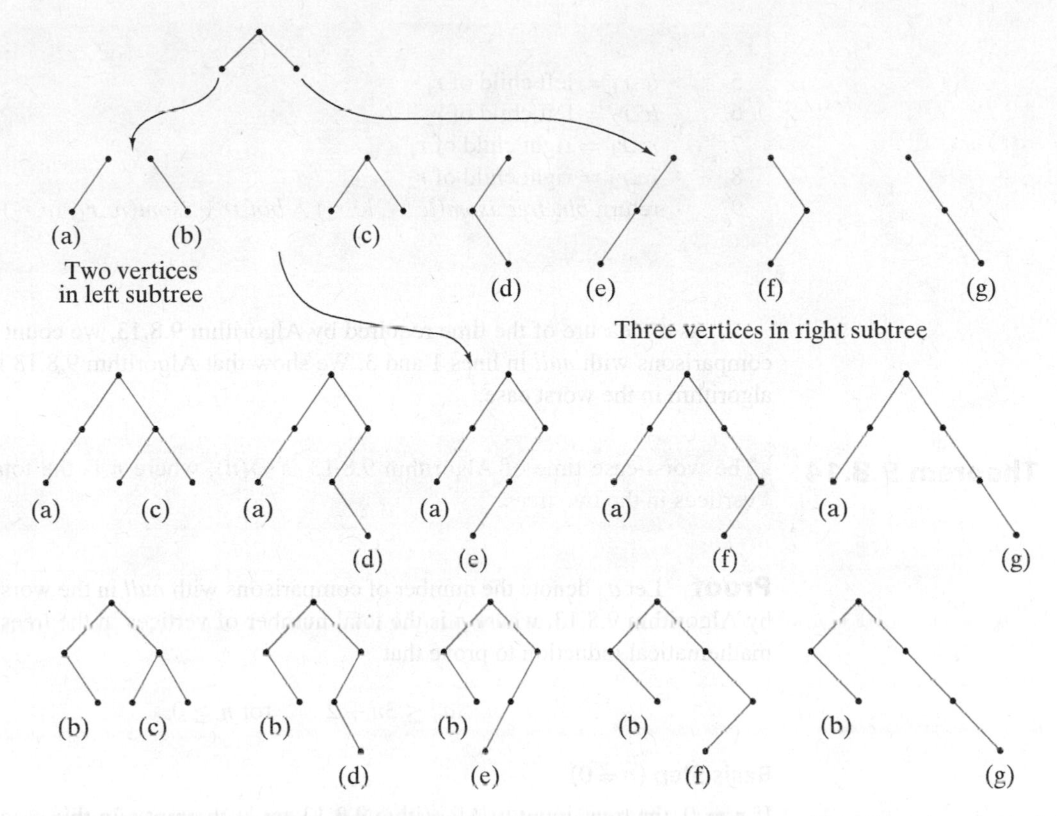


**Figure 9.8.14** The proof of Theorem 9.8.12 for the case $n = 6$ vertices and $k = 2$ vertices in the left subtree.

When discussing graph isomorphisms in Section 8.6, we remarked that there is no efficient method known to decide whether two arbitrary graphs are isomorphic. The situation is different for trees. It is possible to determine in polynomial time whether two arbitrary trees are isomorphic. As a special case, we give a linear-time algorithm to determine whether two binary trees $T_1$ and $T_2$ are isomorphic. The algorithm is based on preorder traversal (see Section 9.6). We first check that each of $T_1$ and $T_2$ is nonempty, after which we check that the left subtrees of $T_1$ and $T_2$ are isomorphic and that the right subtrees of $T_1$ and $T_2$ are isomorphic.

**Algorithm 9.8.13** **Testing Whether Two Binary Trees Are Isomorphic**

**Go Online**
For a C program implementing this algorithm, see `bit.ly/2PiaaxA`

Input: The roots $r_1$ and $r_2$ of two binary trees. (If the first tree is empty, $r_1$ has the special value *null*. If the second tree is empty, $r_2$ has the special value *null*.)

Output: true, if the trees are isomorphic
false, if the trees are not isomorphic

```
    bin_tree_isom(r1, r2) {
1.      if (r1 == null ∧ r2 == null)
2.          return true
        // now one or both of r1 or r2 is not null
3.      if (r1 == null ∨ r2 == null)
4.          return false
        // now neither of r1 or r2 is null
```

```
5.        lc_r1 = left child of r1
6.        lc_r2 = left child of r2
7.        rc_r1 = right child of r1
8.        rc_r2 = right child of r2
9.        return bin_tree_isom(lc_r1, lc_r2) ∧ bin_tree_isom(rc_r1, rc_r2)
     }
```

As a measure of the time required by Algorithm 9.8.13, we count the number of comparisons with *null* in lines 1 and 3. We show that Algorithm 9.8.13 is a linear-time algorithm in the worst case.

**Theorem 9.8.14** The worst-case time of Algorithm 9.8.13 is $\Theta(n)$, where $n$ is the total number of vertices in the two trees.

**Proof** Let $a_n$ denote the number of comparisons with *null* in the worst case required by Algorithm 9.8.13, where $n$ is the total number of vertices in the trees input. We use mathematical induction to prove that

$$a_n \leq 3n + 2 \qquad \text{for } n \geq 0.$$

**Basis Step ($n = 0$)**

If $n = 0$, the trees input to Algorithm 9.8.13 are both empty. In this case, there are two comparisons with *null* at line 1, after which the procedure returns. Thus $a_0 = 2$ and the inequality holds when $n = 0$.

**Inductive Step**

Assume that $a_k \leq 3k + 2$ when $k < n$. We must show that $a_n \leq 3n + 2$.

We first find an upper bound for the number of comparisons in the worst case when the total number of vertices in the trees input to the procedure is $n > 0$ and neither tree is empty. In this case, there are four comparisons at lines 1 and 3. Let $L$ denote the sum of the numbers of vertices in the two left subtrees of the trees input and let $R$ denote the sum of the numbers of vertices in the two right subtrees of the trees input. Then at line 9 there are at most $a_L + a_R$ additional comparisons. Therefore, at most $4 + a_L + a_R$ comparisons are required in the worst case. By the inductive assumption,

$$a_L \leq 3L + 2 \quad \text{and} \quad a_R \leq 3R + 2. \tag{9.8.1}$$

Now

$$2 + L + R = n \tag{9.8.2}$$

because the vertices comprise the two roots, the vertices in the left subtrees, and the vertices in the right subtrees. Combining (9.8.1) and (9.8.2), we obtain

$$4 + a_L + a_R \leq 4 + (3L + 2) + (3R + 2) = 3(2 + L + R) + 2 = 3n + 2.$$

If either tree is empty, four comparisons are required at lines 1 and 3, after which the procedure returns. Thus, whether one of the trees is empty or not, at most $3n + 2$ comparisons are required in the worst case. Therefore, $a_n \leq 3n + 2$, and the Inductive Step is complete. We conclude that the worst-case time of Algorithm 9.8.13 is $O(n)$.

If $n$ is even, say $n = 2k$, one can use induction to show (see Exercise 24) that when two $k$-vertex isomorphic binary trees are input to Algorithm 9.8.13, the number of comparisons is equal to $3n + 2$. Using this result, one can show (see Exercise 25) that if $n$ is odd, say $n = 2k + 1$, when the two binary trees shown in Figure 9.8.15 are input to Algorithm 9.8.13, the number of comparisons is equal to $3n + 1$. Thus the worst-case time of Algorithm 9.8.13 is $\Omega(n)$.

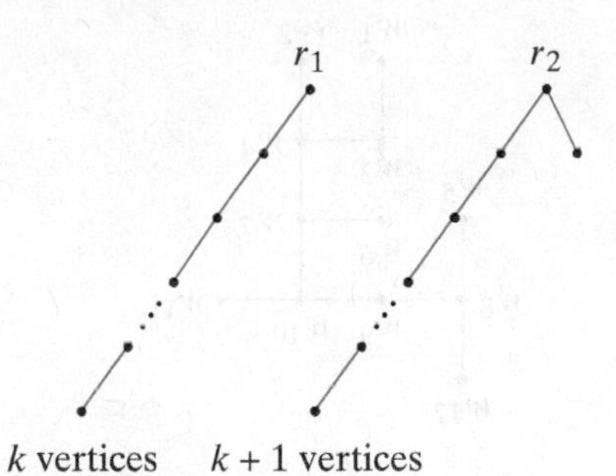


**Figure 9.8.15** Two binary trees that give worst-case run time $3n + 1$ for Algorithm 9.8.13 when $n = 2k + 1$ is odd.

Since the worst-case time is $O(n)$ and $\Omega(n)$, the worst-case time of Algorithm 9.8.13 is $\Theta(n)$. ◀

[Aho] gives an algorithm whose worst-case time is linear in the number of vertices that determines whether two arbitrary (not necessarily binary) rooted trees are isomorphic.

## 9.8 Review Exercises

1. What does it mean for two free trees to be isomorphic?
2. What does it mean for two rooted trees to be isomorphic?
3. What does it mean for two binary trees to be isomorphic?
4. How many $n$-vertex, nonisomorphic binary trees are there?
5. Describe a linear-time algorithm to test whether two binary trees are isomorphic.

## 9.8 Exercises

*In Exercises 1–6, determine whether each pair of free trees is isomorphic. If the pair is isomorphic, specify an isomorphism. If the pair is not isomorphic, give an invariant that one tree satisfies but the other does not.*

1.

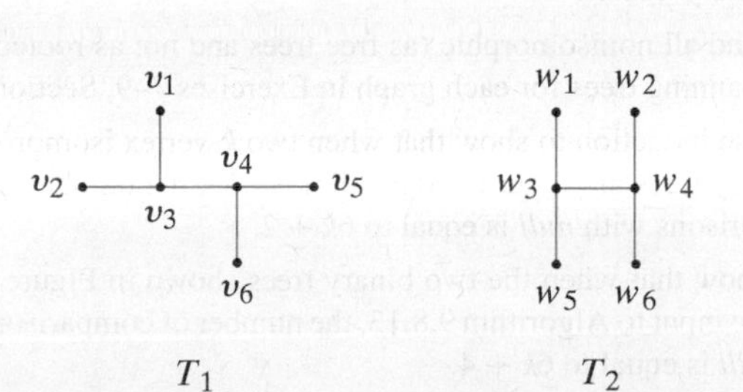


2.

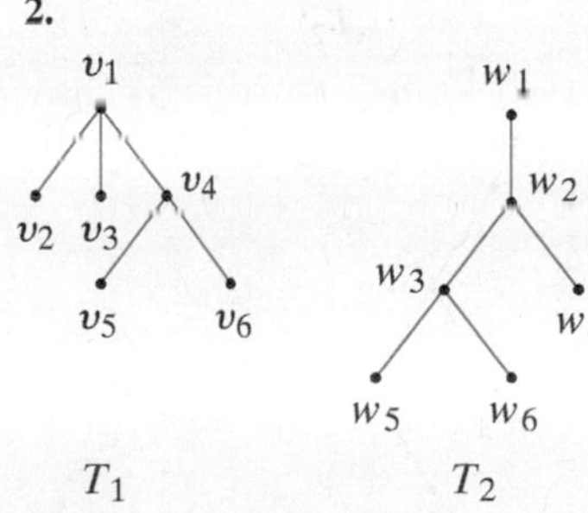


3. $T_1$ as in Exercise 1

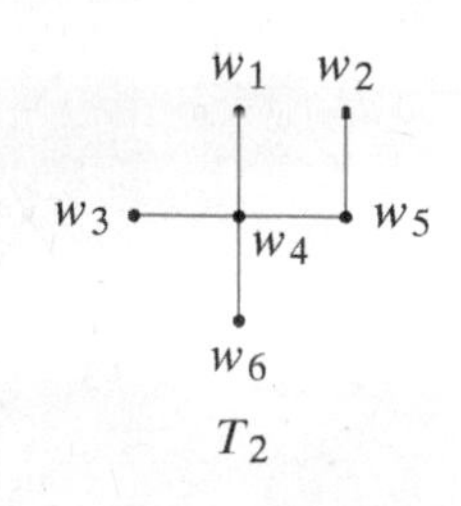

**4.**

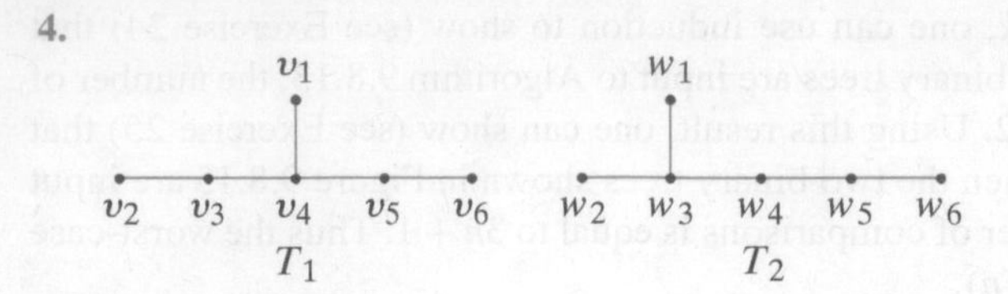


**5.**

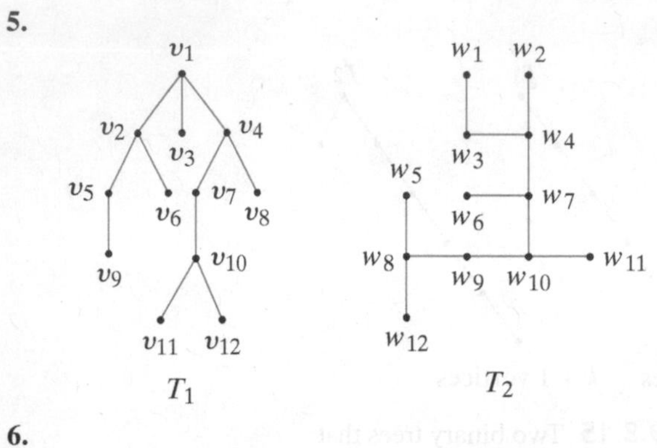


**6.**

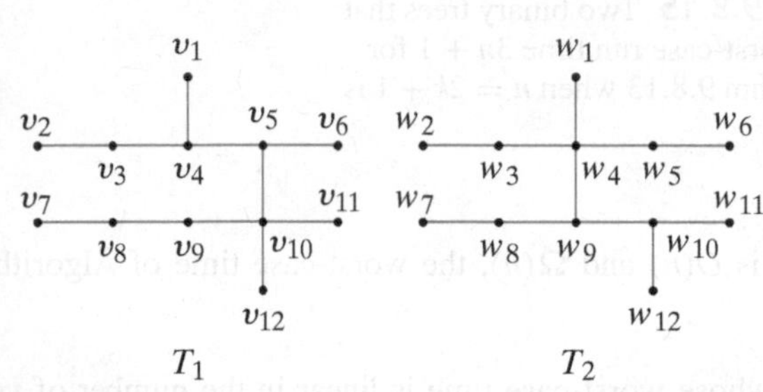


*In Exercises 7–9, determine whether each pair of rooted trees is isomorphic. If the pair is isomorphic, specify an isomorphism. If the pair is not isomorphic, give an invariant that one tree satisfies but the other does not. Also, determine whether the trees are isomorphic as free trees.*

**7.**

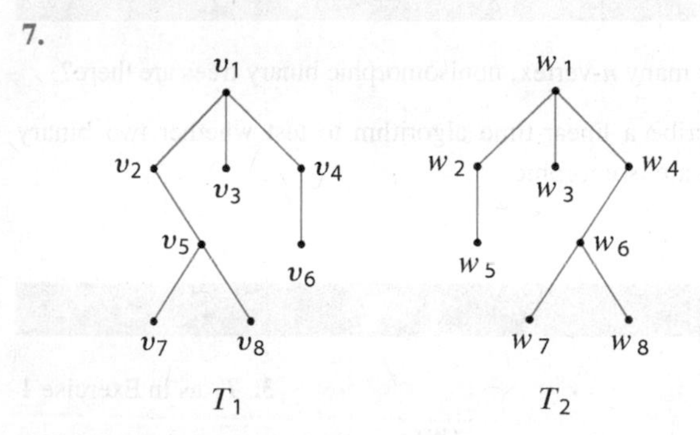


**8.**

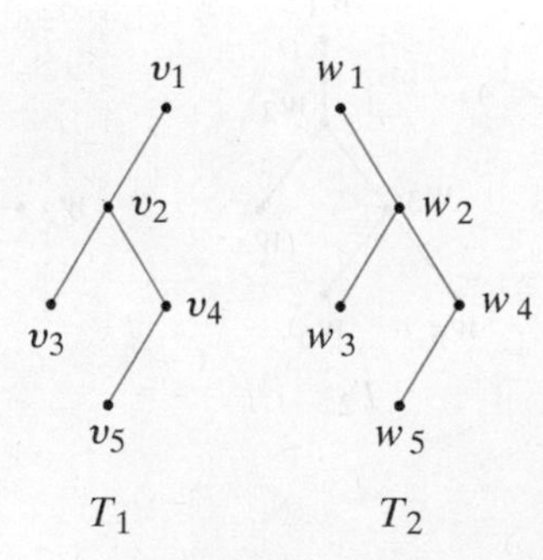


**9.** $T_1$ and $T_2$ as in Exercise 2

*In Exercises 10–12, determine whether each pair of binary trees is isomorphic. If the pair is isomorphic, specify an isomorphism. If the pair is not isomorphic, give an invariant that one tree satisfies but the other does not. Also, determine whether the trees are isomorphic as free trees or as rooted trees.*

**10.** $T_1$ and $T_2$ as in Exercise 8

**11.**

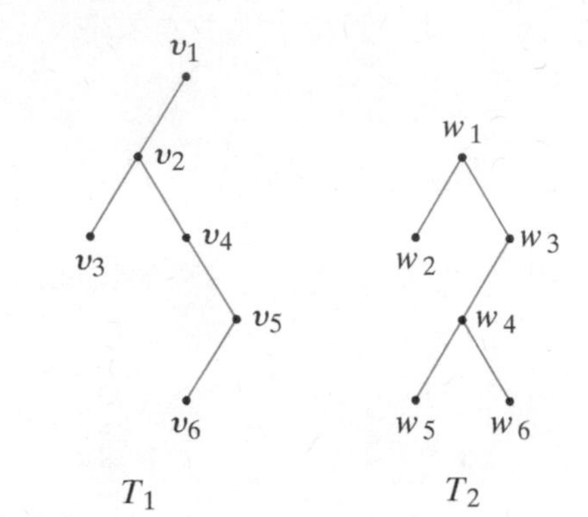


**12.**

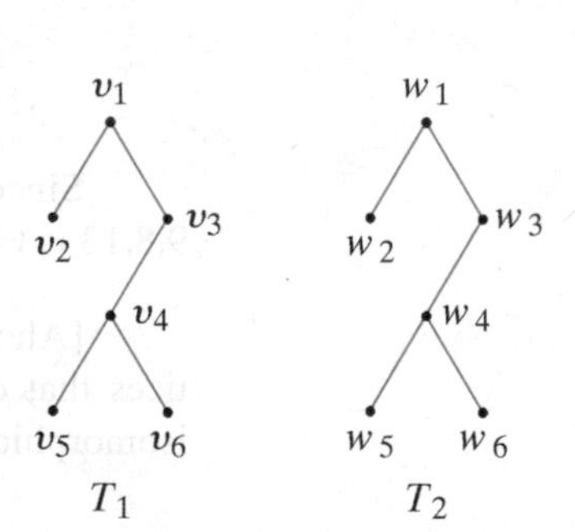


**13.** Draw all nonisomorphic free trees having three vertices.

**14.** Draw all nonisomorphic free trees having four vertices.

**15.** Draw all nonisomorphic free trees having six vertices.

**16.** Draw all nonisomorphic rooted trees having three vertices.

**17.** Draw all nonisomorphic rooted trees having five vertices.

**18.** Draw all nonisomorphic binary trees having two vertices.

**19.** Draw all nonisomorphic binary trees having four vertices.

**20.** Draw all nonisomorphic full binary trees having seven vertices. (A full binary tree is a binary tree in which each internal vertex has two children.)

**21.** Draw all nonisomorphic full binary trees having nine vertices.

**22.** Find a formula for the number of nonisomorphic $n$-vertex full binary trees.

**23.** Find all nonisomorphic (as free trees and not as rooted trees) spanning trees for each graph in Exercises 7–9, Section 9.3.

**24.** Use induction to show that when two $k$-vertex isomorphic binary trees are input to Algorithm 9.8.13, the number of comparisons with *null* is equal to $6k + 2$.

**25.** Show that when the two binary trees shown in Figure 9.8.15 are input to Algorithm 9.8.13, the number of comparisons with *null* is equal to $6k + 4$.

26. Write an algorithm to generate an $n$-vertex random binary tree.

*In Exercises 27–33, $C_1, C_2, \ldots$ denotes the sequence of Catalan numbers. Let $X_1$ denote the set of nonisomorphic full binary trees having n terminal vertices, $n \geq 2$, and let $X_2$ denote the set of nonisomorphic full binary trees having $n+1$ terminal vertices, $n \geq 1$, with one terminal vertex designated as "marked."*

27. Given an $(n-1)$-vertex binary tree $T$, $n \geq 2$, construct a binary tree from $T$ by adding a left child to every vertex in $T$ that does not have a left child, and adding a right child to every vertex in $T$ that does not have a right child. (A terminal vertex will add both a left and right child.) Show that this mapping is one-to-one and onto from the set of all nonisomorphic $(n-1)$-vertex binary trees to $X_1$. Conclude that $|X_1| = C_{n-1}$ for all $n \geq 2$.

28. Show that $|X_2| = (n+1)C_n$ for all $n \geq 1$.

*Given a tree $T \in X_1$, for each vertex v in T, we construct two trees in $X_2$ as follows. One tree in $X_2$ is obtained by inserting two new children of v—one is a new left child, which is marked, and the other is the root of the original subtree in T rooted at v. The other tree in $X_2$ is obtained by inserting two new children of v—one is a new right child, which is marked, and the other is the root of the original subtree in T rooted at v. Let $X_T$ denote the set of all such trees constructed. This construction is due to Ira Gessel and was forwarded to the author by Arthur Benjamin.*

29. Show that $|X_T| = 2(2n-1)$ for all $T \in X_1$.

30. Show that if $T_1$ and $T_2$ are distinct trees in $X_1$, then $X_{T_1} \cap X_{T_2} = \varnothing$.

31. Show that $\bigcup_{T \in X_1} X_T = X_2$.

32. Use Exercises 29–31 to show that $(n+1)C_n = 2(2n-1)C_{n-1}$ for all $n \geq 2$. Exercise 26, Section 7.1, asked for a proof of this identity using the explicit formula for $C_n$. These exercises show a way to prove the identity without using the explicit formula for $C_n$.

33. Use Exercise 32 to give another derivation of the explicit formula for the $n$th Catalan number $C_n = C(2n, n)/(n+1)$. (See also Example 6.2.22.)

34. An *ordered tree* is a tree in which the order of the children is taken into account. For example, the ordered trees

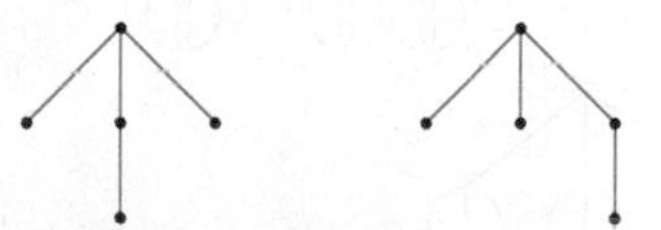

are *not* isomorphic. Show that the number of nonisomorphic ordered trees with $n$ edges is equal to $C_n$, the $n$th Catalan number. *Hint:* Consider a preorder traversal of an ordered tree in which 1 means down and 0 means up.

35. [*Project*] Report on the formulas for the number of nonisomorphic free trees and for the number of nonisomorphic rooted trees with $n$ vertices (see [Deo]).

## 9.9 Game Trees†

**Go Online**
For more on game trees, see
bit.ly/2q58dGw

Trees are useful in the analysis of games such as tic-tac-toe, chess, and checkers, in which players alternate moves. In this section we show how trees can be used to develop game-playing strategies. This kind of approach is used in the development of many computer programs that allow human beings to play against computers or even computers against computers.

As an example of the general approach, consider a version of the game of nim. Initially, there are $n$ piles, each containing a number of identical tokens. Players alternate moves. A move consists of removing one or more tokens from any one pile. The player who removes the last token loses. As a specific case, consider an initial distribution consisting of two piles: one containing three tokens and one containing two tokens. All possible move sequences can be listed in a **game tree** (see Figure 9.9.1). The first player is represented by a box and the second player is represented by a circle. Each vertex shows a particular position in the game. In our game, the initial position is shown as $\binom{3}{2}$. A path represents a sequence of moves. If a position is shown in a square, it is the first player's move; if a position is shown in a circle, it is the second player's move. A terminal vertex represents the end of the game. In nim, if the terminal vertex is a circle, the first player removed the last token and lost the game. If the terminal vertex is a box, the second player lost.

The analysis begins with the terminal vertices. We label each terminal vertex with the value of the position to the first player. If the terminal vertex is a circle, since the first player lost, this position is worthless to the first player and we assign it the value 0 (see Figure 9.9.2). If the terminal vertex is a box, since the first player won, this position

†This section can be omitted without loss of continuity.

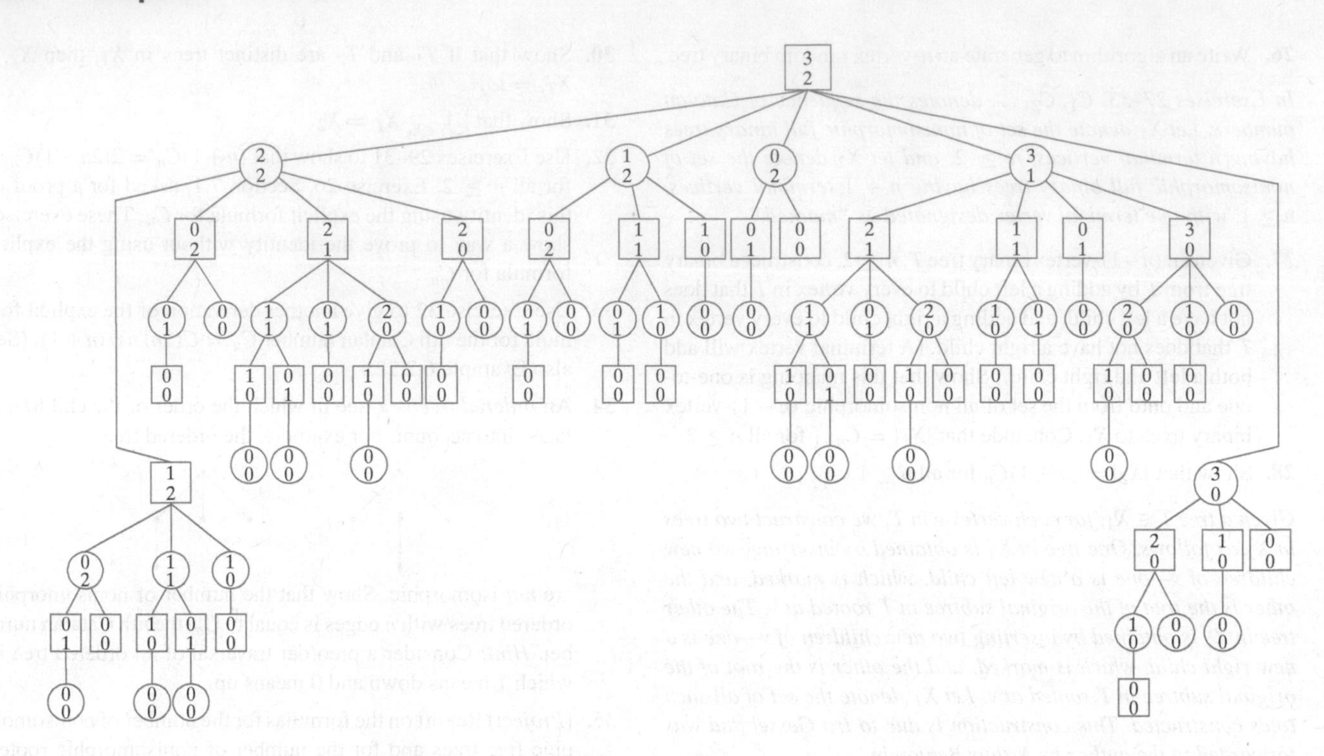

**Figure 9.9.1** A game tree for nim. The initial distribution is two piles of three and two tokens, respectively.

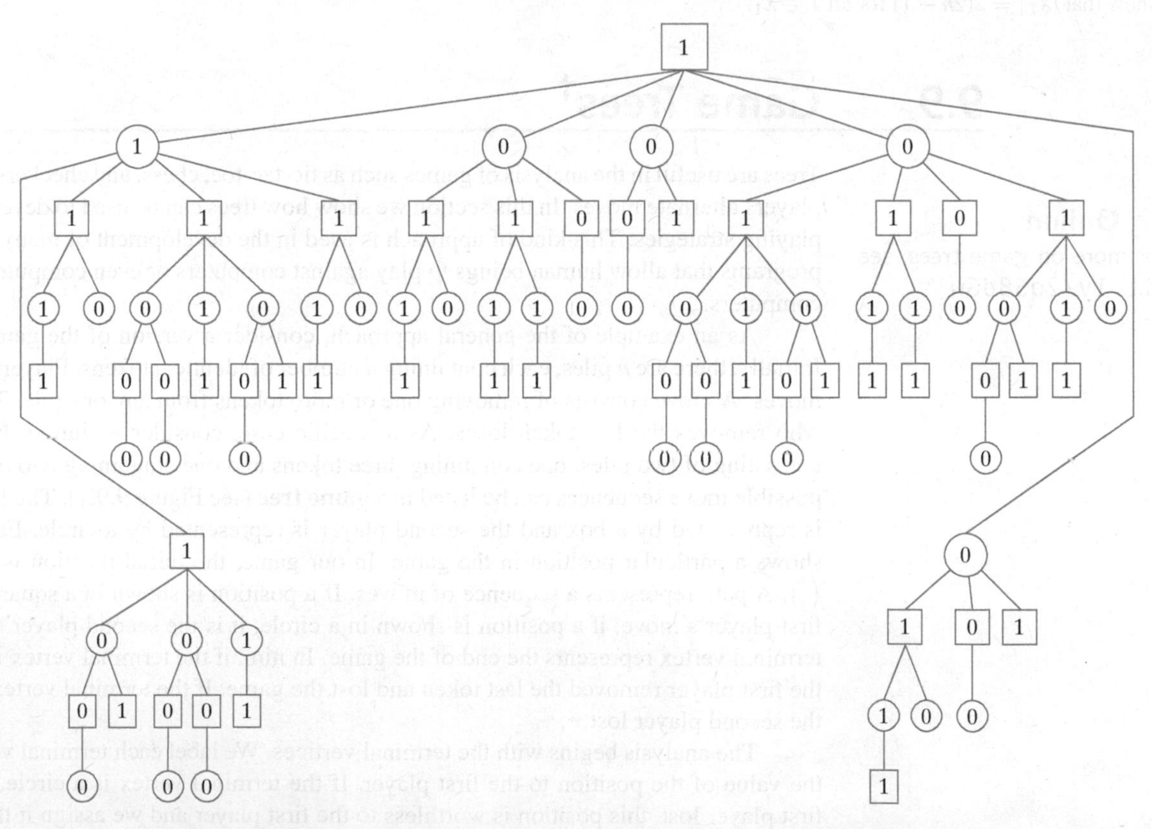

**Figure 9.9.2** The game tree of Figure 9.9.1 showing the values of all vertices.

is valuable to the first player and we label it with a value greater than 0, say 1 (see Figure 9.9.2). At this point, all terminal vertices have been assigned values.

Now, consider the problem of assigning values to the internal vertices. Suppose, for example, that we have an internal box, all of whose children have been assigned a value. For example, if we have the situation shown in Figure 9.9.3, the first player (box) should move to the position represented by vertex $B$, since this position is the most valuable. In other words, box moves to a position represented by a child with the maximum value. We assign this maximum value to the box vertex.

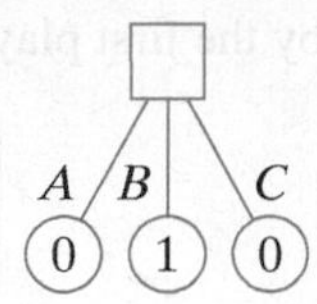


**Figure 9.9.3** The first player (box) should move to position $B$ since it is most valuable. This maximum value (1) is assigned to the box.

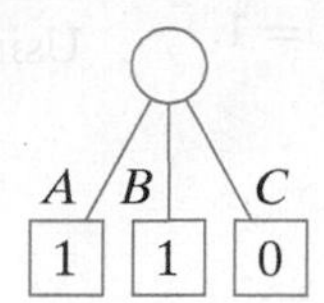


**Figure 9.9.4** The second player (circle) should move to position $C$ since it is least valuable (to box). This minimum value (0) is assigned to the circle.

Consider the situation from the second (circle) player's point of view. Suppose that we have the situation shown in Figure 9.9.4. Circle should move to the position represented by vertex $C$, since this position is least valuable to box and therefore most valuable to circle. In other words, circle moves to a position represented by a child with the minimum value. We assign this minimum value to the circle vertex. The process by which circle seeks the minimum of its children and box seeks the maximum of its children is called the **minimax procedure.**

Working upward from the terminal vertices and using the minimax procedure, we can assign values to all of the vertices in the game tree (see Figure 9.9.2). These numbers represent the value of the game, at any position, to the first player. Notice that the root in Figure 9.9.2, which represents the original position, has a value of 1. This means that the first player can always win the game by using an optimal strategy. This optimal strategy is contained in the game tree: The first player always moves to a position that maximizes the value of the children. No matter what the second player does, the first player can always move to a vertex having value 1. Ultimately, a terminal vertex having value 1 is reached where the first player wins the game.

Many interesting games, such as chess, have game trees so large that it is not feasible to use a computer to generate the entire tree. Nevertheless, the concept of a game tree is still useful for analyzing such games.

When using a game tree we should use a depth-first search. If the game tree is so large that it is not feasible to reach a terminal vertex, we limit the level to which depth-first search is carried out. The search is said to be an ***n*-level search** if we limit the search to $n$ levels below the given vertex. Since the vertices at the lowest level may not be terminal vertices, some method must be found to assign them a value. This is where the specifics of the game must be dealt with. An **evaluation function** $E$ is constructed that assigns each possible game position $P$ the value $E(P)$ of the position to the first player. After the vertices at the lowest level are assigned values by using the function $E$, the minimax procedure can be applied to generate the values of the other vertices. We illustrate these concepts with an example.

**Example 9.9.1** Apply the minimax procedure to find the value of the root in tic-tac-toe using a two-level, depth-first minimax search. Use the evaluation function $E$, which assigns a position the value $NX - NO$ where $NX$ (respectively, $NO$) is the number of rows, columns, or

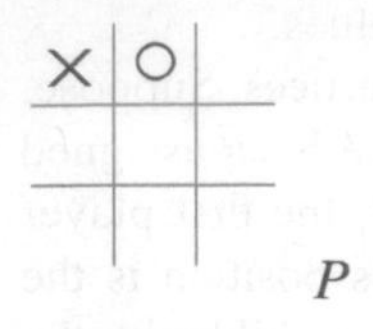


**Figure 9.9.5** The value of position $P$ is $E(P) = NX - NO = 2 - 1 = 1$.

diagonals containing an X (respectively, O) that X (respectively, O) might complete. For example, position $P$ of Figure 9.9.5 has $NX = 2$, since X might complete the column or the diagonal, and $NO = 1$, since O can complete only a column. Therefore, $E(P) = 2 - 1 = 1$.

**SOLUTION** In Figure 9.9.6, we have drawn the game tree for tic-tac-toe to level 2. We have omitted symmetric positions. We first assign to the vertices at level 2 the values given by $E$ (see Figure 9.9.7). Next, we compute circle's values by minimizing over the children. Finally, we compute the value of the root by maximizing over the children. Using this analysis, the first move by the first player would be to the center square.

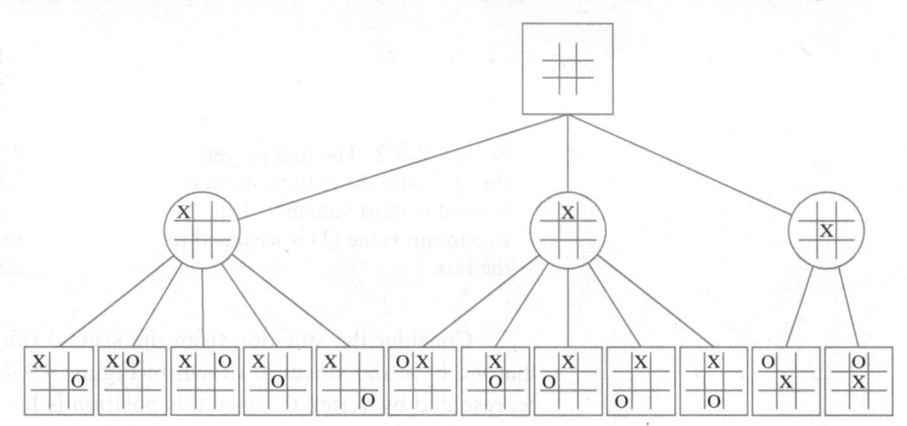

**Figure 9.9.6** The game tree for tic-tac-toe to level 2 with symmetric positions omitted.

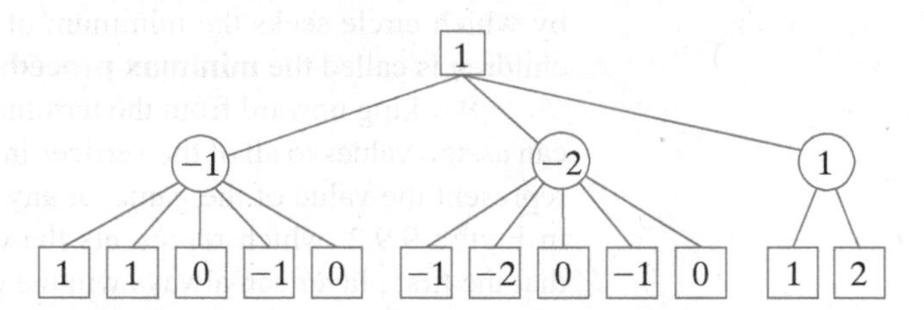


**Figure 9.9.7** The game tree of Figure 9.9.6 with the values of all vertices shown.

◀

Evaluation of a game tree, or even a part of a game tree, can be a time-consuming task, so any technique that reduces the effort is welcomed. The most general technique is called **alpha-beta pruning.** In general, alpha-beta pruning allows us to bypass many vertices in a game tree yet still find the value of a vertex. The value obtained is the same as if we had evaluated all the vertices.

As an example, consider the game tree in Figure 9.9.8. Suppose that we want to evaluate vertex $A$ using a two-level, depth-first search. We evaluate children left to right. We begin at the lower left by evaluating the vertices $E$, $F$, and $G$. The values shown are obtained from an evaluation function. Vertex $B$ is 2, the minimum of its children. At this point, we know that the value $x$ of $A$ must be at least 2, since the value of $A$ is the maximum of its children; that is,

$$x \geq 2. \tag{9.9.1}$$

This lower bound for $A$ is called an **alpha value** of $A$. The next vertices to be evaluated are $H$, $I$, and $J$. When $I$ evaluates to 1, we know that the value $y$ of $C$ cannot exceed 1, since the value of $C$ is the minimum of its children; that is

$$y \leq 1. \tag{9.9.2}$$

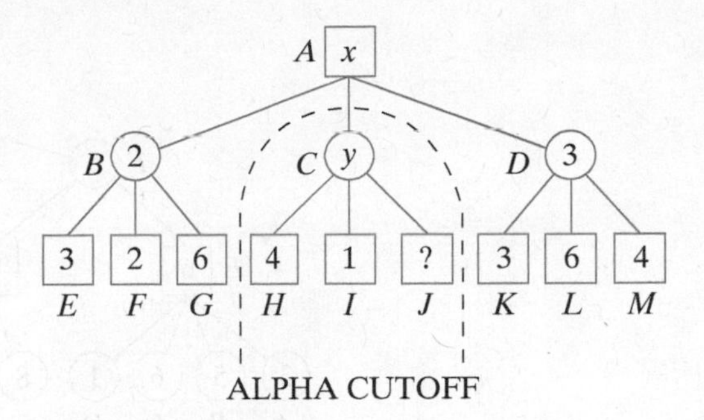


**Figure 9.9.8** Evaluating vertex $A$ using a two-level, depth-first search with alpha-beta pruning. An alpha cutoff occurs at vertex $C$ when vertex $I$ is evaluated since $I$'s value (1) is less than or equal to the current lower-bound estimate (2) for vertex $A$.

It follows from (9.9.1) and (9.9.2) that whatever the value of $y$ is, it will not affect the value of $x$; thus we need not concern ourselves further with the subtree rooted at vertex $C$. We say that an **alpha cutoff** occurs. We next evaluate the children of $D$ and then $D$ itself. Finally, we find that the value of $A$ is 3.

To summarize, an alpha cutoff occurs at a box vertex $v$ when a grandchild $w$ of $v$ has a value less than or equal to the alpha value of $v$. The subtree whose root is the parent of $w$ may be deleted (pruned). This deletion will not affect the value of $v$. An alpha value for a vertex $v$ is only a lower bound for the value of $v$. The alpha value of a vertex is dependent on the current state of the search and changes as the search progresses.

Similarly, a **beta value** of a circle vertex is an upper bound for $v$. A **beta cutoff** occurs at a circle vertex $v$ when a grandchild $w$ of $v$ has a value greater than or equal to the beta value of $v$. The subtree whose root is the parent of $w$ may be pruned. This deletion will not affect the value of $v$. A beta value for a vertex $v$ is only an upper bound for the value of $v$. The beta value of a vertex is dependent on the current state of the search and changes as the search progresses.

**Example 9.9.2** Evaluate the root of the tree of Figure 9.9.9 using depth-first search with alpha-beta pruning. Assume that children are evaluated left to right. For each vertex whose value is computed, write the value in the vertex. Place a check by the root of each subtree that is pruned. The value of each terminal vertex is written under the vertex.

**SOLUTION** We begin by evaluating vertices $A$, $B$, $C$, and $D$ (see Figure 9.9.10). Next, we find that the value of $E$ is 6. This results in a beta value of 6 for $F$. Next, we evaluate

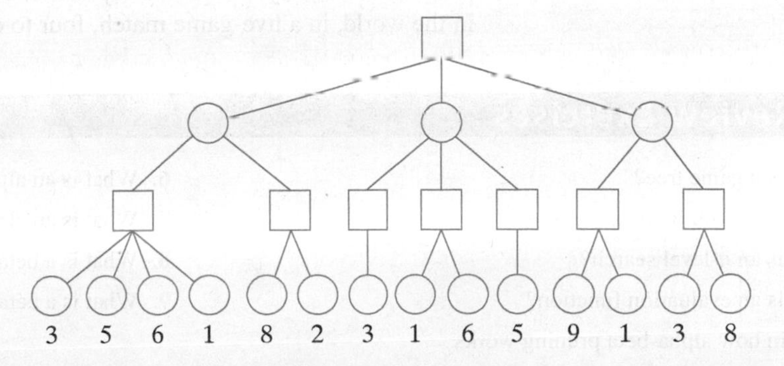


**Figure 9.9.9** The game tree for Example 9.9.2.

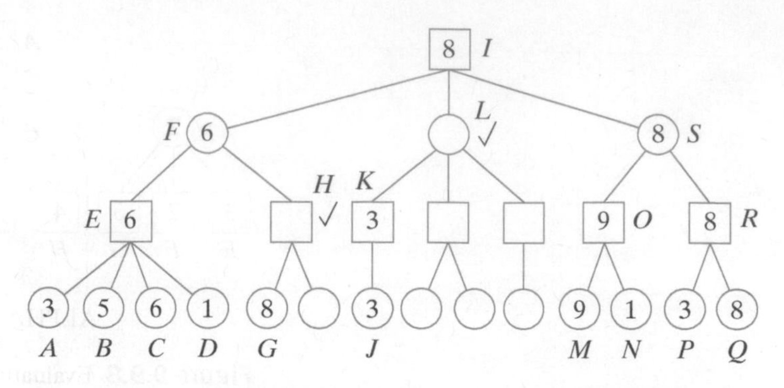


Figure 9.9.10 Evaluating the root of the game tree of Figure 9.9.9 using depth-first search with alpha-beta pruning. Checked vertices are roots of subtrees that are pruned. The values of vertices that are evaluated are written inside the vertices.

vertex $G$. Since its value is 8 and 8 exceeds the beta value of $F$, we obtain a beta cutoff and prune the subtree with root $H$. The value of $F$ is 6. This results in an alpha value of 6 for $I$. Next, we evaluate vertices $J$ and $K$. Since the value 3 of $K$ is less than the alpha value 6 of $I$, an alpha cutoff occurs and the subtree with root $L$ may be pruned. Next, we evaluate $M$, $N$, $O$, $P$, $Q$, $R$, and $S$. No further pruning is possible. Finally, we determine that the root $I$ has value 8. ◀

It has been shown (see [Pearl]) that for game trees in which every parent has $n$ children and in which the terminal values are randomly ordered, for a given amount of time, the alpha-beta procedure permits a search depth $4/3$ greater than the pure minimax procedure, which evaluates every vertex. [Pearl] also shows that for such game trees, the alpha-beta procedure is optimal.

Other techniques have been combined with alpha-beta pruning to facilitate the search of a game tree. One idea is to order the children of the vertices to be evaluated so that the most promising moves are examined first (see Exercises 23–26). Another idea is to allow a variable-depth search in which the search backtracks when it reaches an unpromising position as measured by some function.

**Go Online**
The Chinook site is at
`bit.ly/2q58dGw`

Some game-playing programs have been incredibly successful. The best chess, Go, backgammon, and checkers programs play at a level comparable to the best human players. The world checkers champion is a program named Chinook developed by a team from the University of Alberta. In 2007, they proved that Chinook can never lose. The best an opponent can hope for is a draw. In 1997 the IBM chess program, Deep Blue, defeated Garry Kasparov, who had been world champion since 1985, in a six-game match. Deep Blue won two games, drew three, and lost one. In 2016, AlphaGo, a program created by Google's Deep Mind, defeated Lee Sedol, one of the best Go players in the world, in a five-game match, four to one.

## 9.9 Review Exercises

1. What is a game tree?
2. What is the minimax procedure?
3. What is an $n$-level search?
4. What is an evaluation function?
5. Explain how alpha-beta pruning works.
6. What is an alpha value?
7. What is an alpha cutoff?
8. What is a beta value?
9. What is a beta cutoff?

## 9.9 Exercises

1. Draw the complete game tree for a version of nim in which the initial position consists of one pile of six tokens and a turn consists of taking one, two, or three tokens. Assign values to all vertices so that the resulting tree is analogous to Figure 9.9.2. Assume that the last player to take a token loses. Will the first or second player, playing an optimal strategy, always win? Describe an optimal strategy for the winning player.

2. Draw the complete game tree for nim in which the initial position consists of two piles of three tokens each. Omit symmetric positions. Assume that the last player to take a token loses. Assign values to all vertices so that the resulting tree is analogous to Figure 9.9.2. Will the first or second player, playing an optimal strategy, always win? Describe an optimal strategy for the winning player.

3. Draw the complete game tree for nim in which the initial position consists of two piles, one containing three tokens and the other containing two tokens. Assume that the last player to take a token wins. Assign values to all vertices so that the resulting tree is analogous to Figure 9.9.2. Will the first or second player, playing an optimal strategy, always win? Describe an optimal strategy for the winning player.

4. Draw the complete game tree for nim in which the initial position consists of two piles of three tokens each. Omit symmetric positions. Assume that the last player to take a token wins. Assign values to all vertices so that the resulting tree is analogous to Figure 9.9.2. Will the first or second player, playing an optimal strategy, always win? Describe an optimal strategy for the winning player.

5. Draw the complete game tree for the version of nim described in Exercise 1. Assume that the last person to take a token wins. Assign values to all vertices so that the resulting tree is analogous to Figure 9.9.2. Will the first or second player, playing an optimal strategy, always win? Describe an optimal strategy for the winning player.

6. Give an example of a (possibly hypothetical) complete game tree in which a terminal vertex is 1 if the first player won and 0 if the first player lost having the following properties: There are more 0's than 1's among the terminal vertices, but the first player can always win by playing an optimal strategy.

*Exercises 7 and 8 refer to nim and nim′. Nim is the game using n piles of tokens as described in this section in which the last player to move loses. Nim′ is the game using n piles of tokens as described in this section except that the last player to move wins. We fix n piles each with a fixed number of tokens. We assume that at least one pile has at least two tokens.*

★7. Show that the first player can always win nim if and only if the first player can always win nim′.

★8. Given a winning strategy for a particular player for nim, describe a winning strategy for this player for nim′.

*Evaluate each vertex in each game tree. The values of the terminal vertices are given.*

9.

6 12 2 8 6 10 16 1 7 20

10.

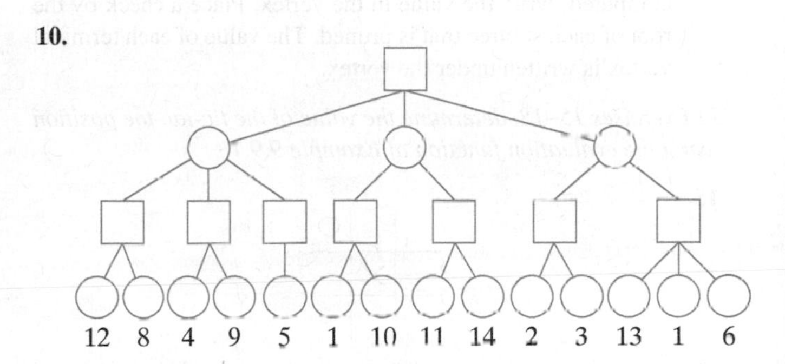


11.

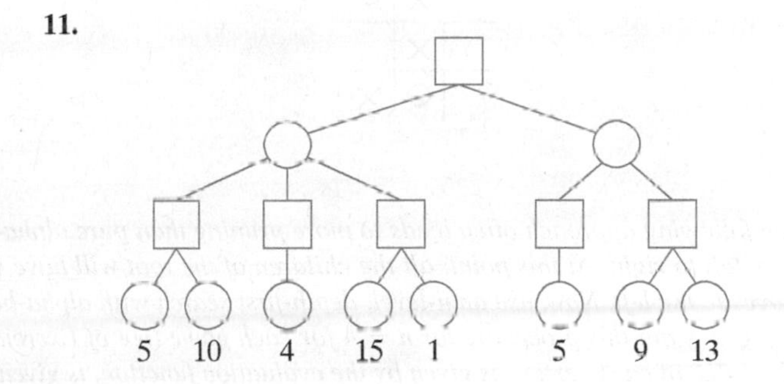


12.

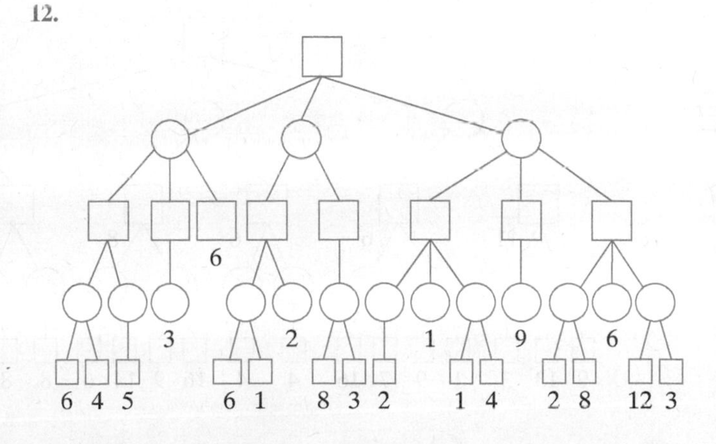

**13.**

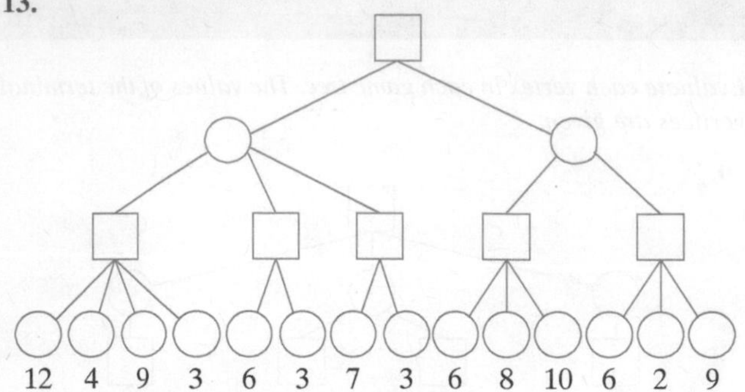


**14.** Evaluate the root of each of the trees of Exercises 9–13 using a depth-first search with alpha-beta pruning. Assume that children are evaluated left to right. For each vertex whose value is computed, write the value in the vertex. Place a check by the root of each subtree that is pruned. The value of each terminal vertex is written under the vertex.

*In Exercises 15–18, determine the value of the tic-tac-toe position using the evaluation function of Example 9.9.1.*

**15.**

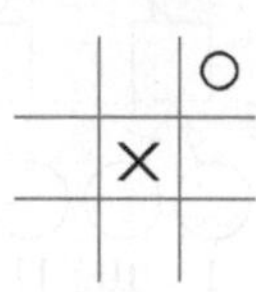

**16.**

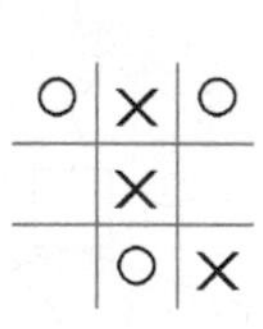

**17.**

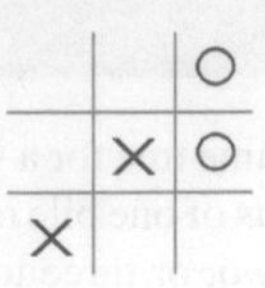

**18.**

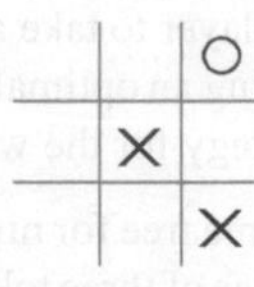

**19.** Assume that the first player moves to the center square in tic-tac-toe. Draw a two-level game tree, with the root having an X in the center square. Omit symmetric positions. Evaluate all the vertices using the evaluation function of Example 9.9.1. Where will O move?

★**20.** Would a two-level search program based on the evaluation function $E$ of Example 9.9.1 play a perfect game of tic-tac-toe? If not, can you alter $E$ so that a two-level search program will play a perfect game of tic-tac-toe?

**21.** Write an algorithm that evaluates vertices of a game tree to level $n$ using depth-first search. Assume the existence of an evaluation function $E$.

★**22.** Write an algorithm that evaluates the root of a game tree using an $n$-level, depth-first search with alpha-beta pruning. Assume the existence of an evaluation function $E$.

*The following approach often leads to more pruning than pure alpha-beta minimax. First, perform a two-level search. Evaluate children from left to right. At this point, all the children of the root will have values. Next, order the children of the root with the most promising moves to the left. Now, use an n-level, depth-first search with alpha-beta pruning. Evaluate children from left to right.*

*Carry out this procedure for $n = 4$ for each game tree of Exercises 23–25. Place a check by the root of each subtree that is pruned. The value of each vertex, as given by the evaluation function, is given under the vertex.*

**23.**

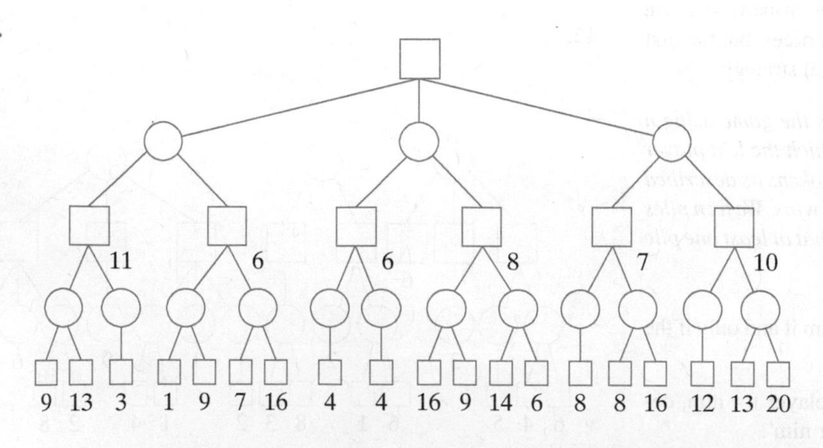

**24.**

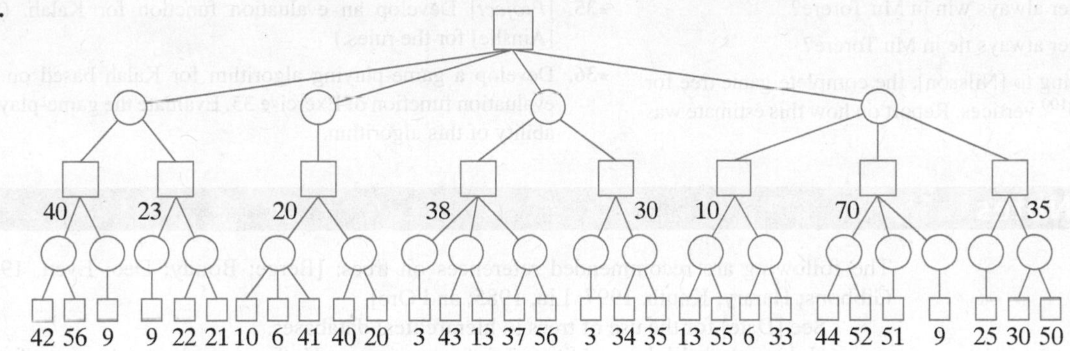


**25.**

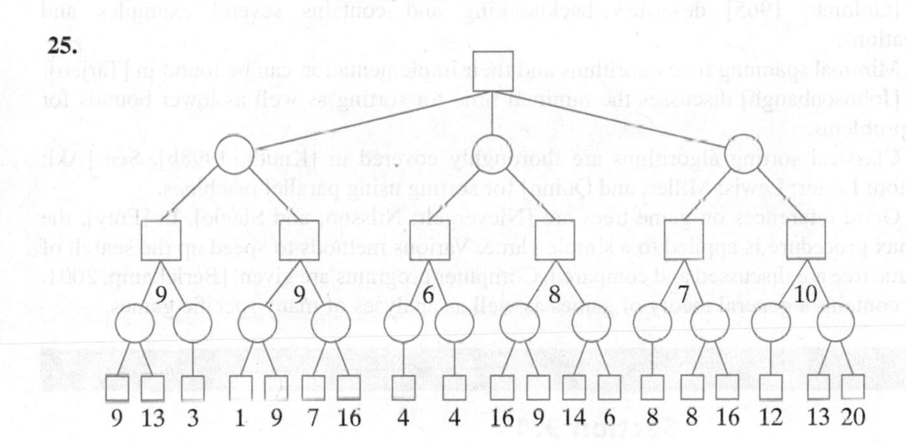


**26.** Write an algorithm to carry out the procedure described before Exercise 23.

*Consider the following simplified version of Jai Alai (see [Moser]) in which there are n players numbered 1, 2, ..., n. When the game begins, the players are aligned in order 1, ..., n. The game begins with the first player (number 1) playing the next player (number 2). When a player wins, that player plays the next player in line. (So if player 1 beats player 2, player 1 next plays player 3.) A losing player goes to the end of the line. (So, again, if player 1 beats player 2, player 2 goes to the end of the line and the line becomes 1, 3, 4, ..., n, 2.) One point is given to each winner in the first $n - 1$ rounds, and two points are given to each winner in the subsequent rounds. The first player to score at least $n - 1$ points wins the game. An extensive analysis of Jai Alai is given in [Skiena].*

**27.** Draw the complete game tree for $n = 3$.

**28.** Draw the complete game tree for $n = 4$.

**★29.** Prove that the height of the game tree is $n(n-1)/2$.

*Mu Torere is a two-person game played by the Maoris (see [Bell]). The board is an eight-pointed star with a circular area in the center known as the putahi.*

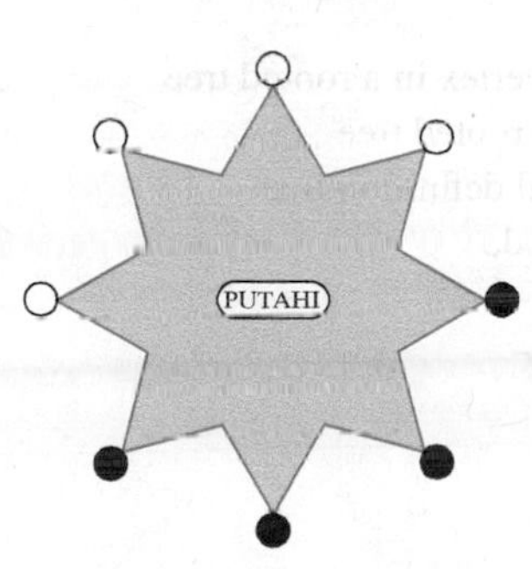


*The first player has four black tokens and the second player has four white tokens. The initial position is shown. A player who cannot make a move loses. Players alternate moves. At most one token can occupy a point of the star or the putahi. A move consists of*

*(a) Moving to an adjacent point*

*(b) Moving from the putahi to a point*

*(c) Moving from a point to the putahi provided that one or both of the adjacent points contain the opponent's pieces*

**★30.** Develop an evaluation function for Mu Torere.

**★31.** Combine the evaluation function in Exercise 30 with a two-level search of the game tree to obtain a game-playing algorithm for Mu Torere. Evaluate the game-playing ability of this algorithm.

★**32.** Can the first player always win in Mu Torere?

★**33.** Can the first player always tie in Mu Torere?

**34.** [*Project*] According to [Nilsson], the complete game tree for chess has over $10^{100}$ vertices. Report on how this estimate was obtained.

★**35.** [*Project*] Develop an evaluation function for Kalah. (See [Ainslie] for the rules.)

★**36.** Develop a game-playing algorithm for Kalah based on the evaluation function of Exercise 35. Evaluate the game-playing ability of this algorithm.

## Chapter 9 Notes

The following are recommended references on trees: [Berge; Bondy; Deo; Even, 1979; Gibbons; Harary; Knuth, 1997; Liu, 1985; and Ore].

See [Date] for the use of trees in hierarchical databases.

[Johnsonbaugh] has additional information on Huffman codes and a proof that Algorithm 9.1.9 constructs an optimal Huffman tree.

[Golomb, 1965] describes backtracking and contains several examples and applications.

Minimal spanning tree algorithms and their implementation can be found in [Tarjan].

[Johnsonbaugh] discusses the minimal time for sorting as well as lower bounds for other problems.

Classical sorting algorithms are thoroughly covered in [Knuth, 1998b]. See [Akl; Leighton; Lester; Lewis; Miller; and Quinn] for sorting using parallel machines.

Good references on game trees are [Nievergelt; Nilsson; and Slagle]. In [Frey], the minimax procedure is applied to a simple game. Various methods to speed up the search of the game tree are discussed and compared. Computer programs are given. [Berlekamp, 2001, 2003] contains a general theory of games as well as analyses of many specific games.

## Chapter 9 Review

### Section 9.1

1. Free tree
2. Rooted tree
3. Level of a vertex in a rooted tree
4. Height of a rooted tree
5. Hierarchical definition tree
6. Huffman code

### Section 9.2

7. Parent
8. Ancestor
9. Child
10. Descendant
11. Sibling
12. Terminal vertex
13. Internal vertex
14. Subtree
15. Acyclic graph
16. Alternative characterizations of trees (Theorem 9.2.3)

### Section 9.3

17. Spanning tree
18. A graph has a spanning tree if and only if it is connected.
19. Breadth-first search
20. Depth-first search
21. Backtracking

### Section 9.4

22. Minimal spanning tree
23. Prim's Algorithm to find a minimal spanning tree
24. Greedy algorithm

### Section 9.5

25. Binary tree
26. Left child in a binary tree
27. Right child in a binary tree
28. Full binary tree
29. If $T$ is full binary tree with $i$ internal vertices, then $T$ has $i + 1$ terminal vertices and $2i + 1$ total vertices.
30. If a binary tree of height $h$ has $t$ terminal vertices, then $\lg t \leq h$.
31. Binary search tree
32. Algorithm to construct a binary search tree

### Section 9.6

33. Preorder traversal
34. Inorder traversal
35. Postorder traversal
36. Prefix form of an expression (Polish notation)
37. Infix form of an expression
38. Postfix form of an expression (reverse Polish notation)
39. Tree representation of an expression

## Section 9.7

**40.** Decision tree

**41.** The height of a decision tree that represents an algorithm is proportional to the worst-case time of the algorithm.

**42.** Any sorting algorithm requires at least $\Omega(n \lg n)$ comparisons in the worst case to sort $n$ items.

## Section 9.8

**43.** Isomorphic free trees

**44.** Isomorphic rooted trees

**45.** Isomorphic binary trees

**46.** The Catalan number $C(2n, n)/(n+1)$ is equal to the number of nonisomorphic binary trees with $n$ vertices.

**47.** Linear-time algorithm (Algorithm 9.8.13) to test whether two binary trees are isomorphic

## Section 9.9

**48.** Game tree

**49.** Minimax procedure

**50.** $n$-level search

**51.** Evaluation function

**52.** Alpha-beta pruning

**53.** Alpha value

**54.** Alpha cutoff

**55.** Beta value

**56.** Beta cutoff

# Chapter 9 Self-Test

*Exercises 1–4 refer to the following tree.*

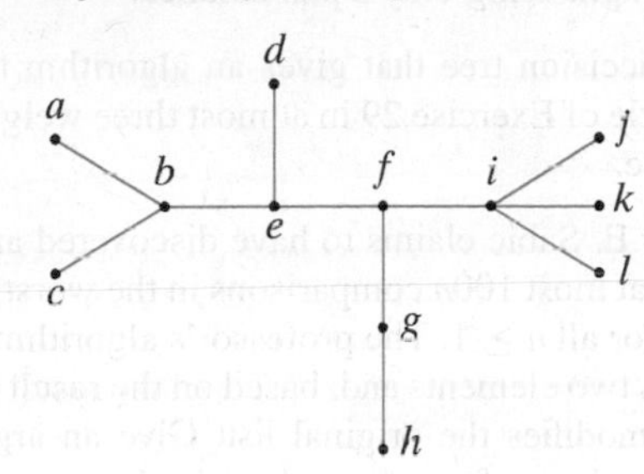


**1.** Draw the tree as a rooted tree with root $c$.

**2.** Find the level of every vertex in the tree rooted at $c$.

**3.** Find the height of the tree rooted at $c$.

**4.** Draw the tree as a rooted tree with root $f$. Find

(a) The parent of $a$.

(b) The children of $b$.

(c) The terminal vertices.

(d) The subtree rooted at $e$.

**5.** Draw a binary tree with exactly two left children and one right child.

**6.** A full binary tree has 15 internal vertices. How many terminal vertices does it have?

*Exercises 7–9 refer to the following binary tree.*

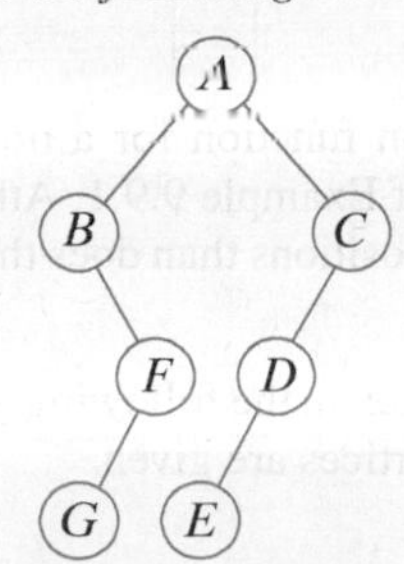


**7.** List the order in which the vertices are processed using preorder traversal.

**8.** List the order in which the vertices are processed using inorder traversal.

**9.** List the order in which the vertices are processed using postorder traversal.

**10.** Find a minimal spanning tree for the following graph.

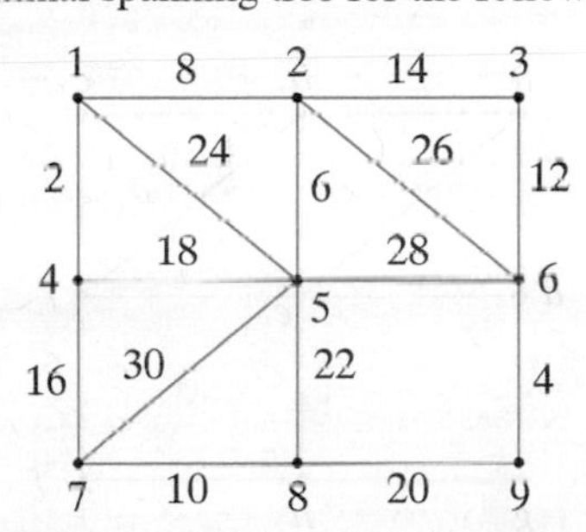


**11.** In what order are the edges added by Prim's Algorithm for the graph of Exercise 10 if the initial vertex is 1?

**12.** In what order are the edges added by Prim's Algorithm for the graph of Exercise 10 if the initial vertex is 6?

**13.** Give an example of the use of the greedy method that does not lead to an optimal algorithm.

**14.** Construct an optimal Huffman code for the set of letters in the table.

| *Letter* | *Frequency* |
|---|---|
| *A* | 5 |
| *B* | 8 |
| *C* | 5 |
| *D* | 12 |
| *E* | 20 |
| *F* | 10 |

*Answer true or false in Exercises 15–17 and explain your answer.*

**15.** If $T$ is a tree with six vertices, $T$ must have five edges.

**16.** If $T$ is a rooted tree with six vertices, the height of $T$ is at most 5.

**17.** An acyclic graph with eight vertices has seven edges.

**18.** Represent the prefix expression $-*E/BD-CA$ as a binary tree. Also write the postfix form and the fully parenthesized infix form of the expression.

*Answer true or false in Exercises 19 and 20 and explain your answer.*

**19.** If $T_1$ and $T_2$ are isomorphic as rooted trees, then $T_1$ and $T_2$ are isomorphic as free trees.

**20.** If $T_1$ and $T_2$ are rooted trees that are isomorphic as free trees, then $T_1$ and $T_2$ are isomorphic as rooted trees.

**21.** Place the words

WORD PROCESSING PRODUCES CLEAN MANUSCRIPTS BUT NOT NECESSARILY CLEAR PROSE

in the order in which they appear, in a binary search tree.

**22.** Explain how we would look for MORE in the binary search tree of Exercise 21.

**23.** Use breadth-first search (Algorithm 9.3.6) with the vertex ordering *eachgbdfi* to find a spanning tree for the following graph.

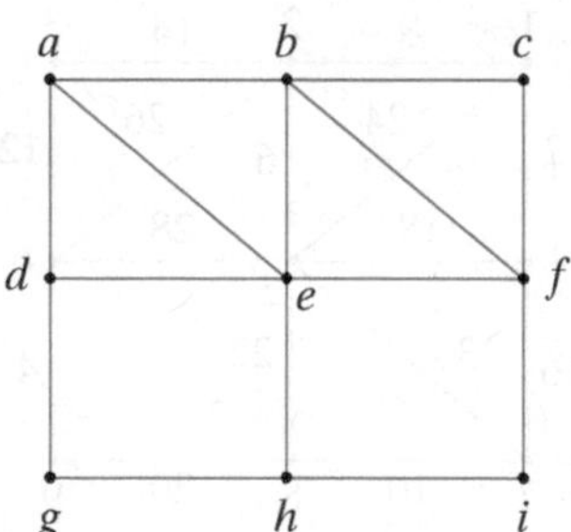


**24.** Use depth-first search (Algorithm 9.3.7) with the vertex ordering *eachgbdfi* to find a spanning tree for the graph of Exercise 23.

**25.** Use breadth-first search (Algorithm 9.3.6) with the vertex ordering *fdehagbci* to find a spanning tree for the graph of Exercise 23.

**26.** Use depth-first search (Algorithm 9.3.7) with the vertex ordering *fdehagbci* to find a spanning tree for the graph of Exercise 23.

**27.** Determine whether the free trees are isomorphic. If the trees are isomorphic, give an isomorphism. If the trees are not isomorphic, give an invariant the trees do not share.

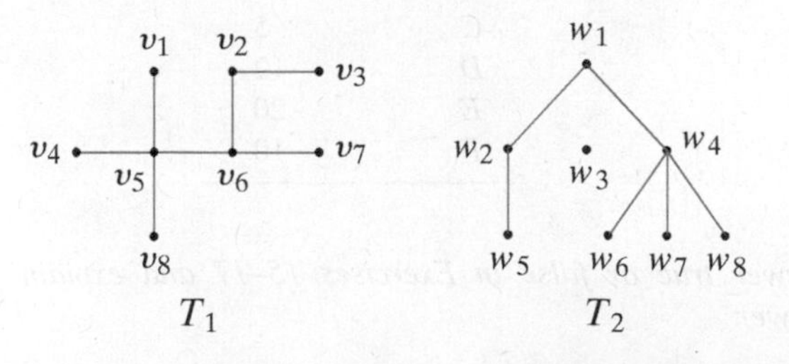


**28.** Determine whether the rooted trees are isomorphic. If the trees are isomorphic, give an isomorphism. If the trees are not isomorphic, give an invariant the trees do not share.

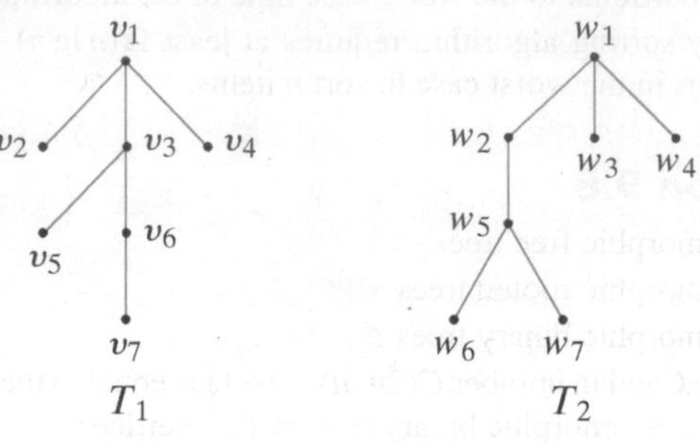


**29.** Six coins are identical in appearance, but one coin is either heavier or lighter than the others, which all weigh the same. Prove that at least three weighings are required in the worst case to identify the bad coin and determine whether it is heavy or light using only a pan balance.

**30.** Draw a decision tree that gives an algorithm to solve the coin puzzle of Exercise 29 in at most three weighings in the worst case.

**31.** Professor E. Sabic claims to have discovered an algorithm that uses at most $100n$ comparisons in the worst case to sort $n$ items, for all $n \geq 1$. The professor's algorithm repeatedly compares two elements and, based on the result of the comparison, modifies the original list. Give an argument that shows that the professor must be mistaken.

**32.** The *binary insertion sort* algorithm sorts an array of size $n$ as follows. If $n = 1$, 2, or 3, the algorithm uses an optimal sort. If $n > 3$, the algorithm sorts $s_1, \ldots, s_n$ in the following way. First, $s_1, \ldots, s_{n-1}$ is recursively sorted. Then binary search is used to determine the correct position for $s_n$, after which $s_n$ is inserted in its correct position. Determine the number of comparisons used by binary insertion sort in the worst case for $n = 4, 5, 6$. Does any algorithm use fewer comparisons for $n = 4, 5, 6$?

**33.** Find the value of the tic-tac-toe position using the evaluation function of Example 9.9.1.

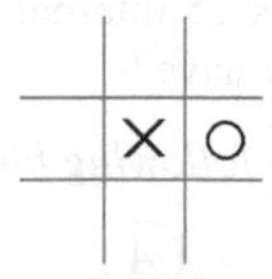


**34.** Give an evaluation function for a tic-tac-toe position different from that of Example 9.9.1. Attempt to discriminate more among the positions than does the evaluation function of Example 9.9.1.

**35.** Evaluate each vertex in the following game tree. The values of the terminal vertices are given.

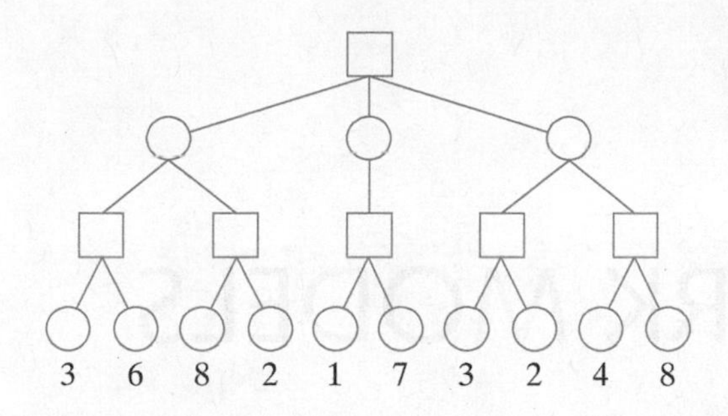


36. Evaluate the root of the tree of Exercise 35 using the minimax procedure with alpha-beta pruning. Assume that the children are evaluated left to right. For each vertex whose value is computed, write the value in the vertex. Place a check by the root of each subtree that is pruned.

## Chapter 9 Computer Exercises

1. Write a program that tests if a graph is a tree.
2. Write a program that, given the adjacency matrix of a tree and a vertex $v$, draws the tree rooted at $v$ using a computer graphics display.
3. Write a program that, given a frequency table for characters, constructs an optimal Huffman code.
4. Write a program that encodes and decodes text given a Huffman code.
5. Compute a table of characters and frequencies by sampling some text. Use your program of Exercise 3 to generate an optimal Huffman code. Use your program of Exercise 4 to encode some sample text. Compare the number of bits used to encode the text using the Huffman code with the number of bits used to encode the text in ASCII.
6. Write a program that, given a tree $T$, computes the eccentricity of each vertex in $T$ and finds the center(s) of $T$.
7. Write a program that, given a rooted tree and a vertex $v$,
   (a) Finds the parent of $v$.
   (b) Finds the ancestors of $v$.
   (c) Finds the children of $v$.
   (d) Finds the descendants of $v$.
   (e) Finds the siblings of $v$.
   (f) Determines whether $v$ is a terminal vertex.
8. Write a program that finds a spanning tree in a graph.
9. Write a program that determines whether a graph is connected.
10. Write a program that finds the components of a graph.
11. Write a program to solve the $n$-queens problem.
12. Write a backtracking program to determine whether two graphs are isomorphic.
13. Write a backtracking program that determines whether a graph can be colored with $n$ colors and, if it can be colored with $n$ colors, produces a coloring.
14. Write a backtracking program that determines whether a graph has a Hamiltonian cycle and, if there is a Hamiltonian cycle, finds it.
15. Write a program that, given a graph $G$ and a spanning tree for $G$, computes the fundamental cycle matrix of $G$.
16. Implement Prim's Algorithm as a program.
17. Implement Kruskal's Algorithm (given before Exercise 20, Section 9.4) as a program.
18. Write a program that accepts strings and puts them into a binary search tree.
19. Write a program that constructs all $n$-vertex binary trees.
20. Write a program that generates a random $n$-vertex binary tree.
21. Implement preorder, postorder, and inorder tree traversals as programs.
22. Implement tournament sort as a program.
23. Implement Algorithm 9.8.13, which tests whether two binary trees are isomorphic, as a program.
24. Write a program to generate the complete game tree for nim in which the initial position consists of two piles of four tokens each. Assume that the last player to take a token loses.
25. Implement the minimax procedure as a program.
26. Implement the minimax procedure with alpha-beta pruning as a program.
27. Implement the method of playing tic-tac-toe in Example 9.9.1 as a program.
28. Write a program that plays a perfect game of tic-tac-toe.
29. [*Project*] Develop a computer program to play a game that has relatively simple rules. Suggested games are Cribbage, Othello, The Mill, Battleship, and Kalah.

# Chapter 10

# NETWORK MODELS



In this chapter we discuss network models, which make use of directed graphs. The major portion of the chapter is devoted to the problem of maximizing the flow through a network. The network might be a transportation network through which commodities flow, a pipeline network through which oil flows, a computer network through which data flow, or any number of other possibilities. In each case the problem is to find a maximal flow. Many other problems, which on the surface seem not to be flow problems, can, in fact, be modeled as network flow problems.

Maximizing the flow in a network is a problem that belongs both to graph theory and to operations research. The traveling salesperson problem furnishes another example of a problem in graph theory and operations research. **Operations research** studies the very broad category of problems of optimizing the performance of a system. Typical problems studied in operations research are network problems, allocation of resources problems, and personnel assignment problems.

## 10.1 Introduction

Consider the directed graph in Figure 10.1.1, which represents an oil pipeline network. Oil is unloaded at the dock $a$ and pumped through the network to the refinery $z$. The vertices $b$, $c$, $d$, and $e$ represent intermediate pumping stations. The directed edges represent subpipelines of the system and show the direction the oil can flow. The labels on the edges show the capacities of the subpipelines. The problem is to find a way to maximize the flow from the dock to the refinery and to compute the value of this maximum flow. Figure 10.1.1 provides an example of a **transport network.**

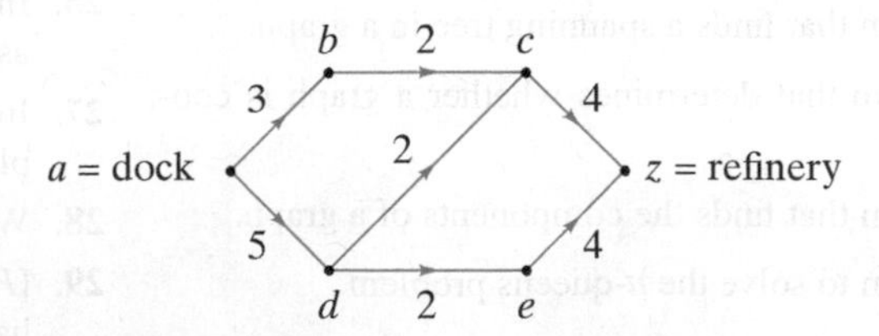


**Figure 10.1.1** A transport network.

**Definition 10.1.1 ▶** A *transport network* (or more simply *network*) is a simple, weighted, directed graph satisfying:

(a) A designated vertex, the *source*, has no incoming edges.

(b) A designated vertex, the *sink*, has no outgoing edges.

(c) The weight $C_{ij}$ of the directed edge $(i, j)$, called the *capacity* of $(i, j)$, is a nonnegative number. ◀

**Example 10.1.2** The graph of Figure 10.1.1 is a transport network. The source is vertex $a$ and the sink is vertex $z$. The capacity of edge $(a, b)$, $C_{ab}$, is 3 and the capacity of edge $(b, c)$, $C_{bc}$, is 2. ◀

Throughout this chapter, if $G$ is a network, we will denote the source by $a$ and the sink by $z$.

A **flow** in a network assigns a flow in each directed edge that does not exceed the capacity of that edge. Moreover, it is assumed that the flow into a vertex $v$, which is neither the source nor the sink, is equal to the flow out of $v$. The next definition makes these ideas precise.

**Definition 10.1.3 ▶** Let $G$ be a transport network. Let $C_{ij}$ denote the capacity of the directed edge $(i, j)$. A *flow* $F$ in $G$ assigns each directed edge $(i, j)$ a nonnegative number $F_{ij}$ such that:

(a) $F_{ij} \leq C_{ij}$.

(b) For each vertex $j$, which is neither the source nor the sink,

$$\sum_i F_{ij} = \sum_i F_{ji}. \tag{10.1.1}$$

[In a sum such as (10.1.1), unless specified otherwise, the sum is assumed to be taken over all vertices $i$. Also, if $(i, j)$ is not an edge, we set $F_{ij} = 0$.]

We call $F_{ij}$ the *flow in edge* $(i, j)$. For any vertex $j$, we call

$$\sum_i F_{ij}$$

the *flow into* $j$ and we call

$$\sum_i F_{ji}$$

the *flow out of* $j$. ◀

The property expressed by equation (10.1.1) is called **conservation of flow**. In the oil-pumping example of Figure 10.1.1, conservation of flow means that oil is neither used nor supplied at pumping stations $b$, $c$, $d$, and $e$.

**Example 10.1.4** The assignments,

$$F_{ab} = 2, \quad F_{bc} = 2, \quad F_{cz} = 3, \quad F_{ad} = 3,$$
$$F_{dc} = 1, \quad F_{de} = 2, \quad F_{ez} = 2,$$

define a flow for the network of Figure 10.1.1. For example, the flow into vertex $d$, $F_{ad} = 3$, is the same as the flow out of vertex $d$, $F_{dc} + F_{de} = 1 + 2 = 3$. ◀

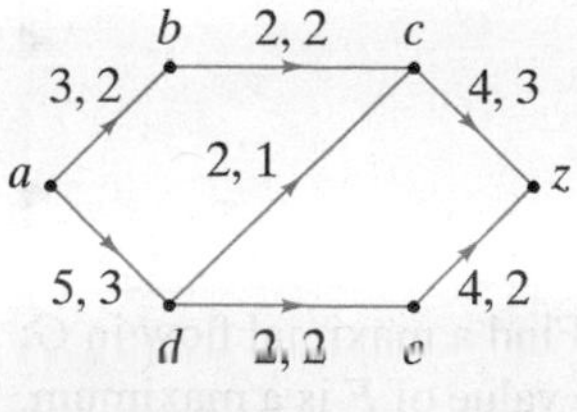

**Figure 10.1.2** Flow in a network. Edges are labeled $x, y$ to indicate capacity $x$ and flow $y$.

In Figure 10.1.2 we have redrawn the network of Figure 10.1.1 to show the flow of Example 10.1.4. An edge $e$ is labeled "$x, y$" if the capacity of $e$ is $x$ and the flow in $e$ is $y$. This notation will be used throughout this chapter.

Notice that in Example 10.1.4, the flow out of the source a, $F_{ab} + F_{ad}$, is the same as the flow into the sink $z$, $F_{cz} + F_{ez}$; both values are 5. The next theorem shows that it is always true that the flow out of the source equals the flow into the sink.

**Theorem 10.1.5** Given a flow $F$ in a network, the flow out of the source $a$ equals the flow into the sink $z$; that is,

$$\sum_i F_{ai} = \sum_i F_{iz}.$$

**Proof** Let $V$ be the set of vertices. We have

$$\sum_{j\in V}\left(\sum_{i\in V} F_{ij}\right) = \sum_{j\in V}\left(\sum_{i\in V} F_{ji}\right),$$

since each double sum is

$$\sum_{e\in E} F_e,$$

where $E$ is the set of edges. Now

$$\begin{aligned}
0 &= \sum_{j\in V}\left(\sum_{i\in V} F_{ij} - \sum_{i\in V} F_{ji}\right) \\
&= \left(\sum_{i\in V} F_{iz} - \sum_{i\in V} F_{zi}\right) + \left(\sum_{i\in V} F_{ia} - \sum_{i\in V} F_{ai}\right) \\
&\quad + \sum_{\substack{j\in V\\ j\neq a,z}}\left(\sum_{i\in V} F_{ij} - \sum_{i\in V} F_{ji}\right) \\
&= \sum_{i\in V} F_{iz} - \sum_{i\in V} F_{ai}
\end{aligned}$$

since $F_{zi} = 0 = F_{ia}$, for all $i \in V$, and (Definition 10.1.3b)

$$\sum_{i\in V} F_{ij} - \sum_{i\in V} F_{ji} = 0 \qquad \text{if } j \in V - \{a, z\}.$$ ◀

In light of Theorem 10.1.5, we can state the following definition.

**Definition 10.1.6 ▶** Let $F$ be a flow in a network $G$. The value

$$\sum_i F_{ai} = \sum_i F_{iz}$$

is called the *value of the flow F*. ◀

**Example 10.1.7** The value of the flow in the network of Figure 10.1.2 is 5. ◀

The problem for a transport network $G$ may be stated: Find a maximal flow in $G$; that is, among all possible flows in $G$, find a flow $F$ so that the value of $F$ is a maximum. In the next section we give an algorithm that efficiently solves this problem. We conclude this section by giving additional examples.

**Example 10.1.8** **A Pumping Network** Figure 10.1.3 represents a pumping network in which water for two cities, $A$ and $B$, is delivered from three wells, $w_1$, $w_2$, and $w_3$. The capacities of the intermediate systems are shown on the edges. Vertices $b$, $c$, and $d$ represent intermediate pumping stations. Model this system as a transport network.

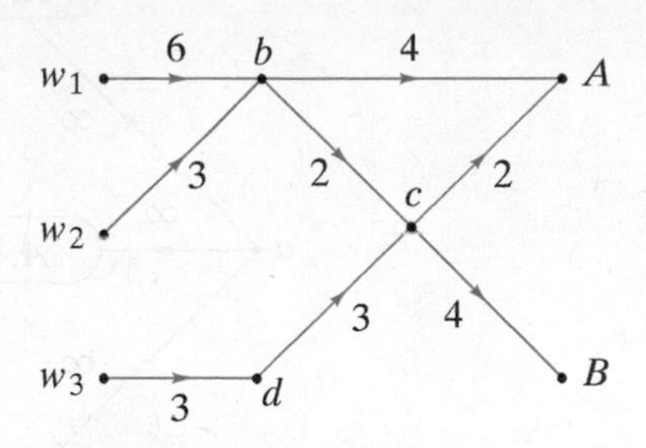


**Figure 10.1.3** A pumping network. Water for cities $A$ and $B$ is delivered from wells $w_1$, $w_2$, and $w_3$. Capacities are shown on the edges.

**SOLUTION** To obtain a designated source and sink, we can obtain an equivalent transport network by tying together the sources into a **supersource** and tying together the sinks into a **supersink** (see Figure 10.1.4). In Figure 10.1.4, $\infty$ represents an unlimited capacity.

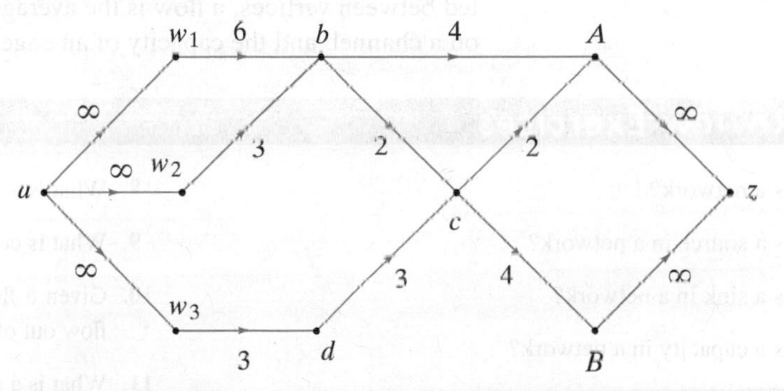


**Figure 10.1.4** The network of Figure 10.1.3 with a designated source and sink. ◀

**Example 10.1.9** **A Traffic Flow Network** It is possible to go from city $A$ to city $C$ directly or by going through city $B$. During the period 6:00 P.M. to 7:00 P.M., the average trip times are

| | |
|---|---|
| $A$ to $B$ | 15 minutes |
| $B$ to $C$ | 30 minutes |
| $A$ to $C$ | 30 minutes. |

The maximum capacities of the routes are

| | |
|---|---|
| $A$ to $B$ | 3000 vehicles |
| $B$ to $C$ | 2000 vehicles |
| $A$ to $C$ | 4000 vehicles. |

Represent the flow of traffic from $A$ to $C$ during the period 6:00 P.M. to 7:00 P.M. as a network.

**SOLUTION** A vertex will represent a city at a particular time (see Figure 10.1.5). An edge connects $X, t_1$ to $Y, t_2$ if we can leave city $X$ at $t_1$ P.M. and arrive at city $Y$ at $t_2$ P.M.

The capacity of an edge is the capacity of the route. Edges of infinite capacity connect $A, t_1$, to $A, t_2$, and $B, t_1$ to $B, t_2$ to indicate that any number of cars can wait at city $A$ or city $B$. Finally, we introduce a supersource and supersink.

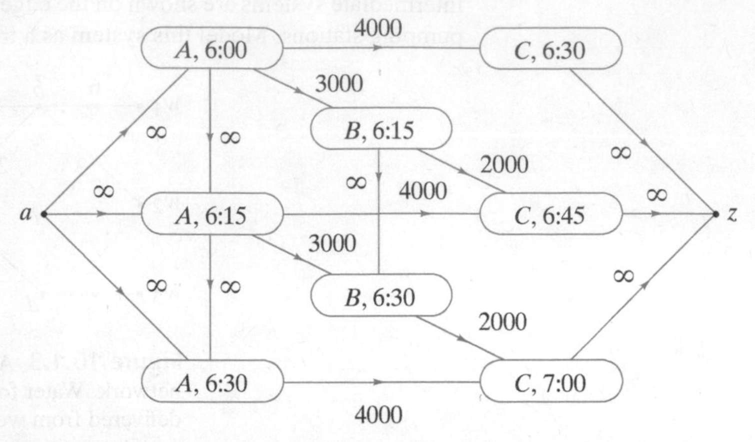


**Figure 10.1.5** A network that represents the flow of traffic from city $A$ to city $C$ during the period 6:00 P.M. to 7:00 P.M.

Variants of the network flow problem have been used in the design of efficient computer networks (see [Jones; Kleinrock]). In a model of a computer network, a vertex is a message or switching center, an edge represents a channel on which data can be transmitted between vertices, a flow is the average number of bits per second being transmitted on a channel, and the capacity of an edge is the capacity of the corresponding channel.

## 10.1 Review Exercises

1. What is a network?
2. What is a source in a network?
3. What is a sink in a network?
4. What is a capacity in a network?
5. What is a flow in a network?
6. What is a flow in an edge?
7. What is a flow into a vertex?
8. What is a flow out of a vertex?
9. What is conservation of flow?
10. Given a flow in a network, what is the relation between the flow out of the source and the flow into the sink?
11. What is a supersource?
12. What is a supersink?

## 10.1 Exercises

*In Exercises 1–3, fill in the missing edge flows so that the result is a flow in the given network. Determine the values of the flows.*

1.

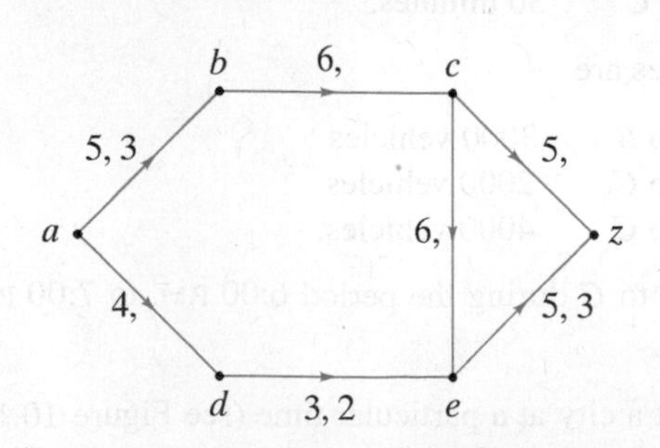


2.

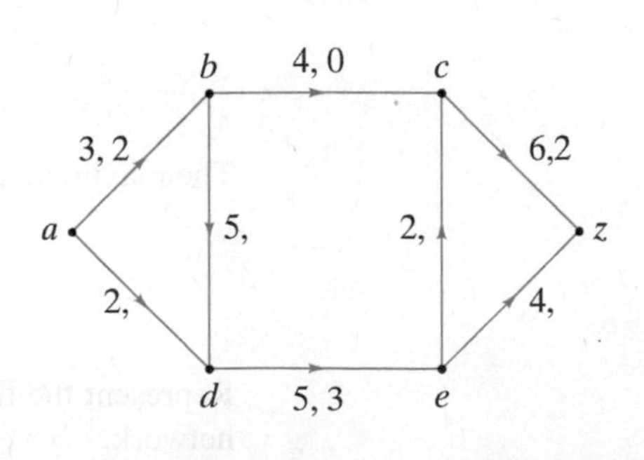

**3.**

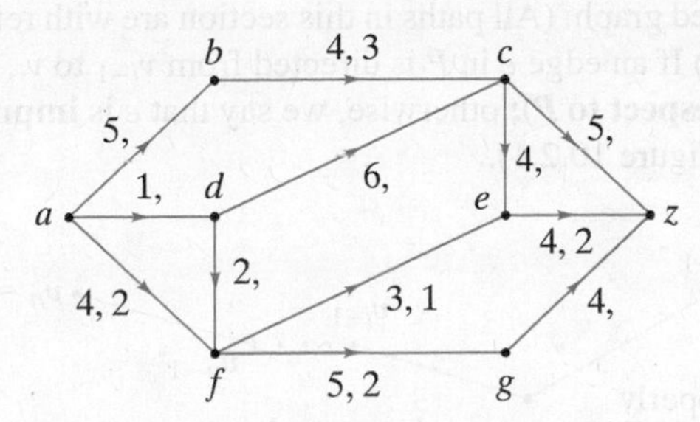


**4.** The following graph represents a pumping network in which oil for three refineries, $A$, $B$, and $C$, is delivered from three wells, $w_1$, $w_2$, and $w_3$. The capacities of the intermediate systems are shown on the edges. Vertices $b$, $c$, $d$, $e$, and $f$ represent intermediate pumping stations. Model this system as a network.

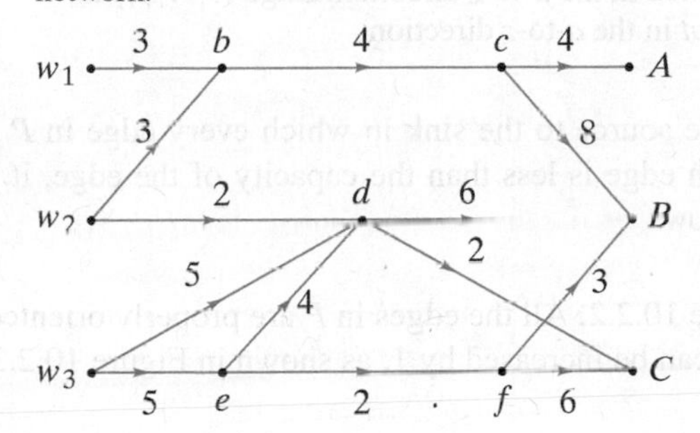


**5.** Model the system of Exercise 4 as a network assuming that well $w_1$ can pump at most 2 units, well $w_2$ at most 4 units, and well $w_3$ at most 7 units.

**6.** Model the system of Exercise 5 as a network assuming, in addition to the limitations on the wells, that city $A$ requires 4 units, city $B$ requires 3 units, and city $C$ requires 4 units.

**7.** Model the system of Exercise 6 as a network assuming, in addition to the limitations on the wells and the requirements by the cities, that the intermediate pumping station $d$ can pump at most 6 units.

**8.** There are two routes from city $A$ to city $D$. One route passes through city $B$ and the other route passes through city $C$. During the period 7:00 A.M. to 8:00 A.M., the average trip times are

| | |
|---|---|
| $A$ to $B$ | 30 minutes |
| $A$ to $C$ | 15 minutes |
| $B$ to $D$ | 15 minutes |
| $C$ to $D$ | 15 minutes. |

The maximum capacities of the routes are

| | |
|---|---|
| $A$ to $B$ | 1000 vehicles |
| $A$ to $C$ | 3000 vehicles |
| $B$ to $D$ | 4000 vehicles |
| $C$ to $D$ | 2000 vehicles. |

Represent the flow of traffic from $A$ to $D$ during the period 7:00 A.M. to 8:00 A.M. as a network.

**9.** In the system shown, we want to maximize the flow from $a$ to $z$. The capacities are shown on the edges. The flow between two vertices, neither of which is $a$ or $z$, can be in either direction. Model this system as a network.

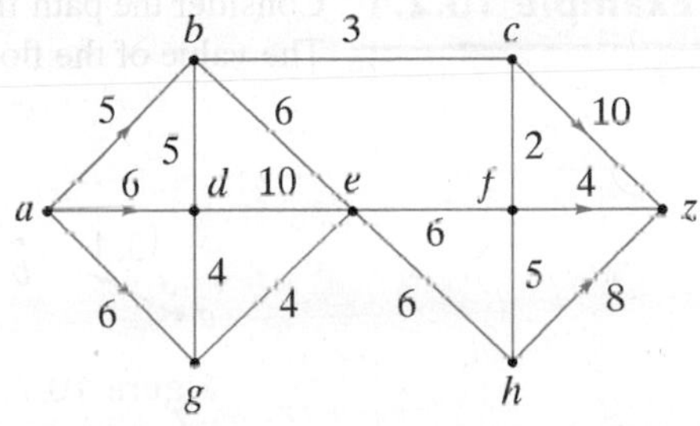


**10.** Give an example of a network with exactly two maximal flows, where each $F_{ij}$ is a nonnegative integer.

**11.** What is the maximum number of edges that an $n$-vertex network can have?

## 10.2 A Maximal Flow Algorithm

If $G$ is a transport network, a **maximal flow** in $G$ is a flow with maximum value. In general, there will be several flows having the same maximum value. In this section we give an algorithm for finding a maximal flow. The basic idea is simple—start with some initial flow and iteratively increase the value of the flow until no more improvement is possible. The resulting flow will then be a maximal flow.

We can take the initial flow to be the one in which the flow in each edge is zero. To increase the value of a given flow, we must find a path from the source to the sink and increase the flow along this path.

It is helpful at this point to introduce some terminology. Throughout this section, $G$ denotes a network with source $a$, sink $z$, and capacity $C$. Momentarily, consider the edges of $G$ to be undirected and let

$$P = (v_0, v_1, \ldots, v_n), \qquad v_0 = a, \qquad v_n = z,$$

be a path from $a$ to $z$ in this undirected graph. (All paths in this section are with reference to the underlying undirected graph.) If an edge $e$ in $P$ is directed from $v_{i-1}$ to $v_i$, we say that $e$ is **properly oriented (with respect to $P$)**; otherwise, we say that $e$ is **improperly oriented (with respect to $P$)** (see Figure 10.2.1).

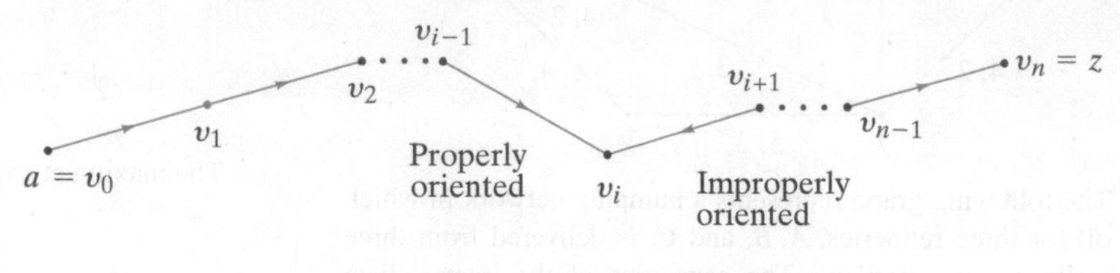


**Figure 10.2.1** Properly and improperly oriented edges. Edge $(v_{i-1}, v_i)$ is properly oriented because it is oriented in the $a$-to-$z$ direction. Edge $(v_i, v_{i+1})$ is improperly oriented because it is *not* in the $a$-to-$z$ direction.

If we can find a path $P$ from the source to the sink in which every edge in $P$ is properly oriented and the flow in each edge is less than the capacity of the edge, it is possible to increase the value of the flow.

**Example 10.2.1** Consider the path from $a$ to $z$ in Figure 10.2.2. All the edges in $P$ are properly oriented. The value of the flow in this network can be increased by 1, as shown in Figure 10.2.3.

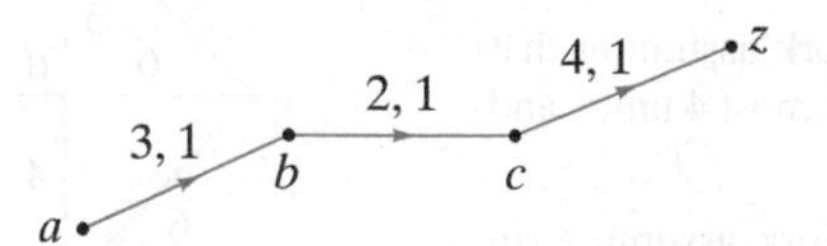


**Figure 10.2.2** A path all of whose edges are properly oriented.

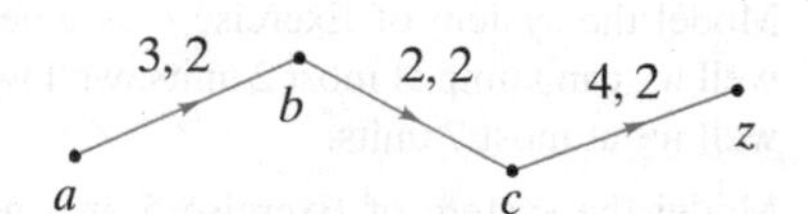


**Figure 10.2.3** After increasing the flow of Figure 10.2.2 by 1. ◀

It is also possible to increase the flow in certain paths from the source to the sink in which we have properly and improperly oriented edges. Let $P$ be a path from a to $z$ and let $x$ be a vertex in $P$ that is neither $a$ nor $z$ (see Figure 10.2.4). There are four possibilities for the orientations of the edges $e_1$ and $e_2$ incident on $x$. In case (a), both edges are properly oriented. In this case, if we increase the flow in each edge by $\Delta$, the flow into $x$ will still equal the flow out of $x$. In case (b), if we increase the flow in $e_2$ by $\Delta$, we must *decrease* the flow in $e_1$ by $\Delta$ so that the flow into $x$ will still equal the flow out of $x$. Case (c) is similar to case (b), except that we increase the flow in $e_1$ by $\Delta$ and decrease the flow in $e_2$ by $\Delta$. In case (d), we decrease the flow in both edges by $\Delta$. In every case, the resulting edge assignments give a flow. Of course, to carry out

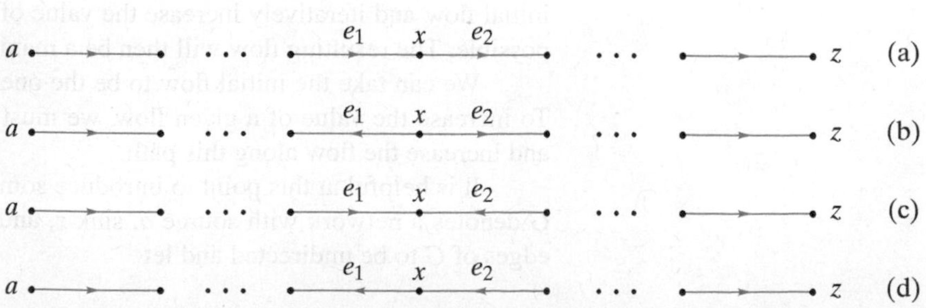

**Figure 10.2.4** The four possible orientations of the edges incident on $x$.

these alterations, we must have flow less than capacity in a properly oriented edge and a nonzero flow in an improperly oriented edge.

**Example 10.2.2** Consider the path from $a$ to $z$ in Figure 10.2.5. Edges $(a, b)$, $(c, d)$, and $(d, z)$ are properly oriented and edge $(c, b)$ is improperly oriented. We decrease the flow by 1 in the improperly oriented edge $(c, b)$ and increase the flow by 1 in the properly oriented edges $(a, b)$, $(c, d)$, and $(d, z)$ (see Figure 10.2.6). The value of the new flow is 1 greater than that of the original flow.

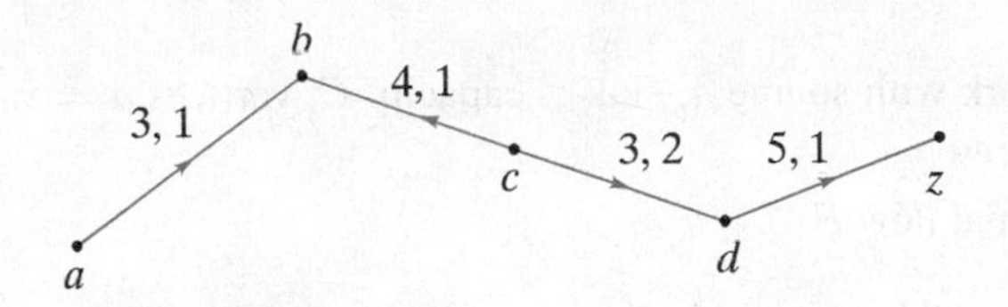


**Figure 10.2.5** A path with an improperly oriented edge: $(c, b)$.

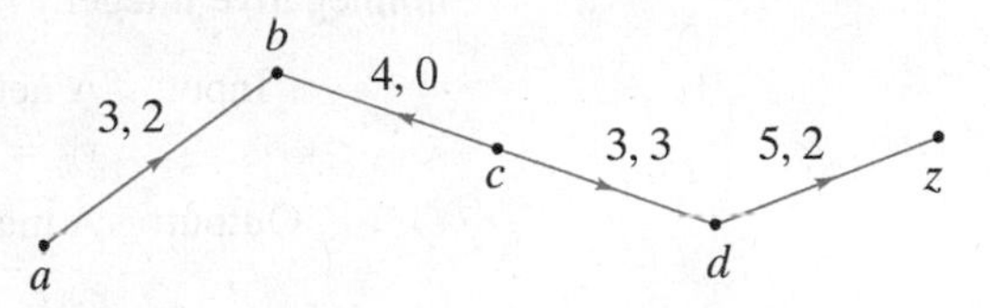


**Figure 10.2.6** After increasing the flow of Figure 10.2.5 by 1. ◀

We summarize the method of Examples 10.2.1 and 10.2.2 as a theorem.

**Theorem 10.2.3** Let $P$ be a path from $a$ to $z$ in a network $G$ satisfying the following conditions:

(*a*) For each properly oriented edge $(i, j)$ in $P$,

$$F_{ij} < C_{ij}.$$

(*b*) For each improperly oriented edge $(i, j)$ in $P$,

$$0 < F_{ij}.$$

Let

$$\Delta = \min X,$$

where $X$ consists of the numbers $C_{ij} - F_{ij}$, for properly oriented edges $(i, j)$ in $P$, and $F_{ij}$, for improperly oriented edges $(i, j)$ in $P$. Define

$$F^*_{ij} = \begin{cases} F_{ij} & \text{if } (i, j) \text{ is not in } P \\ F_{ij} + \Delta & \text{if } (i, j) \text{ is properly oriented in } P \\ F_{ij} - \Delta & \text{if } (i, j) \text{ is not properly oriented in } P. \end{cases}$$

Then $F^*$ is a flow whose value is $\Delta$ greater than the value of $F$.

**Go Online**

For more on the maximal flow algorithm, see `bit.ly/20341ks`

**Proof** (See Figures 10.2.2, 10.2.3, 10.2.5, and 10.2.6.) The argument that $F^*$ is a flow is given just before Example 10.2.2. Since the edge $(a, v)$ in $P$ is increased by $\Delta$, the value of $F^*$ is $\Delta$ greater than the value of $F$. ◀

In the next section we show that if there are no paths satisfying the conditions of Theorem 10.2.3, the flow is maximal. Thus it is possible to construct an algorithm based on Theorem 10.2.3. The outline is as follows:

1. Start with a flow (e.g., the flow in which the flow in each edge is 0).
2. Search for a path satisfying the conditions of Theorem 10.2.3. If no such path exists, stop; the flow is maximal.

3. Increase the flow through the path by $\Delta$, where $\Delta$ is defined as in Theorem 10.2.3, and go to line 2.

In the formal algorithm, we search for a path satisfying the conditions of Theorem 10.2.3 while simultaneously keeping track of the quantities $C_{ij} - F_{ij}$, $F_{ij}$.

**Algorithm 10.2.4** **Finding a Maximal Flow in a Network**

This algorithm finds a maximal flow in a network. The capacity of each edge is a nonnegative integer.

Input: A network with source $a$, sink $z$, capacity $C$, vertices $a = v_0, \ldots, v_n = z$, and $n$

Output: A maximal flow $F$

```
      max_flow(a, z, C, v, n) {
        // v's label is (predecessor(v), val(v))
        // start with zero flow
 1.     for each edge (i, j)
 2.       F_ij = 0
 3.     while (true) {
          // remove all labels
 4.       for i = 0 to n {
 5.         predecessor(v_i) = null
 6.         val(v_i) = null
 7.       }
          // label a
 8.       predecessor(a) = —
 9.       val(a) = ∞
          // U is the set of unexamined, labeled vertices
10.       U = {a}
          // continue until z is labeled
11.       while (val(z) == null) {
12.         if (U == ∅) // flow is maximal
13.           return F
14.         choose v in U
15.         U = U − {v}
16.         Δ = val(v)
17.         for each edge (v, w) with val(w) == null
18.           if (F_vw < C_vw) {
19.             predecessor(w) = v
20.             val(w) = min{Δ, C_vw − F_vw}
21.             U = U ∪ {w}
22.           }
23.         for each edge (w, v) with val(w) == null
24.           if (F_wv > 0) {
25.             predecessor(w) = v
26.             val(w) = min{Δ, F_wv}
27.             U = U ∪ {w}
28.           }
29.       } // end while (val(z) == null) loop
          // find path P from a to z on which to revise flow
```

```
30.        w_0 = z
31.        k = 0
32.        while (w_k ¬= a) {
33.            w_{k+1} = predecessor(w_k)
34.            k = k + 1
35.        }
36.        P = (w_{k+1}, w_k, ..., w_1, w_0)
37.        Δ = val(z)
38.        for i = 1 to k + 1 {
39.            e = (w_i, w_{i-1})
40.            if (e is properly oriented in P)
41.                F_e = F_e + Δ
42.            else
43.                F_e = F_e − Δ
44.        }
45.     } // end while (true) loop
    }
```

A proof that Algorithm 10.2.4 terminates is left as an exercise (Exercise 19). If the capacities are allowed to be nonnegative rational numbers, the algorithm also terminates; however, if arbitrary nonnegative real capacities are allowed and we permit the edges in line 17 to be examined in any order, the algorithm may not terminate (see [Ford, pp. 21–22]).

Algorithm 10.2.4 is often referred to as the **labeling procedure.** We will illustrate the algorithm with two examples.

**Example 10.2.5** In this discussion, if vertex $v$ satisfies

$$predecessor(v) = p \quad \text{and} \quad val(v) = t,$$

we show $v$'s label on the graph as $(p, t)$.

At lines 1 and 2, we initialize the flow to 0 in each edge (see Figure 10.2.7). Next, at lines 4–7 we set all labels to *null*. Then, at lines 8 and 9 we label vertex $a$ as $(-, \infty)$. At line 10 we set $U = \{a\}$. We then enter the while loop (line 11).

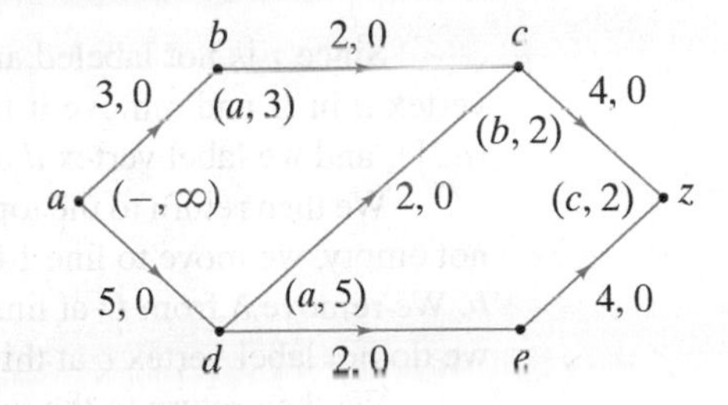


**Figure 10.2.7** After the first labeling. Vertex $v$ is labeled $(predecessor(v), val(v))$.

Since $z$ is not labeled and $U$ is not empty, we move to line 14, where we choose vertex $a$ in $U$ and remove it from $U$ at line 15. At this point, $U = \varnothing$. We set $\Delta$ to $\infty$ [$= val(a)$] at line 16. At line 17 we examine the edges $(a, b)$ and $(a, d)$ since neither $b$ nor $d$ is labeled. For edge $(a, b)$ we have $F_{ab} = 0 < C_{ab} = 3$. At lines 19 and 20, we label vertex $b$ as $(a, 3)$ since $predecessor(b) = a$ and

$$val(b) = \min\{\Delta, 3 - 0\} = \min\{\infty, 3 - 0\} = 3.$$

At line 21, we add $b$ to $U$. Similarly, we label vertex $d$ as $(a, 5)$ and add $d$ to $U$. At this point, $U = \{b, d\}$.

We then return to the top of the while loop (line 11). Since $z$ is not labeled and $U$ is not empty, we move to line 14, where we choose a vertex in $U$. Suppose that we choose $b$. We remove $b$ from $U$ at line 15. We set $\Delta$ to 3 [$= val(b)$] at line 16. At line 17 we examine edge $(b, c)$. At lines 19 and 20 we label vertex $c$ as $(b, 2)$ since $predecessor(c) = b$ and

$$val(c) = \min\{\Delta, 2 - 0\} = \min\{3, 2 - 0\} = 2.$$

At line 21 we add $c$ to $U$. At this point, $U = \{c, d\}$.

We then return to the top of the while loop (line 11). Since $z$ is not labeled and $U$ is not empty, we move to line 14, where we choose a vertex in $U$. Suppose that we choose $c$. We remove $c$ from $U$ at line 15. We set $\Delta$ to 2 [$= val(c)$] at line 16. At line 17 we examine edge $(c, z)$. At lines 19 and 20 we label vertex $z$ as $(c, 2)$. At line 21, we add $z$ to $U$. At this point, $U = \{d, z\}$.

We then return to the top of the while loop (line 11). Since $z$ is labeled, we proceed to line 30. At lines 30–36, by following predecessors from $z$, we find the path $P = (a, b, c, z)$ from $a$ to $z$. At line 37 we set $\Delta$ to 2. Since each edge in $P$ is properly oriented, at line 41 we increase the flow in each edge in $P$ by $\Delta = 2$ to obtain Figure 10.2.8.

We then return to the top of the while loop (line 3). Next, at lines 4–7 we set all labels to *null*. Then, at lines 8 and 9 we label vertex $a$ as $(-, \infty)$ (see Figure 10.2.8). At line 10 we set $U = \{a\}$. We then enter the while loop (line 11).

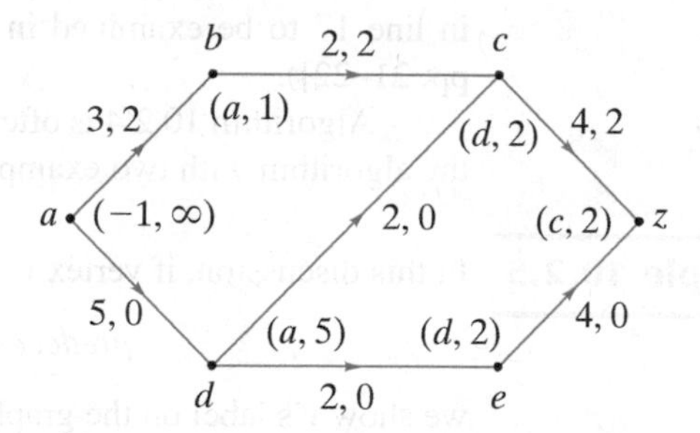


**Figure 10.2.8** After increasing the flow on path $(a, b, c, z)$ by 2 and the second labeling.

Since $z$ is not labeled and $U$ is not empty, we move to line 14, where we choose vertex $a$ in $U$ and remove it from $U$ at line 15. At lines 19 and 20 we label vertex $b$ as $(a, 1)$, and we label vertex $d$ as $(a, 5)$. We add $b$ and $d$ to $U$ so that $U = \{b, d\}$.

We then return to the top of the while loop (line 11). Since $z$ is not labeled and $U$ is not empty, we move to line 14, where we choose a vertex in $U$. Suppose that we choose $b$. We remove $b$ from $U$ at line 15. At line 17 we examine edge $(b, c)$. Since $F_{bc} = C_{bc}$, we do not label vertex $c$ at this point. Now $U = \{d\}$.

We then return to the top of the while loop (line 11). Since $z$ is not labeled and $U$ is not empty, we move to line 14, where we choose vertex $d$ in $U$ and remove it from $U$ at line 15. At lines 19 and 20 we label vertex $c$ as $(d, 2)$ and we label vertex $e$ as $(d, 2)$. We add $c$ and $e$ to $U$ so that $U = \{c, e\}$.

We then return to the top of the while loop (line 11). Since $z$ is not labeled and $U$ is not empty, we move to line 14, where we choose a vertex in $U$. Suppose that we choose $c$ in $U$ and remove it from $U$ at line 15. At lines 19 and 20 we label vertex $z$ as $(c, 2)$. We add $z$ to $U$ so that $U = \{z, e\}$.

We then return to the top of the while loop (line 11). Since $z$ is labeled, we proceed to line 30. At line 36 we find that $P = (a, d, c, z)$. Since each edge in $P$ is properly oriented, at line 41 we increase the flow in each edge in $P$ by $\Delta = 2$ to obtain Figure 10.2.9.

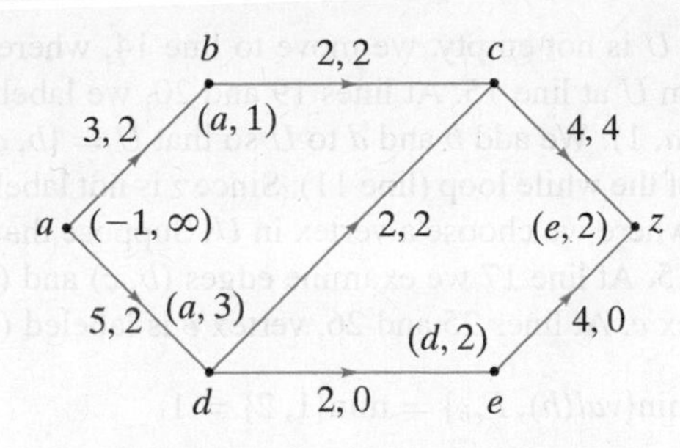


**Figure 10.2.9** After increasing the flow on path $(a, d, c, z)$ by 2 and the third labeling.

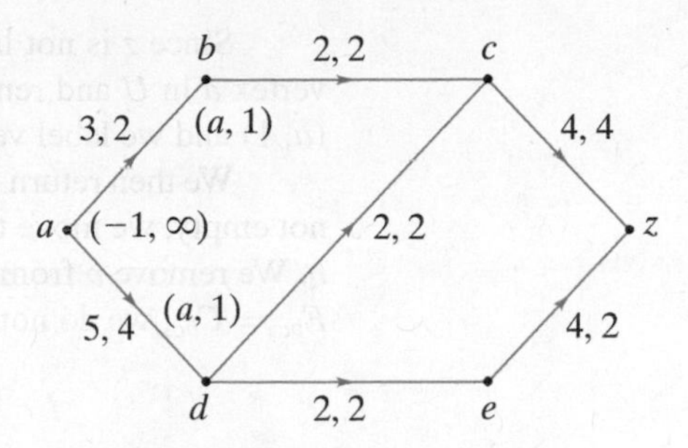


**Figure 10.2.10** After increasing the flow on path $(a, d, e, z)$ by 2 and the fourth and final labeling. The flow is maximal.

You should check that the next iteration of the algorithm produces the labeling shown in Figure 10.2.9. Increasing the flow by $\Delta = 2$ produces Figure 10.2.10.

We then return to the top of the while loop (line 3). Next, at lines 4–7 we set all labels to *null*. Then, at lines 8 and 9 we label vertex $a$ as $(-, \infty)$ (see Figure 10.2.10). At line 10 we set $U = \{a\}$. We then enter the while loop (line 11).

Since $z$ is not labeled and $U$ is not empty, we move to line 14, where we choose vertex $a$ in $U$ and remove it from $U$ at line 15. At lines 19 and 20 we label vertex $b$ as $(a, 1)$ and we label vertex $d$ as $(a, 1)$. We add $b$ and $d$ to $U$ so that $U = \{b, d\}$.

We then return to the top of the while loop (line 11). Since $z$ is not labeled and $U$ is not empty, we move to line 14, where we choose a vertex in $U$. Suppose that we choose $b$. We remove $b$ from $U$ at line 15. At line 17, we examine edge $(b, c)$. Since $F_{bc} = C_{bc}$, we do not label vertex $c$. Now $U = \{d\}$.

We then return to the top of the while loop (line 11). Since $z$ is not labeled and $U$ is not empty, we move to line 14, where we choose vertex $d$ in $U$ and remove it from $U$ at line 15. At line 17 we examine edges $(d, c)$ and $(d, e)$. Since $F_{dc} = C_{dc}$ and $F_{de} = C_{de}$, we do not label either vertex $c$ or vertex $e$. Now $U = \varnothing$.

We then return to the top of the while loop (line 11). Since $z$ is not labeled, we move to line 12. Since $U$ is empty, the algorithm terminates. The flow of Figure 10.2.10 is maximal. ◀

Our last example shows how to modify Algorithm 10.2.4 to generate a maximal flow from a given flow.

**Example 10.2.6** Replace the zero flow in lines 1 and 2 of Algorithm 10.2.4 with the flow of Figure 10.2.11 and then find a maximal flow.

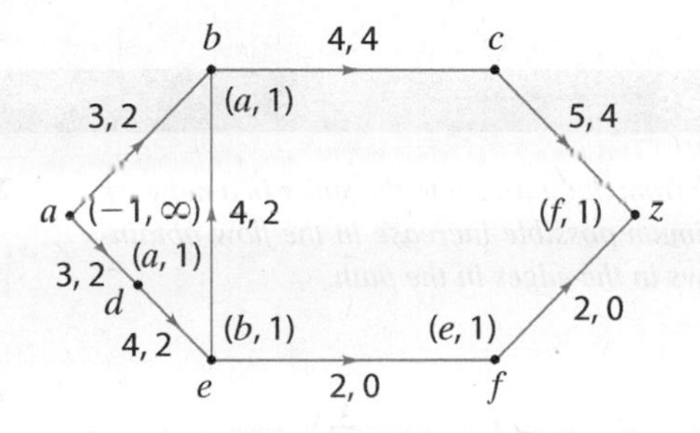


**Figure 10.2.11** After labeling.

**SOLUTION** After initializing the given flow, we move to lines 4–7, where we set all labels to *null*. Then, at lines 8 and 9 we label vertex $a$ as $(-, \infty)$ (see Figure 10.2.11). At line 10 we set $U = \{a\}$. We then enter the while loop (line 11).

Since $z$ is not labeled and $U$ is not empty, we move to line 14, where we choose vertex $a$ in $U$ and remove it from $U$ at line 15. At lines 19 and 20, we label vertex $b$ as $(a, 1)$ and we label vertex $d$ as $(a, 1)$. We add $b$ and $d$ to $U$ so that $U = \{b, d\}$.

We then return to the top of the while loop (line 11). Since $z$ is not labeled and $U$ is not empty, we move to line 14, where we choose a vertex in $U$. Suppose that we choose $b$. We remove $b$ from $U$ at line 15. At line 17 we examine edges $(b, c)$ and $(e, b)$. Since $F_{bc} = C_{bc}$, we do not label vertex $c$. At lines 25 and 26, vertex $e$ is labeled $(b, 1)$ since

$$val(e) = \min\{val(b), F_{eb}\} = \min\{1, 2\} = 1.$$

We then return to the top of the while loop (line 11). We ultimately label $z$ (see Figure 10.2.11), and at line 36 we find the path $P = (a, b, e, f, z)$. Edges $(a, b)$, $(e, f)$, and $(f, z)$ are properly oriented, so the flow in each is increased by 1. Since edge $(e, b)$ is improperly oriented, its flow is decreased by 1. We obtain the flow of Figure 10.2.12.

Another iteration of the algorithm produces the maximal flow shown in Figure 10.2.13.

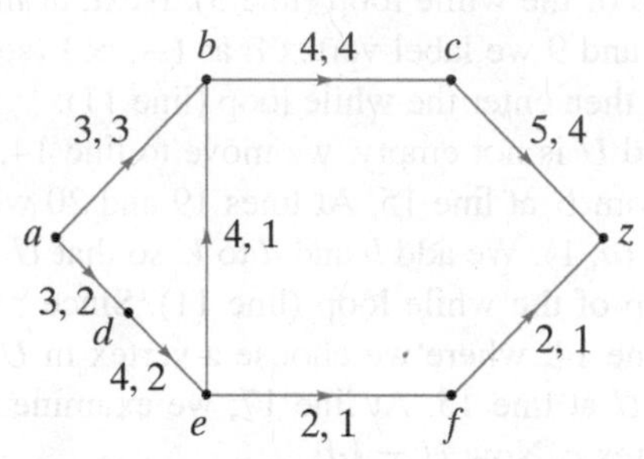


**Figure 10.2.12** After increasing the flow on path $(a, b, e, f, z)$ by 1. Notice that edge $(e, b)$ is improperly oriented and so has its flow *decreased* by 1.

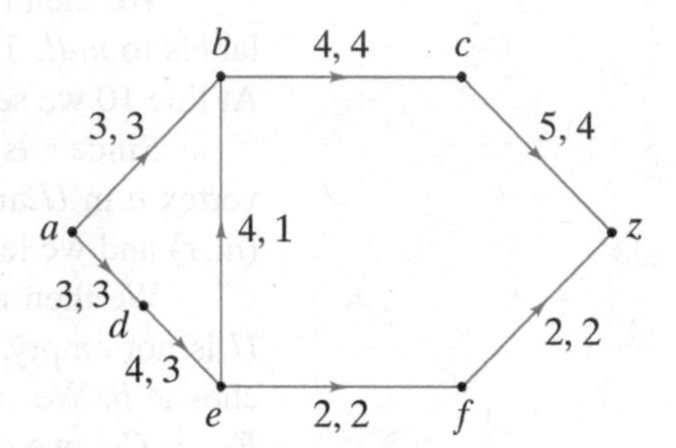


**Figure 10.2.13** After increasing the flow on path $(a, d, e, f, z)$ by 1. The flow is maximal.

◀

## 10.2 Review Exercises

1. What is a maximal flow?
2. What is a properly oriented edge with respect to a path?
3. What is an improperly oriented edge with respect to a path?
4. When can we increase the flow in a path from the source to the sink?
5. Explain how to increase the flow under the conditions of Exercise 4.
6. Explain how to find a maximal flow in a network.

## 10.2 Exercises

*In Exercises 1–3, a path from the source a to the sink z in a network is given. Find the maximum possible increase in the flow obtainable by altering the flows in the edges in the path.*

1.

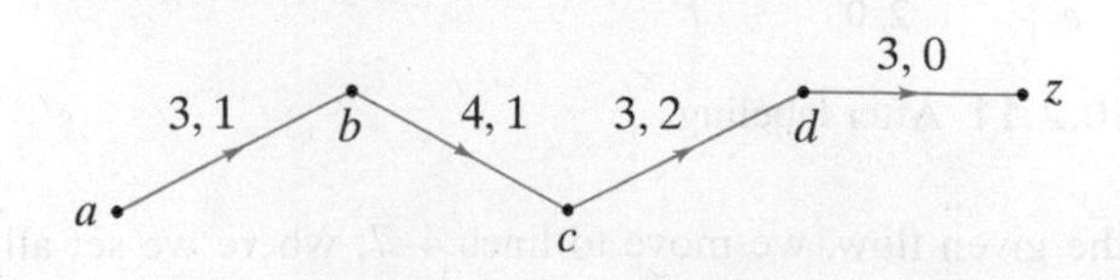


2.

3.

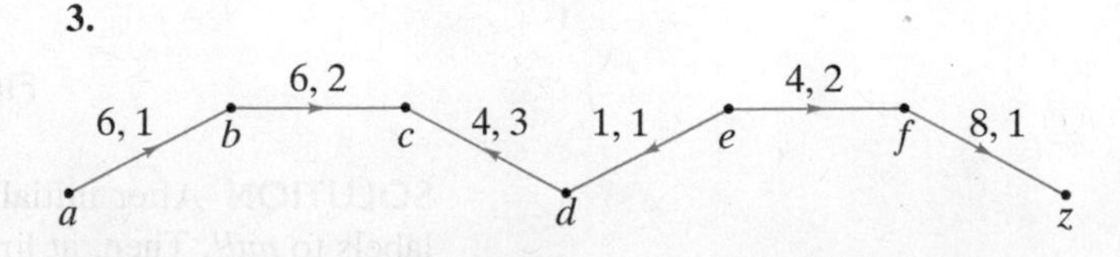

*In Exercises 4–12, use Algorithm 10.2.4 to find a maximal flow in each network.*

**4.** Figure 10.1.4

**5.** Figure 10.1.5

**6.**

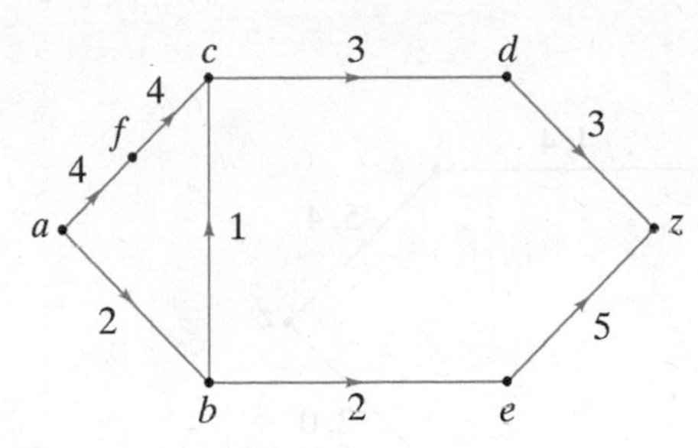


**7.** Exercise 5, Section 10.1

**8.** Exercise 6, Section 10.1

**9.** Exercise 7, Section 10.1

**10.** Exercise 8, Section 10.1

**11.** Exercise 9, Section 10.1

**12.**

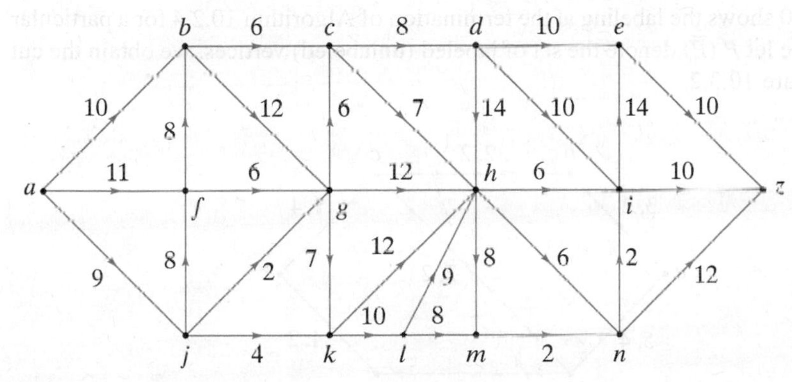


*In Exercises 13–18, find a maximal flow in each network starting with the flow given.*

**13.** Figure 10.1.2

**14.** Exercise 1, Section 10.1

**15.** Exercise 2, Section 10.1

**16.** Exercise 3, Section 10.1

**17.** Figure 10.1.4 with flows

$$F_{a,w_1} = 2, \quad F_{w_1,b} = 2, \quad F_{bA} = 0, \quad F_{cA} = 0,$$
$$F_{Az} = 0, \quad F_{a,w_2} = 0, \quad F_{w_2,b} = 0, \quad F_{bc} = 2,$$
$$F_{cB} = 4, \quad F_{Bz} = 4, \quad F_{a,w_3} = 2, \quad F_{w_3,d} = 2,$$
$$F_{dc} = 2.$$

**18.** Figure 10.1.4 with flows

$$F_{a,w_1} = 1, \quad F_{w_1,b} = 1, \quad F_{bA} = 4,$$
$$F_{cA} = 2, \quad F_{Az} = 6, \quad F_{a,w_2} = 3,$$
$$F_{w_2,b} = 3, \quad F_{bc} = 0, \quad F_{cB} = 1,$$
$$F_{Bz} = 1, \quad F_{a,w_3} = 3, \quad F_{w_3,d} = 3,$$
$$F_{dc} = 3.$$

**19.** Show that Algorithm 10.2.4 terminates.

## 10.3 The Max Flow, Min Cut Theorem

In this section we show that at the termination of Algorithm 10.2.4, the flow in the network is maximal. Along the way we will define and discuss cuts in networks.

Let $G$ be a network and consider the flow $F$ at the termination of Algorithm 10.2.4. Some vertices are labeled and some are unlabeled. Let $P$ ($\overline{P}$) denote the set of labeled (unlabeled) vertices. (Recall that $\overline{P}$ denotes the complement of $P$.) Then the source $a$ is in $P$ and the sink $z$ is in $\overline{P}$. The set $S$ of edges $(v, w)$, with $v \in P$ and $w \in \overline{P}$, is called a **cut,** and the sum of the capacities of the edges in $S$ is called the **capacity of the cut.** We will see that this cut has a minimum capacity and, since a minimal cut corresponds to a maximal flow (Theorem 10.3.9), the flow $F$ is maximal. We begin with the formal definition of cut.

Throughout this section, $G$ is a network with source $a$ and sink $z$. The capacity of edge $(i, j)$ is $C_{ij}$.

**Definition 10.3.1** ▶ A *cut* $(P, \overline{P})$ in $G$ consists of a set $P$ of vertices and the complement $\overline{P}$ of $P$, with $a \in P$ and $z \in \overline{P}$. ◀

**Example 10.3.2** Consider the network $G$ of Figure 10.3.1. If we let $P = \{a, b, d\}$, then $\overline{P} = \{c, e, f, z\}$ and $(P, \overline{P})$ is a cut in $G$. As shown, we sometimes indicate a cut by drawing a dashed line to partition the vertices.

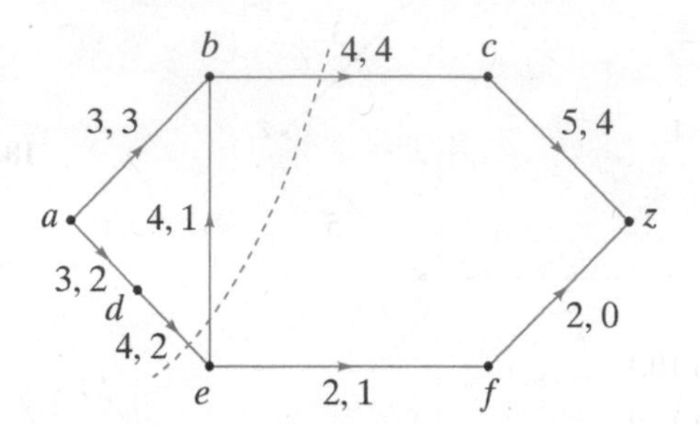


**Figure 10.3.1** A cut in a network. The dashed line divides the vertices into sets $P = \{a, b, d\}$ and $\overline{P} = \{c, e, f, z\}$ producing the cut $(P, \overline{P})$. ◀

**Example 10.3.3** Figure 10.2.10 shows the labeling at the termination of Algorithm 10.2.4 for a particular network. If we let $P$ ($\overline{P}$) denote the set of labeled (unlabeled) vertices, we obtain the cut shown in Figure 10.3.2.

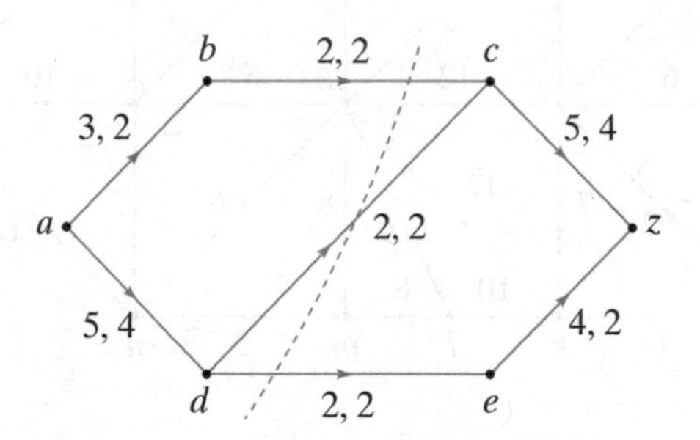


**Figure 10.3.2** A network at termination of Algorithm 10.2.4. The cut $(P, \overline{P})$, $P = \{a, b, d\}$, is obtained by letting $P$ be the set of labeled vertices. ◀

We next define the capacity of a cut.

**Definition 10.3.4** ▶ The *capacity of the cut* $(P, \overline{P})$ is the number

$$C(P, \overline{P}) = \sum_{i \in P} \sum_{j \in \overline{P}} C_{ij}.$$

◀

**Example 10.3.5** The capacity of the cut of Figure 10.3.1 is $C_{bc} + C_{de} = 8$. ◀

**Example 10.3.6** The capacity of the cut of Figure 10.3.2 is $C_{bc} + C_{dc} + C_{de} = 6$. ◀

The next theorem shows that the capacity of any cut is always greater than or equal to the value of any flow.

**Theorem 10.3.7** Let $F$ be a flow in $G$ and let $(P, \overline{P})$ be a cut in $G$. Then the capacity of $(P, \overline{P})$ is greater than or equal to the value of $F$; that is,

$$\sum_{i \in P} \sum_{j \in \overline{P}} C_{ij} \geq \sum_{i} F_{ai}. \qquad \textbf{(10.3.1)}$$

(The notation $\sum_i$ means the sum over all vertices i.)

**Proof** Note that

$$\sum_{j \in P} \sum_{i \in P} F_{ji} = \sum_{j \in P} \sum_{i \in P} F_{ij},$$

since either side of the equation is merely the sum of $F_{ij}$ over all $i, j \in P$.

Now

$$\begin{aligned}
\sum_{i} F_{ai} &= \sum_{j \in P} \sum_{i} F_{ji} - \sum_{j \in P} \sum_{i} F_{ij} \\
&= \sum_{j \in P} \sum_{i \in P} F_{ji} + \sum_{j \in P} \sum_{i \in \overline{P}} F_{ji} - \sum_{j \in P} \sum_{i \in P} F_{ij} - \sum_{j \in P} \sum_{i \in \overline{P}} F_{ij} \\
&= \sum_{j \in P} \sum_{i \in \overline{P}} F_{ji} - \sum_{j \in P} \sum_{i \in \overline{P}} F_{ij} \leq \sum_{j \in P} \sum_{i \in \overline{P}} F_{ji} \leq \sum_{j \in P} \sum_{i \in \overline{P}} C_{ji}.
\end{aligned}$$

◀

**Example 10.3.8** In Figure 10.3.1, the value 5 of the flow is less than the capacity 8 of the cut. ◀

A **minimal cut** is a cut having minimum capacity.

**Theorem 10.3.9** **Max Flow, Min Cut Theorem**

Let $F$ be a flow in $G$ and let $(P, \overline{P})$ be a cut in $G$. If equality holds in (10.3.1), then the flow is maximal and the cut is minimal. Moreover, equality holds in (10.3.1) if and only if

(a) $F_{ij} = C_{ij}$ for $i \in P, j \in \overline{P}$

and

(b) $F_{ij} = 0$ for $i \in \overline{P}, j \in P$.

**Proof** The first statement follows immediately.

The proof of Theorem 10.3.7 shows that equality holds precisely when

$$\sum_{j \in P} \sum_{i \in \overline{P}} F_{ij} = 0 \quad \text{and} \quad \sum_{j \in P} \sum_{i \in \overline{P}} F_{ji} = \sum_{j \in P} \sum_{i \in \overline{P}} C_{ji};$$

thus the last statement is also true. ◀

**Example 10.3.10** In Figure 10.3.2, the value of the flow and the capacity of the cut are both 6; therefore, the flow is maximal and the cut is minimal. ◀

We can use Theorem 10.3.9 to show that Algorithm 10.2.4 produces a maximal flow.

**Theorem 10.3.11** At termination, Algorithm 10.2.4 produces a maximal flow. Moreover, if $P$ (respectively, $\overline{P}$) is the set of labeled (respectively, unlabeled) vertices at the termination of Algorithm 10.2.4, the cut $(P, \overline{P})$ is minimal.

**Proof** Let $P$ ($\overline{P}$) be the set of labeled (unlabeled) vertices of $G$ at the termination of Algorithm 10.2.4. Consider an edge $(i, j)$, where $i \in P, j \in \overline{P}$. Since $i$ is labeled, we must have

$$F_{ij} = C_{ij};$$

otherwise, we would have labeled $j$ at lines 19 and 20. Now consider an edge $(j, i)$, where $j \in \overline{P}, i \in P$. Since $i$ is labeled, we must have

$$F_{ji} = 0;$$

otherwise, we would have labeled $j$ at lines 25 and 26. By Theorem 10.3.9, the flow at the termination of Algorithm 10.2.4 is maximal and the cut $(P, \overline{P})$ is minimal. ◀

## 10.3 Review Exercises

1. What is a cut in a network?
2. What is the capacity of a cut?
3. What is the relation between the capacity of any cut and the value of any flow?
4. What is a minimal cut?
5. State the max flow, min cut theorem.
6. Explain how the max flow, min cut theorem proves that the algorithm of Section 10.2 correctly finds a maximal flow in a network.

## 10.3 Exercises

*In Exercises 1–3, find the capacity of the cut $(P, \overline{P})$. Also, determine whether the cut is minimal.*

1. $P = \{a, d\}$ for Exercise 1, Section 10.1
2. $P = \{a, d, e\}$ for Exercise 2, Section 10.1
3. $P = \{a, b, c, d\}$ for Exercise 3, Section 10.1

*In Exercises 4–16, find a minimal cut in each network.*

4. Figure 10.1.1
5. Figure 10.1.4
6. Figure 10.1.5
7. Exercise 1, Section 10.1
8. Exercise 2, Section 10.1
9. Exercise 3, Section 10.1
10. Exercise 4, Section 10.1
11. Exercise 5, Section 10.1
12. Exercise 6, Section 10.1
13. Exercise 7, Section 10.1
14. Exercise 8, Section 10.1
15. Exercise 9, Section 10.1
16. Exercise 12, Section 10.2

*Exercises 17–22 refer to a network $G$ that, in addition to having nonnegative integer capacities $C_{ij}$, has nonnegative integer minimal edge flow requirements $m_{ij}$. That is, a flow $F$ must satisfy*

$$m_{ij} \leq F_{ij} \leq C_{ij}$$

*for all edges $(i, j)$.*

17. Give an example of a network $G$, in which $m_{ij} \leq C_{ij}$ for all edges $(i, j)$, for which no flow exists.

*Define*

$$C(\overline{P}, P) = \sum_{i \in \overline{P}} \sum_{j \in P} C_{ij},$$

$$m(P, \overline{P}) = \sum_{i \in P} \sum_{j \in \overline{P}} m_{ij}, \qquad m(\overline{P}, P) = \sum_{i \in \overline{P}} \sum_{j \in P} m_{ij}.$$

18. Show that the value $V$ of any flow satisfies
$$m(P, \overline{P}) - C(\overline{P}, P) \leq V \leq C(P, \overline{P}) - m(\overline{P}, P)$$
for any cut $(P, \overline{P})$.
19. Show that if a flow exists in $G$, a maximal flow exists in $G$ with value
$$\min\{C(P, \overline{P}) - m(\overline{P}, P) \mid (P, \overline{P}) \text{ is a cut in } G\}.$$
20. Assume that $G$ has a flow $F$. Develop an algorithm for finding a maximal flow in $G$.
21. Show that if a flow exists in $G$, a minimal flow exists in $G$ with value
$$\max\{m(P, \overline{P}) - C(\overline{P}, P) \mid (P, \overline{P}) \text{ is a cut in } G\}.$$
22. Assume that $G$ has a flow $F$. Develop an algorithm for finding a minimal flow in $G$.
23. True or false? If $F$ is a flow in a network $G$ and $(P, \overline{P})$ is a cut in $G$ and the capacity of $(P, \overline{P})$ exceeds the value of the flow, $F$, then the cut $(P, \overline{P})$ is not minimal and the flow $F$ is not maximal. If true, prove it; otherwise, give a counterexample.

## 10.4 Matching

In this section we consider the problem of matching elements in one set to elements in another set. We will see that this problem can be reduced to finding a maximal flow in a network. We begin with an example.

**Example 10.4.1** Suppose that four persons $A$, $B$, $C$, and $D$ apply for five jobs $J_1$, $J_2$, $J_3$, $J_4$, and $J_5$. Suppose that applicant $A$ is qualified for jobs $J_2$ and $J_5$; applicant $B$ is qualified for jobs $J_2$, and $J_5$; applicant $C$ is qualified for jobs $J_1$, $J_3$, $J_4$, and $J_5$; and applicant $D$ is qualified for jobs $J_2$ and $J_5$. Is it possible to find a job for each applicant?

The situation can be modeled by the graph of Figure 10.4.1. The vertices represent the applicants and the jobs. An edge connects an applicant to a job for which the applicant is qualified. We can show that it is not possible to match a job to each applicant by considering applicants $A$, $B$, and $D$, who are qualified for jobs $J_2$ and $J_5$. If $A$ and $B$ are assigned a job, none remains for $D$. Therefore, no assignments exist for $A$, $B$, $C$, and $D$.

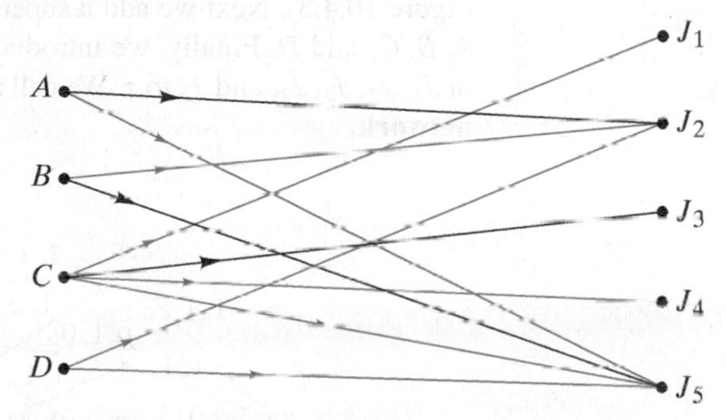


**Figure 10.4.1** Applicants $(A, B, C, D)$ and jobs $(J_1, J_2, J_3, J_4, J_5)$. An edge connects an applicant to a job for which the applicant is qualified. The black lines show a maximal matching (i.e., the maximum number of applicants have jobs). ◀

In Example 10.4.1 a matching consists of finding jobs for qualified persons. A maximal matching finds jobs for the maximum number of persons. A maximal matching for the graph of Figure 10.4.1 is shown with black lines. A complete matching finds jobs for everyone. We showed that the graph of Figure 10.4.1 has no complete matching. The formal definitions follow.

**Definition 10.4.2 ▶** Let $G$ be a directed, bipartite graph with disjoint vertex sets $V$ and $W$ in which the edges are directed from vertices in $V$ to vertices in $W$. (Any vertex in $G$ is either in $V$ or in $W$.) A *matching* for $G$ is a set of edges $E$ with no vertices in common. A *maximal matching* for $G$ is a matching $E$ in which $E$ contains the maximum number of edges. A *complete matching* for $G$ is a matching $E$ having the property that if $v \in V$, then $(v, w) \in E$ for some $w \in W$. ◀

**Example 10.4.3** The matching for the graph of Figure 10.4.2, shown with black lines, is a maximal matching and a complete matching.

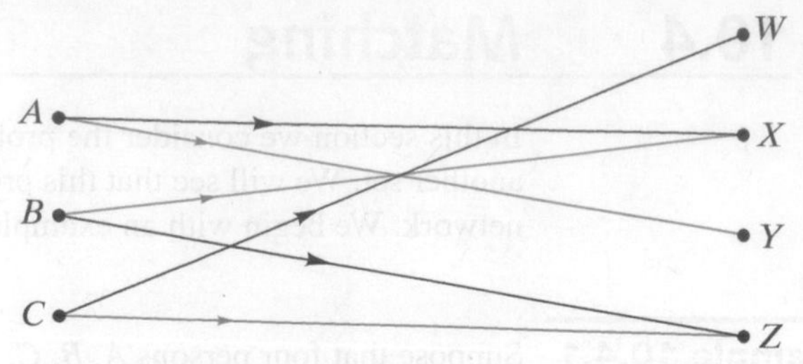


Figure 10.4.2 The black lines show a maximal matching (the maximum number of edges are used) and a complete matching (each of $A$, $B$, and $C$ is matched).

In the next example we illustrate how the matching problem can be modeled as a network problem.

**Example 10.4.4** **A Matching Network** Model the matching problem of Example 10.4.1 as a network.

SOLUTION We first assign each edge in the graph of Figure 10.4.1 capacity 1 (see Figure 10.4.3). Next we add a supersource $a$ and edges of capacity 1 from $a$ to each of $A, B, C$, and $D$. Finally, we introduce a supersink $z$ and edges of capacity 1 from each of $J_1, J_2, J_3, J_4$, and $J_5$ to $z$. We call a network such as that of Figure 10.4.3 a **matching network.**

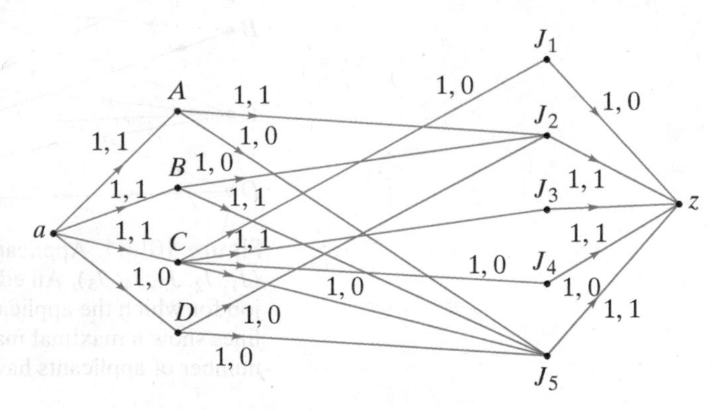


Figure 10.4.3 The matching problem (Figure 10.4.1) as a matching network.

The next theorem relates matching networks and flows.

**Theorem 10.4.5** Let $G$ be a directed, bipartite graph with disjoint vertex sets $V$ and $W$ in which the edges are directed from vertices in $V$ to vertices in $W$. (Any vertex in $G$ is either in $V$ or in $W$.)

(a) A flow in the matching network gives a matching in $G$. The vertex $v \in V$ is matched with the vertex $w \in W$ if and only if the flow in edge $(v, w)$ is 1.

(b) A maximal flow corresponds to a maximal matching.

(c) A flow whose value is $|V|$ corresponds to a complete matching.

**Proof** Let $a$ ($z$) represent the source (sink) in the matching network, and suppose that a flow is given.

Suppose that the edge $(v, w)$, $v \in V$, $w \in W$, has flow 1. The only edge into vertex $v$ is $(a, v)$. This edge must have flow 1; thus the flow into vertex $v$ is 1. Since the flow out of $v$ is also 1, the only edge of the form $(v, x)$ having flow 1 is $(v, w)$. Similarly, the only edge of the form $(x, w)$ having flow 1 is $(v, w)$. Therefore, if $E$ is the set of edges of the form $(v, w)$ having flow 1, the members of $E$ have no vertices in common; thus $E$ is a matching for $G$.

Parts (b) and (c) follow from the fact that the number of vertices in $V$ matched is equal to the value of the corresponding flow. ◀

Since a maximal flow gives a maximal matching, Algorithm 10.2.4 applied to a matching network produces a maximal matching. In practice, the implementation of Algorithm 10.2.4 can be simplified by using the adjacency matrix of the graph (see Exercise 11).

**Example 10.4.6** The matching of Figure 10.4.1 is represented as a flow in Figure 10.4.3. Since the flow is maximal, the matching is maximal. ◀

Next, we turn to the existence of a complete matching in a directed, bipartite graph $G$ with vertex sets $V$ and $W$. If $S \subseteq V$, we let

$$R(S) = \{w \in W \mid v \in S \text{ and } (v, w) \text{ is an edge in } G\}.$$

**Go Online**
For more on the Marriage Problem, see
bit.ly/20341ks

Suppose that $G$ has a complete matching. If $S \subseteq V$, we must have $|S| \leq |R(S)|$. It turns out that if $|S| \leq |R(S)|$ for all subsets $S$ of $V$, then $G$ has a complete matching. This result was first given by the English mathematician Philip Hall and is known as **Hall's Marriage Theorem,** since if $V$ is a set of men and $W$ is a set of women and edges exist from $v \in V$ to $w \in W$ if $v$ and $w$ are compatible, the theorem gives a condition under which each man can marry a compatible woman.

**Theorem 10.4.7** **Hall's Marriage Theorem**

Let $G$ be a directed, bipartite graph with disjoint vertex sets $V$ and $W$ in which the edges are directed from vertices in $V$ to vertices in $W$. (Any vertex in $G$ is either in $V$ or in $W$.) There exists a complete matching in $G$ if and only if

$$|S| \leq |R(S)| \qquad \text{for all } S \subseteq V. \tag{10.4.1}$$

**Proof** We have already pointed out that if there is a complete matching in G, condition (10.4.1) holds.

Suppose that condition (10.4.1) holds. Let $n = |V|$ and let $(P, \overline{P})$ be a minimal cut in the matching network. If we can show that the capacity of this cut is $n$, a maximal flow would have value $n$. The matching corresponding to this maximal flow would be a complete matching.

The argument is by contradiction. Assume that the capacity of the minimal cut $(P, \overline{P})$ is less than $n$. The capacity of this cut is the number of edges in the set

$$E = \{(x, y) \mid x \in P, y \in \overline{P}\}$$

(see Figure 10.4.4). A member of $E$ is one of the three types:

Type I: $(a, v)$, $v \in V$;
Type II: $(v, w)$, $v \in V$, $w \in W$;
Type III: $(w, z)$, $w \in W$.

We will estimate the number of edges of each type.

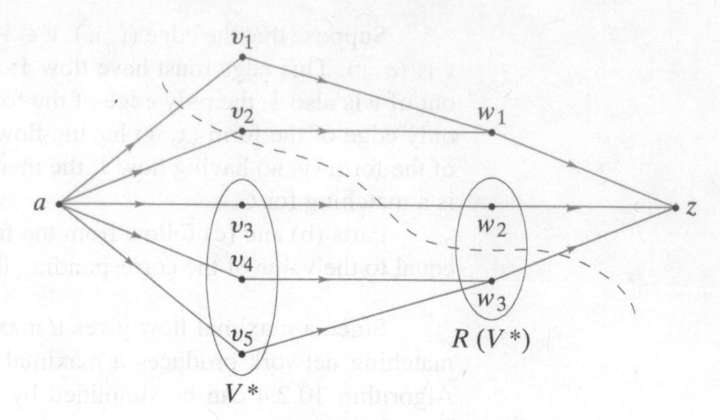


**Figure 10.4.4** The proof of Theorem 10.4.7. $V = \{v_1, v_2, v_3, v_4, v_5\}$; $n = |V| = 5$; $W = \{w_1, w_2, w_3\}$; the cut is $(P, \overline{P})$, where $P = \{a, v_3, v_4, v_5, w_3\}$; $V^* = V \cap P = \{v_3, v_4, v_5\}$; $R(V^*) = \{w_2, w_3\}$; $W_1 = R(V^*) \cap P = \{w_3\}$; $W_2 = R(V^*) \cap \overline{P} = \{w_2\}$; $E = \{(a, v_1), (a, v_2), (v_3, w_2), (w_3, z)\}$. The capacity of the cut is $|E| = 4 < n$. The type I edges are $(a, v_1)$ and $(a, v_2)$. Edge $(v_3, w_2)$ is the only type II edge, and edge $(w_3, z)$ is the only type III edge.

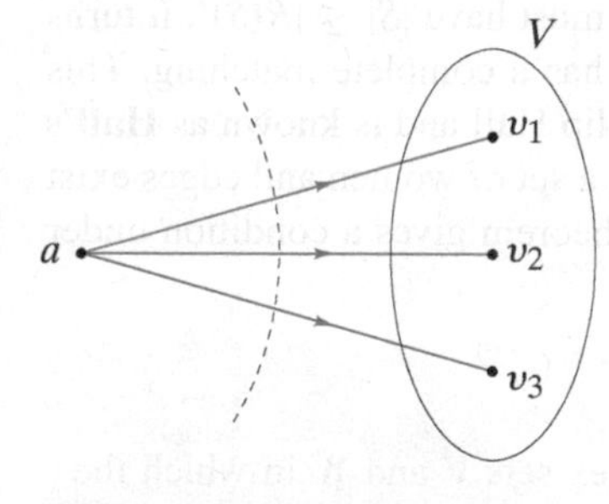


**Figure 10.4.5** The proof of Theorem 10.4.7 for $n = 3$. If $V \subseteq \overline{P}$, as shown the capacity of the cut is $n$. Since we are assuming that the capacity of the cut is less than $n$, this case cannot occur. Therefore, $V \cap P$ is nonempty.

If $V \subseteq \overline{P}$, the capacity of the cut is $n$ (see Figure 10.4.5); thus $V^* = V \cap P$ is nonempty. It follows that there are $n - |V^*|$ edges in $E$ of type I.

We partition $R(V^*)$ into the sets

$$W_1 = R(V^*) \cap P \quad \text{and} \quad W_2 = R(V^*) \cap \overline{P}.$$

Then there are at least $|W_1|$ edges in $E$ of type III. Thus there are less than

$$n - (n - |V^*|) - |W_1| = |V^*| - |W_1|$$

edges of type II in $E$. Since each member of $W_2$ contributes at most one type II edge,

$$|W_2| < |V^*| - |W_1|.$$

Thus

$$|R(V^*)| = |W_1| + |W_2| < |V^*|,$$

which contradicts (10.4.1). Therefore, a complete matching exists. ◀

**Example 10.4.8** For the graph in Figure 10.4.1, if $S = \{A, B, D\}$, we have $R(S) = \{J_2, J_5\}$ and

$$|S| = 3 > 2 = |R(S)|.$$

By Theorem 10.4.7, there is not a complete matching for the graph of Figure 10.4.1. ◀

**Example 10.4.9** There are $n$ computers and $n$ disk drives. Each computer is compatible with $m > 0$ disk drives and each disk drive is compatible with $m$ computers. Is it possible to match each computer with a compatible disk drive?

**SOLUTION** Let $V$ be the set of computers and $W$ be the set of disk drives. An edge exists from $v \in V$ to $w \in W$ if $v$ and $w$ are compatible. Notice that every vertex has degree $m$. Let $S = \{v_1, \ldots, v_k\}$ be a subset of $V$. Then there are $km$ edges from the set $S$. If $R(S) = \{w_1, \ldots, w_j\}$, then $R(S)$ receives at most $jm$ edges from $S$. Therefore, $km \leq jm$. Now $|S| = k \leq j = |R(S)|$. By Theorem 10.4.7 there is a complete matching. Thus it is possible to match each computer with a compatible disk drive. ◄

## 10.4 Review Exercises

1. What is a matching?
2. What is a maximal matching?
3. What is a complete matching?
4. What is the relation between flows and matchings?
5. State Hall's Marriage Theorem.

## 10.4 Exercises

1. Show that the flow in Figure 10.4.3 is maximal by exhibiting a minimal cut whose capacity is 3.
2. Find the flow that corresponds to the matching of Figure 10.4.2. Show that this flow is maximal by exhibiting a minimal cut whose capacity is 3.

*Exercises 3–7 refer to Figure 10.4.1, where we reverse the direction of the edges so that edges are directed from jobs to applicants.*

3. What does a matching represent?
4. What does a maximal matching represent?
5. Show a maximal matching.
6. What does a complete matching represent?
7. Is there a complete matching? If there is a complete matching, show one. If there is not a complete matching, explain why none exists.
8. Applicant $A$ is qualified for jobs $J_1$ and $J_4$; $B$ is qualified for jobs $J_2$, $J_3$, and $J_6$; $C$ is qualified for jobs $J_1$, $J_3$, $J_5$, and $J_6$; $D$ is qualified for jobs $J_1$, $J_3$, and $J_4$; and $E$ is qualified for jobs $J_1$, $J_3$, and $J_6$.
   (a) Model this situation as a matching network.
   (b) Use Algorithm 10.2.4 to find a maximal matching.
   (c) Is there a complete matching?
9. Applicant $A$ is qualified for jobs $J_1$, $J_2$, $J_4$, and $J_5$; $B$ is qualified for jobs $J_1$, $J_4$, and $J_5$; $C$ is qualified for jobs $J_1$, $J_4$, and $J_5$; $D$ is qualified for jobs $J_1$ and $J_5$; $E$ is qualified for jobs $J_2$, $J_3$, and $J_5$; and $F$ is qualified for jobs $J_4$ and $J_5$. Answer parts (a)–(c) of Exercise 8 for this situation.
10. Applicant $A$ is qualified for jobs $J_1$, $J_2$, and $J_4$; $B$ is qualified for jobs $J_3$, $J_4$, $J_5$, and $J_6$; $C$ is qualified for jobs $J_1$ and $J_5$; $D$ is qualified for jobs $J_1$, $J_3$, $J_4$, and $J_8$; $E$ is qualified for jobs $J_1$, $J_2$, $J_4$, $J_6$, and $J_8$; $F$ is qualified for jobs $J_4$ and $J_6$; and $G$ is qualified for jobs $J_3$, $J_5$, and $J_7$. Answer parts (a)–(c) of Exercise 8 for this situation.
11. Five students, $V$, $W$, $X$, $Y$, and $Z$, are members of four committees, $C_1$, $C_2$, $C_3$, and $C_4$. The members of $C_1$ are $V$, $X$, and $Y$; the members of $C_2$ are $X$ and $Z$; the members of $C_3$ are $V$, $Y$, and $Z$; and the members of $C_4$ are $V$, $W$, $X$, and $Z$. Each committee is to send a representative to the administration. No student can represent two committees.
    (a) Model this situation as a matching network.
    (b) What is the interpretation of a maximal matching?
    (c) What is the interpretation of a complete matching?
    (d) Use Algorithm 10.2.4 to find a maximal matching.
    (e) Is there a complete matching?
12. Show that by a suitable ordering of the vertices, the adjacency matrix of a bipartite graph can be written

$$\begin{pmatrix} 0 & A \\ A^T & 0 \end{pmatrix},$$

where 0 is a matrix consisting only of 0's and $A^T$ is the transpose of the matrix $A$.

*In Exercises 13–15, G is a bipartite graph, A is the matrix of Exercise 12, and F is a flow in the associated matching network. Label each entry in A that represents an edge with flow 1.*

13. What kind of labeling corresponds to a matching?
14. What kind of labeling corresponds to a complete matching?
15. What kind of labeling corresponds to a maximal matching?
16. Restate Algorithm 10.2.4, applied to a matching network, in terms of operations on the matrix $A$ of Exercise 12.

*Let G be a directed, bipartite graph with disjoint vertex sets V and W in which the edges are directed from vertices in V to vertices in W. (Any vertex in G is either in V or in W.) We define the deficiency of G as*

$$\delta(G) = \max\{|S| - |R(S)| \mid S \subseteq V\}.$$

17. Show that $G$ has a complete matching if and only if $\delta(G) = 0$.
★18. Show that the maximum number of vertices in $V$ that can be matched with vertices in $W$ is $|V| - \delta(G)$.
19. True or false? Any matching is contained in a maximal matching. If true, prove it; if false, give a counterexample.

## Problem-Solving Corner — Matching

### Problem

Let $G$ be a directed, bipartite graph with disjoint vertex sets $V$ and $W$ in which the edges are directed from vertices in $V$ to vertices in $W$. (Any vertex in $G$ is either in $V$ or in $W$.) Let $M_W$ denote the maximum degree that occurs among vertices in $W$, and let $m_V$ denote the minimum degree that occurs among vertices in $V$. Show that if $0 < M_W \le m_V$, then $G$ has a complete matching.

### Attacking the Problem

Hall's Marriage Theorem (Theorem 10.4.7) says that a directed, bipartite graph with disjoint vertex sets $V$ and $W$ has a complete matching if and only if $|S| \le |R(S)|$ for all $S \subseteq V$. Thus a possible way to solve the problem is to show that the given condition $M_W \le m_V$ implies that $|S| \le |R(S)|$ for all $S \subseteq V$.

### Finding a Solution

Our goal is to prove that if $M_W \le m_V$, then $|S| \le |R(S)|$ for all $S \subseteq V$. Let's start with an example graph $G$ for which $M_W \le m_V$; in fact, let's arrange to have $M_W = 2$ and $m_V = 3$:

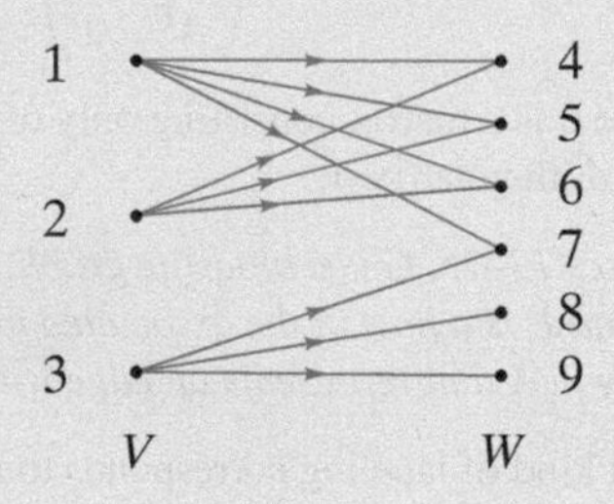


Consider an example subset $S = \{1, 3\}$ of $V$ and the edges incident on vertices in $S$:

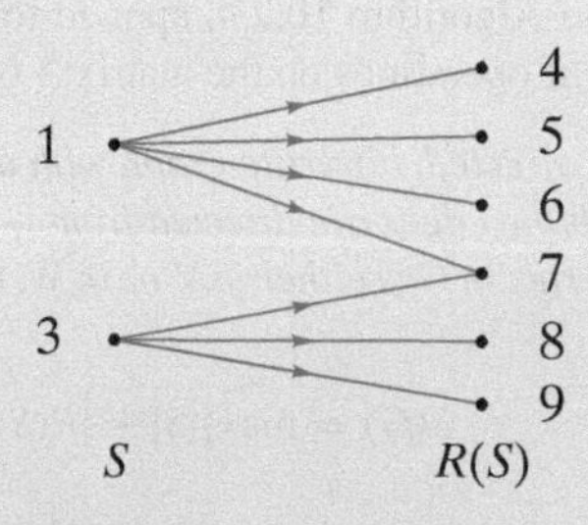


The fact that the minimum degree of vertices in $V$ is 3 means that for *any* subset $S$ of $V$, *at least* three edges are incident on each vertex in $S$. In general, there will be at least $3|S| = m_V|S|$ edges incident on vertices in $S$. In our example, $3|S| = 6$, but there are actually seven edges incident on vertices in $S$. The expression $m_V|S|$ is always a *lower bound* for the number of edges incident on vertices in $S$.

The fact that the maximum degree of vertices in $W$ is 2 means that for *any* subset $S$ of $V$, *at most* two edges are incident on each vertex in $R(S)$. In general, there will be at most $2|R(S)| = M_W|R(S)|$ edges incident on vertices in $R(S)$. In our example, $2|R(S)| = 12$, but there are actually 10 edges incident on vertices in $R(S)$. Since the edges incident on vertices in $S$ are a subset of the edges incident on vertices in $R(S)$, the expression $M_W|R(S)|$ is always an *upper bound* for the number of edges incident on vertices in $S$.

We have two ways of estimating the number of edges incident on vertices in $S$. The first way, using $S$, gives us a lower bound $m_V|S|$ on the number of edges, and the second way, using $R(S)$, gives us an upper bound $M_W|R(S)|$ on the number of edges. Comparing these estimates gives us the inequality

$$m_V|S| \le M_W|R(S)|.$$

We can't deduce $|S| \le |R(S)|$, but we haven't used the hypothesis

$$M_W \le m_V$$

yet! Combining the last two inequalities, we have

$$m_V|S| \le M_W|R(S)| \le m_V|R(S)|.$$

If we now cancel $m_V$ from both ends of the inequality, we obtain

$$|S| \le |R(S)|,$$

which is exactly the inequality we wanted to prove.

### Formal Solution

Let $S \subseteq V$. Each vertex in $S$ is incident on at least $m_V|S|$ edges; thus there are at least $m_V|S|$ edges incident on vertices in $S$. Each vertex in $R(S)$ is incident on at most $M_W$ edges; thus there are at most $M_W|R(S)|$ edges incident on vertices in $R(S)$. It follows that $m_V|S| \le M_W|R(S)|$. Since $M_W \le m_V$, $|R(S)|M_W \le |R(S)|m_V$. Therefore, $m_V|S| \le m_V|R(S)|$, and $|S| \le |R(S)|$. By Theorem 10.4.7, $G$ has a complete matching.

### Summary of Problem-Solving Techniques

- Look at example graphs.
- When looking at examples, it's a good idea to assign distinct values to the parameters in the

problem so you can keep them straight. (In our example we set $M_W = 2$ and $m_V = 3$.)

- Try to reduce given conditions to those in a useful theorem. (We reduced the conditions given in this problem to the conditions given in Hall's Marriage Theorem.)
- An inequality can sometimes be proved by estimating the size of some set in two different ways. If one estimate gives an upper bound $M$ and another gives a lower bound $m$, it follows that $m \leq M$.

## Comments

The last summarized problem-solving technique provides a method of proving an inequality. In a similar way, an equality can sometimes be proved by counting the number of elements in some set in two different ways. If one way of counting gives $c_1$ and the other way of counting gives $c_2$, it follows that $c_1 = c_2$. These techniques are widely used and their usefulness cannot be overemphasized. For example, in Section 6.2 we derived a formula for $C(n, r)$ by counting the number of $r$-permutations of an $n$-element set in two different ways.

## Exercise

1. Give an example of a bipartite graph $G$ that has a complete matching but does not satisfy the condition $M_W \leq m_V$.

# Chapter 10 Notes

General references that contain sections on network models are [Berge; Deo; Liu, 1968, 1985; and Tucker]. The classic work on networks is [Ford]; many of the results on networks, especially the early results, are due to Ford and Fulkerson, the authors of this book. [Tarjan] discusses network flow algorithms and implementation details.

See [Bachelis] for a nice direct proof of Hall's Marriage Theorem (Theorem 10.4.7) using mathematical induction.

The problem of finding a maximal flow in a network $G$, with source $a$, sink $z$, and capacities $C_{ij}$, may be rephrased as follows:

$$\text{maximize} \sum_j F_{aj}$$

subject to

$$0 \leq F_{ij} \leq C_{ij} \qquad \text{for all } i, j,$$

$$\sum_i F_{ij} = \sum_i F_{ji} \qquad \text{for all } j.$$

Such a problem is an example of a **linear programming problem.** In a linear programming problem, we want to maximize (or minimize) a linear expression, such as $\sum_j F_{aj}$, subject to linear inequality and equality constraints, such as $0 \leq F_{ij} \leq C_{ij}$ and $\sum_i F_{ij} = \sum_i F_{ji}$. Although the **simplex algorithm** is normally an efficient way to solve a general linear programming problem, network transport problems are usually more efficiently solved using Algorithm 10.2.4. See [Hillier] for an exposition of the simplex algorithm.

Suppose that for each edge $(i, j)$ in a network $G$, $c_{ij}$ represents the cost of the flow of one unit through edge $(i, j)$. Suppose that we want a maximal flow, with minimal cost

$$\sum_i \sum_j c_{ij} F_{ij}.$$

This problem, called the **transportation problem,** is again a linear programming problem and, as with the maximal flow problem, a specific algorithm can be used to obtain a solution that is, in general, more efficient than the simplex algorithm (see [Hillier]).

## Chapter 10 Review

### Section 10.1

1. (Transport) network
2. Source
3. Sink
4. Capacity
5. Flow in a network
6. Flow in an edge
7. Flow into a vertex
8. Flow out of a vertex
9. Conservation of flow
10. Given a flow $F$ in a network, the flow out of the source equals the flow into the sink. This common value is called the value of the flow $F$.
11. Supersource
12. Supersink

### Section 10.2

13. Maximal flow
14. Properly oriented edge with respect to a path
15. Improperly oriented edge with respect to a path
16. How to increase the flow in a path from the source to the sink when:
    (a) for each properly oriented edge the flow is less than the capacity and
    (b) each improperly oriented edge has positive flow (see Theorem 10.2.3)
17. How to find a maximal flow in a network (Algorithm 10.2.4)

### Section 10.3

18. Cut in a network
19. Capacity of a cut
20. The capacity of any cut is greater than or equal to the value of any flow (Theorem 10.3.7).
21. Minimal cut
22. Max flow, min cut theorem (Theorem 10.3.9)
23. At the termination of the maximal flow algorithm, Algorithm 10.2.4, the set of labeled vertices defines a minimal cut.

### Section 10.4

24. Matching
25. Maximal matching
26. Complete matching
27. Matching network
28. Relationship between flows and matchings (Theorem 10.4.5)
29. Hall's Marriage Theorem (Theorem 10.4.7)

## Chapter 10 Self-Test

*Exercises 1–4 refer to the following network. The capacities are shown on the edges.*

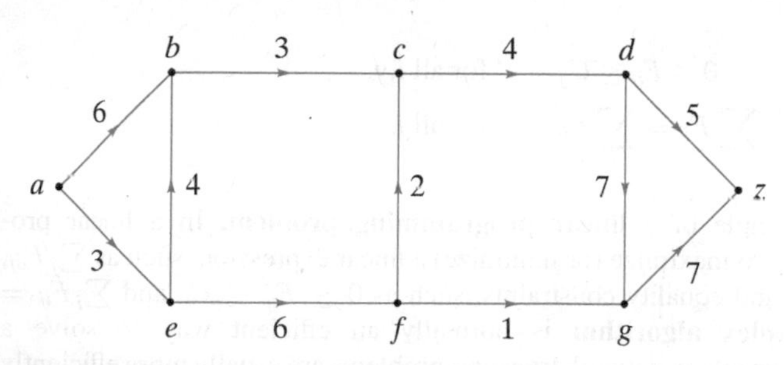


1. Explain why

$$F_{a,e}=2, \quad F_{e,b}=2, \quad F_{b,c}=3,$$
$$F_{c,d}=3, \quad F_{d,z}=3, \quad F_{a,b}=1,$$

with all other $F_{x,y}=0$, is a flow.

2. What is the flow into $b$?
3. What is the flow out of $c$?
4. What is the value of the flow $F$?
5. For the flow of Exercise 1, find a path from $a$ to $z$ satisfying the following: (a) for each properly oriented edge the flow is less than the capacity and (b) each improperly oriented edge has positive flow.
6. By modifying only the flows in the edges of the path of Exercise 5, find a flow with a larger value than $F$.
7. Use Algorithm 10.2.4 to find a maximal flow in the network of Exercise 1 (beginning with the flow in which the flow in each edge is equal to zero).
8. Use Algorithm 10.2.4 to find a maximal flow in the following network (beginning with the flow in which the flow in each edge is equal to zero).

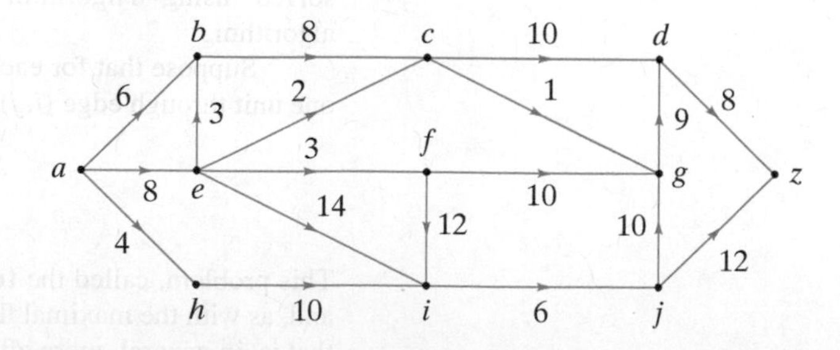

9. In each of parts (a)–(d), answer true if the statement is true for every network; otherwise, answer false.

(a) If the capacity of a cut in a network is equal to $Ca$, then the value of any flow is less than or equal to $Ca$.

(b) If the capacity of a cut in a network is equal to $Ca$, then the value of any flow is greater than or equal to $Ca$.

(c) If the capacity of a cut in a network is equal to $Ca$, then the value of some flow is greater than or equal to $Ca$.

(d) If the capacity of a cut in a network is equal to $Ca$, then the value of some flow is less than or equal to $Ca$.

10. Find the capacity of the cut $(P, \overline{P})$ in the network of Exercise 1, where $P = \{a, b, e, f\}$.

11. Is the cut $(P, \overline{P})$, $P = \{a, b, e, f\}$, in the network of Exercise 1 minimal? Explain.

12. Find a minimal cut in the network of Exercise 8.

*Exercises 13–16 refer to the following situation. Applicant A is qualified for jobs $J_2$, $J_4$, and $J_5$; applicant B is qualified for jobs $J_1$ and $J_3$; applicant C is qualified for jobs $J_1$, $J_3$, and $J_5$; and applicant D is qualified for jobs $J_3$ and $J_5$.*

13. Model the situation as a matching network.

14. Use Algorithm 10.2.4 to find a maximal matching.

15. Is there a complete matching?

16. Find a minimal cut in the matching network.

## Chapter 10 Computer Exercises

**1.** Write a program that accepts as input a network with a given flow and outputs all possible paths from the source to the sink on which the flow can be increased.

**2.** Implement Algorithm 10.2.4 that finds a maximal flow in a network as a program. Have the program output the minimal cut as well as the maximal flow.

**3.** Write a program that computes the deficiency of a network.

Chapter 11

# BOOLEAN ALGEBRAS AND COMBINATORIAL CIRCUITS



**Go Online**
Biographies of Boole and Shannon are at bit.ly/2S7L960

Several definitions honor the nineteenth-century mathematician George Boole—Boolean algebra, Boolean function, Boolean expression, and Boolean ring—to name a few. Boole is one of the persons in a long historical chain who were concerned with formalizing and mechanizing the process of logical thinking. In fact, in 1854 Boole wrote a book entitled *The Laws of Thought.* One of Boole's contributions was the development of a theory of logic using symbols instead of words. For a discussion of Boole's work, see [Hailperin].

Almost a century after Boole's work, it was observed, especially by C. E. Shannon in 1938 (see [Shannon]), that Boolean algebra could be used to analyze electrical circuits. Thus Boolean algebra became an indispensable tool for the analysis and design of electronic computers in the succeeding decades. We explore the relationship of Boolean algebra to circuits throughout this chapter.

## 11.1 Combinatorial Circuits

**Go Online**
For more on combinatorial circuits, see bit.ly/2S7L960

In a digital computer, there are only two possibilities, written 0 and 1, for the smallest, indivisible object. All programs and data are ultimately reducible to combinations of bits. A variety of devices have been used throughout the years in digital computers to store bits. Electronic circuits allow these storage devices to communicate with each other. A bit in one part of a circuit is transmitted to another part of the circuit as a voltage. Thus two voltage levels are needed—for example, a high voltage can communicate 1 and a low voltage can communicate 0.

In this section we discuss **combinatorial circuits.** The output of a combinatorial circuit is uniquely defined for every combination of inputs. A combinatorial circuit has no memory; previous inputs and the state of the system do not affect the output of a combinatorial circuit. Circuits for which the output is a function, not only of the inputs, but also of the state of the system, are called **sequential circuits** and are considered in Chapter 12.

Combinatorial circuits can be constructed using solid-state devices, called **gates,** which are capable of switching voltage levels (bits). We will begin by discussing AND, OR, and NOT gates.

**Definition 11.1.1 ▶** An *AND gate* receives inputs $x_1$ and $x_2$, where $x_1$ and $x_2$ are bits, and produces output denoted $x_1 \wedge x_2$, where

$$x_1 \wedge x_2 = \begin{cases} 1 & \text{if } x_1 = 1 \text{ and } x_2 = 1 \\ 0 & \text{otherwise.} \end{cases}$$

An AND gate is drawn as shown in Figure 11.1.1.

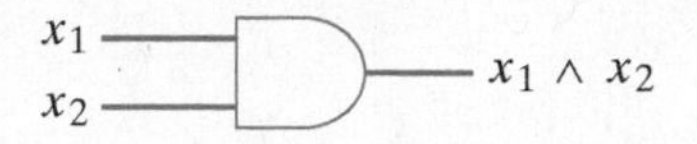


**Figure 11.1.1** AND gate. ◀

**Definition 11.1.2 ▶** An *OR gate* receives inputs $x_1$ and $x_2$, where $x_1$ and $x_2$ are bits, and produces output denoted $x_1 \vee x_2$, where

$$x_1 \vee x_2 = \begin{cases} 1 & \text{if } x_1 = 1 \text{ or } x_2 = 1 \\ 0 & \text{otherwise.} \end{cases}$$

An OR gate is drawn as shown in Figure 11.1.2.

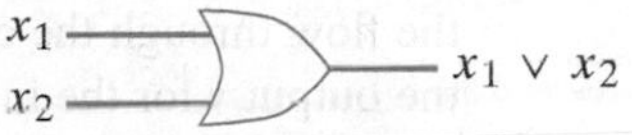

**Figure 11.1.2** OR gate. ◀

**Definition 11.1.3 ▶** A *NOT gate* (or *inverter*) receives input $x$, where $x$ is a bit, and produces output denoted $\overline{x}$, where

$$\overline{x} = \begin{cases} 1 & \text{if } x = 0 \\ 0 & \text{if } x = 1. \end{cases}$$

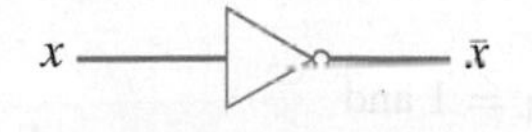


**Figure 11.1.3** NOT gate.

A NOT gate is drawn as shown in Figure 11.1.3. ◀

The **logic table** of a combinatorial circuit lists all possible inputs together with the resulting outputs.

**Example 11.1.4** Following are the logic tables for the basic AND, OR, and NOT circuits (Figures 11.1.1–11.1.3).

| $x_1$ | $x_2$ | $x_1 \wedge x_2$ |
|---|---|---|
| 1 | 1 | 1 |
| 1 | 0 | 0 |
| 0 | 1 | 0 |
| 0 | 0 | 0 |

| $x_1$ | $x_2$ | $x_1 \vee x_2$ |
|---|---|---|
| 1 | 1 | 1 |
| 1 | 0 | 1 |
| 0 | 1 | 1 |
| 0 | 0 | 0 |

| $x$ | $\overline{x}$ |
|---|---|
| 1 | 0 |
| 0 | 1 |

We note that performing the operation AND (OR) is the same as taking the minimum (maximum) of the two bits $x_1$ and $x_2$. ◀

**Example 11.1.5** The circuit of Figure 11.1.4 is an example of a combinatorial circuit since the output $y$ is uniquely defined for each combination of inputs $x_1$, $x_2$, and $x_3$.

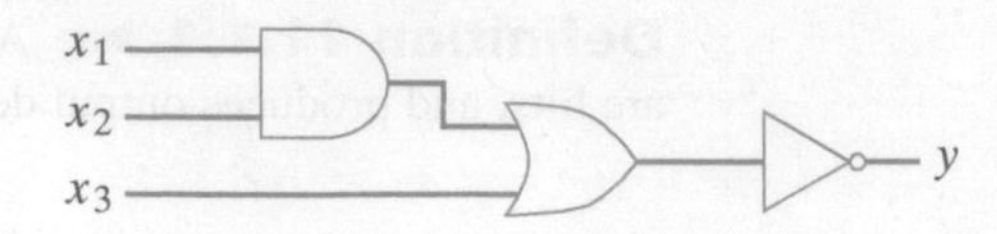


**Figure 11.1.4** A combinatorial circuit.

The logic table for this combinatorial circuit follows.

| $x_1$ | $x_2$ | $x_3$ | $y$ |
|---|---|---|---|
| 1 | 1 | 1 | 0 |
| 1 | 1 | 0 | 0 |
| 1 | 0 | 1 | 0 |
| 1 | 0 | 0 | 1 |
| 0 | 1 | 1 | 0 |
| 0 | 1 | 0 | 1 |
| 0 | 0 | 1 | 0 |
| 0 | 0 | 0 | 1 |

Notice that all possible combinations of values for the inputs $x_1, x_2$, and $x_3$ are listed. For a given set of inputs, we can compute the value of the output $y$ by tracing the flow through the circuit. For example, the fourth line of the table gives the value of the output $y$ for the input values $x_1 = 1,\ x_2 = 0,$ and $x_3 = 0$. If $x_1 = 1$ and $x_2 = 0$, the output from the AND gate is 0 (see Figure 11.1.5). Since $x_3 = 0$, the inputs to the OR gate are both 0. Therefore, the output of the OR gate is 0. Since the input to the NOT gate is 0, it produces output $y = 1$.

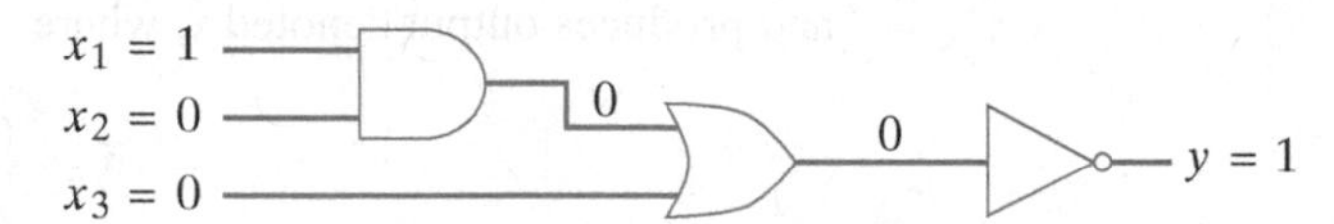


**Figure 11.1.5** The circuit of Figure 11.1.4 when $x_1 = 1$ and $x_2 = x_3 = 0$. ◀

**Example 11.1.6** The circuit of Figure 11.1.6 is not a combinatorial circuit, because the output $y$ is not uniquely defined for each combination of inputs $x_1$ and $x_2$. For example, suppose that $x_1 = 1$ and $x_2 = 0$. If the output of the AND gate is 0, then $y = 0$. On the other hand, if the output of the AND gate is 1, then $y = 1$. Such a circuit might be used to store one bit.

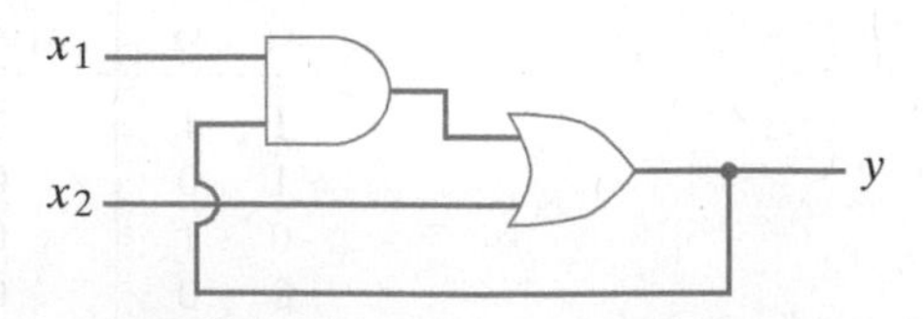


**Figure 11.1.6** A circuit that is not a combinatorial circuit. ◀

**Example 11.1.7** Individual combinatorial circuits may be interconnected. The combinatorial circuits $C_1$, $C_2$, and $C_3$ of Figure 11.1.7 may be combined, as shown, to obtain the combinatorial circuit $C$. ◀

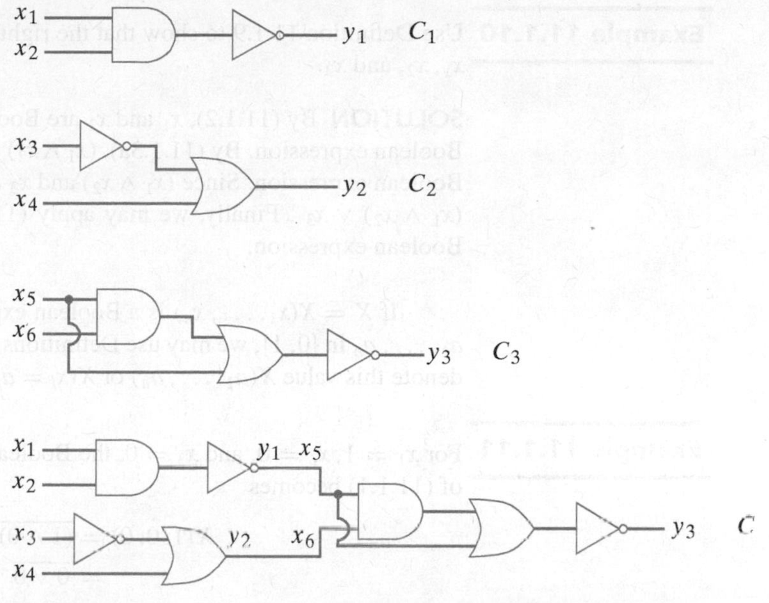

Figure 11.1.7 Combinatorial circuit $C$ is obtained by interconnecting the combinatorial circuits $C_1$, $C_2$, and $C_3$.

**Example 11.1.8** A combinatorial circuit with one output, such as that in Figure 11.1.4, can be represented by an expression using the symbols $\wedge$, $\vee$, and $^-$. We follow the flow of the circuit symbolically. First, $x_1$ and $x_2$ are ANDed (see Figure 11.1.8), which produces output $x_1 \wedge x_2$. This output is then ORed with $x_3$ to produce output $(x_1 \wedge x_2) \vee x_3$. This output is then NOTed. Thus the output $y$ may be

$$y = \overline{(x_1 \wedge x_2) \vee x_3}. \tag{11.1.1}$$

Expressions such as (11.1.1) are called **Boolean expressions.**

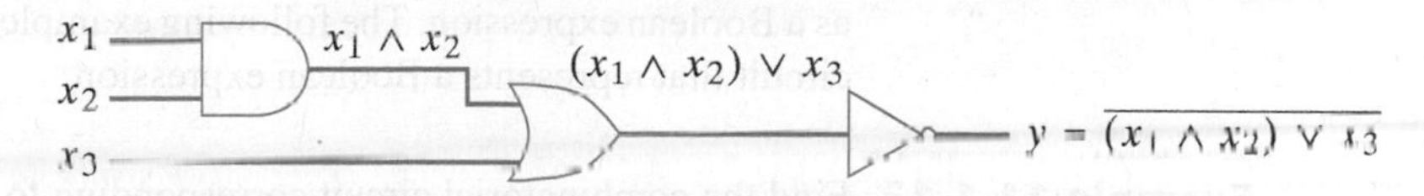


Figure 11.1.8 Representation of a combinatorial circuit by a Boolean expression.

**Definition 11.1.9 ▶** *Boolean expressions* in the symbols $x_1, \ldots, x_n$ are defined recursively as follows.

$$0, 1, x_1, \ldots, x_n \tag{11.1.2}$$

are Boolean expression. If $X_1$ and $X_2$ are Boolean expressions, then

$$\text{(a) } (X_1), \qquad \text{(b) } \overline{X}_1, \qquad \text{(c) } X_1 \vee X_2, \qquad \text{(d) } X_1 \wedge X_2 \tag{11.1.3}$$

are Boolean expressions.

If $X$ is a Boolean expression in the symbols $x_1, \ldots, x_n$, we sometimes write

$$X = X(x_1, \ldots, x_n).$$

Either symbol $x$ or $\overline{x}$ is called a *literal.*

**Example 11.1.10** Use Definition 11.1.9 to show that the right side of (11.1.1) is a Boolean expression in $x_1, x_2$, and $x_3$.

SOLUTION By (11.1.2), $x_1$ and $x_2$ are Boolean expressions. By (11.1.3d), $x_1 \wedge x_2$ is a Boolean expression. By (11.1.3a), $(x_1 \wedge x_2)$ is a Boolean expression. By (11.1.2), $x_3$ is a Boolean expression. Since $(x_1 \wedge x_2)$ and $x_3$ are Boolean expressions, by (11.1.3c), so is $(x_1 \wedge x_2) \vee x_3$. Finally, we may apply (11.1.3b) to conclude that $\overline{(x_1 \wedge x_2) \vee x_3}$ is a Boolean expression. ◀

If $X = X(x_1, \ldots, x_n)$ is a Boolean expression and $x_1, \ldots, x_n$ are assigned values $a_1, \ldots, a_n$ in $\{0, 1\}$, we may use Definitions 11.1.1–11.1.3 to compute a value for $X$. We denote this value $X(a_1, \ldots, a_n)$ or $X(x_i = a_i)$.

**Example 11.1.11** For $x_1 = 1, x_2 = 0$, and $x_3 = 0$, the Boolean expression $X(x_1, x_2, x_3) = \overline{(x_1 \wedge x_2) \vee x_3}$ of (11.1.1) becomes

$$\begin{aligned} X(1,0,0) &= \overline{(1 \wedge 0) \vee 0} \\ &= \overline{0 \vee 0} && \text{since } 1 \wedge 0 = 0 \\ &= \overline{0} && \text{since } 0 \vee 0 = 0 \\ &= 1 && \text{since } \overline{0} = 1. \end{aligned}$$

We have again computed the fourth row of the table in Example 11.1.5. ◀

In a Boolean expression in which parentheses are not used to specify the order of operations, we assume that $\wedge$ is evaluated before $\vee$.

**Example 11.1.12** For $x_1 = 0, x_2 = 0$, and $x_3 = 1$, the value of the Boolean expression $x_1 \wedge x_2 \vee x_3$ is

$$x_1 \wedge x_2 \vee x_3 = 0 \wedge 0 \vee 1 = 0 \vee 1 = 1.$$ ◀

Example 11.1.8 showed how to represent a combinatorial circuit with one output as a Boolean expression. The following example shows how to construct a combinatorial circuit that represents a Boolean expression.

**Example 11.1.13** Find the combinatorial circuit corresponding to the Boolean expression

$$(x_1 \wedge (\overline{x}_2 \vee x_3)) \vee x_2$$

and write the logic table for the circuit obtained.

SOLUTION We begin with the expression $\overline{x}_2 \vee x_3$ in the innermost parentheses. This expression is converted to a combinatorial circuit, as shown in Figure 11.1.9.

The output of this circuit is ANDed with $x_1$ to produce the circuit drawn in Figure 11.1.10. Finally, the output of this circuit is ORed with $x_2$ to give the desired circuit drawn in Figure 11.1.11. The logic table follows.

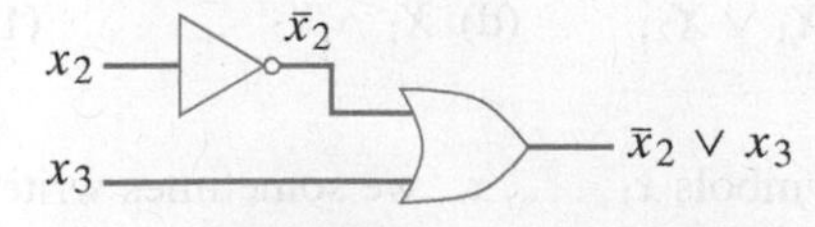


**Figure 11.1.9** The combinatorial circuit corresponding to the Boolean expression $\overline{x}_2 \vee x_3$.

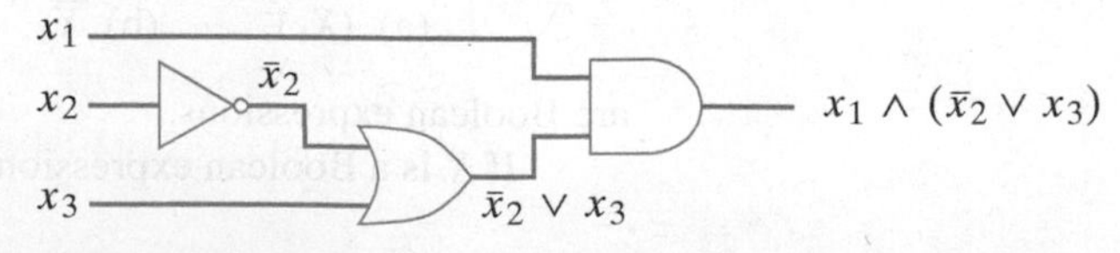


**Figure 11.1.10** The combinatorial circuit corresponding to the Boolean expression $x_1 \wedge (\overline{x}_2 \vee x_3)$.

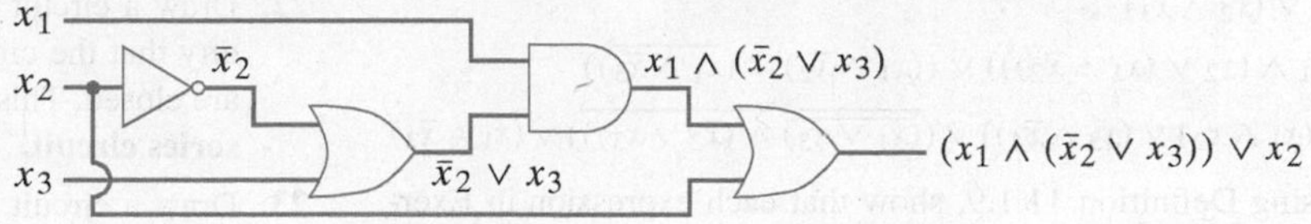

**Figure 11.1.11** The combinatorial circuit corresponding to the Boolean expression $(x_1 \wedge (\overline{x}_2 \vee x_3)) \vee x_2$.

| $x_1$ | $x_2$ | $x_3$ | $(x_1 \wedge (\overline{x}_2 \vee x_3)) \vee x_2$ |
|---|---|---|---|
| 1 | 1 | 1 | 1 |
| 1 | 1 | 0 | 1 |
| 1 | 0 | 1 | 1 |
| 1 | 0 | 0 | 1 |
| 0 | 1 | 1 | 1 |
| 0 | 1 | 0 | 1 |
| 0 | 0 | 1 | 0 |
| 0 | 0 | 0 | 0 |

## 11.1 Review Exercises

1. What is a combinatorial circuit?
2. What is a sequential circuit?
3. What is an AND gate?
4. What is an OR gate?
5. What is a NOT gate?
6. What is an inverter?
7. What is a logic table of a combinatorial circuit?
8. What is a Boolean expression?
9. What is a literal?

## 11.1 Exercises

*In Exercises 1–6, write the Boolean expression that represents the combinatorial circuit, write the logic table, and write the output of each gate symbolically as in Figure 11.1.8.*

1.

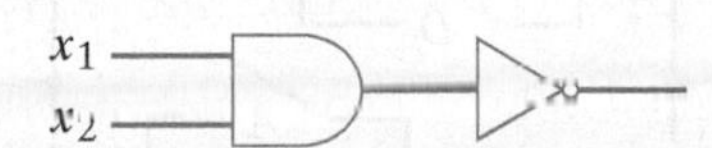


2.

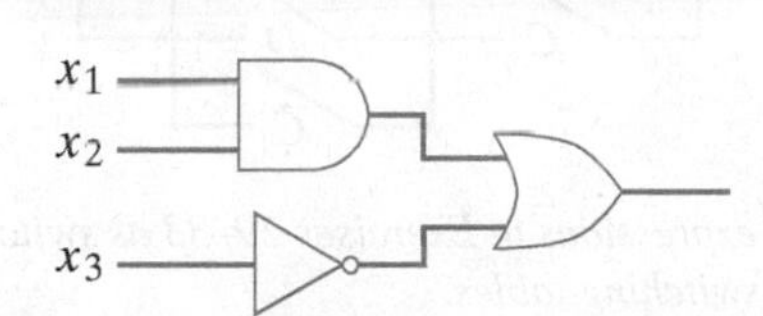


3.

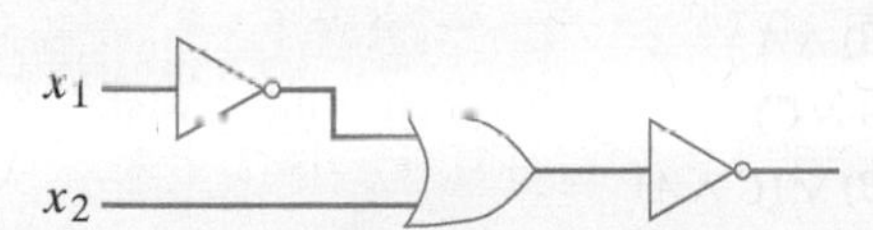


4.

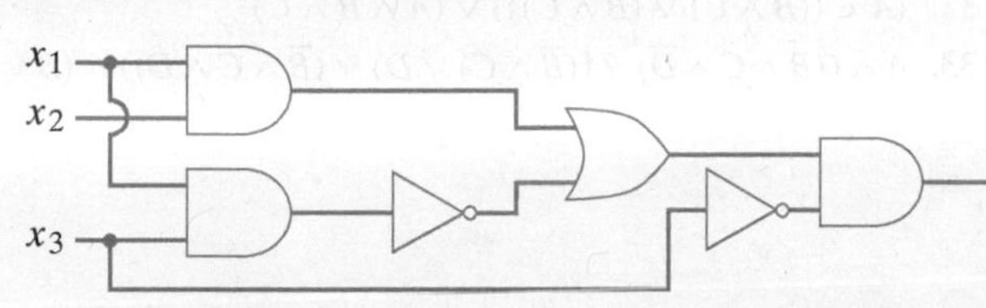


5.

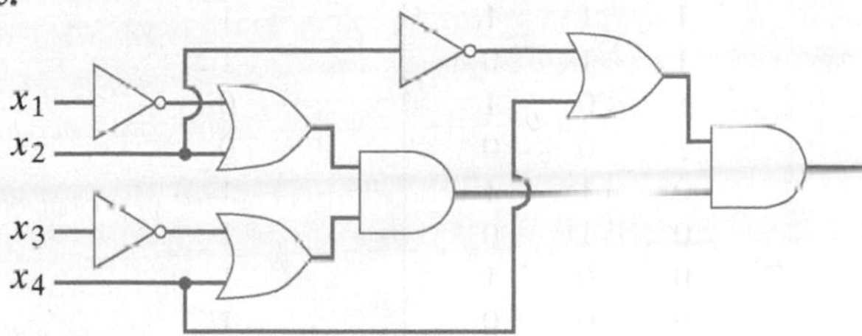


6. The circuit at the bottom of Figure 11.1.7.

*Exercises 7–9 refer to the circuit*

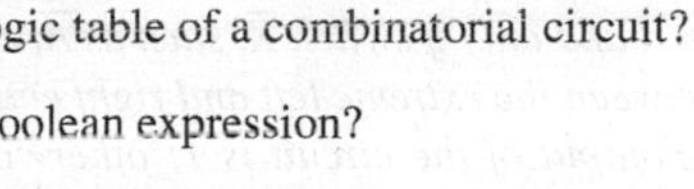

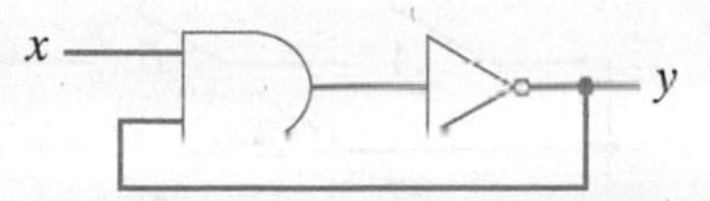


7. Show that this circuit is not a combinatorial circuit.
8. Show that if $x = 0$, the output $y$ is uniquely determined.
9. Show that if $x = 1$, the output $y$ is undetermined.

*In Exercises 10–14, find the value of the Boolean expressions for*

$$x_1 = 1, \quad x_2 = 1, \quad x_3 = 0, \quad x_4 = 1.$$

10. $\overline{x_1 \wedge x_2}$

11. $(x_1 \wedge \overline{x}_2) \vee (x_1 \vee \overline{x}_3)$

12. $x_1 \vee (\overline{x}_2 \wedge x_3)$
13. $(x_1 \wedge (x_2 \vee (x_1 \wedge \overline{x}_2))) \vee ((x_1 \wedge \overline{x}_2) \vee \overline{(x_1 \wedge \overline{x}_3)})$
14. $(((x_1 \wedge x_2) \vee (x_3 \wedge \overline{x}_4)) \vee \overline{((\overline{x_1 \vee x_3}) \wedge (\overline{x}_2 \vee x_3))}) \vee (x_1 \wedge \overline{x}_3)$
15. Using Definition 11.1.9, show that each expression in Exercises 10–14 is a Boolean expression.

*In Exercises 16–20, tell whether the given expression is a Boolean expression. If it is a Boolean expression, use Definition 11.1.9 to show that it is.*

16. $x_1 \wedge (x_2 \vee x_3)$
17. $(x_1)$
18. $x_1 \wedge \overline{x}_2 \vee x_3$
19. $((x_1 \wedge x_2) \vee \overline{x}_3$
20. $((x_1))$
21. Find the combinatorial circuit corresponding to each Boolean expression in Exercises 10–14 and write the logic table.

*A switching circuit is an electrical network consisting of switches each of which is open or closed. An example is given in Figure 11.1.12. If switch X is open (closed), we write X = 0 (X = 1). Switches labeled with the same letter, such as B in Figure 11.1.12, are either all open or all closed. Switch X, such as A in Figure 11.1.12, is open if and only if switch $\overline{X}$, such as $\overline{A}$, is closed. If current can flow between the extreme left and right ends of the circuit, we say that the output of the circuit is 1; otherwise, we say that the output of the circuit is 0. A switching table gives the output of the circuit for all values of the switches. The switching table for Figure 11.1.12 is as follows:*

| A | B | C | *Circuit Output* |
|---|---|---|---|
| 1 | 1 | 1 | 1 |
| 1 | 1 | 0 | 1 |
| 1 | 0 | 1 | 0 |
| 1 | 0 | 0 | 0 |
| 0 | 1 | 1 | 1 |
| 0 | 1 | 0 | 1 |
| 0 | 0 | 1 | 1 |
| 0 | 0 | 0 | 1 |

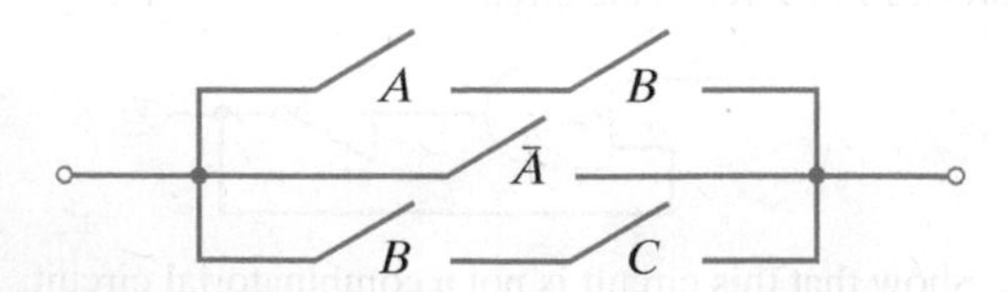

**Figure 11.1.12** A switching circuit.

22. Draw a circuit with two switches $A$ and $B$ having the property that the circuit output is 1 precisely when both $A$ and $B$ are closed. This configuration is labeled $A \wedge B$ and is called a **series circuit.**
23. Draw a circuit with two switches $A$ and $B$ having the property that the circuit output is 1 precisely when either $A$ or $B$ is closed. This configuration is labeled $A \vee B$ and is called a **parallel circuit.**
24. Show that the circuit of Figure 11.1.12 can be represented symbolically as
$$(A \wedge B) \vee \overline{A} \vee (B \wedge C).$$

*Represent each circuit in Exercises 25–29 symbolically and give its switching table.*

25.

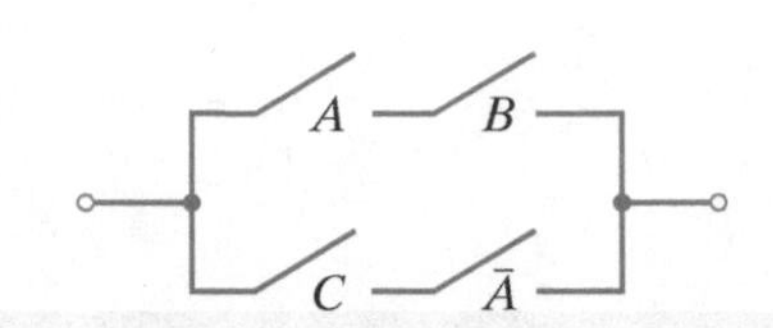


26.

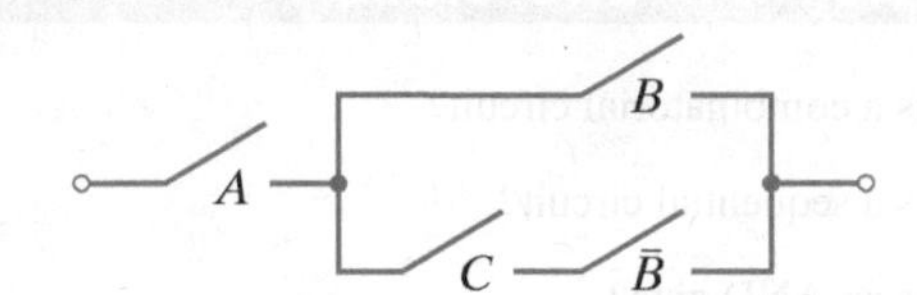


27.

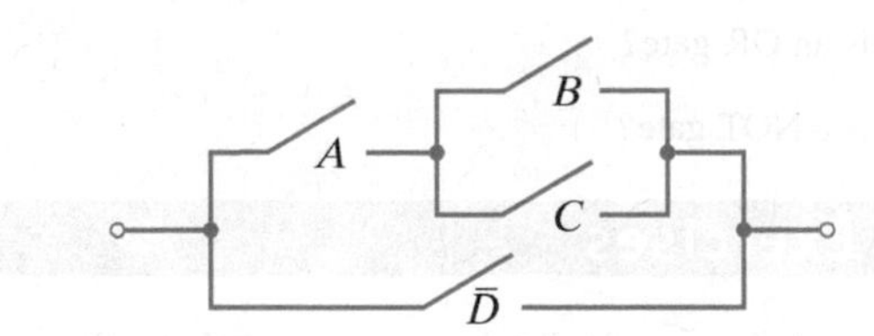


28.

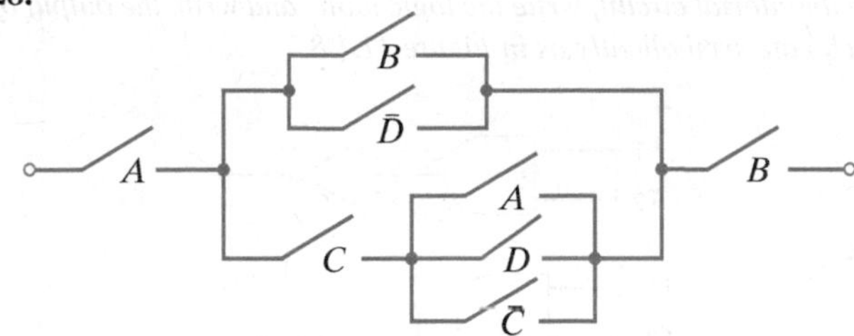


*Represent the expressions in Exercises 29–33 as switching circuits and write the switching tables.*

29. $(A \vee \overline{B}) \wedge A$
30. $A \vee (\overline{B} \wedge C)$
31. $(\overline{A} \wedge B) \vee (C \wedge A)$
32. $(A \wedge ((B \wedge \overline{C}) \vee (\overline{B} \wedge C))) \vee (\overline{A} \wedge B \wedge C)$
33. $A \wedge ((B \wedge C \wedge \overline{D}) \vee ((\overline{B} \wedge C) \vee D) \vee (\overline{B} \wedge \overline{C} \wedge D)) \wedge (B \vee \overline{D})$

## 11.2 Properties of Combinatorial Circuits

In the preceding section we defined two binary operators $\wedge$ and $\vee$ on $Z_2 = \{0, 1\}$ and a unary operator $^-$ on $Z_2$. (Throughout the remainder of this chapter we let $Z_2$ denote the set $\{0, 1\}$.) We saw that these operators could be implemented in circuits as gates. In this section we discuss some properties of the system consisting of $Z_2$ and the operators $\wedge$, $\vee$, and $^-$.

**Theorem 11.2.1** If $\wedge$, $\vee$, and $^-$ are as in Definitions 11.1.1–11.1.3, then the following properties hold.

(a) Associative laws:

$$(a \vee b) \vee c = a \vee (b \vee c)$$
$$(a \wedge b) \wedge c = a \wedge (b \wedge c) \qquad \text{for all } a, b, c \in Z_2.$$

(b) Commutative laws:

$$a \vee b = b \vee a, \qquad a \wedge b = b \wedge a \qquad \text{for all } a, b \in Z_2.$$

(c) Distributive laws:

$$a \wedge (b \vee c) = (a \wedge b) \vee (a \wedge c)$$
$$a \vee (b \wedge c) = (a \vee b) \wedge (a \vee c) \qquad \text{for all } a, b, c \in Z_2.$$

(d) Identity laws:

$$a \vee 0 = a, \qquad a \wedge 1 = a \qquad \text{for all } a \in Z_2.$$

(e) Complement laws:

$$a \vee \overline{a} = 1, \qquad a \wedge \overline{a} = 0 \qquad \text{for all } a \in Z_2.$$

**Proof** The proofs are straightforward verifications. We shall prove the first distributive law only and leave the other equations as exercises (see Exercises 16 and 17).

We must show that

$$a \wedge (b \vee c) = (a \wedge b) \vee (a \wedge c) \qquad \text{for all } a, b, c \in Z_2. \qquad \textbf{(11.2.1)}$$

We simply evaluate both sides of (11.2.1) for all possible values of $a$, $b$, and $c$ in $Z_2$ and verify that in each case we obtain the same result. The table gives the details.

| $a$ | $b$ | $c$ | $a \wedge (b \vee c)$ | $(a \wedge b) \vee (a \wedge c)$ |
|---|---|---|---|---|
| 1 | 1 | 1 | 1 | 1 |
| 1 | 1 | 0 | 1 | 1 |
| 1 | 0 | 1 | 1 | 1 |
| 1 | 0 | 0 | 0 | 0 |
| 0 | 1 | 1 | 0 | 0 |
| 0 | 1 | 0 | 0 | 0 |
| 0 | 0 | 1 | 0 | 0 |
| 0 | 0 | 0 | 0 | 0 |

◀

**Example 11.2.2** By using Theorem 11.2.1, show that the combinatorial circuits of Figure 11.2.1 have identical outputs for given identical inputs.

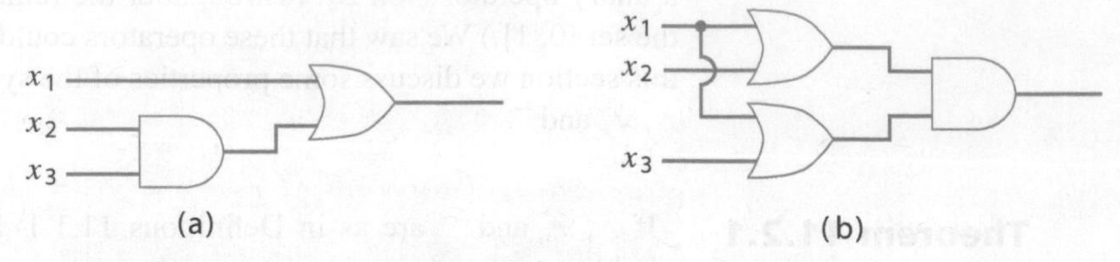


**Figure 11.2.1** The combinatorial circuits (a) and (b) have identical outputs for given identical inputs and are said to be equivalent.

**SOLUTION** The Boolean expressions representing the circuits are, respectively,

$$x_1 \vee (x_2 \wedge x_3), \qquad (x_1 \vee x_2) \wedge (x_1 \vee x_3).$$

By Theorem 11.2.1(c),

$$a \vee (b \wedge c) = (a \vee b) \wedge (a \vee c) \qquad \text{for all } a, b, c \in Z_2. \tag{11.2.2}$$

But (11.2.2) says that the combinatorial circuits of Figure 11.2.1 have identical outputs for given identical inputs. ◀

Arbitrary Boolean expressions are defined to be equal if they have the same values for all possible assignments of bits to the literals.

**Definition 11.2.3** ▶ Let

$$X_1 = X_1(x_1, \ldots, x_n) \qquad \text{and} \qquad X_2 = X_2(x_1, \ldots, x_n)$$

be Boolean expressions. We define $X_1$ to be *equal* to $X_2$ and write $X_1 = X_2$ if

$$X_1(a_1, \ldots, a_n) = X_2(a_1, \ldots, a_n) \qquad \text{for all } a_i \in Z_2.$$ ◀

**Example 11.2.4** Show that

$$\overline{(x \vee y)} = \overline{x} \wedge \overline{y}. \tag{11.2.3}$$

**SOLUTION** According to Definition 11.2.3, equation (11.2.3) holds if the equation is true for all choices of $x$ and $y$ in $Z_2$. Thus we may simply construct a table listing all possibilities to verify (11.2.3).

| $x$ | $y$ | $\overline{(x \vee y)}$ | $\overline{x} \wedge \overline{y}$ |
|---|---|---|---|
| 1 | 1 | 0 | 0 |
| 1 | 0 | 0 | 0 |
| 0 | 1 | 0 | 0 |
| 0 | 0 | 1 | 1 |

◀

If we define a relation $R$ on a set of Boolean expressions by the rule $X_1 R X_2$ if $X_1 = X_2$, $R$ is an equivalence relation. Each equivalence class consists of a set of Boolean expressions any one of which is equal to any other.

Because of the associative laws, Theorem 11.2.1(a), we can unambiguously write

$$a_1 \vee a_2 \vee \cdots \vee a_n \tag{11.2.4}$$

or

$$a_1 \wedge a_2 \wedge \cdots \wedge a_n \tag{11.2.5}$$

for $a_i \in Z_2$. The combinatorial circuit corresponding to (11.2.4) is drawn as in Figure 11.2.2, and the combinatorial circuit corresponding to (11.2.5) is drawn as in Figure 11.2.3.

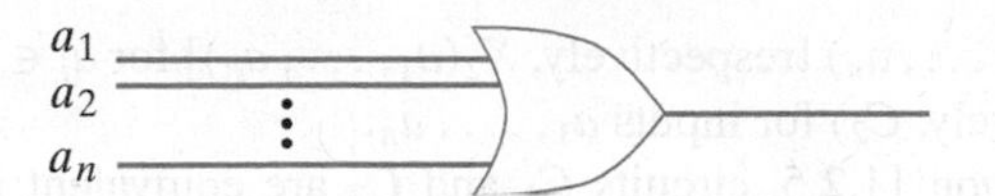


**Figure 11.2.2** An $n$-input OR gate.

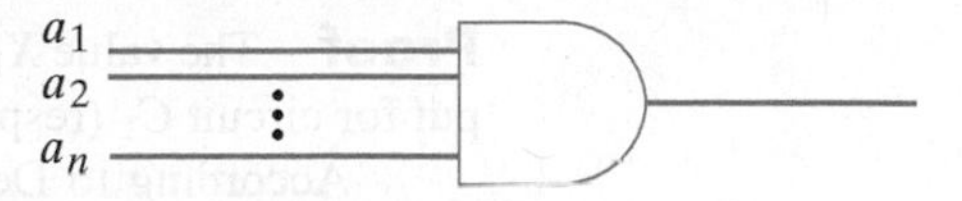


**Figure 11.2.3** An $n$-input AND gate.

The properties listed in Theorem 11.2.1 hold for a variety of systems. Any system satisfying these properties is called a **Boolean algebra.** Abstract Boolean algebras are examined in Section 11.3.

Having defined equality of Boolean expressions, we define equivalence of combinatorial circuits.

**Definition 11.2.5 ▶** We say that two combinatorial circuits, each having inputs $x_1, \ldots, x_n$ and a single output, are *equivalent* if, whenever the circuits receive the same inputs, they produce the same outputs. ◀

**Example 11.2.6** The combinatorial circuits of Figures 11.2.4 and 11.2.5 are equivalent since, as shown, they have identical logic tables.

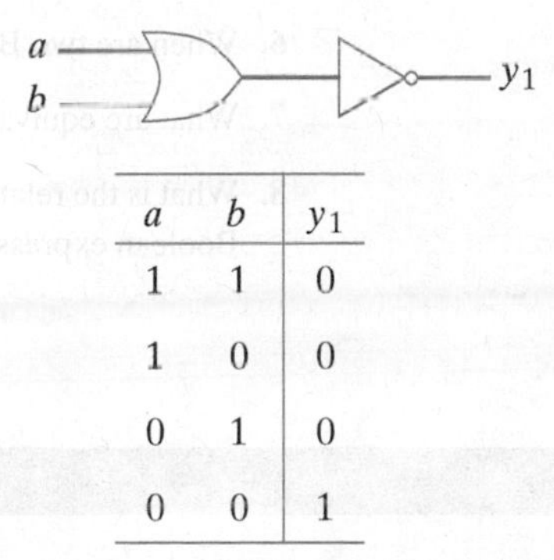


| $a$ | $b$ | $y_1$ |
|---|---|---|
| 1 | 1 | 0 |
| 1 | 0 | 0 |
| 0 | 1 | 0 |
| 0 | 0 | 1 |

**Figure 11.2.4** A combinatorial circuit and its logic table.

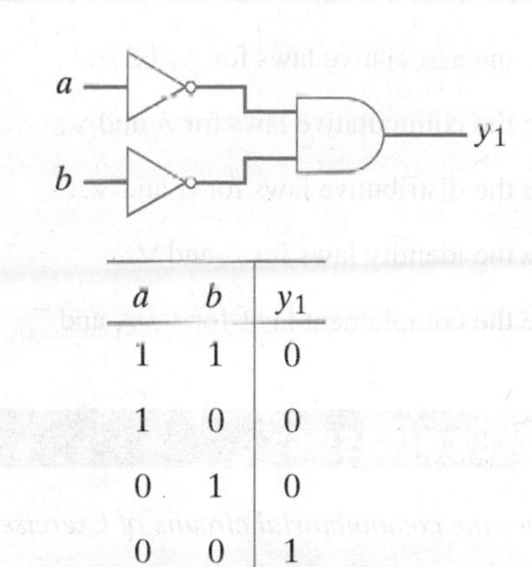


| $a$ | $b$ | $y_1$ |
|---|---|---|
| 1 | 1 | 0 |
| 1 | 0 | 0 |
| 0 | 1 | 0 |
| 0 | 0 | 1 |

**Figure 11.2.5** A combinatorial circuit and its logic table, which is identical to the logic table of Figure 11.2.4. The circuits of Figures 11.2.4 and 11.2.5 are said to be equivalent because they have identical logic tables. ◀

If we define a relation $R$ on a set of combinatorial circuits by the rule $C_1 \, R \, C_2$ if $C_1$ and $C_2$ are equivalent (in the sense of Definition 11.2.5), $R$ is an equivalence relation. Each equivalence class consists of a set of mutually equivalent combinatorial circuits.

Example 11.2.6 shows that equivalent circuits may not have the same number of gates. In general, it is desirable to use as few gates as possible to minimize the cost of the components.

It follows immediately from the definitions that combinatorial circuits are equivalent if and only if the Boolean expressions that represent them are equal.

**Theorem 11.2.7** Let $C_1$ and $C_2$ be combinatorial circuits represented, respectively, by the Boolean expressions $X_1 = X_1(x_1, \ldots, x_n)$ and $X_2 = X_2(x_1, \ldots, x_n)$. Then $C_1$ and $C_2$ are equivalent if and only if $X_1 = X_2$.

**Proof** The value $X_1(a_1, \ldots, a_n)$ [respectively, $X_2(a_1, \ldots, a_n)$] for $a_i \in Z_2$ is the output for circuit $C_1$ (respectively, $C_2$) for inputs $a_1, \ldots, a_n$.

According to Definition 11.2.5, circuits $C_1$ and $C_2$ are equivalent if and only if they have the same outputs $X_1(a_1, \ldots, a_n)$ and $X_2(a_1, \ldots, a_n)$ for all possible inputs $a_1, \ldots, a_n$. Thus circuits $C_1$ and $C_2$ are equivalent if and only if

$$X_1(a_1, \ldots, a_n) = X_2(a_1, \ldots, a_n) \qquad \text{for all values } a_i \in Z_2. \tag{11.2.6}$$

But by Definition 11.2.3, equation (11.2.6) holds if and only if $X_1 = X_2$. ◀

**Example 11.2.8** In Example 11.2.4 we showed that

$$\overline{(x \vee y)} = \overline{x} \wedge \overline{y}.$$

By Theorem 11.2.7, the combinatorial circuits (Figures 11.2.4 and 11.2.5) corresponding to these expressions are equivalent. ◀

## 11.2 Review Exercises

1. State the associative laws for $\wedge$ and $\vee$.
2. State the commutative laws for $\wedge$ and $\vee$.
3. State the distributive laws for $\wedge$ and $\vee$.
4. State the identity laws for $\wedge$ and $\vee$.
5. State the complement laws for $\wedge$, $\vee$, and $^{-}$.
6. When are two Boolean expressions equal?
7. What are equivalent combinatorial expressions?
8. What is the relation between combinatorial expressions and the Boolean expressions that represent them?

## 11.2 Exercises

*Show that the combinatorial circuits of Exercises 1–5 are equivalent.*

1.

$x_1$, $x_2$ → $y_1$

(a)

$x_1$, $x_2$ → $y_1$

(b)

2.

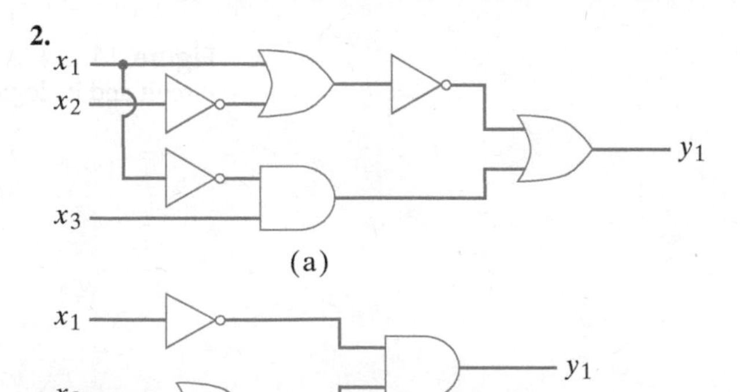

3.

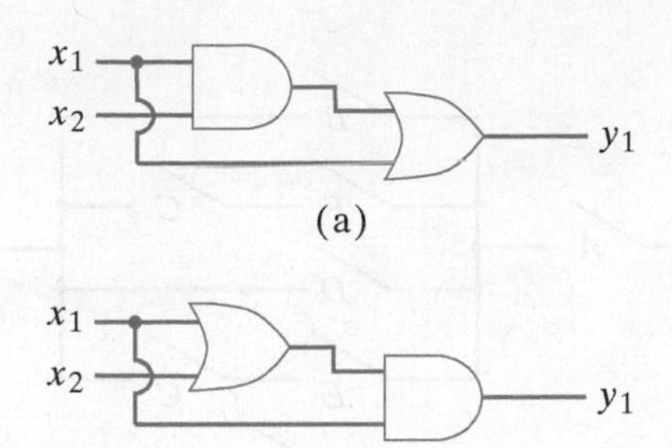


4.

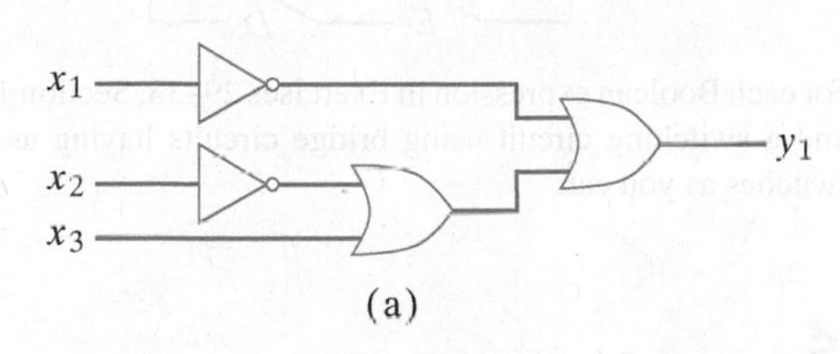


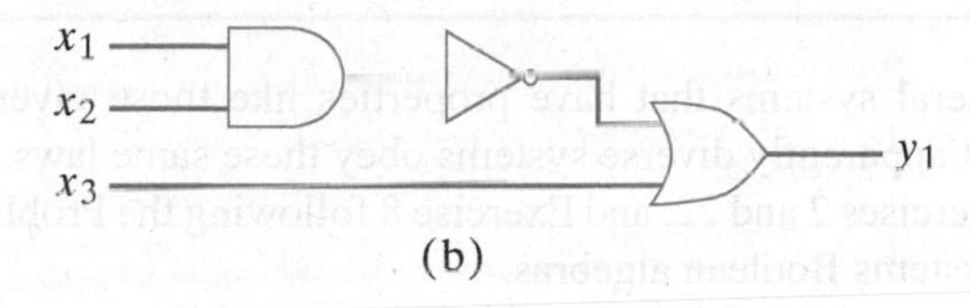


5.

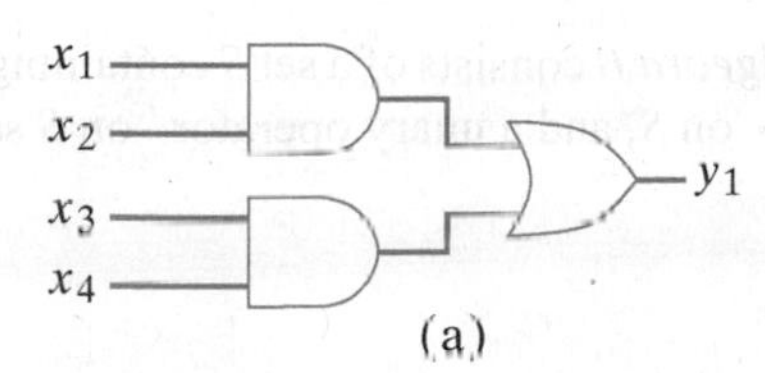


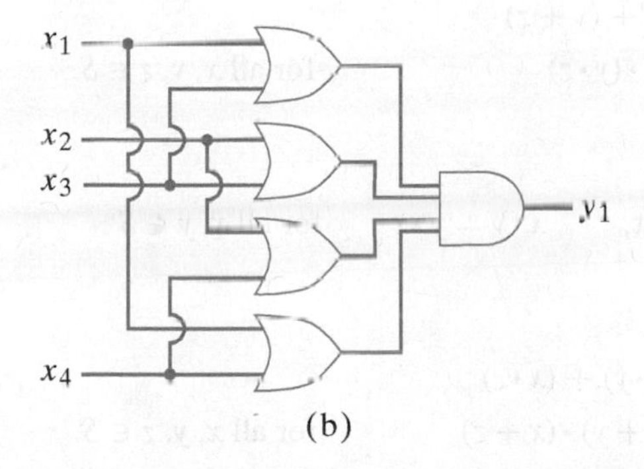


*Verify the equations in Exercises 6–10.*

6. $x_1 \vee x_1 = x_1$

7. $x_1 \wedge \overline{x}_2 = \overline{(\overline{x}_1 \vee x_2)}$

8. $x_1 \vee (x_1 \wedge x_2) = x_1$

9. $x_1 \wedge \overline{(x_2 \wedge x_3)} = (x_1 \wedge \overline{x}_2) \vee (x_1 \wedge \overline{x}_3)$

10. $(x_1 \vee x_2) \wedge (x_3 \vee x_4) = (x_3 \wedge x_1) \vee (x_3 \wedge x_2) \vee (x_4 \wedge x_1) \vee (x_4 \wedge x_2)$

*Prove or disprove the equations in Exercises 11–15.*

11. $\overline{\overline{x}} = x$

12. $\overline{x}_1 \wedge \overline{x}_2 = x_1 \vee x_2$

13. $\overline{x}_1 \wedge ((x_2 \wedge x_3) \vee (x_1 \wedge x_2 \wedge x_3)) = x_2 \wedge x_3$

14. $\overline{((\overline{x}_1 \wedge x_2) \vee (x_1 \wedge \overline{x}_3))} = (x_1 \vee \overline{x}_2) \wedge (x_1 \vee \overline{x}_3)$

15. $(x_1 \vee x_2) \wedge (\overline{x}_3 \vee x_4) \wedge (x_3 \wedge \overline{x}_2) = 0$

16. Prove the second statement of Theorem 11.2.1(c).

17. Prove Theorem 11.2.1, parts (a), (b), (d), and (e).

*We say that two switching circuits are equivalent if the Boolean expressions that represent them are equal.*

18. Show that the switching circuits are equivalent.

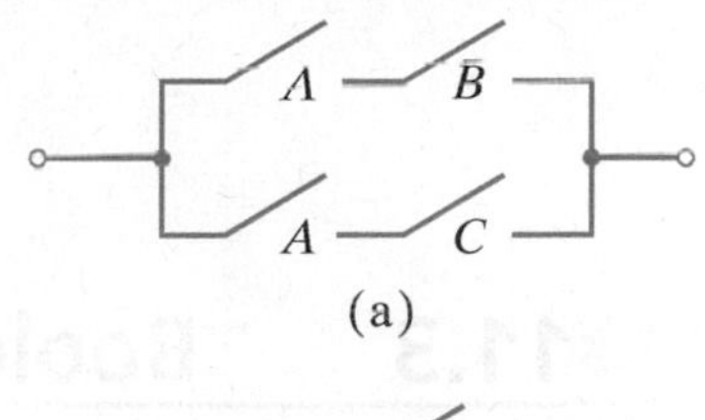


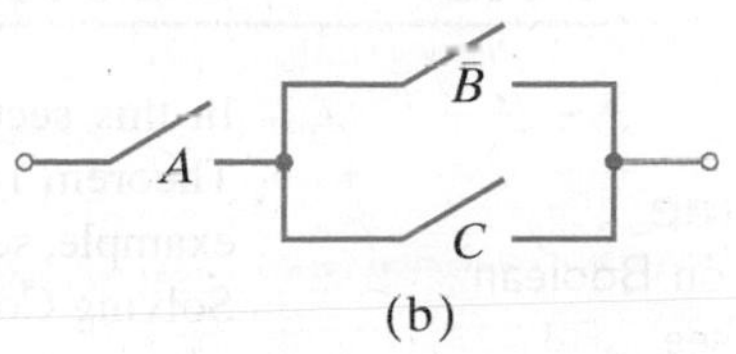


19. For each switching circuit in Exercises 25–28, Section 11.1, find an equivalent switching circuit using parallel and series circuits having as few switches as you can.

20. For each Boolean expression in Exercises 29–33, Section 11.1, find a switching circuit using parallel and series circuits having as few switches as you can.

*A bridge circuit is a switching circuit, such as the one shown here, that uses nonparallel and nonseries circuits.*

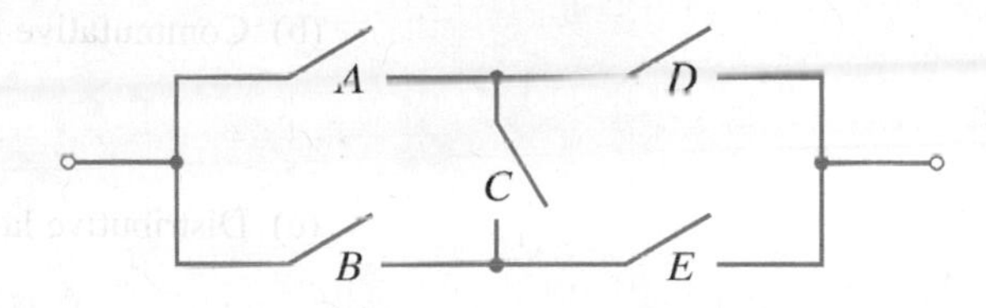


*For each switching circuit, find an equivalent switching circuit using bridge circuits having as few switches as you can.*

21.

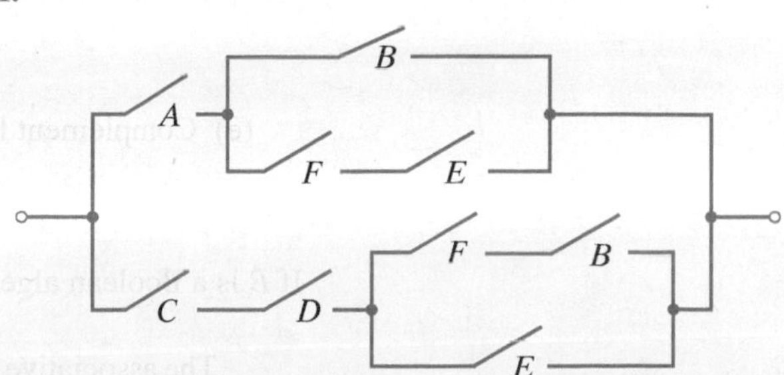

22.

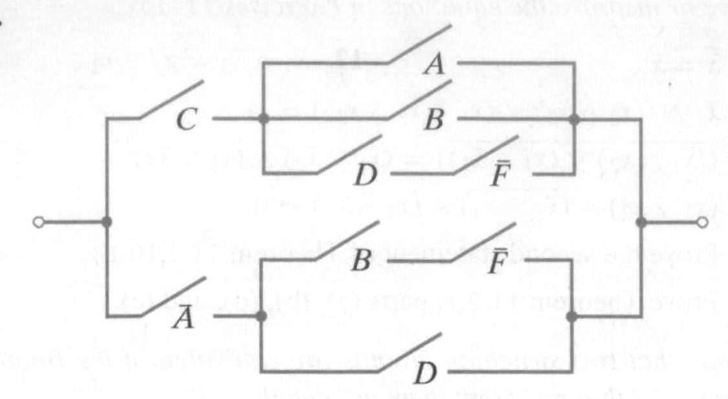


★23.

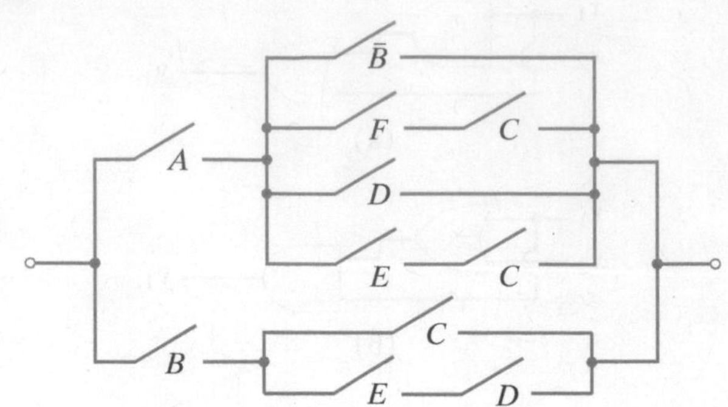


**24.** For each Boolean expression in Exercises 29–33, Section 11.1, find a switching circuit using bridge circuits having as few switches as you can.

## 11.3 Boolean Algebras

**Go Online**
For more on Boolean algebras, see
bit.ly/2S7L960

In this section we consider general systems that have properties like those given in Theorem 11.2.1. We will see that apparently diverse systems obey these same laws (for example, see Example 11.3.2, Exercises 2 and 22, and Exercise 8 following the Problem-Solving Corner). We call such systems Boolean algebras.

**Definition 11.3.1 ▶** A *Boolean algebra B* consists of a set $S$ containing distinct elements 0 and 1, binary operators $+$ and $\cdot$ on $S$, and a unary operator $'$ on $S$ satisfying the following laws.

(a) Associative laws:

$$(x + y) + z = x + (y + z)$$
$$(x \cdot y) \cdot z = x \cdot (y \cdot z) \qquad \text{for all } x, y, z \in S.$$

(b) Commutative laws:

$$x + y = y + x, \qquad x \cdot y = y \cdot x \qquad \text{for all } x, y \in S.$$

(c) Distributive laws:

$$x \cdot (y + z) = (x \cdot y) + (x \cdot z)$$
$$x + (y \cdot z) = (x + y) \cdot (x + z) \qquad \text{for all } x, y, z \in S.$$

(d) Identity laws:

$$x + 0 = x, \qquad x \cdot 1 = x \qquad \text{for all } x \in S.$$

(e) Complement laws:

$$x + x' = 1, \qquad x \cdot x' = 0 \qquad \text{for all } x \in S.$$

If $B$ is a Boolean algebra, we write $B = (S, +, \cdot, ', 0, 1)$. ◀

The associative laws can be omitted from Definition 11.3.1 since they follow from the other laws (see Exercise 24).

**Example 11.3.2** By Theorem 11.2.1, $(Z_2, \vee, \wedge, \bar{}, 0, 1)$ is a Boolean algebra. (We are letting $Z_2$ denote the set $\{0, 1\}$.) The operators $+$, $\cdot$, $'$ in Definition 11.3.1 are $\vee$, $\wedge$, $\bar{}$, respectively. ◀

As is the standard custom, we will usually abbreviate $a \cdot b$ as $ab$. We also assume that $\cdot$ is evaluated before $+$. This allows us to eliminate some parentheses. For example, we can write $(xy) + z$ more simply as $xy + z$.

Several comments are in order concerning Definition 11.3.1. In the first place, 0 and 1 are merely symbolic names and, in general, have nothing to do with the numbers 0 and 1. This same comment applies to $+$ and $\cdot$, which merely denote binary operators and, in general, have nothing to do with ordinary addition and multiplication.

**Example 11.3.3** Let $U$ be a universal set and let $S = \mathcal{P}(U)$, the power set of $U$. If we define the following operations

$$X + Y = X \cup Y, \qquad X \cdot Y = X \cap Y, \qquad X' = \overline{X}$$

on $S$, then $(S, \cup, \cap, \bar{}, \varnothing, U)$ is a Boolean algebra. The empty set $\varnothing$ plays the role of 0 and the universal set $U$ plays the role of 1. If we let $X$, $Y$, and $Z$ be subsets of $S$, properties (a)–(e) of Definition 11.3.1 become the following properties of sets (see Theorem 1.1.22):

(a′) $(X \cup Y) \cup Z = X \cup (Y \cup Z)$
$\quad\;\; (X \cap Y) \cap Z = X \cap (Y \cap Z)$ for all $X, Y, Z \in \mathcal{P}(U)$.

(b′) $X \cup Y = Y \cup X, \quad X \cap Y = Y \cap X$ for all $X, Y \in \mathcal{P}(U)$.

(c′) $X \cap (Y \cup Z) = (X \cap Y) \cup (X \cap Z)$
$\quad\;\; X \cup (Y \cap Z) = (X \cup Y) \cap (X \cup Z)$ for all $X, Y, Z \in \mathcal{P}(U)$.

(d′) $X \cup \varnothing = X, \quad X \cap U = X$ for every $X \in \mathcal{P}(U)$.

(e′) $X \cup \overline{X} = U, \quad X \cap \overline{X} = \varnothing$ for every $X \in \mathcal{P}(U)$. ◀

At this point we will deduce several other properties of Boolean algebras. We begin by showing that the element $x'$ in Definition 11.3.1(e) is unique.

**Theorem 11.3.4** In a Boolean algebra, the element $x'$ of Definition 11.3.1(e) is unique. Specifically, if $x + y = 1$ and $xy = 0$, then $y = x'$.

**Proof**

| | |
|---|---|
| $y = y1$ | Definition 11.3.1(d) |
| $= y(x + x')$ | Definition 11.3.1(e) |
| $= yx + yx'$ | Definition 11.3.1(c) |
| $= xy + yx'$ | Definition 11.3.1(b) |
| $= 0 + yx'$ | Given |
| $= xx' + yx'$ | Definition 11.3.1(e) |
| $= x'x + x'y$ | Definition 11.3.1(b) |
| $= x'(x + y)$ | Definition 11.3.1(c) |
| $= x'1$ | Given |
| $= x'$ | Definition 11.3.1(d) |

◀

**Definition 11.3.5** ▶ In a Boolean algebra, we call the element $x'$ the *complement* of $x$. ◀

We can now derive several additional properties of Boolean algebras.

**Theorem 11.3.6** Let $B = (S, +, \cdot, ', 0, 1)$ be a Boolean algebra. The following properties hold.

(a) Idempotent laws:

$$x + x = x, \qquad xx = x \qquad \text{for all } x \in S.$$

(b) Bound laws:

$$x + 1 = 1, \qquad x0 = 0 \qquad \text{for all } x \in S.$$

(c) Absorption laws:

$$x + xy = x, \qquad x(x + y) = x \qquad \text{for all } x, y \in S.$$

(d) Involution law:

$$(x')' = x \qquad \text{for all } x \in S.$$

(e) 0 and 1 laws:

$$0' = 1, \qquad 1' = 0.$$

(f) De Morgan's laws for Boolean algebras:

$$(x + y)' = x'y', \qquad (xy)' = x' + y' \qquad \text{for all } x, y \in S.$$

**Proof** We will prove (b) and the first statement of parts (a), (c), and (f) and leave the others as exercises (see Exercises 18–20).

(a)

| | |
|---|---|
| $x = x + 0$ | Definition 11.3.1(d) |
| $= x + (xx')$ | Definition 11.3.1(e) |
| $= (x + x)(x + x')$ | Definition 11.3.1(c) |
| $= (x + x)1$ | Definition 11.3.1(e) |
| $= x + x$ | Definition 11.3.1(d) |

(b)

| | |
|---|---|
| $x + 1 = (x + 1)1$ | Definition 11.3.1(d) |
| $= (x + 1)(x + x')$ | Definition 11.3.1(e) |
| $= x + 1x'$ | Definition 11.3.1(c) |
| $= x + x'1$ | Definition 11.3.1(b) |
| $= x + x'$ | Definition 11.3.1(d) |
| $= 1$ | Definition 11.3.1(e) |
| $x0 = x0 + 0$ | Definition 11.3.1(d) |
| $= x0 + xx'$ | Definition 11.3.1(e) |
| $= x(0 + x')$ | Definition 11.3.1(c) |
| $= x(x' + 0)$ | Definition 11.3.1(b) |
| $= xx'$ | Definition 11.3.1(d) |
| $= 0$ | Definition 11.3.1(e) |

(c)

| | |
|---|---|
| $x + xy = x1 + xy$ | Definition 11.3.1(d) |
| $= x(1 + y)$ | Definition 11.3.1(c) |
| $= x(y + 1)$ | Definition 11.3.1(b) |
| $= x1$ | Part (b) |
| $= x$ | Definition 11.3.1(d) |

(f) If we show that

$$(x + y)(x'y') = 0 \tag{11.3.1}$$

and

$$(x + y) + x'y' = 1, \tag{11.3.2}$$

it will follow from Theorem 11.3.4 that $x'y' = (x + y)'$. Now

$$\begin{aligned}
(x + y)(x'y') &= (x'y')(x + y) && \text{Definition 11.3.1(b)}\\
&= (x'y')x + (x'y')y && \text{Definition 11.3.1(c)}\\
&= x(x'y') + (x'y')y && \text{Definition 11.3.1(b)}\\
&= (xx')y' + x'(y'y) && \text{Definition 11.3.1(a)}\\
&= (xx')y' + x'(yy') && \text{Definition 11.3.1(b)}\\
&= 0y' + x'0 && \text{Definition 11.3.1(e)}\\
&= y'0 + x'0 && \text{Definition 11.3.1(b)}\\
&= 0 + 0 && \text{Part (b)}\\
&= 0 && \text{Definition 11.3.1(d)}
\end{aligned}$$

Therefore, (11.3.1) holds.

Next we verify (11.3.2).

$$\begin{aligned}
(x + y) + x'y' &= ((x + y) + x')((x + y) + y') && \text{Definition 11.3.1(c)}\\
&= ((y + x) + x')((x + y) + y') && \text{Definition 11.3.1(b)}\\
&= (y + (x + x'))(x + (y + y')) && \text{Definition 11.3.1(a)}\\
&= (y + 1)(x + 1) && \text{Definition 11.3.1(e)}\\
&= 1 \cdot 1 && \text{Part (b)}\\
&= 1 && \text{Definition 11.3.1(d)}
\end{aligned}$$

By Theorem 11.3.4, $x'y' = (x + y)'$. ◀

**Example 11.3.7** As explained in Example 11.3.3, if $U$ is a set, $\mathcal{P}(U)$ can be considered a Boolean algebra. Therefore, De Morgan's laws, which for sets may be stated

$$\overline{(X \cup Y)} = \overline{X} \cap \overline{Y}, \quad \overline{(X \cap Y)} = \overline{X} \cup \overline{Y} \quad \text{for all } X, Y \in \mathcal{P}(U),$$

hold. These equations may be verified directly (see Theorem 1.1.22), but Theorem 11.3.6 shows that they are a consequence of other laws. ◀

The reader has surely noticed that equations involving elements of a Boolean algebra come in pairs. For example, the identity laws [Definition 11.3.1(d)] are

$$x + 0 = x, \qquad x1 = x.$$

Such pairs are said to be **dual.**

**Definition 11.3.8 ▶** The *dual* of a statement involving Boolean expressions is obtained by replacing 0 by 1, 1 by 0, + by ·, and · by +. ◀

**Example 11.3.9** The dual of $(x + y)' = x'y'$ is $(xy)' = x' + y'$. ◀

Each condition in the definition of a Boolean algebra (Definition 11.3.1) includes its dual. Therefore, we have the following result.

**Theorem 11.3.10** The dual of a theorem about Boolean algebras is also a theorem.

**Proof** Suppose that $T$ is a theorem about Boolean algebras. Then there is a proof $P$ of $T$ involving only the definitions of a Boolean algebra (Definition 11.3.1). Let $P'$ be the sequence of statements obtained by replacing every statement in $P$ by its dual. Then $P'$ is a proof of the dual of $T$. ◀

**Example 11.3.11** The dual of

$$x + x = x \tag{11.3.3}$$

is

$$xx = x. \tag{11.3.4}$$

We proved (11.3.3) earlier [see the proof of Theorem 11.3.6(a)]. If we write the dual of each statement in the proof of (11.3.3), we obtain the following proof of (11.3.4):

$$\begin{aligned} x &= x1 \\ &= x(x + x') \\ &= xx + xx' \\ &= xx + 0 \\ &= xx. \end{aligned}$$

◀

**Example 11.3.12** The proofs given in Theorem 11.3.6 of the two statements of part (b) are dual to each other. ◀

## 11.3 Review Exercises

1. Define *Boolean algebra.*
2. What are the idempotent laws for Boolean algebras?
3. What are the bound laws for Boolean algebras?
4. What are the absorption laws for Boolean algebras?
5. What is the involution law for Boolean algebras?
6. What are the 0/1 laws for Boolean algebras?
7. What are De Morgan's laws for Boolean algebras?
8. How is the dual of a Boolean expression obtained?
9. What can we say about the dual of a theorem about Boolean algebras?

## 11.3 Exercises

1. Verify properties (a′)–(e′) of Example 11.3.3.
2. Let $S = \{1, 2, 3, 6\}$. Define

$$x + y = \text{lcm}(x, y), \qquad x \cdot y = \gcd(x, y), \qquad x' = \frac{6}{x}$$

for $x, y \in S$ (lcm and gcd denote, respectively, the least common multiple and the greatest common divisor). Show that $(S, +, \cdot, ', 1, 6)$ is a Boolean algebra.

3. $S = \{1, 2, 4, 8\}$. Define $+$ and $\cdot$ as in Exercise 2 and define $x' = 8/x$. Show that $(S, +, \cdot, ', 1, 8)$ is not a Boolean algebra.

*Let* $S_n = \{1, 2, \dots, n\}$. *Define*

$$x + y = \max\{x, y\}, \qquad x \cdot y = \min\{x, y\}.$$

4. Show that parts (a)–(c) of Definition 11.3.1 hold for $S_n$.
5. Show that it is possible to define 0, 1 and $'$ so that $(S_n, +, \cdot, ', 0, 1)$ is a Boolean algebra if and only if $n = 2$.
6. Rewrite the conditions of Theorem 11.3.6 for sets as in Example 11.3.3.
7. Interpret Theorem 11.3.4 for sets as in Example 11.3.3.

*Write the dual of each statement in Exercises 8–14.*

8. $(x + y)(x + 1) = x + xy + y$
9. $(x' + y')' = xy$
10. If $x + y = x + z$ and $x' + y = x' + z$, then $y = z$.
11. $xy' = 0$ if and only if $xy = x$.
12. $x = 0$ if and only if $y = xy' + x'y$ for all $y$.
13. If $x + y = 0$, then $x = 0 = y$.
14. $x + x(y + 1) = x$
15. Prove the statements of Exercises 8–14.
16. Prove the duals of the statements of Exercises 8–14.
17. Write the dual of Theorem 11.3.4. How does the dual relate to Theorem 11.3.4 itself?
18. Prove the second statements of parts (a), (c), and (f) of Theorem 11.3.6.
19. Prove the second statements of parts (a), (c), and (f) of Theorem 11.3.6 by dualizing the proofs of the first statements given in the text.

**20.** Prove Theorem 11.3.6, parts (d) and (e).

**⋆21.** Deduce part (a) of Definition 11.3.1 from parts (b)–(e) of Definition 11.3.1.

**22.** Let $U$ be the set of positive integers. Let $S$ be the collection of subsets $X$ of $U$ with either $X$ or $\overline{X}$ finite. Show that $(S, \cup, \cap, ^{-}, \varnothing, U)$ is a Boolean algebra.

**⋆23.** Let $n$ be a positive integer. Let $S$ be the set of all divisors of $n$, including 1 and $n$. Define $+$ and $\cdot$ as in Exercise 2 and define $x' = n/x$. What conditions must $n$ satisfy so that $(S, +, \cdot, ', 1, n)$ is a Boolean algebra?

**⋆24.** Show that the associative laws follow from the other laws of Definition 11.3.1.

# Problem-Solving Corner | Boolean Algebras

## Problem

Let $(S, +, \cdot, ', 0, 1)$ be a Boolean algebra and let $A$ be a subset of $S$. Show that $(A, +, \cdot, ', 0, 1)$ is a Boolean algebra if and only if $1 \in A$ and $xy' \in A$ for all $x, y \in A$.

## Attacking the Problem

Since the given statement is an "if and only if" statement, there are two statements to be proved:

If $(A, +, \cdot, ', 0, 1)$ is a Boolean algebra, then $1 \in A$ and $xy' \in A$ for all $x, y \in A$. **(1)**

If $1 \in A$ and $xy' \in A$ for all $x, y \in A$, then $(A, +, \cdot, ', 0, 1)$ is a Boolean algebra. **(2)**

To prove (1), we can use the laws as specified by the definition of "Boolean algebra" (Definition 11.3.1) and the laws derived in Theorem 11.3.6 that elements of a Boolean algebra must obey. To prove that $(A, +, \cdot, ', 0, 1)$ is a Boolean algebra, we will verify that the laws specified by Definition 11.3.1 are satisfied. Before reading on, you should review Definition 11.3.1 and Theorem 11.3.6.

## Finding a Solution

First let's try to prove (1). We assume that $(A, +, \cdot, ', 0, 1)$ is a Boolean algebra and prove that

- $1 \in A$

and

- $xy' \in A$ for all $x, y \in A$.

Definition 11.3.1 says that a Boolean algebra contains 1. Since $(A, +, \cdot, ', 0, 1)$ is a Boolean algebra, $1 \in A$.

Now suppose that $x, y \in A$. Definition 11.3.1 says that $'$ is a unary operator on $A$. This means that $y' \in A$. Definition 11.3.1 also says that $\cdot$ is a binary operator on $A$. This means that $xy' \in A$. This completes the proof of (1).

Now let's try to prove (2). This time we assume that $1 \in A$ and $xy' \in A$ for all $x, y \in A$ and try to prove that $(A, +, \cdot, ', 0, 1)$ is a Boolean algebra. According to Definition 11.3.1, we must prove that

$A$ contains distinct elements 0 and 1. **(3)**

$+$ and $\cdot$ are binary operators on $A$. **(4)**

$'$ is a unary operator on $A$. **(5)**

The associative laws hold. **(6)**

The commutative laws hold. **(7)**

The distributive laws hold. **(8)**

The identity laws hold. **(9)**

The complement laws hold. **(10)**

$A$ contains 1 by assumption. To prove (3), we must show that $0 \in A$. We have only two assumptions about $A$: $1 \in A$ and if $x, y \in A$, then $xy' \in A$. All we can do at this point is combine these assumptions; that is, take $x = y = 1$ and examine the conclusion: $11' \in A$. Now Theorem 11.3.6(e) [applied to the Boolean algebra $(S, +, \cdot, ', 0, 1)$] says that $1' = 0$. Substituting for $1'$, we know now that $10 \in A$. But Theorem 11.3.6(b) says that for any $x$, $x0 = 0$. Thus $10 = 0$ is in $A$. Success! $A$ contains 1 and 0. 0 and 1 are distinct because they are elements of the Boolean algebra $(S, +, \cdot, ', 0, 1)$. Therefore, (3) is proved.

To prove (4), we must show that $+$ and $\cdot$ are binary operators on $A$; that is, if $x, y \in A$, then $x + y$ and $xy$ are in $A$. Consider proving that $\cdot$ is a binary operator on $A$. What we know is that if $x, y \in A$, then $xy' \in A$, which is close to what we want to prove. If we could somehow replace $y'$ by $y$ in the expression $xy'$, we could conclude that $xy \in A$. What we would like to do is assume that $x, y \in A$, then deduce

$$x, y' \in A, \qquad \textbf{(11)}$$

and then conclude that

$$xy = xy'' \in A.$$

To deduce (11), we need to show if $y \in A$, then $y' \in A$. But this is (5). Detour! Let's work on (5).

We will assume that $y \in A$ and try to prove that $y' \in A$. If we could get rid of that pesky $x$ (in the hypothesis $x, y \in A$ implies $xy' \in A$), we would have exactly what we want. We can effectively eliminate $x$ by taking $x = 1$ since $1y = y$. Formally, we argue as follows. Let $y$ be in $A$. Since $1 \in A, y' = 1y' \in A$. [$y' = 1y'$ by Definition 11.3.1(b) and 11.3.1(d).] We have proved (5).

Now back to (4). Let $x, y \in A$. By the just proved (5), $y' \in A$. By the given condition, $xy = xy'' \in A$. [$y = y''$ by Theorem 11.3.6(d).] We have proved that $\cdot$ is a binary operator on $A$.

De Morgan's laws [Theorem 11.3.6(f)], in effect, allow us to interchange $+$ and $\cdot$, so we can use them to prove that if $x, y \in A$, then $x + y \in A$. Formally we argue as follows. Suppose that $x, y \in A$. By (5), $x'$ and $y'$ are both in $A$. Since we have already proved that $\cdot$ is a binary operator on $A$, $x'y' \in A$. By (5), $(x'y')' \in A$. By De Morgan's laws [Theorem 11.3.6(f)] and Theorem 11.3.6(d), $x+y = x''+y'' = (x'y')' \in A$. Therefore, $+$ is a binary operator on $A$. We have proved (4).

The next statement to prove is (6), which is to verify the associative laws

$$(x + y) + z = x + (y + z),$$
$$(xy)z = x(yz) \qquad \text{for all } x, y, z \in A.$$

Now $(S, +, \cdot, ', 0, 1)$ is a Boolean algebra and so the associative laws hold in $S$. Since $A$ is a subset of $S$, the associative laws surely hold in $A$. Thus (6) holds. For the same reason, properties (7) through (10) also hold in $A$. Therefore, $(A, +, \cdot, ', 0, 1)$ is a Boolean algebra.

## Formal Solution

Suppose that $(A, +, \cdot, ', 0, 1)$ is a Boolean algebra. Then $1 \in A$. Suppose that $x, y \in A$. Then $y' \in A$. Thus $xy' \in A$.

Now suppose that $1 \in A$ and $xy' \in A$ for all $x, y \in A$. Taking $x = y = 1$, we obtain $0 = 11' \in A$. Taking $x = 1$, we obtain $y' = 1y' \in A$. Thus $'$ is a unary operator on $A$. Replacing $y$ by $y'$, we obtain $xy = xy'' \in A$. Thus $\cdot$ is a binary operator on $A$. Now $x + y = x'' + y'' = (x'y')' \in A$. Thus $+$ is a binary operator on $A$. Parts a–e of Definition 11.3.1 automatically hold in $A$ since they hold in $S$. Therefore $(A, +, \cdot, ', 0, 1)$ is a Boolean algebra.

## Summary of Problem-Solving Techniques

- When trying to construct a proof, write out carefully what is assumed and what is to be proved.
- When trying to construct a proof, look at closely related definitions and theorems.
- To prove that something is a Boolean algebra, go directly to the definition (Definition 11.3.1).
- Consider proving statements in an order different from that given. In this problem, it was easier to prove statement (5) before proving statement (4).
- Try various substitutions for the variables in a universally quantified statement. (After all, "universally quantified" means that the statement holds true *for all* values.) By taking $x = y = 1$ in the statement

$$xy' \in A \qquad \text{for all } x, y \in A,$$

we were able to prove that $0 \in A$.

## Exercises

*Use the following definitions for Exercises 1–8. For $a < b$, let*

$$[a, b) = \{x \in \mathbf{R} \mid a \le x < b\},$$

*where, as usual, **R** denotes the set of real numbers. Define a subset $X$ of **R** to be* open relative to the $[a, b)$ definition *if for every $x \in X$, there exists $a, b \in \mathbf{R}$ with $a < b$ and $x \in [a, b) \subseteq X$. Henceforth, we abbreviate "open relative to the $[a, b)$ definition" to "open." Let **R** be the universal set. Define a set $X \subseteq \mathbf{R}$ to be* clopen *if $X$ and $\overline{X}$ are both open. Let $\mathcal{A}$ be the set of clopen subsets of **R**.*

1. Prove that $\mathbf{R}$ and $\varnothing$ are in $\mathcal{A}$.
2. Prove that $[1, 2) \cup [3, 5) \in \mathcal{A}$.
3. Prove that $X = \{x \in \mathbf{R} \mid 3 < x < 10\}$ is open but not clopen.
4. Prove that if $X \in \mathcal{A}$, then $\overline{X} \in \mathcal{A}$.
5. Prove that if $X$ and $Y$ are open, then $X \cup Y$ is open.
6. Prove that if $X$ and $Y$ are open, then $X \cap Y$ is open.
7. Prove that if $X, Y \in \mathcal{A}$, then $X \cap Y \in \mathcal{A}$.
8. Prove that $(\mathcal{A}, \cup, \cap, ^{-}, \varnothing, \mathbf{R})$ is a Boolean algebra.
9. Can we prove Exercises 1 and 4–8 if we replace $[a, b)$, $a < b$, by $\{x \in \mathbf{R} \mid a \le x \le b\}$?

## 11.4 Boolean Functions and Synthesis of Circuits

A circuit is constructed to carry out a specified task. If we want to construct a combinatorial circuit, the problem can be given in terms of inputs and outputs. For example, suppose that we want to construct a combinatorial circuit to compute the **exclusive-OR** of $x_1$ and $x_2$. We can state the problem by listing the inputs and outputs that define the exclusive-OR. This is equivalent to giving the desired logic table.

**Definition 11.4.1 ▶** The *exclusive-OR* of $x_1$ and $x_2$ written $x_1 \oplus x_2$ is defined by Table 11.4.1. ◀

**TABLE 11.4.1 ■ The exclusive-OR**

| $x_1$ | $x_2$ | $x_1 \oplus x_2$ |
|---|---|---|
| 1 | 1 | 0 |
| 1 | 0 | 1 |
| 0 | 1 | 1 |
| 0 | 0 | 0 |

A logic table, with one output, is a function, The domain is the set of inputs and the range is the set of outputs. For the exclusive-OR function given in Table 11.4.1, the domain is the set $\{(1, 1), (1, 0), (0, 1), (0, 0)\}$ and the range is the set $Z_2 = \{0, 1\}$.

If we could develop a formula for the exclusive-OR function of the form

$$x_1 \oplus x_2 = X(x_1, x_2),$$

where $X$ is a Boolean expression, we could solve the problem of constructing the combinatorial circuit. We could merely construct the circuit corresponding to $X$.

Functions that can be represented by Boolean expressions are called **Boolean functions.**

**Definition 11.4.2 ▶** Let $X(x_1, \ldots, x_n)$ be a Boolean expression. A function $f$ of the form

$$f(x_1, \ldots, x_n) = X(x_1, \ldots, x_n)$$

is called a *Boolean function.* ◀

**Example 11.4.3** The function $f: Z_2^3 \rightarrow Z_2$ defined by

$$f(x_1, x_2, x_3) = x_1 \wedge (\overline{x}_2 \vee x_3)$$

is a Boolean function. The inputs and outputs are given in the following table.

| $x_1$ | $x_2$ | $x_3$ | $f(x_1, x_2, x_3)$ |
|---|---|---|---|
| 1 | 1 | 1 | 1 |
| 1 | 1 | 0 | 0 |
| 1 | 0 | 1 | 1 |
| 1 | 0 | 0 | 1 |
| 0 | 1 | 1 | 0 |
| 0 | 1 | 0 | 0 |
| 0 | 0 | 1 | 0 |
| 0 | 0 | 0 | 0 |

◀

In the next example we show how an arbitrary function $f: Z_2^n \rightarrow Z_2$ can be realized as a Boolean function.

**Example 11.4.4** Show that the function $f$ given by the following table is a Boolean function.

| $x_1$ | $x_2$ | $x_3$ | $f(x_1, x_2, x_3)$ |
|---|---|---|---|
| 1 | 1 | 1 | 1 |
| 1 | 1 | 0 | 0 |
| 1 | 0 | 1 | 0 |
| 1 | 0 | 0 | 1 |
| 0 | 1 | 1 | 0 |
| 0 | 1 | 0 | 1 |
| 0 | 0 | 1 | 0 |
| 0 | 0 | 0 | 0 |

**SOLUTION** Consider the first row of the table and the combination

$$x_1 \wedge x_2 \wedge x_3. \tag{11.4.1}$$

Notice that if $x_1 = x_2 = x_3 = 1$, as indicated in the first row of the table, then (11.4.1) is 1. The values of $x_i$ given by any other row of the table give (11.4.1) the value 0. Similarly, for the fourth row of the table we may construct the combination

$$x_1 \wedge \overline{x}_2 \wedge \overline{x}_3. \tag{11.4.2}$$

Expression (11.4.2) has the value 1 for the values of $x_i$ given by the fourth row of the table, whereas the values of $x_i$ given by any other row of the table give (11.4.2) the value 0.

The procedure is clear. We consider a row $R$ of the table where the output is 1. We then form the combination $x_1 \wedge x_2 \wedge x_3$ and place a bar over each $x_i$ whose value is 0 in row $R$. The combination formed is 1 if and only if the $x_i$ have the values given in row $R$. Thus, for row 6, we obtain the combination

$$\overline{x}_1 \wedge x_2 \wedge \overline{x}_3. \tag{11.4.3}$$

Next, we OR the terms (11.4.1)–(11.4.3) to obtain the Boolean expression

$$(x_1 \wedge x_2 \wedge x_3) \vee (x_1 \wedge \overline{x}_2 \wedge \overline{x}_3) \vee (\overline{x}_1 \wedge x_2 \wedge \overline{x}_3). \tag{11.4.4}$$

We claim that $f(x_1, x_2, x_3)$ and (11.4.4) are equal. To verify this, first suppose that $x_1$, $x_2$, and $x_3$ have values given by a row of the table for which $f(x_1, x_2, x_3) = 1$. Then one of (11.4.1)–(11.4.3) is 1, so the value of (11.4.4) is 1. On the other hand, if $x_1, x_2, x_3$ have values given by a row of the table for which $f(x_1, x_2, x_3) = 0$, all of (11.4.1)–(11.4.3) are 0, so the value of (11.4.4) is 0. Thus $f$ and the Boolean expression (11.4.4) agree on $Z_2^3$; therefore,

$$f(x_1, x_2, x_3) = (x_1 \wedge x_2 \wedge x_3) \vee (x_1 \wedge \overline{x}_2 \wedge \overline{x}_3) \vee (\overline{x}_1 \wedge x_2 \wedge \overline{x}_3),$$

as claimed. ◀

After one more definition, we will show that the method of Example 11.4.4 can be used to represent any function $f\colon Z_2^n \to Z_2$.

**Definition 11.4.5 ▶** A *minterm* in the symbols $x_1, \ldots, x_n$ is a Boolean expression of the form

$$y_1 \wedge y_2 \wedge \cdots \wedge y_n,$$

where each $y_i$ is either $x_i$ or $\overline{x}_i$. ◀

**Theorem 11.4.6** If $f\colon Z_2^n \to Z_2$, then $f$ is a Boolean function. If $f$ is not identically zero, let $A_1, \dots, A_k$ denote the elements $A_i$ of $Z_2^n$ for which $f(A_i) = 1$. For each $A_i = (a_1, \dots, a_n)$, set

$$m_i = y_1 \wedge \cdots \wedge y_n,$$

where

$$y_j = \begin{cases} x_j & \text{if } a_j = 1 \\ \overline{x}_j & \text{if } a_j = 0. \end{cases}$$

Then

$$f(x_1, \dots, x_n) = m_1 \vee m_2 \vee \cdots \vee m_k. \tag{11.4.5}$$

**Proof** If $f(x_1, \dots, x_n) = 0$ for all $x_i$, then $f$ is a Boolean function, since 0 is a Boolean expression.

Suppose that $f$ is not identically zero. Let $m_i(a_1, \dots, a_n)$ denote the value obtained from $m_i$ by replacing each $x_j$ with $a_j$. It follows from the definition of $m_i$ that

$$m_i(A) = \begin{cases} 1 & \text{if } A = A_i \\ 0 & \text{if } A \neq A_i. \end{cases}$$

Let $A \in Z_2^n$. If $A = A_i$ for some $i \in \{1, \dots, k\}$, then $f(A) = 1$, $m_i(A) = 1$, and

$$m_1(A) \vee \cdots \vee m_k(A) = 1.$$

On the other hand, if $A \neq A_i$ for any $i \in \{1, \dots, k\}$, then $f(A) = 0$, $m_i(A) = 0$ for $i = 1, \dots, k$, and

$$m_1(A) \vee \cdots \vee m_k(A) = 0.$$

Therefore, (11.4.5) holds. ◀

**Definition 11.4.7 ▶** The representation (11.4.5) of a Boolean function $f\colon Z_2^n \to Z_2$ is called the *disjunctive normal form* of the function $f$. ◀

**Example 11.4.8** Design a combinatorial circuit that computes the exclusive-OR of $x_1$ and $x_2$.

**SOLUTION** The logic table for the exclusive-OR function $x_1 \oplus x_2$ is given in Table 11.4.1. The disjunctive normal form of this function is

$$x_1 \oplus x_2 = (x_1 \wedge \overline{x}_2) \vee (\overline{x}_1 \wedge x_2). \tag{11.4.6}$$

The combinatorial circuit corresponding to (11.4.6) is given in Figure 11.4.1.

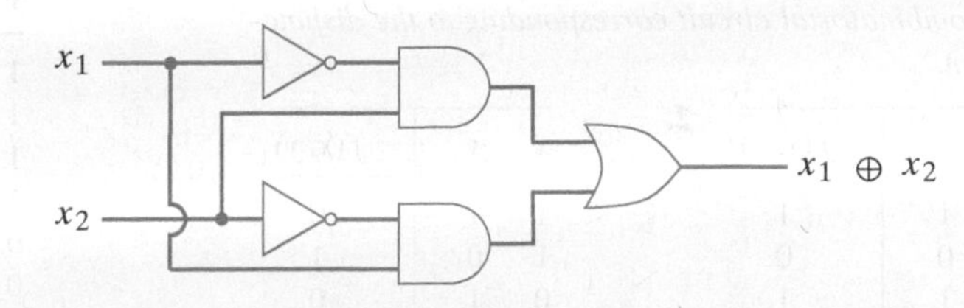


**Figure 11.4.1** A combinatorial circuit for the exclusive-OR. ◀

Suppose that a function is given by a Boolean expression such as

$$f(x_1, x_2, x_3) = (x_1 \vee x_2) \wedge x_3$$

and we wish to find the disjunctive normal form of $f$. We could write the logic table for $f$ and then use Theorem 11.4.6. Alternatively, we can deal directly with the Boolean expression by using the definitions and results of Sections 11.2 and 11.3. We begin by distributing the term $x_3$ as follows:

$$(x_1 \vee x_2) \wedge x_3 = (x_1 \wedge x_3) \vee (x_2 \wedge x_3).$$

Although this represents the Boolean expression as a combination of terms of the form $y \wedge z$, it is not in disjunctive normal form, since each term does not contain all of the symbols $x_1, x_2$, and $x_3$. However, this is easily remedied, as follows:

$$\begin{aligned}(x_1 \wedge x_3) \vee (x_2 \wedge x_3) &= (x_1 \wedge x_3 \wedge 1) \vee (x_2 \wedge x_3 \wedge 1)\\ &= (x_1 \wedge x_3 \wedge (x_2 \vee \overline{x}_2)) \vee (x_2 \wedge x_3 \wedge (x_1 \vee \overline{x}_1))\\ &= (x_1 \wedge x_2 \wedge x_3) \vee (x_1 \wedge \overline{x}_2 \wedge x_3)\\ &\quad \vee (x_1 \wedge x_2 \wedge x_3) \vee (\overline{x}_1 \wedge x_2 \wedge x_3)\\ &= (x_1 \wedge x_2 \wedge x_3) \vee (x_1 \wedge \overline{x}_2 \wedge x_3)\\ &\quad \vee (\overline{x}_1 \wedge x_2 \wedge x_3).\end{aligned}$$

This expression is the disjunctive normal form of $f$.

Theorem 11.4.6 has a dual. In this case the function $f$ is expressed as

$$f(x_1, \ldots, x_n) = M_1 \wedge M_2 \wedge \cdots \wedge M_k. \qquad \textbf{(11.4.7)}$$

Each $M_i$ is of the form

$$y_1 \vee \cdots \vee y_n, \qquad \textbf{(11.4.8)}$$

where $y_j$ is either $x_j$ or $\overline{x}_j$. A term of the form (11.4.8) is called a **maxterm** and the representation of $f$ (11.4.7) is called the **conjunctive normal form**. Exercises 24–28 explore maxterms and the conjunctive normal form in more detail.

**Go Online**
For more on conjunctive normal form, see
`bit.ly/2S7L960`

## 11.4 Review Exercises

1. Define *exclusive-OR*.
2. What is a Boolean function?
3. What is a minterm?
4. What is the disjunctive normal form of a Boolean function?
5. How can one obtain the disjunctive normal form of a Boolean function?
6. What is a maxterm?
7. What is the conjunctive normal form of a Boolean function?

## 11.4 Exercises

*In Exercises 1–10, find the disjunctive normal form of each function and draw the combinatorial circuit corresponding to the disjunctive normal form.*

**1.**

| $x$ | $y$ | $f(x, y)$ |
|---|---|---|
| 1 | 1 | 1 |
| 1 | 0 | 0 |
| 0 | 1 | 1 |
| 0 | 0 | 1 |

**2.**

| $x$ | $y$ | $f(x, y)$ |
|---|---|---|
| 1 | 1 | 0 |
| 1 | 0 | 1 |
| 0 | 1 | 0 |
| 0 | 0 | 1 |

**3.**

| $x$ | $y$ | $z$ | $f(x, y, z)$ |
|---|---|---|---|
| 1 | 1 | 1 | 1 |
| 1 | 1 | 0 | 1 |
| 1 | 0 | 1 | 0 |
| 1 | 0 | 0 | 1 |
| 0 | 1 | 1 | 0 |
| 0 | 1 | 0 | 0 |
| 0 | 0 | 1 | 1 |
| 0 | 0 | 0 | 1 |

**4.**

| $x$ | $y$ | $z$ | $f(x, y, z)$ |
|---|---|---|---|
| 1 | 1 | 1 | 1 |
| 1 | 1 | 0 | 1 |
| 1 | 0 | 1 | 0 |
| 1 | 0 | 0 | 1 |
| 0 | 1 | 1 | 1 |
| 0 | 1 | 0 | 1 |
| 0 | 0 | 1 | 0 |
| 0 | 0 | 0 | 0 |

**5.**

| $x$ | $y$ | $z$ | $f(x, y, z)$ |
|---|---|---|---|
| 1 | 1 | 1 | 0 |
| 1 | 1 | 0 | 1 |
| 1 | 0 | 1 | 1 |
| 1 | 0 | 0 | 1 |
| 0 | 1 | 1 | 1 |
| 0 | 1 | 0 | 1 |
| 0 | 0 | 1 | 1 |
| 0 | 0 | 0 | 0 |

**6.**

| $x$ | $y$ | $z$ | $f(x, y, z)$ |
|---|---|---|---|
| 1 | 1 | 1 | 1 |
| 1 | 1 | 0 | 1 |
| 1 | 0 | 1 | 1 |
| 1 | 0 | 0 | 0 |
| 0 | 1 | 1 | 0 |
| 0 | 1 | 0 | 1 |
| 0 | 0 | 1 | 1 |
| 0 | 0 | 0 | 1 |

**7.**

| $x$ | $y$ | $z$ | $f(x, y, z)$ |
|---|---|---|---|
| 1 | 1 | 1 | 1 |
| 1 | 1 | 0 | 0 |
| 1 | 0 | 1 | 0 |
| 1 | 0 | 0 | 1 |
| 0 | 1 | 1 | 0 |
| 0 | 1 | 0 | 0 |
| 0 | 0 | 1 | 0 |
| 0 | 0 | 0 | 1 |

**8.**

| $x$ | $y$ | $z$ | $f(x, y, z)$ |
|---|---|---|---|
| 1 | 1 | 1 | 0 |
| 1 | 1 | 0 | 0 |
| 1 | 0 | 1 | 0 |
| 1 | 0 | 0 | 1 |
| 0 | 1 | 1 | 1 |
| 0 | 1 | 0 | 1 |
| 0 | 0 | 1 | 1 |
| 0 | 0 | 0 | 0 |

**9.**

| $w$ | $x$ | $y$ | $z$ | $f(w, x, y, z)$ |
|---|---|---|---|---|
| 1 | 1 | 1 | 1 | 1 |
| 1 | 1 | 1 | 0 | 0 |
| 1 | 1 | 0 | 1 | 1 |
| 1 | 1 | 0 | 0 | 0 |
| 1 | 0 | 1 | 1 | 0 |
| 1 | 0 | 1 | 0 | 0 |
| 1 | 0 | 0 | 1 | 0 |
| 1 | 0 | 0 | 0 | 1 |
| 0 | 1 | 1 | 1 | 1 |
| 0 | 1 | 1 | 0 | 0 |
| 0 | 1 | 0 | 1 | 0 |
| 0 | 1 | 0 | 0 | 0 |
| 0 | 0 | 1 | 1 | 1 |
| 0 | 0 | 1 | 0 | 0 |
| 0 | 0 | 0 | 1 | 0 |
| 0 | 0 | 0 | 0 | 0 |

**10.**

| $w$ | $x$ | $y$ | $z$ | $f(w, x, y, z)$ |
|---|---|---|---|---|
| 1 | 1 | 1 | 1 | 0 |
| 1 | 1 | 1 | 0 | 0 |
| 1 | 1 | 0 | 1 | 1 |
| 1 | 1 | 0 | 0 | 1 |
| 1 | 0 | 1 | 1 | 1 |
| 1 | 0 | 1 | 0 | 1 |
| 1 | 0 | 0 | 1 | 0 |
| 1 | 0 | 0 | 0 | 1 |
| 0 | 1 | 1 | 1 | 0 |
| 0 | 1 | 1 | 0 | 1 |
| 0 | 1 | 0 | 1 | 1 |
| 0 | 1 | 0 | 0 | 1 |
| 0 | 0 | 1 | 1 | 0 |
| 0 | 0 | 1 | 0 | 1 |
| 0 | 0 | 0 | 1 | 0 |
| 0 | 0 | 0 | 0 | 1 |

*In Exercises 11–20, find the disjunctive normal form of each function using algebraic techniques. (We abbreviate $a \wedge b$ as $ab$.)*

**11.** $f(x, y) = x \vee xy$

**12.** $f(x, y) = (x \vee y)(\bar{x} \vee \bar{y})$

**13.** $f(x, y, z) = x \vee y(x \vee \bar{z})$

**14.** $f(x, y, z) = (yz \vee x\bar{z})(\overline{x\bar{y} \vee z})$

**15.** $f(x, y, z) = x \vee (\bar{y} \vee (x\bar{y} \vee x\bar{z}))$

**16.** $f(x, y, z) = (\bar{x}y \vee \overline{xz})(\overline{x \vee yz})$

**17.** $f(x, y, z) = (x \vee \bar{x}y \vee \bar{x}y\bar{z})(xy \vee \overline{xz})(y \vee xy\bar{z})$

**18.** $f(w, x, y, z) = wy \vee (w\bar{y} \vee z)(x \vee \bar{w}z)$

**19.** $f(x, y, z) = (\bar{x}y \vee \overline{xz})(\overline{xyz \vee y\bar{z}})(x\bar{y}z \vee x\bar{y} \vee x\overline{yz} \vee \overline{xyz})$

**20.** $f(w, x, y, z) = (\overline{wx}\bar{y}z \vee x\bar{y}\bar{z})(\overline{\overline{wy}z \vee xy\bar{z} \vee yxz})(\bar{w}z \vee xy \vee \bar{w}\bar{y}z \vee xy\bar{z} \vee \bar{x}yz)$

**21.** How many Boolean functions are there from $Z_2^n$ into $Z_2$?

*Let F denote the set of all functions from* $Z_2^n$ *into* $Z_2$. *Define*

$$\begin{aligned} (f \vee g)(x) &= f(x) \vee g(x) && x \in Z_2^n \\ (f \wedge g)(x) &= f(x) \wedge g(x) && x \in Z_2^n \\ \overline{f}(x) &= \overline{f(x)} && x \in Z_2^n \\ 0(x) &= 0 && x \in Z_2^n \\ 1(x) &= 1 && x \in Z_2^n. \end{aligned}$$

**22.** How many elements does $F$ have?

**23.** Show that $(F, \vee, \wedge, \bar{}, 0, 1)$ is a Boolean algebra.

**24.** By dualizing the procedure of Example 11.4.4, explain how to find the conjunctive normal form of a Boolean function from $Z_2^n$ into $Z_2$.

**25.** Find the conjunctive normal form of each function in Exercises 1–10.

**26.** By using algebraic methods, find the conjunctive normal form of each function in Exercises 11–20.

**27.** Show that if $m_1 \vee \cdots \vee m_k$ is the disjunctive normal form of $f(x_1, \ldots, x_n)$, then $\overline{m}_1 \wedge \cdots \wedge \overline{m}_k$ is the conjunctive normal form of $\overline{f(x_1, \ldots, x_n)}$.

**28.** Using the method of Exercise 27, find the conjunctive normal form of $\overline{f}$ for each function $f$ of Exercises 1–10.

**29.** Show that the disjunctive normal form (11.4.5) is unique; that is, show that if we have a Boolean function

$$f(x_1, \ldots, x_n) = m_1 \vee \cdots \vee m_k = m_1' \vee \cdots \vee m_j',$$

where each $m_i$, $m_i'$ is a minterm, then $k = j$ and the subscripts on the $m_i'$ may be permuted so that $m_i = m_i'$ for $i = 1, \ldots, k$.

## 11.5 Applications

**Go Online**

For more on logic design, see bit.ly/2S7L960

In the preceding section we showed how to design a combinatorial circuit using AND, OR, and NOT gates that would compute an arbitrary function from $Z_2^n$ into $Z_2$, where $Z_2 = \{0, 1\}$. In this section we consider using other kinds of gates to implement a circuit. We also consider the problem of efficient design. We will conclude by looking at several useful circuits having multiple outputs. Throughout this section, we write $ab$ for $a \wedge b$.

Before considering alternatives to AND, OR, and NOT gates, we must give a precise definition of "gate."

**Definition 11.5.1** ▶ A *gate* is a function from $Z_2^n$ into $Z_2$. ◀

**Example 11.5.2** The AND gate is the function $\wedge$ from $Z_2^2$ into $Z_2$ defined as in Definition 11.1.1. The NOT gate is the function $\bar{}$ from $Z_2$ into $Z_2$ defined as in Definition 11.1.3. ◀

We are interested in gates that allow us to construct arbitrary combinatorial circuits.

**Definition 11.5.3** ▶ A set of gates $\{g_1, \ldots, g_k\}$ is said to be *functionally complete* if, given any positive integer $n$ and a function $f$ from $Z_2^n$ into $Z_2$, it is possible to construct a combinatorial circuit that computes $f$ using only the gates $g_1, \ldots, g_k$. ◀

**Example 11.5.4** Theorem 11.4.6 shows that the set of gates {AND, OR, NOT} is functionally complete. ◀

It is an interesting fact that we can eliminate either AND or OR from the set {AND, OR, NOT} and still obtain a functionally complete set of gates.

**Theorem 11.5.5** The sets of gates

{AND, NOT} {OR, NOT}

are functionally complete.

**Proof** We will show that the set of gates {AND, NOT} is functionally complete and leave the problem of showing that the other set is functionally complete for the exercises (see Exercise 1).

We have

$$x \vee y = \overline{\overline{x}} \vee \overline{\overline{y}} \qquad \text{involution law}$$
$$= \overline{\overline{x}\,\overline{y}} \qquad \text{De Morgan's law.}$$

Therefore, an OR gate can be replaced by one AND gate and three NOT gates. (The combinatorial circuit is shown in Figure 11.5.1.)

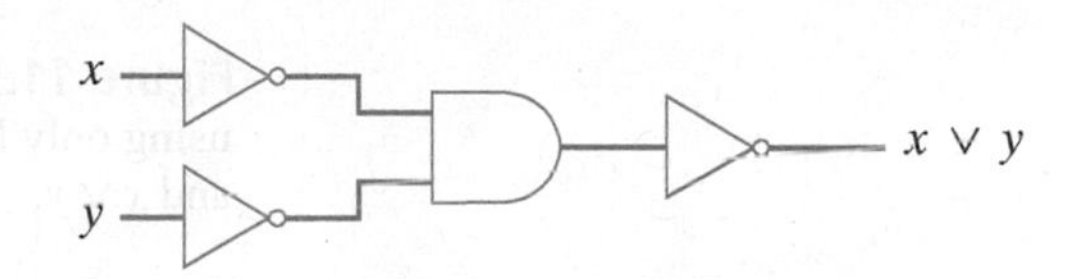


**Figure 11.5.1** A combinatorial circuit using only AND and NOT gates that computes $x \vee y$.

Given any function $f: Z_2^n \rightarrow Z_2$, by Theorem 11.4.6 we can construct a combinatorial circuit $C$ using AND, OR, and NOT gates that computes $f$. But Figure 11.5.1 shows that each OR gate can be replaced by AND and NOT gates. Therefore, the circuit $C$ can be modified so that it consists only of AND and NOT gates. Thus the set of gates {AND, NOT} is functionally complete. ◀

Although none of AND, OR, or NOT singly forms a functionally complete set (see Exercises 2–4), it is possible to define a new gate that by itself forms a functionally complete set.

**Definition 11.5.6 ▶** A *NAND gate* receives inputs $x_1$ and $x_2$, where $x_1$ and $x_2$ are bits, and produces output denoted $x_1 \uparrow x_2$, where

$$x_1 \uparrow x_2 = \begin{cases} 0 & \text{if } x_1 = 1 \text{ and } x_2 = 1 \\ 1 & \text{otherwise.} \end{cases}$$

A NAND gate is drawn as shown in Figure 11.5.2. ◀

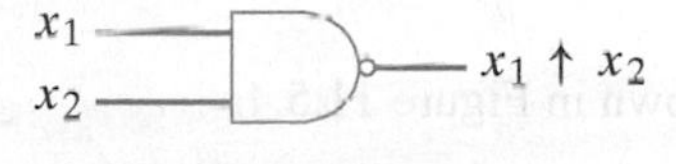


**Figure 11.5.2** NAND gate.

Many basic circuits used in digital computers today are built from NAND gates.

**Theorem 11.5.7** The set {NAND} is a functionally complete set of gates.

**Proof** First we observe that $x \uparrow y = \overline{xy}$. Therefore,

$$\overline{x} = \overline{xx} = x \uparrow x \tag{11.5.1}$$

$$x \vee y = \overline{\overline{x}\,\overline{y}} = \overline{x} \uparrow \overline{y} = (x \uparrow x) \uparrow (y \uparrow y). \tag{11.5.2}$$

Equations (11.5.1) and (11.5.2) show that both OR and NOT can be written in terms of NAND. By Theorem 11.5.5, the set {OR, NOT} is functionally complete. It follows that the set {NAND} is also functionally complete. ◀

**Example 11.5.8** Design combinatorial circuits using NAND gates to compute the functions $f_1(x) = \overline{x}$ and $f_2(x, y) = x \vee y$.

**SOLUTION** The combinatorial circuits, derived from equations (11.5.1) and (11.5.2), are shown in Figure 11.5.3.

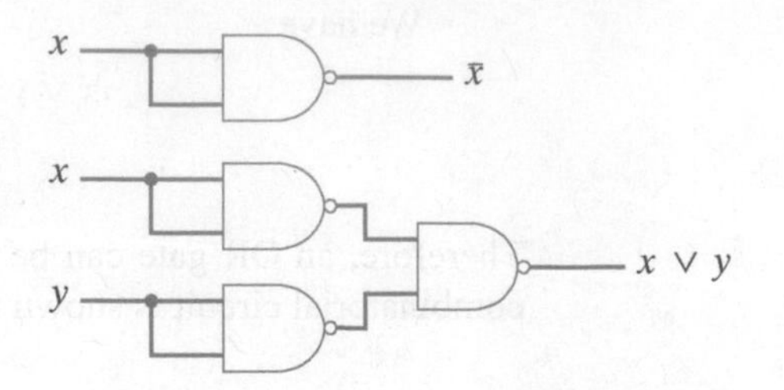


**Figure 11.5.3** Combinatorial circuits using only NAND gates that compute $\overline{x}$ and $x \vee y$. ◀

Consider the problem of designing a combinatorial circuit using AND, OR, and NOT gates to compute the function $f$.

| $x$ | $y$ | $z$ | $f(x, y, z)$ |
|---|---|---|---|
| 1 | 1 | 1 | 1 |
| 1 | 1 | 0 | 1 |
| 1 | 0 | 1 | 0 |
| 1 | 0 | 0 | 1 |
| 0 | 1 | 1 | 0 |
| 0 | 1 | 0 | 0 |
| 0 | 0 | 1 | 0 |
| 0 | 0 | 0 | 0 |

The disjunctive normal form of $f$ is

$$f(x, y, z) = xyz \vee xy\overline{z} \vee x\overline{y}\,\overline{z}. \qquad \textbf{(11.5.3)}$$

The combinatorial circuit corresponding to (11.5.3) is shown in Figure 11.5.4.

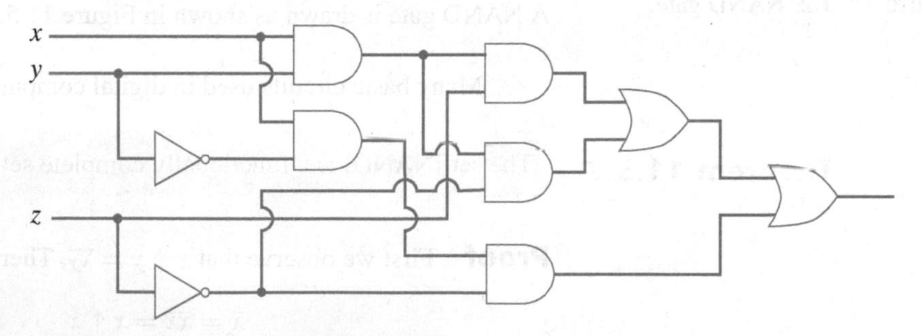


**Figure 11.5.4** A combinatorial circuit that computes $f(x, y, z) = xyz \vee xy\overline{z} \vee x\overline{y}\,\overline{z}$.

The circuit in Figure 11.5.4 has nine gates. As we will show, it is possible to design a circuit having fewer gates. The problem of finding the best circuit is called the **minimization problem.** There are many definitions of "best."

To find a simpler combinatorial circuit equivalent to that in Figure 11.5.4, we attempt to simplify the Boolean expression (11.5.3) that represents it. The equations

$$Ea \vee E\overline{a} = E \tag{11.5.4}$$

$$E = E \vee Ea, \tag{11.5.5}$$

where $E$ represents an arbitrary Boolean expression, are useful in simplifying Boolean expressions.

Equation (11.5.4) may be derived as follows:

$$Ea \vee E\overline{a} = E(a \vee \overline{a}) = E1 = E$$

using the properties of Boolean algebras. Equation (11.5.5) is essentially the absorption law [Theorem 11.3.6(c)].

Using (11.5.4) and (11.5.5), we may simplify (11.5.3) as follows:

$$\begin{aligned} xyz \vee xy\overline{z} \vee x\overline{y}\,\overline{z} &= xy \vee x\overline{y}\,\overline{z} && \text{by (11.5.4)} \\ &= xy \vee xy\overline{z} \vee x\overline{y}\,\overline{z} && \text{by (11.5.5)} \\ &= xy \vee x\overline{z}. && \text{by (11.5.4).} \end{aligned}$$

A further simplification,

$$xy \vee x\overline{z} = x(y \vee \overline{z}), \tag{11.5.6}$$

is possible using the distributive law [Definition 11.3.1(c)]. The combinatorial circuit corresponding to (11.5.6), which requires only three gates, is shown in Figure 11.5.5.

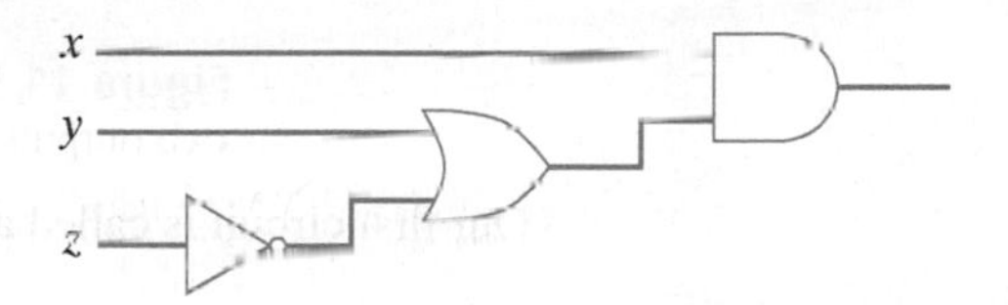


**Figure 11.5.5** A three-gate combinatorial circuit equivalent to that of Figure 11.5.4.

**Example 11.5.9** The combinatorial circuit in Figure 11.4.1 uses five AND, OR, and NOT gates to compute the exclusive-OR $x \oplus y$ of $x$ and $y$. Design a circuit that computes $x \oplus y$ using fewer AND, OR, and NOT gates.

**SOLUTION** Unfortunately, (11.5.4) and (11.5.5) do not help us simplify the disjunctive normal form $x\overline{y} \vee \overline{x}y$ of $x \oplus y$. Thus we must experiment with various Boolean rules until we produce an expression that requires fewer than five gates. One solution is provided by the expression $(x \vee y)\overline{xy}$ whose implementation requires only four gates. This combinatorial circuit is shown in Figure 11.5.6.

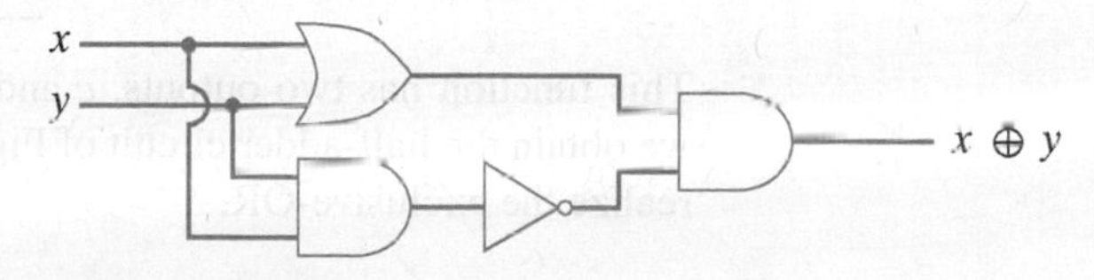


**Figure 11.5.6** A four-gate combinatorial circuit that computes the exclusive-OR $x \oplus y$ of $x$ and $y$. ◀

The set of gates available determines the minimization problem. Since the state of technology determines the available gates, the minimization problem changes through time. In the 1950s, the typical problem was to minimize circuits consisting of AND, OR, and NOT gates. Solutions such as the Quine–McCluskey method and the method of Karnaugh maps were provided. The reader is referred to [Mendelson] for the details of these methods.

Advances in solid-state technology have made it possible to manufacture very small components, called **integrated circuits,** which are themselves entire circuits. Thus circuit design today consists of combining basic gates such as AND, OR, NOT, and NAND gates and integrated circuits to compute the desired functions. Boolean algebra remains an essential tool, as a glance at a book on logic design such as [McCalla] will show.

We conclude this section by considering several useful combinatorial circuits having multiple outputs. A circuit with $n$ outputs can be characterized by $n$ Boolean expressions, as the next example shows.

**Example 11.5.10** Write two Boolean expressions to describe the combinatorial circuit of Figure 11.5.7.

SOLUTION The output $y_1$ is described by the expression $y_1 = \overline{ab}$, and $y_2$ is described by the expression $y_2 = bc \vee \overline{ab}$.

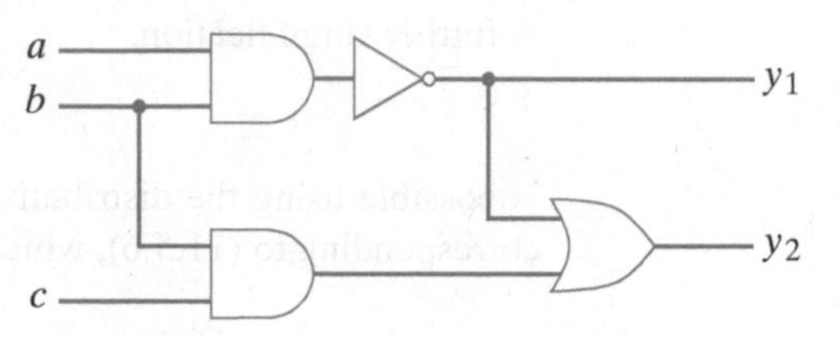


**Figure 11.5.7** A combinatorial circuit with two outputs. ◀

Our first circuit is called a **half adder.**

**Definition 11.5.11 ▶** A *half adder* accepts as input two bits $x$ and $y$ and produces as output the binary sum $cs$ of $x$ and $y$. The term $cs$ is a two-bit binary number. We call $s$ the sum bit and $c$ the carry bit. ◀

**Example 11.5.12** **Half-Adder Circuit** Design a half-adder combinatorial circuit.

SOLUTION The table for the half-adder circuit is as follows:

| $x$ | $y$ | $c$ | $s$ |
|---|---|---|---|
| 1 | 1 | 1 | 0 |
| 1 | 0 | 0 | 1 |
| 0 | 1 | 0 | 1 |
| 0 | 0 | 0 | 0 |

This function has two outputs, $c$ and $s$. We observe that $c = xy$ and $s = x \oplus y$. Thus we obtain the half-adder circuit of Figure 11.5.8. We used the circuit of Figure 11.5.6 to realize the exclusive-OR.

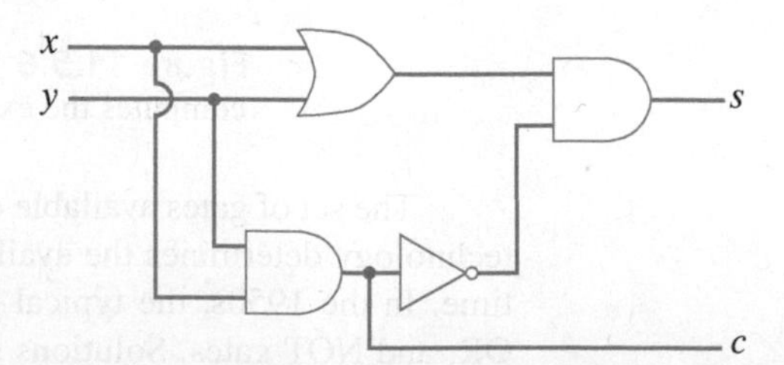


**Figure 11.5.8** A half-adder circuit. ◀

A **full adder** sums three bits and is useful for adding two bits and a third carry bit from a previous addition.

**Definition 11.5.13 ▶** A *full adder* accepts as input three bits $x$, $y$, and $z$ and produces as output the binary sum $cs$ of $x$, $y$, and $z$. The term $cs$ is a two-bit binary number. ◀

**Example 11.5.14 Full-Adder Circuit** Design a full-adder combinatorial circuit.

**SOLUTION** The table for the full-adder circuit is as follows:

| $x$ | $y$ | $z$ | $c$ | $s$ |
|---|---|---|---|---|
| 1 | 1 | 1 | 1 | 1 |
| 1 | 1 | 0 | 1 | 0 |
| 1 | 0 | 1 | 1 | 0 |
| 1 | 0 | 0 | 0 | 1 |
| 0 | 1 | 1 | 1 | 0 |
| 0 | 1 | 0 | 0 | 1 |
| 0 | 0 | 1 | 0 | 1 |
| 0 | 0 | 0 | 0 | 0 |

Checking the eight possibilities, we see that $s = x \oplus y \oplus z$; hence we can use two exclusive-OR circuits to compute $s$.

To compute $c$, we first find the disjunctive normal form

$$c = xyz \vee xy\overline{z} \vee x\overline{y}z \vee \overline{x}yz \tag{11.5.7}$$

of $c$. Next, we use (11.5.4) and (11.5.5) to simplify (11.5.7) as follows:

$$\begin{aligned} xyz \vee xy\overline{z} \vee x\overline{y}z \vee \overline{x}yz &= xy \vee x\overline{y}z \vee \overline{x}yz \\ &= xy \vee xyz \vee x\overline{y}z \vee \overline{x}yz \\ &= xy \vee xz \vee \overline{x}yz \\ &= xy \vee xz \vee xyz \vee \overline{x}yz \\ &= xy \vee xz \vee yz \end{aligned}$$

Additional gates can be eliminated by writing $c = xy \vee z(x \vee y)$. We obtain the full-adder circuit given in Figure 11.5.9.

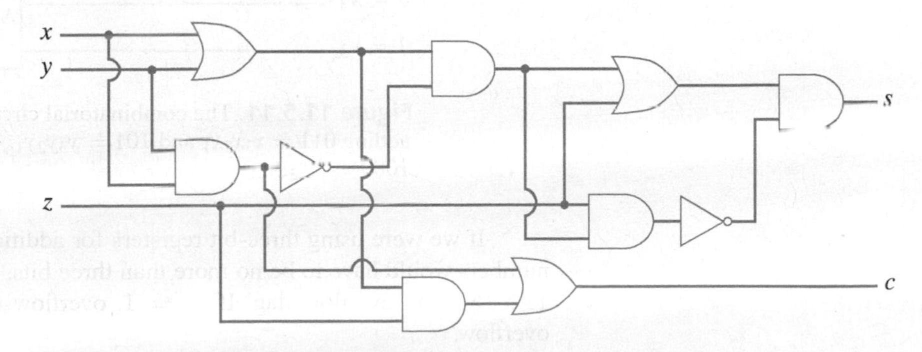


**Figure 11.5.9** A full-adder circuit. ◀

Our last example shows how we may use half-adder and full-adder circuits to construct a circuit to add binary numbers.

**Example 11.5.15 A Circuit to Add Binary Numbers** Using half-adder and full-adder circuits, design a combinatorial circuit that computes the sum of two three-bit numbers.

**SOLUTION** We will let $M = x_3x_2x_1$ and $N = y_3y_2y_1$ denote the numbers to be added and let $z_4z_3z_2z_1$ denote the sum. The circuit that computes the sum of $M$ and $N$ is drawn in Figure 11.5.10. It is an implementation of the standard algorithm for adding numbers since the "carry bit" is indeed *carried* into the next binary addition.

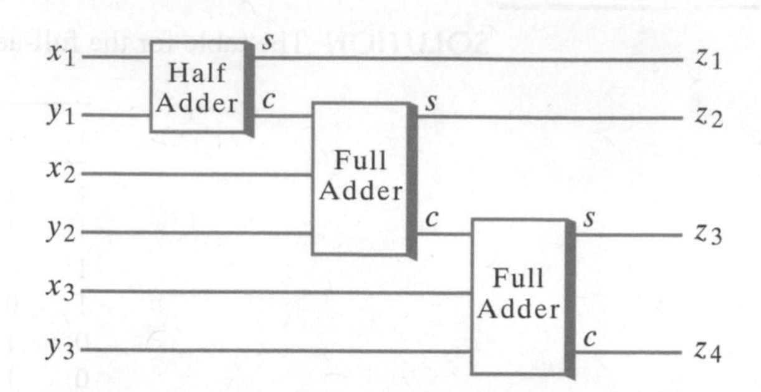


**Figure 11.5.10** A combinatorial circuit that computes the sum of two three-bit numbers.

Figure 11.5.11 shows how the circuit adds $011 = x_3x_2x_1$ and $101 = y_3y_2y_1$. The bits $x_1 = 1$ and $y_1 = 1$ are input to the half adder. Since $1 + 1 = 10$, the sum bit $s = 0 = z_1$ and the carry bit $c = 1$. Next the carry bit (1) and the bits $x_2 = 1$ and $y_2 = 0$ are input to the (middle) full adder. Since $1 + 1 + 0 = 10$, the sum bit $s = 0 = z_2$ and the carry bit $c = 1$. At the last step, the carry bit (1) and the bits $x_3 = 0$ and $y_3 = 1$ are input to the (bottom) full adder. Since $1 + 0 + 1 = 10$, the sum bit $s = 0 = z_3$ and the carry bit $c = 1 = z_4$. The sum $011 + 101 = 1000 = z_4z_3z_2z_1$.

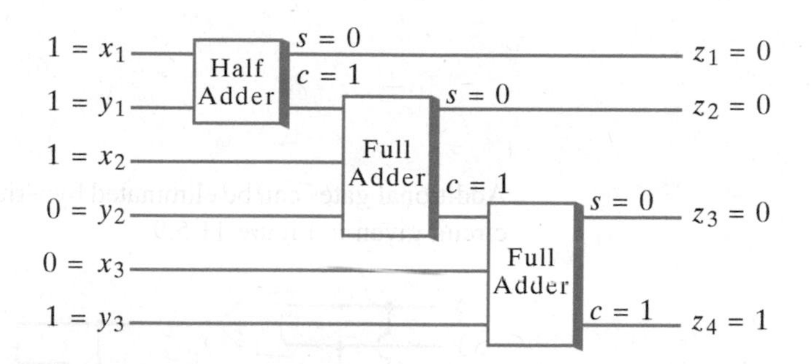


**Figure 11.5.11** The combinatorial circuit of Figure 11.5.10 adding $011 = x_3x_2x_1$ and $101 = y_3y_2y_1$. The sum is $1000 = z_4z_3z_2z_1$.

◀

If we were using three-bit registers for addition, so that the sum of two three-bit numbers would have to be no more than three bits, we could use the $z_4$ bit in Example 11.5.15 as an overflow flag. If $z_4 = 1$, overflow occurred; if $z_4 = 0$, there was no overflow.

In the next chapter (Example 12.1.3), we discuss a sequential circuit that makes use of a primitive internal memory to add binary numbers.

## 11.5 Review Exercises

1. What is a gate?
2. What is a functionally complete set of gates?
3. Give examples of functionally complete sets of gates.
4. What is a NAND gate?
5. Is the set {NAND} functionally complete?
6. What is the minimization problem?
7. What is an integrated circuit?
8. Describe a half-adder circuit.
9. Describe a full-adder circuit.

## 11.5 Exercises

1. Show that the set of gates {OR, NOT} is functionally complete.

*Show that each set of gates in Exercises 2–5 is not functionally complete.*

2. {AND}
3. {NOT}
4. {OR}
5. {AND, OR}
6. Draw a circuit using only NAND gates that computes $xy$.
7. Write $xy$ using only $\uparrow$.
8. Prove or disprove: $x \uparrow (y \uparrow z) = (x \uparrow y) \uparrow z$, for all $x, y, z \in Z_2$.

*Write Boolean expressions to describe the multiple output circuits in Exercises 9–11.*

9.

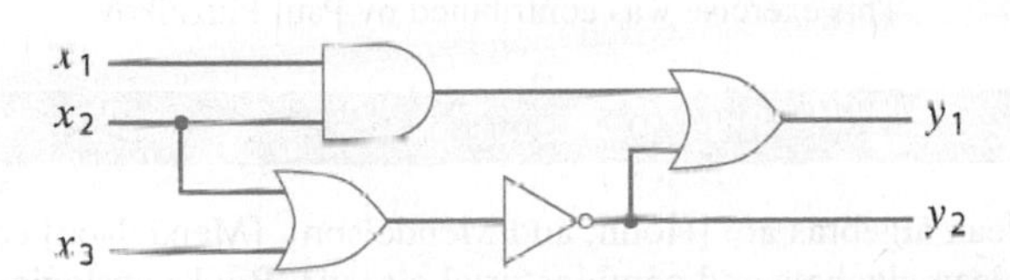


10.

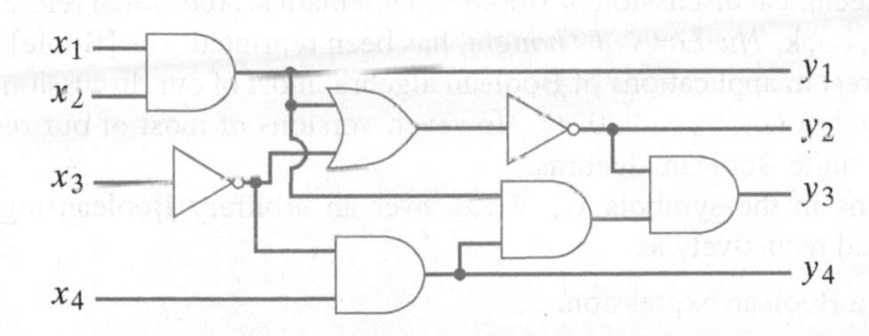


11.

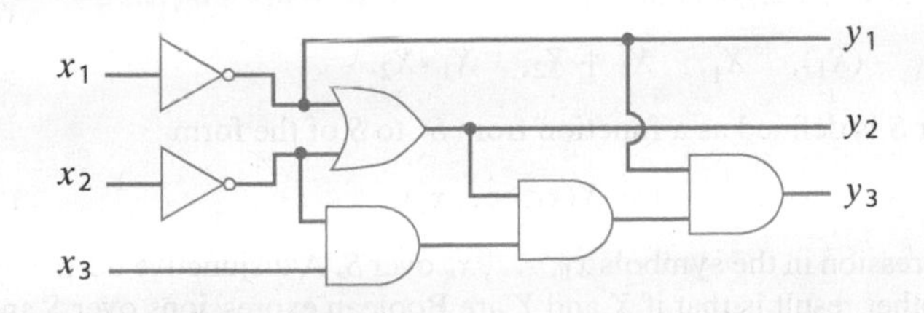


12. Design circuits using only NAND gates to compute the functions of Exercises 1–10, Section 11.4.
13. Can you reduce the number of NAND gates used in any of your circuits for Exercise 12?
14. Design circuits using as few AND, OR, and NOT gates as you can to compute the functions of Exercises 1–10, Section 11.4.
15. Design a half-adder circuit using only NAND gates.

★16. Design a half-adder circuit using five NAND gates.

*A NOR gate receives inputs $x_1$ and $x_2$, where $x_1$ and $x_2$ are bits, and produces output denoted $x_1 \downarrow x_2$, where*

$$x_1 \downarrow x_2 = \begin{cases} 0 & \text{if } x_1 = 1 \text{ or } x_2 = 1 \\ 1 & \text{otherwise.} \end{cases}$$

17. Write $xy$, $x \vee y$, $\bar{x}$, and $x \uparrow y$ in terms of $\downarrow$.
18. Write $x \downarrow y$ in terms of $\uparrow$.
19. Write the logic table for the NOR function.
20. Show that the set of gates {NOR} is functionally complete.
21. Design circuits using only NOR gates to compute the functions of Exercises 1–10, Section 11.4.
22. Can you reduce the number of NOR gates used in any of your circuits for Exercise 21?
23. Design a half-adder circuit using only NOR gates.

★24. Design a half-adder circuit using five NOR gates.

25. Design a circuit with three inputs that outputs 1 precisely when two or three inputs have value 1.
26. Design a circuit that multiplies the binary numbers $x_2x_1$ and $y_2y_1$. The output will be of the form $z_4z_3z_2z_1$.
27. A **2's module** is a circuit that accepts as input two bits $b$ and FLAGIN and outputs bits $c$ and FLAGOUT. If FLAGIN = 1, then $c = \bar{b}$ and FLAGOUT = 1. If FLAGIN = 0 and $b = 1$, then FLAGOUT = 1. If FLAGIN = 0 and $b = 0$, then FLAGOUT = 0. If FLAGIN = 0, then $c = b$. Design a circuit to implement a 2's module.

*The 2's complement of a binary number can be computed by using the following algorithm.*

### Algorithm 11.5.16
*Finding the 2's Complement*

This algorithm computes the 2's complement $C_N C_{N-1} \cdots C_2 C_1$ of the binary number $M = B_N B_{N-1} \cdots B_2 B_1$. The number $M$ is scanned from right to left and the bits are copied until 1 is found. Thereafter, if $B_i = 0$, we set $C_i = 1$ and if $B_i = 1$, we set $C_i = 0$. The flag $F$ indicates whether a 1 has been found ($F$ = **true**) or not ($F$ = **false**).

Input: $B_N B_{N-1} \cdots B_1$

Output: $C_N C_{N-1} \cdots C_1$

```
twos_complement(B) {
    F = false
    i = 1
    while (¬F ∧ i ≤ N) {
        C_i = B_i
        if (B_i == 1)
            F = true
        i = i + 1
    }
    while (i ≤ N) {
        C_i = B_i ⊕ 1
        i = i + 1
    }
    return C
}
```

*Find the 2's complement of the numbers in Exercises 28–30 using Algorithm 11.5.16.*

**28.** 101100 **29.** 11011 **30.** 011010110

**31.** Using 2's modules, design a circuit that computes the 2's complement $y_3y_2y_1$ of the three-bit binary number $x_3x_2x_1$.

★**32.** Let $*$ be a binary operator on a set $S$ containing 0 and 1. Write a set of axioms for $*$, modeled after rules that NAND satisfies, so that if we define

$$\overline{x} = x * x$$
$$x \vee y = (x * x) * (y * y)$$
$$x \wedge y = (x * y) * (x * y),$$

then $(S, \vee, \wedge, \overline{\ }, 0, 1)$ is a Boolean algebra.

★**33.** Let $*$ be a binary operator on a set $S$ containing 0 and 1. Write a set of axioms for $*$, modeled after rules that NOR satisfies, and definitions for $\overline{\ }$, $\vee$, and $\wedge$ so that $(S, \vee, \wedge, \overline{\ }, 0, 1)$ is a Boolean algebra.

★**34.** Show that $\{\rightarrow\}$ is functionally complete (see Definition 1.3.3).

★**35.** Let $B(x, y)$ be a Boolean expression in the variables $x$ and $y$ that uses only the operator $\leftrightarrow$ (see Definition 1.3.8).

(a) Show that if $B$ contains an even number of $x$'s, the values of $B(\overline{x}, y)$ and $B(x, y)$ are the same for all $x$ and $y$.

(b) Show that if $B$ contains an odd number of $x$'s, the values of $B(\overline{x}, y)$ and $\overline{B(x, y)}$ are the same for all $x$ and $y$.

(c) Use parts (a) and (b) to show that $\{\leftrightarrow\}$ is *not* functionally complete.

This exercise was contributed by Paul Pluznikov.

## Chapter 11 Notes

General references on Boolean algebras are [Hohn; and Mendelson]. [Mendelson] contains over 150 references on Boolean algebras and combinatorial circuits. Books on logic design include [Kohavi; McCalla; and Ward].

[Hailperin] gives a technical discussion of Boole's mathematics. Additional references are also provided. Boole's book, *The Laws of Thought,* has been reprinted (see [Boole]).

Because of our interest in applications of Boolean algebra, most of our discussion was limited to the Boolean algebra $(Z_2, \vee, \wedge, \overline{\ }, 0, 1)$. However, versions of most of our results remain valid for arbitrary, finite Boolean algebras.

**Boolean expressions** in the symbols $x_1, \ldots, x_n$ over an arbitrary Boolean algebra $(S, +, \cdot, ', 0, 1)$ are defined recursively as

- For each $s \in S$, $s$ is a Boolean expression.
- $x_1, \ldots, x_n$ are Boolean expressions.

If $X_1$ and $X_2$ are Boolean expressions, so are

$$(X_1), \quad X_1', \quad X_1 + X_2, \quad X_1 \cdot X_2.$$

A **Boolean function** over $S$ is defined as a function from $S^n$ to $S$ of the form

$$f(x_1, \ldots, x_n) = X(x_1, \ldots, x_n),$$

where $X$ is a Boolean expression in the symbols $x_1, \ldots, x_n$ over $S$. A disjunctive normal form can be defined for $f$. Another result is that if $X$ and $Y$ are Boolean expressions over $S$ and

$$X(x_1, \ldots, x_n) = Y(x_1, \ldots, x_n)$$

for all $x_i \in S$, then $Y$ is derivable from $X$ using the definition (Definition 11.3.1) of a Boolean algebra. Other results are that any finite Boolean algebra has $2^n$ elements and that if two Boolean algebras both have $2^n$ elements they are essentially the same. It follows that any finite Boolean algebra is essentially Example 11.3.3, the Boolean algebra of subsets of a finite, universal set $U$. The proofs of these results can be found in [Mendelson].

## Chapter 11 Review

### Section 11.1

1. Combinatorial circuit
2. Sequential circuit
3. AND gate
4. OR gate
5. NOT gate (inverter)
6. Logic table of a combinatorial circuit
7. Boolean expression
8. Literal

### Section 11.2

9. Properties of $\wedge$, $\vee$, and $\bar{}$: associative laws; commutative laws; distributive laws; identity laws; complement laws (see Theorem 11.2.1)
10. Equal Boolean expressions
11. Equivalent combinatorial expressions
12. Combinatorial expressions are equivalent if and only if the Boolean expressions that represent them are equal.

### Section 11.3

13. Boolean algebra
14. $x'$: Complement of $x$
15. Properties of Boolean algebras: Idempotent laws; bound laws; absorption laws; involution law; 0 and 1 laws; De Morgan's laws
16. Dual of statement involving Boolean expressions
17. The dual of a theorem about Boolean algebras is also a theorem.

### Section 11.4

18. Exclusive-OR
19. Boolean function
20. Minterm: $y_1 \wedge y_2 \wedge \cdots \wedge y_n$, where each $y_i$ is $x_i$ or $\overline{x}_i$
21. Disjunctive normal form
22. How to write a Boolean function in disjunctive normal form (Theorem 11.4.6)
23. Maxterm: $y_1 \vee y_2 \vee \cdots \vee y_n$, where each $y_i$ is $x_i$ or $\overline{x}_i$
24. Conjunctive normal form

### Section 11.5

25. Gate
26. Functionally complete set of gates
27. The sets of gates {AND, NOT} and {OR, NOT} are functionally complete.
28. NAND gate
29. The set {NAND} is a functionally complete set of gates.
30. Minimization problem
31. Integrated circuit
32. Half-adder circuit
33. Full-adder circuit

## Chapter 11 Self-Test

1. Write a Boolean expression that represents the combinatorial circuit and write the logic table.

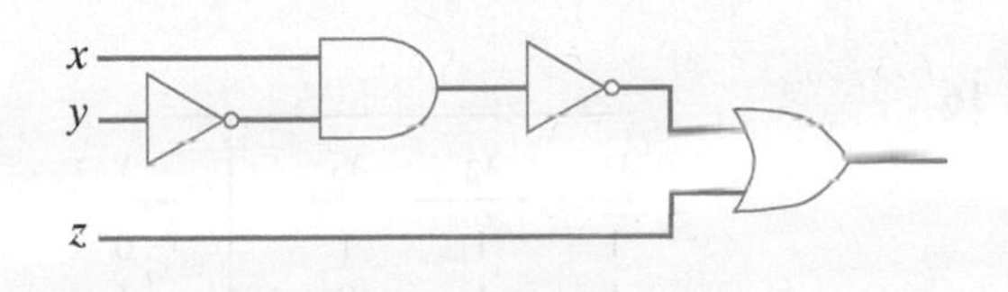


*Are the combinatorial circuits in Exercises 2 and 3 equivalent? Explain.*

2.

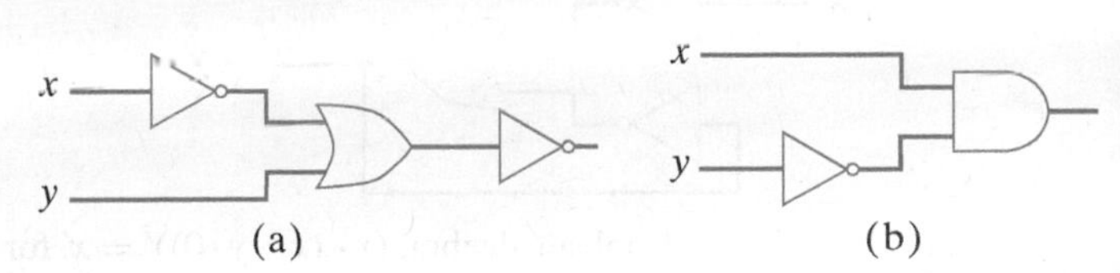

**3.**

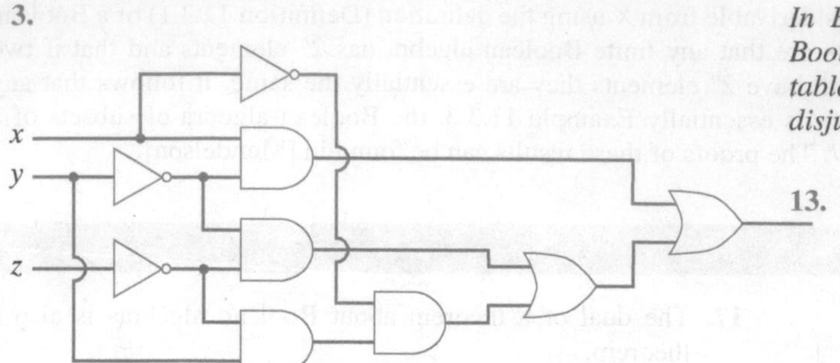


(a)

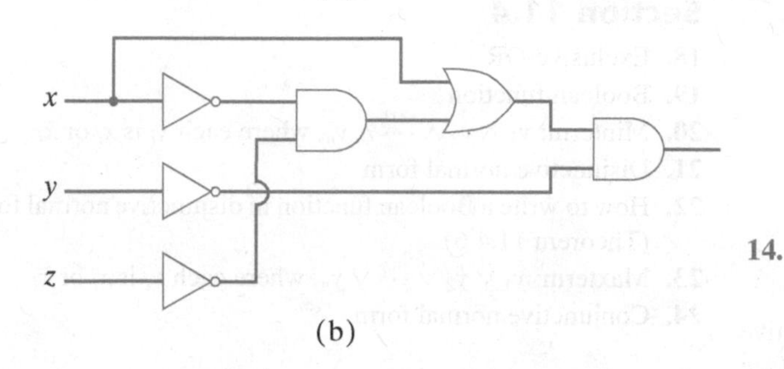


(b)

**4.** If $U$ is a universal set and $S = \mathcal{P}(U)$, the power set of $U$, then

$$(S, \cup, \cap, {}^{-}, \varnothing, U)$$

is a Boolean algebra. State the bound and absorption laws for this Boolean algebra.

**5.** Find the value of the Boolean expression

$$(x_1 \wedge x_2) \vee (\overline{x}_2 \wedge x_3)$$

if $x_1 = x_2 = 0$ and $x_3 = 1$.

**6.** Find a combinatorial circuit corresponding to the Boolean expression of Exercise 5.

*Prove or disprove the equations in Exercises 7 and 8.*

**7.** $(x \wedge y) \vee (\overline{x} \wedge z) \vee (\overline{x} \wedge y \wedge \overline{z}) = y \vee (\overline{x} \wedge z)$

**8.** $(x \wedge y \wedge z) \vee (\overline{x \vee z}) = (x \wedge z) \vee (\overline{x} \wedge \overline{z})$

**9.** Show that the following circuit is not a combinatorial circuit.

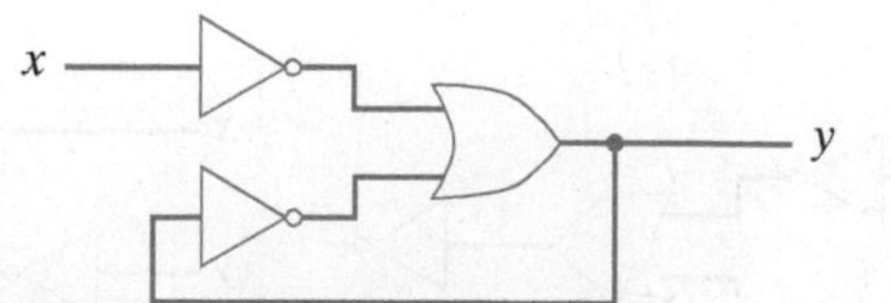


**10.** Prove that in any Boolean algebra, $(x \cdot (x + y \cdot 0))' = x'$ for all $x$ and $y$.

**11.** Write the dual of the statement of Exercise 10 and prove it.

**12.** Let $U$ be the set of positive integers. Let $S$ be the collection of finite subsets of $U$. Why does $(S, \cup, \cap, {}^{-}, \varnothing, U)$ fail to be a Boolean algebra?

*In Exercises 13–16, find the disjunctive normal form of a Boolean expression having a logic table the same as the given table and draw the combinatorial circuit corresponding to the disjunctive normal form.*

**13.**

| $x_1$ | $x_2$ | $x_3$ | $y$ |
|---|---|---|---|
| 1 | 1 | 1 | 0 |
| 1 | 1 | 0 | 0 |
| 1 | 0 | 1 | 0 |
| 1 | 0 | 0 | 0 |
| 0 | 1 | 1 | 0 |
| 0 | 1 | 0 | 0 |
| 0 | 0 | 1 | 0 |
| 0 | 0 | 0 | 0 |

**14.**

| $x_1$ | $x_2$ | $x_3$ | $y$ |
|---|---|---|---|
| 1 | 1 | 1 | 0 |
| 1 | 1 | 0 | 1 |
| 1 | 0 | 1 | 0 |
| 1 | 0 | 0 | 1 |
| 0 | 1 | 1 | 0 |
| 0 | 1 | 0 | 0 |
| 0 | 0 | 1 | 0 |
| 0 | 0 | 0 | 0 |

**15.**

| $x_1$ | $x_2$ | $x_3$ | $y$ |
|---|---|---|---|
| 1 | 1 | 1 | 1 |
| 1 | 1 | 0 | 0 |
| 1 | 0 | 1 | 0 |
| 1 | 0 | 0 | 1 |
| 0 | 1 | 1 | 0 |
| 0 | 1 | 0 | 0 |
| 0 | 0 | 1 | 0 |
| 0 | 0 | 0 | 1 |

**16.**

| $x_1$ | $x_2$ | $x_3$ | $y$ |
|---|---|---|---|
| 1 | 1 | 1 | 0 |
| 1 | 1 | 0 | 1 |
| 1 | 0 | 1 | 0 |
| 1 | 0 | 0 | 1 |
| 0 | 1 | 1 | 1 |
| 0 | 1 | 0 | 0 |
| 0 | 0 | 1 | 1 |
| 0 | 0 | 0 | 0 |

17. Write the logic table for the circuit

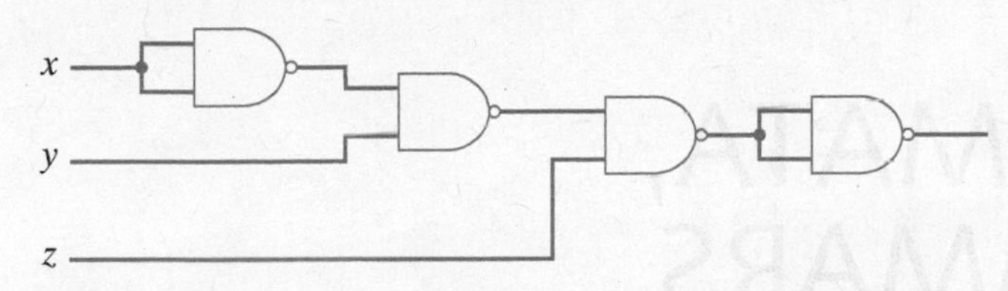


18. Find a Boolean expression in disjunctive normal form for the circuit of part (a) of Exercise 3. Use algebraic methods to simplify the disjunctive normal form. Draw the circuit corresponding to the simplified expression.

19. Design a circuit using only NAND gates to compute $x \oplus y$.

20. Design a full-adder circuit that uses two half adders and one OR gate.

## Chapter 11 Computer Exercises

1. Write a program that inputs a Boolean expression in $x$ and $y$ and prints the logic table of the expression.

2. Write a program that inputs a Boolean expression in $x$, $y$, and $z$ and prints the logic table of the expression.

3. Write a program that outputs the disjunctive normal form of a Boolean expression $p(x, y)$.

4. Write a program that outputs the conjunctive normal form of a Boolean expression $p(x, y)$.

5. Write a program that outputs the disjunctive normal form of a Boolean expression $p(x, y, z)$.

6. Write a program that outputs the conjunctive normal form of a Boolean expression $p(x, y, z)$.

7. Write a program that computes the two's complement of an $n$-bit binary number.

# Chapter 12

# AUTOMATA, GRAMMARS, AND LANGUAGES



In Chapter 11 we discussed combinatorial circuits in which the output depended only on the input. These circuits have no memory. In this chapter we begin by discussing circuits in which the output depends not only on the input but also on the state of the system at the time the input is introduced. The state of the system is determined by previous processing. In this sense, these circuits have memory. Such circuits are called *sequential circuits* and are obviously important in computer design.

Finite-state machines are abstract models of machines with a primitive internal memory. A finite-state automaton is a special kind of finite-state machine that is closely linked to a particular type of language. In the latter part of this chapter, we will discuss finite-state machines, finite-state automata, and languages in some detail.

## 12.1 Sequential Circuits and Finite-State Machines

**Go Online**
For more on sequential circuits and finite-state machines, see
bit.ly/2CAd76a

Operations within a digital computer are carried out at discrete intervals of time. Output depends on the state of the system as well as on the input. We will assume that the state of the system changes only at time $t = 0, 1, \ldots$. A simple way to introduce sequencing in circuits is to introduce a **unit time delay.**

**Definition 12.1.1 ▶** A *unit time delay* accepts as input a bit $x_t$ at time $t$ and outputs $x_{t-1}$, the bit received as input at time $t - 1$. The unit time delay is drawn as shown in Figure 12.1.1.

**Figure 12.1.1** Unit time delay. ◀

As an example of the use of the unit time delay, we discuss the **serial adder.**

**Definition 12.1.2 ▶** A *serial adder* accepts as input two binary numbers

$$x = 0x_N x_{N-1} \cdots x_0 \qquad \text{and} \qquad y = 0y_N y_{N-1} \cdots y_0$$

and outputs the sum $z_{N+1} z_N \cdots z_0$ of $x$ and $y$. The numbers $x$ and $y$ are input sequentially in pairs, $x_0, y_0; \ldots; x_N, y_N; 0, 0$. The sum is output $z_0, z_1, \ldots, z_{N+1}$. ◀

**Example 12.1.3** **Serial-Adder Circuit** A circuit, using a unit time delay, that implements a serial adder is shown in Figure 12.1.2. Show how the serial adder computes the sum of $x = 010$ and $y = 011$.

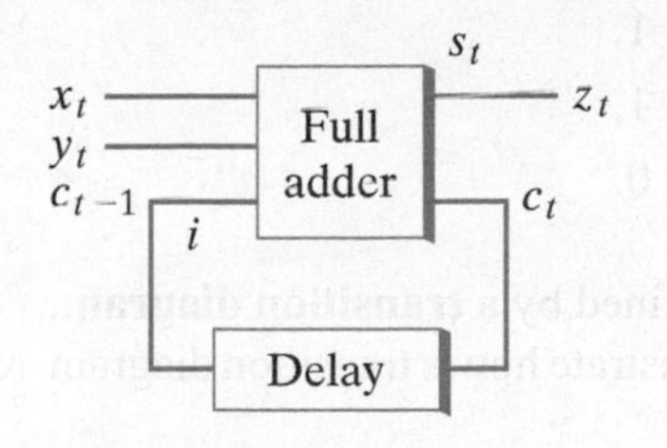


**Figure 12.1.2** A serial-adder circuit.

**SOLUTION** We begin by setting $x_0 = 0$ and $y_0 = 1$. (We assume that at this instant $i = 0$. This can be arranged by first setting $x = y = 0$.) The state of the system is shown in Figure 12.1.3(a). Next, we set $x_1 = y_1 = 1$. The unit time delay sends $i = 0$ as the third bit to the full adder. The state of the system is shown in Figure 12.1.3(b). Finally, we set $x_2 = y_2 = 0$. This time the unit time delay sends $i = 1$ as the third bit to the full adder. The state of the system is shown in Figure 12.1.3(c). We obtain the sum $z = 101$.

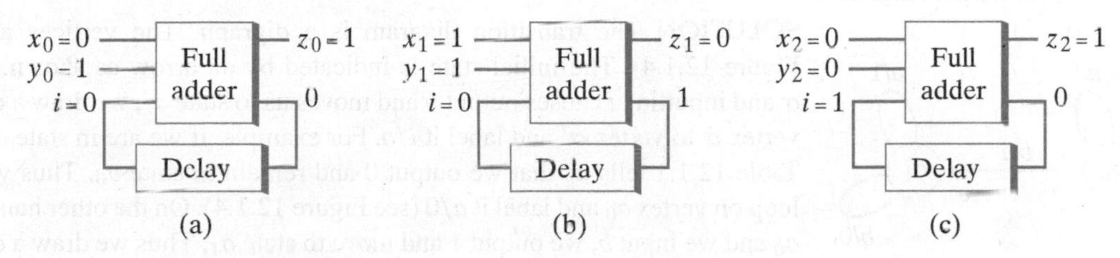


**Figure 12.1.3** Computing $010 + 011$ with the serial-adder circuit. ◀

A **finite-state machine** is an abstract model of a machine with a primitive internal memory.

**Definition 12.1.4 ▶** A *finite-state machine* $M$ consists of

(a) A finite set $\mathcal{I}$ of *input symbols*.

(b) A finite set $\mathcal{O}$ of *output symbols*.

(c) A finite set $\mathcal{S}$ of *states*.

(d) A *next-state function* $f$ from $\mathcal{S} \times \mathcal{I}$ into $\mathcal{S}$.

(e) An *output function* $g$ from $\mathcal{S} \times \mathcal{I}$ into $\mathcal{O}$.

(f) An *initial state* $\sigma \in \mathcal{S}$.

We write $M = (\mathcal{I}, \mathcal{O}, \mathcal{S}, f, g, \sigma)$. ◀

**Example 12.1.5** Let $\mathcal{I} = \{a, b\}$, $\mathcal{O} = \{0, 1\}$, and $\mathcal{S} = \{\sigma_0, \sigma_1\}$. Define the pair of functions $f\colon \mathcal{S} \times \mathcal{I} \to \mathcal{S}$ and $g\colon \mathcal{S} \times \mathcal{I} \to \mathcal{O}$ by the rules given in Table 12.1.1.

**TABLE 12.1.1** ■

| | $f$ | | $g$ | |
|---|---|---|---|---|
| $\mathcal{S}$ \ $\mathcal{I}$ | $a$ | $b$ | $a$ | $b$ |
| $\sigma_0$ | $\sigma_0$ | $\sigma_1$ | 0 | 1 |
| $\sigma_1$ | $\sigma_1$ | $\sigma_1$ | 1 | 0 |

Then $M = (\mathcal{I}, \mathcal{O}, \mathcal{S}, f, g, \sigma_0)$ is a finite-state machine.

Table 12.1.1 is interpreted to mean

$$f(\sigma_0, a) = \sigma_0 \qquad g(\sigma_0, a) = 0,$$
$$f(\sigma_0, b) = \sigma_1 \qquad g(\sigma_0, b) = 1,$$
$$f(\sigma_1, a) = \sigma_1 \qquad g(\sigma_1, a) = 1,$$
$$f(\sigma_1, b) = \sigma_1 \qquad g(\sigma_1, b) = 0.$$

◀

The next-state and output functions can also be defined by a **transition diagram.** Before formally defining a transition diagram, we will illustrate how a transition diagram is constructed.

**Example 12.1.6** Draw the transition diagram for the finite-state machine of Example 12.1.5.

SOLUTION The transition diagram is a digraph. The vertices are the states (see Figure 12.1.4). The initial state is indicated by an arrow as shown. If we are in state $\sigma$ and inputting $i$ causes output $o$ and moves us to state $\sigma'$, we draw a directed edge from vertex $\sigma$ to vertex $\sigma'$ and label it $i/o$. For example, if we are in state $\sigma_0$, and we input $a$, Table 12.1.1 tells us that we output 0 and remain in state $\sigma_0$. Thus we draw a directed loop on vertex $\sigma_0$ and label it $a/0$ (see Figure 12.1.4). On the other hand, if we are in state $\sigma_0$ and we input $b$, we output 1 and move to state $\sigma_1$. Thus we draw a directed edge from $\sigma_0$ to $\sigma_1$ and label it $b/1$. By considering all such possibilities, we obtain the transition diagram of Figure 12.1.4. ◀

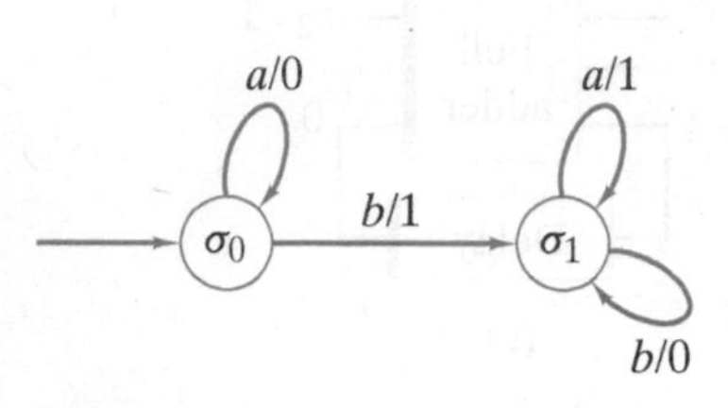


**Figure 12.1.4** A transition diagram.

**Definition 12.1.7 ▶** Let $M = (\mathcal{I}, \mathcal{O}, \mathcal{S}, f, g, \sigma)$ be a finite-state machine. The *transition diagram* of $M$ is a digraph $G$ whose vertices are the members of $\mathcal{S}$. An arrow designates the initial state $\sigma$. A directed edge $(\sigma_1, \sigma_2)$ exists in $G$ if there exists an input $i$ with $f(\sigma_1, i) = \sigma_2$. In this case, if $g(\sigma_1, i) = o$, the edge $(\sigma_1, \sigma_2)$ is labeled $i/o$. ◀

We can regard the finite-state machine $M = (\mathcal{I}, \mathcal{O}, \mathcal{S}, f, g, \sigma)$ as a simple computer. We begin in state $\sigma$, input a string over $\mathcal{I}$, and produce a string of output.

**Definition 12.1.8 ▶** Let $M = (\mathcal{I}, \mathcal{O}, \mathcal{S}, f, g, \sigma)$ be a finite-state machine. An *input string* for $M$ is a string over $\mathcal{I}$. The string

$$y_1 \cdots y_n$$

is the *output string* for $M$ corresponding to the input string

$$\alpha = x_1 \cdots x_n$$

if there exist states $\sigma_0, \ldots, \sigma_n \in \mathcal{S}$ with

$$\sigma_0 = \sigma$$
$$\sigma_i = f(\sigma_{i-1}, x_i) \qquad \text{for } i = 1, \ldots, n;$$
$$y_i = g(\sigma_{i-1}, x_i) \qquad \text{for } i = 1, \ldots, n.$$

◀

**Example 12.1.9** Find the output string corresponding to the input string

$$aababba \qquad \textbf{(12.1.1)}$$

for the finite-state machine of Example 12.1.5.

SOLUTION Initially, we are in state $\sigma_0$. The first symbol input is $a$. We locate the outgoing edge in the transition diagram of $M$ (Figure 12.1.4) from $\sigma_0$ labeled $a/x$, which tells us that if $a$ is input, $x$ is output. In our case, 0 is output. The edge points to the next state, $\sigma_0$. Next, $a$ is input again. As before, we output 0 and remain in state $\sigma_0$. Next, $b$ is input. In this case, we output 1 and change to state $\sigma_1$. Continuing in this way, we find that the output string is

$$0011001. \tag{12.1.2}$$

◀

**Example 12.1.10** **A Serial-Adder Finite-State Machine** Design a finite-state machine that performs serial addition.

SOLUTION We will represent the finite-state machine by its transition diagram. Since the serial adder accepts pairs of bits, the input set is {00, 01, 10, 11} and the output set is {0, 1}.

Given an input $xy$, we take one of two actions: Either we add $x$ and $y$, or we add $x$, $y$, and 1, depending on whether the carry bit was 0 or 1. Thus there are two states, which we will call $C$ (carry) and $NC$ (no carry). The initial state is $NC$. At this point, we can draw the vertices and designate the initial state in our transition diagram (see Figure 12.1.5).

Next, we consider the possible inputs at each vertex. For example, if 00 is input to $NC$, we should output 0 and remain in state $NC$. Thus $NC$ has a loop labeled 00/0. As another example, if 11 is input to $C$, we compute $1+1+1=11$. In this case we output 1 and remain in state $C$. Thus $C$ has a loop labeled 11/1. As a final example, if we are in state $NC$ and 11 is input, we should output 0 and move to state $C$. By considering all possibilities, we arrive at the transition diagram of Figure 12.1.6.

NC C

Figure 12.1.5 Two states for the serial-adder finite-state machine.

00/0 01/1 01/0 11/0 NC C 10/0 00/1 10/1 11/1

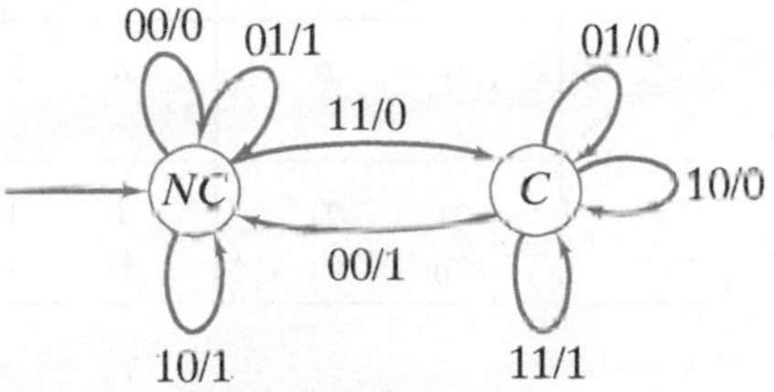

Figure 12.1.6 A finite-state machine that performs serial addition. ◀

**Example 12.1.11** **The *SR* Flip-Flop** A **flip-flop** is a basic component of digital circuits since it serves as a one-bit memory cell. The ***SR* flip-flop** (or **set-reset flip-flop**) can be defined by the table

| $S$ | $R$ | $Q$ |
|---|---|---|
| 1 | 1 | Not allowed |
| 1 | 0 | 1 |
| 0 | 1 | 0 |
| 0 | 0 | $\begin{cases} 1 & \text{if S was last equal to 1} \\ 0 & \text{if R was last equal to 1} \end{cases}$ |

The $SR$ flip-flop "remembers" whether $S$ or $R$ was last equal to 1. (If $Q=1$, $S$ was last equal to 1; if $Q=0$, $R$ was last equal to 1.) We can model the $SR$ flip-flop as a finite-state machine by defining two states: "$S$ was last equal to 1" and "$R$ was last equal to 1" (see Figure 12.1.7). We define the input to be the new values of $S$ and $R$; the notation $sr$ means that $S=s$ and $R=r$. We define $Q$ to be the output. We have arbitrarily designated the initial state as "$S$ was last equal to 1." A sequential circuit implementation of the $SR$ flip-flop is shown in Figure 12.1.8.

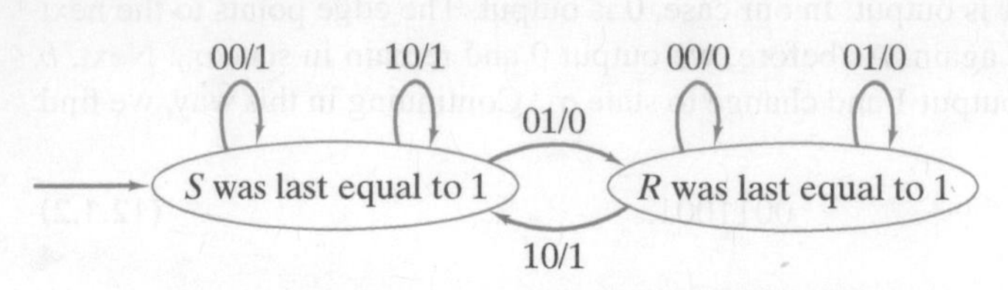


Figure 12.1.7 The *SR* flip-flop as a finite-state machine.

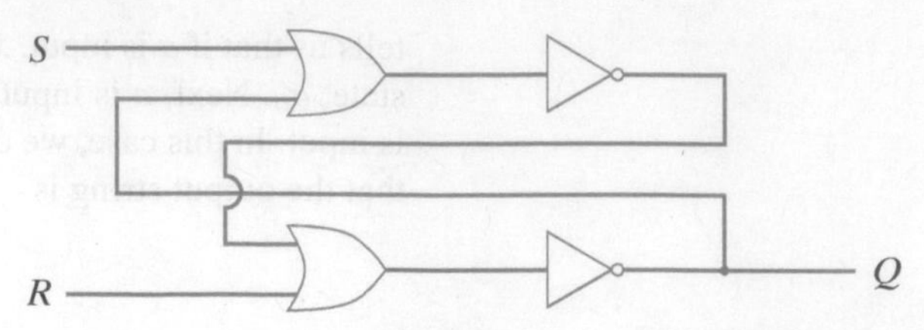


Figure 12.1.8 A sequential circuit implementation of the *SR* flip-flop.

## 12.1 Review Exercises

1. What is a unit time delay?
2. What is a serial adder?
3. Define *finite-state machine.*
4. What is a transition diagram?
5. What is the *SR* flip-flop?

## 12.1 Exercises

*In Exercises 1–5, draw the transition diagram of the finite-state machine* $(\mathcal{I}, \mathcal{O}, \mathcal{S}, f, g, \sigma_0)$.

**1.** $\mathcal{I} = \{a, b\}$, $\mathcal{O} = \{0, 1\}$, $\mathcal{S} = \{\sigma_0, \sigma_1\}$

| $\mathcal{S}$ \ $\mathcal{I}$ | $f$: $a$ | $f$: $b$ | $g$: $a$ | $g$: $b$ |
|---|---|---|---|---|
| $\sigma_0$ | $\sigma_1$ | $\sigma_1$ | 1 | 1 |
| $\sigma_1$ | $\sigma_0$ | $\sigma_1$ | 0 | 1 |

**2.** $\mathcal{I} = \{a, b\}$, $\mathcal{O} = \{0, 1\}$, $\mathcal{S} = \{\sigma_0, \sigma_1\}$

| $\mathcal{S}$ \ $\mathcal{I}$ | $f$: $a$ | $f$: $b$ | $g$: $a$ | $g$: $b$ |
|---|---|---|---|---|
| $\sigma_0$ | $\sigma_1$ | $\sigma_0$ | 0 | 0 |
| $\sigma_1$ | $\sigma_0$ | $\sigma_0$ | 1 | 1 |

**3.** $\mathcal{I} = \{a, b\}$, $\mathcal{O} = \{0, 1\}$, $\mathcal{S} = \{\sigma_0, \sigma_1, \sigma_2\}$

| $\mathcal{S}$ \ $\mathcal{I}$ | $f$: $a$ | $f$: $b$ | $g$: $a$ | $g$: $b$ |
|---|---|---|---|---|
| $\sigma_0$ | $\sigma_1$ | $\sigma_1$ | 0 | 1 |
| $\sigma_1$ | $\sigma_2$ | $\sigma_1$ | 1 | 1 |
| $\sigma_2$ | $\sigma_0$ | $\sigma_0$ | 0 | 0 |

**4.** $\mathcal{I} = \{a, b, c\}$, $\mathcal{O} = \{0, 1\}$, $\mathcal{S} = \{\sigma_0, \sigma_1, \sigma_2\}$

| $\mathcal{S}$ \ $\mathcal{I}$ | $f$: $a$ | $f$: $b$ | $f$: $c$ | $g$: $a$ | $g$: $b$ | $g$: $c$ |
|---|---|---|---|---|---|---|
| $\sigma_0$ | $\sigma_0$ | $\sigma_1$ | $\sigma_2$ | 0 | 1 | 0 |
| $\sigma_1$ | $\sigma_1$ | $\sigma_1$ | $\sigma_0$ | 1 | 1 | 1 |
| $\sigma_2$ | $\sigma_2$ | $\sigma_1$ | $\sigma_0$ | 1 | 0 | 0 |

**5.** $\mathcal{I} = \{a, b, c\}$, $\mathcal{O} = \{0, 1, 2\}$, $\mathcal{S} = \{\sigma_0, \sigma_1, \sigma_2, \sigma_3\}$

| $\mathcal{S}$ \ $\mathcal{I}$ | $f$: $a$ | $f$: $b$ | $f$: $c$ | $g$: $a$ | $g$: $b$ | $g$: $c$ |
|---|---|---|---|---|---|---|
| $\sigma_0$ | $\sigma_1$ | $\sigma_0$ | $\sigma_2$ | 1 | 1 | 2 |
| $\sigma_1$ | $\sigma_0$ | $\sigma_2$ | $\sigma_2$ | 2 | 0 | 0 |
| $\sigma_2$ | $\sigma_3$ | $\sigma_3$ | $\sigma_0$ | 1 | 0 | 1 |
| $\sigma_3$ | $\sigma_1$ | $\sigma_1$ | $\sigma_0$ | 2 | 0 | 2 |

*In Exercises 6–10, find the sets* $\mathcal{I}$, $\mathcal{O}$, *and* $\mathcal{S}$, *the initial state, and the table defining the next-state and output functions for each finite-state machine.*

**6.**

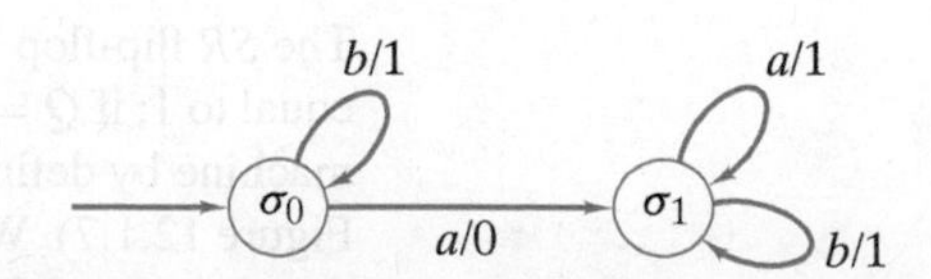

**7.**

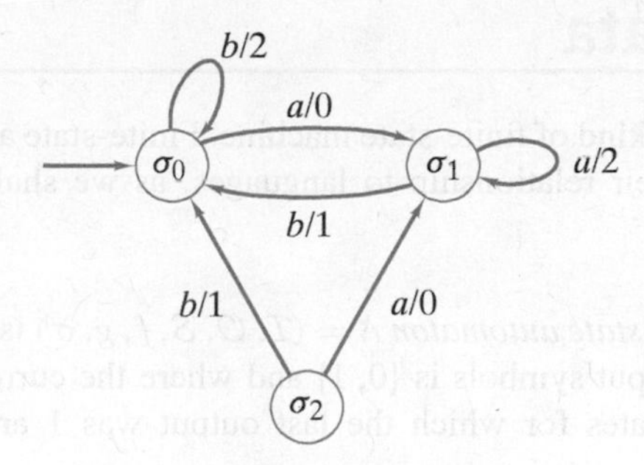


**8.**

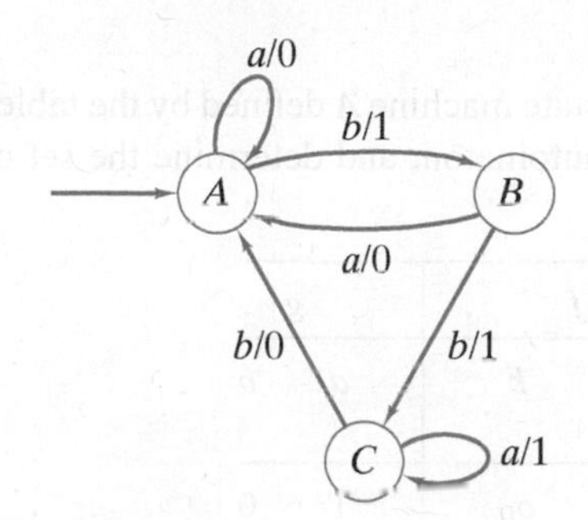


**9.**

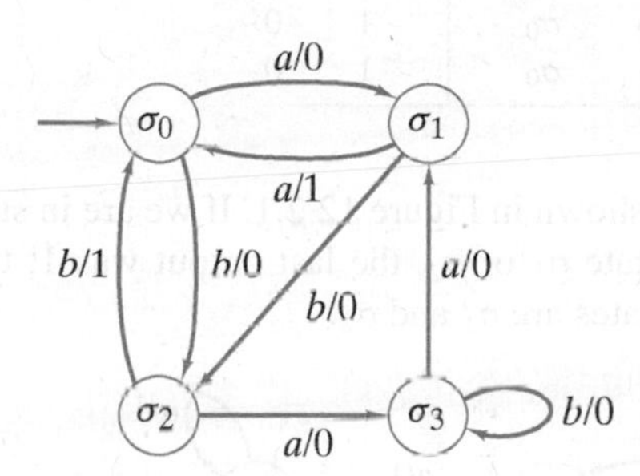


**10.**

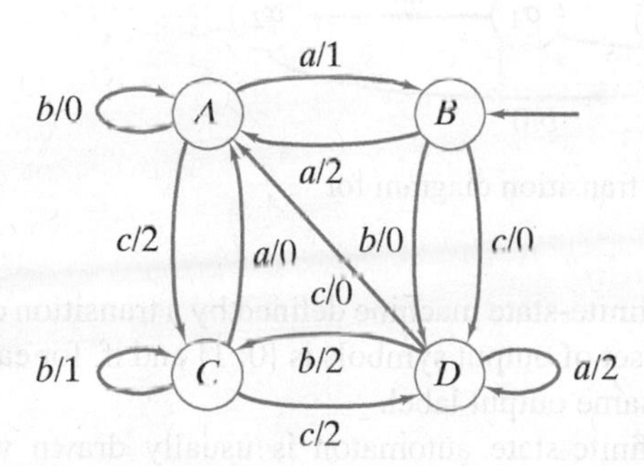


*In Exercises 11–20, find the output string for the given input string and finite-state machine.*

**11.** *abba*; Exercise 1

**12.** *abba*; Exercise 2

**13.** *aabbaba*; Exercise 3

**14.** *aabbcc*; Exercise 4

**15.** *aabaab*; Exercise 5

**16.** *aaa*; Exercise 6

**17.** *baaba*; Exercise 7

**18.** *aabbabaab*; Exercise 8

**19.** *bbababbabaaa*; Exercise 9

**20.** *cacbccbaabac*; Exercise 10

*In Exercises 21–26, design a finite-state machine having the given properties. The input is always a bit string.*

**21.** Outputs 1 if an even number of 1's have been input; otherwise, outputs 0

**22.** Outputs 1 if two or more 1's are input; otherwise, outputs 0

**23.** Outputs 1 if $k$ 1's have been input, where $k$ is a multiple of 3; otherwise, outputs 0

**24.** Outputs 1 whenever it sees 101; otherwise, outputs 0

**25.** Outputs 1 when it sees the first 0 and until it sees another 0; thereafter, outputs 0; in all other cases, outputs 0

**26.** Outputs 1 when it sees 101 and thereafter; otherwise, outputs 0

**27.** Let $\alpha = x_1 \cdots x_n$ be a bit string. Let $\beta = y_1 \cdots y_n$, where

$$y_i = \begin{cases} a & \text{if } x_i = 0 \\ b & \text{if } x_i = 1 \end{cases}$$

for $i = 1, \ldots, n$. Let $\gamma = y_n \cdots y_1$.

Show that if $\gamma$ is input to the finite-state machine of Figure 12.1.4, the output is the 2's complement of $\alpha$ (see Algorithm 11.5.16 for a description of 2's complement).

★**28.** Show that there is no finite-state machine that receives a bit string and outputs 1 whenever the number of 1's input equals the number of 0's input and outputs 0 otherwise.

★**29.** Show that there is no finite-state machine that performs serial multiplication. Specifically, show that there is no finite-state machine that inputs binary numbers $X = x_1 \cdots x_n$, $Y = y_1 \cdots y_n$, as the sequence of two-bit numbers

$$x_n y_n, \quad x_{n-1} y_{n-1}, \quad \ldots, \quad x_1 y_1, \quad 00, \quad \ldots, \quad 00,$$

where there are $n$ 00's, and outputs $z_{2n}, \ldots, z_1$, where $Z = z_1 \cdots z_{2n} = XY$.

*Example:* If there is such a machine, to multiply 101 × 1001 we would input 11,00,10,01,00,00,00,00. The first pair 11 is the pair of rightmost bits (10$\underline{1}$, 100$\underline{1}$); the second pair 00 is the next pair of bits (1$\underline{0}$1, 10$\underline{0}$1); and so on. We pad the input string with four pairs of 00's—the length of the longest number 1001 to be multiplied. Since 101 × 1001 = 101101, it is alleged that we obtain the output shown in the following table.

| *Input* | *Output* |
|---|---|
| 11 | 1 |
| 00 | 0 |
| 10 | 1 |
| 01 | 1 |
| 00 | 0 |
| 00 | 1 |
| 00 | 0 |
| 00 | 0 |

## 12.2 Finite-State Automata

A **finite-state automaton** is a special kind of finite-state machine. Finite-state automata are of special interest because of their relationship to languages, as we shall see in Section 12.5.

**Go Online**
A finite-state automaton simulator is at
`bit.ly/2CAd76a`

**Definition 12.2.1 ▶** A *finite-state automaton* $A = (\mathcal{I}, \mathcal{O}, \mathcal{S}, f, g, \sigma)$ is a finite-state machine in which the set of output symbols is $\{0, 1\}$ and where the current state determines the last output. Those states for which the last output was 1 are called *accepting states*. ◀

**Example 12.2.2** Draw the transition diagram of the finite-state machine $A$ defined by the table. The initial state is $\sigma_0$. Show that $A$ is a finite-state automaton, and determine the set of accepting states.

| $\mathcal{S}$ \ $\mathcal{I}$ | $f$: $a$ | $f$: $b$ | $g$: $a$ | $g$: $b$ |
|---|---|---|---|---|
| $\sigma_0$ | $\sigma_1$ | $\sigma_0$ | 1 | 0 |
| $\sigma_1$ | $\sigma_2$ | $\sigma_0$ | 1 | 0 |
| $\sigma_2$ | $\sigma_2$ | $\sigma_0$ | 1 | 0 |

**SOLUTION** The transition diagram is shown in Figure 12.2.1. If we are in state $\sigma_0$, the last output was 0. If we are in either state $\sigma_1$ or $\sigma_2$, the last output was 1; thus $A$ is a finite-state automaton. The accepting states are $\sigma_1$ and $\sigma_2$.

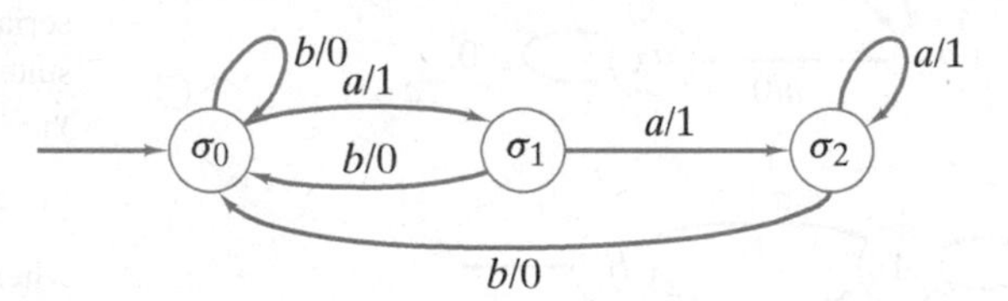


**Figure 12.2.1** The transition diagram for Example 12.2.2. ◀

Example 12.2.2 shows that the finite-state machine defined by a transition diagram will be a finite-state automaton if the set of output symbols is $\{0, 1\}$ and if, for each state $\sigma$, all incoming edges to $\sigma$ have the same output label.

The transition diagram of a finite-state automaton is usually drawn with the accepting states in double circles and the output symbols omitted. When the transition diagram of Figure 12.2.1 is redrawn in this way, we obtain the transition diagram of Figure 12.2.2.

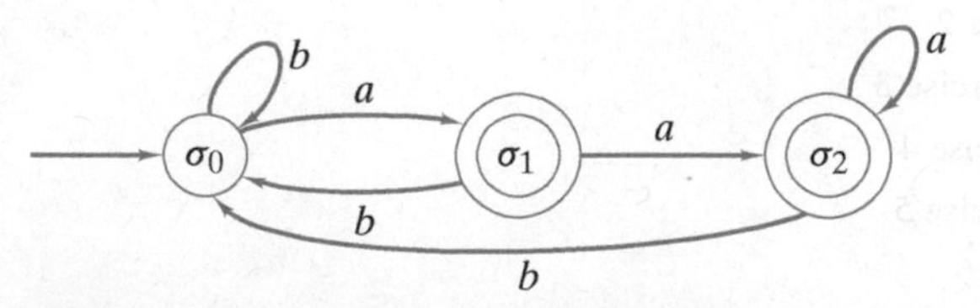


**Figure 12.2.2** The transition diagram of Figure 12.2.1 redrawn with accepting states in double circles and output symbols omitted.

**Example 12.2.3** Draw the transition diagram of the finite-state automaton of Figure 12.2.3 as a transition diagram of a finite-state machine.

**SOLUTION** Since $\sigma_2$ is an accepting state, we label all its incoming edges with output 1 (see Figure 12.2.4). The states $\sigma_0$ and $\sigma_1$ are not accepting, so we label all their incoming edges with output 0. We obtain the transition diagram of Figure 12.2.4.

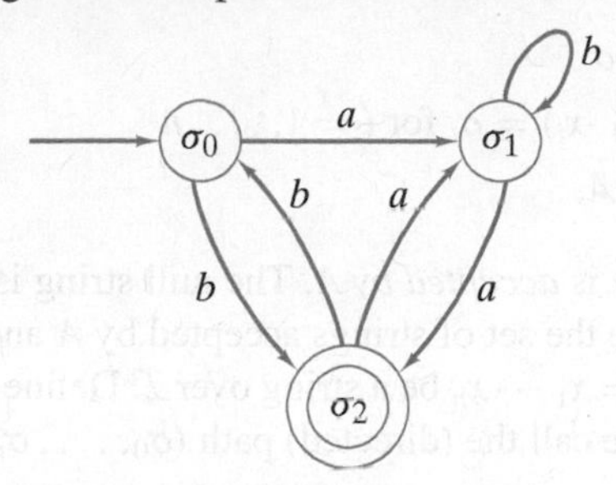


**Figure 12.2.3** A finite-state automaton.

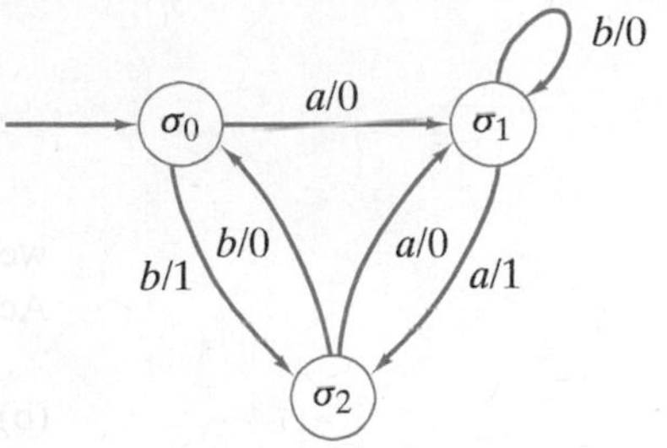

**Figure 12.2.4** The finite-state automaton of Figure 12.2.3 redrawn as a transition diagram of a finite-state machine. ◀

As an alternative to Definition 12.2.1, we can regard a finite-state automaton $A$ as consisting of

1. A finite set $\mathcal{I}$ of *input symbols*
2. A finite set $\mathcal{S}$ of *states*
3. A *next-state function* $f$ from $\mathcal{S} \times \mathcal{I}$ into $\mathcal{S}$
4. A subset $\mathcal{A}$ of $\mathcal{S}$ of *accepting states*
5. An *initial state* $\sigma \in \mathcal{S}$.

If we use this characterization, we write $A = (\mathcal{I}, \mathcal{S}, f, \mathcal{A}, \sigma)$.

**Example 12.2.4** The transition diagram of the finite-state automaton $A = (\mathcal{I}, \mathcal{S}, f, \mathcal{A}, \sigma)$, where

$$\mathcal{I} = \{a, b\}, \quad \mathcal{S} = \{\sigma_0, \sigma_1, \sigma_2\}, \quad \mathcal{A} = \{\sigma_2\}, \quad \sigma = \sigma_0,$$

and $f$ is given by the following table

| $\mathcal{I}$ / $\mathcal{S}$ | $f$: $a$ | $f$: $b$ |
|---|---|---|
| $\sigma_0$ | $\sigma_0$ | $\sigma_1$ |
| $\sigma_1$ | $\sigma_0$ | $\sigma_2$ |
| $\sigma_2$ | $\sigma_0$ | $\sigma_2$ |

is shown in Figure 12.2.5.

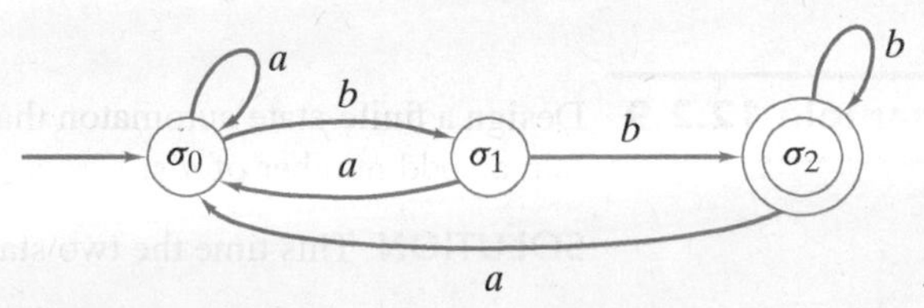


**Figure 12.2.5** The transition diagram for Example 12.2.4. ◀

If a string is input to a finite-state automaton, we will end at either an accepting or a nonaccepting state. The status of this final state determines whether the string is **accepted** by the finite-state automaton.

**Definition 12.2.5 ▶** Let $A = (\mathcal{I}, \mathcal{S}, f, \mathcal{A}, \sigma)$ be a finite-state automaton. Let $\alpha = x_1 \cdots x_n$ be a string over $\mathcal{I}$. If there exist states $\sigma_0, \ldots, \sigma_n$ satisfying

(a) $\sigma_0 = \sigma$

(b) $f(\sigma_{i-1}, x_i) = \sigma_i$ for $i = 1, \ldots, n$

(c) $\sigma_n \in \mathcal{A}$,

we say that $\alpha$ is *accepted by* $A$. The null string is accepted if and only if $\sigma \in \mathcal{A}$. We let $\text{Ac}(A)$ denote the set of strings accepted by $A$ and we say that $A$ *accepts* $\text{Ac}(A)$.

Let $\alpha = x_1 \cdots x_n$ be a string over $\mathcal{I}$. Define states $\sigma_0, \ldots, \sigma_n$ by conditions (a) and (b) above. We call the (directed) path $(\sigma_0, \ldots, \sigma_n)$ the path *representing* $\alpha$ in $A$. ◀

It follows from Definition 12.2.5 that if the path $P$ represents the string $\alpha$ in a finite-state automaton $A$, then $A$ accepts $\alpha$ if and only if $P$ ends at an accepting state.

**Example 12.2.6** Is the string *abaa* accepted by the finite-state automaton of Figure 12.2.2?

**SOLUTION** We begin at state $\sigma_0$. When $a$ is input, we move to state $\sigma_1$. When $b$ is input, we move to state $\sigma_0$. When $a$ is input, we move to state $\sigma_1$. Finally, when the last symbol $a$ is input, we move to state $\sigma_2$. The path $(\sigma_0, \sigma_1, \sigma_0, \sigma_1, \sigma_2)$ represents the string *abaa*. Since the final state $\sigma_2$ is an accepting state, the string *abaa* is accepted by the finite-state automaton of Figure 12.2.2. ◀

**Example 12.2.7** Is the string $\alpha = abbabba$ accepted by the finite-state automaton of Figure 12.2.3?

**SOLUTION** The path representing $\alpha$ terminates at $\sigma_1$. Since $\sigma_1$ is not an accepting state, the string $\alpha$ is not accepted by the finite-state automaton of Figure 12.2.3. ◀

We next give two examples illustrating design problems.

**Example 12.2.8** Design a finite-state automaton that accepts precisely those strings over $\{a, b\}$ that contain no $a$'s.

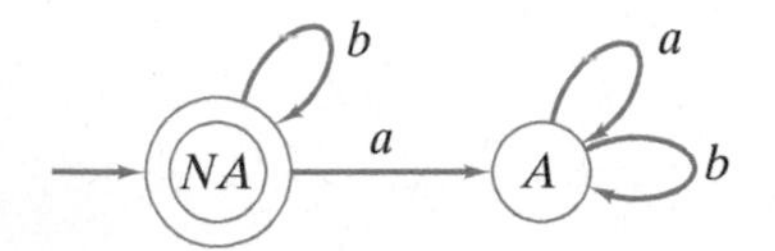


**Figure 12.2.6** A finite-state automaton that accepts precisely those strings over $\{a, b\}$ that contain no $a$'s.

**SOLUTION** The idea is to use two states:

$A$: An $a$ was found.

$NA$: No $a$'s were found.

The state $NA$ is the initial state and the only accepting state. It is now a simple matter to draw the edges (see Figure 12.2.6). Notice that the finite-state automaton correctly accepts the null string. ◀

**Example 12.2.9** Design a finite-state automaton that accepts precisely those strings over $\{a, b\}$ that contain an odd number of $a$'s.

**SOLUTION** This time the two states are

$E$: An even number of $a$'s was found.

$O$: An odd number of $a$'s was found.

The initial state is $E$ and the accepting state is $O$. We obtain the transition diagram shown in Figure 12.2.7. ◀

A finite-state automaton is essentially an algorithm to decide whether or not a given string is accepted. As an example, we convert the transition diagram of Figure 12.2.7 to an algorithm.

**Algorithm 12.2.10**

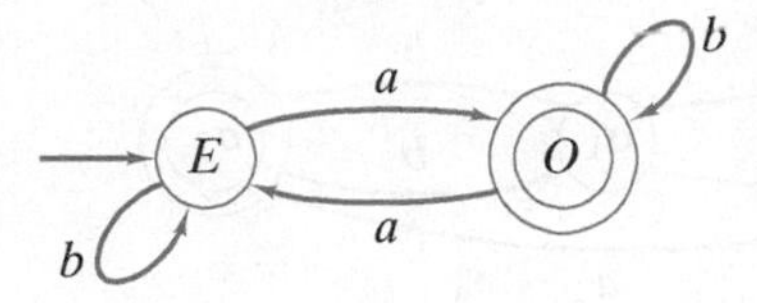


**Figure 12.2.7** A finite-state automaton that accepts precisely those strings over $\{a, b\}$ that contain an odd number of $a$'s.

This algorithm determines whether a string over $\{a, b\}$ is accepted by the finite-state automaton whose transition diagram is given in Figure 12.2.7.

Input: $n$, the length of the string ($n = 0$ designates the null string); $s_1 s_2 \cdots s_n$, the string

Output: "Accept" if the string is accepted
"Reject" if the string is not accepted

```
fsa(s, n) {
  state = 'E'
  for i = 1 to n {
    if (state == 'E' ∧ s_i == 'a')
      state = 'O'
    if (state == 'O' ∧ s_i == 'a')
      state = 'E'
  }
  if (state == 'O')
    return "Accept"
  else
    return "Reject"
}
```

If two finite-state automata accept precisely the same strings, we say that the automata are **equivalent.**

**Definition 12.2.11 ▶** The finite-state automata $A$ and $A'$ are *equivalent* if $\text{Ac}(A) = \text{Ac}(A')$. ◀

**Example 12.2.12** It can be verified that the finite-state automata of Figures 12.2.6 and 12.2.8 are equivalent (see Exercise 33).

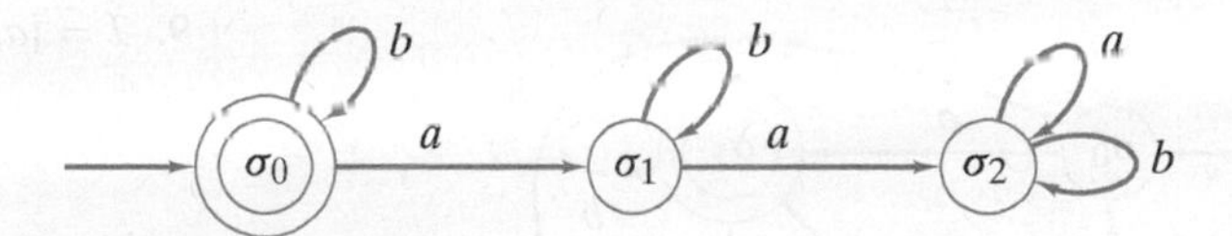


**Figure 12.2.8** A finite-state automaton equivalent to that in Figure 12.2.6.

◀

If we define a relation $R$ on a set of finite-state automata by the rule $A\ R\ A'$ if $A$ and $A'$ are equivalent (in the sense of Definition 12.2.11), $R$ is an equivalence relation. Each equivalence class consists of a set of mutually equivalent finite-state automata.

## 12.2 Review Exercises

1. Define *finite-state automaton.*
2. What does it mean for a string to be accepted by a finite-state automaton?
3. What are equivalent finite-state automata?

## 12.2 Exercises

*In Exercises 1–3, show that each finite-state machine is a finite-state automaton and redraw the transition diagram as the diagram of a finite-state automaton.*

**1.**

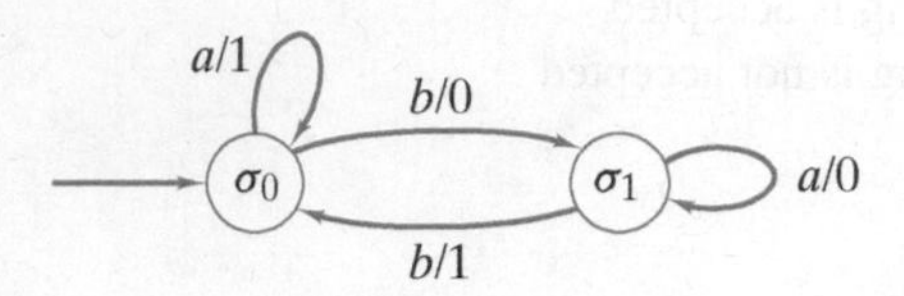


**2.**

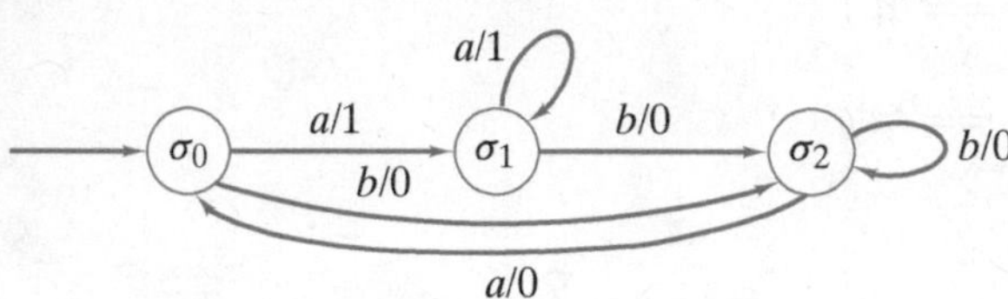


**3.**

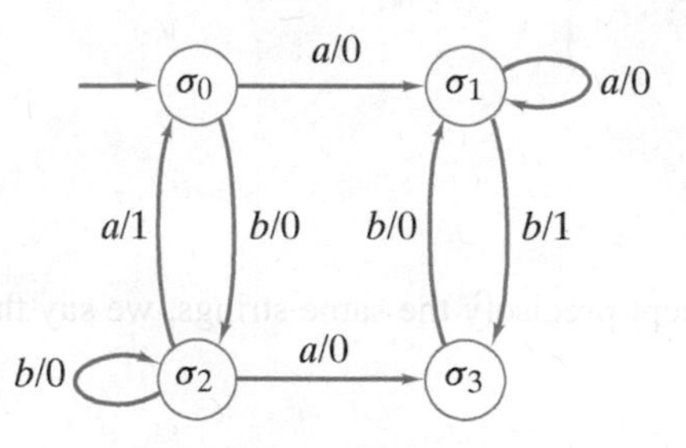


*In Exercises 4–6, redraw the transition diagram of the finite-state automaton as the transition diagram of a finite-state machine.*

4.

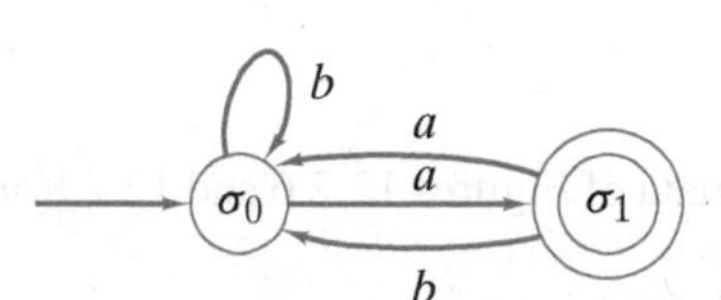


**5.**

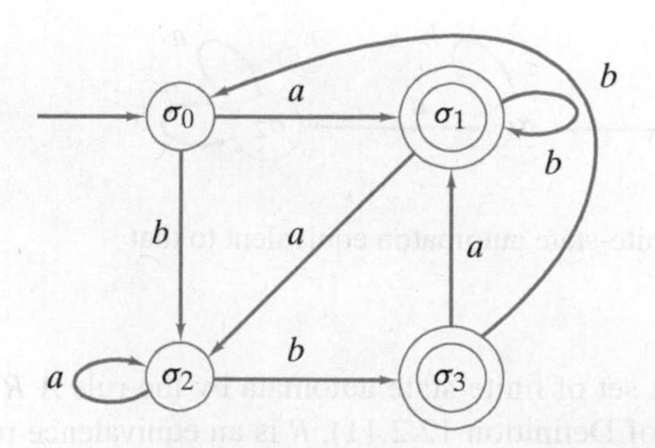


**6.**

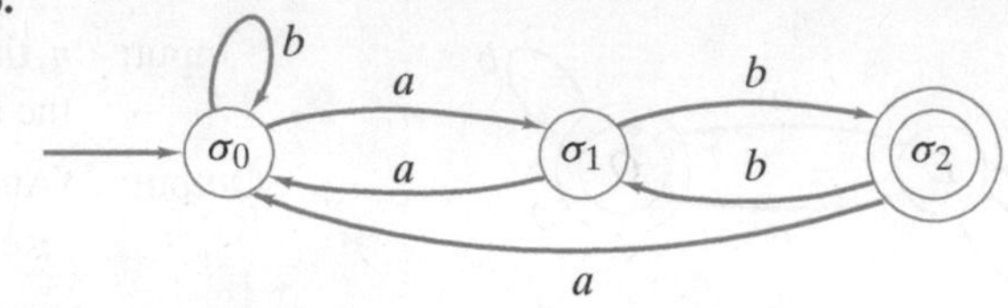


*In Exercises 7–9, draw the transition diagram of the finite-state automaton* $(\mathcal{I}, \mathcal{S}, f, \mathcal{A}, \sigma_0)$.

**7.** $\mathcal{I} = \{a, b\}$, $\mathcal{S} = \{\sigma_0, \sigma_1, \sigma_2\}$, $\mathcal{A} = \{\sigma_0\}$

| $\mathcal{S}$ \\ $\mathcal{I}$ | $f$: $a$ | $b$ |
|---|---|---|
| $\sigma_0$ | $\sigma_1$ | $\sigma_0$ |
| $\sigma_1$ | $\sigma_2$ | $\sigma_0$ |
| $\sigma_2$ | $\sigma_0$ | $\sigma_2$ |

**8.** $\mathcal{I} = \{a, b, c\}$, $\mathcal{S} = \{\sigma_0, \sigma_1, \sigma_2, \sigma_3\}$, $\mathcal{A} = \{\sigma_0, \sigma_2\}$

| $\mathcal{S}$ \\ $\mathcal{I}$ | $f$: $a$ | $b$ | $c$ |
|---|---|---|---|
| $\sigma_0$ | $\sigma_1$ | $\sigma_0$ | $\sigma_2$ |
| $\sigma_1$ | $\sigma_0$ | $\sigma_3$ | $\sigma_0$ |
| $\sigma_2$ | $\sigma_3$ | $\sigma_2$ | $\sigma_0$ |
| $\sigma_3$ | $\sigma_1$ | $\sigma_0$ | $\sigma_1$ |

**9.** $\mathcal{I} = \{a, b\}$, $\mathcal{S} = \{\sigma_0, \sigma_1, \sigma_2\}$, $\mathcal{A} = \{\sigma_0, \sigma_2\}$

| $\mathcal{S}$ \\ $\mathcal{I}$ | $f$: $a$ | $b$ |
|---|---|---|
| $\sigma_0$ | $\sigma_1$ | $\sigma_1$ |
| $\sigma_1$ | $\sigma_0$ | $\sigma_2$ |
| $\sigma_2$ | $\sigma_0$ | $\sigma_1$ |

10. For each finite-state automaton in Exercises 1–6, find the sets $\mathcal{I}$, $\mathcal{S}$, and $\mathcal{A}$, the initial state, and the table defining the next-state function.

11. Which of the finite-state machines of Exercises 1–10, Section 12.1, are finite-state automata?

12. What must the table of a finite-state machine $M$ look like in order for $M$ to be a finite-state automaton?

*In Exercises 13–17, determine whether the given string is accepted by the given finite-state automaton.*

13. *abbaa*; Figure 12.2.2
14. *abbaa*; Figure 12.2.3
15. *aabaabb*; Figure 12.2.5
16. *aaababbab*; Exercise 5
17. *aaabbbaab*; Exercise 6

18. Show that a string $\alpha$ over $\{a, b\}$ is accepted by the finite-state automaton of Figure 12.2.2 if and only if $\alpha$ ends with $a$.

19. Show that a string $\alpha$ over $\{a, b\}$ is accepted by the finite-state automaton of Figure 12.2.5 if and only if $\alpha$ ends with $bb$.

★20. Characterize the strings accepted by the finite-state automata of Exercises 1–9.

*In Exercises 21–31, draw the transition diagram of a finite-state automaton that accepts the given set of strings over $\{a, b\}$.*

21. Even number of $a$'s
22. At least one $b$
23. Exactly one $b$
24. Exactly two $a$'s
25. Contains $m$ $a$'s, where $m$ is a multiple of 3
26. At least two $a$'s
27. Starts with *baa*
★28. Contains *abba*
29. Every $b$ is followed by $a$
★30. Ends with *aba*
★31. Starts with *ab* and ends with *baa*

32. Write algorithms, similar to Algorithm 12.2.10, that decide whether or not a given string is accepted by the finite-state automata of Exercises 1–9.

33. Give a formal argument to show that the finite-state automata of Figures 12.2.6 and 12.2.8 are equivalent.

34. Let $L$ be a finite set of strings over $\{a, b\}$. Show that there is a finite-state automaton that accepts $L$.

35. Let $L$ be the set of strings accepted by the finite-state automaton of Exercise 5. Let $S$ denote the set of all strings over $\{a, b\}$. Design a finite-state automaton that accepts $S - L$.

36. Let $L_i$ be the set of strings accepted by the finite-state automaton $A_i = (\mathcal{I}, \mathcal{S}_i, f_i, \mathcal{A}_i, \sigma_i)$, $i = 1, 2$. Let

$$A = (\mathcal{I}, \mathcal{S}_1 \times \mathcal{S}_2, f, \mathcal{A}, \sigma),$$

where

$$\begin{aligned} f((S_1, S_2), x) &= (f_1(S_1, x), f_2(S_2, x)) \\ \mathcal{A} &= \{(A_1, A_2) \mid A_1 \in \mathcal{A}_1 \text{ and } A_2 \in \mathcal{A}_2\} \\ \sigma &= (\sigma_1, \sigma_2). \end{aligned}$$

Show that $\text{Ac}(A) = L_1 \cap L_2$.

37. Let $L_i$ be the set of strings accepted by the finite-state automaton $A_i = (\mathcal{I}, \mathcal{S}_i, f_i, \mathcal{A}_i, \sigma_i)$, $i = 1, 2$. Let

$$A = (\mathcal{I}, \mathcal{S}_1 \times \mathcal{S}_2, f, \mathcal{A}, \sigma),$$

where

$$\begin{aligned} f((S_1, S_2), x) &= (f_1(S_1, x), f_2(S_2, x)) \\ \mathcal{A} &= \{(A_1, A_2) \mid A_1 \in \mathcal{A}_1 \text{ or } A_2 \in \mathcal{A}_2\} \\ \sigma &= (\sigma_1, \sigma_2). \end{aligned}$$

Show that $\text{Ac}(A) = L_1 \cup L_2$.

*In Exercises 38–42, let $L_i = \text{Ac}(A_i)$, $i = 1, 2$. Draw the transition diagrams of the finite-state automata that accept $L_1 \cap L_2$ and $L_1 \cup L_2$.*

38. $A_1$ given by Exercise 4; $A_2$ given by Exercise 6

39. $A_1$ given by Exercise 4; $A_2$ given by Exercise 5

40. $A_1$ given by Exercise 6; $A_2$ given by Exercise 5

41. $A_1$ given by Exercise 5; $A_2$ given by Exercise 5

42. $A_1$ given by Figure 12.5.7, Section 12.5; $A_2$ given by Exercise 5

## 12.3 Languages and Grammars

According to *Merriam-Webster's Collegiate® Dictionary*, language is "the words, their pronunciation, and the methods of combining them used and understood by a community."[†] Such languages are often called **natural languages** to distinguish them from **formal languages,** which are used to model natural languages and to communicate with computers. The rules of a natural language are very complex and difficult to characterize completely. On the other hand, it is possible to specify completely the rules by which certain formal languages are constructed. We begin with the definition of a formal language.

[†]By Permission. From *Merriam-Webster's Collegiate® Dictionary*, 11th Edition ©2016 Merriam Webster, Inc. (www.Merriam-Webster.com).

**Definition 12.3.1 ▶** Let $A$ be a finite set. A *(formal) language* $L$ over $A$ is a subset of $A^*$, the set of all strings over $A$. ◀

**Example 12.3.2** Let $A = \{a, b\}$. The set $L$ of all strings over $A$ containing an odd number of $a$'s is a language over $A$. As we saw in Example 12.2.9, $L$ is precisely the set of strings over $A$ accepted by the finite-state automaton of Figure 12.2.7. ◀

One way to define a language is to give a list of rules that the language is assumed to obey.

**Definition 12.3.3 ▶** A *phrase-structure grammar* (or, simply, *grammar*) $G$ consists of

(a) A finite set $N$ of *nonterminal symbols*

(b) A finite set $T$ of *terminal symbols* where $N \cap T = \varnothing$

(c) A finite subset $P$ of $[(N \cup T)^* - T^*] \times (N \cup T)^*$, called the set of *productions*

(d) A *starting symbol* $\sigma \in N$.

We write $G = (N, T, P, \sigma)$. ◀

A production $(A, B) \in P$ is usually written

$$A \to B.$$

Definition 12.3.3(c) states that in the production $A \to B$, $A \in (N \cup T)^* - T^*$ and $B \in (N \cup T)^*$; thus $A$ must include at least one nonterminal symbol, whereas $B$ can consist of any combination of nonterminal and terminal symbols.

**Example 12.3.4** Let

$$N = \{\sigma, S\}$$
$$T = \{a, b\}$$
$$P = \{\sigma \to b\sigma, \sigma \to aS, S \to bS, S \to b\}.$$

Then $G = (N, T, P, \sigma)$ is a grammar. ◀

Given a grammar $G$, we can construct a language $L(G)$ from $G$ by using the productions to derive the strings that make up $L(G)$. The idea is to start with the starting symbol and then repeatedly use productions until a string of terminal symbols is obtained. The language $L(G)$ is the set of all such strings obtained. Definition 12.3.5 gives the formal details.

**Definition 12.3.5 ▶** Let $G = (N, T, P, \sigma)$ be a grammar.

If $\alpha \to \beta$ is a production and $x\alpha y \in (N \cup T)^*$, we say that $x\beta y$ is *directly derivable* from $x\alpha y$ and write

$$x\alpha y \Rightarrow x\beta y.$$

If $\alpha_i \in (N \cup T)^*$ for $i = 1, \ldots, n$, and $\alpha_{i+1}$ is directly derivable from $\alpha_i$ for $i = 1, \ldots, n-1$, we say that $\alpha_n$ is *derivable from* $\alpha_1$ and write

$$\alpha_1 \Rightarrow \alpha_n.$$

We call

$$\alpha_1 \Rightarrow \alpha_2 \Rightarrow \cdots \Rightarrow \alpha_n$$

the *derivation of* $\alpha_n$ *(from* $\alpha_1$*)*. By convention, any element of $(N \cup T)^*$ is derivable from itself.

The *language generated by G*, written $L(G)$, consists of all strings over $T$ derivable from $\sigma$. ◀

**Example 12.3.6** Let $G$ be the grammar of Example 12.3.4.

The string $abSbb$ is directly derivable from $aSbb$, written

$$aSbb \Rightarrow abSbb,$$

by using the production $S \rightarrow bS$.

The string $bbab$ is derivable from $\sigma$, written

$$\sigma \Rightarrow bbab.$$

The derivation is

$$\sigma \Rightarrow b\sigma \Rightarrow bb\sigma \Rightarrow bbaS \Rightarrow bbab.$$

The only derivations from $\sigma$ are

$$\begin{aligned}
\sigma &\Rightarrow b\sigma \\
&\;\vdots \\
&\Rightarrow b^n\sigma && n \geq 0 \\
&\Rightarrow b^n aS \\
&\;\vdots \\
&\Rightarrow b^n ab^{m-1}S \\
&\Rightarrow b^n ab^m && n \geq 0, \quad m \geq 1.
\end{aligned}$$

Thus $L(G)$ consists of the strings over $\{a, b\}$ containing precisely one $a$ that end with $b$. ◀

An alternative way to state the productions of a grammar is by using **Backus normal form** (or **Backus–Naur form** or **BNF**). In BNF the nonterminal symbols typically begin with "<" and end with ">." The production $S \rightarrow T$ is written $S ::= T$. Productions of the form

**Go Online**
For more on Backus normal form, see
bit.ly/2CAd76a

$$S ::= T_1, \quad S ::= T_2, \quad \ldots, \quad S ::= T_n$$

may be combined as

$$S ::= T_1 \mid T_2 \mid \cdots \mid T_n.$$

The bar "|" is read "or."

**Example 12.3.7** **A Grammar for Integers** An integer is defined as a string consisting of an optional sign (+ or −) followed by a string of digits (0 through 9). The following grammar generates all integers.

$$\begin{gathered}
<\text{digit}> ::= 0 \mid 1 \mid 2 \mid 3 \mid 4 \mid 5 \mid 6 \mid 7 \mid 8 \mid 9 \\
<\text{integer}> ::= <\text{signed integer}> \mid <\text{unsigned integer}> \\
<\text{signed integer}> ::= + <\text{unsigned integer}> \mid - <\text{unsigned integer}> \\
<\text{unsigned integer}> ::= <\text{digit}> \mid <\text{digit}><\text{unsigned integer}>
\end{gathered}$$

The starting symbol is $<$ integer $>$.

For example, the derivation of the integer −901 is

$$\begin{aligned}
<\text{integer}> &\Rightarrow\ <\text{signed integer}>\\
&\Rightarrow -<\text{unsigned integer}>\\
&\Rightarrow -<\text{digit}><\text{unsigned integer}>\\
&\Rightarrow -<\text{digit}><\text{digit}><\text{unsigned integer}>\\
&\Rightarrow -<\text{digit}><\text{digit}><\text{digit}>\\
&\Rightarrow -9<\text{digit}><\text{digit}>\\
&\Rightarrow -90<\text{digit}>\\
&\Rightarrow -901.
\end{aligned}$$

In the notation of Definition 12.3.3, this language consists of

1. The set $N = \{<\text{digit}>, <\text{integer}>, <\text{signed integer}>, <\text{unsigned integer}>\}$ of nonterminal symbols
2. The set $T = \{0, 1, 2, 3, 4, 5, 6, 7, 8, 9, +, -\}$ of terminal symbols
3. The productions

$$\begin{aligned}
<\text{digit}> &\rightarrow 0, \ldots, <\text{digit}> \rightarrow 9\\
<\text{integer}> &\rightarrow\ <\text{signed integer}>\\
<\text{integer}> &\rightarrow\ <\text{unsigned integer}>\\
<\text{signed integer}> &\rightarrow +<\text{unsigned integer}>\\
<\text{signed integer}> &\rightarrow -<\text{unsigned integer}>\\
<\text{unsigned integer}> &\rightarrow\ <\text{digit}>\\
<\text{unsigned integer}> &\rightarrow\ <\text{digit}><\text{unsigned integer}>
\end{aligned}$$

4. The starting symbol $<\text{integer}>$. ◀

Computer languages, such as Java and C++, are typically specified in BNF. Example 12.3.7 shows how an integer constant in a computer language might be specified in BNF.

Grammars are classified according to the types of productions that define the grammars.

**Definition 12.3.8 ▶** Let $G$ be a grammar and let $\lambda$ denote the null string.

(a) If every production is of the form

$$\alpha A\beta \rightarrow \alpha\delta\beta, \qquad \text{where } \alpha, \beta \in (N \cup T)^*, \quad A \in N, \quad \delta \in (N \cup T)^* - \{\lambda\}, \tag{12.3.1}$$

we call $G$ a *context-sensitive* (or *type* 1) *grammar.*

(b) If every production is of the form

$$A \rightarrow \delta, \qquad \text{where } A \in N, \quad \delta \in (N \cup T)^*, \tag{12.3.2}$$

we call $G$ a *context-free* (or *type* 2) *grammar.*

(c) If every production is of the form

$$A \rightarrow a \text{ or } A \rightarrow aB \text{ or } A \rightarrow \lambda, \qquad \text{where } A, B \in N, \quad a \in T,$$

we call $G$ a *regular* (or *type* 3) *grammar.* ◀

According to (12.3.1), in a context-sensitive grammar, we may replace $A$ by $\delta$ if $A$ is in the context of $\alpha$ and $\beta$. In a context-free grammar, (12.3.2) states that we may replace $A$ by $\delta$ anytime. A regular grammar has especially simple substitution rules: We replace a nonterminal symbol by a terminal symbol, by a terminal symbol followed by a nonterminal symbol, or by the null string.

Notice that a regular grammar is a context-free grammar and that a context-free grammar with no productions of the form $A \to \lambda$ is a context-sensitive grammar.

Some definitions allow $a$ to be replaced by a string of terminals in Definition 12.3.8(c); however, it can be shown (see Exercise 32) that the two definitions produce the same languages.

**Example 12.3.9** The grammar $G$ defined by

$$T = \{a, b, c\}, \quad N = \{\sigma, A, B, C, D, E\},$$

with productions

$$\begin{array}{lllll} \sigma \to aAB, & \sigma \to aB, & A \to aAC, & A \to aC, & B \to Dc, \\ D \to b, & CD \to CE, & CE \to DE, & DE \to DC, & Cc \to Dcc \end{array}$$

and starting symbol $\sigma$ is context-sensitive. For example, the production $CE \to DE$ says that we can replace $C$ by $D$ if $C$ is followed by $E$, and the production $Cc \to Dcc$ says that we can replace $C$ by $Dc$ if $C$ is followed by $c$.

We can derive $DC$ from $CD$ since

$$CD \Rightarrow CE \Rightarrow DE \Rightarrow DC.$$

The string $a^3b^3c^3$ is in $L(G)$, since we have

$$\begin{aligned} \sigma &\Rightarrow aAB \Rightarrow aaACB \Rightarrow aaaCCDc \Rightarrow aaaDCCc \Rightarrow aaaDCDcc \\ &\Rightarrow aaaDDCcc \Rightarrow aaaDDDccc \Rightarrow aaabbbccc. \end{aligned}$$

It can be shown (see Exercise 33) that

$$L(G) = \{a^nb^nc^n \mid n = 1, 2, \ldots\}.$$

◀

It is natural to allow the language $L(G)$ to inherit a property of a grammar $G$. The next definition makes this concept precise.

**Definition 12.3.10 ▶** A language $L$ is *context-sensitive* (respectively, *context-free*, *regular*) if there is a context-sensitive (respectively, context-free, regular) grammar $G$ with $L = L(G)$. ◀

**Example 12.3.11** According to Example 12.3.9, the language

$$L = \{a^nb^nc^n \mid n = 1, 2, \ldots\}$$

is context-sensitive. It can be shown (see [Hopcroft]) that there is no context-free grammar $G$ with $L = L(G)$; hence $L$ is not a context-free language. ◀

**Example 12.3.12** The grammar $G$ defined by

$$T = \{a, b\} \text{ and } N = \{\sigma\}$$

with productions

$$\sigma \to a\sigma b \text{ and } \sigma \to ab$$

and starting symbol $\sigma$ is context-free. The only derivations of $\sigma$ are

$$\begin{aligned}\sigma &\Rightarrow a\sigma b\\ &\vdots\\ &\Rightarrow a^{n-1}\sigma b^{n-1}\\ &\Rightarrow a^{n-1}abb^{n-1} = a^n b^n.\end{aligned}$$

Thus $L(G)$ consists of the strings over $\{a, b\}$ of the form $a^n b^n$, $n = 1, 2 \ldots$. This language is context-free. In Section 12.5 (see Example 12.5.6), we will show that $L(G)$ is not regular. ◀

It follows from Examples 12.3.11 and 12.3.12 that the set of context-free languages that do not contain the null string is a proper subset of the set of context-sensitive languages and that the set of regular languages is a proper subset of the set of context-free languages. It can also be shown that there are languages that are not context-sensitive.

**Example 12.3.13** The grammar $G$ defined in Example 12.3.4 is regular. Thus the language

$$L(G) = \{b^n ab^m \mid n = 0, 1, \ldots; m = 1, 2, \ldots\}$$

it generates is regular. ◀

**Example 12.3.14** The grammar of Example 12.3.7 is context-free but not regular. However, if we change the productions to

$$\begin{aligned}< \text{integer} > &::= + < \text{unsigned integer} >| - < \text{unsigned integer} >|\\ &\quad 0 < \text{digits} >| \; 1 < \text{digits} >| \cdots | \; 9 < \text{digits} >\\ < \text{unsigned integer} > &::= 0 < \text{digits} >| \; 1 < \text{digits} >| \cdots | \; 9 < \text{digits} >\\ < \text{digits} > &::= 0 < \text{digits} >| \; 1 < \text{digits} >| \cdots | \; 9 < \text{digits} >| \; \lambda,\end{aligned}$$

the resulting grammar is regular. Since the language generated is unchanged, it follows that the set of strings representing integers is a regular language. ◀

Example 12.3.14 motivates the following definition.

**Definition 12.3.15 ▶** Grammars $G$ and $G'$ are *equivalent* if $L(G) = L(G')$. ◀

**Example 12.3.16** The grammars of Examples 12.3.7 and 12.3.14 are equivalent. ◀

If we define a relation $R$ on a set of grammars by the rule $GRG'$ if $G$ and $G'$ are equivalent (in the sense of Definition 12.3.15), $R$ is an equivalence relation. Each equivalence class consists of a set of mutually equivalent grammars.

We close this section by briefly introducing another kind of grammar that can be used to generate fractal curves.

**Go Online**
For more on Lindenmayer grammars, see
bit.ly/2CAd76a

**Definition 12.3.17 ▶** A *context-free interactive Lindenmayer grammar* consists of

(a) A finite set $N$ of *nonterminal symbols*

(b) A finite set $T$ of *terminal symbols* where $N \cap T = \varnothing$

(c) A finite set $P$ of *productions* $A \to B$, where $A \in N \cup T$ and $B \in (N \cup T)^*$

(d) A *starting symbol* $\sigma \in N$. ◀

The difference between a context-free interactive Lindenmayer grammar and a context-free grammar is that a context-free interactive Lindenmayer grammar allows productions of the form $A \to B$, where $A$ is a terminal or a nonterminal. (In a context-free grammar, $A$ must be a nonterminal.)

The rules for deriving strings in a context-free interactive Lindenmayer grammar are different from the rules for deriving strings in a phrase-structure grammar (see Definition 12.3.5). In a context-free interactive Lindenmayer grammar, to derive the string $\beta$ from the string $\alpha$, *all* symbols in $\alpha$ must be replaced *simultaneously*. The formal definition follows.

**Definition 12.3.18** ▶ Let $G = (N, T, P, \sigma)$ be a context-free interactive Lindenmayer grammar. If

$$\alpha = x_1 \cdots x_n$$

and there are productions

$$x_i \to \beta_i$$

in $P$, for $i = 1, \ldots, n$, we write

$$\alpha \Rightarrow \beta_1 \cdots \beta_n$$

and say that $\beta_1 \cdots \beta_n$ is *directly derivable* from $\alpha$. If $\alpha_{i+1}$ is directly derivable from $\alpha_i$ for $i = 1, \ldots, n-1$, we say that $\alpha_n$ is *derivable from* $\alpha_1$ and write

$$\alpha_1 \Rightarrow \alpha_n.$$

We call

$$\alpha_1 \Rightarrow \alpha_2 \Rightarrow \cdots \Rightarrow \alpha_n$$

the *derivation of* $\alpha_n$ (*from* $\alpha_1$). The *language generated by* $G$, written $L(G)$, consists of all strings over $T$ derivable from $\sigma$. ◀

**Example 12.3.19** **The von Koch Snowflake**

Let

$$N = \{D\}$$

$$T = \{d, +, -\}$$

$$P = \{D \to D - D + + D - D, D \to d, + \to +, - \to -\}.$$

We regard $G(N, T, P, D)$ as a context-free Lindenmayer grammar. As an example of a derivation from $D$, we have

$$D \Rightarrow D - D + + D - D \Rightarrow d - d + + d - d.$$

Thus $d - d + + d - d \in L(G)$.

**Go Online**

A biography of von Koch is at bit.ly/2CAd7Ga

We now impose a meaning on the strings in $L(G)$. We interpret the symbol $d$ as a command to draw a straight line of a fixed length in the current direction; we interpret $+$ as a command to turn right by 60°; and we interpret $-$ as a command to turn left by 60°. If we begin at the left and the first move is horizontal to the right, when the string $d - d + + d - d$ is interpreted, we obtain the curve shown in Figure 12.3.1(a).

The next-longest string in $L(G)$ is

$$d - d + + d - d - d - d + + d - d + + d - d + + d - d - d - d + + d - d,$$

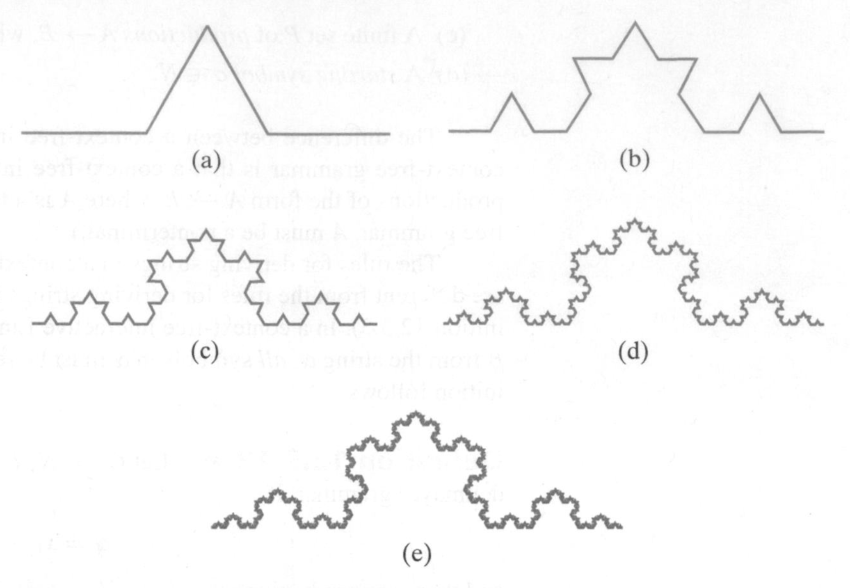


**Figure 12.3.1** von Koch snowflakes.

whose derivation is

$$\begin{aligned} D &\Rightarrow D-D++D-D \\ &\Rightarrow D-D++D-D-D-D++D-D++D \\ &\quad -D++D-D-D-D++D-D \\ &\Rightarrow d-d++d-d-d-d++d-d++d \\ &\quad -d++d-d-d-d++d-d. \end{aligned}$$

No shorter string is possible because *all* symbols must be replaced simultaneously using productions (Definition 12.3.18). If we replace some $D$'s by $d$ and other $D$'s by $D-D++D-D$, we cannot derive any string from the resulting string, let alone a terminal string, since $d$ does not occur on the left side of any production.

When the string

$$d-d++d-d-d-d++d-d++d-d++d-d-d-d++d-d$$

is interpreted, we obtain the curve shown in Figure 12.3.1(b).

The curves obtained by interpreting the next-longest strings in $L(G)$ are shown in Figure 12.3.1(c)–(e). These curves are known as **von Koch snowflakes.** ◀

Curves such as the von Koch snowflake are called **fractal curves** (see [Peitgen]). A characteristic of fractal curves is that a part of the whole resembles the whole. For example, as shown in Figure 12.3.2, when the part of the von Koch snowflake indicated is extracted and enlarged, it resembles the original.

Context-free and context-sensitive interactive Lindenmayer grammars were invented in 1968 by A. Lindenmayer (see [Lindenmayer]) to model the growth of plants. As Example 12.3.19 suggests, these grammars can be used in computer graphics to generate images (see [Prusinkiewicz 1986, 1988; Smith]). It can be shown (see [Wood, p. 503]) that the class of languages generated by context-sensitive Lindenmayer grammars is exactly the same as the class of languages generated by phrase-structure grammars.

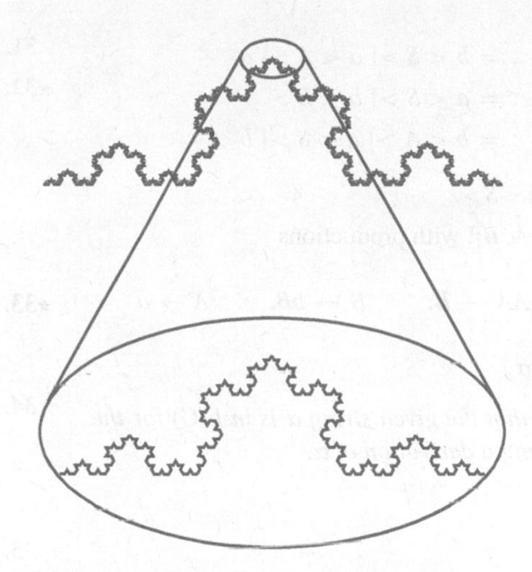

**Figure 12.3.2** The fractal nature of the von Koch snowflake. When the top part of the von Koch snowflake is extracted and enlarged, it resembles the original.

## 12.3 Review Exercises

1. Contrast natural and formal languages.
2. Define *phrase-structure grammar.*
3. What is a directly derivable string?
4. What is a derivable string?
5. What is a derivation?
6. What is the language generated by a grammar?
7. What is Backus normal form?
8. Define *context-sensitive grammar.*
9. Define *context-free grammar.*
10. Define *regular grammar.*
11. Which grammar is equivalent to a type 1 grammar?
12. Which grammar is equivalent to a type 2 grammar?
13. Which grammar is equivalent to a type 3 grammar?
14. Define *context-sensitive language.*
15. Define *context-free language.*
16. Define *regular language.*
17. What is a context-free interactive Lindenmayer grammar?
18. How is the von Koch snowflake generated?
19. What is a fractal curve?

## 12.3 Exercises

*In Exercises 1–6, determine whether the given grammar is context-sensitive, context-free, regular, or none of these. Give all characterizations that apply.*

**1.** $T = \{a, b\}$, $N = \{\sigma, A\}$, with productions

$$\sigma \to b\sigma, \quad \sigma \to aA, \quad A \to a\sigma,$$
$$A \to bA, \quad A \to a, \quad \sigma \to b,$$

and starting symbol $\sigma$.

**2.** $T = \{a, b\}$, $N = \{\sigma, A, B\}$, with productions

$$\sigma \to A, \quad \sigma \to AAB, \quad Aa \to ABa,$$
$$A \to aa, \quad Bb \to ABb, \quad AB \to ABB,$$
$$B \to b,$$

and starting symbol $\sigma$.

**3.** $T = \{a, b, c\}$, $N = \{\sigma, A, B\}$, with productions

$$\sigma \to AB, \quad AB \to BA, \quad A \to aA,$$
$$B \to Bb, \quad A \to a, \quad B \to b,$$

and starting symbol $\sigma$.

**4.** $T = \{a, b, c\}$, $N = \{\sigma, A, B\}$, with productions

$$\sigma \to BAB, \quad \sigma \to ABA, \quad A \to AB,$$
$$B \to BA, \quad A \to aA, \quad A \to ab,$$
$$B \to b,$$

and starting symbol $\sigma$.

5.
$$<S> ::= b<S> \mid a<A> \mid a$$
$$<A> ::= a<S> \mid b<B>$$
$$<B> ::= b<A> \mid a<S> \mid b$$

with starting symbol $<S>$.

6. $T = \{a, b\}, N = \{\sigma, A, B\}$, with productions

$$\sigma \to AA\sigma, \quad AA \to B, \quad B \to bB, \quad A \to a$$

and starting symbol $\sigma$.

*In Exercises 7–11, show that the given string $\alpha$ is in $L(G)$ for the given grammar $G$ by giving a derivation of $\alpha$.*

7. *bbabbab*, Exercise 1
8. *aabbaab*, Exercise 2
9. *abab*, Exercise 3
10. *abbbaabab*, Exercise 4
11. *abaabbabba*, Exercise 5
12. Write the grammars of Examples 12.3.4 and 12.3.9 and Exercises 1–4 and 6 in BNF.

★13. Let $G$ be the grammar of Exercise 1. Show that $\alpha \in L(G)$ if and only if $\alpha$ is nonnull and contains an even number of $a$'s.

★14. Let $G$ be the grammar of Exercise 5. Characterize $L(G)$.

*In Exercises 15–24, write a grammar that generates the strings having the given property.*

15. Strings over $\{a, b\}$ starting with $a$
16. Strings over $\{a, b\}$ containing $ba$
17. Strings over $\{a, b\}$ ending with $ba$

★18. Strings over $\{a, b\}$ not ending with $ab$

19. Integers with no leading 0's
20. Floating-point numbers (numbers such as .294, 89., 67.284)
21. Exponential numbers (numbers including floating-point numbers and numbers such as 6.9E3, 8E12, 9.6E–4, 9E–10)
22. All strings over $\{a, b\}$
23. Boolean expressions in $X_1, \ldots, X_n$
24. Strings $x_1 \cdots x_n$ over $\{a, b\}$ with $x_1 \cdots x_n = x_n \cdots x_1$

*Each grammar in Exercises 25–31 is proposed as generating the set $L$ of strings over $\{a, b\}$ that contain equal numbers of $a$'s and $b$'s. If the grammar generates $L$, prove that it does so. If the grammar does not generate $L$, give a counterexample and prove that your counterexample is correct. In each grammar, $S$ is the starting symbol.*

25. $S \to aSb \mid bSa \mid \lambda$
26. $S \to aB \mid bA \mid \lambda, B \to b \mid bA, A \to a \mid aB$
27. $S \to aSb \mid bSa \mid SS \mid \lambda$
28. $S \to abS \mid baS \mid aSb \mid bSa \mid \lambda$
29. $S \to aB \mid bA, A \to a \mid SA, B \to b \mid SB$
30. $S \to aSb \mid bSa \mid abS \mid baS \mid Sab \mid Sba \mid \lambda$
31. $S \to aSbS \mid bSaS \mid \lambda$

★32. Let $G$ be a grammar and let $\lambda$ denote the null string. Show that if every production is of the form

$$A \to \alpha \text{ or } A \to \alpha B \text{ or } A \to \lambda,$$
$$\text{where } A, B \in N, \quad \alpha \in T^* - \{\lambda\},$$

there is a regular grammar $G'$ with $L(G) = L(G')$.

★33. Let $G$ be the grammar of Example 12.3.9. Show that

$$L(G) = \{a^n b^n c^n \mid n = 1, 2, \ldots\}.$$

34. Show that the language

$$\{a^n b^n c^k \mid n, k \in \{1, 2, \ldots\}\}$$

is a context-free language.

35. Let

$$N = \{S, D\}$$
$$T = \{d, +, -\}$$
$$P = \{S \to D + D + D + D,$$
$$D \to D + D - D - DD + D + D - D \mid d,$$
$$+ \to +, - \to -\}.$$

Regard $G = (N, T, P, S)$ as a context-free Lindenmayer grammar. Interpret the symbol $d$ as a command to draw a straight line of a fixed length in the current direction; interpret + as a command to turn right by 90°; and interpret − as a command to turn left by 90°. Generate the two smallest strings in $L(G)$ and draw the corresponding curves. These curves are known as *quadratic Koch islands*.

★36. The following figure

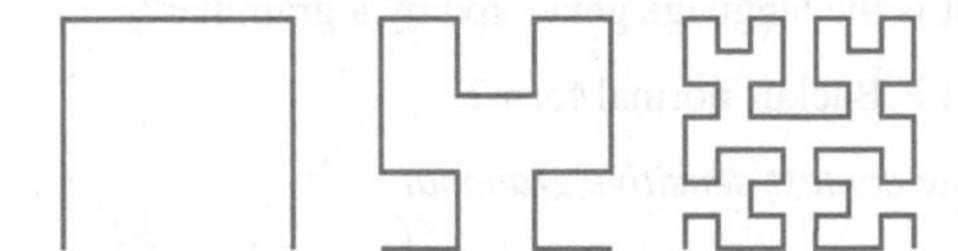

shows the first three stages of the *Hilbert curve*. Define a context-free Lindenmayer grammar that generates strings that when appropriately interpreted generate the Hilbert curve.

37. This exercise assumes familiarity with musical notation.

Pictures such as those in Exercise 36 that consist of only horizontal and vertical lines can be interpreted as music. An arbitrary starting note is chosen. Thereafter, when we follow the curve, the length of a horizontal segment determines the duration of the note, and the length of a vertical segment tells how to change the pitch. Following an $n$-unit vertical line up is interpreted as changing the pitch by moving up $n$ half-steps. Following an $n$-unit vertical line down is interpreted as changing the pitch by moving down $n$ half-steps.

Write out the music corresponding to the second figure in Exercise 36. Assume that we start at the lower-left of the figure and that C is the first note. Assume also that the first horizontal segment is two units long, which is interpreted as a quarter note.

For more on mathematics and music, see [Harkleroad].

## 12.4 Nondeterministic Finite-State Automata

In this section and the next, we show that regular grammars and finite-state automata are essentially the same in that either is a specification of a regular language. We begin with an example that illustrates how we can convert a finite-state automaton to a regular grammar.

**Example 12.4.1** Write the regular grammar given by the finite-state automaton of Figure 12.2.7.

**SOLUTION** The terminal symbols are the input symbols $\{a, b\}$. The states $E$ and $O$ become the nonterminal symbols. The initial state $E$ becomes the starting symbol. The productions correspond to the directed edges. If there is an edge labeled $x$ from $S$ to $S'$, we write the production $S \to xS'$. In our case, we obtain the productions

$$E \to bE, \qquad E \to aO, \qquad O \to aE, \qquad O \to bO. \tag{12.4.1}$$

In addition, if $S$ is an accepting state, we include the production $S \to \lambda$. In our case, we obtain the additional production

$$O \to \lambda. \tag{12.4.2}$$

Then the grammar $G = (N, T, P, E)$, with $N = \{O, E\}$, $T = \{a, b\}$, and $P$ consisting of the productions (12.4.1) and (12.4.2), generates the language $L(G)$, which is the same as the set of strings accepted by the finite-state automaton of Figure 12.2.7. ◀

**Theorem 12.4.2** Let $A$ be a finite-state automaton given as a transition diagram. Let $\sigma$ be the initial state. Let $T$ be the set of input symbols and let $N$ be the set of states. Let $P$ be the set of productions

$$S \to xS'$$

if there is an edge labeled $x$ from $S$ to $S'$ and

$$S \to \lambda$$

if $S$ is an accepting state. Let $G$ be the regular grammar

$$G = (N, T, P, \sigma).$$

Then the set of strings accepted by $A$ is equal to $L(G)$.

**Proof** First we show that $\text{Ac}(A) \subseteq L(G)$. Let $\alpha \in \text{Ac}(A)$. If $\alpha$ is the null string, then $\sigma$ is an accepting state. In this case, $G$ contains the production $\sigma \to \lambda$. The derivation

$$\sigma \Rightarrow \lambda \tag{12.4.3}$$

shows that $\alpha \in L(G)$.

Now suppose $\alpha \in \text{Ac}(A)$ and $\alpha$ is not the null string. Then $\alpha = x_1 \cdots x_n$ for some $x_i \in T$. Since $\alpha$ is accepted by $A$, there is a path $(\sigma, S_1, \ldots, S_n)$, where $S_n$ is an accepting state, with edges successively labeled $x_1, \ldots, x_n$. It follows that $G$ contains the productions

$$\sigma \to x_1 S_1$$
$$S_{i-1} \to x_i S_i \qquad \text{for } i = 2, \ldots, n.$$

Since $S_n$ is an accepting state, $G$ also contains the production $S_n \to \lambda$. The derivation

$$\begin{aligned} \sigma &\Rightarrow x_1 S_1 \\ &\Rightarrow x_1 x_2 S_2 \\ &\;\vdots \\ &\Rightarrow x_1 \cdots x_n S_n \\ &\Rightarrow x_1 \cdots x_n \end{aligned} \qquad \textbf{(12.4.4)}$$

shows that $\alpha \in L(G)$.

We complete the proof by showing that $L(G) \subseteq \text{Ac}(A)$. Suppose that $\alpha \in L(G)$. If $\alpha$ is the null string, $\alpha$ must result from the derivation (12.4.3) since a derivation that starts with any other production would yield a nonnull string. Thus the production $\sigma \to \lambda$ is in the grammar. Therefore, $\sigma$ is an accepting state in $A$. It follows that $\alpha \in \text{Ac}(A)$.

Now suppose $\alpha \in L(G)$ and $\alpha$ is not the null string. Then $\alpha = x_1 \cdots x_n$ for some $x_i \in T$. It follows that there is a derivation of the form (12.4.4). If, in the transition diagram, we begin at $\sigma$ and trace the path $(\sigma, S_1, \ldots, S_n)$, we can generate the string $\alpha$. The last production used in (12.4.4) is $S_n \to \lambda$; thus the last state reached is an accepting state. Therefore, $\alpha$ is accepted by $A$, so $L(G) \subseteq \text{Ac}(A)$. The proof is complete. ◀

Next, we consider the reverse situation. Given a regular grammar $G$, we want to construct a finite-state automaton $A$ so that $L(G)$ is precisely the set of strings accepted by $A$. It might seem, at first glance, that we can simply reverse the procedure of Theorem 12.4.2. However, the next example shows that the situation is a bit more complex.

**Example 12.4.3** Consider the regular grammar defined by $T = \{a, b\}$ and $N = \{\sigma, C\}$, with productions

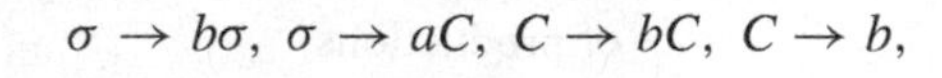

$$\sigma \to b\sigma,\ \sigma \to aC,\ C \to bC,\ C \to b,$$

and starting symbol $\sigma$.

The nonterminals become states with $\sigma$ as the initial state. For each production of the form $S \to xS'$, we draw an edge from state $S$ to state $S'$ and label it $x$. The productions

**Figure 12.4.1** The graph corresponding to the productions $\sigma \to b\sigma$, $\sigma \to aC$, $C \to bC$.

$$\sigma \to b\sigma,\ \sigma \to aC,\ \text{and } C \to bC$$

give the graph shown in Figure 12.4.1. The production $C \to b$ is equivalent to the two productions

$$C \to bF \text{ and } F \to \lambda,$$

where $F$ is an additional nonterminal symbol. The productions

$$\sigma \to b\sigma, \qquad \sigma \to aC, \qquad C \to bC, \qquad C \to bF$$

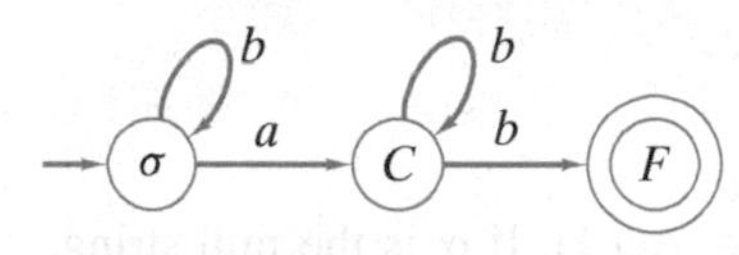


**Figure 12.4.2** The nondeterministic finite-state automaton corresponding to the grammar $\sigma \to b\sigma$, $\sigma \to aC$, $C \to bC$, $C \to b$.

give the graph shown in Figure 12.4.2. The production $F \to \lambda$ tells us that $F$ should be an accepting state (see Figure 12.4.2). ◀

Unfortunately, the graph of Figure 12.4.2 is not a finite-state automaton. There are several problems. Vertex $C$ has no outgoing edge labeled $a$, and vertex $F$ has no outgoing edges at all. Also, vertex $C$ has two outgoing edges labeled $b$. A diagram such as that of Figure 12.4.2 defines another kind of automaton called a **nondeterministic finite-state automaton.** The reason for the word "nondeterministic" is that when we are in a state where there are multiple outgoing edges all having the same label $x$, if $x$ is input the situation is nondeterministic—we have a choice of next states. For example, if in Figure 12.4.2 we are in state $C$ and $b$ is input, we have a choice of next states—we can either remain in state $C$ or go to state $F$.

**Definition 12.4.4 ▶** A *nondeterministic finite-state automaton A* consists of

(a) A finite set $\mathcal{I}$ of *input symbols*

(b) A finite set $\mathcal{S}$ of *states*

(c) A *next-state function f* from $\mathcal{S} \times \mathcal{I}$ into $\mathcal{P}(\mathcal{S})$

(d) A subset $\mathcal{A}$ of $\mathcal{S}$ of *accepting states*

(e) An *initial state* $\sigma \in \mathcal{S}$.

We write $A = (\mathcal{I}, \mathcal{S}, f, \mathcal{A}, \sigma)$. ◀

The only difference between a nondeterministic finite-state automaton and a finite-state automaton is that in a finite-state automaton the next-state function takes us to a uniquely defined state, whereas in a nondeterministic finite-state automaton the next-state function takes us to a set of states.

**Example 12.4.5** For the nondeterministic finite-state automaton of Figure 12.4.2, we have

$$\mathcal{I} = \{a, b\}, \qquad \mathcal{S} = \{\sigma, C, F\}, \qquad \mathcal{A} = \{F\}.$$

The initial state is $\sigma$ and the next-state function $f$ is given by

| $\mathcal{S}$ \ $\mathcal{I}$ | $f$: $a$ | $b$ |
|---|---|---|
| $\sigma$ | $\{C\}$ | $\{\sigma\}$ |
| $C$ | $\varnothing$ | $\{C, F\}$ |
| $F$ | $\varnothing$ | $\varnothing$ |

◀

We draw the transition diagram of a nondeterministic finite-state automaton similarly to that of a finite-state automaton. We draw an edge from state $S$ to each state in the set $f(S, x)$ and label each $x$.

**Example 12.4.6** The transition diagram of the nondeterministic finite-state automaton

$$\mathcal{I} = \{a, b\}, \qquad \mathcal{S} = \{\sigma, C, D\}, \qquad \mathcal{A} = \{C, D\}$$

with initial state $\sigma$ and next-state function

| $\mathcal{S}$ \ $\mathcal{I}$ | $f$: $a$ | $b$ |
|---|---|---|
| $\sigma$ | $\{\sigma, C\}$ | $\{D\}$ |
| $C$ | $\varnothing$ | $\{C\}$ |
| $D$ | $\{C, D\}$ | $\varnothing$ |

is shown in Figure 12.4.3.

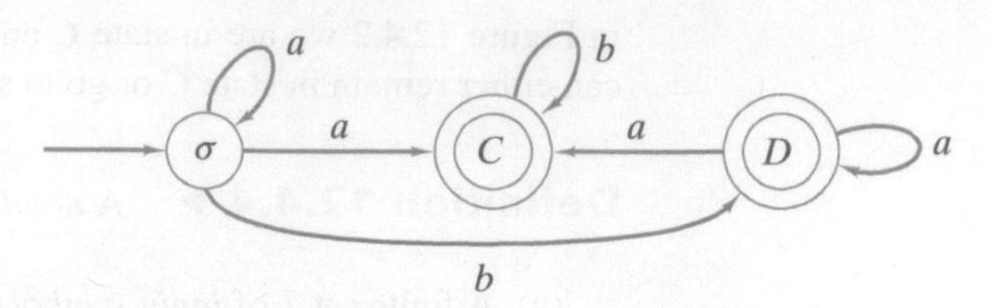


**Figure 12.4.3** The transition diagram of the nondeterministic finite-state automaton of Example 12.4.6.

A string $\alpha$ is accepted by a nondeterministic finite-state automaton $A$ if there is some path representing $\alpha$ in the transition diagram of $A$ beginning at the initial state and ending in an accepting state. The formal definition follows.

**Definition 12.4.7** ▶ Let $A = (\mathcal{I}, \mathcal{S}, f, \mathcal{A}, \sigma)$ be a nondeterministic finite-state automaton. The null string is *accepted* by $A$ if and only if $\sigma \in \mathcal{A}$. If $\alpha = x_1 \cdots x_n$ is a nonnull string over $\mathcal{I}$ and there exist states $\sigma_0, \ldots, \sigma_n$ satisfying the following conditions:

(a) $\sigma_0 = \sigma$

(b) $\sigma_i \in f(\sigma_{i-1}, x_i)$ for $i = 1, \ldots, n$

(c) $\sigma_n \in \mathcal{A}$,

we say that $\alpha$ is *accepted* by $A$. We let Ac($A$) denote the set of strings accepted by $A$ and we say that $A$ *accepts* Ac($A$).

If $A$ and $A'$ are nondeterministic finite-state automata and $\text{Ac}(A) = \text{Ac}(A')$, we say that $A$ and $A'$ are *equivalent*.

If $\alpha = x_1 \cdots x_n$ is a string over $\mathcal{I}$ and there exist states $\sigma_0, \ldots, \sigma_n$ satisfying conditions (a) and (b), we call the path $(\sigma_0, \ldots, \sigma_n)$ a *path representing* $\alpha$ in $A$.

**Example 12.4.8** The string $\alpha = bbabb$ is accepted by the nondeterministic finite-state automaton of Figure 12.4.2, since the path $(\sigma, \sigma, \sigma, C, C, F)$, which ends at an accepting state, represents $\alpha$. Notice that the path $P = (\sigma, \sigma, \sigma, C, C, C)$ also represents $\alpha$ but that $P$ does not end at an accepting state. Nevertheless, the string $\alpha$ is accepted because there is at least one path representing $\alpha$ that ends at an accepting state. A string $\beta$ will fail to be accepted if no path represents $\beta$ or every path representing $\beta$ ends at a nonaccepting state.

**Example 12.4.9** The string $\alpha = aabaabbb$ is accepted by the nondeterministic finite-state automaton of Figure 12.4.3. The reader should locate the path representing $\alpha$, which ends at state $C$.

**Example 12.4.10** The string $\alpha = abba$ is not accepted by the nondeterministic finite-state automaton of Figure 12.4.3. Starting at $\sigma$, when we input $a$, there are two choices: Go to $C$ or remain at $\sigma$. If we go to $C$, when we input two $b$'s, our moves are determined and we remain at $C$. But now when we input the final $a$, there is no edge along which to move. On the other hand, suppose that when we input the first $a$, we remain at $\sigma$. Then, when we input $b$, we move to $D$. But now when we input the next $b$, there is no edge along which to move. Since there is no path representing $\alpha$ in Figure 12.4.3, the string $\alpha$ is not accepted by the nondeterministic finite-state automaton of Figure 12.4.3.

We formulate the construction of Example 12.4.3 as a theorem.

**Theorem 12.4.11** Let $G = (N, T, P, \sigma)$ be a regular grammar. Let

$$\mathcal{I} = T$$
$$\mathcal{S} = N \cup \{F\}, \text{ where } F \notin N \cup T$$
$$f(S, x) = \{S' \mid S \to xS' \in P\} \cup \{F \mid S \to x \in P\}$$
$$\mathcal{A} = \{F\} \cup \{S \mid S \to \lambda \in P\}.$$

Then the nondeterministic finite-state automaton $A = (\mathcal{I}, \mathcal{S}, f, \mathcal{A}, \sigma)$ accepts precisely the strings $L(G)$.

**Proof** The proof is essentially the same as the proof of Theorem 12.4.2 and is therefore omitted. ◀

**Example 12.4.12** Consider the regular grammar $G$ defined by $T = \{a, b\}$ and $N = \{S\}$ with productions

$$S \to \lambda, \; S \to b, \text{ and } S \to aS,$$

and starting symbol $S$. Construct the nondeterministic finite-state automaton given by Theorem 12.4.11 that accepts the strings $L(G)$.

**Figure 12.4.4** The transition diagram corresponding to the regular grammar of Example 12.4.12.

SOLUTION The set $\mathcal{I}$ of input symbols is the same as the set $\{a, b\}$ of terminal symbols. The set $\mathcal{S}$ of states is the set $\{S, F\}$ of nonterminal symbols together with a new state $F$, which is also an accepting state. The initial state is $S$, the starting symbol.

The production $S \to \lambda$ means that $S$ is an accepting state. Thus the set $\mathcal{A}$ of accepting states is $\{S, F\}$. The production $S \to b$ yields a transition from state $S$ to state $F$ on input symbol $b$. The production $S \to aS$ yields a transition from state $S$ to state $S$ on input symbol $a$. We obtain the transition diagram shown in Figure 12.4.4. ◀

It may seem that a nondeterministic finite-state automaton is a more general concept than a finite-state automaton; however, in the next section we will show that given a nondeterministic finite-state automaton $A$, we can construct a finite-state automaton that is equivalent to $A$.

## 12.4 Review Exercises

1. Given a finite-state automaton $A$, how can we construct a regular grammar $G$ so that the set of strings accepted by $A$ is equal to the language generated by $G$?
2. What is a nondeterministic finite-state automaton?
3. What does it mean for a string to be accepted by a nondeterministic finite-state automaton?
4. What are equivalent nondeterministic finite-state automata?
5. Given a regular grammar $G$, how can we construct a nondeterministic finite-state automaton $A$ so that the language generated by $G$ is equal to the set of strings accepted by $A$?

## 12.4 Exercises

*In Exercises 1–5, draw the transition diagram of the nondeterministic finite-state automaton $(\mathcal{I}, \mathcal{S}, f, \mathcal{A}, \sigma_0)$.*

1. $\mathcal{I} = \{a, b\}$, $\mathcal{S} = \{\sigma_0, \sigma_1, \sigma_2\}$, $\mathcal{A} = \{\sigma_0\}$

| $\mathcal{S}$ \ $\mathcal{I}$ | $a$ | $b$ |
|---|---|---|
| $\sigma_0$ | $\varnothing$ | $\{\sigma_1, \sigma_2\}$ |
| $\sigma_1$ | $\{\sigma_2\}$ | $\{\sigma_0, \sigma_1\}$ |
| $\sigma_2$ | $\{\sigma_0\}$ | $\varnothing$ |

**2.** $\mathcal{I} = \{a, b\}$, $\mathcal{S} = \{\sigma_0, \sigma_1, \sigma_2, \sigma_3\}$, $\mathcal{A} = \{\sigma_1\}$

| $\mathcal{S}$ \ $\mathcal{I}$ | $a$ | $b$ |
|---|---|---|
| $\sigma_0$ | $\varnothing$ | $\{\sigma_3\}$ |
| $\sigma_1$ | $\{\sigma_1, \sigma_2\}$ | $\{\sigma_3\}$ |
| $\sigma_2$ | $\varnothing$ | $\{\sigma_0, \sigma_1, \sigma_3\}$ |
| $\sigma_3$ | $\varnothing$ | $\varnothing$ |

**3.** $\mathcal{I} = \{a, b\}$, $\mathcal{S} = \{\sigma_0, \sigma_1, \sigma_2\}$, $\mathcal{A} = \{\sigma_0, \sigma_1\}$

| $\mathcal{S}$ \ $\mathcal{I}$ | $a$ | $b$ |
|---|---|---|
| $\sigma_0$ | $\{\sigma_1\}$ | $\{\sigma_0, \sigma_2\}$ |
| $\sigma_1$ | $\varnothing$ | $\{\sigma_2\}$ |
| $\sigma_2$ | $\{\sigma_1\}$ | $\varnothing$ |

**4.** $\mathcal{I} = \{a, b, c\}$, $\mathcal{S} = \{\sigma_0, \sigma_1, \sigma_2\}$, $\mathcal{A} = \{\sigma_0\}$

| $\mathcal{S}$ \ $\mathcal{I}$ | $a$ | $b$ | $c$ |
|---|---|---|---|
| $\sigma_0$ | $\{\sigma_1\}$ | $\varnothing$ | $\varnothing$ |
| $\sigma_1$ | $\{\sigma_0\}$ | $\{\sigma_2\}$ | $\{\sigma_0, \sigma_2\}$ |
| $\sigma_2$ | $\{\sigma_0, \sigma_1, \sigma_2\}$ | $\{\sigma_0\}$ | $\{\sigma_0\}$ |

**5.** $\mathcal{I} = \{a, b, c\}$, $\mathcal{S} = \{\sigma_0, \sigma_1, \sigma_2, \sigma_3\}$, $\mathcal{A} = \{\sigma_0, \sigma_3\}$

| $\mathcal{S}$ \ $\mathcal{I}$ | $a$ | $b$ | $c$ |
|---|---|---|---|
| $\sigma_0$ | $\{\sigma_1\}$ | $\{\sigma_0, \sigma_1, \sigma_3\}$ | $\varnothing$ |
| $\sigma_1$ | $\{\sigma_2, \sigma_3\}$ | $\varnothing$ | $\varnothing$ |
| $\sigma_2$ | $\varnothing$ | $\{\sigma_0, \sigma_3\}$ | $\{\sigma_1, \sigma_2\}$ |
| $\sigma_3$ | $\varnothing$ | $\varnothing$ | $\{\sigma_0\}$ |

*For each nondeterministic finite-state automaton in Exercises 6–10, find the sets $\mathcal{I}$, $\mathcal{S}$, and $\mathcal{A}$, the initial state, and the table defining the next-state function.*

**6.**

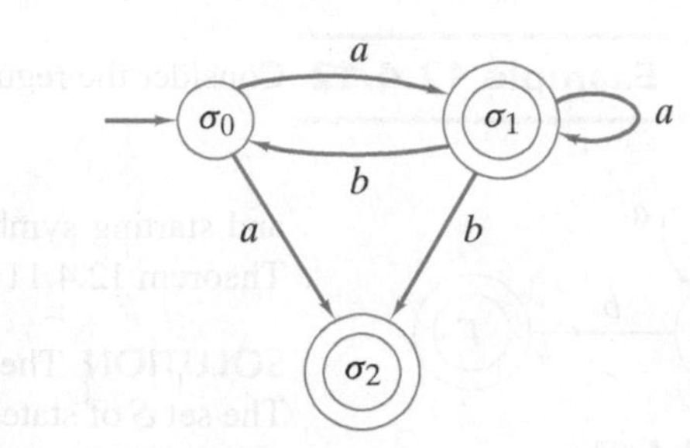


**7.**

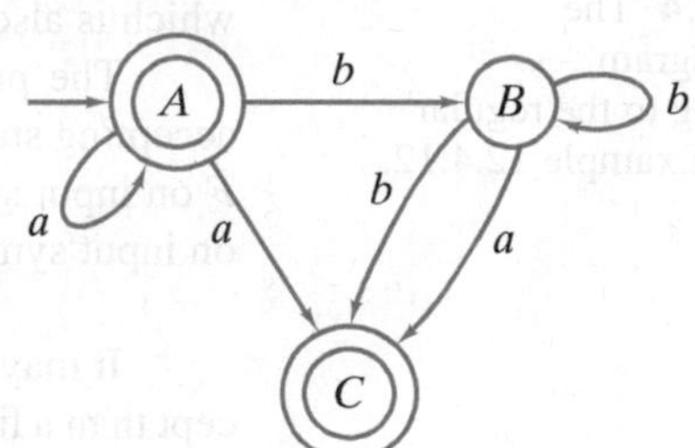


**8.**

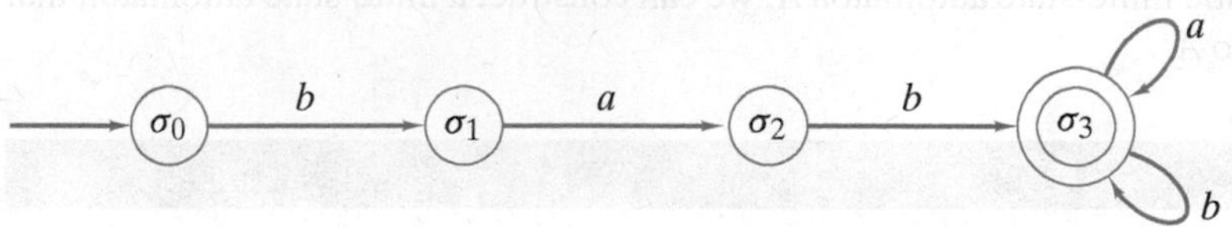


**9.**

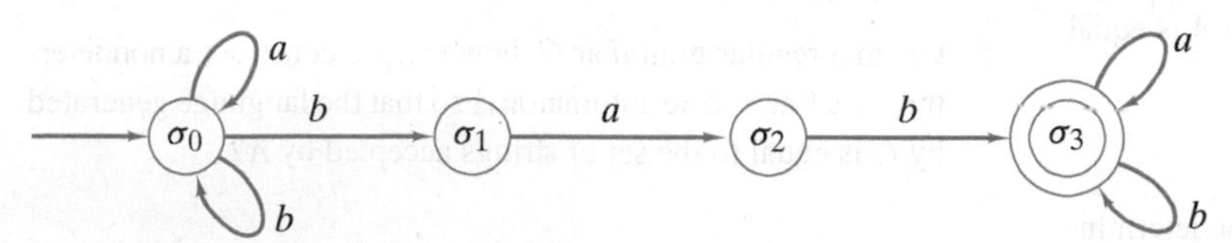


**10.**

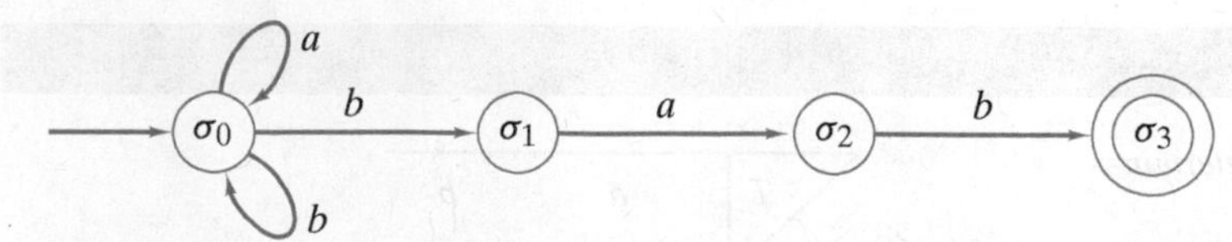


**11.** Write the regular grammars given by the finite-state automata of Exercises 4–9, Section 12.2.

**12.** Is the string *bbabbb* accepted by the nondeterministic finite-state automaton of Figure 12.4.2? Prove your answer.

**13.** Represent the grammars of Exercises 1 and 5, Section 12.3, and Example 12.3.14 by nondeterministic finite-state automata.

**14.** Is the string *bbabab* accepted by the nondeterministic finite-state automaton of Figure 12.4.2? Prove your answer.

15. Is the string *aaabba* accepted by the nondeterministic finite-state automaton of Figure 12.4.3? Prove your answer.
16. Show that a string $\alpha$ over $\{a, b\}$ is accepted by the nondeterministic finite-state automaton of Figure 12.4.2 if and only if $\alpha$ contains exactly one $a$ and ends with $b$.
17. Is the string *aaaab* accepted by the nondeterministic finite-state automaton of Figure 12.4.3? Prove your answer.
18. Characterize the strings accepted by the nondeterministic finite-state automaton of Figure 12.4.3.
19. Show that the strings accepted by the nondeterministic finite-state automaton of Exercise 10 are precisely those strings over $\{a, b\}$ that end *bab*.
★20. Characterize the strings accepted by the nondeterministic finite-state automata of Exercises 1–9.

*Design nondeterministic finite-state automata that accept the strings over $\{a, b\}$ having the properties specified in Exercises 21–29.*

21. Starting either *abb* or *ba*
22. Containing either *abb* or *ba*
23. Ending either *abb* or *ba*
★24. Containing *bab* and *bb*
25. Starting with *abb* and ending with *ab*
26. Having each *b* preceded and followed by an *a*
★27. Starting with *ab* but not ending with *ab*
28. Not containing *ba* or *bbb*
★29. Not containing *abba* or *bbb*
30. Write regular grammars that generate the strings of Exercises 21–29.

## 12.5 Relationships Between Languages and Automata

In the preceding section we showed (Theorem 12.4.2) that if $A$ is a finite-state automaton, there exists a regular grammar $G$, with $L(G) = \text{Ac}(A)$. As a partial converse, we showed (Theorem 12.4.11) that if $G$ is a regular grammar, there exists a nondeterministic finite-state automaton $A$ with $L(G) = \text{Ac}(A)$. In this section we show (Theorem 12.5.4) that if $G$ is a regular grammar, there exists a finite-state automaton $A$ with $L(G) = \text{Ac}(A)$. This result will be deduced from Theorem 12.4.11 by showing that any nondeterministic finite-state automaton can be converted to an equivalent finite-state automaton (Theorem 12.5.3). We will first illustrate the method by an example.

**Example 12.5.1** Find a finite-state automaton equivalent to the nondeterministic finite-state automaton of Figure 12.4.2.

**SOLUTION** The set of input symbols is unchanged. The states consist of all subsets

$$\varnothing, \quad \{\sigma\}, \quad \{C\}, \quad \{F\}, \quad \{\sigma, C\}, \quad \{\sigma, F\}, \quad \{C, F\}, \quad \{\sigma, C, F\}$$

of the original set $S = \{\sigma, C, F\}$ of states. The initial state is $\{\sigma\}$. The accepting states are all subsets

$$\{F\}, \quad \{\sigma, F\}, \quad \{C, F\}, \quad \{\sigma, C, F\}$$

of $S$ that contain an accepting state of the original nondeterministic finite-state automaton.

An edge is drawn from $X$ to $Y$ and labeled $x$ if $X = \varnothing = Y$ or if

$$\bigcup_{S \in X} f(S, x) = Y.$$

We obtain the finite-state automaton of Figure 12.5.1. The states

$$\{\sigma, F\}, \quad \{\sigma, C\}, \quad \{\sigma, C, F\}, \quad \{F\},$$

which can never be reached, can be deleted. Thus we obtain the simplified, equivalent finite-state automaton of Figure 12.5.2.

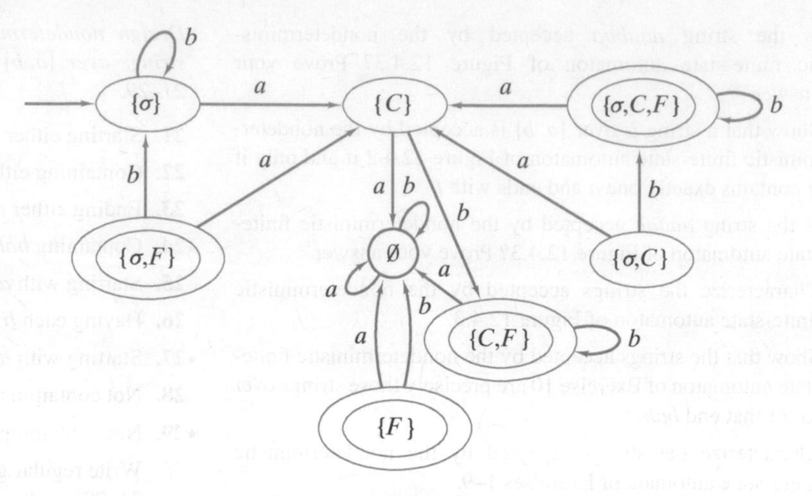


**Figure 12.5.1** A finite-state automaton equivalent to the nondeterministic finite-state automaton of Figure 12.4.2.

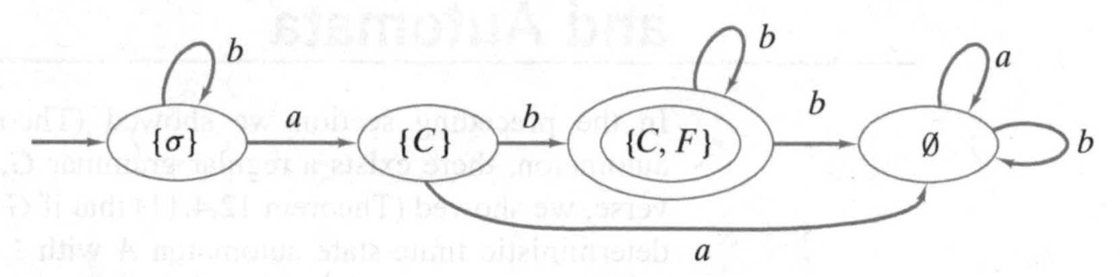


**Figure 12.5.2** A simplified version of Figure 12.5.1 (with unreachable states deleted).

**Example 12.5.2** The finite-state automaton equivalent to the nondeterministic finite-state automaton of Example 12.4.6 is shown in Figure 12.5.3.

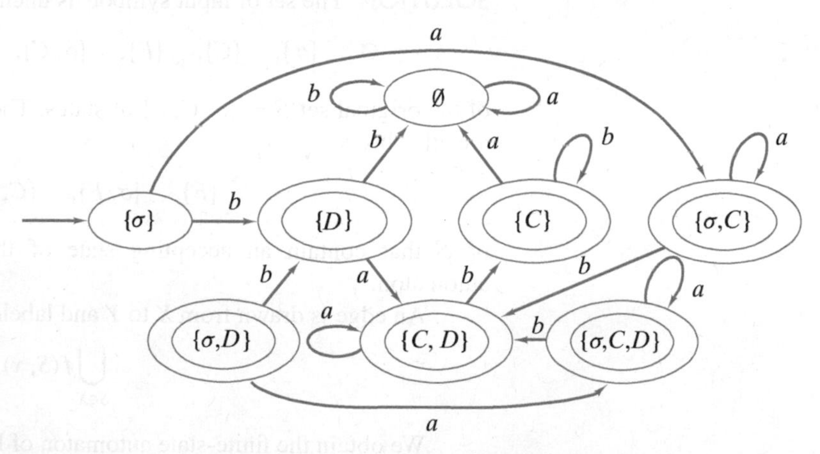


**Figure 12.5.3** A finite-state automaton equivalent to the nondeterministic finite-state automaton of Example 12.4.6. ◀

We now formally justify the method of Examples 12.5.1 and 12.5.2.

**Theorem 12.5.3** Let $A = (\mathcal{I}, \mathcal{S}, f, \mathcal{A}, \sigma)$ be a nondeterministic finite-state automaton. Let

(a) $\mathcal{S}' = \mathcal{P}(\mathcal{S})$

(b) $\mathcal{I}' = \mathcal{I}$

(c) $\sigma' = \{\sigma\}$

(d) $\mathcal{A}' = \{X \subseteq \mathcal{S} \mid X \cap \mathcal{A} \neq \varnothing\}$

(e) $f'(X, x) = \begin{cases} \varnothing & \text{if } X = \varnothing \\ \bigcup_{S \in X} f(S, x) & \text{if } X \neq \varnothing. \end{cases}$

Then the finite-state automaton $A' = (\mathcal{I}', \mathcal{S}', f', \mathcal{A}', \sigma')$ is equivalent to $A$.

**Proof** Suppose that the string $\alpha = x_1 \cdots x_n$ is accepted by $A$. Then there exist states $\sigma_0, \ldots, \sigma_n \in \mathcal{S}$ with

$$\sigma_0 = \sigma$$

$$\sigma_i \in f(\sigma_{i-1}, x_i) \qquad \text{for } i = 1, \ldots, n$$

$$\sigma_n \in \mathcal{A}.$$

Set $Y_0 = \{\sigma_0\}$ and

$$Y_i = f'(Y_{i-1}, x_i) \qquad \text{for } i = 1, \ldots, n.$$

Since

$$Y_1 = f'(Y_0, x_1) = f'(\{\sigma_0\}, x_1) = f(\sigma_0, x_1),$$

it follows that $\sigma_1 \in Y_1$. Now

$$\sigma_2 \in f(\sigma_1, x_2) \subseteq \bigcup_{S \in Y_1} f(S, x_2) = f'(Y_1, x_2) = Y_2.$$

Again,

$$\sigma_3 \in f(\sigma_2, x_3) \subseteq \bigcup_{S \in Y_2} f(S, x_3) = f'(Y_2, x_3) = Y_3.$$

The argument may be continued (formally, we would use induction) to show that $\sigma_n \in Y_n$. Since $\sigma_n$ is an accepting state in $A$, $Y_n$ is an accepting state in $A'$. Thus, in $A'$, we have

$$f'(\sigma', x_1) = f'(Y_0, x_1) = Y_1$$

$$f'(Y_1, x_2) = Y_2$$

$$\vdots$$

$$f'(Y_{n-1}, x_n) = Y_n.$$

Therefore, $\alpha$ is accepted by $A'$.

Now suppose that the string $\alpha = x_1 \cdots x_n$ is accepted by $A'$. Then there exist subsets $Y_0, \ldots, Y_n$ of $\mathcal{S}$ such that

$$Y_0 = \sigma' = \{\sigma\}$$

$$f'(Y_{i-1}, x_i) = Y_i \qquad \text{for } i = 1, \ldots, n;$$

also, there exists a state $\sigma_n \in Y_n \cap \mathcal{A}$. Since

$$\sigma_n \in Y_n = f'(Y_{n-1}, x_n) = \bigcup_{S \in Y_{n-1}} f(S, x_n),$$

there exists $\sigma_{n-1} \in Y_{n-1}$ with $\sigma_n \in f(\sigma_{n-1}, x_n)$. Similarly, since

$$\sigma_{n-1} \in Y_{n-1} = f'(Y_{n-2}, x_{n-1}) = \bigcup_{S \in Y_{n-2}} f(S, x_{n-1}),$$

there exists $\sigma_{n-2} \in Y_{n-2}$ with $\sigma_{n-1} \in f(\sigma_{n-2}, x_{n-1})$. Continuing, we obtain

$$\sigma_i \in Y_i \qquad \text{for } i = 0, \ldots, n,$$

with

$$\sigma_i \in f(\sigma_{i-1}, x_i) \qquad \text{for } i = 1, \ldots, n.$$

In particular,

$$\sigma_0 \in Y_0 = \{\sigma\}.$$

Thus $\sigma_0 = \sigma$, the initial state in $A$. Since $\sigma_n$ is an accepting state in $A$, the string $\alpha$ is accepted by $A$. ◀

The next theorem summarizes these results and those of the preceding section.

**Theorem 12.5.4** A language $L$ is regular if and only if there exists a finite-state automaton that accepts precisely the strings in $L$.

**Proof** This theorem restates Theorems 12.4.2, 12.4.11, and 12.5.3. ◀

**Example 12.5.5** Find a finite-state automaton $A$ that accepts precisely the strings generated by the regular grammar $G$ having productions

$$\sigma \to b\sigma, \qquad \sigma \to aC, \qquad C \to bC, \qquad C \to b.$$

The starting symbol is $\sigma$, the set of terminal symbols is $\{a, b\}$, and the set of nonterminal symbols is $\{\sigma, C\}$.

**SOLUTION** The nondeterministic finite-state automaton $A'$ that accepts $L(G)$ is shown in Figure 12.4.2. A finite-state automaton equivalent to $A'$ is shown in Figure 12.5.1, and an equivalent simplified finite-state automaton $A$ is shown in Figure 12.5.2. The finite-state automaton $A$ accepts precisely the strings generated by $G$. ◀

We close this section by giving some applications of the methods and theory we have developed.

**Example 12.5.6** **A Language That Is Not Regular** Show that the language

$$L = \{a^n b^n \mid n = 1, 2, \ldots\}$$

is not regular.

**SOLUTION** If $L$ is regular, there exists a finite-state automaton $A$ such that $\text{Ac}(A) = L$. Suppose that $A$ has $k$ states. The string $\alpha = a^k b^k$ is accepted by $A$. Consider the path $P$, which represents $\alpha$. Since there are $k$ states, some state $\sigma$ is revisited on the part of the path representing $a^k$. Thus there is a cycle $C$, all of whose edges are labeled $a$, that contains $\sigma$. We change the path $P$ to obtain a path $P'$ as follows. When we arrive at $\sigma$ in $P$, we follow $C$. After returning to $\sigma$ on $C$, we continue on $P$ to the end. If the length of $C$ is $j$, the path $P'$ represents the string $\alpha' = a^{j+k} b^k$. Since $P$ and $P'$ end at the same state

$\sigma'$ and $\sigma'$ is an accepting state, $\alpha'$ is accepted by $A$. This is a contradiction, since $\alpha'$ is not of the form $a^n b^n$. Therefore, $L$ is not regular. ◀

**Example 12.5.7** Let $L$ be the set of strings accepted by the finite-state automaton $A$ of Figure 12.5.4. Construct a finite-state automaton that accepts the strings

$$L^R = \{x_n \cdots x_1 \mid x_1 \cdots x_n \in L\}.$$

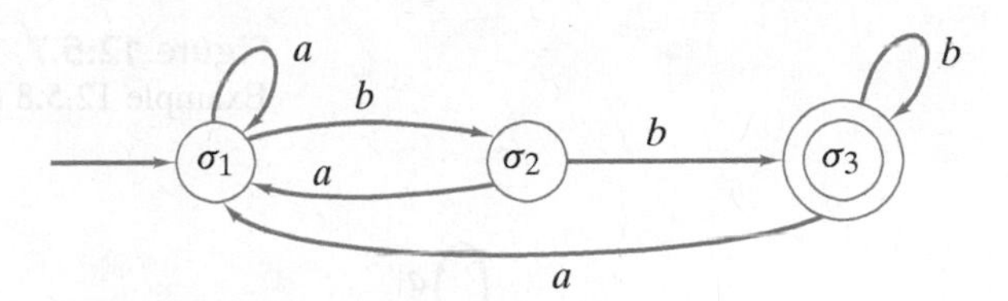


**Figure 12.5.4** The finite-state automaton for Example 12.5.7 that accepts $L$.

**SOLUTION** We want to convert $A$ to a finite-state automaton that accepts $L^R$. The string $\alpha = x_1 \cdots x_n$ is accepted by $A$ if there is a path $P$ in $A$ representing $\alpha$ that starts at $\sigma_1$ and ends at $\sigma_3$. If we start at $\sigma_3$ and trace $P$ in reverse, we end at $\sigma_1$ and process the edges in the order $x_n, \ldots, x_1$. Thus we need only reverse all arrows in Figure 12.5.4 and make $\sigma_3$ the starting state and $\sigma_1$ the accepting state (see Figure 12.5.5). The result is a nondeterministic finite-state automaton that accepts $L^R$.

After finding an equivalent finite-state automaton and eliminating the unreachable states, we obtain the equivalent finite-state automaton of Figure 12.5.6.

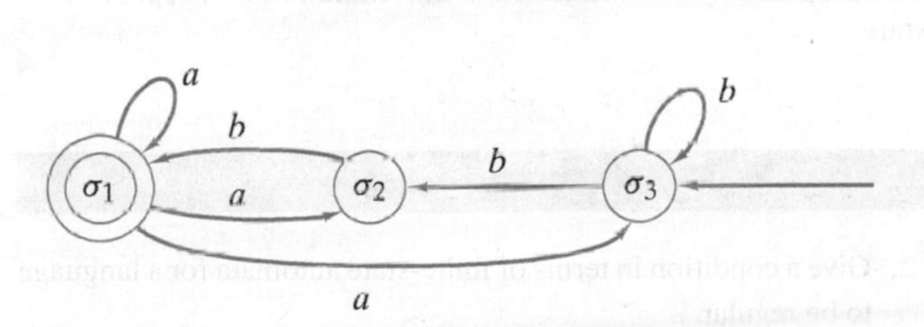


**Figure 12.5.5** A nondeterministic finite-state automaton that accepts $L^R$.

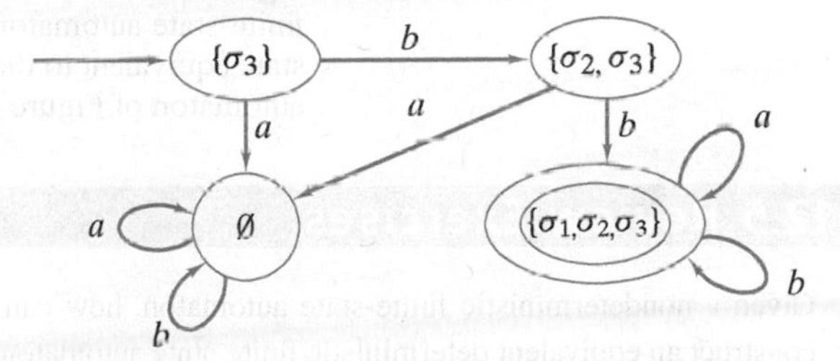


**Figure 12.5.6** A finite-state automaton that accepts $L^R$.

◀

**Example 12.5.8** Let $L$ be the set of strings accepted by the finite-state automaton $A$ of Figure 12.5.7. Construct a nondeterministic finite-state automaton that accepts the strings

$$L^R = \{x_n \cdots x_1 \mid x_1 \cdots x_n \in L\}.$$

**SOLUTION** If $A$ had only one accepting state, we could use the procedure of Example 12.5.7 to construct the desired nondeterministic finite-state automaton. Thus we first construct a nondeterministic finite-state automaton equivalent to $A$ with one accepting state. To do this we introduce an additional state $\sigma_5$. Then we arrange for paths terminating at $\sigma_3$ or $\sigma_4$ to optionally terminate at $\sigma_5$ (see Figure 12.5.8). The desired nondeterministic finite-state automaton is obtained from Figure 12.5.8 by the method of Example 12.5.7 (see Figure 12.5.9). Of course, if desired, we could construct an equivalent finite-state automaton.

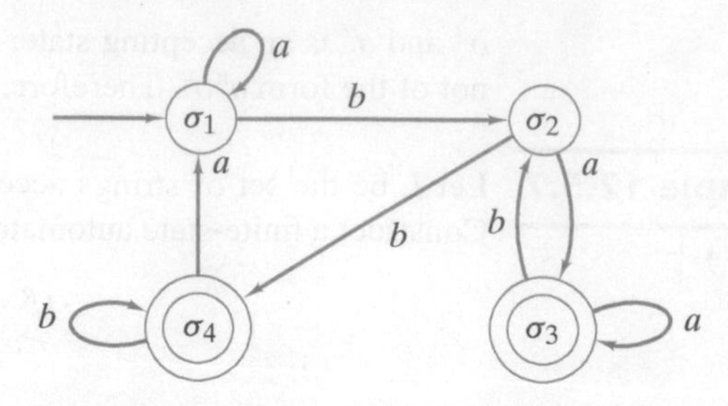


**Figure 12.5.7** The finite-state automaton for Example 12.5.8 that accepts $L$.

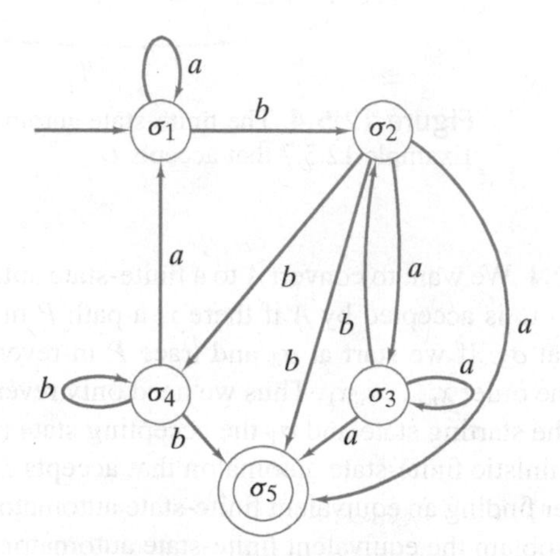


**Figure 12.5.8** A nondeterministic finite-state automaton with one accepting state equivalent to the finite-state automaton of Figure 12.5.7.

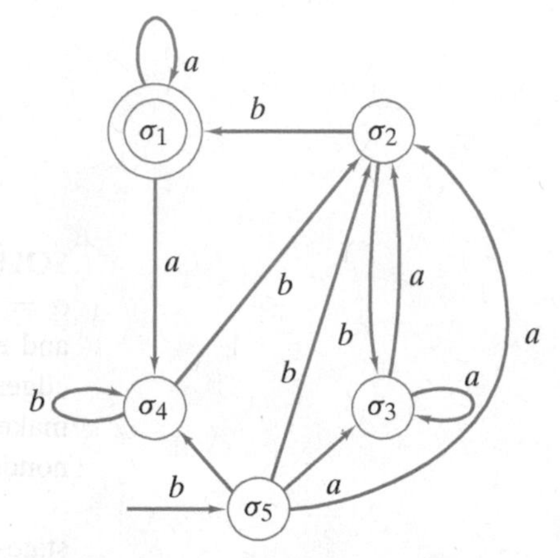


**Figure 12.5.9** A nondeterministic finite-state automaton that accepts $L^R$.

◀

## 12.5 Review Exercises

1. Given a nondeterministic finite-state automaton, how can we construct an equivalent deterministic finite-state automaton?
2. Give a condition in terms of finite-state automata for a language to be regular.

## 12.5 Exercises

1. Find the finite-state automata equivalent to the nondeterministic finite-state automata of Exercises 1–10, Section 12.4.

*In Exercises 2–6, find finite-state automata that accept the strings generated by the regular grammars.*

2. Grammar of Exercise 1, Section 12.3
3. Grammar of Exercise 5, Section 12.3
4. 

$$
\begin{aligned}
<S> &::= a<A> \mid a<B> \\
<A> &::= a<B> \mid b<S> \mid b \\
<B> &::= b<S> \mid b
\end{aligned}
$$

with starting symbol $<S>$

5. 

$$
\begin{aligned}
<S> &::= a<S> \mid a<A> \mid b<C> \mid a \\
<A> &::= b<A> \mid a<C> \\
<B> &::= a<S> \mid a \\
<C> &::= a<B> \mid a<C>
\end{aligned}
$$

with starting symbol $<S>$

6. 

$$
\begin{aligned}
<S> &::= a<A> \mid a<B> \\
<A> &::= b<S> \mid b \\
<B> &::= a<B> \mid a<C> \\
<C> &::= a<S> \mid b<A> \mid a<C> \mid a
\end{aligned}
$$

with starting symbol $<S>$

7. Find finite-state automata that accept the strings of Exercises 21–29, Section 12.4.

8. By eliminating unreachable states from the finite-state automaton of Figure 12.5.3, find a simpler, equivalent finite-state automaton.

9. Show that the nondeterministic finite-state automaton of Figure 12.5.5 accepts a string $\alpha$ over $\{a, b\}$ if and only if $\alpha$ begins *bb*.

★10. Characterize the strings accepted by the nondeterministic finite-state automata of Figures 12.5.7 and 12.5.9.

*In Exercises 11–21, find a nondeterministic finite-state automaton that accepts the given set of strings. If $S_1$ and $S_2$ are sets of strings, we let*

$$S_1^+ = \{u_1u_2\cdots u_n \mid u_i \in S_1, n \in \{1, 2, \ldots\}\};$$
$$S_1S_2 = \{uv \mid u \in S_1, v \in S_2\}.$$

11. $Ac(A)^R$, where $A$ is the automaton of Exercise 4, Section 12.2

12. $Ac(A)^R$, where $A$ is the automaton of Exercise 5, Section 12.2

13. $Ac(A)^R$, where $A$ is the automaton of Exercise 6, Section 12.2

14. $Ac(A)^+$, where $A$ is the automaton of Exercise 4, Section 12.2

15. $Ac(A)^+$, where $A$ is the automaton of Exercise 5, Section 12.2

16. $Ac(A)^+$, where $A$ is the automaton of Exercise 6, Section 12.2

17. $Ac(A)^+$, where $A$ is the automaton of Figure 12.5.7

18. $Ac(A_1)Ac(A_2)$, where $A_1$ is the automaton of Exercise 4, Section 12.2, and $A_2$ is the automaton of Exercise 6, Section 12.2

19. $Ac(A_1)Ac(A_2)$, where $A_1$ is the automaton of Exercise 5, Section 12.2, and $A_2$ is the automaton of Exercise 6, Section 12.2

20. $Ac(A_1)Ac(A_1)$, where $A_1$ is the automaton of Exercise 5, Section 12.2

21. $Ac(A_1)Ac(A_2)$, where $A_1$ is the automaton of Figure 12.5.7 and $A_2$ is the automaton of Exercise 6, Section 12.2

22. Find a regular grammar that generates the language $L^R$, where $L$ is the language generated by the grammar of Exercise 5, Section 12.3.

23. Find a regular grammar that generates the language $L^+$, where $L$ is the language generated by the grammar of Exercise 5, Section 12.3.

24. Let $L_1$ (respectively, $L_2$) be the language generated by the grammar of Exercise 5, Section 12.3 (respectively, Example 12.5.5). Find a regular grammar that generates the language $L_1L_2$.

★25. Show that the set

$$L = \{x_1 \cdots x_n \mid x_1 \cdots x_n = x_n \cdots x_1\}$$

of strings over $\{a, b\}$ is not a regular language.

26. Show that if $L_1$ and $L_2$ are regular languages over $\mathcal{I}$ and $S$ is the set of all strings over $\mathcal{I}$, then each of $S - L_1$, $L_1 \cup L_2$, $L_1 \cap L_2$, $L_1^+$, and $L_1L_2$ is a regular language.

★27. Show, by example, that there are context-free languages $L_1$ and $L_2$ such that $L_1 \cap L_2$ is not context-free.

★28. Prove or disprove: If $L$ is a regular language, so is

$$\{u^n \mid u \in L, n \in \{1, 2, \ldots\}\}.$$

## Chapter 12 Notes

General references on automata, grammars, and languages are [Carroll; Cohen; Davis; Hopcroft; Kelley; McNaughton; Sudkamp; and Wood].

The systematic development of fractal geometry was begun by Benoit B. Mandelbrot (see [Mandelbrot, 1977, 1982]).

A finite-state machine has a primitive internal memory in the sense that it remembers which state it is in. By permitting an external memory on which the machine can read and write data, we can define more powerful machines. Other enhancements are achieved by allowing the machine to scan the input string in either direction and by allowing the machine to alter the input string. It is then possible to characterize the classes of machines that accept context-free languages, context-sensitive languages, and languages generated by phrase-structure grammars.

**Turing machines** form a particularly important class of machines. Like a finite-state machine, a Turing machine is always in a particular state. The input string to a Turing machine is assumed to reside on a tape that is infinite in both directions. A Turing machine scans one character at a time and after scanning a character, the machine either halts or does some, none, or all of the following: alter the character; move one position left or right; change states. In particular, the input string can be changed. A Turing machine $T$ accepts a string $\alpha$ if, when $\alpha$ is input to $T$, $T$ halts in an accepting state. It can be shown that a language $L$ is generated by a phrase-structure grammar if and only if there is a Turing machine that accepts $L$.

The real importance of Turing machines results from the widely held belief that any function that can be computed by some, perhaps hypothetical, digital computer can be computed by some Turing machine. This last assertion is known as **Turing's hypothesis** or **Church's thesis.** Church's thesis implies that a Turing machine is the correct abstract model of a digital computer. These ideas also yield the following formal definition of algorithm. An **algorithm** is a Turing machine that, given an input string, eventually stops.

## Chapter 12 Review

### Section 12.1

1. Unit time delay
2. Serial adder
3. Finite-state machine
4. Input symbol
5. Output symbol
6. State
7. Next-state function
8. Output function
9. Initial state
10. Transition diagram
11. Input and output strings for a finite-state machine
12. *SR* flip-flop

### Section 12.2

13. Finite-state automaton
14. Accepting state
15. String accepted by a finite-state automaton
16. Equivalent finite-state automata

### Section 12.3

17. Natural language
18. Formal language
19. Phrase-structure grammar
20. Nonterminal symbol
21. Terminal symbol
22. Production
23. Starting symbol
24. Directly derivable string
25. Derivable string
26. Derivation
27. Language generated by a grammar
28. Backus normal form (= Backus–Naur form = BNF)
29. Context-sensitive grammar (= type 1 grammar)
30. Context-free grammar (= type 2 grammar)
31. Regular grammar (= type 3 grammar)
32. Context-sensitive language
33. Context-free language
34. Regular language
35. Context-free interactive Lindenmayer grammar
36. von Koch snowflake
37. Fractal curves

### Section 12.4

38. Given a finite-state automaton $A$, how to construct a regular grammar $G$, such that the set of strings accepted by $A$ is equal to the language generated by $G$ (see Theorem 12.4.2)
39. Nondeterministic finite-state automaton
40. String accepted by a nondeterministic finite-state automaton
41. Equivalent nondeterministic finite-state automata
42. Given a regular grammar $G$, how to construct a nondeterministic finite-state automaton $A$ such that the language generated by $G$ is equal to the set of strings accepted by $A$ (see Theorem 12.4.11)

### Section 12.5

43. Given a nondeterministic finite-state automaton, how to construct an equivalent deterministic finite-state automaton (see Theorem 12.5.3)
44. A language $L$ is regular if and only if there exists a finite-state automaton that accepts the strings in $L$.

## Chapter 12 Self-Test

1. Draw the transition diagram of the finite-state machine $(\mathcal{I}, \mathcal{O}, \mathcal{S}, f, g, \sigma_0)$, where $\mathcal{I} = \{a, b\}$, $\mathcal{O} = \{0, 1\}$, and $\mathcal{S} = \{\sigma_0, \sigma_1\}$.

| | $f$ | | $g$ | |
|---|---|---|---|---|
| $\mathcal{S}$ \ $\mathcal{I}$ | $a$ | $b$ | $a$ | $b$ |
| $\sigma_0$ | $\sigma_1$ | $\sigma_0$ | 0 | 1 |
| $\sigma_1$ | $\sigma_0$ | $\sigma_1$ | 1 | 0 |

2. For the finite-state machine of Exercise 1, find the output string for the input string *bbaa*.
3. Draw the transition diagram of the finite-state automaton $(\mathcal{I}, \mathcal{S}, f, \mathcal{A}, S)$, where $\mathcal{I} = \{0, 1\}$, $\mathcal{S} = \{S, A, B\}$, and $\mathcal{A} = \{A\}$.

| | $f$ | |
|---|---|---|
| $\mathcal{S}$ \ $\mathcal{I}$ | 0 | 1 |
| $S$ | $A$ | $S$ |
| $A$ | $S$ | $B$ |
| $B$ | $A$ | $S$ |

4. Is the string 11010 accepted by the finite-state automaton of Exercise 3?
5. Is the grammar

$$S \to aSb, \quad S \to Ab, \quad A \to aA, \quad A \to b, \quad A \to \lambda$$

context-sensitive, context-free, regular, or none of these? Give all characterizations that apply.

6. Show that the string $\alpha = aaaabbbb$ is in the language generated by the grammar of Exercise 5 by giving a derivation of $\alpha$.

7. Characterize the language generated by the grammar of Exercise 5.

8. Draw the transition diagram of the nondeterministic finite-state automaton $(\mathcal{I}, \mathcal{S}, f, \mathcal{A}, \sigma_0)$, where $\mathcal{I} = \{a, b\}$, $\mathcal{S} = \{\sigma_0, \sigma_1, \sigma_2\}$, and $\mathcal{A} = \{\sigma_2\}$.

| $\mathcal{S}$ \ $\mathcal{I}$ | $a$ | $b$ |
|---|---|---|
| $\sigma_0$ | $\{\sigma_0\}$ | $\{\sigma_2\}$ |
| $\sigma_1$ | $\{\sigma_0, \sigma_1\}$ | $\varnothing$ |
| $\sigma_2$ | $\{\sigma_2\}$ | $\{\sigma_0, \sigma_1\}$ |

9. Find a finite-state automaton equivalent to the nondeterministic finite-state automaton of Exercise 8.

10. Find the sets $\mathcal{I}$, $\mathcal{O}$, and $\mathcal{S}$, the initial state, and the table defining the next-state and output functions for the following finite-state machine.

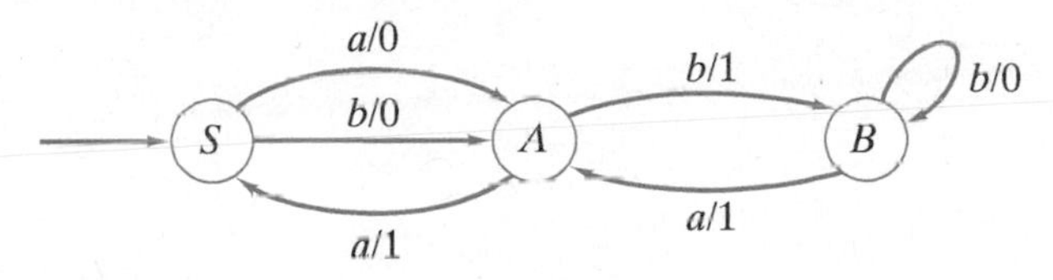


11. Design a finite-state machine whose input is a bit string that outputs 0 when it sees 001 and thereafter; otherwise, it outputs 1.

12. Draw the transition diagram of a finite-state automaton that accepts the set of strings over $\{0, 1\}$ that contain an even number of 0's and an odd number of 1's.

13. Characterize the set of strings accepted by the following finite-state automaton.

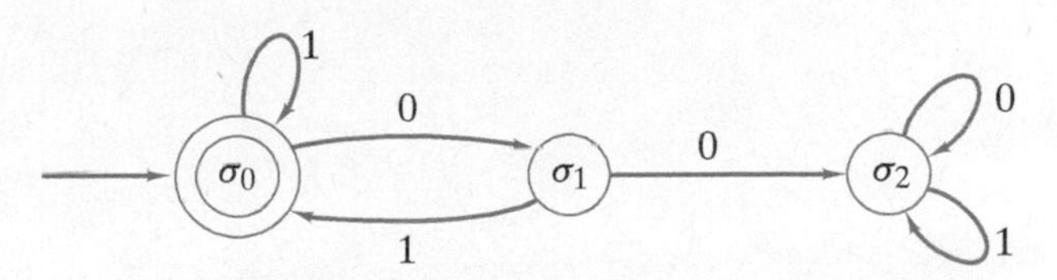


14. Write a grammar that generates all nonnull strings over $\{0, 1\}$ having an equal number of 0's and 1's.

15. Find the sets $\mathcal{I}$, $\mathcal{S}$, and $\mathcal{A}$, the initial state, and the table defining the next-state function for the nondeterministic finite-state automaton shown below.

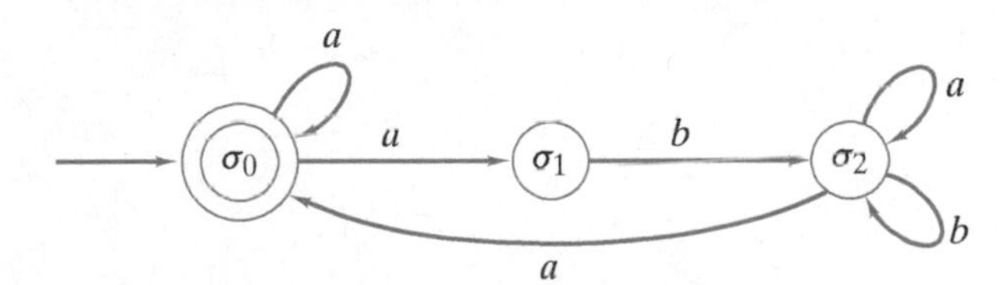


16. Is the string *aabaaba* accepted by the nondeterministic finite-state automaton of Exercise 15? Explain.

17. Design a nondeterministic finite-state automaton that accepts all strings over $\{0, 1\}$ that begin 01 and contain 110.

18. Find a finite-state automaton equivalent to the nondeterministic finite-state automaton of Exercise 15.

19. Explain how to construct a nondeterministic finite-state automaton that accepts the language

$$L_1L_2 = \{\alpha\beta \mid \alpha \in L_1, \beta \in L_2\},$$

given finite-state automata that accept regular languages $L_1$ and $L_2$.

20. Prove that any regular language that does not contain the null string is accepted by a nondeterministic finite-state automaton with exactly one accepting state. Give an example to show that this statement is false for arbitrary regular languages (i.e., if we allow the null string as a member of the regular language).

## Chapter 12 Computer Exercises

1. Write a program that simulates an arbitrary finite-state machine. The program should initially receive as input the next-state function, the output function, and the initial state. The program should then accept strings, simulate the action of the finite-state machine, and output the string produced by the finite-state machine.

2. Write a program that simulates an arbitrary finite-state automaton. The program should initially receive as input the next-state function, the set of accepting states, and the initial state. The program should then accept strings, simulate the action of the finite-state automaton, and print messages indicating whether the strings are accepted.

3. Write a program to draw a fractal given the context-free Interactive Lindenmayer grammar that generates it (see Section 12.3).

4. Report on the Knuth–Morris–Pratt algorithm (see [Johnsonbaugh]) that determines whether a string contains a particular substring. This algorithm makes use of a finite-state automaton.

# Appendix A

# MATRICES

It is a common practice to organize data into rows and columns. In mathematics, such an array of data is called a **matrix.** In this appendix we summarize some definitions and elementary properties of matrices. We begin with the definition of "matrix."

**Definition A.1** ▶ A *matrix*

$$A = \begin{pmatrix} a_{11} & a_{12} & \dots & a_{1n} \\ a_{21} & a_{22} & \dots & a_{2n} \\ \vdots & \vdots & & \vdots \\ a_{m1} & a_{m2} & \dots & a_{mn} \end{pmatrix} \tag{A.1}$$

is a rectangular array of data.

If $A$ has $m$ rows and $n$ columns, we say that the *size* of $A$ is $m$ by $n$ (written $m \times n$). ◀

We will often abbreviate equation (A.1) to $A = (a_{ij})$. In this equation, $a_{ij}$ denotes the element of $A$ appearing in the $i$th row and $j$th column.

**Example A.2** The matrix

$$A = \begin{pmatrix} 2 & 1 & 0 \\ -1 & 6 & 14 \end{pmatrix}$$

has two rows and three columns, so its size is $2 \times 3$. If we write $A = (a_{ij})$, we would have, for example,

$$a_{11} = 2, \qquad a_{21} = -1, \qquad a_{13} = 0.$$ ◀

**Definition A.3** ▶ Two matrices $A$ and $B$ are *equal,* written $A = B$, if they are the same size and their corresponding entries are equal. ◀

**Example A.4** Determine $w$, $x$, $y$, and $z$ so that

$$\begin{pmatrix} x+y & y \\ w+z & w-z \end{pmatrix} = \begin{pmatrix} 5 & 2 \\ 4 & 6 \end{pmatrix}.$$

**SOLUTION** According to Definition A.3, since the matrices are the same size, they will be equal provided that

$$x + y = 5 \qquad y = 2$$
$$w + z = 4 \qquad w - z = 6.$$

Solving these equations, we obtain

$$w = 5, \qquad x = 3, \qquad y = 2, \qquad z = -1.$$ ◀

We describe next some operations that can be performed on matrices. The **sum** of two matrices is obtained by adding the corresponding entries. The **scalar product** is obtained by multiplying each entry in the matrix by a fixed number.

**Definition A.5** ▶ Let $A = (a_{ij})$ and $B = (b_{ij})$ be two $m \times n$ matrices. The *sum* of $A$ and $B$ is defined as

$$A + B = (a_{ij} + b_{ij}).$$

The *scalar product* of a number $c$ and a matrix $A = (a_{ij})$ is defined as

$$cA = (ca_{ij}).$$

If $A$ and $B$ are matrices, we define $-A = (-1)A$ and $A - B = A + (-B)$. ◀

**Example A.6** If

$$A = \begin{pmatrix} 4 & 2 \\ -1 & 0 \\ 6 & -2 \end{pmatrix}, \qquad B = \begin{pmatrix} 1 & -3 \\ 4 & 4 \\ -1 & -3 \end{pmatrix},$$

then

$$A + B = \begin{pmatrix} 5 & -1 \\ 3 & 4 \\ 5 & -5 \end{pmatrix}, \qquad 2A = \begin{pmatrix} 8 & 4 \\ -2 & 0 \\ 12 & -4 \end{pmatrix}, \qquad -B = \begin{pmatrix} -1 & 3 \\ -4 & -4 \\ 1 & 3 \end{pmatrix}.$$

◀

Multiplication of matrices is another important matrix operation.

**Definition A.7** ▶ Let $A = (a_{ij})$ be an $m \times n$ matrix and let $B = (b_{jk})$ be an $n \times l$ matrix. The *matrix product* of $A$ and $B$ is defined as the $m \times l$ *matrix*

$$AB = (c_{ik}),$$

where

$$c_{ik} = \sum_{j=1}^{n} a_{ij} b_{jk}.$$ ◀

To multiply the matrix $A$ by the matrix $B$, Definition A.7 requires that the number of columns of $A$ be equal to the number of rows of $B$.

**Example A.8** Let

$$A = \begin{pmatrix} 1 & 6 \\ 4 & 2 \\ 3 & 1 \end{pmatrix}, \qquad B = \begin{pmatrix} 1 & 2 & -1 \\ 4 & 7 & 0 \end{pmatrix}.$$

The matrix product $AB$ is defined since the number of columns of $A$ is the same as the number of rows of $B$; both are equal to 2. Entry $c_{ik}$ in the product $AB$ is obtained by using the $i$th row of $A$ and the $k$th column of $B$. For example, the entry $c_{31}$ will be computed using the third row

$$(3 \quad 1)$$

of $A$ and the first column

$$\begin{pmatrix} 1 \\ 4 \end{pmatrix}$$

of $B$. We then multiply, consecutively, each element in the third row of $A$ by each element in the first column of $B$ and then sum to obtain

$$3 \cdot 1 + 1 \cdot 4 = 7.$$

Since the number of columns of $A$ is the same as the number of rows of $B$, the elements pair up correctly. Proceeding in this way, we obtain the product

$$AB = \begin{pmatrix} 25 & 44 & -1 \\ 12 & 22 & -4 \\ 7 & 13 & -3 \end{pmatrix}.$$

◀

**Example A.9** The matrix product

$$\begin{pmatrix} a & b \\ c & d \end{pmatrix} \begin{pmatrix} x \\ y \end{pmatrix} \quad \text{is} \quad \begin{pmatrix} ax + by \\ cx + dy \end{pmatrix}.$$

◀

**Definition A.10 ▶** Let $A$ be an $n \times n$ matrix. If $m$ is a positive integer, the $m$th *power* of $A$ is defined as the matrix product

$$A^m = \underbrace{A \cdots A}_{m\ A\text{'s}}.$$

◀

**Example A.11** If

$$A = \begin{pmatrix} 1 & -3 \\ -2 & 4 \end{pmatrix},$$

then

$$A^2 = AA = \begin{pmatrix} 1 & -3 \\ -2 & 4 \end{pmatrix} \begin{pmatrix} 1 & -3 \\ -2 & 4 \end{pmatrix} = \begin{pmatrix} 7 & -15 \\ -10 & 22 \end{pmatrix}$$

$$A^4 = AAAA = A^2A^2 = \begin{pmatrix} 7 & -15 \\ -10 & 22 \end{pmatrix} \begin{pmatrix} 7 & -15 \\ -10 & 22 \end{pmatrix} = \begin{pmatrix} 199 & -435 \\ -290 & 634 \end{pmatrix}.$$

◀

## Exercises

1. Compute the sum

$$\begin{pmatrix} 2 & 4 & 1 \\ 6 & 9 & 3 \\ 1 & -1 & 6 \end{pmatrix} + \begin{pmatrix} a & b & c \\ d & e & f \\ g & h & i \end{pmatrix}.$$

*In Exercises 2–8, let*

$$A = \begin{pmatrix} 1 & 6 & 9 \\ 0 & 4 & -2 \end{pmatrix}, \quad B = \begin{pmatrix} 4 & 1 & -2 \\ -7 & 6 & 1 \end{pmatrix}$$

*and compute each expression.*

2. $A + B$
3. $B + A$
4. $-A$
5. $3A$
6. $-2B$
7. $2B + A$
8. $B - 6A$

*In Exercises 9–13, compute the products.*

9. $$\begin{pmatrix} 1 & 2 & 3 \\ -1 & 2 & 3 \\ 0 & 1 & 4 \end{pmatrix} \begin{pmatrix} 2 & 8 \\ -1 & 1 \\ 6 & 0 \end{pmatrix}$$

**10.** $\begin{pmatrix} 1 & 6 \\ -8 & 2 \\ 4 & 1 \end{pmatrix} \begin{pmatrix} 4 & 1 \\ 7 & -6 \end{pmatrix}$

**11.** $A^2$, where $A = \begin{pmatrix} 1 & -2 \\ 6 & 2 \end{pmatrix}$

**12.** $\begin{pmatrix} 2 & -4 & 6 & 1 & 3 \end{pmatrix} \begin{pmatrix} 1 \\ 3 \\ -2 \\ 6 \\ 4 \end{pmatrix}$

**13.** $\begin{pmatrix} 2 & 4 & 1 \\ 6 & 9 & 3 \\ 1 & -1 & 6 \end{pmatrix} \begin{pmatrix} a & b \\ c & d \\ e & f \end{pmatrix}$

**14.**

(a) Give the size of each matrix.

$$A = \begin{pmatrix} 1 & 4 & 6 \\ 0 & 1 & 7 \end{pmatrix}, \quad B = \begin{pmatrix} 1 & 4 & 7 \\ 8 & 2 & 1 \\ 0 & 1 & 6 \end{pmatrix},$$

$$C = \begin{pmatrix} 4 & 2 \\ 0 & 0 \\ 2 & 9 \end{pmatrix}$$

(b) Using the matrices of part (a), decide which of the products

$$A^2, \quad AB, \quad BA, \quad AC, \quad CA, \quad AB^2,$$
$$BC, \quad CB, \quad C^2$$

are defined and then compute these products.

**15.** Determine $x$, $y$, and $z$ so that the equation

$$\begin{pmatrix} x+y & 3x+y \\ x+z & x+y-2z \end{pmatrix} = \begin{pmatrix} -1 & 1 \\ 9 & -17 \end{pmatrix}$$

holds.

**16.** Determine $w$, $x$, $y$, and $z$ so that the equation

$$\begin{pmatrix} 2 & 1 & -1 & 7 \\ 6 & 8 & 0 & 3 \end{pmatrix} \begin{pmatrix} x & 2x \\ y & -y+z \\ x+w & w-2y+x \\ z & z \end{pmatrix} = -\begin{pmatrix} 45 & 46 \\ 3 & 87 \end{pmatrix}$$

holds.

**17.** Define the $n \times n$ matrix $I_n = (a_{ij})$ by

$$a_{ij} = \begin{cases} 1 & \text{if } i = j \\ 0 & \text{if } i \neq j. \end{cases}$$

The matrix $I_n$ is called the $n \times n$ **identity matrix.**

Show that if $A$ is an $n \times n$ matrix (such a matrix is called a **square matrix**), then

$$AI_n = A = I_nA.$$

*An $n \times n$ matrix $A$ is said to be invertible if there exists an $n \times n$ matrix $B$ satisfying*

$$AB = I_n = BA.$$

*(The matrix $I_n$ is defined in Exercise 17.)*

**18.** Show that the matrix

$$\begin{pmatrix} 2 & 1 \\ 1 & 1 \end{pmatrix}$$

is invertible.

**★19.** Show that the matrix

$$\begin{pmatrix} a & b \\ c & d \end{pmatrix}$$

is invertible if and only if $ad - bc \neq 0$.

**20.** Suppose that we want to solve the system

$$AX = C,$$

where

$$A = \begin{pmatrix} a_{11} & a_{12} \\ a_{21} & a_{22} \end{pmatrix}$$
$$X = \begin{pmatrix} x \\ y \end{pmatrix}$$
$$C = \begin{pmatrix} c_1 \\ c_2 \end{pmatrix}$$

for $x$ and $y$.

Show that if $A$ is invertible, the system has a solution.

**21.** The **transpose** of a matrix $A = (a_{ij})$ is the matrix $A^T = (a'_{ji})$, where $a'_{ji} = a_{ij}$. *Example*:

$$\begin{pmatrix} 1 & 3 \\ 4 & 6 \end{pmatrix}^T = \begin{pmatrix} 1 & 4 \\ 3 & 6 \end{pmatrix}.$$

If $A$ and $B$ are $m \times k$ and $k \times n$ matrices, respectively, show that

$$(AB)^T = B^T A^T.$$

# Appendix B

# ALGEBRA REVIEW

In this appendix, we review basic algebra: rules for combining and simplifying expressions; fractions; exponents; factoring; quadratic equations; inequalities; and logarithms. For a more extensive treatment of basic algebra, see [Bleau; Lial; Sullivan].

## Grouping

Terms with a common symbol can be combined:

$$ac + bc = (a + b)c, \qquad ac - bc = (a - b)c.$$

Technically, these equations are known as **distributive laws.**

**Example B.1** $2x + 3x = (2 + 3)x = 5x$ ◀

The distributive laws, rewritten as

$$a(b + c) = ab + ac, \qquad a(b - c) = ab - ac,$$

can be used to simplify expressions.

**Example B.2** $2(x + 1) = 2x + 2 \cdot 1 = 2x + 2$ ◀

**Example B.3** $2(x + 1) + 2(x - 1) = 2x + 2 + 2x - 2 = 4x$ ◀

## Fractions

Formulas useful for adding, subtracting, and multiplying fractions are given as Theorem B.4.

**Theorem B.4** **Combining Fractions**

(a) $\dfrac{a}{c} + \dfrac{b}{c} = \dfrac{a+b}{c}$

(b) $\dfrac{a}{c} - \dfrac{b}{c} = \dfrac{a-b}{c}$

(c) $\dfrac{a}{c} + \dfrac{b}{d} = \dfrac{ad+bc}{cd}$

(d) $\dfrac{a}{c} - \dfrac{b}{d} = \dfrac{ad - bc}{cd}$

(e) $\dfrac{a}{c} \cdot \dfrac{b}{d} = \dfrac{ab}{cd}$

**Example B.5** Using Theorem B.4(a), we obtain

$$\frac{x-1}{2} + \frac{x+1}{2} = \frac{(x-1)+(x+1)}{2} = \frac{2x}{2} = x.$$ ◀

**Example B.6** Using Theorem B.4(b), we obtain

$$\frac{x-1}{2} - \frac{x+1}{2} = \frac{(x-1)-(x+1)}{2} = \frac{-2}{2} = -1.$$ ◀

**Example B.7** Using Theorem B.4(c), we obtain

$$\frac{x-1}{2} + \frac{x+1}{3} = \frac{3(x-1)+2(x+1)}{2 \cdot 3} = \frac{5x-1}{6}.$$ ◀

**Example B.8** Using Theorem B.4(d), we obtain

$$\frac{x-1}{2} - \frac{x+1}{3} = \frac{3(x-1)-2(x+1)}{2 \cdot 3} = \frac{x-5}{6}.$$ ◀

**Example B.9** Using Theorem B.4(e), we obtain

$$\frac{2}{x} \cdot \frac{4}{y} = \frac{8}{xy}.$$ ◀

## Exponents

If $n$ is a positive integer and $a$ is a real number, we define $a^n$ as

$$a^n = \underbrace{a \cdot a \cdots a}_{n\ a\text{'s}}.$$

If $a$ is a nonzero real number, we define $a^0 = 1$. If $n$ is a negative integer and $a$ is a nonzero real number, we define $a^n$ as

$$a^n = \frac{1}{a^{-n}}.$$

**Example B.10** If $a$ is a real number, $a^4 = a \cdot a \cdot a \cdot a$. As a specific example, $2^4 = 2 \cdot 2 \cdot 2 \cdot 2 = 16$. If $a$ is a nonzero real number,

$$a^{-4} = \frac{1}{a^4}.$$

As a specific example,

$$2^{-4} = \frac{1}{2^4} = \frac{1}{16}.$$ ◀

If $a$ is a positive real number and $n$ is a positive integer, we define $a^{1/n}$ to be the positive number $b$ satisfying $b^n = a$. We call $b$ the *nth root* of $a$.

**Example B.11** $3^{1/4}$ to nine significant digits is 1.316074013 because $(1.316074013)^4$ is approximately 3. ◀

If $a$ is a positive real number, $m$ is an integer, and $n$ is a positive integer, we define

$$a^{m/n} = (a^{1/n})^m.$$

The preceding equation defines $a^q$ for all positive real numbers $a$ and *rational numbers* $q$. (Recall that a rational number is a number that is the quotient of integers.)

**Example B.12** Since $3^{1/4}$ to nine significant digits is 1.316074013,

$$3^{9/4} = (1.316074013)^9 = 11.84466612.$$

The decimal values are approximations. ◀

If $a$ is a positive real number, the definition of $a^x$ can be extended to include all real numbers $x$ (rational or irrational). The following theorem lists five important laws of exponents.

**Theorem B.13** **Laws of Exponents**

Let $a$ and $b$ be positive real numbers, and let $x$ and $y$ be real numbers. Then

(a) $a^{x+y} = a^x a^y$

(b) $(a^x)^y = a^{xy}$

(c) $\dfrac{a^x}{a^y} = a^{x-y}$

(d) $a^x b^x = (ab)^x$

(e) $\dfrac{a^x}{b^x} = \left(\dfrac{a}{b}\right)^x$.

**Example B.14** Let $a = 3$, $x = 2$, and $y = 4$. Then $a^x = 9$, $a^y = 81$, and $a^{x+y} = 3^{2+4} = 729$. Now

$$a^{x+y} = 729 = 9 \cdot 81 = a^x a^y,$$

which illustrates Theorem B.13(a). ◀

**Example B.15** Let $a = 3$, $x = 2$, and $y = 4$. Then $a^x = 9$ and $a^{xy} = 3^8 = 6561$. Now

$$(a^x)^y = 9^4 = 6561 = a^{xy},$$

which illustrates Theorem B.13(b). ◀

**Example B.16** Let $a = 3$, $x = 2$, and $y = 4$. Then $a^x = 9$, $a^y = 81$, and $a^{x-y} = 3^{-2} = 1/9$. Now

$$\frac{a^x}{a^y} = \frac{9}{81} = \frac{1}{9} = a^{x-y},$$

which illustrates Theorem B.13(c). ◀

**Example B.17** Let $a = 3$, $b = 4$, and $x = 2$. Then $a^x = 9$, $b^x = 16$, and $(ab)^x = 12^2 = 144$. Now

$$a^x b^x = 9 \cdot 16 = 144 = (ab)^x,$$

which illustrates Theorem B.13(d). ◀

**Example B.18** Let $a = 3$, $b = 4$, and $x = 2$. Then $a^x = 9$, $b^x = 16$, and

$$\left(\frac{a}{b}\right)^x = \left(\frac{3}{4}\right)^2 = \frac{9}{16}.$$

Now

$$\frac{a^x}{b^x} = \frac{9}{16} = \left(\frac{a}{b}\right)^x,$$

which illustrates Theorem B.13(e). ◀

**Example B.19** $2^x 2^x = 2^{x+x} = 2^{2x} = (2^2)^x = 4^x$ ◀

## Factoring

We may use the equation

$$(x + b)(x + d) = x^2 + (b + d)x + bd$$

to factor an expression of the form $x^2 + c_1 x + c_2$.

**Example B.20** Factor $x^2 + 3x + 2$.

**SOLUTION** We look for integer constants in the factorization. According to the previous equation, $x^2 + 3x + 2$ factors as $(x + b)(x + d)$, where $b + d = 3$ and $bd = 2$. If $bd = 2$ and $b$ and $d$ are integers, the only choices for $b$ and $d$ are 1, 2 and $-1, -2$. We find that $b = 1$ and $d = 2$ satisfy both $b + d = 3$ and $bd = 2$. Thus

$$x^2 + 3x + 2 = (x + 1)(x + 2).$$ ◀

Special cases of

$$(x + b)(x + d) = x^2 + (b + d)x + bd$$

are

$$\begin{aligned}(x + b)^2 &= x^2 + 2bx + b^2\\ (x - b)^2 &= x^2 - 2bx + b^2\\ (x + b)(x - b) &= x^2 - b^2.\end{aligned}$$

**Example B.21** Using the equation $(x + b)^2 = x^2 + 2bx + b^2$, we have

$$(x + 9)^2 = x^2 + 18x + 81.$$ ◀

**Example B.22** Factor $x^2 - 36$.

**SOLUTION** Since $36 = 6^2$, we have

$$x^2 - 36 = (x + 6)(x - 6).$$ ◀

We may use the equation

$$(ax + b)(cx + d) = (ac)x^2 + (ad + bc)x + bd$$

to factor an expression of the form $c_0x^2 + c_1x + c_2$.

**Example B.23** Factor $6x^2 - x - 2$.

**SOLUTION** We look for integer constants in the factorization. Using the preceding notation, we must have

$$ac = 6, \qquad ad + bc = -1, \qquad bd = -2.$$

Since $ac = 6$, the possibilities for $a$ and $c$ are

$$1, 6 \quad 2, 3 \quad -1, -6 \quad -2, -3.$$

Since $bd = -2$, the only possibilities for $b$ and $d$ are $1, -2$ and $-1, 2$. Since we must also have $ad + bc = -1$, we find that $a = 2$, $b = 1$, $c = 3$, and $d = -2$ provide a solution. Therefore, the factorization is

$$6x^2 - x - 2 = (2x + 1)(3x - 2).$$ ◀

**Example B.24** Show that

$$\left[\frac{n(n+1)}{2}\right]^2 + (n+1)^3 = \left[\frac{(n+1)(n+2)}{2}\right]^2.$$

**SOLUTION** We show how the left side of the equation can be rewritten as the right side of the equation. By Theorem B.13(d) and (e), we have

$$\left[\frac{n(n+1)}{2}\right]^2 + (n+1)^3 = \frac{n^2(n+1)^2}{4} + (n+1)^3.$$

Since $(n+1)^2$ is a common factor of the right side of this equation, we may write

$$\frac{n^2(n+1)^2}{4} + (n+1)^3 = (n+1)^2\left[\frac{n^2}{4} + (n+1)\right].$$

Since

$$\frac{n^2}{4} + (n+1) = \frac{n^2 + 4n + 4}{4} = \frac{(n+2)^2}{4},$$

it follows that

$$(n+1)^2\left[\frac{n^2}{4} + (n+1)\right] = (n+1)^2\left[\frac{(n+2)^2}{4}\right] = \left[\frac{(n+1)(n+2)}{2}\right]^2.$$ ◀

## Solving a Quadratic Equation

A **quadratic equation** is an equation of the form

$$ax^2 + bx + c = 0, \ a \neq 0.$$

A **solution** is a value for $x$ that satisfies the equation.

**Example B.25** The value $x = -3$ is a solution of the quadratic equation

$$2x^2 + 2x - 12 = 0$$

because

$$2(-3)^2 + 2(-3) - 12 = 2 \cdot 9 - 6 - 12 = 18 - 18 = 0.$$ ◀

If a quadratic equation can be easily factored, its solutions may be readily obtained.

**Example B.26** Solve the quadratic equation

$$3x^2 - 10x + 8 = 0.$$

**SOLUTION** We may factor $3x^2 - 10x + 8$ as

$$3x^2 - 10x + 8 = (x - 2)(3x - 4).$$

For this expression to be equal to zero, either $x-2$ or $3x-4$ must equal zero. If $x-2=0$, we must have $x = 2$. If $3x - 4 = 0$, we must have $x = 4/3$. Thus the solutions of the given quadratic equation are

$$x = 2 \quad \text{and} \quad x = \frac{4}{3}.$$ ◀

The solutions of a quadratic equation can *always* be obtained from the **quadratic formula.**

**Theorem B.27** **Quadratic Formula**

The solutions of

$$ax^2 + bx + c = 0, \; a \neq 0,$$

are

$$x = \frac{-b \pm \sqrt{b^2 - 4ac}}{2a}.$$

**Example B.28** The quadratic formula gives the solutions of $x^2 - x - 1 = 0$ as

$$x = \frac{-(-1) \pm \sqrt{(-1)^2 - 4 \cdot 1 \cdot (-1)}}{2 \cdot 1} = \frac{1 \pm \sqrt{1+4}}{2} = \frac{1 \pm \sqrt{5}}{2}.$$

Thus the solutions are

$$x = \frac{1 + \sqrt{5}}{2} \quad \text{and} \quad x = \frac{1 - \sqrt{5}}{2}.$$ ◀

## Inequalities

If $a$ is **less than** $b$, we write $a < b$. If $a$ is **less than or equal** to $b$, we write $a \leq b$. If $a$ is **greater than** $b$, we write $a > b$. If $a$ is **greater than or equal** to $b$, we write $a \geq b$.

**Example B.29** Suppose that $a = 2$, $b = 8$, $c = 2$. We have

$$a < b,\ b > a,\ a \leq b,\ b \geq a,\ a \leq c,\ a \geq c.$$ ◀

Important laws of inequalities are given as Theorem B.30.

**Theorem B.30** **Laws of Inequalities**

(a) If $a < b$ and $c$ is any number whatsoever, then $a + c < b + c$.

(b) If $a \leq b$ and $c$ is any number whatsoever, then $a + c \leq b + c$.

(c) If $a > b$ and $c$ is any number whatsoever, then $a + c > b + c$.

(d) If $a \geq b$ and $c$ is any number whatsoever, then $a + c \geq b + c$.

(e) If $a < b$ and $c > 0$, then $ac < bc$.

(f) If $a \leq b$ and $c > 0$, then $ac \leq bc$.

(g) If $a < b$ and $c < 0$, then $ac > bc$.

(h) If $a \leq b$ and $c < 0$, then $ac \geq bc$.

(i) If $a > b$ and $c > 0$, then $ac > bc$.

(j) If $a \geq b$ and $c > 0$, then $ac \geq bc$.

(k) If $a > b$ and $c < 0$, then $ac < bc$.

(l) If $a \geq b$ and $c < 0$, then $ac \leq bc$.

(m) If $a < b$ and $b < c$, then $a < c$.

(n) If $a < b$ and $b \leq c$, then $a < c$.

(o) If $a \leq b$ and $b < c$, then $a < c$.

(p) If $a \leq b$ and $b \leq c$, then $a \leq c$.

(q) If $a > b$ and $b > c$, then $a > c$.

(r) If $a > b$ and $b \geq c$, then $a > c$.

(s) If $a \geq b$ and $b > c$, then $a > c$.

(t) If $a \geq b$ and $b \geq c$, then $a \geq c$.

**Example B.31** Solve the inequality $x - 5 < 6$.

**SOLUTION** By Theorem B.30(a), we may add 5 to both sides of the inequality to obtain the solution

$$x < 11.$$ ◀

**Example B.32** Solve the inequality

$$3x + 4 < x + 10.$$

**SOLUTION** By Theorem B.30(a), we may add $-x$ to both sides of the inequality to obtain $2x + 4 < 10$. Again, by Theorem B.30(a), we may add $-4$ to both sides of the inequality to obtain $2x < 6$. Finally, we may use Theorem B.30(e) to multiply both sides of the inequality by $1/2$ and obtain the solution $x < 3$. ◀

**Example B.33** Show that if $n > 2m$ and $m > 2p$, then $n > 4p$.

SOLUTION We may use Theorem B.30(i) to multiply both sides of $m > 2p$ by 2 to obtain $2m > 4p$. Since $n > 2m$, we may use Theorem B.30(q) to obtain $n > 4p$. ◀

**Example B.34** Show that

$$\frac{n+2}{n+1} < \frac{4(n+1)^2}{(2n+1)^2}$$

for every positive integer $n$.

SOLUTION Since $(n+1)(2n+1)^2$ is positive, by Theorem B.30(e),

$$(n+1)(2n+1)^2 \cdot \frac{n+2}{n+1} < (n+1)(2n+1)^2 \cdot \frac{4(n+1)^2}{(2n+1)^2},$$

which can be rewritten as

$$(2n+1)^2(n+2) < (n+1)4(n+1)^2.$$

Expanding each side of the inequality, we obtain

$$4n^3 + 12n^2 + 9n + 2 < 4n^3 + 12n^2 + 12n + 4.$$

By Theorem B.30(a), we may add $-4n^3 - 12n^2 - 9n - 2$ to both sides of the inequality to obtain $0 < 3n + 2$. This last inequality is true for all positive integers $n$ because the right side is always at least 5. Since the steps are reversible (i.e., beginning with $0 < 3n + 2$ we can obtain the original inequality using Theorem B.30), we have proved the given inequality. ◀

## Logarithms

Throughout this subsection, $b$ is a positive real number not equal to 1. If $x$ is a positive real number, the **logarithm to the base $b$ of $x$** is the exponent to which $b$ must be raised to obtain $x$. We denote the logarithm to the base $b$ of $x$ as $\log_b x$. Thus if we let $y = \log_b x$, the definition states that $b^y = x$.

**Example B.35** We have $\log_2 8 = 3$ because $2^3 = 8$. ◀

**Example B.36** Given $2^{2^x} = n$, where $n$ is a positive integer, solve for $x$.

SOLUTION Let lg denote the logarithm to the base 2. Then from the definition of logarithm, $2^x = \lg n$. Again, from the definition of logarithm, $x = \lg(\lg n)$. ◀

The following theorem lists important laws of logarithms.

**Theorem B.37** **Laws of Logarithms**

Suppose that $b > 0$ and $b \neq 1$. Then

(a) $b^{\log_b x} = x$

(b) $\log_b(xy) = \log_b x + \log_b y$

(c) $\log_b\left(\dfrac{x}{y}\right) = \log_b x - \log_b y$

(d) $\log_b(x^y) = y\log_b x$

(e) If $a > 0$ and $a \neq 1$, we have $\log_a x = \dfrac{\log_b x}{\log_b a}$.

(f) If $b > 1$ and $x > y > 0$, then $\log_b x > \log_b y$.

Theorem B.37(e) is known as the **change-of-base formula for logarithms.** If we know how to compute logarithms to the base $b$, we can perform the computation on the right side of the equation to obtain the logarithm to the base $a$. Theorem B.37(f) says that if $b > 1$, $\log_b(x)$ is an increasing function.

**Example B.38** Let $b = 2$ and $x = 8$. Then $\log_b x = 3$. Now

$$b^{\log_b x} = 2^3 = 8 = x,$$

which illustrates Theorem B.37(a). ◀

**Example B.39** Let $b = 2$, $x = 8$, and $y = 16$. Then $\log_b x = 3$, $\log_b y = 4$, and $\log_b(xy) = \log_2 128 = 7$. Now

$$\log_b(xy) = 7 = 3 + 4 = \log_b x + \log_b y,$$

which illustrates Theorem B.37(b). ◀

**Example B.40** Let $b = 2$, $x = 8$, and $y = 16$. Then $\log_b x = 3$, $\log_b y = 4$, and

$$\log_b\left(\frac{x}{y}\right) = \log_2\frac{1}{2} = -1.$$

Now

$$\log_b\left(\frac{x}{y}\right) = -1 = \log_b x - \log_b y,$$

which illustrates Theorem B.37(c). ◀

**Example B.41** Let $b = 2$, $x = 4$, and $y = 3$. Then $\log_b x = 2$ and

$$\log_b(x^y) = \log_2 64 = 6.$$

Now

$$\log_b(x^y) = 6 = 3 \cdot 2 = y\log_b x,$$

which illustrates Theorem B.37(d). ◀

**Example B.42** Suppose that we have a calculator with a logarithm key that computes logarithms to the base 10 but does not have a key that computes logarithms to the base 2. Use Theorem B.37(e) to compute $\log_2 40$.

**SOLUTION** Using our calculator, we compute

$$\log_{10}40 = 1.602060, \quad \log_{10}2 = 0.301030.$$

Theorem B.37(e) now gives

$$\log_2 40 = \frac{\log_{10}40}{\log_{10}2} = \frac{1.602060}{0.301030} = 5.321928.$$

◀

**Example B.43** Show that if $k$ and $n$ are positive integers satisfying

$$2^{k-1} < n < 2^k,$$

then

$$k - 1 < \lg n < k,$$

where lg denotes the logarithm to the base 2.

**SOLUTION** By Theorem B.37(f), the logarithm function is increasing. Therefore,

$$\lg(2^{k-1}) < \lg n < \lg(2^k).$$

By Theorem B.37(d),

$$\lg(2^{k-1}) = (k-1)\lg 2.$$

Since $\lg 2 = \log_2 2 = 1$, we have

$$\lg(2^{k-1}) = (k-1)\lg 2 = k - 1.$$

Similarly, $\lg(2^k) = k$. The given inequality now follows. ◀

## Exercises

*In Exercises 1–3, simplify the given expression by combining like terms.*

**1.** $8x - 12x$

**2.** $8x + 3a - 4y - 9a$

**3.** $6(a+b) - 8(a-b)$

*In Exercises 4–6, combine the given fractions.*

**4.** $\frac{8x-4b}{3} + \frac{7x+b}{3}$ **5.** $\frac{8x-4b}{2} - \frac{7x+b}{4}$

**6.** $\frac{8x-4b}{3} \cdot \frac{7x+b}{3}$

**7.** Show that

$$\frac{1}{n} - \frac{1}{n+1} = \frac{1}{n(n+1)}.$$

Use this fact to show that

$$\sum_{i=1}^{n} \frac{1}{i(i+1)} = \frac{n}{n+1}.$$

*Find the value of each expression in Exercises 8–13 without using a calculator.*

**8.** $3^4$ **9.** $3^{-4}$ **10.** $(-3)^4$

**11.** $(-3)^{-4}$ **12.** $1^{10}$ **13.** $1000^0$

**14.** Which expressions are equal?

(a) $3^4 3^{10}$ (b) $(3^4)^{10}$ (c) $3^{14}$

(d) $4^3 10^3$ (e) $2^3 20^3$ (f) $3^{40}$

(g) $2187^2$

**15.** Show that $5^n + 4 \cdot 5^n = 5^{n+1}$ for every positive integer $n$.

*In Exercises 16–24, expand the given expression.*

**16.** $(x+3)(x+5)$ **17.** $(x-3)(x+4)$

**18.** $(2x+3)(3x-4)$ **19.** $(x+4)^2$

**20.** $(x-4)^2$ **21.** $(3x+4)^2$

**22.** $(x-2)(x+2)$ **23.** $(x+a)(x-a)$

**24.** $(2x-3)(2x+3)$

*In Exercises 25–36, factor the given expression.*

**25.** $x^2 + 6x + 5$ **26.** $x^2 - 3x - 10$

**27.** $x^2 + 6x + 9$ **28.** $x^2 - 8x + 16$

**29.** $x^2 - 81$ **30.** $x^2 - 4b^2$

**31.** $2x^2 + 11x + 5$ **32.** $6x^2 + x - 15$

**33.** $4x^2 - 12x + 9$

**34.** $4x^2 - 9$

**35.** $9a^2 - 4b^2$

**36.** $12x^2 - 50x + 50$

**37.** Show that

$$(n+1)! + (n+1)(n+1)! = (n+2)!$$

for every positive integer $n$.

**38.** Show that

$$\frac{n(n+1)(2n+1)}{6} + (n+1)^2 = \frac{(n+1)(n+2)(2n+3)}{6}$$

for every positive integer $n$.

**39.** Show that

$$\frac{n}{2n+1} + \frac{1}{(2n+1)(2n+3)} = \frac{n+1}{2n+3}$$

for every positive integer $n$.

**40.** Show that

$$7(3 \cdot 2^{n-1} - 4 \cdot 5^{n-1}) - 10(3 \cdot 2^{n-2} - 4 \cdot 5^{n-2}) = 3 \cdot 2^n - 4 \cdot 5^n$$

for every positive integer $n$.

**41.** Simplify $2r(n-1)r^{n-1} - r^2(n-2)r^{n-2}$.

*In Exercises 42–44, solve the quadratic equation.*

**42.** $x^2 - 6x + 8 = 0$

**43.** $6x^2 - 7x + 2 = 0$

**44.** $2x^2 - 4x + 1 = 0$

*In Exercises 45–47, solve the given inequality.*

**45.** $2x + 3 \le 9$

**46.** $2x - 8 > 3x + 1$

**47.** $\dfrac{x-3}{6} < \dfrac{4x+3}{2}$

**48.** Show that $\sum_{i=1}^{n} i \le n^2$.

**49.** Show that

$$(1+ax)(1+x) \ge 1 + (a+1)x$$

for any $x$ and $a \ge 0$.

**50.** Show that

$$\left(\frac{3}{2}\right)^{n-2}\left(\frac{5}{2}\right) > \left(\frac{3}{2}\right)^{n}$$

for every integer $n \ge 2$.

**51.** Show that

$$\frac{2n+1}{(n+2)n^2} > \frac{2}{(n+1)^2}$$

for every positive integer $n$.

**52.** Show that $6n^2 < 6n^2 + 4n + 1$ for every positive integer $n$.

**53.** Show that $6n^2 + 4n + 1 \le 11n^2$ for every positive integer $n$.

*Find the value of each expression in Exercises 54–58 without using a calculator (lg means $\log_2$).*

**54.** $\lg 64$

**55.** $\lg \frac{1}{128}$

**56.** $\lg 2$

**57.** $2^{\lg 10}$

**58.** $\lg 2^{1000}$

*Given that* $\lg 3 = 1.584962501$ *and* $\lg 5 = 2.321928095$, *find the value of each expression in Exercises 59–63 (lg means* $\log_2$*).*

**59.** $\lg 6$

**60.** $\lg 30$

**61.** $\lg 59049$

**62.** $\lg 0.6$

**63.** $\lg 0.0375$

*Use a calculator with a logarithm key to find the value of each expression in Exercises 64–67.*

**64.** $\log_5 47$

**65.** $\log_7 0.30881$

**66.** $\log_9 8.888^{100}$

**67.** $\log_{10}(\log_{10} 1054)$

*In Exercises 68–70, use a calculator with a logarithm key to solve for x.*

**68.** $5^x = 11$

**69.** $5^{2x}6^x = 811$

**70.** $5^{11^x} = 10^{100}$

**71.** Show that $x^{\log_b y} = y^{\log_b x}$.

# Appendix C PSEUDOCODE

In this appendix, we describe the pseudocode used in this book.

We let = denote the **assignment operator.** In pseudocode, $x = y$ means "copy the value of $y$ into $x$" or, equivalently, "replace the current value of $x$ by the value of $y$." When $x = y$ is executed, the value of $y$ is unchanged.

**Example C.1** Suppose that the value of $x$ is 5 and the value of $y$ is 10. After

$$x = y$$

the value of $x$ is 10 and the value of $y$, which is unchanged, is also 10. ◀

In the **if statement**

if (*condition*)
    *action*

if *condition* is true, *action* is executed and control passes to the statement following *action*. If *condition* is false, *action* is not executed and control passes immediately to the statement following *action*. As shown, we use indentation to identify the statements that make up *action*.

**Example C.2** Suppose that the value of $x$ is 5, the value of $y$ is 10, and the value of $z$ is 15. Consider the segment

```
if (y > x)
    z = x
y = z
```

Since $y > x$ is true,

$$z = x$$

executes, and the value of $z$ is set to 5. Next

$$y = z$$

executes and the value of $y$ is set to 5. Now each of the variables $x$, $y$, and $z$ has value 5. ◀

We show reserved words (e.g., if) in the regular typeface and user-chosen words (e.g., variables such as $x$) in italics.

We use the usual arithmetic operators $+$, $-$, $*$ (for multiplication), and / as well as the relational operators $==$ (equals), $\neg=$ (not equal), $<$, $>$, $\leq$, and $\geq$ and the logical operators $\wedge$ (and), $\vee$ (or), and $\neg$ (not). We will use $==$ to denote the equality operator and $=$ to denote the assignment operator. We will sometimes use less formal statements when to do otherwise would obscure the meaning. (*Example:* Choose an element $x$ in $S$.)

**Example C.3** Suppose that the value of $x$ is 5, the value of $y$ is 10, and the value of $z$ is 15. Consider the segment

```
if (y == x)
    z = x
y = z
```

Since $y\ == x$ is false,

$$z = x$$

does *not* execute. Next

$$y = z$$

executes, and the value of $y$ is set to 15. Now the value of $x$ is 5, and each of the variables $y$ and $z$ has value 15. ◀

In an if statement, if *action* consists of multiple statements, we enclose them in braces. An example of a multiple-statement *action* in an if statement is

```
if (x ≥ 0) {
    x = x + 1
    a = b + c
}
```

An alternative form of the if statement is the **if else statement.** In the if else statement

```
if (condition)
    action1
else
    action2
```

if *condition* is true, *action1* (but not *action2*) is executed and control passes to the statement following *action2*. If *condition* is false, *action2* (but not *action1*) is executed and control passes to the statement following *action2*. If *action1* or *action2* consists of multiple statements, they are enclosed in braces.

**Example C.4** Suppose that the value of $x$ is 5, the value of $y$ is 10, and the value of $z$ is 15. Consider the segment

```
if (y ¬= x)
    y = x
else
    z = x
a = z
```

Since $y \neg= x$ is true,

$$y = x$$

executes and $y$ is set to 5. The statement

$$z = x$$

does *not* execute. Next

$$a = z$$

executes and $a$ is set to 15. Now the value of each of $x$ and $y$ is equal to 5, and the value of each of $a$ and $z$ is equal to 15. ◀

**Example C.5** Suppose that the value of $x$ is 5, the value of $y$ is 10, and the value of $z$ is 15. Consider the segment

```
if (y < x)
    y = x
else
    z = x
a = z
```

Since $y < x$ is false,

$$y = x$$

does *not* execute. Rather,

$$z = x$$

executes and $z$ is set to 5. Next

$$a = z$$

executes and $a$ is set to 5. Now the value of each of $x$, $z$, and $a$ is equal to 5, and the value of $y$ is equal to 10. ◀

Two slash marks // signal the beginning of a **comment,** which then extends to the end of the line. Comments help the reader understand the code but are not executed.

**Example C.6** In the segment

```
if (x ≥ 0) { // if x is nonnegative, update x and a
    x = x + 1
    a = b + c
}
```

the part

// if $x$ is nonnegative, update $x$ and $a$

is a comment. The segment executes as if it were written

```
if (x ≥ 0) {
    x = x + 1
    a = b + c
}
```

◀

The **while loop** is written

```
while (condition)
    action
```

If *condition* is true, *action* is executed and this sequence is repeated; that is, if *condition* is true, *action* is executed again. This sequence is repeated until *condition* becomes false. Then control passes immediately to the statement following *action*. If *action* consists of multiple statements, we enclose them in braces.

**Example C.7** Let $s_1, \ldots, s_n$ be a sequence. After the segment

```
large = s_1
i = 2
while (i ≤ n) {
    if (s_i > large)
        large = s_i
    i = i + 1
}
```

executes, *large* is equal to the largest element in the sequence. The idea is to step through the sequence and save the largest value seen so far in *large*. ◀

In Example C.7 we stepped through a sequence by using the variable $i$ that took on the integer values 2 through $n$. This kind of loop is so common that a special loop, called the **for loop,** is often used instead of the while loop. The form of the **for loop** is

```
for var = init to limit
    action
```

As in the previous if statement and while loop, if *action* consists of multiple statements, we enclose them in braces. When the for loop is executed, *action* is executed for values of *var* from *init* to *limit*. More precisely, *init* and *limit* are expressions that have integer values. The variable *var* is first set to the value *init*. If $var \le limit$, we execute *action* and then add 1 to *var*. The process is then repeated. Repetition continues until $var > limit$. Notice that if $init > limit$, *action* will not be executed at all.

**Example C.8** The segment of Example C.7 may be rewritten using a for loop as

```
large = s_1
for i = 2 to n
    if (s_i > large)
        large = s_i
```

◀

A **function** is a unit of code that can receive input, perform computations, and produce output. The **parameters** describe the data, variables, and so on that are input to and output from the function. The syntax is

```
function name(parameters separated by commas) {
    code for performing computations
}
```

**Example C.9** The following function, named *max1*, finds the largest of the numbers $a$, $b$, and $c$. The parameters $a$, $b$, and $c$ are *input* parameters (i.e., they are assigned values before the function executes), and the parameter $x$ is an *output* parameter (i.e., the function will assign $x$ a value—namely, the largest of the numbers $a$, $b$, and $c$).

```
max1(a, b, c, x) {
   x = a
   if (b > x) // if b is larger than x, update x
      x = b
   if (c > x) // if c is larger than x, update x
      x = c
}
```

The **return statement**

```
return x
```

terminates a function and returns the value of $x$ to the invoker of the function. The return statement

```
return
```

(without the $x$) simply terminates a function. If there is no return statement, the function terminates just before the closing brace.

**Example C.10** The function of Example C.9 can be rewritten using the return statement as

```
max2(a, b, c) {
   x = a
   if (b > x) // if b is larger than x, update x
      x = b
   if (c > x) // if c is larger than x, update x
      x = c
   return x
}
```

Rather than using a parameter for the largest of $a$, $b$, and $c$, *max2* returns the largest value. ◀

We use the functions *print* and *println* for output. The function *println* adds a newline after printing its argument(s) (which causes the next output to occur flush left on the next line); otherwise, the functions are the same. The operator + concatenates strings. Strings are delimited by double quotation marks (i.e., "and"). If exactly one of +'s operands is a string, the other argument is converted to a string, after which the concatenation occurs. The concatenation operator is useful in the print functions.

**Example C.11** The segment

```
for i = 1 to n
   println(s_i)
```

prints the values in the sequence $s_1, \ldots, s_n$, one per line. ◀

**Example C.12** The segment

for $i = 1$ to $n$
    *print*($s_i$ + " ")
*println*()

prints the values in the sequence $s_1, \ldots, s_n$, separated by space, on one line. The sequence values are followed by a newline. ◀

## Exercises

1. Show how the segment in Example C.7 finds the largest element in the sequence

$$s_1 = 2, \quad s_2 = 3, \quad s_3 = 8, \quad s_4 = 6.$$

2. Show how the segment in Example C.7 finds the largest element in the sequence

$$s_1 = 8, \quad s_2 = 8, \quad s_3 = 4, \quad s_4 = 1.$$

3. Show how the segment in Example C.7 finds the largest element in the sequence

$$s_1 = 1, \quad s_2 = 1, \quad s_3 = 1, \quad s_4 = 1.$$

4. Show how the function *max1* of Example C.9 finds the largest of the numbers $a = 4$, $b = -3$, and $c = 5$.
5. Show how the function *max1* of Example C.9 finds the largest of the numbers $a = b = 4$ and $c = 2$.
6. Show how the function *max1* of Example C.9 finds the largest of the numbers $a = b = c = 8$.
7. Write a function that returns the minimum of $a$ and $b$.
8. Write a function that returns the maximum of $a$ and $b$.
9. Write a function that swaps the values $a$ and $b$. (Here $a$ and $b$ will be both input and output parameters.)
10. Write a function that prints all odd numbers between 1 and $n$, inclusive.
11. Write a function that prints all negative numbers, one per line, that occur in the sequence of numbers $s_1, \ldots, s_n$.
12. Write a function that prints all indexes of values in the sequence $s_1, \ldots, s_n$ that are equal to the value *val*.
13. Write a function that returns the product of the sequence of numbers $s_1, s_2, \ldots, s_n$.
14. Write a function that prints every other value in the sequence $s_1, s_2, \ldots, s_n$ (i.e., $s_1, s_3, s_5, \ldots$) one value per line.

# REFERENCES


AGARWAL, M., N. SAXENA, and N. KAYAL, "PRIMES is in P," *Ann. Math.*, 160 (2004) 781–793.

AHO, A., J. HOPCROFT, and J. ULLMAN, *Data Structures and Algorithms*, Addison-Wesley, Reading, Mass., 1983.

AIGNER, M. and G. M. ZIEGLER, *Proofs from THE BOOK*, 3rd ed., Springer-Verlag, Berlin, 2004.

AINSLIE, T., *Ainslie's Complete Hoyle*, Simon and Schuster, 1975.

AKL, S. G., *The Design and Analysis of Parallel Algorithms*, Prentice Hall, Englewood Cliffs, N.J., 1989.

APPEL, K. and W. HAKEN, "Every planar map is four-colorable," *Illinois J. Math.*, 21 (1977), 429–567.

APPLEGATE, D. L., R. E. BIXBY, V. CHVÁTAL, and W. J. COOK, *The Traveling Salesman Problem: A Computational Study*, Princeton University Press, Princeton, N.J., 2006.

BAASE, S. and A. VAN GELDER, *Computer Algorithms: Introduction to Design and Analysis*, 3rd ed., Addison-Wesley, Reading, Mass., 2000.

BABAI, L. and T. KUCERA, "Canonical labelling of graphs in linear average time," *Proc. 20th Symposium on the Foundations of Computer Science*, 1979, 39–46.

BACHELIS, G. F., "A short proof of Hall's theorem on SDRs," *Amer. Math. Monthly*, 109 (2002), 473–474.

BAIN, V., "An algorithm for drawing the *n*-cube," *College Math. J.*, 29 (1998), 320–322.

BARKER, S. F., *The Elements of Logic*, 5th ed., McGraw-Hill, New York, 1989.

BELL, R. C., *Board and Table Games from Many Civilizations*, rev. ed., Dover, New York, 1979.

BENJAMIN, A. T. and J. J. QUINN, *Proofs that Really Count: The Art of Combinatorial Proof*, Mathematical Association of America, Washington, D.C., 2003.

BENTLEY, J., *Programming Pearls*, 2nd ed., Addison-Wesley, Reading, Mass., 2000.

BERGE, C., *Graphs and Hypergraphs*, North-Holland, Amsterdam, 1979.

BERLEKAMP, E. R., J. H. CONWAY, and R. K. GUY, *Winning Ways*, Vol. 1, 2nd ed., A. K. Peters, New York, 2001.

BERLEKAMP, E. R., J. H. CONWAY, and R. K. GUY, *Winning Ways*, Vol. 2, 2nd ed., A. K. Peters, New York, 2003.

BILLINGSLEY, P., *Probability and Measure*, 3rd ed., Wiley, New York, 1995.

BLEAU B. L., M. CLEMENS, and G. CLEMENS, *Forgotten Algebra*, 4th ed., Barron's, Hauppauge, N.Y., 2013.

BONDY, J. A. and U. S. R. MURTY, *Graph Theory with Applications*, American Elsevier, New York, 1976.

BOOLE, G., *The Laws of Thought*, reprinted by Dover, New York, 1951.

BRASSARD, G. and P. BRATLEY, *Fundamentals of Algorithms*, Prentice Hall, Upper Saddle River, N.J., 1996.

BRUALDI, R. A., *Introductory Combinatorics*, 5th ed., Prentice Hall, Upper Saddle River, N.J., 2004.

CARMONY, L., "Odd pie fights," *Math. Teacher*, 72 (1979), 61–64.

CARROLL, J. and D. LONG, *Theory of Finite Automata*, Prentice Hall, Englewood Cliffs, N.J., 1989.

CHARTRAND, G., L. LESNIAK, and P. ZHANG, *Graphs and Digraphs*, 5th ed., Chapman and Hall/CRC, Boca Raton, Fla., 2010.

CHRYSTAL, G., *Textbook of Algebra*, Vol. II, 7th ed., Chelsea, New York, 1964.

CHU, I. P. and R. JOHNSONBAUGH, "Tiling deficient boards with trominoes," *Math. Mag.*, 59 (1986), 34–40.

CODD, E. F., "A relational model of data for large shared databanks," *Comm. ACM*, 13 (1970), 377–387.

COHEN, D. I. A., *Introduction to Computer Theory*, 2nd ed., Wiley, New York, 1997.

COPI, I. M., C. COHEN and K. MCMAHON, *Introduction to Logic*, 14th ed., Pearson, Upper Saddle River, N.J., 2011.

COPPERSMITH, D. and S. WINOGRAD, "Matrix multiplication via arithmetic progressions," *J. Symbolic Comput.*, 9 (1990), 251–280.

CORMEN, T. H., C. E. LEISERSON, R. L. RIVEST, and C. STEIN, *Introduction to Algorithms*, 3rd ed., MIT Press, Cambridge, Mass., 2009.

CULL, P. and E. F. ECKLUND, JR., "Towers of Hanoi and analysis of algorithms," *Amer. Math. Monthly*, 92 (1985), 407–420.

D'ANGELO, J. P. and D. B. WEST, *Mathematical Thinking: Problem Solving and Proofs*, 2nd ed., Prentice Hall, Upper Saddle River, N.J., 2000.

DATE, C. J., *An Introduction to Database Systems*, 8th ed., Addison-Wesley, Reading, Mass., 2004.

DAVIS, M. D., R. SIGAL, and E. J. WEYUKER, *Computability, Complexity, and Languages*, 2nd ed., Academic Press, San Diego, 1994.

DE BERG, M., M. VAN KREVELD, M. OVERMARS, and O. SCHWARZKOPF, *Computational Geometry*, 2nd rev. ed., Springer, Berlin, 2008.

DEO, N., *Graph Theory and Applications to Engineering and Computer Science*, Prentice Hall, Englewood Cliffs, N.J., 1974.

DIJKSTRA, E. W., "A note on two problems in connexion with graphs," *Numer. Math.*, 1 (1959), 260–271.

DIJKSTRA, E. W., "Cooperating sequential processes," in *Programming Languages*, F. Genuys, ed., Academic Press, New York, 1968.

EDELSBRUNNER, H., *Algorithms in Combinatorial Geometry*, Springer-Verlag, New York, 1987.

EDGAR, W. J., *The Elements of Logic*, SRA, Chicago, 1989.

ENGLISH, E. and S. HAMILTON, "Network security under siege, the timing attack," *Computer* (March 1996), 95–97.

EVEN, S., *Algorithmic Combinatorics*, Macmillan, New York, 1973.

EVEN, S., *Graph Algorithms*, Computer Science Press, Rockville, Md., 1979.

EZEKIEL, M., "The cobweb theorem," *Quart. J. Econom.*, 52 (1938), 255–280.

FORD, L. R., JR., and D. R. FULKERSON, *Flows in Networks*, Princeton University Press, Princeton, N.J., 1962.

FOWLER, P. A., "The Königsberg bridges—250 years later," *Amer. Math. Monthly*, 95 (1988), 42–43.

FREY, P., "Machine-problem solving—Part 3: The alpha-beta procedure," *BYTE*, 5 (November 1980), 244–264.

FUKUNAGA, K., *Introduction to Statistical Pattern Recognition*, 2nd ed., Academic Press, New York, 1990.

GALLIER, J. H., *Logic for Computer Science*, Harper & Row, New York, 1986.

GARDNER, M., *Mathematical Puzzles and Diversions*, Simon and Schuster, 1959.

GARDNER, M., "A new kind of cipher that would take millions of years to break," *Sci. Amer.* (February 1977), 120–124.

GARDNER, M., *Mathematical Circus*, Mathematical Association of America, Washington, 1992.

GENESERETH, M. R. and N. J. NILSSON, *Logical Foundations of Artificial Intelligence*, Morgan Kaufmann, Los Altos, Calif., 1987.

GHAHRAMANI, S., *Fundamentals of Probability*, 3rd ed., Prentice Hall, Upper Saddle River, N.J., 2005.

GIBBONS, A., *Algorithmic Graph Theory*, Cambridge University Press, Cambridge, 1985.

GOLDBERG, S., *Introduction to Difference Equations*, Wiley, New York, 1958.

GOLOMB, S. W., "Checker boards and polyominoes," *Amer. Math. Monthly*, 61 (1954), 675–682.

GOLOMB, S. and L. BAUMERT, "Backtrack programming," *J. ACM*, 12 (1965), 516–524.

GOSE, E., R. JOHNSONBAUGH, and S. JOST, *Pattern Recognition and Image Analysis*, Prentice Hall, Upper Saddle River, N.J., 1996.

GRAHAM, R. L., "An efficient algorithm for determining the convex hull of a finite planar set," *Info. Proc. Lett.*, 1 (1972), 132–133.

GRAHAM, R. L., D. E. KNUTH, and O. PATASHNIK, *Concrete Mathematics: A Foundation for Computer Science*, 2nd ed., Addison-Wesley, Reading, Mass., 1994.

GRIES, D., *The Science of Programming*, Springer-Verlag, New York, 1981.

HAILPERIN, T., "Boole's algebra isn't Boolean algebra," *Math. Mag.*, 54 (1981), 137–184.

HALMOS, P. R., *Naive Set Theory*, Springer-Verlag, New York, 1974.

HARARY, F., *Graph Theory*, Addison-Wesley, Reading, Mass., 1969.

HARKLEROAD, L., *The Math Behind the Music*, Cambridge University Press, New York, and Mathematical Association of America, Washington, D.C., 2006.

HELL, P., "Absolute retracts in graphs," in *Graphs and Combinatorics*, R. A. Bari and F. Harary, eds., Lecture Notes in Mathematics, Vol. 406, Springer-Verlag, New York, 1974.

HILLIER, F. S. and G. J. LIEBERMAN, *Introduction to Operations Research*, 9th ed., McGraw-Hill, New York, 2017.

HINZ, A. M., "The Tower of Hanoi," *Enseignement Math.*, 35 (1989), 289–321.

HOHN, F., *Applied Boolean Algebra*, 2nd ed., Macmillan, New York, 1966.

HOLTON, D. A. and J. SHEEHAN, *The Petersen Graph*, Cambridge University Press, 1993.

HOPCROFT, J. E., R. MOTWANI, and J. D. ULLMAN, *Introduction to Automata Theory, Languages, and Computation*, 3rd ed., Addison-Wesley, Boston, 2007.

HU, T. C., *Combinatorial Algorithms*, Addison-Wesley, Reading, Mass., 1982.

JACOBS, H. R., *Geometry*, 3rd ed., W. H. Freeman, San Francisco, 2003.

JOHNSONBAUGH, R. and M. SCHAEFER, *Algorithms*, Prentice Hall, Upper Saddle River, N.J., 2004.

JONES, R. H. and N. C. STEELE, *Mathematics in Communication Theory*, Ellis Horwood, Chichester, England, 1989.

KELLEY, D., *Automata and Formal Languages*, Prentice Hall, Upper Saddle River, N.J., 1995.

KELLY, D. G., *Introduction to Probability*, Prentice Hall, Upper Saddle River, N.J., 1994.

KLEINROCK, L., *Queueing Systems*, Vol. 2: *Computer Applications*, Wiley, New York, 1976.

KLINE, M., *Mathematical Thought from Ancient to Modern Times*, Oxford University Press, New York, 1972.

KNUTH, D. E., "Algorithms," *Sci. Amer.* (April 1977), 63–80.

KNUTH, D. E., "Algorithmic thinking and mathematical thinking," *Amer. Math. Monthly*, 92 (1985), 170–181.

KNUTH, D. E., *The Art of Computer Programming*, Vol. 1: *Fundamental Algorithms*, 3rd ed., Addison-Wesley, Reading, Mass., 1997.

KNUTH, D. E., *The Art of Computer Programming*, Vol. 2: *Seminumeric Algorithms*, 3rd ed., Addison-Wesley, Reading, Mass., 1998a.

KNUTH, D. E., *The Art of Computer Programming*, Vol. 3: *Sorting and Searching*, 2nd ed., Addison-Wesley, Reading, Mass., 1998b.

KÖBLER, J., U. SCHÖNING, and J. TORÁN, *The Graph Isomorphism Problem: Its Structural Complexity*, Birkhäuser Verlag, Basel, Switzerland, 1993.

KOHAVI, Z., *Switching and Finite Automata Theory*, 2nd ed., McGraw-Hill, New York, 1978.

KÖNIG, D., *Theorie der endlichen und unendlichen Graphen*, Akademische Verlags-gesellschaft, Leipzig, 1936. (Reprinted in 1950 by Chelsea, New York.) (English translation: *Theory of Finite and Infinite Graphs*, Birkhäuser Boston, Cambridge, Mass., 1990.)

KRANTZ, S. G., *Techniques of Problem Solving*, American Mathematical Society, Providence, R.I., 1997.

KROENKE, D. M. and D. J. AUER, *Database Processing: Fundamentals, Design and Implementation*, 14th ed., Pearson, Upper Saddle River, N.J., 2016.

KRUSE, R. L. and A. RYBA, *Data Structures and Program Design in C++*, Prentice Hall, Upper Saddle River, N.J., 1999.

KUROSAKA, R. T., "A ternary state of affairs," *BYTE*, 12 (February 1987), 319–328.

LEIGHTON, F. T., *Introduction to Parallel Algorithms and Architectures*, Morgan Kaufmann, San Mateo, Calif., 1992.

LESTER, B. P., *The Art of Parallel Programming*, Prentice Hall, Upper Saddle River, N.J., 1993.

LEWIS, T. G. and H. EL-REWINI, *Introduction to Parallel Computing*, Prentice Hall, Upper Saddle River, N.J., 1992.

LIAL, M. L., E. J. HORNSBY, D. I. SCHNEIDER, and C. J. DANIELS, *College Algebra*, 12th ed., Pearson, Upper Saddle River, N.J., 2017.

LINDENMAYER, A., "Mathematical models for cellular interaction in development," Parts I and II, *J. Theoret. Biol.*, 18 (1968), 280–315.

LIPSCHUTZ, S., *Schaum's Outline of Theory and Problems of Set Theory and Related Topics*, 2nd ed., McGraw-Hill, New York, 1998.

LIU, C. L., *Introduction to Combinatorial Mathematics*, McGraw-Hill, New York, 1968.

LIU, C. L., *Elements of Discrete Mathematics*, 2nd ed., McGraw-Hill, New York, 1985.

MANBER, U., *Introduction to Algorithms*, Addison-Wesley, Reading, Mass., 1989.

MANDELBROT, B. B., *Fractals: Form, Chance, and Dimension*, W. H. Freeman, San Francisco, 1977.

MANDELBROT, B. B., *The Fractal Geometry of Nature*, W. H. Freeman, San Francisco, 1982.

MARTIN, G. E., *Polyominoes: A Guide to Puzzles and Problems in Tiling*, Mathematical Association of America, Washington, D.C., 1991.

MCCALLA, T. R., *Digital Logic and Computer Design*, Merrill, New York, 1992.

MCNAUGHTON, R., *Elementary Computability, Formal Languages, and Automata*, Prentice Hall, Englewood Cliffs, N.J., 1982.

MENDELSON, E., *Boolean Algebra and Switching Circuits*, Schaum, New York, 1970.

MILLER, R. and L. BOXER, *A Unified Approach to Sequential and Parallel Algorithms*, Prentice Hall, Upper Saddle River, N.J., 2000.

MITCHISON, G. J., "Phyllotaxis and the Fibonacci series," *Science*, 196 (1977), 270–275.

MOSER, L. K., "A mathematical analysis of the game of Jai Alai," *Amer. Math. Monthly*, 89 (1982), 292–300.

NADLER, M. and E. P. SMITH, *Pattern Recognition Engineering*, Wiley, New York, 1993.

NANDI, R. and A. GUEDES, "Graph isomorphism applied to fingerprint matching," http://euler.mat.ufrgs.br/~trevisan/workgraph/regina.pdf.

NAYLOR, M., "Golden, $\sqrt{2}$, and $\pi$ flowers: A spiral story," *Math. Mag.*, 75 (2002), 163–172.

NEWMAN, J. R., "Leonhard Euler and the Koenigsberg bridges," *Sci. Amer.* (July 1953), 66–70.

NIEVERGELT, J., J. C. FARRAR, and E. M. REINGOLD, *Computer Approaches to Mathematical Problems*, Prentice Hall, Englewood Cliffs, N.J., 1974.

NILSSON, N. J., *Problem-Solving Methods in Artificial Intelligence*, McGraw-Hill, New York, 1971.

NIVEN, I., *Mathematics of Choice*, Mathematical Association of America, Washington, D.C., 1965.

NIVEN, I., and H. S. ZUCKERMAN, *An Introduction to the Theory of Numbers*, 5th ed., Wiley, New York, 1991.

NYHOFF, L. R., *C++: An Introduction to Data Structures*, Prentice Hall, Upper Saddle River, N.J., 1999.

ORE, O., *Graphs and Their Uses*, Mathematical Association of America, Washington, D.C., 1963.

PEARL, J., "The solution for the branching factor of the alpha-beta pruning algorithm and its optimality," *Comm. ACM*, 25 (1982), 559–564.

PEITGEN, H. and D. SAUPE, eds., *The Science of Fractal Images*, Springer-Verlag, New York, 1988.

PFLEEGER, C. P. and S. L. PFLEEGER, *Security in Computing*, 5th ed., Prentice Hall, Upper Saddle River, N.J., 2016.

POŚA, L., "Hamiltonian circuits in random graphs," *Discrete Math.*, 14 (1976) 359–364.

PREPARATA, F. P. and S. J. HONG, "Convex hulls of finite sets of points in two and three dimensions," *Comm. ACM*, 20 (1977), 87–93.

PREPARATA, F. P. and M. I. SHAMOS, *Computational Geometry*, Springer-Verlag, New York, 1985.

Problem 1186, *Math. Mag.*, 58 (1985), 112–114.

PRODINGER, H. and R. TICHY, "Fibonacci numbers of graphs," *Fibonacci Quarterly*, 20 (1982), 16–21.

PRUSINKIEWICZ, P., "Graphical applications of L-systems," *Proc. of Graphics Interface 1986—Vision Interface* (1986), 247–253.

PRUSINKIEWICZ, P. and J. HANAN, "Applications of L-systems to computer imagery," in *Graph Grammars and Their Application to Computer Science; Third International Workshop*, H. Ehrig, M. Nagl, A. Rosenfeld, and G. Rozenberg, eds., Springer-Verlag, New York, 1988.

QUINN, M. J., *Designing Efficient Algorithms for Parallel Computers*, McGraw-Hill, New York, 1987.

READ, R. C. and D. G. CORNEIL, "The graph isomorphism disease," *J. Graph Theory*, 1 (1977), 339–363.

REINGOLD, E., J. NIEVERGELT, and N. DEO, *Combinatorial Algorithms*, Prentice Hall, Englewood Cliffs, N.J., 1977.

RIORDAN, J., *An Introduction to Combinatorial Analysis*, Wiley, New York, 1958.

ROBERTS, F. S. and B. TESMAN, *Applied Combinatorics*, 2nd ed., Prentice Hall, Upper Saddle River, N.J., 2005.

ROBINSON, J. A., "A machine-oriented logic based on the resolution principle," *J. ACM*, 12 (1965), 23–41.

ROSS, S. M., *A First Course in Probability*, 9th ed., Prentice Hall, Upper Saddle River, N.J., 2014.

ROZANOV, Y. A., *Probability Theory: A Concise Course*, Dover, New York, 1969.

SAAD, Y. and M. H. SCHULTZ, "Topological properties of hypercubes," *IEEE Trans. Computers*, 37 (1988), 867–872.

SCHUMER, P., "The Josephus problem: Once more around," *Math. Mag.*, 75 (2002), 12–17.

SCHWENK, A. J., "Which rectangular chessboards have a knight's tour?" *Math. Mag.*, 64 (1991), 325–332.

SEIDEL, R., "A convex hull algorithm optimal for points in even dimensions," M.S. thesis, Tech. Rep. 81–14, Dept. of Comp. Sci., Univ. of British Columbia, Vancouver, Canada, 1981.

SHANNON, C. E., "A symbolic analysis of relay and switching circuits," *Trans. Amer. Inst. Electr. Engrs.*, 47 (1938), 713–723.

SIGLER, L., *Fibonacci's Liber Abaci*, Springer-Verlag, New York, 2003.

SKIENA, S., *Calculated Bets: Computers, Gambling, and Mathematical Modeling to Win*, Mathematical Association of America, Washington, D.C., 2001.

SLAGLE, J. R., *Artificial Intelligence: The Heuristic Programming Approach*, McGraw-Hill, New York, 1971.

SMITH, A. R., "Plants, fractals, and formal languages," *Computer Graphics*, 18 (1984), 1–10.

SOLOW, D., *How to Read and Do Proofs*, 6th ed., Wiley, New York, 2014.

STANDISH, T. A., *Data Structures in Java*, Addison-Wesley, Reading, Mass., 1998.

STOLL, R. R., *Set Theory and Logic*, Dover, New York, 1979.

SUDKAMP, T. A., *Languages and Machines: An Introduction to the Theory of Computer Science*, 3rd ed., Addison-Wesley, Reading, Mass., 2006.

SULLIVAN, M., *College Algebra*, 10th ed., Prentice Hall, Upper Saddle River, N.J., 2016.

TARJAN, R. E., *Data Structures and Network Algorithms*, Society for Industrial and Applied Mathematics, Philadelphia, 1983.

TAUBES, G., "Small army of code-breakers conquers a 129-digit giant," *Science*, 264 (1994), 776–777.

TUCKER, A., *Applied Combinatorics*, 6th ed., Wiley, New York, 2012.

ULLMAN, J. D. and J. D. WIDOM, *A First Course in Database Systems*, 3rd ed., Prentice Hall, Upper Saddle River, N.J., 2008.

VILENKIN, N. Y., *Combinatorics*, Academic Press, New York, 1971.

WAGON, S., "Fourteen proofs of a result about tiling a rectangle," *Amer. Math. Monthly*, 94 (1987), 601–617. (Reprinted in R. K. Guy and R. E. Woodrow, eds., *The Lighter Side of Mathematics*, Mathematical Association of America, Washington, D.C., 1994, 113–128.)

WARD, S. A. and R. H. HALSTEAD, JR., *Computation Structures*, MIT Press, Cambridge, Mass., 1990.

WEST, D., *Introduction to Graph Theory*, 2nd ed., Prentice Hall, Upper Saddle River, N.J., 2001.

WILLETT P., "Matching of chemical and biological structures using subgraph and maximal common subgraph isomorphism algorithms," *IMA Volumes in Mathematics and its Applications*, 108 (1999), 11–38.

WILSON, R. J., *Introduction to Graph Theory*, 5th ed., Addison-Wesley, Reading, Mass., 2010.

WONG, D. F. and C. L. LIU, "A new algorithm for floorplan design," 23rd Design Automation Conference, (1986), 101–107.

WOOD, D., *Theory of Computation*, Harper & Row, New York, 1987.

WOS, L., R. OVERBEEK, E. LUSK, and J. BOYLE, *Automated Reasoning*, Prentice Hall, Englewood Cliffs, N.J., 1984.

# HINTS/SOLUTIONS TO Selected Exercises

## Section 1.1 Review

**1.** A set is a collection of objects.

**2.** A set may be defined by listing the elements in it. For example, $\{1, 2, 3, 4\}$ is the set consisting of the integers 1, 2, 3, 4. A set may also be defined by listing a property necessary for membership. For example,

$$\{x \mid x \text{ is a positive, real number}\}$$

defines the set consisting of the positive, real numbers.

**3.**

| *Set* | *Description* | *Examples of Members* |
|---|---|---|
| **Z** | Integers | $-3, 2$ |
| **Q** | Rational numbers | $-3/4, 2.13074$ |
| **R** | Real numbers | $-2.13074, \sqrt{2}$ |
| $\mathbf{Z}^+$ | Positive integers | $2, 10$ |
| $\mathbf{Q}^+$ | Positive rational numbers | $3/4, 2.13074$ |
| $\mathbf{R}^+$ | Positive real numbers | $2.13074, \sqrt{2}$ |
| $\mathbf{Z}^-$ | Negative integers | $-12, -10$ |
| $\mathbf{Q}^-$ | Negative rational numbers | $-3/8, -2.13074$ |
| $\mathbf{R}^-$ | Negative real numbers | $-2.13074, -\sqrt{2}$ |
| $\mathbf{Z}^{nonneg}$ | Nonnegative integers | $0, 3$ |
| $\mathbf{Q}^{nonneg}$ | Nonnegative rational numbers | $0, 3.13074$ |
| $\mathbf{R}^{nonneg}$ | Nonnegative real numbers | $0, \sqrt{3}$ |

**4.** The cardinality of $X$ (i.e., the number of elements in $X$)

**5.** $x \in X$ **6.** $x \notin X$ **7.** $\varnothing$

**8.** Sets $X$ and $Y$ are equal if they have the same elements. Set equality is denoted $X = Y$.

**9.** Prove that for every $x$, if $x$ is in $X$, then $x$ is in $Y$, and if $x$ is in $Y$, then $x$ is in $X$.

**10.** Prove one of the following: (a) There exists $x$ such that $x \subset X$ and $x \notin Y$. (b) There exists $x$ such that $x \notin X$ and $x \in Y$.

**11.** $X$ is a subset of $Y$ if every element of $X$ is an element of $Y$. $X$ is a subset of $Y$ is denoted $X \subseteq Y$.

**12.** To prove that $X$ is a subset of $Y$, let $x$ be an arbitrary element of $X$ and prove that $x$ is in $Y$.

**13.** Find $x$ such that $x$ is in $X$, but $x$ is not in $Y$.

**14.** $X$ is a proper subset of $Y$ if $X \subseteq Y$ and $X \neq Y$. $X$ is a proper subset of $Y$ is denoted $X \subset Y$.

**15.** To prove that $X$ is a proper subset of $Y$, prove that $X$ is a subset of $Y$ and find $x$ in $Y$ such that $x$ is not in $X$.

**16.** The power set of $X$ is the collection of all subsets of $X$. It is denoted $\mathcal{P}(X)$.

**17.** $X$ *union* $Y$ is the set of elements that belong to either $X$ or $Y$ or both. It is denoted $X \cup Y$.

**18.** The union of $\mathcal{S}$ is the set of elements that belong to at least one set in $\mathcal{S}$. It is denoted $\cup \mathcal{S}$.

**19.** $X$ *intersect* $Y$ is the set of elements that belong to both $X$ and $Y$. It is denoted $X \cap Y$.

**20.** The intersection of $\mathcal{S}$ is the set of elements that belong to every set in $\mathcal{S}$. It is denoted $\cap \mathcal{S}$.

**21.** $X \cap Y = \varnothing$

**22.** A collection of sets $\mathcal{S}$ is pairwise disjoint if, whenever $X$ and $Y$ are distinct sets in $\mathcal{S}$, $X$ and $Y$ are disjoint.

**23.** The difference of $X$ and $Y$ is the set of elements that are in $X$ but not in $Y$. It is denoted $X - Y$.

**24.** A universal set is a set that contains all of the sets under discussion.

**25.** The complement of $X$ is $U - X$, where $U$ is a given universal set. The complement of $X$ is denoted $\overline{X}$.

**26.** A Venn diagram provides a pictorial view of sets. In a Venn diagram, a rectangle depicts a universal set, and subsets of the universal set are drawn as circles. The inside of a circle represents the members of that set.

**27.**

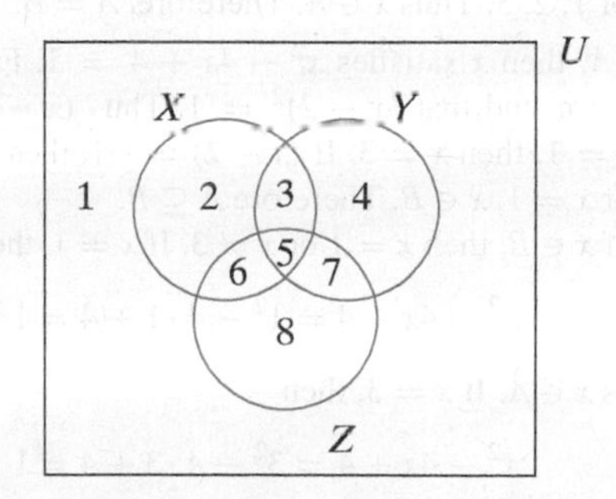

Region 1 represents elements in none of $X$, $Y$, or $Z$. Region 2 represents elements in $X$, but in neither $Y$ nor $Z$. Region 3 represents elements in $X$ and $Y$, but not in $Z$. Region 4 represents elements in $Y$, but in neither $X$ nor $Z$. Region 5 represents elements in $X$, $Y$, and $Z$. Region 6 represents elements in $X$ and $Z$, but not in $Y$. Region 7 represents elements in $Y$ and $Z$, but not in $X$. Region 8 represents elements in $Z$, but in neither $X$ nor $Y$.

**28.** $(A \cup B) \cup C = A \cup (B \cup C)$, $(A \cap B) \cap C = A \cap (B \cap C)$

**29.** $A \cup B = B \cup A$, $A \cap B = B \cap A$

**30.** $A \cap (B \cup C) = (A \cap B) \cup (A \cap C)$, $A \cup (B \cap C) = (A \cup B) \cap (A \cup C)$

**31.** $A \cup \varnothing = A$, $A \cap U = A$

**32.** $A \cup \overline{A} = U$, $A \cap \overline{A} = \varnothing$

**33.** $A \cup A = A$, $A \cap A = A$

**34.** $A \cup U = U$, $A \cap \varnothing = \varnothing$

**35.** $A \cup (A \cap B) = A$, $A \cap (A \cup B) = A$

**36.** $\overline{\overline{A}} = A$ **37.** $\overline{\varnothing} = U$, $\overline{U} = \varnothing$

**38.** $\overline{(A \cup B)} = \overline{A} \cap \overline{B}$, $\overline{(A \cap B)} = \overline{A} \cup \overline{B}$

**39.** A collection $\mathcal{S}$ of nonempty subsets of $X$ is a partition of $X$ if every element in $X$ belongs to exactly one member of $\mathcal{S}$.

**40.** The Cartesian product of $X$ and $Y$ is the set of all ordered pairs $(x, y)$ where $x \in X$ and $y \in Y$. It is denoted $X \times Y$.

**41.** The Cartesian product of $X_1, X_2, \ldots, X_n$ is the set of all $n$-tuples $(x_1, x_2, \ldots, x_n)$ where $x_i \in X_i$ for $i = 1, \ldots, n$. It is denoted $X_1 \times X_2 \times \cdots \times X_n$.

## Section 1.1

**1.** $\{1, 2, 3, 4, 5, 7, 10\}$ **4.** $\{2, 3, 5\}$

**7.** $\varnothing$ **10.** $U$

**13.** $\{6, 8\}$

**16.** $\{1, 2, 3, 4, 5, 7, 10\}$

**17.** $Y$ is the set of odd, positive integers.

**20.** $\{2, 4\}$

**23.** $\{n \in \mathbf{Z}^+ \mid n = 2 \text{ or } n = 4 \text{ or } n \geq 6\}$

**26.** $\{2n + 1 \mid n \in \mathbf{Z}^+ \text{ and } n \geq 3\}$

**28.** 0 **31.** 5

**32.** If $x \in A$, then $x$ is one of 3, 2, 1. Thus $x \in B$. If $x \in B$, then $x$ is one of 1, 2, 3. Thus $x \in A$. Therefore, $A = B$.

**35.** If $x \in A$, then $x$ satisfies $x^2 - 4x + 4 = 1$. Factoring $x^2 - 4x + 4$, we find that $(x - 2)^2 = 1$. Thus $(x - 2) = \pm 1$. If $(x - 2) = 1$, then $x = 3$. If $(x - 2) = -1$, then $x = 1$. Since $x = 3$ or $x = 1$, $x \in B$. Therefore $A \subseteq B$.

If $x \in B$, then $x = 1$ or $x = 3$. If $x = 1$, then

$$x^2 - 4x + 4 = 1^2 - 4 \cdot 1 + 4 = 1$$

and thus $x \in A$. If $x = 3$, then

$$x^2 - 4x + 4 = 3^2 - 4 \cdot 3 + 4 = 1$$

and again $x \in A$. Therefore $B \subseteq A$. We conclude that $A = B$.

**36.** Since $1 \in A$, but $1 \notin B$, $A \neq B$.

**39.** Note that $A = B \cap C = \{2, 4\}$. Since $1 \in B$, but $1 \notin A$, $A \neq B$.

**40.** Equal **43.** Not equal

**44.** Let $x \in A$. Then $x = 1$ or $x = 2$. In either case, $x \in B$. Therefore $A \subseteq B$.

**47.** First note that $B = \mathbf{Z}^+$. Now let $x \in A$. Then $x = 2n$ for some $n \in \mathbf{Z}^+$. Since $2 \in \mathbf{Z}^+$, $2n \in \mathbf{Z}^+ = B$. Therefore $A \subseteq B$.

**48.** Since $3 \in A$, but $3 \notin B$, $A$ is not a subset of $B$.

**51.** Since $3 \in A$, but $3 \notin B$, $A$ is not a subset of $B$.

**52.**

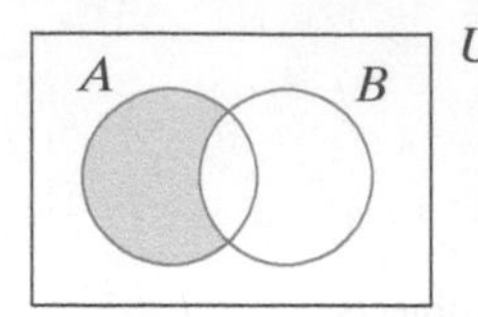


**55.** Same as Exercise 52

**58.**

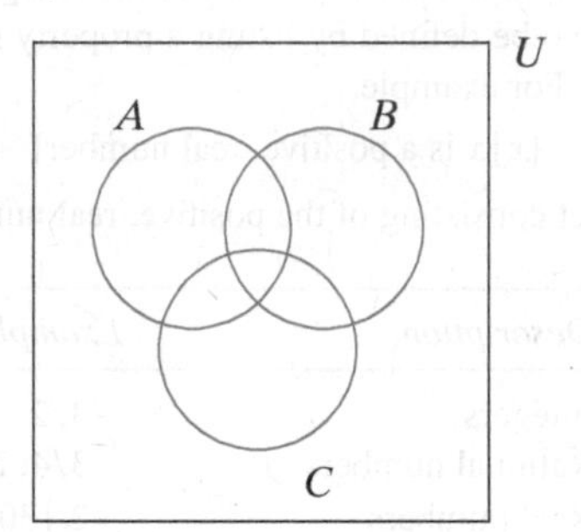


**60.** The shaded area represents the beverage, which has great taste *and* is less filling.

**61.** 10 **64.** 64 **66.** 4

**68.** $\{(1, a), (1, b), (1, c), (2, a), (2, b), (2, c)\}$

**71.** $\{(a, a), (a, b), (a, c), (b, a), (b, b), (b, c), (c, a), (c, b), (c, c)\}$

**72.** $\{(1, a, \alpha), (1, a, \beta), (2, a, \alpha), (2, a, \beta)\}$

**75.** $\{(a, 1, a, \alpha), (a, 2, a, \alpha), (a, 1, a, \beta), (a, 2, a, \beta)\}$

**76.** The entire $xy$-plane

**78.** Parallel horizontal lines spaced one unit apart. There is a lowest line [passing through $(0, 0)$] but the lines continue indefinitely above the lowest line.

**81.** Parallel planes stacked one above another one unit apart. The planes continue indefinitely in both directions [above and below the origin $(0, 0, 0)$].

**83.** $\{\{1\}\}$

**86.** $\{\{a, b, c, d\}\}$, $\{\{a, b, c\}, \{d\}\}$,
$\{\{a, b, d\}, \{c\}\}$, $\{\{a, c, d\}, \{b\}\}$, $\{\{b, c, d\}, \{a\}\}$,
$\{\{a, b\}, \{c\}, \{d\}\}$, $\{\{a, c\}, \{b\}, \{d\}\}$,
$\{\{a, d\}, \{b\}, \{c\}\}$,
$\{\{b, c\}, \{a\}, \{d\}\}$, $\{\{b, d\}, \{a\}, \{c\}\}$, $\{\{c, d\}, \{a\}, \{b\}\}$,
$\{\{a, b\}, \{c, d\}\}$, $\{\{a, c\}, \{b, d\}\}$, $\{\{a, d\}, \{b, c\}\}$,
$\{\{a\}, \{b\}, \{c\}, \{d\}\}$

**87.** True **90.** True

**93.** $\varnothing, \{a\}, \{b\}, \{a, b\}$. All but $\{a, b\}$ are proper subsets.

**96.** $2^n - 1$

**97.** $A \subseteq B$

**100.** $B \subseteq A$

**101.** $\{1, 4, 5\}$

**104.** The center of the circle

## Section 1.2 Review

**1.** A proposition is a sentence that is either true or false, but not both.

**2.** The truth table of a proposition $P$ made up of the individual propositions $p_1, \ldots, p_n$ lists all possible combinations of truth values for $p_1, \ldots, p_n$, T denoting true and F denoting false, and for each such combination lists the truth value of $P$.

**3.** The conjunction of propositions $p$ and $q$ is the proposition $p$ *and* $q$. It is denoted $p \wedge q$.

**4.**

| $p$ | $q$ | $p \wedge q$ |
|---|---|---|
| T | T | T |
| T | F | F |
| F | T | F |
| F | F | F |

**5.** The disjunction of propositions $p$ and $q$ is the proposition $p$ *or* $q$. It is denoted $p \vee q$.

**6.**

| $p$ | $q$ | $p \vee q$ |
|---|---|---|
| T | T | T |
| T | F | T |
| F | T | T |
| F | F | F |

**7.** The negation of proposition $p$ is the proposition *not* $p$. It is denoted $\neg p$.

**8.**

| $p$ | $\neg p$ |
|---|---|
| T | F |
| F | T |

## Section 1.2

**1.** Is a proposition. Negation: $2 + 5 \neq 19$

**5.** Not a proposition; it is a question.

**8.** Not a proposition; it is a command.

**11.** Not a proposition; it is a description of a mathematical expression (i.e., $p - q$, where $p$ and $q$ are primes).

**13.** Ten heads were not obtained. (Alternative: At least one tail was obtained.)

**16.** No heads were obtained. (Alternative: Ten tails were obtained.)

**17.** True

**20.** True

**23.**

| $p$ | $q$ | $p \wedge \neg q$ |
|---|---|---|
| T | T | F |
| T | F | T |
| F | T | F |
| F | F | F |

**26.**

| $p$ | $q$ | $(p \wedge q) \wedge \neg q$ |
|---|---|---|
| T | T | F |
| T | F | F |
| F | T | F |
| F | F | F |

**29.**

| $p$ | $q$ | $(p \vee q) \wedge (\neg p \vee q) \wedge (p \vee \neg q) \wedge (\neg p \vee \neg q)$ |
|---|---|---|
| T | T | F |
| T | F | F |
| F | T | F |
| F | F | F |

**31.** $p \wedge q$; false

**34.** Lee does not take computer science.

**37.** Lee takes computer science or Lee does not take mathematics.

**40.** You play football and you miss the midterm exam.

**43.** It is not the case that you play football or you miss the midterm exam, or you pass the course.

**45.** Today is Monday or it is raining.

**48.** (Today is Monday and it is raining) and it is not the case that (it is hot or today is Monday).

**50.** $\neg p$

**53.** $\neg p \wedge \neg q$

**56.** $p \wedge \neg q$

**59.** $\neg p \wedge \neg r \wedge \neg q$

**61.** $p \wedge r$

**64.** $(p \vee q) \wedge \neg r$

**68.** Inclusive-or: To enter Utopia, you must show a driver's license or a passport or both. Exclusive-or: To enter Utopia, you must show a driver's license or a passport but not both. Exclusive-or is the intended meaning.

**71.** Inclusive-or: The car comes with a cupholder that heats or cools your drink or both. Exclusive-or: The car comes with a cupholder that heats or cools your drink but not both. Exclusive or is the intended meaning.

**74.** Inclusive-or: The meeting will be canceled if fewer than 10 persons sign up or at least 3 inches of snow falls or both. Exclusive-or: The meeting will be canceled if fewer than 10 persons sign up or at least 3 inches of snow falls but not both. Inclusive-or is the intended meaning.

**75.** No, assuming the interpretation: It shall be unlawful for any person to keep more than three [3] dogs and more than three [3] cats upon his property within the city. A judge ruled that the ordinance was "vague." Presumably, the intended meaning was: "It shall be unlawful for any person to keep more than three [3] dogs *or* more than three [3] cats upon his property within the city."

**76.** "national park" "north dakota" OR "south dakota"

## Section 1.3 Review

**1.** If $p$ and $q$ are propositions, the conditional proposition is the proposition if $p$ then $q$. It is denoted $p \to q$.

**2.**

| $p$ | $q$ | $p \to q$ |
|---|---|---|
| T | T | T |
| T | F | F |
| F | T | T |
| F | F | T |

**3.** In the conditional proposition $p \to q$, $p$ is the hypothesis.

**4.** In the conditional proposition $p \to q$, $q$ is the conclusion.

**5.** In the conditional proposition $p \to q$, $q$ is a necessary condition.

**6.** In the conditional proposition $p \to q$, $p$ is a sufficient condition.

**7.** The converse of $p \to q$ is $q \to p$.

**8.** If $p$ and $q$ are propositions, the biconditional proposition is the proposition $p$ if and only if $q$. It is denoted $p \leftrightarrow q$.

**9.**

| $p$ | $q$ | $p \leftrightarrow q$ |
|---|---|---|
| T | T | T |
| T | F | F |
| F | T | F |
| F | F | T |

**10.** If the propositions $P$ and $Q$ are made up of the propositions $p_1, \ldots, p_n$, $P$ and $Q$ are logically equivalent provided that given any truth values of $p_1, \ldots, p_n$, either $P$ and $Q$ are both true or $P$ and $Q$ are both false.

**11.** $\neg(p \vee q) \equiv \neg p \wedge \neg q$, $\neg(p \wedge q) \equiv \neg p \vee \neg q$

**12.** The contrapositive of $p \to q$ is $\neg q \to \neg p$.

## Section 1.3

**1.** If Joey studies hard, then he will pass the discrete mathematics exam.

**4.** If Katrina passes discrete mathematics, then she will take the algorithms course.

**7.** If you inspect the aircraft, then you have the proper security clearance.

**10.** If the program is readable, then it is well structured.

**12.** (For Exercise 1) If Joey passes the discrete mathematics exam, then he studied hard.

**14.** True **17.** False **20.** False

**23.** True **26.** True **29.** True

**32.** True **33.** True **36.** False

**39.** True **42.** True **43.** $p \to q$

**46.** $q \leftrightarrow (p \wedge \neg r)$

**47.** $p \to q$

**50.** $q \leftrightarrow (p \wedge r)$

**53.** If today is Monday, then it is raining.

**56.** It is not the case that today is Monday or it is raining if and only if it is hot.

**59.** Let $p$: $4 < 6$ and $q$: $9 > 12$.
Given statement: $p \to q$; false.
Converse: $q \to p$; if $9 > 12$, then $4 < 6$; true.
Contrapositive: $\neg q \to \neg p$; if $9 \le 12$, then $4 \ge 6$; false.

**62.** Let $p$: $|4| < 3$ and $q$: $-3 < 4 < 3$.
Given statement: $q \to p$; true.
Converse: $p \to q$; if $|4| < 3$, then $-3 < 4 < 3$; true.
Contrapositive:
$\neg p \to \neg q$, if $|4| \ge 3$, then $-3 \ge 4$ or $4 \ge 3$; true.

**63.** $P \not\equiv Q$ **66.** $P \not\equiv Q$

**69.** $P \not\equiv Q$ **72.** $P \not\equiv Q$

**73.** Pat will not use the treadmill and will not lift weights.

**76.** To make chili, you do not need red pepper or you do not need onions.

**77.**

| $p$ | $q$ | $p$ *impl* $q$ | $q$ *impl* $p$ |
|---|---|---|---|
| T | T | T | T |
| T | F | F | F |
| F | T | F | F |
| F | F | T | T |

Since $p$ *impl* $q$ is true precisely when $q$ *impl* $p$ is true, $p$ *impl* $q \equiv q$ *impl* $p$.

**80.**

| $p$ | $q$ | $p \to q$ | $\neg p \vee q$ |
|---|---|---|---|
| T | T | T | T |
| T | F | F | F |
| F | T | T | T |
| F | F | T | T |

Since $p \to q$ is true precisely when $\neg p \vee q$ is true, $p \to q \equiv \neg p \vee q$.

## Section 1.4 Review

**1.** Deductive reasoning refers to the process of drawing a conclusion from a sequence of propositions.

**2.** In the argument $p_1, p_2, \ldots, p_n / \therefore q$, the hypotheses are $p_1, p_2, \ldots, p_n$.

**3.** "Premise" in another name for hypothesis.

**4.** In the argument $p_1, p_2, \ldots, p_n / \therefore q$, the conclusion is $q$.

**5.** The argument $p_1, p_2, \ldots, p_n / \therefore q$ is valid provided that if $p_1$ and $p_2$ and ... and $p_n$ are all true, then $q$ must also be true.

**6.** An invalid argument is an argument that is not valid.

**7.** $p \to q$
$p$
$\therefore q$

**8.** $p \to q$
$\neg q$
$\therefore \neg p$

**9.** $p$
$\therefore p \vee q$

**10.** $p \wedge q$
$\therefore p$

**11.** $p$
$q$
$\therefore p \wedge q$

**12.** $p \to q$
$q \to r$
$\therefore p \to r$

**13.** $p \vee q$
$\neg p$
$\therefore q$

## Section 1.4

**1.** Valid $p \to q$
$p$
$\therefore q$

**4.** Invalid $(p \vee r) \to q$
$q$
$\therefore \neg p \to r$

**6.** Valid $(p \to s) \wedge (q \to r)$
$p \vee q$
$\therefore r \vee s$

**9.** Valid $p \to (r \vee s)$
$q \to r$
$\neg s$
$p$
$\therefore r \vee q$

**10.** Valid. If 4 megabytes is better than no memory at all, then we will buy a new computer. If 4 megabytes is better than no memory at all, then we will buy more memory. Therefore, if 4 megabytes is better than no memory at all, then we will buy a new computer and we will buy more memory.

**13.** Invalid. If we will not buy a new computer, then 4 megabytes is not better than no memory at all. We will buy a new computer. Therefore, 4 megabytes is better than no memory at all.

**15.** Valid. If the for loop is faulty or the while loop is faulty, then the hardware is unreliable or the output is correct. The for loop is faulty. The hardware is not unreliable. Therefore the output is correct.

**18.** Valid If the for loop is faulty, then the while loop is faulty. If the while loop is faulty, then the hardware is unreliable. The hardware is not unreliable. If the output is correct, then the hardware is unreliable. Therefore the output is not correct.

**20.** Invalid **23.** Invalid

**26.** An analysis of the argument must take into account the fact that "nothing" is being used in two very different ways.

**27.** Addition

**30.** Let $p$ denote the proposition "there is gas in the car," let $q$ denote the proposition "I go to the store," and let $r$ denote the proposition "I get a soda." Then the hypotheses are as follows:

$$p \to q$$
$$q \to r$$
$$p$$

From $p \to q$ and $q \to r$, we may use the hypothetical syllogism to conclude $p \to r$. From $p \to r$ and $p$, we may use modus ponens to conclude $r$. Since $r$ represents the proposition "I get a soda," we conclude that the conclusion does follow from the hypotheses.

**33.** We construct a truth table for all the propositions involved:

| $p$ | $q$ | $p \to q$ | $\neg q$ | $\neg p$ |
|---|---|---|---|---|
| T | T | T | F | F |
| T | F | F | T | F |
| F | T | T | F | T |
| F | F | T | T | T |

We observe that whenever the hypotheses $p \to q$ and $\neg q$ are true, the conclusion $\neg p$ is also true; therefore, the argument is valid.

**36.** We construct a truth table for all the propositions involved:

| $p$ | $q$ | $p \wedge q$ |
|---|---|---|
| T | T | T |
| T | F | F |
| F | T | F |
| F | F | F |

We observe that whenever the hypotheses $p$ and $q$ are true, the conclusion $p \wedge q$ is also true; therefore, the argument is valid.

## Section 1.5 Review

**1.** If $P(x)$ is a statement involving the variable $x$, we call $P$ a propositional function if for each $x$ in the domain of discourse, $P(x)$ is a proposition.

**2.** A domain of discourse for a propositional function $P$ is a set $D$ such that $P(x)$ is defined for every $x$ in $D$.

**3.** A universally quantified statement is a statement of the form for all $x$ in the domain of discourse, $P(x)$.

**4.** A counterexample to the statement $\forall x\, P(x)$ is a value of $x$ for which $P(x)$ is false.

**5.** An existentially quantified statement is a statement of the form for some $x$ in the domain of discourse, $P(x)$.

**6.** $\neg(\forall x\, P(x))$ and $\exists x \neg P(x)$ have the same truth values. $\neg(\exists x\, P(x))$ and $\forall x \neg P(x)$ have the same truth values.

**7.** To prove that the universally quantified statement $\forall x\, P(x)$ is true, show that for every $x$ in the domain of discourse, the proposition $P(x)$ is true.

**8.** To prove that the existentially quantified statement $\exists x\, P(x)$ is true, find one value of $x$ in the domain of discourse for which the proposition $P(x)$ is true.

**9.** To prove that the universally quantified statement $\forall x\, P(x)$ is false, find one value of $x$ in the domain of discourse for which the proposition $P(x)$ is false.

**10.** To prove that the existentially quantified statement $\exists x\, P(x)$ is false, show that for every $x$ in the domain of discourse, the proposition $P(x)$ is false.

**11.** $\forall x P(x)$
$\therefore P(d)$ if $d \in D$

**12.** $P(d)$ for every $d \in D$
$\therefore \forall x\, P(x)$

**13.** $\exists x P(x)$
$\therefore P(d)$ for some $d \in D$

**14.** $P(d)$ for some $d \in D$
$\therefore \exists x\, P(x)$

## Section 1.5

**1.** Is a propositional function. The domain of discourse could be taken to be all integers.

**4.** Is a propositional function. The domain of discourse is the set of all movies.

**7.** 11 divides 77. True.

**10.** For every positive integer $n$, $n$ divides 77. False.

**13.** There exists $n$ such that $n$ does not divide 77. True.

**16.** True **19.** False **22.** False

**25.** $P(1) \wedge P(2) \wedge P(3) \wedge P(4)$

**28.** $P(1) \vee P(2) \vee P(3) \vee P(4)$

**31.** $P(2) \wedge P(3) \wedge P(4)$

**32.** Every student is taking a math course.

**35.** Some student is not taking a math course.

**38.** (For Exercise 32) $\exists x \neg P(x)$. Some student is not taking a math course.

**39.** Every professional athlete plays soccer. False.

**42.** Either someone does not play soccer or some soccer player is a professional athlete. True.

**45.** Everyone is a professional athlete and plays soccer. False.

**47.** (For Exercise 39) $\exists x(P(x) \wedge \neg Q(x))$. Someone is a professional athlete and does not play soccer.

**48.** $\forall x(P(x) \rightarrow Q(x))$

**51.** $\exists x(P(x) \wedge Q(x))$

**52.** (For Exercise 48) $\exists x(P(x) \wedge \neg Q(x))$. Some accountant does not own a Porsche.

**53.** False. A counterexample is $x = 0$.

**56.** True. The value $x = 2$ makes $(x > 1) \rightarrow (x^2 > x)$ true.

**59.** (For Exercise 53) $\exists x(x^2 \leq x)$. There exists $x$ such that $x^2 \leq x$.

**61.** The literal meaning is: No man cheats on his wife. The intended meaning is: Some man does not cheat on his wife. Let $P(x)$ denote the statement "$x$ is a man," and $Q(x)$ denote the statement "$x$ cheats on his wife." Symbolically, the clarified statement is $\exists x(P(x) \wedge \neg Q(x))$.

**64.** The literal meaning is: No environmental problem is a tragedy. The intended meaning is: Some environmental problem is not a tragedy. Let $P(x)$ denote the statement "$x$ is an environmental problem," and $Q(x)$ denote the statement "$x$ is a tragedy." Symbolically, the clarified statement is $\exists x(P(x) \wedge \neg Q(x))$.

**67.** The literal meaning is: Everything is not sweetness and light. The intended meaning is: Not everything is sweetness and light. Let $P(x)$ denote the statement "$x$ is sweetness and light." Symbolically, the clarified statement is $\exists x \neg P(x)$.

**70.** The literal meaning is: No circumstance is right for a formal investigation. The intended meaning is: Some circumstance is not right for a formal investigation. Let $P(x)$ denote the statement "$x$ is a circumstance," and $Q(x)$ denote the statement "$x$ is right for a formal investigation." Symbolically, the clarified statement is $\exists x(P(x) \wedge \neg Q(x))$.

**72.** (a)

| $p$ | $q$ | $p \rightarrow q$ | $q \rightarrow p$ |
|---|---|---|---|
| T | T | T | T |
| T | F | F | T |
| F | T | T | F |
| F | F | T | T |

One of $p \rightarrow q$ or $q \rightarrow p$ is true since in each row, one of the last two entries is true.

(b) The statement, "All integers are positive or all positive numbers are integers," which is false, in symbols is

$$(\forall x(I(x) \rightarrow P(x))) \vee (\forall x(P(x) \rightarrow I(x))).$$

This is *not* the same as the given statement

$$\forall x((I(x) \rightarrow P(x)) \vee (P(x) \rightarrow I(x))),$$

which is true. The ambiguity results from attempting to distribute $\forall$ across the *or*.

**75.** Universal instantiation

**76.** Let $P(x)$ denote the propositional function "$x$ has a graphing calculator," and let $Q(x)$ denote the propositional function "$x$ understands the trigonometric functions." The hypotheses are $\forall x\, P(x)$ and $\forall x(P(x) \rightarrow Q(x))$. By universal instantiation, we have $P(\text{Ralphie})$ and $P(\text{Ralphie}) \rightarrow Q(\text{Ralphie})$. The modus ponens rule of inference now gives $Q(\text{Ralphie})$, which represents the proposition "Ralphie understands the trigonometric functions." We conclude that the conclusion does follow from the hypotheses.

**79.** By definition, the proposition $\forall x\, P(x)$ is true when $P(x)$ is true for all $x$ in the domain of discourse. We are given that $P(d)$ is true for any $d$ in the domain of discourse $D$. Therefore, $\forall x\, P(x)$ is true.

## Section 1.6 Review

**1.** For every $x$ and for every $y$, $P(x, y)$. Let the domain of discourse be $X \times Y$. The statement is true if, for every $x \in X$ and for every $y \in Y$, $P(x, y)$ is true. The statement is false if there is at least one $x \in X$ and at least one $y \in Y$ such that $P(x, y)$ is false.

**2.** For every $x$, there exists $y$ such that $P(x, y)$. Let the domain of discourse be $X \times Y$. The statement is true if, for every $x \in X$, there is at least one $y \in Y$ for which $P(x, y)$ is true. The statement is false if there is at least one $x \in X$ such that $P(x, y)$ is false for every $y \in Y$.

**3.** There exists $x$ such that for every $y$, $P(x, y)$. Let the domain of discourse be $X \times Y$. The statement is true if there is at least one $x \in X$ such that $P(x, y)$ is true for every $y \in Y$. The statement is false if, for every $x \in X$, there is at least one $y \in Y$ such that $P(x, y)$ is false.

**4.** There exists $x$ and there exists $y$ such that $P(x, y)$. Let the domain of discourse be $X \times Y$. The statement is true if there is at least one $x \in X$ and at least one $y \in Y$ such that $P(x, y)$ is true. The statement is false if, for every $x \in X$ and for every $y \in Y$, $P(x, y)$ is false.

**5.** Let $P(x, y)$ be the propositional function "$x \leq y$" with domain of discourse $\mathbf{Z} \times \mathbf{Z}$. Then $\forall x \exists y P(x, y)$ is true since, for every integer $x$, there exists an integer $y$ (e.g., $y = x$) such that $x \leq y$ is true. On the other hand, $\exists x \forall y P(x, y)$ is false. For every integer $x$, there exists an integer $y$ (e.g., $y = x - 1$) such that $x \leq y$ is false.

**6.** $\exists x \exists y \neg P(x, y)$

**7.** $\exists x \forall y \neg P(x, y)$

**8.** $\forall x \exists y \neg P(x, y)$

**9.** $\forall x \forall y \neg P(x, y)$

**10.** Given a quantified propositional function, you and your opponent, whom we call Farley, play a logic game. Your goal is to try to make the propositional function true, and Farley's goal is to try to make it false. The game begins with the first (left) quantifier. If the quantifier is $\forall$, Farley chooses a value for that variable; if the quantifier is $\exists$, you choose a value for that variable. The game continues with the second quantifier. After values are chosen for all the variables, if the propositional function is true, you win; if it is false, Farley wins. If you can always win regardless of how Farley chooses values for the variables, the quantified propositional function is true, but if Farley can choose values for the variables so that you cannot win, the quantified propositional function is false.

## Section 1.6

**1.** Everyone is taller than everyone.

**4.** Someone is taller than someone.

**5.** (For Exercise 1) In symbols: $\exists x \exists y \neg T_1(x, y)$. In words: Someone is not taller than someone.

**6.** (Exercise 1) False. Garth is not taller than Garth. (Exercise 2) False. Erin is taller than no one. (Exercise 3) False. No one is taller than everyone (himself, in particular). (Exercise 4) True. Garth is taller than Erin.

**9.** (Exercise 1) False. Garth is not taller than Pat. (Exercise 2) False. Garth is not taller than Pat, Sandy, or Gale. (Exercise 3) True. Marty is taller than Pat, Sandy, and Gale. (Exercise 4) True. Marty is taller than Pat.

**12.** (Exercise 1) False. Dale is not taller than Dale. (Exercise 2) False. Erin is not taller than Dale. (Exercise 3) False. No one is taller than Dale. (Exercise 4) False. No one is taller than Dale.

**15.** (Exercise 1) False. Dale is not taller than Dale. (Exercise 2) True. Dale is taller than Garth. (Exercise 3) False. Dale is not taller than Dale. (Exercise 4) True. Dale is taller than Garth.

**18.** (Exercise 1) False. Pat is not taller than Marty. (Exercise 2) True. Pat, Sandy, and Gale are all taller than Erin. (Exercise 3) False. Marty is taller than Pat, Sandy, and Gale. (Exercise 4) True. Pat is taller than Erin.

**21.** (Exercise 1) False. Pat is not taller than Pat. (Exercise 2) False. Pat is not taller than everyone. (Exercise 3) Pat is not taller than Pat; Sandy is not taller than Pat; and Gale is not taller than Pat. (Exercise 4) False. None of Pat, Sandy, or Gale is taller than everyone.

**22.** Everyone is taller than or the same height as everyone.

**25.** Someone is taller than or the same height as someone.

**26.** (For Exercise 22) In symbols: $\exists x \exists y \neg T_2(x, y)$. In words: Someone is shorter than someone.

**27.** (For Exercises 22, 6, and 1) False. Erin is not taller than or the same height as Garth.

**28.** For any two people, if they are distinct, the first is taller than the second.

**31.** There are two distinct people and the first is taller than the second.

**32.** (For Exercise 28) In symbols: $\exists x \exists y \neg T_3(x, y)$. In words: There are two distinct people and the first is shorter than or the same height as the second.

**33.** (For Exercises 28, 6, and 1) False. Erin and Garth are distinct persons, but Erin is not taller than Garth.

**34.** $\exists x \forall y L(x, y)$. True (think of a saint).

**37.** $\forall x \exists y L(x, y)$. True (according to Dean Martin's song, "Everybody Loves Somebody Sometime").

**38.** (For Exercise 34) Everyone does not love someone. $\forall x \exists y \neg L(x, y)$

**39.** $\exists y A(\text{Brit}, y)$

**42.** $\forall y \exists x A(x, y)$

**43.** False

**46.** True

**47.** (For Exercise 43) $\exists x \exists y \neg P(x, y)$ or $\exists x \exists y (x < y)$

**48.** False. A counterexample is $x = 2, y = 0$.

**51.** True. Take $x = y = 0$.

**54.** False. A counterexample is $x = y = 2$.

**57.** True. Take $x = 1, y = \sqrt{8}$.

**60.** True. Take $x = 0$. Then for all $y$, $x^2 + y^2 \geq 0$.

**63.** True. For any $x$, if we set $y = x - 1$, the conditional proposition, if $x < y$, then $x^2 < y^2$, is true because the hypothesis is false.

**66.** (For Exercise 48) $\exists x \exists y (x^2 \geq y + 1)$

**69.** (For Exercise 48) Since both quantifiers are $\forall$, Farley chooses values for both $x$ and $y$. Since Farley can choose values that make $x^2 < y + 1$ false (e.g., $x = 2$, $y = 0$), Farley can win the game. Therefore, the proposition is false.

**72.** Since the first two quantifiers are $\forall$, Farley chooses values for both $x$ and $y$. The last quantifier is $\exists$, so you choose a value for $z$. Farley can choose values (e.g., $x = 1, y = 2$) so that no matter which value you choose for $z$, the expression

$$(x < y) \rightarrow ((z > x) \wedge (z < y))$$

is false. Since Farley can choose values for the variables so that you cannot win, the quantified statement is false.

**74.** $\forall x \exists y P(x, y)$ must be true. Since $\forall x \forall y P(x, y)$ is true, regardless of which value of $x$ is selected, $P(x, y)$ is true *for all* $y$. Thus for any $x$, $P(x, y)$ is true *for any* particular $y$.

**77.** $\forall x \forall y P(x, y)$ might be false. Let $P(x, y)$ denote the expression $x \le y$. If the domain of discourse is $\mathbf{Z}^+ \times \mathbf{Z}^+$, $\exists x \forall y P(x, y)$ is true; however, $\forall x \forall y P(x, y)$ is false.

**80.** $\forall x \forall y P(x, y)$ might be false. Let $P(x, y)$ denote the expression $x \le y$. If the domain of discourse is $\mathbf{Z}^+ \times \mathbf{Z}^+$, $\exists x \exists y P(x, y)$ is true; however, $\forall x \forall y P(x, y)$ is false.

**83.** $\forall x \exists y P(x, y)$ might be true. Let $P(x, y)$ denote the expression $x \le y$. If the domain of discourse is $\mathbf{Z}^+ \times \mathbf{Z}^+$, $\forall x \exists y P(x, y)$ is true; however, $\forall x \forall y P(x, y)$ is false.

**86.** $\forall x \forall y P(x, y)$ must be false. Since $\forall x \exists y P(x, y)$ is false, there exists $x$, say $x = x'$, such that for all $y$, $P(x, y)$ is false. Choose $y = y'$ in the domain of discourse. Then $P(x', y')$ is false. Therefore $\forall x \forall y P(x, y)$ is false.

**89.** $\forall x \forall y P(x, y)$ must be false. Since $\exists x \forall y P(x, y)$ is false, for every $x$ there exists $y$ such that $P(x, y)$ is false. Choose $x = x'$ in the domain of discourse. For this choice of $x$, there exists $y = y'$ such that $P(x', y')$ is false. Therefore $\forall x \forall y P(x, y)$ is false.

**92.** $\forall x \forall y P(x, y)$ must be false. Since $\exists x \exists y P(x, y)$ is false, for every $x$ and for every $y$, $P(x, y)$ is false. Choose $x = x'$ and $y = y'$ in the domain of discourse. For these choices of $x$ and $y$, $P(x', y')$ is false. Therefore $\forall x \forall y P(x, y)$ is false.

**95.** $\exists x \neg(\forall y P(x, y))$ is not logically equivalent to $\neg(\forall x \exists y P(x, y))$. Let $P(x, y)$ denote the expression $x < y$. If the domain of discourse is $\mathbf{Z} \times \mathbf{Z}$, $\exists x \neg(\forall y P(x, y))$ is true; however, $\neg(\forall x \exists y P(x, y))$ is false.

**98.** $\exists x \exists y \neg P(x, y)$ is not logically equivalent to $\neg(\forall x \exists y P(x, y))$. Let $P(x, y)$ denote the expression $x < y$. If the domain of discourse is $\mathbf{Z} \times \mathbf{Z}$, $\exists x \exists y \neg P(x, y)$ is true; however, $\neg(\forall x \exists y P(x, y))$ is false.

**99.** $\forall \varepsilon > 0 \exists \delta > 0 \forall x ((0 < |x - a| < \delta) \rightarrow (|f(x) - L| < \varepsilon))$

## Chapter 1 Self-Test

1. [Section 1.1] $\varnothing$
2. [Section 1.2] False
3. [Section 1.3] If Leah gets an A in discrete mathematics, then Leah studies hard.
4. [Section 1.3] Converse: If Leah studies hard, then Leah gets an A in discrete mathematics. Contrapositive: If Leah does not study hard, then Leah does not get an A in discrete mathematics.
5. [Section 1.1] $A \subseteq B$
6. [Section 1.1] Yes
7. [Section 1.4] The argument is invalid. If $p$ and $r$ are true and $q$ is false, the hypotheses are true, but the conclusion is false.
8. [Section 1.3] True
9. [Section 1.4] Let

$p$: The Skyscrapers win.
$q$: I'll eat my hat.
$r$: I'll be quite full.

Then the argument symbolically is

$$\begin{array}{l} p \rightarrow q \\ q \rightarrow r \\ \hline \therefore r \rightarrow p \end{array}$$

The argument is invalid. If $p$ and $q$ are false and $r$ is true, the hypotheses are true, but the conclusion is false.

10. [Section 1.5] The statement is not a proposition. The truth value cannot be determined without knowing what "the team" refers to.
11. [Section 1.5] The statement is a propositional function. When we substitute a particular team for the variable "team," the statement becomes a proposition.
12. [Section 1.6] $\exists x \forall y \neg K(x, y)$
13. [Section 1.6] $\forall x \exists y K(x, y)$; everybody knows somebody.
14. [Section 1.1] Since $|A| = 3$ and $|\mathcal{P}(A)| = 2^3 = 8$, $|\mathcal{P}(A) \times A| = 8 \cdot 3 = 24$.
15. [Section 1.2]

| $p$ | $q$ | $r$ | $\neg(p \wedge q) \vee (p \vee \neg r)$ |
|---|---|---|---|
| T | T | T | T |
| T | T | F | T |
| T | F | T | T |
| T | F | F | T |
| F | T | T | T |
| F | T | F | T |
| F | F | T | T |
| F | F | F | T |

16. [Section 1.2] I take hotel management and either I do not take recreation supervision or I take popular culture.
17. [Section 1.2] $p \vee (q \wedge \neg r)$
18. [Section 1.5] For all positive integers $n$, $n$ and $n + 2$ are prime. The proposition is false. A counterexample is $n = 7$.
19. [Section 1.5] For some positive integer $n$, $n$ and $n + 2$ are prime. The proposition is true. For example, if $n = 5$, $n$ and $n + 2$ are prime.
20. [Section 1.4] Hypothetical syllogism
21. [Section 1.4] Let

$p$: The Council approves the funds.
$q$: New Atlantic gets the Olympic Games.
$r$: New Atlantic builds a new stadium.
$s$: The Olympic Games are canceled.

Then the argument symbolically is

$$\begin{array}{l} p \rightarrow q \\ q \rightarrow r \\ \neg r \\ \hline \therefore \neg p \vee s \end{array}$$

From $p \rightarrow q$ and $q \rightarrow r$, we may use the hypothetical syllogism to conclude $p \rightarrow r$. From $p \rightarrow r$ and $\neg r$, we may use the modus tollens to conclude $\neg p$. We may then use addition to conclude $\neg p \lor s$.

**22.** [Section 1.6] The statement is true. For every $x$, there exists $y$, namely the *cube root* of $x$, such that $x = y^3$. In words: Every real number has a cube root.

**23.** [Section 1.6]

$$\begin{aligned}\neg(\forall x \exists y \forall z P(x, y, z)) &\equiv \exists x \neg(\exists y \forall z P(x, y, z))\\ &\equiv \exists x \forall y \neg(\forall z P(x, y, z))\\ &\equiv \exists x \forall y \exists z \neg P(x, y, z)\end{aligned}$$

**24.** [Section 1.3] $(\neg r \lor q) \to \neg q$

## Section 2.1 Review

**1.** A mathematical system consists of axioms, definitions, and undefined terms.

**2.** An axiom is a proposition that is assumed to be true.

**3.** A definition creates a new concept in terms of existing ones.

**4.** An undefined term is a term that is not explicitly defined but rather is implicitly defined by the axioms.

**5.** A theorem is a proposition that has been proved to be true.

**6.** A proof is an argument that establishes the truth of a theorem.

**7.** A lemma is a theorem that is usually not too interesting in its own right but is useful in proving another theorem.

**8.** A direct proof assumes that the hypotheses are true and then, using the hypotheses as well as other axioms, definitions, and previously derived theorems, shows directly that the conclusion is true.

**9.** An integer $n$ is even if there exists an integer $k$ such that $n = 2k$.

**10.** An integer $n$ is odd if there exists an integer $k$ such that $n = 2k + 1$.

**11.** Within a proof, a proof of an auxiliary result is called a subproof.

**12.** To disprove the universally quantified statement $\forall x P(x)$, we need to find one member $x$ in the domain of discourse that makes $P(x)$ false.

## Section 2.1

**1.** If three points are not collinear, then there is exactly one plane that contains them.

**4.** If $x$ is a nonnegative real number and $n$ is a positive integer, $x^{1/n}$ is the nonnegative number $y$ satisfying $y^n = x$.

**7.** Let $m$ and $n$ be even integers. Then there exist $k_1$ and $k_2$ such that $m = 2k_1$ and $n = 2k_2$. Now

$$m + n = 2k_1 + 2k_2 = 2(k_1 + k_2).$$

Therefore $m + n$ is even.

**10.** Let $m$ and $n$ be odd integers. Then there exist $k_1$ and $k_2$ such that $m = 2k_1 + 1$ and $n = 2k_2 + 1$. Now

$$\begin{aligned}mn = (2k_1 + 1)(2k_2 + 1) &= 4k_1k_2 + 2k_1 + 2k_2 + 1\\ &= 2(2k_1k_2 + k_1 + k_2) + 1.\end{aligned}$$

Therefore $mn$ is odd.

**13.** Let $x$ and $y$ be rational numbers. Then there exist integers $m_1, n_1, m_2, n_2$ such that $x = m_1/n_1$ and $y = m_2/n_2$. Now

$$x + y = \frac{m_1}{n_1} + \frac{m_2}{n_2} = \frac{m_1n_2 + m_2n_1}{n_1n_2}.$$

Since $m_1n_2 + m_2n_1$ and $n_1n_2$ are integers, $x + y$ is a rational number.

**16.** If $m = 3k_1 + 1$ and $n = 3k_2 + 1$, then

$$\begin{aligned}mn = (3k_1 + 1)(3k_2 + 1) &= 9k_1k_2 + 3k_1 + 3k_2 + 1\\ &= 3(3k_1k_2 + k_1 + k_2) + 1.\end{aligned}$$

**18.** From the definition of max, it follows that $d \ge d_1$ and $d \ge d_2$. From $x \ge d$ and $d \ge d_1$, we may derive $x \ge d_1$ from a previous theorem (the second theorem of Example 2.1.5). From $x \ge d$ and $d \ge d_2$, we may derive $x \ge d_2$ from the same previous theorem. Therefore, $x \ge d_1$ and $x \ge d_2$.

**21.** Let $x \in X \cap Y$. From the definition of "intersection," we conclude that $x \in X$. Therefore $X \cap Y \subseteq X$.

**24.** Let $x \in X \cap Z$. From the definition of "intersection," we conclude that $x \in X$ and $x \in Z$. Since $X \subseteq Y$ and $x \in X$, $x \in Y$. Since $x \in Y$ and $x \in Z$, from the definition of "intersection," we conclude that $x \in Y \cap Z$. Therefore $X \cap Z \subseteq Y \cap Z$.

**27.** Let $x \in Y$. From the definition of "union," we conclude that $x \in X \cup Y$. Since $X \cup Y = X \cup Z$, $x \in X \cup Z$. From the definition of "union," we conclude that $x \in X$ or $x \in Z$. If $x \in Z$, we conclude that $Y \subseteq Z$. If $x \in X$, from the definition of "intersection," we conclude that $x \in X \cap Y$. Since $X \cap Y = X \cap Z$, $x \in X \cap Z$. Therefore $x \in Z$, and again $Y \subseteq Z$.

The argument that $Z \subseteq Y$ is the same as that for $Y \subseteq Z$ with the roles of $Y$ and $Z$ reversed. Thus $Y = Z$.

**30.** Since $X \in \mathcal{P}(X)$, $X \in \mathcal{P}(Y)$. Therefore $X \subseteq Y$.

**33.** False. If $X = \{1, 2\}$ and $Y = \{2, 3\}$, then $X$ is not a subset of $Y$ since $1 \in X$, but $1 \notin Y$. Also, $Y$ is not a subset of $X$ since $3 \in Y$, but $3 \notin X$.

**36.** False. Let $X = \{1, a\}$, $Y = \{1, 2, 3\}$, and $Z = \{3\}$. Then $Y - Z = \{1, 2\}$ and $(X \cup Y) - (X \cup Z) = \{2\}$.

**39.** False. Let $X = Y = \{1\}$ and $U = \{1, 2\}$. Then $\overline{X \cap Y} = \{2\}$, which is not a subset of $X$.

**42.** False. Let $X = Y = \{1\}$ and $U = \{1, 2\}$. Then $\overline{X \times Y} = \{(1, 2), (2, 1), (2, 2)\}$ and $\overline{X} \times \overline{Y} = \{(2, 2)\}$.

**45.** False. Let $X = \{1, 2\}$, $Y = \{1\}$ and $Z = \{2\}$. Then $X \cap (Y \times Z) = \varnothing$ and $(X \cap Y) \times (X \cap Z) = \{(1, 2)\}$.

**46.** We prove only $(A \cup B) \cup C = A \cup (B \cup C)$. Let $x \in (A \cup B) \cup C$. Then $x \in A \cup B$ or $x \in C$. If $x \in A \cup B$, then $x \in A$ or $x \in B$. Thus $x \in A$ or $x \in B$ or $x \in C$. If $x \in A$, then $x \in A \cup (B \cup C)$. If $x \in B$ or $x \in C$, then $x \in B \cup C$. Again, $x \in A \cup (B \cup C)$.

Now suppose that $x \in A \cup (B \cup C)$. Then $x \in A$ or $x \in B \cup C$. If $x \in B \cup C$, then $x \in B$ or $x \in C$. Thus $x \in A$ or $x \in B$ or $x \in C$. If $x \in A$ or $x \in B$, then $x \in A \cup B$. Thus

$x \in (A \cup B) \cup C$. If $x \in C$, again $x \in (A \cup B) \cup C$. Therefore $(A \cup B) \cup C = A \cup (B \cup C)$.

**49.** We prove only $A \cup \varnothing = A$. Let $x \in A \cup \varnothing$. Then $x \in A$ or $x \in \varnothing$. But $x \notin \varnothing$, so $x \in A$.

Now suppose that $x \in A$. Then $x \in A \cup \varnothing$. Therefore $A \cup \varnothing = A$.

**52.** We prove only $A \cup U = U$. By definition, any set is a subset of the universal set, so $A \cup U \subseteq U$.

If $x \in U$, then $x \in A \cup U$. Thus $U \subseteq A \cup U$. Therefore $A \cup U = U$.

**55.** We prove only $\overline{\varnothing} = U$. By definition, any set is a subset of the universal set, so $\overline{\varnothing} \subseteq U$.

Now suppose that $x \in U$. Then $x \notin \varnothing$ (by the definition of "empty set"). Thus $x \in \overline{\varnothing}$ and $U \subseteq \overline{\varnothing}$. Therefore $\overline{\varnothing} = U$.

**57.** Let $x \in A \triangle B$. Then $x \in A \cup B$ and $x \notin A \cap B$. Since $x \in A \cup B$, $x \in A$ or $x \in B$. Since $x \notin A \cap B$, $x \notin A$ or $x \notin B$. If $x \in A$, then $x \notin B$. Thus $x \in A - B$; hence $x \in (A - B) \cup (B - A)$. If $x \in B$, then $x \notin A$. Thus $x \in B - A$ and again $x \in (A - B) \cup (B - A)$. Therefore $A \triangle B \subseteq (A - B) \cup (B - A)$.

Now suppose that $x \in (A - B) \cup (B - A)$. Then $x \in A - B$ or $x \in B - A$. If $x \in A - B$, then $x \in A$ and $x \notin B$. Thus $x \in A \cup B$ and $x \notin A \cap B$. Therefore $x \in (A \cup B) - (A \cap B) = A \triangle B$. If $x \in B - A$, then $x \in B$ and $x \notin A$. Then $x \in A \cup B$ and $x \notin A \cap B$. Again $x \in (A \cup B) - (A \cap B) = A \triangle B$. Therefore $(A - B) \cup (B - A) \subseteq A \triangle B$. We have proved that $(A - B) \cup (B - A) = A \triangle B$.

**60.** False. Let $A = \{1, 2, 3\}$, $B = \{2, 3, 4\}$, and $C = \{1, 2, 4\}$. Then $A \triangle (B \cup C) = \{4\}$ and $(A \triangle B) \cup (A \triangle C) = \{1, 4\} \cup \{3, 4\} = \{1, 3, 4\}$.

**63.** True. Using Example 2.1.11,, we find that

$$\begin{aligned} A \cap (B \triangle C) &= A \cap [(B \cup C) - (B \cap C)] \\ &= [A \cap (B \cup C)] - [A \cap (B \cap C)]. \end{aligned}$$

Using the distributive law and observing that $(A \cap B) \cap (A \cap C) = A \cap (B \cap C)$, we find that

$$\begin{aligned} (A \cap B) \triangle (A \cap C) &= [(A \cap B) \cup (A \cap C)] - [(A \cap B) \cap (A \cap C)] \\ &= [A \cap (B \cup C)] - [A \cap (B \cap C)]. \end{aligned}$$

Therefore

$$A \cap (B \triangle C) = (A \cap B) \triangle (A \cap C).$$

## Section 2.2 Review

1. A proof by contradiction assumes that the hypotheses are true and that the conclusion is false and then, using the hypotheses and the negated conclusion as well as other axioms, definitions, and previously derived theorems, derives a contradiction.
2. Example 2.2.1
3. "Indirect proof" is another name for proof by contradiction.
4. To prove $p \to q$, proof by contrapositive proves the equivalent statement $\neg q \to \neg p$.
5. Example 2.2.5
6. Instead of proving
$$(p_1 \vee p_2 \vee \cdots \vee p_n) \to q,$$
in proof by cases, we prove
$$(p_1 \to q) \wedge (p_2 \to q) \wedge \cdots \wedge (p_n \to q).$$
7. Example 2.2.6
8. Proof of equivalence shows that two or more statements are all true or all false.
9. Example 2.2.11
10. If the statements are $p$, $q$, and $r$, we can show that they are equivalent by proving that $p \to q$, $q \to r$, and $r \to p$ are all true.
11. A proof of $\exists x\, P(x)$ is called an existence proof.
12. An existence proof of $\exists x\, P(x)$ that exhibits an element $a$ of the domain of discourse that makes $P(a)$ true is called a constructive proof.
13. Example 2.2.12
14. A proof of $\exists x\, P(x)$ that does not exhibit an element $a$ of the domain of discourse that makes $P(a)$ true, but rather proves $\exists x\, P(x)$ some other way (e.g., using proof by contradiction), is called a nonconstructive proof.
15. Example 2.2.14

## Section 2.2

**1.** Suppose, by way of contradiction, that $x$ is rational. Then there exist integers $p$ and $q$ such that $x = p/q$. Now $x^2 = p^2/q^2$ is rational, which is a contradiction.

**4.** Suppose, by way of contradiction, that $n$ is even. Then there exists $k$ such that $n = 2k$. Now $n^2 = 2(2k^2)$; thus $n^2$ is even, which is a contradiction.

**7.** Suppose, by way of contradiction, that $\sqrt[3]{2}$ is rational. Then there exist integers $p$ and $q$ such that $\sqrt[3]{2} = p/q$. We assume that the fraction $p/q$ is in lowest terms so that $p$ and $q$ are not both even. Cubing $\sqrt[3]{2} = p/q$ gives $2 = p^3/q^3$, and multiplying by $q^3$ gives $2q^3 = p^3$. It follows that $p^3$ is even. An argument like that in Example 2.2.1 shows that $p$ is even. Therefore, there exists an integer $k$ such that $p = 2k$. Substituting $p = 2k$ into $2q^3 = p^3$ gives $2q^3 = (2k)^3 = 8k^3$. Canceling 2 gives $q^3 = 4k^3$. Therefore $q^3$ is even and, thus, $q$ is even. Thus $p$ and $q$ are both even, which contradicts our assumption that $p$ and $q$ are not both even. Therefore, $\sqrt[3]{2}$ is irrational.

**10.** Since the integers increase without bound, there exists $n \in \mathbf{Z}$ such that $1/(b - a) < n$. Therefore $1/n < b - a$. Choose $m \in \mathbf{Z}$ as large as possible satisfying $m/n \le a$. Then, by the choice of $m$, $a < (m + 1)/n$. Also

$$\frac{m+1}{n} = \frac{m}{n} + \frac{1}{n} < a + (b - a) = b.$$

Therefore $x = (m + 1)/n$ is a rational number satisfying $a < x < b$.

**13.** True. Let $a = b = 2$. Then $a$ and $b$ are rational numbers and $a^b = 4$ is also rational. This is a constructive existence proof.

**16.** A team can score any number of points greater than or equal to two. To prove this, let $n$ be an integer greater than or equal to 2. If $n$ is even, say $n = 2k$, if the team scores $k$ safeties, its total score is $2k = n$. If $n$ is odd, say $n = 2k + 1$, if the team scores one field goal and $k - 1$ safeties, its total score is $3 + 2(k-1) = 2k+1 = n$. (Note that $k \geq 1$, so the prescribed scoring is possible.)

**19.** True. We give a proof by contradiction. Suppose that $(X-Y) \cap (Y-X)$ is nonempty. Then there exists $x \in (X-Y) \cap (Y-X)$. Thus $x \in X - Y$ and $x \in Y - X$. Since $x \in X - Y$, $x \in X$ and $x \notin Y$. Since $x \in Y - X$, $x \in Y$ and $x \notin X$. We now have $x \in X$ and $x \notin X$, which is a contradiction. Therefore $(X-Y) \cap (Y-X) = \varnothing$.

**22.** Suppose, by way of contradiction, that no two bags contain the same number of coins. Suppose that we arrange the bags in increasing order of the number of coins that they contain. Then the first bag contains at least one coin; the second bag contains at least two coins; and so on. Thus the total number of coins is at least

$$1 + 2 + 3 + \cdots + 9 = 45.$$

This contradicts the hypothesis that there are 40 coins. Thus if 40 coins are distributed among nine bags so that each bag contains at least one coin, at least two bags contain the same number of coins.

**25.** Let $i$ be the greatest integer for which $s_i$ is positive. Since $s_1$ is positive and the set of indexes $1, 2, \ldots, n$ is finite, such an $i$ exists. Since $s_n$ is negative, $i < n$. Now $s_{i+1}$ is equal to either $s_i + 1$ or $s_i - 1$. If $s_{i+1} = s_i + 1$, then $s_{i+1}$ is a positive integer (since $s_i$ is a positive integer). This contradicts the fact that $i$ is the *greatest* integer for which $s_i$ is positive. Therefore, $s_{i+1} = s_i - 1$. Again, if $s_i - 1$ is a positive integer, we have contradiction. Therefore, $s_{i+1} = s_i - 1 = 0$.

**27.** We use proof by contradiction and assume the negation of the conclusion

$$\neg\exists i(s_i \leq A).$$

By the generalized De Morgan's laws for logic, this latter statement is equivalent to

$$\forall i(s_i > A).$$

Thus we assume

$$s_1 > A$$
$$s_2 > A$$
$$\vdots$$
$$s_n > A.$$

Adding these inequalities yields

$$s_1 + s_2 + \cdots + s_n > nA.$$

Dividing by $n$ gives

$$\frac{s_1 + s_2 + \cdots + s_n}{n} > A,$$

which contradicts the hypothesis. Therefore, there exists $i$ such that $s_i \leq A$.

**30.** Since $s_i \neq s_j$, either $s_i \neq A$ or $s_j \neq A$. By changing the notation, if necessary, we may assume that $s_i \neq A$. Either $s_i < A$ or $s_i > A$. If $s_i < A$, the proof is complete; so assume that $s_i > A$. We show that there exists $k$ such that $s_k < A$. Suppose, by way of contradiction, that $s_m \geq A$ for all $m$, that is,

$$s_1 \geq A$$
$$s_2 \geq A$$
$$\vdots$$
$$s_n \geq A.$$

Adding these inequalities yields

$$s_1 + s_2 + \cdots + s_i + \cdots + s_n > nA$$

since $s_i > A$. Dividing by $n$ gives

$$\frac{s_1 + s_2 + \cdots + s_n}{n} > A,$$

which is a contradiction. Therefore there exists $k$ such that $s_k < A$.

**32.** Notice that if $n \geq 2$ and $m \geq 1$,

$$2m + 5n^2 \geq 2m + 5 \cdot 2^2 > 20,$$

so the only possible solution is if $n = 1$. However, if $n = 1$,

$$2m + 5n^2 = 2m + 5,$$

which is odd being the sum of an even integer and an odd integer. Thus this sum cannot equal 20. Therefore, $2m + 5n^2 = 20$ has no solution in positive integers.

**35.** We claim that if $n$ and $n+1$ are consecutive integers, one is odd and one is even. Suppose that $n$ is odd. Then there exists $k$ such that $n = 2k + 1$. Now $n + 1 = 2k + 2 = 2(k+1)$, which is even. If $n$ is even, there exists $k$ such that $n = 2k$. Now $n + 1 = 2k + 1$, which is odd. Since one of $n$ and $n+1$ is even and the other is odd, their product is even (see Exercise 11, Section 2.1).

**37.** We consider four cases: $x \geq 0, y \geq 0$, $x < 0, y \geq 0$; $x \geq 0, y < 0$; and $x < 0, y < 0$.

First assume that $x \geq 0$ and $y \geq 0$. Then $xy \geq 0$ and $|xy| = xy = |x||y|$. Next assume that $x < 0$ and $y \geq 0$. Then $xy \leq 0$ and $|xy| = -xy = (-x)(y) = |x||y|$. Next assume that $x \geq 0$ and $y < 0$. Then $xy \leq 0$ and $|xy| = -xy = (x)(-y) = |x||y|$. Finally assume that $x < 0$ and $y < 0$. Then $xy > 0$ and $|xy| = xy = (-x)(-y) = |x||y|$.

**39.** We consider three cases: $x > 0$, $x = 0$, and $x < 0$. If $x > 0$, $|x| = x$ and $\operatorname{sgn}(x) = 1$. Therefore,

$$|x| = x = 1 \cdot x = \operatorname{sgn}(x)x.$$

If $x = 0$, $|x| = 0$ and $\operatorname{sgn}(x) = 0$. Therefore,

$$|x| = 0 = 0 \cdot 0 = \operatorname{sgn}(x)x.$$

If $x < 0$, $|x| = -x$ and $\operatorname{sgn}(x) = -1$. Therefore,

$$|x| = -x = -1 \cdot x = \operatorname{sgn}(x)x.$$

In every case, we have $|x| = \operatorname{sgn}(x)x$.

**42.** We consider two cases: $x \geq y$ and $x < y$.

If $x \geq y$,

$$\max\{x, y\} = x \quad \text{and} \quad \min\{x, y\} = y.$$

Therefore,

$$\max\{x, y\} + \min\{x, y\} = x + y.$$

If $x < y$,

$$\max\{x, y\} = y \quad \text{and} \quad \min\{x, y\} = x.$$

Therefore,

$$\max\{x, y\} + \min\{x, y\} = y + x = x + y.$$

In either case,

$$\max\{x, y\} + \min\{x, y\} = x + y.$$

**46.** We first prove that if $n$ is even, then $n + 2$ is even. Assume that $n$ is even. Then there exists $k$ such that $n = 2k$. Now $n + 2 = 2k + 2 = 2(k + 1)$ is even.

We next prove that if $n + 2$ is even, then $n$ is even. Assume that $n + 2$ is even. Then there exists $k$ such that $n + 2 = 2k$. Now $n = 2k - 2 = 2(k - 1)$ is even.

**48.** We first prove that if $A \subseteq B$, then $\overline{B} \subseteq \overline{A}$. Assume that $A \subseteq B$. Let $x \in \overline{B}$. Then $x \notin B$. If $x \in A$, then $x \in B$, which is not the case. Therefore $x \notin A$. Therefore $x \in \overline{A}$. Thus $\overline{B} \subseteq \overline{A}$.

We next prove that if $\overline{B} \subseteq \overline{A}$, then $A \subseteq B$. Assume that $\overline{B} \subseteq \overline{A}$. From the first part of the proof, we can deduce $\overline{\overline{A}} \subseteq \overline{\overline{B}}$. Since $\overline{\overline{A}} = A$ and $\overline{\overline{B}} = B$, $A \subseteq B$.

**51.** We first show that if $(a, b) = (c, d)$, then $a = c$ and $b = d$. Assume that $(a, b) = (c, d)$. Then

$$\{\{a\}, \{a, b\}\} = \{\{c\}, \{c, d\}\}. \qquad (*)$$

First suppose that $a \neq b$. Then the set on the left-hand side of equation (*) contains two distinct sets: $\{a\}$ and $\{a, b\}$. Thus the set on the right-hand side of equation (*) also contains two distinct sets: $\{c\}$ and $\{c, d\}$. Therefore $c \neq d$. (If $c = d$, $\{c, d\} = \{c, c\} = \{c\}$.) Since $a \neq b$ and $c \neq d$, we must have

$$\{a\} = \{c\} \quad \text{and} \quad \{a, b\} = \{c, d\}.$$

It follows that $a = c$ and $b = d$.

Now suppose that $a = b$. Then

$$\{\{a\}, \{a, b\}\} = \{\{a\}, \{a, a\}\} = \{\{a\}, \{a\}\} = \{\{a\}\}.$$

Thus the set on the left-hand side of equation (*) contains one set. Therefore the set on the right-hand side of equation (*) also contains one set. We must have $c = d$; otherwise the set on the right-hand side of equation (*) would contain two distinct sets. Thus

$$\{\{c\}, \{c, d\}\} = \{\{c\}\}.$$

Equation (*) now becomes

$$\{\{a\}\} = \{\{c\}\}.$$

It follows that $a = c$ and, hence, $b = d$. We have shown that if $(a, b) = (c, d)$, then $a = c$ and $b = d$.

If $a = c$ and $b = d$, then

$$(a, b) = \{\{a\}, \{a, b\}\} = \{\{c\}, \{c, d\}\} = (c, d).$$

The proof is complete.

**52.** [(a) $\rightarrow$ (b)] Assume that $n$ is odd. Then there exists $k'$ such that $n = 2k' + 1$. Since $n = 2(k' + 1) - 1$, taking $k = k' + 1$, we have $n = 2k - 1$.

[(b) $\rightarrow$ (c)] Assume that there exists $k$ such that $n = 2k - 1$. Then

$$n^2 + 1 = (2k - 1)^2 + 1 = (4k^2 - 4k + 1) + 1 = 2(2k^2 - 2k + 1).$$

Therefore $n^2 + 1$ is even.

[(c) $\rightarrow$ (a)] We prove the contrapositive: If $n$ is even, then $n^2 + 1$ is odd.

Suppose that $n$ is even. Then there exists $k$ such that $n = 2k$. Now

$$n^2 + 1 = (2k)^2 + 1 = 4k^2 + 1 = 2(2k^2) + 1.$$

Therefore $n^2 + 1$ is odd.

## Section 2.3 Review

1. $p \vee q$, $\neg p \vee r / \therefore q \vee r$
2. A clause consists of terms separated by *or*'s, where each term is a variable or the negation of a variable.
3. A proof by resolution proceeds by repeatedly applying the rule in Exercise 1 to pairs of statements to derive new statements until the conclusion is derived.

## Section 2.3

**1.**

| $p$ | $q$ | $r$ | $p \vee q$ | $\neg p \vee r$ | $q \vee r$ |
|---|---|---|---|---|---|
| T | T | T | T | T | T |
| T | T | F | T | F | T |
| T | F | T | T | T | T |
| T | F | F | T | F | F |
| F | T | T | T | T | T |
| F | T | F | T | T | T |
| F | F | T | F | T | T |
| F | F | F | F | T | F |

**2.** 1. $\neg p \vee q \vee r$
2. $\neg q$
3. $\neg r$
4. $\neg p \vee r$     From 1 and 2
5. $\neg p$     From 3 and 4

**5.** First we note that $p \rightarrow q$ is logically equivalent to $\neg p \vee q$. We now argue as follows:

1. $\neg p \vee q$
2. $p \vee q$
3. $q$     From 1 and 2

**7.** (For Exercise 2)

| | | |
|---|---|---|
| 1. | $\neg p \vee q \vee r$ | Hypothesis |
| 2. | $\neg q$ | Hypothesis |
| 3. | $\neg r$ | Hypothesis |
| 4. | $p$ | Negation of conclusion |
| 5. | $\neg p \vee r$ | From 1 and 2 |
| 6. | $\neg p$ | From 3 and 5 |

Now 4 and 6 combine to give a contradiction.

## Section 2.4 Review

**1.** Suppose that we have propositional function $S(n)$ whose domain of discourse is the set of positive integers. Suppose that $S(1)$ is true and, for all $n \geq 1$, if $S(n)$ is true, then $S(n+1)$ is true. Then $S(n)$ is true for every positive integer $n$.

**2.** We first verify that $S(1)$ is true (Basis Step). We then assume that $S(n)$ is true and prove that $S(n+1)$ is true (Inductive Step).

**3.** $\dfrac{n(n+1)}{2}$

**4.** The geometric sum is the sum

$$a + ar^1 + ar^2 + \cdots + ar^n.$$

It is equal to

$$\frac{a(r^{n+1} - 1)}{r - 1}.$$

## Section 2.4

**1. Basis Step** $1 = 1^2$

**Inductive Step** Assume true for $n$.

$$1 + \cdots + (2n-1) + (2n+1) = n^2 + 2n + 1 = (n+1)^2$$

**4. Basis Step** $1^2 = (1 \cdot 2 \cdot 3)/6$

**Inductive Step** Assume true for $n$.

$$\begin{aligned} 1^2 + \cdots + n^2 + (n+1)^2 &= \frac{n(n+1)(2n+1)}{6} + (n+1)^2 \\ &= \frac{(n+1)(n+2)(2n+3)}{6} \end{aligned}$$

**7. Basis Step** $1/(1 \cdot 3) = 1/3$

**Inductive Step** Assume true for $n$.

$$\begin{aligned} &\frac{1}{1 \cdot 3} + \cdots + \frac{1}{(2n-1)(2n+1)} + \frac{1}{(2n+1)(2n+3)} \\ &= \frac{n}{2n+1} + \frac{1}{(2n+1)(2n+3)} = \frac{n+1}{2n+3} \end{aligned}$$

**11. Basis Step** $\cos x = \dfrac{\cos[(x/2) \cdot 2] \sin(x/2)}{\sin(x/2)}$

**Inductive Step** Assume true for $n$. Then

$$\begin{aligned} &\cos x + \cdots + \cos nx + \cos(n+1)x \\ &= \frac{\cos[(x/2)(n+1)] \sin(nx/2)}{\sin(x/2)} + \cos(n+1)x. \quad (*) \end{aligned}$$

We must show that the right-hand side of $(*)$ is equal to

$$\frac{\cos[(x/2)(n+2)] \sin[(n+1)x/2]}{\sin(x/2)}.$$

This is the same as showing that [after multiplying by the term $\sin(x/2)$]

$$\begin{aligned} &\cos\left[\frac{x}{2}(n+1)\right] \sin\frac{nx}{2} + \cos(n+1)x \sin\frac{x}{2} \\ &= \cos\left[\frac{x}{2}(n+2)\right] \sin\left[\frac{(n+1)x}{2}\right]. \end{aligned}$$

If we let $\alpha = (x/2)(n+1)$ and $\beta = x/2$, we must show that

$$\cos\alpha \sin(\alpha - \beta) + \cos 2\alpha \sin\beta = \cos(\alpha + \beta) \sin\alpha.$$

This last equation can be verified by reducing each side to terms involving $\alpha$ and $\beta$.

**13. Basis Step** $1/2 \leq 1/2$

**Inductive Step** Assume true for $n$.

$$\begin{aligned} \frac{1 \cdot 3 \cdot 5 \cdots (2n-1)(2n+1)}{2 \cdot 4 \cdot 6 \cdots (2n)(2n+2)} &\geq \frac{1}{2n} \cdot \frac{2n+1}{2n+2} \\ &= \frac{2n+1}{2n} \cdot \frac{1}{2n+2} \geq \frac{1}{2n+2} \end{aligned}$$

**16. Basis Step ($n = 4$)** $2^4 = 16 \geq 16 = 4^2$

**Inductive Step** Assume true for $n$.

$$\begin{aligned} (n+1)^2 = n^2 + 2n + 1 &\leq 2^n + 2n + 1 \\ &\leq 2^n + 2^n \quad \text{by Exercise 15} \\ &= 2^{n+1} \end{aligned}$$

**19.** $r^0 + r^1 + \cdots + r^n = \dfrac{1 - r^{n+1}}{1 - r} < \dfrac{1}{1 - r}$

**22. Basis Step** $7^1 - 1 = 6$ is divisible by 6.

**Inductive Step** Suppose that 6 divides $7^n - 1$. Now

$$7^{n+1} - 1 = 7 \cdot 7^n - 1 = 7^n - 1 + 6 \cdot 7^n.$$

Since 6 divides both $7^n - 1$ and $6 \cdot 7^n$, it divides their sum, which is $7^{n+1} - 1$.

**25. Basis Step** $3^1 + 7^1 - 2 = 8$ is divisible by 8.

**Inductive Step** Suppose that 8 divides $3^n + 7^n - 2$. Now

$$3^{n+1} + 7^{n+1} - 2 = 3(3^n + 7^n - 2) + 4(7^n + 1).$$

By the inductive assumption, 8 divides $3^n + 7^n - 2$. We can use mathematical induction to show that 2 divides $7^n + 1$ for all $n \geq 1$ (the argument is similar to that given in the hint for Exercise 22). It then follows that 8 divides $4(7^n + 1)$. Since 8 divides both $3(3^n + 7^n - 2)$ and $4(7^n + 1)$, it divides their sum, which is $3^{n+1} + 7^{n+1} - 2$.

**28.** We prove the assertion using induction on $n$. The Basis Step is $n = 1$. In this case, there is one subset of $\{1\}$ with an even number of elements, namely, $\varnothing$. Since $2^{n-1} = 2^0 = 1$, the assertion is true when $n = 1$.

Assume that the number of subsets of $\{1, \ldots, n\}$ containing an even number of elements is $2^{n-1}$. We must prove that the number of subsets of $\{1, \ldots, n+1\}$ containing an even number of elements is $2^n$.

Let $E_1, \ldots, E_{2^{n-1}}$ denote the subsets of $\{1, 2, \ldots, n\}$ containing an even number of elements. Since there are $2^n$ subsets of $\{1, 2, \ldots, n\}$ altogether and $2^{n-1}$ contain an even number of elements, there are $2^n - 2^{n-1} = 2^{n-1}$ subsets of $\{1, \ldots, n\}$ that contain an odd number of elements. Denote these as $O_1, \ldots, O_{2^{n-1}}$. Now $E_1, \ldots, E_{2^{n-1}}$ are the subsets of $\{1, \ldots, n+1\}$ containing an even number of elements that do not contain $n+1$, and

$$O_1 \cup \{n+1\}, \ldots, O_{2^{n-1}} \cup \{n+1\}$$

are the subsets of $\{1, \ldots, n+1\}$ containing an even number of elements that contain $n+1$. Thus there are $2^{n-1} + 2^{n-1} = 2^n$ subsets of $\{1, \ldots, n+1\}$ that contain an even number of elements. The Inductive Step is complete.

**30.** At the Inductive Step when the $(n+1)$st line is added, because of the assumptions, the line will intersect each of the other $n$ lines. Now, imagine traveling along the $(n+1)$st line. Each time we pass through one of the original regions, it is divided into two regions.

**33.**

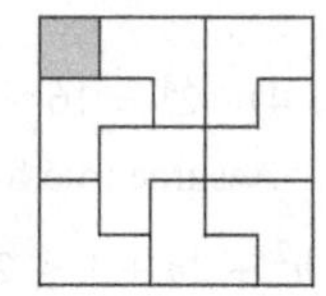

**37.** We denote the square in row $i$, column $j$ by $(i, j)$. Then, by symmetry, we need only consider $7 \times 7$ boards with squares $(i, j)$ removed where $i \le j \le 4$. The solution when square $(1, 1)$ is removed is shown in the following figure.

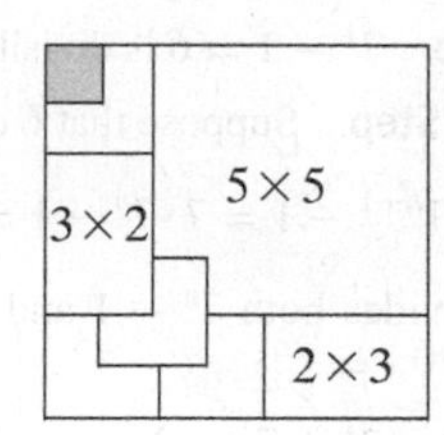


Not all trominoes of the tiling are shown. By Exercise 36, the 3×2 subboards have tilings. By Exercise 33, the 5×5 subboard with a corner square removed has a tiling. Essentially the same figure gives tilings if square (1, 2) or (2, 2) is deleted.

A similar argument gives tilings for the remaining cases.

**40. Basis Step ($n = 1$)** The board is a tromino.

**Inductive Step** Assume that any $2^n \times 2^n$ deficient board can be tiled with trominoes. We must prove that any $2^{n+1} \times 2^{n+1}$ deficient board can be tiled with trominoes.

Given a $2^{n+1} \times 2^{n+1}$ deficient board, we divide the board into four $2^n \times 2^n$ subboards as shown in Figure 2.4.6. By the inductive assumption, we can tile the subboard containing the missing square. The three remaining subboards form a $2^n \times 2^n$ L-shape, which may be tiled using Exercise 39. Thus, the $2^{n+1} \times 2^{n+1}$ deficient board is tiled. The inductive step is complete.

**41.** Number the squares as shown:

| | | | |
|---|---|---|---|
| 1 | 2 | 3 | 1 |
| 2 | 3 | 1 | 2 |
| 3 | 1 | 2 | 3 |
| 1 | 2 | 3 | 1 |

Notice that each tromino covers exactly one 1, one 2, and one 3. Therefore, if there is a tiling, five 2's are covered. Since five trominoes are required, the missing square cannot be a 2. Similarly the missing square cannot be a 3.

The same argument applied to

| | | | |
|---|---|---|---|
| 1 | 3 | 2 | 1 |
| 2 | 1 | 3 | 2 |
| 3 | 2 | 1 | 3 |
| 1 | 3 | 2 | 1 |

shows that the only possibility for the missing square is a corner. Such a board can be tiled:

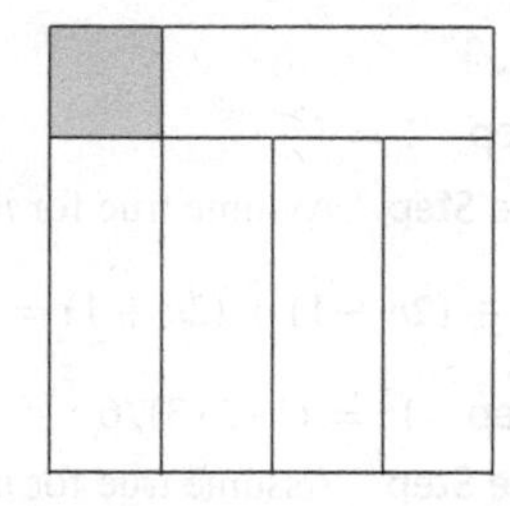

**45.** We prove that $pow = a^{i-1}$ is a loop invariant for the while loop. Just before the while loop begins executing, $i = 1$ and $pow = 1$, so $pow = a^{1-1}$. We have proved the Basis Step.

Assume that $pow = a^{i-1}$. If $i \le n$ (so that the loop body executes again), $pow$ becomes

$$pow * a = a^{i-1} * a = a^i,$$

and $i$ becomes $i+1$. We have proved the Inductive Step. Therefore $pow = a^{i-1}$ is an invariant for the while loop.

The while loop terminates when $i = n + 1$. Because $pow = a^{i-1}$ is an invariant, at this point $pow = a^n$.

**49.** (a) $S_1 = 0 \ne 2$;

$$\begin{aligned} 2 + \cdots + 2n + 2(n+1) &= S_n + 2n + 2 \\ &= (n+2)(n-1) + 2n + 2 \\ &= (n+3)n = S_{n+1}. \end{aligned}$$

(b) We must have $S'_n = S'_{n-1} + 2n$; thus

$$
\begin{aligned}
S'_n &= S'_{n-1} + 2n \\
&= [S'_{n-2} + 2(n-1)] + 2n \\
&= S'_{n-2} + 2n + 2(n-1) \\
&= S'_{n-3} + 2n + 2(n-1) + 2(n-2) \\
&\vdots \\
&= S'_1 + 2[n + (n-1) + \cdots + 2] \\
&= C' + 2\left[\frac{n(n+1)}{2} - 1\right] \\
&= n^2 + n + C.
\end{aligned}
$$

**53.** If $n = 2$, each person throws a pie at the other and there are no survivors.

**56.** The statement is false. In

1 2 3 4 5

1 and 5 are farthest apart, but neither is a survivor.

**58.** Let $x$ and $y$ be points in $X \cap Y$. Then $x$ is in $X$ and $y$ is in $X$. Since $X$ is convex, the line segment from $x$ to $y$ is in $X$. Similarly, the line segment from $x$ to $y$ is in $Y$. Therefore, the line segment from $x$ to $y$ is in $X \cap Y$. Thus, $X \cap Y$ is convex.

**61.** Let $x_1, \ldots, x_n$ denote the $n$ points, and let $X_i$ be the circle of radius 1 centered at $x_i$. Apply Helly's Theorem to $X_1, \ldots, X_n$.

**63.** 1

**66.** **Basis Step ($i = 1$)** Since 2 is eliminated, 1 survives. Thus $J(2) = 1$.

**Inductive Step** Assume true for $i$. Now suppose that $2^{i+1}$ persons are arranged in a circle. We begin by eliminating $2, 4, 6, \ldots, 2^{i+1}$. We then have $2^i$ persons arranged in a circle, and, beginning with 1, we eliminate the second person, then the fourth person, and so on. By the inductive assumption, 1 survives. Therefore,

$$J(2^{i+1}) = J(2^i) = 1.$$

**69.** The greatest power of 2 less than or equal to 100,000 is $2^{16}$. Thus, in the notation of Exercise 67, $n = 100{,}000$, $i = 16$, and

$$j = n - 2^i = 100{,}000 - 2^{16} = 100{,}000 - 65{,}536 = 34{,}464.$$

By Exercise 67,

$$J(100{,}000) = J(n) = 2j + 1 = 2 \cdot 34{,}464 + 1 = 68{,}929.$$

**70.**

$$
\begin{aligned}
b_1 + b_2 + \cdots + b_n &= (a_2 - a_1) + (a_3 - a_2) \\
&\quad + \cdots + (a_{n+1} - a_n) \\
&= -a_1 + a_{n+1} = a_{n+1} - a_1
\end{aligned}
$$

since $a_2, \ldots, a_n$ cancel.

**73.** Let

$$a_n = \frac{1}{n}.$$

Then

$$\Delta a_n = a_{n+1} - a_n = \frac{1}{n+1} - \frac{1}{n} = \frac{-1}{n(n+1)}.$$

Let $b_n = \Delta a_n$. By Exercise 70,

$$
\begin{aligned}
\frac{-1}{1 \cdot 2} + \cdots + \frac{-1}{n(n+1)} &= \Delta a_1 + \cdots + \Delta a_n \\
&= b_1 + \cdots + b_n \\
&= a_{n+1} - a_1 \\
&= \frac{1}{n+1} - 1 = \frac{-n}{n+1}.
\end{aligned}
$$

Multiplying by $-1$ yields the desired formula.

## Section 2.5 Review

**1.** Suppose that we have a propositional function $S(n)$ whose domain of discourse is the set of integers greater than or equal to $n_0$. Suppose that $S(n_0)$ is true; and for all $n > n_0$, if $S(k)$ is true for all $k$, $n_0 \le k < n$, then $S(n)$ is true. Then $S(n)$ is true for every integer $n \ge n_0$.

**2.** Every nonempty set of nonnegative integers has a least element.

**3.** If $d$ and $n$ are integers, $d > 0$, there exist unique integers $q$ (quotient) and $r$ (remainder) satisfying $n = dq + r$, $0 \le r < d$.

## Section 2.5

**1.** **Basis Steps ($n = 6, 7$)** We can make six cents postage by using three 2-cent stamps. We can make seven cents postage by using one 7-cent stamp.

**Inductive Step** We assume that $n \ge 8$ and postage of $k$ cents or more can be achieved by using only 2-cent and 7-cent stamps for $6 \le k < n$. By the inductive assumption, we can make postage of $n - 2$ cents. We may add a 2-cent stamp to make $n$ cents postage.

**4.** **Basis Step ($n = 4$)** We can make four cents postage by using two 2-cent stamps.

**Inductive Step** We assume that we can make $n$ cents postage, and we prove that we can make $n + 1$ cents postage.

If among the stamps that make $n$ cents postage there is at least one 5-cent stamp, we replace one 5-cent stamp by three 2-cent stamps to make $n + 1$ cents postage. If there are no 5-cent stamps among the stamps that make $n$ cents postage, there are at least two 2-cent stamps (because $n \ge 4$). We replace two 2-cent stamps by one 5-cent stamp to make $n + 1$ cents postage.

**7.** In the Inductive Step, we must have $k = \lfloor n/2 \rfloor \ge 3$. Since this inequality fails for $n = 4, 5$, the Basis Steps are $n = 3, 4, 5$.

**10.** $c_2 = 4, c_3 = 9, c_4 = 20, c_5 = 29$

**12.** $c_2 = 2, c_3 = 3, c_4 = 12, c_5 = 13$

**15.** Notice that

$$
\begin{aligned}
c_0 &= 0 \\
c_1 &= c_0 + 3 = 3 \\
c_2 &= c_1 + 3 = 6
\end{aligned}
$$

$$c_3 = c_1 + 3 = 6$$
$$c_4 = c_2 + 3 = 9.$$

Thus the assertion $c_n \le 2n$ fails for $n = 4$.

In the Inductive Step, we must have $k = \lfloor n/2 \rfloor \ge 3$. Since this inequality fails for $n = 4, 5$, the Basis Steps are $n = 3, 4, 5$. In the fallacious proof, only the case $n = 3$ was proved in the Basis Steps. In fact, since the statement is false for $n = 4$, the Basis Steps $n = 3, 4, 5$ cannot be proved.

**17.** $q = 5,\ r = 2$

**20.** $q = -1,\ r = 2$

**23.** $\frac{5}{6} = \frac{1}{2} + \frac{1}{3} = \frac{1}{2} + \frac{1}{4} + \frac{1}{12}$

**27.** We may assume that $p/q > 1$. Choose the largest integer $n$ satisfying

$$\frac{1}{1} + \frac{1}{2} + \cdots + \frac{1}{n} \le \frac{p}{q}.$$

(The previous Problem-Solving Corner shows that the sum

$$\frac{1}{1} + \frac{1}{2} + \cdots + \frac{1}{n}$$

is unbounded; so such an $n$ exists.) If we obtain an equality, $p/q$ is in Egyptian form, so suppose that

$$\frac{1}{1} + \frac{1}{2} + \cdots + \frac{1}{n} < \frac{p}{q}. \qquad (*)$$

Set

$$D = \frac{p}{q} - \left(\frac{1}{1} + \cdots + \frac{1}{n}\right).$$

Clearly, $D > 0$. Since $n$ is the largest integer satisfying $(*)$,

$$\frac{1}{1} + \frac{1}{2} + \cdots + \frac{1}{n} + \frac{1}{n+1} \ge \frac{p}{q}.$$

Thus

$$\begin{aligned} D &= \frac{p}{q} - \left(\frac{1}{1} + \cdots + \frac{1}{n}\right) \\ &\le \left(\frac{1}{1} + \cdots + \frac{1}{n} + \frac{1}{n+1}\right) - \left(\frac{1}{1} + \cdots + \frac{1}{n}\right) \\ &= \frac{1}{n+1}. \end{aligned}$$

In particular, $D < 1$. By Exercise 25, $D$ may be written in Egyptian form:

$$D = \frac{1}{n_1} + \cdots + \frac{1}{n_k},$$

where the $n_i$ are distinct. Since

$$\frac{1}{n_i} \le D \le \frac{1}{1+n}, \qquad \text{for } i = 1, \ldots, k,$$

$n < n + 1 \le n_i$ for $i = 1, \ldots, k$. It follows that

$$1, 2, \ldots, n, n_1, \ldots, n_k$$

are distinct. Thus

$$\frac{p}{q} = D + \frac{1}{1} + \cdots + \frac{1}{n} = \frac{1}{n_1} + \cdots + \frac{1}{n_k} + \frac{1}{1} + \cdots + \frac{1}{n}$$

is represented in Egyptian form.

**28.** In this solution, the induction is over the set $X$ of odd integers $n > 5$, where 3 divides $n^2 - 1$. Such an induction can be justified by considering the first statement to be about the smallest integer in $X$, the second statement to be about the second smallest integer in $X$, and so on.

**Basis Steps ($n = 7, 11$)** Exercise 37, Section 2.4, gives a solution if $n = 7$.

If $n = 11$, enclose the missing square in a corner $7 \times 7$ subboard (see the following figure). Tile this subboard using the result of Exercise 37, Section 2.4. Tile the two $6 \times 4$ subboards using the result of Exercise 36, Section 2.4. Tile the $5 \times 5$ subboard with a corner square missing using the result of Exercise 33, Section 2.4. Thus the $11 \times 11$ board can be tiled.

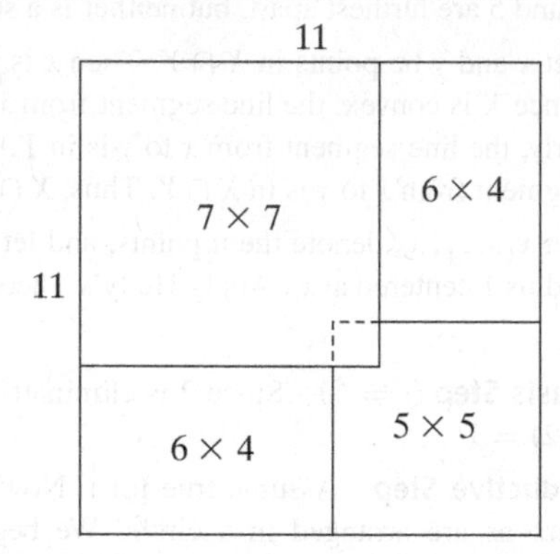


**Inductive Step** Suppose that $n > 11$ and assume that if $k < n$, $k$ is odd, $k > 5$, and 3 divides $k^2 - 1$, then a $k \times k$ deficient board can be tiled with trominoes.

Consider an $n \times n$ deficient board. Enclose the missing square in a corner $(n-6) \times (n-6)$ subboard. By the inductive assumption, this board can be tiled with trominoes. Tile the two $6 \times (n - 7)$ subboards using the result of Exercise 36, Section 2.4. Tile the deficient $7 \times 7$ subboard using the result of Exercise 37, Section 2.4. The $n \times n$ board is tiled, and the inductive step is complete.

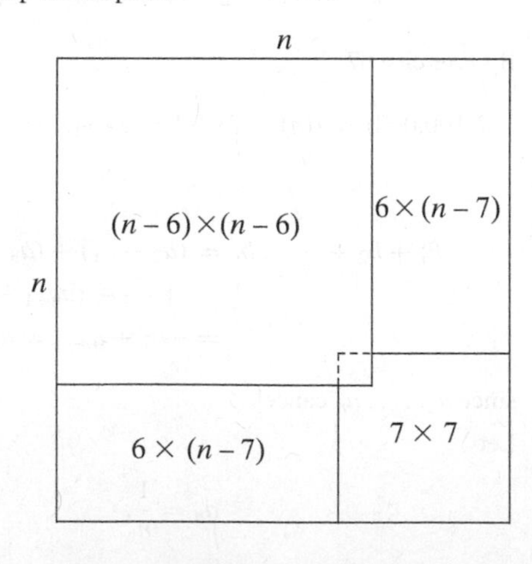

**31.**

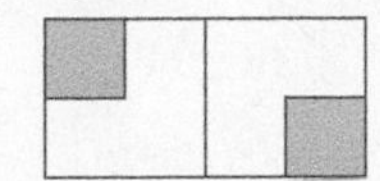

**36.** Let $X$ be a nonempty set of nonnegative integers. We must prove that $X$ has a least element. Using mathematical induction, we prove that for all $n \geq 0$, if $X$ contains an element less than or equal to $n$, then $X$ has a least element. Notice that this proves that $X$ has a least element. (Since $X$ is nonempty, $X$ contains an integer $n$. Now $X$ contains an element less than or equal to $n$; so it follow that $X$ has a least element.)

**Basis Step ($n = 0$)** If $X$ contains an element less than or equal to 0, then $X$ contains 0 since $X$ consists of nonnegative integers. In this case, 0 is the least element in $X$.

**Inductive Step** Now we assume that if $X$ contains an element less than or equal to $n$, then $X$ has a least element. We must prove that if $X$ contains an element less than or equal to $n+1$, then $X$ has a least element.

Suppose that $X$ contains an element less than or equal to $n+1$. We consider two cases: $X$ contains an element less than or equal to $n$, and $X$ does not contain an element less than or equal to $n$. If $X$ contains an element less than or equal to $n$, by the inductive assumption, $X$ has a least element. If $X$ does not contain an element less than or equal to $n$, since $X$ contains an element less than or equal to $n+1$, $X$ must contain $n+1$, which is the least element in $X$. The inductive step is complete.

## Chapter 2 Self-Test

**1.** [Section 2.1] Axioms are statements that are assumed to be true. Definitions are used to create new concepts in terms of existing ones.

**2.** [Section 2.2] In a direct proof, the negated conclusion is not assumed, whereas in a proof by contradiction, the negated conclusion is assumed.

**3.** [Section 2.2] Suppose that if four teams play seven games, no pair of teams plays at least two times; or, equivalently, if four teams play seven games, each pair of teams plays at most one time. If the teams are $A$, $B$, $C$, and $D$ and each pair of teams plays at most one time, the most games that can be played are:

$A$ and $B$; $A$ and $C$; $A$ and $D$; $B$ and $C$; $B$ and $D$; $C$ and $D$.

Thus at most six games can be played. This is a contradiction. Therefore, if four teams play seven games, some pair of teams plays at least two times.

**4.** [Section 2.1] Since $x$ and $y$ are rational numbers, there exist integers $m_1, n_1, m_2, n_2$ such that $x = m_1/n_1$ and $y = m_2/n_2$. Since $y \neq 0$, $m_2 \neq 0$. Now

$$\frac{x}{y} = \frac{m_1/n_1}{m_2/n_2} = \frac{m_1 n_2}{n_1 m_2}.$$

Since $x/y$ is the quotient of integers, it is rational.

**5.** [Section 2.2] We consider two cases: $a \leq b$ and $a > b$. In each of these two cases, we consider the two cases: $b \leq c$ and $b > c$.

First suppose that $a \leq b$. If $b \leq c$, then

$$\min\{\min\{a, b\}, c\} = \min\{a, c\} = a = \min\{a, b\} = \min\{a, \min\{b, c\}\}.$$

If $b > c$, then

$$\min\{\min\{a, b\}, c\} = \min\{a, c\} = \min\{a, \min\{b, c\}\}.$$

In either case,

$$\min\{\min\{a, b\}, c\} = \min\{a, \min\{b, c\}\}.$$

Now suppose that $a > b$. If $b \leq c$, then

$$\min\{\min\{a, b\}, c\} = \min\{b, c\} = b = \min\{a, b\} = \min\{a, \min\{b, c\}\}.$$

If $b > c$, then

$$\min\{\min\{a, b\}, c\} = \min\{b, c\} = c = \min\{a, c\} = \min\{a, \min\{b, c\}\}.$$

In either case,

$$\min\{\min\{a, b\}, c\} = \min\{a, \min\{b, c\}\}.$$

Therefore, for all $a, b, c$,

$$\min\{\min\{a, b\}, c\} = \min\{a, \min\{b, c\}\}.$$

*In Exercises 6–9, only the Inductive Step is given.*

**6.** [Section 2.4] $2 + 4 + \cdots + 2n + 2(n+1) = n(n+1) + 2(n+1) = (n+1)(n+2)$

**7.** [Section 2.4]

$$2^2 + 4^2 + \cdots + (2n)^2 + [2(n+1)]^2 = \frac{2n(n+1)(2n+1)}{3} + [2(n+1)]^2 = \frac{2(n+1)(n+2)[2(n+1)+1]}{3}$$

**8.** [Section 2.4]

$$\frac{1}{2!} + \frac{2}{3!} + \cdots + \frac{n}{(n+1)!} + \frac{n+1}{(n+2)!} = 1 - \frac{1}{(n+1)!} + \frac{n+1}{(n+2)!} = 1 - \frac{1}{(n+2)!}$$

**9.** [Section 2.4]

$$\begin{aligned} 2^{n+2} &= 2 \cdot 2^{n+1} < 2[1 + (n+1)2^n] = 2 + (n+1)2^{n+1} \\ &= 1 + [1 + (n+1)2^{n+1}] \\ &< 1 + [2^{n+1} + (n+1)2^{n+1}] \\ &= 1 + (n+2)2^{n+1} \end{aligned}$$

**10.** [Section 2.1] Suppose that $m$ and $m-n$ are odd. Then there exist integers $k_1$ and $k_2$ such that $m = 2k_1+1$ and $m-n = 2k_2+1$. Now

$$n = m - (m-n) = (2k_1 + 1) - (2k_2 + 1) = 2(k_1 - k_2).$$

Therefore $n$ is even.

**11.** [Section 2.2] [(a) $\rightarrow$ (b)] We prove the contrapositive: If $A \cap \overline{B} \neq \varnothing$, then $A$ is not a subset of $B$. Since $A \cap \overline{B} \neq \varnothing$, there exists $x$ with $x \in A$ and $x \in \overline{B}$. Thus there exists $x$ with $x \in A$ and $x \notin B$. Therefore $A$ is not a subset of $B$.

[(b) $\rightarrow$ (c)] If $x \in B$, then $x \in A \cup B$. Therefore $B \subseteq A \cup B$.

Let $x \in A \cup B$. We must show that $x \in B$. Now $x \in A$ or $x \in B$. If $x \in B$, this part of the proof is complete; so suppose that $x \in A$. Since $A \cap \overline{B} = \varnothing$, $x \notin \overline{B}$. Thus $x \in B$ and $A \cup B \subseteq B$. Therefore $A \cup B = B$.

[(c) $\rightarrow$ (a)] Let $x \in A$. Then $x \in A \cup B$. Since $A \cup B = B$, $x \in B$. Therefore $A \subseteq B$.

**12.** [Section 2.1] We first prove that $X \subseteq Z$. Let $x \in X$. Since $X \subseteq Y$, $x \in Y$. Since $Y \subset Z$, $x \in Z$. Therefore $X \subseteq Z$.

We now show that $X$ is a *proper* subset of $Z$. Since $Y \subset Z$, there exists $z \in Z$ such that $z \notin Y$. Now $z \notin X$, because if it were, we would have $z \in Y$. Therefore $X \subset Z$.

**13.** [Section 2.5] $q = 9$, $r = 2$

**14.** [Section 2.5] $c_2 = 2$, $c_3 = 3$, $c_4 = 8$, $c_5 = 9$.

**15.** [Section 2.5] **Basis Step ($n = 1$)** $c_1 = 0 \leq 0 = 1 \lg 1$

**Inductive Step**

$$\begin{aligned} c_n &= 2c_{\lfloor n/2 \rfloor} + n \\ &\leq 2\lfloor n/2 \rfloor \lg \lfloor n/2 \rfloor + n \\ &\leq 2(n/2)\lg(n/2) + n \\ &= n(\lg n - 1) + n = n \lg n \end{aligned}$$

**16.** [Section 2.5] Let $X$ be a nonempty set of nonnegative integers that has an upper bound. We must show that $X$ contains a largest element.

Let $Y$ be the set of integer upper bounds for $X$. By assumption, $Y$ is nonempty. Since $X$ consists of nonnegative integers, $Y$ also consists of nonnegative integers. By the Well-Ordering Property, $Y$ has a least element, say $n$. Since $Y$ consists of upper bounds for $X$, $k \leq n$ for every $k$ in $X$. Suppose, by way of contradiction, that $n$ is not in $X$. Then $k \leq n - 1$ for every $k$ in $X$. Thus, $n - 1$ is an upper bound for $X$, which is a contradiction. Therefore, $n$ is in $X$. Since $k \leq n$ for every $k$ in $X$, $n$ is the largest element in $X$.

**17.** [Section 2.3]

$$\begin{aligned} (p \vee q) \rightarrow r &\equiv \neg(p \vee q) \vee r \\ &\equiv \neg p \neg q \vee r \\ &\equiv (\neg p \vee r)(\neg q \vee r) \end{aligned}$$

**18.** [Section 2.3]

$$\begin{aligned} (p \vee \neg q) \rightarrow \neg rs &\equiv \neg(p \vee \neg q) \vee \neg rs \\ &\equiv \neg pq \vee \neg rs \\ &\equiv (\neg p \vee \neg r)(\neg p \vee s)(q \vee \neg r)(q \vee s) \end{aligned}$$

**19.** [Section 2.3]

1. $\neg p \vee q$
2. $\neg q \vee \neg r$
3. $p \vee \neg r$
4. $\neg p \vee \neg r$ From 1 and 2
5. $\neg r$ From 3 and 4

**20.** [Section 2.3]

1. $\neg p \vee q$
2. $\neg q \vee \neg r$
3. $p \vee \neg r$
4. $r$ Negation of conclusion
5. $\neg p \vee \neg r$ From 1 and 2
6. $\neg r$ From 3 and 5

Now 4 and 6 give a contradiction.

## Section 3.1 Review

**1.** Let $X$ and $Y$ be sets. A function $f$ from $X$ to $Y$ is a subset of the Cartesian product $X \times Y$ having the property that for each $x \in X$, there is exactly one $y \in Y$ with $(x, y) \in f$.

**2.** In an arrow diagram of the function $f$, there is an arrow from $i$ to $j$ if $(i, j) \in f$.

**3.** The graph of a function $f$, whose domain and codomain are subsets of the real numbers, consists of the points in the plane that correspond to the elements in $f$.

**4.** A set $S$ of points in the plane defines a function when each vertical line intersects at most one point of $S$.

**5.** The remainder when $x$ is divided by $y$

**6.** A hash function takes a data item to be stored or retrieved and computes the first choice for a location for the item.

**7.** A collision occurs for a hash function $H$ if $H(x) = H(y)$ but $x \neq y$.

**8.** When a collision occurs, a collision resolution policy determines an alternative location for one of the data items.

**9.** Pseudorandom numbers are numbers that appear random even though they are generated by a program.

**10.** A linear congruential random number generator uses a formula of the form

$$x_n = (ax_{n-1} + c) \bmod m.$$

Given the pseudorandom number $x_{n-1}$, the next pseudorandom number $x_n$ is given by the formula. A "seed" is used as the first pseudorandom number in the sequence. As an example, the formula

$$x_n = (7x_{n-1} + 5) \bmod 11$$

with seed 3 gives a sequence that begins 3, 4, 0, 5, ....

**11.** The floor of $x$ is the greatest integer less than or equal to $x$. It is denoted $\lfloor x \rfloor$.

**12.** The ceiling of $x$ is the least integer greater than or equal to $x$. It is denoted $\lceil x \rceil$.

**13.** A function $f$ from $X$ to $Y$ is said to be one-to-one if for each $y \in Y$, there is at most one $x \in X$ with $f(x) = y$. The function $\{(a, 1), (b, 3), (c, 0)\}$ is one-to-one. If a function from $X$ to $Y$ is one-to-one, each element in $Y$ in its arrow diagram will have at most one arrow pointing to it.

**14.** A function $f$ from $X$ to $Y$ is said to be onto $Y$ if the range of $f$ is $Y$. The function $\{(a, 1), (b, 3), (c, 0)\}$ is onto $\{0, 1, 3\}$. If a function from $X$ to $Y$ is onto $Y$, each element in $Y$ in its arrow diagram will have at least one arrow pointing to it.

**15.** A bijection is a function that is one-to-one and onto. The function of Exercises 13 and 14 is one-to-one and onto $\{0, 1, 3\}$.

**16.** If $f$ is a one-to-one, onto function from $X$ to $Y$, the inverse function is

$$f^{-1} = \{(y, x) \mid (x, y) \in f\}.$$

If $f$ is the function of Exercises 13 and 14, we have

$$f^{-1} = \{(1, a), (3, b), (0, c)\}.$$

Given the arrow diagram for a one-to-one, onto function $f$ from $X$ to $Y$, we can obtain the arrow diagram for $f^{-1}$ by reversing the direction of each arrow.

**17.** Suppose that $g$ is a function from $X$ to $Y$ and $f$ is a function from $Y$ to $Z$. The composition function from $X$ to $Z$ is defined as

$$f \circ g = \{(x, z) \mid (x, y) \in g \text{ and } (y, z) \in f \text{ for some } y \in Y\}.$$

If $g = \{(1, 2), (2, 2)\}$ and $f = \{(2, a)\}$, $f \circ g = \{(1, a), (2, a)\}$. Given the arrow diagrams for functions $g$ from $X$ to $Y$ and $f$ from $Y$ to $Z$, we can obtain the arrow diagram of $f \circ g$ by drawing an arrow from $x \in X$ to $z \in Z$ provided that there are arrows from $x$ to some $y \in Y$ and from $y$ to $z$.

**18.** A binary operator on $X$ is a function from $X \times X$ to $X$. The addition operator $+$ is a binary operator on the set of integers.

**19.** A unary operator on $X$ is a function from $X$ to $X$. The minus operator $-$ is a unary operator on the set of integers.

## Section 3.1

**1.** Valid

**4.** Invalid. The check digit should be 5.

**7.** When 82 is transposed to 28, the sum becomes

$$8+6+9+0+6+5+4+8+2+3+5+5+8+6+1 = 76$$

and the check digit becomes 4.

**8.** It is a function from $X$ to $Y$; domain $= X$, range $= \{a, b, c\}$; it is neither one-to-one nor onto. Its arrow diagram is

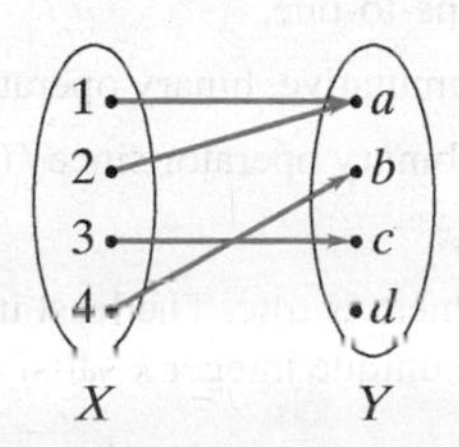


**11.** It is not a function (from $X$ to $Y$).

**13.**

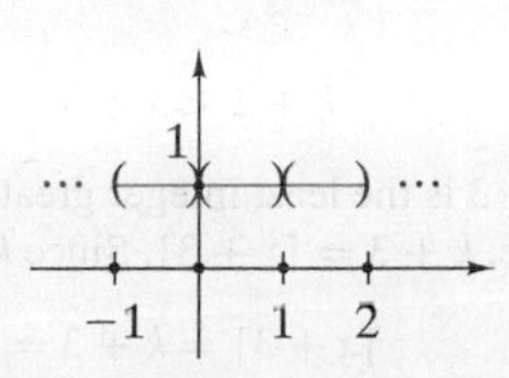


**16.**

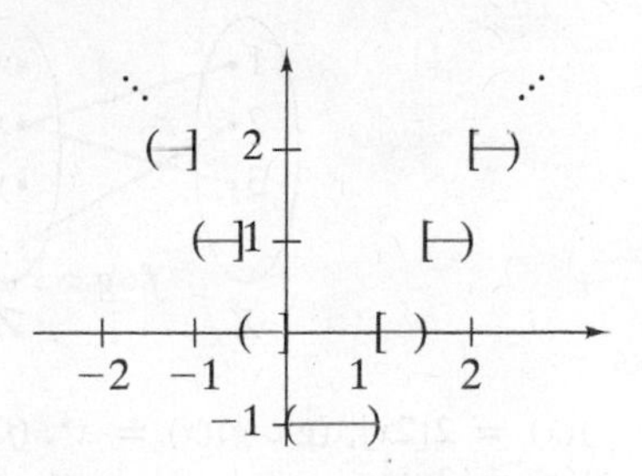


**17.** The function $f$ is both one-to-one and onto. To prove that $f$ is one-to-one, suppose that $f(n) = f(m)$. Then $n + 1 = m + 1$. Thus $n = m$. Therefore $f$ is one-to-one.

To prove that $f$ is onto, let $m$ be an integer. Then $f(m - 1) = (m - 1) + 1 = m$. Therefore $f$ is onto.

**20.** The function $f$ is neither one-to-one nor onto. Since $f(-1) = |-1| = 1 = f(1)$, $f$ is not one-to-one. Since $f(n) \geq 0$ for all $n \in \mathbf{Z}$, $f(n) \neq -1$ for all $n \in \mathbf{Z}$. Therefore $f$ is not onto.

**23.** The function $f$ is not one-to-one, but it is onto. Since $f(2, 1) = 2 - 1 = 1 = 3 - 2 = f(3, 2)$, $f$ is not one-to-one. Suppose that $k \in \mathbf{Z}$. Then $f(k, 0) = k - 0 = k$. Therefore $f$ is onto.

**26.** The function $f$ is neither one-to-one nor onto. Since $f(2, 1) = 2^2 + 1^2 = 1^2 + 2^2 = f(1, 2)$, $f$ is not one-to-one. Since $f(m, n) \geq 0$ for all $m, n \in \mathbf{Z}$, $f(m, n) \neq -1$ for all $m, n \in \mathbf{Z}$. Therefore $f$ is not onto.

**29.** Suppose that $f(a, b) = f(c, d)$. Then $2^a 3^b = 2^c 3^d$. We claim that $a = c$. If not, either $a > c$ or $a < c$. We assume that $a > c$. (The argument is the same if $a < c$.) We may then cancel $2^c$ from both sides of $2^a 3^b = 2^c 3^d$ to obtain $2^{a-c} 3^b = 3^d$. Since $a - c > 0$, $2^{a-c} 3^b$ is even. Since $3^d$ is odd, we have a contradiction. Therefore $a = c$.

We may now cancel $2^a$ from both sides of $2^a 3^b = 2^c 3^d$ to obtain $3^b = 3^d$. An argument like that in the preceding paragraph shows that $b = d$. Since $a = c$ and $b = d$, $f$ is one-to-one.

Since $f(m, n) \neq 5$ for all $m, n \in \mathbf{Z}^+$, $f$ is not onto. [Note that $f(m, n) \geq 6$ for all $m, n \in \mathbf{Z}^+$.]

**30.** $f$ is both one-to-one and onto.

**33.** $f$ is both one-to-one and onto.

**36.** Define a function $f$ from $\{1, 2, 3, 4\}$ to $\{a, b, c, d, e\}$ as

$$f = \{(1, a), (2, c), (3, b), (4, d)\}.$$

Then $f$ is one-to-one, but not onto.

**39.** $\forall x_1 \forall x_2 ((f(x_1) = f(x_2)) \rightarrow (x_1 = x_2))$

**42.**
$$\begin{aligned} \neg(\forall y \in Y \exists x \in X (f(x) = y)) &\equiv \exists y \in Y \neg(\exists x \in X (f(x) = y)) \\ &\equiv \exists y \in Y\, \forall x \in X \neg(f(x) = y) \\ &\equiv \exists y \in Y\, \forall x \in X (f(x) \neq y) \end{aligned}$$

**43.** $f^{-1}(y) = (y - 2)/4$

**46.** $f^{-1}(y) = 1/(y - 3)$

**49.** $f \circ g = \{(1, x), (2, z), (3, x)\}$

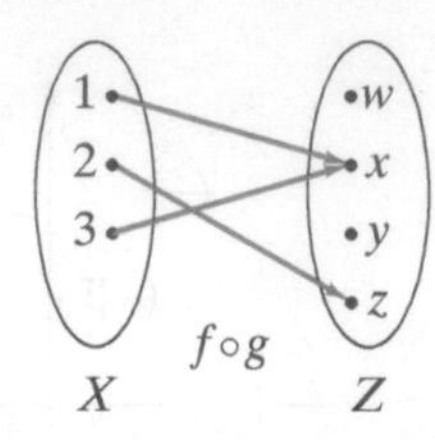


**52.** $(f \circ f)(x) = 2\lfloor 2x \rfloor$, $(g \circ g)(x) = x^4$, $(f \circ g)(x) = \lfloor 2x^2 \rfloor$, $(g \circ f)(x) = \lfloor 2x \rfloor^2$

**54.** Let $g(x) = \log_2 x$ and $h(x) = x^2 + 2$. Then $f(x) = (g \circ h)(x)$.

**57.** Let $g(x) = 2x$ and $h(x) = \sin x$. Then $f(x) = (g \circ h)(x)$.

**60.** $f = \{(-5, 25), (-4, 16), (-3, 9), (-2, 4), (-1, 1), (0, 0), (1, 1), (2, 4), (3, 9), (4, 16), (5, 25)\}$. $f$ is neither one-to-one nor onto. We omit the arrow diagram of $f$.

**63.** $f = \{(0, 0), (1, 4), (2, 3), (3, 2), (4, 1)\}$; $f$ is one-to-one and onto. The arrow diagram of $f$ is

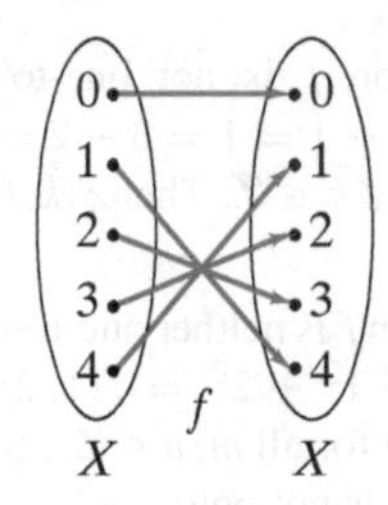


**66.** The ISBN's check digit is invalid. It should be 8.

**69.** Valid

**72.** If $i$ is an untripled digit, $i$ contributes the value $i$ to the sum. Thus if an untripled digit is changed, the sum will change by an amount $a$ satisfying $0 < |a| < 10$. It follows that the check digit will change. If a tripled digit is changed, the following table shows that the sum mod 10 will again change by an amount $a$ satisfying $0 < |a| < 10$:

| *Digit* | *Digit Times* 3 | *Contribution to Sum* mod 10 |
|---|---|---|
| 0 | 0 | 0 |
| 1 | 3 | 3 |
| 2 | 6 | 6 |
| 3 | 9 | 9 |
| 4 | 12 | 2 |
| 5 | 15 | 5 |
| 6 | 18 | 8 |
| 7 | 21 | 1 |
| 8 | 24 | 4 |
| 9 | 27 | 7 |

Again, the check digit will change.

*In the solutions to Exercises 73 and 76, a : b means "store item a in cell b."*

**73.** 53 : 9, 13 : 2, 281 : 6, 743 : 7, 377 : 3, 20 : 10, 10 : 0, 796 : 4

**76.** 714 : 0, 631 : 6, 26 : 5, 373 : 1, 775 : 8, 906 : 13, 509 : 2, 2032 : 7, 42 : 4, 4 : 3, 136 : 9, 1028 : 10

**79.** During a search, if we stop the search at an empty cell, we may not find the item even if it is present. The cell may be empty because an item was deleted. One solution is to mark deleted cells and consider them nonempty during a search.

**80.** False. Take $g = \{(1, a), (2, b)\}$ and $f = \{(a, z), (b, z)\}$.

**83.** True. Let $z \in Z$. Since $f$ is onto, there exists $y \in Y$ such that $f(y) = z$. Since $g$ is onto, there exists $x \in X$ such that $g(x) = y$. Now $f(g(x)) = f(y) = z$. Therefore $f \circ g$ is onto.

**86.** True. Suppose that $g(x_1) = g(x_2)$. Then $f(g(x_1)) = f(g(x_2))$. Since $f \circ g$ is one-to-one, $x_1 = x_2$. Therefore, $g$ is one-to-one.

**88.** $g(S) = \{a\}$, $g(T) = \{a, c\}$, $g^{-1}(U) = \{1\}$, $g^{-1}(V) = \{1, 2, 3\}$

**93.** No. Let $f(x) = x$ and $g(x) = x^2$. Then
$$E_1(f) = f(1) = 1 = g(1) = E_1(g).$$

**96.** 101

**99.** Suppose that $S(Y_1) = s_1 s_2 s_3 = S(Y_2)$. Now $a \in Y_1$ if and only if $s_1 = 1$ if and only if $a \in Y_2$. Also $b \in Y_1$ if and only if $s_2 = 1$ if and only if $b \in Y_2$. Also $c \in Y_1$ if and only if $s_3 = 1$ if and only if $c \in Y_2$. It follows that $Y_1 = Y_2$ and $S$ is one-to-one.

**101.** If $x \in X \cap Y$, $C_{X \cap Y}(x) = 1 = 1 \cdot 1 = C_X(x)C_Y(x)$. If $x \notin X \cap Y$, then $C_{X \cap Y}(x) = 0$. Since either $x \notin X$ or $x \notin Y$, either $C_X(x) = 0$ or $C_Y(x) = 0$. Thus $C_X(x)C_Y(x) = 0 = C_{X \cap Y}(x)$.

**104.** If $x \in X - Y$, then
$$C_{X-Y}(x) = 1 = 1 \cdot [1 - 0] = C_X(x)[1 - C_Y(x)].$$
If $x \notin X - Y$, then either $x \notin X$ or $x \in Y$. In case $x \notin X$,
$$C_{X-Y}(x) = 0 = 0 \cdot [1 - C_Y(x)] = C_X(x)[1 - C_Y(x)].$$
In case $x \in Y$,
$$C_{X-Y}(x) = 0 = C_X(x)[1 - 1] = C_X(x)[1 - C_Y(x)].$$
Thus the equation holds for all $x \in U$.

**107.** $f$ is onto by definition. Suppose that $f(X) = f(Y)$. Then $C_X(x) = C_Y(x)$, for all $x \in U$. Suppose that $x \in X$. Then $C_X(x) = 1$. Thus $C_Y(x) = 1$. Therefore, $x \in Y$. This argument shows that $X \subseteq Y$. Similarly, $Y \subseteq X$. Therefore $X = Y$ and $f$ is one-to-one.

**109.** $f$ is a commutative, binary operator.

**112.** $f$ is not a binary operator since $f(x, 0)$ is not defined.

**114.** $g(x) = -x$

**117.** The statement is true. The least integer greater than or equal to $x$ is the unique integer $k$ satisfying
$$k - 1 < x \leq k.$$
Now
$$k + 2 < x + 3 \leq k + 3.$$
Thus, $k + 3$ is the least integer greater than or equal to $x + 3$. Therefore, $k + 3 = \lceil x + 3 \rceil$. Since $k = \lceil x \rceil$, we have
$$\lceil x + 3 \rceil = k + 3 = \lceil x \rceil + 3.$$

**120.** If $n$ is an odd integer, $n = 2k + 1$ for some integer $k$. Now

$$\frac{n^2}{4} = \frac{(2k+1)^2}{4} = \frac{4k^2+4k+1}{4} = k^2 + k + \frac{1}{4}.$$

Since $k^2 + k$ is an integer,

$$\left\lfloor \frac{n^2}{4} \right\rfloor = k^2 + k.$$

The result now follows because

$$\left(\frac{n-1}{2}\right)\left(\frac{n+1}{2}\right) = \left[\frac{(2k+1)-1}{2}\right]\left[\frac{(2k+1)+1}{2}\right] = \frac{2k(2k+2)}{4} = \frac{4k^2+4k}{4} = k^2 + k.$$

**123.** Let $k = \lceil x \rceil$. Then $k - 1 < x \le k$ and $2x \le 2k$. Thus $\lceil 2x \rceil \le 2k = 2\lceil x \rceil$. Now $\lceil x \rceil = k < x + 1$. Therefore $2\lceil x \rceil < 2x + 2 \le \lceil 2x \rceil + 2$, so $2\lceil x \rceil - 2 < \lceil 2x \rceil$. Therefore $2\lceil x \rceil - 1 \le \lceil 2x \rceil$.

**126.** April, July

## Section 3.2 Review

**1.** A sequence is a function in which the domain is a subset of integers.

**2.** If $s_n$ denotes element $n$ of the sequence, we call $n$ the index of the sequence.

**3.** A sequence $s$ is increasing if for all $i$ and $j$ in the domain of $s$, if $i < j$, then $s_i < s_j$.

**4.** A sequence $s$ is decreasing if for all $i$ and $j$ in the domain of $s$, if $i < j$, then $s_i > s_j$.

**5.** A sequence $s$ is nonincreasing if for all $i$ and $j$ in the domain of $s$, if $i < j$, then $s_i \ge s_j$.

**6.** A sequence $s$ is nondecreasing if for all $i$ and $j$ in the domain of $s$, if $i < j$, then $s_i \le s_j$.

**7.** Let $\{s_n\}$ be a sequence defined for $n = m, m+1, \ldots$, and let $n_1, n_2, \ldots$ be an increasing sequence whose values are in the set $\{m, m+1, \ldots\}$. We call the sequence $\{s_{n_k}\}$ a subsequence of $\{s_n\}$.

**8.** $a_m + a_{m+1} + \cdots + a_n$ **9.** $a_m a_{m+1} \cdots a_n$

**10.** A string over $X$ is a finite sequence of elements from $X$.

**11.** The null string is the string with no elements.

**12.** $X^*$ is the set of all strings over $X$.

**13.** $X^+$ is the set of all nonnull strings over $X$.

**14.** The length of a string $\alpha$ is the number of elements in $\alpha$. It is denoted $|\alpha|$.

**15.** The concatenation of strings $\alpha$ and $\beta$ is the string consisting of $\alpha$ followed by $\beta$. It is denoted $\alpha\beta$.

**16.** A string $\beta$ is a substring of the string $\alpha$ if there are strings $\gamma$ and $\delta$ with $\alpha = \gamma\beta\delta$.

## Section 3.2

**1.** $c$ **2.** $c$

**3.** $cddcdc$

**5.** Increasing, decreasing, nonincreasing, nondecreasing

**8.** None of (a)–(d) **31.** 52

**32.** 52 **33.** No

**34.** No **35.** No

**36.** Yes **45.** 12

**46.** 23 **47.** 7

**48.** 46 **49.** 1

**50.** 3 **51.** 3

**52.** 21 **53.** No

**54.** No **55.** No

**56.** Yes **73.** 15

**74.** 155 **75.** $2n + 3(n-1)n/2$

**76.** Yes **77.** No

**78.** No **79.** Yes

**88.** 2/9

**91.** $z_4 = a_3 + a_4 = 2/9 + 3/64 = 155/576$

**94.** Yes. We must show that

$$a_n = \frac{n-1}{n^2(n-2)^2} > \frac{n}{(n+1)^2(n-1)^2} = a_{n+1} \quad \text{for all } n \ge 3.$$

The preceding inequality is equivalent, successively, to

$$(n-1)^2(n+1)^2(n-1) > n^3(n-2)^2$$
$$(n^2-1)^2(n-1) > n^3(n-2)^2$$
$$(n^4-2n^2+1)(n-1) > n^3(n^2-4n+4)$$
$$3n^4 + 2n^2 + n - 1 > 6n^3.$$

Since $3n^4 > 6n^3$ and $2n^2 + n - 1 > 0$ for all $n > 3$,

$$3n^4 + 2n^2 + n - 1 > 3n^4 > 6n^3 \quad \text{for all } n > 3.$$

By plugging in $n = 3$ into the inequality $3n^4 + 2n^2 + n - 1 > 6n^3$, we can verify that the inequality also holds for $n = 3$. Thus

$$3n^4 + 2n^2 + n - 1 > 6n^3 \quad \text{for all } n \ge 3$$

and hence the equivalent inequality $a_n > a_{n+1}$ also holds for all $n \ge 3$. Therefore $a$ is decreasing.

**97.** Since $a_i > 0$ for all $i$, $z_{n+1} > z_n$ for all $n$. Thus $z$ is increasing.

**100.** Yes, $z$ is nondecreasing. See the solution to Exercise 97.

**101.** We have $n < 4n$ and $n \cdot 4n$ is a perfect square.

**105.** Since $6 = 2 \cdot 3$, the $a$-sequence must contain a multiple of 3. The smallest multiple of 3 that can contribute to a perfect square is $4 \cdot 3 = 12$. Since $6 < 8 < 12$ and $6 \cdot 8 \cdot 12$ is a perfect square, $s_6 = 12$.

**109.** 1, 3, 5, 7, 9, 11, 13 **110.** 1, 5, 9, 13, 17, 21, 25

**111.** $n_k = 2k - 1$ **112.** $s_{n_k} = 4k - 3$

**117.** 88 **118.** 1140

**119.** 48 **120.** 3168

**137.** $b_1 = 1, b_2 = 2, b_3 = 3, b_4 = 4, b_5 = 5, b_6 = 126$

**140.** Let $s_0 = 0$. Then

$$\sum_{k=1}^{n} a_k b_k = \sum_{k=1}^{n} (s_k - s_{k-1}) b_k$$
$$= \sum_{k=1}^{n} s_k b_k - \sum_{k=1}^{n} s_{k-1} b_k$$
$$= \sum_{k=1}^{n} s_k b_k - \sum_{k=1}^{n} s_k b_{k+1} + s_n b_{n+1}$$
$$= \sum_{k=1}^{n} s_k (b_k - b_{k+1}) + s_n b_{n+1}.$$

**143.** 00, 01, 10, 11

**146.** 000, 010, 001, 011, 100, 110, 101, 111, 00, 01, 11, 10, 0, 1, $\lambda$

**149. Basis Step (n = 1)** In this case, $\{1\}$ is the only nonempty subset of $\{1\}$, so the sum is

$$\frac{1}{1} = 1 = n.$$

**Inductive Step** Assume that the statement is true for $n$. We divide the subsets of

$$\{1, \dots, n, n+1\}$$

into two classes:

$C_1$ = class of nonempty subsets that do not contain $n + 1$
$C_2$ = class of subsets that contain $n + 1$.

By the inductive assumption,

$$\sum_{C_1} \frac{1}{n_1 \cdots n_k} = n.$$

Since a set in $C_2$ consists of $n + 1$ together with a subset (empty or nonempty) of $\{1, \dots, n\}$,

$$\sum_{C_2} \frac{1}{(n+1) n_1 \cdots n_k} = \frac{1}{n+1} + \frac{1}{n+1} \sum_{C_1} \frac{1}{n_1 \cdots n_k}.$$

[The term $1/(n + 1)$ results from the subset $\{n + 1\}$.] By the inductive assumption,

$$\frac{1}{n+1} + \frac{1}{n+1} \sum_{C_1} \frac{1}{n_1 \cdots n_k} = \frac{1}{n+1} + \frac{1}{n+1} \cdot n = 1.$$

Therefore,

$$\sum_{C_2} \frac{1}{(n+1) n_1 \cdots n_k} = 1.$$

Finally,

$$\sum_{C_1 \cup C_2} \frac{1}{n_1 \cdots n_k} = \sum_{C_1} \frac{1}{n_1 \cdots n_k} + \sum_{C_2} \frac{1}{(n+1) n_1 \cdots n_k} = n+1.$$

**151.** Since $x_1 \le x \le x_n$, $|x - x_1| = x - x_1$ and $|x - x_n| = x_n - x$. Thus

$$\sum_{i=1}^{n} |x - x_i| = |x - x_1| + \sum_{i=2}^{n-1} |x - x_i| + |x - x_n|$$
$$= (x - x_1) + \sum_{i=2}^{n-1} |x - x_i| + (x_n - x)$$
$$= \sum_{i=2}^{n-1} |x - x_i| + (x_n - x_1).$$

**154.** Using Exercise 4, Section 2.4, we have

$$\sum_{i=1}^{n} \sum_{j=1}^{n} (i - j)^2 = \sum_{i=1}^{n} \sum_{j=1}^{n} (i^2 - 2ij + j^2)$$
$$= \sum_{i=1}^{n} \sum_{j=1}^{n} i^2 - 2 \sum_{i=1}^{n} \sum_{j=1}^{n} ij + \sum_{i=1}^{n} \sum_{j=1}^{n} j^2$$
$$= \sum_{j=1}^{n} \sum_{i=1}^{n} i^2 - 2 \sum_{i=1}^{n} i \sum_{j=1}^{n} j + \sum_{i=1}^{n} \sum_{j=1}^{n} j^2$$
$$= \sum_{j=1}^{n} \frac{n(n+1)(2n+1)}{6} - 2 \left[ \frac{n(n+1)}{2} \right]^2$$
$$+ \sum_{i=1}^{n} \frac{n(n+1)(2n+1)}{6}$$
$$= n \left[ \frac{n(n+1)(2n+1)}{6} \right] - \frac{n^2(n+1)^2}{2}$$
$$+ n \left[ \frac{n(n+1)(2n+1)}{6} \right]$$
$$= 2n \left[ \frac{n(n+1)(2n+1)}{6} \right] - \frac{n^2(n+1)^2}{2}$$
$$= \frac{n^2(n+1)[2(2n+1) - 3(n+1)]}{6}$$
$$= \frac{n^2(n+1)[n-1]}{6} = \frac{n^2(n^2-1)}{6}.$$

**155.** The function $f$ is one-to-one. Suppose that $f(\alpha) = f(\beta)$. Then $\alpha ab = \beta ab$. Thus $\alpha = \beta$.

The function $f$ is not onto. Since $|f(\alpha)| \ge 2$ for all $\alpha \in X^*$, $f(\alpha) \ne \lambda$ for all $\alpha \in X^*$.

**158.** Let $\alpha = \lambda$. Then $\alpha \in L$ and the first rule states that $ab = a\alpha b \in L$. Now $\beta = ab \in L$ and the first rule states that $aabb = a\beta b \in L$. Now $\gamma = aabb \in L$ and the first rule states that $aaabbb = a\gamma b \in L$.

**161.** We use strong induction on the length $n$ of $\alpha$ to show that if $\alpha \in L$, $\alpha$ has an equal number of $a$'s and $b$'s.

The Basis Step is $n = 0$. In this case, $\alpha$ is the null string, which has an equal number of $a$'s and $b$'s.

We turn now to the Inductive Step. We assume any string in $L$ of length $k < n$ has an equal number of $a$'s and $b$'s. We must show that any string in $L$ of length $n$ has an equal number of $a$'s and $b$'s. Let $\alpha \in L$ and suppose that $|\alpha| = n > 0$. Now $\alpha$ is in $L$ because of either rule 1 or rule 2.

Suppose that $\alpha$ is in $L$ because of rule 1. In this case, $\alpha = a\beta b$ or $\alpha = b\beta a$, where $\beta \in L$. Since $|\beta| < n$, by the inductive hypothesis $\beta$ has an equal number of $a$'s and $b$'s. Since $\alpha = a\beta b$ or $\alpha = b\beta a$, $\alpha$ also has an equal number of $a$'s and $b$'s.

Suppose that $\alpha$ is in $L$ because of rule 2. In this case, $\alpha = \beta\gamma$, where $\beta \in L$ and $\gamma \in L$. Since $|\beta| < n$ and $|\gamma| < n$, by the inductive hypothesis $\beta$ and $\gamma$ each have equal numbers of $a$'s and $b$'s. Since $\alpha = \beta\gamma$, $\alpha$ also has an equal number of $a$'s and $b$'s. The proof by induction is complete.

## Section 3.3 Review

**1.** A binary relation from a set $X$ to a set $Y$ is a subset of the Cartesian product $X \times Y$.

**2.** In a digraph of a relation on $X$, vertices represent the elements of $X$ and directed edges from $x$ to $y$ represent the elements $(x, y)$ in the relation.

**3.** A relation $R$ on a set $X$ is reflexive if $(x, x) \in R$ for every $x \in X$. The relation $\{(1, 1), (2, 2)\}$ is a reflexive relation on $\{1, 2\}$. The relation $\{(1, 1)\}$ is not a reflexive relation on $\{1, 2\}$.

**4.** A relation $R$ on a set $X$ is symmetric if for all $x, y \in X$, if $(x, y) \in R$, then $(y, x) \in R$. The relation $\{(1, 2), (2, 1)\}$ is a symmetric relation on $\{1, 2\}$. The relation $\{(1, 2)\}$ is not a symmetric relation on $\{1, 2\}$.

**5.** A relation $R$ on a set $X$ is antisymmetric if for all $x, y \in X$, if $(x, y) \in R$ and $(y, x) \in R$, then $x = y$. The relation $\{(1, 2)\}$ is an antisymmetric relation on $\{1, 2\}$. The relation $\{(1, 2), (2, 1)\}$ is not an antisymmetric relation on $\{1, 2\}$.

**6.** A relation $R$ on a set $X$ is transitive if for all $x, y, z \in X$, if $(x, y)$ and $(y, z) \in R$, then $(x, z) \in R$. The relation $\{(1, 2), (2, 3), (1, 3)\}$ is a transitive relation on $\{1, 2, 3\}$. The relation $\{(1, 2), (2, 1)\}$ is not a transitive relation on $\{1, 2\}$.

**7.** A relation $R$ on a set $X$ is a partial order if $R$ is reflexive, antisymmetric, and transitive. The relation

$$\{(1, 1), (2, 2), (3, 3), (1, 2), (2, 3), (1, 3)\}$$

is a partial order on $\{1, 2, 3\}$.

**8.** If $R$ is a relation from $X$ to $Y$, the inverse of $R$ is the relation from $Y$ to $X$:

$$R^{-1} = \{(y, x) \mid (x, y) \in R\}.$$

The inverse of the relation $\{(1, 2), (1, 3)\}$ is $\{(2, 1), (3, 1)\}$.

**9.** Let $R_1$ be a relation from $X$ to $Y$ and $R_2$ be a relation from $Y$ to $Z$. The composition of $R_1$ and $R_2$ is the relation from $X$ to $Z$

$$R_2 \circ R_1 = \{(x, z) \mid (x, y) \in R_1 \text{ and } (y, z) \in R_2 \text{ for some } y \in Y\}.$$

The composition of the relations

$$R_1 = \{(1, 2), (1, 3), (2, 2)\}$$

and

$$R_2 = \{(2, 1), (2, 3), (1, 4)\}$$

is

$$R_2 \circ R_1 = \{(1, 1), (1, 3), (2, 1), (2, 3)\}.$$

## Section 3.3

**1.** {(8840, Hammer), (9921, Pliers), (452, Paint), (2207, Carpet)}

**4.** $\{(a, a), (b, b)\}$

**5.**

| | |
|---|---|
| $a$ | 6 |
| $b$ | 2 |
| $a$ | 1 |
| $c$ | 1 |

**8.**

| | |
|---|---|
| Mercury | 1 |
| Venus | 2 |
| Earth | 3 |
| Mars | 4 |
| Jupiter | 5 |
| Saturn | 6 |
| Uranus | 7 |
| Neptune | 8 |

**9.**

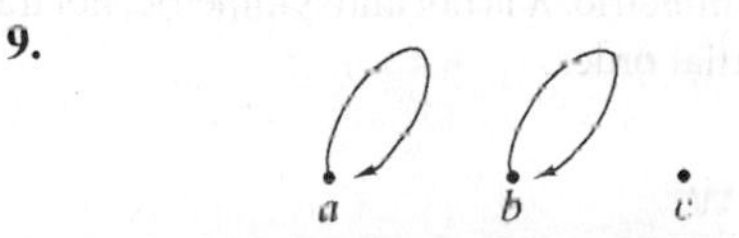


**12.**

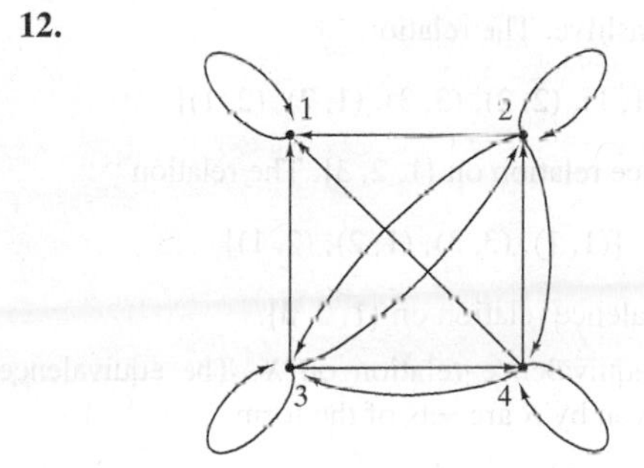


**13.** $\{(a, b), (a, c), (b, a), (b, d), (c, c), (c, d)\}$

**16.** $\{(b, c), (c, b), (d, d)\}$

**17.** (Exercise 1) {(Hammer, 8840), (Pliers, 9921), (Paint, 452), (Carpet, 2207)}

**18.** $\{(1, 1), (1, 4), (2, 2), (2, 5), (3, 3), (4, 1), (4, 4), (5, 2), (5, 5)\}$

**20.** $R = R^{-1} = \{(1, 1), (1, 2), (1, 3), (1, 4), (1, 5), (2, 1), (2, 2), (2, 3), (2, 4), (3, 1), (3, 2), (3, 3), (4, 1), (4, 2), (5, 1)\}$

**23.** Antisymmetric

**24.** Symmetric, antisymmetric, transitive

**27.** Antisymmetric

**30.** Reflexive, symmetric, antisymmetric, transitive, partial order

**33.** Antisymmetric

**35.** Reflexive, antisymmetric, transitive, partial order

**38.** Reflexive: Suppose that $(x_1, x_2)$ is in $X_1 \times X_2$. Since $R_i$ is reflexive, $x_1 R_1 x_1$ and $x_2 R_2 x_2$. Thus $(x_1, x_2) R (x_1, x_2)$.

Antisymmetric: Suppose that $(x_1, x_2) R (x'_1, x'_2)$ and $(x'_1, x'_2) R (x_1, x_2)$. Then $x_1 R_1 x'_1$ and $x'_1 R_1 x_1$. Since $R_1$ is antisymmetric, $x_1 = x'_1$. Similarly, $x_2 = x'_2$. Therefore $(x_1, x_2) = (x'_1, x'_2)$ and $R$ is antisymmetric.

Transitivity is proved similarly.

**40.** $\{(1, 1), (2, 2), (3, 3), (4, 4), (1, 2), (2, 3), (2, 1), (3, 2)\}$

**43.** $\{(1, 1), (1, 2), (2, 1), (2, 2)\}$

**45.** True. Let $(x, y), (y, z) \in R^{-1}$. Then $(z, y), (y, x) \in R$. Since $R$ is transitive, $(z, x) \in R$. Thus $(x, z) \in R^{-1}$. Therefore, $R^{-1}$ is transitive.

**48.** True. We must show that $(x, x) \in R \circ S$ for all $x \in X$. Let $x \in X$. Since $R$ and $S$ are reflexive, $(x, x) \in R$ and $(x, x) \in S$. Therefore, $(x, x) \in R \circ S$ and $R \circ S$ is reflexive.

**51.** True. Let $(x, y) \in R \cap S$. Then $(x, y) \in R$ and $(x, y) \in S$. Since $R$ and $S$ are symmetric, $(y, x) \in R$ and $(y, x) \in S$. Therefore, $(y, x) \in R \cap S$ and $R \cap S$ is symmetric.

**54.** False. Let $R = \{(1, 2)\}$, $S = \{(2, 1)\}$.

**57.** True. Suppose that $(x, y), (y, x) \in R^{-1}$. Then $(y, x), (x, y) \in R$. Since $R$ is antisymmetric, $y = x$. Therefore $R^{-1}$ is antisymmetric.

**59.** $R$ is reflexive and symmetric. $R$ is not antisymmetric, not transitive, and not a partial order.

## Section 3.4 Review

**1.** An equivalence relation is a relation that is reflexive, symmetric, and transitive. The relation

$$\{(1, 1), (2, 2), (3, 3), (1, 2), (2, 1)\}$$

is an equivalence relation on $\{1, 2, 3\}$. The relation

$$\{(1, 1), (3, 3), (1, 2), (2, 1)\}$$

is not an equivalence relation on $\{1, 2, 3\}$.

**2.** Let $R$ be an equivalence relation on $X$. The equivalence classes of $X$ given by $R$ are sets of the form

$$\{x \in X \mid xRa\},$$

where $a \in X$.

**3.** If $R$ is an equivalence relation on $X$, the equivalence classes partition $X$. Conversely, if $\mathcal{S}$ is a partition of $X$ and we define $xRy$ to mean that for some $S \in \mathcal{S}$, both $x$ and $y$ belong to $S$, then $R$ is an equivalence relation.

## Section 3.4

**1.** Equivalence relation: $[1] = [3] = \{1, 3\}$, $[2] = \{2\}$, $[4] = \{4\}$, $[5] = \{5\}$

**4.** Equivalence relation: $[1] = [3] = [5] = \{1, 3, 5\}$, $[2] = \{2\}$, $[4] = \{4\}$

**7.** Not an equivalence relation (neither transitive nor reflexive)

**10.** Equivalence relation: $[1] = [3] = [5] = \{1, 3, 5\}$, $[2] = [4] = \{2, 4\}$

**11.** The relation is an equivalence relation.

**14.** The relation is not an equivalence relation. It is neither reflexive nor symmetric.

**17.** $\{(1, 1), (1, 2), (2, 1), (2, 2), (3, 3), (3, 4), (4, 3), (4, 4)\}$, $[1] = [2] = \{1, 2\}$, $[3] = [4] = \{3, 4\}$

**20.** $\{(1, 1), (1, 2), (1, 3), (2, 1), (2, 2), (2, 3), (3, 1), (3, 2), (3, 3), (4, 4)\}$, $[1] = [2] = [3] = \{1, 2, 3\}$, $[4] = \{4\}$

**24.** $\{1\}, \{1, 3\}, \{1, 4\}, \{1, 3, 4\}$

**26.** [Part (b)]
{San Francisco, San Diego, Los Angeles}, {Pittsburgh, Philadelphia}, {Chicago}

**28.** $R = \{(x, x) \mid x \in X\}$

**31.** Suppose that $x R y$. Since $R$ is reflexive, $y R y$. Taking $z = y$ in the given condition, we have $y R x$. Therefore $R$ is symmetric.

Now suppose that $x R y$ and $y R z$. The given condition tells us that $z R x$. Since $R$ is symmetric, $x R z$. Therefore $R$ is transitive. Since $R$ is reflexive, symmetric, and transitive, $R$ is an equivalence relation.

**34.** [Part (b)]
$(1, 1), (1, 2), (1, 3), (1, 4), (1, 5), (1, 6), (1, 7), (1, 8), (1, 9), (1, 10), (2, 1), (3, 1), (4, 1), (5, 1), (6, 1), (7, 1), (8, 1), (9, 1), (10, 1)$

**37.** (a) We show symmetry only. Let $(x, y) \in R_1 \cap R_2$. Then $(x, y) \in R_1$ and $(x, y) \in R_2$. Since $R_1$ and $R_2$ are symmetric, $(y, x) \in R_1$ and $(y, x) \in R_2$. Thus $(y, x) \in R_1 \cap R_2$ and, therefore, $R_1 \cap R_2$ is symmetric.

(b) $A$ is an equivalence class of $R_1 \cap R_2$ if and only if there are equivalence classes $A_1$ of $R_1$ and $A_2$ of $R_2$ such that $A = A_1 \cap A_2$.

**40.** [Part (b)] Torus

**43.** If $x \in X$, then $x \in f^{-1}(f(\{x\}))$. Thus $\cup\{S \mid S \in \mathcal{S}\} = X$. Suppose that

$$a \in f^{-1}(\{y\}) \cap f^{-1}(\{z\})$$

for some $y, z \in Y$. Then $f(a) = y$ and $f(a) = z$. Thus $y = z$. Therefore, $\mathcal{S}$ is a partition of $X$. The equivalence relation that generates this partition is given in Exercise 41.

**46.** Suppose, by way of contradiction, that $a \in [b]$. Then $(a, b) \in R$. Since $R$ is symmetric, $(b, a) \in R$. Since $R$ is transitive, $(b, b) \in R$, which is a contradiction. Therefore $[b] = \varnothing$.

**49.** Since $R$ is not transitive, there exist $(a, b), (b, c) \in R$, but $(a, c) \notin R$. Then $a \in [b]$, $b \in [c]$, and $a \notin [c]$. Since $R$ is reflexive, $b \in [b]$. Therefore $[b] \cap [c] \neq \varnothing$, but $[b] \neq [c]$. Thus the collection of pseudo equivalence classes does not partition $X$.

**53.** $\rho(R_1) = \{(1, 1), (2, 2), (3, 3), (4, 4), (1, 2), (3, 4), (4, 2)\}$
$\sigma(R_1) = \{(1, 1), (2, 1), (1, 2), (3, 4), (4, 3), (4, 2), (2, 4)\}$
$\tau(R_1) = \{(1, 1), (1, 2), (3, 4), (4, 2), (3, 2)\}$
$\tau(\sigma(\rho(R_1))) = \{(x, y) \mid x, y \in \{1, 2, 3, 4\}\}$

**56.** Let $(x, y), (x, z) \in \tau(R)$. Then $(x, y) \in R^m$ and $(y, z) \in R^n$. Thus $(x, z) \in R^{m+n}$. Therefore, $(x, z) \in \tau(R)$ and $\tau(R)$ is transitive.

**59.** Suppose that $R$ is transitive. If $(x, y) \in \tau(R) = \cup\{R^n\}$, then there exist $x = x_0, \ldots, x_n = y \in X$ such that $(x_{i-1}, x_i) \in R$ for $i = 1, \ldots, n$. Since $R$ is transitive, it follows that $(x, y) \in R$. Thus $R \supseteq \tau(R)$. Since we always have $R \subseteq \tau(R)$, it follows that $R = \tau(R)$.

Suppose that $\tau(R) = R$. By Exercise 56, $\tau(R)$ is transitive. Therefore, $R$ is transitive.

**60.** True. Let $D = \{(x, x) \mid x \in X\}$. Then, by definition, $\rho(R) = R \cup D$, where $R$ is any relation on $X$. Now

$$\begin{aligned}\rho(R_1 \cup R_2) = (R_1 \cup R_2) \cup D &= (R_1 \cup D) \cup (R_2 \cup D)\\ &= \rho(R_1) \cup \rho(R_2).\end{aligned}$$

**63.** False. Let $R_1 = \{(1, 2), (2, 3)\}, R_2 = \{(1, 3), (3, 4)\}$.

**66.** True. Using the notation and hint for Exercise 60,

$$\rho(\tau(R_1)) = \iota(R_1) \cup D$$

and

$$\tau(\rho(R_1)) = \tau(R_1 \cup D).$$

So we must show that $\tau(R_1) \cup D = \tau(R_1 \cup D)$.

We first note that if $A \subseteq B$, then $\tau(A) \subseteq \tau(B)$. Now $R_1 \subseteq R_1 \cup D$. Therefore, $\tau(R_1) \subseteq \tau(R_1 \cup D)$. Also, $D \subseteq R_1 \cup D$. Therefore, $D = \tau(D) \subseteq \tau(R_1 \cup D)$. It follows that $\tau(R_1) \cup D \subseteq \tau(R_1 \cup D)$.

Since $R_1 \subseteq \tau(R_1)$, $R_1 \cup D \subseteq \tau(R_1) \cup D$. By the note in the preceding paragraph, we have $\tau(R_1 \cup D) \subseteq \tau(\tau(R_1) \cup D)$. Since $\tau(R_1) \cup D$ is transitive, $\tau(\tau(R_1) \cup D) = \tau(R_1) \cup D$ (Exercise 59). Therefore, $\tau(R_1 \cup D) \subseteq \tau(R_1) \cup D$.

**67.** A set is equivalent to itself by the identity function.

If $X$ is equivalent to $Y$, there is a one-to-one, onto function $f$ from $X$ to $Y$. Now $f^{-1}$ is a one-to-one, onto function from $Y$ to $X$.

If $X$ is equivalent to $Y$, there is a one-to-one, onto function $f$ from $X$ to $Y$. If $Y$ is equivalent to $Z$, there is a one-to-one, onto function $g$ from $Y$ to $Z$. Now $g \circ f$ is a one-to-one, onto function from $X$ to $Z$.

## Section 3.5 Review

**1.** To obtain the matrix of a relation from $X$ to $Y$, we label the rows with the elements of $X$ and the columns with the elements of $Y$. We then set the entry in row $x$ and column $y$ to 1 if $xRy$ and to 0 otherwise.

**2.** A relation is reflexive if and only if its matrix has 1's on the main diagonal.

**3.** A relation is symmetric if and only if its matrix $A$ satisfies the following: For all $i$ and $j$, the $ij$th entry of $A$ is equal to the $ji$th entry of $A$.

**4.** See the paragraph following the proof of Theorem 3.5.6.

**5.** The matrix of the relation $R_2 \circ R_1$ is obtained by replacing each nonzero term in $A_1A_2$ by 1.

## Section 3.5

**1.**
$$\begin{array}{c} \\ 1 \\ 2 \\ 3 \end{array}\begin{array}{c} \begin{array}{cccc} \alpha & \beta & \Sigma & \delta \end{array} \\ \begin{pmatrix} 0 & 0 & 0 & 1 \\ 1 & 0 & 1 & 0 \\ 0 & 1 & 1 & 0 \end{pmatrix} \end{array}$$

**4.**
$$\begin{array}{c} \\ 1 \\ 2 \\ 3 \\ 4 \\ 5 \end{array}\begin{array}{c} \begin{array}{ccccc} 1 & 2 & 3 & 4 & 5 \end{array} \\ \begin{pmatrix} 0 & 1 & 0 & 0 & 0 \\ 0 & 0 & 1 & 0 & 0 \\ 0 & 0 & 0 & 1 & 0 \\ 0 & 0 & 0 & 0 & 1 \\ 0 & 0 & 0 & 0 & 0 \end{pmatrix} \end{array}$$

**8.** $R = \{(a, w), (a, y), (c, y), (d, w), (d, x), (d, y), (d, z)\}$

**11.** The test is, whenever the $ij$th entry is 1, $i \neq j$, then the $ji$th entry is *not* 1.

**14.** (For Exercise 8)

$$\begin{array}{c} \\ w \\ x \\ y \\ z \end{array}\begin{array}{c} \begin{array}{cccc} a & b & c & d \end{array} \\ \begin{pmatrix} 1 & 0 & 0 & 1 \\ 0 & 0 & 0 & 1 \\ 1 & 0 & 1 & 1 \\ 0 & 0 & 0 & 1 \end{pmatrix} \end{array}$$

**16.** (a) $A_1 = \begin{pmatrix} 1 & 1 \\ 1 & 0 \\ 1 & 0 \end{pmatrix}$

(b) $A_2 = \begin{pmatrix} 0 & 1 & 0 \\ 1 & 1 & 1 \end{pmatrix}$

(c) $A_1A_2 = \begin{pmatrix} 1 & 2 & 1 \\ 0 & 1 & 0 \\ 0 & 1 & 0 \end{pmatrix}$

(d) We change each nonzero entry in part (c) to 1 to obtain

$$A_1A_2 = \begin{pmatrix} 1 & 1 & 1 \\ 0 & 1 & 0 \\ 0 & 1 & 0 \end{pmatrix}.$$

(e) $\{(1, b), (1, a), (1, c), (2, b), (3, b)\}$

**19.** Each column that contains 1 in row $x$ corresponds to an element of the equivalence class containing $x$.

**21.** Suppose that the $ij$th entry of $A$ is 1. Then the $ij$th entry of either $A_1$ or $A_2$ is 1. Thus either $(i, j) \in R_1$ or $(i, j) \in R_2$. Therefore, $(i, j) \in R_1 \cup R_2$. Now suppose that $(i, j) \in R_1 \cup R_2$. Then the $ij$th entry of either $A_1$ or $A_2$ is 1. Therefore, the $ij$th entry of $A$ is 1. It follows that $A$ is the matrix of $R_1 \cup R_2$.

**25.** Each row must contain exactly one 1 for the relation to be a function.

## Section 3.6 Review

**1.** An $n$-ary relation is a set of $n$-tuples.

**2.** A database management system is a program that helps users access information in a database.

**3.** A relational database represents data as tables and provides ways to manipulate the tables.

4. A single attribute or combination of attributes for a relation is a key if the values of the attributes uniquely define an $n$-tuple.
5. A query is a request for information from a database.
6. The selection operator chooses certain $n$-tuples from a relation. The choices are made by giving conditions on the attributes (see Example 3.6.3).
7. The project operator chooses specified columns from a relation. In addition, duplicates are eliminated (see Example 3.6.4).
8. The join operation on relations $R_1$ and $R_2$ begins by examining all pairs of tuples, one from $R_1$ and one from $R_2$. If the join condition is satisfied, the tuples are combined to form a new tuple. The join condition specifies a relationship between an attribute in $R_1$ and an attribute in $R_2$ (see Example 3.6.5).

## Section 3.6

1. {(1089, Suzuki, Zamora), (5620, Kaminski, Jones), (9354, Jones, Yu), (9551, Ryan, Washington), (3600, Beaulieu, Yu), (0285, Schmidt, Jones), (6684, Manacotti, Jones)}

5. EMPLOYEE [Name]
Suzuki, Kaminski, Jones, Ryan, Beaulieu, Schmidt, Manacotti

8. BUYER [Name]
United Supplies, ABC Unlimited, JCN Electronics, Danny's, Underhanded Sales, DePaul University

11. TEMP := BUYER [Part No = 20A8]
TEMP [Name]
Underhanded Sales, Danny's, ABC Unlimited

14. TEMP1 := BUYER [Name = Danny's]
TEMP2 := TEMP1 [Part No = Part No] SUPPLIER
TEMP2 [Dept]
04, 96

17. TEMP1 := BUYER [Name = JCN Electronics]
TEMP2 := TEMP1 [Part No = Part No] SUPPLIER
TEMP3 := TEMP2 [Dept = Dept] DEPARTMENT
TEMP4 := TEMP3 [Manager = Manager] EMPLOYEE
TEMP4 [Name]
Kaminski, Schmidt, Manacotti

22. Let $R_1$ and $R_2$ be two $n$-ary relations. Suppose that the set of elements in the $i$th column of $R_1$ and the set of elements in the $i$th column of $R_2$ come from a common domain for $i = 1, \ldots, n$. The *union* of $R_1$ and $R_2$ is the $n$-ary relation $R_1 \cup R_2$.

TEMP1 := DEPARTMENT [Dept = 23]
TEMP2 := DEPARTMENT [Dept = 96]
TEMP3 := TEMP1 *union* TEMP2
TEMP4 := TEMP3 [Manager = Manager] EMPLOYEE
TEMP4 [Name]
Kaminski, Schmidt, Manacotti, Suzuki

## Chapter 3 Self-Test

1. [Section 3.1] $f$ is not one-to-one. $f$ is onto.
2. [Section 3.2]
(a) $b_5 = 35,\ b_{10} = 120$
(b) $(n+1)^2 - 1$
(c) Yes
(d) No
3. [Section 3.3] Reflexive, symmetric, transitive
4. [Section 3.3] Symmetric
5. [Section 3.4] Yes. It is reflexive, symmetric, and transitive.
6. [Section 3.4] $[3] = \{3, 4\}$. There are two equivalence classes.
7. [Section 3.1] $x = y = 2.3$
8. [Section 3.2]
(a) 14
(b) 18
(c) 192
(d) $a_{n_k} = 4k$
9. [Section 3.1] Define $f$ from $X = \{1, 2\}$ to $\{3\}$ by $f(1) = f(2) = 3$. Define $g$ from $\{1\}$ to $X$ by $g(1) = 1$.
10. [Section 3.2]
(a) *ccddccccdd*
(b) *cccddccddc*
(c) 5 (d) 20
11. [Section 3.1] ($a : b$ means "store item $a$ in cell $b$.") 1 : 1, 784 : 4, 18 : 5, 329 : 6, 43 : 7, 281 : 8, 620 : 9, 1141 : 10, 31 : 11, 684 : 12
12. [Section 3.4] $\{(a, a), (b, b), (b, d), (b, e), (d, b), (d, d), (d, e), (e, b), (e, d), (e, e), (c, c)\}$
13. [Section 3.3] All counterexample relations are on $\{1, 2, 3\}$.
(a) False. $R = \{(1, 1)\}$.
(b) True (c) True
(d) False. $R = \{(1, 1)\}$.
14. [Section 3.2] $\sum_{k=-1}^{n-2} (n - k - 2)r^{k+2}$
15. [Section 3.3] $R = \{(1, 1), (2, 2), (3, 3), (4, 4), (1, 2), (2, 1), (2, 3)\}$
16. [Section 3.4]
(a) $R$ is reflexive because any eight-bit string has the same number of zeros as itself.
$R$ is symmetric because, if $s_1$ and $s_2$ have the same number of zeros, then $s_2$ and $s_1$ have the same number of zeros.
To see that $R$ is transitive, suppose that $s_1$ and $s_2$ have the same number of zeros and that $s_2$ and $s_3$ have the same number of zeros. Then $s_1$ and $s_3$ have the same number of zeros. Therefore, $R$ is an equivalence relation.
(b) There are nine equivalence classes.

(c) 11111111, 01111111, 00111111, 00011111, 00001111, 00000111, 00000011, 00000001, 00000000

**17.** [Section 3.5] $\begin{pmatrix} 1 & 0 \\ 1 & 1 \\ 0 & 1 \end{pmatrix}$

**18.** [Section 3.5] $\begin{pmatrix} 1 & 1 & 0 \\ 1 & 0 & 1 \end{pmatrix}$

**19.** [Section 3.5] $\begin{pmatrix} 1 & 0 \\ 2 & 1 & 1 \\ 1 & 0 & 1 \end{pmatrix}$

**20.** [Section 3.5] $\begin{pmatrix} 1 & 1 & 0 \\ 1 & 1 & 1 \\ 1 & 0 & 1 \end{pmatrix}$

**21.** [Section 3.6]

ASSIGNMENT [Team]

Blue Sox, Mutts, Jackalopes

**22.** [Section 3.6]

PLAYER [Name, Age]

Johnsonbaugh, 22; Glover, 24; Battey, 18; Cage, 30; Homer, 37; Score, 22; Johnsonbaugh, 30; Singleton, 31

**23.** [Section 3.6]

TEMP1 := PLAYER [Position = p]

TEMP2 := TEMP1 [ID Number = PID] ASSIGNMENT

TEMP2 [Team]

Mutts, Jackalopes

**24.** [Section 3.6]

TEMP1 := PLAYER [Age ≥ 30]

TEMP2 := TEMP1 [ID Number = PID] ASSIGNMENT

TEMP2 [Team]

Blue Sox, Mutts

## Section 4.1 Review

**1.** An algorithm is a step-by-step method of solving a problem.

**2.** Input—the algorithm receives input. Output—the algorithm produces output. Precision—the steps are precisely stated. Determinism—the intermediate results of each step of execution are unique and are determined only by the inputs and the results of the preceding steps. Finiteness—the algorithm terminates; that is, it stops after finitely many instructions have been executed. Correctness—the output produced by the algorithm is correct; that is, the algorithm correctly solves the problem. Generality—the algorithm applies to a set of inputs.

**3.** A trace of an algorithm is a simulation of execution of the algorithm.

**4.** The advantages of pseudocode over ordinary text are that pseudocode has more precision, structure, and universality. It is often readily converted to computer code.

**5.** An algorithm is made up of one or more pseudocode functions.

## Section 4.1

**3.** The algorithm does not receive input (but, logically, it needs none). If some even number greater than 2 is not the sum of two prime numbers, the algorithm will stop and output "no". If every even number greater than 2 is the sum of two prime numbers, lines 2 and 3 become an infinite loop—in this case, the algorithm will not terminate and, therefore, will not produce output. The algorithm lacks precision; in order to execute line 2, we need to know how to check whether $n$ is the sum of two primes. The algorithm does have the determinism property. The algorithm *may* lack the finiteness property. We have already noted that if every even number greater than 2 is the sum of two prime numbers (which is currently unsettled), the algorithm will not terminate. The algorithm is not general; that is, it does not apply to a set of inputs. Rather, it applies to *one* set of inputs—namely, the empty set.

**6.** Input: $s, n$

Output: *small*, the smallest value in the sequence $s$

```
min(s, n) {
    small = s_1
    for i = 2 to n
        if (s_i < small) // smaller value found
            small = s_i
    return small
}
```

**9.** Input: $s, n$

Output: *small* (smallest), *large* (largest)

```
small_large(s, n, small, large) {
    small = large = s_1
    for i = 2 to n {
        if (s_i < small)
            small = s_i
        if (s_i > large)
            large = s_i
    }
}
```

**12.** Input: $s, n$

Output: *sum*

```
seq_sum(s, n) {
    sum = 0
    for i = 1 to n
        sum = sum + s_i
    return sum
}
```

**15.** Input: $s, n$

Output: $s$ (in reverse)

```
reverse(s, n) {
    i = 1
    j = n
    while (i < j) {
```

```
    swap(s_i, s_j)
    i = i + 1
    j = j − 1
  }
}
```

**18.** Input: $A$ (an $n \times n$ matrix of a relation $R$), $n$

Output: true, if $R$ is reflexive; false, if $R$ is not reflexive

```
is_reflexive(A, n) {
  for i = 1 to n
    if (A_ii == 0)
      return false
  return true
}
```

**21.** Input: $A$ (an $n \times n$ matrix of a relation $R$), $n$

Output: true, if $R$ is antisymmetric; false, if $R$ is not antisymmetric

```
is_reflexive(A, n) {
  for i = 1 to n − 1
    for j = i + 1 to n
      if (A_ij == 1 ∧ A_ji == 1)
        return false
  return true
}
```

**24.** Input: $A$ (an $m \times k$ matrix of a relation $R_1$), $B$ (a $k \times n$ matrix of a relation $R_2$), $m$, $k$, $n$

Output: $C$ (the $m \times n$ matrix of the relation $R_2 \circ R_1$)

```
comp_relation(A, B, m, k, n, C) {
  // first compute the matrix product AB
  for i = 1 to m
    for j = 1 to n {
      C_ij = 0
      for t = 1 to k
        C_ij = C_ij + A_it B_tj
  }
  // replace each nonzero entry in C by 1
  for i = 1 to m
    for j = 1 to n
      if (C_ij > 0)
        C_ij = 1
}
```

## Section 4.2 Review

1. Finding web pages containing key words is a searching problem. The programs that perform the search are called *search engines*. Finding medical records in a hospital is another searching problem. Such a search may be carried out by people or by a computer.

2. We are given text $t$ and we want to find the first occurrence of pattern $p$ in $t$ or determine that $p$ does not occur in $t$.

3. We use the notation in the solution to Exercise 2. Determine whether $p$ is in $t$ starting at index 1 in $t$. If so, stop. Otherwise, determine whether $p$ is in $t$ starting at index 2 in $t$. If so, stop. Continue in this way until finding $p$ in $t$ or determining that $p$ cannot be in $t$. In the latter case, the search can be terminated when the index in $t$ is so large that there are not enough characters remaining in $t$ to accommodate $p$.

4. Sorting a sequence $s$ means to rearrange the data so that $s$ is in order (nonincreasing order or nondecreasing order).

5. The entries in a book's index are sorted in increasing order, thus making it easy to quickly locate an entry in the index.

6. To sort $s_1, \ldots, s_n$ using insertion sort, first insert $s_2$ in $s_1$ so that $s_1, s_2$ is sorted. Next, insert $s_3$ in $s_1, s_2$ so that $s_1, s_2, s_3$ is sorted. Continue until inserting $s_n$ in $s_1, \ldots, s_{n-1}$ so that the entire sequence $s_1, \ldots, s_n$ is sorted.

7. The time required by an algorithm is the number of steps to termination. The space required by an algorithm is the amount of storage required by the input, local variables, and so on.

8. Knowing or being able to estimate the time and space required by an algorithm gives an indication of how the algorithm will perform for input of various sizes when run on a computer. Knowing or being able to estimate the time and space required by two or more algorithms that solve the same problem makes it possible to compare the algorithms.

9. Many practical problems are too difficult to be solved efficiently, and compromises either in generality or correctness are necessary.

10. When a randomized algorithm executes, at some points it makes random choices.

11. The requirement that the intermediate results of each step of execution be uniquely defined and depend only on the inputs and results of the preceding steps is violated.

12. To shuffle $s_1, \ldots, s_n$, first swap $s_1$ and a randomly chosen element in $s_1, \ldots, s_n$. Next, swap $s_2$ and a randomly chosen element in $s_2, \ldots, s_n$. Continue until swapping $s_{n-1}$ and a randomly chosen element in $s_{n-1}, s_n$.

13. We might generate random arrangements of sequences to use as input to test or time a sorting program.

## Section 4.2

1. First $i$ and $j$ are set to 1. The while loop then compares $t_1 \cdots t_4 =$ "bala" with $p =$ "bala". Since the comparison succeeds, the algorithm returns $i = 1$ to indicate that $p$ was found in $t$ starting at index 1 in $t$.

4. First 20 is inserted in

| 34 |
|---|

Since $20 < 34$, 34 must move one position to the right

| | 34 |
|---|---|

Now 20 is inserted

| 20 | 34 |
|---|---|

Since $144 > 34$, it is immediately inserted to 34's right

| 20 | 34 | 144 |
|---|---|---|

Since $55 < 144$, 144 must move one position to the right

| 20 | 34 | | 144 |
|---|---|---|---|

Since $55 > 34$, 55 is now inserted

| 20 | 34 | 55 | 144 |
|---|---|---|---|

The sequence is now sorted.

**7.** Since each element is greater than or equal to the element to its left, the element is always inserted in its original position.

**8.** We first swap $a_i$ and $a_j$, where $i = 1$ and $j = rand(1, 5) = 5$. After the swap we have

| 135 | 57 | 72 | 101 | 34 |
|---|---|---|---|---|
| ↑ $i$ | | | | ↑ $j$ |

We next swap $a_i$ and $a_j$, where $i = 2$ and $j = rand(2, 5) = 4$. After the swap we have

| 135 | 101 | 72 | 57 | 34 |
|---|---|---|---|---|
| | ↑ $i$ | | ↑ $j$ | |

We next swap $a_i$ and $a_j$, where $i = 3$ and $j = rand(3, 5) = 3$. The sequence is unchanged.

We next swap $a_i$ and $a_j$, where $i = 4$ and $j = rand(4, 5) = 5$. After the swap we have

| 135 | 101 | 72 | 34 | 57 |
|---|---|---|---|---|
| | | | ↑ $i$ | ↑ $j$ |

**11.** Yes. The following values for *rand* sort the input in increasing order:

$$rand(1, 5) = 3, \quad rand(2, 5) = 2, \quad rand(3, 5) = 5, \quad rand(4, 5) = 5.$$

**13.** The while loop tests whether $p$ occurs at index $i$ in $t$. If $p$ does occur at index $i$ in $t$, $t_{i+j-1}$ will be equal to $p_j$ for all $j = 1, \ldots, m$. Thus $j$ becomes $m + 1$ and the algorithm returns $i$. If $p$ does not occur at index $i$ in $t$, $t_{i+j-1}$ will not be equal to $p_j$ for some $j$. In this case the while loop terminates (without executing return $i$).

Now suppose that $p$ occurs in $t$ and its first occurrence is at index $i$ in $t$. As noted in the previous paragraph, the algorithm correctly returns $i$, the smallest index in $t$ where $p$ occurs.

If $p$ does not occur in $t$, then the while loop terminates for every $i$ and $i$ increments in the for loop. Therefore, the for loop runs to completion, and the algorithm correctly returns 0 to indicate that $p$ was not found in $t$.

**16.** Input: $s$ (the sequence $s_1, \ldots, s_n$), $n$, and *key*

Output: $i$ (the index of the last occurrence of *key* in $s$, or 0 if *key* is not in $s$)

```
reverse_linear_search(s, n, key) {
    i = n
    while (i ≥ 1) {
        if (s_i == key)
            return i
        i = i − 1
    }
    return 0
}
```

**19.** We measure the time of the algorithm by counting the number of comparisons ($t_{i+j-1} == p_j$) in the while loop.

No comparisons will be made if $n - m + 1 \le 0$. In the remainder of this solution, we assume that $n - m + 1 > 0$.

If $p$ is in $t$, $m$ comparisons must be performed to verify that $p$ is, in fact, in $t$. We can guarantee that exactly $m$ comparisons are performed if $p$ is at index 1 in $t$.

If $p$ is not in $t$, at least one comparison must be performed for each $i$. We can guarantee that exactly one comparison is performed for each $i$ if the first character in $p$ does not occur in $t$. In this case, $n - m + 1$ comparisons are made.

If $m < n - m + 1$, the best case is that $p$ is at index 1 in $t$. If $n - m + 1 < m$, the best case is that the first character in $p$ does not occur in $t$. If $m = n - m + 1$, either situation is the best case.

**22.** Input: $s$ (the sequence $s_1, \ldots, s_n$) and $n$

Output: $s$ (sorted in nondecreasing order)

```
selection_sort(s, n) {
    for i = 1 to n − 1 {
        // find smallest in s_i, ..., s_n
        small_index = i
        for j = i + 1 to n
            if (s_j < s_small_index)
                small_index = j
        swap(s_i, s_small_index)
    }
}
```

## Section 4.3 Review

**1.** Analysis of algorithms refers to the process of deriving estimates for the time and space needed to execute algorithms.

**2.** The worst-case time for input of size $n$ of an algorithm is the maximum time needed to execute the algorithm among all inputs of size $n$.

**3.** The best-case time for input of size $n$ of an algorithm is the minimum time needed to execute the algorithm among all inputs of size $n$.

**4.** The average-case time for input of size $n$ of an algorithm is the average time needed to execute the algorithm over some finite set of inputs all of size $n$.

**5.** $f(n) = O(g(n))$ if there exists a positive constant $C_1$ such that $|f(n)| \le C_1|g(n)|$ for all but finitely many positive integers $n$. This notation is called the big oh notation.

**6.** Except for constants and a finite number of exceptions, $f$ is bounded above by $g$.

**7.** $f(n) = \Omega(g(n))$ if there exists a positive constant $C_2$ such that $|f(n)| \ge C_2|g(n)|$ for all but finitely many positive integers $n$. This notation is called the omega notation.

**8.** Except for constants and a finite number of exceptions, $f$ is bounded below by $g$.

**9.** $f(n) = \Theta(g(n))$ if $f(n) = O(g(n))$ and $f(n) = \Omega(g(n))$. This notation is called the theta notation.

**10.** Except for constants and a finite number of exceptions, $f$ is bounded above and below by $g$.

## Section 4.3

**1.** $\Theta(n)$ **4.** $\Theta(n^2)$ **7.** $\Theta(n^2)$

**10.** $\Theta(n)$ **13.** $\Theta(n \lg n)$ **14.** $\Theta(n^2)$

**17.** $\Theta(n)$ **20.** $\Theta(n^2)$ **23.** $\Theta(n^3)$

**26.** $\Theta(n \lg n)$ **29.** $\Theta(1)$

**32.** When $n = 1$, we obtain

$$1 = A + B + C.$$

When $n = 2$, we obtain

$$3 = 4A + 2B + C.$$

When $n = 3$, we obtain

$$6 = 9A + 3B + C.$$

Solving this system for $A$, $B$, $C$, we obtain

$$A = B = \frac{1}{2}, \qquad C = 0.$$

We obtain the formula

$$1 + 2 + \cdots + n = \frac{n^2}{2} + \frac{n}{2} + 0 = \frac{n(n+1)}{2},$$

which can be proved using mathematical induction (see Section 2.4).

**34.** $n! = n(n-1)\cdots 2 \cdot 1 \le n \cdot n \cdots n = n^n$

**37.** Since $n = 2^{\lg n}$, $n^{n+1} = (2^{\lg n})^{n+1} = 2^{(n+1)\lg n}$. Thus, it suffices to show that $(n+1)\lg n \le n^2$ for all $n \ge 1$. A proof by induction shows that $n \le 2^{n-1}$ for all $n \ge 1$. Thus, $\lg n \le n-1$ for all $n \ge 1$. Therefore,

$$(n+1)\lg n \le (n+1)(n-1) = n^2 - 1 < n^2 \quad \text{for all } n \ge 1.$$

**40.** Since $f(n) = O(g(n))$, there exist constants $C' > 0$ and $N$ such that

$$f(n) \le C'g(n) \quad \text{for all } n \ge N.$$

Let

$$C = \max\{C', f(1)/g(1), f(2)/g(2), \ldots, f(N)/g(N)\}.$$

For $n \le N$,

$$f(n)/g(n) \le \max\{f(1)/g(1), f(2)/g(2), \ldots, f(N)/g(N)\} \le C.$$

For $n \ge N$,

$$f(n) \le C'g(n) \le Cg(n).$$

Therefore, $f(n) \le Cg(n)$ for all $n$.

**43.** False. If the statement were true, we would have $n^n \le C2^n$ for some constant $C$ and for all sufficiently large $n$. The preceding inequality may be rewritten as

$$\left(\frac{n}{2}\right)^n \le C$$

for some constant $C$ and for all sufficiently large $n$. Since $(n/2)^n$ becomes arbitrarily large as $n$ becomes large, we cannot have $n^n \le C2^n$ for some constant $C$ and for all sufficiently large $n$.

**46.** True

**49.** True

**52.** False. If $n! \ge C(n+1)!$ for all $n \ge K$, then, canceling $n!$, $1 \ge C(n+1)$ for all $n \ge K$. But $1 \ge C(n+1)$ is false if $n > \max\{1/C - 1, K\}$.

**54.** True

**57.** True

**60.** True

**62.** False. A counterexample is $f(n) = n$ and $g(n) = 2n$.

**65.** True

**68.** False. A counterexample is $f(n) = 1$ and $g(n) = 1/n$.

**69.** $f(n) \ne O(g(n))$ means that for every positive constant $C$, $|f(n)| > C|g(n)|$ for infinitely many positive integers $n$.

**72.** We first find nondecreasing positive functions $f_0$ and $g_0$ such that for infinitely many $n$, $f_0(n) = n^2$ and $g_0(n) = n$. This implies that $f_0(n) \ne O(g_0(n))$. Our functions also satisfy $f_0(n) = n$ and $g_0(n) = n^2$ for infinitely many $n$ [obviously different $n$ than those for which $f_0(n) = n^2$ and $g_0(n) = n$]. This implies that $g_0(n) \ne O(f_0(n))$. If we then set $f(n) = f_0(n) + n$ and $g(n) = g_0(n) + n$, we obtain increasing positive functions for which $f(n) \ne O(g(n))$ and $g(n) \ne O(f(n))$.

We begin by setting $f_0(2) = 2$ and $g_0(2) = 2^2$. Then

$$f_0(n) = n, \quad g_0(n) = n^2, \qquad \text{if } n = 2.$$

Because $g_0$ is nondecreasing, the least $n$ for which we may have $g_0(n) = n$ is $n = 2^2$. So we define $f_0(2^2) = 2^4$ and $g_0(2^2) = 2^2$. Then

$$f_0(n) = n^2, \quad g_0(n) = n, \qquad \text{if } n = 2^2.$$

The preceding discussion motivates defining

$$f_0(2^{2^k}) = \begin{cases} 2^{2^k} & \text{if } k \text{ is even} \\ 2^{2^{k+1}} & \text{if } k \text{ is odd} \end{cases}$$

$$g_O(2^{2^k}) = \begin{cases} 2^{2^{k+1}} & \text{if } k \text{ is even} \\ 2^{2^k} & \text{if } k \text{ is odd.} \end{cases}$$

Suppose that $n = 2^{2^k}$. If $k$ is odd, $f_O(n) = n^2$ and $g_O(n) = n$; if $k$ is even, $f_O(n) = n$ and $g_O(n) = n^2$. Now $f_O$ and $g_O$ are defined only for $n = 2^{2^k}$, but they are nondecreasing on this domain. To extend their domains to the set of positive integers, we may simply define $f_O(1) = g_O(1) = 1$ and make them constant on sets of the form $\{i \mid 2^{2^k} \le i < 2^{2^{k+1}}\}$.

**76.** No

**78.** (a) The sum of the areas of the rectangles below the curve is equal to

$$\frac{1}{2} + \frac{1}{3} + \cdots + \frac{1}{n}.$$

This area is less than the area under the curve, which is equal to

$$\int_1^n \frac{1}{x}\,dx = \log_e n.$$

The given inequality now follows immediately.

(b) The sum of the areas of the rectangles whose bases are on the $x$-axis and whose tops are above the curve is equal to

$$1 + \frac{1}{2} + \cdots + \frac{1}{n-1}.$$

Since this area is greater than the area under the curve, the given inequality follows immediately.

(c) Part (a) shows that

$$1 + \frac{1}{2} + \cdots + \frac{1}{n} = O(\log_e n).$$

Since $\log_e n = \Theta(\lg n)$ (see Example 4.3.6),

$$1 + \frac{1}{2} + \cdots + \frac{1}{n} = O(\lg n).$$

Similarly, we can conclude from part (b) that

$$1 + \frac{1}{2} + \cdots + \frac{1}{n} = \Omega(\lg n).$$

Therefore,

$$1 + \frac{1}{2} + \cdots + \frac{1}{n} = \Theta(\lg n).$$

**80.** Replacing $a$ by $b$ in the sum yields

$$\frac{b^{n+1} - a^{n+1}}{b - a} = \sum_{i=0}^{n} a^i b^{n-i} < \sum_{i=0}^{n} b^i b^{n-i}$$

$$= \sum_{i=0}^{n} b^n = (n+1)b^n.$$

**83.** By Exercise 81, the sequence $\{(1 + 1/n)^n\}_{n=1}^{\infty}$ is increasing. Therefore

$$2 = \left(1 + \frac{1}{1}\right)^1 \le \left(1 + \frac{1}{n}\right)^n$$

for every positive integer $n$. Exercise 82 shows that

$$\left(1 + \frac{1}{n}\right)^n < 4$$

for every positive integer $n$. Taking logs to the base 2, we obtain

$$1 = \lg 2 \le \lg\left(1 + \frac{1}{n}\right)^n < \lg 4 = 2.$$

Since

$$\lg\left(1 + \frac{1}{n}\right)^n = n \lg\left(1 + \frac{1}{n}\right) = n \lg\left(\frac{n+1}{n}\right)$$
$$= n[\lg(n+1) - \lg n],$$

we have

$$1 \le n[\lg(n+1) - \lg n] < 2.$$

Dividing by $n$ gives the desired inequality.

**86.** Replacing $b$ by $a$ in the sum yields

$$\frac{b^{n+1} - a^{n+1}}{b - a} = \sum_{i=0}^{n} a^i b^{n-i} > \sum_{i=0}^{n} a^i a^{n-i}$$

$$= \sum_{i=0}^{n} a^n = (n+1)a^n.$$

**89.** By Exercise 88, the sequence $\{(1 + 1/n)^{n+1}\}_{n=1}^{\infty}$ is decreasing. Since $(1 + 1/n)^{n+1} = 4$, when $n = 1$,

$$4 \ge \left(1 + \frac{1}{n}\right)^{n+1} = \left(\frac{n+1}{n}\right)^{n+1}.$$

Taking logs to the base 2, we obtain

$$2 = \lg 4 \ge \lg\left(\frac{n+1}{n}\right)^{n+1} = (n+1)\lg\left(\frac{n+1}{n}\right)$$
$$= (n+1)[\lg(n+1) - \lg n].$$

Dividing by $n + 1$ gives the desired inequality.

**91.** True. Since $\lim_{n\to\infty} f(n)/g(n) = 0$, taking $\varepsilon = 1$, there exists $N$ such that

$$\left|\frac{f(n)}{g(n)}\right| < 1, \quad \text{for all } n \ge N.$$

Therefore, for all $n \ge N$, $|f(n)| < |g(n)|$ and $f(n) = O(g(n))$.

**94.** True. Let $d = |c|$. Since $\lim_{n\to\infty} |f(n)|/|g(n)| = d > 0$, taking $\varepsilon = d/2$, there exists $N$ such that

$$\left|\frac{|f(n)|}{|g(n)|} - d\right| < d/2, \quad \text{for all } n \ge N.$$

This last inequality may be written

$$-\frac{d}{2} < \frac{|f(n)|}{|g(n)|} - d < \frac{d}{2}, \quad \text{for all } n \ge N,$$

or

$$\frac{d}{2} < \frac{|f(n)|}{|g(n)|} < \frac{3d}{2}, \quad \text{for all } n \ge N,$$

or

$$\frac{d}{2}|g(n)| < |f(n)| < \frac{3d}{2}|g(n)|, \quad \text{for all } n \geq N.$$

Therefore, $f(n) = \Theta(g(n))$.

**99.** Multiply both sides of the inequality in Exercise 98 by $\lg e$ and use the change-of-base formula for logarithms.

## Section 4.4 Review

**1.** An algorithm that contains a recursive function

**2.** A function that invokes itself

**3.**

```
factorial(n) {
   if (n == 0)
      return 1
   return n * factorial(n − 1)
}
```

**4.** The original problem is divided into two or more subproblems. Solutions are then found for the subproblems (usually by further subdivision). These solutions are then combined in order to obtain a solution to the original problem.

**5.** In a base case, a solution is obtained directly, that is, without a recursive call.

**6.** If a recursive function had no base case, it would continue to call itself and never terminate.

**7.** $f_1 = 1, f_2 = 1, f_n = f_{n-1} + f_{n-2}$ for $n \geq 3$

**8.** $f_1 = 1, f_2 = 1, f_3 = 2, f_4 = 3$

## Section 4.4

**1.** (a) At line 2, since $4 \neq 0$, we proceed to line 4. The algorithm is invoked with input 3.

(b) At line 2, since $3 \neq 0$, we proceed to line 4. The algorithm is invoked with input 2.

(c) At line 2, since $2 \neq 0$, we proceed to line 4. The algorithm is invoked with input 1.

(d) At line 2, since $1 \neq 0$, we proceed to line 4. The algorithm is invoked with input 0.

(e) At lines 2 and 3, since $0 = 0$, we return 1.

Execution resumes in part (d) at line 4 after computing $0!$ $(= 1)$. We return $0! \cdot 1 = 1$.

Execution resumes in part (c) at line 4 after computing $1!$ $(= 1)$. We return $1! \cdot 2 = 2$.

Execution resumes in part (b) at line 4 after computing $2!$ $(= 2)$. We return $2! \cdot 3 = 6$.

Execution resumes in part (a) at line 4 after computing $3!$ $(= 6)$. We return $3! \cdot 4 = 24$.

**4.** We use induction on $i$, where $n = 2^i$. The Basis Step is $i = 1$. In this case, the board is a tromino $T$. The algorithm correctly tiles the board with $T$ and returns. Thus the algorithm is correct for $i = 1$.

Now assume that if $n = 2^i$, the algorithm is correct. Let $n = 2^{i+1}$. The algorithm divides the board into four $(n/2) \times (n/2)$ subboards. It then places one right tromino in the center as in Figure 2.4.6. It considers each of the squares covered by the center tromino as missing. It then tiles the four subboards, and, by the inductive assumption, these subboards are correctly tiled. Therefore, the $n \times n$ board is correctly tiled. The Inductive Step is complete. The algorithm is correct.

**7.** The proof is by strong induction on $n$. The Basis Steps $(n = 1, 2)$ are readily verified.

Assume that the algorithm is correct for all $k < n$. We must show that the algorithm is correct for $n > 2$. Since $n > 2$, the algorithm executes the return statement

$$\text{return } walk(n-1) + walk(n-2)$$

By the inductive assumption, the values of $walk(n-1)$ and $walk(n-2)$ are correctly computed by the algorithm. Since

$$walk(n) = walk(n-1) + walk(n-2),$$

the algorithm returns the correct value of $walk(n)$.

**10.** (a) Input: $n$

Output: $2 + 4 + \cdots + 2n$

```
1. sum(s, n) {
2.    if (n == 1)
3.       return 2
4.    return sum(n − 1) + 2n
5. }
```

(b) **Basis Step ($n = 1$)** If $n$ is equal to 1, we correctly return 2.

**Inductive Step** Assume that the algorithm correctly computes the sum when the input is $n - 1$. Now suppose that the input to this algorithm is $n > 1$. At line 2, since $n \neq 1$, we proceed to line 4, where we invoke this algorithm with input $n - 1$. By the inductive assumption, the value returned, $sum(n-1)$, is equal to

$$2 + \cdots + 2(n-1).$$

At line 4, we then return

$$sum(n-1) + 2n = 2 + \cdots + 2(n-1) + 2n,$$

which is the correct value.

**13.** Input: The sequence $s_1, \ldots, s_n$ and the length $n$ of the sequence

Output: The maximum value in the sequence

```
find_max(s, n) {
   if (n == 1)
      return s1
   x = find_max(s, n − 1)
   if (x > sn)
      return x
   else
      return sn
}
```

We prove that the algorithm is correct using induction on $n$. The base case is $n = 1$. If $n = 1$, the only item in the sequence is $s_1$ and the algorithm correctly returns it.

Assume that the algorithm computes the maximum for input of size $n - 1$, and suppose that the algorithm receives input of size $n$. By assumption, the recursive call

$x = find\_max(s, n - 1)$

correctly computes $x$ as the maximum value in the sequence $s_1, \ldots, s_{n-1}$. If $x$ is greater than $s_n$, the maximum value in the sequence $s_1, \ldots, s_n$ is $x$—the value returned by the algorithm. If $x$ is not greater than $s_n$, the maximum value in the sequence $s_1, \ldots, s_n$ is $s_n$—again, the value returned by the algorithm. In either case, the algorithm correctly computes the maximum value in the sequence. The Inductive Step is complete, and we have proved that the algorithm is correct.

**16.** To list all of the ways that a robot can walk $n$ meters, set $s$ to the null string and invoke this algorithm.

Input: $n$, $s$ (a string)

Output: All the ways the robot can walk $n$ meters. Each method of walking $n$ meters includes the extra string $s$ in the list.

```
list_walk1(n, s) {
   if (n == 1) {
      println(s + "take one step of length 1")
      return
   }
   if (n == 2) {
      println(s + "take two steps of length 1")
      println(s + "take one step of length 2")
      return
   }
   s' = s + "take one step of length 2" // concatenation
   list_walk1(n − 2, s')
   s' = s + "take one step of length 1" // concatenation
   list_walk1(n − 1, s')
}
```

**18.** After one month, there is still just one pair because a pair does not become productive until after one month. Therefore, $a_1 = 1$. After two months, the pair alive in the beginning becomes productive and adds one additional pair. Therefore, $a_2 = 2$. The increase in pairs of rabbits $a_n - a_{n-1}$ from month $n - 1$ to month $n$ is due to each pair alive in month $n - 2$ producing an additional pair. That is, $a_n - a_{n-1} = a_{n-2}$. Since $\{a_n\}$ satisfies the same recurrence relation as $\{f_n\}$, $a_1 = f_2$, and $a_2 = f_3$, $a_n = f_{n+1}$, $n \geq 1$.

**21. Basis Step (n = 2)**

$$f_2^2 = 1 = 1 \cdot 2 - 1 = f_1 f_3 + (-1)^3$$

**Inductive Step**

$$\begin{aligned} f_n f_{n+2} + (-1)^{n+2} &= f_n(f_{n+1} + f_n) + (-1)^{n+2} \\ &= f_n f_{n+1} + f_n^2 + (-1)^{n+2} \\ &= f_n f_{n+1} + f_{n-1} f_{n+1} + (-1)^{n+1} + (-1)^{n+2} \\ &= f_{n+1}(f_n + f_{n-1}) = f_{n+1}^2 \end{aligned}$$

**24. Basis Step (n = 1)** $f_1^2 = 1^2 = 1 = 1 \cdot 1 = f_1 f_2$

**Inductive Step** $\displaystyle\sum_{k=1}^{n+1} f_k^2 = \sum_{k=1}^{n} f_k^2 + f_{n+1}^2 = f_n f_{n+1} + f_{n+1}^2$

$$= f_{n+1}(f_n + f_{n+1}) = f_{n+1} f_{n+2}$$

**27.** We use strong induction.

**Basis Steps (n = 6,7)** $f_6 = 8 > 7.59 = (3/2)^5$. $f_7 = 13 > 11.39 = (3/2)^6$.

**Inductive Step**

$$\begin{aligned} f_n = f_{n-1} + f_{n-2} &> \left(\frac{3}{2}\right)^{n-2} + \left(\frac{3}{2}\right)^{n-3} \\ &= \left(\frac{3}{2}\right)^{n-1}\left[\left(\frac{3}{2}\right)^{-1} + \left(\frac{3}{2}\right)^{-2}\right] \\ &= \left(\frac{3}{2}\right)^{n-1}\left[\frac{16}{9}\right] > \left(\frac{3}{2}\right)^{n-1} \end{aligned}$$

**30.** We use strong induction on $n$.

**Basis Step (n = 1)** $1 = f_1$

**Inductive Step** Suppose that $n \geq 2$ and that every positive integer less than $n$ can be expressed as the sum of distinct Fibonacci numbers, no two of which are consecutive. Let $f_{k_1}$ be the largest Fibonacci number satisfying $n \geq f_{k_1}$. If $n = f_{k_1}$, then $n$ is trivially the sum of distinct Fibonacci numbers, no two of which are consecutive. Suppose that $n > f_{k_1}$. By the inductive assumption, $n - f_{k_1}$ can be expressed as the sum of distinct Fibonacci numbers $f_{k_2} > f_{k_3} > \cdots > f_{k_m}$, no two of which are consecutive:

$$n - f_{k_1} = \sum_{i=2}^{m} f_{k_i}.$$

Now $n$ is expressed as the sum of Fibonacci numbers:

$$n = \sum_{i=1}^{m} f_{k_i}. \qquad (*)$$

We next show that $f_{k_1} > f_{k_2}$, so that, in particular, $n$ is the sum of *distinct* Fibonacci numbers.

Notice that $f_{k_2} < n$. Since $f_{k_1}$ is the largest Fibonacci number satisfying $n \geq f_{k_1}$, $f_{k_2} \leq f_{k_1}$. If $f_{k_2} = f_{k_1}$,

$$n \geq f_{k_1} + f_{k_2} > f_{k_1} + f_{k_1 - 1} = f_{k_1 + 1}.$$

This last inequality contradicts the choice of $f_{k_1}$ as the largest Fibonacci number satisfying $n \geq f_{k_1}$. Therefore $f_{k_1} > f_{k_2}$.

The only Fibonacci numbers in the sum (*) that might be consecutive are $f_{k_1}$ and $f_{k_2}$. If they are consecutive, we may also write (*) as

$$\begin{aligned} n &= \sum_{i=1}^{m} f_{k_i} \\ &= f_{k_1} + f_{k_2} + \sum_{i=3}^{m} f_{k_i} \end{aligned}$$

$$= f_{k_1} + f_{k_1-1} + \sum_{i=3}^{m} f_{k_i}$$

$$= f_{k_1+1} + \sum_{i=3}^{m} f_{k_i}.$$

Now $f_{k_1+1} \leq n$ and $f_{k_1+1} > f_{k_1}$. This contradicts the choice of $f_{k_1}$ as the largest Fibonacci number satisfying $n \geq f_{k_1}$. The inductive step is complete.

**33.** Using the formula $f_k f_{k+2} - f_{k+1}^2 = (-1)^{k+1}$ from Exercise 21, we obtain

$$1 + \sum_{k=1}^{n} \frac{(-1)^{k+1}}{f_k f_{k+1}} = 1 + \sum_{k=1}^{n} \frac{f_k f_{k+2} - f_{k+1}^2}{f_k f_{k+1}}$$

$$= 1 + \sum_{k=1}^{n} \left( \frac{f_{k+2}}{f_{k+1}} - \frac{f_{k+1}}{f_k} \right)$$

$$= 1 + \left( \frac{f_3}{f_2} - \frac{f_2}{f_1} \right) + \left( \frac{f_4}{f_3} - \frac{f_3}{f_2} \right)$$

$$+ \cdots + \left( \frac{f_{n+2}}{f_{n+1}} - \frac{f_{n+1}}{f_n} \right)$$

$$= 1 + \frac{f_{n+2}}{f_{n+1}} - \frac{f_2}{f_1} = \frac{f_{n+2}}{f_{n+1}}.$$

**36. Basis Step (n = 4)**

$$\sum_{k=1}^{4} (-1)^k k f_k = -f_1 + 2f_2 - 3f_3 + 4f_4 = -1 + 2 - 6 + 12 = 7$$

and

$$(-1)^4(4f_3 + f_1) - 2 = 8 + 1 - 2 = 7.$$

**Inductive Step**

$$\sum_{k=1}^{n+1} (-1)^k k f_k = \sum_{k=1}^{n} (-1)^k k f_k + (-1)^{n+1}(n+1) f_{n+1}$$

$$= (-1)^n (n f_{n-1} + f_{n-3}) - 2 + (-1)^{n+1}(n+1) f_{n+1}$$

$$= (-1)^{n+1}[(n+1) f_{n+1} - n f_{n-1} - f_{n-3}] - 2$$

$$= (-1)^{n+1}[(n+1)(f_n + f_{n-1}) - n f_{n-1} - f_{n-3}] - 2$$

$$= (-1)^{n+1}[(n+1) f_n + f_{n-1} - f_{n-3}] - 2$$

$$= (-1)^{n+1}[(n+1) f_n + f_{n-2}] - 2.$$

The last step is justified by the equality $f_{n-1} = f_{n-2} + f_{n-3}$.

## Chapter 4 Self-Test

**1.** [Section 4.1] At line 2, we set *large* to 12. At line 3, since $b > large$ $(3 > 12)$ is false, we move to line 5. At line 5, since $c > large$ $(0 > 12)$ is false, we move to line 7, where we return *large* (12), the maximum of the given values.

**2.** [Section 4.1] If the set $S$ is an infinite set, the algorithm will not terminate, so it lacks the finiteness and output properties. Line 1 is not precisely stated since *how* to list the subsets of $S$ and their sums is not specified; thus the algorithm lacks the precision property. The order of the subsets listed in line 1 depends on the method used to generate them, so the algorithm lacks the determinism property. Since line 2 depends on the order of the subsets generated in line 1, the determinism property is lacking here as well.

**3.** [Section 4.2] The while loop first tests whether "110" occurs in $t$ at index 1. Since "110" does not occur in $t$ at index 1, the algorithm next tests whether "110" occurs in $t$ at index 2. Since "110" does occur in $t$ at index 2, the algorithm returns the value 2.

**4.** [Section 4.2] First 64 is inserted in

| 44 |
|---|

Since $64 > 44$, it is immediately inserted to 44's right

| 44 | 64 |
|---|---|

Next 77 is inserted. Since $77 > 64$, it is immediately inserted to 64's right

| 44 | 64 | 77 |
|---|---|---|

Next 15 is inserted. Since $15 < 77$, 77 must move one position to the right

| 44 | 64 | | 77 |
|---|---|---|---|

Since $15 < 64$, 64 must move one position to the right

| 44 | | 64 | 77 |
|---|---|---|---|

Since $15 < 44$, 44 must move one position to the right

| | 44 | 64 | 77 |
|---|---|---|---|

Now 15 is inserted

| 15 | 44 | 64 | 77 |
|---|---|---|---|

Finally 3 is inserted. Since $3 < 77$, 77 must move one position to the right

| 15 | 44 | 64 | | 77 |
|---|---|---|---|---|

Since $3 < 64$, 64 must move one position to the right

| 15 | 44 | | 64 | 77 |
|---|---|---|---|---|

Since $3 < 44$, 44 must move one position to the right

| 15 | | 44 | 64 | 77 |
|---|---|---|---|---|

Since $3 < 15$, 15 must move one position to the right

| | 15 | 44 | 64 | 77 |
|---|---|---|---|---|

Now 3 is inserted

| 3 | 15 | 44 | 64 | 77 |
|---|---|---|---|---|

The sequence is sorted.

**5.** [Section 4.2] We first swap $a_i$ and $a_j$, where $i = 1$ and $j = rand(1, 5) = 1$. The sequence is unchanged.

We next swap $a_i$ and $a_j$, where $i = 2$ and $j = rand(2, 5) = 3$. After the swap we have

| 5 | 2 | 51 | 44 | 96 |
|---|---|---|---|---|

We next swap $a_i$ and $a_j$, where $i = 3$ and $j = rand(3, 5) = 5$. After the swap we have

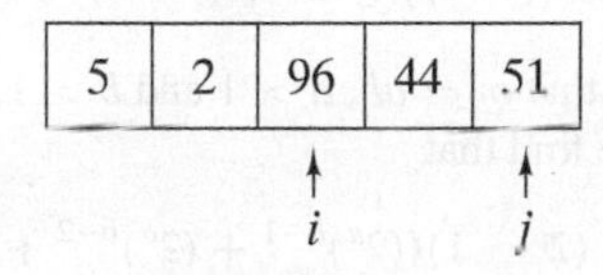

We next swap $a_i$ and $a_j$, where $i = 4$ and $j = rand(4, 5) = 5$. After the swap we have

| 5 | 2 | 96 | 51 | 44 |
|---|---|---|---|---|

**6.** [Section 4.1]

```
sort(a, b, c, x, y, z) {
   x = a
   y = b
   z = c
   if (y < x)
      swap(x, y)
   if (z < x)
      swap(x, z)
   if (z < y)
      swap(y, z)
}
```

**7.** [Section 4.2]

```
repeaters(s, n) {
   i = 1
   while (i < n) {
      if (s_i == s_{i+1})
         println(s_i)
      // skip to next element not equal to s_i
      j = i
      while (i < n ∧ s_i == s_j)
         i = i + 1
   }
}
```

**8.** [Section 4.1]

```
test_distinct(a, b, c) {
   if (a == b ∨ a == c ∨ b == c)
      return false
   return true
}
```

**9.** [Section 4.3]

Input: $A$ and $B$ ($n \times n$ matrices) and $n$

Output: true (if $A = B$); false (if $A \neq B$)

```
equal_matrices(A, B, n) {
   for i = 1 to n
      for j = 1 to n
         if (A_ij ¬= B_ij)
            return false
   return true
}
```

The worst-case time is $\Theta(n^2)$.

**10.** [Section 4.3] $\Theta(n^3)$

**11.** [Section 4.3] $\Theta(n^4)$

**12.** [Section 4.3] $\Theta(n^2)$

**13.** [Section 4.4] Since $n \neq 2$, we proceed immediately to line 6, where we divide the board into four $4 \times 4$ boards. At line 7, we rotate the board so that the missing square is in the upper-left quadrant. At line 8, we place one tromino in the center. We then proceed to lines 9–12, where we call the algorithm to tile the subboards. We obtain the tiling:

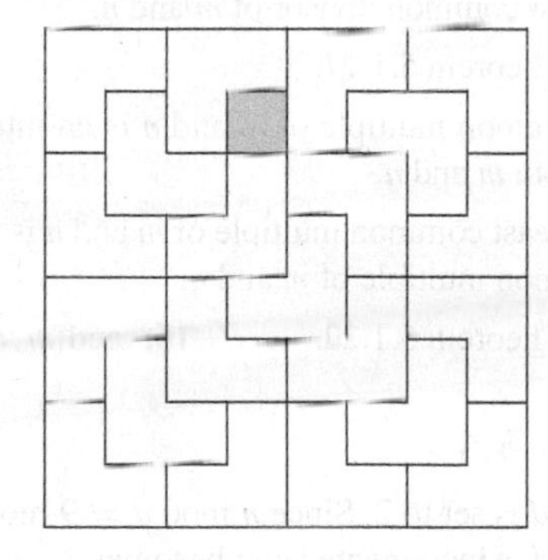

**14.** [Section 4.4] $t_4 = 3$, $t_5 = 5$

**15.** [Section 4.4]

Input: $n$, an integer greater than or equal to 1

Output: $t_n$

```
   tribonacci(n) {
1.    if (n == 1 ∨ n == 2 ∨ n == 3)
2.       return 1
3.    return tribonacci(n − 1) + tribonacci(n − 2)
            + tribonacci(n − 3)
   }
```

**16.** [Section 4.4] **Basis Steps (n = 1, 2, 3)** If $n = 1, 2, 3$, at lines 1 and 2 we return the correct value, 1. Therefore, the algorithm is correct in these cases.

**Inductive Step** Assume that $n > 3$ and that the algorithm correctly computes $t_k$, if $k < n$. Since $n > 3$, we proceed to line 3. We then call this algorithm to compute $t_{n-1}$, $t_{n-2}$, and $t_{n-3}$. By the inductive assumption, the values computed are correct. The algorithm then computes $t_{n-1} + t_{n-2} + t_{n-3}$. But the formula shows that this value is equal to $t_n$. Therefore, the algorithm returns the correct value for $t_n$.

## Section 5.1 Review

1. We say that $d$ divides $n$ if there exists an integer $q$ satisfying $n = dq$.
2. If $d$ divides $n$, we say that $d$ is a divisor of $n$.
3. If $d$ divides $n$, $n = dq$, we call $q$ the quotient.
4. An integer greater than 1 whose only positive divisors are itself and 1 is called prime.
5. An integer greater than 1 that is not prime is called composite.
6. If $n$ is composite, it must have a divisor $d$ satisfying $2 \le d \le \lfloor\sqrt{n}\rfloor$ (see Theorem 5.1.7).
7. Algorithm 5.1.8 does not run in time polynomial in the *size* of the input.
8. Any integer greater than 1 can be written as a product of primes. Moreover, if the primes are written in nondecreasing order, the factorization is unique.
9. See the proof of Theorem 5.1.12.
10. A common divisor of $m$ and $n$, not both zero, is an integer that divides both $m$ and $n$.
11. The greatest common divisor of $m$ and $n$, not both zero, is the largest common divisor of $m$ and $n$.
12. See Theorem 5.1.17.
13. A common multiple of $m$ and $n$ is an integer that is divisible by both $m$ and $n$.
14. The least common multiple of $m$ and $n$ is the smallest positive common multiple of $m$ and $n$.
15. See Theorem 5.1.22.
16. $\gcd(m, n) \cdot \operatorname{lcm}(m, n) = mn$

## Section 5.1

1. First $d$ is set to 2. Since $n \bmod d = 9 \bmod 2 = 1$ is not equal to 0, $d$ is incremented and becomes 3.

   Now $n \bmod d = 9 \bmod 3$ equals 0, so the algorithm returns $d = 3$ to indicate that $n = 9$ is composite and 3 is a divisor of 9.

4. When $d$ is set to $2, \ldots, 6$, $n \bmod d$ is not equal to zero. However, when $d$ becomes 7, $n \bmod d = 637 \bmod 7$ equals 0, so the algorithm returns $d = 7$ to indicate that $n = 637$ is composite and 7 is a divisor of 637.

7. First $d$ is set to 2. Since $n \bmod d = 3738 \bmod 2$ equals 0, the algorithm returns $d = 2$ to indicate that $n = 3738$ is composite and 2 is a divisor of 3738.

9. 47
12. 17
15. 1
18. 20
21. 13
24. $3^2 \cdot 7^3 \cdot 11$
25. (For Exercise 14) 25

28. Since $d$ divides $m$, there exists $q$ such that $m = dq$. Multiplying by $n$ gives $mn = d(qn)$. Therefore, $d$ divides $mn$ (with quotient $qn$).

31. Since $a$ divides $b$, there exists $q_1$ such that $b = aq_1$. Since $b$ divides $c$, there exists $q_2$ such that $c = bq_2$. Now

$$c = bq_2 = (aq_1)q_2 = a(q_1q_2).$$

Therefore, $a$ divides $c$ (with quotient $q_1q_2$).

36. **Basis Step (n = 1)** $\prod_{i=0}^{0} F_i = F_0 = 3 = 5 - 2 = F_1 - 2.$
**Inductive Step** Assume true for $n$. Now

$$\prod_{i=0}^{n} F_i = F_n \left(\prod_{i=0}^{n-1} F_i\right) = F_n(F_n - 2)$$
$$= (2^{2^n} + 1)(2^{2^n} - 1) = 2^{2^{n+1}} - 1 = F_{n+1} - 2.$$

39. We use the identity

$$C^n - 1 = (C - 1)(C^{n-1} + C^{n-2} + \cdots + C + 1).$$

Now suppose that $m = ab$, $a > 1$ and $b > 1$. Taking $C = 2^a$ and $n = b$, we find that

$$(2^a)^b - 1 = (2^a - 1)((2^a)^{b-1} + (2^a)^{b-2} + \cdots + 2^a + 1).$$

Thus $2^a - 1$ divides $(2^a)^b - 1 = 2^m - 1$. Since $a > 1$, $2^a - 1 > 1$. Therefore $2^m - 1$ is composite.

40. Since 2 is not the product of two or more elements in $E$, it is $E$-prime.

43. Since $12 = 2 \cdot 6$, it is $E$-composite.

46. Notice that $2 \cdot m$, where $m$ is odd, is $E$-prime since it is *not* the product of two or more elements in $E$. It follows that the set of $E$-primes is infinite.

49. The $E$-prime 18 divides $6 \cdot 6$ ($36 = 2 \cdot 18$), but 18 does not divide 6.

## Section 5.2 Review

1. $\sum_{i=0}^{n} d_i 10^i$
2. $\sum_{i=0}^{n} b_i 2^i$
3. $\sum_{i=0}^{n} h_i 16^i$
4. $\lfloor 1 + \lg n \rfloor$
5. Perform the computation $\sum_{i=0}^{n} b_i 2^i$ in decimal.
6. Divide the number to be converted to binary by 2. The remainder gives the 1's bit. Divide the quotient by 2. The remainder gives the 2's bit. Continue.
7. Perform the computation $\sum_{i=0}^{n} h_i 16^i$ in decimal.
8. Divide the number to be converted to hexadecimal by 16. The remainder gives the number of 1's. Divide the quotient by 16. The remainder gives the number of 16's. Continue.
9. Use the ordinary algorithm for adding decimal numbers to add binary numbers—except replace the decimal addition table by the binary addition table.
10. Use the ordinary algorithm for adding decimal numbers to add hexadecimal numbers—except replace the decimal addition table by the hexadecimal addition table.

**11.** Let

$$n = \sum_{i=0}^{m} b_i 2^i$$

be the binary expansion of $n$. Using repeated squaring, compute $a^1, a^2, a^4, a^8, \ldots, a^{b_m}$. Then

$$a^n = a^{\sum_{i=0}^{m} b_i 2^i} = \prod_{i=0}^{m} a^{b_i 2^i}.$$

**12.** Proceed as described in the solution to Exercise 11, only use the formula

$$ab \bmod z = [(a \bmod z)(b \bmod z)] \bmod z.$$

## Section 5.2

**1.** 6 **4.** 7 **7.** 1585

**10.** 163 **11.** 9 **14.** 32

**17.** 100010 **20.** 110010000 **23.** 11000

**26.** 1001000 **29.** 58 **32.** 2563

**35.** (For Exercise 11) 9

**38.** FE **41.** 3DBF9

**43.** 2010 cannot represent a number in binary because 2 is an illegal symbol in binary. 2010 could represent a number in either decimal or hexadecimal.

**45.** 51 **48.** 4570

**51.** (For Exercise 11) 11 **54.** (For Exercise 45) 33

**57.** 9450 cannot represent a number in binary because 9, 4, and 5 are illegal symbols in binary. 9450 cannot represent a number in octal because 9 is an illegal symbol in octal. 9450 represents a number in either decimal or hexadecimal.

**59.** The algorithm begins by setting *result* to 1 and $x$ to $a$. Since $n = 16 > 0$, the body of the while loop executes. Since $n$ mod 2 is not equal to 1, *result* is not modified. $x$ becomes $a^2$, and $n$ becomes 8.

Since $n = 8 > 0$, the body of the while loop executes. Since $n$ mod 2 is not equal to 1, *result* is not modified. $x$ becomes $a^4$, and $n$ becomes 4.

Since $n = 4 > 0$, the body of the while loop executes. Since $n$ mod 2 is not equal to 1, *result* is not modified. $x$ becomes $a^8$, and $n$ becomes 2.

Since $n = 2 > 0$, the body of the while loop executes. Since $n$ mod 2 is not equal to 1, *result* is not modified. $x$ becomes $a^{16}$, and $n$ becomes 1.

Since $n = 1 > 0$, the body of the while loop executes. Since $n$ mod 2 is equal to 1, *result* becomes $result * x = 1 * a^{16} = a^{16}$. $x$ becomes $a^{32}$, and $n$ becomes 0.

Since $n = 0$ is not greater than 0, the while loop terminates. The algorithm returns *result*, which is equal to $a^{16}$.

**62.** The algorithm begins by setting *result* to 1 and $x$ to $a \bmod z = 5 \bmod 21 = 5$. Since $n = 10 > 0$, the body of the while loop executes. Since $n$ mod 2 is not equal to 1, *result* is not modified. $x$ is set to $x * x \bmod z = 25 \bmod 21 = 4$, and $n$ is set to 5.

Since $n = 5 > 0$, the body of the while loop executes. Since $n$ mod 2 is equal to 1, *result* is set to $(result * x) \bmod z = 4 \bmod 21 = 4$. $x$ is set to $x * x \bmod z = 16 \bmod 21 = 16$, and $n$ is set to 2.

Since $n = 2 > 0$, the body of the while loop executes. Since $n$ mod 2 is not equal to 1, *result* is not modified. $x$ is set to $x * x \bmod z = 256 \bmod 21 = 4$, and $n$ is set to 1.

Since $n = 1 > 0$, the body of the while loop executes. Since $n$ mod 1 is equal to 1, *result* is set to $(result * x) \bmod z = 16 \bmod 21 = 16$. $x$ is set to $x * x \bmod z = 16 \bmod 21 = 16$, and $n$ is set to 0.

Since $n = 0$ is not greater than 0, the while loop terminates. The algorithm returns *result*, which is equal to $a^n \bmod z = 5^{10} \bmod 21 = 16$.

**65.** If $m_k$ is the highest power of 2 that divides $m$, then $m = 2^{m_k} p$, where $p$ is odd. Similarly, if $n_k$ is the highest power of 2 that divides $n$, then $n = 2^{n_k} q$, where $q$ is odd. Now $mn = 2^{m_k + n_k} pq$. Since $pq$ is odd, $m_k + n_k$ is the highest power of 2 that divides $mn$, and the result follows.

## Section 5.3 Review

**1.** See Algorithm 5.3.3.

**2.** If $a$ is a nonnegative integer, $b$ is a positive integer, and $r = a \bmod b$, then $\gcd(a, b) = \gcd(b, r)$.

**3.** $a \geq f_{n+2}$ and $b \geq f_{n+1}$

**4.** $\log_{3/2} 2m/3$

**5.** Write the nonzero remainders as found by the Euclidean algorithm in the form

$$r = n - dq$$

in the order in which the Euclidean algorithm computes them. Substitute the formula for the next-to-last remainder into the last equation. Call the resulting equation $E_1$. Substitute the second-to-last formula for the remainder into $E_1$. Call the resulting equation $E_2$. Substitute the third-to-last formula for the remainder into $E_2$. Continue until the first formula for the remainder is substituted into the last $E_k$ equation.

**6.** $s$ is the inverse of $n$ mod $z$ if $ns \bmod z = 1$.

**7.** Find numbers $s'$ and $t'$ such that $s'n + t'\phi = 1$. Set $s = s' \bmod \phi$.

## Section 5.3

**1.** 90 mod 60 = 30; 60 mod 30 = 0; so gcd(60, 90) = 30.

**4.** 825 mod 315 = 195; 315 mod 195 = 120; 195 mod 120 = 75; 120 mod 75 = 45; 75 mod 45 = 30; 45 mod 30 = 15; 30 mod 15 = 0; so gcd(825, 315) = 15.

**7.** 4807 mod 2091 = 625; 2091 mod 625 = 216; 625 mod 216 = 193; 216 mod 193 = 23; 193 mod 23 = 9; 23 mod 9 = 5; 9 mod 5 = 4; 5 mod 4 = 1; 4 mod 1 = 0; so gcd(2091, 4807) = 1.

**10.** 490256 mod 337 = 258; 337 mod 258 = 79; 258 mod 79 = 21; 79 mod 21 = 16; 21 mod 16 = 5; 16 mod 5 = 1; 5 mod 1 = 0; so gcd(490256, 337) = 1.

**13.** (For Exercise 10) The nonzero remainders in the order they are computed by the Euclidean algorithm are

$$490256 \bmod 337 = 258$$
$$337 \bmod 258 = 79$$
$$258 \bmod 79 = 21$$
$$79 \bmod 21 = 16$$
$$21 \bmod 16 = 5$$
$$16 \bmod 5 = 1.$$

Writing these equations in the form $r = n - dq$, where $r$ is the remainder and $q$ is the quotient, yields

$$258 = 490256 - 337 \cdot 1454$$
$$79 = 337 - 258 \cdot 1$$
$$21 = 258 - 79 \cdot 3$$
$$16 = 79 - 21 \cdot 3$$
$$5 = 21 - 16 \cdot 1$$
$$1 = 16 - 5 \cdot 3.$$

Substituting the next-to-last formula for 5 into the last equation yields

$$1 = 16 - (21 - 16 \cdot 1) \cdot 3 = 16 \cdot 4 - 21 \cdot 3.$$

Substituting the second-to-last formula for 16 into the previous equation yields

$$1 = (79 - 21 \cdot 3)4 - 21 \cdot 3 = 79 \cdot 4 - 21 \cdot 15.$$

Substituting the third formula for 21 into the previous equation yields

$$1 = 79 \cdot 4 - (258 - 79 \cdot 3)15 = 79 \cdot 49 - 258 \cdot 15.$$

Substituting the second formula for 79 into the previous equation yields

$$1 = (337 - 258)49 - 258 \cdot 15 = 337 \cdot 49 - 258 \cdot 64.$$

Finally, substituting the first formula for 258 into the previous equation yields

$$1 = 337 \cdot 49 - (490256 - 337 \cdot 1454)64$$
$$= 337 \cdot 93105 - 490256 \cdot 64.$$

Thus, if we set $s = -64$ and $t = 93105$,

$$s \cdot 490256 + t \cdot 337 = \gcd(490256, 337) = 1.$$

**16.**

```
gcd_recurs(a, b) {
   make a largest
   if (a < b)
      swap(a, b)
   return gcd_recurs1(a, b)
}

gcd_recurs1(a, b) {
   if (b == 0)
      return a
   r = a mod b
   return gcd_recurs1(b, r)
}
```

**19.**

```
gcd_subtract(a, b) {
   while (true) {
      // make a largest
      if (a < b)
         swap(a, b)
      if (b == 0)
         return a
      a = a - b
   }
}
```

**22.** By Theorem 5.3.5, a pair $a, b,\ a > b$, would require $n$ modulus operations when input to the Euclidean Algorithm only if $a \geq f_{n+2}$ and $b \geq f_{n+1}$. Now $f_{29} = 514229$, $f_{30} = 832040$, and $f_{31} = 1346269$. Thus, no pair can require more than 28 modulus operations in the worst case because 29 modulus operations would require one member of the pair to exceed 1000000. The pair 514229, 832040 itself requires 28 modulus operations.

**25.** We prove the statement by induction on $n$.
**Basis Step (n = 1)** $\gcd(f_1, f_2) = \gcd(1, 1) = 1$
**Inductive Step** Assume that $\gcd(f_n, f_{n+1}) = 1$. Now

$$\gcd(f_{n+1}, f_{n+2}) = \gcd(f_{n+1}, f_{n+1} + f_n) = \gcd(f_{n+1}, f_n) = 1.$$

We use Exercise 18 with $a = f_{n+1} + f_n$ and $b = f_{n+1}$ to justify the second equality.

**29.** If $m = 1$, the result is immediate, so we assume that $m > 1$.

Suppose that $f$ is one-to-one and onto. Since $m > 1$, there exists $x$ such that $f(x) = nx \bmod m = 1$. Thus there exists $q$ such that

$$nx = mq + 1.$$

Let $g$ be the greatest common divisor of $m$ and $n$. Then $g$ divides both $m$ and $n$ and also $nx - mq = 1$. Therefore, $g = 1$.

Now suppose that $\gcd(m, n) = 1$. By Theorem 5.3.7, there exist $s$ and $t$ such that

$$1 = sm + tn.$$

Let $k \in X$. Then

$$k = msk + ntk.$$

Therefore,

$$(ntk) \bmod m = (k - msk) \bmod m = k \bmod m = k.$$

We may argue as in the Computing an Inverse Modulo an Integer subsection that if we set $x = tk \bmod m$, then $f(x) = (ntk) \bmod m$. Therefore, $f$ is onto. Since $f$ is a function from $X$ to $X$, $f$ is also one-to-one.

**30.** If $a \neq 0$, $a = 1 \cdot a + 0 \cdot b > 0$. In this case, $a \in X$. Similarly, if $b \neq 0$, $b \in X$.

**33.** Suppose that $g$ does not divide $a$. Then $a = qg + r$, $0 < r < g$. Since $g \in X$, there exist $s$ and $t$ such that $g = sa + tb$. Now

$$r = a - qg = a - q(sa + tb) = (1 - qs)a + (-qt)b.$$

Therefore $r \in X$. Since $g$ is the least element in $X$ and $0 < r < g$, we have a contradiction. Therefore, $g$ divides $a$. Similarly, $g$ divides $b$.

**35.** $\gcd(3, 2) = \gcd(2, 1) = \gcd(1, 0) = 1$, $s = 2$

**38.** $\gcd(47, 11) = \gcd(11, 3) = \gcd(3, 2) = \gcd(2, 1) = \gcd(1, 0) = 1, s = 30$

**41.** $\gcd(243, 100) = \gcd(100, 43) = \gcd(43, 14) = \gcd(14, 1) = \gcd(1, 0) = 1, s = 226$

**42.** We argue by contradiction. Suppose that 6 has an inverse modulo 15; that is, suppose that there exists $s$ such that $6s \bmod 15 = 1$. Then there exists $q$ such that

$$15 - 6sq = 1.$$

Since 3 divides 15 and 3 divides $6sq$, 3 divides 1. We have obtained the desired contradiction. Thus, 6 does not have an inverse modulo 15.

That 6 does not have an inverse modulo 15 does not contradict the result preceding Example 5.3.9. In order to guarantee that $n$ has an inverse modulo $\phi$, the result preceding Example 5.3.9 requires that $\gcd(n, \phi) = 1$. In this exercise, $\gcd(6, 15) = 3$.

## Section 5.4 Review

**1.** Cryptology is the study of systems for secure communications.

**2.** A cryptosystem is a system for secure communications.

**3.** To encrypt a message is to transform the message so that only an authorized recipient can reconstruct it.

**4.** To decrypt a message is to transform an encrypted message so that it can be read.

**5.** Compute $c = a^n \bmod z$ and send $c$.

**6.** Compute $c^s \bmod z$. $z$ is chosen as the product of primes $p$ and $q$. $s$ satisfies $ns \bmod (p-1)(q-1) = 1$.

**7.** The security of the RSA encryption system relies mainly on the fact that at present there is no efficient algorithm known for factoring integers.

## Section 5.4

**1.** FKKGEJAIMWQ

**4.** BUSHWHACKED

**7.** 53

**9.** $z = pq = 17 \cdot 23 = 391$

**12.** $c = a^n \bmod z = 101^{31} \bmod 391 = 186$

**14.** $z = pq = 59 \cdot 101 = 5959$

**17.** $c = a^n \bmod z = 584^{41} \bmod 5959 = 3237$

## Chapter 5 Self-Test

**1.** [Section 5.1] For $d = 2, \ldots, 6$, 539 mod $d$ is not equal to zero, so $d$ increments. When $d = 7$, 539 mod $d$ equals zero, so the algorithm returns $d = 7$ to indicate that 539 is composite and 7 is a divisor of 539.

**2.** [Section 5.2] 150

**3.** [Section 5.1] $539 = 7^2 \cdot 11$

**4.** [Section 5.3] $\gcd(480, 396) = \gcd(396, 84) = \gcd(84, 60) = \gcd(60, 24) = \gcd(24, 12) = \gcd(12, 0) = 12$

**5.** [Section 5.1] $7^2 \cdot 13^2$

**6.** [Section 5.1] $2 \cdot 5^2 \cdot 7^4 \cdot 13^4 \cdot 17$

**7.** [Section 5.2] 110101110, 1AE

**8.** [Section 5.3] Since

$$\log_{3/2} \frac{2(100{,}000{,}000)}{3} = \log_{3/2} 100^4 + \log_{3/2} \frac{2}{3}$$
$$= (4 \log_{3/2} 100) - 1$$
$$= 4(11.357747) - 1 = 44.430988,$$

an upper bound for the number of modulus operations required by the Euclidean algorithm for integers in the range 0 to 100,000,000 is 44.

**9.** [Section 5.2] The algorithm begins by setting *result* to 1 and $x$ to $a$. Since $n = 30 > 0$, the body of the while loop executes. Since $n \bmod 2$ is not equal to 1, *result* is not modified. $x$ becomes $a^2$, and $n$ becomes 15.

Since $n = 15 > 0$, the body of the while loop executes. Since $n \bmod 2$ is equal to 1, *result* becomes $result * x = 1 * a^2 = a^2$. $x$ becomes $a^4$, and $n$ becomes 7.

Since $n = 7 > 0$, the body of the while loop executes. Since $n \bmod 2$ is equal to 1, *result* becomes $result * x = a^2 * a^4 = a^6$. $x$ becomes $a^8$, and $n$ becomes 3.

Since $n = 3 > 0$, the body of the while loop executes. Since $n \bmod 2$ is equal to 1, *result* becomes $result * x = a^6 * a^8 = a^{14}$. $x$ becomes $a^{16}$, and $n$ becomes 1.

Since $n = 1 > 0$, the body of the while loop executes. Since $n \bmod 2$ is equal to 1, *result* becomes $result * x = a^{14} * a^{16} = a^{30}$. $x$ becomes $a^{32}$, and $n$ becomes 0.

Since $n = 0$ is not greater than 0, the while loop terminates. The algorithm returns *result*, which is equal to $a^{30}$.

**10.** [Section 5.2] The algorithm begins by setting *result* to 1 and $x$ to $a \bmod z = 50 \bmod 11 = 6$. Since $n = 30 > 0$, the body of the while loop executes. Since $n \bmod 2$ is not equal to 1, *result* is not modified. $x$ is set to $x * x \bmod z = 36 \bmod 11 = 3$, and $n$ is set to 15.

Since $n = 15 > 0$, the body of the while loop executes. Since $n \bmod 2$ is equal to 1, *result* is set to $(result*x) \bmod z = 3 \bmod 11 = 3$. $x$ is set to $x * x \bmod z = 9 \bmod 11 = 9$, and $n$ is set to 7.

Since $n = 7 > 0$, the body of the while loop executes. Since $n \bmod 2$ is equal to 1, *result* is set to $(result*x) \bmod z = 27 \bmod 11 = 5$. $x$ is set to $x * x \bmod z = 81 \bmod 11 = 4$, and $n$ is set to 3.

Since $n = 3 > 0$, the body of the while loop executes. Since $n \bmod 2$ is equal to 1, *result* is set to $(result*x) \bmod z = 20 \bmod 11 = 9$. $x$ is set to $x * x \bmod z = 16 \bmod 11 = 5$, and $n$ is set to 1.

Since $n = 1 > 0$, the body of the while loop executes. Since $n \bmod 2$ is equal to 1, *result* is set to $(result*x) \bmod z = 45 \bmod 11 = 1$. $x$ is set to $x * x \bmod z = 25 \bmod 11 = 3$, and $n$ is set to 0.

Since $n = 0$ is not greater than 0, the while loop terminates. The algorithm returns *result*, which is equal to $a^n \bmod z = 50^{30} \bmod 11 = 1$.

**11.** [Section 5.3] The nonzero remainders in the order they are computed by the Euclidean algorithm are

$$480 \bmod 396 = 84$$
$$396 \bmod 84 = 60$$
$$84 \bmod 60 = 24$$
$$60 \bmod 24 = 12.$$

Writing these equations in the form $r = n - dq$, where $r$ is the remainder and $q$ is the quotient, yields

$$84 = 480 - 396 \cdot 1$$
$$60 = 396 - 84 \cdot 4$$
$$24 = 84 - 60 \cdot 1$$
$$12 = 60 - 24 \cdot 2.$$

Substituting the next-to-last formula for 24 into the last equation yields

$$12 = 60 - 24 \cdot 2 = 60 - (84 - 60) \cdot 2 = 3 \cdot 60 - 2 \cdot 84.$$

Substituting the second formula for 60 into the previous equation yields

$$12 = 3 \cdot (396 - 84 \cdot 4) - 2 \cdot 84 = 3 \cdot 396 - 14 \cdot 84.$$

Finally, substituting the first formula for 84 into the previous equation yields

$$12 = 3 \cdot 396 - 14 \cdot (480 - 396) = 17 \cdot 396 - 14 \cdot 480.$$

Thus, if we set $s = 17$ and $t = -14$,

$$s \cdot 396 + t \cdot 480 = \gcd(396, 480) = 12.$$

**12.** [Section 5.3] The nonzero remainders in the order they are computed by the Euclidean algorithm are

$$425 \bmod 196 = 33$$
$$196 \bmod 33 = 31$$
$$33 \bmod 31 = 2$$
$$31 \bmod 2 = 1.$$

Writing these equations in the form $r = n - dq$, where $r$ is the remainder and $q$ is the quotient, yields

$$33 = 425 - 196 \cdot 2$$
$$31 = 196 - 33 \cdot 5$$
$$2 = 33 - 31 \cdot 1$$
$$1 = 31 - 2 \cdot 15.$$

Substituting the next-to-last formula for 2 into the last equation yields

$$1 = 31 - (33 - 31) \cdot 15 = 16 \cdot 31 - 15 \cdot 33.$$

Substituting the second formula for 31 into the previous equation yields

$$1 = 16 \cdot (196 - 33 \cdot 5) - 15 \cdot 33 = 16 \cdot 196 - 95 \cdot 33.$$

Finally, substituting the first formula for 33 into the previous equation yields

$$1 = 16 \cdot 196 - 95 \cdot (425 - 196 \cdot 2) = 206 \cdot 196 - 95 \cdot 425.$$

Thus, if we set $s' = 206$ and $t' = -95$,

$$s' \cdot 196 + t' \cdot 425 = \gcd(196, 425) = 1.$$

Thus $s = s' \bmod 425 = 206 \bmod 425 = 206$.

**13.** [Section 5.4] $z = pq = 13 \cdot 17 = 221$, $\phi = (p-1)(q-1) = 12 \cdot 16 = 192$

**14.** [Section 5.4] $s = 91$

**15.** [Section 5.4] $c = a^n \bmod z = 144^{19} \bmod 221 = 53$

**16.** [Section 5.4] $a = c^s \bmod z = 28^{91} \bmod 221 = 63$

## Section 6.1 Review

**1.** If an activity can be constructed in $t$ successive steps and step 1 can be done in $n_1$ ways, step 2 can be done in $n_2$ ways, ..., and step $t$ can be done in $n_t$ ways, then the number of different possible activities is $n_1 \cdot n_2 \cdots n_t$. As an example, if there are two choices for an appetizer and four choices for a main dish, the total number of dinners is $2 \cdot 4 = 8$.

**2.** Suppose that $X_1, \ldots, X_t$ are sets and that the $i$th set $X_i$ has $n_i$ elements. If $\{X_1, \ldots, X_t\}$ is a pairwise disjoint family, the number of possible elements that can be selected from $X_1$ or $X_2$ or ... or $X_t$ is $n_1 + n_2 + \cdots + n_t$. As an example, suppose that within a set of strings, two start with $a$ and four start with $b$. Then $2 + 4 = 6$ start with either $a$ or $b$.

**3.** $|X \cup Y| = |X| + |Y| - |X \cap Y|$. An example of the use of the Inclusion-Exclusion Principle for two sets is provided by Example 6.1.14.

## Section 6.1

**1.** $2 \cdot 4$ **4.** $8 \cdot 4 \cdot 5$ **7.** $6^2$

**10.** $5 \cdot 4 \cdot 3$ **13.** $6 + 12 + 9$ **16.** $m + n$

**19.** $1 + 1$

**22.** There are $5 \cdot 4 \cdot 3$ strings that begin with $AC$ and $5 \cdot 4 \cdot 3$ strings that begin with $DB$. Thus the answer is $5 \cdot 4 \cdot 3 + 5 \cdot 4 \cdot 3$.

**25.** Since there are three kinds of cabs, two kinds of cargo beds, and five kinds of engines, the correct number of ways to personalize the big pickups is $3 \cdot 2 \cdot 5 = 30$, not 32.

**26.** 3: (1, 3), (2, 2), (3, 1), where $(b, r)$ means the blue die shows $b$ and the red die shows $r$.

**29.** 6: (2, 1), (2, 2), (2, 3), (2, 4), (2, 5), (2, 6), where $(b, r)$ means the blue die shows $b$ and the red die shows $r$.

**32.** Since each die can show any one of five values, by the Multiplication Principle there are $5 \cdot 5$ outcomes in which neither die shows 2.

**34.** $10 \cdot 5$ **37.** $2^4$ **40.** 8

**43.** $2^4$. (Once the first four bits are assigned values, the last four bits are determined.)

**44.** $5 \cdot 4 \cdot 3$ **47.** $3 \cdot 4 \cdot 3$ **50.** $5^3$

**53.** $4 \cdot 3$ **56.** $5^3 - 4^3$ **58.** $200 - 5 + 1$

**61.** 40

**64.** One one-digit number contains 7. The distinct two-digit numbers that contain 7 are $17, 27, \ldots, 97$ and $70, 71, \ldots, 76, 78, 79$. There are 18 of these. The distinct three-digit numbers that contain 7 are 107 and $1xy$, where $xy$ is one of the two-digit numbers listed above. The answer is $1+18+19$.

**67.** $5 + (8 + 7 + \cdots + 1) + (7 + 6 + \cdots + 1)$

**70.** $10!$ **73.** $(3!)(5!)(2!)(3!)$

**77.** $2^{10}$ **80.** $2^{14}(2^{16} - 2)$

**83.** We count the number of $n \times n$ matrices that represent symmetric relations on an $n$-element set. Example 6.1.7 showed that there are $n^2 - n$ entries off the main diagonal. Of these, half, $(n^2 - n)/2$, are above the main diagonal. Since there are $n$ entries on the main diagonal, there are

$$\frac{n^2 - n}{2} + n = \frac{n^2 + n}{2}$$

entries on or above the main diagonal. These entries can be assigned values arbitrarily and there are $2^{(n^2+n)/2}$ ways to assign these values. Because the relation is symmetric, once these values are assigned, the values below the main diagonal are determined. (Entry $ij$ is 1 if and only if entry $ji$ is 1.) Therefore there are $2^{(n^2+n)/2}$ symmetric relations on an $n$-element set.

**86.** We count the number of $n \times n$ matrices that represent reflexive and antisymmetric relations on an $n$-element set.

Since the relation is reflexive, the main diagonal must consist of 1's. For $i$ and $j$ satisfying $1 \le i < j \le n$, we can assign the entries in row $i$, column $j$ and row $j$, column $i$ in three ways:

| *Row i, Column j* | *Row j, Column i* |
|---|---|
| 0 | 0 |
| 1 | 0 |
| 0 | 1 |

Since there are $(n^2 - n)/2$ values of $i$ and $j$ satisfying $1 \le i < j \le n$, we can assign the off-diagonal values in $3^{(n^2-n)/2}$ ways. Therefore there are $3^{(n^2-n)/2}$ reflexive and antisymmetric relations on an $n$-element set.

**89.** The truth table of an $n$-variable function has $2^n$ rows since each variable can be either T or F. Each row of the table can assign the function the value T or F. Since there are $2^n$ rows, the function value assignments can be made in $2^{2^n}$ ways. Therefore there are $2^{2^n}$ truth tables for an $n$-variable function.

**92.** By the Inclusion-Exclusion Principle, the total number of possibilities = number of strings that begin 100 + number of strings that have the fourth bit 1 − number of strings that begin 100 and have the fourth bit 1, so the answer is $2^5 + 2^7 - 2^4$.

**95.** By the Inclusion-Exclusion Principle, the total number of possibilities = number in which Connie is chairperson + number in which Alice is an officer − number in which Connie is chairperson and Alice is an officer, so the answer is $5 \cdot 4 + 3 \cdot 5 \cdot 4 - 2 \cdot 4$.

**99.** Let $F$ be the set of students taking French, let $B$ be the set of students taking business, and let $M$ be the set of students taking music. We are given that $|F \cap B \cap M| = 10$, $|F \cap B| = 36$, $|F \cap M| = 20$, $|B \cap M| = 18$, $|F| = 65$, $|B| = 76$, and $|M| = 63$. By Exercise 98,

$$\begin{aligned} |F \cup B \cup M| &= |F| + |B| + |M| - |F \cap B| - |F \cap M| \\ &\quad - |B \cap M| + |F \cap B \cap M| \\ &= 65 + 76 + 63 - 36 - 20 - 18 + 10 = 140. \end{aligned}$$

Thus 140 students are taking French or business or music. Since there are 191 students, $191 - 140 = 51$ are not any of the three courses.

**102.** Let $X$ be the set of integers between 1 and 10,000 that are multiples of 3, let $Y$ be the set of integers between 1 and 10,000 that are multiples of 5, and let $Z$ be the set of integers between 1 and 10,000 that are multiples of 11.

A multiple of 3 is of the form $3k$ for some integer $k$, so a multiple of 3 between 1 and 10,000 satisfies

$$1 \le 3k \le 10{,}000.$$

Dividing by 3, we obtain

$$0.333\ldots = \frac{1}{3} \le k \le \frac{10{,}000}{3} = 3333.333\ldots.$$

Thus the multiples of 3 between 1 and 10,000 correspond to the values $k = 1, 2, \ldots, 3333$. Therefore there are 3333 multiples of 3 between 1 and 10,000. Similarly, there are 2000 multiples of 5 between 1 and 10,000 and 909 multiples of 11 between 1 and 10,000. Therefore $|X| = 3333$, $|Y| = 2000$, and $|Z| = 909$.

A number that is a multiple of 3 and 5 is a multiple of 15. Arguing as in the previous paragraph, we find that there are 666 multiples of 3 and 5 between 1 and 10,000. Similarly, there are 303 multiples of 3 and 11 between 1 and 10,000, there are 181 multiples of 5 and 11 between 1 and 10,000, and there are 60 multiples of 3, 5, and 11 between 1 and 10,000. Therefore $|X \cap Y| = 666$, $|X \cap Z| = 303$, $|Y \cap Z| = 181$, and $|X \cap Y \cap Z| = 60$. By Exercise 98,

$$\begin{aligned} |X \cup Y \cup Z| &= |X| + |Y| + |Z| - |X \cap Y| - |X \cap Z| \\ &\quad - |Y \cap Z| + |X \cap Y \cap Z| \\ &= 3333 + 2000 + 909 - 666 - 303 - 181 \\ &\quad + 60 = 5152. \end{aligned}$$

Therefore there are 5152 integers between 1 and 10,000 that are multiples of 3 or 5 or 11 or any combination thereof.

## Section 6.2 Review

**1.** An ordering of $x_1, \ldots, x_n$

**2.** There are $n!$ permutations of an $n$-element set. There are $n$ ways to choose the first item, $n - 1$ ways to choose the second item, and so on. Therefore, the total number of permutations is

$$n(n-1)\cdots 2\cdot 1 = n!.$$

**3.** An ordering of $r$ elements selected from $x_1, \ldots, x_n$

**4.** There are $n(n-1)\cdots(n-r+1)$ $r$-permutations of an $n$-element set. There are $n$ ways to choose the first item, $n-1$ ways to choose the second item, ..., and $n-r+1$ ways to choose the $r$th element. Therefore, the total number of $r$-permutations is

$$n(n-1)\cdots(n-r+1).$$

**5.** $P(n, r)$

**6.** An $r$-element subset of $\{x_1, \ldots, x_n\}$

**7.** There are

$$\frac{n!}{(n-r)!r!}$$

$r$-combinations of an $n$-element set.

There are $P(n, r)$ ways to select an $r$-permutation of an $n$-element set. This $r$-permutation can also be constructed by first choosing an $r$-combination [$C(n, r)$ ways] and then ordering it [$r!$ ways]. Therefore, $P(n, r) = C(n, r)r!$. Thus

$$C(n,r) = \frac{P(n,r)}{r!} = \frac{n(n-1)\cdots(n-r+1)}{r!} = \frac{n!}{(n-r)!r!}.$$

**8.** $C(n, r)$

## Section 6.2

**1.** $4! = 24$

**4.** *abc, acb, bac, bca, cab, cba, abd, adb, bad, bda, dab, dba, acd, adc, cad, cda, dac, dca, bcd, bdc, cbd, cdb, dbc, dcb*

**7.** $P(11, 3) = 11\cdot 10\cdot 9$ **10.** $3!$

**13.** $4!$ contain the substring *AE* and $4!$ contain the substring *EA*; therefore, the total number is $2\cdot 4!$.

**16.** We first count the number $N$ of strings that contain either the substring *AB* or the substring *BE*. The answer to the exercise will be: Total number of strings $- N$ or $5! - N$.

According to Exercise 71, Section 6.1, the number of strings that contain *AB* or *BE* = number of strings that contain *AB* + number of strings that contain *BE* − number of strings that contain *AB* and *BE*. A string contains *AB* and *BE* if and only if it contains *ABE* and the number of such strings is $3!$. The number of strings that contain *AB* = number of strings that contain *BE* = $4!$. Thus the number of strings that contain *AB* or *BE* is $4!+4!-3!$. The solution to the exercise is

$$5! - (2\cdot 4! - 3!).$$

**19.** $8!P(9, 5) = 8!(9\cdot 8\cdot 7\cdot 6\cdot 5)$

**21.** $10!$

**24.** Fix a seat for a Jovian. There are $7!$ arrangements for the remaining Jovians. For each of these arrangements, we can place the Martians in five of the eight in-between positions, which can be done in $P(8, 5)$ ways. Thus there are $7!P(8, 5)$ such arrangements.

**25.** $C(4, 3) = 4$ **28.** $C(11, 3)$

**31.** $C(17, 0) + C(17, 1) + C(17, 2) + C(17, 3) + 4$

**33.** $C(13, 5)$

**36.** A committee that has at most one man has exactly one man or no men. There are $C(6, 1)C(7, 3)$ committees with exactly one man. There are $C(7, 4)$ committees with no men. Thus the answer is $C(6, 1)C(7, 3) + C(7, 4)$.

**39.** $C(10, 4)C(12, 3)C(4, 2)$

**42.** First, we count the number of eight-bit strings with no two 0's in a row. We divide this problem into counting the number of such strings with exactly eight 1's, with exactly seven 1's, and so on.

There is one eight-bit string with no two 0's in a row that has exactly eight 1's. Suppose that an eight-bit string with no two 0's in a row has exactly seven 1's. The 0 can go in any one of eight positions; thus there are eight such strings. Suppose that an eight-bit string with no two 0's in a row has exactly six 1's. The two 0's must go in two of the blanks shown:

$$\_1\_1\_1\_1\_1\_1\_.$$

Thus the two 0's can be placed in $C(7, 2)$ ways. Thus there are $C(7, 2)$ such strings. Similarly, there are $C(6, 3)$ eight-bit strings with no two 0's in a row that have exactly five 1's and there are $C(5, 4)$ eight-bit strings with no two 0's in a row that have exactly four 1's in a row. If a string has less than four 1's, it will have two 0's in a row. Therefore, the number of eight-bit strings with no two 0's in a row is

$$1 + 8 + C(7, 2) + C(6, 3) + C(5, 4).$$

Since there are $2^8$ eight-bit strings, there are

$$2^8 - [1 + 8 + C(7, 2) + C(6, 3) + C(5, 4)]$$

eight-bit strings that contain at least two 0's in a row.

**43.** $1\cdot 48$ (The four aces can be chosen in one way and the fifth card can be chosen in 48 ways.)

**46.** First, we count the number of hands containing cards in spades and hearts. Since there are 26 spades and hearts, there are $C(26, 5)$ ways to select five cards from among these 26. However, $C(13, 5)$ contain only spades and $C(13, 5)$ contain only hearts. Therefore, there are

$$C(26, 5) - 2C(13, 5)$$

ways to select five cards containing cards in spades and hearts.

Since there are $C(4, 2)$ ways to select two suits, the number of hands containing cards of exactly two suits is

$$C(4, 2)[C(26, 5) - 2C(13, 5)].$$

**49.** There are nine consecutive patterns: A2345, 23456, 34567, 45678, 56789, 6789T, 789TJ, 89TJQ, 9TJQK. Corresponding to the four possible suits, there are four ways for each pattern to occur. Thus there are $9\cdot 4$ hands that are consecutive and of the same suit.

**52.** $C(52, 13)$

**55.** $1 \cdot C(48, 9)$ (Select the aces, then select the nine remaining cards.)

**58.** There are $C(13, 4)C(13, 4)C(13, 4)C(13, 1)$ hands that contain four spades, four hearts, four diamonds, and one club. Since there are four ways to select the three suits to have four cards each, there are $4C(13, 4)^3C(13, 1)$ hands that contain four cards of three suits and one card of the fourth suit.

**60.** $2^{10}$ **63.** $2^9$ **65.** $C(50, 4)$

**68.** $C(50, 4) - C(46, 4)$ (Total number − number with no defectives)

**72.** Order the $2n$ items. The first item can be paired in $2n - 1$ ways. The next (not yet selected) item can be paired in $2n - 3$ ways, and so on.

**73.** A list of votes where Wright is never behind Upshaw and each receives $r$ votes is a string of $r$ $W$'s and $r$ $U$'s where, reading the string from left to right, the number of $W$'s is always greater than or equal to the number of $U$'s. Such a string can also be considered a path of the type described in Example 6.2.23, where $W$ is a move right and $U$ is a move up. Example 6.2.23 proved that there are $C_r$ such paths. Therefore, the number of ways the votes can be counted in which Wright is never behind Upshaw is $C_r$.

**78.** By Exercise 77, $k$ vertical steps can occur in $C(k, \lceil k/2 \rceil)$ ways, since, at any point, the number of up steps is greater than or equal to the number of down steps. Then, $n - k$ horizontal steps can be inserted among the $k$ vertical steps in $C(n, k)$ ways. Since each horizontal step can occur in two ways, the number of paths containing exactly $k$ vertical steps that never go strictly below the $x$-axis is

$$C(k, \lceil k/2 \rceil)C(n, k)2^{n-k},$$

Summing over all $k$, we find that the total number of paths is

$$\sum_{k=0}^{n} C(k, \lceil k/2 \rceil)C(n, k)2^{n-k}.$$

**85.** The solution counts *ordered* hands.

**87.** Once—when we choose the five slots with 0's and 1's for the remaining slots.

**92.** Use Theorems 3.4.1 and 3.4.8.

**95.** Note that

$$\frac{n-i}{k-i} \ge \frac{n}{k},$$

for $i = 0, 1, \ldots, k - 1$. Therefore,

$$\begin{aligned} C(n, k) = \frac{n!}{(n-k)!k!} &= \frac{n(n-1)\cdots(n-k+1)}{k(k-1)\cdots 1} \\ &= \frac{n}{k}\frac{n-1}{k-1}\cdots\frac{n-k+1}{1} \\ &\ge \frac{n}{k}\frac{n}{k}\cdots\frac{n}{k} \\ &= \left(\frac{n}{k}\right)^k. \end{aligned}$$

Also,

$$C(n, k) = \frac{n(n-1)\cdots(n-k+1)}{k!} \le \frac{nn\cdots n}{k!} = \frac{n^k}{k!}.$$

## Section 6.3 Review

**1.** $n!/(n_1!\cdots n_t!)$. The formula derives from the Multiplication Principle. We first assign positions to the $n_1$ items of type 1, which can be done in $C(n, n_1)$ ways. Having made these assignments, we next assign positions to the $n_2$ items of type 2, which can be done in $C(n - n_1, n_2)$ ways, and so on. The number of orderings is then

$$C(n, n_1)C(n - n_1, n_2)\cdots C(n - n_1 - \cdots - n_{t-1}, n_t),$$

which, after applying the formula for $C(n, k)$ and simplification, gives $n!/(n_1!\cdots n_t!)$.

**2.** $C(k + t - 1, t - 1)$. The formula is obtained by considering $k + t - 1$ slots and $k + t - 1$ symbols consisting of $k$ ×'s and $t - 1$ |'s. Each placement of these symbols into the slots determines a selection. The number $n_1$ of ×'s between the first and second | represents $n_1$ copies of the first element in the set. The number $n_2$ of ×'s between the second and third | represents $n_2$ copies of the second element in the set, and so on. Since there are $C(k + t - 1, t - 1)$ ways to select the positions of the |'s, there are also $C(k + t - 1, t - 1)$ selections.

## Section 6.3

**1.** $5!$ **4.** $6!/(4!2!)$

**7.** Permute one token with four $S$'s and other tokens with one letter each from among *ALEPERON*, which can be done in $9!/2!$ ways.

**10.** $C(6 + 6 - 1, 6 - 1)$

**13.** Each such route can be designated by a string of $i$ $X$'s, $j$ $Y$'s, and $k$ $Z$'s, where an $X$ means move one unit in the $x$-direction, a $Y$ means move one unit in the $y$-direction, and a $Z$ means move one unit in the $z$-direction. There are

$$\frac{(i + j + k)!}{i!j!k!}$$

such strings.

**17.** $10!/(5! \cdot 3! \cdot 2!)$

**18.** $C(10 + 3 - 1, 10)$ **21.** $C(9 + 2 - 1, 9)$

**24.** Four, since the possibilities are (0, 0), (2, 1), (4, 2), and (6, 3), where the pair $(r, g)$ designates $r$ red and $g$ green balls.

**25.** $C(15 + 3 - 1, 15)$ **28.** $C(13 + 2 - 1, 13)$

**31.** $C(12 + 4 - 1, 12)$
$- [C(7 + 4 - 1, 7) + C(6 + 4 - 1, 6)$
$+ C(3 + 4 - 1, 3) + C(2 + 4 - 1, 2)$
$- C(1 + 4 - 1, 1)]$

**36.** $52!/(13!)^4$ **39.** $C(20, 5)$ **42.** $C(20, 5)^2$

**45.** $C(15 + 6 - 1, 15)$ **48.** $C(10 + 12 - 1, 10)$

**51.** Apply the result of Example 6.3.9 to the inner $k - 1$ nested loops of that example. Next, write out the number of iterations for $i_1 = 1$; then $i_1 = 2$; and so on. By Example 6.3.9, this sum is equal to $C(k + n - 1, k)$.

## Section 6.4 Review

**1.** Let $\alpha = s_1 \cdots s_p$ and $\beta = t_1 \cdots t_q$ be strings over $\{1, 2, \ldots, n\}$. Then $\alpha$ is lexicographically less than $\beta$ if either $p < q$ and $s_i = t_i$ for all $i = 1, \ldots, p$; or for some $i$, $s_i \neq t_i$ and for the smallest such $i$, we have $s_i < t_i$.

**2.** Given a string $s_1 \ldots s_r$, which represents the $r$-combination $\{s_1, \ldots, s_r\}$, to find the following string $t_1 \ldots t_r$, find the rightmost element $s_m$ that is not at its maximum value. ($s_r$'s maximum value is $n$, $s_{r-1}$'s maximum value is $n - 1$, etc.) Then set $t_i = s_i$ for $i = 1, \ldots, m - 1$; set $t_m = s_m + 1$; and set $t_{m+1} \cdots t_r = (s_m + 2)(s_m + 3) \cdots$. Begin with the string $12 \cdots r$.

**3.** Given a string $s$, which represents a permutation, to find the following string, find the rightmost digit $d$ of $s$ whose right neighbor exceeds $d$. Find the rightmost element $r$ that satisfies $d < r$. Swap $d$ and $r$. Finally, reverse the substring to the right of $d$'s original position. Begin with the string $12 \cdots n$.

## Section 6.4

**1.** 1357 **4.** 12435

**7.** (For Exercise 1) At lines 8–12, we find the rightmost $s_m$ not at its maximum value. In this case, $m = 4$. At line 14, we increment $s_m$. This makes the last digit 7. Since $m$ is the rightmost position, at lines 16 and 17, we do nothing. The next combination is 1357.

**9.** 123, 124, 125, 126, 134, 135, 136, 145, 146, 156, 234, 235, 236, 245, 246, 256, 345, 346, 356, 456

**12.** 12, 21 **14.** 12379

**17.** None, since each combination has the digits in increasing order.

**20.** 123456897

**23.** A permutation that ends 1234 begins with a permutation of $\{5, 6, 7, 8, 9\}$, and there are 5! of these; therefore, the answer is 5!.

**24.** Input: $r, n$

Output: A list of all $r$-combinations of $\{1, 2, \ldots, n\}$ in increasing lexicographic order

```
r_comb(r, n) {
   s0 = −1
   for i = 1 to r
      si = i
   println(s1, ..., sn)
   while (true) {
      m = r
      max_val = n
      while (sm == max_val) {
         m = m − 1
         max_val = max_val − 1
      }
      if (m == 0)
         return
      sm = sm + 1
      for j = m + 1 to r
         sj = sj−1 + 1
      println(s1, ..., sn)
   }
}
```

**27.** Input: $s_1, \ldots, s_r$ (an $r$-combination of $\{1, \ldots, n\}$), $r$, and $n$

Output: $s_1, \ldots, s_r$, the next $r$-combination (The first $r$-combination follows the last $r$-combination.)

```
next_comb(s, r, n) {
   s0 = n + 1 // dummy value
   m = r
   max_val = n
   // loop test always fails if m = 0
   while (sm == max_val) {
      // find rightmost element not at its maximum value
      m = m − 1
      max_val = max_val − 1
   }
   if (m == 0) { // last r-combination detected
      s1 = 0
      m = 1
   }
   // increment rightmost element
   sm = sm + 1
   // rest of elements are successors of sm
   for j = m + 1 to r
      sj = sj−1 + 1
}
```

**29.** Input: $s_1, \ldots, s_r$ (an $r$-combination of $\{1, \ldots, n\}$), $r$, and $n$

Output: $s_1, \ldots, s_r$, the previous $r$-combination (The last $r$-combination precedes the first $r$-combination.)

```
prev_comb(s, r, n) {
   s0 = n // dummy value
   // find rightmost element at least
   // 2 larger than its left neighbor
   m = r
   // loop test always fails if m = 1
   while (sm − sm−1 == 1)
      m = m − 1
   sm = sm − 1
   if (m == 1 ∧ s1 == 0)
      m = 0
   // set elements to right of index m to max values
   for j = m + 1 to r
      sj = n + j − r
}
```

**31.** Input: $r, s_k, s_{k+1}, \ldots, s_n$, a string $\alpha$, $k$, and $n$

Output: A list of all $r$-combinations of $\{s_k, s_{k+1}, \ldots, s_n\}$ each prefixed by $\alpha$ [To list all $r$-combinations of $\{s_1, s_2, \ldots, s_n\}$, invoke this function as $r\_comb2(r, s, 1, n, \lambda)$, where $\lambda$ is the null string.]

```
r_comb2(r, s, k, n, α) {
  if (r == 0) {
    println(α)
    return
  }
  if (k == n) {
    println(α, s_n)
    return
  }
  β = α + " " + s_k
  // print r-combinations containing s_k
  r_comb2(r − 1, s, k + 1, n, β)
  // print r-combinations not containing s_k
  if (r ≤ n − k)
    r_comb2(r, s, k + 1, n, α)
}
```

## Section 6.5 Review

1. An experiment is a process that yields an outcome.
2. An event is an outcome or combination of outcomes from an experiment.
3. The sample space is the event consisting of all possible outcomes.
4. The number of outcomes in the event divided by the number of outcomes in the sample space

## Section 6.5

1. (H, 1), (H, 2), (H, 3), (H, 4), (H, 5), (H, 6), (T, 1), (T, 2), (T, 3), (T, 4), (T, 5), (T, 6)
4. (H, 1), (H, 2), (H, 3)
5. (1, 1), (1, 3), (1, 5), (2, 2), (2, 4), (2, 6), (3, 1), (3, 3), (3, 5), (4, 2), (4, 4), (4, 6), (5, 1), (5, 3), (5, 5), (6, 2), (6, 4), (6, 6)
8. Three dice are rolled.
11. 1/6
14. 1/52
17. 4/36
20. $C(90, 4)/C(100, 4)$
23. $1/10^3$
26. $1/[C(50, 5) \cdot 36]$
28. $\dfrac{4 \cdot C(13, 5) \cdot 3 \cdot C(13, 4)C(13, 2)^2}{C(52, 13)}$
30. $1/2^{10}$
33. $C(10, 5)/2^{10}$
34. $2^{10}/3^{10}$
37. 1/5!
38. $10/C(12, 3)$
41. 18/38
44. 2/38
45. 1/3
49. 1/4
52. The possibilities are: A (correct), B (incorrect), C (incorrect); and A (incorrect), B (correct), C (incorrect). In the first case, if the student stays with A, the answer will be correct; but if the student switches to B, the answer will be incorrect. In the second case, if the student stays with A, the answer will be incorrect; but if the student switches to B, the answer will be correct. Thus, the probability of a correct answer is 1/2.
55. There are $C(10+3-1, 3-1)$ ways to distribute 10 compact discs to Mary, Ivan, and Juan. If each receives at least two compact discs, we must distribute the remaining six discs, and there are $C(4+3-1, 3-1)$ ways to do this. Thus the probability that each person receives at least two discs is
$$\frac{C(4+3-1, 3-1)}{C(10+3-1, 3-1)}.$$

## Section 6.6 Review

1. A probability function $P$ assigns to each outcome $x$ in a sample space $S$ a number $P(x)$ so that
$$0 \le P(x) \le 1, \quad \text{for all } x \in S$$
and
$$\sum_{x \in S} P(x) = 1.$$
2. $P(x) = 1/n$, where $n$ is the size of the sample space.
3. The probability of $E$ is
$$P(E) = \sum_{x \in E} P(x).$$
4. $P(E) + P(\overline{E}) = 1$
5. $E_1$ or $E_2$ (or both)
6. $E_1$ and $E_2$
7. $P(E_1 \cup E_2) = P(E_1) + P(E_2) - P(E_1 \cap E_2)$. $P(E_1) + P(E_2)$ equals $P(x)$ for all $x \in E_1$ plus $P(x)$ for all $x \in E_2$, which is equal to $P(E_1) + P(E_2)$, except that $P(x)$, for $x \in E_1 \cap E_2$, is counted twice. The formula now follows.
8. Events $E_1$ and $E_2$ are mutually exclusive if $E_1 \cap E_2 = \varnothing$.
9. If we roll two dice, the events "roll doubles" and "the sum is odd" are mutually exclusive.
10. $P(E_1 \cup E_2) = P(E_1) + P(E_2)$. This formula follows from the formula of Exercise 7 because $P(E_1 \cap E_2) = 0$
11. $E$ *given* $F$ is the event $E$ given that event $F$ occurred.
12. $E \mid F$
13. $P(E \mid F) = P(E \cap F)/P(F)$
14. Events $E$ and $F$ are independent if $P(E \cap F) = P(E)P(F)$.
15. If we roll two dice, the events "get an odd number on the first die" and "get an even number on the second die" are independent.
16. Pattern recognition places items into various classes based on features of the items.

**17.** Suppose that the possible classes are $C_1, \ldots, C_n$. Suppose further that each pair of classes is mutually exclusive and each item to be classified belongs to one of the classes. For a feature set $F$, we have

$$P(C_j \mid F) = \frac{P(F \mid C_j)P(C_j)}{\sum_{i=1}^{n} P(F \mid C_i)P(C_i)}.$$

The equation

$$P(C_j \mid F) = \frac{P(C_j \cap F)}{P(F)} = \frac{P(F \mid C_j)P(C_j)}{P(F)}$$

follows from the definition of conditional probability. The proof is completed by showing that

$$P(F) = \sum_{i=1}^{n} P(F \mid C_i)P(C_i),$$

which follows from the fact that each pair of classes is mutually exclusive and each item to be classified belongs to one of the classes.

## Section 6.6

**1.** $1/8$

**4.** $P(1) = P(3) = 3/10$, $P(2) = P(4) = P(5) = P(6) = 1/10$

**7.** $1 - 3/10 = 7/10$

**10.** $P(2) = P(4) = P(6) = 1/12$. $P(1) = P(3) = P(5) = 3/12$.

**13.** $1 - (1/4)$

**14.** $3(1/12)^2 + 3(3/12)^2$

**17.** Let $E$ denote the event "sum is 6," and let $F$ denote the event "at least one die shows 2." Then

$$P(E \cap F) = P((2,4)) + P((4,2)) = 2\left(\frac{1}{12}\right)^2 = \frac{2}{144},$$

and

$$\begin{aligned} P(F) &= P((1,2)) + P((2,1)) + P((2,2)) + P((2,3)) \\ &\quad + P((2,4)) + P((2,5)) + P((2,6)) + P((3,2)) \\ &\quad + P((4,2)) + P((5,2)) + P((6,2)) \\ &= \left(\frac{3}{12}\right)\left(\frac{1}{12}\right) + \left(\frac{1}{12}\right)\left(\frac{3}{12}\right) + \left(\frac{1}{12}\right)^2 \\ &\quad + \left(\frac{1}{12}\right)\left(\frac{3}{12}\right) + \left(\frac{1}{12}\right)^2 + \left(\frac{1}{12}\right)\left(\frac{3}{12}\right) \\ &\quad + \left(\frac{1}{12}\right)^2 + \left(\frac{3}{12}\right)\left(\frac{1}{12}\right) + \left(\frac{1}{12}\right)^2 \\ &\quad + \left(\frac{3}{12}\right)\left(\frac{1}{12}\right) + \left(\frac{1}{12}\right)^2 = \frac{23}{144}. \end{aligned}$$

Therefore,

$$P(E \mid F) = \frac{P(E \cap F)}{P(F)} = \frac{\frac{2}{144}}{\frac{23}{144}} = \frac{2}{23}.$$

**20.** (T,1), (T,2), (T,3), (T,4), (T,5), (T,6), (H,3)

**23.** Yes

**25.** $C(90, 6)/C(100, 6)$

**28.** $1/2^4$

**31.** $\dfrac{\frac{1}{2^4}}{\frac{2^4-1}{2^4}} = \dfrac{1}{15}$

**34.** Let $E_1$ denote the event "children of both sexes," and let $E_2$ denote the event "at most one boy." Then

$$P(E_1) = \frac{14}{16}, \quad P(E_2) = \frac{5}{16}, \quad \text{and} \quad P(E_1 \cap E_2) = \frac{4}{16}.$$

Now

$$P(E_1 \cap E_2) = \frac{1}{4} \neq \frac{35}{128} = P(E_1)P(E_2).$$

Therefore, the events $E_1$ and $E_2$ are not independent.

**37.** $1/2^{10}$

**40.** $1 - (1/2^{10})$

**43.** Let $E$ denote the event "four or five or six heads," and let $F$ denote the event "at least one head." Then

$$\begin{aligned} P(E \cap F) &= \frac{C(10,4)}{2^{10}} + \frac{C(10,5)}{2^{10}} + \frac{C(10,6)}{2^{10}} \\ &= \frac{210 + 252 + 210}{2^{10}} = 0.65625. \end{aligned}$$

Since $P(F) = 1 - (1/2^{10}) = 0.999023437$,

$$P(E \mid F) = \frac{P(E \cap F)}{P(F)} = \frac{0.65625}{0.999023437} = 0.656891495.$$

**46.** If $k > n/2$, the probability is 1. (If none of your numbers was drawn, there are $k$ numbers drawn and, together with your $k$ numbers that are distinct from those drawn, we account for $k + k > n$ numbers, which is impossible.)

Suppose then that $k \leq n/2$. The probability that none of your numbers was drawn is $C(n-k, k)/C(n, k)$. Thus the probability that at least one of your numbers was drawn is $1 - C(n-k, k)/C(n, k)$.

**47.** The probability that neither Maya or Chloe wins a prize is $C(68, 20)/C(70, 20)$. Thus the probability that at least one of them wins a prize is $1 - C(68, 20)/C(70, 20)$.

**50.** The numbers drawn include Maya's number, but not Chloe's. Assigning Maya's number and leaving out Chloe's leaves 68 remaining numbers from which to draw 19 numbers. Thus the answer is $C(68, 19)/C(70, 20)$.

**53.** Let $E$ be the event "at least one person has a birthday on April 1." Then $\overline{E}$ is the event "no one has a birthday on April 1." Now

$$P(E) = 1 - P(\overline{E}) = 1 - \frac{364 \cdot 364 \cdots 364}{365 \cdot 365 \cdots 365} = 1 - \left(\frac{364}{365}\right)^n.$$

**57.** Let $E_1$ denote the event "over 350 pounds," and let $E_2$ denote the event "bad guy." Then

$$\begin{aligned} P(E_1 \cup E_2) &= P(E_1) + P(E_2) - P(E_1 \cap E_2) \\ &= \frac{35}{90} + \frac{20}{90} - \frac{15}{90} = \frac{40}{90}. \end{aligned}$$

**59.** $P(A) = 0.55$, $P(D) = 0.10$, $P(N) = 0.35$

**62.** $P(B) = P(B \mid A)P(A) + P(B \mid D)P(D) + P(B \mid N)P(N)$
$$= (0.10)(0.55) + (0.30)(0.10) + (0.30)(0.35) = 0.19$$

**63.** We require that
$$P(H \mid Pos) = 0.5 = \frac{(0.95)P(H)}{(0.95)P(H) + (0.02)(1 - P(H))}.$$
Solving for $P(H)$ gives $P(H) = .0206$.

**66.** Yes. Suppose that $E$ and $F$ are independent, that is, $P(E)P(F) = P(E \cap F)$. Now
$$\begin{aligned} P(\overline{E})P(\overline{F}) &= (1 - P(E))(1 - P(F)) \\ &= 1 - P(E) - P(F) + P(E)P(F) \\ &= 1 - P(E) - P(F) + P(E \cap F). \end{aligned}$$
By De Morgan's law for sets,
$$\overline{E} \cap \overline{F} = \overline{E \cup F};$$
thus,
$$\begin{aligned} P(\overline{E} \cap \overline{F}) &= P(\overline{E \cup F}) \\ &= 1 - P(E \cup F) \\ &= 1 - [P(E) + P(F) - P(E \cap F)] \\ &= 1 - P(E) - P(F) + P(E \cap F). \end{aligned}$$
Therefore,
$$P(\overline{E})P(\overline{F}) = P(\overline{E} \cap \overline{F}),$$
and $\overline{E}$ and $\overline{F}$ are independent.

**69.** Let $E_i$ be the event "runner completes the marathon on attempt $i$." The error in the reasoning is assuming that $P(E_2) = 1/3 = P(E_3)$. In fact, $P(E_2) \neq 1/3 \neq P(E_3)$ because, if the runner completes the marathon, it is not run again. Although $P(E_1) = 1/3$,
$$\begin{aligned} P(E_2) &= P(\text{fail on attempt 1 and succeed on attempt 2}) \\ &= P(\text{fail on attempt 1})P(\text{succeed on attempt 2}) \\ &= \frac{2}{3} \cdot \frac{1}{3} = \frac{2}{9}. \end{aligned}$$
Similarly,
$$P(E_3) = \frac{2}{3} \cdot \frac{2}{3} \cdot \frac{1}{3} = \frac{4}{27}.$$
Thus, the probability of completing the marathon is
$$\begin{aligned} P(E_1 \cup E_2 \cup E_3) &= P(E_1) + P(E_2) + P(E_3) \\ &= \frac{1}{3} + \frac{2}{9} + \frac{4}{27} = \frac{19}{27} = 0.704, \end{aligned}$$
which means that there is about a 70 percent chance that the runner will complete the marathon—not exactly a virtual certainty!

## Section 6.7 Review

**1.** If $a$ and $b$ are real numbers and $n$ is a positive integer, then
$$(a + b)^n = \sum_{k=0}^{n} C(n, k)a^{n-k}b^k.$$

**2.** In the expansion of
$$(a + b)^n = \underbrace{(a + b)(a + b) \cdots (a + b)}_{n \text{ factors}},$$
the term $a^{n-k}b^k$ arises from choosing $b$ from $k$ factors and $a$ from the other $n - k$ factors, which can be done in $C(n, k)$ ways. Summing over all $k$ gives the Binomial Theorem.

**3.** Pascal's triangle is an arrangement of the binomial coefficients in triangular form. The border consists of 1's, and any interior value is the sum of the two numbers above it:

1
1 1
1 2 1
1 3 3 1
1 4 6 4 1
1 5 10 10 5 1
⋮

**4.** $C(n, 0) = C(n, n) = 1$, for all $n \geq 0$; and $C(n + 1, k) = C(n, k - 1) + C(n, k)$, for all $1 \leq k \leq n$

## Section 6.7

**1.** $x^4 + 4x^3y + 6x^2y^2 + 4xy^3 + y^4$

**3.** $C(11, 7)x^4y^7$

**6.** 5,987,520

**9.** $C(7, 3) + C(5, 2)$, since
$$\begin{aligned} (a + \sqrt{ax} + x)^2(a + x)^5 &= [(a + x) + \sqrt{ax}]^2(a + x)^5 \\ &= (a + x)^7 + 2\sqrt{ax}(a + x)^6 + ax(a + x)^5. \end{aligned}$$

**10.** $C(10 + 3 - 1, 10)$

**13.** 1 8 28 56 70 56 28 8 1

**17.** [Inductive Step only] Assume that the theorem is true for $n$.
$$\begin{aligned} (a + b)^{n+1} &= (a + b)(a + b)^n \\ &= (a + b)\sum_{k=0}^{n} C(n, k)a^{n-k}b^k \\ &= \sum_{k=0}^{n} C(n, k)a^{n+1-k}b^k \\ &\quad + \sum_{k=0}^{n} C(n, k)a^{n-k}b^{k+1} \\ &= \sum_{k=0}^{n} C(n, k)a^{n+1-k}b^k \\ &\quad + \sum_{k=1}^{n+1} C(n, k - 1)a^{n+1-k}b^k \\ &= C(n, 0)a^{n+1}b^0 + \sum_{k=1}^{n} C(n, k)a^{n+1-k}b^k \\ &\quad + C(n, n)a^0b^{n+1} \end{aligned}$$

$$+\sum_{k=1}^{n} C(n, k-1)a^{n+1-k}b^k$$
$$= C(n+1, 0)a^{n+1}b^0$$
$$+\sum_{k=1}^{n}[C(n, k) + C(n, k-1)]a^{n+1-k}b^k$$
$$+ C(n+1, n+1)a^0b^{n+1}$$
$$= C(n+1, 0)a^{n+1}b^0$$
$$+\sum_{k=1}^{n} C(n+1, k)a^{n+1-k}b^k$$
$$+ C(n+1, n+1)a^0b^{n+1}$$
$$= \sum_{k=0}^{n+1} C(n+1, k)a^{n+1-k}b^k$$

**20.** The number of solutions in nonnegative integers of

$$x_1 + x_2 + \cdots + x_{k+2} = n - k$$

is $C(k+2+n-k-1, n-k) = C(n+1, k+1)$. The number of solutions is also the number of solutions $C(k+1+n-k-1, n-k) = C(n, k)$ with $x_{k+2} = 0$ plus the number of solutions

$$C(k+1+n-k-1-1, n-k-1) = C(n-1, k)$$

with $x_{k+2} = 1$ plus $\cdots$ plus the number of solutions $C(k+1+0-1, 0) = C(k, k)$ with $x_{k+2} = n-k$. The result now follows.

**23.** Take $a = 1$ and $b = 2$ in the Binomial Theorem.

**26.** $x^3 + 3x^2y + 3x^2z + 3xy^2 + 6xyz + 3xz^2 + y^3 + 3y^2z + 3yz^2 + z^3$

**32.** Set $a = 1$ and $b = x$ and replace $n$ by $n-1$ in the Binomial Theorem to obtain

$$(1+x)^{n-1} = \sum_{k=0}^{n-1} C(n-1, k)x^k.$$

Now multiply by $n$ to obtain

$$n(1+x)^{n-1} = n\sum_{k=0}^{n-1} C(n-1, k)x^k$$
$$= n\sum_{k=1}^{n} C(n-1, k-1)x^{k-1}$$
$$= \sum_{k=1}^{n} \frac{n(n-1)!}{(n-k)!\,(k-1)!}x^{k-1}$$
$$= \sum_{k=1}^{n} \frac{n!}{(n-k)!\,k!}kx^{k-1}$$
$$= \sum_{k=1}^{n} C(n, k)kx^{k-1}.$$

**35.** The solution is by induction on $k$. We omit the Basis Step. Assume that the statement is true for $k$. After $k$ iterations, we obtain the sequence defined by

$$a'_j = \sum_{i=0}^{k-1} a_{i+j}\frac{B_i}{2^n}.$$

Let $B'_0, \ldots, B'_k$ denote the row after $B_0, \ldots, B_{k-1}$ in Pascal's triangle. Smoothing $a'$ by $c$ to obtain $a''$ yields

$$a''j = \frac{1}{2}(a'_j + a'_{j+1})$$
$$= \frac{1}{2^{n+1}}\left(\sum_{i=0}^{k-1} a_{i+j}B_i + \sum_{i=0}^{k-2} a_{i+j+1}B_i\right)$$
$$= \frac{1}{2^{n+1}}\left(a_jB_0 + \sum_{i=1}^{k-1} a_{i+j}B_i + \sum_{i=0}^{k-2} a_{i+j+1}B_i + a_{k+j}B_{k-1}\right)$$
$$= \frac{1}{2^{n+1}}\left(a_jB_0 + \sum_{i=1}^{k-1} a_{i+j}B_i + \sum_{i=1}^{k-1} a_{i+j}B_{i-1} + a_{k+j}B_{k-1}\right)$$
$$= \frac{1}{2^{n+1}}\left(a_jB'_0 + \sum_{i=1}^{k-1} a_{i+j}B'_i + a_{k+j}B'_k\right)$$
$$= \frac{1}{2^{n+1}}\sum_{i=0}^{k} a_{i+j}B'_i,$$

and the Inductive Step is complete.

**38.** [Inductive Step only] Notice that

$$C(n+1, i)^{-1} + C(n+1, i+1)^{-1} = \frac{n+2}{n+1}C(n, i)^{-1}.$$

Now

$$\sum_{i=1}^{n+1} C(n+1, i)^{-1}$$
$$= \frac{1}{2}\left(\sum_{i=1}^{n+1} C(n+1, i)^{-1} + \sum_{i=0}^{n} C(n+1, i+1)^{-1}\right)$$
$$= \frac{1}{2}\left(C(n+1, 1)^{-1} + \frac{n+2}{n+1}\sum_{i=1}^{n} C(n, i)^{-1} + C(n+1, n+1)^{-1}\right)$$
$$= \frac{1}{2}\left(\frac{n+2}{n+1} + \frac{n+2}{2^n}\sum_{i=0}^{n-1}\frac{2^i}{i+1}\right)$$
$$= \frac{n+2}{2^{n+1}}\sum_{i=0}^{n}\frac{2^i}{i+1}.$$

**41.** We use the Hint, which follows from the Binomial Theorem applied to $(j+1)^p$. By Exercise 40, $p$ divides $C(p, i)$ for all $i$, $1 \le i \le p-1$. Therefore $p$ divides $(j+1)^p - j^p - 1$ for all $j$, $1 \le j \le a-1$. Thus $p$ divides

$$\sum_{j=1}^{a-1}[(j+1)^p - j^p - 1] = [2^p - 1^p - 1] + [3^p - 2^p - 1] + \cdots + [a^p - (a-1)^p - 1]$$
$$= a^p - 1^p - (a-1)\cdot 1 = a^p - a.$$

## Section 6.8 Review

1. *First Form:* If $n$ pigeons fly into $k$ pigeonholes and $k < n$, some pigeonhole contains at least two pigeons.

   *Second Form:* If $f$ is a function from a finite set $X$ to a finite set $Y$ and $|X| > |Y|$, then $f(x_1) = f(x_2)$ for some $x_1, x_2 \in X$, $x_1 \neq x_2$.

   *Third Form:* Let $f$ be a function from a finite set $X$ to a finite set $Y$. Suppose that $|X| = n$ and $|Y| = m$. Let $k = \lceil n/m \rceil$. Then there are at least $k$ values $a_1, \ldots, a_k \in X$ such that

$$f(a_1) = f(a_2) = \cdots = f(a_k).$$

2. *First Form:* If 20 persons (pigeons) go into six rooms (pigeonholes), then some room contains at least two persons.

   *Second Form:* In the previous example, let $X$ be the set of persons, and let $Y$ be the set of rooms. If $p$ is a person, define a function $f$ by letting $f(p)$ be the room in which person $p$ is located. Then for some distinct persons $p_1$ and $p_2$, $f(p_1) = f(p_2)$; that is, the distinct persons $p_1$ and $p_2$ are in the same room.

   *Third Form:* Let $X$, $Y$, and $f$ be as in the last example. Then there are at least $\lceil 20/6 \rceil = 4$ persons $p_1, p_2, p_3, p_4$ such that

$$f(p_1) = f(p_2) = f(p_3) = f(p_4);$$

   that is, there are at least four persons in the same room.

## Section 6.8

1. Let the five cards be the pigeons and the four suits be the pigeonholes. Assign each card (pigeon) to its suit (pigeonhole). By the Pigeonhole Principle, some pigeonhole (suit) will contain at least two pigeons (cards), that is, at least two cards are of the same suit.

4. Let the 35 students be the pigeons and the 24 letters of the alphabet be the pigeonholes. Assign each student (pigeon) the first letter of the first name (pigeonhole). By the Pigeonhole Principle, some pigeonhole (letter) will contain at least two pigeons (students); that is, at least two students have first names that start with the same letter.

7. Let the 13 persons be the pigeons and the $12 = 3 \cdot 4$ possible names be the pigeonholes. Assign each person (pigeon) that person's name (pigeonhole). By the Pigeonhole Principle, some pigeonhole (name) will contain at least two pigeons (persons); that is, at least two persons have the same first and last names.

10. Yes. Connect processors 1 and 2, 2 and 3, 2 and 4, 3 and 4. Processor 5 is not connected to any processors. Now only processors 3 and 4 are directly connected to the same number of processors.

13. Let $a_i$ denote the position of the $i$th unavailable item. Consider

$$a_1, \ldots, a_{30}; \quad a_1 + 3, \ldots, a_{30} + 3; \quad a_1 + 6, \ldots, a_{30} + 6.$$

   These 90 numbers range in value from 1 to 86. By the second form of the Pigeonhole Principle, two of these numbers are the same. If $a_i = a_j + 3$, two are three apart. If $a_i = a_j + 6$, two are six apart. If $a_i + 3 = a_j + 6$, two are three apart.

18. $n + 1$

19. Suppose that $k \leq m/2$. Clearly, $k \geq 1$. Since $m \leq 2n + 1$,

$$k \leq \frac{m}{2} \leq n + \frac{1}{2} < n + 1.$$

   Suppose that $k > m/2$. Then

$$m - k < m - \frac{m}{2} = \frac{m}{2} < n + 1.$$

   Because $m$ is the largest element in $X$, $k < m$. Thus $k + 1 \leq m$ and so $1 \leq m - k$. Therefore, the range of $a$ is contained in $\{1, \ldots, n\}$.

20. The second form of the Pigeonhole Principle applies.

21. Suppose that $a_i = a_j$. Then either $i \leq m/2$ and $j > m/2$ or $j \leq m/2$ and $i > m/2$. We may assume that $i \leq m/2$ and $j > m/2$. Now

$$i + j = a_i + m - a_j = m.$$

31. When we divide $a$ by $b$, the possible remainders are $0, 1, \ldots, b - 1$. Consider what happens after $b$ divisions.

35. We suppose that the board has three rows and seven columns. We call two squares in one column that are the same color a *colorful pair*. By the Pigeonhole Principle, each column contains at least one colorful pair. Thus the board contains seven colorful pairs, one in each column. Again by the Pigeonhole Principle, at least four of these seven colorful pairs are the same color, say red. Since there are three pairs of rows and four red colorful pairs, a third application of the Pigeonhole Principle shows that at least two columns contain red colorful pairs in the same rows. These colorful pairs determine a rectangle whose four corner squares are red.

38. Suppose that it is possible to mark $k$ squares in the upper-left $k \times k$ subgrid and $k$ squares in the lower-right $k \times k$ subgrid so that no two marked squares are in the same row, column, or diagonal of the $2k \times 2k$ grid. Then the $2k$ marked squares are contained in $2k - 1$ diagonals. One diagonal begins at the top left square and runs to the bottom right square; $k - 1$ diagonals begin at the $k - 1$ squares immediately to the right of the top left square and run parallel to the first diagonal described; and $k - 1$ diagonals begin at the $k - 1$ squares immediately under the top left square and run parallel to the others described. By the first form of the Pigeonhole Principle, some diagonal contains two marked squares. This contradiction shows that it is impossible to mark $k$ squares in the upper-left $k \times k$ subgrid and $k$ squares in the lower-right $k \times k$ subgrid so that no two marked squares are in the same row, column, or diagonal of the $2k \times 2k$ grid.

41. Since there are $n + 1$ numbers in $X$, but only $n$ distinct odd parts, the odd parts of two numbers must coincide. Thus one of these two numbers divides the other.

44. We use the fact that the sum of the interior angles of a planar polygon with $n$ sides is $(n-1)180$ degrees. Now suppose that at most two interior angles are less than 180 degrees. Then at

least $n - 2$ interior angles are greater than or equal to 180 degrees. Thus the sum of the interior angles and the other angles is greater than $(n - 2)180$ degrees, which is a contradiction.

## Chapter 6 Self-Test

1. [Section 6.1] $2^4$
2. [Section 6.2] $6!/(3!\,3!) = 20$
3. [Section 6.2] We construct the strings by a three-step process. First, we choose positions for $A$, $C$, and $E$ [$C(6, 3)$ ways]. Next, we place $A$, $C$, and $E$ in these positions. We can place $C$ one way (last), and we can place $A$ and $E$ two ways ($AE$ or $EA$). Finally, we place the remaining three letters (3! ways). Therefore, the total number of strings is $C(6, 3) \cdot 2 \cdot 3!$.
4. [Section 6.3] $8!/(3!2!)$
5. [Section 6.3] We count the number of strings in which no $I$ appears before any $L$ and then subtract from the total number of strings.

   We construct strings in which no $I$ appears before any $L$ by a two-step process. First, we choose positions for $N$, $O$, and $S$; then we place the $I$'s and $L$'s. We can choose positions for $N$, $O$, and $S$ in $8 \cdot 7 \cdot 6$ ways. The $I$'s and $L$'s can then be placed in only one way because the $L$'s must come first. Thus there are $8 \cdot 7 \cdot 6$ strings in which no $I$ appears before any $L$.

   Exercise 4 shows that there are $8!/(3!\,2!)$ strings formed by ordering the letters *ILLINOIS*. Therefore, there are
$$\frac{8!}{3!\,2!} - 8 \cdot 7 \cdot 6$$
   strings formed by ordering the letters *ILLINOIS* in which some $I$ appears before some $L$.
6. [Section 6.1] $6 \cdot 9 \cdot 7 + 6 \cdot 9 \cdot 4 + 6 \cdot 7 \cdot 4 + 9 \cdot 7 \cdot 4$
7. [Section 6.1] $2^n - 2$
8. [Section 6.2] Two suits can be chosen in $C(4, 2)$ ways. We can choose three cards of one suit in $C(13, 3)$ ways and we can choose three cards of the other suit in $C(13, 3)$ ways. Therefore, the total number of hands is $C(4, 2)C(13, 3)^2$.
9. [Section 6.7]
$$\begin{aligned}(s - r)^4 &= C(4, 0)s^4 + C(4, 1)s^3(-r)\\ &\quad + C(4, 2)s^2(-r)^2 + C(4, 3)s(-r)^3\\ &\quad + C(4, 4)(-r)^4\\ &= s^4 - 4s^3r + 6s^2r^2 - 4sr^3 + r^4\end{aligned}$$
10. [Section 6.8] Let the 15 individual socks be the pigeons and let the 14 types of pairs be the pigeonholes. Assign each sock (pigeon) to its type (pigeonhole). By the Pigeonhole Principle, some pigeonhole will contain at least two pigeons (the matched socks).
11. [Section 6.1] $6 \cdot 5 \cdot 4 \cdot 3 + 6 \cdot 5 \cdot 4 \cdot 3 \cdot 2$
12. [Section 6.2] We must select either three or four defective discs. Thus the total number of selections is $C(5, 3)C(95, 1) + C(5, 4)$.
13. [Section 6.3] $12!/(3!)^4$
14. [Section 6.7] $2^3 \cdot 8!/(3!\,1!\,4!)$
15. [Section 6.7] If we set $a = 2$ and $b = -1$ in the Binomial Theorem, we obtain
$$1 = 1^n = [2 + (-1)]^n = \sum_{k=0}^{n} C(n, k)2^{n-k}(-1)^k.$$
16. [Section 6.3] $C(11 + 4 - 1, 4 - 1)$
17. [Section 6.8] There are $3 \cdot 2 \cdot 3 = 18$ possible names for the 19 persons. We can consider the assignment of names to people to be that of assigning pigeonholes to the pigeons. By the Pigeonhole Principle, some name is assigned to at least two persons.
18. [Section 6.7] $C(n, 1) = n$
19. [Section 6.8] Let $a_i$ denote the position of the $i$th available item. The 220 numbers
$$a_1, \ldots, a_{110}; \qquad a_1 + 19, \ldots, a_{110} + 19$$
   range from 1 to 219. By the Pigeonhole Principle, two are the same.
20. [Section 6.8] Each point has an $x$-coordinate that is either even or odd and a $y$-coordinate that is either even or odd. Since there are four possibilities and there are five points, by the Pigeonhole Principle at least two points, $p_i = (x_i, y_i)$ and $p_j = (x_j, y_j)$ have

   - Both $x_i$ and $x_j$ even or both $x_i$ and $x_j$ odd.

   *and*

   - Both $y_i$ and $y_j$ even or both $y_i$ and $y_j$ odd.

   Therefore, $x_i + x_j$ is even and $y_i + y_j$ is even. In particular, $(x_i + x_j)/2$ and $(y_i + y_j)/2$ are integers. Thus the midpoint of the pair $p_i$ and $p_j$ has integer coordinates.
21. [Section 6.4] 12567
22. [Section 6.4] 234567
23. [Section 6.4] 6427153
24. [Section 6.4] 631245
25. [Section 6.5] 1/4
26. [Section 6.6] $P(H) = 5/6$, $P(T) = 1/6$
27. [Section 6.6] Let $S$ denote the event "children of both sexes," and let $G$ denote the event "at most one girl." Then
$$P(S) = \frac{6}{8} = \frac{3}{4}$$
$$P(G) = \frac{4}{8} = \frac{1}{2}$$
$$P(S \cap G) = \frac{3}{8}.$$
   Therefore,
$$P(S)P(G) = \frac{3}{4} \cdot \frac{1}{2} = \frac{3}{8} = P(S \cap G),$$
   and $S$ and $G$ are independent.
28. [Section 6.5] 5/36
29. [Section 6.5] $\dfrac{C(7, 5)C(31 - 7, 2)}{C(31, 7)} = \dfrac{21 \cdot 276}{2629575} = 0.002204158$

**30.** [Section 6.6] Let $J$ denote the event "Joe passes," and let $A$ denote the event "Alicia passes." Then

$$\begin{aligned} P(\text{Joe fails}) &= P(\overline{J}) = 1 - P(J) = 0.25 \\ P(\text{both pass}) &= P(J \cap A) = P(J)P(A) \\ &= (0.75)(0.80) = 0.6 \\ P(\text{both fail}) &= P(\overline{J} \cap \overline{A}) = P(\overline{J \cup A}) \\ &= 1 - P(J \cup A) \\ &= 1 - [P(J) + P(A) - P(J \cap A)] \\ &= 1 - [0.75 + 0.80 - 0.6] = 0.05 \\ P(\text{at least one passes}) &= 1 - P(\text{both fail}) \\ &= 1 - 0.05 = 0.95. \end{aligned}$$

**31.** [Section 6.6] Let $B$ denote the event "bug present," and let $T$, $R$, and $J$ denote the events "Trisha (respectively, Roosevelt, José) wrote the program." Then

$$\begin{aligned} P(J \mid B) &= \frac{P(B \mid J)P(J)}{P(B \mid J)P(J) + P(B \mid T)P(T) + P(B \mid R)P(R)} \\ &= \frac{(0.05)(0.25)}{(0.05)(0.25) + (0.03)(0.30) + (0.02)(0.45)} \\ &= 0.409836065. \end{aligned}$$

**32.** [Section 6.5] $\dfrac{4 \cdot C(13, 6) \cdot 3 \cdot C(13, 5) \cdot 2 \cdot C(13, 2)}{C(52, 13)}$

## Section 7.1 Review

**1.** A recurrence relation defines the $n$th term of a sequence in terms of certain of its predecessors.

**2.** An initial condition for a sequence is an explicitly given value for a particular term in the sequence.

**3.** Compound interest is interest on interest. If a person invests $d$ dollars at $p$ percent compounded annually and we let $A_n$ be the amount of money earned after $n$ years, the recurrence relation

$$A_n = \left(1 + \frac{p}{100}\right) A_{n-1}$$

together with the initial condition $A_0 = d$ defines the sequence $\{A_n\}$.

**4.** The Tower of Hanoi puzzle consists of three pegs mounted on a board and disks of various sizes with holes in their centers. Only a disk of smaller diameter can be placed on a disk of larger diameter. Given all the disks stacked on one peg, the problem is to transfer the disks to another peg by moving one disk at a time.

**5.** If there is one disk, move it and stop. If there are $n > 1$ disks, recursively move $n - 1$ disks to an empty peg. Move the largest disk to the remaining empty peg. Recursively move $n - 1$ disks on top of the largest disk.

**6.** We assume that at time $n$, the quantity $q_n$ sold at price $p_n$ is given by the equation $p_n = a - bq_n$, where $a$ and $b$ are positive parameters. We also assume that $p_n = kq_{n+1}$, where $k$ is another positive parameter. If we graph the price and quantity over time, the graph resembles a cobweb (see, e.g., Figure 7.1.5).

**7.** Ackermann's function $A(m, n)$ is defined by the recurrence relations

$$\begin{aligned} A(m, 0) &= A(m - 1, 1) & m &\ge 1 \\ A(m, n) &= A(m - 1, A(m, n - 1)) & m &\ge 1, n \ge 1 \end{aligned}$$

and initial conditions

$$A(0, n) = n + 1 \qquad n \ge 0.$$

## Section 7.1

**1.** $a_n = a_{n-1} + 4$; $a_1 = 3$

**4.** $A_n = (1.14)A_{n-1}$ **5.** $A_0 = 2000$

**6.** $A_1 = 2280, A_2 = 2599.20, A_3 = 2963.088$

**7.** $A_n = (1.14)^n 2000$

**8.** We must have $A_n = 4000$ or $(1.14)^n 2000 = 4000$ or $(1.14)^n = 2$. Taking the logarithm of both sides, we must have $n \log 1.14 = \log 2$. Thus

$$n = \frac{\log 2}{\log 1.14} = 5.29.$$

**18.** We count the number of $n$-bit strings not containing the pattern 000.

- Begin with 1. In this case, if the remaining $(n - 1)$-bit string does not contain 000, neither will the $n$-bit string. There are $S_{n-1}$ such $(n - 1)$-bit strings.
- Begin with 0. There are two cases to consider.
  1. Begin with 01. In this case, if the remaining $(n - 2)$-bit string does not contain 000, neither will the $n$-bit string. There are $S_{n-2}$ such $(n-2)$-bit strings.
  2. Begin with 00. Then the third bit must be a 1 and if the remaining $(n - 3)$-bit string does not contain 000, neither will the $n$ bit string. There are $S_{n-3}$ such $(n - 3)$-bit strings.

Since the cases are mutually exclusive and cover all $n$-bit strings ($n > 3$) not containing 000, we have $S_n = S_{n-1} + S_{n-2} + S_{n-3}$ for $n > 3$. $S_1 = 2$ (there are two 1-bit strings), $S_2 = 4$ (there are four 2-bit strings), and $S_3 = 7$ (there are eight 3-bit strings but one of them is 000).

**19.** There are $S_{n-1}$ $n$-bit strings that begin 1 and do not contain the pattern 00 and there are $S_{n-2}$ $n$ bit strings that begin 0 (since the second bit must be 1) and do not contain the pattern 00. Thus $S_n = S_{n-1} + S_{n-2}$. Initial conditions are $S_1 = 2$, $S_2 = 3$.

**22.** $S_1 = 2, S_2 = 4, S_3 = 7, S_4 = 12$

**25.** $C_3 = 5, C_4 = 14, C_5 = 42$

**28.** We first prove that if $n \ge 5$, then $C_n$ is not prime. Suppose, by way of contradiction, that $C_n$ is prime for some $n \ge 5$. By Exercise 27, $n + 2 < C_n$. Thus $C_n$ does not divide $n + 2$. By Exercise 26,

$$(n + 2)C_{n+1} = (4n + 2)C_n.$$

Thus $C_n$ divides $(n+2)C_{n+1}$. By Exercise 27, Section 5.3, $C_n$ divides either $n+2$ or $C_{n+1}$. Since $C_n$ does not divide $n+2$, $C_n$ divides $C_{n+1}$. Therefore there exists an integer $k \geq 1$ satisfying $C_{n+1} = kC_n$. Thus

$$(n+2)kC_n = (4n+2)C_n.$$

Canceling $C_n$, we obtain

$$(n+2)k = (4n+2).$$

If $k=1$, the preceding equation becomes $n+2 = 4n+2$, and thus $n=0$, which contradicts the fact that $n \geq 5$. Similarly, if $k=2$, then $n=1$, and if $k=3$, then $n=4$, both of which contradict the fact that $n \geq 5$. If $k \geq 4$,

$$4n+2 = k(n+2) \geq 4(n+2) = 4n+8.$$

Therefore $0 \geq 6$. Thus $k$ does not exist. This contradiction shows that if $n \geq 5$, $C_n$ is not prime.

Directly checking $n = 0, 1, 2, 3, 4$ shows that only $C_2 = 2$ and $C_3 = 5$ are prime.

**31.** Let $P_n$ denote the number of ways to divide a convex $(n+2)$-sided polygon, $n \geq 1$, into triangles by drawing $n-1$ lines through the corners that do not intersect in the interior of the polygon. We note that $P_1 = 1$.

Suppose that $n > 1$ and consider a convex $(n+2)$-sided polygon (see the following figure).

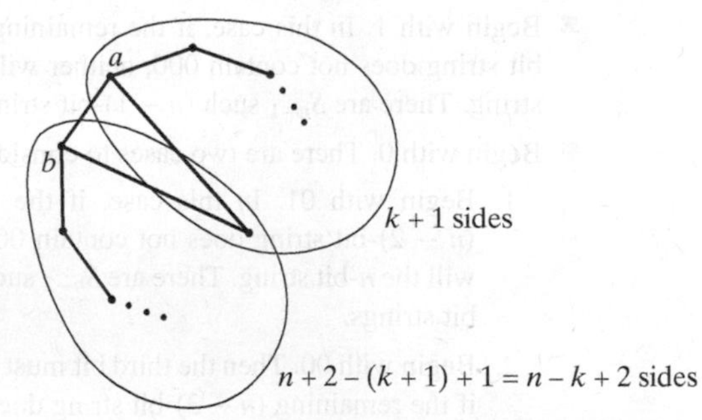


We choose one edge $ab$ and construct a partition of the polygon by a two-step procedure. First we select a triangle to which side $ab$ belongs. This triangle divides the original polygon into two polygons: one having $k+1$ sides, for some $k$ satisfying $1 \leq k \leq n$; and the other having $n-k+2$ sides (see the preceding figure). By definition, the $(k+1)$-sided polygon can be partitioned in $P_{k-1}$ ways and the $(n-k+2)$-sided polygon can be partitioned in $P_{n-k}$ ways. (For the degenerate cases $k=1$ and $k=n$, we set $P_0 = 1$.) Therefore, the total number of ways to partition the $(n+2)$-sided polygon is

$$P_n = \sum_{k=1}^{n} P_{k-1}P_{n-k}.$$

Since the sequence $P_1, P_2, \ldots$ satisfies the same recurrence relation as the Catalan sequence $C_1, C_2, \ldots$ and $P_0 = P_1 = 1 = C_0 = C_1$, it follows that $P_n = C_n$ for all $n \geq 1$.

**36.** [For $n=3$]

Step 1—move disk 3 from peg 1 to peg 3.
Step 2—move disk 2 from peg 1 to peg 2.
Step 3—move disk 3 from peg 3 to peg 2.
Step 4—move disk 1 from peg 1 to peg 3.
Step 5—move disk 3 from peg 2 to peg 1.
Step 6—move disk 2 from peg 2 to peg 3.
Step 7—move disk 3 from peg 1 to peg 3.

**38.** Let $\alpha$ and $\beta$ be the angles shown in Figure 7.1.6. The geometry of the situation shows that the price tends to stabilize if and only if $\alpha + \beta > 180°$. This last condition holds if and only if $-\tan\beta < \tan\alpha$. Since $b = -\tan\beta$ and $k = \tan\alpha$, we conclude that the price stabilizes if and only if $b < k$.

**40.** $A(2,2) = 7, A(2,3) = 9$ **43.** $A(3,n) = 2^{n+3} - 3$

**46.** If $m = 0$,

$$\begin{aligned} A(m, n+1) &= A(0, n+1) \\ &= n+2 > n+1 \\ &= A(0, n) = A(m, n). \end{aligned}$$

The last inequality follows from Exercise 44.

**47.** We use induction on $z$. For the Basis Step ($z=0$), we have $AO(3, y, 0) = 1 = y^0$. For the Inductive Step, assume the result for $z$. We then have $AO(3, y, z+1) = AO(2, y, y^z) = y \cdot y^z = y^{z+1}$.

**49.** Use Exercises 41 and 42.

**52.** We prove the statement by using induction on $x$. The inductive step will itself require induction on $y$.

Exercise 49 shows that the equation is true for $x = 0, 1, 2$ and for all $y$.

**Basis Step ($x=2$)** See Exercise 49.

**Inductive Step (Case $x$ implies case $x+1$)** Assume that $x \geq 2$ and

$$A(x, y) = AO(x, 2, y+3) - 3 \qquad \text{for all } y \geq 0.$$

We must prove that

$$A(x+1, y) = AO(x+1, 2, y+3) - 3 \qquad \text{for all } y \geq 0.$$

We establish this last equation by induction on $y$.

**Basis Step ($y=0$)** We must prove that

$$A(x+1, 0) = AO(x+1, 2, 3) - 3.$$

Now

$$\begin{aligned} &AO(x+1, 2, 3) - 3 \\ &= AO(x, 2, AO(x+1, 2, 2)) - 3 && \text{by definition} \\ &= AO(x, 2, 4) - 3 && \text{by Exercise 51} \\ &= A(x, 1) && \text{by the inductive assumption on } x \\ &= A(x+1, 0) && \text{by (7.1.12).} \end{aligned}$$

**Inductive Step (Case $y$ implies case $y+1$)** Assume that

$$A(x+1, y) = AO(x+1, 2, y+3) - 3.$$

We must prove that

$$A(x+1, y+1) = AO(x+1, 2, y+4) - 3.$$

Now

$$\begin{aligned}
&AO(x+1, 2, y+4) - 3 \\
&= AO(x, 2, AO(x+1, 2, y+3)) - 3 \quad \text{by definition} \\
&= AO(x, 2, A(x+1, y) + 3) - 3 \quad \text{by the inductive assumption on } y \\
&= A(x, A(x+1, y)) \quad \text{by the inductive assumption on } x \\
&= A(x+1, y+1) \quad \text{by (7.1.13).}
\end{aligned}$$

**55.** Suppose that we have $n$ dollars. If we buy orange juice the first day, we have $n-1$ dollars left, which may be spent in $R_{n-1}$ ways. Similarly, if the first day we buy milk or beer, there are $R_{n-2}$ ways to spend the remaining dollars. Since these cases are disjoint, $R_n = R_{n-1} + 2R_{n-2}$.

**58.** $S_3 = 1/2$, $S_4 = 3/4$

**60.** A function $f$ from $X = \{1, \ldots, n\}$ into $X$ will be denoted $(i_1, i_2, \ldots, i_n)$, which means that $f(k) = i_k$. The problem then is to count the number of ways to select $i_1, \ldots, i_n$ so that if $i$ occurs, so do $1, 2, \ldots, i-1$.

We shall count the number of such functions having exactly $j$ 1's. Such functions can be constructed in two steps: Pick the positions for the $j$ 1's; then place the other numbers. There are $C(n, j)$ ways to place the 1's. The remaining numbers must be selected so that if $i$ appears, so do $1, \ldots, i-1$. There are $F_{n-j}$ ways to select the remaining numbers, since the remaining numbers must be selected from $\{2, \ldots, n\}$. Thus there are $C(n, j)F_{n-j}$ functions of the desired type having exactly $j$ 1's. Therefore, the total number of functions from $X$ into $X$ having the property that if $i$ is in the range of $f$, then so are $1, \ldots, i-1$, is

$$\sum_{j=1}^{n} C(n, j)F_{n-j} = \sum_{j=1}^{n} C(n, n-j)F_{n-j} = \sum_{j=0}^{n-1} C(n, j)F_j.$$

**63.** $\{u_n\}$ is not a recurrence relation because, if $n$ is odd and greater than 1, $u_n$ is defined in terms of the *successor* $u_{3n+1}$. $u_i$, for $2 \le i \le 7$, is equal to one. As examples,

$$u_2 = u_1 = 1$$

$$u_3 = u_{10} = u_5 = u_{16} = u_8 = u_4 = u_2 = 1.$$

**66.** Use equation (6.7.4) to write

$$S(k, n) = \sum_{i=1}^{n} S(k-1, i).$$

**69.** We use the terminology of Exercise 90, Section 6.2. Choose one of $n+1$ people, say $P$. There are $s_{n,j-1}$ ways for $P$ to sit alone. (Seat the other $n$ people at the other $k-1$ tables.) Next we count the number of arrangements in which $P$ is not alone. Seat everyone but $P$ at $k$ tables. This can be done in $s_{n,k}$ ways. Now $P$ can be seated to the right of someone in $n$ ways. Thus there are $ns_{n,k}$ arrangements in which $P$ is not alone. The recurrence relation now follows.

**72.** Let $A_n$ denote the amount at the end of $n$ years and let $i$ be the interest rate expressed as a decimal. The discussion following Example 7.1.3 shows that $A_n = (1+i)^n A_0$. The value of $n$ required to double the amount satisfies

$$2A_0 = (1+i)^n A_0 \quad \text{or} \quad 2 = (1+i)^n.$$

If we take the natural logarithm (logarithm to the base $e$) of both sides of this equation, we obtain

$$\ln 2 = n \ln(1+i).$$

Thus

$$n = \frac{\ln 2}{\ln(1+i)}.$$

Since $\ln 2 = 0.6931472\ldots$ and $\ln(1+i)$ is approximately equal to $i$ for small values of $i$, $n$ is approximately equal to $0.69\ldots/i$, which, in turn, is approximately equal to $70/r$.

**74.** 1, 3, 2; 2, 3, 1; $E_3 = 2$

**77.** We count the number of rise/fall permutations of $1, \ldots, n$ by considering how many have $n$ in the second, fourth, $\ldots$, positions.

Suppose that $n$ is in the second position. Since any of the remaining numbers is less than $n$, any of them may be placed in the first position. Thus we may select the number to be placed in the first position in $C(n-1, 1)$ ways and, after selecting it, we may arrange it in $E_1 = 1$ way. The last $n-2$ positions can be filled in $E_{n-2}$ ways since any rise/fall permutation of the remaining $n-2$ numbers gives a rise/fall permutation of $1, \ldots, n$. Thus the number of rise/fall permutations of $1, \ldots, n$ with $n$ in the second position is $C(n-1, 1)E_1E_{n-2}$.

Suppose that $n$ is in the fourth position. We may select numbers to be placed in the first three positions in $C(n-1, 3)$ ways. After selecting the three items, we may arrange them in $E_3$ ways. The last $n-4$ numbers can be arranged in $E_{n-4}$ ways. Thus the number of rise/fall permutations of $1, \ldots, n$ with $n$ in the fourth position is $C(n-1, 3)E_3E_{n-4}$.

In general, the number of rise/fall permutations of $1, \ldots, n$ with $n$ in the $(2j)$th position is

$$C(n-1, 2j-1)E_{2j-1}E_{n-2j}.$$

Summing over all $j$ gives the desired recurrence relation.

## Section 7.2 Review

**1.** Use the recurrence relation to write the $n$th term in terms of certain of its predecessors. Then successively use the recurrence relation to replace each of the resulting terms by certain of their predecessors. Continue until an explicit formula is obtained.

**2.** An $n$th-order, linear homogeneous recurrence relation with constant coefficients is a recurrence relation of the form

$$a_n = c_1a_{n-1} + c_2a_{n-2} + \cdots + c_ka_{n-k}.$$

**3.** $a_n = 6a_{n-1} - 8a_{n-2}$

**4.** To solve

$$a_n = c_1a_{n-1} + c_2a_{n-2},$$

first solve the equation

$$t^2 = c_1 t + c_2$$

for $t$. Suppose that the roots are $t_1$ and $t_2$ and that $t_1 \neq t_2$. Then the general solution is of the form

$$a_n = bt_1^n + dt_2^n,$$

where $b$ and $d$ are constants. The values of the constants can be obtained from the initial conditions.

If $t_1 = t_2 = t$, the general solution is of the form

$$a_n = bt^n + dnt^n,$$

where again $b$ and $d$ are constants. The values of the constants can again be obtained from the initial conditions.

## Section 7.2

**1.** Yes; order 1

**4.** No

**7.** No

**10.** Yes; order 3

**11.** $a_n = 2(-3)^n$

**15.** $a_n = 2^{n+1} - 4^n$

**18.** $a_n = (2^{2-n} + 3^n)/5$

**21.** $a_n = 2(-4)^n + 3n(-4)^n$

**24.** $R_n = [(-1)^n + 2^{n+1}]/3$

**28.** Let $d_n$ denote the deer population at time $n$. The initial condition is $d_0 = 0$. The recurrence relation is

$$d_n = 100n + 1.2d_{n-1}, \quad n > 0.$$

$$\begin{aligned}
d_n &= 100n + 1.2d_{n-1} = 100n + 1.2[100(n-1) + 1.2d_{n-2}] \\
&= 100n + 1.2 \cdot 100(n-1) + 1.2^2 d_{n-2} \\
&= 100n + 1.2 \cdot 100(n-1) \\
&\quad + 1.2^2[100(n-2) + 1.2d_{n-3}] \\
&= 100n + 1.2 \cdot 100(n-1) \\
&\quad + 1.2^2 \cdot 100(n-2) + 1.2^3 d_{n-3} \\
&\ \ \vdots \\
&= \sum_{i=0}^{n-1} 1.2^i \cdot 100(n-i) + 1.2^n d_0 \\
&= \sum_{i=0}^{n-1} 1.2^i \cdot 100(n-i) \\
&= 100n \sum_{i=0}^{n-1} 1.2^i - 1.2 \cdot 100 \sum_{i=1}^{n-1} i \cdot 1.2^{i-1} \\
&= \frac{100n(1.2^n - 1)}{1.2 - 1} \\
&\quad - 120 \frac{(n-1)1.2^n - n1.2^{n-1} + 1}{(1.2-1)^2}, \qquad n > 0.
\end{aligned}$$

**29.** 1/2

**32.** 0

**35.** $n/C$

**38.** Set $b_n = a_n/n!$ to obtain $b_n = -2b_{n-1} + 3b_{n-2}$. Solving gives $a_n = n!\, b_n = (n!/4)[5 - (-3)^n]$.

**41.** We establish the inequality by using induction on $n$.

The base cases $n = 1$ and $n = 2$ are left to the reader. Now assume that the inequality is true for values less than $n + 1$. Then

$$\begin{aligned}
f_{n+2} &= f_{n+1} + f_n \\
&\geq \left(\frac{1+\sqrt{5}}{2}\right)^{n-1} + \left(\frac{1+\sqrt{5}}{2}\right)^{n-2} \\
&= \left(\frac{1+\sqrt{5}}{2}\right)^{n-2} \left(\frac{1+\sqrt{5}}{2} + 1\right) \\
&= \left(\frac{1+\sqrt{5}}{2}\right)^{n-2} \left(\frac{1+\sqrt{5}}{2}\right)^{2} \\
&= \left(\frac{1+\sqrt{5}}{2}\right)^{n}
\end{aligned}$$

and the Inductive Step is complete.

**43.** $a_n = b2^n + d4^n + 1$

**46.** $a_n = b/2^n + d3^n - (4/3)2^n$

**49.** The argument is identical to that given in Theorem 7.2.11.

**52.** Recursively invoking this algorithm to move the $n - k_n$ disks at the top of peg 1 to peg 2 takes $T(n - k_n)$ moves. Moving the $k_n$ disks on peg 1 to peg 4 requires $2^{k_n} - 1$ moves (see Example 7.2.4). Recursively invoking this algorithm to move the $n - k_n$ disks on peg 2 to peg 4 again takes $T(n - k_n)$ moves. The recurrence relation now follows.

**55.** From the inequality

$$\frac{k_n(k_n+1)}{2} \leq n,$$

we can deduce $k_n \leq \sqrt{2n}$. Since

$$n - k_n \leq \frac{k_n(k_n+1)}{2},$$

it follows that $r_n \leq k_n$. Therefore,

$$\begin{aligned}
T(n) &= (k_n + r_n - 1)2^{k_n} + 1 \\
&< 2k_n 2^{k_n} + 1 \\
&\leq 2\sqrt{2n}2^{\sqrt{2n}} + 1 \\
&= O(4^{\sqrt{n}}).
\end{aligned}$$

## Section 7.3 Review

**1.** Let $b_n$ denote the time required for input of size $n$. Simulate the execution of the algorithm and count the time required by the various steps. Then $b_n$ is equal to the sum of the times required by the various steps.

**2.** Selection sort selects the largest element, places it last, and then recursively sorts the remaining sequence.

**3.** $\Theta(n^2)$

**4.** Binary search examines the middle item in the sequence. If the middle item is the desired item, binary search terminates. Otherwise, binary search compares the middle item with the desired item. If the desired item is less than the middle item, binary search recursively searches in the left half of the sequence. If the desired item is greater than the middle item, binary search recursively searches in the right half of the sequence. The input must be sorted.

**5.** If $a_n$ is the worst-case time for input of size $n$, $a_n = 1+a_{\lfloor n/2 \rfloor}$.

**6.** $\Theta(\lg n)$

**7.** Merge maintains two pointers to elements in the two input sequences. Initially the pointers reference the first elements in the sequences. Merge copies the smaller element to the output and moves the pointer to the next element in the sequence that contains the element just copied. It then repeats this process. When a pointer moves off the end of one of the sequences, merge concludes by copying the rest of the other sequence to the output. Both input sequences must be sorted.

**8.** $\Theta(n)$, where $n$ is the sum of the lengths of the input sequences

**9.** Merge sort first divides the input into two nearly equal parts. It then recursively sorts each half and merges the halves to produce sorted output.

**10.** $a_n = a_{\lfloor n/2 \rfloor} + a_{\lfloor (n+1)/2 \rfloor} + n - 1$

**11.** If the input size is a power of two, the size is always divisible by 2 and the floors vanish.

**12.** An arbitrary input size falls between two powers of two. Since we know the worst-case time when the input size is a power of two, we may bound the worst-case time for input of arbitrary size by the worst-case times for inputs whose sizes are the powers of two that bound it.

**13.** $\Theta(n \lg n)$

## Section 7.3

**1.** At line 2, since $i > j$ $(1 > 5)$ is false, we proceed to line 4, where we set $k$ to 3. At line 5, since *key* ('*G*') is not equal to $s_3$ ('*J*'), we proceed to line 7. At line 7, $key < s_k$ ('*G*' < '*J*') is true, so at line 8 we set $j$ to 2. We then invoke this algorithm with $i = 1, j = 2$ to search for *key* in

$$s_1 = \text{'}C\text{'}, \qquad s_2 = \text{'}G\text{'}.$$

At line 2, since $i > j$ $(1 > 2)$ is false, we proceed to line 4, where we set $k$ to 1. At line 5, since *key* ('*G*') is not equal to $s_1$ ('*C*'), we proceed to line 7. At line 7, $key < s_k$ ('*G*' < '*C*') is false, so at line 10 we set $i$ to 2. We then invoke this algorithm with $i = j = 2$ to search for *key* in

$$s_2 = \text{'}G\text{'}.$$

At line 2, since $i > j$ $(2 > 2)$ is false, we proceed to line 4, where we set $k$ to 2. At line 5, since *key* ('*G*') is equal to $s_2$ ('*G*'), we return 2, the index of *key* in the sequence $s$.

**4.** At line 2, since $i > j$ $(1 > 5)$ is false, we proceed to line 4, where we set $k$ to 3. At line 5, since *key* ('*Z*') is not equal to $s_3$ ('*J*'), we proceed to line 7. At line 7, $key < s_k$ ('*Z*' < '*J*') is false, so at line 10 we set $i$ to 4. We then invoke this algorithm with $i = 4, j = 5$ to search for *key* in

$$s_4 = \text{'}M\text{'}, \qquad s_5 = \text{'}X\text{'}.$$

At line 2, since $i > j$ $(4 > 5)$ is false, we proceed to line 4, where we set $k$ to 4. At line 5, since *key* ('*Z*') is not equal to $s_4$ ('*M*') we proceed to line 7. At line 7, $key < s_k$ ('*Z*' < '*M*') is false, so at line 10 we set $i$ to 5. We then invoke this algorithm with $i = j = 5$ to search for *key* in

$$s_5 = \text{'}X\text{'}.$$

At line 2, since $i > j$ $(5 > 5)$ is false, we proceed to line 4, where we set $k$ to 5. At line 5, since *key* ('*Z*') is not equal to $s_5$ ('*X*'), we proceed to line 7. At line 7, $key < s_k$ ('*Z*' < '*X*') is false, so at line 10 we set $i$ to 6. We then invoke this algorithm with $i = 6, j = 5$.

At line 2, since $i > j$ $(6 > 5)$ is true, we return 0 to indicate that we failed to find *key*.

**7.** Consider the input 10, 4, 2 and $key = 10$.

**10.** The idea is to repeatedly divide the sequence as nearly as possible into two parts and retain the part that might contain the key. Only after obtaining a subsequence of length 1 or 2, do we test whether the subsequence contains the key. The following algorithm implements this design.

```
binary_search_nonrecurs(s, n, key) {
  i = 1
  j = n
  // the body of the loop executes only if the subsequence
  // s_i, ..., s_j has length greater than or equal to 3
  while (i < j − 1) {
    k = ⌊(i + j)/2⌋
    if (s_k < key)
      i = k + 1
    else
      j = k
  }
  for k = i to j
    if (s_k == key)
      return k
  return 0
}
```

We first prove that if a sequence of length $n$ is input to the while loop, where $n$ is a power of 2, say $n = 2^m$, $m \geq 2$, the loop iterates $m - 1$ times. The proof is by induction on $m$. The Basis Step is $m = 2$. In this case $n = 4$. Assuming that $i = 1$ and $n = 4$, in the while loop, $k$ is first set to 2. Then either $i$ is set to 3 or $j$ is set to 2. Thus the loop does not execute again. Therefore the loop iterates $1 = m - 1$ time. The Basis Step is complete.

Now suppose that if a sequence of length $n = 2^m$ is input to the while loop, the loop iterates $m - 1$ times. Suppose that $n = 2^{m+1}$. Assuming that $i = 1$ and $n = 2^{m+1}$, in the while loop, $k$ is first set to $2^m$. Then either $i$ is set to $2^m + 1$ or $j$ is set to $2^m$. Thus at the next iteration of the loop, a sequence of length $2^m$ is processed. By the inductive assumption, the loop iterates an additional $m - 1$ times. Therefore the loop iterates a total of $m$ times. The Inductive Step is complete.

Next we prove that if a sequence of length $n$, where $n$ satisfies $2^{m-1} < n \le 2^m$, $m \ge 2$, is input to the while loop, the loop iterates at most $m - 1$ times. The proof is by induction on $m$. The Basis Step is $m = 2$. In this case we have $2 < n \le 4$. Thus $n$ is either 3 or 4. In the preceding paragraphs, we proved that if $n = 4$ the loop iterates one time. If $n = 3$, it is easy to check that the loop iterates one time. The Basis Step is complete.

Now assume that if a sequence of length $n$, where $n$ satisfies $2^{m-1} < n \le 2^m$, $m \ge 2$, is input to the while loop, the loop iterates at most $m - 1$ times. Suppose that $n$ satisfies $2^m < n \le 2^{m+1}$. When $n$ is even, the sequence is divided evenly and the next sequence processed by the loop has length $n/2$. Since $n/2$ satisfies $2^{m-1} < n/2 \le 2^m$, by the inductive assumption the loop iterates at most $m - 1$ more times. When $n$ is odd, the sequence is divided into two parts—one part of length $(n - 1)/2$ and the other of length $(n+1)/2$. Since $n$ is odd, $2^m < n < 2^{m+1}$. Therefore $2^m < n+1 \le 2^{m+1}$. Thus $2^{m-1} < (n+1)/2 \le 2^m$. In this case, the inductive assumption tells us that the loop iterates at most $m - 1$ more times. We also have $2^m \le n-1 < 2^{m+1}$ and $2^{m-1} \le (n-1)/2 < 2^m$. If $2^{m-1} < (n-1)/2$, we may use the inductive assumption to conclude that the loop iterates at most $m - 1$ more times. If $2^{m-1} = (n-1)/2$, we may use the result proved just after the algorithm to conclude that the loop iterates $m-2$ more times. In every case the loop iterates at most $m - 1$ more times. Together with the first iteration, we conclude that if $n$ satisfies $2^m < n \le 2^{m+1}$, the while loop iterates at most $m$ times. The Inductive Step is complete.

Suppose that $n$ satisfies $2^{m-1} < n \le 2^m$. Then the while loop iterates at most $m - 1$ times. This accounts for $m-1$ tests of the form $s_k < key$. At the for loop, either $i = j$ or $i = j+1$. Thus there are at most two additional comparisons (of the form $s_k == key$). Thus if $n$ satisfies $2^{m-1} < n \le 2^m$, the algorithm uses at most $m+1$ comparisons. Since $2^{m-1} < n \le 2^m$, $m - 1 < \lg n \le m$. Therefore $\lceil \lg n \rceil = m$. Thus the algorithm uses at most $1 + m = 1 + \lceil \lg n \rceil$ comparisons.

**13.** The algorithm is not correct. If $s$ is a sequence of length 1, $s_1 = 9$, and $key = 8$, the algorithm does not terminate.

**16.** The algorithm is correct. The worst-case time is $\Theta(\log n)$.

**18.** Algorithm B is superior if $2 \le n \le 15$. (For $n = 1$ and $n = 16$, the algorithms require equal numbers of comparisons.)

**21.** Suppose that the sequences are $a_1, \ldots, a_n$ and $b_1, \ldots, b_n$. (a) $a_1 < b_1 < a_2 < b_2 < \cdots$ (b) $a_n < b_1$

**24.** 11

**28.** 4,1; 4,3; 4,2; 3,2

**31.** 4,3,2,1

**34.** Using the notation in the hint, every pair $p_i, p_j$ is an inversion in either $p$ or $p^R$ but not both. Since there are $n(n-1)/2$ total pairs $p_i, p_j$, the total number of inversions in $p$ and $p^R$ is $n(n-1)/2$. Since there are $n!$ permutations of $n$ items, adding the inversions in $p$ and $p^R$ for all permutations $p$ gives $n!n(n-1)/2$. But in this computation, each inversion was counted twice [once in the pair $p$, $p^R$ and once in the pair $p^R, (p^R)^R = p^R, p$]. Therefore, the total number of inversions in all of the permutations of $\{1, \ldots, n\}$ is $n!n(n-1)/4$.

**37.** The idea is to use a divide-and-conquer approach. Suppose that we divide the input $p_1, \ldots, p_n$ into two (approximately) equal parts: $p_1, \ldots, p_k$ and $p_{k+1}, \ldots, p_n$, where $k = \lfloor (n+1)/2 \rfloor$. We can then recursively count the number of inversions in the first part and the number of inversions in the second part. The remaining inversions to count are those of the form $p_i, p_j$, where index $i$ is in the first part and index $j$ is in the second part. The total number of inversions is, then, the sum of the number of inversions in the first part, the number of inversions in the second part, and the number of inversions of the form $p_i, p_j$, where index $i$ is in the first part and index $j$ is in the second part. We can count the latter type of inversions if the first and second parts are sorted (after counting the inversions in each part prior to sorting).

Let $i = 1$ and $j = k+1$. If $p_i < p_j$, since each half is sorted, $p_i < p_m$ for all indexes $m$ in the second half. Therefore $p_i$ is not in any inversion where the other element is from the first second half. In this case, we need not consider $p_i$ further. We can increment $i$ and repeat this process. If $p_i > p_j$, then $p_i, p_j$ is an inversion. Furthermore, since the first half is sorted, $p_m > p_j$ for all indexes $m$ in the first half. Therefore $p_m, p_j$ is an inversion for all indexes $m$ in the first half. We can count $k$ inversions. We need not consider $p_j$ further. We can increment $j$ and repeat this process.

If we copy each element we need not consider further, we simply have the merge algorithm together with counting inversions. To effect the sorting, we can use merge sort. So if we use merge sort, together with the enhanced merge, we have our $\Theta(n \lg n)$ algorithm.

**39.** Algorithm 7.3.11 computes $a^n$ by using the formula $a^n = a^m a^{n-m}$.

**40.** $b_n = b_{\lfloor n/2 \rfloor} + b_{\lfloor (n+1)/2 \rfloor} + 1$, $b_1 = 0$

**41.** $b_2 = 1$, $b_3 = 2$, $b_4 = 3$ **42.** $b_n = n - 1$

**43.** We prove the formula by using mathematical induction. The Basis Step, $n = 1$, has already been established.

Assume that $b_k = k - 1$ for all $k < n$. We show that $b_n = n - 1$. Now

$$b_n = b_{\lfloor n/2 \rfloor} + b_{\lfloor (n+1)/2 \rfloor} + 1$$

$$= \left\lfloor \frac{n}{2} \right\rfloor - 1 + \left\lfloor \frac{n+1}{2} \right\rfloor - 1 + 1$$

by the inductive assumption

$$= \left\lfloor \frac{n}{2} \right\rfloor + \left\lfloor \frac{n+1}{2} \right\rfloor - 1 = n - 1.$$

**56.** If $n = 1$, then $i = j$ and we return before reaching line 6b, 10, or 14. Therefore, $b_1 = 0$. If $n = 2$, then $j = i + 1$. There is one comparison at line 6b and we return before reaching line 10 or 14. Therefore, $b_2 = 1$.

**57.** $b_3 = 3$, $b_4 = 4$

**58.** When $n > 2$, $b_{\lfloor (n+1)/2 \rfloor}$ comparisons are required for the first recursive call and $b_{\lfloor n/2 \rfloor}$ comparisons are required for the second recursive call. Two additional comparisons are required at lines 10 and 14. The recurrence relation now follows.

**59.** Suppose that $n = 2^k$. Then (7.3.12) becomes

$$b_{2^k} = 2b_{2^{k-1}} + 2.$$

Now

$$\begin{aligned} b_{2^k} &= 2b_{2^{k-1}} + 2 \\ &= 2[2b_{2^{k-2}} + 2] + 2 \\ &= 2^2 b_{2^{k-2}} + 2^2 + 2 = \cdots \\ &= 2^{k-1} b_{2^1} + 2^{k-1} + 2^{k-2} + \cdots + 2 \\ &= 2^{k-1} + 2^{k-1} + \cdots + 2 \\ &= 2^{k-1} + 2^k - 2 \\ &= n - 2 + \frac{n}{2} = \frac{3n}{2} - 2. \end{aligned}$$

**60.** We use the following fact, which can be verified by considering the cases $x$ even and $x$ odd:

$$\left\lceil \frac{3x}{2} - 2 \right\rceil + \left\lceil \frac{3(x+1)}{2} - 2 \right\rceil = 3x - 2 \text{ for } x = 1, 2, \ldots.$$

Let $a_n$ denote the number of comparisons required by the algorithm in the worst case. The cases $n = 1$ and $n = 2$ may be directly verified. (The case $n = 2$ is the Basis Step.)

**Inductive Step** Assume that $a_k \le \lceil (3k/2) - 2 \rceil$ for $2 \le k < n$. We must show that the inequality holds for $k = n$.

If $n$ is odd, the algorithm partitions the array into subclasses of sizes $(n-1)/2$ and $(n+1)/2$. Now

$$\begin{aligned} a_n &= a_{(n-1)/2} + a_{(n+1)/2} + 2 \\ &< \left\lceil \frac{(3/2)(n-1)}{2} - 2 \right\rceil \\ &\quad + \left\lceil \frac{(3/2)(n+1)}{2} - 2 \right\rceil + 2 \\ &= \frac{3(n-1)}{2} - 2 + 2 = \frac{3n}{2} - \frac{3}{2} \\ &= \left\lceil \frac{3n}{2} - 2 \right\rceil. \end{aligned}$$

The case $n$ even is treated similarly.

**69.** $\Theta(n)$

**70.** If $n = 1$, *sort* just returns; therefore, all of the zeros precede all of the ones. The Basis Step is proved.

Assume that for input of size $n - 1$, after *sort* is invoked all of the zeros precede all of the ones. Suppose that *sort* is invoked with input of size $n$. If the first element is a one, it is swapped with the last element. *sort* is then called recursively on the first $n-1$ elements. By the inductive assumption, within the first $n - 1$ elements all of the zeros precede all of the ones. Since the last element is a one, all of the zeros precede all of the ones for all $n$ elements. If the first element is a zero, *sort* is called recursively on the last $n-1$ elements. By the inductive assumption, within the last $n - 1$ elements all of the zeros precede all of the ones. Since the first element is a zero, all of the zeros precede all of the ones for all $n$ elements. In either case, *sort* does produce as output a rearranged version of the input sequence in which all of the zeros precede all of the ones, and the Inductive Step is complete.

**75.** If $n = 2^k$,

$$a_{2^k} = 3a_{2^{k-1}} + 2^k,$$

so

$$\begin{aligned} a_n = a_{2^k} &= 3a_{2^{k-1}} + 2^k \\ &= 3[3a_{2^{k-2}} + 2^{k-1}] + 2^k \\ &= 3^2 a_{2^{k-2}} + 3 \cdot 2^{k-1} + 2^k \\ &\;\;\vdots \\ &= 3^k a_{2^0} + 3^{k-1} \cdot 2^1 + 3^{k-2} \cdot 2^2 + \cdots \\ &\quad + 3 \cdot 2^{k-1} + 2^k \\ &= 3^k + 2(3^k - 2^k) \qquad (*) \\ &= 3 \cdot 3^k - 2 \cdot 2^k \\ &= 3 \cdot 3^{\lg n} - 2n. \end{aligned}$$

Line $(*)$ results from the equation

$$(a-b)(a^{k-1}b^0 + a^{k-2}b^1 + \cdots + a^1 b^{k-2} + a^0 b^{k-1}) = a^k - b^k$$

with $a = 3$ and $b = 2$.

**77.** $b_n = b_{\lfloor (1+n)/2 \rfloor} + b_{\lfloor n/2 \rfloor} + 3$

**80.** $b_n = 4n - 3$

**83.** We will show that $b_n \le b_{n+1}$, $n = 1, 2, \ldots$. We have the recurrence relation

$$b_n = b_{\lfloor (1+n)/2 \rfloor} + b_{\lfloor n/2 \rfloor} + c_{\lfloor (1+n)/2 \rfloor, \lfloor n/2 \rfloor}.$$

**Basis Step** $b_2 = 2b_1 + c_{1,1} \ge 2b_1 \ge b_1$

**Inductive Step** Assume that the statement holds for $k < n$. In case $n$ is even, we have $b_n = 2b_{n/2} + c_{n/2, n/2}$; so

$$\begin{aligned} b_{n+1} &= b_{(n+2)/2} + b_{n/2} + c_{(n+2)/2, n/2} \\ &\ge b_{n/2} + b_{n/2} + c_{n/2, n/2} = b_n. \end{aligned}$$

The case $n$ is odd is similar.

**85.**

```
ex85(s, i, j) {
    if (i == j)
        return
    m = ⌊(i + j)/2⌋
    ex85(s, i, m)
    ex85(s, m + 1, j)
    combine(s, i, m, j)
}
```

**88.** We prove the inequality by using mathematical induction.

**Basis Step** $a_1 = 0 \le 0 = b_1$

**Inductive Step** Assume that $a_k \le b_k$ for $k < n$. Then

$$\begin{aligned} a_n &\le a_{\lfloor n/2 \rfloor} + a_{\lfloor (n+1)/2 \rfloor} + 2\lg n \\ &\le b_{\lfloor n/2 \rfloor} + b_{\lfloor (n+1)/2 \rfloor} + 2\lg n = b_n. \end{aligned}$$

**91.** Let $c = a_1$. If $n$ is a power of $m$, say $n = m^k$, then

$$\begin{aligned} a_n = a_{m^k} &= a_{m^{k-1}} + d \\ &= [a_{m^{k-2}} + d] + d \\ &= a_{m^{k-2}} + 2d \\ &\vdots \\ &= a_{m^0} + kd = c + kd. \end{aligned}$$

An arbitrary value of $n$ falls between two powers of $m$, say

$$m^{k-1} < n \le m^k.$$

This last inequality implies that

$$k - 1 < \log_m n \le k.$$

Since the sequence $a$ is nondecreasing,

$$a_{m^{k-1}} \le a_n \le a_{m^k}.$$

Now

$$\begin{aligned} \Omega(\log_m n) = c + (-1 + \log_m n)d &\le c + (k-1)d \\ &= a_{m^{k-1}} \le a_n \end{aligned}$$

and

$$\begin{aligned} a_n \le a_{m^k} &= c + kd \\ &\le c + (1 + \log_m n)d = O(\log_m n). \end{aligned}$$

Thus $a_n = \Theta(\log_m n)$. By Example 4.3.6, $a_n = \Theta(\lg n)$.

## Section 7.4 Review

1. Computational geometry is concerned with the design and analysis of algorithms to solve geometry problems.
2. Given $n$ points in the plane, find a closest pair.
3. Compute the distance between each pair of points and choose the minimum distance.
4. Find a vertical line $l$ that divides the points into two nearly equal parts. Then recursively solve the problem for each of the parts. Let $\delta_L$ be the distance between a closest pair in the left part, and let $\delta_R$ be the distance between a closest pair in the right part. Let $\delta = \min\{\delta_L, \delta_R\}$. Then examine the points that lie within a vertical strip of width $2\delta$ centered about $l$. Order the points in this strip in increasing order of the $y$-coordinates and examine the points in this order. Compute the distance between each point $p$ and the following seven points. Anytime there is a pair whose distance is less than $\delta$, update $\delta$. At the conclusion, $\delta$ is the distance between a closest pair.
5. The worst-case time of the brute-force algorithm is $\Theta(n^2)$. The worst-case time of the divide-and-conquer algorithm is $\Theta(n \lg n)$.

## Section 7.4

**1.** The 16 points sorted by $x$-coordinate are (1, 2), (1, 5), (1, 9), (3, 7), (3, 11), (5, 4), (5, 9), (7, 6), (8, 4), (8, 7), (8, 9), (11, 3), (11, 7), (12, 10), (14, 7), (17, 10), so the dividing point is (7, 6). We next find $\delta_L = \sqrt{8}$, the minimum distance among the left-side points (1, 2), (1, 5), (1, 9), (3, 7), (3, 11), (5, 4), (5, 9), (7, 6), and $\delta_R = 2$, the minimum distance among the right-side points (8, 4), (8, 7), (8, 9), (11, 3), (11, 7), (12, 10), (14, 7), (17, 10). Thus $\delta = \min\{\delta_L, \delta_R\} = 2$. The points, sorted by $y$-coordinate in the vertical strip, are (8, 4), (7, 6), (8, 7), (8, 9). In this case we compare each point in the strip to all the following points. The distances from (8, 4) to (7, 6), (8, 7), (8, 9) are not less than 2, so $\delta$ is not updated at this point. The distance from (7, 6) to (8, 7) is $\sqrt{2}$, so $\delta$ is updated to $\sqrt{2}$. The distances from (7, 6) to (8, 9) and from (8, 7) to (8, 9) are greater than $\sqrt{2}$, so $\delta$ remains $\sqrt{2}$. Therefore, the distance between the closest pair is $\sqrt{2}$.

**4.** Consider the extreme case when all of the points are on the vertical line.

**7.**

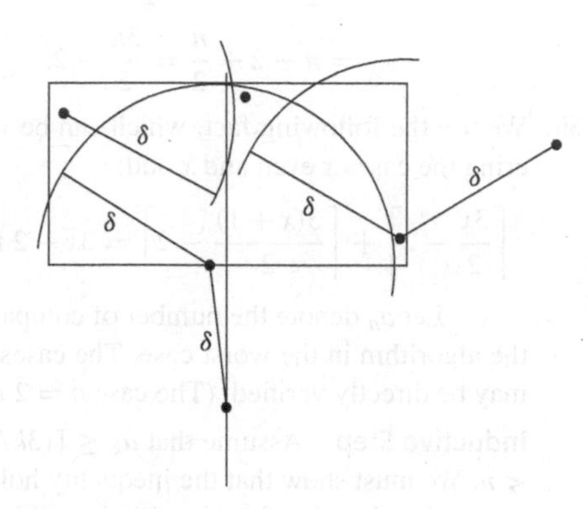


**10.** Let $B$ be either of the left or right $\delta \times \delta$ squares that make up the $\delta \times 2\delta$ rectangle (see Figure 7.4.2). We argue by contradiction and assume that $B$ contains four or more points. We partition $B$ into four $\delta/2 \times \delta/2$ squares as shown in Figure 7.4.3. Then each of the four squares contains at most one point, and therefore exactly one point. Subsequently we refer to these four squares as the subsquares of $B$.

The figure

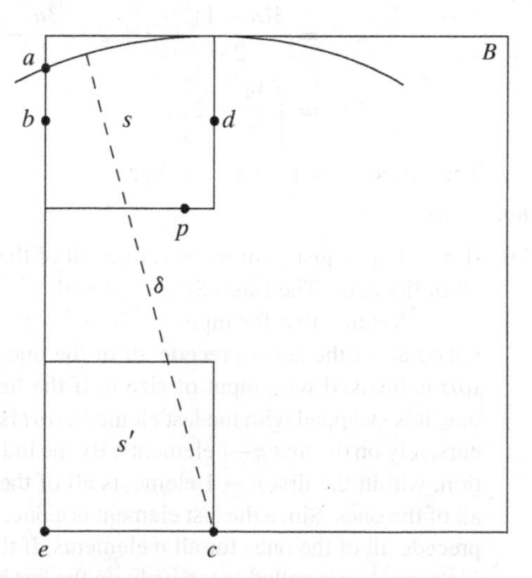


shows the following construction. We reduce the size of the subsquares, if possible, so that

- Each subsquare contains one point.
- The subsquares are the same size.
- The subsquares are as small as possible.

Since at least one point is not in a corner of $B$, the subsquares do not collapse to points and so at least one point is on a side of a subsquare $s$ interior to $B$. We choose such a point and call it $p$. We select a subsquare $s'$ nearest $p$. We label the two corner points of $s'$ on the side farthest from $p$, $e$ and $c$. We draw a circle of radius $\delta$ with center at $c$ and let $a$ be the (noncorner) point where this circle meets the side of $s$. Note that this circle meets a side of $s$ in a noncorner point. Choose a point $b$ in $s$ on the same side as $a$ between $a$ and $e$. Let $d$ be the corresponding point on the opposite side of $s$. Now the length of the diameter of rectangle $R = bdce$ is less than $\delta$; hence, $R$ contains at most one point. This is contradiction since $R$ contains $p$ and the point in $s'$. Therefore, $B$ contains at most three points.

**13.** In addition to $p.x$ and $p.y$, we assume that each point $p$ has another field $p.side$, which we use to indicate whether $p$ is on the left side or the right side when the points are divided into two nearly equal parts. The extra argument, *label*, to *rec_find_all_2δ_once* sets $p.side$ to *label* for all points $p$.

```
find_all_2δ_once(p,n) {
   δ = closest_pair(p, n) // original procedure
   if (δ > 0) {
      sort p_1, ..., p_n by x-coordinate
      rec_find_all_2δ_once(p, 1, n, δ, λ) // λ = empty string
   }
}
rec_find_all_2δ_once(p, i, j, δ, label) {
   if (j − i < 3) {
      sort p_i, ..., p_j by y-coordinate
      directly find and output all distinct pairs less than
         2δ apart
      for k = i to j
         p_k.side = label
      return
   }
   k = ⌊(i + j)/2⌋
   l = p_k.x
   rec_find_all_2δ_once(p, i, k, δ, L)
   rec_find_all_2δ_once(p, k + 1, j, δ, R)
   merge p_i, ..., p_k and p_{k+1}, ..., p_j by y-coordinate
   t = 0
   for k = i to j
      if (p_k.x > l − 2 * δ ∧ p_k.x < l + 2 * δ) {
         t = t + 1
         v_t = p_k
      }
   for k = 1 to t − 1
      for s = k + 1 to min{t, k + 31}
         if (dist(v_k, v_s) < 2 * δ ∧ v_k.side ¬= v_s.side)
            println(v_k + " " + v_s)
   for k = i to j
      p_k.side = label
}
```

**16.** We show that for each point $p$, there are at most 6 distinct points whose distance to $p$ is $\delta$. It will then follow that the number of pairs $\delta$ apart is less than or equal to $6n$.

We argue by contradiction. Suppose that some point $p$ has 7 distinct neighbors $p_1, \ldots, p_7$ whose distance to $p$ is $\delta$. Then we have the situation

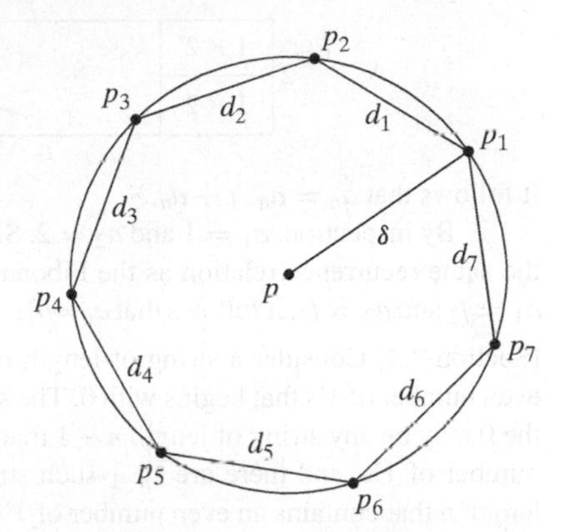


Let $C$ be the circumference of this circle. Since each $d_i$ is at least $\delta$, we have

$$2\pi\delta = C > \sum_{i=1}^{7} d_i \geq 7\delta.$$

Therefore, $\pi > 7/2 = 3.5$, which is a contradiction. (A more careful estimate shows that for each point $p$, there are at most 5 distinct points whose distance to $p$ is $\delta$.)

**18.** The inequality is proved by induction. The Basis Steps are $n = 2, 3$, which are immediate. For the Inductive Step, we have

$$a_n \leq a_{\lfloor n/2 \rfloor} + a_{\lfloor (n+1)/2 \rfloor} + cn \leq t_{\lfloor n/2 \rfloor} + t_{\lfloor (n+1)/2 \rfloor} + cn = t_n.$$

**21.** $a_n \leq t_n = \Theta(n \lg n)$

## Chapter 7 Self-Test

1. [Section 7.1] (a) 3,5,8,12 (b) $a_1 = 3$ (c) $a_n = a_{n-1} + n$
2. [Section 7.2] Yes
3. [Section 7.1] $A_n = (1.17)A_{n-1}$, $A_0 = 4000$
4. [Section 7.1] Let $X$ be an $n$-element set and choose $x \in X$. Let $k$ be a fixed integer, $0 \leq k \leq n - 1$. We can select a $k$-element subset $Y$ of $X - \{x\}$ in $C(n-1, k)$ ways. Having done this, we can partition $Y$ in $P_k$ ways. This partition together with $X - Y$ partitions $X$. Since all partitions of $X$ can be generated in this way, we obtain the desired recurrence relation.
5. [Section 7.2] $a_n = 2(-2)^n - 4n(-2)^n$
6. [Section 7.2] $a_n = 3 \cdot 5^n + (-2)^n$
7. [Section 7.1] If the first domino is placed as shown, there are $a_{n-1}$ ways to cover the $2 \times (n-1)$ board that remains.

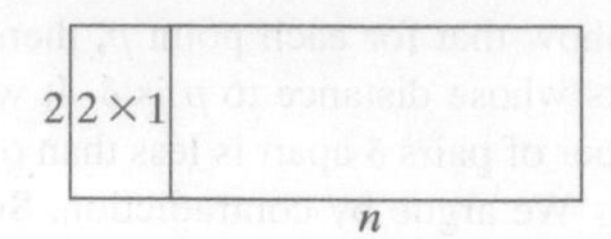


If the first two dominoes are placed as shown, there are $a_{n-2}$ ways to cover the $2 \times (n-2)$ board that remains.

1×2
2
1×2
n

It follows that $a_n = a_{n-1} + a_{n-2}$.

By inspection, $a_1 = 1$ and $a_2 = 2$. Since $\{a_n\}$ satisfies the same recurrence relation as the Fibonacci sequence and $a_1 = f_2$ and $a_2 = f_3$, it follows that $a_i = f_{i+1}$ for $i = 1, 2, \ldots$.

**8.** [Section 7.2] Consider a string of length $n$ that contains an even number of 1's that begins with 0. The string that follows the 0 may be any string of length $n-1$ that contains an even number of 1's, and there are $c_{n-1}$ such strings. A string of length $n$ that contains an even number of 1's that begins with 2 can be followed by any string of length $n-1$ that contains an even number of 1's, and there are $c_{n-1}$ such strings. A string of length $n$ that contains an even number of 1's that begins with 1 can be followed by any string of length $n-1$ that contains an odd number of 1's. Since there are $3^{n-1}$ strings altogether of length $n-1$ and $c_{n-1}$ of these contain an even number of 1's, there are $3^{n-1} - c_{n-1}$ strings of length $n-1$ that contain an odd number of 1's. It follows that

$$c_n = 2c_{n-1} + 3^{n-1} - c_{n-1} = c_{n-1} + 3^{n-1}.$$

An initial condition is $c_1 = 2$, since there are two strings (0 and 2) that contain an even number (namely, zero) of 1's.

We may solve the recurrence relation by iteration:

$$\begin{aligned} c_n &= c_{n-1} + 3^{n-1} = c_{n-2} + 3^{n-2} + 3^{n-1} \\ &\vdots \\ &= c_1 + 3^1 + 3^2 + \cdots + 3^{n-1} \\ &= 2 + \frac{3^n - 3}{3 - 1} = \frac{3^n + 1}{2}. \end{aligned}$$

**9.** [Section 7.3] $b_n = b_{n-1} + 1$, $b_0 = 0$

**10.** [Section 7.3] $b_1 = 1$, $b_2 = 2$, $b_3 = 3$

**11.** [Section 7.3] $b_n = n$

**12.** [Section 7.3] $n(n+1)/2 = O(n^2)$. The given algorithm is faster than the straightforward technique and is, therefore, preferred.

**13.** [Section 7.4] The 18 points sorted by the $x$-coordinate are (1, 8), (2, 2), (3, 13), (4, 4), (4, 8), (6, 10), (7, 1), (7, 7), (7, 13), (10, 1), (10, 5), (10, 9), (10, 13), (12, 3), (13, 8), (14, 5), (16, 4), (16, 10), so the dividing point is (7, 13). We next find $\delta_L = \sqrt{8}$, the minimum distance among the left-side points (1, 8), (2, 2), (3, 13), (4, 4), (4, 8), (6, 10), (7, 1), (7, 7), (7, 13), and $\delta_R = \sqrt{5}$, the minimum distance among the right-side points (10, 1), (10, 5), (10, 9), (10, 13), (12, 3), (13, 8), (14, 5), (16, 4), (16, 10). Thus $\delta = \min\{\delta_L, \delta_R\} = \sqrt{5}$. The points, sorted by $y$-coordinate in the vertical strip, are (7, 1), (7, 7), (6, 10), (7, 13). In this case we compare each point in the strip to all the following points. Since no pair is closer than $\sqrt{5}$, the algorithm does not update $\delta$. Therefore, the distance between the closest pair is $\sqrt{5}$.

**14.** [Section 7.4] If we replace "three" by "two," when there are three points, the algorithm would be called recursively with inputs of sizes 1 and 2. But a set consisting of one point has no pair—let alone a closest pair.

**15.** [Section 7.4] Each $\delta/2 \times \delta/2$ box contains at most one point, so there are at most four points in the lower half of the rectangle.

**16.** [Section 7.4] $\Theta(n(\lg n)^2)$

## Section 8.1 Review

**1.** An undirected graph consists of a set $V$ of vertices and a set $E$ of edges such that each edge $e \in E$ is associated with an unordered pair of vertices.

**2.** Friendship can be modeled by an undirected graph by letting the vertices denote the people and placing an edge between two people if they are friends.

**3.** A directed graph consists of a set $V$ of vertices and a set $E$ of edges such that each edge $e \in E$ is associated with an ordered pair of vertices.

**4.** Precedence can be modeled by a directed graph by letting the vertices denote the tasks and placing a directed edge from task $t_i$ to task $t_j$ if $t_i$ must be completed before $t_j$.

**5.** If edge $e$ is associated with vertices $v$ and $w$, $e$ is said to be incident on $v$ and $w$.

**6.** If edge $e$ is associated with vertices $v$ and $w$, $v$ and $w$ are said to be incident on $e$.

**7.** If edge $e$ is associated with vertices $v$ and $w$, $v$ and $w$ are said to be adjacent.

**8.** Parallel edges are edges that are incident on the same pair of vertices.

**9.** An edge incident on a single vertex is called a loop.

**10.** A vertex that is not incident on any edge is called an isolated vertex.

**11.** A simple graph is a graph with neither loops nor parallel edges.

**12.** A weighted graph is a graph with numbers assigned to the edges.

**13.** A map with distances can be modeled as a weighted graph. The vertices are the cities, the edges are the roads between the cities, and the numbers on the edges are the distances between the cities.

**14.** The length of a path in a weighted graph is the sum of the weights of its edges.

**15.** A similarity graph has a dissimilarity function $s$ where $s(v, w)$ measures the dissimilarity of vertices $v$ and $w$.

**16.** The $n$-cube has $2^n$ vertices labeled $0, 1, \ldots, 2^n - 1$. An edge connects two vertices if the binary representation of their labels differs in exactly one bit.

**17.** A serial computer executes one instruction at a time.

**18.** A serial algorithm executes one instruction at a time.

**19.** A parallel computer can execute several instructions at a time.

**20.** A parallel algorithm can execute several instructions at a time.

**21.** The complete graph on $n$ vertices has one edge between each distinct pair of vertices. It is denoted $K_n$.

**22.** A graph $G = (V, E)$ is bipartite if there exist subsets $V_1$ and $V_2$ (either possibly empty) of $V$ such that $V_1 \cap V_2 = \varnothing$, $V_1 \cup V_2 = V$, and each edge in $E$ is incident on one vertex in $V_1$ and one vertex in $V_2$.

**23.** The complete bipartite graph on $m$ and $n$ vertices has disjoint vertex sets $V_1$ with $m$ vertices and $V_2$ with $n$ vertices in which the edge set consists of all edges of the form $(v_1, v_2)$ with $v_1 \in V_1$ and $v_2 \in V_2$.

## Section 8.1

**1.** The graph is an undirected, simple graph.

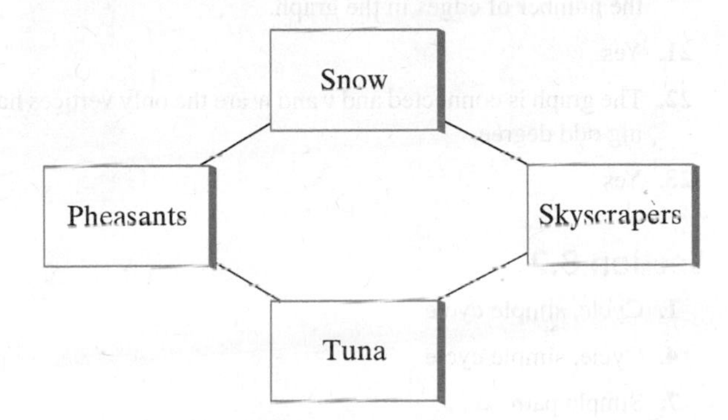


**4.** The graph is a directed, nonsimple graph.

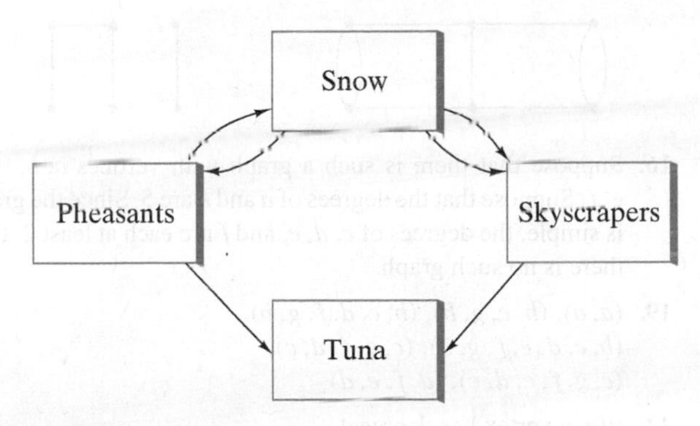


**5.** Since an odd number of edges touch some vertices ($c$ and $d$), there is no path from $a$ to $a$ that passes through each edge exactly one time.

**8.** $(a, c, e, b, c, d, e, f, d, b, a)$

**11.** $V = \{v_1, v_2, v_3, v_4\}$. $E = \{e_1, e_2, e_3, e_4, e_5, e_6\}$. $e_1$ and $e_6$ are parallel edges. $e_5$ is a loop. There are no isolated vertices. $G$ is not a simple graph. $e_1$ is incident on $v_1$ and $v_2$.

**14.**

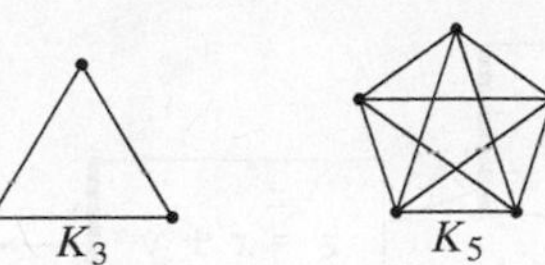


**17.** Bipartite. $V_1 = \{v_1, v_2, v_5\}$, $V_2 = \{v_3, v_4\}$.

**20.** Not bipartite

**23.** Bipartite. $V_1 = \{v_1\}$, $V_2 = \{v_2, v_3\}$.

**24.**

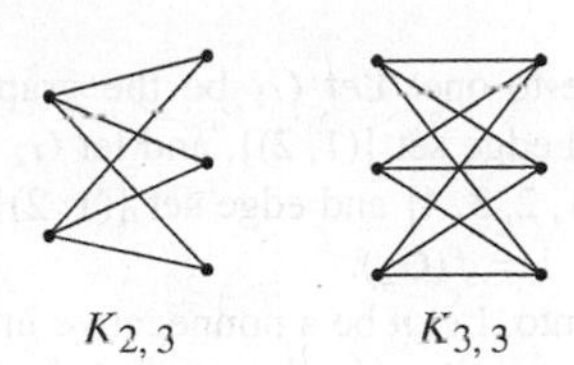


**27.** $(b, c, a, d, e)$

**32.** Two classes

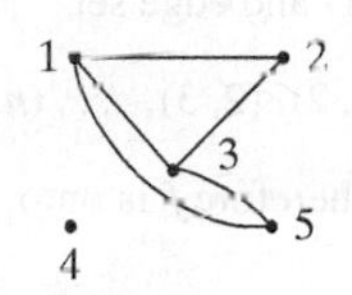


**37.**

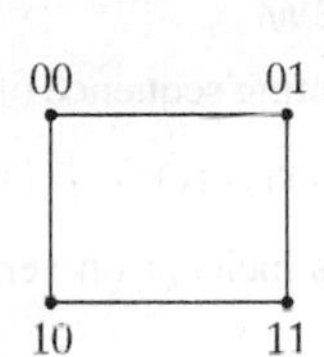


**40.** $n$

**43.**

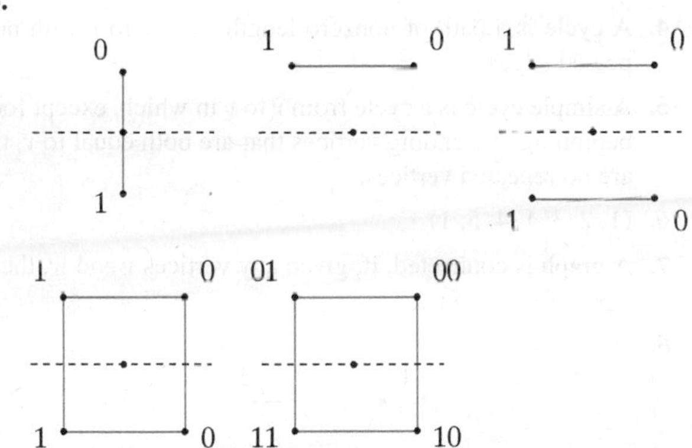


**46.**

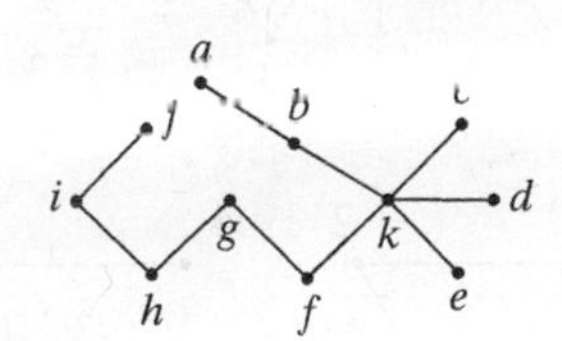


**49.**

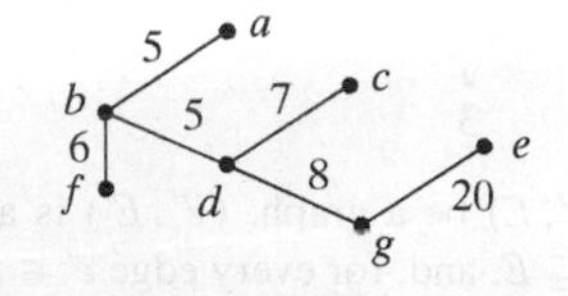

**50.**

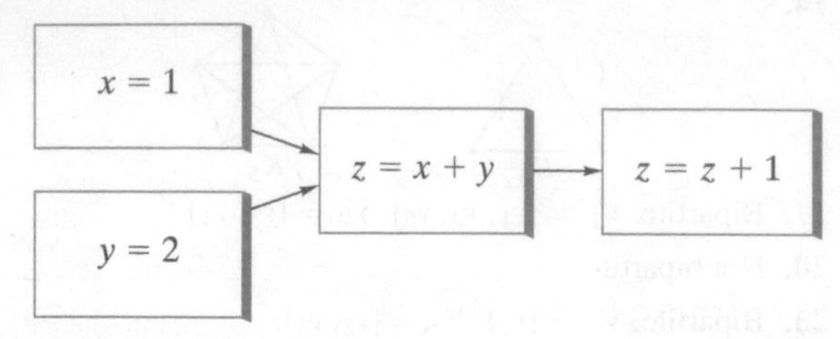


**53.** $f$ is not one-to-one. Let $G_1$ be the graph with vertex set $\{1, 2, 3\}$ and edge set $\{(1, 2)\}$, and let $G_2$ be the graph with vertex set $\{1, 2, 3, 4\}$ and edge set $\{(1, 2)\}$. Then $G_1 \neq G_2$, but $f(G_1) = 1 = f(G_2)$.

$f$ is onto. Let $n$ be a nonnegative integer. If $n = 0$, let $G$ be the graph with vertex set $\{1, 2, 3\}$ and edge set $\varnothing$. Then $f(G) = 0 = n$. If $n > 0$, let $G$ be the graph with vertex set $\{1, 2, \ldots, n, n + 1\}$ and edge set

$$\{(1, 2), (2, 3), \ldots, (n, n + 1)\}.$$

Then $f(G) = n$. Therefore $f$ is onto.

## Section 8.2 Review

**1.** A path is an alternating sequence of vertices and edges

$$(v_0, e_1, v_1, e_2, v_2, \ldots, v_{n-1}, e_n, v_n),$$

in which edge $e_i$ is incident on vertices $v_{i-1}$ and $v_i$ for $i = 1, \ldots, n$.

**2.** A simple path is a path with no repeated vertices.

**3.** (1, 2, 3, 1)

**4.** A cycle is a path of nonzero length from $v$ to $v$ with no repeated edges.

**5.** A simple cycle is a cycle from $v$ to $v$ in which, except for the beginning and ending vertices that are both equal to $v$, there are no repeated vertices.

**6.** (1, 2, 3, 1, 4, 5, 1)

**7.** A graph is connected, if, given any vertices $v$ and $w$, there is a path from $v$ to $w$.

**8.**

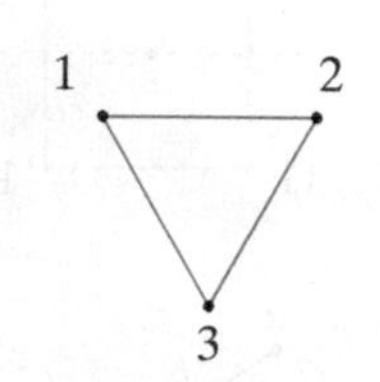


**9.**

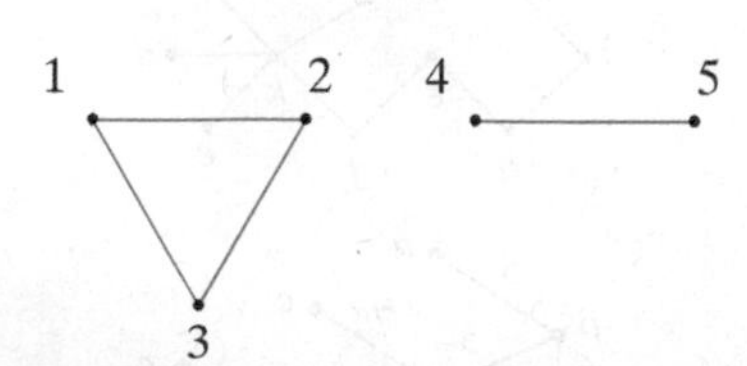


**10.** Let $G = (V, E)$ be a graph. $(V', E')$ is a subgraph of $G$ if $V' \subseteq V$, $E' \subseteq E$, and, for every edge $e' \in E'$, if $e'$ is incident on $v'$ and $w'$, then $v', w' \in V'$.

**11.** The graph of Exercise 8 is a subgraph of the graph of Exercise 9.

**12.** Let $G$ be a graph and let $v$ be a vertex in $G$. The subgraph $G'$ of $G$ consisting of all edges and vertices in $G$ that are contained in some path beginning at $v$ is called the component of $G$ containing $v$.

**13.** The graph of Exercise 8 is a component of the graph of Exercise 9.

**14.** One

**15.** The degree of vertex $v$ is the number of edges incident on $v$.

**16.** An Euler cycle in a graph $G$ is a cycle that includes all of the edges and all of the vertices of $G$.

**17.** A graph $G$ has an Euler cycle if and only if $G$ is connected and the degree of every vertex is even.

**18.** The graph of Exercise 8 has the Euler cycle (1, 2, 3, 1).

**19.** The graph of Exercise 9 does not have an Euler cycle because it is not connected.

**20.** The sum of the degrees of the vertices in a graph equals twice the number of edges in the graph.

**21.** Yes

**22.** The graph is connected and $v$ and $w$ are the only vertices having odd degree.

**23.** Yes

## Section 8.2

**1.** Cycle, simple cycle

**4.** Cycle, simple cycle

**7.** Simple path

**10.** **13.**

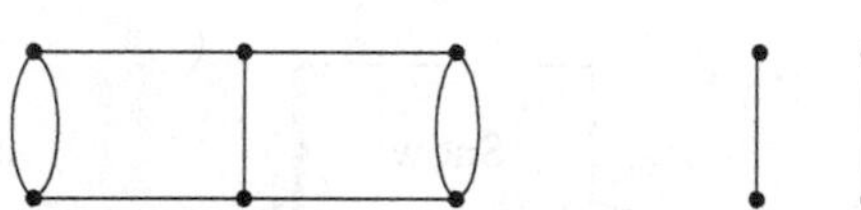

**16.** Suppose that there is such a graph with vertices $a, b, c, d, e, f$. Suppose that the degrees of $a$ and $b$ are 5. Since the graph is simple, the degrees of $c$, $d$, $e$, and $f$ are each at least 2; thus there is no such graph.

**19.** $(a, a)$, $(b, c, g, b)$, $(b, c, d, f, g, b)$, $(b, c, d, e, f, g, b)$, $(c, g, f, d, c)$, $(c, g, f, e, d, c)$, $(d, f, e, d)$

**22.** Every vertex has degree 4.

**24.** $G_1 = (\{v_1\}, \varnothing)$

$G_2 = (\{v_2\}, \varnothing)$

$G_3 = (\{v_1, v_2\}, \varnothing)$

$G_4 = (\{v_1, v_2\}, \{e_1\})$

**27.** There are 17 subgraphs. **28.** No Euler cycle

**31.** No Euler cycle

**34.** For

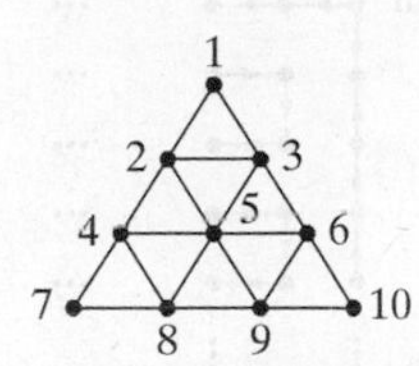


an Euler cycle is (10, 9, 6, 5, 9, 8, 5, 4, 8, 7, 4, 2, 5, 3, 2, 1, 3, 6, 10). The method generalizes.

**37.** $m = n = 2$ or $m = n = 1$

**39.** $d$ and $e$ are the only vertices of odd degree.

**43.** The argument is similar to that of the proof of Theorem 8.2.23.

**46.** True. In the path, for all repeated $a$,

$$(\ldots, a, \ldots, b, a, \ldots)$$

eliminate $a, \ldots, b$.

**48.** Suppose that $e = (v, w)$ is in a cycle. Then there is a path $P$ from $v$ to $w$ not including $e$. Let $x$ and $y$ be vertices in $G - \{e\}$. Since $G$ is connected, there is a path $P'$ in $G$ from $v$ to $w$. Replace any occurrence of $e$ in $P'$ by $P$. The resulting path from $v$ to $w$ lies in $G - \{e\}$. Therefore, $G - \{e\}$ is connected.

**51.** The union of all connected subgraphs containing $G'$ is a component.

**54.** Let $G$ be a simple, disconnected graph with $n$ vertices having the maximum number of edges. Show that $G$ has two components. If one component has $i$ vertices, show that the components are $K_i$ and $K_{n-i}$. Use Exercise 11, Section 8.1, to find a formula for the number of edges in $G$ as a function of $i$. Show that the maximum occurs when $i = 1$.

**56.**

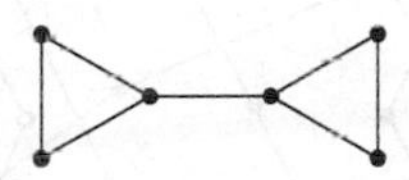

**59.** Modify the proofs of Theorems 8.2.17 and 8.2.18.

**62.** Use Exercises 59 and 61.

**65.** We first count the number of paths $(v_0, v_1, \ldots, v_k)$ of length $k \geq 1$. The first vertex $v_0$ may be chosen in $n$ ways. Each subsequent vertex may be chosen in $n - 1$ ways (since it must be different from its predecessor). Thus the number of paths of length $k$ is $n(n-1)^k$.

The number of paths of length $k$, $1 \leq k \leq n$, is

$$\sum_{k=1}^{n} n(n-1)^k = n(n-1)\frac{(n-1)^k - 1}{(n-1) - 1}$$

$$= \frac{n(n-1)[(n-1)^k - 1]}{n-2}.$$

**69.** If $v$ is a vertex in $V$, the path consisting of $v$ and no edges is a path from $v$ to $v$; thus $vRv$ for every vertex $v$ in $V$. Therefore, $R$ is reflexive.

Suppose that $vRw$. Then there is a path $(v_0, \ldots, v_n)$, where $v_0 = v$ and $v_n = w$. Now $(v_n, \ldots, v_0)$ is a path from $w$ to $v$, and thus $wRv$. Therefore, $R$ is symmetric.

Suppose that $vRw$ and $wRx$. Then there is a path $P_1$ from $v$ to $w$ and a path $P_2$ from $w$ to $x$. Now $P_1$ followed by $P_2$ is a path from $v$ to $x$, and thus $vRx$. Therefore, $R$ is transitive.

Since $R$ is reflexive, symmetric, and transitive on $V$, $R$ is an equivalence relation on $V$.

**71.** 2

**74.** Let $s_n$ denote the number of paths of length $n$ from $v_1$ to $v_1$. We show that the sequences $s_1, s_2, \ldots$ and $f_1, f_2, \ldots$ satisfy the same recurrence relation, $s_1 = f_2$, and $s_2 = f_3$, from which it follows that $s_n = f_{n+1}$ for $n \geq 1$.

If $n = 1$, there is one path of length 1 from $v_1$ to $v_1$, namely the loop on $v_1$; thus, $s_1 = f_2$.

If $n = 2$, there are two paths of length 2 from $v_1$ to $v_1$: $(v_1, v_1, v_1)$ and $(v_1, v_2, v_1)$; thus, $s_2 = f_3$.

Assume that $n > 2$. Consider a path of length $n$ from $v_1$ to $v_1$. The path must begin with the loop $(v_1, v_1)$ or the edge $(v_1, v_2)$.

If the path begins with the loop, the remainder of the path must be a path of length $n - 1$ from $v_1$ to $v_1$. Since there are $s_{n-1}$ such paths, there are $s_{n-1}$ paths of length $n$ from $v_1$ to $v_1$ that begin $(v_1, v_1, \ldots)$.

If the path begins with the edge $(v_1, v_2)$, the next edge in the path must be $(v_2, v_1)$. The remainder of the path must be a path of length $n - 2$ from $v_1$ to $v_1$. Since there are $s_{n-2}$ such paths, there are $s_{n-2}$ paths of length $n$ from $v_1$ to $v_1$ that begin $(v_1, v_2, v_1, \ldots)$.

Since any path of length $n > 2$ from $v_1$ to $v_1$ begins with the loop $(v_1, v_1)$ or the edge $(v_1, v_2)$, it follows that

$$s_n = s_{n-1} + s_{n-2}.$$

Because the sequences $s_1, s_2, \ldots$ and $f_1, f_2, \ldots$ satisfy the same recurrence relation, $s_1 = f_2$, and $s_2 = f_3$, it follows that $s_n = f_{n+1}$ for $n \geq 1$.

**76.** Suppose that every vertex has an out edge. Choose a vertex $v_0$. Follow an edge out of $v_0$ to a vertex $v_1$. (By assumption, such an edge exists.) Continue to follow an edge out of $v_i$ to a vertex $v_{i+1}$. Since there are a finite number of vertices, we will eventually return to a previously visited vertex. At this point, we will have discovered a cycle, which is a contradiction. Therefore, a dag has at least one vertex with no out edges.

## Section 8.3 Review

**1.** A Hamiltonian cycle in a graph $G$ is a cycle that contains each vertex in $G$ exactly once, except for the starting and ending vertex that appears twice.

**2.** The graph of Figure 8.3.9 has a Hamiltonian cycle and an Euler cycle. The Hamiltonian and Euler cycles are the graph itself.

**3.** The graph of Figure 8.3.2 has a Hamiltonian cycle, but not an Euler cycle. The Hamiltonian cycle is shown in Figure 8.3.3.

The graph does not have an Euler cycle, because all of the vertices have odd degree.

4. The graph

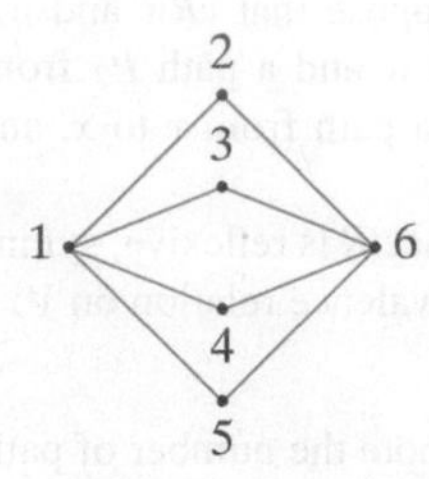


has an Euler cycle because it is connected and every vertex has even degree. It does not have a Hamiltonian cycle. To prove that it does not have a Hamiltonian cycle, we argue by contradiction. Suppose that the graph has a Hamiltonian cycle. Then, because vertices 2, 3, 4, and 5 all have degree 2, all the edges in the graph would have to be included in a Hamiltonian cycle. Since the graph itself is not a cycle, we have a contradiction.

5. The graph consisting of two vertices and no edges has neither a Hamiltonian cycle nor an Euler cycle because it is not connected.

6. The traveling salesperson problem is: Given a weighted graph $G$, find a minimum-length Hamiltonian cycle in $G$. The Hamiltonian cycle problem simply asks for a Hamiltonian cycle—any Hamiltonian cycle will do. The traveling salesperson problem asks not just for a Hamiltonian cycle, but for one of minimum length.

7. A simple cycle

8. A Gray code is a sequence $s_1, s_2, \ldots, s_{2^n}$, where each $s_i$ is a string of $n$ bits, satisfying the following:
   - Every $n$-bit string appears somewhere in the sequence.
   - $s_i$ and $s_{i+1}$ differ in exactly one bit, $i = 1, \ldots, 2^n - 1$.
   - $s_{2^n}$ and $s_1$ differ in exactly one bit.

9. See Theorem 8.3.6.

10. See the discussion preceding Algorithm 8.3.10.

## Section 8.3

1. $(d, a, e, b, c, h, g, f, j, i, d)$

3. We would have to eliminate two edges each at $b$, $d$, $i$, and $k$, leaving $19 - 8 = 11$ edges. A Hamiltonian cycle would have 12 edges.

6. $(a, b, c, j, i, m, k, d, e, f, l, g, h, a)$

9.

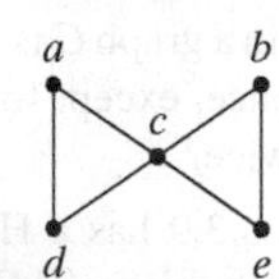


12. If $n$ is even and $m > 1$ or if $m$ is even and $n > 1$, there is a Hamiltonian cycle. The sketch shows the solution in case $n$ is even.

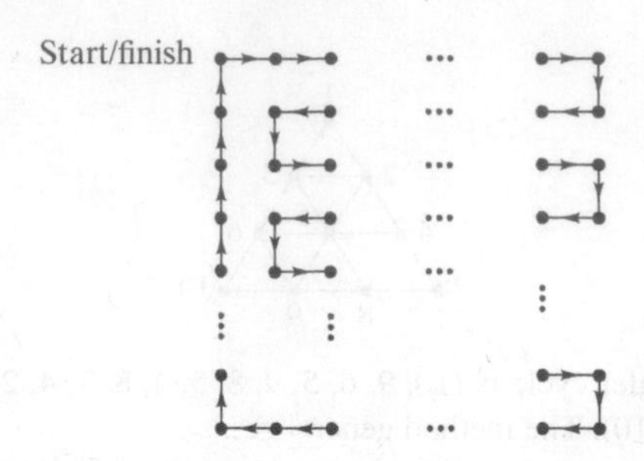


If $n = 1$ or if $m = 1$, there is no cycle and, in particular, there is no Hamiltonian cycle. Suppose that $n$ and $m$ are both odd and that the graph has a Hamiltonian cycle. Since there are $nm$ vertices, this cycle has $nm$ edges; therefore, the Hamiltonian cycle contains an odd number of edges. However, we note that in a Hamiltonian cycle, there must be as many "up" edges as "down" edges and as many "left" edges as "right" edges. Thus a Hamiltonian cycle must have an even number of edges. This contradiction shows that if $n$ and $m$ are both odd, the graph does not have a Hamiltonian cycle.

15. When $m = n$ and $n > 1$

18. Any cycle $C$ in the $n$-cube has even length since the vertices in $C$ alternate between an even and an odd number of 1's.

Suppose that the $n$-cube has a simple cycle of length $m$. We just observed that $m$ is even. Now $m > 0$, by definition. Since the $n$-cube is a simple graph, $m \neq 2$. Therefore, $m \geq 4$.

Now suppose that $m \geq 4$ and $m$ is even. Let $G$ be the first $m/2$ members of the Gray code $G_{n-1}$. Then $0G$, $1G^R$ describes a simple cycle of length $m$ in the $n$-cube.

21.

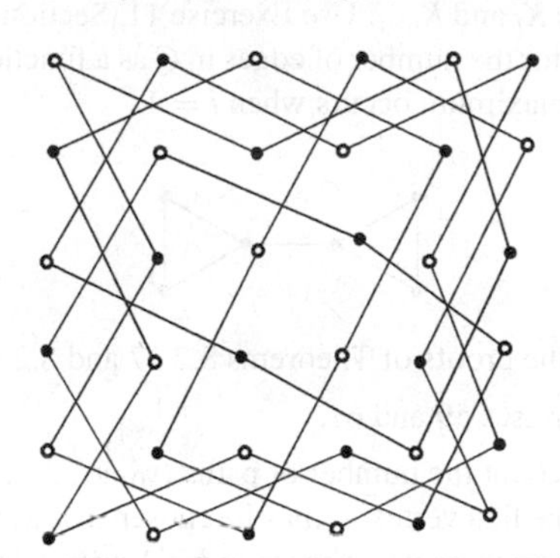

25. Suppose that the algorithm begins $(a, b, c, e, f)$, which will then not extend. Since $f$ is adjacent to $a$, the path will be modified to $(a, f, e, c, b)$. This path cannot be extended. Since $b$ is adjacent to $a$, the path will be modified to $(a, b, c, e, f)$. We are now in an infinite loop.

28. If the first choices are $(b, c, d, e)$, the algorithm will not terminate.

32. For all $n \geq 3$, the algorithm always finds a Hamiltonian cycle in $K_n$.

35. No. Let $(v_1, \ldots, v_n, v_1)$ be a Hamiltonian cycle. If the choices are $v_1, \ldots, v_n$, the algorithm finds a Hamiltonian cycle.

**37.** Yes. If $(v_1, \ldots, v_{n-1}, v_n)$, $v_1 = v_n$, is a Hamiltonian cycle, $(v_1, \ldots, v_{n-1})$ is a Hamiltonian path.

**40.** Yes, $(a, b, d, g, m, l, h, i, j, e, f, k, c)$

**43.** Yes, $(i, j, g, h, e, d, c, b, a, f)$

**46.** Yes, $(a, c, d, f, g, e, b)$

## Section 8.4 Review

**1.** Label the start vertex 0 and all other vertices $\infty$. Let $T$ be the set of all vertices. Choose $v \in T$ with minimum label and remove $v$ from $T$. For each $x \in T$ adjacent to $v$, relabel $x$ with the minimum of its current label and the label of $v + w(v, x)$, where $w(v, x)$ is the weight of edge $(v, x)$. Repeat if $z \notin T$.

**2.** See Example 8.4.2.

**3.** See the proof of Theorem 8.4.3.

## Section 8.4

**1.** 7; $(a, b, c, f)$

**4.** 7; $(b, c, f, j)$

**6.** An algorithm can be modeled after Example 8.4.2.

**9.** Modify Algorithm 8.4.1 so that it begins by assigning the weight $\infty$ to each nonexistent edge. The algorithm then continues as written. At termination, $L(z)$ will be equal to $\infty$ if there is no path from $a$ to $z$.

## Section 8.5 Review

**1.** Order the vertices and label the rows and columns of a matrix with the ordered vertices. The entry in row $i$, column $j$, $i \neq j$, is the number of edges incident on $i$ and $j$. If $i = j$, the entry is twice the number of loops incident on $i$. The resulting matrix is the adjacency matrix of the graph.

**2.** The $ij$th entry in $A^n$ is equal to the number of paths of length $n$ from vertex $i$ to vertex $j$.

**3.** Order the vertices and edges and label the rows of a matrix with the vertices and the columns with the edges. The entry in row $v$ and column $e$ is 1 if $e$ is incident on $v$ and 0 otherwise. The resulting matrix is the incidence matrix of the graph.

## Section 8.5

**1.**

$$\begin{array}{c} \\ a \\ b \\ c \\ d \\ e \end{array} \begin{array}{c} \begin{array}{ccccc} a & b & c & d & e \end{array} \\ \begin{pmatrix} 0 & 1 & 1 & 1 & 1 \\ 1 & 0 & 1 & 0 & 0 \\ 1 & 1 & 0 & 1 & 1 \\ 1 & 0 & 1 & 0 & 1 \\ 1 & 0 & 1 & 1 & 0 \end{pmatrix} \end{array}$$

**4.**

$$\begin{array}{c} \\ v_1 \\ v_2 \\ v_3 \\ v_4 \\ v_5 \\ v_6 \end{array} \begin{array}{c} \begin{array}{cccccc} v_1 & v_2 & v_3 & v_4 & v_5 & v_6 \end{array} \\ \begin{pmatrix} 0 & 1 & 1 & 0 & 0 & 0 \\ 1 & 0 & 1 & 0 & 0 & 0 \\ 1 & 1 & 0 & 0 & 0 & 0 \\ 0 & 0 & 0 & 0 & 0 & 0 \\ 0 & 0 & 0 & 0 & 0 & 1 \\ 0 & 0 & 0 & 0 & 1 & 0 \end{pmatrix} \end{array}$$

**7.**

$$\begin{array}{c} \\ a \\ b \\ c \\ d \\ e \end{array} \begin{array}{c} \begin{array}{cccccccc} x_1 & x_2 & x_3 & x_4 & x_5 & x_6 & x_7 & x_8 \end{array} \\ \begin{pmatrix} 1 & 0 & 1 & 0 & 0 & 1 & 0 & 0 \\ 1 & 1 & 0 & 0 & 0 & 0 & 0 & 0 \\ 0 & 1 & 0 & 1 & 1 & 0 & 1 & 0 \\ 0 & 0 & 0 & 1 & 0 & 1 & 0 & 1 \\ 0 & 0 & 1 & 0 & 0 & 0 & 1 & 1 \end{pmatrix} \end{array}$$

**10.**

$$\begin{array}{c} \\ 1 \\ 2 \\ 3 \\ 4 \\ 5 \\ 6 \\ 7 \end{array} \begin{array}{c} \begin{array}{cccccccc} e_1 & e_2 & e_3 & e_4 & e_5 & e_6 & e_7 & e_8 \end{array} \\ \begin{pmatrix} 1 & 0 & 0 & 0 & 0 & 0 & 0 & 0 \\ 1 & 1 & 0 & 1 & 1 & 1 & 0 & 0 \\ 0 & 1 & 1 & 0 & 0 & 0 & 0 & 0 \\ 0 & 0 & 1 & 1 & 0 & 0 & 0 & 0 \\ 0 & 0 & 0 & 0 & 1 & 0 & 1 & 0 \\ 0 & 0 & 0 & 0 & 0 & 1 & 1 & 1 \\ 0 & 0 & 0 & 0 & 0 & 0 & 0 & 1 \end{pmatrix} \end{array}$$

**13.**

**16.**

**19.** [For $K_5$]

$$\begin{pmatrix} 4 & 3 & 3 & 3 & 3 \\ 3 & 4 & 3 & 3 & 3 \\ 3 & 3 & 4 & 3 & 3 \\ 3 & 3 & 3 & 4 & 3 \\ 3 & 3 & 3 & 3 & 4 \end{pmatrix}$$

**22.** The graph is not connected.

**24.**

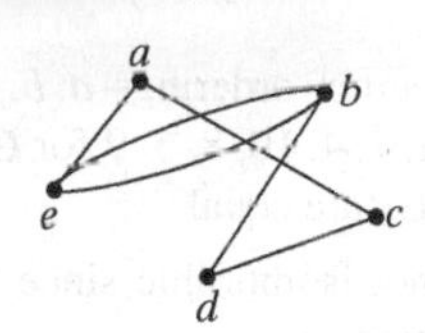


**27.** $G$ is not connected.

**28.** Because of the symmetry of the graph, if $v$ and $w$ are vertices in $K_5$, there is the same number of paths of length $n$ from $v$ to $v$ as there is from $w$ to $w$. Thus all the diagonal elements of $A^n$ are equal. Similarly, all the off-diagonal elements of $A^n$ are equal.

**31.** If $n \geq 2$,

$$\begin{aligned} d_n &= 4a_{n-1} && \text{by Exercise 29} \\ &= 4\left(\frac{1}{5}\right)[4^{n-1} + (-1)^n] && \text{by Exercise 30.} \end{aligned}$$

The formula can be directly verified for $n = 1$.

## Section 8.6 Review

1. Graphs $G_1$ and $G_2$ are isomorphic if there is a one-to-one, onto function $f$ from the vertices of $G_1$ to the vertices of $G_2$ and a one-to-one, onto function $g$ from the edges of $G_1$ to the edges of $G_2$, so that an edge $e$ is incident on $v$ and $w$ in $G_1$ if and only if the edge $g(e)$ is incident on $f(v)$ and $f(w)$ in $G_2$.

2. The following graphs

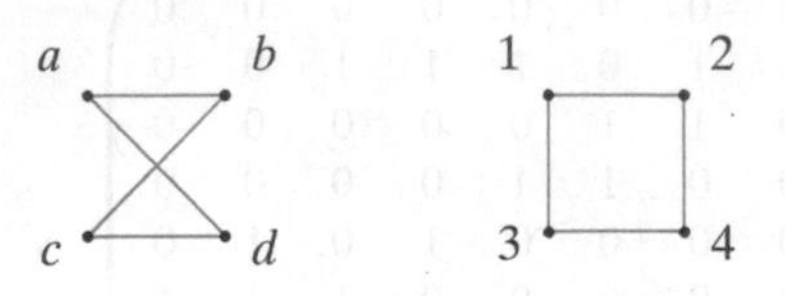


are isomorphic. An isomorphism is given by $f(a) = 1$, $f(b) = 2$, $f(c) = 4$, $f(d) = 3$, and $g(a, b) = (1, 2)$, $g(b, c) = (2, 4)$, $g(c, d) = (4, 3)$, $g(d, a) = (3, 1)$.

3. The following graphs

are not isomorphic; the first graph has two vertices, but the second graph has three vertices.

4. A property $P$ is an invariant if, whenever $G_1$ and $G_2$ are isomorphic graphs, if $G_1$ has property $P$, then $G_2$ also has property $P$.

5. To show that two graphs are not isomorphic, find an invariant that one graph has and the other does not have.

6. Two graphs are isomorphic if and only if for some orderings of their vertices, their adjacency matrices are equal.

7. A rectangular array of vertices

## Section 8.6

1. Relative to the vertex orderings $a, b, c, d, e, f, g$ for $G_1$, and $1, 3, 5, 7, 2, 4, 6$ for $G_2$, the adjacency matrices of $G_1$ and $G_2$ are equal.

4. Relative to the vertex orderings $a, b, c, d, e, f, g, h, i, j$ for $G_1$, and $5, 6, 1, 2, 7, 4, 10, 8, 3, 9$ for $G_2$, the adjacency matrices of $G_1$ and $G_2$ are equal.

7. The graphs are not isomorphic since they do not have the same number of vertices.

10. The graphs are not isomorphic since $G_1$ has a simple cycle of length 3 and $G_2$ does not.

13. The graphs are not isomorphic. The edge $(1, 4)$ in $G_2$ has $\delta(1) = 3$ and $\delta(4) = 3$ but there is no such edge in $G_1$ (see also Exercise 26).

17. All are isomorphic to $K_2$.

20. If $r = 2$, all are isomorphic to

There are no connected, simple, 3-regular, 5-vertex graphs since the number of vertices having odd degree is even.

If $r = 4$, all are isomorphic to $K_5$.

*In Exercises 22–28, we use the notation of Definition 8.6.1.*

22. If $(v_0, v_1, \dots, v_k)$ is a simple cycle of length $k$ in $G_1$, then $(f(v_0), f(v_1), \dots, f(v_k))$ is a simple cycle of length $k$ in $G_2$. [The vertices $f(v_i)$, $i = 1, \dots, k-1$, are distinct, since $f$ is one-to-one.]

25. In the hint to Exercise 22, we showed that if $C = (v_0, v_1, \dots, v_k)$ is a simple cycle of length $k$ in $G_1$, then $(f(v_0), f(v_1), \dots, f(v_k))$, which here we denote $f(C)$, is a simple cycle of length $k$ in $G_1$. Let $C_1, C_2, \dots, C_n$ denote the $n$ simple cycles of length $k$ in $G_1$. Then $f(C_1)$, $f(C_2), \dots, f(C_n)$ are $n$ simple cycles of length $k$ in $G_2$. Moreover, since $f$ is one-to-one, $f(C_1), f(C_2), \dots f(C_n)$ are distinct.

28. The property is an invariant. If $(v_0, v_1, \dots, v_n)$ is an Euler cycle in $G_1$, then, since $g$ is onto, $(f(v_0), f(v_1), \dots, f(v_n))$ is an Euler cycle in $G_2$.

31.

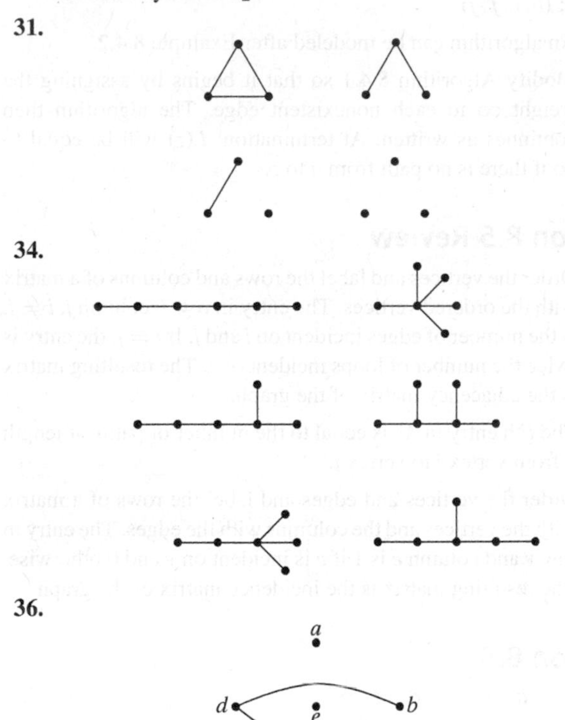


39.

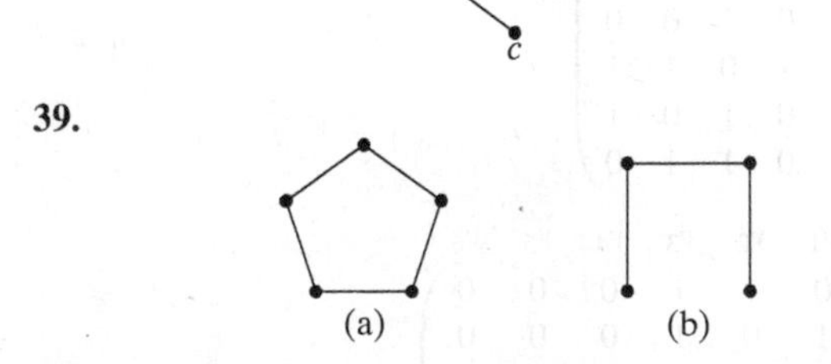


42. Define $g((v, w)) = (f(v), f(w))$.

43. $f(a) = 1, f(b) = 2, f(c) = 3, f(d) = 2$

46. $f(a) = 1, f(b) = 2, f(c) = 3, f(d) = 1$

## Section 8.7 Review

1. A graph that can be drawn in the plane without its edges crossing
2. A contiguous region
3. $f = e - v + 2$
4. Edges of the form $(v, v_1)$ and $(v, v_2)$, where $v$ has degree 2 and $v_1 \neq v_2$
5. Given edges of the form $(v, v_1)$ and $(v, v_2)$, where $v$ has degree 2 and $v_1 \neq v_2$, a series reduction deletes vertex $v$ and replaces $(v, v_1)$ and $(v, v_2)$ by $(v_1, v_2)$.
6. Two graphs are homeomorphic if they can be reduced to isomorphic graphs by performing a sequence of series reductions.
7. A graph is planar if and only if it does not contain a subgraph homeomorphic to $K_5$ or $K_{3,3}$.

## Section 8.7

1.

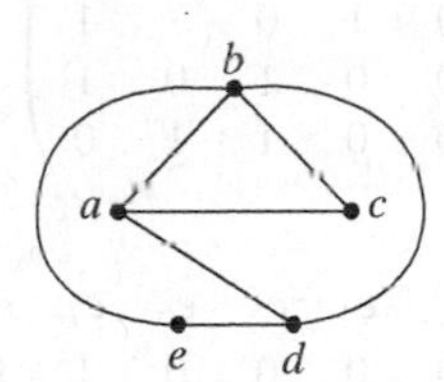


4.

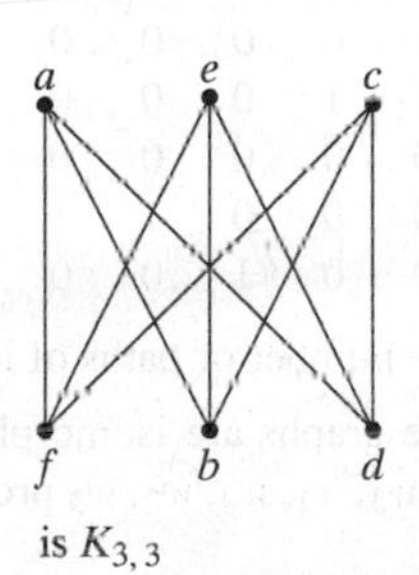


is $K_{3,3}$

6. Planar

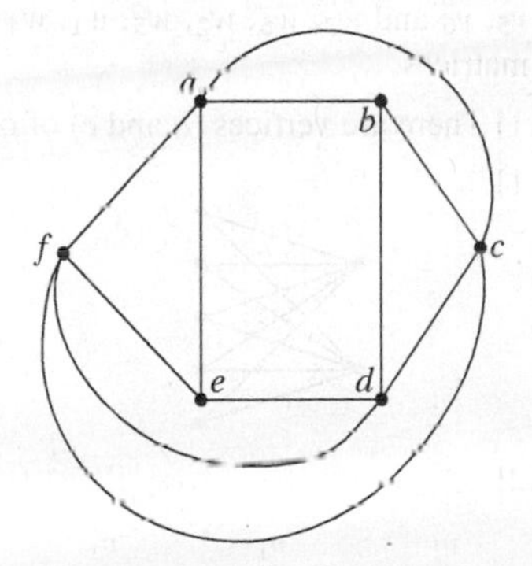


9. $2e = 2 + 2 + 2 + 3 + 3 + 3 + 4 + 4 + 5$, so $e = 14$. $f = e - v + 2 = 14 - 9 + 2 = 7$

12. A graph with five or fewer vertices and a vertex of degree 2 is homeomorphic to a graph with four or fewer vertices. Such a graph cannot contain a homeomorphic copy of $K_{3,3}$ or $K_5$.

15. If $K_5$ is planar, $e \leq 3v - 6$ becomes $10 \leq 3 \cdot 5 - 6 = 9$.

19.

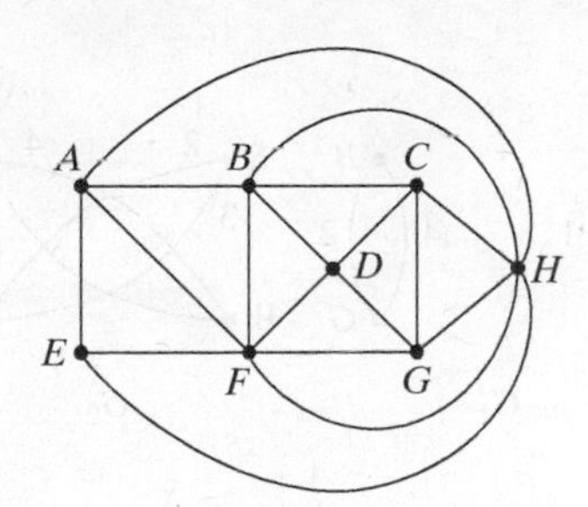


23.

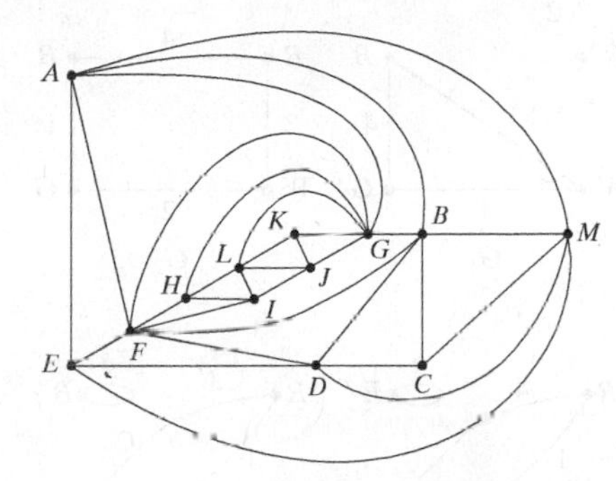


26.

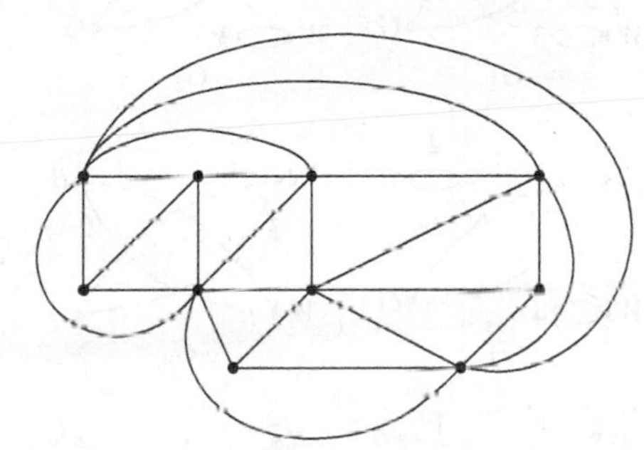

29. It contains

32. Assume that $G$ does not have a vertex of degree 5. Show that $2e \geq 6v$. Now use Exercise 13 to deduce a contradiction.

## Section 8.8 Review

1. Instant Insanity consists of four cubes each of whose faces is painted one of the four colors, red, white, blue, or green. The problem is to stack the cubes, one on top of the other, so that whether the stack is viewed from front, back, left, or right, one sees all four colors.
2. Draw a graph $G$, where the vertices represent the four colors and an edge labeled $i$ connects two vertices if the opposing faces of cube $i$ have those colors. Find two graphs where
   - Each vertex has degree 2.
   - Each cube represents an edge exactly once in each graph.
   - The graphs have no edges in common.

   One graph represents the front/back stacking, and the other represents the left/right stacking.

## Section 8.8

**1.**

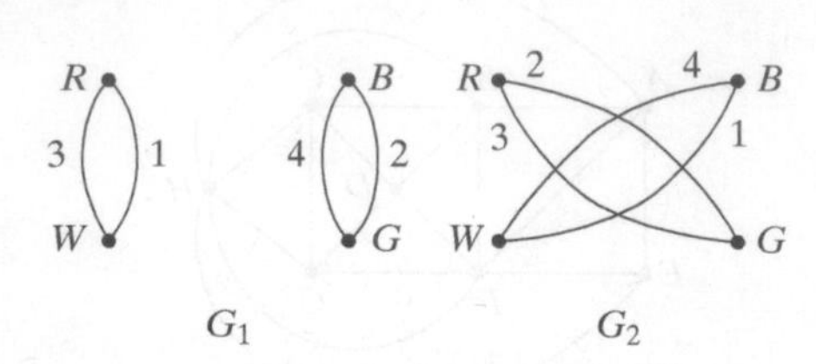


**4.**

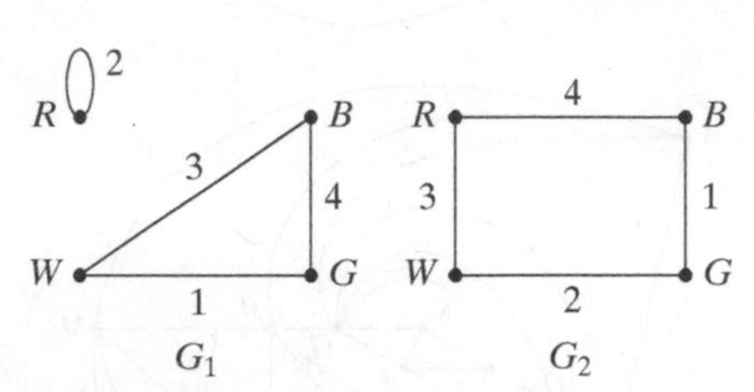


**7.** (a)

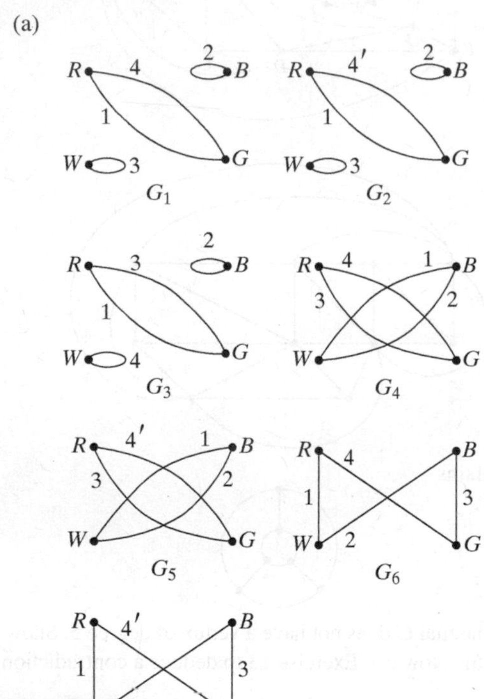


(b) Solutions are $G_1, G_5$; $G_1, G_7$; $G_2, G_4$; $G_2, G_6$; $G_3, G_6$; and $G_3, G_7$.

**13.** One edge can be chosen in $C(2 + 4 - 1, 2) = 10$ ways. The three edges labeled 1 can be chosen in $C(3 + 10 - 1, 3) = 220$ ways. Thus the total number of graphs is $220^4$.

**15.**

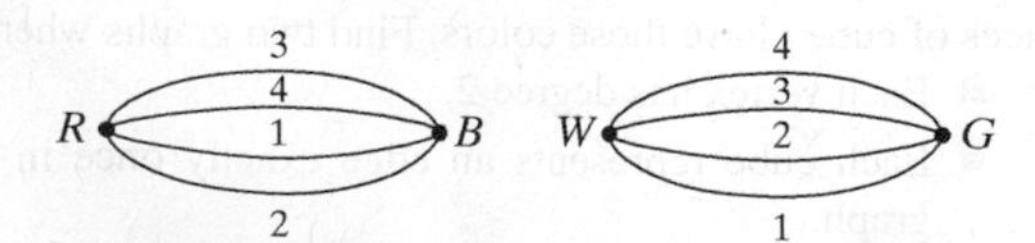


**19.** According to Exercise 14, not counting loops, every vertex must have degree at least 4. In Figure 8.8.5, not counting loops, vertex $W$ has degree 3 and, therefore, Figure 8.8.5 does not have a solution to the modified version of Instant Insanity. Figure 8.8.3 gives a solution to regular Instant Insanity for Figure 8.8.5.

## Chapter 8 Self-Test

**1.** [Section 8.1] $V = \{v_1, v_2, v_3, v_4\}$. $E = \{e_1, e_2, e_3\}$. $e_1$ and $e_2$ are parallel edges. There are no loops. $v_1$ is an isolated vertex. $G$ is not a simple graph. $e_3$ is incident on $v_2$ and $v_4$. $v_2$ is incident on $e_1$, $e_2$, and $e_3$.

**2.** [Section 8.2] It is a cycle.

**3.** [Section 8.3] $(v_1, v_2, v_3, v_4, v_5, v_7, v_6, v_1)$

**4.** [Section 8.5]

| | $v_1$ | $v_2$ | $v_3$ | $v_4$ | $v_5$ | $v_6$ | $v_7$ |
|---|---|---|---|---|---|---|---|
| $v_1$ | 0 | 1 | 0 | 0 | 0 | 1 | 0 |
| $v_2$ | 1 | 0 | 1 | 1 | 0 | 1 | 1 |
| $v_3$ | 0 | 1 | 0 | 1 | 0 | 0 | 0 |
| $v_4$ | 0 | 1 | 1 | 0 | 1 | 0 | 0 |
| $v_5$ | 0 | 0 | 0 | 1 | 0 | 1 | 1 |
| $v_6$ | 1 | 1 | 0 | 0 | 1 | 0 | 1 |
| $v_7$ | 0 | 1 | 0 | 0 | 1 | 1 | 0 |

**5.** [Section 8.5]

| | $e_1$ | $e_2$ | $e_3$ | $e_4$ | $e_5$ | $e_6$ | $e_7$ | $e_8$ | $e_9$ | $e_{10}$ | $e_{11}$ |
|---|---|---|---|---|---|---|---|---|---|---|---|
| $v_1$ | 1 | 0 | 0 | 0 | 0 | 0 | 1 | 0 | 0 | 0 | 0 |
| $v_2$ | 1 | 1 | 0 | 1 | 1 | 1 | 0 | 0 | 0 | 0 | 0 |
| $v_3$ | 0 | 1 | 1 | 0 | 0 | 0 | 0 | 0 | 0 | 0 | 0 |
| $v_4$ | 0 | 0 | 1 | 1 | 0 | 0 | 0 | 0 | 0 | 1 | 0 |
| $v_5$ | 0 | 0 | 0 | 0 | 0 | 0 | 0 | 1 | 1 | 1 | 0 |
| $v_6$ | 0 | 0 | 0 | 0 | 0 | 1 | 1 | 0 | 1 | 0 | 1 |
| $v_7$ | 0 | 0 | 0 | 0 | 1 | 0 | 0 | 1 | 0 | 0 | 1 |

**6.** [Section 8.5] The number of paths of length 3 from $v_2$ to $v_3$

**7.** [Section 8.6] The graphs are isomorphic. The orderings $v_1$, $v_2$, $v_3$, $v_4$, $v_5$ and $w_3$, $w_1$, $w_4$, $w_2$, $w_5$ produce equal adjacency matrices.

**8.** [Section 8.6] The graphs are isomorphic. The orderings $v_1$, $v_2$, $v_3$, $v_4$, $v_5$, $v_6$ and $w_3$, $w_6$, $w_2$, $w_5$, $w_1$, $w_4$ produce equal adjacency matrices.

**9.** [Section 8.1] There are vertices ($a$ and $e$) of odd degree.

**10.** [Section 8.1]

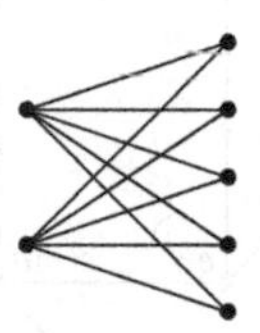

**11.** [Section 8.2]

$v_1$: $e_1$, $e_2$; $v_2$ $v_3$ — $v_1$: $e_1$, $e_3$; $v_2$ $v_4$ — $v_1$: $e_2$, $e_3$; $v_3$ $v_4$

$v_1$: $e_1$, $e_2$; $v_2$ $v_3$ $v_4$ — $v_1$: $e_1$, $e_3$; $v_2$ $v_3$ $v_4$ — $v_1$: $e_2$, $e_3$; $v_2$ $v_3$ $v_4$

**12.** [Section 8.2]

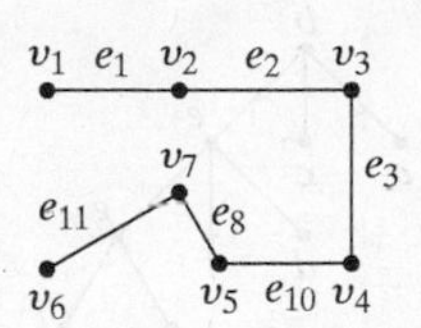


**13.** [Section 8.3] (000, 001, 011, 010, 110, 111, 101, 100, 000)

**14.** [Section 8.3] A Hamiltonian cycle would have seven edges. Suppose that the graph has a Hamiltonian cycle. We would have to eliminate three edges at vertex $b$ and one edge at vertex $f$. This leaves $10 - 4 = 6$ edges, not enough for a Hamiltonian cycle. Therefore, the graph does not have a Hamiltonian cycle.

**15.** [Section 8.5] No. Each edge is incident on at least one vertex.

**16.** [Section 8.1] If we let $V_1$ denote the set of vertices containing an even number of 1's and $V_2$ the set of vertices containing an odd number of 1's, each edge is incident on one vertex in $V_1$ and one vertex in $V_2$. Therefore, the $n$-cube is bipartite.

**17.** [Section 8.2] No. There are vertices of odd degree.

**18.** [Section 8.3] In a minimum-weight Hamiltonian cycle, every vertex must have degree 2. Therefore, edges $(a, b)$, $(a, j)$, $(j, i)$, $(i, h)$, $(g, f)$, $(f, e)$, and $(e, d)$ must be included. We cannot include edge $(b, h)$ or we will complete a cycle. This implies that we must include edges $(h, g)$ and $(b, c)$. Since vertex $g$ now has degree 2, we cannot include edge $(c, g)$ or $(g, d)$. Thus we must include $(c, d)$. This is a Hamiltonian cycle and the argument shows that it is unique. Therefore, it is minimal.

**19.** [Section 8.6]

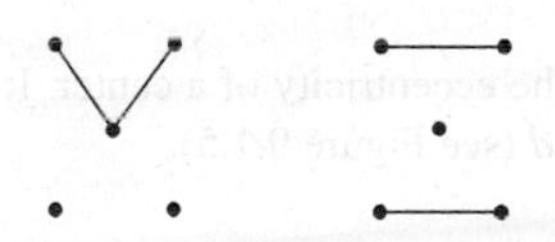

**20.** [Section 8.6]

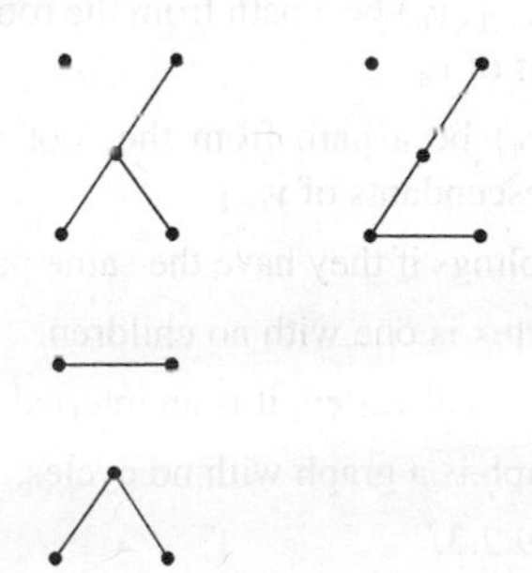

**21.** [Section 8.4] 9

**22.** [Section 8.4] 11

**23.** [Section 8.4] $(a, e, f, i, g, z)$

**24.** [Section 8.4] 12

**25.** [Section 8.7] The graph is planar:

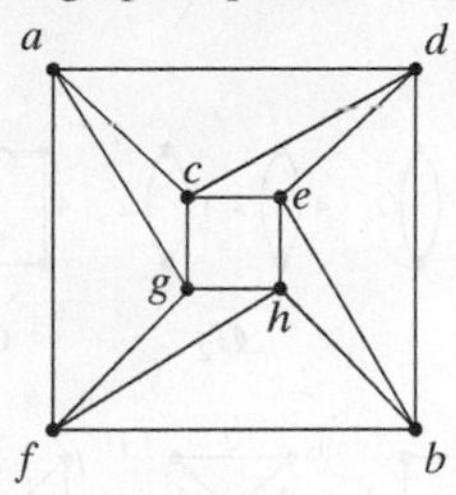


**26.** [Section 8.7] The graph is not planar; the following subgraph is homeomorphic to $K_5$:

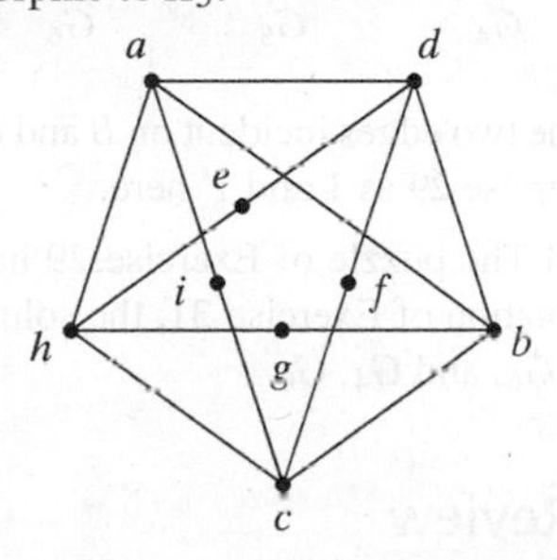


**27.** [Section 8.7] A simple, planar, connected graph with $e$ edges and $v$ vertices satisfies $e \leq 3v - 6$ (see Exercise 13, Section 8.7). If $e = 31$ and $v = 12$, the inequality is not satisfied, so such a graph cannot be planar.

**28.** [Section 8.7] For $n = 1, 2, 3$, it is possible to draw the $n$-cube in the plane without having any of its edges cross:

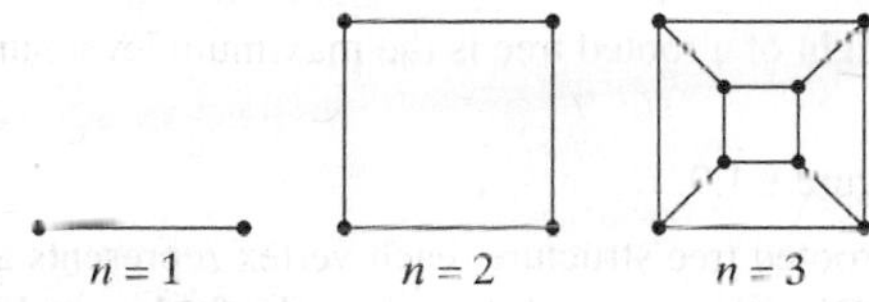


We argue by contradiction to show that the 4-cube is not planar. Suppose that the 4-cube is planar. Since every cycle has at least four edges, each face is bounded by at least four edges. Thus the number of edges that bound faces is at least $4f$. In a planar graph, each edge belongs to at most two bounding cycles. Therefore, $2e \geq 4f$. Using Euler's formula for graphs, we find that

$$2e \geq 4(e - v + 2).$$

For the 4-cube, we have $e = 32$ and $v = 16$, so Euler's formula becomes

$$64 = 2 \cdot 32 \geq 4(32 - 16 + 2) = 72,$$

which is a contradiction. Therefore, the 4-cube is not planar. The $n$-cube, for $n > 4$, is not planar since it contains the 4-cube.

**29.** [Section 8.8]

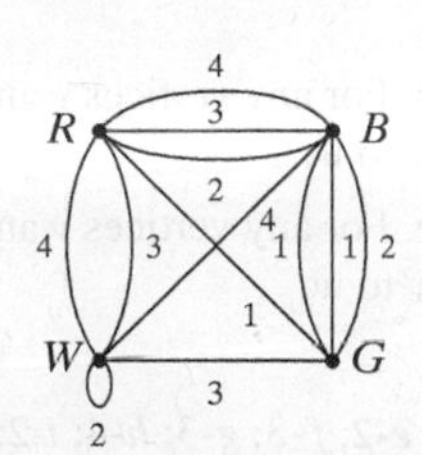

**30.** [Section 8.8] See the hints for Exercises 31 and 32.

**31.** [Section 8.8]

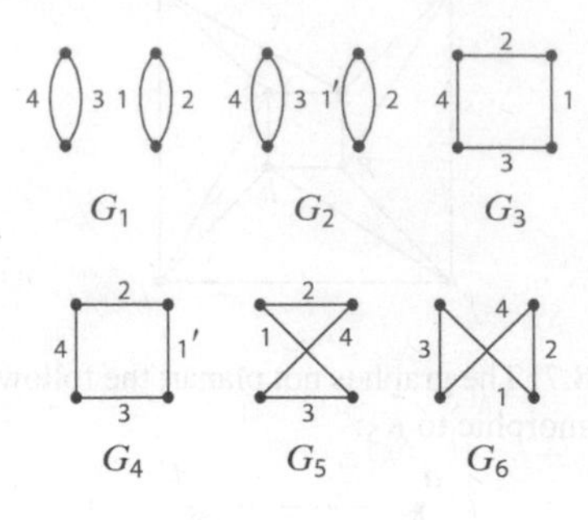


We denote the two edges incident on $B$ and $G$ labeled 1 in the graph of Exercise 29 as 1 and $1'$ here.

**32.** [Section 8.8] The puzzle of Exercise 29 has four solutions. Using the notation of Exercise 31, the solutions are $G_1$, $G_5$; $G_2$, $G_5$; $G_3$, $G_6$; and $G_4$, $G_6$.

## Section 9.1 Review

**1.** A free tree $T$ is a simple graph satisfying the following: If $v$ and $w$ are vertices in $T$, there is a unique simple path from $v$ to $w$.

**2.** A rooted tree is a tree in which a particular vertex is designated the root.

**3.** The level of a vertex $v$ is the length of the simple path from the root to $v$.

**4.** The height of a rooted tree is the maximum level number that occurs.

**5.** See Figure 9.1.9.

**6.** In the rooted tree structure, each vertex represents a file or a folder. Directly under a folder $f$ are the folders and files contained in $f$.

**7.** A Huffman code can be defined by a rooted tree. The code for a particular character is obtained by following the simple path from the root to that character. Each edge is labeled with 0 or 1, and the sequence of bits encountered on the simple path is the code for that character.

**8.** Suppose that there are $n$ frequencies. If $n = 2$, build the tree shown in Figure 9.1.11 and stop. Otherwise, let $f_i$ and $f_j$ denote the smallest frequencies, and replace them in the list by $f_i + f_j$. Recursively construct an optimal Huffman coding tree using the modified list. In the tree that results, add two edges to a vertex labeled $f_i + f_j$, and label the added vertices $f_i$ and $f_j$.

## Section 9.1

**1.** The graph is a tree. For any vertices $v$ and $w$, there is a unique simple path from $v$ to $w$.

**4.** The graph is a tree. For any vertices $v$ and $w$, there is a unique simple path from $v$ to $w$.

**7.** $n = 1$

**8.** $a$-1; $b$-1; $c$-1; $d$-1; $e$-2; $f$-3; $g$-3; $h$-4; $i$-2; $j$-3; $k$-0

**11.** Height = 4

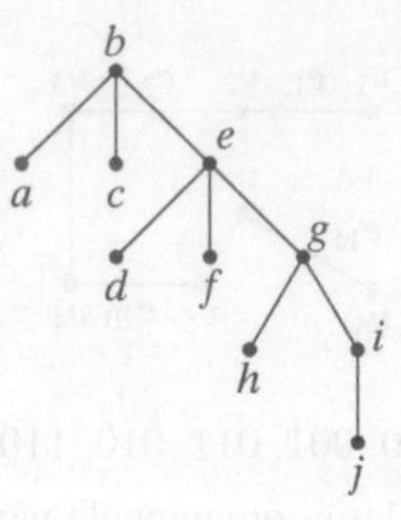


**14.** PEN

**17.** SALAD

**18.** 0111100010

**21.** 011000010010000 1111

**24.**

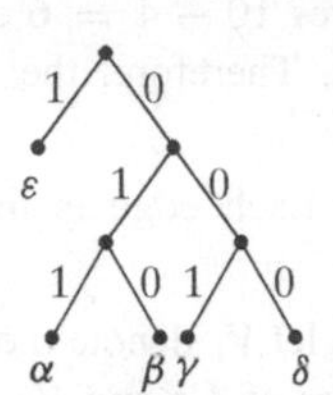


**27.** Another tree is shown in the hint for Exercise 24.

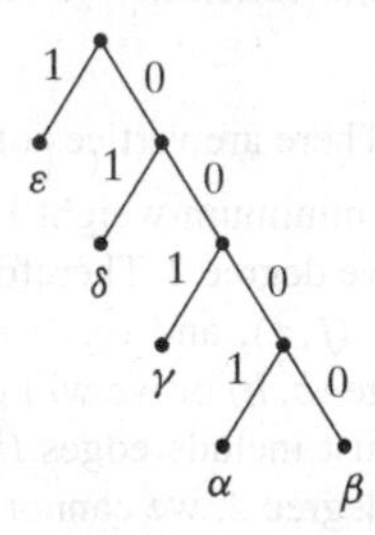


**32.** Let $T$ be a tree. Root $T$ at some arbitrary vertex. Let $V$ be the set of vertices on even levels and let $W$ be the set of vertices on odd levels. Since each edge is incident on a vertex in $V$ and a vertex in $W$, $T$ is a bipartite graph.

**35.** $e$, $g$

**38.** The radius is the eccentricity of a center. It is not necessarily true that $2r = d$ (see Figure 9.1.5).

## Section 9.2 Review

**1.** Let $(v_0, \ldots, v_{n-1}, v_n)$ be a path from the root $v_0$ to $v_n$. We call $v_{n-1}$ the parent of $v_n$.

**2.** Let $(v_0, \ldots, v_n)$ be a path from the root $v_0$ to $v_n$. We call $(v_i, \ldots, v_n)$ descendants of $v_{i-1}$.

**3.** $v$ and $w$ are siblings if they have the same parent.

**4.** A terminal vertex is one with no children.

**5.** If $v$ is not a terminal vertex, it is an internal vertex.

**6.** An acyclic graph is a graph with no cycles.

**7.** See Theorem 9.2.3.

## Section 9.2

**1.** Kronos

**4.** Apollo, Athena, Hermes, Heracles

**7.** $b$; $d$

**10.** $e, f, g, j$; $j$

**13.** $a, b, c, d, e$

**17.** They are siblings.

**22.**

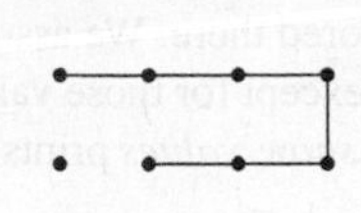

**25.**

**27.** A single vertex is a "cycle" of length 0.

**30.** Each component of a forest is connected and acyclic and, therefore, a tree.

**33.** Suppose that $G$ is connected. Add parallel edges until the resulting graph $G^*$ has $n - 1$ edges. Since $G^*$ is connected and has $n - 1$ edges, by Theorem 9.2.3, $G^*$ is acyclic. But adding an edge in parallel introduces a cycle. Contradiction.

**36.**

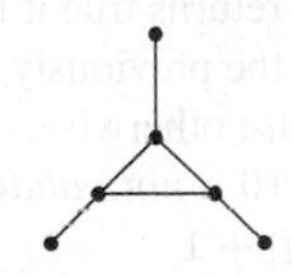

## Section 9.3 Review

1. A tree $T$ is a spanning tree of a graph $G$ if $T$ is a subgraph of $G$ that contains all of the vertices of $G$.
2. A graph $G$ has a spanning tree if and only if $G$ is connected.
3. Select an ordering of the vertices. Select the first vertex and label it the root. Let $T$ consist of this single vertex and no edges. Add to the tree all edges incident on this single vertex that do not produce a cycle when added to the tree. Also add the vertices incident on these edges. Repeat this procedure with the vertices on level 1, then those on level 2, and so on.
4. Select an ordering of the vertices. Select the first vertex and label it the root. Add an edge incident on this vertex to the tree, and add the additional vertex $v$ incident on this edge. Next add an edge incident on $v$ that does not produce a cycle when added to the tree, and add the additional vertex incident on this edge. Repeat this process. If, at any point, we cannot add an edge incident on a vertex $w$, we backtrack to the parent $p$ of $w$ and try to add an edge incident on $p$. When we finally backtrack to the root and cannot add more edges, depth first search concludes.
5. Depth-first search

## Section 9.3

**1.**

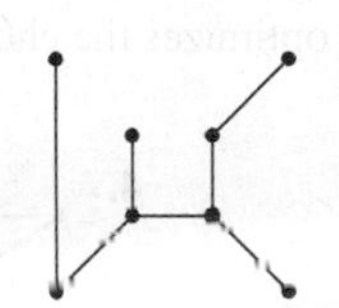

**4.** The path $(h, f, e, g, b, d, c, a)$

**7.**

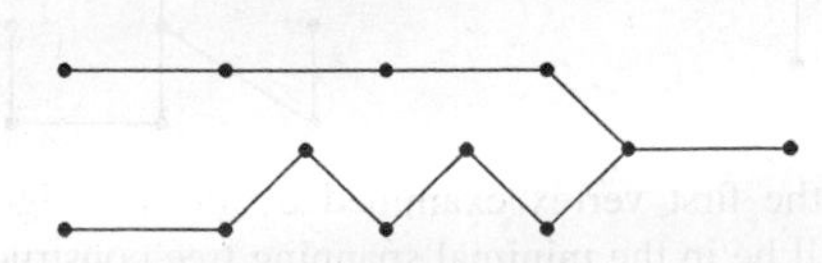

**10.** The two-queens problem clearly has no solution. For the three-queens problem, by symmetry, the only possible first column positions are upper left and second from top. If the first move is first column, upper left, the second move must be to the bottom of the second column. Now no move is possible for the third column. If the first move is first column, second from top, there is no move possible in column two. Therefore, there is no solution to the three-queens problem.

**13.**

| | | | | |
|---|---|---|---|---|
| | | × | | |
| × | | | | |
| | | | × | |
| | × | | | |
| | | | | × |

| | | | | |
|---|---|---|---|---|
| | | | × | |
| × | | | | |
| | | × | | |
| | | | | × |
| | × | | | |

**17.** False. Consider $K_4$.

**20.** First, show that the graph $T$ constructed is a tree. Now use induction on the level of $T$ to show that $T$ contains all the vertices of $G$.

**23.** Suppose that $x$ is incident on vertices $a$ and $b$. Removing $x$ from $T$ produces a disconnected graph with two components, $U$ and $V$. Vertices $a$ and $b$ belong to different components—say, $a \in U$ and $b \in V$. There is a path $P$ from $a$ to $b$ in $T'$. As we move along $P$, at some point we encounter an edge $y = (v, w)$ with $v \in U$, $w \in V$. Since adding $y$ to $T - \{x\}$ produces a connected graph, $(T - \{x\}) \cup \{y\}$ is a spanning tree. Clearly, $(T' - \{y\}) \cup \{x\}$ is a spanning tree.

**26.** Suppose that $T$ has $n$ vertices. If an edge is added to $T$, the resulting graph $T'$ is connected. If $T'$ were acyclic, $T'$ would be a tree with $n$ edges and $n$ vertices. Thus $T'$ contains a cycle. If $T'$ contains two or more cycles, we would be able to produce a connected graph $T''$ by deleting two or more edges from $T'$. But now $T''$ would be a tree with $n$ vertices and fewer than $n - 1$ edges—an impossibility.

**27.**

$$\begin{array}{r} \\ (abca) \\ (acda) \\ (acdb) \\ (bcdeb) \end{array} \begin{array}{c} \begin{array}{cccccccc} e_1 & e_2 & e_6 & e_5 & e_3 & e_4 & e_7 & e_8 \end{array} \\ \begin{pmatrix} 1 & 0 & 0 & 0 & 1 & 1 & 0 & 0 \\ 0 & 1 & 0 & 0 & 1 & 0 & 0 & 1 \\ 0 & 0 & 1 & 0 & 0 & 1 & 0 & 1 \\ 0 & 0 & 0 & 1 & 0 & 1 & 1 & 1 \end{pmatrix} \end{array}$$

**30.** Input: A graph $G = (V, E)$ with $n$ vertices

Output: true if $G$ is connected
false if $G$ is not connected

```
is_connected(V, E) {
    T = bfs(V, E)
    // T = (V', E') is the spanning tree returned by bfs
    if (|V'| == n)
        return true
    else
        return false
}
```

**33.**

```
bfs_track_parent(V, E, parent) {
    S = (v1)
    // set v1's parent to 0 to indicate that v1 has no parent
    parent(v1) = 0
```

```
    V' = {v1}
    E' = Ø
    while (true) {
      for each x ∈ S, in order,
        for each y ∈ V − V', in order
          if ((x, y) is an edge) {
            add edge (x, y) to E' and y to V'
            parent(y) = x
          }
      if (no edges were added)
        return T
      S = children of S ordered consistently with the original
          vertex ordering
    }
}
```

**34.**

```
print_parents(V, parent) {
  for each v ∈ V
    println(v, parent(v))
}
```

**37.** An algorithm can be obtained by modifying the four-queens algorithm. The array *row* is replaced by the array *p*, which is the permutation. A conflict for $p(k)$ now means that for some $i < k$, $p(i) = p(k)$, that is, the value $p(k)$ has already been assigned. To obtain *all* of the permutations, when we find a permutation, we print it and continue (whereas in the four-queens algorithm, being content with one solution, we terminated the algorithm).

```
perm(n) {
  k = 1
  p(1) = 0
  while (k > 0) {
    p(k) = p(k) + 1
    while (p(k) ≤ n ∧ p(k) conflicts)
      p(k) = p(k) + 1
    if (p(k) ≤ n)
      if (k == n)
        println(p)
      else {
        k = k + 1
        p(k) = 0
      }
    else
      k = k − 1
  }
}
```

**40.** The idea of the backtracking algorithm is to scan the grid (we chose to scan top to bottom, left to right), skipping positions where numbers were preassigned, and, at the next available position, we try 1, then 2, then 3, and so on, until we find a legal value (i.e., a value that does not conflict within its $3\times 3$ subsquare, within its row, or within its column). If such a value is found, we continue with the next available position. If no such value can be found, we backtrack to the last position where we assigned a value; if that value was $i$, we try $i+1$, $i+2$ and so on.

In the following algorithm, the value $s(i, j)$ is the value in row $i$, column $j$, or 0 if no value is stored there. We assume that initially all values in $s$ are set to 0, except for those values that are specified in the puzzle. Finally, *show_values* prints the array $s$.

```
sudoku(s) {
  i = 0
  j = 1
  // advance advances i and j to the next position in which
  // a value is not specified. It proceeds down a column first.
  advance(i, j)
  while (i ≥ 1 ∧ j ≥ 1) {
    // search for a legal value
    s(i, j) = s(i, j) + 1
    // not_valid(i, j) returns true if the value s(i, j)
    // conflicts with the previously chosen and specified
    // values, and false otherwise.
    while (s(i, j) < 10 ∧ not_valid(i, j))
      s(i, j) = s(i, j) + 1
    // if no value found, backtrack
    if (s(i, j) == 10) {
      s(i, j) = 0
      // retreat moves i and j to the previous position in
      // which a value is not specified. It proceeds up a
      // column first.
      retreat(i, j)
    }
    else
      advance(i, j) // sets j to 10 if advanced off board
    if (j == 10) {
      // Solution!
      show_values()
      return
    }
  }
}
```

## Section 9.4 Review

1. A minimal spanning tree is a spanning tree with minimum weight.
2. Prim's Algorithm builds a minimal spanning tree by iteratively adding edges. The algorithm begins with a single vertex. Then at each iteration, it adds to the current tree a minimum-weight edge that does not complete a cycle.
3. A greedy algorithm optimizes the choice at each iteration.

## Section 9.4

**1.**

**4.**

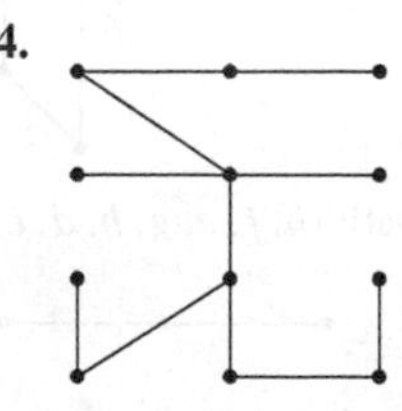

**10.** If $v$ is the first vertex examined by Prim's Algorithm, the edge will be in the minimal spanning tree constructed by the algorithm.

**13.** Suppose that $G$ has two minimal spanning trees $T_1$ and $T_2$. Then, there exists an edge $x$ in $T_1$ that is not in $T_2$. By Exercise 23, Section 9.3, there exists an edge $y$ in $T_2$ that is not in $T_1$ such that $T_3 = (T_1 - \{x\}) \cup \{y\}$ and $T_4 = (T_2 - \{y\}) \cup \{x\}$ are spanning trees. Since $x$ and $y$ have different weights, either $T_3$ or $T_4$ has weight less than $T_1$. This is a contradiction.

**14.** False

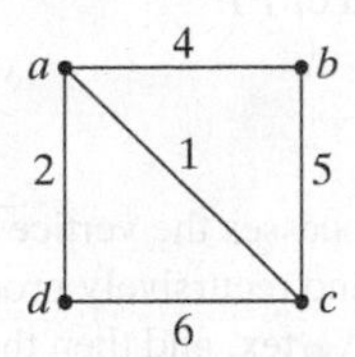


**16.** False. Consider $K_5$ with the weight of every edge equal to 1.

**20.** Input: The edges $E$ of an $n$-vertex, connected, weighted graph. If $e$ is an edge, $w(e)$ is equal to the weight of $e$; if $e$ is not an edge, $w(e)$ is equal to $\infty$ (a value greater than any actual weight).

Output: A minimal spanning tree.

```
kruskal(E, w, n) {
  V' = ∅
  E' = ∅
  T' = (V', E')
  while (|E'| < n − 1) {
    among all edges that if added to T' would not
        complete a cycle, choose e = (v_i, v_j) of
        minimum weight
    E' = E' ∪ {e}
    V' = V' ∪ {v_i, v_j}
    T' = (V', E')
  }
  return T'
}
```

**23.** Terminate Kruskal's Algorithm after $k$ iterations. This groups the data into $n - k$ classes.

**27.** We show that $a_1 = 7$ and $a_2 = 3$ provide a solution. We use induction on $n$ to show that the greedy solution gives an optimal solution for $n \geq 1$. The cases $n = 1, 2, \ldots, 8$ may be verified directly.

We first show that if $n \geq 9$, there is an optimal solution containing at least one 7. Let $S'$ be an optimal solution. Suppose that $S'$ contains no 7's. Since $S'$ contains at most two 1's (since $S'$ is optimal), $S'$ contains at least three 3's. We replace three 3's by one 7 and two 1's to obtain a solution $S$. Since $|S| = |S'|$, $S$ is optimal.

If we remove a 7 from $S$, we obtain a solution $S^*$ to the $(n - 7)$-problem. If $S^*$ were not optimal, $S$ could not be optimal. Thus $S^*$ is optimal. By the inductive assumption, the greedy solution $GS^*$ to the $(n - 7)$-problem is optimal, so $|S^*| = |GS^*|$. Notice that 7 together with $GS^*$ is the greedy solution $GS$ to the $n$-problem. Since $|GS| = |S|$, $GS$ is optimal.

**30.** Suppose that the greedy algorithm is optimal for all denominations less than $a_{m-1} + a_m$. We use induction on $n$ to show that the greedy algorithm is optimal for all $n$. We may assume that $n \geq a_{m-1} + a_m$.

Consider an optimal solution $\mathcal{S}$ for $n$. First suppose that $\mathcal{S}$ uses at least one $a_m$ coin. The solution, $\mathcal{S}$ with one $a_m$ coin removed, is optimal for $n - a_m$. (If there was a solution for $n - a_m$ using fewer coins, we could add one $a_m$ coin to it to obtain a solution for $n$ using fewer coins than $\mathcal{S}$, which is impossible.) By the inductive assumption, the greedy solution for $n - a_m$ is optimal. If we add one $a_m$ coin to the greedy solution for $n - a_m$, we obtain a solution $\mathcal{G}$ for $n$ that uses the same number of coins as $\mathcal{S}$. Therefore, $\mathcal{G}$ is optimal. But $\mathcal{G}$ is also greedy because the greedy solution begins by removing one $a_m$ coin.

Now suppose that $\mathcal{S}$ does not use an $a_m$ coin. Let $i$ be the largest index such that $\mathcal{S}$ uses an $a_i$ coin. The solution, $\mathcal{S}$ with one $a_i$ coin removed, is optimal for $n - a_i$. By the inductive assumption, the greedy solution for $n - a_i$ is optimal. Now

$$n \geq a_{m-1} + a_m \geq a_i + a_m,$$

so $n - a_i \geq a_m$. Therefore, the greedy solution uses at least one $a_m$ coin. Thus there is an optimal solution for $n - a_i$ that uses an $a_m$ coin. If we add one $a_i$ coin to this optimal solution, we obtain an optimal solution for $n$ that uses an $a_m$ coin. The argument in the preceding paragraph can now be repeated to show that the greedy solution is optimal.

## Section 9.5 Review

**1.** A binary tree is a rooted tree in which each vertex has either no children, one child, or two children.

**2.** A left child of vertex $v$ is a child designated as "left."

**3.** A right child of vertex $v$ is a child designated as "right."

**4.** A full binary tree is a binary tree in which each vertex has either two children or zero children.

**5.** $i + 1$

**6.** $2i + 1$

**7.** If a binary tree of height $h$ has $t$ terminal vertices, then $\lg t \leq h$.

**8.** A binary search tree is a binary tree $T$ in which data are associated with the vertices. The data are arranged so that, for each vertex $v$ in $T$, each data item in the left subtree of $v$ is less than the data item in $v$, and each data item in the right subtree of $v$ is greater than the data item in $v$.

**9.** See Figures 9.5.4 and 9.5.5.

**10.** Insert the first data item in a vertex and label it the root. Insert the next data items in the tree according to the following steps. Begin at the root. If the data item to be added is less than the data item at the current vertex, move to the left child and repeat; otherwise, move to the right child and repeat. If there is no child, create one, put an edge incident on it and the last vertex visited, and store the data item in the added vertex.

## Section 9.5

**1.** Example 9.5.5 showed that $n-1$ games are played. Since there are two choices for the winner of each game, the tournament can unfold in $2^{n-1}$ ways.

**4.** No. Based on past performance, it is likely that certain teams will defeat other teams. Someone knowledgeable about basketball will take this into account. For example, through 2016 a number 16 seed has never defeated a number 1 seed.

**5.**

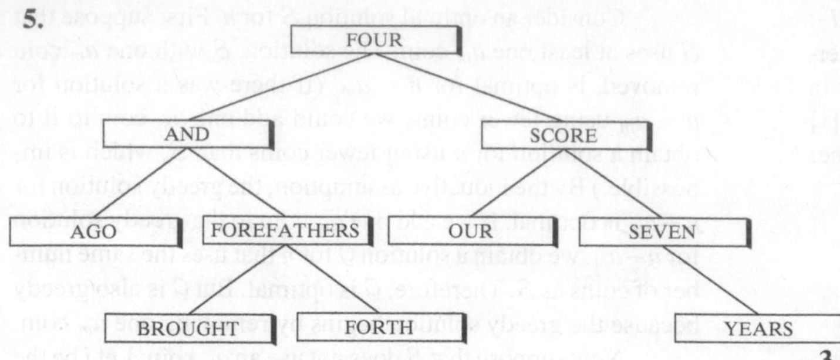


**8.** False. Consider

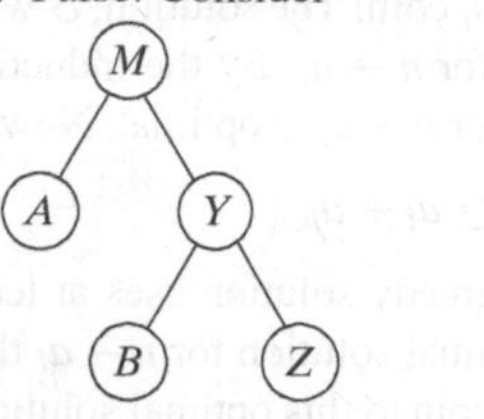


**9.**

**12.** $mi + 1$, $(m-1)i + 1$

**15.** $t - 1$

**18.** Balanced

**21.** Balanced

**22.** A tree of height 0 has one vertex, so $N_0 = 1$. In a balanced binary tree of height 1, the root must have at least one child. If the root has exactly one child, the number of vertices will be minimized. Therefore, $N_1 = 2$. In a balanced binary tree of height 2, there must be a path from the root to a terminal vertex of length 2. This accounts for three vertices. But for the tree to be balanced, the root must have two children. Therefore, $N_2 = 4$.

**25.** Suppose that there are $n$ vertices in a balanced binary tree of height $h$. Then

$$n \geq N_h = f_{h+3} - 1 > \left(\frac{3}{2}\right)^{h+2} - 1,$$

for $h \geq 3$. The equality comes from Exercise 24 and the last inequality comes from Exercise 27, Section 4.4. Therefore,

$$n + 1 > \left(\frac{3}{2}\right)^{h+2}.$$

Taking the logarithm to the base 3/2 of each side, we obtain

$$\log_{3/2}(n+1) > h + 2.$$

Therefore,

$$h < [\log_{3/2}(n+1)] - 2 = O(\lg n).$$

## Section 9.6 Review

**1.** Preorder traversal processes the vertices of a binary tree by beginning at the root and recursively processing the current vertex, the vertex's left subtree, and then the vertex's right subtree.

**2.** Input: *PT*, the root of a binary tree

Output: Dependent on how "process" is interpreted

```
preorder (PT) {
  if (PT == null)
    return
  process PT
  l = left child of PT
  preorder(l)
  r = right child of PT
  preorder (r)
}
```

**3.** Inorder traversal processes the vertices of a binary tree by beginning at the root and recursively processing the vertex's left subtree, the current vertex, and then the vertex's right subtree.

**4.** Input: *PT*, the root of a binary tree

Output: Dependent on how "process" is interpreted

```
inorder (PT) {
  if (PT == null)
    return
  l = left child of PT
  inorder(l)
  process PT
  r = right child of PT
  inorder (r)
}
```

**5.** Postorder traversal processes the vertices of a binary tree by beginning at the root and recursively processing the vertex's left subtree, the vertex's right subtree, and then the current vertex.

**6.** Input: *PT*, the root of a binary tree

Output: Dependent on how "process" is interpreted

```
postorder (PT) {
  if (PT == null)
    return
  l = left child of PT
  postorder (l)
  r = right child of PT
  postorder (r)
  process PT
}
```

**7.** In the prefix form of an expression, an operator precedes its operands.

**8.** Polish notation

**9.** In the infix form of an expression, an operator is between its operands.

**10.** In the postfix form of an expression, an operator follows its operands.

**11.** Reverse Polish notation

**12.** No parentheses are needed.

**13.** In a tree representation of an expression, the internal vertices represent operators, and the operators operate on the subtrees.

## Section 9.6

**1.** *preorder* *inorder* *postorder*
*ABDCE* *BDAEC* *DBECA*

**4.** *preorder* *inorder* *postorder*
*ABCDE* *EDCBA* *EDCBA*

**6.**

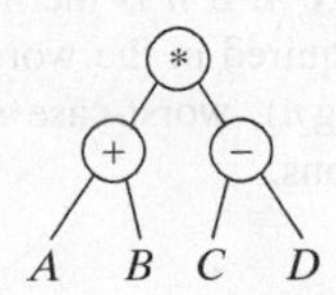

prefix: $* + AB - CD$
postfix: $AB + CD - *$

**9.**

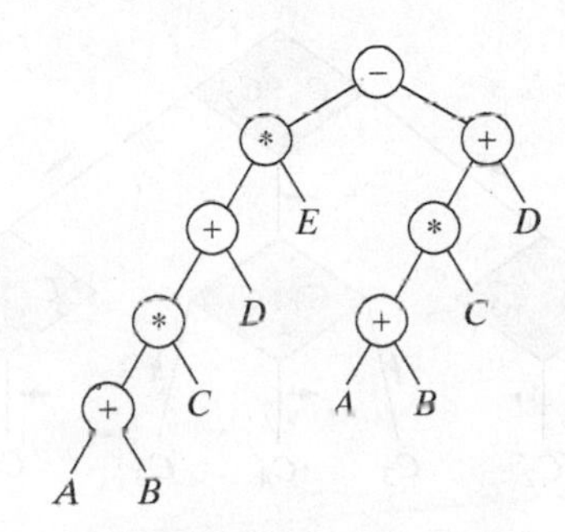

prefix: $- * + * +ABCDE + * + ABCD$
postfix: $AB + C * D + E * AB + C * D + -$

**11.**

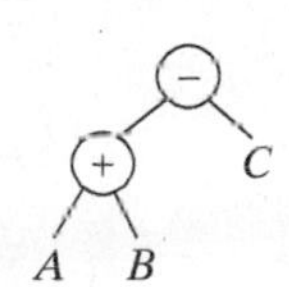

prefix: $- + ABC$
usual infix: $A + B - C$
parened infix: $((A + B) - C)$

**14.**

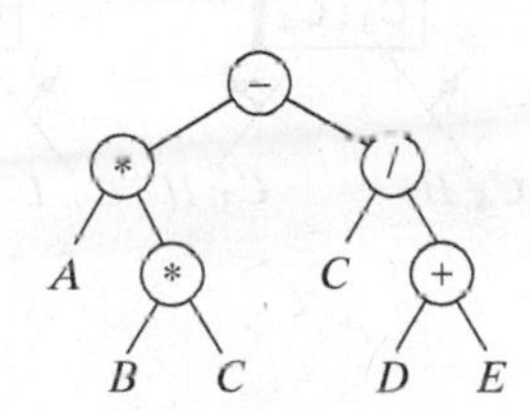

prefix: $- * A * BC/C + DE$
usual infix: $A * B * C - C/(D + E)$
parened infix: $((A * (B * C)) - (C/(D + E)))$

**16.** $-4$

**19.** 0

**22.**

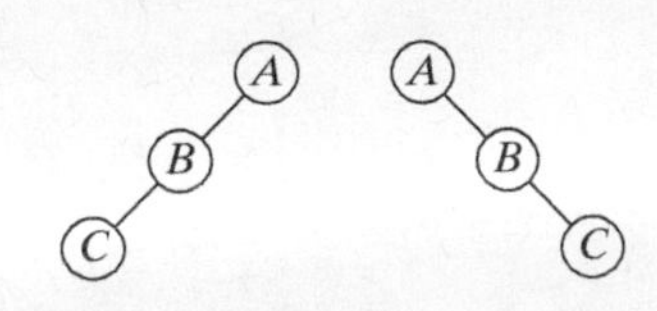

**25.**

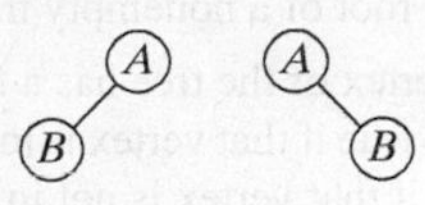

**28.** Input: *PT*, the root of a binary tree
Output: *PT*, the root of the modified binary tree

```
swap_children (PT) {
  if (PT == null)
    return
  swap the left and right children of PT
  l = left child of PT
  swap_children(l)
  r = right child of PT
  swap_children(r)
}
```

**31.** If $T$ is a binary tree, we let $post(T)$ denote the order in which the vertices of $T$ are visited under postorder traversal. We let $revpost(T)$ denote the reverse of $post(T)$. We prove by induction on the number of nodes in a tree $T$ that the order in which *funnyorder* visits the nodes of $T$ is $revpost(T)$.

The assertion is evident if $T$ has no nodes. Thus the basis step is proved.

Now assume that the order in which *funnyorder* visits the nodes of a tree $T'$ having fewer than $n$ nodes is $revpost(T')$. Let $T$ be an $n$-node tree. We must prove that the order in which *funnyorder* visits the nodes of $T$ is $revpost(T)$.

Let $T_1$ be the left subtree of $T$, let $T_2$ be the right subtree of $T$, and let $r$ be the root of $T$. By the inductive assumption, the order in which *funnyorder* visits the nodes of $T_1$ is $revpost(T_1)$, and the order in which *funnyorder* visits the nodes of $T_2$ is $revpost(T_2)$. The pseudocode shows that the order in which *funnyorder* visits the nodes of $T$ is

$$r, revpost(T_2), revpost(T_1).$$

The reverse of this list is

$$post(T_1), post(T_2), r,$$

which is the order in which postorder visits the nodes of $T$. The inductive step is complete.

**32.** Define an *initial segment* of a string to be the first $i \geq 1$ characters for some $i$. Define $r(x) = 1$, for $x = A, B, \ldots, Z$; and $r(x) = -1$, for $x = +, -, *, /$. If $x_1 \cdots x_n$ is a string over $\{A, \ldots, Z, +, -, *, /\}$, define

$$r(x_1 \cdots x_n) = r(x_1) + \cdots + r(x_n).$$

Then a string $s$ is a postfix string if and only if $r(s) = 1$ and $r(s') \geq 1$, for all initial segments $s'$ of $s$.

**35.** Let $G$ be the graph with vertex set $\{1, 2, \ldots, n\}$ and edge set

$$\{(1, i) \mid i = 2, \ldots n\}.$$

The $\{1\}$ is a vertex cover of $G$ of size 1.

**38.** Input: *PT*, the root of a nonempty tree

Output: Each vertex of the tree has a field *in_cover* that is set to true if that vertex is in the vertex cover or to false if that vertex is not in the vertex cover.

```
tree_cover(PT) {
    flag = false
    ptr = first child of PT
    while (ptr != null) {
        tree_cover(ptr)
        if (in_cover of ptr == false)
            flag = true
        ptr = next sibling of ptr
    }
    in_cover of PT = flag
}
```

## Section 9.7 Review

**1.** A decision tree is a binary tree in which the internal vertices contain questions with two possible answers, the edges are labeled with answers to the questions, and the terminal vertices represent decisions. If we begin at the root, answer each question, and follow the appropriate edges, we will eventually arrive at a terminal vertex that represents a decision.

**2.** The worst-case time of an algorithm is proportional to the height of the decision tree that represents the algorithm.

**3.** A decision tree that represents a sorting algorithm has $n!$ terminal vertices corresponding to the $n!$ possible permutations of input of size $n$. If $h$ is the height of the tree, then $h$ comparisons are required in the worst case. Since $\lg n! \leq h$ and $\lg n! = \Theta(n \lg n)$, worst-case sorting requires at least $\Omega(n \lg n)$ comparisons.

## Section 9.7

**1.**

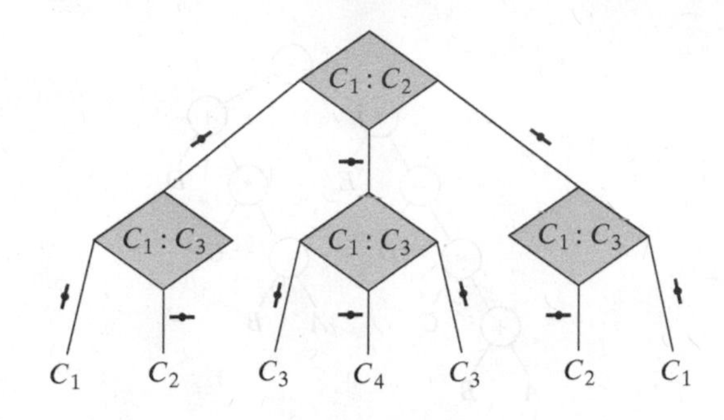


**4.** In this graph only, if the left pan is heavier, go right.

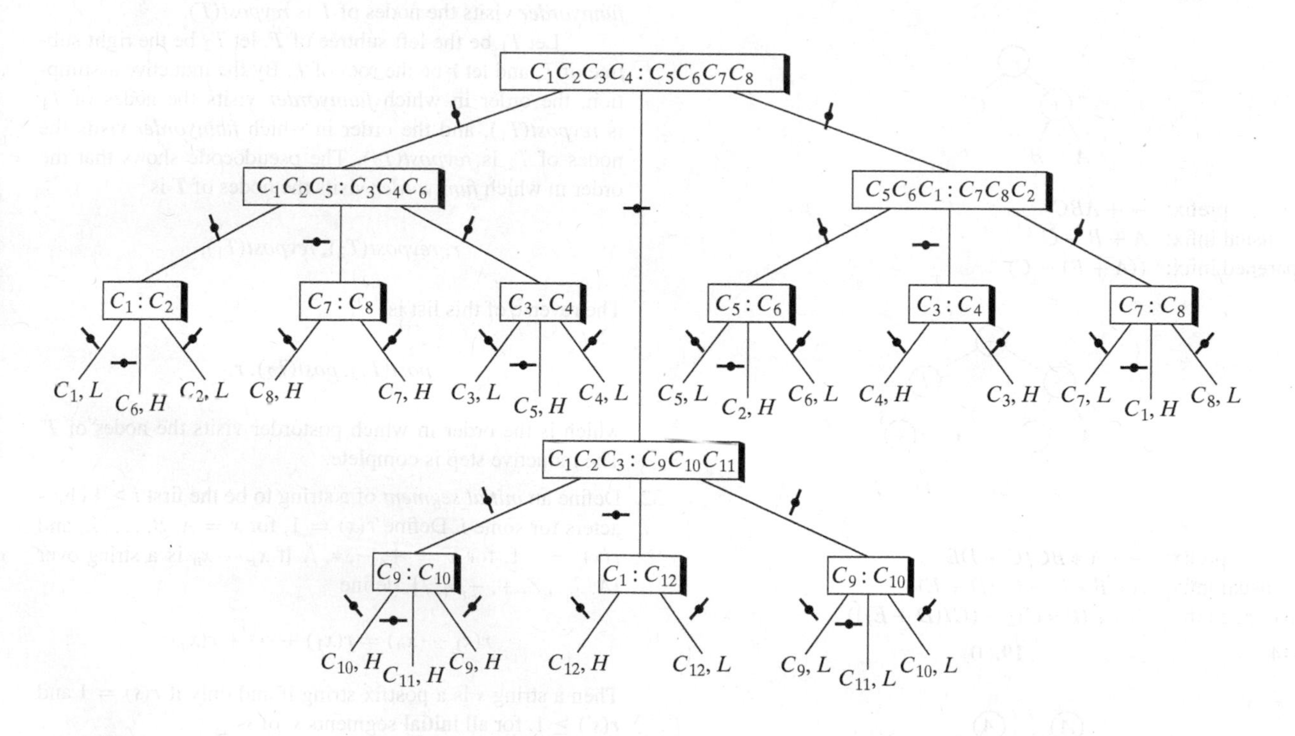

**7.** There are 28 possible outcomes to the fourteen-coins puzzle. A tree of height 3 has at most 27 terminal vertices; thus at least four weighings are required in the worst case. In fact, there is an algorithm that uses four weighings in the worst case: We begin by weighing four coins against four coins. If the coins do not balance, we proceed as in the solution given for Exercise 4 (for the twelve-coins puzzle). In this case, at most three weighings are required. If the coins do balance, we disregard these coins; our problem then is to find the bad coin from among the remaining six coins. The six-coins puzzle can be solved in at most three weighings in the worst case, which, together with the initial weighing, requires four weighings in the worst case.

**9.** Let $f(n)$ denote the number of weighings needed to solve the $n$-coin problem in the worst case. Let $T$ be the decision tree that represents this algorithm for input of size $n$, and let $h$ denote the height of $T$. Then the algorithm requires $h$ weighings in the worst case, so $h = f(n)$. Since there are $n - 1$ possible outcomes, $T$ has at least $n-1$ terminal vertices. By the analog of Theorem 9.5.6 for "trinary" trees, $\log_3(n - 1) \leq h = f(n)$.

**12.** The decision tree analysis shows that at least $\lceil \lg 5! \rceil = 7$ comparisons are required to sort five items in the worst case. The following algorithm sorts five items using at most seven comparisons in the worst case.

Given the sequence $a_1, \ldots, a_5$, we first sort $a_1, a_2$ (one comparison) and then $a_3, a_4$ (one comparison). (We assume now that $a_1 < a_2$ and $a_3 < a_4$.) We then compare $a_2$ and $a_4$. Let us assume that $a_2 < a_4$. (The case $a_2 > a_4$ is symmetric, and for this reason that part of the algorithm is omitted.) At this point we know that

$$a_1 < a_2 < a_4 \quad \text{and} \quad a_3 < a_4.$$

Next we determine where $a_5$ belongs among $a_1$, $a_2$, and $a_4$ by first comparing $a_5$ with $a_2$. If $a_5 < a_2$, we next compare $a_5$ with $a_1$; but if $a_5 > a_2$, we next compare $a_5$ with $a_4$. In either case, two additional comparisons are required. At this point, $a_1, a_2, a_4, a_5$ is sorted. Finally, we insert $a_3$ in its proper place. If we first compare $a_3$ with the second-smallest item among $a_1, a_2, a_4, a_5$, only one additional comparison will be required, for a total of seven comparisons. To justify this last statement, we note that the following arrangements are possible after we insert $a_5$ in its correct position:

$$a_5 < a_1 < a_2 < a_4$$
$$a_1 < a_5 < a_2 < a_4$$
$$a_1 < a_2 < a_5 < a_4$$
$$a_1 < a_2 < a_4 < a_5.$$

If $a_3$ is less than the second item, only one additional comparison is needed (with the first item) to locate the correct position for $a_3$. If $a_3$ is greater than the second item, at most one additional comparison is needed to locate the correct position for $a_3$. In the first three cases, we need only compare $a_3$ with either $a_2$ or $a_5$ to find the correct position for $a_3$ since we already know that $a_3 < a_4$. In the fourth case, if $a_3$ is greater than $a_2$, we know that it goes between $a_2$ and $a_4$.

**14.** We can consider the numbers as contestants and the internal vertices as winners where the larger value wins.

**17.** Suppose we have an algorithm that finds the largest value among $x_1, \ldots, x_n$. Let $x_1, \ldots, x_n$ be the vertices of a graph. An edge exists between $x_i$ and $x_j$ if the algorithm compares $x_i$ and $x_j$. The graph must be connected. The least number of edges necessary to connect $n$ vertices is $n - 1$.

**20.** By Exercise 16, tournament sort requires $2^k - 1$ comparisons to find the largest element. By Exercise 18, tournament sort requires $k$ comparisons to find the second largest element. Similarly, tournament sort requires at most $k$ comparisons to find the third-largest, at most $k$ comparisons to find the fourth-largest, and so on. Thus the total number of comparisons is at most

$$\begin{aligned}[2^k - 1] + (2^k - 1)k &\leq 2^k + k2^k \\ &\leq k2^k + k2^k \\ &= 2 \cdot 2^k k = 2n \lg n.\end{aligned}$$

## Section 9.8 Review

**1.** Free trees $T_1$ and $T_2$ are isomorphic if there is a one-to-one, onto function $f$ from the vertex set of $T_1$ to the vertex set of $T_2$ satisfying the following: Vertices $v_i$ and $v_j$ are adjacent in $T_1$ if and only if the vertices $f(v_i)$ and $f(v_j)$ are adjacent in $T_2$.

**2.** Let $T_1$ be a rooted tree with root $r_1$ and let $T_2$ be a rooted tree with root $r_2$. Then $T_1$ and $T_2$ are isomorphic if there is a one-to-one, onto function $f$ from the vertex set of $T_1$ to the vertex set of $T_2$ satisfying the following:

(a) $v_i$ and $v_j$ are adjacent in $T_1$ if and only if $f(v_i)$ and $f(v_j)$ are adjacent in $T_2$.

(b) $f(r_1) = f(r_2)$.

**3.** Let $T_1$ be a binary tree with root $r_1$ and let $T_2$ be a binary tree with root $r_2$. Then $T_1$ and $T_2$ are isomorphic if there is a one-to-one, onto function $f$ from the vertex set of $T_1$ to the vertex set of $T_2$ satisfying the following:

(a) $v_i$ and $v_j$ are adjacent in $T_1$ if and only if $f(v_i)$ and $f(v_j)$ are adjacent in $T_2$.

(b) $f(r_1) = f(r_2)$.

(c) $v$ is a left child of $w$ in $T_1$ if and only if $f(v)$ is a left child of $f(w)$ in $T_2$.

(d) $v$ is a right child of $w$ in $T_1$ if and only if $f(v)$ is a right child of $f(w)$ in $T_2$.

**4.** $C(2n, n)/(n + 1)$

**5.** Given binary trees $T_1$ and $T_2$, we first check whether either is empty (in which case it is immediate whether they are isomorphic). If both are nonempty, we first check whether the left subtrees are isomorphic and then whether the right subtrees are isomorphic. $T_1$ and $T_2$ are isomorphic if and only if their left and right subtrees are isomorphic.

## Section 9.8

**1.** Isomorphic. $f(v_1) = w_1, f(v_2) = w_5, f(v_3) = w_3, f(v_4) = w_4, f(v_5) = w_2, f(v_6) = w_6$.

**4.** Not isomorphic. $T_2$ has a simple path of length 2 from a vertex of degree 1 to a vertex of degree 1, but $T_1$ does not.

**7.** Isomorphic as rooted trees. $f(v_1)=w_1$, $f(v_2)=w_4$, $f(v_3)=w_3$, $f(v_4)=w_2$, $f(v_5)=w_6$, $f(v_6)=w_5$, $f(v_7)=w_7$, $f(v_8)=w_8$. Also isomorphic as free trees.

**10.** Not isomorphic as binary trees. The root of $T_1$ has a left child but the root of $T_2$ does not. Isomorphic as rooted trees and as free trees.

**13.**

**16.**

**19.**

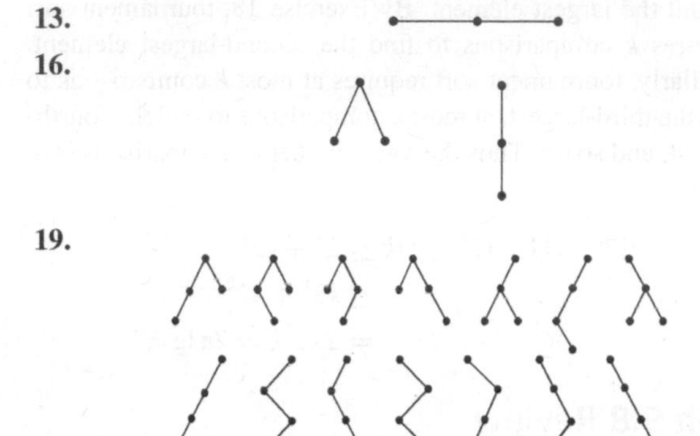

**22.** Let $b_n$ denote the number of nonisomorphic, $n$-vertex full binary trees. Since every full binary tree has an odd number of vertices, $b_n = 0$ if $n$ is even. We show that if $n = 2i + 1$ is odd, $b_n = C_i$, where $C_i$ denotes the $i$th Catalan number.

The last equation follows from the fact that there is a one-to-one, onto function from the set of $i$-vertex binary trees to the set of $(2i + 1)$-vertex full binary trees. Such a function may be constructed as follows. Given an $i$-vertex binary tree, at every terminal vertex we add two children. At every vertex with one child, we add an additional child. Since the tree that is obtained has $i$ internal vertices, there are $2i+1$ vertices total (Theorem 9.5.4). The tree constructed is a full binary tree. Notice that this function is one-to-one. Given a $(2i + 1)$-vertex full binary tree $T'$, if we eliminate all the terminal vertices, we obtain an $i$-vertex binary tree $T$. The image of $T$ is $T'$. Therefore, the function is onto.

**25.** There are four comparisons at lines 1 and 3. By Exercise 24, the call *bin_tree_isom*$(lc\_r_1, lc\_r_2)$ requires $6(k-1)+2$ comparisons. The call *bin_tree_isom*$(rc\_r_1, rc\_r_2)$ requires four comparisons. Thus the total number of comparisons is

$$4 + 6(k-1) + 2 + 4 = 6k + 4.$$

**27.** Let $T^*$ denote the tree constructed. Then $T^*$ is a full binary tree. Each vertex in $T$ becomes an internal vertex in $T^*$. Since we added only terminal vertices, the original $n - 1$ vertices in $T$ are the only internal vertices in $T^*$. By Theorem 9.5.4, $T^*$ has $n$ terminal vertices. Therefore $T^* \in X_1$. We leave it to the reader to check that this mapping is a bijection. By Theorem 9.8.12, there are $C_{n-1}$ $(n-1)$-vertex binary trees. Therefore $|X_1| = C_{n-1}$.

**29.** By Theorem 9.5.4, a tree in $X_1$ has $n - 1$ internal vertices and $2n - 1$ total vertices. Thus we may choose the vertex $v$ in $2n - 1$ ways and the vertex to mark (left or right) in 2 ways. Therefore $|X_T| = 2(2n - 1)$.

**33.** Using iteration, we have

$$\begin{aligned}
C_n &= \frac{2(2n-1)}{n+1} C_{n-1} \\
&= \frac{2(2n-1)}{n+1}\frac{2(2n-3)}{n} C_{n-2} \\
&= \frac{2^2(2n-1)(2n-3)}{(n+1)n} C_{n-2} \\
&= \frac{2^3(2n-1)(2n-3)(2n-5)}{(n+1)n(n-1)} C_{n-3} \\
&\;\;\vdots \\
&= \frac{2^{n-1}(2n-1)(2n-3)\cdots 3}{(n+1)n(n-1)\cdots 3} C_1 \\
&= \frac{1}{n+1}\left[\frac{2^n(2n-1)(2n-3)\cdots 3}{n(n-1)\cdots 3\cdot 2}\right] \\
&= \frac{1}{n+1}\left[\frac{2^n n!(2n-1)(2n-3)\cdots 3}{n!n!}\right] \\
&= \frac{1}{n+1}\left\{\frac{[(2n)(2n-2)\cdots 2][(2n-1)(2n-3)\cdots 3]}{n!n!}\right\} \\
&= \frac{1}{n+1}\frac{(2n)!}{n!n!} = \frac{1}{n+1}C(2n, n).
\end{aligned}$$

## Section 9.9 Review

**1.** In a game tree, each vertex shows a particular position in the game. In particular, the root shows the initial configuration of the game. The children of a vertex show all possible responses by a player to the position shown in the vertex.

**2.** In the minimax procedure, values are first assigned to the terminal vertices in a game tree. Then, working from the bottom up, the value of a circle is set to the minimum of the values of its children, and the value of a box is set to the maximum of the values of its children.

**3.** A search that terminates $n$ levels below the given vertex.

**4.** An evaluation function assigns to each possible game position the value of the position to the first player.

**5.** Alpha-beta pruning deletes (prunes) parts of the game tree and thus omits evaluating parts of it when the minimax procedure is applied. Alpha-beta pruning works as follows. Suppose that a box vertex $v$ is known to have a value of at least $x$. When a grandchild $w$ of $v$ has a value of at most $x$, the subtree whose root is the parent of $w$ is deleted. Similarly, suppose that a circle vertex $v$ is known to have a value of at most $x$. When a grandchild $w$ of $v$ has a value of at least $x$, the subtree whose root is the parent of $w$ is deleted.

**6.** An alpha value is a lower bound for a box vertex.

**7.** An alpha cutoff occurs at a box vertex when a grandchild $w$ of $v$ has a value less than or equal to the alpha value of $v$.

**8.** A beta value is an upper bound for a circle vertex.

**9.** A beta cutoff occurs at a circle vertex when a grandchild $w$ of $v$ has a value greater than or equal to the beta value of $v$.

## Section 9.9

1.

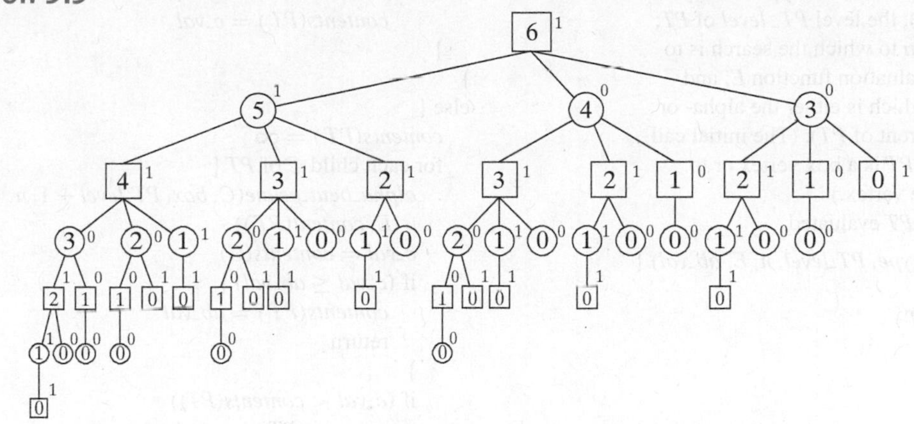

The first player always wins. The winning strategy is to first take one token; then, whatever the second player does, leave one token.

4. The second player always wins. If two piles remain, leave piles with equal numbers of tokens. If one pile remains, take it.

7. Suppose that the first player can win in nim. The first player can always win in nim′ by adopting the following strategy: Play nim′ exactly like nim unless the move would leave an odd number of singleton piles and no other pile. In this case, leave an even number of piles.

Suppose that the first player can always win in nim′. The first player can always win in nim by adopting the following strategy: Play nim exactly like nim′ unless the move would leave an even number of singleton piles and no other pile. In this case, leave an odd number of piles.

9.

12. The value of the root is 3.

14. (For Exercise 11)

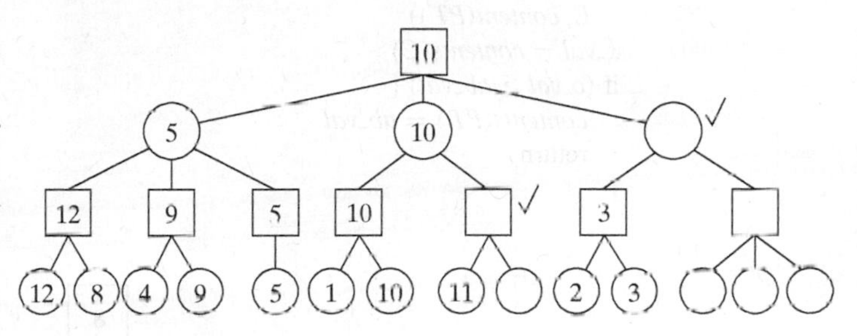


15. $3 - 2 = 1$

18. $4 - 1 = 3$

19.

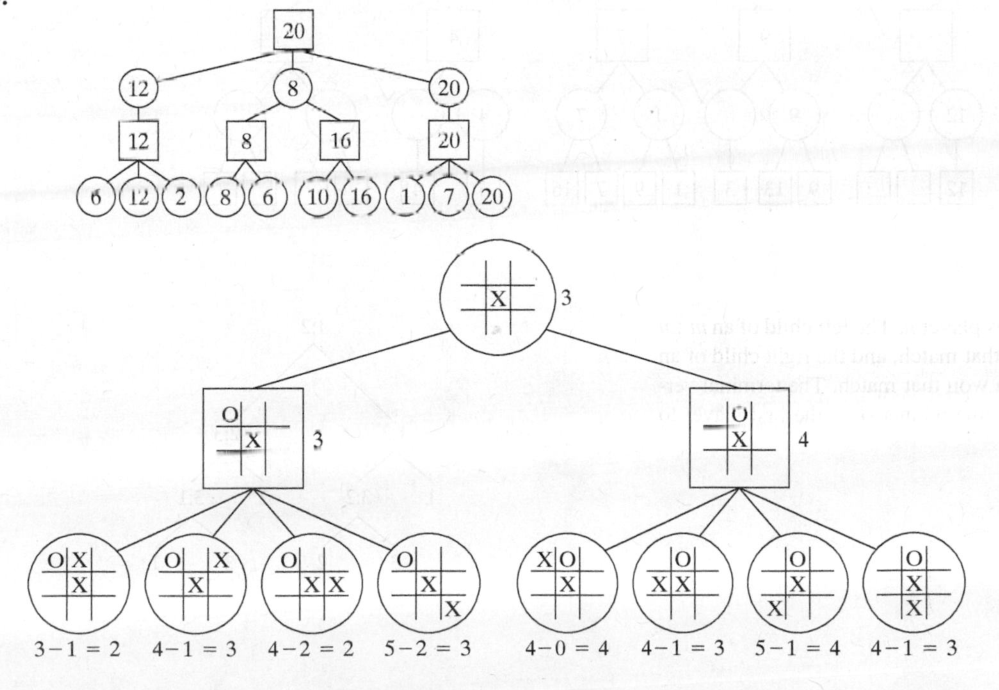


O will move to a corner.

**22.** Input: The root *PT* of a game tree; the type *PT_type* of *PT* (*box* or *circle*); the level *PT_level* of *PT*; the maximum level *n* to which the search is to be conducted; an evaluation function *E*; and a number *ab_val* (which is either the alpha- or beta-value of the parent of *PT*). (The initial call sets *ab_val* to $\infty$ if *PT* is a box vertex or to $-\infty$ if *PT* is a circle vertex.)

Output: The game tree with *PT* evaluated

```
alpha_beta_prune(PT, PT_type, PT_level, n, E, ab_val) {
  if (PT_level == n) {
    contents(PT) = E(PT)
    return
  }
  if (PT_type == box) {
    contents(PT) = −∞
    for each child C of PT {
      alpha_beta_prune(C, circle, PT_level + 1, n,
        E, content(PT))
      c_val = contents(C)
      if (c_val ≥ ab_val) {
        contents(PT) = ab_val
        return
      }
      if (c_val > contents(PT))
        contents(PT) = c_val
    }
  }
  else {
    contents(PT) = ∞
    for each child C of PT {
      alpha_beta_prune(C, box, PT_level + 1, n,
        E, content(PT))
      c_val = contents(C)
      if (c_val ≤ ab_val) {
        contents(PT) = ab_val
        return
      }
      if (c_val < contents(PT))
        contents(PT) = c_val
    }
  }
}
```

**23.** We first obtain the values 6, 6, 7 for the children of the root. Then we order the children of the root with the rightmost child first and use the alpha-beta procedure to obtain

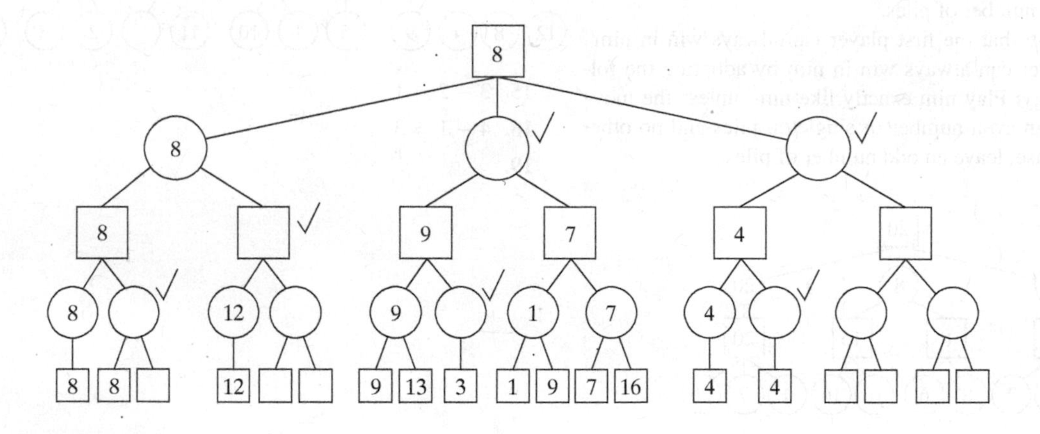

**27.** $m : n$ means player $m$ plays player $n$. The left child of an $m : n$ vertex means that $m$ won that match, and the right child of an $m : n$ vertex means that $n$ won that match. The terminal vertices show who was the game winner (i.e., the first player to score at least 2 points).

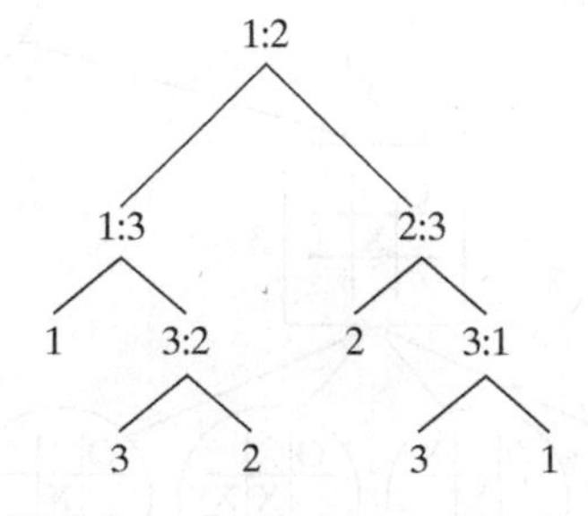

## Chapter 9 Self-Test

**1.** [Section 9.1]

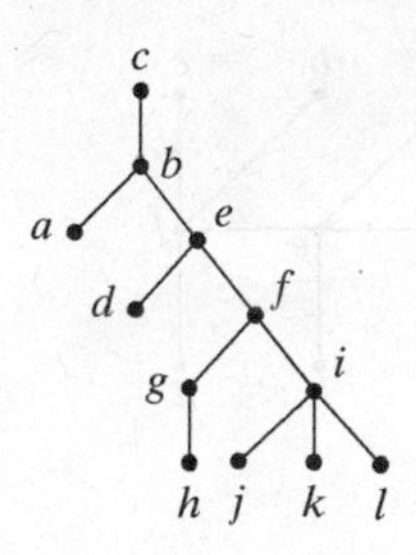


**2.** [Section 9.1] $a$-2, $b$-1, $c$-0, $d$-3, $e$-2, $f$-3, $g$-4, $h$-5, $i$-4, $j$-5, $k$-5, $l$-5

**3.** [Section 9.1] 5

**4.** [Section 9.2]

(a) $b$

(b) $a, c$

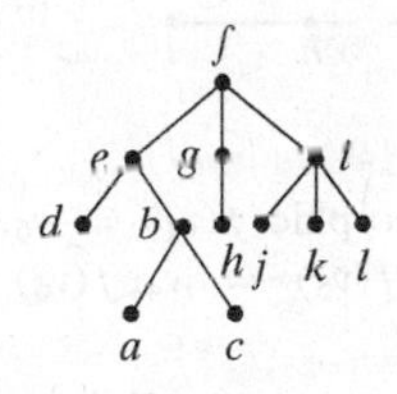


(c) $d, a, c, h, j, k, l$

(d)

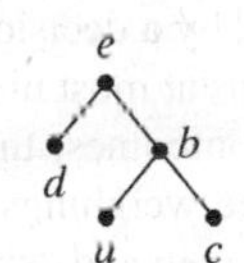


**5.** [Section 9.5]

**6.** [Section 9.5] 16

**7.** [Section 9.6] $ABFGCDE$

**8.** [Section 9.6] $BGFAEDC$

**9.** [Section 9.6] $GFBEDCA$

**10.** [Section 9.4]

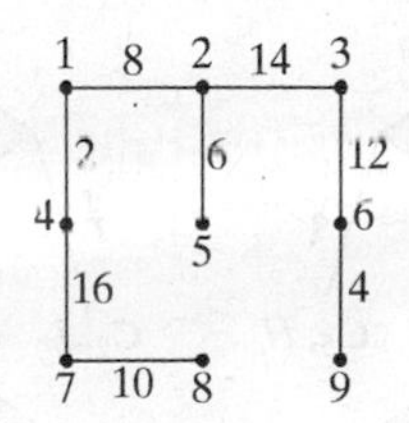


**11.** [Section 9.4] (1, 4), (1, 2), (2, 5), (2, 3), (3, 6), (6, 9), (4, 7), (7, 8)

**12.** [Section 9.4] (6, 9), (3, 6), (2, 3), (2, 5), (1, 2), (1, 4), (4, 7), (7, 8)

**13.** [Section 9.4] Consider a "shortest-path algorithm" in which at each step we select an available edge having minimum weight incident on the most recently added vertex (see the discussion preceding Theorem 9.4.5).

**14.** [Section 9.1]

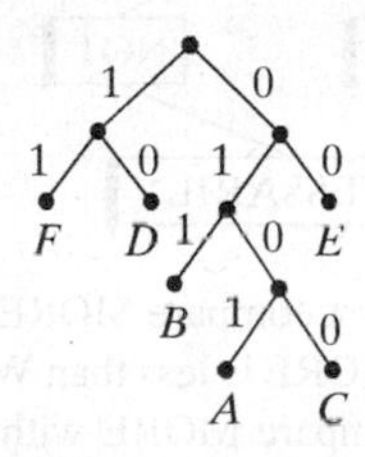


**15.** [Section 9.2] True. See Theorem 9.2.3.

**16.** [Section 9.2] True. A tree of height 6 or more must have seven or more vertices.

**17.** [Section 9.2] False.

**18.** [Section 9.6]

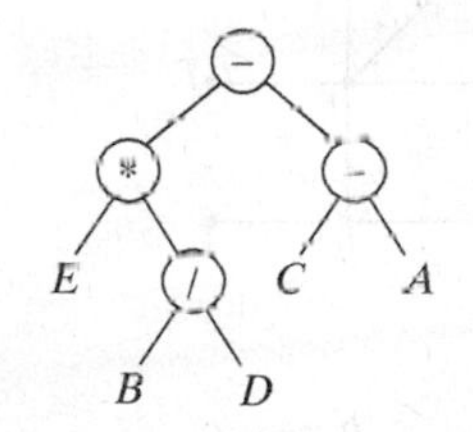


postfix: $EBD/*CA--$

parened infix: $((E*(B/D))-(C-A))$

**19.** [Section 9.8] True. If $f$ is an isomorphism of $T_1$ and $T_2$ as rooted trees, $f$ is also an isomorphism of $T_1$ and $T_2$ as free trees.

**20.** [Section 9.8] False.

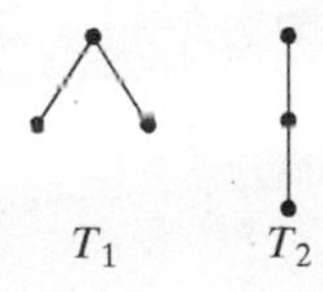

**21.** [Section 9.5]

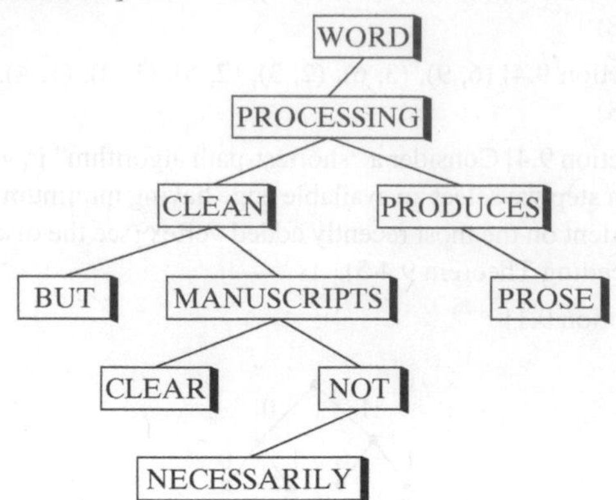


**22.** [Section 9.5] We first compare MORE with the word WORD in the root. Since MORE is less than WORD, we go to the left child. Next, we compare MORE with PROCESSING. Since MORE is less than PROCESSING, we go to the left child. Since MORE is greater than CLEAN, we go to the right child. Since MORE is greater than MANUSCRIPTS, we go to the right child. Since MORE is less than NOT, we go to the left child. Since MORE is less than NECESSARILY, we attempt to go to the left child. Since there is no left child, we conclude that MORE is not in the tree.

**23.** [Section 9.3]

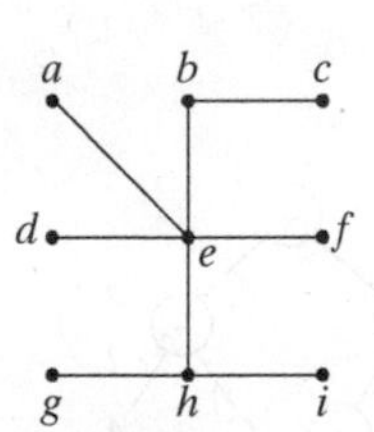


**24.** [Section 9.3]

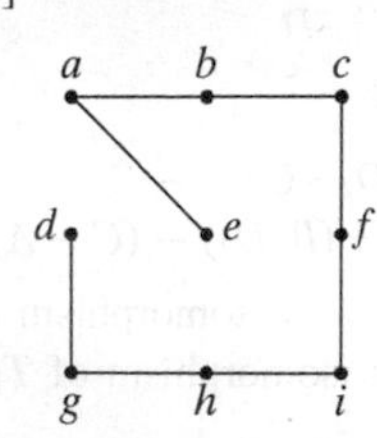


**25.** [Section 9.3]

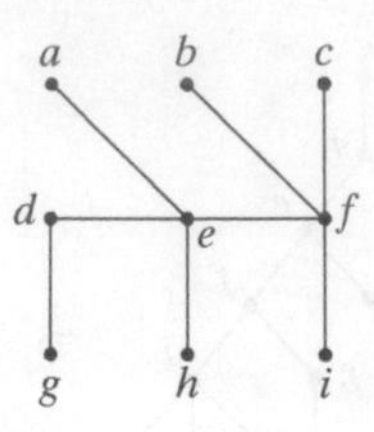


**26.** [Section 9.3]

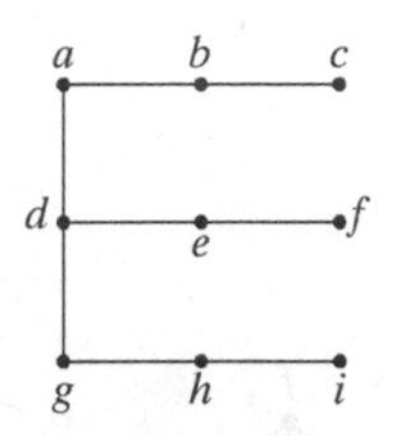


**27.** [Section 9.8] Isomorphic. $f(v_1) = w_6, f(v_2) = w_2, f(v_3) = w_5, f(v_4) = w_7, f(v_5) = w_4, f(v_6) = w_1, f(v_7) = w_3, f(v_8) = w_8$.

**28.** [Section 9.8] Not isomorphic. $T_1$ has a vertex ($v_3$) on level 1 of degree 3, but $T_2$ does not.

**29.** [Section 9.7] An algorithm that requires at most two weighings can be represented by a decision tree of height at most 2. However, such a tree has at most nine terminal vertices. Since there are 12 possible outcomes, there is no such algorithm. Therefore, at least three weighings are required in the worst case to identify the bad coin and determine whether it is heavy or light.

**30.** [Section 9.7]

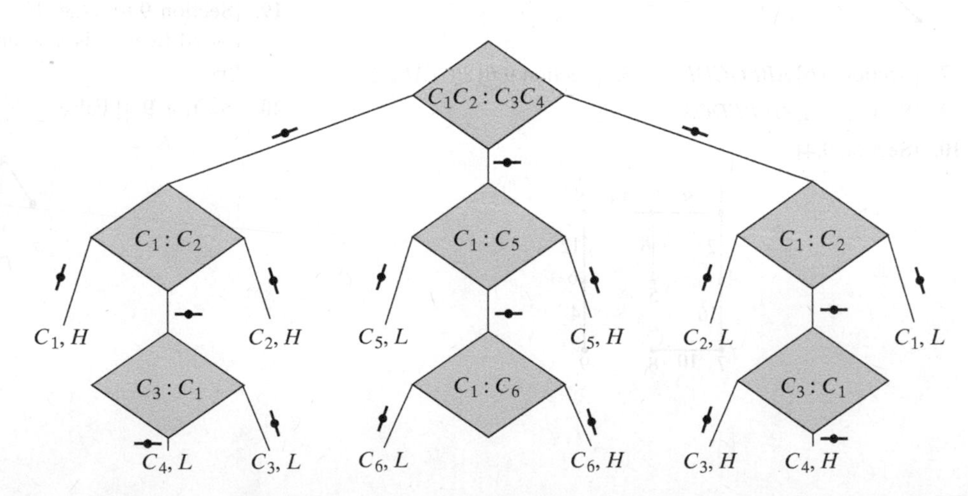

**31.** [Section 9.7] According to Theorem 9.7.3, any sorting algorithm requires at least $Cn \lg n$ comparisons in the worst case. Since Professor Sabic's algorithm uses at most $100n$ comparisons, we must have $Cn \lg n \le 100n$ for all $n \ge 1$. If we cancel $n$, we obtain $C \lg n \le 100$ for all $n \ge 1$, which is false. Therefore, the professor does not have a sorting algorithm that uses at most $100n$ comparisons in the worst case for all $n \ge 1$.

**32.** [Section 9.7] In the worst case, three comparisons are required to sort three items using an optimal sort (see Example 9.7.2).

If $n = 4$, binary insertion sort sorts three items (three comparisons—worst case) and then inserts the fourth item in the sorted three-item list (two comparisons—worst case) for a total of five comparisons in the worst case.

If $n = 5$, binary insertion sort sorts four items (five comparisons—worst case) and then inserts the fifth item in the sorted four-item list (three comparisons—worst case) for a total of eight comparisons in the worst case.

If $n = 6$, binary insertion sort sorts five items (eight comparisons—worst case) and then inserts the sixth item in the sorted five-item list (three comparisons—worst case) for a total of eleven comparisons in the worst case.

The decision tree analysis shows that any algorithm requires at least five comparisons in the worst case to sort four items. Thus binary insertion sort is optimal if $n = 4$.

The decision tree analysis shows that any algorithm requires at least seven comparisons in the worst case to sort five items. It is possible, in fact, to sort five items using seven comparisons in the worst case. Thus binary insertion sort is not optimal if $n = 5$.

The decision tree analysis shows that any algorithm requires at least ten comparisons in the worst case to sort six items. It is possible, in fact, to sort six items using ten comparisons in the worst case. Thus binary insertion sort is not optimal if $n = 6$.

**33.** [Section 9.9] $3 - 1 = 2$

**34.** [Section 9.9] Let each row, column, or diagonal that contains one X and two blanks count 1. Let each row, column, or diagonal that contains two X's and one blank count 5. Let each row, column, or diagonal that contains three X's count 100. Let each row, column, or diagonal that contains one O and two blanks count $-1$. Let each row, column, or diagonal that contains two O's and one blank count $-5$. Let each row, column, or diagonal that contains three O's count $-100$. Sum the values obtained.

**35.** [Section 9.9]

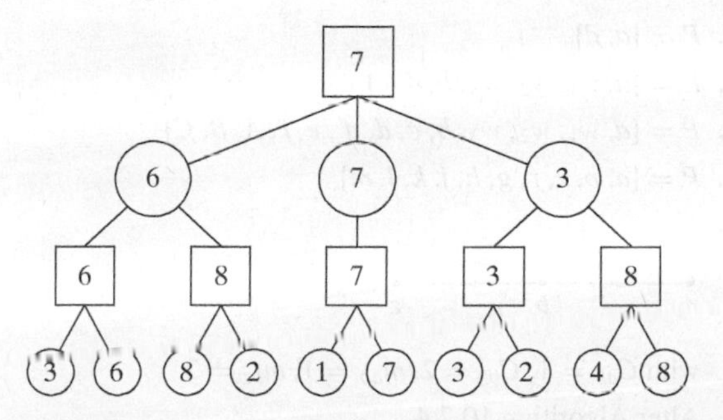


**36.** [Section 9.9]

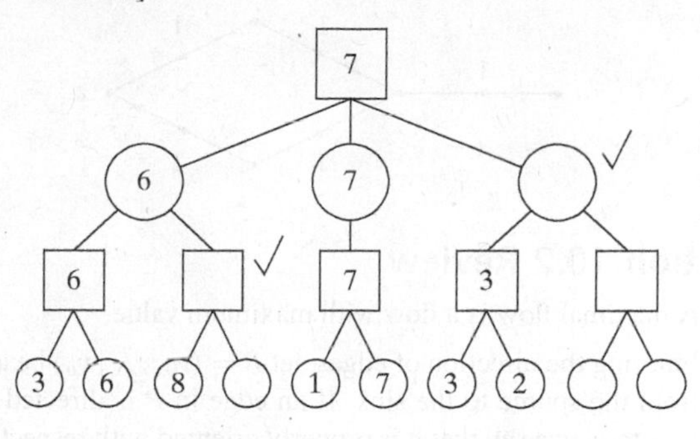


## Section 10.1 Review

1. A network is a simple, weighted, directed graph with a designated vertex having no incoming edges, a designated vertex having no outgoing edges, and nonnegative weights.
2. A source is a vertex with no incoming edges.
3. A sink is a vertex with no outgoing edges.
4. The weight of an edge is called its capacity.
5. A flow assigns each edge a nonnegative number that does not exceed the capacity of the edge such that for each vertex $v$, which is neither the source nor the sink, the flow into $v$ equals the flow out of $v$.
6. The flow in an edge is the nonnegative number assigned to it as in Exercise 5.
7. If $F_{ij}$ is the flow in edge $(i, j)$, the flow into vertex $j$ is $\sum_i F_{ij}$.
8. If $F_{ij}$ is the flow in edge $(i, j)$, the flow out of vertex $i$ is $\sum_j F_{ij}$.
9. Conservation of flow refers to the equality of the flow into and out of a vertex.
10. They are equal.
11. If a network has multiple sources, they can be tied together into a single vertex called the supersource.
12. If a network has multiple sinks, they can be tied together into a single vertex called the supersink.

## Section 10.1

1. $(b, c)$ is 6, 3; $(a, d)$ is 4, 2; $(c, e)$ is 6, 1; $(c, z)$ is 5, 2. The value of the flow is 5.

4. Add edges $(a, w_1)$, $(a, w_2)$, $(a, w_3)$, $(A, z)$, $(B, z)$, and $(C, z)$ each having capacity $\infty$.

7.

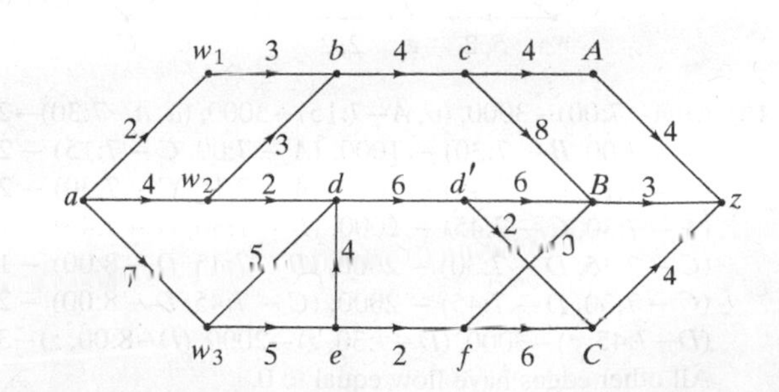

**10.**

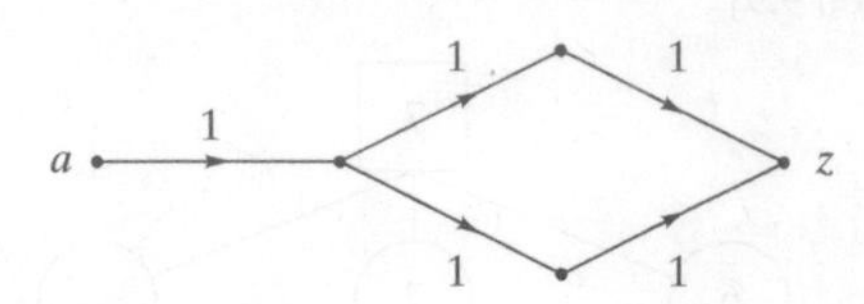


## Section 10.2 Review

1. A maximal flow is a flow with maximum value.
2. Ignoring the direction of edges, let $P = (v_0, \ldots, v_n)$ be a path from the source to the sink. If an edge in $P$ is directed from $v_{i-1}$ to $v_i$, we say that it is properly oriented with respect to $P$.
3. Ignoring the direction of edges, let $P = (v_0, \ldots, v_n)$ be a path from the source to the sink. If an edge in $P$ is directed from $v_i$ to $v_{i-1}$, we say that it is improperly oriented with respect to $P$.
4. We can increase the flow in a path when every properly oriented edge is under capacity and every improperly oriented edge has positive flow.
5. Let $\Delta$ be the minimum of the numbers $C_{ij} - F_{ij}$, for properly oriented edges $(i, j)$ in the path, and $F_{ij}$, for improperly oriented edges $(i, j)$ in the path. Then the flow can be increased by $\Delta$ by adding $\Delta$ to the flow in each properly oriented edge and by subtracting $\Delta$ from the flow in each improperly oriented edge.
6. Start with a flow (e.g., assign each edge flow zero). Search for a path as described in Exercise 4. Increase the flow in such a path as described in Exercise 5.

## Section 10.2

**1.** 1

**4.** $(a, w_1) - 6$, $(a, w_2) - 0$, $(a, w_3) - 3$, $(w_1, b) - 6$, $(w_2, b) - 0$, $(w_3, d) - 3$, $(d, c) - 3$, $(b, c) - 2$, $(b, A) - 4$, $(c, A) - 2$, $(c, B) - 3$, $(A, z) - 6$, $(B, z) - 3$

**7.**

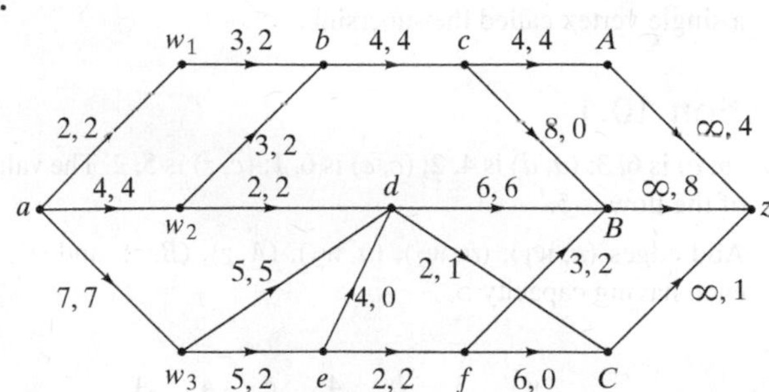


**10.** $(a, A-7{:}00)-3000$, $(a, A-7{:}15)-3000$, $(a, A-7{:}30)-2000$, $(A-7{:}00, B-7{:}30) - 1000$, $(A-7{:}00, C-7{:}15) - 2000$, $(A-7{:}15, B-7{:}45) - 1000$, $(A-7{:}15, C-7{:}30) - 2000$, $(A-7{:}30, C-7{:}45) - 2000$, $(B-7{:}30, D-7{:}45) - 1000$, $(C-7{:}15, D-7{:}30) - 2000$, $(B-7{:}45, D-8{:}00) - 1000$, $(C-7{:}30, D-7{:}45) - 2000$, $(C-7{:}45, D-8{:}00) - 2000$, $(D-7{:}45, z)-3000$, $(D-7{:}30, z)-2000$, $(D-8{:}00, z)-3000$. All other edges have flow equal to 0.

**13.**

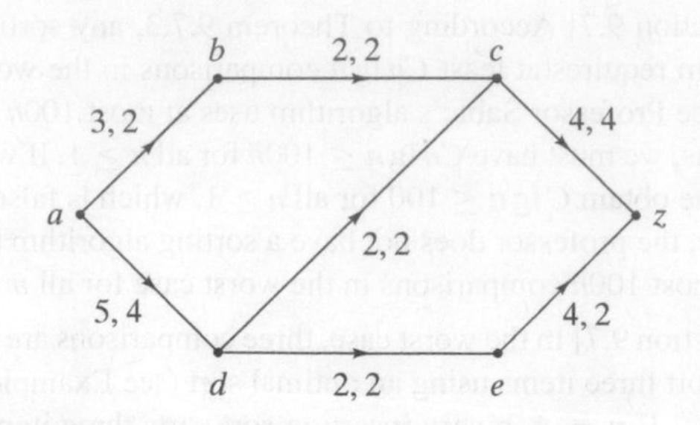


**16.** The maximum flow is 9.

**19.** Suppose that the sum of the capacities of the edges incident on $a$ is $U$. Each iteration of Algorithm 10.2.5 increases the flow by 1. Since the flow cannot exceed $U$, eventually the algorithm must terminate.

## Section 10.3 Review

1. A cut in a network consists of a set $P$ of vertices and the complement $\overline{P}$ of $P$, where the source is in $P$ and the sink is in $\overline{P}$.
2. The capacity of a cut $(P, \overline{P})$ is the number
$$\sum_{i \in P} \sum_{j \in \overline{P}} C_{ij}.$$
3. The capacity of any cut is greater than or equal to the value of any flow.
4. A minimal cut is a cut having minimum capacity.
5. If the value of a flow equals the capacity of a cut, then the flow is maximal and the cut is minimal. The value of a flow $F$ equals the capacity of a cut $(P, \overline{P})$ if and only if $F_{ij} = C_{ij}$ for all $i \in P$, $j \in \overline{P}$, and $F_{ij} = 0$ for all $i \in \overline{P}$, $j \in P$.
6. Let $P$ be the set of labeled vertices, and let $\overline{P}$ be the set of unlabeled vertices at the termination of Algorithm 10.2.4. It can be shown that the conditions
   - $F_{ij} = C_{ij}$ for all $i \in P$, $j \in \overline{P}$
   - $F_{ij} = 0$ for all $i \in \overline{P}$, $j \in P$

   of Exercise 5 hold. Thus the flow is maximal.

## Section 10.3

**1.** 8; minimal

**4.** $P = \{a, b, d\}$

**7.** $P = \{a, d\}$

**10.** $P = \{a, w_1, w_2, w_3, b, d, e\}$

**13.** $P = \{a, w_1, w_2, w_3, b, c, d, d', e, f, A, B, C\}$

**16.** $P = \{a, b, c, f, g, h, j, k, l, m\}$

**17.**

1, 1   2, 1

a — b — z

with $C_{ab} = 1$, $C_{bz} = 2$, $m_{ab} = 1$, $m_{bz} = 2$.

**20.** Alter Algorithm 10.2.4.

**23.** False. Consider the flow

$$a \xrightarrow{1,1} b \xrightarrow{2,1} z$$

and the cut $P = \{a, b\}$.

## Section 10.4 Review

*In the solutions to Exercises 1–5, G is a directed, bipartite graph with disjoint vertex sets V and W in which the edges are directed from V to W.*

1. A matching for $G$ is a set of edges with no vertices in common.
2. A maximal matching for $G$ is a matching containing the maximum number of edges.
3. A complete matching for $G$ is a matching $E$ having the property that if $v \in V$, then $(v, w) \in E$ for some $w \in W$.
4. Add a supersource $a$ and edges from $a$ to each vertex in $V$. Add a supersink $z$ and edges from each vertex in $W$ to $z$. Assign all edges capacity 1. We call the resulting network a matching network. Then, a flow in the matching network gives a matching in $G$ [$v$ is matched with $w$ if and only if the flow in edge $(v, w)$ is 1]; a maximal flow corresponds to a maximal matching; and a flow whose value is $|V|$ corresponds to a complete matching.
5. If $S \subseteq V$, let

$$R(S) = \{w \in W \mid v \in S \text{ and } (v, w) \text{ is an edge in } G\}.$$

Hall's Marriage Theorem states that there exists a complete matching in $G$ if and only if $|S| \leq |R(S)|$ for all $S \subseteq V$.

## Section 10.4

1. $P = \{a, A, B, D, J_2, J_5\}$
3. Finding qualified persons for jobs
6. Finding qualified persons for all jobs
9. All unlabeled edges are 1, 0. There is no complete matching.

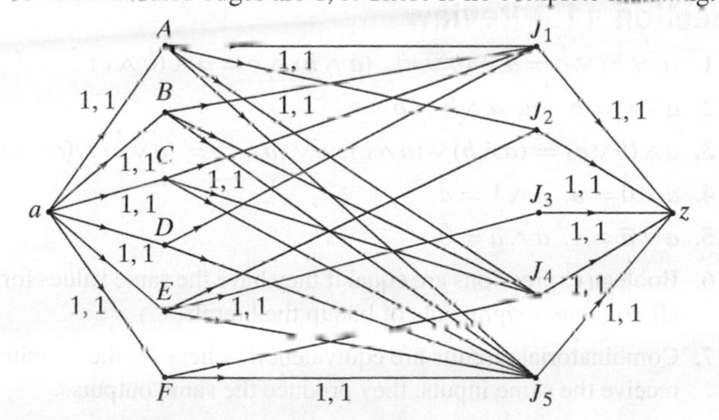


13. Each row and column has at most one label.
17. If $\delta(G) = 0$, then $|S| - |R(S)| \leq 0$, for all $S \subseteq V$. By Theorem 10.4.7, $G$ has a complete matching.

If $G$ has a complete matching, then $|S| - |R(S)| \leq 0$, for all $S \subseteq V$, so $\delta(G) \leq 0$. If $S = \varnothing$, $|S| - |R(S)| = 0$, so $\delta(G) = 0$.

## Chapter 10 Self-Test

1. [Section 10.1] In each edge, the flow is less than or equal to the capacity and, except for the source and sink, the flow into each vertex $v$ is equal to the flow out of $v$.
2. [Section 10.1] 3
3. [Section 10.1] 3
4. [Section 10.1] 3
5. [Section 10.2] $(a, b, e, f, g, z)$
6. [Section 10.2] Change the flows to $F_{a,b} = 2$, $F_{e,b} = 1$, $F_{e,f} = 1$, $F_{f,g} = 1$, $F_{g,z} = 1$.
7. [Section 10.2] $F_{a,b} = 3$, $F_{b,c} = 3$, $F_{c,d} = 4$, $F_{d,z} = 4$, $F_{a,e} = 2$, $F_{e,f} = 2$, $F_{f,c} = 2$, $F_{f,g} = 1$, $F_{g,z} = 1$, and all other edge flows zero.
8. [Section 10.2] $F_{a,b} = 0$, $F_{b,c} = 5$, $F_{c,d} = 5$, $F_{d,z} = 8$, $F_{e,b} = 3$, $F_{g,d} = 3$, $F_{a,e} = 8$, $F_{e,f} = 3$, $F_{f,g} = 3$, $F_{a,h} = 4$, $F_{e,i} = 2$, $F_{j,z} = 6$, $F_{h,i} = 4$, $F_{i,j} = 6$, and all other edge flows zero.
9. [Section 10.3] a—True, b—False, c—False, d—True
10. [Section 10.3] 6
11. [Section 10.3] No. The capacity of $(P, \overline{P})$ is 6, but the capacity of $(P', \overline{P'})$, $P' = \{a, b, c, e, f\}$, is 5.
12. [Section 10.3] $P = \{a, b, c, e, f, g, h, i\}$
13. [Section 10.4]

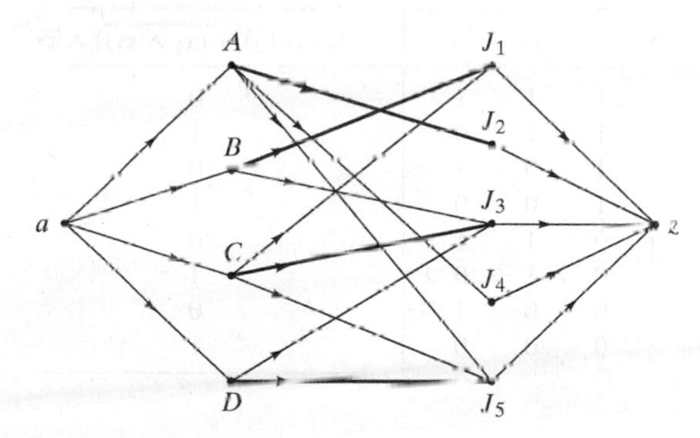


14. [Section 10.4] See the solution to Exercise 13.
15. [Section 10.4] $A - J_2$, $B - J_1$, $C - J_3$, $D - J_5$ is a complete matching.
16. [Section 10.4] $P = \{a\}$

## Section 11.1 Review

1. A combinatorial circuit is a circuit in which the output is uniquely defined for every combination of inputs.
2. A sequential circuit is a circuit in which the output is a function of the input and state of the system.
3. An AND gate receives input $x_1$ and $x_2$, where $x_1$ and $x_2$ are bits, and produces output 1 if $x_1$ and $x_2$ are both 1, and 0 otherwise.
4. An OR gate receives input $x_1$ and $x_2$, where $x_1$ and $x_2$ are bits, and produces output 0 if $x_1$ and $x_2$ are both 0, and 1 otherwise.

5. A NOT gate receives input $x$, where $x$ is a bit, and produces output 1 if $x$ is 0, and 0 if $x$ is 1.

6. An inverter is a NOT gate.

7. A logic table of a combinatorial circuit lists all possible inputs together with the resulting outputs.

8. Boolean expressions in the symbols $x_1, \ldots, x_n$ are defined recursively as follows. $0, 1, x_1, \ldots, x_n$ are Boolean expressions. If $X_1$ and $X_2$ are Boolean expressions, then $(X_1)$, $\overline{X_1}$, $X_1 \vee X_2$, and $X_1 \wedge X_2$ are Boolean expressions.

9. A literal is the symbol $x$ or $\overline{x}$ that appears in a Boolean expression.

## Section 11.1

1. $\overline{x_1 \wedge x_2}$

| $x_1$ | $x_2$ | $\overline{x_1 \wedge x_2}$ |
|---|---|---|
| 1 | 1 | 0 |
| 1 | 0 | 1 |
| 0 | 1 | 1 |
| 0 | 0 | 1 |

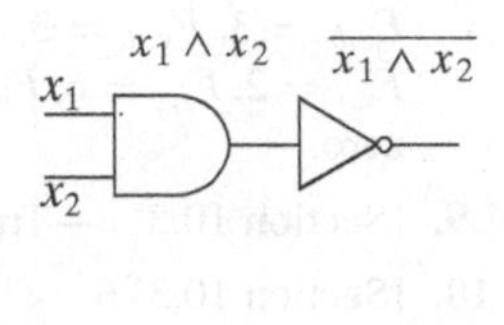


4.

| $x_1$ | $x_2$ | $x_3$ | $((x_1 \wedge x_2) \vee \overline{(x_1 \wedge x_3)}) \wedge \overline{x}_3$ |
|---|---|---|---|
| 1 | 1 | 1 | 0 |
| 1 | 1 | 0 | 1 |
| 1 | 0 | 1 | 0 |
| 1 | 0 | 0 | 1 |
| 0 | 1 | 1 | 0 |
| 0 | 1 | 0 | 1 |
| 0 | 0 | 1 | 0 |
| 0 | 0 | 0 | 1 |

7. If $x = 1$, the output $y$ is undetermined: Suppose that $x = 1$ and $y = 0$. Then the input to the AND gate is 1, 0. Thus the output of the AND gate is 0. Since this is then NOTed, $y = 1$. Contradiction. Similarly, if $x = 1$ and $y = 1$, we obtain a contradiction.

10. 0

13. 1

16. Is a Boolean expression. $x_1$, $x_2$, and $x_3$ are Boolean expressions by (11.1.2). $x_2 \vee x_3$ is a Boolean expression by (11.1.3c). $(x_2 \vee x_3)$ is a Boolean expression by (11.1.3a). $x_1 \wedge (x_2 \vee x_3)$ is a Boolean expression by (11.1.3d).

19. Not a Boolean expression

22.

A B

25. $(A \wedge B) \vee (C \wedge \overline{A})$

| $A$ | $B$ | $C$ | $(A \wedge B) \vee (C \wedge \overline{A})$ |
|---|---|---|---|
| 1 | 1 | 1 | 1 |
| 1 | 1 | 0 | 1 |
| 1 | 0 | 1 | 0 |
| 1 | 0 | 0 | 0 |
| 0 | 1 | 1 | 1 |
| 0 | 1 | 0 | 0 |
| 0 | 0 | 1 | 1 |
| 0 | 0 | 0 | 0 |

27. $(A \wedge (C \vee (D \wedge C))) \vee (B \wedge (\overline{D} \vee (C \wedge A) \vee \overline{C}))$

29.

| $A$ | $B$ | $(A \vee \overline{B}) \wedge A$ |
|---|---|---|
| 1 | 1 | 1 |
| 1 | 0 | 1 |
| 0 | 1 | 0 |
| 0 | 0 | 0 |

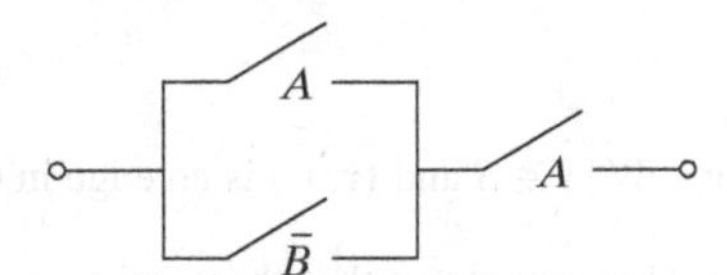


32.

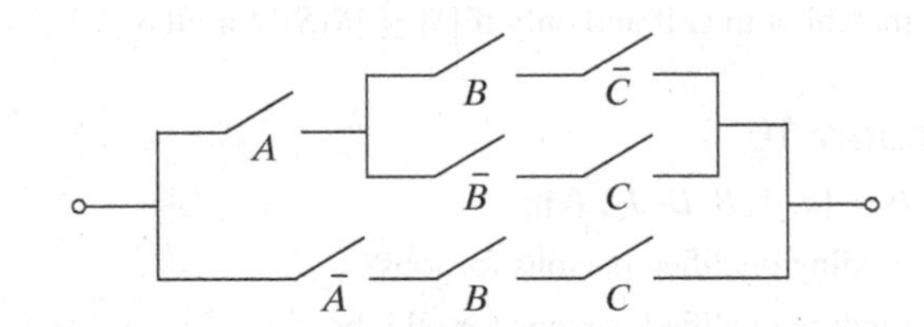


## Section 11.2 Review

1. $(a \vee b) \vee c = a \vee (b \vee c)$, $(a \wedge b) \wedge c = a \wedge (b \wedge c)$

2. $a \vee b = b \vee a$, $a \wedge b = b \wedge a$

3. $a \wedge (b \vee c) = (a \wedge b) \vee (a \wedge c)$, $a \vee (b \wedge c) = (a \vee b) \wedge (a \vee c)$

4. $a \vee 0 = a$, $a \wedge 1 = a$

5. $a \vee \overline{a} = 1$, $a \wedge \overline{a} = 0$

6. Boolean expressions are equal if they have the same values for all possible assignments of bits to the literals.

7. Combinatorial circuits are equivalent if, whenever the circuits receive the same inputs, they produce the same outputs.

8. Let $C_1$ and $C_2$ be combinatorial circuits represented, respectively, by the Boolean expressions $X_1$ and $X_2$. Then $C_1$ and $C_2$ are equivalent if and only if $X_1 = X_2$.

## Section 11.2

**1.**

| $x_1$ | $x_2$ | $\overline{x_1 \wedge x_2}$ | $\overline{x}_1 \vee \overline{x}_2$ |
|---|---|---|---|
| 1 | 1 | 0 | 0 |
| 1 | 0 | 1 | 1 |
| 0 | 1 | 1 | 1 |
| 0 | 0 | 1 | 1 |

**4.**

| $x_1$ | $x_2$ | $x_3$ | $\overline{x}_1 \vee (\overline{x}_2 \vee x_3)$ | $\overline{(x_1 \wedge x_2)} \vee x_3$ |
|---|---|---|---|---|
| 1 | 1 | 1 | 1 | 1 |
| 1 | 1 | 0 | 0 | 0 |
| 1 | 0 | 1 | 1 | 1 |
| 1 | 0 | 0 | 1 | 1 |
| 0 | 1 | 1 | 1 | 1 |
| 0 | 1 | 0 | 1 | 1 |
| 0 | 0 | 1 | 1 | 1 |
| 0 | 0 | 0 | 1 | 1 |

**6.**

| $x_1$ | $x_1 \vee x_1$ |
|---|---|
| 1 | 1 |
| 0 | 0 |

**9.**

| $x_1$ | $x_2$ | $x_3$ | $x_1 \wedge \overline{(x_2 \wedge x_3)}$ | $(x_1 \wedge \overline{x}_2) \vee (x_1 \wedge \overline{x}_3)$ |
|---|---|---|---|---|
| 1 | 1 | 1 | 0 | 0 |
| 1 | 1 | 0 | 1 | 1 |
| 1 | 0 | 1 | 1 | 1 |
| 1 | 0 | 0 | 1 | 1 |
| 0 | 1 | 1 | 0 | 0 |
| 0 | 1 | 0 | 0 | 0 |
| 0 | 0 | 1 | 0 | 0 |
| 0 | 0 | 0 | 0 | 0 |

**11.**

| $x$ | $\overline{\overline{x}}$ |
|---|---|
| 1 | 1 |
| 0 | 0 |

**14.** False. Take $x_1 = 1, x_2 = 1, x_3 = 0$.

**16.**

| $a$ | $b$ | $c$ | $a \vee (b \wedge c)$ | $(a \vee b) \wedge (a \vee c)$ |
|---|---|---|---|---|
| 1 | 1 | 1 | 1 | 1 |
| 1 | 1 | 0 | 1 | 1 |
| 1 | 0 | 1 | 1 | 1 |
| 1 | 0 | 0 | 1 | 1 |
| 0 | 1 | 1 | 1 | 1 |
| 0 | 1 | 0 | 0 | 0 |
| 0 | 0 | 1 | 0 | 0 |
| 0 | 0 | 0 | 0 | 0 |

**18.** The Boolean expressions that represent the circuits are $(A \wedge \overline{B}) \vee (A \wedge C)$ and $A \wedge (\overline{B} \vee C)$. The expressions are equal by Theorem 11.2.1(c). Therefore, the switching circuits are equivalent.

**21.**

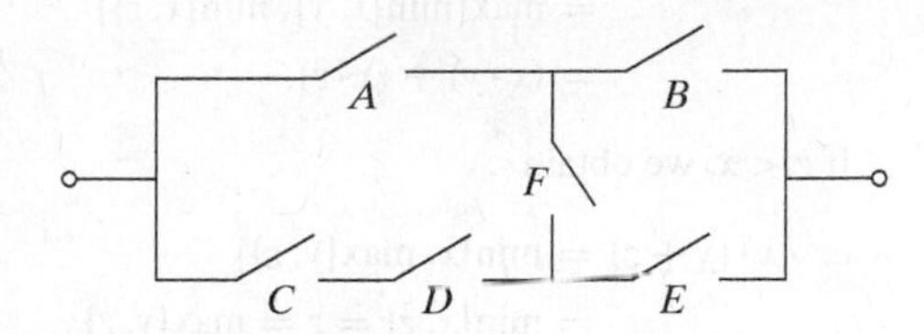


## Section 11.3 Review

**1.** A Boolean algebra consists of a set $S$ containing distinct elements 0 and 1, binary operators $+$ and $\cdot$, and a unary operator $'$ on $S$ satisfying the associative, commutative, distributive, identity, and complement laws.

**2.** $x + x = x, xx = x$

**3.** $x + 1 = 1, x0 = 0$

**4.** $x + xy = x, x(x + y) = x$

**5.** $(x')' = x$

**6.** $0' = 1, 1' = 0$

**7.** $(x + y)' = x'y', (xy)' = x' + y'$

**8.** The dual of a Boolean expression is obtained by replacing 0 by 1, 1 by 0, $+$ by $\cdot$, and $\cdot$ by $+$.

**9.** The dual of a theorem about Boolean algebras is also a theorem.

## Section 11.3

**2.** One can show that the associative and distributive laws hold for lcm and gcd directly. The commutative law clearly holds. To see that the identity laws hold, note that

$$\text{lcm}(x, 1) = x \quad \text{and} \quad \gcd(x, 6) = x.$$

Since

$$\text{lcm}(x, 6/x) = 6 \quad \text{and} \quad \gcd(x, 6/x) = 1,$$

the complement laws hold. Therefore, $(S, +, \cdot, ', 1, 6)$ is a Boolean algebra.

**4.** We show only

$$x \cdot (x + z) = (x \cdot y) + (x \cdot z) \quad \text{for all } x, y, z \in S_n.$$

Now

$$x \cdot (y + z) = \min\{x, \max\{y, z\}\}$$
$$(x \cdot y) + (x \cdot z) = \max\{\min\{x, y\}, \min\{x, z\}\}.$$

We assume that $y \le z$. (The argument is similar if $y > z$.) There are three cases to consider: $x < y$; $y \le x \le z$; and $z < x$.

If $x < y$, we obtain

$$\begin{aligned} x \cdot (y + z) &= \min\{x, \max\{y, z\}\} \\ &= \min\{x, z\} = x = \max\{x, x\} \\ &= \max\{\min\{x, y\}, \min\{x, z\}\} \\ &= (x \cdot y) + (x \cdot z). \end{aligned}$$

If $y \le x \le z$, we obtain

$$\begin{aligned} x \cdot (y+z) &= \min\{x, \max\{y, z\}\} \\ &= \min\{x, z\} = x = \max\{y, x\} \\ &= \max\{\min\{x, y\}, \min\{x, z\}\} \\ &= (x \cdot y) + (x \cdot z). \end{aligned}$$

If $z < x$, we obtain

$$\begin{aligned} x \cdot (y+z) &= \min\{x, \max\{y, z\}\} \\ &= \min\{x, z\} = z = \max\{y, z\} \\ &= \max\{\min\{x, y\}, \min\{x, z\}\} \\ &= (x \cdot y) + (x \cdot z). \end{aligned}$$

**7.** If $X \cup Y = U$ and $X \cap Y = \varnothing$, then $Y = \overline{X}$.

**8.** $xy + x0 = x(x + y)y$

**11.** $x + y' = 1$ if and only if $x + y = x$.

**14.** $x(x + y0) = x$

**15.** (For Exercise 13)

$$\begin{aligned} 0 = x + y &= (x + x) + y \\ &= x + (x + y) = x + 0 = x \end{aligned}$$

Similarly, $y = 0$.

**18.** [For part (c)]

$$\begin{aligned} x(x + y) &= (x + 0)(x + y) \\ &= x + 0y = x + y0 = x + 0 = x \end{aligned}$$

**21.** First, show that if $ba = ca$ and $ba' = ca'$, then $b = c$. Now take $a = x$, $b = x + (y + z)$, and $c = (x + y) + z$ and use this result.

**23.** If the prime $p$ divides $n$, $p^2$ does not divide $n$.

## Section 11.4 Review

**1.** The exclusive-OR of $x_1$ and $x_2$ is 0 if $x_1 = x_2$, and 1 otherwise.

**2.** A Boolean function is a function of the form

$$f(x_1, \ldots, x_n) = X(x_1, \ldots, x_n),$$

where $X$ is a Boolean expression.

**3.** A minterm is a Boolean expression of the form

$$y_1 \wedge y_2 \wedge \cdots \wedge y_n,$$

where each $y_i$ is either $x_i$ or $\overline{x_i}$.

**4.** The disjunctive normal form of a not identically zero Boolean function $f$ is

$$f(x_1, \ldots, x_n) = m_1 \vee m_2 \vee \cdots \vee m_k,$$

where each $m_i$ is a minterm.

**5.** Let $A_1, \ldots, A_k$ denote the elements $A_i$ of $Z_2^n$ for which $f(A_i) = 1$. For each $A_i = (a_1, \ldots, a_n)$, set $m_i = y_1 \wedge \cdots \wedge y_n$, where $y_j = x_j$ if $a_j = 1$, and $y_j = \overline{x_j}$ if $a_j = 0$. Then

$$f(x_1, \ldots, x_n) = m_1 \vee m_2 \vee \cdots \vee m_k.$$

**6.** A maxterm is a Boolean expression of the form

$$y_1 \vee y_2 \vee \cdots \vee y_n,$$

where each $y_i$ is either $x_i$ or $\overline{x_i}$.

**7.** The conjunctive normal form of a not identically one Boolean function $f$ is

$$f(x_1, \ldots, x_n) = m_1 \wedge m_2 \wedge \cdots \wedge m_k,$$

where each $m_i$ is a maxterm.

## Section 11.4

*In these hints, $a \wedge b$ is written $ab$.*

**1.** $xy \vee \overline{x}y \vee \overline{x}\,\overline{y}$

**4.** $xyz \vee xy\overline{z} \vee x\overline{y}\,\overline{z} \vee \overline{x}yz \vee \overline{x}y\overline{z}$

**7.** $xyz \vee x\overline{y}\,\overline{z} \vee \overline{x}\,\overline{y}\,\overline{z}$

**10.** $wx\overline{y}z \vee wx\overline{y}\,\overline{z} \vee w\overline{x}yz \vee w\overline{x}y\overline{z} \vee w\overline{x}\,\overline{y}\,\overline{z}$
$\vee \overline{w}xy\overline{z} \vee \overline{w}x\overline{y}z \vee \overline{w}x\overline{y}\,\overline{z} \vee \overline{w}\,\overline{x}y\overline{z} \vee \overline{w}\,\overline{x}\,\overline{y}\,\overline{z}$

**11.** $xy \vee x\overline{y}$

**14.** $xy\overline{z}$

**17.** $xyz \vee \overline{x}yz \vee xy\overline{z} \vee \overline{x}y\overline{z}$

**20.** 0

**22.** $2^{2^n}$

**25.** (For Exercise 3)

$$(\overline{x} \vee y \vee \overline{z})(x \vee \overline{y} \vee \overline{z})(x \vee \overline{y} \vee z)$$

**28.** (For Exercise 3)

$$(\overline{x} \vee \overline{y} \vee \overline{z})(\overline{x} \vee \overline{y} \vee z)(\overline{x} \vee y \vee z)(x \vee y \vee \overline{z})(x \vee y \vee z)$$

## Section 11.5 Review

**1.** A gate is a function from $Z_2^n$ into $Z_2$.

**2.** A set of gates $G$ is functionally complete if, given any positive integer $n$ and a function $f$ from $Z_2^n$ into $Z_2$, it is possible to construct a combinatorial circuit that computes $f$ using only the gates in $G$.

**3.** {AND, OR, NOT}

**4.** A NAND gate receives input $x_1$ and $x_2$, where $x_1$ and $x_2$ are bits, and produces output 0 if $x_1$ and $x_2$ are both 1, and 1 otherwise.

**5.** Yes

**6.** The problem of finding the best circuit

**7.** Small components that are themselves entire circuits

**8.** See Figure 11.5.8.

**9.** See Figure 11.5.9.

## Section 11.5

**1.** AND can be expressed in terms of OR and NOT: $xy = \overline{\overline{x} \vee \overline{y}}$.

**2.** A combinatorial circuit consisting only of AND gates would always output 0 when all inputs are 0.

**5.** We use induction on $n$ to show that there is no $n$-gate combinatorial circuit consisting of only AND and OR gates that computes $f(x) = \overline{x}$.

If $n = 0$, the input $x$ equals the output $x$, and so it is impossible for a 0-gate circuit to compute $f$. The Basis Step is proved.

Suppose that there is no $n$-gate combinatorial circuit consisting of only AND and OR gates that computes $f$. Consider an $(n+1)$-gate combinatorial circuit consisting of only AND and OR gates. The input $x$ first arrives at either an AND or an OR gate. Suppose that $x$ first arrives at an AND gate. (The argument is similar if $x$ first arrives at an OR gate and is omitted.) Because the circuit is a combinatorial circuit, the other input to the AND gate is either $x$ itself, the constant 1, or the constant 0. If both inputs to the AND gate are $x$ itself, then the output of the AND gate is equal to the input. In this case, the behavior of the circuit is unchanged if we remove the AND gate and connect $x$ to what was the output line of the AND gate. But we now have an equivalent $n$-gate circuit, which, by the inductive hypothesis, cannot compute $f$. Thus the $(n+1)$-gate circuit cannot compute $f$.

If the other input to the AND gate is the constant 1, the output of the AND gate is again equal to the input and we can argue as in the previous case that the $(n+1)$-gate circuit cannot compute $f$.

If the other input to the AND gate is the constant 0, the AND gate always outputs 0 and, so, changing the value of $x$ does not affect the output of the circuit. In this case, the circuit cannot compute $f$. The Inductive Step is complete. Therefore, no $n$-gate combinatorial circuit consisting of only AND and OR gates can compute $f(x) = \bar{x}$. Thus {AND, OR} is not functionally complete.

**6.** 

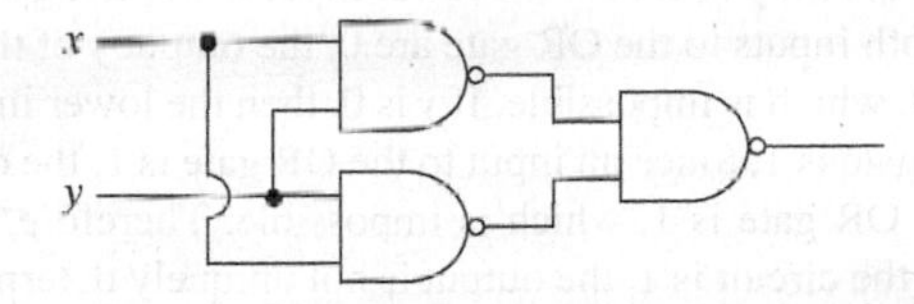


**9.** $y_1 = x_1x_2 \vee \overline{(x_2 \vee x_3)}$; $y_2 = \overline{x_2 \vee x_3}$

**12.** (For Exercise 3) The dnf may be simplified to $xy \vee x\bar{z} \vee \bar{x}\,\bar{y}$ and then rewritten as $x(y \vee \bar{z}) \vee \bar{x}\,\bar{y} = (x\overline{\bar{y}z}) \vee \bar{x}\,\bar{y} = \overline{\overline{x\overline{\bar{y}z}}\;\overline{\bar{x}\,\bar{y}}}$, which gives the circuit

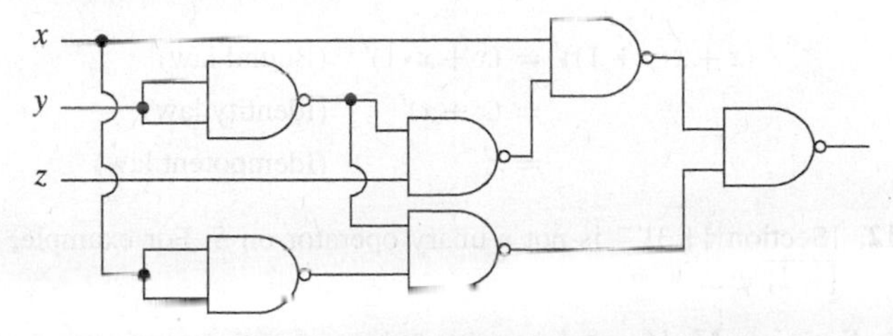


**15.**

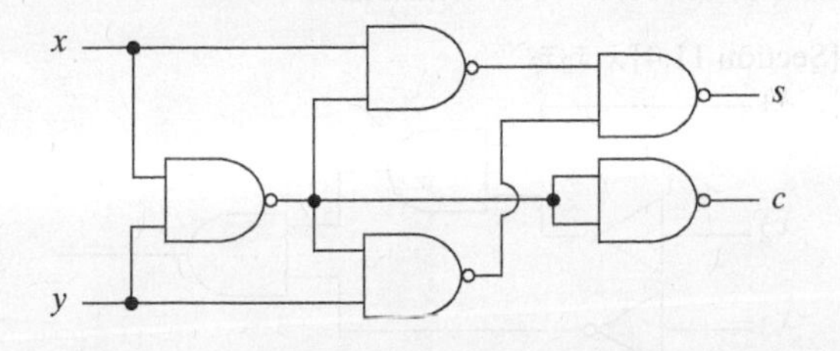


**17.** $xy = (x \downarrow x) \downarrow (y \downarrow y)$
$x \vee y = (x \downarrow y) \downarrow (x \downarrow y)\ \bar{x} = x \downarrow x$
$x \uparrow y = [(x \downarrow x) \downarrow (y \downarrow y)] \downarrow [(x \downarrow x) \downarrow (y \downarrow y)]$

**20.** Since

$$\bar{x} = x \downarrow x, \quad x \vee y = (x \downarrow y) \downarrow (x \downarrow y),$$

and {NOT, OR} is functionally complete, {NOR} is functionally complete.

**23.**

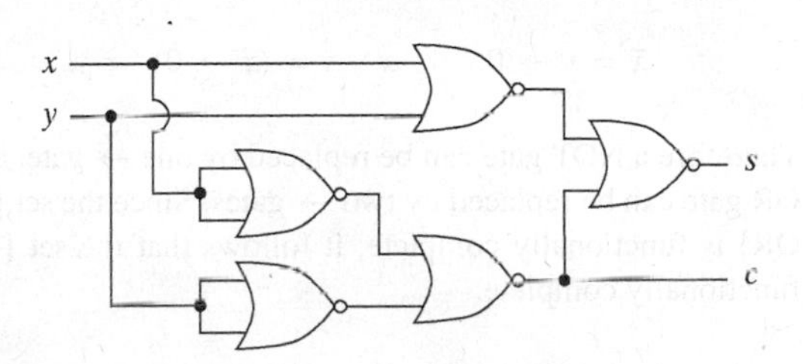


**25.** The logic table is

| $x$ | $y$ | $z$ | *Output* |
|---|---|---|---|
| 1 | 1 | 1 | 1 |
| 1 | 1 | 0 | 1 |
| 1 | 0 | 1 | 1 |
| 1 | 0 | 0 | 0 |
| 0 | 1 | 1 | 1 |
| 0 | 1 | 0 | 0 |
| 0 | 0 | 1 | 0 |
| 0 | 0 | 0 | 0 |

**27.** The logic table is

| $b$ | FLAGIN | $c$ | FLAGOUT |
|---|---|---|---|
| 1 | 1 | 0 | 1 |
| 1 | 0 | 1 | 1 |
| 0 | 1 | 1 | 1 |
| 0 | 0 | 0 | 0 |

Thus $c = b \oplus$ FLAGIN and FLAGOUT $= b \vee$ FLAGIN. We obtain the circuit

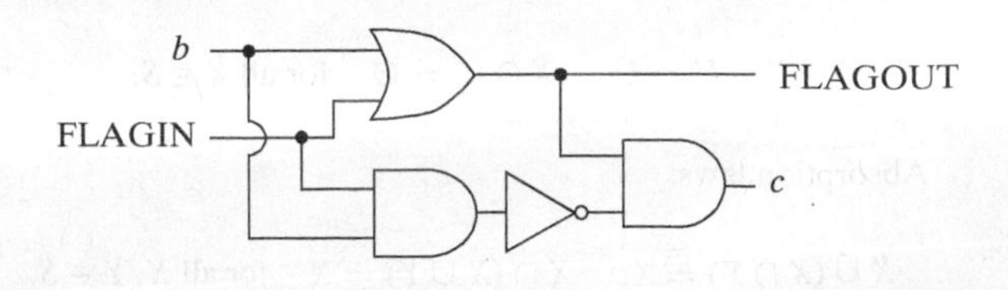


**28.** 010100

**31.**

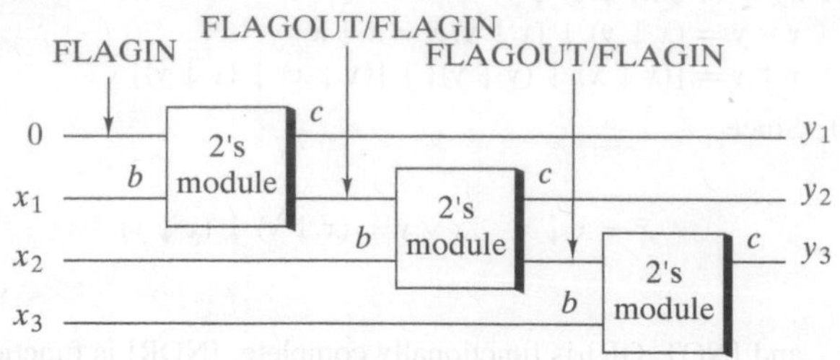


**34.** Writing the truth tables shows that

$$\overline{x} = x \to 0, \qquad x \vee y = (x \to 0) \to y.$$

Therefore a NOT gate can be replaced by one → gate, and an OR gate can be replaced by two → gates. Since the set {NOT, OR} is functionally complete, it follows that the set {→} is functionally complete.

## Chapter 11 Self-Test

**1.** [Section 11.1]

| $x$ | $y$ | $z$ | $\overline{(x \wedge \overline{y}) \vee z}$ |
|---|---|---|---|
| 1 | 1 | 1 | 1 |
| 1 | 1 | 0 | 1 |
| 1 | 0 | 1 | 1 |
| 1 | 0 | 0 | 0 |
| 0 | 1 | 1 | 1 |
| 0 | 1 | 0 | 1 |
| 0 | 0 | 1 | 1 |
| 0 | 0 | 0 | 1 |

**2.** [Section 11.2] The circuits are equivalent. The logic table for either circuit is

| $x$ | $y$ | *Output* |
|---|---|---|
| 1 | 1 | 0 |
| 1 | 0 | 1 |
| 0 | 1 | 0 |
| 0 | 0 | 0 |

**3.** [Section 11.2] The circuits are not equivalent. If $x = 0$, $y = 1$, and $z = 0$, the output of circuit (a) is 1, but the output of circuit (b) is 0.

**4.** [Section 11.3] Bound laws:

$$X \cup U = U, \quad X \cap \varnothing = \varnothing \quad \text{for all } X \in S.$$

Absorption laws:

$$X \cup (X \cap Y) = X, \quad X \cap (X \cup Y) = X \quad \text{for all } X, Y \in S.$$

**5.** [Section 11.1] 1

**6.** [Section 11.1]

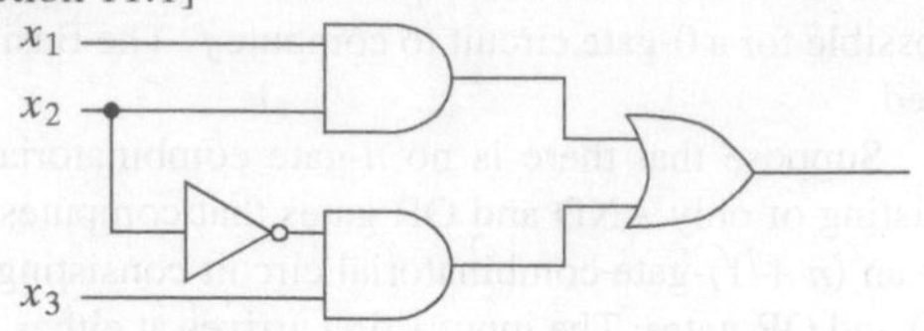


**7.** [Section 11.2] The equation is true. The logic table for either expression is

| $x$ | $y$ | $z$ | *Value* |
|---|---|---|---|
| 1 | 1 | 1 | 1 |
| 1 | 1 | 0 | 1 |
| 1 | 0 | 1 | 0 |
| 1 | 0 | 0 | 0 |
| 0 | 1 | 1 | 1 |
| 0 | 1 | 0 | 1 |
| 0 | 0 | 1 | 1 |
| 0 | 0 | 0 | 0 |

**8.** [Section 11.2] The equation is false. If $x = 1$, $y = 0$, and $z = 1$, then

$$(x \wedge y \wedge z) \vee \overline{(x \vee z)} = 0,$$

but

$$(x \wedge z) \vee (\overline{x} \wedge \overline{z}) = 1.$$

**9.** [Section 11.1] Suppose that $x$ is 1. Then the upper input to the OR gate is 0. If $y$ is 1, then the lower input to the OR gate is 0. Since both inputs to the OR gate are 0, the output $y$ of the OR gate is 0, which is impossible. If $y$ is 0, then the lower input to the OR gate is 1. Since an input to the OR gate is 1, the output $y$ of the OR gate is 1, which is impossible. Therefore, if the input to the circuit is 1, the output is not uniquely determined. Thus the circuit is not a combinatorial circuit.

**10.** [Section 11.3]
$$\begin{aligned}(x(x+y\cdot 0))' &= (x(x+0))' && \text{(Bound law)}\\ &= (x\cdot x)' && \text{(Identity law)}\\ &= x' && \text{(Idempotent law)}\end{aligned}$$

**11.** [Section 11.3] Dual: $(x + x(y+1))' = x'$

$$\begin{aligned}(x + x(y+1))' &= (x + x\cdot 1)' && \text{(Bound law)}\\ &= (x+x)' && \text{(Identity law)}\\ &= x' && \text{(Idempotent law)}\end{aligned}$$

**12.** [Section 11.3] $^{-}$ is not a unary operator on $S$. For example, $\overline{\{1, 2\}} \notin S$.

*In Exercises 13–16, $a \wedge b$ is written $ab$.*

**13.** [Section 11.4] $x_1\overline{x}_2\overline{x}_3$

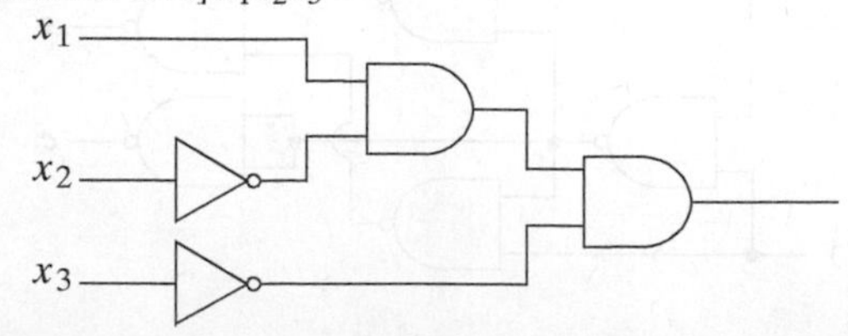

**14.** [Section 11.4] $x_1x_2\bar{x}_3 \vee x_1\bar{x}_2\bar{x}_3$

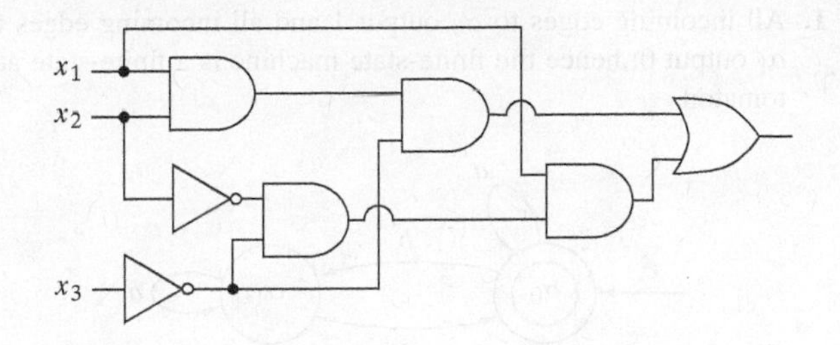


**15.** [Section 11.4] $x_1x_2x_3 \vee x_1\bar{x}_2\bar{x}_3 \vee \bar{x}_1\bar{x}_2\bar{x}_3$

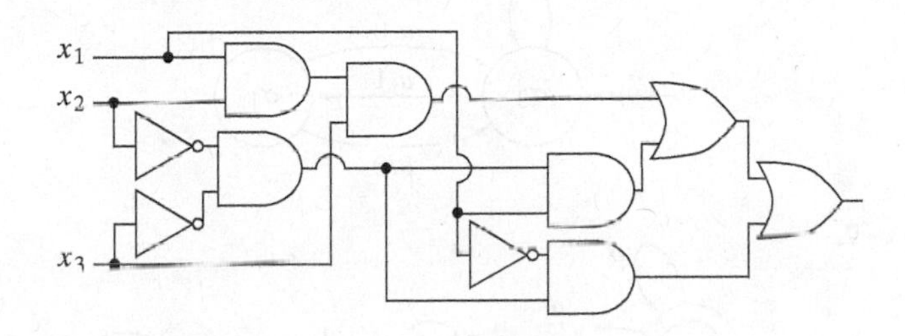


**16.** [Section 11.4] $x_1x_2\bar{x}_3 \vee x_1\bar{x}_2\bar{x}_3 \vee \bar{x}_1x_2x_3 \vee \bar{x}_1\bar{x}_2x_3$

**17.** [Section 11.5]

| $x$ | $y$ | $z$ | *Output* |
|---|---|---|---|
| 1 | 1 | 1 | 1 |
| 1 | 1 | 0 | 0 |
| 1 | 0 | 1 | 1 |
| 1 | 0 | 0 | 0 |
| 0 | 1 | 1 | 0 |
| 0 | 1 | 0 | 0 |
| 0 | 0 | 1 | 1 |
| 0 | 0 | 0 | 0 |

**18.** [Section 11.5] Disjunctive normal form: $x\bar{y}z \vee x\bar{y}\bar{z} \vee \bar{x}y\bar{z} \vee \bar{x}\bar{y}\bar{z}$

$$(x\bar{y}z \vee x\bar{y}\bar{z}) \vee \bar{x}y\bar{z} \vee \bar{x}\bar{y}\bar{z} = x\bar{y} \vee (\bar{x}y\bar{z} \vee \bar{x}\bar{y}\bar{z})$$
$$= x\bar{y} \vee \bar{x}\bar{z}$$

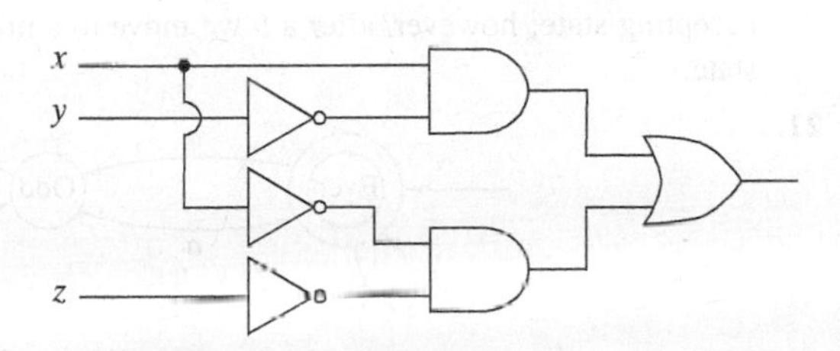


**19.** [Section 11.5]

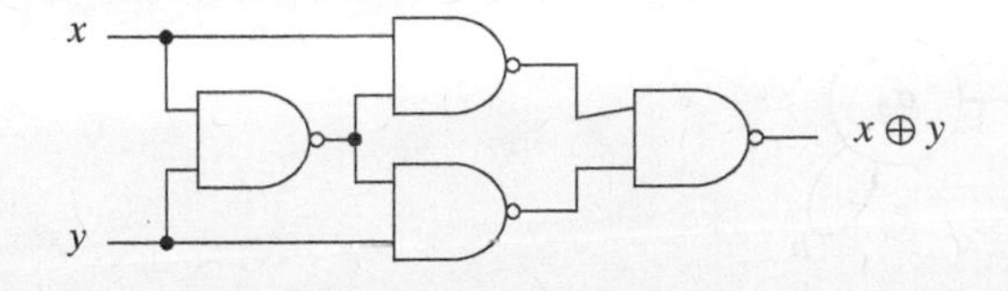


**20.** [Section 11.5]

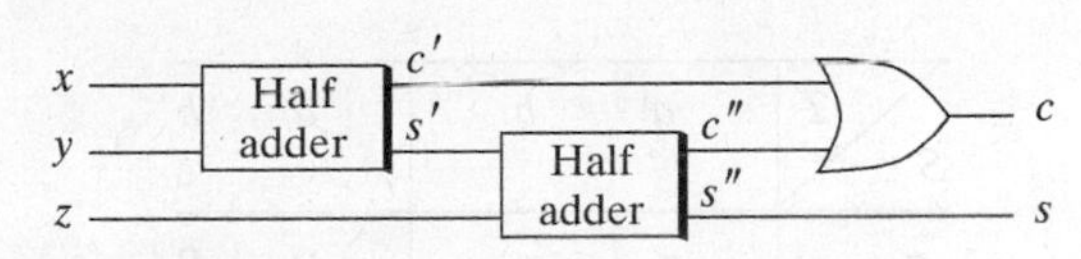


## Section 12.1 Review

1. A unit time delay accepts as input a bit $x_t$ at time $t$ and outputs $x_{t-1}$, the bit received as input at time $t - 1$.
2. A serial adder inputs two binary numbers and outputs their sum.
3. A finite-state machine consists of a finite set $\mathcal{I}$ of input symbols, a finite set $\mathcal{O}$ of output symbols, a finite set $\mathcal{S}$ of states, a next-state function $f$ from $\mathcal{S} \times \mathcal{I}$ into $\mathcal{S}$, an output function $g$ from $\mathcal{S} \times \mathcal{I}$ into $\mathcal{O}$, and an initial state $\sigma \in \mathcal{S}$.
4. Let $M = (\mathcal{I}, \mathcal{O}, \mathcal{S}, f, g, \sigma)$ be a finite-state machine. The transition diagram of $M$ is a digraph $G$ whose vertices are the states. An arrow designates the initial state. A directed edge $(\sigma_1, \sigma_2)$ exists in $G$ if there exists an input $i$ with $f(\sigma_1, i) = \sigma_2$. In this case, if $g(\sigma_1, i) = o$, the edge $(\sigma_1, \sigma_2)$ is labeled $i/o$.
5. The *SR* flip-flop is defined by the table

| *S* | *R* | *Q* |
|---|---|---|
| 1 | 1 | Not allowed |
| 1 | 0 | 1 |
| 0 | 1 | 0 |
| 0 | 0 | 1 *ifSwaslastequalto*1<br>0 *ifRwaslastequalto*1 |

## Section 12.1

**1.**

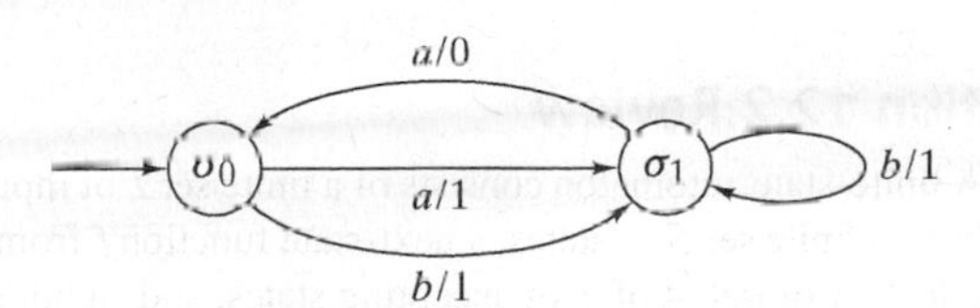


**4.**

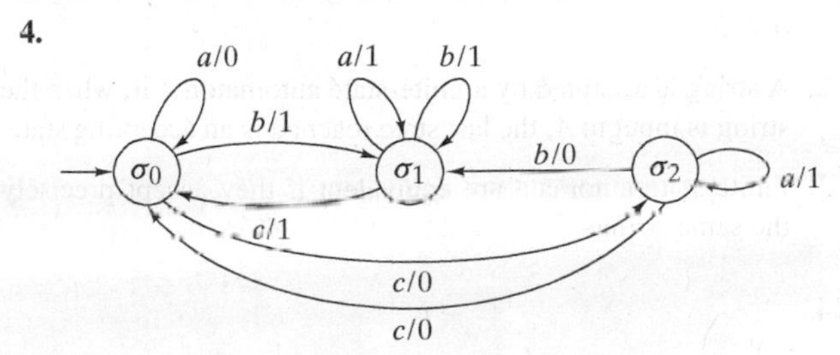


**6.** $\mathcal{I} = \{a, b\}$; $\mathcal{O} = \{0, 1\}$; $\mathcal{S} = \{\sigma_0, \sigma_1\}$; initial state $= \sigma_0$

| $\mathcal{S}$ \ $\mathcal{I}$ | $a$ | $b$ | $a$ | $b$ |
|---|---|---|---|---|
| $\sigma_0$ | $\sigma_1$ | $\sigma_0$ | 0 | 1 |
| $\sigma_1$ | $\sigma_1$ | $\sigma_1$ | 1 | 1 |

9. $\mathcal{I} = \{a, b\}$; $\mathcal{O} = \{0, 1\}$; $\mathcal{S} = \{\sigma_0, \sigma_1, \sigma_2, \sigma_3\}$; initial state $= \sigma_0$

| $\mathcal{S}$ \ $\mathcal{I}$ | $a$ | $b$ | $a$ | $b$ |
|---|---|---|---|---|
| $\sigma_0$ | $\sigma_1$ | $\sigma_2$ | 0 | 0 |
| $\sigma_1$ | $\sigma_0$ | $\sigma_2$ | 1 | 0 |
| $\sigma_2$ | $\sigma_3$ | $\sigma_0$ | 0 | 1 |
| $\sigma_3$ | $\sigma_1$ | $\sigma_3$ | 0 | 0 |

11. 1110 14. 001110

17. 20210 20. 020022201020

21.

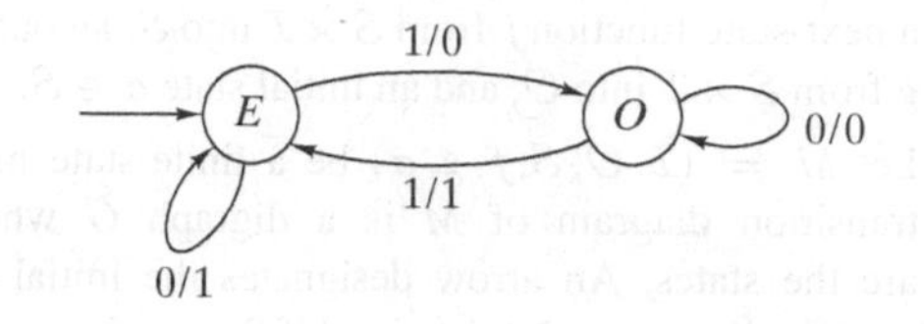


24.

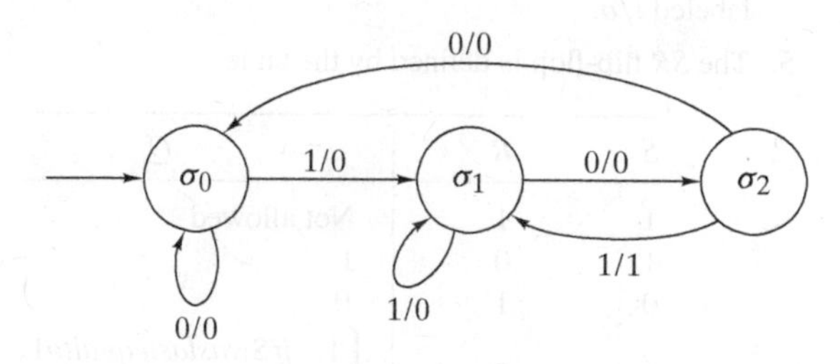


27. When $\gamma$ is input, the machine outputs $x_n, x_{n-1}, \ldots$ until $x_i = 1$. Thereafter, it outputs $\overline{x}_i$. However, according to Algorithm 11.5.16, this is the 2's complement of $\alpha$.

## Section 12.2 Review

1. A finite-state automaton consists of a finite set $\mathcal{I}$ of input symbols, a finite set $\mathcal{S}$ of states, a next-state function $f$ from $\mathcal{S} \times \mathcal{I}$ into $\mathcal{S}$, a subset $\mathcal{A}$ of $\mathcal{S}$ of accepting states, and an initial state $\sigma \in \mathcal{S}$.

2. A string is accepted by a finite-state automaton $A$ if, when the string is input to $A$, the last state reached is an accepting state.

3. Finite-state automata are equivalent if they accept precisely the same strings.

24.

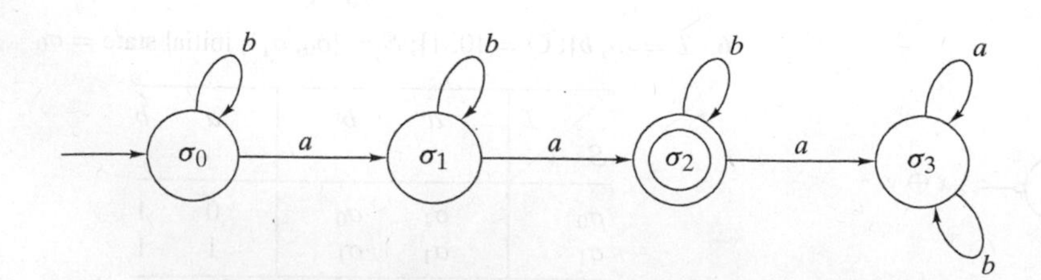


## Section 12.2

1. All incoming edges to $\sigma_0$ output 1 and all incoming edges to $\sigma_1$ output 0; hence the finite-state machine is a finite-state automaton.

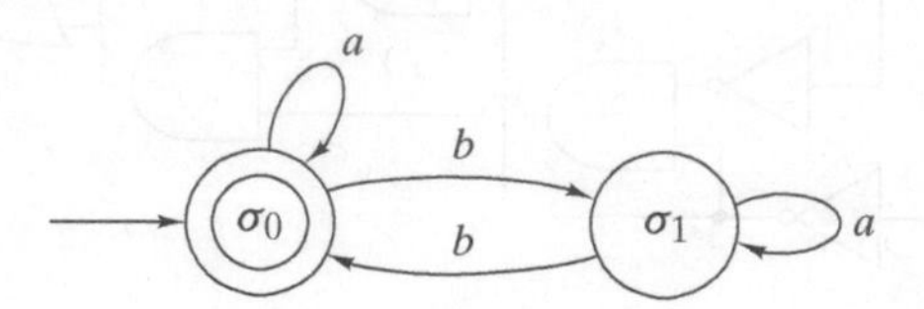


4.

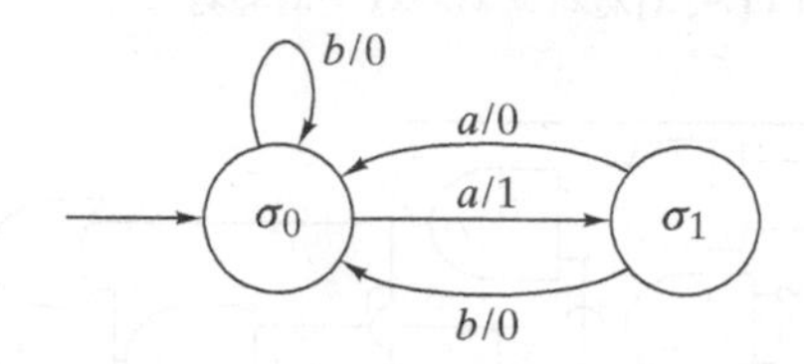


7.



10. (For Exercise 1) $\mathcal{I} = \{a, b\}$; $\mathcal{S} = \{\sigma_0, \sigma_1\}$; $\mathcal{A} = \{\sigma_0\}$; initial state $= \sigma_0$

| $\mathcal{S}$ \ $\mathcal{I}$ | $a$ | $b$ |
|---|---|---|
| $\sigma_0$ | $\sigma_0$ | $\sigma_1$ |
| $\sigma_1$ | $\sigma_1$ | $\sigma_0$ |

13. Accepted

16. Accepted

18. No matter which state we are in, after an $a$ we move to an accepting state; however, after a $b$ we move to a nonaccepting state.

21.

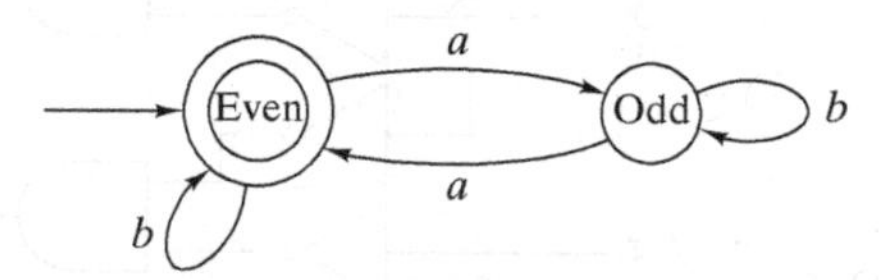

**27.**

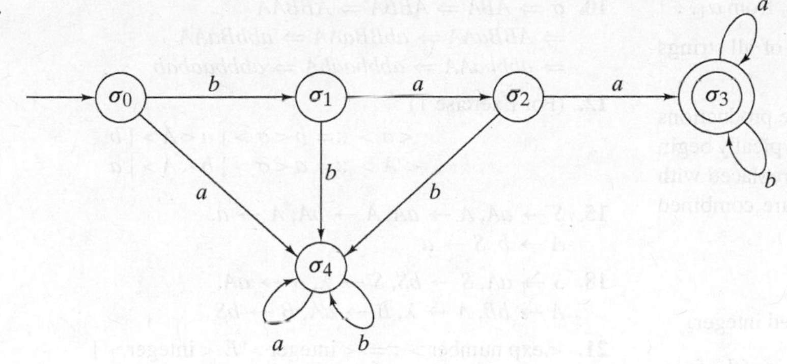

**30.**

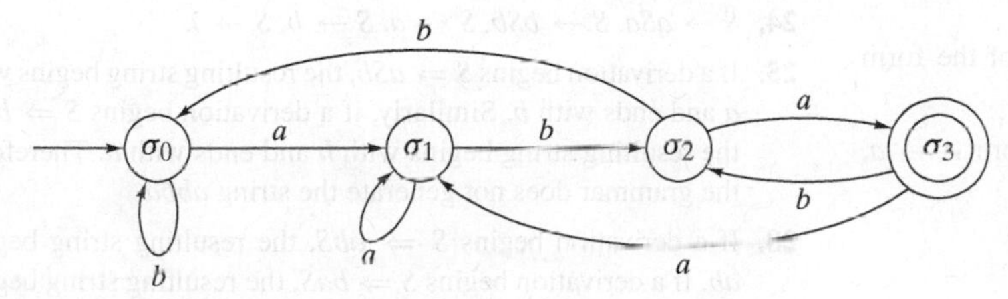

**32.** (For Exercise 1) This algorithm determines whether a string over $\{a, b\}$ is accepted by the finite-state automaton whose transition diagram is given in Exercise 1.

Input: $n$, the length of the string ($n = 0$ designates the null string); $s_1 \cdots s_n$, the string
Output: "Accept" if the string is accepted
"Reject" if the string is not accepted

```
ex32(s, n) {
  state = σ0
  for i = 1 to n {
    if (state == σ0 ∧ s_i == 'b')
      state = σ1
    if (state == σ1 ∧ s_i == 'b')
      state = σ0
  }
  if (state == σ0)
    return "Accept"
  else
    return "Reject"
}
```

**35.** Make each accepting state nonaccepting and each nonaccepting state accepting.

**38.** Using the construction given in Exercises 36 and 37, we obtain the following finite-state automaton that accepts $L_1 \cap L_2$. (We designate the states in Exercise 6 with primes.)

The finite-state automaton that accepts $L_1 \cup L_2$ is the same as the finite-state automaton that accepts $L_1 \cap L_2$ except that the set of accepting states is

$$\{(\sigma_1, \sigma_0'), \quad (\sigma_1, \sigma_1'), \quad (\sigma_1, \sigma_2'), \quad (\sigma_0, \sigma_2')\}.$$

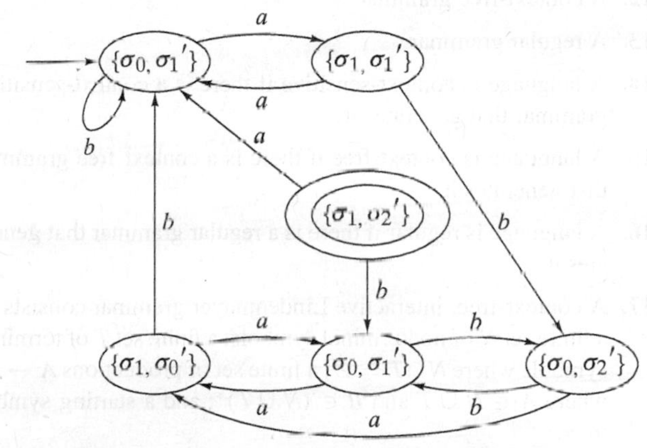

**41.** Use the construction of Exercises 36 and 37.

## Section 12.3 Review

**1.** A "natural language" refers to ordinary written and spoken words and combinations of words. A "formal language" is an artificial language consisting of a specified set of strings. Formal languages are used to model natural languages and to communicate with computers.

**2.** A phrase-structure grammar consists of a finite set $N$ of nonterminal symbols, a finite set $T$ of terminal symbols where $N \cap T = \varnothing$, a finite subset of $[(N \cup T)^* - T^*] \times (N \cup T)^*$ called the set of productions, and a starting symbol in $N$.

**3.** If $\alpha \to \beta$ is a production and $x\alpha y \in (N \cup T)^*$, we say that $x\beta y$ is directly derivable from $x\alpha y$.

**4.** If $\alpha_i \in (N \cup T)^*$ for $i = 1, \ldots, n$, and $\alpha_{i+1}$ is directly derivable from $\alpha_i$ for $i = 1, \ldots, n-1$, we say that $\alpha_n$ is derivable from $\alpha_1$ and write $\alpha_1 \Rightarrow \alpha_n$.

5. We call $\alpha_1 \Rightarrow \alpha_2 \Rightarrow \cdots \Rightarrow \alpha_n$ a derivation of $\alpha_n$ from $\alpha_1$.

6. The language generated by a grammar consists of all strings in terminals derivable from the start symbol.

7. Backus normal form (BNF) is a way to write the productions of a grammar. In BNF the nonterminal symbols typically begin with "⟨" and end with "⟩". Also the arrow $\rightarrow$ is replaced with ::=. Productions with the same left-hand side are combined using the bar "|". An example is

$$\langle\text{signed integer}\rangle ::= \quad + \langle\text{unsigned integer}\rangle \mid - \langle\text{unsigned integer}\rangle$$

8. In a context-sensitive grammar, every production is of the form $\alpha A \beta \rightarrow \alpha \delta \beta$, where $\alpha, \beta \in (N \cup T)^*$, $A \in N$, and $\delta \in (N \cup T)^* - \{\lambda\}$.

9. In a context-free grammar, every production is of the form $A \rightarrow \delta$, where $A \in N$ and $\delta \in (N \cup T)^*$.

10. In a regular grammar, every production is of the form $A \rightarrow a$, $A \rightarrow aB$, or $A \rightarrow \lambda$, where $A, B \in N$ and $a \in T$.

11. A context-sensitive grammar

12. A context-free grammar

13. A regular grammar

14. A language is context-sensitive if there is a context-sensitive grammar that generates it.

15. A language is context-free if there is a context-free grammar that generates it.

16. A language is regular if there is a regular grammar that generates it.

17. A context-free, interactive Lindenmayer grammar consists of a finite set $N$ of nonterminal symbols; a finite set $T$ of terminal symbols where $N \cap T = \varnothing$; a finite set of productions $A \rightarrow B$, where $A \in N \cup T$ and $B \in (N \cup T)^*$; and a starting symbol in $N$.

18. The von Koch snowflake is generated by the context-free, interactive Lindenmayer grammar

$$N = \{D\}$$
$$T = \{d, +, -\}$$
$$P = \{D \rightarrow D - D + + D - D, D \rightarrow d, + \rightarrow +, - \rightarrow -\}.$$

$d$ means "draw a straight line of a fixed length in the current direction," + means "turn right by 60°," and − means "turn left by 60°."

19. Fractal curves are characterized by having a part of the whole curve resemble the whole.

## Section 12.3

1. Regular, context-free, context-sensitive

4. Context-free, context-sensitive

7. $\sigma \Rightarrow b\sigma \Rightarrow bb\sigma \Rightarrow bbaA \Rightarrow bbabA \Rightarrow bbabbA \Rightarrow bbabba\sigma \Rightarrow bbabbab$

10. $\sigma \Rightarrow ABA \Rightarrow ABBA \Rightarrow ABBAA \Rightarrow ABBaAA \Rightarrow abBBaAA \Rightarrow abbBaAA \Rightarrow abbbaAA \Rightarrow abbbaabA \Rightarrow abbbaabab$

12. (For Exercise 1)

$$<\sigma> ::= b<\sigma> \mid a<A> \mid b$$
$$<A> ::= a<\sigma> \mid b<A> \mid a$$

15. $S \rightarrow aA, A \rightarrow aA, A \rightarrow bA, A \rightarrow a, A \rightarrow b, S \rightarrow a$

18. $S \rightarrow aA, S \rightarrow bS, S \rightarrow \lambda, A \rightarrow aA, A \rightarrow bB, A \rightarrow \lambda, B \rightarrow aA, B \rightarrow bS$

21. <exp number> ::= <integer> $E$ <integer> |
<float number> |
<float number> $E$ <integer>

24. $S \rightarrow aSa, S \rightarrow bSb, S \rightarrow a, S \rightarrow b, S \rightarrow \lambda$

25. If a derivation begins $S \Rightarrow aSb$, the resulting string begins with $a$ and ends with $b$. Similarly, if a derivation begins $S \Rightarrow bSa$, the resulting string begins with $b$ and ends with $a$. Therefore, the grammar does not generate the string *abba*.

28. If a derivation begins $S \Rightarrow abS$, the resulting string begins $ab$. If a derivation begins $S \Rightarrow baS$, the resulting string begins $ba$. If a derivation begins $S \Rightarrow aSb$, the resulting string starts with $a$ and ends with $b$. If a derivation begins $S \Rightarrow bSa$, the resulting string begins with $b$ and ends with $a$. Therefore, the grammar does not generate the string *aabbabba*.

31. The grammar does generate $L$, the set of all strings over $\{a, b\}$ with equal numbers of $a$'s and $b$'s.

Any string generated by the grammar has equal numbers of $a$'s and $b$'s since whenever any of the productions are used in a derivation, equal numbers of $a$'s and $b$'s are added to the string.

To prove the converse, we consider an arbitrary string $\alpha$ in $L$, and we use induction on the length $|\alpha|$ of $\alpha$ to show that $\alpha$ is generated by the grammar. The Basis Step is $|\alpha| = 0$. In this case, $\alpha$ is the null string, and $S \Rightarrow \lambda$ is a derivation of $\alpha$.

Let $\alpha$ be a nonnull string, and suppose that any string in $L$ whose length is less than $|\alpha|$ is generated by the grammar. We first consider the case that $\alpha$ starts with $a$. Then $\alpha$ can be written $\alpha = a\alpha_1 b\alpha_2$, where $\alpha_1$ and $\alpha_2$ have equal numbers of $a$'s and $b$'s. By the inductive hypothesis, there are derivations $S \Rightarrow \alpha_1$ and $S \Rightarrow \alpha_2$ of $\alpha_1$ and $\alpha_2$. But now

$$S \Rightarrow aSbS \Rightarrow a\alpha_1 b\alpha_2$$

is a derivation of $\alpha$. Similarly, if $\alpha$ starts with $b$, there is a derivation of $\alpha$. The Inductive Step is finished, and the proof is complete.

32. Replace each production

$$A \rightarrow x_1 \cdots x_n B,$$

where $n > 1$, $x_i \in T$, and $B \in N$, with the productions

$$A \rightarrow x_1 A_1$$
$$A_1 \rightarrow x_2 A_2$$
$$\vdots$$
$$A_{n-1} \rightarrow x_n B,$$

where $A_1, \ldots, A_{n-1}$ are additional nonterminal symbols.

**35.** $S \Rightarrow D+D+D+D \Rightarrow d+d+d+d$

$$\begin{aligned}S &\Rightarrow D+D+D+D\\ &\Rightarrow D+D-D-DD+D+D-D\\ &\quad +D+D-D-DD+D+D-D\\ &\quad +D+D-D-DD+D+D-D\\ &\quad +D+D-D-DD+D+D-D\\ &\Rightarrow d+d-d-dd+d+d-d\\ &\quad +d+d-d-dd+d+d-d\\ &\quad +d+d-d-dd+d+d-d\\ &\quad +d+d-d-dd+d+d-d\end{aligned}$$

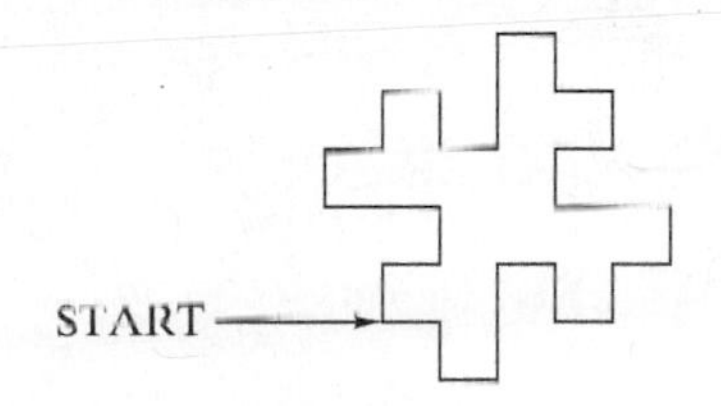


## Section 12.4 Review

**1.** Let $\sigma$ be the start state, let $T$ be the set of input symbols, and let $N$ be the set of states. Let $P$ be the set of productions $S \to xS'$, if there is an edge labeled $x$ from $S$ to $S'$, and $S \to \lambda$ if $S$ is an accepting state. Let $G$ be the regular grammar $(N, T, P, \sigma)$. Then the set of strings accepted by $A$ is equal to $L(G)$.

**2.** A nondeterministic finite-state automaton consists of a finite set $\mathcal{I}$ of input symbols, a finite set $\mathcal{S}$ of states, a next-state function $f$ from $\mathcal{S} \times \mathcal{I}$ into $\mathcal{P}(\mathcal{S})$, a subset $\mathcal{A}$ of $\mathcal{S}$ of accepting states, and an initial state $\sigma \in \mathcal{S}$.

**3.** A string $\alpha$ is accepted by a nondeterministic finite-state automaton $A$ if there is some path representing $\alpha$ in the transition diagram of $A$ beginning at the initial state and ending in an accepting state.

**4.** Nondeterministic finite-state automata are equivalent if they accept precisely the same strings.

**5.** Let $G = (N, T, P, \sigma)$ be a regular grammar. The finite-state automaton $A$ is constructed as follows. The set of input symbols is $T$. The set of states is $N$ together with an additional state $F \notin N \cup T$. The next-state function $f$ is defined as

$$f(S, x) = \{S' \mid S \to xS' \in P\} \cup \{F \mid S \to x \in P\}.$$

The set of accepting states is $F$ together with all $S$ for which $S \to \lambda$ is a production. Then $A$ accepts precisely the strings $L(G)$.

## Section 12.4

**1.**

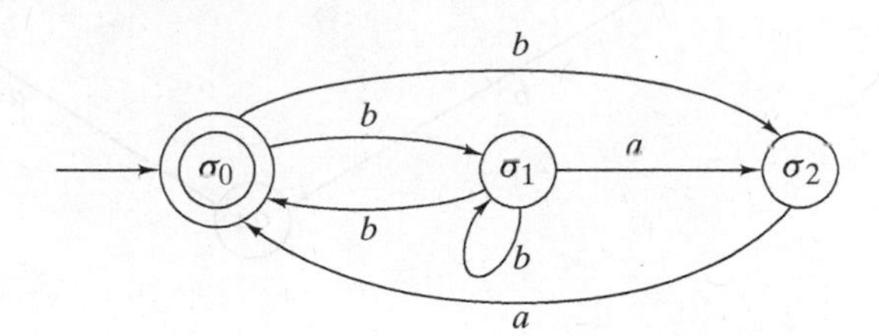


**4.**

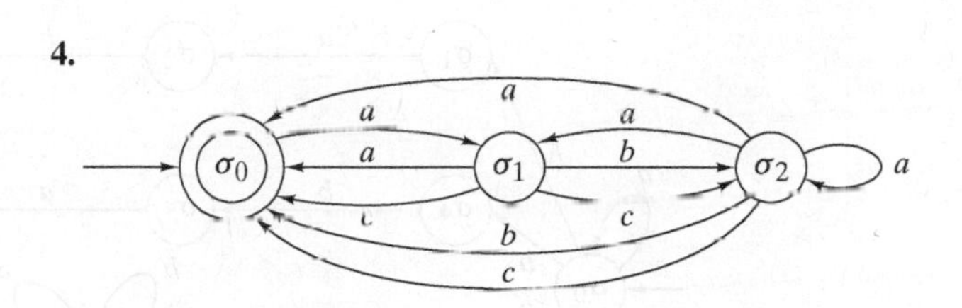


**6.** $\mathcal{I} = \{a, b\}$; $\mathcal{S} = \{\sigma_0, \sigma_1, \sigma_2\}$; $\mathcal{A} = \{\sigma_1, \sigma_2\}$; initial state $= \sigma_0$

| $\mathcal{S}$ \ $\mathcal{I}$ | $a$ | $b$ |
|---|---|---|
| $\sigma_0$ | $\{\sigma_1, \sigma_2\}$ | $\varnothing$ |
| $\sigma_1$ | $\{\sigma_1\}$ | $\{\sigma_0, \sigma_2\}$ |
| $\sigma_2$ | $\varnothing$ | $\varnothing$ |

**9.** $\mathcal{I} = \{a, b\}$; $\mathcal{S} = \{\sigma_0, \sigma_1, \sigma_2, \sigma_3\}$; $\mathcal{A} = \{\sigma_3\}$; initial state $= \sigma_0$

| $\mathcal{S}$ \ $\mathcal{I}$ | $a$ | $b$ |
|---|---|---|
| $\sigma_0$ | $\{\sigma_0\}$ | $\{\sigma_0, \sigma_1\}$ |
| $\sigma_1$ | $\{\sigma_2\}$ | $\varnothing$ |
| $\sigma_2$ | $\varnothing$ | $\{\sigma_3\}$ |
| $\sigma_3$ | $\{\sigma_3\}$ | $\{\sigma_3\}$ |

**11.** (For Exercise 6) $N = \{\sigma_0, \sigma_1, \sigma_2\}$, $T = \{a, b\}$,

$$\sigma_0 \to a\sigma_1, \quad \sigma_0 \to b\sigma_0, \quad \sigma_1 \to a\sigma_0, \quad \sigma_1 \to b\sigma_2,$$
$$\sigma_2 \to b\sigma_1, \quad \sigma_2 \to a\sigma_0, \quad \sigma_2 \to \lambda$$

**14.** No. For the first three characters, *bba*, the moves are determined and we end at $C$. From $C$, no edge contains an $a$; therefore, *bbabab* is not accepted.

**17.** Yes. The path $(\sigma, \sigma, \sigma, \sigma, C, C)$, which represents the string *aaaab*, ends at $C$, which is an accepting state.

**21.**

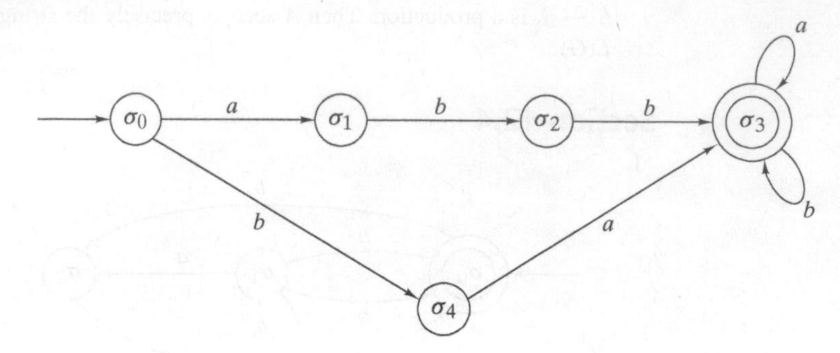


**24.**

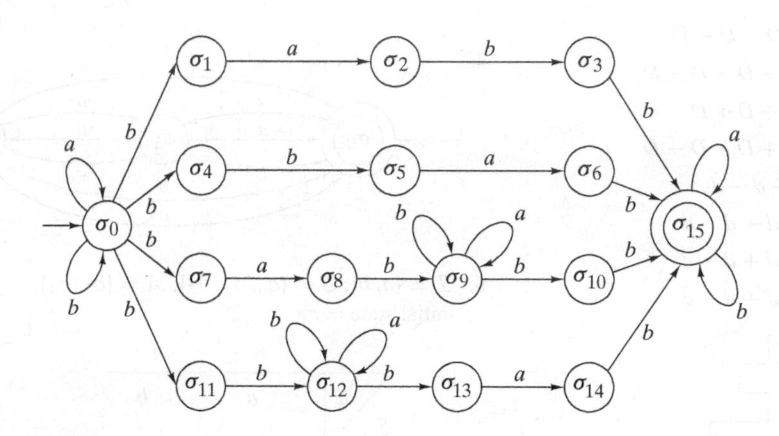

**27.**

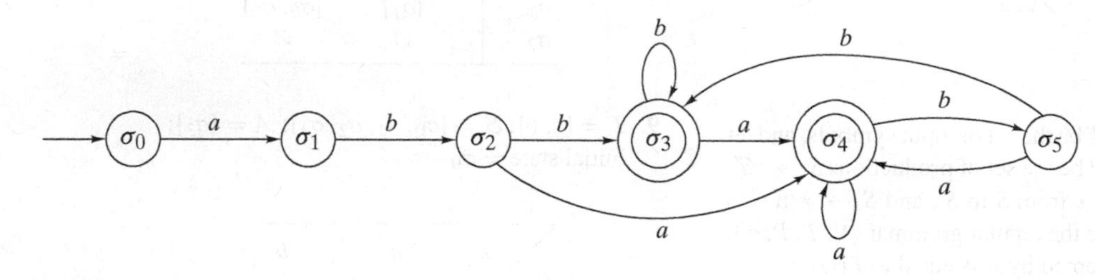


**30.** (For Exercise 21) $\sigma_0 \to a\sigma_1$, $\sigma_0 \to b\sigma_4$, $\sigma_1 \to b\sigma_2$, $\sigma_2 \to b\sigma_3$, $\sigma_3 \to a\sigma_3$, $\sigma_3 \to b\sigma_3$, $\sigma_4 \to a\sigma_3$, $\sigma_3 \to \lambda$

## Section 12.5 Review

**1.** Let $A = (\mathcal{I}, \mathcal{S}, f, \mathcal{A}, \sigma)$ be a nondeterministic finite-state automaton. An equivalent deterministic finite-state automaton can be constructed as follows. The set of states is the power set of $\mathcal{S}$. The set of input symbols is $\mathcal{I}$ (unchanged). The start symbol is $\{\sigma\}$ (essentially unchanged). The set of accepting states consists of all subsets of $\mathcal{S}$ that contain at least one accepting state of $A$. The next state function is defined by the rule

$$f'(X, x) = \begin{cases} \varnothing & \text{if } X = \varnothing \\ \bigcup_{S \in X} f(S, x) & \text{if } X \neq \varnothing. \end{cases}$$

**2.** A language $L$ is regular if and only if there exists a finite-state automaton that accepts precisely the strings in $L$.

## Section 12.5

**1.** (For Exercise 1)

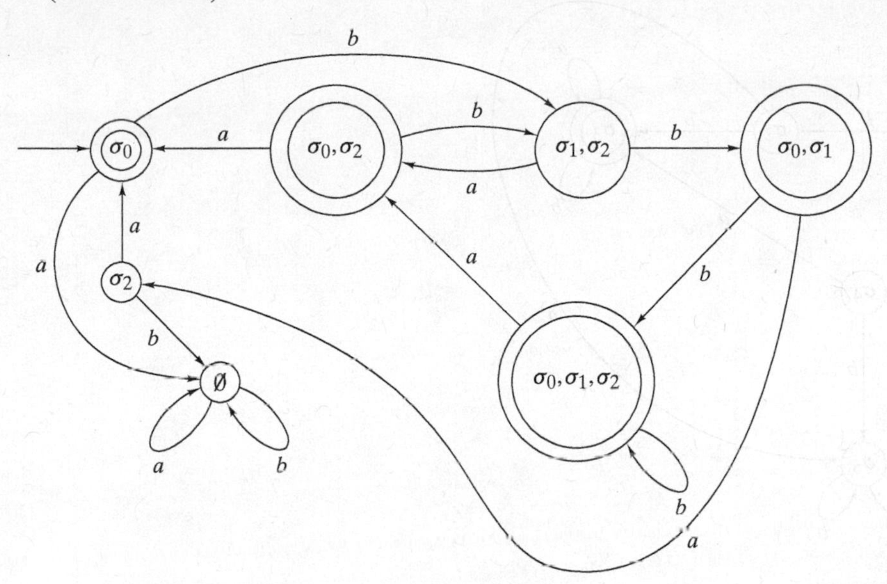


**2.**

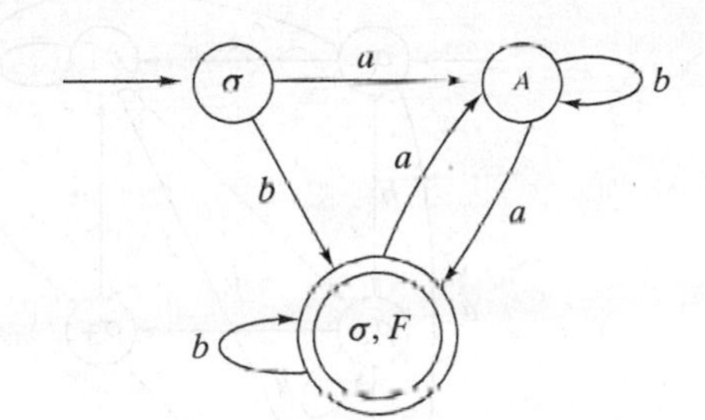


**5.**

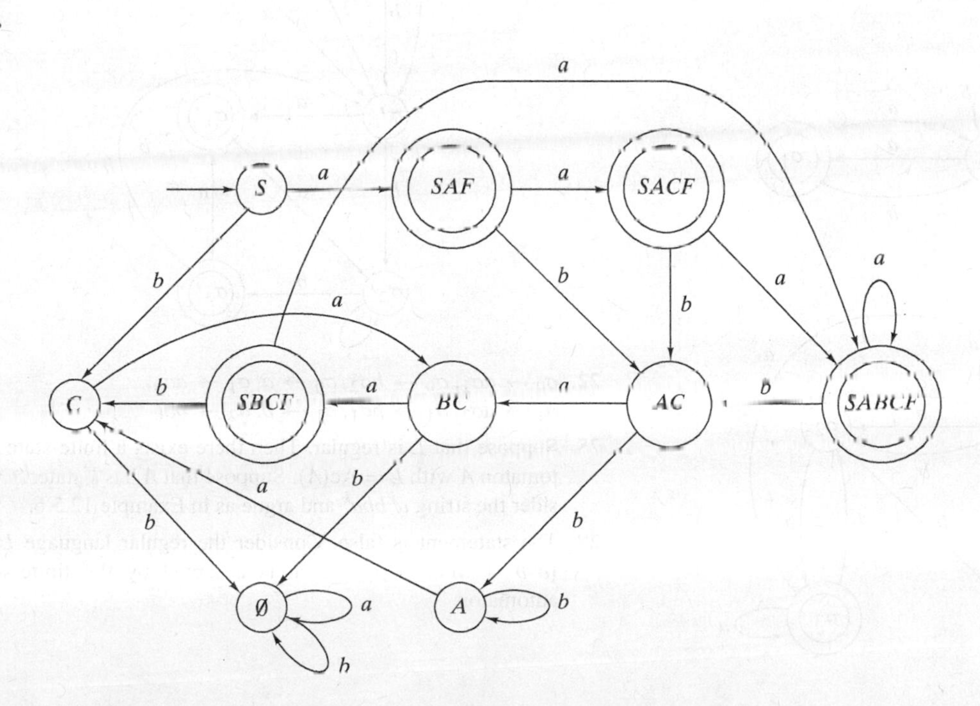

**7.** (For Exercise 21)

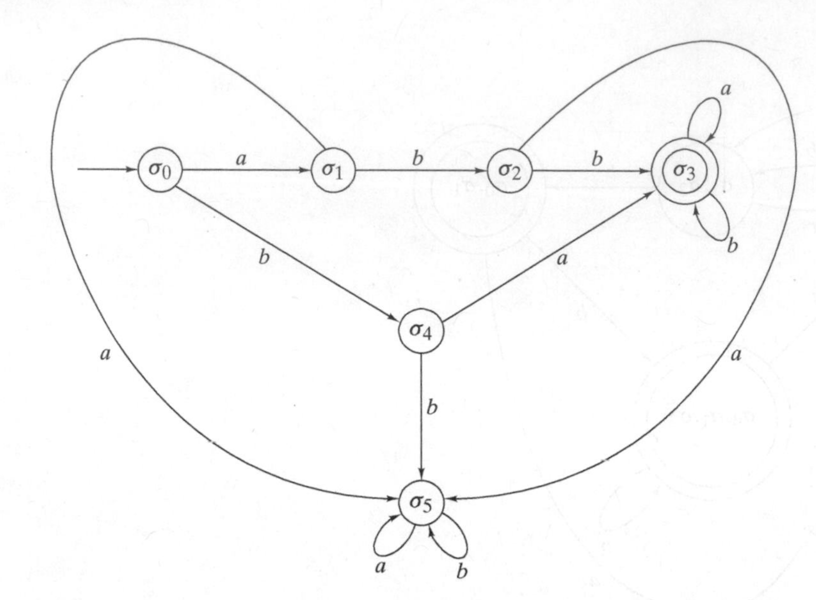


**10.** Figure 12.5.7 accepts the string $ba^n$, $n \geq 1$, and strings that end $b^2$ or $aba^n$, $n \geq 1$. Using Example 12.5.8, we see that Figure 12.5.9 accepts the string $a^n b$, $n \geq 1$, and strings that start $b^2$ or $a^n ba$, $n \geq 1$.

**11.**

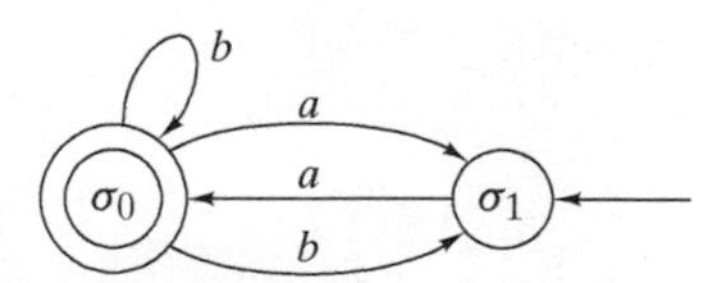


**14.**

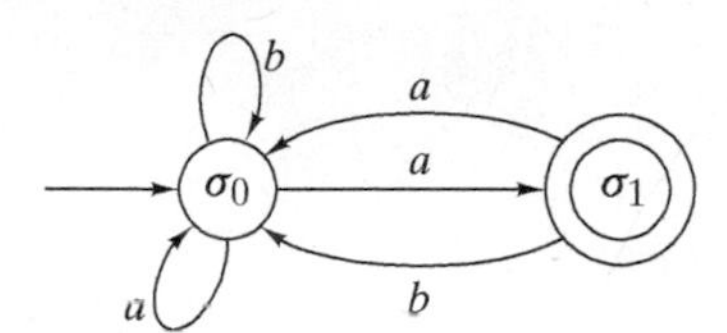


**17.**

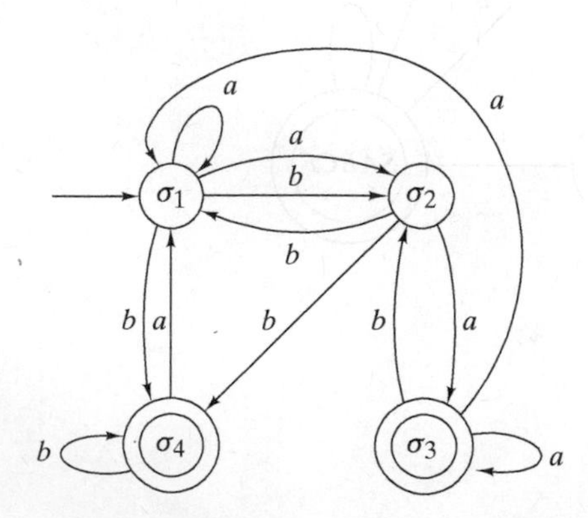


**20.**

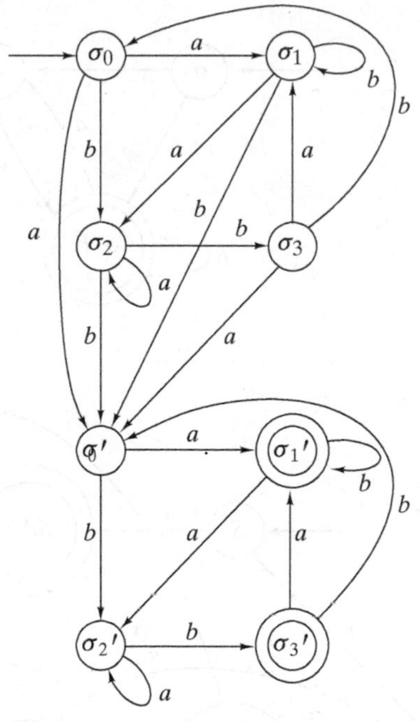


**22.** $\sigma_0 \to a\sigma_1$, $\sigma_0 \to b\sigma_2$, $\sigma_0 \to a$, $\sigma_1 \to a\sigma_0$, $\sigma_1 \to a\sigma_2$, $\sigma_1 \to b\sigma_1$, $\sigma_1 \to b$, $\sigma_2 \to b\sigma_0$

**25.** Suppose that $L$ is regular. Then there exists a finite-state automaton $A$ with $L = \text{Ac}(A)$. Suppose that $A$ has $k$ states. Consider the string $a^k bba^k$ and argue as in Example 12.5.6.

**28.** The statement is false. Consider the regular language $L = \{a^n b \mid n \geq 0\}$, which is accepted by the finite-state automaton

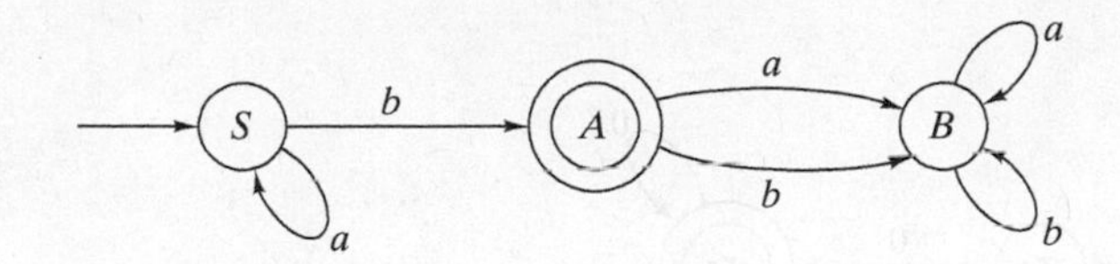


The language

$$L' = \{u^n \mid u \in L, n \in \{1, 2, \dots\}\}$$

is not regular. Suppose that $L'$ is regular. Then there is a finite-state automaton $A$ that accepts $L'$. In particular, $A$ accepts $a^n b$ for every $n$. It follows that for sufficiently large $n$, the path representing $a^n b$ contains a cycle of length $k < n$. Since $A$ accepts $a^n b a^n b$, $A$ also accepts $a^{n+k} b a^n b$, which is a contradiction.

## Chapter 12 Self-Test

**1.** [Section 12.1]

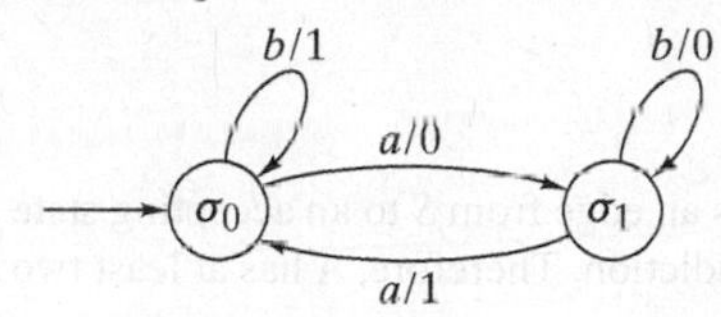


**2.** [Section 12.1] 1101

**3.** [Section 12.2]

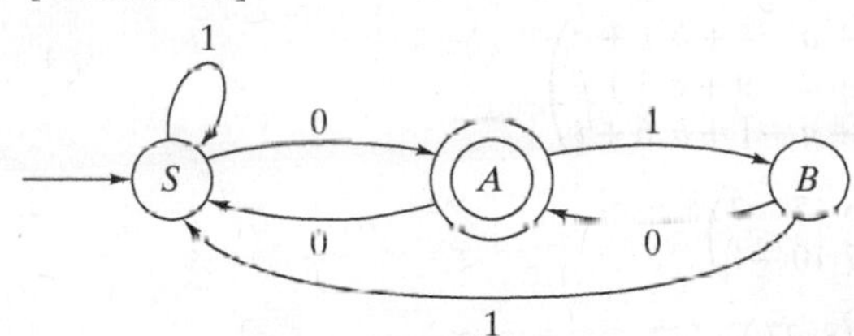


**4.** [Section 12.2] Yes

**5.** [Section 12.3] Context-free

**6.** [Section 12.3] $S \Rightarrow aSb \Rightarrow aaSbb \Rightarrow aaaSbbb \Rightarrow aaaAbbbb \Rightarrow aaaaAbbbb \Rightarrow aaaabbbb$

**7.** [Section 12.3] $a^i b^j, j \le 2 + i, j \ge 1, i \ge 0$

**8.** [Section 12.4]

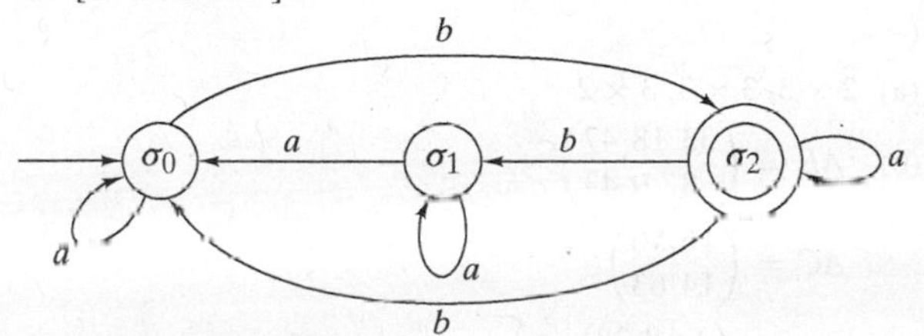


**9.** [Section 12.5]

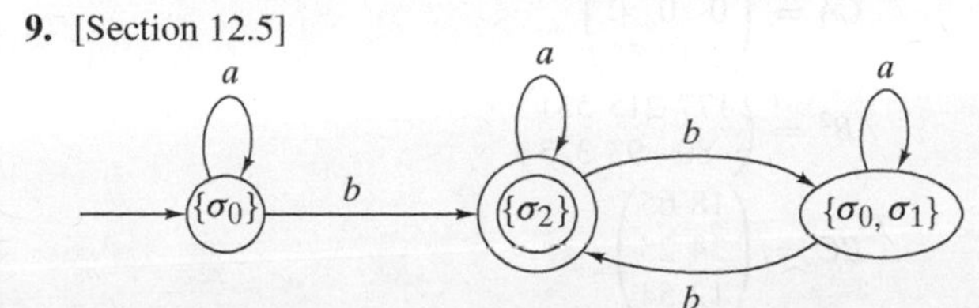


**10.** [Section 12.1] $\mathcal{I} = \{a, b\}$; $\mathcal{O} = \{0, 1\}$; $\mathcal{S} = \{S, A, B\}$; initial state = $S$

| $\mathcal{S}$ \ $\mathcal{I}$ | $f$: $a$ | $f$: $b$ | $g$: $a$ | $g$: $b$ |
|---|---|---|---|---|
| $S$ | $A$ | $A$ | 0 | 0 |
| $A$ | $S$ | $B$ | 1 | 1 |
| $B$ | $A$ | $B$ | 1 | 0 |

**11.** [Section 12.1]

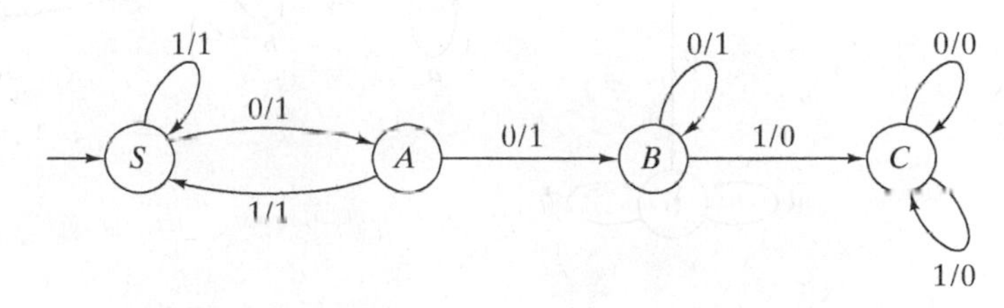


**12.** [Section 12.2]

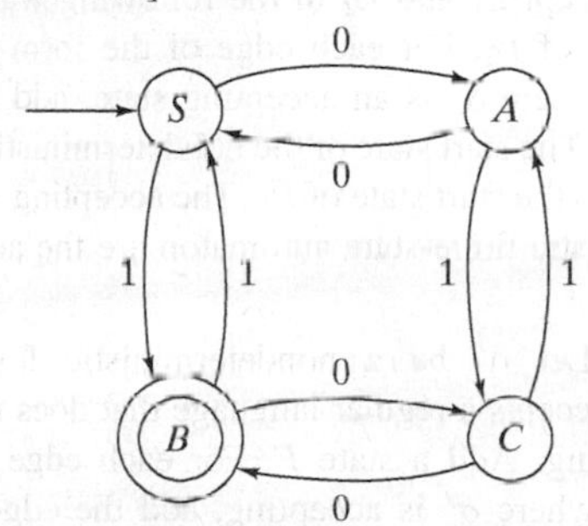


**13.** [Section 12.2] Every 0 is followed by a 1.

**14.** [Section 12.3] $S \to ASB$, $S \to AB$, $AB \to BA$, $BA \to AB$, $A \to a$, $B \to b$

**15.** [Section 12.4] $\mathcal{I} = \{a, b\}$; $\mathcal{S} = \{\sigma_0, \sigma_1, \sigma_2\}$; $\mathcal{A} = \{\sigma_0\}$; initial state = $\sigma_0$

| $\mathcal{S}$ \ $\mathcal{I}$ | $a$ | $b$ |
|---|---|---|
| $\sigma_0$ | $\{\sigma_0, \sigma_1\}$ | $\varnothing$ |
| $\sigma_1$ | $\varnothing$ | $\{\sigma_2\}$ |
| $\sigma_2$ | $\{\sigma_0, \sigma_2\}$ | $\{\sigma_2\}$ |

**16.** [Section 12.4] Yes, since the path

$$(\sigma_0, \sigma_0, \sigma_1, \sigma_2, \sigma_2, \sigma_2, \sigma_2, \sigma_0)$$

represents $aababa$ and $\sigma_0$ is an accepting state.

**17.** [Section 12.4]

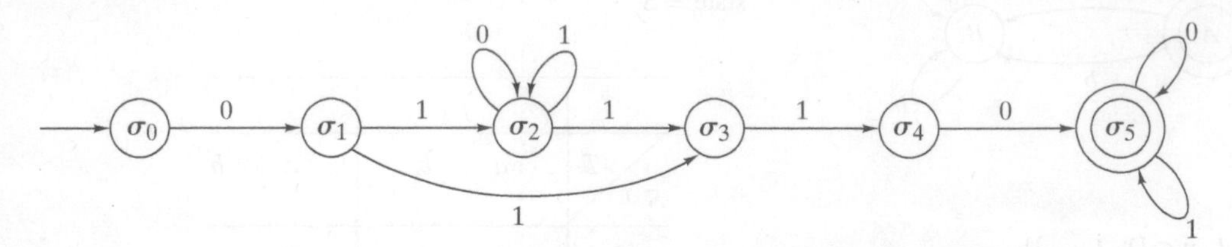


**18.** [Section 12.5]

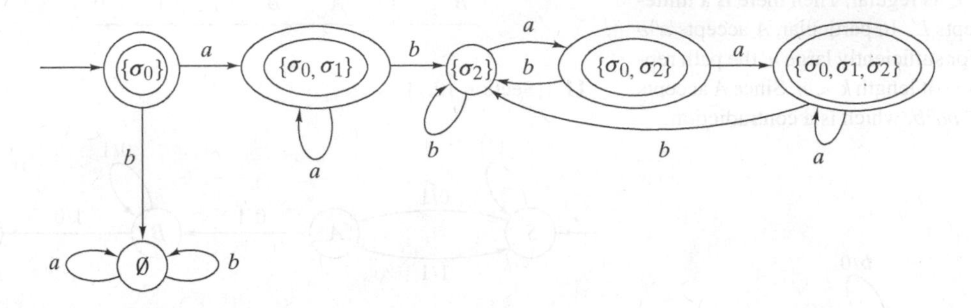


**19.** [Section 12.5] Combine the nondeterministic finite-state automata that accept $L_1$ and $L_2$ in the following way. Let $S$ be the start state of $L_2$. For each edge of the form $(S_1, S_2)$ labeled $a$ in $L_1$ where $S_2$ is an accepting state, add an edge $(S_1, S)$ labeled $a$. The start state of the nondeterministic finite-state automaton is the start state of $L_1$. The accepting states of the nondeterministic finite-state automaton are the accepting states of $L_2$.

**20.** [Section 12.5] Let $A'$ be a nondeterministic finite-state automaton that accepts a regular language that does not contain the null string. Add a state $F$. For each edge, $(\sigma, \sigma')$ labeled $a$ in $A'$ where $\sigma'$ is accepting, add the edge $(\sigma, F)$ labeled $a$. Make $F$ the only accepting state. The resulting nondeterministic finite-state automaton $A$ has one accepting state. We claim that $\text{Ac}(A) = \text{Ac}(A')$.

We show that $\text{Ac}(A) \subseteq \text{Ac}(A')$. [The argument that $\text{Ac}(A') \subseteq \text{Ac}(A)$ is similar and omitted.] Suppose that $\alpha \in \text{Ac}(A)$. There is a path

$$(\sigma_0, \sigma_1, \ldots, \sigma_{n-1}, \sigma_n)$$

that represents $\alpha$ in $A$, with $\sigma_n$ an accepting state. Since $\alpha \neq \lambda$, there is a last symbol $a$ in $\alpha$. Thus the edge $(\sigma_{n-1}, \sigma_n)$ is labeled $a$. Now the path

$$(\sigma_0, \sigma_1, \ldots, \sigma_{n-1}, F)$$

represents $\alpha$ in $A'$ and terminates in an accepting state. Therefore, $\alpha \in \text{Ac}(A')$.

To see that the statement is false for an arbitrary regular language, consider the regular language

$$L = \{\lambda\} \cup \{0^i \mid i \text{ is odd }\}$$

and a nondeterministic finite-state automaton $A$ with start state $S$ that accepts $L$. Since $\lambda \in L$, $S$ is an accepting state. If $S$ has a loop labeled 0, then $A$ accepts all strings of 0's; therefore, there is no loop at $S$ labeled 0. Since $0 \in L$ and there is no loop at $S$, there is an edge from $S$ to an accepting state $S' \neq S$, which is a contradiction. Therefore, $A$ has at least two accepting states.

## Appendix A

**1.** $\begin{pmatrix} 2+a & 4+b & 1+c \\ 6+d & 9+e & 3+f \\ 1+g & -1+h & 6+i \end{pmatrix}$

**2.** $\begin{pmatrix} 5 & 7 & 7 \\ -7 & 10 & -1 \end{pmatrix}$

**5.** $\begin{pmatrix} 3 & 18 & 27 \\ 0 & 12 & -6 \end{pmatrix}$

**8.** $\begin{pmatrix} -2 & -35 & -56 \\ -7 & -18 & 13 \end{pmatrix}$

**9.** $\begin{pmatrix} 18 & 10 \\ 14 & -6 \\ 23 & 1 \end{pmatrix}$

**12.** $(-4)$

**14.** (a) $2 \times 3$, $3 \times 3$, $3 \times 2$

(b) $AB = \begin{pmatrix} 33 & 18 & 47 \\ 8 & 9 & 43 \end{pmatrix}$

$AC = \begin{pmatrix} 16 & 56 \\ 14 & 63 \end{pmatrix}$

$CA = \begin{pmatrix} 4 & 18 & 38 \\ 0 & 0 & 0 \\ 2 & 17 & 75 \end{pmatrix}$

$AB^2 = \begin{pmatrix} 177 & 215 & 531 \\ 80 & 93 & 323 \end{pmatrix}$

$BC = \begin{pmatrix} 18 & 65 \\ 34 & 25 \\ 12 & 54 \end{pmatrix}$

**17.** Let $A = (b_{ij})$, $I_n = (a_{jk})$, $AI_n = (c_{ik})$. Then

$$c_{ik} = \sum_{j=1}^{n} b_{ij}a_{jk} = b_{ik}a_{kk} = b_{ik}.$$

Therefore, $AI_n = A$. Similarly, $I_nA = A$.

**20.** The solution is $X = A^{-1}C$.

## Appendix B

**1.** $-4x$

**4.** $\dfrac{15x - 3b}{3} = 5x - b$

**7.** $\dfrac{1}{n} - \dfrac{1}{n+1} = \dfrac{n+1-n}{n(n+1)} = \dfrac{1}{n(n+1)}$

We may use this equation to compute $\sum_{i=1}^{n} \frac{1}{i(i+1)}$ as follows:

$$\begin{aligned}
&\sum_{i=1}^{n} \frac{1}{i(i+1)} \\
&= \sum_{i=1}^{n} \frac{1}{i} - \frac{1}{i+1} \\
&= \left(1 - \frac{1}{2}\right) + \left(\frac{1}{2} - \frac{1}{3}\right) + \cdots + \left(\frac{1}{n-1} - \frac{1}{n}\right) \\
&\quad + \left(\frac{1}{n} - \frac{1}{n+1}\right) \\
&= 1 - \frac{1}{n+1} = \frac{n+1-1}{n+1} = \frac{n}{n+1}.
\end{aligned}$$

**8.** 81

**11.** 1/81

**14.** (a), (c), and (g) are equal. (b) and (f) are equal. (d) and (e) are equal.

**16.** $x^2 + 8x + 15$

**19.** $x^2 + 8x + 16$

**22.** $x^2 - 4$

**25.** $(x + 5)(x + 1)$

**28.** $(x - 4)^2$

**31.** $(2x + 1)(x + 5)$

**34.** $(2x + 3)(2x - 3)$

**37.** $(n + 1)! + (n + 1)(n + 1)! = (n + 1)![1 + (n + 1)] = (n + 1)!(n + 2) = (n + 2)!$

**40.**

$$\begin{aligned}
&7(3\cdot 2^{n-1} - 4\cdot 5^{n-1}) - 10(3\cdot 2^{n-2} - 4\cdot 5^{n-2}) \\
&= 2^{n-2}(7\cdot 3\cdot 2 - 10\cdot 3) + 5^{n-2}(-7\cdot 4\cdot 5 + 10\cdot 4) \\
&= 2^{n-2}\cdot 12 + 5^{n-2}(-100) \\
&= 2^{n-2}(2^2\cdot 3) - 5^{n-2}(5^2\cdot 4) \\
&= 3\cdot 2^n - 4\cdot 5^n
\end{aligned}$$

**42.** Factoring gives $(x - 4)(x - 2) = 0$, which has solutions $x = 4, 2$.

**45.** $2x \leq 6$, $x \leq 3$

**48.** $i \leq n$ for $i = 1, \ldots, n$. Summing these inequalities, we obtain

$$\sum_{i=1}^{n} i \leq n\cdot n = n^2.$$

**51.** Multiply by $(n + 2)n^2(n + 1)^2$ to get

$$(2n + 1)(n + 1)^2 > 2(n + 2)n^2$$

or

$$2n^3 + 5n^2 + 4n + 1 > 2n^3 + 4n^2$$

or

$$n^2 + 4n + 1 > 0,$$

which is true if $n \geq 1$.

**54.** 6

**57.** 10

**59.** 2.584962501

**62.** −0.736965594

**64.** 2.392231208

**67.** 0.480415248

**68.** 1.489896102

**71.** Let $u = \log_b y$ and $v = \log_b x$. By definition, $b^u = y$ and $b^v = x$. Now

$$x^{\log_b y} = x^u = (b^v)^u = b^{vu} = (b^u)^v = y^v = y^{\log_b x}.$$

## Appendix C

**1.** First *large* is set to 2 and $i$ is set to 2. Since $i \leq n$ is true, the body of the while loop executes. Since $s_i >$ *large* is true, *large* is set to 3. $i$ is set to 3 and the while loop executes again.

Since $i \leq n$ is true, the body of the while loop executes. Since $s_i >$ *large* is true, *large* is set to 8. $i$ is set to 4 and the while loop executes again.

Since $i \leq n$ is true, the body of the while loop executes. Since $s_i >$ *large* is false, the value of *large* does not change. $i$ is set to 5 and the while loop executes again.

Since $i \leq n$ is false, the while loop terminates. The value of *large* is 8, the largest element in the sequence.

**4.** First $x$ is set to 4. Since $b > x$ is false, $x = b$ is *not* executed. Since $c > x$ is true, $x = c$ executes, and $x$ is set to 5. Thus $x$ is the largest of the numbers $a$, $b$, and $c$.

**7.**

```
min(a, b) {
   if (a < b)
      return a
   else
      return b
}
```

**10.**

```
odds(n) {
   i = 1
   while (i ≤ n) {
      println(i)
      i = i + 2
   }
}
```

**13.**

```
product(s, n) {
   partial_product = 1
   for i = 1 to n
      partial_product = partial_product * s_i
   return partial_product
}
```

# INDEX

## A

**H**

# List of Symbols

## Boolean Algebras and Circuits

| | |
|---|---|
| $x \vee y$ | $x$ or $y$ (1 if $x$ or $y$ is 1, 0 otherwise); page 533 |
| $x \wedge y$ | $x$ and $y$ (1 if $x$ and $y$ are 1, 0 otherwise); page 533 |
| $x \oplus y$ | exclusive-OR of $x$ and $y$ (0 if $x = y$, 1 otherwise); page 551 |
| $\overline{x}$ | not $x$ (0 if $x$ is 1, 1 if $x$ is 0); page 533 |
| $x \downarrow y$ | $x$ NOR $y$ (0 if $x$ or $y$ is 1, 1 otherwise); page 563 |
| $x \uparrow y$ | $x$ NAND $y$ (0 if $x$ and $y$ are 1, 1 otherwise); page 557 |

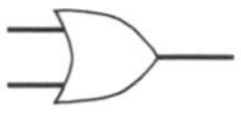

OR gate; page 533

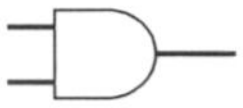

AND gate; page 533

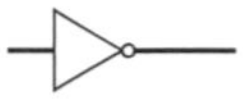

NOT gate (inverter); page 533

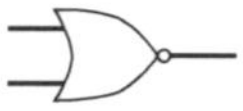

NOR gate; page 563

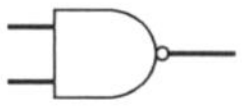

NAND gate; page 557

## Strings, Grammars, and Languages

| | |
|---|---|
| $\lambda$ | null string; page 135 |
| $\lvert s \rvert$ | length of the string $s$; page 135 |
| $st$ | concatenation of strings $s$ and $t$ ($s$ followed by $t$); page 135 |
| $a^n$ | $aa \cdots a$ ($n$ $a$'s); page 135 |
| $X^*$ | set of all strings over $X$; page 135 |
| $X^+$ | set of all nonnull strings over $X$; page 135 |
| $\alpha \to \beta$ | production in a grammar; page 580 |
| $\alpha \Rightarrow \beta$ | $\beta$ is derivable from $\alpha$; page 580 |
| $\alpha_1 \Rightarrow \alpha_2 \Rightarrow \cdots \Rightarrow \alpha_n$ | derivation of $\alpha_n$ from $\alpha_1$; page 580 |
| $L(G)$ | language generated by the grammar $G$; page 581 |
| $S ::= T$ | Backus normal form (BNF); page 581 |
| $S ::= T_1 \mid T_2$ | $S ::= T_1$, $S ::= T_2$; page 581 |
| $\text{Ac}(A)$ | set of strings accepted by $A$; page 576 |

## Matrices

| | |
|---|---|
| $(a_{ij})$ | matrix with entries $a_{ij}$; page 605 |
| $A = B$ | matrices $A$ and $B$ are equal ($A$ and $B$ are the same size and their corresponding entries are equal); page 605 |
| $A + B$ | matrix sum; page 606 |
| $cA$ | scalar product; page 606 |
| $-A$ | $(-1)A$; page 606 |
| $A - B$ | $A + (-B)$; page 606 |
| $AB$ | matrix product; page 606 |
| $A^n$ | matrix product $AA \cdots A$ ($n$ $A$'s); page 607 |